# ANNUAL REVIEW OF BIOCHEMISTRY

# ANNUAL REVIEW OF BIOCHEMISTRY

VOLUME 60, 1991

CHARLES C. RICHARDSON, *Editor*
Harvard Medical School

JOHN N. ABELSON, *Associate Editor*
California Institute of Technology

ALTON MEISTER, *Associate Editor*
Cornell University Medical College

CHRISTOPHER T. WALSH, *Associate Editor*
Harvard Medical School

ANNUAL REVIEWS INC.    4139 EL CAMINO WAY    P.O. BOX 10139    PALO ALTO, CALIFORNIA 94303-0897

ANNUAL REVIEWS INC.
Palo Alto, California, USA

*International Standard Serial Number: 0066–4154*
*International Standard Book Number: 0–8243–0860-3*
*Library of Congress Catalog Card Number: 32-25093*

Annual Review and publication titles are registered trademarks of Annual Reviews Inc.

∞ The paper used in this publication meets the minimum requirements of American National Standard for Information Sciences—Permanence of Paper for Printed Library Materials, ANSI Z39.48-1984.

Annual Reviews Inc. and the Editors of its publications assume no responsibility for the statements expressed by the contributors to this *Review*.

Typesetting by Kachina Typesetting Inc., Tempe, Arizona; John Olson, President
Typesetting Coordinator, Janis Hoffman

PRINTED AND BOUND IN THE UNITED STATES OF AMERICA

*Annual Review of Biochemistry*
*Volume 60, 1991*

# CONTENTS

vi    CONTENTS    (*Continued*)

# SOME RELATED ARTICLES IN OTHER *ANNUAL REVIEWS*

From the *Annual Review of Biophysics and Biophysical Chemistry*, Volume 20 (1991)

> *Beyond Gene Sequencing: Analysis of Protein Structure with Mass Spectrometry*, C. Fenselau
> *Biochemistry of Genetic Recombination: Energetics and Mechanism of DNA Strand Exchange*, S. C. Kowalczykowski
> *Chloramphenicol Acetyltransferase*, W. V. Shaw and A. G. W. Leslie
> *Structure and Function of a Lipoprotein: Lipovitellin*, L. J. Banaszak, W. Sharrock, and P. Timmins
> *Structure and Function of Retroviral Proteases*, P. M. D. Fitzgerald and J. P. Springer
> *The Zipper-Like Folding of Collagen Triple Helices and the Effects of Mutations that Disrupt the Zipper*, J. Engel and D. J. Prockop

From the *Annual Review of Cell Biology*, Volume 7 (1991)

> *Replication and Transcription of Vertebrate Mitochondrial DNA*, D. A. Clayton
> *Mechanisms and Functions of Cell Death*, R. E. Ellis, J. Yuan, and H. R. Horvitz
> *Cell and Substrate Adhesion Molecules in* Drosophila, M. Hortsch and C. S. Goodman
> *Centromeres: an Integrated Protein/DNA Complex Required for Chromosome Movement*, I. Schulman and K. S. Bloom
> *Activation of Replication Origins Within Yeast Chromosomes*, W. L. Fangman and B. J. Brewer
> *Biochemical Mechanisms of Constitutive and Regulated Pre-mRNA Splicing*, M. R. Green
> *Ubiquitination*, D. Finley and V. Chau

From the *Annual Review of Genetics*, Volume 24 (1990)

> *Genetic Analysis of Protein Export in* Escherichia coli, P. J. Schatz and J. Beckwith
> *Phage T4 Introns: Self-Splicing and Mobility*, M. Belfort

From the *Annual Review of Immunology*, Volume 9 (1991)

> *Regulation of Human B-Cell Activation and Adhesion*, E. A. Clark and P. J. L. Lane
> *Interactions between Immunogenic Peptides and MHC Proteins*, J. B. Rothbard and M. L. Gefter

From the *Annual Review of Medicine,* Volume 42 (1991)

> *Growth Factors and Cytokines in Bone Cell Metabolism,* E. Canalis, T. L. McCarthy, and M. Centrella
> *Lipoprotein Transport Gene Abnormalities Underlying Coronary Heart Disease Susceptibility,* J. L. Breslow
> *Growth Factors and Wound Healing: Platelet-Derived Growth Factor as a Model Cytokine,* T. F. Deuel, R. S. Kawahara, T. A. Mustoe, and G. F. Pierce

From the *Annual Review of Microbiology,* Volume 45 (1991)

> *Chaperone-Assisted Assembly and Molecular Architecture of Adhesive Pili,* S. J. Hultgren, S. Normark, and S. N. Abraham
> *Gene Amplification in* Leishmania, S. M. Beverly
> *Proper and Improper Folding of Proteins in the Cellular Environment,* B. Nilsson and S. Anderson
> *Serine β-Lactamases and Penicillin-Binding Proteins,* J.-M. Ghuysen

From the *Annual Review of Neuroscience,* Volume 14 (1991)

> *Cellular and Molecular Biology of the Presynaptic Nerve Terminal,* W. S. Trimble, M. Linial, and R. H. Scheller
> *Molecular Biology of Serotonin Receptors,* D. Julius
> *Molecular Approaches to Hereditary Diseases of the Nervous System: Huntington's Disease as a Paradigm,* N. S. Wexler, E. A. Rose, and D. E. Housman

From the *Annual Review of Nutrition,* Volume 11 (1991)

> *Phosphorus Magnetic Resonance Spectroscopy in Nutritional Research,* Z. Argov and B. Chance
> *Current Issues in Fructose Metabolism,* R. R. Henry, P. A. Crapo, and A. W. Thorburn
> *Biosynthesis of Prostaglandins,* W. E. M. Lands

From the *Annual Review of Physiology,* Volume 53 (1991)

> *Receptor G Protein Signaling in Yeast,* K. J. Blumer and J. Thorner
> *Regulators of Angiogenesis,* M. Klagsbrun and P. A. D'Amore

From the *Annual Review of Plant Physiology and Plant Molecular Biology,* Volume 42 (1991)

> *Plant Lipoxygenase: Structure and Function,* J. N. Siedow
> *Chlorophyll Fluorescence and Photosynthesis: the Basics,* G. H. Krause and E. Weis
> *Glycerolipid Synthesis: Biochemistry and Regulation,* J. Browse and C. Somerville

For the convenience of readers, a detachable order form/envelope is bound into the back of this volume.

Herman M Kalckar

Annu. Rev. Biochem. 1991. 60:1–37

# 50 YEARS OF BIOLOGICAL RESEARCH—FROM OXIDATIVE PHOSPHORYLATION TO ENERGY REQUIRING TRANSPORT REGULATION

## Herman M. Kalckar*

Department of Biochemistry, Harvard University, Cambridge, Massachusetts 02138

KEY WORDS: ATP, molecular enzyme defects in infants, galactose chemotactic mutants, energized down-regulation.

## CONTENTS

* The Editorial Committee was saddened to learn that Dr. Kalckar died at the age of 83 on May 17, 1991.

0066-4154/91/0701-0001$02.00

## My School and Family

This autobiography, requested by the Editorial Committee of the *Annual Review of Biochemistry*, has, I hope, mobilized my wit sufficiently to make the accidental reader dwell for more than a few moments and to raise curiosity about my intellectual pursuits in Copenhagen, where I grew up.

Copenhagen at that time was often referred to as "The Athens of the North." Although I was an erratic pupil in school and occasionally felt miserable, I have many interesting and rewarding memories from the old-fashioned school I attended, which, according to modern standards, especially for athletic activities, must be considered very modest. Its name was "Østre Borgerdyd Skole," which literally meant "the school for civic merits in the Eastern (Østre) borough of Copenhagen." I suppose that "merits" were meant to be applied to a larger part of the world than that quiet, attractive part of the Danish capital in which the school was located (the three Kalckar sons, of which I was the middle one, lived in easy walking distance from the school).

The Athenian flavor of "Borgerdyd Skolen" could not help making itself felt. The retired "Rector" (headmaster), J. L. Heiberg, was a world-renowned scholar in Greek, having translated Archimedes from Latin to Greek with great precision before the original Greek inscriptions were discovered. Our physics teacher, H. C. Christiansen, from the heather-covered, austere region of Western Jutland, was a formidable and passionately devoted teacher in mathematical physics. He had written a very concentrated, and in many ways very fine, textbook in physics, in which calculus played a prominent role. We were a little scared of him, because his temper would flare up if we tried to cut corners in our preparations and tried instead to memorize the text and formulas from his book. Christiansen insisted in no uncertain terms that physics and calculus, vector analysis, etc can only be grasped by using pencil and paper over and over again in order to find the right polarity in the fine network of a creative scientific argument. I dare say that we got the Ørsted-Ampere rules right.

I have allocated so much space to this alert teacher because I happened later to choose biological research and teaching as my occupation. Scientific arguments in pursuit of "truth," or I prefer the term "high fidelity," require a special kind of alertness.

Although our education in biology in school was largely systematic and somewhat static, it was brightened greatly by some extraordinary demonstrations in human physiology by a world-famous "Jutlandian" zoo-physiologist, August Krogh, professor at the University of Copenhagen and Nobel Laureate of 1923 for his description of capillary blood flow and its regulation. Krogh was, to my knowledge, the only physiologist in the 1920s who took an active interest in introducing the principles of human physiology to Danish high school boys. Krogh had constructed the microtonometer for measuring oxygen tension in small samples of blood as well as a respiratory spirometer for metabolic measurements. The spirometer and the ergometer bicycle were very popular at these school demonstrations.

Although I was perhaps the most restless of three, very different, brothers, the relatively modest Kalckar home where I grew up was an oasis for me. Here most of my interest in the humanistic disciplines developed and thrived. My mother, Bertha Rosalie, born Melchior, read French (Flaubert, Proust, etc) and German (Goethe, Heine, Lessing, etc) and spoke both languages. English came home at a slower rate (see below). Both my parents insisted that I treat the Danish language with care, reverence, and love. We belonged to a middle-class, Jewish-Danish family—Danish for several generations. We were mainly free-thinkers; however, my mother was attached to the Jewish faith and thought that it behooved us to acquire some religious education under the auspices of the main synagogue in Copenhagen and learn a minimum of Hebrew. My father, Ludvig Kalckar, took interest in reading Jewish and Yiddish humorists. All of us, however, responded most to Danish humor: cartoons, limericks, prose, and especially the theater.

My father, born in 1860, attended the world premiere of Ibsen's "A Doll's House" at the Royal Theater in November 1879. He wrote an enthusiastic report about the event. Although he was a businessman by necessity rather than by choice, my father carried on his profession as a consultant and broker most conscientiously. During World War I, he could easily have earned 10 to 20 times more than he did, had he not stubbornly refused to deal with stocks. He said in his quiet, unassuming way that he did not care to earn money on sunken ships and lost lives, and he "meant business." He always insisted, in his shy fashion, that this was a personal issue, and that he enjoyed seeing his relatives and friends socially, speculators and nonspeculators alike. So in spite of my father's late evening hours in his office, we did not grow rich. My mother was very patient and understood his principles. During the war she helped the International Red Cross in its relief efforts for prisoners of war and their families.

In contrast to its experience in World War II, Denmark was rather lucky during and after the World War I, almost too lucky in terms of gluttony, be it profiteering at the stock market or in the wild export of milk products.

We were very interested in the history of Denmark's relations with its neighbors, especially Germany. Germany under the Kaiser was not felt to be a trustworthy neighbor. Like so many other Danes, neither of my parents liked the Kaiser. My father had an extensive library on the history of Slesvig and Holstein, the disputed border provinces annexed by Bismarck in 1864. In 1918–1920, the popular, nationalistic Danish King Christian X wanted both provinces returned to Denmark. My Francophile mother fully supported this idea. In contrast, my father felt that restraint was more important than blind nationalism. He supported the return of North Slesvig, but felt unable to make a judgment in regard to the southern provinces. Fortunately, Denmark was blessed with a democratic statesman, Danish by birth and by sentiment, yet very familiar with the populations of both North and South Slesvig. His name was Hans Peter Hansen-Nørremølle, and he and his equally wise friend, H. V. Claussen, approached people unfailingly as human beings, regardless of whether they were from North or South Slesvig.

The verdict of their fair and patient inquiry was that Denmark should concentrate its efforts on North Slesvig, where the pro-Danes were in the majority, and not on South Slesvig, where the majority of the population remained pro-German. The liberal Danish government at that time respected this genuinely democratic recommendation, although it was not universally popular at the time. In general, the liberal Danes at that time showed great and truly democratic foresight. Imagine the abysmal conditions in Denmark under the Nazi occupation 1940–1945 if it had annexed the Slesvig-Holstein provinces. The real Nazi terror in Denmark started only three years after the invasion. I may be wrong in my analysis of these factors; after all why should the brave Norwegians have been exposed over five years to Nazi Terror? Both countries were lucky to have their staunch Kings at that time, Christian X and the heroic Haakon VII of Norway.

## My Two Great Teachers

Before discussing my debut in research on biological phosphorylations in 1935, it seems fitting to pay my special respects to my two great teachers, Ejnar Lundsgaard and Fritz Lipmann. Fritz Lipmann's memorial words about Ejnar Lundsgaard in 1968 are so beautifully chosen that I have selected excerpts from them:

> Word has come from Copenhagen of the death, during the last days of 1968, of Ejnar Lundsgaard. In these times of rush and short recall, I am particularly keen to refresh the memory of the scientific community about his truly great discovery which is now slipping into the background. Personally and scientifically, I encountered Lundsgaard during my period of maturation and I owe him much. When I moved to Copenhagen in 1932 and stayed until 1939, we saw each other a great deal. In the fall of 1967, his friends and colleagues went to Copenhagen to celebrate the 40th anniversary of the discovery of what he called the a-lactacid contraction of iodoacetate-poisoned muscle. When this startling news reached us in Meyerhof's laboratory, it was very upsetting to our group, which looked upon glycolytic lactic acid as the link between metabolic energy generation and muscle contraction.
>
> The interest in iodinated organic compounds which might have metabolic actions similar to thyroxin induced Lundsgaard to study iodinated acetic acid. Its injection into animals, however, yielded a rather unexpected effect: for a few minutes the animal behaved quite normally, but suddenly it turned over and its muscles became rigid. Such rigor was dependent on prior muscle activity, since denervated or curarized muscles did not respond.

But when rigor developed, the expected burst of lactic acid was missing. Lundsgaard elegantly solved the puzzle. He showed that: (i) iodoacetate inhibited glycolysis; and (ii) poisoned muscle performed a limited number of normal contractions at the expense of dephosphorylation of creatine phosphate, then newly found in need of a function. The rigor mortis-like condition developed when the limited supply of creatine phosphate was exhausted.

Lundsgaard had discovered that the muscle machine can be driven by phosphate-bond energy, and he shrewdly realized that this type of energy was "nearer," as he expressed it, to the conversion of metabolic energy into mechanical energy than lactic acid. He was right, because it soon developed that the glycolytic reaction is a feeder of phosphate-bond energy and not of acid. On the way, Lundsgaard also provided an enormously useful tool for studying enzyme mechanisms. Iodoacetate has become one of the standard reagents for SH-blocking in enzymes. Thus, iodoacetate inhibits glycolysis because it blocks the functional SH in phosphoglyceraldehyde dehydrogenase.

In the middle thirties, Lundsgaard became professor of physiology at Copenhagen University and trained many biochemists and physicians. Herman Kalckar was one of his graduate students. And, even though I did not formally work with him, I consider myself his pupil. My subsequent work was profoundly influenced by his discoveries, which changed our concepts of metabolic energy transformation.

## First Descriptions of Oxidative Phosphorylations in Tissue

During my studies in physiology and pathology at the University of Copenhagen, my interest was captivated by a few revolutionary developments in cell physiology that forced us all to reinterpret the role of lactic acid in mammalian c,lls. Otto Warburg found that tumor cells were unable to suppress their high generation of lactic acid formation even if they were supplied with excess oxygen. Ejnar Lundsgaard discovered that muscle contractions could occur without the formation of lactic acid, and Fritz Lipmann was able to interpret the Pasteur reaction. Lundsgaard found that frog muscles that have been exposed to iodoacetate are only able to perform a limited number of twitches; the phosphocreatine splits into phosphate and creatine. Lipmann was the first to believe in Lundsgaard's discovery as well as his interpretation. A few years later, in 1933, Lipmann was invited to Copenhagen. I was indeed lucky to have the good fortune of having Lundsgaard and Lipmann become my mentors.

In 1934 Lundsgaard became chief of the physiology department. His time was therefore more restricted, and so Fritz Lipmann became my foremost mentor. Lipmann encouraged me to study the newer literature on carbohydrate and phosphate metabolism in isolated mammalian tissue extracts or tissue particle preparations. For example, Hans Krebs seemed to have had much luck in disclosing different types of respiratory metabolism in tissue particle preparations from a variety of organs, especially pigeon breast muscle. About the same time, he formulated the tricarboxylic cycle, later called the Krebs tricarboxylic cycle.

For a variety of reasons, I elected to study the metabolism of phosphate and carbohydrate in extracts of kidney cortex. One might argue that pigeon breast muscle, bristling with enzymes catalyzing the Krebs cycle, would be superior. In order to obtain this tissue, however, I would have had to train myself to cut off the heads of pigeons, and I greatly preferred to avoid killing animals, probably from a lack of courage. In any case I was lucky to obtain fresh mammalian kidney from the perfusion experiments on anesthetized cats and rabbits, conducted at least once a week in the physiology department.

My studies on phosphorylation and respiration in kidney cortex extracts turned out to be rewarding. In the presence of oxygen (aeration in Warburg manometer vessels), inorganic phosphate was captured and esterified to various substrates such as glucose or glycerol. A special gift of 5'adenylic acid (5'AMP) from the late Dr. Pawel Ostern of Lwow, Poland gave me additional valuable insight. The kidney extracts converted the 5'AMP to adenosine triphosphate (ATP). When phosphatase activity was arrested by the addition of sodium fluoride, the P/O ratio approached 1. Under anaerobic conditions, no phosphorylation was detectable,[1] even in preparations well "spiked" with fluoride. I also noted that addition of dicarboxylic acids, like fumarate or succinate, stimulated oxygen consumption as well as phosphorylation. I never managed to get P/O ratios higher than 1.5, however, and then only if succinate was added to the cortex preparation (1).

Oxidative phosphorylation remained at about the same order of magnitude when I merely added fumarate or succinate to the extract, i.e. in the absence of glucose. This observation seemed very puzzling, and it induced me to pursue this particular aspect. I found that fumarate (or malate) in the absence of glucose not only stimulated respiration, but also gave rise to an accumulation of a phosphoric ester with properties different from those of hexose phosphate or glycerophosphate. What was the nature of this phosphoric ester? In pursuing this question, I once more received encouragement and advice from Lipmann.

In 1934–1935, work by Gustav Embden opened up new perspectives regarding phosphoric esters and their role in glycolysis; this new panorama was further expanded with the discovery of *phosphoenolpyruvic acid* (PEP) by Lohmann & Meyerhof. It was customary at that time to use the stability of phosphoric esters in acid or alkali as a first criterion of the nature of the different esters. The properties of the phosphoric ester formed in my fumarate experiments did not correspond to those of ATP. Since PEP was able to serve as a phosphate donor, and even able to catalyze the conversion of ADP to ATP, PEP attracted my special interest. PEP was known to be chemically

---

[1]The awareness that respiration stabilizes P-ester levels was expressed in 1932 not only by Lundsgaard but independently by Engelhardt in his work on nucleated avian erothrocytes, to which he added redox dyes.

*Figure 1*    V. A. Belitser, H. M. Kalckar, June 1960, Kiev

dephosphorylated rapidly at room temperature in alkaline iodine and even more rapidly by mercuric chloride. I used these highly specific criteria and found that the ester formed from fumarate or malate fulfilled all of these criteria. PEP seemed indeed to be formed in the cortex preparations from oxidation of malate in the presence of inorganic phosphate.

This type of generation of PEP could scarcely have originated from hexose phosphate, since the presence of sodium fluoride barred not only phosphatase activity but, as shown by Lohmann & Meyerhof, enolase activity as well; in this way the conversion of 2-phosphoglycerate to PEP was arrested. I therefore believed that with fumarate or malate in the experiments, PEP must have originated from one of the dicarboxylic acids, including oxaloacetate, all members of the Szent-Gyorgyi-Krebs di-tricarboxylic acid pathway. I published my new observations in *Nature* (3).

Lipmann's response to my discovery of PEP generation from dicarboxylic acids in respiring kidney cortex dispersions was unreservedly positive. As I mentioned, in spite of my repeated efforts, I never succeeded in obtaining P/O ratios significantly above 1.5. The further development in this direction we owe to V. A. Belitser in the USSR; in an extensive paper on oxidative phosphorylation he showed that pigeon breast muscle and rabbit heart muscle, in the presence of oxygen, are able to bring about phosphorylation of creatine to phosphocreatine. Belitser & Tsybakova showed that in the presence of arsenous acid and fluoride, the oxidation of succinate to fumarate did in fact generate P/O ratios as high as 3–4. The paper appeared shortly before the outbreak of World War II in 1939 (2). Much later, in June 1960, I had the privilege to meet Belitser in Kiev (see photo, p. 7), when the National Academy of Sciences sent me to the USSR Academy (Akademi NAYK) at the beginning of the scholarly exchange between the two academies.

In 1938, I was an active participant at the International Congress of Physiology in Zürich. I presented some of my recent findings on the oxidative phosphorylation of glycerol and the generation of PEP from dicarboxylic acids (3). My text and communication were presented in German, since I did not yet trust my ability to lecture in English. However, Professor R. A. Peters from the University of Oxford, who was chairman of the session, and Dorothy and Joseph Needham, University of Cambridge, were actively interested in some of the metabolic aspects of my topic, and they asked questions. I therefore had to switch to English, and enjoyed a very reassuring dialogue. I can still see and hear the very tall Joseph Needham rising and introducing himself to the audience: "Joseph Needham, Caius ("Kee-ees") College, Cambridge—can I ask a few questions?" My English apparently sufficed.

## Plans for Leaving Denmark

While I was trying to finish my PhD thesis, we lost my younger brother Fritz, just six months after his exciting journey to California in 1937. He died suddenly in a "status epilepticus," which, according to his California friends, he had been close to a couple of times in 1937. His fatal attack happened barely a year before the effective treatment of epilepsy became available in the United States and Denmark. My mother never got over this loss. The Bohr family, who also loved Fritz dearly, rendered deep personal comfort. We were touched by Niels Bohr's speech at the modest cremation ceremony and by his request to bring with him young Aage Bohr, who had enjoyed a warm friendship with Fritz. This all happened in January 1938. My PhD thesis became dedicated to the memory of Fritz Kalckar.

Although I admittedly was depressed during 1938, I had much encouragement and inspiration as well. Lundsgaard and Lipmann were formidable mentors and generous personalities. Funding for my research came from three sources: the University of Copenhagen, financed by the Ministry of Education; the Carlsberg Foundation; and the Rockefeller Foundation.

My personal salary for teaching, during the last years as assistant professor, came from the University of Copenhagen. The medical students who came to my afternoon lectures in physiology numbered 30 to 40, only about 20–30% of those who showed up at the professor's morning lectures. Yet I felt that I had the cream of the crop of medical students and that we generated a sophisticated dialogue. Our texts were Lundsgaard's excellent textbook interwoven occasionally with the British classic, "Starling's."

In the late 1930s the defense for a "Dr. Med." degree was a highly ceremonial event, whether at Danish, Swedish, or Norwegian universities. In my case, the ceremony took place in the old original section of "København" under the auspices of the almost 500-year-old university, in the middle of January 1939. My formal defense of my thesis was not as formidable as Professor Lundsgaard's concentrated and stylish opposition, but once he was over the critique of the language (mainly my casual Danish text and some of the tables), the scientific dialogue proceeded very well indeed. Lundsgaard agreed with Lipmann that the thesis revealed several new and important features regarding a relatively new field, metabolic biology, barely 20 years old.

During 1938 I came to know a most interesting scholarly chemist and biochemist, Kaj Linderstrøm-Lang at the Carlsberg Laboratory. Lang attracted scores of bright young biochemists, especially from Great Britain and the United States. Among the latter I came to know one of Lang's favorite American scholars, Rollin Hotchkiss from the Rockefeller Institute. By 1937 he and Lang had already developed protein chemistry along fundamentally new lines.

Through Lipmann, Lang had developed a strong interest in the bioenergetics of protein synthesis and wanted to learn more about phosphorylation systems. Lang fully agreed with Lipmann that the postulate that peptidases catalyzed synthesis as well as hydrolysis of peptides in proteins was not the answer to the problem of cellular protein synthesis. I was mainly a listener to these fascinating discussions and cherished the ideas that transphosphorylations and ATP might well be involved in cellular protein synthesis. Lipmann's visionary ideas began to take shape in 1938 via his bold experiment in which he generated ATP from his "homemade" acetylphosphate, catalyzed by a bacterial enzyme. This was performed in early 1939, after I had left Denmark, and only six months before the outbreak of World War II. It was Lipmann's last performance in Denmark during his last months there.

In 1938, Dr. Warren Weaver, Director of the Rockefeller Foundation, visited Copenhagen. My candidacy for a research fellowship was being discussed. I was appointed a Rockefeller Research Fellow in January 1939, shortly after my defense of my PhD degree at the University of Copenhagen. Now came the important problem: Which laboratory in the United States would stimulate and widen my horizon in general biology and bioenergetics? Linderstrøm-Lang was very familiar with Cal Tech's Departments of Chemistry, with Linus Pauling and Charles Coryell, and Biology, with the great geneticists T. H. Morgan, A. Sturtevant, and Calvin Bridges. Bridges had once shown my brother the *Drosophila* laboratory. Lang also knew the biochemist there, Henry Borsook, who was interested in thermodynamics and the bioenergetics of peptide formation. Borsook still operated on the old-fashioned notion of reversal of peptidase activity in order to reclaim polypeptides and knew as yet nothing about phosphorylations. I followed Lang's recommendation and decided on Cal Tech.

## California Institute of Technology and Writing a Review on ATP

In January 1939 I visited London for the first time. My purpose was to visit Dr. Robert Robison, an outstanding biochemist and the Editor of the *Biochemical Journal*. He was instrumental in publishing my fumarate-PEP paper (4) in the journal. Although I had decided to spend most of the year 1939–1940 at Cal Tech's Biology Department, I wanted to make a brief stop at Washington University in St. Louis, where Carl and Gerty Cori worked and where in 1936 they had discovered the first phosphorolytic fission with isolation of glucose-1-phosphate, generated from glycogen and catalyzed by a muscle enzyme, which they now called phosphorylase. I arrived in the bleak and sooty city of St. Louis in February 1939, and was well received by the Coris. Unexpectedly, they were actively interested in my articles on oxidative phosphorylation but were unable to repeat the experiments. Gerty wanted me to make a few demonstrations, but Carl cut it short and simply asked a few questions, and so did I. I saw that the Coris used the old-fashioned Meyerhof extract technique, using test tubes. I explained to them that kidney cortex extracts show a very high oxygen consumption, and in order to see oxidative phosphorylation I always aerated the cortex extracts by shaking them in Warburg vessels.

A bright young researcher in the Cori department, Dr. Sidney Colowick, witnessed my discussion with the Coris. He was asked to compare my technique with theirs. Colowick soon confirmed that the simple shaking technique was indeed needed in order to observe oxidative phosphorylation in the cortex preparations. The Coris wrote to me with the good news and with their best "thanks." This response may well have prepared the way for Carl Cori's later generous invitation to come to their lab when my first-year fellowship was running out.

On March 1, 1939 I arrived in Pasadena. I was received by a young student who soon became my good friend, David Bonner, originally from Salt Lake City. My other friend there was Max Delbrück, whom I had met briefly in Copenhagen at Bohr's Institute. I should later share many interesting reunions with Delbrück.

Thanks to the Bonner brothers, James and the younger David, I managed to find a bungalow shortly after the arrival of my wife Vibeke. We all agreed that the Cal Tech faculty club, "Atheneum," was nice but too expensive as a domicile for a research fellow. The Bonners were also most helpful in introducing me to a superb auto mechanic, José Plannes (of Basque origin); José was THE auto mechanic all the young Cal Tech scholars turned to then. José helped me superbly too. A Ford coupe, vintage 1934, was for sale and I bought it. It held up, not only on desert roads but in the mountains as well. After I received my driver's license, I had the nerve in early April to drive Dr. Hugo Theorell from Stockholm the whole way up to Mt. Wilson observatory, which he wanted to see. The snow was just beginning to melt.

Through Theorell I met Linus Pauling, who later came to play a very positive role in my career. I gave seminars and began to write an extensive review, much inspired by one of Dr. Linus Pauling's former coworkers, Dr. Charles Coryell, who was very interested in the energetics of phosphoryl donors like phosphocreatine and ATP. Pauling's ideas about resonance had greatly influenced Coryell's thinking, and I incorporated some of this into the review. I did not know then about the intriguing review that Lipmann had started, for which he was collecting thermal data. Lipmann's review was to have a great impact in biochemical circles.

I was lucky in the summer of 1939 to be enrolled in a combined lab and lecture course taught by an unusual microbiologist, Dr. Cornelius B. van Niel, who, perhaps more than anybody else I had then encountered, sensed what he called "The Unity of Life." van Niel was Dutch and a graduate of Delft University, Biotechnological School, a highly distinguished school. There he wrote his PhD thesis about a new type of sugar fermentation catalyzed by certain bacteria that generated propionic acid and carbon dioxide. van Niel reminded me that a Danish cheese expert, Orla Jensen, had first sniffed the existence of such a type of fermentation when he found out that the gas in the "eyes" of Swiss cheese was pure carbon dioxide. I had not paid attention to these matters while in Denmark. Perhaps worse, I was ignorant of some important work at the Carlsberg Lab. O. Winge had made great basic contributions in yeast genetics by describing the existence of diploid yeast as well as monoploid. Clearly my fixation on the four "big L's" had made me overlook the great tradition of Danish genetics at that time, founded by Winge's great teacher and researcher, Wilhelm Johannsen.

Although the van Niel Lab was very near the Pacific coast (in Pacific Grove, California), we discussed rather sparingly the role of the microorganisms in the ocean, except algae and the chemosynthetic microorganisms. Photosynthesis was, of course, broadly covered and various postulates critically discussed. These discussions inspired me to expand my biochemistry review that I had started at Cal Tech.

The autumn of 1939 was unusually hot, even for Pasadena, in stark contrast to the climate of Pacific Grove. Fortunately, the hospitable Physics Professor,

C. C. Lauritsen, and his wife invited us over to their swimming pool. November was sunny and pleasant, as was December 1939. My wife and I were invited to a real traditional American Christmas dinner at the Morgan's (T. H. and Lilian) in the stylish old house built and designed by the founder of The Kerckhoff Institute for the Biological Sciences. The founder was of course Dr. Thomas Hunt Morgan himself, and the multidisciplined institute had largely served as my scientific home base. Morgan asked about Denmark and more specifically about Niels Bohr.

In early 1940 I said farewell to my friends at Cal Tech, and Linus Pauling generously offered to get my extensive review on bioenergetics published in *Chemical Review,* provided I could manage to send him a finished script by the end of 1940.

We sold our car and left California in January 1940 and travelled more or less comfortably by train to the Midwest. I sat up most of the night on the train to St. Louis writing page after page of the review for Linus Pauling. After a brief couple of days in colorful New Orleans, we continued north, arriving late February 1940 in the damp and smoky St. Louis. It was a depressing city, but there I had the benefit of working in a lab that had developed a new type of enzymology, which now had become my main interest.

I had partly lost my enthusiasm for pursuing P/O ratios in various tissue particle preparations. Yet Carl Cori and his able coworkers had managed to obtain good P/O ratios in heart tissue dispersions, keeping in mind my warning against using unshaken test tubes for trying to obtain P/O ratios in tissue dispersions. Admittedly, by now I would rather have learned enzymology "a la Cori." Nevertheless, Carl persuaded me first to participate in collaborative work with him and Sidney Colowick on P/O ratios. Later, Sidney and I tried to delve into enzymology on our own. Thus began some of my happiest months in research.

While writing these lines I realized that W. A. Engelhardt and M. N. Lyubimova discovered 50 years ago that myosin was endowed with ATP-ase activity (5). In my big composition entitled "The Nature of Energetic Coupling on Biological Synthesis" (6), I took the liberty to make a speculative sketch from the Engelhardt discovery on the possibility that phosphorylation and dephosphorylation of myosin are "synonymous" with contraction and relaxation of myosin.

More important than my speculative efforts of 1939 on the ATP-ase report, however, was Engelhardt's experimental design, which was based on a hypothesis that expressed much the same ideas, albeit with experimental data. The interesting article by Engelhardt came to light in the *Yale Journal of Biology & Medicine* 15:21–38 (1942–1943), thanks to Paul Talalay's translation from the Russian [Engelhardt, W. A. 1941. *Adv. Contemp. Biochem.* (USSR) 14:172–90; cf. Kalckar, H. M. 1969. *Biological Phosphorylations,*

*Development of Concepts,* pp. 444–87. Engelwood Cliffs, NJ: Prentice Hall)]. In this masterful article, Engelhardt studied the changes of extensibility of isolated myosin fibers from muscle preparations upon the addition of ATP. A torsion balance with a reflecting mirror was used for recording. Engelhardt found that addition of ATP to freshly made myosin fibers brought about a specific increase in the extensibility of the myosin fiber, with a splitting of ATP to ADP. Engelhardt wisely refrained from interpreting whether this mechano-chemical effect was a contraction or a relaxation of the myosin fiber. In any case, this work opened the way for new discoveries in the life sciences.

## *Resumption of Laboratory Projects*

Only 25 years old, Dr. Sidney Colowick had achieved wisdom as well as a warm sense of humor and brilliant analytical insight. Colowick and I were interested in studying the enzymes of transphosphorylations. Our first attempt centered around a further fractionation and purification of yeast hexokinase. We soon found that the semipurified hexokinase preparations catalyzed the transfer of only one mole of P from ATP per hexose phosphate formed, and ATP was converted to ADP; in contrast, in crude hexokinase preparations, ATP was converted to 5'AMP, and 2 moles of hexose phosphate were formed.

Although we did not have the luck to find the suspected factor in yeast preparations, we did have luck testing extracts from skeletal muscle. Through a simple and effective purification of muscle extracts (merely by acidification with HCl and heating the acidified extract to 90°C for one to two minutes, then neutralizing partly to pH 6.5 and spinning), we obtained a clear supernatant (containing only 1% of the original protein content). This fraction turned out to contain the complementary activity. In other words, the mixing of the fractionated yeast preparation and the heat-resistant muscle fraction brought about the catalysis of formation of two hexose phosphates per ATP consumed, and the appearance of one 5'AMP. We called the active complementary fraction from muscle, "myokinase" (7).

At the suggestion of Dr. Marvin Johnson, University of Wisconsin, we pursued the following research problem: Does myokinase catalyze an "ADP dismutation" in which two ADP molecules are converted to one ATP and one 5'AMP? In order to prove this option, we needed to expand our analytical methods. Meanwhile Pearl Harbor forced the United States into the War, and Sidney Colowick was "drafted" for defense-related research in early 1942.

I was therefore left alone to find analytic methods to prove the existence of an enzymic "ADP dismutation." For this difficult task I used fractionation by means of barium acetate and analytical determinations of ammonia and of pentose as well as phosphorus (7a). The specific 5'AMP deaminase (described by Gerhard Schmidt in 1928) served well for this purpose. The analytical data supported the ADP dismutation scheme. The enzyme was later called "adenylate kinase." In 1946 I was able to reconfirm this enzymic ADP dismutase by UV spectrophotometric shifts.

In 1943 I received an invitation from Dr. Oliver H. Lowry to join a research institute of biochemistry, partly supported by the city of New York, partly from private funds. The division was headed by Oliver Lowry. The section on nutrition was headed by Dr. Otto A. Bessey (an old associate of C. G. King, President of the Nutrition Foundation). The names of the Committee of Overseers I scarcely remember, except that I do indeed remember Dr. Michael Heidelberger, then at Columbia University Medical College.

The biochemistry lab was very well equipped. For me the number one attraction was a Beckman UV spectrophotometer, which soon became my "Stradivarius." Playing this instrument, I came to develop a new field, the enzymatic synthesis of purine nucleosides (8). One of my favorite "tunes" was the metabolism of inosine, a hypoxanthine riboside. Why did I select this odd nucleoside? Because I soon found out that addition of xanthine oxidase gave rise to a highly conspicuous spectral change from 2480 to 2900 Å, since xanthine oxidase converted the hypoxanthine via xanthine to uric acid. I soon suspected that I was essentially dealing with a genuine phosphorylase reaction (8).

In pursuing this interesting problem, Oliver Lowry, a master analyst in the field of phosphate metabolism, came to play a leading role. He and his coworker, Jeanne Lopez, first tried to develop a better method for analyzing phosphocreatine. They succeeded and, in addition, via the same method (which operates at pH 4), they demonstrated that phophorolysis of inosine and guanosine gave rise to the formation of an acid-labile phosphoric ester, ribose-1-phosphate (9). Incubating this labile ester with hypoxanthine or guanine in the presence of the "new" phosphorylase gave rise to the resynthesis of inosine or guanosine. Thirty to 40 years later, it was found that inborn defects in the phosphorolytic fission of inosine to hypoxanthine and phosphate interfere in the conversion of T-cells to B-cells.

## Reunion with Linderstrøm-Lang in New York

Contact with Denmark until 1943 had been possible provided the letters reported about "neutral" events, since all the letters were opened and censored by the German occupation army. Denmark was now, as were so many other European countries, in the hands of the SS army and the Gestapo (joined by a small but vicious band of Danish Nazis). I received a laconic message that my mother had died in her home near the end of August 1943. It was very disturbing news, so much more so because I did not know what happened to my older brother or to my brave friends who had actively resisted the Nazi infestation in Denmark. I learned before my return that my brother Emil had been saved in Sweden.

In late 1945 the great Kaj Linderstrøm-Lang came to New York, invited by Warren Weaver, director of the Rockefeller Foundation. After a joint meeting, I had the rich opportunity to spend time alone with Lang. He told me in his low-keyed way about his encounters with the Gestapo and their brutalities, which led to the loss of many lives, including members of his own family. He knew about my mother's death, and he expressed that moving, warm empathy that so characterized him. Then we came to talk about the future. He had followed my scientific projects from my bioenergetics paper in early 1941 in the *Chemical Review* to my most recent paper on nucleoside phosphorolysis, which had just appeared as a full paper in the *Federation Proceedings, 1945.* He expressed the hope that I eventually would return to Copenhagen, but at present he agreed with me that I should continue to pursue my highly promising career in the United States.

Curiously enough, during our joint meeting (without Warren Weaver's knowledge), a very zealous assistant manager for fellowships of The Rockefeller Foundation expressed his personal feelings in a letter addressed to me that a former Rockefeller Foundation Fellow (even though the fellowship had terminated in early 1941) still ought to return to his homeland, i.e. Denmark. Fortunately, Ollie Lowry, Lang, and Weaver found this message under my circumstances absurd. I did not let myself be "bamboozled" by such a schoolmasterish message. My future decisions were to be influenced by other forces.

In 1945 I went to Cold Spring Harbor for Max Delbrück's first laboratory course in bacteriophage genetics. We were only four or five "students," among them my friend Rollin Hotchkiss from the Rockefeller Institute. We were all working happily more than ten hours a day reading plaques (lysed bacterial hosts, by various *Escherichia coli* strains). *E. coli* strains that were resistant to a phage mutant would be read on the plate as a "carpet" without any plaques. Delbrück was with us in the lab until well after midnight.

Late in 1945 there arrived an important letter from Ejnar Lundsgaard in Copenhagen with a generous offer to expand my research position at the University of Copenhagen. Keeping Lang's words in mind, I responded warmly in favor of returning to Denmark.

## Return to the University of Copenhagen

I returned to Denmark on a nice Norwegian ship, after first anchoring in Bergen, an ancient city, which probably was not right then in the mood to call itself an old "Hansa City" after all its bitter fight against five years of Nazi tyranny. After Bergen, we finally reached Copenhagen in April 1946. Among the survivors of the dark years was my older brother Emil (rescued by friends in Sweden), his wife, and their two children, Jørgen (10 years) and Birgit (7 years). I shall not dwell anymore on the loss of my mother and her brother.

My scientific life was rapidly restored, thanks to help from many quarters. First of all, my mentor of 12 years ago, Professor Ejnar Lundsgaard, saw to it that a sizable unit of his institute was set aside for my research. He initiated a promotion in my rank at the University and found starting funds for the continuation of my research. I had managed in advance to obtain some American financial support from The Rockefeller Foundation, and thanks to Dr. Thomas Jukes, from Lederle Laboratories. This together with support from the Carlsberg Foundation enabled me to purchase equipment for my lab, which I called the "Cytofysiologisk Institute."

In 1947, I was lucky enough to be approached by a young student, Hans Klenow ("studiosis magistrarum," as he was assigned, since he came from the graduate school, University of Copenhagen): Klenow was interested to learn and participate in research projects. Another bright "stud. mag.," Niels Ole Kjeldgaard, joined us. As a starting point in the new lab, we investigated the intriguing enzymological puzzles of the activities observed in two oxidases: xanthine oxidase and xanthopterin oxidase. This project soon disclosed quite a number of interesting observations, which we monitored not only by spectrophotometry but also by fluorometry. We isolated xanthine and pterine oxidase from cows' milk by using a milk centrifuge of Danish make (10). Our preference was to collect the enzyme from the whey fraction and purify the enzyme by ammonium sulfate fractionation.

Most folic acid preparations exerted a marked inhibitory effect on xanthine and xanthopterine oxidase, due apparently to an inhibitory factor in most pteroyl glutamate acid preparations (PGA). Much of this was resolved when my friend Tom Jukes sent us a PGA preparation of particularly high purity that seemed free of most of the inhibitory effect on the two oxidases. In addition, Oliver Lowry identified the oxidase inhibitor to be 2-amino-4-hydroxy-6-formylpteridine (i.e. a 6-aldehyde) as the strong inhibitor of the oxidases. This finding was further supported by our observations that incubation with xanthine oxidase catalyzed the 6-aldehyde group to a 6-carboxyl group, a derivative that did not exert any inhibition of the oxidases.

Aside from its value as an exercise in biochemical methods, I cannot say that this project deserved to be further pursued, and soon Klenow as well as Kjeldgaard showed their own great abilities in independent research. I was longing to resume work on the nucleoside phosphorylase project, and was lucky to be approached by a brilliant young PhD postdoctoral biochemist on a National Institutes of Health fellowship. His name was Morris E. Friedkin.

## Young American Biologists Choose Copenhagen Laboratories

Young Dr. Morris Friedkin from Lehninger's laboratory at the University of Chicago wanted to come to the University of Copenhagen to spend a year with me on his NIH postdoctoral Fellowship, 1949–1950. This association was to become one of the culminations in my own career. Friedkin had followed my work on inosine and guanosine phosphorolysis from 1947, and he was also aware of the Manson-Lampen abstract on phosphorolysis of deoxynucleosides from thymus preparations.

Friedkin and I preferred first to monitor the phosphorolysis of guanosine-deoxyriboside (DR) by enzymatic oxidation, observing the spectral shift as I had described in 1947. The enzymatic phosphorolysis of guanine-DR yielded a highly acid-labile phosphoric ester, which reacted with hypoxanthine in the presence of liver nucleoside phosphorylase to form hypoxanthine-DR. The highly acid labile ester was rapidly dephosphorylated at pH 4 at room temperature and subsequently lost its ability to generate purine-DR (11).

Upon enzymatic phosphorolysis (at pH 8) of guanine-DR, Friedkin then isolated a deoxyribosyl-1-phosphate as the crystalline salt of the organic base, cyclohexylamine. The salt contained no inorganic phosphate unless subjected to hydrolysis at pH 4. Hypoxanthine-DR was synthesized when the cyclohexamide salt of DR-1-phosphate was incubated with hypoxanthine in the presence of nucleoside phosphorylase. The formation of hypoxanthine-DR was shown by spectrophotometric methods. It indicated that the enzymatic synthesis of hypoxanthine-DR was greatly favored. The same was observed in the case of deoxy-guanosine. This is illustrated in Figure 2 (12). Friedkin and I were fortunate to have the excellent Danish microbiologist E. Hoff-Jørgensen join us. He showed that the microbiologic assay (with *Thermobacterium acidophilus* R 26) confirmed our enzymatic data for deoxy-guanosine (13).

On my visit to Chicago in early 1950 I was invited to a meeting on bacteriophage with Max Delbrück and Salvadore Luria. They were running the meeting at a hectic tempo. I was introduced to a shy young man from Indiana University. His name was James Dewey Watson. Both Delbrück and Luria recommended to young Watson that he use his upcoming fellowship to visit Dr. Ole Maaløe and me in Copenhagen, since we were supposedly attuned to phage biology and biochemistry. The research topic had actually been initiated by an able virologist, Alfred Hershey. Hershey's game plan was to label T-phages via their *E. coli* host with $^{32}$P and $^{35}$S to see whether the two isotopes were transmitted to the progeny. $^{32}$P was supposed to represent DNA and $^{35}$S, protein. It turned out that 40–50% of either isotope was transmitted to the progeny.

When I learned that Maaløe and Watson would also examine various T-phages by labeling them with $^{32}$P, I wondered whether I could not enrich their program by providing them with $^{14}$C-labeled adenine.

Thanks to the generous help from Professor Alexander Todd (head of the University Chemistry Laboratory, Cambridge) and his young coworker, Dr. Malcolm Clark, I was able to provide our lab with highly labeled [8-$^{14}$C]–adenine. Actually, Clark and I worked together, and half a year before

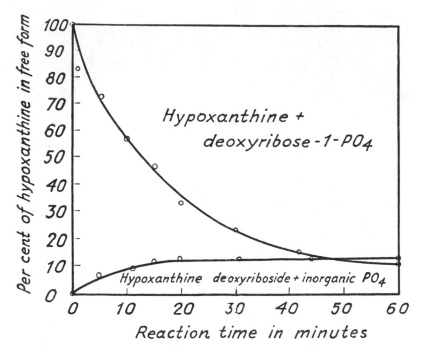

*Figure 2*  Equilibrium between deoxyribose-1-phosphate plus hypoxanthine and hypoxanthine deoxyriboside plus phosphate. Abscissa: time. Ordinate: per cent hypoxanthine in free form.

Watson arrived we had [8-$^{14}$C] adenine ready for him and Maaløe. Todd let us publish alone and helped us to get our findings published in the British *Journal of the Chemical Society,* which he found needed some useful publications in the field of "radio-chemistry" (14).

A hard-working and deeply original American microbiologist, Walter McNutt, wanted to come to our Copenhagen Institute in 1950. Within a few months, McNutt found that dialyzed enzyme preparations from *Lactobacillus helveticus* catalyzed a seemingly related reaction of fission of deoxynucleosides (DR), albeit in the total absence of phosphate. Moreover, this type of DR transfer, as McNutt showed, operated between pyrimidine and purine DRs. This bacterial enzyme was classified as a *trans-N*-glycosidase (15).

The $^{14}$C-labeled adenine was mobilized to try to solve a dispute, the enzymic interaction of deoxyribosides between adenine-DR and hypoxanthine-DR. The *L. helveticus* was the enzyme source with which McNutt had demonstrated the existence of *trans-N*-glycosidases between purines and pyrimidines. In the case of adenine-DR reacting with hypoxanthine or vice versa, however, two options were plausible, and in this case the use of a $^{14}$C adenine turned out to be critical. The enzyme might catalyze 1. a transamination or 2. a transglycosidation.

1. Transamination would catalyze the following reaction: $^{14}$C adenine + hypoxanthine-DR $<= = =>$ $^{14}$C hypoxanthine + adenine-DR.
2. Trans-$N$-glycosidase would catalyze the reaction: $^{14}$C adenine + hypoxanthine-DR $<= = =>$ hypoxanthine + $^{14}$C adenine-DR.

The experimental outcome supported Reaction 2, i.e. the transglycosidase reaction (16).

The second occasion for which our [8-$^{14}$C]adenine was put to use was when Watson and Maaløe were studying the transmission of nucleic acid to the progeny of a T-phage in a purine requiring $E.$ $coli$ host. It was found that the progeny of the T-phages had recovered 40–50% of the $^{32}$P label as well as 40–50% of the $^{14}$C adenine label (17).

## Otto Warburg in Denmark

In the spring of 1952, Kaj Linderstrøm-Lang at the Carlsberg Laboratory in Copenhagen invited Otto Warburg as a special guest lecturer at Carlsberg. After all, thanks to the great Warburg, Denmark had been fortunate enough to claim two great biologists, Albert Fischer, a pioneering tissue culturist, and Fritz Lipmann, a leading genius in biochemistry, both of whom he inspired. Besides, Hans Klenow and I had been Otto Warburg's personal guests in Berlin-Dahlen in 1950. Warburg was pleased to accept the invitation from the Carlsberg Laboratory in the spring of 1952. He gave two lectures, concentrating on his work on crystalline metabolic enzymes. Lang was impressed by the depth and style of the Warburg lectures, which barely consumed one hour. (Fortunately, photosynthesis was not one of the topics selected, since Warburg was disinclined to discuss quantum yields less than perfect, namely one!, i.e. 1.0!)

After delivering a second lecture on the kinetics of the crystalline enolase for a wider circle of Danish scholars, Warburg expressed a warm interest in seeing more of Denmark. There was, of course, a variety of choices, but to my mind the superior one was Elsinore and Hamlet's Castle. My good friend from Pacific Grove (a fellow student in C. B. van Niel's second microbiology course in 1939), Andrew Benson, happened to be a visitor in Copenhagen, and he joined Warburg and myself on the little excursion to Elsinore.

The excursion turned out better than anticipated. It was a crystal-clear, windy, and sunny day. Warburg seemed to like the scenery very much. In fact he seemed moved to open up and ask me, "Why do I encounter so much alienation from British and American biochemists?" I tried to answer this difficult question by quoting his wise and warm admirer, Hans H. Weber, who, like so many of us, treasured the monumental Warburgian prose, but occasionally worried about some of his opinionated footnotes. "Be honest with me," Warburg said, "give me one example." "O.k.," I said. "Why did you call the late Sir Frederic Gowland Hopkins a Romantic?" "Did I really?"

Warburg said. "Yes, in one of your footnotes," I answered. Warburg assured me that he had always been an admirer of Gowland Hopkins, "The generous 'Dean' of British Biochemistry" and the former President of the Royal Society, who undoubtedly had voted for Warburg as a Foreign Associate in 1934. In any case, Warburg's open confession was remarkable; perhaps Hamlet's castle played its part. I believe that I transmitted this Warburgian adventure to Carl and Gerty Cori, old friends of Warburg. It might also interest other believers in the Warburgian enzymological approach, such as Arthur Kornberg and Eugene Kennedy.

## The Galactose Pathway and UDP-Glucose

By 1951, Leloir and coworkers in Buenos Aires had already described the complex nature of the biochemical conversion of glucose phosphate to galactose phosphate. This seemingly so simple positional "swap" of a hydroxyl of the C-4 group of glucose with the hydrogen on the same carbon, thus forming galactose, was found to require at least one coenzyme, UDPGlc or UDPGal. We suggested that such nucleotides might be formed by an interaction between galactose-1-phosphate (Gal-1-P) and UDPGlc, and by a transfer between Glc-1-P and UTP, as summarized in Table 1. In 1952 we initiated our studies on the complex galactose metabolism in eukaryotic cells. These studies took place in our research unit of cell physiology at the University of Copenhagen. Along with Agnete Munch-Petersen and visiting research fellows, among them Evelyn Smith-Mills from Scotland, we conducted studies on the enzymatic interactions of UTP, UDPGlc, Glc-1-P, and Gal-1-P. The enzymes involved were prepared from the same type of yeast as the one used

**Table 1**   The galactose pathway towards glucose metabolites

| Reaction | Enzyme | EC number |
|---|---|---|
| 1. Gal + ATP → Gal-1-P + ADP | Galactokinase | 2.7.1.6 |
| 2. Gal-1-P + UDPGlc ↔ UDPGal + Glc-1-P | P-Gal-uridylyl transferase | 2.7.7.12 |
| 3. UDPGal ↔ UDPGlc | UDPGal-4-epimerase | 5.1.3.2 |
| Sum: | | |
| Gal + ATP → Glc-1-P + ADP | | |
| UDP Gal formation from glucose | | |
| 4. Glc + ATP → Glc-6-P + ADP | Hexokinase | 2.7.1.1 |
| 5. Glc-6-P ↔ Glc-1-P | Phosphoglucomutase | 5.4.2.2 |
| 6. Glc-1-P + UTP ↔ UDPGlc + PPi | PP-uridylyl transferase | 2.7.7.9 |
| 7. UDPGlc ↔ UDPGal[a] | UDPGal-4-epimerase | 5.1.3.2 |
| Sum: | | |
| Glc + ATP + UTP → UDPGal + ADP + PPi | | |

[a] UDPGal is donor of galactosyl units to various low- or high-molecular-weight acceptors; the transfer is catalyzed by various galactosyl transferases.
Sources: (20, 21)

by the Leloir group, i.e. the yeast *Saccharomyces fragilis* preinduced by galactose. We were able to detect two types of "uridyl transferase" in yeast:

1. Gal-1-P + UDPG $\leftrightarrow$ Glc-1-P + UDPGal (Gal-1-P uridyl transferase) (Ref. 18)

2. Glc-1-P + UTP $\leftrightarrow$ UDPGlc + PP (Pyro P uridyl transferase) (Ref. 19)

## Description of a Human Enzyme Defect

In 1953 my contact with biochemical research in the United States was renewed in, let me stress, a very creative way via "The Visiting Scientists Program" at the National Institutes of Health in Bethesda. Thanks to Dr. Bernard L. Horecker, chief of research at the "National Institute of Metabolic Diseases," as it was called in 1953, I received a generous invitation to conduct basic medical research in a field of biochemistry of my own choice. In 1954, I had not only Horecker's encouragement and support but also that of Dr. Hans De Witt Stetten, incoming Institute director. Stetten, as well as Horecker, soon became my trusted friends during these years.

At NIH I was able to pursue further research on the enzymology of the UDP-glucose system, thanks to an enlightened and active group of biochemists and pharmacologists there. I joined forces with Drs. Jack L. Strominger and Elizabeth S. Maxwell. For the first opus we had Dr. Julius Axelrod (from the National Institute of Mental Health) as an additional participant. The joint papers described a novel dehydrogenase pathway from liver by which UDP-glucose (UDPG), through a two-step dehydrogenation of the free 6-hydroxyl group of the glucosyl moiety, could be converted to UDP-glucuronic acid ("active glucoronic acid"). Here I merely emphasize a few aspects of this reaction in connection with our ongoing studies of galactose metabolism. UDP-glucose dehydrogenase turned out to be highly specific for UDP-glucose; it required NAD as a coenzyme and was inhibited strongly by NADH. In my further studies on galactose metabolism, I used this specific and sensitive spectrophotometric enzyme test for UDPGlc, catalyzing the reduction of 2 moles of NAD per UDPGlc oxidized (i.e. UDPGlc + 2 NAD $\rightarrow$ UDP-glucuronic acid + 2 NADH). In human hemolysates, however, UDPGlc was only consumed if Gal-1-P also was present.

Several research groups had clearly demonstrated the presence of active galactokinase in hemolysates from normal as well as galactosemic subjects, i.e. incubation of galactose and ATP brought about a marked accumulation of Gal-1-P. The view on galactosemia in 1952 seemed to center around defects in the so-called "Galactowaldenase," now named "epimerase," which supposedly was needed to convert Gal-1-P to Glc-1-P. The latter was sketched in an oversimplified way,

Gal-1-P↔Glc-1-P,

without taking into account whether UDPGlc was involved or not.

My own view on the question of the defective step in galactosemia differed decisively from this formulation, and for several reasons. Some of my reservations arose from numerous reported clinical observations on galactosemic infants. The development of toxic symptoms, such as cataracts of the lens or mental retardation, did not develop in galactosemic infants, provided they were administered, as early as possible, a lactose-free diet. Soybean formulas, containing sucrose instead of lactose, serve as a telling example. In short, galactosemic infants kept on a galactose-free diet seem to have an excellent chance of normal development.

It was known that the central nervous system, and especially the mammalian brain, contains large amounts of galactolipids and that almost half of the galactose is deposited after birth, i.e. converted from UDPG to UDPGal. Yet, normal brain function is actually better preserved in galactosemic infants kept on galactose-free diets. A defect of the enzyme Gal-1-P uridyl transferase would not interfere with the synthesis of galactolipids, even on a galactose-free diet. In this case a normally functioning 4-epimerase would secure a normal conversion from glucose via UDPGlc to UDPGal and hence an unhindered formation of galactosyl compounds from glucose (Reactions 4–7, Table 1).

The disease symptoms observable in galactosemic infants fed lactose may rather be due to a "traffic jam," i.e. an accumulation of Gal-1-P that is unable to be converted to UDPGal due to the absence of the enzyme Gal-1-P uridyl transferase. Our first priority, therefore, was to compare the levels of the Gal-1-P uridyl transferase enzyme levels to hemolysates from galactosemic and nongalactosemic subjects.

Although galactosemia is a rare disease, it was possible for us, thanks to the skill and efforts of Dr. Kurt Isselbacher (with additional help from Dr. Josephine Kety and Dr. Elisabeth P. Anderson), to acquire blood samples from eight subjects with galactosemia (and more than a dozen subjects who were either normal or afflicted with unrelated diseases). We found that the galactosemic patients specifically lacked Gal-1-P uridyl transferase (20, 21) but did have normal levels of the other uridyl transferase, which reacts with pyrophosphate.[2]

We wanted to ascertain that the type of galactosemia that we had been studying was an exclusive defect in the enzyme Gal-1-P uridyl transferase, and did not affect UDPGal-4-epimerase. The absence of the latter in hemoly-

---

[2]This was further corroborated by Dr. Kurt Isselbacher in tests on white blood cells and on a few trace amounts of liver biopsies.

**Table 2**   Quantitative comparison of activities of enzymes of galactose metabolism in three classes of subjects

| Classes of subjects | Average activities, $\mu$M conversion/hour/gram cells (37°C) (lysates of erythrocytes) | | | |
|---|---|---|---|---|
| | Galactokinase | Gal-1-P uridyl transferase | UDP-Gal-4-epimerase | G-1-P, PP-uridyl transferase |
| Nongalactosemics | 0·10 | 4·8 | 0·32 | >10·0 |
| Galactosemics | 0·08 | <0·02 | 0·35 | >10·0 |
| Galactosemic carriers | | 2·9 | | |

sates from galactosemic as well as normal subjects was clarified through the following sequence of events in our laboratory, all within the same lucky year of 1956. Dr. Elizabeth Maxwell in our group was able to purify UDPGal-4-epimerase (abbreviated: epimerase) from calf liver. It turned out that the purified epimerase required NAD for its activity and was strongly inhibited by NADH. The absence of epimerase activity in normal as well as in galactosemic hemolysates in earlier experiments was due to an unanticipated need for addition of NAD to the hemolysates in order to overcome the inhibition of the NADH present in all human hemolysates (22).

In 1957 Dr. H. N. Kirkman joined our group. Kirkman was an MD interested in biochemistry. As a pediatrician, he was also interested in the field of inborn errors of metabolism in humans. He developed improved methods for studying the kinetics of the enzyme Gal-1-P uridyl transferase. This enabled him to demonstrate partial enzyme defects in galactosemic carriers in siblings or parents of galactosemic subjects (totaling 15 carriers). Table 2 summarizes these records as well as records from nongalactosemic families (23, 24[3]).

Liz Maxwell and I decided that the chemical name "Waldenase" should not be used anymore in the context of the enzymatic reversible conversion of UDP-galactose to UDP-glucose. The enzyme catalyzing this reaction should be called "UDPGal-UDPG epimerase" (25).

## Gal *Mutants in* E. coli

In 1957 Dr. Kiyoshi Kurahashi and I were lucky enough to obtain a valuable gift from Dr. E. M. Lederberg: some samples from the Morse-Lederberg collection of *E. coli* galactose mutants. Kurahashi and an able French microbiologist, Dr. Huguette de Robichon-Szulmajster (from Gif sur Yvette), were

[3]Reports from Great Britain and Switzerland about a human hereditary UDP-Gal epimerase defect that does not respond favorably to a galactose-free diet were recorded around 1970.

able to classify the *Gal* mutants and to study their physiology just before an international biochemistry meeting in Tokyo, Japan. Some of the *Gal* mutants were highly defective in the enzyme Gal-1-P uridyl transferase, the type we had singled out as the defective enzyme in the children afflicted with the "inborn error, galactosemia." Indeed, the *E. coli* with a defective Gal-1-P uridyl transferase enzyme, if grown in media containing galactose, turned out to lag in rate of growth. The effect of galactose on this mutant was especially noteworthy when the growth rate of the mutant was monitored in a minimum medium, such as ammonia mineral glycerol medium, with or without galactose. In the presence of galactose, growth was very slow and was soon arrested (26).

## *Japan—India—Israel*

At the International Symposium on Enzyme Research in Tokyo, September–October 1957, I had the opportunity to present our joint work (H.M.K. Huguette de Robichon-Szulmajster, and Kiyoshi Kurahashi) on mutants of Gal-1-P uridylyl transferase in man and microorganism (46). After the symposium I flew to India, where Professor J. B. S. Haldane, now a resident of India, had invited me to stay for a week at the Indian Statistical Institute.

Professor Haldane received me at the airport and brought me safely out of Calcutta to the Statistical Institute. Haldane was greatly interested in our work on hereditary galactosemia. His own concern about galactose evolved around the consumption of the so-called wrinkled peas, which are stuffed with a galactose-containing trisaccharide, stachyose (containing fructose, glucose, and galactose). Could this type of undeveloped pea (which Mendel had discovered to be recessive) prove to be toxic? What would happen if stachyose peas were offered to a galactosemic subject? Later the verdict came from a friend, Professor G. Semenza in Zürich: Because the human intestine is unable to hydrolyze stachyose, galactose would not be released.

Haldane had a broad view of the diversities of evolution. One of his favorite illustrations was Alkaptonuria. "All insects have Alkaptonuria. They have phenolic acid around which they use to tar and blacken their cuticle. Alkaptonuria in man merely results in black ear cartilages."

After my visit with Haldane, I stopped at several places in mid-India (Sarnath, Delhi, and Bombay). I was invited to Thrombay by Professor Bhaba, head of the Atomic station. Bhaba was a friend of my brother, Fritz Kalckar, from the Niels Bohr Institute in Copenhagen.

Finally, my last invitation was to The Weizmann Institute of Sciences with Drs. Ephraim Katchalski and Michael Sela. They were then both heavily engaged in work on a novel type of immunology. They were making synthetic polypeptides, conferring them with specific antigenic markers, like DNP, from which, after injection into rabbits, they could harvest specific antibodies to DNP. My friend, Dr. Edgar Haber, happened to be visiting Michael Sela at the same time.

## Johns Hopkins University Years

In 1958 I was invited by Dr. William D. McElroy, Scientific Director of The McCollum Pratt Laboratories of Biochemistry at Johns Hopkins University, to assume a full professorship. Among the advanced graduate students, I had the great luck to have Michael Yarmolinsky and Elke Jordan join me, and we were all enthusiastic to have Dr. K. Kurahashi join us too.

We developed methods to analyze the levels of the enzymes of the "galactose sequence." We were puzzled by the fact that the galacto kinase-negative mutant (*Gal-2* mutant) invariably also showed high, "constitutive," levels of the next enzyme of the Gal sequence, i.e. Gal-1-P uridyl transferase ("Gal-3"). In contrast, if galactokinase was operating normally in the *E. coli,* the Gal-1-P uridyltransferase levels were low and depended on induction from outside, i.e. by addition of galactose to the growth medium, in order to increase the levels of both enzymes (27, 28). This involved response by the *E. coli* Gal sequence continued to occupy my mind. Ten years later, it eventually awakened my great interest in galactose chemotaxis (discussed below).

In April 1959 I received a warm personal congratulation from Johns Hopkins University President Milton Eisenhower. I had been elected to be a Member of the US National Academy of Sciences (NAS) in Washington, D.C. Curiously enough, I was the only biochemist elected. It was said that the Academy President, Detlev Bronk, an outstanding biophysicist in his own right, was hesitant to crowd the Academy with too many biochemists. I definitely felt that Carl and Gerty Cori had convinced Detlev Bronk that I was a deserving choice. Perhaps my academicianship in the Royal Danish Academy of 1948 was a contributing factor in my favor.

Be that as it may, a call was extended for a young academician to participate in a scientific exchange between the NAS and the USSR Academy NAYK. I volunteered immediately and was accepted. Hans Stetten at NIH encouraged me to tell the Russians part of our hereditary galactosemia story. Dr. Bentley Glass at Hopkins (also elected an NAS member) supported this idea. Thanks to the skill and culture of my friend Dr. Simon Black, this became a reality. Black had taught himself Russian, and when in Russia travelled without a guide on modest third-class trains with native Russians. I arrived in Moscow with his Russian slides ($2\frac{1}{2}'' \times 3\frac{1}{4}''$) in my pocket for my lecture there.

## My Solo Visit to the USSR

On June 1, 1959, I flew to Paris and from there flew with "Aeroflot" directly to Moscow; the jet was not more than half filled. In Moscow I was met by a young Scandinavian-looking student from Akademy NAYK, Michkael Kritsky, who turned out to be an interesting and attractive travel companion. Kritsky was a 24-year-old graduate student in biochemistry at one of the academy institutes in Moscow under Prof. Belozorsky, a highly regarded biochemist in the field of nucleic acids, whom I met later.

At the Bach Institute, I had the opportunity to meet several distinguished and active biochemists in my own field. Among them was Wladimir A. Engelhardt, a grand old friend and scholar, whom I had recently seen at Johns Hopkins University during his visit to the Biology Department there. Engelhardt was studying membrane ATP-ases and cation transport. In a few months he was due to move to a new academy institute. His wife, M. N. Lyubimova, was studying ATP-ases in various filamentous proteins—spermosin, mimosin (from mimosa plants), and filaments in trypanosomes; all these filamentous proteins seem to contract when exposed to ATP. My lecture, entitled "Nasledstvennaya Galacktotsemia," was to take place in Academician A. L. Oparin's division. Oparin presided and Dr. V. Gorkin translated paragraph by paragraph into Russian. In this way almost two hours elapsed, in spite of the Russian slides, which I duly presented. The audience included junior as well as senior biologists and biochemists (Engelhardt, among the latter). The topic of inborn errors of metabolism in man seemed to interest the audience greatly.

I had arranged for a 20-minute question period, if necessary. Five minutes turned out to suffice, since only four questions were raised, either philosophical or physiological. I then decided to give a brief summary about the X-linked Glc-6-P dehydrogenase defects. The audience was surprised to learn about the high incidence in their own country of Favism (especially among oriental Jews). This is a sex-linked hereditary Glc-6-P dehydrogenase defect in red blood cells that leads to hemolytic anemia.

During this visit to Moscow, I had the pleasure to meet the great biochemist and polymer chemist, Ephraim K. Katchalski, from Israel. Our subsequent friendship resulted in several invitations to visit The Weizmann Institute of Sciences in Israel. I did not participate actively in the polymer symposium in Moscow. I learned, however, that this discipline was considered important in the USSR, and that during the Krushchev regime at the time, Jewish scientists were held in high esteem, several of whom, such as N. N. Semenov, occupied leading positions.

My guide Kritsky and I then travelled by train to Leningrad (the night express was called "Red Arrow" and was comfortable). The light nights in Leningrad encouraged us to take long evening walks near the Neva. The gracious city had a spirit of European urbanity. At the Hermitage there was an exhibition of the ancient Crimean Gold. Yet, for me, the unforgettable sight there was of the French paintings, culminating with the magnificent painting by Henri Matisse, "Les Danseuses."

From Leningrad a jet carried us nonstop to Kiev. My main interest there was to meet V. A. Belitser, who, as I have mentioned, in 1939 first described in detail a reliable system that generated oxidative phosphorylation. I later received permission from the USSR reprint office to have his 1939 article with Tsibakova translated to English and reprinted in my monograph of 1969. My immediate problem in 1959, however, was to find Belitser. I did find him. He was most cordial and seemed to enjoy very much to meet me and to be photographed with me (see photo of both of us on p. 7). We spent about half an hour together. I then travelled on to Armenia (see Ref. 46 for my further trip to Armenia and Yerevan).

The year 1959 was, in a way, with the U-2 incident, a tragic year. President Eisenhower and his wise advisor, George Kistiakovsky of Harvard, were constantly fed lies by the generals in the "Defense"—nay! "War"—Department. Later, back in Moscow at the International Congress of Biochemistry in 1961, the mood was mixed. In all the turmoil, a few of us were lucky enough to hear Marshall Nirenberg disclose publicly the first breaking of the genetic code. UUU stands for phenylalanine. My friend and gentle humorist, the biochemist, Rollin Hotchkiss, coined the moving event in 1961 "the U-3 incident."

## Massachusetts General Hospital

In 1960 I received a distinguished invitation that pleased me very much from Dr. Paul C. Zamecnik, chairman of The Huntington Laboratories at Massachusetts General Hospital (MGH) and Professor of Medicine at Harvard Medical School (HMS), and the new chairman of The Department of Biological Chemistry at HMS, Dr. Eugene P. Kennedy. Professor Carl Cori, who had become an adviser to MGH for the last couple of years, added his voice in strong support. The invitation was to fill the position left open since Fritz Lipmann had accepted the Professorship at the Rockefeller Institute in 1957. Dr. Oliver Cope, chairman and Professor of Surgery at MGH, and Dr. George Packer Berry, Dean of Harvard Medical School, had carefully looked into the possibility of financing and restoring a new professorship at MGH. Moreover, thanks to Dr. Walter Bauer, Professor of Medicine at HMS and MGH, and his friendship with Sir Henry Dale, chairman of The Wellcome Trust in London, further generous support of the new professorship was offered from London. I was pleased and honored to accept this new professorship.

In Boston, I met two old friends from the Carlsberg Lab in Copenhagen, Mahlon Hoagland and Paul Plesner, and their guest, a most exciting, beautiful lady with serene poise, who also came from Copenhagen. Thirty years ago I had met and admired Agnete Fridericia. Agnete had lost her husband, Svend Laursen, in 1960. Her friends and colleagues respected and admired more than ever Agnete's courage in the face of the agony, and her ability to get through it with her two young children. Agnete's and my cheerful conversations during the 1960s, most of it in Danish, changed my life. We married in 1968.

BACTERIAL MUTANTS AFFECTING THE TOPOGRAPHY OF THE CELL WALL
In starting my research at MGH, I was lucky enough, through the encouragement of Dr. Roger Jeanloz, head of the Carbohydrate Unit at MGH, to have Dr. A. M. Rapin join our unit in early 1962. Moreover, Dr. T. A. Sundararajan from India also joined my unit at the same time. Already this joint work opened up the discovery of a new type of *E. coli* Gal mutant, which we called *Gal 23*. Because *Gal 23* contained both K. T. and E. enzymes, we wondered whether the new Gal mutant might be unable to synthesize UDPG (i.e. UDPG pyrophosphorylase). We were able to substantiate this idea. We found that induction with galactose (or D-fucose) in minimum medium or in broth failed to increase the low levels of UDPG synthetase (29).

At the same time and completely independently, Fukasawa, Jokura & Kurahashi in Tokyo also isolated an *E. coli* mutant unable to synthesise UDP-glucose. Dr. Hiroshi Nikaido, who had recently joined our lab at MGH, became a guiding spirit in the study of bacterial cell walls. Nikaido analyzed the lipopolysaccharides in *Salmonella enterides,* the so-called O-antigens. From his collaboration with Fukasawa on the UDP-galactose-epimerase-less *Salmonella* mutants, and adopting the concept of a core and side chains and the data from the epimerase-less mutants, Nikaido invoked a theory in which galactose forms a link between the core of the side chains. Missing the galactose link elicited the lack of incorporation of a number of core sugars, as well as those of the side chain (i.e. mannose, rhamnose, tyvelose, etc) (30). Clearly, Nikaido was a great observer. I was fortunate to have him visit our lab and to be able to obtain funds for him as an independent researcher in our department at MGH.

LEARNING PROTEIN BIOCHEMISTRY AND BIOPHYSICS IN AN UNANTICIPATED WAY   In 1958, E. Maxwell, R. Szulmajster, and I had observed that a highly purified yeast enzyme, UDP galactose epimerase with NAD(H) bound to the enzyme, showed a conspicuously intense blue fluorescence. In our lab at MGH, Robert A. Darrow and his coworkers found that purified epimerase from yeast is asymmetrical in the sense that one dimer enzyme protein carries only one bound NAD, not two bound NAD (31, 32). At about the same time, Creighton & Yanofsky revealed a related type of asymmetry in tryptophan synthetase. Later, when Dr. Othmar Gabriel joined our lab, he found a related type of asymmetry in *E. coli* TDPG oxidoreductase (33).

We had studied the effects of NAD and NADH on the conformation of alcohol dehydrogenase at the Carlsberg Laboratory, where we found that the addition of NAD or NADH exerted a tightening effect on the apo-enzyme in terms of changes in optical rotatory dispersion (ORD) and circular dichroism (CD). In our joint work, Professor Martin Ottesen had recommended additional studies of conformational changes of the apo-enzyme, such as tritium exchange methods combined with sephedex fractionation.

A. U. Bertland studied the changes in fluorescence parameters as well as changes in ORD and CD. He found that if the purified yeast epimerase was incubated with 5' UMP and D-galactose or 5' UMP and L-arabinose for 2 to 3 hours, a 30-fold fluorescence enhancement (based on the mol. fluorescence of free NADH) ensued. The "superfluorescent" epimerase lost more than 90% of the catalytic activity (34). A quantum yield of 3% for the fluorescence of free NADH and an enhancement factor of 30 for the epimerase-bound [NADH] would amount to a quantum yield of nearly 90% for the bound NAD (35). The fluorescence of the bound [NADH] showed high polarization, not only through the excitation band (3400 Å) to the emission band (4400 Å), but also from the aromatic region (2800 Å). Since the S. fragiles epimerase contains 40 tyrosines but only 3–4 tryptophans per dimer (see summary by Kalckar et al 1969), the energy transfer from the tyrosine rings to the reduced nicotine amide ring is remarkably high.

In the discussion at the Copenhagen Symposium (cf. Kalckar et al. 1969), Dr. Howard Schachman had noted that one of the slides carried vector signs and wondered whether we were also following the polarization of fluorescence. I was pleased to answer his question affirmatively. Bertland had indeed been studying the polarization of fluorescence and did find polarization of fluorescence from the aromatic band as well as from the excitation and emission bands (36). The weird device by which Bertland managed to obtain catalytically active epimerase (NAD) with quenched fluorescence was also described at the Symposium. Prolonged storage at high dilution (0.1–0.3 mg protein per ml buffer) would suffice to accomplish the transformation. I should add that epimerase reduced by sodium borohydride did not respond in this way but remained generally unresponsive as NADH, which remained catalytically inactive. As mentioned above, the reductive inhibition of yeast epimerase by $NaBH_4$ is only short-lived. Bertland, Y. Seyama, and I found, however, that the further addition of 5'UMP would transform it into a long-lived, reductive inhibition (37).

A HOME-SPUN TRANSITION TO GAL CHEMOTAXIS    In 1964, Henry C. P. Wu approached me regarding graduate work to obtain a PhD degree, which he eventually achieved in 1969. In 1966, another brilliant young biochemist, Dr. Winfried Boos from the University of Konstanz, Bodensee, also joined our laboratory.

We again concentrated our efforts on trying to understand the regulation of the Gal operon. To make a long story short, *GalK,i* releases galactose from its cellular UDPGal and seems unable to recapture it from the medium. In contrast, *GalK,c* seems to possess a galactose "scavenger mechanism" sufficiently effective to maintain the induced state of *GalK,c*.

Winfried Boos then developed biochemical methods to isolate the Gal scavenger from either of the GalK strains. By a skillful use of polyacrylamide gel electrophoresis of cell extracts, obtained by osmotic shock devices, he found that the fractionated shock fluid from *GalK,c* showed a specific high-affinity Gal-binding protein, whereas shock fluid from *GalK,i* did not. The Gal-binding protein was considered a periplasmic protein, residing in the space between the outer and inner membranes (38).

Shortly afterwards, while listening to a fascinating lecture by Julius Adler from the University of Wisconsin on galactose chemotaxis in various *E. coli* mutants, I became excited by certain parallels. Adler had found a galactose chemotactic receptor located in the periplasmic space of *E. coli*. Chemotaxis was then only partly characterized, but it was suspected to be related to one of the most primitive sensory responses known in nature.

I decided that Boos and Adler should exchange the Gal mutants that had scored positive (+) or negative (−) in their respective tests. For what purpose? I had exhilarated myself with the notion that the *GalK,c* strain would score positive in Adler's chemotactic test and that the *GalK,i* strain would score negative in the chemotactic test. Unfortunately, due to my habitual type of "daydreaming," as Boos used to call it, my monomaniacal mumbling, "soliloquizing," was obviously not deciphered by either Adler or Boos. Finally, after a good deal of cajolery, both friends began to understand me. They did finally exchange their bacterial mutants for executing their respective tests. The finding was in fact that only the strain with high-affinity Gal binders *(GalK,c)* scored positive in Gal chemotaxis, and this was also reflected in Boos's tests of Gal binding versus Gal chemotaxis. The receptors in Gal chemotaxis seemed indeed identical with the Gal binder.

As Boos mused in his generous and thoughtful chapter in the "Festschrift" to my seventy-fifth birthday, "I feel obliged to tell you on this occasion that it was really Herman Kalckar who first perceived the idea that a transport-related binding protein is chemoreceptor in *E. coli*, even though the published record may suggest otherwise." In an article in *Scientific American*, Adler (39) also recalled our fruitful encounter of 1970.

Since one of my favorite pupils, Dr. Paul Plesner, recently challenged me to try to express some of my philosophical thoughts regarding my long trot along the trail of biochemistry, I shall try to construct a philosophical parable to the story that I have just finished. The grand old British biochemist, Sir Frederic Gowland Hopkins, ventured to answer a question posed to him through the philosophy of Alfred North Whitehead, "Has the modern biochemist, in analyzing the organism into parts, so departed from reality that his studies no longer have biological meaning?" In a dialogue with himself, Hopkins wrote: "So long as his analysis involves the isolation of events, and not merely of substances, he is not in danger of such departure." Boos, Wu, Adler, and I actually managed in our own way, somehow, to make a transition from biochemical analysis to cellular events, like endogenous induction and sugar chemotaxis. Our case, that a transport-related protein might also be used as a chemoreceptor, could only be postulated on account of the recognized specificity of mutational steps, selecting mutants by two widely different criteria. Adler and his group had been seeking mutants in chemotaxis, while we were looking for mutants in galactose metabolism. Now both fields came together. This makes it feasible to "isolate" two different types of events and thus to disclose one of the capricious ways of the economy of Nature.

My old friend, the late Max Delbrück, who had always teased me about my mumbling, seemed to have enjoyed the mutual Gal taxis adventure, and he invited me in 1970 to give a lecture at Cal Tech, Biology Division. This happened also to be the first "Jean Weigle Lecture," which was later published in *Science* (40).

TUMOR BIOLOGY AND HEXOSE TRANSPORT REGULATION     Some time during my stay at MGH and Harvard Medical School, I became captivated by a new important field that had originated through observations on nutrient transport into tumor cell cultures. It was possible to transform a fibroblast culture into a sarcoma culture by exposure to a virus. This was first explored on amino acid transport by Foster & Pardee in 1969 and by Isselbacher in 1972 on glucose as well as on amino acids.

My own interest in tumor biology was aroused in 1973, and I was lucky to be invited to visit Dr. S. Hakomori at the University of Washington, Seattle, who had initiated active studies on tumor biology. We studied the biology of glucose and galactose transport in nontransformed and polyoma-transformed hamster cell cultures. We found enhancement of transport, of glucose as well as galactose, in the transformed cultures (41), and also morphological changes in the transformed cultures. These changes may well have represented the loss of a surface protein, as described in transformed cells by R. Hynes the same year. The surface protein that is present in normal cultures Hynes named "fibronectin."

The same year my interest in general regulation of hexose transport was alerted by some exciting observations by Harold Amos and his coworkers in the Microbiology Department at Harvard Medical School. Their studies focused on regulation of glucose transport into primary chick embryo fibroblasts. They found persistently that depriving normal chick embryo fibroblasts (CEF) of glucose brought about a dramatic enhancement of the rate of influx of glucose, as compared with tests on glucose-fed cultures. As Dr. Amos was happy about my enthusiasm, and he was as always concerned with the future of his young coworkers, he asked Dr. C. W. Christopher if he would be interested to join my unit at MGH. At that time, although I was retired from Harvard, I was fortunate to have a unit in Dr. Zamecnik's department and to be well funded. Moreover, I had my able coworker, Donna B. Ullrey, in my unit, and so in 1975 we had the pleasure of having Christopher join us.

Hexose uptake by hamster fibroblast cell cultures was increased 5–10-fold by either substituting D-fructose for glucose or by completely omitting D-glucose from the culture medium for 24 hours. Conversely, when cyclohex-imide was present for 24 hours in media containing glucose, up to 20-fold decreases in hexose uptake were observed. These decreases in uptake activity were only observed over a narrow range of cycloheximide (CHX) con-centrations, however. After extended exposure to low concentration of c CHX (0.05 to 10 $\mu$g/ml), the uptake by the fed cells decreased in parallel with inhibition of protein synthesis, whereas at high concentrations of CHX (>50 $\mu$g/ml), uptake was increased. The effect of CHX on the uptake mechanism was found to be biphasic with respect to the concentration of CHX (42). Cells deprived of glucose and maintained in the presence of cycloheximide did not show decreases in uptake activity. In separate experiments, the high uptake rates of glucose-starved cells could be decreased by the addition of glucose to the glucose-free medium. The reversal was complete in 6 to 8 hours. The analog of glucose, 2-deoxy-D-glucose, did not promote the time-dependent decrease, thus suggesting that the 6-phosphoester of glucose is not an in-hibitor of transport. In addition, when cycloheximide was added at the same time as glucose, there was no decrease in uptake for at least 12 hours. It was proposed that the turnover of components of hexose uptake systems could account for part of the control of hexose transport.

## Japan 1977

Once more, in September–October 1977, my wife and I were the guests of the hospitable Japan Science Council. Dr. Yousuko Seyama, who had been in my laboratory in 1972–1973, was our host in Tokyo, and Dr. Kiyoshi Kurahashi was our host in Osaka. This was also the year we visited Hiroshima. The joint Japanese-American Atom Bomb Committee (to whom I was introduced by a letter from Dr. Alexander Leaf, MGH) invited me to summarize my article on stromtium fallout and my article in *Nature* about collection of children's teeth, a study that was organized by the Dental School of Washington University in St. Louis (see Ref. 46).

A unique contribution to peace was initiated by the president of the University of Hiroshima, Dr. Orata Osada, with the book, *Children of the*

*A-Bomb—A Testament of the Surviving Boys and Girls of Hiroshima.* Dr. Osada, in spite of his own radiation sickness, had managed to collect hundreds of letters from the schoolchildren of Hiroshima. One of the translators of this unique book, Dr. Jean Clark Dan, a marine biologist and a mother of five children, was an old friend of mine and sent me the English translation by her and her friend Ruth S. Morgan. The testaments of the schoolchildren, from grammar school to junior high school, are completely free of any propaganda and only tell what they experienced. It was finally possible, after several years, to get this book distributed through a Canadian publishing house. No US publisher, as far as I know, was willing to distribute the book.

After a few more stops in southern Japan we returned to Tokyo where our friends helped us get ready for our trip to China.

## China 1977

Thanks to the help of Dr. Joseph Needham and to his friendship with Kuo Mo Jo, a prominent Chinese historian, I was invited to Academia Sinica in Peking and Shanghai for two weeks in late November 1977. I lectured in both places.

Just prior to our arrival, the outskirts of Peking had been hit by an earthquake. The epicenter was in Tiensin. A couple of days later, I was invited by the director, Dr. Pei Shih Shang, to visit the Biophysics Institute of Academia Sinica, Peking. Dr. Shang, 77 years old, was a member of the standing committee of the party central committee. He did not speak English, but since he had been in Freiburg, Germany in 1927, he still spoke German, and I had not forgotten my German from Denmark, so our conversations in German went quite well. In view of the prevalence of earthquakes in China one would expect that the Biophysics Institute would have had some seismographic monitors. They did not have such luxuries, however. They used pigeons as monitors, a method they called "bio-seismography." I was told that the feet of pigeons have some sensitive "corpuscles"; these sensors are called the "Herbst sensors." They are supposed to transmit mechanical vibrations into increased spike frequency. I am not sure that I followed the pigeon "tale."

Obviously the large staff had to deal with a lot of biophysical topics such as radioactive isotopes and crystallography especially of insulin crystals, which they received from Shanghai Institute of Biochemistry. The equipment was protected against the late November chill, but the director and the senior or junior scholars had merely their old overcoats on.

After a week in Peking we flew to Shanghai where I visited the Shanghai Institute of Biochemistry, Academia Sinica. The director, Wang Ying-lai, was born in 1907 in Fukian Province, China. After graduating from Nanking University, he earned his PhD at Cambridge University in 1941. He remained in Cambridge for four years of postdoctoral research on cytochromes with

David Keilin at the Molteno Institute. After his return to China, he published in the 1950s procedures for the isolation of succinic dehydrogenase, which received world-wide use. In 1958 he was appointed director of the newly founded Institute of Biochemistry of the Academia Sinica in Shanghai. It was under his administration that the Institute undertook the major effort in basic research that culminated in the total synthesis of insulin. Wang Ying-lai has represented the Chinese biochemists internationally and was a visitor in the United States in 1975.

With the changes in administration that accompanied the Cultural Revolution, Wang was appointed the senior Vice-Chairman of the Revolutionary Committee of the Shanghai Institute of Biochemistry. At the time fundamental studies on peptide synthesis included research on the structure-function relationships of insulin and methods of synthesis of oxytocin, vasopressin, and glucagon. Selected publications were credited to groups rather than to individual authors: *The Total Synthesis of Crystalline Insulin,* Institute of Biochemistry (Shanghai), Institute of Organic Chemistry (Shanghai), and Department of Chemistry.

I was lucky to attend a biochemistry meeting in which Professor Wang requested one of his coworkers, Dr. Pan Xiu Jia, to report on an unusual hormone project undertaken by the biochemistry lab on the behalf of the fresh-water fish farmers. Dr. Pan gave her report in Chinese, which was then translated into English. The first step was to sneak one D-amino-acid, D-alanine, into a nonapeptide of L-amino acids of a hypothalamic hormone luteinizing releasing hormone (LRH). Dr. Pan brought this unique mosaic hormone analogue to the Shanghai cooperative fresh-water fish farm in order to study the possible spawning response of the silver carp at the temperature of the farm, 25°. The injection of a few micrograms of the mosaic D-analogue per kilogram carp elicited a dramatic spawning response that lasted for weeks, since no antibodies against the mosaic hormone were formed. The D–amino acid protected the hormone against hydrolysis.

I also paid a visit to the Shanghai Institute for Physiology and Acupuncture. The Institute tried to map the neural paths of acupuncture response. I found a visit to the Da Hua Hospital most interesting. There I observed the following: Female 20–30 years old—acupuncture for appendectomy: 2–3 needles in right ear (120 vibrations per minute) + 1 needle in right foot (200 vibrations per minute). Male 60–70 years old—acupuncture for gastrectomy: pretreatment with luminal, 4 needles in right ear, and 2 + 2 needles in each foot.

I stayed for 30–45 minutes during these operations, which were surprisingly peaceful. Afterwards I was the only one of ten foreign observers who volunteered to have the acupuncture needle inserted in my hand between the thumb and index finger; with the vibrations it felt as if my hand "grew" bigger and bigger within 10 minutes. This method is used by pregnant women giving birth, according to Ms. Wang, our intelligent guide, who had had it during the birth of her child.

Before we left China, Agnete and I were invited by a lady superintendent of a junior high school in Peking to attend a class. The teacher was a young, spirited lady who taught vector analysis. The usual standard pictures of Mao, Lenin, and Marx were on the wall, but the blackboard was bristling with mathematical formulas. The rapport between students and teacher seemed excellent.

A few days later we left China for Tokyo, and, with the help of our Japanese friends, from there on to United States.

## Boston University

After my second retirement from MGH, my friend and colleague, Dr. John M. Buchanan of MIT, approached his colleague, Dr. Norman Lichtin, Chairman of the Chemistry Department at Boston University Graduate School, who invited me as visiting professor. Thanks to the skill and energy of my assistant, Donna Ullrey, we performed innovative research in biochemistry there for more than ten years.

As mentioned above, feeding D-glucose to Chinese hamster lung fibroblast down-regulates its own transport system. This effect depends on glucose metabolism, since a mutant that lacks phosphoglucose isomerase depends specifically on glucose metabolism (43). In collaboration with Dr. Paul Plesner, we found that the mutant was unable to keep its ATP at normal levels in glucose media, whereas mannose as a hexose source was a competent energy source (44). Later, Ullrey and I found that the rare sugar D-allose is much more effective than glucose in controlling the glucose transport system (45); it is also phosphorylated (46).

## Summary

In 1930 adenosine triphosphate appeared in the literature from W. A. Engelhardt's work on avian erythrocytes. This was an early example of oxidative phosphorylation in intact cells, and it required methylene blue and oxygen. Both Belitser and I realized that the use of Warburg manometers for aeration was critical in order to generate oxidative phosphorylation of glucose in tissue preparations. Test tube techniques did not work.

In 1956 we were able to describe a human type of diabetes called "galactose diabetes," in which consumption of human or cows' milk provokes mental retardation. Replacement of human or cows' milk products with "vegetable milk" formula in early infancy can prevent retardation. We determined that the disease results from a defect of galactose-one-phosphate uridylyltransferase, a hereditary enzyme. This type of enzyme defect, if discovered and treated in early infancy, is a benign molecular disease.

Regulation of transport systems in mammalian cell cultures are frequently complex energized systems. Perhaps my greatest surprise in this regard was the mere fact that an all-cis "odd" hexose-D-allose turned out to be a highly intense down-regulator of the hexose transport system. Additions of inhibitors of oxidative phosphorylation (such as oligomycin or di-nitrophenol) arrested the allose-mediated down-regulation. We have reason to suspect that the strong down-regulator is a phosphorylated form of D-allose.

Thus ends my story about oxidative energized biological phosphorylation systems.

ACKNOWLEDGMENTS

I thank Dr. Rollin D. Hotchkiss, Mrs. Agnete Kalckar, Dr. Eugene Kennedy, and Donna B. Ullrey.

## Literature Cited

1. Kalckar, H. M. 1937. *Enzymologia* 2:47–53
2. Belitser, V. A., Tsibakova, E. T. 1939. *Biokhimya* 4:516–34
3. Kalckar, H. M. 1938. *Nature* 142: 871
4. Kalckar, H. M. 1939. *Biochem. J.* 33:531–41
5. Engelhardt, W. A., Lyubimova, M. N. 1941. *Nature* 144:668–72
6. Kalckar, H. M. 1941. *Chem. Rev.* 28:71–178
7. Colowick, S. P., Kalckar, H. M. 1943. *J. Biol. Chem.* 148:117–26
7a. Kalckar, H. M. 1943. *J. Biol. Chem.* 148:127–37
8. Kalckar, H. M. 1945. *Fed. Proc.* 4:248–52
9. Lowry, O. H., Lopez, J. 1946. *J. Biol. Chem.* 162:421–28
10. Kalckar, H. M., Klenow, H., Kjeldgaard, N. O. 1950. *Biochem. Biophys. Acta* 5:575–85
11. Friedkin, M., Kalckar, H. M. 1950. *J. Biol. Chem.* 184:437–48
12. Friedkin, M. 1950. *J. Biol. Chem.* 184:449–59
13. Hoff-Jørgensen, E., Friedkin, M., Kalckar, H. M. 1950. *J. Biol. Chem.* 184:461–64
14. Clark, V. M., Kalckar, H. M. 1950. *J. Chem. Soc.*, pp. 1029–30
15. McNutt, W. S. 1950. *Biochem. J.* 50:384–97
16. Kalckar, H. M., McNutt, W. S., Hoff-Jørgensen, E. 1952. *Biochem. J.* 50:397–400
17. Watson, J. D., Maaløe, O. 1953. *Biochim. Biophys. Acta* 10:432–42
18. Munch-Petersen, A., Kalckar, H. M., Cutolo, E., Smith, E. B. 1952. *Nature* 172:1936–39
19. Kalckar, H. M., Braganca, B., Munch-Petersen, A. 1952. *Nature* 172:2–4
20. Kalckar, H. M., Anderson, E. P., Isselbacher, K. J. 1956. *Biochim. Biophys. Acta* 20:262–68
21. Kalckar, H. M. 1985. *BioEssays* 3:134–37
22. Maxwell, E. S. 1957. *J. Biol. Chem.* 229:139–51
23. Kirkman, H. N., Kalckar, H. M. 1958. *Ann. NY Acad. Sci.* 75:274–78
24. Kirkman, H. N., Bynum, E. 1959. *Ann. Hum. Genet.* 29:117–26
25. Kalckar, H. M., Maxwell, E. S. 1958. *Physiol. Rev.* 38:77–84
26. Kalckar, H. M., de Robichon-Szulmaister, H., Kurahashi, K. 1958. *Proc. Int. Symp. Enzymol. Chem.*, Tokyo-Kyoto-Maruzen, pp. 52–56
27. Kalckar, H. M., Kurahashi, K., Jordan, E. 1959. *Proc. Natl. Acad. Sci. USA* 45:1775–86
28. Jordan, E., Yarmolinsky, M. B., Kalckar, H. M. 1962. *Proc. Natl. Acad. Sci. USA* 48:32–40
29. Sundararajan, T. A., Rapin, A. M., Kalckar, H. M. 1962. *Proc. Natl. Acad. Sci. USA* 48:2187–93
30. Nikaido, H. 1962. In *The Specificity of Cell Surfaces,* ed. B. D. Davis, L. Warren, pp. 67–71. Englewood Cliffs, NJ: Prentice-Hall
31. Darrow, R. A., Creveling, C. R. 1964. *J. Biol. Chem.* 239:362–63
32. Darrow, R. A., Rodstrom, R. 1968. *Biochemistry* 7:1645–54

33. Wang, S. F., Gabriel, O. 1969. *J. Biol. Chem.* 244:3430
34. Gabriel, O., Kalckar, H. M., Darrow, R. A. 1975. In *Subunit Enzymes*, ed. K. E. Ebner, pp. 85–135. New York: Dekker
35. Bhaduri, A., Christensen, A., Kalckar, H. M. 1965. *Biochem. Biophys. Res. Commun.* 21:632–37
36. Kalckar, H. M., Bertland, A. U., Johansen, J. T., Ottesen, M. 1969. *The Role of Nucleotides for the Function and Conformation of Enzymes. First Benzon Symp. 1968–1969*, ed. H. M. Kalckar, H. Klenow, A. Munch-Petersen, M. Ottesen, J. Hess-Thaysen, pp. 247–75. Copenhagen: Munksgaard
37. Bertland, A. U., Seyama, Y., Kalckar, H. M. 1971. *Biochemistry* 10:1545–51
38. Boos, W. 1969. *Eur. J. Biochem.* 10:66–73
39. Adler, J. 1976. *Sci. Am.* 234:40–42
40. Kalckar, H. M. 1971. *Science* 174:557–65
41. Kalckar, H. M., Ullrey, D. B., Kijomota, S., Hakomori, S. 1973. *Proc. Natl. Acad. Sci. USA* 70:839–43
42. Christopher, C. W., Colby, M., Ullrey, D. B. 1976. *J. Cell Physiol.* 89:684–92
43. Ullrey, D. B., Franchi, A., Pouyssegur, J., Kalckar, H. M. 1982. *Proc. Natl. Acad. Sci. USA* 79:3777–79
44. Plesner, P., Ullrey, D. B., Kalckar, H. M. 1985. *Proc. Natl. Acad. Sci. USA* 82:2761–63
45. Ullrey, D. B., Kalckar, H. M. 1986. *Proc. Natl. Acad. Sci. USA* 83:5858–60
46. Kalckar, H. M. 1990. *Compr. Biochem.: Autobiographical Notes from a Nomadic Biochemist.* 37:101–76

*Annu. Rev. Biochem. 1991. 60:39–71*

# PROTEIN-PRIMING OF DNA REPLICATION

## Margarita Salas

Centro de Biología Molecular (CSIC-UAM), Universidad Autónoma, Canto Blanco, 28049 Madrid, Spain.

KEY WORDS:   adenovirus, bacteriophage φ29, DNA polymerase, initiation of replication, terminal protein.

## CONTENTS

## PERSPECTIVES AND SUMMARY

The fact that all known DNA polymerases require a free 3'-OH group for chain elongation raised the problem of how replication is initiated. A general mechanism to initiate replication is the formation of an RNA primer synthesized by specific enzymes, the primases (1); a primer can be used in the case of circular DNA molecules or linear DNAs that are converted either to circular or concatemeric molecules. An alternative mechanism for the initiation of replication is the specific nicking of one of the strands of a circular double-stranded DNA, producing a free 3'-OH group used for elongation (1). In the case of linear DNAs that remain as such for replication, RNA priming cannot be used for the initiation of replication since, after removal of the RNA, it is not possible to fill the gap that would result at the 5'-ends of the newly synthesized DNA chain. Some linear DNA molecules contain a palindromic nucleotide sequence at the 3'-end which allows the formation of a hairpin structure that provides the 3'-OH group for elongation (1, 2). The finding of specific proteins covalently linked to the 5'-ends of viral linear double-stranded DNAs, the so-called terminal proteins (TP), led to the discovery of a new mechanism for the initiation of replication in which the primer, instead of being the 3'-OH group of a nucleotide provided by RNA or DNA molecules, is the OH group of a serine, threonine, or tyrosine residue of the TP.

The protein-priming mechanism of DNA replication, as unravelled from the development of in vitro replication systems with purified proteins, mainly in the case of adenovirus and bacteriophage $\phi$29 DNAs, is summarized in Figure 1. Specific initiation protein(s) (see Table 1) interact with the origin of replication, giving rise to the unwinding of the double-helix to expose a region of single-stranded DNA. A free molecule of the TP forms a complex with a specific DNA polymerase, and this complex interacts with the origins of replication by recognition of the parental TP and specific sequences at either DNA end. In the presence of the dNTP corresponding to the 5'-termini, the DNA polymerase catalyzes the formation of a covalent bond between the dNMP and the OH group of a specific serine, threonine, or tyrosine in the TP. After this initiation step, the same DNA polymerase catalyzes chain elongation by a strand-displacement mechanism, with the concomitant removal of the initiation protein(s) and the binding of SSB protein to the parental single-strand that is being displaced. Replication is continuous without the need of Okazaki fragments. No other accessory proteins seem to be needed to produce full-length DNA synthesis, except a topoisomerase I activity in the case of adenovirus. Indeed, it has been shown that the $\phi$29 DNA polymerase has helicase-like activity and is able to produce strand displacement.

For the replication of the nontemplate (displaced) strand, two models have

**Table 1**    Proteins involved in the replication of TP-containing DNA viruses

| Virus | Priming protein | DNA polymerase | Initiation proteins | SSB | Topoisomerase |
|---|---|---|---|---|---|
| Adenovirus | pTP | Ad pol | NFI, NFIII | DBP | NFII |
| Phage φ29 | p3 | p2 | p6 | p5 | — |
| Phage PRD1 | p8 | p1 | ? | p12 | — |

been proposed. In one, first presented by Daniell (3), complete displacement of the parental strand is assumed to occur with the formation of a "panhandle" structure by hybridization of the self-complementary terminal sequences that provide a replication origin identical to the one present at the ends of double-stranded DNA. In the other, indicated in Figure 1, before displacement synthesis initiated at one end of the DNA is completed, initiation of replication occurs at the other end, giving rise to the so-called type-I molecules with two single-stranded tails. When the two replication forks meet, separation occurs, producing type-II molecules. Replication is completed from the latter molecules with the concomitant removal of the SSB protein; once full-length synthesis has been achieved, the DNA polymerase will be recycled by joining another TP and starting a new initiation event.

In this review I deal not only with the systems that have been shown to replicate by protein-priming, such as adenovirus and bacteriophages φ29, PRD1, and Cp-1, but also with linear double-stranded DNA plasmids that contain a TP and are likely to replicate by protein-priming. I also discuss the hepadnaviruses that contain TP at the 5' end of one of the strands of the circular DNA, which, it has been suggested, may be involved in a special protein-priming mechanism for reverse transcription. In addition, I review the TP-containing RNA viruses and their possible role in the initiation of replication. Recent reviews are available for adenovirus DNA replication (4–7), bacteriophages φ29, PRD1, and Cp-1 DNA replication (8, 9), eukaryotic linear plasmids (10), hepadnavirus replication (11, 12), and TP-containing RNA viruses (12–14).

## ADENOVIRUS DNA REPLICATION

The human adenovirus DNA genome consists of a linear double-stranded DNA of 35–36 kb with a 103–163-bp-long inverted terminal repetition (ITR) (15–17). The finding of circular molecules and concatemers in the DNA isolated from viral particles, and their conversion into linear, unit-length DNA after protease treatment, suggested the existence of protein at the ends of adenovirus DNA (18). In addition, the infectivity of the adenovirus 5 DNA-protein complex was greatly reduced after treatment with pronase (19).

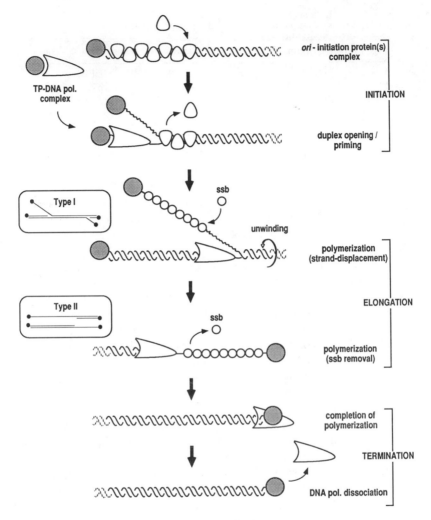

*Figure 1*   Model for protein-priming of DNA replication.

Indeed, a protein of 55,000 daltons was found to be covalently linked at the 5'
ends of adenovirus DNA (20, 21). A model first proposed by Rekosh et al
(20) suggested that a free molecule of the TP could act as a primer for the
initiation of replication by formation of a covalent linkage with the 5' terminal
nucleotide, dCMP, that would provide the 3'OH group needed for elongation
by the DNA polymerase. The TP is synthesized as a precursor of 80,000
daltons, the preterminal protein (pTP) (22). The linkage between both the pTP
and the TP and DNA, readily hydrolyzed in alkali, is a phosphoester bond
between the $\beta$-OH group of a specific serine residue at position 580 of the
pTP and 5' dCMP (22–24).

## Replication in Vivo

Electron microscopic and restriction enzyme analysis (reviewed in 25) of viral replicative intermediates from adenovirus-infected cells indicated that initiation of replication occurs at either end of the molecule and proceeds by a strand-displacement mechanism giving rise to two basic types of replicating molecules. Type-I molecules consisted of unit length linear duplexes with one or more single-stranded branches from the same or different DNA end, and accounted for about 33% of the molecules (26). Type-II replicating molecules consisted of unit-length linear molecules that were partially double-stranded and partially single-stranded, accounting for about 54% of the molecules (26). The two pathways possible for the synthesis of the DNA strands complementary to the displaced parental strands have been summarized already.

## Replication in Vitro

A breakthrough in the study of adenovirus DNA replication was the development by Challberg & Kelly (27) of an in vitro replication system consisting of soluble extracts from the nuclei of HeLa cells infected with adenovirus 5 using as template DNA-protein complex isolated from adeno 5 virions. Type-I molecules similar to the ones observed in vivo were synthesized in the in vitro system (28). The 5'-end of each nascent daughter strand was covalently linked to a protein of 80,000 daltons (22, 29), structurally related to the 55,000-dalton protein covalently linked to the 5'-ends of adenovirus DNA. By using the adeno 2 ts 1 mutant grown at the restrictive temperature, a DNA −80,000 daltons protein complex was isolated that served as a template for adenovirus DNA replication (30, 31). The 80,000-dalton precursor is processed to the 55,000-dalton form during virus maturation by a virus-encoded protease, which is temperature sensitive in the ts 1 mutant.

## Viral Proteins Required for Replication

Two virus-encoded proteins are required for the initiation of adenovirus DNA replication, the 80,000-dalton pTP and a 140,000-dalton DNA polymerase. The genes coding for these two replication proteins map in the E2B region that encodes an mRNA that allows the expression of both proteins after splicing (32). Expression of functional pTP and DNA polymerase also requires sequences that map at position 39 in a minor exon (33). Only plasmids that contain genome coordinates 24.7 to 9.2 and, in addition, the Hind III J fragment that encodes the exon at coordinate 39, can produce full-length, enzymatically active DNA polymerase (34). Similarly, only plasmids that contain DNA from coordinates 29.4 to 23.9 and, in addition, the Hind III J fragment, can express functional pTP (35). Indeed, when 9 bp from the putative exon starting at position 39 were ligated to the main open reading frame coding for the DNA polymerase or for the pTP and the sequences were inserted behind a promoter of vaccinia virus, enzymatically active proteins were synthesized (36).

The pTP and DNA polymerase have been copurified in a functional form from adeno 2-infected HeLa cells (32, 37, 38). The pTP-DNA polymerase complex can be separated by sedimentation in glycerol gradient containing urea; reconstitution experiments indicate that both proteins are required for initiation activity (32, 39, 40). The 140,000-dalton DNA polymerase contains intrinsic $3' \rightarrow 5'$ exonuclease activity on single-stranded DNA (41). Mutations in the pTP have been constructed recently. When two to six amino acids from the amino terminus were removed, the initiation activity gradually decreased and a deletion of 18 amino acids rendered a completely inactive protein (42). Replacement of cysteine at position 8, a residue conserved between the different serotypes, by serine had little effect, whereas replacement of serine 580, the site of covalent attachment of dCMP, by threonine or alanine abolished pTP function.

The third viral protein required for adenovirus DNA replication is the single-stranded DNA binding protein (DBP). The DBP is an early protein synthesized in large amounts in adenovirus-infected cells (43). The protein contains 529 amino acids (44) and binds cooperatively to single-stranded DNA, although binding to double-stranded DNA was also reported (43, 45–47). Extracts from cells infected with the temperature-sensitive mutant H5ts125 were defective in adenovirus DNA replication and could be complemented by purified wild-type DBP (38, 48–54). DBP was also shown to be essential for in vitro adenovirus DNA replication with purified proteins (55, 56). Although DBP is not absolutely required for initiation of DNA replication, it stimulates pTP-dCMP complex formation (57, 58). On the other hand, the study of the elongation reaction catalyzed by the adenovirus DNA polymerase using poly(dT)-oligo(dA) as template-primer showed that the presence of DBP greatly increased the processivity of the polymerase and allowed its translocation through duplex DNA (41, 59).

## Cellular Proteins Required for Replication

Three cellular proteins, nuclear factor I (NF-I), nuclear factor II (NF-II), and nuclear factor III (NF-III), are needed for efficient adenovirus DNA replication in vitro. NF-I stimulates the initiation reaction and becomes essential in the presence of DBP (57, 58). NF-I binds specifically to a region of nucleotides 19–39 at the adenovirus DNA ends (58, 60–65). NF-I has been purified (62, 63, 66) and shown to be indistinguishable from the transcription factor CTF (66). The N-terminal 220 amino acids of NF-I are sufficient for specific DNA binding, protein dimerization, and adenovirus replication, whereas transcription activation requires an additional C-terminal domain (67, 68). In the presence of DBP, the affinity of NF-I for its binding site in the replication origin is increased (69). This is due to the formation of a multimeric protein complex of DBP and double-stranded DNA that enhances the binding of NF-I (70). NF-II, required for the elongation of replicating intermediates of adenovirus DNA to full-length, was isolated from nuclear extracts from uninfected HeLa cells (56). NF-III was shown to stimulate the initiation of adenovirus DNA replication in vitro and to bind to a sequence present at positions 39–50 in the adenovirus origin of replication adjacent to the NF-I binding site (64, 71–73). The sequences recognized by NF-III in the adenovirus origin of replication are similar to promoter and enhancer elements (72–74). Indeed, NF-III is identical to the octamer transcription factor-1 (OTF-1) (75, 76). NF-III has a conserved DNA binding domain (POU domain) present in several transcription factors, and the 160-amino-acid POU domain is enough for the stimulation of adenovirus DNA replication in vitro (C. P. Verrijzer, A. J. Kal, P. C. van der Vliet, personal communication).

## Replication Origin

In addition to the viral DNA-TP complex, plasmids containing the cloned adenovirus terminal sequences are also templates for adenovirus 2 or 5 DNA replication in vitro, although with reduced efficiency, after linearization to locate the terminal sequence at the end of the linear molecule (77, 78). The use of plasmids that contain deletions or point mutations in the adenovirus terminal sequences has shown the existence of three different functional domains for the initiation reaction. The first domain (domain A) consists of the first 18 bp of the viral DNA and constitutes the minimal replication origin, although in the absence of the other two domains its activity is low (79–83). It is very AT-rich and contains the sequence 9-ATAATATACC-18 fully conserved in different adenovirus serotypes (84–87). Mutations in this conserved sequence greatly reduce the template activity, whereas several base pairs in region 1–9 can be mutated without loss of activity (79, 88). The 9–18 conserved region may be involved in the binding of the pTP (89). The second domain (domain B) consists of the DNA sequence between nucleotides 19 and 39 and is the site of NF-I interaction (58, 80, 82, 83). The presence of this domain, which is also homologous among the different adenovirus serotypes, increases the efficiency of initiation about 10-fold. This NF-I site can be inverted or replaced by cellular NF-I binding sites, but its position must be fixed; insertion or deletion of 1 or 2 bp between the minimal origin and domain B abolishes the stimulatory effect of this domain (65, 83, 90). This critical spacing requirement suggests the existence of specific interactions between the factors that recognize the minimal origin and domain B (4). The third domain (domain C) consists of the highly conserved sequence between nucleotides 39 and 51 and corresponds to the NF-III binding site; it is not essential but stimulates the initiation reaction about threefold (64, 71, 83).

The in vitro results defining the adenovirus 2 or 5 origin of replication are in good agreement with in vivo data using linear plasmids containing inverted copies of the adenovirus 2 ITR (91). Deletion analysis showed that molecules containing the terminal 45 bp were fully active (92, 93). Although the three domains contribute to the efficiency of in vivo replication, when the NF-III domain is linked to the minimal replication origin, stimulation of replication is obtained. In addition, the NF-I domain can stimulate replication in the absence of the NF-III domain (94). However, with plasmids containing adenovirus 4 DNA sequences, only the terminal 18 bp of the viral genome were required for efficient replication in vivo (95). Similar results were obtained for the initiation of replication in vitro; addition of either the NF-I site, the NF-III site, or both sites together did not stimulate the initiation of replication (96).

In addition to the factors described before, the in vitro initiation of replication using pronase-treated adenovirus DNA requires the presence of a factor named pL (97). Factor pL is a 5'→3' exonuclease that degrades the 5'-end of the nontemplate strand, thus creating a single-stranded region at the 3'-end of the template strand (98). The initiation of replication of linearized plasmids containing the adenovirus replication origin at the DNA end is also strongly stimulated by factor pL (98). In agreement with this finding, partially duplex oligonucleotides that lack up to 14 nucleotides from the 5'-end of the nontemplate strand are about 20-fold more efficient in initiating replication than are fully duplex oligonucleotides (99). Protein-primed initiation of replication was also observed on different single-stranded DNAs (57, 77, 88, 100). The single-stranded template strand of the adenovirus origin was much better in supporting the initiation reaction than were other single-stranded DNAs, including the nontemplate strand (96, 99). These findings suggest a model in which the 3'-end of the template strand is made single-stranded as a prerequisite for initiation of adenovirus DNA replication (99). With natural single-stranded templates, dCTP is the preferred initiating nucleotide. With templates containing no dG residues, the specific nucleotide requirements for the initiation reaction were changed (88), indicating that the nucleotide specificity for the initiation reaction is determined by the terminal nucleotide at the template strand.

## Replication of the Nontemplate Strand

To provide an insight into the in vitro replication of the nontemplate (displaced) strand, an artificial panhandle molecule was constructed that contained the adeno 2 double-stranded ITR region and a single-stranded loop (101). Such a molecule was an efficient template and could initiate replication by protein-priming; the initiation reaction was stimulated by NF-I and NF-III, and adenovirus DBP was also essential for replication. On the other hand, adenoviruses with nonidentical terminal sequences are viable, giving rise after transfection to DNA molecules with identical termini. These results could be explained by strand-displacement with the formation of a "panhandle" structure and efficient repair of one end with the opposite terminus used as a template (102). Up to 51 bp can be deleted from one end of the adenovirus genome without loss of infectivity (103). This finding was taken as evidence for the existence of the "panhandle" structure as a replicative intermediate in vivo. However, the possibility that in vivo complementary strand synthesis occurs with adenovirus DNA molecules that are replicated from both ends cannot be excluded, if it is taken into account that such replicating molecules have been found (26).

## BACTERIOPHAGE $\phi$29 DNA REPLICATION

The *Bacillus subtilis* phage $\phi$29 genome consists of a linear double-stranded DNA 19,285-bp long (104–106). The first evidence for the existence of a protein attached at the ends of a linear double-stranded DNA was the finding that $\phi$29 DNA could be isolated from the viral particles as circular molecules or concatemers and that these DNA structures could be converted into unit-length linear DNA by treatment with proteolytic enzymes (107). In addition, when $\phi$29 DNA was treated with proteolytic enzymes, the capacity to transfect competent *B. subtilis* was lost (108), and the DNA isolated from a $\phi$29 ts mutant in gene 3 was thermolabile for transfection (109). Indeed, it was shown that a protein of 28,000 daltons, characterized as the product of the viral gene 3 (110), was covalently linked at the two 5'-ends of the viral DNA (110–112). Other *B. subtilis* phages morphologically similar to $\phi$29 that contain linear double-stranded DNA and TP of similar size, classified in three serological classes, are: (*a*) $\phi$29, $\phi$15, PZA, and PZE; (*b*) Nf, M2, and B103; and (*c*) GA-1 (113–116). All the phages of the $\phi$29 family have a short ITR, six nucleotides long (AAAGTA) for $\phi$29, PZA, $\phi$15, and B103 DNAs; eight nucleotides long (AAAGTAAG) for Nf and M2 DNAs; and seven nucleotides long (AAATAGA) for GA-1 DNA. Other sequence similarities among the ends of these phage DNAs have been recently reviewed (8, 9). The amino acid sequence of the TP of phage PZA is almost identical (117) to that of phage $\phi$29 (104, 118), and the amino acid sequence of the TP of phage Nf is approximately 66% homologous with the $\phi$29 and PZA TPs, showing similar hydropathy and secondary structure predictions (119).

The linkage between the $\phi$29 TP and DNA, readily hydrolyzed by alkali treatment, is a phosphoester bond between the OH group of serine residue 232 in the 266 amino acids TP and 5'-dAMP (120, 121). Prediction of the secondary structure in the region around the linking site suggested that the above serine residue is located in a $\beta$ turn preceded by an $\alpha$-helix (121).

## Replication in Vivo

Electron microscopic analysis of the replicative intermediates synthesized in *B. subtilis* infected with φ29 showed the presence of two basic types of replicating molecules (122–124), similar to the ones found in adenovirus-infected cells. Type-I and type-II molecules accounted for about 20% and 55%, respectively, of the total number of molecules (122). Type-I molecules with a single-stranded branch from each DNA end were also seen. The results indicated that replication starts at either DNA end and proceeds by strand-displacement toward the other end.

Genes 1, 2, 3, 5, 6, and 17 were shown to be required for the synthesis of the viral DNA in vivo (125–128); genes 2 and 3 are involved in initiation, whereas genes 5 and 6 are concerned with some elongation step, although the possibility that they also play a role in initiation could not be ruled out (129). In phage M2, three genes involved in the viral DNA replication have been characterized: G, E, and T. Genes G and E correspond to the φ29 genes 2 and 3, respectively (130). Development of φ29 occurs normally in most of the *B. subtilis* replication mutants available (reviewed in 8, 9).

## Replication in Vitro

When extracts from φ29-infected *B. subtilis* were incubated with $[\alpha\text{-}^{32}\text{P}]$ dATP in the presence of φ29 DNA-TP complex as template, a $^{32}\text{P}$-labeled protein with the electrophoretic mobility of p3 was found (131–134). Incubation of the $^{32}\text{P}$-labeled protein with piperidine released 5'-dAMP, indicating the formation of a TP-dAMP covalent complex (131). When extracts from phage M2–infected *B. subtilis* were used, an initiation complex between the M2 TP and 5'-dAMP was also found (130).

## Viral Proteins Essential for the Initiation of Replication

Phage φ29 genes 2 and 3 and phage M2 genes G and E were shown to be essential for the in vitro initiation reaction (130, 135). Phage φ29 genes 2 (136) and 3 (137, 138) were cloned, and both proteins, p2 (139) and p3 (138, 140), were overproduced and highly purified in a functional form.

The purified protein p2, in addition to catalyzing the initiation reaction, was shown to have DNA polymerase activity when assayed with a template-primer (138, 139); moreover, a partially purified DNA polymerase isolated from *B. subtilis* infected with a ts mutant in gene 2 showed a greater heat lability than the one present in the wild-type preparation (141). The φ29 DNA polymerase has 3' to 5' exonuclease (142, 143) that is about 10-fold more active on single- than on double-stranded DNA and has properties expected for an enzyme involved in proofreading (C. Garmendia, A. Bernad, L. Blanco, M. Salas, in preparation). More recently, a pyrophosphorolytic activity has been characterized in the φ29 DNA polymerase (M. A. Blasco, A. Bernad, L. Blanco, M. Salas, submitted).

The in vitro initiation reaction is greatly stimulated by $NH_4^+$ ions (144) because the latter stabilize the formation of a complex between DNA polymerase and TP (145). In fact, the two proteins copurify when extracts from $\phi$29-infected cells are used (146, 147). The purified system consisting of TP and DNA polymerase with $\phi$29 DNA-TP as template gave rise to the synthesis of a high level of full-length $\phi$29 DNA in a very processive way (144, 148). When primed M13 DNA was used as template, the $\phi$29 DNA polymerase synthesized DNA chains greater than 70 kb in a very processive way. These results indicated that the $\phi$29 DNA polymerase is a processive enzyme, able to produce strand-displacement without need of accessory proteins (148).

FUNCTIONAL DOMAINS IN THE $\phi$29 TP    Deletion mutants in the $\phi$29 TP have been constructed, and their priming activity and the capacity to interact with the $\phi$29 DNA polymerase and with DNA have been studied (149–152). The results indicated the existence of two DNA polymerase–binding regions in the TP, located at positions 72–80 and 241–261 (152). On the other hand, the amino acid regions of the TP at positions 13–18, 30–51, and 56–71 are important for DNA binding (152). None of the typical motifs considered to be responsible for sequence-specific DNA binding are found in protein p3. There is a positive cluster of amino acids at the amino end of the TP that contains the sequence KKKY flanked by aromatic and hydrophobic residues. This sequence is deleted in the mutant protein lacking amino acids at positions 30–51, in which no binding to DNA is detected, and it is conserved in the TP of phage Nf (119). Interestingly, a similar amino acid region is present in the $\alpha$-like DNA polymerases (153), where the DNA-binding domain could be located (154).

Site-directed mutagenesis at and near the DNA-linking site of the $\phi$29 TP has been carried out also. Thus, serine 232, involved in the covalent linkage to dAMP, has been changed into threonine. The purified mutant protein was completely inactive in the priming reaction (155). When the serine 232 residue was changed into cysteine, the purified mutant TP had about 0.7% of the priming activity of the wild-type protein (156). Residues leucine 220, serine 223, and serine 226 were independently changed into proline; the purified TP mutants had about 3%, 140%, and 1% of the priming activity of the wild-type protein, respectively (156). All the mutant TPs were able to interact with the $\phi$29 DNA polymerase and with DNA, suggesting that leucine 220 and serine 226, in addition to serine 232, form part of a functional domain involved in the process of initiation of DNA replication. These three amino acid residues are conserved in the TP of phage Nf (119).

Recently, the presence of the amino acid sequence arginine-glycine-aspartic acid (RGD), found in cell-adhesive proteins, has been pointed out in the $\phi$29 and M2 TPs, at positions 256–258 (157). The synthetic peptide RGD, but not RGE, inhibited transfection of $\phi$29 and M2 DNAs, as well as the in vitro initiation reaction.

FUNCTIONAL DOMAINS IN THE $\phi$29 DNA POLYMERASE. HOMOLOGIES WITH PROKARYOTIC AND EUKARYOTIC DNA POLYMERASES    Aphidicolin, phosphonoacetic acid, and the nucleotide analogues butyl-anilino dATP and butyl-phenyldGTP, known inhibitors of the eukaryotic DNA polymerase $\alpha$ (158, 159), inhibit the $\phi$29 DNA polymerase (160, 161). Indeed, phage $\phi$29 aphidicolin-resistant mutants have an altered DNA polymerase with reduced sensitivity to aphidicolin (162). Three main conserved amino acid regions (D--SLYP, K---NS-YG, and YGDTDS for regions 1, 2, and 3, respectively) are present, in the same linear arrangement, in the C-terminal portion of $\alpha$-like DNA polymerases of viral and cellular origin, including those of adenovirus, phages $\phi$29, M2, and PRD1, and the putative DNA polymerases from the TP-containing linear plasmids pGKL1 and pGKL2 from yeast, pAl2 and pCIK1 from fungi, and S1 mitochondrial DNA from maize (153, 161, 163–175). These conserved regions have been proposed to form the catalytic site involved in dNTP binding (176).

Site-directed mutagenesis has been carried out in the phage $\phi$29 DNA polymerase. The Tyr residue at position 254, present in region 1, was changed into Phe (Y254F); the Tyr residue at position 390 from region 2 was changed into Phe or Ser (Y390F and Y390S); and the YCDTD sequence present at positions 454–458 in region 3 was changed as follows: Y454F, C455G, D456G, T457P, and D458G. None of the above mutations reduced the 3'→5' exonuclease activity (177; L. Blanco, A. Bernad, M. A. Blasco, M. Salas, submitted). The initiation activity was greatly reduced by the Y254F mutation, and the nonprocessive elongation was affected by the Y390F mutation; both the initiation and nonprocessive elongation activities were undetectable in the T457P and D458G mutant proteins. The processive elongation was greatly reduced or undetected with all the mutant proteins except the C455G one. It has been proposed that the YGDTDS motif in the $\phi$29 DNA polymerase is involved in the metal binding associated with the dNTP site (177) and that the three regions of amino acid homology are essential for the synthetic activities of the $\phi$29 DNA polymerase, i.e. initiation and polymerization, but not for the 3'→5' exonuclease activity (177; L. Blanco, A. Bernad, M. A. Blasco, M. Salas, submitted). Temperature-sensitive mutants available in $\phi$29 gene 2 have been studied. Mutant ts2 (24) consisted of the change A355V, and the DNA polymerase had a ts phenotype both in the initiation and elongation reactions (178). This mutation is located between regions 1 and 2, in a conserved region present only in protein-primed DNA polymerases (178). Mutant ts2 (98) consisted of the A492V mutation; the ts2 (98) mutant DNA polymerase was rather insensitive to temperature, but its initiation and elongation activities were stimulated by $Mn^{2+}$ ions to a greater extent than the wild-type DNA polymerase (178). This result could be related to the location of the ts2 (98) mutation in a new region of amino acid homology, region 4, proposed to be involved in dNTP binding (L. Blanco, A. Bernad, M. A. Blasco, M. Salas, submitted) and close to the proposed $Me^{2+}$ binding site (177).

Regions 1, 2, and 3 have been also found in *Escherichia coli* DNA polymerase I and other Pol I-like DNA polymerases (L. Blanco, A. Bernad, M. A. Blasco, M. Salas, submitted). Moreover, region 4 is present in all the α-like and Pol I-like DNA polymerases analyzed (L. Blanco, A. Bernad, M. A. Blasco, M. Salas, submitted). In the case of the *E. coli* DNA polymerase I, residue K758 in this region is involved in dNTP binding (179, 180). Site-directed mutagenesis in the corresponding region of the φ29 DNA polymerase has shown that mutation K498R has very high initiation but very poor elongation activity, and mutation K498T has undetectable initiation and elongation activities (M. A. Blasco, A. Bernad, L. Blanco, M. Salas, unpublished results).

On the other hand, three regions of amino acid homology named Exo I, Exo II, and Exo III, containing the critical residues involved in the 3'→5' exonuclease activity in *E. coli* DNA polymerase I (181–183), are located at the amino-terminal end and in the same linear arrangement in several prokaryotic and eukaryotic DNA polymerases including those of phages φ29, M2, and PRD1, adenovirus, and those of the linear plasmids pGKL1, pGKL2, pClK1, pAl2, and S1 (184; L. Blanco, A. Bernad, M. A. Blasco, M. Salas, submitted). The conservation of regions Exo I and Exo II among Pol I and several α-like DNA polymerases has been reported also (154, 185). Site-directed mutagenesis in the φ29 DNA polymerase at amino acid residues Asp12 and Glu14, in region Exo I, and Asp 66 in region Exo II, has shown that none of these mutations affected the initiation or elongation activities, whereas they greatly reduced the 3'→5' exonuclease activity (184). These results indicated that the 3'→5' exonuclease domain of the φ29 DNA polymerase is located at the amino terminus of the protein, as is the case with the Klenow fragment of *E. coli* DNA polymerase I. Given the homology of this region with the other DNA polymerases indicated above, the results reflect a high evolutionary conservation of this catalytic domain.

Altogether, the results with the φ29 DNA polymerase indicate that the latter may have a tridimensional structure similar to that of the Klenow fragment of *E. coli* DNA polymerase I (181); the 3'→5' exonuclease activity is located at the amino end of the protein and the synthetic activities, i.e. initiation and polymerization, at the carboxyl end. These results can be extrapolated to DNA polymerases from TP-containing genomes whose sequence is presently available and, most likely, to the other DNA polymerases from this family.

## Other Viral Proteins Involved in Replication

GENE 1 AND 17 PRODUCTS    The role of the products of genes 1 and 17, required for $\phi$29 DNA replication in vivo has not yet been elucidated. Gene 1, encoded in ORF 6 in the sequence of Yoshikawa & Ito (104), was cloned, and protein p1, of molecular weight 10,000, was overproduced and purified (128). Transfection of *B. subtilis* protoplasts with $\phi$29 DNA-TP complex isolated from mutant sus17 (112), or with recombinant $\phi$29 DNA-TP molecules lacking gene 17, indicated that protein p17 is partially dispensable, probably because some bacterial protein can substitute for protein p17 (186). Comparison of the nucleotide sequence of phages $\phi$29, $\phi$15, and PZA has shown that deletions in the central and carboxy-terminal parts of gene 17 are tolerated and that the only portion that has to be conserved to encode the functional product is the region corresponding to 83 amino acids from the amino-end of the protein (187).

GENE 6 PRODUCT    Gene 6 was cloned and protein p6 was overproduced and highly purified (188). The apparent molecular weight of the purified protein was 23,600, suggesting that the native form of protein p6 is a dimer. Purified protein p6 stimulated the formation of the p3-dAMP initiation complex (188) due to a decrease in the $K_m$ value for dATP (189). Protein p6 also stimulated elongation, but only at a dATP concentration higher than 1 $\mu$M, when transition of the p3-dAMP initiation complex to the first elongated products, p3-$(dAMP)_2$ and p3-$(dAMP)_3$, occurs (145, 189, 190). The effect of protein p6 was not on the rate of elongation but rather on the amount of elongated product, stimulating the transition from initiation to elongation (190).

By gel retardation assays, protein p6 was shown to bind to double-stranded DNA. DNase I footprinting experiments indicated that protein p6 could bind in a cooperative way to fragments containing the left or right terminal sequences of $\phi$29 DNA, producing a characteristic pattern of hypersensitive bands spaced about 24 nucleotides apart along 200–300 bp, flanking protected regions (191, 192). Hydroxyl radical footprinting of protein p6 bound to terminal $\phi$29 DNA fragments gave rise to a specific pattern of protected regions of 3–4 nucleotides regularly spaced every 12 nucleotides (193). These results, together with the fact that binding of protein p6 to circular DNA restrains positive supercoiling (191), support a model in which a DNA right-handed superhelix tightly wraps around a multimeric protein p6 core (193). By deletion analysis, main p6-recognition signals have been located between nucleotides 62 and 125 at the right $\phi$29 DNA end and between nucleotides 46 and 68 at the left end. No sequence similarity exists in these regions, thus protein p6 does not seem to recognize a specific sequence in the DNA but rather a structural feature, which could be bendability (192). The activity of p6 in the initiation of $\phi$29 DNA replication not only requires the formation of the complex, but also its correct positioning relative to the $\phi$29 DNA ends, indicating that other proteins involved in the initiation of $\phi$29 DNA replication recognize, at a precise position, either the p6 core or the DNA conformational change induced by protein p6 (193).

Deletions at the carboxyl and amino ends of protein p6 have been constructed, and their activity in the initiation of $\phi$29 DNA replication and their capacity to interact with the $\phi$29 DNA ends were studied. The 14 carboxy-terminal amino acids of p6 are fully dispensable; the activity of the protein decreased with deletions from 23 to 39 amino acids and was undetectable when 44 amino acids were removed (194). The activity of protein p6 decreased considerably when 5 amino acids were removed from the amino end and was undetectable after deletion of 13 amino acids. Both mutant proteins were unable to interact specifically with the $\phi$29 DNA ends (195).

GENE 5 PRODUCT    Gene 5 was cloned, and protein p5 was overproduced (196), purified, and shown to bind to single-stranded DNA (197). Electron microscopy indicated that protein p5 binds to M13 ssDNA producing a twofold reduction in the DNA length (352). Protein p5 did not affect either the formation of the p3-dAMP initiation complex or the rate of elongation. However, protein p5 greatly stimulated $\phi$29 DNA-TP replication at incubation times where the replication in the absence of p5 leveled off (197). Replication of primed M13 ssDNA was also stimulated by protein p5. Other SSB proteins such as *E. coli* SSB, T4 gp32, adenovirus DBP and human RF-A can functionally substitute for protein p5 (352).

## Replication Origin

The purified replication system was used to show that proteinase K-treated $\phi$29 DNA is not a template for the initiation reaction. However, piperidine-treated $\phi$29 DNA was active, although the activity was reduced 5–10-fold from that obtained with $\phi$29 DNA-TP complex (116). The DNA sequence requirements for the initiation reaction were studied by cloning terminal $\phi$29 DNA fragments from the left and right ends. Fragments released by restriction nuclease digestion, containing the $\phi$29 DNA terminal sequences at the DNA ends, were active templates for the initiation reaction. No template activity was obtained with the circular plasmid or when the terminal sequences were not placed at the DNA end (198). The analysis of deletion derivatives indicated that the minimal origins of replication are located within the terminal 12 bp at each $\phi$29 DNA end (199). Point site–directed and random mutagenesis in those sequences indicated that a change of the second or third A in the $\phi$29 DNA sequence into a C completely abolished the template activity. Changes at positions 4–12 were tolerated to a great extent, suggesting that the sequence requirements at the end-proximal region of the origin of replication are more strict that those at the distal region (199). In spite of the above in vitro results, which suggest the partial dispensability of the parental TP, no replication was obtained in vivo after transfection of *B. subtilis* protoplasts with $\phi$29 DNA molecules lacking the parental TP (186).

In addition to the in vitro template activity of protein-free DNA containing the $\phi$29 DNA sequences corresponding to the terminal 12 bp, a single-stranded DNA of 29 nucleotides, corresponding to the sequence at the 3' right terminus, gave rise to the formation of the TP-dAMP initiation complex with an efficiency about 15% of that obtained with the natural template $\phi$29 DNA-TP (L. Blanco, A. Bernad, M. Salas, in preparation); the preferred nucleotide is, by far, dATP. When the sequence complementary to the latter was used, the template activity was greatly reduced (L. Blanco, A. Bernad, M. Salas, in preparation). Initiation activity was also obtained with unspecific single-stranded DNA such as polydC; the initiation complex formed in this case is TP-dGMP with an efficiency of about 2% with respect to the TP-dAMP complex formed with the natural template (L. Blanco, A. Bernad, M. Salas, in preparation).

Taking advantage of the fact that the initiation reaction is stimulated about 50-fold when $Mn^{2+}$ ions are used instead of $Mg^{2+}$ ions (J. A. Esteban, A. Bernad, M. Salas, L. Blanco, in preparation), some initiation activity in the presence of $Mn^{2+}$ could be detected in the absence of template, although the affinity for the nucleotide was greatly decreased. In this case, the initiation complex could be formed between the TP and any of the nucleotides; the efficiency is about 0.3% of that obtained in the formation of the TP-dAMP complex with the $\phi$29 DNA-TP template (L. Blanco, A. Bernad, M. Salas, in preparation). These results and the ones presented above indicate that the nucleotide specificity for the initiation reaction is provided by the 3' terminal nucleotide at the template strand.

## Replication of the Nontemplate Strand

To test the in vivo replication of the nontemplate DNA strand, recombinant $\phi$29 DNA molecules containing parental TP at only one DNA end were constructed (186). No replication in *B. subtilis* protoplasts was obtained, suggesting that the fully displaced nontemplate DNA strand is not an active template for replication in vivo. In agreement with the in vivo results, when replication of $\phi$29 DNA-TP was studied in the purified in vitro system, a significant amount of type-II replicative intermediates was found at an incubation time at which no synthesis of full-length $\phi$29 DNA was detected (C. Gutiérrez, J. M. Sogo, M. Salas, unpublished results). These results indicate that the appearance of type-II replicative intermediates does not require synthesis of full-length DNA and displacement of the nontemplate strand;

rather the results support the model in which initiation of replication can occur from both DNA ends, while type-II molecules are produced by separation of the two displacement forks when they meet.

## BACTERIOPHAGE PRD1 DNA REPLICATION

Bacteriophage PRD1 is a member of a family of lipid-containing phages (reviewed in 200) that infect a variety of Gram-negative bacteria including *Escherichia coli* and *Salmonella typhimurium*. The PRD1 genome is a linear, double-stranded DNA of about 14,700 bp with a TP linked to the 5' ends (201). Other lipid-containing phages closely related to PRD1 are PR3, PR4, PR5, PR722, and L17 (200). The DNA of all these phages have an ITR 110–111 bp long (202, 203).

Two very early genes, VIII and I, located at the left-hand terminus of PRD1 DNA, are involved in the synthesis of the viral DNA (204). Gene VIII codes for the TP p8 of 259 amino acids (165, 201, 205, 206). Gene I encodes the DNA polymerase p1, of 553 amino acids (165, 166). Gene XII, the third known very early gene, located at the right-hand terminus, encodes protein p12, of 160 amino acids, involved in the shut-off of early protein synthesis and in the synthesis of the viral DNA (204, 207, 208). Protein p12 has been purified and shown to have affinity for single-stranded DNA and, to a lesser extent, for double-stranded DNA. The purified protein p12 stimulates replication when $\phi$29 DNA-TP is used as template (353) to an extent similar to the $\phi$29 single-stranded DNA binding protein p5 (197, 352).

When extracts from PRD1-infected *S. typhimurium* were incubated with $[\alpha\text{-}^{32}P]dGTP$, a labeled protein with the electrophoretic mobility of the TP was found (201). This reaction required genes I and VIII as the only two viral genes (205). Synthesis of full-length PRD1 DNA was obtained, starting at or near the ends of the DNA molecules (209). The covalent linkage between p8 and dGMP was shown to be a phosphoester bond with tyrosine (205). The use of extracts from *E. coli* cells temperature-sensitive for replication genes suggested that the host replication complex may be needed for the initiation reaction (205; reviewed in 9). Gene VIII has been recently cloned and protein p8 was overproduced and purified in a functional form (210). No substantial sequence homology was found between the PRD1 TP and those of $\phi$29 and Nf. However, when the hydropathic profiles of the three TPs were compared they were superimposable at the C-terminal portion, suggesting a functional homology at the carboxyl end of the three proteins. Taking these results into account as well as the fact that the linking site in the $\phi$29 TP is at the serine residue at position 232 (121), it has been proposed that the linking site in the PRD1-TP is the tyrosine residue 226 (206).

## BACTERIOPHAGE Cp-1 DNA REPLICATION

The *Streptococcus pneumoniae* phage Cp-1 contains a linear, double-stranded DNA of about $12 \times 10^6$ daltons with a TP of 28,000 daltons covalently linked to the two 5'-ends (211). The Cp-1–related phages Cp-5 and Cp-7 also have a TP (212). Phage Cp-1, Cp-5, and Cp-7 DNAs have a long ITR of 236, 343, and 347 bp, respectively (213, 214).

Incubation of extracts of Cp-1–infected *S. pneumoniae* with $[\alpha\text{-}^{32}P]$ dATP produced a labeled protein with the electrophoretic mobility of the TP. Treatment of the $^{32}P$-labeled protein with piperidine released 5'dAMP, indicating the formation of a covalent complex between the TP and dAMP. Moreover, the TP-dAMP complex could be elongated in vitro by addition of the four dNTPs (215). The linkage between the TP and dAMP is a phosphoester bond with threonine (215).

## BACTERIOPHAGE HB-3 TP

Bacteriophage HB-3 is a member of a family of temperate phages from *S. pneumoniae* (216). The HB-3 genome consists in a linear, double-stranded DNA of about 40 kb (217). Other members of the same family are phages HB-623 and HB-746. A protein of 23,000 daltons was shown to be covalently linked at the 5'-ends of the DNA of these phages (217). The function of the TP, it has been speculated, may be to protect the DNA molecules during their transport from the membrane to the insertion region of the host bacteria, or it may serve a function during integration. In this relation, the free linear double-stranded T-DNA molecules from *Agrobacterium tumefaciens* become covalently linked through the 5'-ends to the VirD2 protein before these T-DNA molecules are transferred to the genome of the host plant (218, 219). The possible role of these 5'-linked proteins in DNA replication remains to be determined; a model, the invertron, has been recently proposed, suggesting a role in replication and integration (220).

## TERMINAL PROTEINS IN LINEAR PLASMIDS

A variety of linear plasmids have been isolated in the past 10 years from bacteria, fungi, and higher plants. In most of the cases long ITRs have been characterized, and in many of them evidence for the existence of a TP linked at the 5'-ends of the DNA has been reported. Table 2 shows a summary of the different linear plasmids isolated and their sources, as well as of the characterization of TP and ITR. Only relevant features not indicated in the table are described below.

The yeast *Kluyveromyces lactis* contains two linear killer plasmids, pGKL1

**Table 2**  Linear plasmids with TP and ITR

| Linear plasmid | Source | Size | ITR | TP | Refs. |
|---|---|---|---|---|---|
| | Bacteria | | | | |
| pSLA1 | *Streptomyces rochei* | 17 kb | nd | yes | (221, 222) |
| pSLA2 | *Streptomyces rochei* | 17 kb | 614 bp | yes | (221, 222) |
| pSCL | *Streptomyces clavuligerus* | 12 kb | yes | yes | (223) |
| pSCP1 | *Streptomyces coelicolor* | 350 kb | 70 kb | yes | (224–226) |
| | Yeast | | | | |
| pGKL1 | *Kluyveromyces lactis* | 8.9 kb | 202 bp | 28 kDa | (227–231) |
| pGKL2 | *Kluyveromyces lactis* | 13.4 kb | 182–184 bp | 36 kDa | (169, 227–231) |
| pSKL | *Saccharomyces kluyveri* | 14.2 kb | 483 bp | yes | (236) |
| | Fungi | | | | |
| pAI2 | *Ascobolus immersus* | 5.6 kb | ≈700 bp | yes | (237, 238) |
| pMC32 | *Morchella conica* | 6 kb | ≈750 bp | nd | (239) |
| pCF637 | *Ceratocystis fimbriata* | 8.2 kb | nd | yes | (240) |
| pFQ501 | *Ceratocystis fimbriata* | 6 kb | 750 bp | nd | (241) |
| pRS64 | *Rhizoctonia solani* | 2.6 kb | nd | nd | (242) |
| pEM | *Agaricus bitorquis* | 7.3 kb | ≈1 kb | nd | (243) |
| pMPJ | *Agaricus bitorquis* | 3.6 kb | nd | nd | (243) |
| kal DNA | *Neurospora intermedia* | 9 kb | 1361 bp | yes | (244, 245) |
| pFOXC2 | *Fusarium oxysporum* | 1.9 kb | 50 bp | yes | (10, 246) |
| pClK1 | *Claviceps purpurea* | 6.7 kb | 327 bp | yes | (247) |
| pFSC1 | *Fusarium solani* f.sp. cucurbitae | 9.2 kb | 1211 bp | 80 kDa | (248, 249) |
| pFSC2 | *Fusarium solani* f.sp. cucurbitae | 8.3 kb | 1027 bp | 80 kDa | (248, 249) |
| pLPO1 | *Pleurotus ostreatus* | 10 kb | nd | yes | (250) |
| pLPO2 | *Pleurotus ostreatus* | 9.4 kb | nd | yes | (250) |
| pLLE1 | *Lentinus edodes* | 11 kb | nd | yes | (251) |
| | Higher plants | | | | |
| 11.5 kb plasmid | *Brassica campestris* | 11.5 kb | 325 bp | yes | (252) |
| 10.4 kb plasmid | *Beta maritima* | 10.4 kb | nd | nd | (253) |
| N-1 | *Sorghum bicolor* | 5.7 kb | nd | nd | (254, 255) |
| N-2 | *Sorghum bicolor* | 5.3 kb | nd | nd | (254, 255) |
| S-1 | *Zea mays* L. | 6.4 kb | 208 bp | yes | (256, 257, 260) |
| S-2 | *Zea mays* L. | 5.4 kb | 208 bp | yes | (256, 257, 260) |
| n | *Zea mays* L. | 2.35 kb | 170 bp | yes | (258, 259, 260) |

and pGKL2. Plasmid pGKL2, required for the maintenance of pGKL1 in the cell, can replicate in the absence of pGKL1. Both plasmids replicate by a strand-displacement mechanism and both type-I and type-II molecules were found (232). ORF1 in plasmid pGKL1 (161, 233) and ORF2 in pGKL2 (169) contain three regions of amino acid homology that are found at the carboxyl part of α-like DNA polymerases. By analogy with the viral systems it is likely that the TPs of pGKL1 and pGKL2 are plasmid-encoded products. Since functions have been assigned to the four ORFs of pGKL1, it is possible that the TPs for both plasmids are encoded in pGKL2. A DNA-binding protein has been detected in extracts of yeast cells carrying pGKL1 and pGKL2. By deletion mapping and DNase I footprinting the binding occurs within the 202 bp ITR of pGKL1 (D. McNeel, F. Tamanoi, personal communication). This binding protein could be similar in function to protein p6 and factors NF-I and NF-III that bind to the replication origins of phage $\phi$29 and adenovirus, respectively, and stimulate the initiation of replication.

When the pGKL plasmids were transferred into *Saccharomyces cerevisiae*, nonkiller transformants harboring pGKL2 and two new plasmids, F1 and F2, of 7.8 kb and 3.9 kb, respectively, were obtained. F2 was a linear DNA formed by a 5-kb deletion of the right part of pGKL1 and has two different terminal structures. One end has a protein attached at the 5' terminus, whereas the two strands are linked together at the other end forming a hairpin structure. F1 was an inverted dimer of F2, and it was thought to be generated from F2 by protein-priming and elongation by strand-displacement and to contain a TP at each 5'-end of the DNA (234).

By UV-irradiation of *K. lactis* containing pGKL1 and pGKL2, two deleted plasmids of pGKL1 were found, pK192L and pK192S. They are linear and arose by a 6.5-kb deletion of the right part of pGKL1. The pK192L is a palindromic plasmid of 4903 bp, consisting in a unique sequence of 215 bp and ITRs of 2344 bp. The pK192S was suggested to be a fold-back form of pK192L, consisting of 2344 bases of double-stranded DNA and 215 unpaired nucleotides forming a hairpin structure. These findings suggested that pK192L has the terminal structure of pGKL1, which is 202-bp ITRs and TP in both termini, and that pK192S has this structure only at one end (235). The 2.4-kb left terminal fragment of pGKL1 carried in the pK192 plasmids contained only part of the ORF1 gene, encoding the putative DNA polymerase required for the protein-primed replication. Segregation analysis showed that the two plasmids were maintained only in cells carrying pGKL1 with the intact ORF1, as well as pGKL2, whereas plasmids F1 and F2, which contain the complete ORF1 from pGKL1, are stably maintained in cells harboring only pGKL2. These results suggest that, for the replication of the pGKL1-derived plasmids, the presence of the putative DNA polymerase encoded in ORF1 is necessary and, therefore, that the putative DNA polymerase coded by pGKL2 does not function in the replication of pGKL1 (235). If the TPs of pGKL1 and pGKL2 form a specific complex with the corresponding DNA polymerase, this complex is likely to recognize specifically the different terminal nucleotide sequences present in pGKL1 and pGKL2.

Replication of plasmid pAI2 from *Ascobolus immersus* starts at the DNA ends (173). A large ORF spanning 1202 amino acids shares homology with the three conserved regions present in $\alpha$-like DNA polymerases and in putative DNA polymerases from linear plasmids (173). ORF1 from the *Claviceps purpurea* plasmid pClK1 codifies for a protein of 1097 amino acids, which is likely to be a DNA polymerase from the amino acid homology at the three regions conserved in $\alpha$-like DNA polymerases and in the putative DNA polymerases from linear plasmids (174). The gene coding for the TP is unknown. Taking into account that in the case of the $\phi$29 and adenovirus TPs the amino acid involved in the linkage to the DNA is located in a $\beta$-turn preceded by an $\alpha$-helix (121), it is possible that the TP of pClK1 is encoded by 400 amino acids at the amino-terminal part of ORF1. In fact, serine residue 327 of this ORF meets the criteria for a nucleotide-linking site (174). S1 and S2 are linear mitochondrial DNA molecules found in a male sterile cytoplasm of maize, with idential ITRs of 208 bp. S1 and S2 contain a 1462-bp sequence of nearly perfect homology (256, 257). In the linear mitochondrial DNA, n, present in a male fertile cytoplasm (258), of the first 17 nucleotides of the termini, 16 are identical to those of the termini of S1 and S2 DNAs (259). ORF 3 of plasmid S1 codifies for a protein with amino acid homology with the three conserved regions present in $\alpha$-like DNA polymerases (161, 168) and in the putative DNA polymerases encoded by linear plasmids. All these data suggest that the linear plasmids replicate by a protein-priming mechanism.

## HEPADNAVIRUS TP

The most characteristic feature of the hepadnavirus family is the structure of the 3-kb virion DNA. One strand, the minus (noncoding) strand, is complete, with unique 3' and 5' termini. The other, the viral plus strand, is incomplete, with a unique 5'-end and a 3'-end that is heterogeneous (261). The duplex DNA is held in a circular form by a cohesive overlap between the 5'-ends of the two DNA strands (262). Human hepatitis B virus (HBV), ground squirrel hepatitis virus (GSHV), and duck hepatitis B virus (DHBV) were shown to contain a protein covalently linked at the 5'-end of the complete viral DNA minus strand (263–265). The protein was also bound to minus-strand DNA intermediates, as small as 30 bases, isolated from replicating complexes present in infected duck liver, supporting the idea that the protein may function as a primer for synthesis of the minus-strand DNA. The DNA-protein linkage in DHBV most likely involves a phosphoester bond between a tyrosine residue and dGMP, inferred from the fact that the DNA-peptide bond was resistant to alkali treatment under conditions in which phosphodiester bonds to serine or threonine are cleaved (264).

The mechanism by which hepadnavirus DNA replicates involves a reverse transcription step (266). The RNA template for reverse transcription, named pre-genome, is produced by copying the minus-strand DNA in the closed circular double-stranded DNA molecule formed from the virion DNA (266, 267). The formation of the covalently closed double-stranded DNA molecule seems to require the splitting-off of the template-bound protein (reviewed in 12). Initiation of the minus-strand DNA synthesis on the pre-genome is supposed to take place via a protein-priming mechanism (268–270). The minus-strand generated in this way is further used as template for DNA-dependent plus-strand DNA synthesis, which is initiated by a capped oligoribonucleotide (267, 269, 270).

The mammalian hepadnaviruses (271–273) and DHBV (274, 275) have a similar genetic organization. The P-gene encodes the viral DNA polymerase/reverse transcriptase, as well as the genome-linked protein. Different P-gene proteins, presumably covalently bound at a common region and subsequently processed, are bound to the 5'-end of the DHBV minus strand (276). Limited protease treatment cleaved C-terminal P-proteins from the viral DNA, indicating that the TP forms a separate domain encoded in the N-terminal quarter of the 785-amino-acid multifunctional primary gene-P product (277). Functional analysis of a deletion mutant also indicated that a nonessential spacer of ~200 amino acids separates the TP from the polymerase domain, the latter residing in the C-terminal half of the P-gene (277). Recently, the C-terminal part of the hepadnaviral P-proteins was shown to be significantly similar to ribonuclease H of *E. coli* (278), and homology of a region of the P protein of HBV with the polymerase domain of retrovirus reverse transcriptase had been previously shown (279). Thus, the main proteins required for hepadnaviral reverse transcription, the primer, the DNA polymerase/reverse transcriptase, and possibly also the RNase H, seem to be synthesized as a polyprotein precursor initially linked as such to its DNA product. By comparison of N-terminal parts of P-proteins of hepadnaviruses to TPs of picornaviruses, a conserved region was found between positions 61 and 84 of HBV and VPg of human rhinovirus 14. In particular, the Gly-X-Tyr sequence present in picornaviruses at the linking site is also present at the N-terminal region of hepadnaviruses (278).

## GENOME-LINKED PROTEINS OF RNA VIRUSES

Two groups of animal viruses, picornavirus and calicivirus, and five groups of plant viruses, comovirus, luteovirus, nepovirus, potyvirus, and sobemovirus, have been shown to contain a genome-linked viral protein (VPg) attached to the 5'-end of the single-stranded genome RNA. In addition, two animal viruses, infectious pancreatic necrosis virus and drosophila X virus, have double-stranded RNA molecules with VPg (see Table 3; reviewed in 12–14). The presence of VPg in the viral RNA is essential for infectivity in most

**Table 3**   RNA viruses with genome-linked protein (VPg)

| Virus group | Virus | Genome[a] | VPg, size | Linkage | Refs. |
|---|---|---|---|---|---|
| Picornavirus | Poliovirus | ssRNA | 22 aa | Tyr-pU | (280–287) |
| | Encephalomyocarditis virus | ssRNA | ~10,000 and 8,000 | Tyr-pU | (308, 309) |
| | Foot and mouth disease virus | ssRNA | 24, 25, and 26 aa | Tyr-pU | (311–315) |
| | Hepatitis A | ssRNA | 21–23 aa | Tyr-pU? | (316, 317) |
| Calicivirus | Vesicular exanthema virus | ssRNA | ~10,000 | nd | (318, 319) |
| — | Infectious pancreatic necrosis virus | dsRNA (2) | ~110,000 | nd | (320, 321) |
| — | Drosophila X virus | dsRNA (2) | 67,000 | nd | (322, 323) |
| Comovirus | Cowpea mosaic virus | ssRNA (2) | 28aa | Ser-pU | (324–327, 335) |
| | Radish mosaic virus | ssRNA (2) | nd | Ser-pN | (336) |
| | Squash mosaic virus | ssRNA (2) | nd | nd | (337) |
| | Etches Ackerbohn mosaic virus | ssRNA (2) | nd | nd | (337) |
| Luteovirus | Potato leafroll virus | ssRNA | ~7,000 | nd | (338) |
| | Barley yellow dwarf virus | ssRNA | ~17,000 | nd | (339) |
| Nepovirus | Tobacco ringspot virus | ssRNA (2) | ~4,000 | nd | (340–342) |
| | Tomato black ring virus | ssRNA (2) | ~6,000 | nd | (340, 341, 343) |
| | Arabis mosaic virus | ssRNA (2) | ~4,000 | nd | (344) |
| | Strawberry latent ringspot virus | ssRNA (2) | ~4,000 | nd | (344) |
| | Raspberry ringspot virus | ssRNA (2) | ~4,000 | nd | (344) |
| | Cherry leafroll virus | ssRNA (2) | ~3,500 | nd | (345) |
| Potyvirus | Tobacco etch virus | ssRNA | ~6,000 | nd | (346) |
| | Tobacco vein mottling virus | ssRNA | ~24,000 | nd | (347) |
| | Plum pox virus | ssRNA | ~20,000 | nd | (348) |
| Sobemovirus | Southern bean mosaic virus | ssRNA | ~12,000 | nd | (349, 350) |
| | Pea enation mosaic virus | ssRNA | ~17,000 | nd | (351) |

[a] The number of RNA molecules that compose the genome is indicated in parentheses when more than one molecule is present.

cases, except for picornaviruses and comoviruses. Only relevant features not indicated in Table 3 are described below.

VPg was found to be covalently linked to the 5'-end of poliovirus RNA (280, 281), to nascent strands of poliovirus replicative intermediates, and to poly(U) of minus strands (281–283). These findings were consistent with the idea that the protein serves as a primer for the synthesis of both the plus and minus strands of poliovirus RNA. The tyrosine involved in the covalent linkage to dUMP is located at position 3 from the amino-end of VPg (288).

Antibodies against VPg specifically precipitated not only VPg but also the replicase precursor protein P3-1b (286, 288, 289) and other proteins; one of them, named P3-9 or pre-VPg-3, is the amino-terminal cleavage product of P3-1b and contains the sequence of protein VPg at its C-terminus (290, 291).

A poliovirus-specific RNA polymerase isolated from infected cells (292–294), lost after poly(U)-agarose chromatography, was recovered by addition of a host-factor or oligo(U) (295). The product of the replicase reactions, initiated either with host factor or with oligo(U), included full-length (35S) RNA molecules, largely in double-stranded form (296). In addition, in the presence of host factor, molecules approximately twice the length of the template were synthesized resulting from hairpin formation and template-directed elongation (297, 298). The products of this reaction contained newly synthesized plus-strand sequences as well as the expected minus-strand sequences. Products primed with oligo(U) were all of minus-strand polarity. Anti-VPg antibodies inhibited the host-factor-dependent poliovirus RNA synthesis by the poliovirus replicase, but not oligo(U)-primed synthesis. Complexes of VPg-related polypeptides covalently linked to newly made RNA of 50 to 150 nucleotides could be immunoprecipitated by the anti-VPg antibody. These findings implicated the VPg precursors in the de novo initiation of replicase products (299). Two forms of VPg were immunoprecipitated by anti-VPg antibodies from poliovirus-infected cells. One was free VPg and the other was VPg-pUpU (300). Use of a membrane fraction from poliovirus-infected HeLa cells made possible the in vitro synthesis of VPg-pU and VPg-pUpU (301, 302). The latter could be extended to longer RNA molecules, suggesting that the uridylylated protein may serve as a primer for RNA-dependent RNA synthesis (303). Moreover, incubation of VPg-pUpU with nucleoside triphosphates resulted in the formation of VPg-pUUAAAACAGp, which corresponds to the 5' sequence of the poliovirus plus-strand RNA.

An alternative model to that of priming by VPg, a VPg precursor or a uridylylated form of VPg, is that recently proposed by Tobin et al (304) implicating a hairpin formed at the 3'-end of poliovirus RNA as a primer for minus-strand synthesis by the polymerase. This model was based on the observation that minus-strand product RNA synthesized in vitro by the purified polymerase and host factor is covalently linked at one end to the template RNA (297, 305), and that purified terminal uridylyltransferase stimulates the initiation of RNA synthesis by the polymerase in vitro (306, 307). In the template-priming model, a mechanism should exist for the breakage of the phosphodiester bond linking the template and product RNAs and for linking VPg to minus-strand RNA. Tobin et al (304) have recently shown that VPg-linked product RNA was formed in a reaction that required VPg, $Mg^{2+}$, and a poliovirus RNA replication intermediate synthesized in vitro on poliovirus RNA. VPg was linked to the product RNA synthesized in vitro by a phosphoester bond between the 5' terminal phosphate in the RNA and the tyrosine residue in VPg. Thus, the synthesis of VPg-linked RNA requires the breakage of a phosphodiester bond in the RNA and the formation of a new phosphodiester bond between VPg and the product RNA. Because the VPg linkage reaction requires the template-linked product RNA and is inactive on oligo(U)-primed product RNA, the nucleophilic attack by VPg appears to take place at the terminal poly(A)-poly(U) hairpin. This model, although it accounts for the synthesis of the VPg-linked minus-strand RNA, does not explain the synthesis of the plus-strand RNA. It is possible that VPg priming is used for the synthesis of plus-strand RNA.

The VPg linked to the RNA from encephalomyocarditis virus (EMCV) (308) is derived from protein C, which contains the replicase gene (289). A crude membrane-bound replication complex isolated from EMCV-infected cells is able to initiate the synthesis of viral RNA. Both the formation of the primer, VPg-pU, and its utilization for the initiation of RNA chains take place in this system (310). The VPg of foot and mouth disease virus (FMDV) is heterogeneous (312). Sequence data revealed a region of the FMDV genome with a coding potential for three VPgs with 24, 25, and 26 amino acids, respectively, all of them containing a Tyr residue at position 3 from the N-end (313–315). The VPg of hepatitis A virus is 21–23 amino acids long and contains a tyrosine residue as a potential binding site for RNA in position 3 from its N-terminus (316, 317).

The genome of cowpea mosaic virus (CPMV) is composed of two RNA molecules, B-RNA and M-RNA (324, 325). The B-RNA is capable of self-replication in cowpea mesophyll protoplasts and codes for the VPg (328). B-RNA is translated into a 200-kDa polyprotein (329) that is processed to give a 60-kDa polypeptide (330), membrane-bound, which is the precursor of VPg (331, 332), located at the carboxy-terminus (333). Alignment of the 28-amino-acid sequence of CPMV VPg and poliovirus VPg shows eight identical residues and three basic amino acids located at similar positions (333, 334).

## CONCLUSION

The availability of purified in vitro replication systems, such as those of adenovirus and bacteriophage $\phi$29, has greatly contributed to the elucidation of the protein-priming mechanism of DNA replication. In all cases a specific DNA polymerase covalently links the 5' terminal dNMP to the OH group of a specific amino acid residue in the TP. In vitro systems have been developed also for the replication of bacteriophages PRD1 and Cp-1, showing that they use a similar protein-priming mechanism for the initiation of replication. A

TP has been characterized in most of the linear plasmids, and in the cases in which the DNA sequence is available, a putative DNA polymerase has been identified. Although in vitro systems are still lacking, it is very likely that linear plasmids also use the protein-priming mechanism for the initiation of replication. The TP of hepadnaviruses may function as a primer for the synthesis of minus-strand DNA, although direct evidence is still lacking. At the present time there is controversy as to the role of protein primers of the TP (VPg) linked at the 5'-end of RNA viruses, and further work is needed to answer this question.

ACKNOWLEDGMENTS

I thank all the authors who sent preprints and unpublished results of their work and my colleagues in the laboratory for many helpful discussions. I also thank Ms. Carmen Hermoso for efficiently typing the manuscript. Work in the author's laboratory was supported by grants from the National Institutes of Health (5 R01 GM27242-11), Dirección General de Investigación Científica y Técnica (PB87-0323), and Fundación Ramón Areces.

*Literature Cited*

1. Kornberg, A. 1980. *DNA Replication.* San Francisco: Freeman
2. Kornberg, A. 1982. *DNA Replication. Supplement.* San Francisco: Freeman
3. Daniell, E. 1976. *J. Virol.* 19:685–708
4. Kelly, T. J., Wold, M. S., Li, J. 1988. *Adv. Virus Res.* 34:1–42
5. Hay, R. T., Russell, W. C. 1989. *Biochem. J.* 258:3–16
6. Stillman, B. W. 1989. *Annu. Rev. Cell. Biol.* 5:197–245
7. van der Vliet, P. C. 1990. In *The Eukaryotic Nucleus: Molecular Biochemistry and Macromolecular Assemblies,* ed. P. Strauss, S. H. Wilson, pp. 1–32. New York: Telford
8. Salas, M. 1988. *Curr. Top. Microbiol. Immunol.* 136:72–88
9. Salas, M. 1988. In *The Bacteriophages,* ed. R. Calendar, 1:169–91. New York: Plenum
10. Meinhardt, F., Kempken, F., Kämper, J., Esser, K. 1990. *Curr. Genet.* 17:89–95
11. Mason, W. S., Taylor, J. M., Hull, R. 1987. *Adv. Virus Res.* 32:35–96
12. Vartapetian, A. B., Bogdanov, A. A. 1987. *Prog. Nucleic Acids Res. Mol. Biol.* 34:209–51
13. Wimmer, E. 1982. *Cell* 28:199–201
14. Daubert, S. D., Bruening, G. 1984. *Methods Virol.* 8:347–79
15. Steenbergh, P. H., Maat, J., Van Ormondt, H., Sussenbach, J. S. 1977. *Nucleic Acids Res.* 4:4371–89
16. Arrand, J., Roberts, R. J. 1979. *J. Mol. Biol.* 128:577–94
17. Shinagawa, M., Padmanabhan, R. 1979. *Biochem. Biophys. Res. Commun.* 87:671–78
18. Robinson, A. J., Younghusband, H. B., Bellett, A. J. D. 1973. *Virology* 56:54–69
19. Sharp, P. A., Moore, C., Haverty, J. L. 1976. *Virology* 75:442–56
20. Rekosh, D. M. K., Russell, W. C., Bellett, A. J. D. 1977. *Cell* 11:283–95
21. Carusi, E. A. 1977. *Virology* 76:380–94
22. Challberg, M. D., Desiderio, S. V., Kelly, T. J. Jr. 1980. *Proc. Natl. Acad. Sci. USA* 77:5105–9
23. Desiderio, S. V., Kelly, T. J. Jr. 1981. *J. Mol. Biol.* 145:319–37
24. Smart, J. E., Stillman, B. W. 1982. *J. Biol. Chem.* 257:13499–506
25. Challberg, M. D., Kelly, T. J. Jr. 1982. *Annu. Rev. Biochem.* 51:901–34
26. Lechner, R. L., Kelly, T. J. Jr. 1977. *Cell* 12:1007–20
27. Challberg, M. D., Kelly, T. J. Jr. 1979. *Proc. Natl. Acad. Sci. USA* 76:655–59
28. Challberg, M. D., Kelly, T. J. Jr. 1979. *J. Mol. Biol.* 135:999–1012
29. Lichy, J. H., Horwitz, M. S., Hurwitz, J. 1981. *Proc. Natl. Acad. Sci. USA* 78:2678–82

30. Challberg, M. D., Kelly, T. J. Jr. 1981. *J. Virol.* 38:272–77
31. Stillman, B. W., Lewis, J. B., Chow, L. T., Mathews, M. B., Smart, J. E. 1981. *Cell* 23:497–508
32. Stillman, B. W., Tamanoi, F., Mathews, M. B. 1982. *Cell* 31:613–23
33. Shu, L., Pettit, S. C., Engler, J. A. 1988. *Virology* 165:348–56
34. Shu, L., Horwitz, M. S., Engler, J. A. 1987. *Virology* 161:520–26
35. Pettit, S. C., Horwitz, M. S., Engler, J. A. 1988. *J. Virol.* 62:496–500
36. Stunnenberg, H. G., Lange, H., Philipson, L., van Miltenburg, R. T., van der Vliet, P. C. 1988. *Nucleic Acids Res.* 16:2431–44
37. Enomoto, T., Lichy, J. H., Ikeda, J. E., Hurwitz, J. 1981. *Proc. Natl. Acad. Sci. USA* 78:6779–83
38. Ostrove, J. M., Rosenfeld, P., Williams, J., Kelly, T. J. Jr. 1983. *Proc. Natl. Acad. Sci. USA* 80:935–39
39. Lichy, J. H., Field, J., Horwitz, M. S., Hurwitz, J. 1982. *Proc. Natl. Acad. Sci. USA* 79:5225–29
40. Friefeld, B. R., Lichy, J. H., Hurwitz, J., Horwitz, M. S. 1983. *Proc. Natl. Acad. Sci. USA* 80:1589–93
41. Field, J., Gronostajski, R. M., Hurwitz, J. 1984. *J. Biol. Chem.* 259:9487–95
42. Pettit, S. C., Horwitz, M. S., Engler, J. A. 1989. *J. Virol.* 63:5244–50
43. van der Vliet, P. C., Levine, A. J. 1973. *Nature New Biol.* 246:170–74
44. Kruijer, W., van Shaik, F. M. A., Sussenbach, J. S. 1981. *Nucleic Acids Res.* 9:4439–57
45. van der Vliet, P. C., Keegstra, W., Jansz, H. S. 1978. *Eur. J. Biochem.* 86:389–98
46. Fowlkes, D. M., Lord, S. T., Linné, T., Pettersson, U., Philipson, L. 1979. *J. Mol. Biol.* 132:163–80
47. Schechter, N. M., Davies, W., Anderson, C. W. 1980. *Biochemistry* 19:2802–10
48. Horwitz, M. S. 1978. *Proc. Natl. Acad. Sci. USA* 75:4291–95
49. Kaplan, L. M., Ariga, H., Hurwitz, J., Horwitz, M. S. 1979. *Proc. Natl. Acad. Sci. USA* 76:5534–38
50. Ariga, H., Klein, H., Levine, A. J., Horwitz, M. S. 1980. *Virology* 101:307–10
51. Challberg, M. D., Ostrove, J. M., Kelly, T. J. Jr. 1982. *J. Virol.* 41:265–70
52. Friefeld, B. R., Krevolin, M. D., Horwitz, M. S. 1983. *Virology* 124:380–89
53. Van Bergen, B. G. M., van der Vliet, P. C. 1983. *J. Virol.* 42:642–48
54. Prelich, G., Stillman, B. W. 1986. *J. Virol.* 57:883–92
55. Ikeda, J. E., Enomoto, T., Hurwitz, J. 1981. *Proc. Natl. Acad. Sci. USA* 78:884–88
56. Nagata, K., Guggenheimer, R. A., Hurwitz, J. 1983. *Proc. Natl. Acad. Sci. USA* 80:4266–70
57. Nagata, K., Guggenheimer, R. A., Enomoto, T., Lichy, J. H., Hurwitz, J. 1982. *Proc. Natl. Acad. Sci. USA* 79:6438–42
58. de Vries, E., van Driel, W., Tromp, M., van Boom, J., van der Vliet, P. C. 1985. *Nucleic Acids Res.* 13:4935–52
59. Lindenbaum, J. O., Field, J., Hurwitz, J. 1986. *J. Biol. Chem.* 261:10218–27
60. Nagata, K., Guggenheimer, R. A., Hurwitz, J. 1983. *Proc. Natl. Acad. Sci. USA* 80:6177–81
61. Leegwater, P. A. J., van Driel, W., van der Vliet, P. C. 1985. *EMBO J.* 4:1515–21
62. Diffley, J. F. X., Stillman, B. 1986. *Mol. Cell. Biol.* 6:1363–73
63. Rosenfeld, P. J., Kelly, T. J. 1986. *J. Biol. Chem.* 261:1398–408
64. Rosenfeld, P. J., O'Neill, E. A., Wides, R. J., Kelly, T. J. 1987. *Mol. Cell. Biol.* 7:875–86
65. de Vries, E., van Driel, W., van der Heuvel, S. J. L., van der Vliet, P. C. 1987. *EMBO J.* 6:161–68
66. Jones, K. A., Kadonaga, J. T., Rosenfeld, P. J., Kelly, T. J., Tjian, R. 1987. *Cell* 48:79–89
67. Mermod, N., O'Neill, E. A., Kelly, T. J., Tjian, R. 1989. *Cell* 58:741–53
68. Gounari, F., de Francesco, R., Schmitt, J., van der Vliet, P. C., Cortese, R., et al. 1990. *EMBO J.* 9:559–66
69. Cleat, P. H., Hay, R. T. 1989. *EMBO J.* 8:1841–48
70. Stuiver, M. H., van der Vliet, P. C. 1990. *J. Virol.* 65:379–86
71. Pruijn, G. J. M., van Driel, W., van der Vliet, P. C. 1986. *Nature* 322:656–59
72. O'Neill, E. A., Kelly, T. J. 1988. *J. Biol. Chem.* 263:931–37
73. Pruijn, G. J. M., van Driel, W., van Miltenburg, R. T., van der Vliet, P. C. 1987. *EMBO J.* 6:3771–78
74. Pruijn, G. J. M., van Miltenburg, R. T., Claessens, J. A. J., van der Vliet, P. C. 1988. *J. Virol.* 62:3092–3102
75. O'Neill, E. A., Fletcher, C., Burrow, C. R., Heintz, N., Roeder, R. G., et al. 1988. *Science* 241:1210–13
76. Pruijn, G. J. M., van der Vliet, P. C., Dathan, N. A., Mattaj, I. W. 1989. *Nucleic Acids Res.* 17:1845–63
77. Tamanoi, F., Stillman, B. W. 1982. *Proc. Natl. Acad. Sci. USA* 79:2221–25
78. van Bergen, B. G. M., van der Ley, P. A., van Driel, W., van Mansfeld, A. D.

M., van der Vliet, P. C. 1983. *Nucleic Acids Res.* 11:1975–89
79. Tamanoi, F., Stillman, B. W. 1983. *Proc. Natl. Acad. Sci. USA* 80:6446–50
80. Guggenheimer, R. A., Stillman, B. W., Nagata, K., Tamanoi, F., Hurwitz, J. 1984. *Proc. Natl. Acad. Sci. USA* 81: 3069–73
81. Lally, C., Dörper, T., Gröger, W., Antoine, G., Winnacker, E. L. 1984. *EMBO J.* 3:333–37
82. Rawlins, D. R., Rosenfeld, P. J., Wides, R. J., Challberg, M. D., Kelly, T. J. Jr. 1984. *Cell* 37:309–19
83. Wides, R. J., Challberg, M. D., Rawlins, D. R., Kelly, T. J. 1987. *Mol. Cell. Biol.* 7:864–74
84. Tolun, A., Aleström, P., Pettersson, U. 1979. *Cell* 17:705–13
85. Shinagawa, M., Ishiyama, T., Padmanabhan, R., Fujinaga, K., Kamada, M., et al. 1983. *Virology* 125:491–95
86. Aleström, P., Stenlund, A., Li, P., Pettersson, U. 1982. *Gene* 18:193–97
87. Stillman, B. W., Topp, W. C., Engler, J. A. 1982. *J. Virol.* 44:530–37
88. Challberg, M. D., Rawlins, D. R. 1984. *Proc. Natl. Acad. Sci. USA* 81:100–4
89. Rijnders, A. W. M., van Bergen, B. G. M., van der Vliet, P. C., Sussenbach, J. S. 1983. *Nucleic Acids Res.* 11:8777–89
90. Adhya, S., Shneidman, P. S., Hurwitz, J. 1986. *J. Biol. Chem.* 261:3339–46
91. Hay, R. T., Stow, N. D., McDougall, I. M. 1984. *J. Mol. Biol.* 175:493–510
92. Hay, R. T. 1985. *EMBO J.* 4:421–26
93. Wang, K., Pearson, G. D. 1985. *Nucleic Acids Res.* 13:5173–87
94. Hay, R. T., Clark, L., Cleat, P. H., Harris, M. P. G., Robinson, E. C., et al. 1988. *Cancer Cells* 6:71–75
95. Hay, R. T. 1985. *J. Mol. Biol.* 186:129–36
96. Harris, M. P. G., Hay, R. T. 1988. *J. Mol. Biol.* 201:57–67
97. Guggenheimer, R. A., Nagata, K., Kenny, M., Hurwitz, J. 1984. *J. Biol. Chem.* 259:7815–25
98. Kenny, M. K., Balogh, L. A., Hurwitz, J. 1988. *J. Biol. Chem.* 263:9801–8
99. Kenny, M. K., Hurwitz, J. 1988. *J. Biol. Chem.* 263:9809–17
100. Ikeda, J. E., Enomoto, T., Hurwitz, J. 1982. *Proc. Natl. Acad. Sci. USA* 79: 2442–46
101. Leegwater, P. A. J., Rombouts, R. F. A., van der Vliet, P. C. 1988. *Biochim. Biophys. Acta* 951:403–10
102. Lippé, R., Graham, F. L. 1989. *J. Virol.* 63:5133–41
103. Stow, N. D. 1982. *Nucleic Acids Res.* 10:5105–19

104. Yoshikawa, H., Ito, J. 1982. *Gene* 17: 323–35
105. Garvey, K. J., Yoshikawa, H., Ito, J. 1985. *Gene* 40:301–9
106. Vlček, C., Pačes, V. 1986. *Gene* 46: 215–25
107. Ortín, J., Viñuela, E., Salas, M., Vásquez, C. 1971. *Nature New Biol.* 234: 275–77
108. Hirokawa, H. 1972. *Proc. Natl. Acad. Sci. USA* 69:1555–59
109. Yanofsky, S., Kawamura, F., Ito, J. 1976. *Nature* 259:60–63
110. Salas, M., Mellado, R. P., Viñuela, E., Sogo, J. M. 1978. *J. Mol. Biol.* 119:269–91
111. Yehle, C. O. 1978. *J. Virol.* 27:776–83
112. Ito, J. 1978. *J. Virol.* 28:895–904
113. Geiduschek, E. P., Ito, J. 1982. In *The Molecular Biology of the Bacilli*, ed. D. A. Dubnau, 1:203–45. London: Academic
114. Fučik, V., Grunow, R., Grünnerová, H., Hostomský, Z., Zadrazil, S. 1980. In *DNA: Recombination, Interactions and Repair*, ed. S. Zadrazil, J. Sponar, 63:111–18. Oxford: Pergamon
115. Yoshikawa, H., Ito, J. 1981. *Proc. Natl. Acad. Sci. USA* 78:2596–600
116. Gutiérrez, J., Vinós, J., Prieto, I., Méndez, E., Hermoso, J. M., et al. 1986. *Virology* 155:474–83
117. Pačes, V., Vlček, C., Urbánek, P., Hostomský, Z. 1985. *Gene* 38:45–56
118. Escarmís, C., Salas, M. 1982. *Nucleic Acids Res.* 10:5785–98
119. Leavitt, M. C., Ito, J. 1987. *Nucleic Acids Res.* 15:5251–59
120. Hermoso, J. M., Salas, M. 1980. *Proc. Natl. Acad. Sci. USA* 77:6425–28
121. Hermoso, J. M., Méndez, E., Soriano, F., Salas, M. 1985. *Nucleic Acids Res.* 13:7715–28
122. Inciarte, M. R., Salas, M., Sogo, J. M. 1980. *J. Virol.* 28:895–904
123. Harding, N. E., Ito, J. 1980. *Virology* 104:323–38
124. Sogo, J. M., García, J. A., Peñalva, M. A., Salas, M. 1982. *Virology* 116:1–18
125. Talavera, A., Salas, M., Viñuela, E. 1972. *Eur. J. Biochem.* 31:367–71
126. Carrascosa, J. L., Camacho, A., Moreno, F., Jiménez, F., Mellado, R. P., et al. 1976. *Eur. J. Biochem.* 66:229–41
127. Hagen, E. W., Reilly, B. E., Tosi, M. E., Anderson, D. L. 1976. *J. Virol.* 19:501–17
128. Prieto, I., Méndez, E., Salas, M. 1989. *Gene* 77:195–204
129. Mellado, R. P., Peñalva, M. A., Inciarte, M. R., Salas, M. 1980. *Virology* 104:84–96

130. Matsumoto, K., Saito, T., Hirokawa, H. 1983. *Mol. Gen. Genet.* 191:26–30
131. Peñalva, M. A., Salas, M. 1982. *Proc. Natl. Acad. Sci. USA* 79:5522–26
132. Shih, M. F., Watabe, K., Ito, J. 1982. *Biochem. Biophys. Res. Commun.* 105:1031–36
133. Watabe, K., Shih, M. F., Sugino, A., Ito, J. 1982. *Proc. Natl. Acad. Sci. USA* 79:5245–58
134. Shih, M. F., Watabe, K., Yoshikawa, H., Ito, J. 1984. *Virology* 133:56–64
135. Blanco, L., García, J. A., Peñalva, M. A., Salas, M. 1983. *Nucleic Acids Res.* 11:1309–23
136. Blanco, L., García, J. A., Salas, M. 1984. *Gene* 29:33–40
137. García, J. A., Pastrana, R., Prieto, I., Salas, M. 1983. *Gene* 21:65–76
138. Watabe, K., Leusch, M., Ito, J. 1984. *Proc. Natl. Acad. Sci. USA* 81:5374–78
139. Blanco, L., Salas, M. 1984. *Proc. Natl. Acad. Sci. USA* 81:5325–29
140. Prieto, I., Lázaro, J. M., García, J. A., Hermoso, J. M., Salas, M. 1984. *Proc. Natl. Acad. Sci. USA* 81:1639–43
141. Watabe, K., Ito, J. 1983. *Nucleic Acids Res.* 11:8333–42
142. Watabe, K., Leusch, M., Ito, J. 1984. *Biochem. Biophys. Res. Commun.* 123:1019–26
143. Blanco, L., Salas, M. 1985. *Nucleic Acids Res.* 13:1239–49
144. Blanco, L., Salas, M. 1985. *Proc. Natl. Acad. Sci. USA* 82:6404–8
145. Blanco, L., Prieto, I., Gutiérrez, J., Bernad, A., Lázaro, J. M., et al. 1987. *J. Virol.* 61:3983–91
146. Watabe, K., Shih, M. F., Ito, J. 1983. *Proc. Natl. Acad. Sci. USA* 80:4248–52
147. Matsumoto, K., Saito, T., Kim, C. I., Ando, T., Hirokawa, H. 1984. *Mol. Gen. Genet.* 196:381–86
148. Blanco, L., Bernad, A., Lázaro, J. M., Martín, G., Garmendia, C., et al. 1989. *J. Biol. Chem.* 264:8935–40
149. Zaballos, A., Salas, M., Mellado, R. P. 1986. *Gene* 43:106–10
150. Zaballos, A., Mellado, R. P., Salas, M. 1988. *Gene* 63:113–21
151. Zaballos, A., Lázaro, J. M., Méndez, E., Mellado, R. P., Salas, M. 1989. *Gene* 83:187–95
152. Zaballos, A., Salas, M. 1989. *Nucleic Acids Res.* 17:10353–66
153. Tomalski, M. D., Wu, J., Miller, L. K. 1988. *Virology* 167:591–600
154. Reha-Krantz, L. J. 1988. *J. Mol. Biol.* 202:711–24
155. Garmendia, C., Salas, M., Hermoso, J. M. 1988. *Nucleic Acids Res.* 16:5727–40
156. Garmendia, C., Hermoso, J. M., Salas, M. 1990. *Gene* 88:73–79
157. Kobayashi, H., Matsumoto, K., Misawa, S., Miura, K., Hirokawa, H. 1989. *Mol. Gen. Genet.* 220:8–11
158. Huberman, J. A. 1981. *Cell* 23:647–48
159. Khan, N. N., Wright, G. E., Dudycz, L. W., Brown, N. C. 1984. *Nucleic Acids Res.* 12:3695–706
160. Blanco, L., Salas, M. 1986. *Virology* 153:179–87
161. Bernad, A., Zaballos, A., Salas, M., Blanco, L. 1987. *EMBO J.* 6:4219–25
162. Matsumoto, K., Kim, C. I., Urano, S., Ohashi, M., Hirokawa, H. 1986. *Virology* 152:32–38
163. Larder, B. A., Kemp, S. D., Darby, G. 1987. *EMBO J.* 6:169–75
164. Knopf, C. W. 1987. *J. Gen. Virol.* 68:1429–33
165. Savilahti, H., Bamford, D. H. 1987. *Gene* 57:121–30
166. Jung, G., Leavitt, M. C., Hsieh, J. C., Ito, J. 1987. *Proc. Natl. Acad. Sci. USA* 84:8287–91
167. Fukuhara, H. 1987. *Nucleic Acids Res.* 15:10046
168. Kuzmin, E. V., Levchenko, I. V. 1987. *Nucleic Acids Res.* 15:6758
169. Tommasino, M., Ricci, S., Galeotti, C. L. 1988. *Nucleic Acids Res.* 16:5863–78
170. Wong, S. W., Wahl, A. F., Yuan, P. M., Arai, N., Pearson, B. E., et al. 1988. *EMBO J.* 7:37–47
171. Spicer, E. K., Rush, J., Fung, C., Reha-Krantz, L. J., Karam, J. D., et al. 1988. *J. Biol. Chem.* 263:7478–85
172. Pizzagalli, A., Valsasnini, P., Plevani, P., Lucchini, G. 1988. *Proc. Natl. Acad. Sci. USA* 85:3772–76
173. Kempken, F., Meinhardt, F., Esser, K. 1989. *Mol. Gen. Genet.* 218:523–30
174. Oeser, B., Tudzynski, P. 1989. *Mol. Gen. Genet.* 217:132–40
175. Matsumoto, K., Takano, H., Kim, C. I., Hirokawa, H. 1989. *Gene* 84:247–55
176. Gibbs, J. S., Chiou, H. C., Hall, J. D., Mount, D. W., Retondo, M. J., et al. 1985. *Proc. Natl. Acad. Sci. USA* 82:7969–73
177. Bernad, A., Lázaro, J. M., Salas, M., Blanco, L. 1990. *Proc. Natl. Acad. Sci. USA* 87:4610–14
178. Blasco, M. A., Blanco, L., Parés, E., Salas, M., Bernad, A. 1990. *Nucleic Acids Res.* 18:4763–70
179. Pandey, V. N., Stone, K. L., Williams, K. R., Modak, M. J. 1987. *Biochemistry* 26:7744–48
180. Rush, J., Konigsberg, W. H. 1990. *J. Biol. Chem.* 265:4821–27
181. Ollis, D. L., Brick, R., Hamlin, R.,

Xuong, N. G., Steitz, T. A. 1985. *Nature* 313:762–66
182. Derbyshire, V., Freemont, P. S., Sanderson, M. R., Beese, L., Friedman, J. M., et al. 1988. *Science* 240:199–201
183. Freemont, P. S., Friedman, J. M., Beese, L. S., Sanderson, M. R., Steitz, T. A. 1988. *Proc. Natl. Acad. Sci. USA* 85:8924–28
184. Bernad, A., Blanco, L., Lázaro, J. M., Martín, G., Salas, M. 1989. *Cell* 59:219–28
185. Leavitt, M. C., Ito, J. 1989. *Proc. Natl. Acad. Sci. USA* 86:4465–69
186. Escarmís, C., Guirao, D., Salas, M. 1989. *Virology* 169:152–60
187. Benes, V., Arnold, L., Smrt, J., Pačes, V. 1989. *Gene* 75:341–47
188. Pastrana, R., Lázaro, J. M., Blanco, L., García, J. A., Méndez, E., et al. 1985. *Nucleic Acids Res.* 13:3083–100
189. Blanco, L., Gutiérrez, J., Lázaro, J. M., Bernad, A., Salas, M. 1986. *Nucleic Acids Res.* 14:4923–37
190. Blanco, L., Bernad, A., Salas, M. 1988. *J. Virol.* 62:4167–72
191. Prieto, I., Serrano, M., Lázaro, J. M., Salas, M., Hermoso, J. M. 1988. *Proc. Natl. Acad. Sci. USA* 85:314–18
192. Serrano, M., Gutiérrez, J., Prieto, I., Hermoso, J. M., Salas, M. 1989. *EMBO J.* 8:1879–85
193. Serrano, M., Salas, M., Hermoso, J. M. 1990. *Science* 248:1012–16
194. Otero, M. J., Salas, M. 1989. *Nucleic Acids Res.* 17:4567–77
195. Otero, M. J., Lázaro, J. M., Salas, M. 1990. *Gene* 95:25–30
196. Martín, G., Salas, M. 1988. *Gene* 67: 193–201
197. Martín, G., Lázaro, J. M., Méndez, E., Salas, M. 1989. *Nucleic Acids Res.* 17:3663–72
198. Gutiérrez, J., García, J. A., Blanco, L., Salas, M. 1986. *Gene* 43:1–11
199. Gutiérrez, J., Garmendia, C., Salas, M. 1988. *Nucleic Acids Res.* 16:5895–914
200. Mindich, L., Bamford, D. H. 1988. In *The Bacteriophages*, ed. R. Calendar, 2:475–520. New York: Plenum
201. Bamford, D. H., McGraw, T., MacKenzie, G., Mindich, L. 1983. *J. Virol.* 47:311–16
202. Savilahti, H., Bamford, D. H. 1986. *Gene* 49:199–205
203. Gerendasy, D. D., Ito, J. 1987. *J. Virol.* 61:549–96
204. Mindich, L., Bamford, D. H., Goldthwait, C., Laverty, M., MacKenzie, G. 1982. *J. Virol.* 44:1013–20
205. Bamford, D. H., Mindich, L. 1984. *J. Virol.* 50:309–15

206. Hsieh, J. C., Jung, G., Leavitt, M. C., Ito, J. 1987. *Nucleic Acids Res.* 15:8999–9009
207. Pakula, T., Savilahti, H., Bamford, D. H. 1989. *Gene* 85:53–58
208. Gerendasy, D., Ito, J. 1990. *J. Bacteriol.* 172:1889–98
209. Yoo, S. K., Ito, J. 1989. *Virology* 170: 442–49
210. Savilahti, H., Caldentey, J., Bamford, D. H. 1989. *Gene* 85:45–51
211. García, E., Gómez, A., Ronda, C., Escarmís, C., López, R. 1983. *Virology* 128:92–104
212. López, R., Ronda, C., García, P., Escarmís, C., García, E. 1984. *Mol. Gen. Genet.* 197:67–74
213. Escarmís, C., Gómez, A., García, E., Ronda, C., López, R., et al. 1984. *Virology* 133:166–71
214. Escarmís, C., García, P., Méndez, E., López, R., Salas, M., et al. 1985. *Gene* 36:341–48
215. García, P., Hermoso, J. M., García, J. A., García, E., López, R., et al. 1986. *J. Virol.* 58:31–35
216. Bernheimer, H. P. 1977. *Science* 195: 66–68
217. Romero, A., López, R., Lurz, R., García, P. 1990. *J. Virol.* 64:5149–55
218. Herrera-Estrella, A., Chen, Z., van Montagu, M., Wang, K. 1988. *EMBO J.* 7:4055–62
219. Dürrenberger, F., Crameri, A., Hohn, B., Koukolikova-Nicola, Z. 1989. *Proc. Natl. Acad. Sci. USA* 86:9154–58
220. Sakaguchi, K. 1990. *Microbiol. Rev.* 54:66–74
221. Hirochika, H., Sakaguchi, K. 1982. *Plasmid* 7:59–65
222. Hirochika, H., Nakamura, K., Sakaguchi, K. 1984. *EMBO J.* 3:761–66
223. Keen, C. L., Mendelovitz, S., Cohen, G., Aharonowitz, Y., Roy, K. L. 1988. *Mol. Gen. Genet.* 212:172–76
224. Kinashi, H., Shimaji, M. 1987. *J. Antibiot.* 40:913–16
225. Kinashi, H., Shimaji, M., Sakai, A. 1987. *Nature* 328:454–56
226. Kinashi, H., Shimaji, M., Hanafusa, T. 1990. *Abstr. 1990 UCLA Symp. Mol. Biol. Streptomyces. J. Cell. Biochem.*, *Suppl.* 14A, p. 100
227. Hishinuma, F., Nakamura, K., Hirai, K., Nishizawa, R., Gunge, N., et al. 1984. *Nucleic Acids Res.* 12:7581–97
228. Stark, M. J. R., Mileham, A. J., Romanos, M. A., Boyd, A. 1984. *Nucleic Acids Res.* 12:6011–30
229. Sor, F., Wesolowski, M., Fukuhara, H. 1983. *Nucleic Acids Res.* 11:5037–44

230. Kikuchi, Y., Hirai, K., Hishinuma, F. 1984. *Nucleic Acids Res.* 12:5685–92
231. Stam, J. C., Kwakman, J., Meijer, M., Stuitje, A. R. 1986. *Nucleic Acids Res.* 14:6871–84
232. Fujimura, H., Yamada, T., Hishinuma, F., Gunge, N. 1988. *FEMS Microbiol. Lett.* 49:441–44
233. Jung, G., Leavitt, M. C., Ito, J. 1987. *Nucleic Acids Res.* 15:9088
234. Kikuchi, Y., Hirai, K., Gunge, N., Hishinuma, F. 1985. *EMBO J.* 4:1881–86
235. Kitada, K., Gunge, N. 1988. *Mol. Gen. Genet.* 215:46–52
236. Kitada, K., Hishinuma, F. 1987. *Mol. Gen. Genet.* 206:377–81
237. Francou, F. 1981. *Mol. Gen. Genet.* 184:440–44
238. Meinhardt, F., Kempken, F., Esser, K. 1986. *Curr. Genet.* 11:243–46
239. Meinhardt, F., Esser, K. 1987. *Appl. Microbiol. Biotechnol.* 27:276–82
240. Giasson, L., Lalonde, M. 1987. *Curr. Genet.* 11:331–34
241. Normand, P., Simonet, P., Giasson, L., Ravel-Chapins, P., Fortin, J. A., et al. 1987. *Curr. Genet.* 11:335–38
242. Hashiba, T., Homma, Y., Hyakumachi, M., Matsuda, I. 1984. *J. Gen. Microbiol.* 130:2067–70
243. Mohan, M., Meyer, R. J., Anderson, J. B., Horgen, P. A. 1984. *Curr. Genet.* 8:615–19
244. Bertrand, H., Griffiths, A. J. F., Court, D. A., Cheng, C. K. 1986. *Cell* 47:829–37
245. Bertrand, H., Griffiths, A. J. F. 1989. *Genome* 31:155–59
246. Kistler, H. C., Leong, S. A. 1986. *J. Bacteriol.* 167:587–93
247. Düvell, A., Hessberg-Stutzke, H., Oeser, B., Rogmann-Backwinkel, P., Tudzynski, P. 1988. *Mol. Gen. Genet.* 214:128–34
248. Samac, D. A., Leong, S. A. 1988. *Plasmid* 19:57–67
249. Samac, D. A., Leong, S. A. 1989. *Curr. Genet.* 16:187–94
250. Yui, Y., Katayose, Y., Shishido, K. 1988. *Biochim. Biophys. Acta* 951:53–60
251. Katayose, Y., Kajiwara, S., Shishido, K. 1990. *Nucleic Acids Res.* 18:1395–400
252. Turpen, T., Garger, S. J., Marks, M. D., Grill, L. K. 1987. *Mol. Gen. Genet.* 209:227–33
253. Saumitou-Laprade, P., Pannenbecker, G., Maggouta, F., Jean, R., Michaelis, G. 1989. *Curr. Genet.* 16:181–86
254. Pring, D. R., Conde, M. F., Schertz, K.

F., Levings, C. S. III. 1982. *Mol. Gen. Genet.* 186:180–84
255. Chase, C. D., Pring, D. R. 1986. *Plant Mol. Biol.* 6:53–64
256. Paillard, M., Sederoff, R. R., Levings, C. S. III. 1985. *EMBO J.* 4:1125–28
257. Levings, C. S. III, Sederoff, R. R. 1983. *Proc. Natl. Acad. Sci. USA* 80:4055–59
258. Leon, P., Walbot, V., Bedinger, P. 1989. *Nucleic Acids Res.* 17:4089–99
259. Bedinger, P., Hostos, E. L., Leon, P., Walbot, V. 1986. *Mol. Gen. Genet.* 205:206–12
260. Kemble, R. J., Thompson, R. D. 1982. *Nucleic Acids Res.* 10:8181–90
261. Summers, J., O'Connell, A., Millman, I. 1975. *Proc. Natl. Acad. Sci. USA* 72:4597–601
262. Sattler, F., Robinson, W. S. 1979. *J. Virol.* 32:226–33
263. Gerlich, W. H., Robinson, W. S. 1980. *Cell* 21:801–9
264. Molnar-Kimber, K. L., Summers, J., Taylor, J. M., Mason, W. S. 1983. *J. Virol.* 45:165–72
265. Weiser, B., Ganem, D., Seeger, C., Varmus, H. E. 1983. *J. Virol.* 48:1–9
266. Summers, J., Mason, W. S. 1982. *Cell* 29:403–15
267. Buscher, M., Reiser, W., Will, H., Schaller, H. 1985. *Cell* 40:717–24
268. Molnar-Kimber, K. L., Summers, J. W., Mason, S. W. 1984. *J. Virol.* 51:181–91
269. Seeger, C., Ganem, D., Varmus, H. E. 1986. *Science* 232:477–84
270. Lien, J. M., Aldrich, C. E., Mason, W. S. 1986. *J. Virol.* 57:229–36
271. Galibert, F., Chen, T. N., Mandart, E. 1982. *J. Virol.* 41:51–65
272. Seeger, C., Ganem, D., Varmus, H. E. 1984. *J. Virol.* 51:367–75
273. Kodama, K., Ogasawara, N., Yoshikawa, H., Murakami, S. 1985. *J. Virol.* 56:978–86
274. Mandart, E., Kay, A., Galibert, F. 1984. *J. Virol.* 49:782–92
275. Sprengel, R., Kuhn, C., Will, H., Schaller, H. 1985. *J. Med. Virol.* 15:323–33
276. Bosch, V., Bartenschlager, R., Radziwill, G., Schaller, H. 1988. *Virology* 166:475–85
277. Bartenschlager, R., Schaller, H. 1988. *EMBO J.* 7:4185–92
278. Khudyakov, Y. E., Makhov, A. M. 1989. *FEBS Lett.* 243:115–18
279. Toh, H., Hayashida, H., Miyata, T. 1983. *Nature* 305:829–31
280. Lee, Y. F., Nomoto, A., Detjen, B.,

Wimmer, E. 1977. *Proc. Natl. Acad. Sci. USA* 74:59–63

281. Flanegan, J. B., Pettersson, R. F., Ambros, V., Hewlett, M. J., Baltimore, D. 1977. *Proc. Natl. Acad. Sci. USA* 74:961–65

282. Nomoto, A., Detjen, B., Pozzati, R., Wimmer, E. 1977. *Nature* 268:208–13

283. Pettersson, R. F., Ambros, V., Baltimore, D. 1978. *J. Virol.* 27:357–65

284. Rothberg, P. G., Harris, J. R. T., Nomoto, A., Wimmer, E. 1978. *Proc. Natl. Acad. Sci. USA* 75:4868–72

285. Ambros, V., Baltimore, D. 1978. *J. Biol. Chem.* 253:5263–66

286. Kitamura, N., Adler, C. J., Rothberg, P. G., Martinko, J., Nathenson, S. G., et al. 1980. *Cell* 21:295–302

287. Racaniello, V. R., Baltimore, D. 1981. *Proc. Natl. Acad. Sci. USA* 78:4887–91

288. Kitamura, N., Semler, B. L., Rothberg, P. G., Larsen, G. R., Adler, C. J., et al. 1981. *Nature* 291:547–53

289. Pallansch, M. A., Kew, O. M., Palmenberg, A. C., Golini, F., Wimmer, E., et al. 1980. *J. Virol.* 35:414–19

290. Semler, B. L., Anderson, C. W., Hanecak, R., Dorner, L. F., Wimmer, E. 1982. *Cell* 28:405–12

291. Baron, M. H., Baltimore, D. 1982. *Cell* 28:395–404

292. Dasgupta, A., Baron, M. H., Baltimore, D. 1979. *Proc. Natl. Acad. Sci. USA* 76:2679–83

293. Flanegan, J. B., van Dyke, T. A. 1979. *J. Virol.* 32:155–61

294. Tuschall, D. M., Hiebert, E., Flanegan, J. B. 1982. *J. Virol.* 44:209–16

295. Dasgupta, A., Zabel, P., Baltimore, D. 1980. *Cell* 19:423–29

296. Baron, M. H., Baltimore, D. 1982. *J. Biol. Chem.* 257:12359–66

297. Young, D. C., Tuschall, D. M., Flanegan, J. B. 1985. *J. Virol.* 54:256–64

298. Hey, T. D., Richards, O. C., Ehrenfeld, E. 1986. *J. Virol.* 58:790–96

299. Baron, M. H., Baltimore, D. 1982. *Cell* 30:745–52

300. Crawford, N. M., Baltimore, D. 1983. *Proc. Natl. Acad. Sci. USA* 80:7452–55

301. Takegami, T., Kuhn, R. J., Anderson, C. W., Wimmer, E. 1983. *Proc. Natl. Acad. Sci. USA* 80:7447–51

302. Takeda, N., Kuhn, R. J., Yang, C. F., Takegami, T., Wimmer, E. 1986. *J. Virol.* 60:43–53

303. Takegami, T., Semler, B. L., Anderson, C. W., Wimmer, E. 1983. *Virology* 128:33–47

304. Tobin, G. J., Young, D. C., Flanegan, J. B. 1989. *Cell* 59:511–19

305. Young, D. C., Dunn, B. M., Tobin, G. J., Flanegan, J. B. 1986. *J. Virol.* 58:715–23

306. Andrews, N. C., Levin, D., Baltimore, D. 1985. *J. Biol. Chem.* 260:7628–35

307. Andrews, N. C., Baltimore, D. 1986. *Proc. Natl. Acad. Sci. USA* 83:221–25

308. Hruby, D. E., Roberts, W. K. 1978. *J. Virol.* 25:413–15

309. Vartapetian, A. B., Drygin, Yu. F., Chumakov, K. M., Bogdanov, A. A. 1980. *Nucleic Acids Res.* 8:3729–42

310. Vartapetian, A. B., Koonin, E. V., Agol, V. I., Bogdanov, A. A. 1984. *EMBO J.* 3:2593–98

311. Sangar, D. V., Rowlands, D. J., Harris, T. J. R., Brown, F. 1977. *Nature* 268:648–50

312. King, A. M. Q., Sangar, D. V., Harris, T. J. R., Brown, F. 1980. *J. Virol.* 34:627–34

313. Forss, S., Strebel, K., Beck, E., Schaller, H. 1984. *Nucleic Acids Res.* 12:6587–601

314. Carroll, A. R., Rowlands, D. J., Clarke, B. E. 1974. *Nucleic Acids Res.* 12:2461–72

315. Robertson, B. H., Grubman, M. J., Wendell, G. N., Moore, D. M., Welsh, J. D., et al. 1985. *J. Virol.* 54:651–60

316. Weitz, M., Baroudy, B. M., Maloy, W. L., Ticehurst, J. R., Purcell, R. H. 1986. *J. Virol.* 60:124–30

317. Paul, A. V., Tada, H., von der Helm, K., Wissel, T., Kiehn, R., et al. 1987. *Virus Res.* 8:153–71

318. Burroughs, J. N., Brown, F. 1978. *J. Gen. Virol.* 41:443–46

319. Black, D. N., Burroughs, J. N., Harris, T. J. R., Brown, F. 1978. *Nature* 274:614–15

320. Dobos, P. 1976. *Nucleic Acids Res.* 3:1903–24

321. Persson, R. H., MacDonald, R. D. 1982. *J. Virol.* 44:437–43

322. Dobos, P., Hill, B. J., Hallett, R., Kells, D. T. C., Becht, H., et al. 1979. *J. Virol.* 32:593–605

323. Revet, B., Delain, E. 1982. *Virology* 123:29–44

324. Lomonossoff, G., Shanks, M. 1983. *EMBO J.* 2:2253–58

325. Van Wezenbeek, P., Verver, J., Harmsen, J., Vos, P., van Kammen, A. 1983. *EMBO J.* 2:941–46

326. Stanley, J., Rottier, P., Davies, J. W., Zabel, P., van Kammen, A. 1978. *Nucleic Acids Res.* 5:4505–22

327. Daubert, S. D., Bruening, G., Najarian, R. C. 1978. *Eur. J. Biochem.* 92:45–51

328. Stanley, J., Goldbach, R., van Kammen, A. 1980. *Virology* 106:180–82

329. Rezelman, G., Goldbach, R., van Kammen, A. 1980. *J. Virol.* 36:366–73
330. Goldbach, R., Rezelman, G. 1983. *J. Virol.* 46:614–19
331. Zabel, P., Moerman, M., van Straaten, F., Goldbach, R., van Kammen, A. 1982. *J. Virol.* 41:1083–88
332. Goldbach, R., Rezelman, G., Zabel, P., van Kammen, A. 1982. *J. Virol.* 42:630–35
333. Zabel, P., Moerman, M., Lomonossoff, G., Shanks, M., Beyreuther, K. 1984. *EMBO J.* 3:1629–34
334. Wellink, J., Rezelman, G., Goldbach, R., Beyreuther, K. 1986. *J. Virol.* 59:50–58
335. Jaegle, M., Wellink, J., Goldbach, R. 1987. *J. Gen. Virol.* 68:627–32
336. Drygin, Y. F., Sapotsky, M. V., Bogdanov, A. A. 1987. *FEBS Lett.* 215:247–51
337. Daubert, S. D., Bruening, G. 1979. *Virology* 98:246–50
338. Mayo, M. A., Barker, H., Robinson, D. J., Tamada, T., Harrison, B. D. 1982. *J. Gen. Virol.* 59:163–67
339. Murphy, J. F., D'Arcy, C. J., Clark, J. M. Jr. 1989. *J. Gen. Virol.* 70:2253–56
340. Harrison, B. D., Barker, H. 1978. *J. Gen. Virol.* 40:711–15
341. Chu, P. W. G., Boccardo, G., Francki, R. I. B. 1981. *Virology* 109:428–30
342. Mayo, M. A., Barker, H., Harrison, B. D. 1979. *J. Gen. Virol.* 43:735–40
343. Koenig, I., Fritsch, C. 1982. *J. Gen. Virol.* 60:343–53
344. Mayo, M. A., Barker, H., Harrison, B. D. 1982. *J. Gen. Virol.* 59:149–62
345. Hellen, C. U. T., Cooper, J. I. 1987. *J. Gen. Virol.* 68:2913–17
346. Hari, V. 1981. *Virology* 112:391–99
347. Siaw, M. F. E., Shahabuddin, M., Ballard, S., Shaw, J. G., Rhoads, R. E. 1985. *Virology* 142:134–43
348. Riechmann, J. L., Laín, S., García, J. A. 1989. *J. Gen. Virol.* 70:2785–89
349. Ghosh, A., Dasgupta, R., Salerno-Rife, T., Rutgers, T., Kaesberg, P. 1979. *Nucleic Acids Res.* 7:2137–46
350. Ghosh, A., Rutgers, T., Ke-Qiang, M., Kaesberg, P. 1981. *J. Virol.* 39:87–92
351. Reisman, D., de Zoeten, G. A. 1982. *J. Gen. Virol.* 62:187–90
352. Gutiérrez, C., Martin, G., Sogo, J. M., Salas, M. 1991. *J. Biol. Chem.* In press
353. Pakula, T., Caldentey, J., Serrano, M., Gutiérrez, C., Hermoso, J. M., et al. 1990. *Nucleic Acids Res.* 18:6553–57

Annu. Rev. Biochem. 1991. 60:73–99

# PHOSPHOLIPID TRANSFER PROTEINS

## K. W. A. Wirtz

Centre for Biomembranes and Lipid Enzymology, State University of Utrecht, Utrecht, the Netherlands

KEY WORDS:  phosphatidylcholine transfer protein, phosphatidylinositol transfer protein, sterol carrier protein 2, lipid-binding site, lipid metabolism.

## CONTENTS

0066-4154/91/0701-0073$02.00

## INTRODUCTION AND PERSPECTIVES

The transfer of lipid molecules between lipid-protein structures such as membranes and serum lipoproteins is a very common process that is of great importance to a proper functioning of cells and organisms (1, 2). Isotope-labeling studies have indicated that in the cell there is a continuous flux of phospholipids, glycolipids, and cholesterol from the sites of synthesis (i.e. endoplasmic reticulum and Golgi system) to other subcellular membrane structures, including the plasma membrane (3–5). This flux does not necessarily reflect a unidirectional mass transfer of lipids but may also be part of very rapid exchange reactions, as was, for instance, shown for the extensive transfer of radiolabeled phospholipids to mitochondria (6). The reverse flow from the plasma membrane to distinct sites in the cell has been observed mainly by use of fluorescently labeled lipids introduced into the plasma membrane from the medium (7). Apart from these intracellular lipid transfer processes, a continuous and extensive trafficking of cholesterol and phospholipids is known to occur in the circulation between serum lipoproteins, as well as between lipoproteins and blood cells (e.g. erythrocytes, blood platelets) and tissues (1, 8).

The mechanisms of intracellular lipid transfer may be manifold, including bulk lipid flow by way of membrane vesicles, lateral diffusion of lipids within the plane of the membrane through contact sites, spontaneous redistribution between membranes by lipid monomer equilibration, and redistribution controlled by lipid transfer proteins. Given the distinct lipid composition of subcellular organelles, intracellular lipid transfer must be strictly regulated, and must be subject to a sorting mechanism that is still poorly understood (9, 10). As for the lipid transfer processes in the blood, monomer diffusion and collisional interactions most certainly ensure the transport of cholesterol and phospholipids between serum lipoproteins as well as between lipoproteins and cell plasma membranes (8). In addition, lipid transfer proteins have been identified in plasma that have a specific affinity for phospholipids, cholesterol esters, and triglycerides (11). In this instance, ample evidence shows that the protein-dependent transfer of these lipids is directly linked to their enzymatic conversion and, hence, to the metabolism and the interconversion of the various serum lipoprotein classes (12).

In this review, I limit discussion to the phospholipid and cholesterol-transferring proteins that occur in the cell. These proteins have been identified in a great variety of organisms including mammalian tissues, plants, yeast, fungi, and certain bacteria. It has been proposed that the lipid transfer proteins play a role in membrane biogenesis, in the maintenance of the membrane lipid composition, and in various aspects of lipid metabolism. What physiological function these proteins actually fulfill, however, is not yet well established. In addition to phospholipid and cholesterol-transferring proteins, mammalian tissues contain a series of other proteins that bind and transfer hydrophobic ligands, including glycolipids (13, 14), oxysterols (15), fatty acids (16), and vitamin A derivatives such as retinol and retinoic acid (17). Most of these proteins are under intensive investigation.

I refer the reader to two previous comprehensive reviews for information on how the field of phospholipid transfer proteins has developed since their initial discovery in 1968 (18, 19). My aim in the present review is to discuss the most relevant information on phospholipid transfer proteins gathered during the past ten years, with an emphasis on the phosphatidylcholine transfer protein (PC-TP), the phosphatidylinositol transfer protein (PI-TP), and the nonspecific lipid transfer protein (nsL-TP), also referred to as sterol carrier protein 2 ($SCP_2$). For further information on these specific transfer proteins, the reader may also consult some recent reviews (20–24).

## PHOSPHOLIPID TRANSFER PROTEINS IN MAMMALS, PLANTS, AND UNICELLULAR ORGANISMS

### Introductory Remarks

In most studies, the membrane-free cytosol is taken as the starting material for phospholipid transfer proteins. Their activity is usually estimated by measuring the time-dependent transfer of labeled phospholipids from a "donor" membrane to an unlabeled "acceptor" membrane. In the transfer assays, most commonly used, the phospholipids are either radiolabeled or carry a fluorescent fatty acid (e.g. parinaric acid, pyrene-labeled acyl chain) (19, 25, 26). By varying the polar head group of the probe lipids, one is able to establish the activity of the transfer protein of interest. The assay based on the transfer of radiolabeled phospholipids is discontinuous, as the estimation of transfer activity requires the separation of donor and acceptor membranes at the end of the incubation. In general, this kind of assay is used for establishing which different phospholipid transfer activities are present in a particular cytosol. This assay is also commonly used in the purification of phospholipid transfer proteins. Fluorescently labeled phospholipids are applied in a continuous assay, which makes use of the change in the spectral properties of the probe lipids when they are transferred from the donor to the acceptor membrane. Since initial rates of transfer can be accurately measured, these spectrofluorometric assays are very convenient in studies on the kinetic properties and regulatory mechanisms of pure phospholipid transfer proteins.

### Mammalian Tissues

Phospholipid transfer activity was first discovered in the membrane-free cytosol from rat liver (27–29). All mammalian tissues investigated to date contain this activity, which is most commonly estimated by measuring the transfer of phosphatidylcholine (PC), phosphatidylinositol (PI), and phosphatidylethanolamine (PE) (30). In some studies it was also established that the cytosol contains activity for the transfer of phosphatidylserine (PS), phosphatidic acid (PA), and sphingomyelin (SPM) (31, 32). As of to date, the bulk of phospholipid transfer activity in the cytosol can be attributed to three major proteins: (a) PC-TP, which is specific for PC (33, 34); (b) PI-TP, which shows a marked preference for PI but is also able to transfer PC and to a lesser extent phosphatidylglycerol (PG) (35–38); and (c) nsL-TP, which transfers all common diacyl phospholipids as well as glycolipids and cholesterol (32, 39). The latter protein accounts for the PE-transfer activity measured in mammalian tissues. So far, phospholipid transfer proteins with other specificities have not been identified. The three proteins mentioned above have been purified to homogeneity from a number of different tissues (Table 1). Bovine and rat liver PC-TP have molecular weights of 25,000 and 28,000 and isoelectric points of 8.4 and 5.8, respectively. Antibodies raised against either protein are not cross-reactive, thereby suggesting that rat and bovine PC-TP are substantially different. The cross-reactivity of these antibodies with PC-TP from other mammals has not been determined. Bovine PI-TP, with its molecular weight of 33,000 and its isoelectric points of 5.3 and 5.6, closely resembles rat and human PI-TP. This resemblance is reflected in the cross-reactivity of the antibovine PI-TP antibody with rat PI-TP. Similarly, antirat PI-TP antibody is cross-reactive with PI-TP in bovine and human brain cytosol (43). In fact, PI-TP appears strongly conserved, as antibovine PI-TP antibody reacted with a 35–36-kDa protein in the membrane-free cytosol from mammals, birds, reptiles, amphibians, and insects (54). The phospholipid transfer protein from rat lung was reported to have a rather broad substrate specificity with a distinct preference for dipalmitoyl-PC (55). In view of its molecular weight, isoelectric point, and amino-terminal sequence, however, this protein appears closely related, if not identical, to PI-TP (54, 56).

Bovine, rat, and human nsL-TP have a molecular weight of 14,000 and isoelectric points in the order of 8.6 to 9.6 (32, 39, 47–51). Antibodies raised against either protein were cross-reactive. All available evidence, including amino acid compositions, indicates that nsL-TP from these sources is highly conserved. On the other hand, despite having a comparable isoelectric point and molecular weight, nsL-TP from goat liver was reported to have a different amino acid composition (52). Another nsL-TP with distinctly different properties was purified from rat hepatoma 27 (31). This protein has a molecular weight of 11,000, is acidic with an isoelectric point of 5.2, and has, unlike the other nsL-TPs, a pronounced activity toward SPM. This protein has been found in several carcinoma cell lines as well as in fetal rat liver, but it does not occur in adult rat liver (57). These data were taken to indicate

**Table 1**   Properties of phospholipid transfer proteins from mammalian sources and microorganisms

| Protein | Source | Molecular weight | Isoelectric point | Substrate specificity | References |
|---------|--------|------------------|-------------------|-----------------------|------------|
| PC-TP | bovine liver | 24,680 | 5.8 | PC | (33, 40) |
| PC-TP | rat liver | 28,000 | 8.4 | PC | (34, 41) |
| PI-TP | bovine brain | 32,500 | 5.3, 5.6 | PI > PC = PG | (35, 42) |
|  | bovine heart | 33,500 | 5.3, 5.6 | PI > PC >> SPM | (37) |
| PI-TP | rat liver | 35,000 | 5.1, 5.3 | PI > PC | (34) |
|  | rat brain | 36,000 | 4.9, 5.3 | PI > PC | (43) |
| PI-TP | human platelet | 29,000 | 5.6, 5.9 | PI > PC = PG | (38) |
| PI-TP | *Saccharomyces cerevisiae* | 35,000 | 4.6 | PI > PC | (44, 45) |
| PI-TP | *Mucor mucedo* | 24,000 | 5.05 | PI > PC > PE | (46) |
| nsL-TP | rat liver | 13,500 | 8.6 | PC, PI, PE, PS | (39, 47–49) |
|  | bovine liver | 14,000 | 9.55 | cholesterol | (32, 50) |
|  | human liver | 14,000 | n.d. | n.d. | (51) |
|  | goat liver | 12,000 | 8.65 | PC, PI, PE | (52) |
|  | rat hepatoma | 11,200 | 5.2 | PC, PI, PE, SPM | (31) |
| PG-TP | *Rhodopseudomonas* | 27,000 | 5.2 | PG > PC > PE | (53) |

that this hepatoma nsL-TP may belong to the family of carcinoembryonic protein.

## Plants

Phospholipid transfer proteins have been purified from maize seedlings (58), castor bean endosperms (59), and spinach leaves (60). These proteins have a broad phospholipid substrate specificity similar to nsL-TP from mammalian sources. Moreover, castor bean nsL-TP was shown to transfer typical plant lipids such as monogalactosyl- and digalactosyldiacylglycerol. By use of an enzyme immunoassay, it was estimated that nsL-TP accounted for 2% of the proteins in the membrane-free supernatant from maize seedlings (61). nsL-TPs from these three sources have a similar molecular weight of 9000; they are basic, with isoelectric points of around 9 (maize and spinach nsL-TP) to as high as 10.5 (castor bean nsL-TP) (62). Although their occurrence has been reported (63), transfer proteins with a specificity for a single class of phospholipids have not yet been purified from plant material. A transfer protein with a high activity toward PI was purified from filamentous fungi (46).

## Yeast

Transfer activity for all common diacyl phospholipids has been detected in the membrane-free cytosol of the yeast *Saccharomyces cerevisiae* (64). Attempts to purify transfer proteins from yeast cytosol yielded one protein with a molecular weight of 35,000 and an isoelectric point of 4.6 (44). Apart from these molecular properties, this protein has a pronounced similarity to bovine and rat PI-TP, as it transfers only PI and PC. The antibody against bovine PI-TP, however, is not cross-reactive with yeast PI-TP (45). The purification of another closely related transfer protein ($M_r$ 33,400, pI 6.3) was reported (65). Recently, the gene encoding PI-TP was isolated from a yeast genomic library (66). Disruption of the gene by insertion of a LEU2 gene yielded nonviable spores, thereby indicating that a functional PI-TP gene is essential for cell growth. Haploid cells containing the disrupted gene could be rescued by transformation with a plasmid-borne copy of the wild-type gene.

In addition to PI-TP, it appears that yeast cytosol contains a transfer protein that is specific for PS and PE (67). So far, a PS/PE transfer protein has not been detected in mammalian cells. Determination of transfer activity during growth indicated that the level of PC-transfer activity increased as the cells entered the exponential phase (44, 68). In these studies, however, it was emphasized that there was no strict correlation between PC-transfer activity and active membrane biogenesis.

## Bacteria

Phospholipid transfer activity is present in the crude cytosol of *Bacillus subtilis* and of *Rhodopseudomonas sphaeroides* (69, 70). From the latter bacterium, a phospholipid transfer protein was purified with a molecular weight of 27,000 and an isoelectric point of 5.2 (53). Evidence was provided that the purified transfer protein was restricted to the cytoplasm of this gram-negative bacterium, whereas the periplasmic space between outer and cytoplasmic membranes contained an additional phospholipid transfer protein of an estimated molecular weight of 56,000 (71). Determination of the specificities of transfer indicated that the 27-kDa protein preferred phosphatidylglycerol (PG) over PE and PC, whereas the 56-kDa protein appeared to be nonspecific (71). Of the total phospholipid transfer activity, 30 to 40% was detected in the periplasm and the remainder in the cytoplasm. *R. sphaeroides* is a photoheterotrophic bacterium that, in response to incident light, forms a cytoplasmic membrane structure harboring the photosynthetic apparatus (72). There was a clear correlation between light intensity during growth and the level of phospholipid transfer activity in the cells. Cells grown chemoheterotrophically, however, contained substantial levels of phospholipid transfer activity as well (70). This result shows that the occurrence of the phospholipid transfer proteins in *R. sphaeroides* is not strictly linked to the ability of these cells to assemble a photosynthetic apparatus.

# PRIMARY STRUCTURE: LACK OF SEQUENCE HOMOLOGIES

To date, the primary structure of a limited number of mammalian phospho-lipid-transferring proteins has been elucidated: those of bovine liver PC-TP (40), rat brain PI-TP (54), and rat and bovine liver nsL-TP (50, 73). Bovine PC-TP consists of a single polypeptide chain of 213 amino acid residues with $N$-acetyl methionine as the amino terminus. It contains two disulfide bridges and a number of very hydrophobic peptide segments (for a detailed discussion, see Ref. 23). The sequence of rat PI-TP, as deduced from complementary DNA (cDNA) analysis, is composed of 271 amino acid residues. The N-terminal peptide segment (21 residues) was virtually identical with that of rat lung PI-TP determined by Edman degradation (56). It showed extensive homology (about 80%) with the N-terminal sequence of bovine brain PI-TP (54). It was further reported that rat PI-TP cDNA hybridizes with genomic DNA of *Drosophila melanogaster* and a host of higher eukaryotes, thereby suggesting that the primary structure of PI-TP may be strongly conserved (54). Recently, it was recognized that the *SEC14* structural gene of *Saccharomyces cerevisiae,* which encodes for a protein required for the export of secretory proteins from the Golgi complex, was identical to the gene encoding for yeast PI-TP (66, 74, 74a). Surprisingly, the deduced amino acid sequence displayed no sequence homology with rat PI-TP. Furthermore, in contrast to what one would expect from the similarities in phospholipid-binding and transfer characteristics (see below), bovine PC-TP and rat PI-TP have no significant sequence homology. Moreover, it is apparent from hydropathy profiles as well as from predicted structural analyses that the structural design of these two proteins appears to be completely different (23, 54). Computer searches have failed to identify any other proteins with which rat PI-TP and bovine PC-TP share significant sequence homologies. This is also the case for the SEC14 protein (yeast PI-TP) (74).

Bovine nsL-TP has a sequence of 121 amino acid residues with Ser-Ser-Ser- at the amino terminal end and with alanine as the C-terminal residue (50). The sequence of rat nsL-TP (SCP$_2$) is highly homologous (10 substitutions), except that it consists of 123 residues due to an extension with Lys-Leu at the C-terminal end (73, 75). nsL-TP contains a single cysteine residue, modification of which inhibits the activity (76, 77). Certain peptide segments of rat SCP$_2$ (residues 14–67) and of bovine PC-TP (residues 153–205) have a significant (44%) sequence homology (73). Significant homology was also observed between SCP$_2$ (residues 35–115) and the variable domains of the heavy chain of immunoglobulin G. Despite their apparently similar mode of action, there is no sequence homology between mammalian and plant nsL-TP. Sequences elucidated for castor bean (three isoforms), maize, and spinach nsL-TP consist of a chain of 91–93 amino acid residues (78–81). In contrast to mammalian nsL-TP, plant nsL-TPs, even within the same species, display a wide variation in primary structure (a similarity in the order of 30% to 50%) (81).

## PROPERTIES OF THE LIPID-BINDING SITE

### Introductory Remarks

PC-TP from bovine liver was found to contain one molecule of noncovalently bound PC (33, 82). By interacting with a membrane interface, PC-TP may replace this endogenous molecule for another PC molecule present. By this exchange reaction, spin-labeled, photolabeled, and fluorescently labeled PC analogs have been incorporated into PC-TP, thus enabling extensive investigation of the lipid-binding site (see below). Mammalian PI-TPs are less specific than PC-TP, as they have the ability to bind either PI or PC. Due to the one negative charge difference between these lipids, PI-TP may have an isoelectric point of either 5.5 (PI bound) or 5.7 (PC bound) (83). These two forms have been detected during the purification of PI-TP from bovine brain (84), bovine heart (37), rat liver (34), and human platelets (38). On the other hand, only one form of PI-TP (isoelectric point of 5.2) was detected in yeast even though, like mammalian PI-TP, this protein could bind both PI and PC (45).

All available evidence indicates that nsL-TP ($SCP_2$) has lipid-binding characteristics that differ considerably from those of PC-TP and PI-TP. In general, nsL-TP in purified form does not contain a phospholipid or cholesterol molecule (32, 85). Whenever nsL-TP greatly stimulated the transfer of cholesterol from lipid monolayers and bovine heart mitochondria to acceptor membranes, binding of cholesterol to nsL-TP could not be observed (86). On the other hand, it has been reported that nsL-TP forms a one-to-one molar complex with cholesterol upon incubation with adrenal lipid droplets (22). The formation of this complex was confirmed by titrating nsL-TP with dehydroergosterol, which is a fluorescent analog of cholesterol (87). From Scatchard binding analysis of light scattering, fluorescence lifetime, and fluorescence energy transfer data, a 1:1 molar stoichiometry was estimated with a dissociation constant of about 1.4 $\mu$M. Recently, nsL-TP was also shown to be able to bind PC molecules by incubating the protein with vesicles consisting of fluorescently labeled PC. In contrast to results obtained with dehydroergosterol, however, binding was limited to a small proportion of the protein (less than 10%) (77). Moreover, this PC/nsL-TP complex could not be isolated by, for instance, column chromatography, a finding that strongly suggests that the complex only exists in the presence of vesicles. The apparent discrepancy between these observations probably stems from the fact that nsL-TP, in contrast to PC-TP and PI-TP, has a low-affinity lipid-binding site (77). The extent of binding to nsL-TP would then reflect the affinity of the lipid for this binding site relative to its affinity for the membrane.

## Specific Binding Sites for the sn-1- and sn-2 Fatty Acyl Chains

The site on bovine liver PC-TP that accommodates the acyl chains was initially explored by electron-spin resonance spectroscopy with spin-labeled analogs of PC that carry a nitroxide moiety on the *sn*-2 acyl chain (88, 89). From the spectra of the spin-labeled PC/PC-TP complex, it was apparent that the nitroxide moiety was strongly immobilized irrespective of whether the spin-label was close to the carboxy- or methyl-terminal end of the *sn*-2-fatty acyl chain. Moreover, addition of ascorbate did not result in a reduction of the nitroxide moiety, thus indicating that the moiety embedded in the protein was inaccessible from the medium. These data suggested that the 2-acyl chain strongly interacted with a hydrophobic pocket in PC-TP, a notion that then raised the question of whether the 1-acyl chain was accommodated in the same pocket or at another site on the protein.

The first evidence for the occurrence of two independent binding sites came from time-resolved fluorescence studies with PC analogs that carried as a fluorophore the *cis*-parinaroyl moiety at either the *sn*-1-, *sn*-2-, or both *sn*-1- and *sn*-2-positions (1-PnA-PC, 2-PnA-PC, and diPnA-PC) (90). From fluorescence depolarization measurements, it was estimated that the rotational correlation time of the parinaroyl chromophore in the 2-PnA-PC/PC-TP complex was 11 ns. A similar analysis of the 1-PnA-PC/PC-TP complex yielded a correlation time of 26 ns. These very long correlation times indicate that both the 1- and 2-parinaroyl chains are immobilized following the rotation of the protein. The difference in correlation times is compatible with a model in which PC-TP has a prolate ellipsoid shape (axial ratio of 2.50), accommodating the 1-acyl chain in a site parallel to the long symmetry axis and the 2-acyl chain in a site more or less perpendicular (an estimated angle of 60–90°) to this axis (91, 92). The correlation time of 15 ns, estimated for the diPnA-PC/PC-TP complex, is in agreement with this model in which the 1- and 2-parinaroyl chains are independently photoselected and behave as isolated probes (90).

The 1-acyl- and 2-acyl-binding sites of PC-TP were further explored by measuring the binding of fluorescent PC analogs carrying both pyrene-labeled and unlabeled saturated acyl chains of different lengths at either the *sn*-1- or *sn*-2-position (93). The protein preferred the analogs with the saturated acyl chains at the *sn*-1-position and the pyrene-labeled acyl chains at the *sn*-2-position. In general, binding of the positional isomers was much reduced.

Apparently, saturated acyl chains fit well into the 1-acyl binding site, whereas the 2-acyl binding site is more flexible in being able effectively to accommodate the pyrenylacyl chains. Moreover, the binding affinity was markedly dependent on the length of the acyl chains; optimal binding was observed with analogs carrying palmitic acid at the *sn*-1-position and 10-pyrenyldecanoic and 12-pyrenyllauric acid (equivalent in length to acyl chains of 16 and 18 C-atoms) at the *sn*-2-position. This result shows that the characteristics of the 1- and 2-acyl-binding sites are ideally suited for interaction of PC-TP with natural species of PC that usually carry an unsaturated fatty acyl chain at the *sn*-2-position. In agreement with this conclusion, PC-TP has a distinctly higher affinity for 1-palmitoyl, 2-arachidonoyl-PC (16:0/20:4-PC) and 1-palmitoyl, 2-docosahexanoyl-PC (16:0/22:6-PC) than for other naturally occurring PC-species (94).

From the ability to discriminate between positional isomers of pyrenylacyl-labeled PC it was inferred that similar to PC-TP, bovine brain PI-TP has independent binding sites for the 1- and 2-acyl chains (95). In contrast to PC-TP, however, PI-TP prefers PC analogs with a pyrenylacyl chain in the *sn*-1 position. In general, the 2-acyl-binding site of PI-TP appears to be less well suited to accommodate bulky acyl chains, in agreement with the observation that compared to PC-TP, PI-TP has a very low affinity for 16:0/20:4-PC and 16:0/22:6-PC (94). Steady-state and time-resolved fluorescence measurements with 2-parinaroyl- and 2-pyrenylacyl-labeled analogs of PI and PC have clearly indicated that both phospholipids share the same 2-acyl-binding site on PI-TP (83, 95). This would imply that PI-TP has a relatively low affinity for the most abundant species of PI in mammalian tissues, i.e. 1-stearoyl, 2-arachidonoyl-PI. Of course, it remains to be established in what way, if any, the metabolism of this PI species is related to the activity of PI-TP.

In summary, both PC-TP and PI-TP appear to have distinct binding sites for the 1- and 2-fatty-acyl chains. Owing to the different properties of these sites, both proteins can discriminate between molecular phospholipid species as well as between positional isomers of these species. So far, these are the only proteins known that display such a discriminatory behavior in their interaction with phospholipids.

## Use of Photoactivatable Analogs of PC

The 2-acyl-binding site of PC-TP was probed by incorporation of PC analogs that carry the photoactivatable m-diazirinophenoxy-moiety on the methyl-terminal end of the *sn*-2-[1-$^{14}$C]hexanoyl and *sn*-2-[1-$^{14}$C] undecanoyl chain (96). Photolysis resulted in the covalent coupling of the $^{14}$C-acyl chains to amino acid residues that supposedly form part of the 2-acyl-binding site. Sequence analysis of $^{14}$C-labeled peptides obtained from the photolyzed

PC-TP-adduct showed that major sites of coupling were present in the hydrophobic peptide segments Leu51-Tyr-Ala-Tyr-Lys-Val-Phe57 (peptide A) and Val171-Phe-Met-Tyr-Tyr-Phe-Asp177 (peptide B). In peptide A, both acyl chains were almost exclusively coupled to Tyr54; in peptide B, the hexanoyl-chain was preferentially coupled to Tyr175 and Asp177 and the undecanoyl-chain to Val171 and Met173. This shift, as well as the alternating sites of coupling, may indicate that the 2-acyl chain is aligned along peptide B, which has adopted a $\beta$-strand conformation. In support of this, computer analyses have predicted that both peptides A and B are part of very hydrophobic antiparallel $\beta$-strand structures (97). These data suggest that the sn-2 acyl chain is accommodated in a hydrophobic pocket formed by $\beta$-strands. In this model the polar head group of PC would be pointing toward the strongly predicted $\beta$-turn Asn178-Pro-Gly-Gly181, which presumably is located on the protein surface. So far, we have not succeeded in growing crystals of PC-TP to prove that these structural arrangements are correct. On the other hand, elucidation of the tertiary structure of a number of hydrophobic ligand-binding proteins has revealed that $\beta$-sheet structures form the core of the lipid-binding site. So it was demonstrated that palmitate bound to rat intestinal fatty-acid-binding protein was located between two $\beta$-sheets made up of 10 antiparallel $\beta$-strands (98, 99). The acyl chain was almost exclusively surrounded by apolar amino acid residues; the carboxylate group was located in the interior of the protein interacting with polar amino acid residues. A similar $\beta$-sheet structure forms the binding site for vitamin A in the serum-retinol-binding protein (100).

## Specific Recognition Sites for the Polar Head Groups

Given the specific affinity for PC, it is presumed that PC-TP exposes a recognition site for the phosphorylcholine head group at the interface. Modifications of the phosphorylcholine moiety (e.g. partial demethylation, increase of the distance between phosphate and tertiary amine) greatly reduced, or even completely abolished, the transport of the ensuing phospholipid analogs, thereby demonstrating the strict configurational constraints of this site (101). Both electrostatic and hydrophobic requirements have to be met for a lipid molecule to be transferred by PC-TP, however. In this respect, PC-TP transfers neither 1-acyl-lysoPC (101) nor a dimeric form of PC consisting of two dipalmitoyl-PC molecules linked together by a disulfide bond at the methyl terminal ends of the 1-acyl chains (102). Moreover, the presence of these PC derivatives in a membrane bilayer has no effect on the rate at which PC-TP transfers PC. This finding suggests that there is no competition between transferable and nontransferable PC derivatives at the interface. It could also mean that the interaction with the recognition site does not affect the rate-limiting step in the transfer process, which most likely is the desorption of PC from the interface (see below).

The ability of bovine PI-TP to transfer both PI and PC has been taken to indicate that it has independent recognition sites for the phosphorylinositol and the phosphorylcholine head groups (37). From competition-binding experiments it was estimated that PI-TP has a 16-fold higher affinity for PI than for PC (103). A 12–15-fold higher binding affinity was also reported for yeast PI-TP (45). Partial periodate oxidation of the inositol ring and subsequent borohydride reduction yielded PI-derivatives containing two, three, and four hydroxyl groups. These derivatives, as well as PG, were transferred by PI-TP at about one tenth of the rate observed for PI, whereas PA was not transferred at all (42). Thus it appears that the intact inositol ring is a prerequisite for PI to be preferentially transferred. Moreover, PI-TP does not transfer PI-4-phosphate (PIP) and PI-4,5-bisphosphate (PIP$_2$) (104). This finding probably indicates that phosphorylation of the inositol moiety prevents the interaction with the phosphorylinositol-recognition site on PI-TP. On the other hand, the transfer activity of PI-TP was very efficiently inhibited by PIP$_2$ micelles (105). An inhibition of 50% was observed at a molar ratio of PI-TP to PIP$_2$ of $1:7$, a result that suggests that PI-TP has a high affinity for PIP$_2$ micelles. This affinity is in the same order as that observed for profilin, the actin-binding protein that, by interacting with PIP$_2$ micelles, inhibits PIP$_2$ hydrolysis by the phosphoinositide-specific phospholipase C (106).

## CARRIERS OF PHOSPHOLIPIDS BETWEEN MEMBRANES

### Introductory Remarks

Most membrane lipids, including phospholipids, have a finite monomer concentration in the aqueous phase. By being in equilibrium with the interface, monomer lipids have a tendency to redistribute spontaneously between membranes (107–111). In general, spontaneous transfer of natural phospholipids by monomer diffusion proceeds very slowly (half-times in the order of days) and is presumably of little consequence to intracellular lipid transfer (111). This conclusion may be premature, however, as at high vesicle membrane concentrations spontaneous transfer of PC was found to be greatly enhanced due to collisional transfer between closely appositioned vesicles (112). Provided that similar conditions occur in the cell, one cannot exclude the possibility that lipids do spontaneously redistribute between intracellular membranes, and by doing so contribute to important processes like membrane biogenesis. The fact remains that in vitro under most conditions, protein-mediated phospholipid transfer is several orders of magnitude faster than spontaneous transfer. Kinetic studies have shown that by interacting with the membrane interface, phospholipid transfer proteins enhance the lipid desorption reaction, which may be considered the rate-limiting step in the overall phospholipid transfer process (113, 114).

## Mode of Action

One of the fundamental questions that has to be addressed in studying the mode of action of phospholipid transfer proteins is what effect these proteins have on the stability of membranes. Phospholipid monolayer studies have clearly shown that when phospholipid transfer is greatly stimulated, PC-TP and PI-TP do not penetrate into the interface (36, 82). Bovine nsL-TP is somewhat more surface active, but there is, as yet, no evidence that its penetration is a prerequisite for transfer activity (86). In agreement with these observations, transfer proteins do not perturb the membrane bilayer; their action is not accompanied by an enhanced transbilayer movement of phospholipids. These transfer proteins are therefore ideally suited to probe the phospholipids in the outer membrane leaflet and to provide information on membrane phospholipid asymmetry and rates of transbilayer phospholipid movement (for details, see Refs. 24, 115, 116).

For PC-TP to bind a PC-molecule, it is presumed to disturb the membrane lipid packing locally such that the acyl chains of PC may become accommodated in the hydrophobic 1- and 2-acyl binding sites. For this reaction to occur, the energetics of binding to PC-TP are probably of the same order as the energetics that govern the Van der Waals interactions between PC molecules in the bilayer. In this model, an increase of Van der Waals interactions would be reflected in a diminished rate of PC-TP-mediated transfer of PC between membranes. In support of this, the rate constant at which PC bound to PC-TP dissociated from the membrane increased five-fold when the vesicles underwent a phase transition from the gel to the liquid crystalline state (113). Moreover, the apparent activation energy for egg PC transfer to vesicles in the liquid crystalline state was 3–4 times lower than to vesicles in the gel state (117). Van der Waals interactions between PC molecules are less efficient in highly curved single bilayer vesicles than in large multilamellar liposomes (118). In line with the proposed model, PC-TP-mediated transfer of PC was 100 times faster between vesicles than between vesicles and liposomes (119). This effect of curvature was also reflected in the rate constant of dissociation of the PC-TP-interface complex, which was about 100-fold lower for liposomes than for vesicles (120). On the other hand, sizing of the vesicles by molecular sieve chromatography demonstrated that the association constant of the PC-TP-vesicle complex was about 30-fold higher for small vesicles (diameter of 22 nm) than for larger vesicles (diameter of 30 nm) (121). This strongly suggests that subtle changes in vesicle bilayer organization have major effects on the interaction of PC-TP with the interface and, therefore, on the rates of intermembrane PC transfer.

To date, all studies have indicated that PI-TP mediates a strict molecular exchange reaction in the sense that the flux of PI and PC from donor to acceptor membrane equals that in the reverse direction (122, 123). This implies that at the interface, PI-TP may exchange its bound PI molecule for PI or PC; similarly it may exchange its bound PC for PI or PC. Which of these exchange reactions will prevail probably depends on the relative amounts of PI and PC in the membrane. From kinetic analyses it was estimated that PI-TP exchanges its bound PI 20 times faster for PI than for PC, and its bound PC five times faster for PI than for PC (103). The only reason that PI-TP carries PC at all is that in most membranes the PC-pool is large relative to the PI-pool. Amphiphilic amines, like local anesthetics (e.g. dibucaine) and chlorpromazine, effectively inhibit the PI-TP-mediated transfer of both PI and PC (124, 125). At the membrane bilayer, these cationic agents may form ion pairs with the anionic PI and the zwitterionic PC, thereby interfering with the interaction of PI-TP with these phospholipids. In general, an accumulation of these cationic agents at the interface may repulse PI-TP, as its membrane interaction site is thought to be positively charged (105). The incorporation of PE into the acceptor membrane stimulates the transfer activity, whereas the incorporation of SPM has a pronounced inhibitory effect (126, 127). These zwitterionic phospholipids may directly affect the hydration properties of the interface and therefore may affect the ability of PI-TP to interact with the polar head groups of PI and PC (126). Apart from these electrostatic effects, PI-TP appears to be very sensitive toward the hydrocarbon fluidity of the membrane. Highest transfer activities were observed with the most fluid membranes (128). All available evidence indicates that the mode of action of PI-TP is very similar to that of PC-TP. This being the case, it is surprising that PI-TP lacks the extended regions of highly hydrophobic amino acids that were observed for PC-TP (54, 97).

In contrast to PC-TP and PI-TP, nsL-TP (SCP$_2$) does not display any distinct specificity in the lipid transfer reaction. By using fluorescent acyl-labeled PA, PC, and PE analogs in a donor-acceptor vesicle transfer assay, it was shown that the nsL-TP-dependent rates of transfer were highly correlated with the rates of spontaneous transfer in the order PA > PC > PE (114). A similar high correlation was observed for PC species carrying fluorescent pyrenylacyl chains of different lengths (i.e. rates decreased with increasing chain length) and for various oxysterol derivatives (i.e. rates decreased in the order 25-hydroxycholesterol > 7$\alpha$-hydroxycholesterol > 7-ketocholesterol > cholesterol) (86). These studies strongly suggest that by interacting with membranes, nsL-TP nonspecifically lowers the energy barrier for the lipid monomer-membrane interface equilibration reaction. In this model, nsL-TP has no preference for any special kind of lipid; it may just enforce the natural tendency of a lipid molecule to leave the membrane, thereby particularly affecting the desorption of molecules that are perpendicularly displaced from, but still associated with, the lipid bilayer (112, 129). In agreement with this mode of action, highly apolar lipids like cholesteryl oleate and triglycerides are not transferred by nsL-TP (32). In this process, a low-affinity lipid-binding site may be involved, thus giving the lipid molecule the extra push to leave the interface for nsL-TP (77). In this model, binding to nsL-TP is invariably favored by lipids whose hydrophobic interactions in the bilayer are weakest (T. W. J. Gadella, unpublished observation). The lipid-nsL-TP complex formed would then function as an intermediate in the intermembrane lipid transfer reaction (130).

## Net Transfer versus Exchange: Effect on Membrane Properties

Inherent to the one-for-one molecular exchange reaction at the interface, PC-TP transfers as much PC from donor to acceptor membrane as in the opposite direction, irrespective of whether both membranes contain the same molecular species of PC (131). This property has the interesting consequence that a species with a small molecular surface area (e.g. dimyristoyl-PC) can be replaced for one with a larger surface area (e.g. dioleoyl-PC). With vesicles, this type of exchange reaction generates an imbalance between the molecular packing of the outer and inner leaflets and, as a result, enhances the transbilayer movement of PC (132). If similar exchange reactions were carried out with intact human erythrocytes, changes in cell shape were observed (133). Upon introduction of dipalmitoyl-PC, the cells turned echinocytic (membrane bent outward), and upon introduction of dilinoleoyl-PC, stomatocytic (membrane bent inward). These drastic changes in shape were thought to be related to differences in the geometry of the native PC and the PC species newly introduced (134, 135).

Although PC-TP generally catalyzes an exchange reaction, it can also release PC into an interface devoid of this phospholipid. PC-TP can mediate a net mass transfer of PC to vesicles prepared of only PA, PE-PA (80:20 mol %), or total phospholipid from *Escherichia coli* (114, 121, 136, 137). This result implies that PC-TP can dissociate from a membrane surface without a bound PC molecule. The rates of PC transfer in these studies were at least one order of magnitude lower than under conditions in which the acceptor vesicles contained PC, thus suggesting that the insertion of PC by PC-TP into a membrane proceeds faster when PC is there to replace the endogenous PC molecule. This enhanced rate has been interpreted to indicate that PC-TP acts by way of a concerted, one-step catalytic mechanism for PC exchange (137).

Bovine brain PI-TP will mediate a transfer of PI from donor membranes to PC acceptor vesicles that are initially devoid of PI (36, 122, 127). Inherent to its dual specificity, PI-TP may also mediate a net transfer of PC to vesicles that contain PI but lack PC (122). This net flux of PI (or PC) is compensated for by the transfer of an equal amount of PC (or PI) in the opposite direction. So far, there is no evidence that bovine PI-TP can give rise to a net mass transfer of phospholipids, as has been shown for PC-TP (122). In the case of yeast PI-TP, however, a moderate but significant net mass transfer of PC to PE-containing vesicles was observed (45). PI-TP may be involved in the transfer of PI from the site of synthesis, i.e. endoplasmic reticulum, to the plasma membrane, where, as a result of agonist-induced receptor activation, the PI-pool is subject to a high turnover (20). It is possible that the trigger for net PI transfer is the decrease in the molar ratio of PI to PC, which in mammalian membranes is maintained at a value of about 0.1. Kinetic measurements have indicated that, with a 16-fold higher affinity for PI than for PC, PI-TP is ideally suited to maintain this ratio because it inserts PI into the depleted plasma membrane in exchange for PC (103).

In contrast to PC-TP and PI-TP, nsL-TP has a low-affinity lipid-binding site. This implies that in the interaction with the interface, only a small proportion of nsL-TP contains a lipid molecule at any one time. As a result, nsL-TP will mediate a net mass transfer of lipids whenever there exists a chemical lipid gradient between the membranes involved. The net mass transfer is then superimposed on an exchange reaction. This property makes nsL-TP ideally suited to change both the lipid content and the composition of membranes. nsL-TP was shown to facilitate a net mass transfer of phospholipids from vesicles to rat liver mitochondrial and microsomal membranes (116, 138). Under these conditions, the microsomal glucose-6-phosphate phosphohydrolase activity was found to be greatly affected by alterations in the PE and PC content (139). Similarly, nsL-TP can mediate a net mass transfer of cholesterol (49). For instance, the cholesterol-to-phospholipid molar ratios of synaptic plasma membranes and synaptosomes could be drastically changed by incubating these membranes with nsL-TP in the presence of an excess of either PC or cholesterol/PC vesicles (140). A decrease of the cholesterol/phospholipid ratio in these membrane fractions resulted in as much as a 70–100% loss of sodium-dependent $\gamma$-aminobutyric acid (GABA) uptake. Reinsertion of cholesterol into the depleted membranes restored the GABA uptake. These studies demonstrate that, in agreement with a former proposal (141), nsL-TP can be used to investigate systematically the lipid dependence of membrane-bound enzyme activity.

# TISSUE DISTRIBUTION AND INTRACELLULAR LOCALIZATION

From measuring phospholipid transfer activities in the membrane-free cytosol, it became apparent that each rat tissue may have different levels of PC-TP, PI-TP, and nsL-TP (84). This was corroborated by use of immunoassays and immunoblotting techniques. With a radioimmunoassay in which an affinity-purified polyclonal antibody against rat liver PC-TP was used, it was established that PC-TP was most abundant in liver and intestinal mucosa and virtually absent from brain and heart. Intermediate to low levels were found in kidney, spleen, lung, and adrenals (142). In rat liver, 60% of the total PC-TP was present in the supernatant fraction as compared to 40% distributed evenly over the various particulate fractions. The significance of the membrane-associated form of PC-TP is not yet clear. It may well be that the high levels of PC-TP in liver and intestinal mucosa are related to the very active synthesis of PC in these tissues (see below). In addition, levels of PC-TP may be positively correlated with growth rate and cell proliferation. Levels of PC-TP were four times higher in the fast-growing, poorly differentiated Morris hepatoma 7777 than in the slow-growing, well-differentiated Morris hepatoma 9633 (143).

PI-TP occurs in all tissues investigated and is predominantly present in the cytosolic fraction. In line with a previous study (84), it was recently observed that in the rat (out of a total of 13 different tissues), the specific activity for PI-transfer was highest in brain (43). Very low levels of PI-transfer activity were detected in adipose tissue and skeletal muscle. Immunoblots of cytosolic fractions with antibovine brain PI-TP antibody showed that an immunoreactive PI-TP band ($M_r$ 36,000) was present in each rat tissue, except for rat testis, in which one additional immunoreactive band ($M_r$ 41,000) was detected (144). This 41-kDa protein has PI-transfer activity and appears unique to testis. Its relationship with PI-TP remains to be established. Additional studies on rat brain demonstrated that PI-TP is rather evenly distributed among all major anatomical regions, except for the pituitary gland, which has a reduced level (145). Subfractionation of total rat brain, however, showed that PI-transfer activity was highly enriched in synaptosomes and myelin fractions (146). Moreover, PI-transfer activity was found to increase in developing rat brain at the onset of myelination (147–149) and during postnatal development (150). These observations have been taken to indicate that the cytosolic levels of PI-transfer activity may be correlated with active membrane biogenesis and enhanced PI-metabolism.

Affinity-purified polyclonal antibodies against rat liver nsL-TP (SCP$_2$) have been used routinely to determine the levels and subcellular distribution of nsL-TP in rat tissues. By use of a specific enzyme immunoassay, it was established that levels are high in the cytosolic fraction of liver and intestinal mucosa and low in the cytosolic fraction of all other tissues tested (151). Immunoblot analysis of subcellular membrane fractions subsequently showed that in most rat tissues, except for liver, the bulk of nsL-TP is membrane-bound (86, 152). For instance, in adrenals, about 80% of nsL-TP was associated with membranes as compared to 30% in liver (86, 153). Surprisingly, membrane-bound nsL-TP was found to have a higher isoelectric point than the cytosolic form (154). At this point it is not clear what effect membrane-bound nsL-TP has on the transfer of lipids between membranes.

## Involvement of Peroxisomes

Through using electron microscopy with the affinity-purified anti-nsL-TP antibody and protein A-colloidal gold, a dense labeling of peroxisomes was observed in rat liver and in adrenal glands (86, 155–157). Significant labeling was also observed in the cytoplasm, inside mitochondria, and on the cytoplasmic surface of the endoplasmic reticulum. Immunoblot analysis indicated that nsL-TP was highly concentrated in the peroxisomal fraction from liver (156, 157). In addition, an immunoreactive 58-kDA protein was present in the peroxisomes (155–157). Density-gradient analysis confirmed that in rat liver, the 58-kDA protein was peroxisomal (coincided with peroxisomal catalase); in contrast, the bulk of membrane-bound nsL-TP did not coincide with the catalase (154). Similar observations were made in wild-type Chinese hamster ovary cells (158). This suggests that, contrary to what has been stated by others (156, 157), nsL-TP may be primarily bound to a membrane fraction different from peroxisomes. Whether the 58-kDa protein in peroxisomes can be processed to form nsL-TP is not yet known. That the occurrence of nsL-TP is somehow related to peroxisomes, however, was indicated by the recent finding that nsL-TP is deficient in livers of infants with cerebro-hepato-renal (Zellweger) syndrome, a condition in which hepatic peroxisomes are absent (51; G. P. H. Van Heusden, unpublished observation). Similarly, nsL-TP was absent from mutant Chinese hamster ovary cells that lacked peroxisomes (158). In these latter cells, relatively low levels of 58-kDa protein could be observed bound to membranes, which may have been "peroxisomal remnants." On the other hand, pulse-chase experiments in rat hepatoma H-35 cells were interpreted to indicate that the 58-kDa protein is not a precursor of nsL-TP (159). It was further proposed that, given its occurrence in different subcellular organelles, isoforms of nsL-TP may exist with specific targeting sequences (157). In vitro translation studies with mRNA isolated from rat adrenals and liver have provided evidence for the existence of a precursor of nsL-TP with an N-terminal leader sequence of 20 amino acid residues (159,

160). In what subcellular organelle(s) this precursor is posttranslationally processed is not yet known.

Full-length cDNA clones encoding nsL-TP and the 58-kDa protein have not yet been isolated. Recently, a cDNA clone was identified that consisted of 1851 base pairs starting at the 5' end with an open reading frame of 1545 base pairs. The 369-base-pair segment at the 3' end of this open reading frame corresponded exactly to the amino acid sequence of nsL-TP (161). This observation has been independently confirmed by sequence analysis of a cDNA clone of 1431 base pairs (162). The latter clone may encode for proteins of about 15 and 30 kDa, in agreement with open reading frames present at 20 and 150 amino acids upstream from the amino-terminal serine of $SCP_2$ (160, 162). Since the open reading frame of the 1851-base-pair insert (encoding a 55-kDa protein) does not contain an in-frame stop codon toward the 5' end, it is possible that this cDNA is a fragment of the cDNA clone encoding the rat liver 58-kDa protein (161).

# STIMULATION OF LIPID METABOLISM

## Phospholipids

As for their physiological function, lipid transfer proteins may possibly be involved in intracellular lipid metabolism. Studies on fetal rat lung indicated that, together with an increase in choline kinase and cholinephosphotransferase activities, the levels of PC-TP more than doubled two days before term (142). This suggests a possible relationship between PC-TP and the rate of PC synthesis and/or PC secretion in the lung surfactant. Recently, the important observation was made that both bovine and rat liver PC-TP can stimulate the microsomal conversion of diglyceride into PC up to 10-fold (163). In contrast, despite its affinity for PC, rat liver PI-TP had no effect on this reaction. It was suggested that this stimulatory effect was due to a direct interaction of PC-TP with cholinephosphotransferase. Another suggestion was that PC-TP specifically removes the newly synthesized PC molecule from the enzyme (163). This would then prevent the back-reaction by which cholinephosphotransferase cleaves the PC molecule into diglyceride and CDP-choline. This is the first evidence that the synthesis of PC may not only be regulated at the level of CDP-choline (164) but also at the level of PC-TP. Attempts to show that PI-TP has a stimulatory effect on the de novo synthesis of PI have been unsuccessful (43).

In mammalian cells, PS synthase is a microsomal enzyme (165) and PS decarboxylase an inner mitochondrial membrane enzyme (166). The in situ conversion of PS into PE has been interpreted to indicate that PS can be translocated from the endoplasmic reticulum to mitochondria, where it then becomes decarboxylated (167). This translocation-decarboxylation reaction has also been observed in vitro by incubating microsomes that contain [$^{14}$C]PS with intact rat liver mitochondria (168). Decarboxylation of PS was found to be stimulated when the incubation was carried out in the presence of nsL-TP, because nsL-TP enhances the transfer of PS from microsomes to mitochondria (168). The nsL-TP-mediated transfer of PS in vitro does not reflect the dynamic situation in the intact cell. Recently, it was observed that the incorporation of serine into PS and the subsequent conversion into PE was normal in mutant Chinese hamster ovary cells that were deficient in nsL-TP (158). This observation provides direct evidence that in these cells the intracellular translocation of PS from the endoplasmic reticulum to mitochondria occurs independently of nsL-TP.

## Cholesterol

nsL-TP, also designated "sterol carrier protein 2" (SCP$_2$), has been proposed to play an important role in intracellular cholesterol metabolism (for a review, see Ref. 22). So it was shown that in vitro SCP$_2$ activates the microsomal enzymes involved in the conversion of sterol intermediates into cholesterol. Specifically, SCP$_2$ stimulated the conversion of 4,4-dimethyl-$\Delta^8$-cholesterol into a C$_{27}$-sterol (47) as well as the conversion of $\Delta^7$-cholesterol into 7-dehydroxycholesterol (49) and of 7-dehydrocholesterol into cholesterol (47, 49). In these studies, the substrates were introduced into the rat liver microsomes from organic solvents. Their enhanced conversion in the presence of SCP$_2$ suggests that this protein helps to make these substrates available to the enzyme sites. Similar observations have been made for the conversion of exogenous cholesterol into cholesterol ester by microsomal acyl-CoA cholesterol acyltransferase (ACAT). [$^{14}$C]cholesterol introduced into microsomes via solvent dispersion was not effective as a substrate. Only in the presence of SCP$_2$ was a striking enhancement of cholesterol ester formation observed (169). Interestingly, addition of SCP$_2$ to native rat liver microsomes already led to a 1.5- to 2-fold enhanced esterification of endogenous cholesterol (49, 170). This suggests that SCP$_2$ has enlarged the cholesterol pool available to ACAT by reshuffling cholesterol within the microsomal membrane fraction. In agreement with this notion, preincubation of rat liver microsomes with cholesterol-containing vesicles in the presence of SCP$_2$ stimulated the ACAT activity up to five-fold (49, 87, 170). A comparable stimulation of ACAT activity was observed when microsomes from various Morris hepatomas were tested (171). A similar mechanism may account for the observation that SCP$_2$ stimulates the microsomal conversion of cholesterol into 7$\alpha$-hydroxycholesterol, which is the rate-limiting step in bile acid synthesis (172, 173).

## Steroidogenesis

Evidence has become available that $SCP_2$ may play an important role in steroid hormone biosynthesis (for a review, see Ref. 174). The rate-limiting step in this process is the translocation of cholesterol from intracellular stores to the $P_{450}$-cholesterol side-chain cleavage enzyme, the inner mitochondrial membrane protein that catalyzes the conversion of cholesterol to pregnenolone (175–177). $SCP_2$ would then be one of the cytosolic factors involved in the translocation reaction. In vitro $SCP_2$ will transfer cholesterol from adrenal lipid droplets to adrenal mitochondria, giving rise to a greatly enhanced pregnenolone production (85). Just the addition of $SCP_2$ to rat adrenal mitochondria enhanced the conversion of endogenous cholesterol to pregnenolone (85, 178, 179), a result later confirmed with mitochondria isolated from rat tumor Leydig cells (180). This stimulation is in agreement with the observation that addition of $SCP_2$ to adrenal mitochondria shifts cholesterol from the outer to inner membrane by an, as yet, unknown mechanism (181). Moreover, there is some evidence that rat adrenal cytosol contains a factor that potentiates the $SCP_2$ action (179). Pregnenolone production by adrenal mitochondria was enhanced eight-fold after a 15-min stimulation of rats with adrenal corticotropic hormone (ACTH); this treatment had no effect on the cellular levels of $SCP_2$. Since the stimulatory effect on pregnenolone production was inhibited by prior treatment of the rats with cycloheximide, it appears that ACTH affects the synthesis of an essential protein factor different from $SCP_2$ (179). On the other hand, up to a four-fold increase in the synthesis of $SCP_2$ was noted with rat adrenal cortical cells in culture treated for 48 hr with ACTH or dibutyryl cyclic AMP (160). In addition, stimulation of steroidogenesis by treatment of rats with gonadotropin resulted in a five-fold increase in mRNAs encoding $SCP_2$ in the ovaries (162). These studies support the model in which increased $SCP_2$ synthesis is related to an enhanced pregnenolone production.

## ROLE IN MEMBRANE VESICLE FLOW

A continuous and massive flow of membrane vesicles occurs within the cell from the endoplasmic reticulum (ER) through the Golgi complex to the plasma membrane. Secretory and membrane proteins, as well as lipids including cholesterol, sphingomyelin, and glycolipids, may be carried from their site of synthesis to the cell surface by this bulk flow mechanism (182–186). From measuring the secretion of an inert bulk phase marker (i.e. $N$-acyl-Asn-Tyr-Thr-NH$_2$) by cells in culture, it was estimated that as a result of this vesicle flow, about half of the total ER phospholipid was transferred into the Golgi stack every 10 min (187). It was suggested that in order to maintain the integrity of the ER, phospholipids are carried back from the Golgi to the ER as individual molecules through the cytosol, possibly by phospholipid transfer proteins. This would then imply that in situ these proteins carry out a net mass transfer of phospholipids to the ER. Since in vitro these proteins are normally found to catalyze an exchange of phospholipids between membranes, however, one has to propose the presence of specific components in the ER that help these transfer proteins to unload selectively their bound phospholipid molecule.

Recently, it was shown that PI-TP from yeast is identical to the SEC14p, a protein required for the export of yeast secretory proteins from the Golgi complex (66, 74, 74a). From studies on *sec14-1*$^{\text{ts}}$ mutants, it was evident that the SEC14p exerted its effect at a late Golgi stage of secretion (188, 189). Under nonpermissive conditions, an accumulation of Golgi-derived structures was observed, thus suggesting that the SEC14p plays a role in the budding of vesicles from the *trans*-Golgi cisternae. If the physiological role of SEC14p is to transfer PI between membranes, it is possible that the maintenance of the PI/PC ratio in the *trans*-Golgi stack is very important to an uninterrupted vesicle flow. Another possibility is that PI-TP is part of the process involved in phospholipid retrieval from the Golgi to the ER (187). These studies provide the first direct evidence that a phospholipid transfer protein plays an important role in an essential intracellular process, i.e. protein secretion by membrane vesicle flow. On the other hand, the complete lack of sequence homology between rat and yeast PI-TP (54, 66) makes one wonder whether PI-TP plays a role in protein secretion in mammalian cells.

## CONCLUDING REMARKS

Although definite proof is still lacking, it is reasonable to assume that in situ phospholipid transfer proteins are involved in the rapid translocation of phospholipids between intracellular organelles. This rapid translocation may be linked to phospholipid metabolism, the maintenance of the distinct lipid composition of different subcellular membranes and/or, in concert with additional, as yet unidentified, cellular factors, membrane biogenesis. So far, there is no evidence that specific transfer proteins for other phospholipids such as PE, PS, or PA are present in mammalian tissues. This raises the question whether PC-TP and PI-TP have a particular function in the above processes. The recent discovery that PI-TP is essential in membrane vesicle flow in yeast makes clear that research in the physiological role of phospholipid transfer proteins may reveal new and unexpected connections between lipid dynamics and the functioning of the cell.

ACKNOWLEDGMENTS

I acknowledge with thanks my long-time collaborator Mr. J. Westerman, whose dedication and experimental skills have contributed significantly to the field reviewed here.

## Literature Cited

1. Bell, F. P. 1978. *Prog. Lipid Res.* 17:207–43
2. Dawidowicz, E. A. 1987. *Curr. Top. Membr. Transp.* 29:175–202
3. Kaplan, M. R., Simoni, R. D. 1985. *J. Cell Biol.* 101:441–45
4. Kaplan, M. R., Simoni, R. D. 1985. *J. Cell Biol.* 101:446–53
5. Vance, J. E. 1988. *Biochim. Biophys. Acta* 963:10–20
6. Yaffe, M. P., Kennedy, E. P. 1983. *Biochemistry* 22:1497–1507
7. Pagano, R. E., Sleight, R. G. 1985. *Science* 229:1051–57
8. Phillips, M. C., Johnson, W. J., Rothblat, G. H. 1987. *Biochim. Biophys. Acta* 906:223–76
9. Van Meer, G. 1989. *Annu. Rev. Cell Biol.* 5:247–75
10. Voelker, D. R. 1990. *Experientia* 46:569–79
11. Tall, A. R. 1986. *J. Lipid Res.* 27:361–67
12. Morton, R. E. 1990. *Experientia* 46:552–60
13. Sasaki, T. 1990. *Experientia* 46:611–16
14. Brown, R. E., Stephenson, F. A., Markello, T., Barenholz, Y., Thompson, T. E. 1985. *Chem. Phys. Lipids* 38:79–93
15. Taylor, F. R., Kandutsch, A. A. 1985. *Chem. Phys. Lipids* 38:187–94
16. Paulussen, R. J. A., Veerkamp, J. H. 1990. See Ref. 190, pp. 175–226
17. Chytil, F., Ong, D. E. 1984. In *The Retinoids*, ed. M. B. Sporn, A. B. Roberts, D. S. Goodman, 2:89–123. Orlando: Academic
18. Kader, J. C., Douady, D., Mazliak, P. 1982. In *Phospholipids*, ed. J. N. Hawthorne, G. B. Ansell, pp. 279–311. Amsterdam: Elsevier
19. Wirtz, K. W. A. 1982. In *Lipid-Protein Interactions*, ed. P. Jost, O. H. Griffith, 1:151–231. New York: Wiley
20. Helmkamp, G. M. 1986. *J. Bioenerg. Biomembr.* 18:71–91
21. Helmkamp, G. M. 1990. See Ref. 190, pp. 129–74
22. Scallen, T. J., Pastuszyn, A., Noland, B. J., Chanderbhan, R., Kharroubi, A.,

Vahouny, G. V. 1985. *Chem. Phys. Lipids* 38:239–61
23. Wirtz, K. W. A., Teerlink, T., Akeroyd, R. 1985. In *The Enzymes of Biological Membranes*, ed. A. N. Martonosi, 2:111–38. New York: Plenum
24. Wirtz, K. W. A., Op den Kamp, J. A. F., Roelofsen, B. 1986. In *Progress in Protein-Lipid Interactions*, ed. A. Watts, J. J. H. H. M. de Pont, 2:221–65. Amsterdam: Elsevier
25. Wetterau, J. R., Zilversmit, D. B. 1984. *Methods Biochem. Anal.* 30:199–226
26. Somerharju, P. J., Van Paridon, P. A., Wirtz, K. W. A. 1990. See Ref. 190, pp. 21–43
27. Wirtz, K. W. A., Zilversmit, D. B. 1968. *J. Biol. Chem.* 243:3596–3602
28. Akiyama, M., Sakagami, T. 1969. *Biochim. Biophys. Acta* 187:105–12
29. McMurray, W. C., Dawson, R. M. C. 1969. *Biochim. J.* 12:91–108
30. Harvey, M. S., Wirtz, K. W. A., Kamp, H. H., Zegers, B. J. M., Van Deenen, L. L. M. 1973. *Biochim. Biophys. Acta* 323:234–39
31. Dyatlovitskaya, E. V., Timofeeva, N. G., Bergelson, L. 1978. *Eur. J. Biochem.* 82:463–71
32. Crain, R. C., Zilversmit, D. B. 1980. *Biochemistry* 19:1433–39
33. Kamp, H. H., Wirtz, K. W. A., Van Deenen, L. L. M. 1973. *Biochim. Biophys. Acta* 318:313–25
34. Lumb, R. H., Kloosterman, A. D., Wirtz, K. W. A., Van Deenen, L. L. M. 1976. *Eur. J. Biochem.* 69:15–22
35. Helmkamp, G. M., Harvey, M. S., Wirtz, K. W. A., Van Deenen, L. L. M. 1974. *J. Biol. Chem.* 249:6382–89
36. Demel, R. A., Kalsbeek, R., Wirtz, K. W. A., Van Deenen, L. L. M. 1977. *Biochim. Biophys. Acta* 466:10–22
37. DiCorleto, P. E., Warach, J. B., Zilversmit, D. B. 1979. *J. Biol. Chem.* 254:7795–7802
38. George, P. Y., Helmkamp, G. M. 1985. *Biochim. Biophys. Acta* 836:176–84
39. Bloj, B., Zilversmit, D. B. 1977. *J. Biol. Chem.* 252:1613–19

40. Akeroyd, R., Moonen, P., Westerman, J., Puyk, W. C., Wirtz, K. W. A. 1981. *Eur. J. Biochem.* 114:385–91
41. Poorthuis, B. J. H. M., Van der Krift, T. P., Teerlink, T., Akeroyd, R., Hostetler, K. Y., Wirtz, K. W. A. 1980. *Biochim. Biophys. Acta* 600:376–86
42. Somerharju, P. J., Van Paridon, P., Wirtz, K. W. A. 1983. *Biochim. Biophys. Acta* 731:186–95
43. Venuti, S. E., Helmkamp, G. M. 1988. *Biochim. Biophys. Acta* 946:119–28
44. Daum, G., Paltauf, F. 1984. *Biochim. Biophys. Acta* 794:385–91
45. Szolderits, G., Hermetter, A., Paltauf, F., Daum, G. 1989. *Biochim. Biophys. Acta* 986:301–9
46. Grondin, P., Vergnolle, C., Chavant, L., Kader, J. C. 1990. *Int. J. Biochem.* 22:93–98
47. Noland, B. J., Arebalo, R. E., Hansbury, E., Scallen, T. J. 1980. *J. Biol. Chem.* 255:4282–89
48. Poorthuis, B. J. H. M., Glatz, J. F. C., Akeroyd, R., Wirtz, K. W. A. 1981. *Biochim. Biophys. Acta* 665:256–61
49. Traszkos, J. M., Gaylor, J. L. 1983. *Biochim. Biophys. Acta* 751:52–65
50. Westerman, J., Wirtz, K. W. A. 1985. *Biochem. Biophys. Res. Commun.* 127:333–38
51. Van Amerongen, A., Helms, J. B., Van der Krift, T. P., Schutgens, R. B. H., Wirtz, K. W. A. 1987. *Biochim. Biophys. Acta* 919:149–55
52. Basu, J., Kundu, M., Bhattacharya, U., Mazumder, C., Chakrabarti, P. 1988. *Biochim. Biophys. Acta* 959:134–42
53. Tai, S. P., Kaplan, S. 1984. *J. Biol. Chem.* 259:12178–83
54. Dickeson, S. K., Lim, C. N., Schuyler, G. T., Dalton, T. P., Helmkamp, G. M., Yarbrough, L. R. 1989. *J. Biol. Chem.* 264:16557–64
55. Read, R. J., Funkhouser, J. D. 1983. *Biochim. Biophys. Acta* 752:118–26
56. Funkhouser, J. D. 1987. *Biochem. Biophys. Res. Commun.* 145:1310–14
57. Dyatlovitskaya, E. V., Timofeeva, N. G., Yakimenko, E. F., Barsukov, L. I., Muzya, G. I., Bergelson, L. D. 1982. *Eur. J. Biochem.* 123:311–15
58. Douady, D., Grosbois, M., Guerbette, F., Kader, J. C. 1985. *Physiol. Veg.* 23:373–80
59. Watanabe, S., Yamada, M. 1986. *Biochim. Biophys. Acta* 876:116–23
60. Kader, J. C., Julienne, M., Vergnolle, C. 1984. *Eur. J. Biochem.* 139:411–16
61. Grosbois, M., Guerbette, F., Douady, D., Kader, J. C. 1987. *Biochim. Biophys. Acta* 917:162–68

62. Arondel, V., Kader, J. C. 1990. *Experientia* 46:579–85
63. Tanaka, T., Yamada, M. 1982. In *Biochemistry and Metabolism of Plant Lipids*, ed. J. F. G. M. Wintermans, P. J. C. Kuiper, pp. 99–106. Amsterdam: Elsevier
64. Cobon, G. S., Crowfoot, P. D., Murphy, M., Linnane, A. W. 1976. *Biochim. Biophys. Acta* 441:255–63
65. Bozzato, R. P., Tinker, D. O. 1987. *Biochem. Cell. Biol.* 65:195–202
66. Aitken, J. F., Van Heusden, G. P. H., Temkin, M., Dowhan, W. 1990. *J. Biol. Chem.* 265:4711–17
67. Paltauf, F., Daum, G. 1990. See Ref. 190, pp. 279–99
68. Daum, G., Schwelberger, H. G., Paltauf, F. 1986. *Biochim. Biophys. Acta* 879:240–46
69. Lemaresquier, H., Bureau, G., Mazliak, P., Kader, J. C. 1982. *Int. J. Biochem.* 14:71–74
70. Cohen, L. K., Lueking, D. R., Kaplan, S. 1979. *J. Biol. Chem.* 254:721–28
71. Tai, S. P., Kaplan, S. 1985. *J. Bacteriol.* 164:181–86
72. Kaplan, S. 1981. *Photochem. Photobiol.* 34:769–74
73. Pastuszyn, A., Noland, B. J., Bazan, J. F., Fletterick, R. J., Scallen, T. J. 1987. *J. Biol. Chem.* 262:13219–27
74. Bankaitis, V. A., Malehorn, D. E., Emr, S. D., Greene, R. 1989. *J. Cell Biol.* 108:1271–81
74a. Bankaitis, V. A., Aitken, J. R., Cleves, A. E., Dowhan, W. 1990. *Nature* 347:561–62
75. Morris, H. R., Larsen, B. S., Billheimer, J. T. 1988. *Biochem. Biophys. Res. Commun.* 154:476–82
76. Van Amerongen, A., Teerlink, T., Van Heusden, G. P. H., Wirtz, K. W. A. 1985. *Chem. Phys. Lipids* 38:195–204
77. Nichols, J. W. 1987. *J. Biol. Chem.* 262:14172–77
78. Takashima, K., Watanabe, S., Yamada, M., Mamiya, G. 1986. *Biochim. Biophys. Acta* 870:248–55
79. Takashima, K., Watanabe, S., Yamada, M., Suga, T., Mamiya, G. 1988. *Eur. J. Biochem.* 177:241–49
80. Bouillon, P., Drischel, C., Vergnolle, C., Duranton, H., Kader, J. C. 1987. *Eur. J. Biochem.* 166:387–91
81. Tchang, F., This, P., Stiefel, V., Arondel, V., Morch, M. D., et al. 1988. *J. Biol. Chem.* 263:16849–55
82. Demel, R. A., Wirtz, K. W. A., Kamp, H. H., Geurts van Kessel, W. S. M., Van Deenen, L. L. M. 1973. *Nature New Biol.* 246:102–5

83. Van Paridon, P. A., Visser, A. J. W. G., Wirtz, K. W. A. 1987. *Biochim. Biophys. Acta* 898:172–80
84. Helmkamp, G. M., Nelemans, S. A., Wirtz, K. W. A. 1976. *Biochim. Biophys. Acta* 424:168–82
85. Chanderbhan, R., Noland, B. J., Scallen, T. J., Vahouny, G. V. 1982. *J. Biol. Chem.* 257:8928–34
86. Van Amerongen, A., Demel, R. A., Westerman, J., Wirtz, K. W. A. 1989. *Biochim. Biophys. Acta* 1004:36–43
87. Schroeder, F., Butko, P., Nemecz, G., Scallen, T. J. 1990. *J. Biol. Chem.* 265:151–57
88. Devaux, P. F., Moonen, P., Bienvenue, A., Wirtz, K. W. A. 1977. *Proc. Natl. Acad. Sci. USA* 74:1807–10
89. Machida, K., Ohnishi, S. I. 1978. *Biochim. Biophys. Acta* 507:156–64
90. Berkhout, T. A., Visser, A. J. W. G., Wirtz, K. W. A. 1984. *Biochemistry* 23:1505–13
91. Visser, A. J. W. G., Penners, N. H. G., Müller, F. 1983. In *Mobility and Recognition in Cell Biology*, ed. H. Sund, C. Veeger, pp. 137–54. Berlin: de Gruyter
92. Wirtz, K. W. A., Visser, A. J. W. G., Op den Kamp, J. A. F., Roelofsen, B., Van Deenen, L. L. M. 1985. In *Recent Advances in Biological Membrane Studies: Structure and Biogenesis, Oxidation and Energetics*, ed. L. Packer, 91:29–43. New York: Plenum
93. Somerharju, P. J., Van Loon, D., Wirtz, K. W. A. 1987. *Biochemistry* 26:7193–99
94. Kasurinen, J., Van Paridon, P. A., Wirtz, K. W. A., Somerharju, P. J. 1990. *Biochemistry* 29:3548–54
95. Van Paridon, P. A., Gadella, T. W. J., Somerharju, P. J., Wirtz, K. W. A. 1988. *Biochemistry* 27:6208–14
96. Westerman, J., Wirtz, K. W. A., Berkhout, T., Van Deenen, L. L. M., Radhakrishnan, R., Khorana, H. G. 1983. *Eur. J. Biochem.* 132:441–49
97. Akeroyd, R., Lenstra, J. A., Westerman, J., Vriend, G., Wirtz, K. W. A., Van Deenen, L. L. M. 1982. *Eur. J. Biochem.* 121:391–94
98. Sacchettini, J. C., Gordon, J. I., Banaszak, L. J. 1989. *J. Mol. Biol.* 208:327–39
99. Sacchettini, J. C., Gordon, J. I., Banaszak, L. J. 1989. *Proc. Natl. Acad. Sci. USA* 86:7736–40
100. Newcomer, M. E., Liljas, A., Sundelin, J., Rask, L., Peterson, P. A. 1984. *J. Biol. Chem.* 259:5230–31
101. Kamp, H. H., Wirtz, K. W. A., Baer,

P. R., Slotboom, A. J., Rosenthal, A. F., Paltauf, F., Van Deenen, L. L. M. 1977. *Biochemistry* 16:1310–16
102. Runquist, E. A., Helmkamp, G. M. 1988. *Biochim. Biophys. Acta* 940:10–20
103. Van Paridon, P. A., Gadella, T. W. J., Somerharju, P. J., Wirtz, K. W. A. 1987. *Biochim. Biophys. Acta* 903:68–77
104. Schermoly, M. J., Helmkamp, G. M. 1983. *Brain Res.* 268:197–200
105. Van Paridon, P. A., Gadella, T. W. J., Wirtz, K. W. A. 1988. *Biochim. Biophys. Acta* 943:76–86
106. Goldschmidt-Clermont, P. J., Machesky, L. M., Baldassare, J. J., Pollard, T. D. 1990. *Science* 247:1575–78
107. Martin, F. J., MacDonald, R. C. 1976. *Biochemistry* 15:321–27
108. Duckwitz-Peterlein, G., Eilenberger, G., Overath, P. 1977. *Biochim. Biophys. Acta* 448:245–64
109. Roseman, M. A., Thompson, T. E. 1980. *Biochemistry* 19:439–44
110. McLean, L. R., Phillips, M. C. 1981. *Biochemistry* 20:2893–2900
111. McLean, L. R., Phillips, M. C. 1984. *Biochemistry* 23:4624–30
112. Jones, J. D., Thompson, T. E. 1989. *Biochemistry* 28:129–34
113. Bozzato, R. P., Tinker, D. O. 1982. *Can. J. Biochem.* 60:409–18
114. Nichols, J. W., Pagano, R. E. 1983. *J. Biol. Chem.* 258:5368–71
115. Bloj, B., Zilversmit, D. B. 1981. *Mol. Cell. Biochem.* 40:163–72
116. Crain, R. C. 1982. *Lipids* 17:935–43
117. Kasper, A. M., Helmkamp, G. M. 1981. *Biochemistry* 20:146–51
118. Huang, C., Mason, J. T. 1978. *Proc. Natl. Acad. Sci. USA* 75:308–10
119. Machida, K., Ohnishi, S. I. 1980. *Biochim. Biophys. Acta* 596:201–9
120. Wirtz, K. W. A., Vriend, G., Westerman, J. 1979. *Eur. J. Biochem.* 94:215–21
121. Berkhout, T. A., Van den Bergh, C., Mos, H., De Kruijff, B., Wirtz, K. W. A. 1984. *Biochemistry* 23:6894–6900
122. Kasper, A. M., Helmkamp, G. M. 1981. *Biochim. Biophys. Acta* 664:22–32
123. Zborowski, J., Demel, R. A. 1982. *Biochim. Biophys. Acta* 688:381–87
124. Badger, C. R., Helmkamp, G. M. 1982. *Biochim. Biophys. Acta* 692:33–40
125. Mullikin, L. J., Helmkamp, G. M. 1984. *J. Biol. Chem.* 259:2764–68
126. Helmkamp, G. M. 1980. *Biochim. Biophys. Acta* 595:222–34
127. Demel, R. A., Van Bergen, B. G. M.,

Van den Eeden, A. L. G., Zborowski, J., Defize, L. H. K. 1982. *Biochim. Biophys. Acta* 710:264–70

128. Helmkamp, G. M. 1980. *Biochemistry* 19:2050–56
129. Nichols, J. W. 1985. *Biochemistry* 24: 6390–98
130. Nichols, J. W. 1988. *Biochemistry* 27: 1889–96
131. Helmkamp, G. M. 1980. *Biochem. Biophys. Res. Commun.* 97:1091–96
132. De Kruijff, B., Wirtz, K. W. A. 1977. *Biochim. Biophys. Acta* 468:318–26
133. Kuypers, F. A., Berendsen, W., Roelofsen, B., Op den Kamp, J. A. F., Van Deenen, L. L. M. 1984. *J. Cell Biol.* 99:2260–67
134. Op den Kamp, J. A. F., Roelofsen, B., Van Deenen, L. L. M. 1985. *Trends Biochem. Sci.* 10:320–23
135. Christiansson, A., Kuypers, F. A., Roelofsen, B., Op den Kamp, J. A. F., Van Deenen, L. L. M. 1985. *J. Cell Biol.* 101:1455–62
136. Wirtz, K. W. A., Devaux, P. F., Bienvenue, A. 1980. *Biochemistry* 19:3395–99
137. Runquist, E. A., Helmkamp, G. M. 1988. *Biochim. Biophys. Acta* 940:21–32
138. Crain, R. C., Zilversmit, D. B. 1980. *Biochim. Biophys. Acta* 620:37–48
139. Crain, R. C., Zilversmit, D. B. 1981. *Biochemistry* 20:5320–26
140. North, P., Fleischer, S. 1983. *J. Biol. Chem.* 258:1242–53
141. Bergelson, L. D., Barsukov, L. I. 1977. *Science* 197:224–30
142. Teerlink, T., Van der Krift, T. P., Post, M., Wirtz, K. W. A. 1982. *Biochim. Biophys. Acta* 713:61–67
143. Teerlink, T., Poorthuis, B. J. H. M., Van der Krift, T. P., Wirtz, K. W. A. 1981. *Biochim. Biophys. Acta* 665:74–80
144. Thomas, P. J., Wendelburg, B. E., Venuti, S. E., Helmkamp, G. M. 1989. *Biochim. Biophys. Acta* 982:24–30
145. Venuti, S. E., Helmkamp, G. M. 1988. *Neurochem. Int.* 13:531–41
146. Wirtz, K. W. A., Jolles, J., Westerman, J., Neys, F. 1976. *Nature* 260:354–55
147. Brophy, P. J., Aitken, J. W. 1979. *J. Neurochem.* 33:355–56
148. Ruenwongsa, P., Singh, H., Jungalwala, F. B. 1979. *J. Biol. Chem.* 254:9385–93
149. Nyquist, D. A., Helmkamp, G. M. 1989. *Biochim. Biophys. Acta* 987:165–70
150. Carey, E. M., Foster, P. C. 1984. *Biochim. Biophys. Acta* 792:48–58

151. Teerlink, T., Van der Krift, T. P., Van Heusden, G. P. H., Wirtz, K. W. A. 1984. *Biochim. Biophys. Acta* 793:251–59
152. Van Amerongen, A., Van Noort, M., Van Beckhoven, J. R. C. M., Rommerts, F. F. G., Orly, J., Wirtz, K. W. A. 1989. *Biochim. Biophys. Acta* 1001:243–48
153. Chanderbhan, R. F., Kharroubi, A. T., Noland, B. J., Scallen, T. J., Vahouny, G. V. 1986. *Endocrine Res.* 12:351–70
154. Van Heusden, G. P. H., Bos, K., Wirtz, K. W. A. 1990. *Biochim. Biophys. Acta* 1046:315–21
155. Van der Krift, T. P., Leunissen, J., Teerlink, T., Van Heusden, G. P. H., Verkleij, A. J., Wirtz, K. W. A. 1985. *Biochim. Biophys. Acta* 812:387–92
156. Tsuneoka, M., Yamamoto, A., Fujiki, Y., Tashiro, Y. 1988. *J. Biochem.* 104: 560–64
157. Keller, G. A., Scallen, T. J., Clarke, D., Maher, P. A., Krisans, S. K., Singer, S. J. 1989. *J. Cell Biol.* 108:1353–61
158. Van Heusden, G. P. H., Bos, K., Raetz, C. R. H., Wirtz, K. W. A. 1990. *J. Biol. Chem.* 265:4105–10
159. Fujiki, Y., Tsuneoka, M., Tashiro, Y. 1989. *J. Biochem.* 106:1126–31
160. Trzeciak, W. H., Simpson, E. R., Scallen, T. J., Vahouny, G. V., Waterman, M. R. 1987. *J. Biol. Chem.* 262:3713–17
161. Ossendorp, B. C., Van Heusden, G. P. H., Wirtz, K. W. A. 1990. *Biochem. Biophys. Res. Commun.* 168:631–36
162. Billheimer, J. T., Strehl, L. L., Davis, G. L., Strauss, J. F., Davis, L. G. 1990. *DNA Cell Biol.* 9:159–65
163. Khan, Z. U., Helmkamp, G. M. 1990. *J. Biol. Chem.* 265:700–5
164. Pelech, S. L., Vance, D. E. 1984. *Biochim. Biophys. Acta* 779:217–51
165. Dennis, E. A., Kennedy, E. P. 1972. *J. Lipid Res.* 13:263–67
166. Van Golde, L. M. G., Raben, J., Batenburg, J. J., Fleischer, B., Zambrano, F., Fleischer, S. 1974. *Biochim. Biophys. Acta* 360:179–92
167. Voelker, D. R. 1984. *Proc. Natl. Acad. Sci. USA* 81:2669–73
168. Voelker, D. R. 1989. *J. Biol. Chem.* 264:8019–25
169. Gavey, K. L., Noland, B. J., Scallen, T. J. 1981. *J. Biol. Chem.* 256:2993–99
170. Poorthuis, B. J. H. M., Wirtz, K. W. A. 1982. *Biochim. Biophys. Acta* 710:99–105
171. Van Heusden, G. P. H., Van der Krift,

# PHOSPHOLIPID TRANSFER PROTEINS    99

T. P., Hostetler, K. Y., Wirtz, K. W.
A. 1983. *Cancer Res.* 43:4207–10
172. Seltman, H., Diven, W., Rizk, M., Noland, B. J., Chanderbhan, R., et al. 1985. *Biochem. J.* 230:19–24
173. Lidström-Olsson, B., Wikvall, K. 1986. *Biochem. J.* 238:879–84
174. Vahouny, G. V., Chanderbhan, R., Kharroubi, A., Noland, B. J., Pastuszyn, A., Scallen, T. J. 1987. *Adv. Lipid Res.* 22:83–113
175. Privalle, C. T., Crivello, J. F., Jefcoate, C. R. 1983. *Proc. Natl. Acad. Sci. USA* 80:702–6
176. Simpson, E. R. 1979. *Mol. Cell. Endocrinol.* 13:213–27
177. Churchill, P. F., Kimura, T. 1979. *J. Biol. Chem.* 256:10443
178. Vahouny, G. V., Chanderbhan, R., Noland, B. J., Irwin, D., Dennis, P., et al. 1983. *J. Biol. Chem.* 258:11731–37
179. McNamara, B. C., Jefcoate, C. R. 1989. *Arch. Biochem. Biophys.* 275:53–62
180. Van Noort, M., Rommerts, F. F. G., Van Amerongen, A., Wirtz, K. W. A.

1988. *Biochem. Biophys. Res. Commun.* 154:60–65
181. Vahouny, G. V., Dennis, P., Chanderbhan, R., Fiskum, G., Noland, B. J., Scallen, T. J. 1984. *Biochem. Biophys. Res. Commun.* 122:509–15
182. Warren, G. 1985. *Trends Biochem. Sci.* 10:439–43
183. Rothman, J. E. 1987. *Cell* 50:521–22
184. Dawidowicz, E. A. 1987. *Annu. Rev. Biochem.* 56:43–61
185. Sleight, R. G. 1987. *Annu. Rev. Physiol.* 49:193–208
186. Bishop, W. R., Bell, R. M. 1988. *Annu. Rev. Cell Biol.* 4:579–610
187. Wieland, F. T., Gleason, M. L., Serafini, T. A., Rothman, J. E. 1987. *Cell* 50:289–300
188. Novick, P., Field, C., Schekman, R. 1980. *Cell* 21:205–15
189. Schekman, R. 1982. *Trends Biochem. Sci.* 7:243–46
190. Hilderson, H. J., Spener, F., eds. 1990. *Subcellular Biochemistry: Intracellular Membrane Transfer of Lipids*, Vol. 16. New York: Plenum

Annu. Rev. Biochem. 1991. 60:101–24

# THE ENZYMOLOGY OF PROTEIN TRANSLOCATION ACROSS THE *Escherichia coli* PLASMA MEMBRANE

*William Wickner, Arnold J. M. Driessen,[1] and Franz-Ulrich Hartl[2]*

Molecular Biology Institute and Department of Biological Chemistry, University of California at Los Angeles, Los Angeles, California 90024-1570

KEY WORDS:  translocase, secretion, translocation ATPase, leader peptide, protonmotive force.

## CONTENTS

[1]Permanent address: Department of Microbiology, University of Groningen, Kerklaan 30, 9751 NN Haren, The Netherlands.

[2]Permanent address: Institut für Physiologische Chemie, Universität München, Goethestrabe 33, D-8000 München 2, Federal Republic of Germany.

0066-4154/91/0701-0101$02.00

## PERSPECTIVES AND SUMMARY

Converging physiological, genetic, and biochemical studies have established the salient features of preprotein translocation across the plasma membrane of *Escherichia coli*. Translocation is catalyzed by two proteins, a soluble chaperone and a membrane-bound translocase. SecB, the major chaperone for export, forms a complex with preproteins. Complex formation inhibits side-reactions such as aggregation and misfolding and aids preprotein binding to the membrane surface. Translocase consists of functionally linked peripheral and integral membrane protein domains. SecA protein, the peripheral membrane domain of translocase, is the primary receptor for the SecB/preprotein complex. SecA hydrolyzes ATP, promoting cycles of translocation, preprotein release, $\Delta\bar{\mu}_{H^+}$-dependent translocation, and rebinding of the preprotein. The membrane-embedded domain of translocase is the SecY/E protein. It has, as subunits, the SecY and SecE polypeptides. The SecY/E protein stabilizes and activates SecA and participates in binding it to the membrane. SecA recognizes the leader domain of preproteins, whereas both SecA and SecB recognize the mature domain. Many proteins translocate without requiring SecB, and some proteins do not need translocase to assemble into the plasma membrane. Translocation is usually followed by endoproteolytic cleavage by leader peptidase. The availability of virtually every pure protein and cloned gene involved in the translocation process makes *E. coli* the premier organism for the study of translocation mechanisms.

## INTRODUCTION

Cells are divided into multiple compartments, each a membrane or a membrane-bounded aqueous space with a unique set of resident proteins. Compartmentation allows metabolic reactions that would otherwise lead to futile cycles to be separated and regulated and allows metabolic control through selective routing of proteins and other molecules. Eukaryotic cells have more than 20 distinct compartments, whereas *E. coli* has four: cytoplasm, inner (plasma) membrane, periplasm, and outer membrane. During the past two decades, the pathways of protein targeting and translocation out of the cytoplasm have been heavily studied for *E. coli* and for eukaryotic organelles such as the rough endoplasmic reticulum, mitochondria, chloroplast, peroxisome, and nucleus. In each case, the basic questions are the same: What are the sequence elements or structural features in a protein that specify its correct target membrane for translocation? What proteins aid in protein targeting to the membrane? Is translocation via a proteinaceous transport system or through the lipid phase? What are the energetics of this process?

*E. coli* offers unique advantages for the study of these questions. For physiological studies, cells can be grown in a defined minimal medium and pulse-labeled with radioactive amino acids. Translocation across the plasma membrane exposes newly made proteins to externally added protease. These technical advantages have allowed extensive in vivo studies of translocation in this organism, whereas there are fewer studies of in vivo translocation into endoplasmic reticulum or mitochondria. The sophisticated genetics of *E. coli* has allowed the selection of mutants in the genes for both exported proteins and the proteins that facilitate export (1) and studies of their interactions (2). Biochemical studies of in vitro translocation reactions (3–7) have allowed dissection to yield pure components, first for the precursor of M13 coat protein (8) and, more recently, for proteins with more complex translocation requirements (9–13). A rough chronology of these studies is presented in Table 1. In this review, however, we emphasize our current view of the enzymology of preprotein translocation across the plasma membrane (Figure 1) rather than chronology. An excellent review of the genetics of bacterial export has appeared recently (1). We do not review here the export of colicins and toxins by "dedicated" transport systems.

## LEADER PEPTIDES

Periplasmic and outer membrane proteins, as well as some plasma membrane proteins, are made with amino-terminal leader (signal) sequences (14–21) that are removed immediately after translocation. Leader sequences are 16–26 residues long, and comprise a basic amino-terminal region, a central apolar region, and a nonhelical carboxy-terminal domain. Genetic studies have clearly established that the basic character of the amino terminus (22–25) and the apolar composition of the central domain (26–33) are essential for translocation. The conserved features of the last part of the leader, a helix-disrupting glycyl or prolyl residue and small side-chain residues at positions $-1$ and $-3$ with respect to the cleavage site (14, 19), are not needed for translocation (34, 35) but are essential for recognition by leader peptidase (or, for lipoproteins, lipoprotein signal peptidase) (34, 36, 37). Exchange of leader sequences between *E. coli* proteins usually allows continued export

**Table 1**   A chronology of some important findings in bacterial protein export[a]

| Year | Finding | References |
|------|---------|-----------|
| 1969 | Sequence of the first integral membrane protein | 177 |
| 1976 | Fusion proteins used to study secretion | 178 |
| 1977 | Bacterial membrane-bound polysomes | 179, 180, 193, 194 |
| 1977 | In vivo kinetics of protein export | 51 |
| 1978 | Discovery of *E. coli* leader peptidase | 181 |
| 1979 | Loop model of bacterial secretion | 168 |
| 1979–1983 | Translocation is not coupled to translation | 52–56, 182, 183, 195 |
| 1980 | Leader sequence mutations block secretion | 26, 184 |
| 1980 | Discovery of lipoprotein signal peptidase | 185 |
| 1980 | Membrane potential is needed for protein export | 98–101 |
| 1981 | Isolation of *secA* mutants | 196 |
| 1981 | Isolation of *prlA* mutants | 46 |
| 1982 | Isolation of pure precursor proteins | 155, 156, 186 |
| 1982–1983 | Isolation of leader peptidase, lipoprotein signal peptidase, and their genes | 187–191, 197, 198 |
| 1983 | Isolation of *SecB* mutants | 48 |
| 1983 | Isolation of *SecY* mutants | 86 |
| 1983 | Reconstitution of procoat translocation with pure components | 8 |
| 1984 | Interference | 80 |
| 1984 | In vitro Sec-dependent translocation | 5–7 |
| 1984–1986 | Preproteins are unfolded for export | 164, 166 |
| 1985 | ATP is needed for translocation | 105 |
| 1985 | M13 procoat is Sec-independent | 93 |
| 1988 | Interference is via SecB; SecB as chaperone | 10, 49, 57, 64 |
| 1988 | Isolation of SecA protein | 11, 12 |
| 1988 | Isolation of SecB protein | 9, 10 |
| 1988 | Isolation of *SecE* mutants | 88, 89 |
| 1989 | Suppressor directed inactivation | 192 |
| 1989 | SecA is the catalytic element of translocation ATPase | 76 |
| 1989 | Translocation intermediates | 114, 168a |
| 1990 | Purification of SecY/E protein | 13 |
| 1990 | SDI demonstration of SecY and SecE interaction | 2 |

[a] The authors offer their sincere apologies for inadvertant omissions that undoubtedly occur here.

(38, 39), albeit at a reduced rate, thereby suggesting that the leader peptide and the rest of the protein have distinct functions in export.

Two kinds of genetic studies have established that these are the only features of functional leader peptides that are essential. In one study, the apolar domain of a leader was replaced by leucyl residues (29). Normal export ensued, establishing that apolarity is the only important feature of the center of leader peptides. In a similar vein, the addition of several lysyl residues at the amino terminus of an apolar membrane-spanning region can convert it into a functional leader (40). In a second approach to leader function, secretion of pre-invertase was measured in yeast after the normal leader region was completely deleted and replaced by random peptides (41). Approximately one in five random peptides supported translocation. Sequence analysis showed that the only common theme among these peptides was simply their basic and apolar character. How can such a simple, and common, motif serve to direct the selective and efficient process of secretion? Recent studies, reviewed below, suggest that the mature domains of exported proteins also bear some of the information specifying export. In addition, there may be a selective pressure on proteins that remain in the cytosol to avoid leader-like sequences near their amino terminus (41). Nevertheless, the relative roles in leader function of either protein recognition or of physical partitioning into membrane lipids remains an area of active inquiry.

Leader peptides serve a number of distinct functions. The leader regions of pre-MBP (42) and pre-$\beta$-lactamase (43) retard the folding of the mature domain, and this delayed folding may be crucial for interaction with the SecB chaperone (see below). The leader region is directly recognized by SccA (44, 45), SecY (46), and leader peptidase (47). Direct evidence for its interaction with lipid during Sec-dependent translocation is not yet available.

## SEC PROTEIN–DEPENDENT TRANSLOCATION

Genetic and biochemical studies of bacterial secretion have converged, providing a model (Figure 1A) of how proteins are selected for translocation. Export of many precursor proteins involves interaction with a cytosolic chaperone, most commonly SecB (48, 49). Most proteins (50) then use a complex, membrane-bound translocase. Periplasmic and outer membrane preproteins must be cleaved by leader peptidase to leave the outer surface of the inner membrane (35).

### SecB

Preprotein synthesis and membrane transit are not coupled in *E. coli*. (51–56). During the interval between synthesis and translocation, some precursor proteins are stabilized by forming a stoichiometric complex with SecB (57–61), a soluble oligomeric "chaperone" protein. Though SecB is not essential for either in vivo (61a) or in vitro (62) translocation, it prevents nonproductive reactions such as tight protein folding (63, 64), aggregation (58), or misassociation with low-affinity sites on the membrane (65). It has been postulated that chaperones such as SecB might function by association with apolar regions (66), though there is no direct evidence. In any case, SecB clearly associates with the mature domain of precursor proteins:

**A** Sec Dependent Protein Export

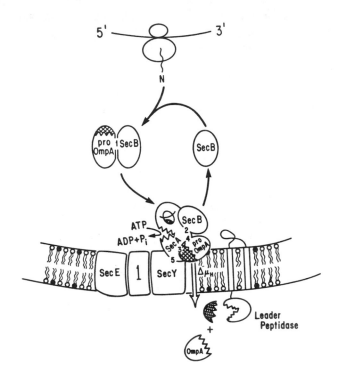

**B** Catalytic Cycle of Preprotein Translocation

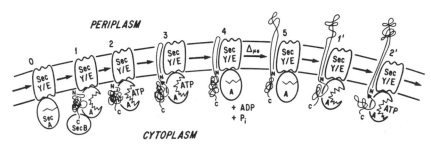

*Figure 1* Sec-dependent protein translocation. Modified from *A*. Hartl et al (65) and *B*. Schiebel et al (84). Small case numbers in part *A* refer to the binding of 1. SecB to the mature domain of a precursor protein, 2. SecB to SecA, 3. SecA to the leader domain, 4. SecA to the mature domain of a precursor protein, and 5. SecA to SecY/E.

1. SecB can bind to mature domains of precursor proteins in vitro (57, 63, 67).
2. Exchange of leader regions between pre-maltose-binding protein, which interacts strongly with SecB, and pre-alkaline phosphatase or pre-ribose-binding protein, which do not, has shown that SecB-dependence is conferred by the mature domain, not the leader (60; D. Collier, S. Strobel, P. J. Bassford Jr., personal communication).
3. Overproduction of a mutant pre-maltose-binding protein that lacks its leader region interferes with the export of wild-type precursor proteins (68, 69). This interference is overcome by overproduction of SecB protein, thus indicating that the interfering maltose-binding protein was titrating the available SecB protein.
4. Regions of interaction with SecB have been genetically defined within mature domains of precursor proteins (E. Altmann and S. Emr, personal communication). Though SecB associates with the mature domain of preproteins, the leader region, by retarding the folding of the mature domain, may strongly influence binding to SecB (64).

There is no evidence that SecB functions as an "unfoldase." Association with SecB prevents pre-maltose-binding protein from folding into its final, stable folded structure (64), which cannot cross the membrane. This does not show, however, that the preprotein is being held in an unfolded state, devoid of stable secondary or tertiary structure, or being actively unfolded once folding occurs. ProOmpA in complex with SecB has been shown to have substantial folded structure (58). In this case, the SecB association shields apolar surfaces of proOmpA and prevents its aggregation.

SecB is the major chaperone for preprotein export, yet other chaperone proteins have a role as well. GroEL selectively facilitates the secretion of pre-$\beta$-lactamase (70) and DnaK can support the secretion of artificial fusions between leader peptides and $\beta$-galactosidase (71). Either GroEL or trigger factor, an abundant cytosolic protein, can stabilize proOmpA for in vitro translocation across membrane vesicles (57). Genetic studies show that the interaction of proOmpA and trigger factor occurs in vivo, but is actually inhibitory for export (72). The biochemical basis for the unique role of SecB, rather than the more abundant GroEL (73) or trigger factor (72), may lie in the specific affinity of SecA for SecB (65). SecB may contribute a major share of the binding affinity of the SecB/preprotein complex for SecA (65). SecB can even stabilize proOmpA while they are both bound to the membrane via SecA (65).

Several salient questions remain about the mechanism of SecB function. How is a 1 : 1 binding stoichiometry achieved by an oligomeric protein? Perhaps the several SecB subunits associate with several different apolar domains of the bound preprotein. What features of the mature domains of precursor proteins are recognized by SecB? Crystals of a SecB/preprotein complex may be essential for clear answers to these questions. Finally, the nanomolar binding affinity of SecB for SecA suggests that it might remain bound to SecA in the high protein concentration of the cell. In this light, even SecB could be viewed as a subunit of the peripheral domain of translocase.

## SecA

SecA is a peripheral membrane protein (12, 74) of 102 kDa (75). It serves as the high-affinity receptor for the SecB/preprotein complex (65). SecA hydrolyzes ATP and couples this energy to translocation (76, 77). SecA function and high-affinity membrane binding require acidic lipids (78; J. Hendrick, W. Wickner, in preparation) and the membrane-embedded SecY/E protein (65). The synergy of action of suppressor mutants in *SecA(prlD)* (45) and *SecY (prlA)* (79) is consistent with the physical interaction of these proteins. SecA and SecB are apparently the sole peripheral membrane proteins with direct roles in translocation (12, 13). The binding of SecA to the membrane strongly enhances its affinity for the SecB/preprotein complex (65). SecA binds this complex by virtue of its affinities for SecB (65) and for both the leader (44, 80) and mature domains of the precursor protein (78). Basic residues of the leader are especially important for its recognition by SecA (81, 82). SecA requires both SecY/E protein and acidic lipids for high-affinity binding to the membrane. This requirement for acidic lipids may be one reason for the requirement for these lipids in the overall translocation process (83). When SecA protein is in contact with each of the essential elements for translocation—a preprotein with an authentic leader and mature domain, the SecY/E protein, and acidic phospholipids (78)—it hydrolyzes ATP. This activated ATPase activity is termed the "translocation ATPase," since it depends on each of the elements needed for successful translocation (76). ATP binding to SecA drives a limited preprotein translocation (84). Model reactions suggest that ATP hydrolysis drives a release of precursor from SecA (84). $\Delta\bar{\mu}_{H^+}$-driven translocation occurs when the preprotein is not bound to SecA nucleotide. SecA, ATP, $\Delta\bar{\mu}_{H^+}$, and SecY/E protein are required for translocation as the protein progresses across the membrane (84). Each preprotein undergoes many cycles of SecA binding, ATP binding to SecA, limited preprotein translocation, ATP hydrolysis, release, $\Delta\bar{\mu}_{H^+}$-driven translocation, and rebinding (84). The progress of the translocation reaction is readily reversible (84). The two sources of metabolic energy for translocation, ATP and $\Delta\bar{\mu}_{H^+}$, govern progress across the membrane at each stage (Figure 1B).

SecA is clearly a complex enzyme with multiple catalytic and regulatory interactions. Its synthesis is regulated by the overall functionality of translocase (85), underscoring its central role in selecting proteins for export. Its amino acid sequence has three potential nucleoside-triphosphate-binding motifs (D. Oliver, personal communication), and the protein can be covalently modified by up to three moles of $N_3$-ATP per mole of SecA (76). Hydrolytic and regulatory roles have not yet been assigned to the individual ATP-binding sites. ATP hydrolysis is regulated by each of the essential components of translocation—SecY/E protein, the preprotein itself, and acidic lipids. Even competitive inhibition of translocation ATPase by synthetic leader peptide is manifest as an increase in the $K_m$ for ATP (44). In turn, ATP hydrolysis promotes preprotein release (84). Upon interaction with lipids, SecA is rendered thermolabile, but is stabilized by its interactions with ATP, SecY/E, and the leader and mature domains of precursor proteins (78). Strikingly, the stabilization of the ATPase activity of SecA by precursor protein can be efficiently mimicked by a mixture of synthetic leader peptide and the mature protein (78), affording enzymologists an opportunity to explore directly the separable specificities of SecA for the leader and mature domains of preproteins.

## SecY/E

Both biochemical (13) and genetic (2) studies have established that the SecY/E protein is the integral membrane protein that supports translocation. It constitutes approximately 0.6% of the protein of the plasma membrane, and comprises three subunits, the SecY polypeptide (86, 87), the SecE polypeptide (88–90), and a polypeptide of unknown gene. The latter subunit, termed band 1, copurifies with the SecY and SecE subunits during either column chromatography or immunoprecipitation (13). Confirmation of its role in translocase, however, will have to await either determination of its role in the mechanism of translocation or the identification and mutation of its gene. Although the transbilayer topology of the SecY and SecE polypeptides is known from genetic studies (90, 91), the number and stoichiometry of each subunit in the SecY/E protein is not yet known. SecY and SecE are exceptionally hydrophobic proteins; the former can even aggregate in the presence of SDS (87) and is not solubilized from the plasma membrane by most common detergents (44). In order to solubilize and maintain the SecY/E protein in a form that was isolable and could be reconstituted with lipid to support functional translocation, it was necessary to employ a combination of detergent and lipid mixed micelles in solutions with lowered water content, achieved by glycerol addition (92). Liposomes bearing the purified SecY/E protein support a translocation reaction that requires SecA, ATP, and an authentic preprotein and responds to an electrochemical potential (13). Equivalent translocation units of inner membrane vesicles and proteoliposomes with purified SecY/E protein have almost equivalent amounts of SecY protein (13), a result that suggests that no other proteins are essential for the basic translocation reaction. Since SecA is bound to the membrane at the SecY/E protein (65), we have defined "preprotein translocase" as the combination of SecY/E protein with its bound SecA. The former constitutes an integral membrane domain of translocase, whereas the latter is its peripheral domain.

There are more questions than answers about the basic functions of the SecY/E protein (92a). Is the energy of ATP hydrolysis transferred between SecY/E and SecA, much as energy is transferred between the integral and peripheral domains of the $F_1F_0$-ATP synthase? The availability of pure SecY/E may allow its assay as a proton-conducting channel, perhaps gated by the preprotein. Does the SecY/E protein serve as a transport system for the preprotein, conducting it through its center or along its surface? The existence in this gene of prlA mutants (prlA = secY), which are strong suppressors of leader sequence mutations (46), indicates that the SecY/E protein interacts directly with the leader peptide region during translocation. This function is now open to direct biochemical test in the reconstituted translocation reaction.

## TRANSLOCATION THAT IS SEC-INDEPENDENT

Though conditionally lethal mutants in the sec genes affect the secretion of most bacterial proteins, certain small proteins such as M13 procoat protein (93), PF3 coat protein (A. Kuhn, personal communication), and honeybee prepromelittin (94) do not require the Sec proteins for translocation. This cannot simply be a question of size, as secretion of the 8100 $M_r$ precursor of heat-stable enterotoxin (prepro-$ST_B$) requires SecA (95) and the MalF protein requires neither SecA nor SecY for its membrane insertion (I. McGovern, B. Traxler, J. Beckwith, personal communication).

M13 procoat protein, the first preprotein found not to require Sec proteins for membrane insertion (93) (Figure 2), has a typical leader sequence (96). Genetic exchange of the leader domains between M13 procoat and proOmpA, a Sec-dependent protein, showed that Sec-dependence was a feature of the mature domain rather than a particular property of the leader region (39). This was also established by point mutations, which showed that each part of the mature domain of the protein is essential for its assembly into the membrane (31, 32). A similar conclusion has been reached for prepromellitin insertion into the endoplasmic reticulum (97). This is in striking contrast to the Sec-dependent preproteins, in which few point mutations in the mature domain are reported to have a severe effect on export.

What is the relationship between the Sec-dependent and Sec-independent pathways of protein translocation? Occam's razor suggests that features of translocation shared between these pathways reflect a shared underlying mechanism. Sec-dependent and Sec-independent proteins have typical N-terminal leader sequences, thereby suggesting a role in all translocation for

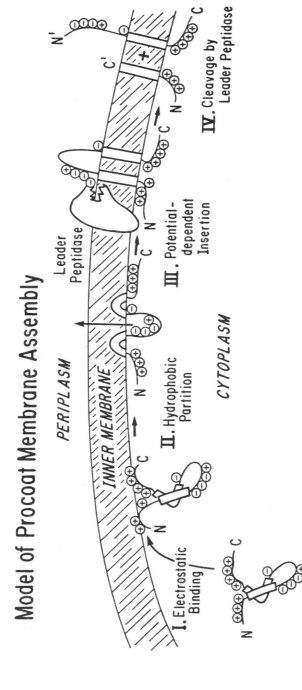

*Figure 2*  The Sec-independent insertion of M13 procoat into the plasma membrane of *E. coli*. From Kuhn et al (32).

the leader region that goes beyond its recognition by SecA. Since Sec-independent translocation requires the membrane electrochemical potential, the potential may also act on the Sec-dependent preproteins per se rather than solely via a postulated proton/protein antiport mechanism of SecY/E.

## ENERGETICS OF EXPORT

Translocation requires two forms of energy, ATP, and $\Delta\bar{\mu}_{H^+}$. In vivo dissipation of $\Delta\bar{\mu}_{H^+}$ by iono- or protonophores arrests translocation, and precursor proteins accumulate within the cell (54, 55, 98–104). Evidence that both ATP (105) and $\Delta\bar{\mu}_{H^+}$ (106, 107) are involved in translocation emerged from in vitro studies. The requirement for ATP is absolute, whereas $\Delta\bar{\mu}_{H^+}$ is strongly stimulatory.

### The Role of $\Delta\bar{\mu}_{H^+}$

The requirement for $\Delta\bar{\mu}_{H^+}$ varies, depending on the precursor protein species (108). Observations that $\Delta\bar{\mu}_{H^+}$ is only stimulatory in vitro, whereas it is clearly required in vivo, led to the suggestion that the role of $\Delta\bar{\mu}_{H^+}$ is indirect (105). Compelling evidence that $\Delta\bar{\mu}_{H^+}$ is an intrinsic part of the translocation mechanism was provided by studies with the reconstituted translocation reaction (13, 84, 109). $\Delta\bar{\mu}_{H^+}$ stimulates translocation in a reaction with purified components that utilizes both the transmembrane electrical potential $\Delta\psi$ and the transmembrane pH gradient $\Delta$pH (109), as it does in vivo (104). $\Delta$pH may perform a dual function in translocation (109), acting as a direct energy source and indirectly modulating the rate of translocation via the effect of the cytosolic pH on the activity of the translocase.

$\Delta\bar{\mu}_{H^+}$ may directly affect the precursor protein, activate SecY/E to promote transport, or both. Insofar as the $\Delta\bar{\mu}_{H^+}$ acts directly on precursor proteins during translocation, the roles of $\Delta$pH and $\Delta\psi$ may be different. $\Delta$pH may drive translocation by promoting the facile deprotonation of positively charged residues prior to membrane transit and their protonation afterward (109). $\Delta\psi$ may have an electrophoretic effect to promote the transfer of regions bearing a net negative charge. Interactions of the N-terminal domain of the leader sequence with the surface of the membrane may be followed by an electrophoretic transmembrane alignment of the internal dipole of the leader in response to $\Delta\psi$ (100). Charged residues provide a barrier to translocation, as illustrated by the strict "positive inside-rule" that guides the membrane topology of polytopic membrane proteins (40, 110, 111). Positively charged residues appear to be significantly more difficult to translocate than negatively charged ones (112). The $\Delta\bar{\mu}_{H^+}$ could serve directly as a driving force for translocation if polypeptide transit requires concurrent proton movement in the opposite direction. Proton transfer by a proton/polypeptide antiport mechanism could be gated in response to the distribution of positive charges in the precursor protein or obey a strict coupling stoichiometry. The SecY/E protein may conduct such proton movements, a biochemical function that can be directly tested with the reconstituted translocation reaction.

## The Role of ATP

ATP promotes translocation solely via its hydrolysis by the SecA protein (76). SecA may use the energy of ATP binding to translocate small domains of the precursor (84), whereas the energy of ATP hydrolysis drives preprotein release (84). $\Delta\bar{\mu}_{H^+}$ appears to have a role very different from ATP hydrolysis (Figure 1B). Although SecA may allow $\Delta\bar{\mu}_{H^+}$-independent translocation (113), $\Delta\bar{\mu}_{H^+}$ has no significant effect on the kinetics of SecA binding and function (109). Early and late stages of proOmpA translocation are reported to differ in their requirement for ATP and $\Delta\bar{\mu}_{H^+}$ (84, 109, 114, 115). The very first stage, a limited translocation that allows processing of the leader peptide, strictly requires ATP (84). Subsequent translocation can be efficiently driven by either SecA and ATP or by $\Delta\bar{\mu}_{H^+}$ alone (84). Studies of the kinetic isotope solvent effect show that the overall rate of translocation is limited by a critical proton transfer reaction (109). This involvement of catalytic protons provides further support for a direct role of $\Delta\bar{\mu}_{H^+}$ in translocation.

## COMPARISONS WITH OTHER PROTEIN TRANSLOCATION REACTIONS

A number of eukaryotic membranes are competent for protein translocation, including the membranes of the endoplasmic reticulum (ER) (116), mitochondria (117, 118), chloroplasts (119), and peroxisomes (120). All protein transport appears to follow certain common principles:
1. Proteins to be transported contain contiguous targeting sequences.
2. The membranes expose receptors or receptor-like components at their cytosolic surfaces.
3. Except in special cases, translocation requires enzymic catalysis and metabolic energy in the form of ATP or GTP and, for some, a membrane potential.
4. Preproteins have to be in a translocation-competent conformation, probably loosely folded, which is maintained by their interaction with molecular chaperones.

Comparison of different translocation reactions (Table 2) suggests general mechanisms of protein translocation and highlights important unresolved questions.

**Table 2** A comparison of bacterial protein export with protein transport into endoplasmic reticulum and mitochondria

| | Bacterial protein export |
|---|---|
| Targeting signals | 16–26 residues; positively charged $N$-terminus followed by central apolar region. Hydrophobic moment along the long axis of the polypeptide chain. |
| Cytosolic chaperones | SecB, GroEL, Hsp70 (Dnak). |
| Recognition | Via peripheral membrane components SecB and SecA. Recognition of SecB/ preprotein complex through affinity of SecA for SecB and for leader and mature parts of the preprotein. Subsequent interaction with phospholipid or SecY/E. Sequential receptor cascade. Bypass of SecA by M13 procoat. |
| Translocation | Across single membrane; requires integral membrane protein SecY/E. |
| Proteolytic processing | Cleavage by membrane integrated enzymes leader peptidase or lipoprotein signal peptidase. |
| Energy requirement | Cytosolic ATP (SecA-ATPase) and $\Delta\bar{\mu}_{H^+}$ required throughout translocation. Slow membrane transit possible without $\Delta\bar{\mu}_{H^+}$, and small parts of polypeptide chain can translocate independently of ATP. |

| | Transport across ER membrane |
|---|---|
| Targeting signals | As for bacterial protein export, ER and bacterial signal sequences are functionally interchangable. |
| Cytosolic chaperones | Hsp70 (SSA) in posttranslational transport in yeast; may be functionally replaced by tightly coupled cotranslational translocation in mammalian cells. |
| Recognition | Via SRP and docking protein in mammalian cells. Sequential receptor model. Bypass of SRP/docking protein by small precursors. |
| Translocation | Across single membrane. Requires "signal sequence receptor" (SSR) in mammalian cells and integral membrane proteins Sec61–63 in yeast. Sequence homology between Sec61 and bacterial SecY. |
| Proteolytic processing | Cleavage by membrane-integrated signal peptidase. Cleavage specificity similar to bacterial leader peptidase. |
| Energy requirement | GTP required for SRP/docking protein. ATP-requirement for release of hsp70 (SSA) in yeast. ATP in the ER lumen for function of hsp70 (BiP) in translocation and folding/assembly of proteins. |

| | Mitochondrial protein import |
|---|---|
| Targeting signals | 10–70 residues; rich in positively charged and hydroxylated residues. Can form amphiphilic $\alpha$-helices. Hydrophobic moment perpendicular to the long axis of the polypeptide chain. |
| Cytosolic chaperones | Hsp70 (SSA). |
| Recognition | Via multiple surface receptors in the outer membrane followed by outer membrane general insertion protein. Converging receptor model. Bypass of surface receptors possible. |
| Translocation | Across two apposed membranes at translocation contact sites. Requires hsp70 (SSC1) in the matrix. |
| Proteolytic processing | Presequences completely translocated into matrix space. Cleavage by metal-dependent, soluble enzyme. |
| Energy requirement | Cytosolic ATP for release of hsp70 (SSA). ATP in the matrix for function of hsp70 (SSC1) and hsp60. $\Delta\psi$ across inner membrane to trigger insertion or translocation of positively charged leader. Membrane transit of mature domain is possible independent of $\Delta\psi$. |

## Targeting and Recognition

The signals required to direct noncytosolic proteins to their target compartments are mostly located at the amino terminus as cleavable presequences; however, proteins having internal targeting sequences are known. In general, the proteolytic cleavage of presequences, performed by specialized proteases (121–123), is not necessary for translocation. There is no sequence similarity among the targeting signals for each compartment. Hence the relevant information must reside in more degenerate features, such as the distribution of charges, hydrophilicity and hydrophobicity, and secondary structure.

Bacterial and ER signals are almost indistinguishable and are functionally interchangeable (124–126). In contrast, the typical mitochondrial matrix-targeting sequence lacks contiguous hydrophobic segments (127) but has several positively charged and hydroxylated residues. These sequences can form amphiphilic $\alpha$-helices, with basic residues exposed on one face and uncharged and hydrophobic residues on the other (16, 17, 128). Amphiphilicity is also a structural feature of all bacterial and ER targeting signals due to the linear arrangement of polar and apolar regions. The structural differences between secretory and mitochondrial targeting signals probably correspond to mechanistic differences of the initial steps of translocation.

Targeting and recognition involve the interaction of preproteins with cytosolic proteins as well as with peripheral and integral membrane components. In eukaryotes such as yeast, cytosolic heat-shock proteins of the hsp70 family may interact with secretory and mitochondrial preproteins, stabilizing them in translocation-competent conformations (129, 130). Chaperone-preprotein interactions are also a common theme in the bacterial cytosol. However, the function of the main prokaryotic secretion chaperone, SecB, is distinguished by its specific interaction with the receptor SecA, an interaction that contributes to the targeting of the preprotein-SecB complex (65). In mammalian cells, preproteins interact via their leader sequences with the signal recognition particle (SRP) (131) as they emerge from ribosomes. This may slow translation until the complex of ribosome, SRP, and nascent chain has bound to the SRP receptor (132), or docking protein (133), at the ER membrane, thus ensuring cotranslational translocation. However, several studies (134–137) have shown that large domains of preproteins can be made prior to initiating successful translocation. SecB and SecA appear to exert part of the functions ascribed to SRP without being structurally related to it. SecA recognizes SecB (65) as well as the leader and mature parts of the preprotein and SecB stabilizes the mature part of the precursor. SRP from dog pancreas can substitute for the stabilizing function of SecB in bacterial protein export in vitro (62). The complex of preprotein and SecB "docks" at its receptor, the "translocase," comprising SecA and membrane- embedded SecY/E protein (65).

Nuclear-encoded mitochondrial proteins encounter their target compartment by directly binding to receptor proteins integrated into the mitochondrial outer membrane (118). An interaction with cytosolic components specific for the targeting sequences seems not to be required. Receptor components have been identified (138–140) that have overlapping specificity. The structural elements in precursor proteins recognized by these receptors are unclear. Receptor-bound precursors converge to a common membrane insertion site, the general insertion protein GIP (141). This component is probably located at contact sites in which outer and inner mitochondrial membranes are closely apposed.

## Translocation and Energy Requirement

Investigators of ER and mitochondrial protein translocation favor the idea of a proteinaceous pore or channel that preproteins are thought to penetrate in extended conformations (142–144). In the case of mitochondria, such a transport protein could span the two membranes at contact sites (145, 146). Alternatively, it might only span the inner membrane and be connected to the apparatus for binding and insertion in the outer membrane. Candidates for the ER translocase are the membrane-integrated products of the SEC61–63 genes in yeast (147) and the 35-kDa signal sequence receptor (SSR) in mammalian ER (148). Preproteins may reach the SSR either directly or via SRP and docking protein. The SRP and docking-protein-independent pathway seems to be followed by small precursor proteins. Interestingly, one of these proteins, prepromellitin, can also be exported across the plasma membrane in E. coli in the absence of SecA and SecY function (94). The Sec61 protein has homology to SecY in cytosolic and membrane-embedded parts (R. Schekman, personal communication), again underscoring the similarity between the bacterial and ER systems.

Bacterial and mitochondrial protein transport require both ATP and the inner membrane electrochemical potential. The requirement for cytosolic ATP is connected to keeping the preprotein in a more open conformation or perhaps even actively unfolding it during translocation. SecA may be such an ATP-driven unfoldase. In the case of mitochondria, cytosolic ATP is probably required for the release of hsp70 from the receptor-bound precursor (118, 149). In striking contrast to the bacterial system, both ER and mitochondria have an additional requirement for ATP on the *trans*-side of the membrane (150–152). The energy input by SecA and ATP on the *cis*-side of the bacterial membrane could replace the functions performed by components that bind to the translocating chain on the *trans*-side in other translocation reactions. As long as the part of the bacterial protein that has not yet translocated is actively unfolded, translocation may be driven by the energy derived both from $\Delta\bar{\mu}_{H^+}$

and from folding in the periplasm. Movement of the polypeptide across the membrane is readily reversible, and ATP-hydrolysis and/or $\Delta\bar{\mu}_{H^+}$ is required to achieve unidirectionality (84). In contrast, reversibility of chain movement has not been observed in mitochondria, perhaps because of the presence of efficient binding proteins on the *trans*-side of the membrane. An additional possibility is that SecD and SecF of the *E. coli* inner membrane support translocation by binding to translocating preproteins as they emerge into the periplasm (153). The mechanism of such an interaction, however, would have to differ from that suggested for hsp70 proteins in the lumen of ER and mitochondria, since ATP is absent from the bacterial periplasm.

The proton-motive force accelerates each stage of the movement of pre-proteins across the bacterial plasma membrane (84, 109). In contrast, the mitochondrial system utilizes the electrical component $\Delta\psi$ only to drive the translocation of the positively charged leader sequence across the inner membrane (154), perhaps by a direct electrophoretic effect of $\Delta\psi$ (negative inside). Notably, bacterial proteins move across the membrane in the opposite direction with respect to the polarity of the membrane potential as in mitochondria. However, the basic N-terminus of the bacterial leader apparently remains on the cytosolic side of the membrane, whereas the basic residues of the mitochondrial leader cross into the matrix.

# CURRENT QUESTIONS IN BACTERIAL EXPORT

## Procoat

How does M13 procoat translocate without Sec-protein-mediated catalysis? If the small size of M13 procoat inherently prevents its stable folding, it would be unnecessary for SecA to catalyze unfolding. As one proposed function of SecB is to shield apolar surfaces, the two apolar regions of procoat may be apposed to each other prior to membrane interactions. The three charged segments of procoat may also function to shield these apolar domains; indeed, the character of each of these segments is important for procoat membrane insertion. However, our enthusiasm for these ideas is tempered by the low solubility of purified procoat (155, 156). A second question is what blocks procoat recognition by SecB and translocase when a mutation is introduced into the mature domain of procoat that prevents its Sec-independent transloca-tion? The answer to this question may lie in the recognition specificity of SecB and SecA for the mature domains of exported proteins.

## Mature Domains

Initial studies of protein translocation focused on the leader sequence; howev-er, simple joining of an authentic leader to a cytosolic protein does not always guarantee its rapid and efficient export (157). Large deletions in the mature region of an exported protein can dramatically lower both the rate and extent of export (158). One apparent constraint on the mature sequences of exported proteins is that lysyl and arginyl residues are excluded from the first few residues after the leader (16, 17, 50, 111, 159, 160). It is not known why these basic residues would interfere with translocation. Acidic amino acids often occur 3–4 residues (one turn of an $\alpha$-helix) from basic residues in exported proteins (161), suggesting a role for charge pairing in translocation. Another possible constraint on mature regions is that they must not fold in a manner that obscures the leader from recognition by SecA or that prevents unfolding. This would lead to the testable proposal that cytosolic proteins, given a leader sequence, would be more readily exported if rendered structurally labile by genetic means. Finally, SecB and/or SecA may recognize features of preprotein charge, polarity, or secondary structure. In any case, the existence of a stable ternary complex among SecB, a preprotein, and membrane-bound SecA (65) suggests that SecB and SecA recognize different aspects of preproteins.

## Unfolding

Only a few studies have been made of the state of preprotein folding. Studies of protease sensitivity (162, 163) show that proteins emerge into the periplasm in an initially unfolded state. Prior to translocation, leader peptidase is far more susceptible to digestion in cell lysates than after it has assumed its final *trans*-membrane form (164). Pre-MBP must remain unfolded to some degree in order to translocate (165, 166), though the degree of secondary or tertiary structure in the export-competent preMBP is not yet known. Trimethoprim, by stabilizing the folded form of a proOmpA/dihydrofolate reductase fusion protein, prevents its translocation (167). ProOmpA, in complex with SecB, has significant secondary and tertiary structure (58). If, as is widely assumed, exported proteins cross the membrane as a simple loop (39, 168), an active unfolding process would be necessary. Although there is no direct evidence, it is possible that SecA could actively unfold preproteins at the expense of ATP hydrolysis.

## Translocation Pathway

Do preproteins cross the membrane through a SecY/E "pore," through the lipid phase, or along a special surface of the SecY/E protein? It has recently been shown that proOmpA can still translocate when its cysteines are joined as a disulfide, forming a protein loop (168a), or when the cysteines are derivatized by coumarin (58) or *N*-ethylmaleimide (84). Since the plasma membrane remains impermeant to protons at all times, it could not have a water-filled pore such as is found in the outer membrane. It seems unlikely to us that a transport protein for preproteins could have the structural flexibility to accommodate a loop or a coumarin and yet remain "proton-tight." The establishment of conditions for accumulating preproteins as translocation intermediates (84, 114) may now allow assessment of their association with SecY/E or accessibility to lipid-embedded probes. The requirements for SecA, ATP hydrolysis, and $\Delta\bar{\mu}_{H^+}$ persist throughout the course of transloca-tion (84, 114, 115), demonstrating that the role of SecA goes beyond simply serving as a receptor.

## Role of Lipid

Although we have emphasized the role of the translocase enzyme in protein export from *E. coli,* the lipid phase also plays a vital, if poorly understood, role. Acidic lipids (78, 83; J. Hendrick, W. Wickner, in preparation) are required for high-affinity SecA binding to the membrane and for the activa-tion of translocation ATPase. Intriguing reports have suggested that features of the fatty acyl phase, such as fluidity, may play a major role in the translocation process (51, 169). A large body of data shows that leader peptides must retain their capacity to partition into lipid to be functional (170, 171). Whether or not they do so during the translocation process remains to be determined.

## Role of $\Delta\bar{\mu}_{H^+}$

Some of the possible roles of $\Delta\bar{\mu}_{H^+}$ in translocation are outlined above. The reconstitution of translocation with all purified components (13) may allow a more quantitative study of the effects of $\Delta\bar{\mu}_{H^+}$.

## Folding in the Periplasm

Newly made proteins must be cleaved by leader peptidase for release into the periplasm (35), and may reside on the outer surface of the plasma membrane for a brief period after membrane transit (160). In other translocation reac-tions, Hsp60 catalyzes protein folding after import into mitochondria (151, 172) and chloroplasts (173). BiP may assist the folding of proteins as they enter the ER (RER) and appears to be required for translocation (150, 174). SecD and SecF, which are largely exposed to the periplasm, might catalyze periplasmic protein folding (175). The rapid rate of protein export in vivo might require SecD and SecF action to "clear" the periplasmic surface of the SecY/E protein for further rounds of translocation. The slower rates and lower extents of in vitro translocation may render detection of the SecD and SecF catalysis difficult.

## Role of Translation

Translocation in *E coli* is clearly not coupled to ongoing polypeptide chain growth. Nevertheless, it is premature to conclude that no distinctions exist between polypeptides that happen to initiate translocation while still attached to the ribosome and those that cross entirely posttranslationally. A major aspect of SecB and SecA function relates to the folded state of the protein, as noted above. The polypeptides that cross cotranslationally may not fold prior to translocation and therefore might be less dependent on possible SecA "unfolding" or SecB chaperone functions. Thus, in SecA- or SecB-deficient strains, the export defect is dramatically relieved by drugs or mutants that selectively slow protein synthesis (165, 176), and the dependence of translocation on SecA or SecB is diminished when only a small domain of the polypeptide, which might not fold as stably, remains to be translocated (84, 114, 115).

## Regulation

The processes of secretion and membrane assembly must now be dovetailed with other elements of what has been called "Project K12," the attempt to understand the interrelationships of the essential systems of one of the simplest free-living organisms. Export can be regulated by the rates of translation, by the levels of ATP and $\Delta\bar{\mu}_{H^+}$, and by the synthesis of SecA. It will, in turn, be interesting to know how the cell measures when sufficient cell-surface growth has occurred to warrant cell division or another round of replication.

## PROSPECTUS

Our current limited understanding of bacterial export provides tantalizing glimpses of how nature has solved some basic problems. Specificity may be achieved by a cascade of receptors, SecB, SecA, and perhaps SecY/E, acting in series to "proofread" the choice of proteins. Leader and mature interactions may prevent final, stable folding that would block export, and the metastable, "export-competent" folded state can be supported by a chaperone. Translocation itself is driven by two distinct energy sources, acting in distinct manners. These ideas, and others, are now accessible to more rigorous biochemical and genetic tests.

The facile interplay between genetics and biochemistry in *E. coli* should allow researchers during the coming decade to obtain a clear view of the mechanisms of bacterial protein export. This knowledge will enrich other areas of inquiry, such as the fundamental problem of protein folding. Progress in understanding the bacterial export reaction may also help to accelerate the parallel studies in eukaryotic and mammalian systems.

## Literature Cited

1. Bieker, K. L., Phillips, G. J., Silhavy, T. 1990. *J. Bioenerget. Biomembr.* 22:291–310
2. Bieker, K. L., Silhavy, T. 1990. *Cell* 61:833–42
3. Chang, C. N., Model, P., Blobel, G. 1979. *Proc. Natl. Acad. Sci. USA* 76:1251–55
4. Chang, C. N., Inouye, H., Model, P., Beckwith, J. 1980. *J. Bacteriol.* 142:726–28
5. Rhoads, D. B., Tai, P. C., Davis, B. 1984. *J. Bacteriol.* 159:63–70
6. Müller, M., Blobel, G. 1984. *Proc. Natl. Acad. Sci. USA* 81:7421–25
7. Müller, M., Blobel, G. 1984. *Proc. Natl. Acad. Sci. USA* 81:7737–41
8. Ohno-Iwashita, Y., Wickner, W. 1983. *J. Biol. Chem.* 258:1895–1900
9. Kumamoto, C. A., Chen, L., Fandl, J., Tai, P. C. 1989. *J. Biol. Chem.* 264:2242–49
10. Weiss, J. B., Ray, P. H., Bassford, P. J. Jr. 1988. *Proc. Natl. Acad. Sci. USA* 85:8978–82
11. Cabelli, R. J., Chen, L., Tai, P. C., Oliver, D. B. 1988. *Cell* 55:683–92
12. Cunningham, K., Lill, R., Crooke, E., Rice, M., Moore, K., Wickner, W., Oliver, D. 1989. *EMBO J.* 8:955–59
13. Brundage, L., Hendrick, J. P., Schiebel, E., Driessen, A. J. M., Wickner, W. 1990. *Cell.* 62:649–57
14. vonHeijne, G. 1983. *Eur. J. Biochem.* 133:17–21
15. Deleted in proof
16. vonHeijne, G. 1986. *EMBO J.* 5:1335–42
17. vonHeijne, G. 1986. *EMBO J.* 5:3021–27
18. vonHeijne, G. 1986. *J. Mol. Biol.* 192:287–90
19. Perlman, D., Halvorson, H. O. 1983. *J. Mol. Biol.* 167:391–409
20. Randall, L. L., Hardy, S. J. S. 1989. *Science* 243:1156–59
21. Gierasch, L. M. 1989. *Biochemistry* 28:923–30
22. Inouye, S., Soberon, X., Franceschini, T., Nakamura, K., Itakura, K., Inouye, M. 1982. *Proc. Natl. Acad. Sci. USA* 79:3438–41
23. Vlasuk, G. P., Inouye, S., Ito, H., Itakura, K., Inouye, M. 1983. *J. Biol. Chem.* 258:7141–48
24. Sasaki, S., Matsuyama, S.-I., Mizushima, S. 1990. *J. Biol. Chem.* 265:4358–63
25. Iino, T., Takahashi, M., Sako, T. 1987. *J. Biol. Chem.* 7412–17
26. Bedouelle, H., Bassford, P. J. Jr., Fowler, A. V., Zabin, I., Beckwith, J., Hofnung, M. 1980. *Nature* 285:78–85
27. Emr, S. D., Silhavy, T. 1982. *J. Cell Biol.* 95:689–96
28. Bankaitis, V. A., Rasmussen, B. A., Bassford, P. J. Jr. 1984. *Cell* 37:243–52
29. Kendall, D. A., Bock, S. C., Kaiser, E. T. 1986. *Nature* 321:706–8
30. Pollitt, S., Inouye, S., Inouye, M. 1985. *J. Biol. Chem.* 260:7965–69
31. Kuhn, A., Kreil, G., Wickner, W. 1986. *EMBO J.* 5:3681–85
32. Kuhn, A., Wickner, W., Kreil, G. 1986. *Nature* 322:335–39
33. Michaelis, S., Hunt, J. F., Beckwith, J. 1986. *J. Bacteriol.* 167:160–67
34. Koshland, D., Sauer, R. T., Botstein, D. 1982. *Cell* 30:903–14
35. Dalbey, R. E., Wickner, W. 1985. *J. Biol. Chem.* 260:15925–31
36. Kuhn, A., Wickner, W. 1985. *J. Biol. Chem.* 260:15914–18
37. Pollitt, S., Inouye, S., Inouye, S. 1986. *J. Biol. Chem.* 261:1835–37
38. Lehnhardt, S., Pollitt, S., Inouye, M. 1987. *J. Biol. Chem.* 262:1716–19
39. Kuhn, A., Kreil, G., Wickner, W. 1987. *EMBO J.* 6:501–5
40. vonHeijne, G. 1989. *Nature* 341:456–58
41. Kaiser, C. A., Preuss, D., Grisafi, P., Botstein, D. 1987. *Science* 235:312–17
42. Park, S., Liu, G., Topping, T. B., Cover, W. H., Randall, L. L. 1988. *Science* 239:1033–35
43. Laminet, A. A., Plückthun, A. 1989. *EMBO J.* 8:1469–77
44. Cunningham, K., Wickner, W. 1989. *Proc. Natl. Acad. Sci. USA* 86:8630–34
45. Fikes, J. D., Bassford, P. J. Jr. 1989. *J. Bacteriol.* 171:402–9
46. Emr, S. D., Hanley-Way, S., Silhavy, T. J. 1981. *Cell* 23:79–88
47. Dierstein, R., Wickner, W. 1985. *J. Biol. Chem.* 260:15919–24
48. Kumamoto, C. A., Beckwith, J. 1983. *J. Bacteriol.* 154:253–60
49. Kumamoto, C. A., Nault, A. K. 1989. *Gene* 75:167–75
50. Liss, L. R., Oliver, D. B. 1986. *J. Biol. Chem.* 261:2299–2303
51. Ito, K., Sato, T., Yura, T. 1977. *Cell* 11:551–59
52. Date, T., Wickner, W. 1981. *J. Virol.* 37:1087–89
53. Koshland, D., Botstein, D. 1980. *Cell* 20:749–60

54. Zimmermann, R., Wickner, W. 1983. *J. Biol. Chem.* 258:3920–25
55. Zimmermann, R., Wickner, W. 1983. *J. Biol. Chem.* 258:12073–80
56. Randall, L. L. 1983. *Cell* 33:231–40
57. Lecker, S., Lill, R., Ziegelhoffer, T., Bassford, P. J. Jr., Kumamoto, C. A., Wickner, W. 1989. *EMBO J.* 8:2703–9
58. Lecker, S., Wickner, W. 1990. *EMBO J.* 9:2309–14
59. Kumamoto, C. A. 1989. *Proc. Natl. Acad. Sci. USA* 86:5320–24
60. Gannon, P. M., Li, P., Kumamoto, C. A. 1989. *J. Bacteriol.* 171:813–18
61. Watanabe, M., Blobel, G. 1989. *Cell* 58:695–705
61a. Kumamoto, C. A., Beckwith, J. 1985. *J. Bacteriol.* 163:267–74
62. Crooke, E., Guthrie, B., Lecker, S., Lill, R., Wickner, W. 1988. *Cell* 54:1003–11
63. Weiss, J. B., Bassford, P. J. Jr. 1990. *J. Bacteriol.* 172:3023–29
64. Liu, G., Topping, T. B., Randall, L. L. 1989. *Proc. Natl. Acad. Sci. USA* 86:9213–17
65. Hartl, F.-U., Lecker, S., Schiebel, E., Hendrick, J. P., Wickner, W. 1990. *Cell* 63:269–79
66. Hemmingsen, S. M., Woolford, C., van der Vies, S. M., Tilly, K., Dennis, D. T., et al. 1988. *Nature* 333:330–34
67. Randall, L. L., Topping, T. B., Hardy, S. J. S. 1990. *Science* 248:860–63
68. Bankaitis, V. A., Bassford, P. J. Jr. 1984. *J. Biol. Chem.* 259:12193–12200
69. Collier, D. N., Bankaitis, V. A., Weiss, J. B., Bassford, P. J. Jr. 1988. *Cell* 53:273–83
70. Bochkareva, E. S., Lissin, N. M., Girshovich, A. S. 1988. *Nature* 336:254–57
71. Phillips, G. J., Silhavy, T. J. 1990. *Nature* 344:882–84
72. Guthrie, B., Wickner, W. 1990. *J. Bacteriol.* 172:5555–62
73. Kusukawa, N., Yura, T., Ueguchi, C., Akiyama, Y., Ito, K. 1989. *EMBO J.* 8:3517–21
74. Oliver, D. B., Beckwith, J. 1982. *Cell* 30:311–19
75. Schmidt, M. G., Rollo, E. E., Grodberg, J., Oliver, D. B. 1988. *J. Bacteriol.* 170:3404–14
76. Lill, R., Cunningham, K., Brundage, L., Ito, K., Oliver, D., Wickner, W. 1989. *EMBO J.* 8:961–66
77. Swidersky, U. E., Hoffschulte, H. K., Müller, M. 1990. *EMBO J.* 9:1777–85
78. Lill, R., Dowhan, W., Wickner, W. 1990. *Cell* 60:271–80
79. Ryan, J. P., Bassford, P. J. Jr. 1985. *J. Biol. Chem.* 260:14832–37
80. Bankaitis, V. A., Bassford, P. J. Jr. 1985. *J. Bacteriol.* 161:169–78
81. Puziss, J. W., Fikes, J. D., Bassford, P. J. Jr. 1989. *J. Bacteriol.* 171:2303–11
82. Akita, M., Sasaki, S., Matsuyama, S.-i., Mizushima, S. 1990. *J. Biol. Chem.* 265:8164–69
83. deVrije, T., deSwart, R. L., Dowhan, W., Tommassen, J., deKruijff, B. 1988. *Nature* 334:173–75
84. Schiebel, E., Driessen, A. J. M., Hartl, F.-U., Wickner, W. 1991. *Cell.* In press
85. Rollo, E. E., Oliver, D. B. 1988. *J. Bacteriol.* 170:3281–82
86. Ito, K., Wittekind, M., Nomura, M., Shiba, K., Yura, T., Miura, A., Nashimoto, H. 1983. *Cell* 32:789–97
87. Ito, K. 1984. *Mol. Gen. Genet.* 197:204–8
88. Riggs, P. D., Derman, A. I., Beckwith, J. 1988. *Genetics* 118:571–79
89. Schatz, P. J., Riggs, P. D., Jacq, A., Fath, M. J., Beckwith, J. 1989. *Genes Dev.* 3:1035–44
90. Schatz, P. J., Riggs, P. D., Jacq, A., Fath, M. J., Beckwith, J. 1990. *EMBO J.* In press
91. Akiyama, Y., Ito, K. 1987. *EMBO J.* 6:3465–70
92. Driessen, A. J. M., Wickner, W. 1990. *Proc. Natl. Acad. Sci. USA* 87:3107–11
92a. Watanabe, M., Nicchitta, C. V., Blobel, G. 1990. *Proc. Natl. Acad. Sci. USA* 87:1960–64
93. Wolfe, P. B., Rice, M., Wickner, W. 1985. *J. Biol. Chem.* 260:1836–41
94. Cobet, W., Mollay, C., Müller, G., Zimmermann, R. 1989. *J. Biol. Chem.* 264:10169–76
95. Kupersztoch, Y. M., Tachias, K., Moomaw, C. R., Dreyfus, L. A., Urban, R., et al. 1990. *J. Bacteriol.* 172:2427–32
96. Sugimoto, K., Sugisaki, H., Okamoto, T., Takanami, M. 1977. *J. Mol. Biol.* 110:487–507
97. Muller, G., Zimmermann, R. 1987. *EMBO J.* 6:2099–2107
98. Date, T., Goodman, J. G., Wickner, W. 1980. *Proc. Natl. Acad. Sci. USA* 77:4669–73
99. Date, T., Zwizinski, C., Ludmerer, S., Wickner, W. 1980. *Proc. Natl. Acad. Sci. USA* 77:827–31
100. Daniels, C. J., Bole, D. G., Quay, S. C., Oxender, D. L. 1981. *Proc. Natl. Acad. Sci. USA* 77:5396–5400
101. Enequist, H. G., Hirst, T. R., Harayama, S., Hardy, S. J. S., Randall,

L. L. 1981. *Eur. J. Biochem.* 116:227–33
102. Pagès, J. M., Lazdunski, C. 1982. *Eur. J. Biochem.* 124:561–66
103. Pagès, J. M., Lazdunski, C. 1982. *FEBS Lett.* 149:51–54
104. Bakker, E. P., Randall, L. L. 1984. *EMBO J.* 3:895–900
105. Chen, L., Tai, P. C. 1985. *Proc. Natl. Acad. Sci. USA* 82:4384–88
106. Geller, B. L., Movva, N. R., Wickner, W. 1986. *Proc. Natl. Acad. Sci. USA* 83:4219–22
107. Yamade, K., Ichihara, S., Mizushima, S. 1987. *J. Biol. Chem.* 262:2358–62
108. Yamada, H., Tokuda, H., Mizushima, S. 1989. *J. Biol. Chem.* 264:1723–28
109. Driessen, A. J. M., Wickner, W. 1991. *Proc. Natl. Acad. Sci. USA.* In press
110. vonHeijne, G., Gravel, Y. 1986. *Eur. J. Biochem.* 174:671–78
111. Dalbey, R. E. 1990. *Trends Biochem. Sci.* 15:253–57
112. Nillson, I., vonHeijne, G. 1990. *Cell* 62:1135–41
113. Yamada, H., Matsuyama, S., Mizushima, S. 1989. *J. Biol. Chem.* 264:18577–81
114. Tani, K., Shiozuka, K., Tokuda, H., Mizushima, S. 1989. *J. Biol. Chem.* 264:18582–88
115. Geller, B. L., Green, H. M. 1989. *J. Biol. Chem.* 264:16465–69
116. Walter, P., Lingappa, V. R. 1986. *Annu. Rev. Cell Biol.* 2:499–516
117. Attardi, G., Schatz, G. 1988. *Annu. Rev. Cell Biol.* 4:289–333
118. Hartl, F.-U., Neupert, W. 1990. *Science* 247:930–38
119. Lubben, T. H., Theg, S. M., Keegstra, K. 1988. *Photosynth. Res.* 17:173–95
120. Lazarow, P. B., Fujiki, Y. 1985. *Annu. Rev. Cell Biol.* 1:489–530
121. Zwizinski, C., Wickner, W. 1980. *J. Biol. Chem.* 255:7973–77
122. Evans, F. A., Gilmore, R., Blobel, G. 1986. *Proc. Natl. Acad. Sci. USA* 83:581–85
123. Hawlitschek, G., Schneider, H., Schmidt, B., Tropschug, M., Hartl, F.-U., Neupert, W. 1988. *Cell* 53:795–806
124. Watts, C., Wickner, W., Zimmermann, R. 1983. *Proc. Natl. Acad. Sci. USA* 80:2809–13
125. Lingappa, V. R., Chaidez, I., Yost, C. S., Hedgpeth, J. 1984. *Proc. Natl. Acad. Sci. USA* 81:456
126. vonHeijne, G. 1988. *Biochim. Biophys. Acta* 947:307
127. Hartl, F.-U., Pfanner, N., Nicholson, D. W., Neupert, W. 1989. *Biochim. Biophys. Acta* 988:1–45

128. Roise, D., Horvath, S. D., Tomich, J. M., Richards, J. H., Schatz, G. 1986. *EMBO J.* 5:1327–34
129. Deshaies, R. J., Koch, B. D., Werner-Washburne, M., Craig, E. A., Schekman, R. 1988. *Nature* 332:800–5
130. Chirico, W. J., Waters, M. G., Blobel, G. 1988. *Nature* 332:805–10
131. Walter, P., Blobel, G. 1982. *Nature* 299:691–98
132. Gilmore, R., Walter, P., Blobel, G. 1982. *J. Cell Biol.* 95:470–77
133. Meyer, D. I., Krause, E., Dobberstein, B. 1982. *Nature* 297:117–20
134. Schlenstedt, G., Gudmundsson, G. H., Boman, H. G., Zimmermann, R. 1990. *J. Biol. Chem.* 265:13960–68
135. Perara, E., Rothman, R. E., Lingappa, V. R. 1986. *Science* 232:348–52
136. Mueckler, M., Lodish, H. 1986. *Nature* 322:549–52
137. Ainger, K. J., Meyer, D. I. 1986. *EMBO J.* 5:951–55
138. Sollner, T., Griffiths, G., Pfaller, R., Pfanner, N., Neupert, W. 1989. *Cell* 59:1061–70
139. Vestweber, D., Brunner, J., Baker, A., Schatz, G. 1989. *Nature* 341:205–9
140. Sollner, T., Pfaller, R., Griffiths, G., Pfanner, N., Neupert, W. 1990. *Cell* 62:107–15
141. Pfaller, R., Steger, H. F., Rassow, J., Pfanner, N., Neupert, W. 1988. *J. Cell Biol.* 107:2483–90
142. Gilmore, R., Blobel, G. 1985. *Cell* 42:497–505
143. Pfanner, N., Hartl, F.-U., Guiard, B., Neupert, W. 1987. *Eur. J. Biochem.* 169:289–93
144. Rassow, J., Guiard, B., Wienhues, U., Herzog, V., Hartl, F.-U., Neupert, W. 1989. *J. Cell Biol.* 109:1421–28
145. Schleyer, M., Neupert, W. 1985. *Cell* 43:339–50
146. Vestweber, D., Schatz, G. 1988. *J. Cell Biol.* 107:2037–43
147. Rothblatt, J. A., Deshaies, R. J., Sanders, S. L., Daum, G., Schekman, R. 1989. *J. Cell Biol.* 109:2641–52
148. Wiedmann, M., Kurzchalia, T. V., Hartmann, E., Rapoport, T. A. 1987. *Nature* 328:830–33
149. Pfanner, N., Tropschug, M., Neupert, W. 1987. *Cell* 49:815–23
150. Rose, M. D., Misra, L. M., Vogel, J. P. 1989. *Cell* 57:1211
151. Ostermann, J., Horwich, A. L., Neupert, W., Hartl, F.-U. 1989. *Nature* 341:125–30
152. Hwang, S. T., Schatz, G. 1989. *Proc. Natl. Acad. Sci. USA* 86:8432
153. Gardel, C., Johnson, K., Jacq, A.,

# 124    WICKNER, DRIESSEN & HARTL

Beckwith, J. 1990. *EMBO J.* 9:3209–16

154. Pfanner, N., Neupert, W. 1985. *EMBO J.* 4:2819–25
155. Silver, P., Watts, C., Wickner, W. 1981. *Cell* 25:341–45
156. Zwizinski, C., Wickner, W. 1982. *EMBO J.* 1:573–78
157. Moreno, F., Fowler, A. V., Hall, M., Silhavy, T. J., Zabin, I., Schwartz, M. 1980. *Nature* 286:356–59
158. Freudl, R., Schwarz, H., Degen, M., Henning, U. 1987. *J. Bacteriol.* 169:66–71
159. Yamane, K., Mizushima, S. 1988. *J. Biol. Chem.* 263:19690–96
160. Summers, R. G., Harris, C. R., Knowles, J. R. 1989. *J. Biol. Chem.* 264:20082–88
161. vonHeijne, G. 1980. *Biochem. Biophys. Res. Commun.* 93:82–86
162. Minsky, A., Summers, R. G., Knowles, J. R. 1986. *Proc. Natl. Acad. Sci. USA* 83:4180–84
163. Sanders, R. L., May, B. K. 1975. *J. Bacteriol.* 123:806–14
164. Wolfe, P. B., Wickner, W. 1984. *Cell* 36:1067–72
165. Kumamoto, C. A., Gannon, P. M. 1988. *J. Biol. Chem.* 263:11554–58
166. Randall, L. L., Hardy, S. J. S. 1986. *Cell* 46:921–28
167. Freudl, R., Schwarz, H., Kramps, S., Hindennach, I., Henning, U. 1988. *J. Biol. Chem.* 263:17084
168. Halegoua, S., Inouye, M. 1979. In *Bacterial Outer Membranes*, ed. M. Inouye, pp. 67–114. New York: Wiley
168a. Tani, K., Tokuda, H., Mizushima, S. 1990. *J. Biol. Chem.* 265:17341–47
169. Pagès, J. M., Piovant, M., Varenne, S., Lazdunski, C. 1978. *Eur. J. Biochem.* 86:589–602
170. Batenburg, A. M., Demel, R. A., Verkleij, A. J., deKruijff, B. 1988. *Biochemistry* 27:5678–85
171. Briggs, M. S., Gierasch, L. M., Zlotnick, A., Lear, J. D., deGrado, W. F. 1985. *Science* 282:1096–99
172. Cheng, M. Y., Hartl, F.-U., Martin, J., Pollock, R. A., Kalousek, F., et al. 1989. *Nature* 337:620–25
173. Musgrove, J. E., Johnson, R. A., Ellis, R. J. 1987. *Eur. J. Biochem.* 163:529–34
174. Vogel, J. P., Misra, L. M., Rose, M. D. 1990. *J. Cell Biol.* 110:1885–95
175. Gardel, C., Benson, S., Hunt, J.,

Michaelis, S., Beckwith, J. 1987. *J. Bacteriol.* 169:1286–90
176. Lee, C. A., Beckwith, J. 1986. *J. Bacteriol.* 166:878–83
177. Asbeck, F., Beyreuther, K., Kohler, H., vonWettstein, G., Braunitzer, G. 1969. *Hoppe-Seyler's Z. Physiol. Chem.* 350:1047–55
178. Silhavy, T. J., Casadaban, M. J., Shuman, H. A., Beckwith, J. A. 1976. *Proc. Natl. Acad. Sci. USA* 73:3423–27
179. Randall, U. L., Hardy, S. J. S. 1977. *Eur. J. Biochem.* 75:43–53
180. Smith, W. P., Tai, P.-C., Thompson, R. C., Davis, B. D. 1977. *Proc. Natl. Acad. Sci. USA* 74:2830–34
181. Chang, C. N., Blobel, G., Model, P. 1978. *Proc. Natl. Acad. Sci. USA* 75:361–65
182. Ito, K., Mandel, G., Wickner, W. 1979. *Proc. Natl. Acad. Sci. USA* 76:1199–1203
183. Koshland, D., Botstein, D. 1982. *Cell* 30:893–902
184. Emr, S. D., Hedgpeth, J., Clement, J. M., Silhavy, T. J., Hofnung, M. 1980. *Nature* 285:82–85
185. Hussain, M., Ichihara, S., Mizushima, S. 1980. *J. Biol. Chem.* 255:3707–12
186. Ito, K. 1982. *J. Biol. Chem.* 257:9895–97
187. Wolfe, P. B., Wickner, W., Goodman, J. 1983. *J. Biol. Chem.* 258:12073–80
188. Tokunaga, M., Loranger, J. M., Chang, S.-Y., Regue, M., Chang, S., Wu, H. C. 1985. *J. Biol. Chem.* 260:5610–15
189. Yu, F., Yamada, H., Daishima, K., Mizushima, S. 1984. *FEBS Lett.* 173:264–68
191. Innis, M. A., Tokunaga, M., Williams, M. E., Loranger, J. M., Chang, S.-Y., et al. 1984. *Proc. Natl. Acad. Sci. USA* 81:3708–12
192. Bieker, K. L., Silhavy, T. 1989. *Proc. Natl. Acad. Sci. USA* 86:968–72
193. Smith, W. P., Tai, P.-C., Davis, B. D. 1978. *Proc. Natl. Acad. Sci. USA* 75:814–17
194. Cancedda, R., Schlesinger, M. J. 1974. *J. Bacteriol.* 117:290–301
195. Crowles, I., Gamon, K., Henning, U. 1981. *Eur. J. Biochem.* 113:375–80
196. Oliver, D. B., Beckwith, J. 1981. *Cell* 25:765–72
197. Wolfe, P. B., Silver, P., Wickner, W. 1982. *J. Biol. Chem.* 257:7898–7902
198. Yamagata, H., Daishima, K., Mizushima, S. 1983. *FEBS Lett.* 158:301–4

tion of the DNA is mediated by specialized phage enzymes called terminases, which bind to specific nucleotide sequences in the phage genomes called *pac* or *cos* sites. Cutting, catalyzed by the same enzyme, takes place in the vicinity of the binding sites, a process that may or may not require the presence of the procapsid, depending on the particular phage. Packaging involves the translocation of the DNA into the procapsid, a function that may also be the responsibility of the terminase. Packaging requires ATP hydrolysis and results in the condensation of the DNA about 30- to 200-fold inside the capsid (6, 7).

The elucidation of the main features of λ DNA maturation and packaging had a tremendous practical impact in modern molecular biology, since it allowed the development of techniques for in vitro packaging, which in turn opened the doors to the cloning of eukaryotic genes.

In this review I have tried to summarize the present state of knowledge of the three major processes mentioned: recognition, cutting, and packaging for phage λ. An excellent comparative review on DNA maturation and packaging is that of Black (6). Recent reviews on aspects of λ DNA maturation include those of Feiss (8) and Becker & Murialdo (9); the latest article covering all aspects of the subject was published in 1983 (10). I have concentrated more on the dynamic aspects of DNA maturation and packaging and have left untouched the important areas of DNA condensation and structure in the capsid. Reviews in this field include those of Black (6, 6a) and Earnshaw & Casjens (11).

During its lytic life cycle, λ DNA alternates between a covalently closed, circular (ring) and an open, linear (rod) molecule. The mature chromosome is of unique sequence, 48,502 nucleotide pairs long (12). Its 5'-terminal ends protrude as single-stranded chains of 12 nucleotides. The sequences of the ends are complementary and they can cohere to form a ring (13–15). The mature molecule is present in the capsid of the virion in a condensed form (11). It is converted into a ring within minutes following injection into a cell (16–20).

Closure of the molecule upon entry to the host cell (16, 18–20) is a relatively simple process. The complementary single-stranded ends anneal in a reaction that is thermodynamically favored and is greatly accelerated by some positional effect that restricts diffusion of the ends, keeping them in close proximity (21–23). Annealing is then followed by the formation of phosphodiester bonds, a reaction catalyzed by the host DNA ligase (24). On the other hand, the opposite process, the opening of the DNA molecule, is normally coupled to its encapsidation and it is extremely complex.

Cutting and packaging take place when capsid precursors known as pro-capsids or proheads interact with newly replicated DNA molecules known as concatemers (Figures 1 and 2) (25–27). The concatemers are the product of

DNA replication, which proceeds in two stages. The first stage takes place when, upon injection into the host and circularization and covalent closure, λ DNA undergoes several rounds of θ-replication in which rings generate daughter rings. At later times during phage vegetative growth, σ or rolling-circle replication takes over in which individual rings give rise to linear concatemers of λ DNA (Figure 1) (28–38). These concatemers, which in the absence of packaging may be up to 10 chromosomal units long, are the normal substrates for packaging into procapsids (37–39). Prophages and polymeric rings, which are formed by recombination between monomeric rings, are also substrates for cleavage and packaging (37, 40–46) (Figure 1).

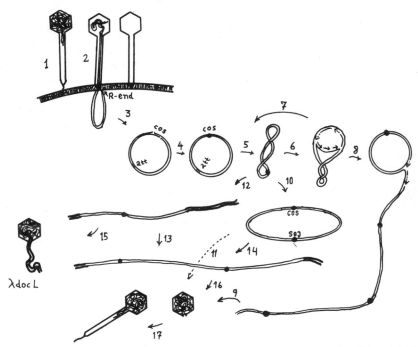

*Figure 1*    DNA replication cycle of λ. Sizes and shapes are purely schematic. Thin and thick lines represent λ and *E. coli* DNA strands, respectively. *cos* is represented by a dot and *att* by three cross lines between the DNA lines. 1. Phage adsorption; 2. DNA injection; 3. circulariza-tion; 4. ligation; 5. supercoiling introduced by gyrase; 6. θ replication; 7. daughter circles may go to a second round of θ replication; 8. switch to σ replication and production of concatemers; 9. cutting and packaging; 17. addition of tails. Recombination of monomeric circles generates dimeric circles (10), which can be packaged (11). Integration of a monomeric circle generates a monolysogen (12), which, in the absence of excision, is packaged to form a defective particle called λdocL (15). Dilysogens can be formed by integration of a second circle into a monoly-sogen (13) or by integration of a dimer (14). The DNA bracketed by the two *cos*'s can, in the absence of excision, be packaged (16).

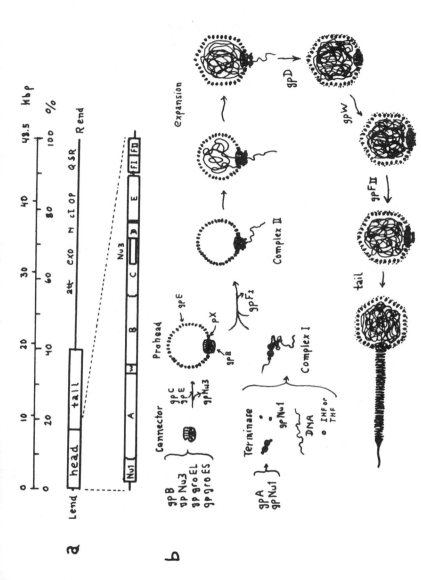

*Figure 2* (a) Genetic map of the λ chromosome. The size of the genes in the left arm are from Ref. 12. Only a few genes in the right arm are labeled for general orientation. The lower part shows an expanded view of the head genetic region. (b) Pathway of λ head morphogenesis. The figure is purely schematic and is not intended to represent molecular shapes.

These DNA molecules and proheads interact, and unit size DNA molecules are cut and inserted into the cavity of the prohead. The enzyme that catalyzes cutting and encapsidation is called λ DNA terminase (40, 47–49), and its site of action in the DNA is called *cos*, for *co*hesive end *s*ite (42, 50, 51). During these processes, two nicks, staggered 12 nucleotides apart on opposite strands of the duplex, are introduced at *cos*. In this way the monomeric chromosomal units are cut from the replicative oligomers and the unique single-stranded ends of the mature molecule are regenerated. Cleavage is then completed by the separation of the cohesive ends, a reaction that has been observed in vitro (52).

The terminase of λ is composed of two subunits (53–57), both of which recognize specific sequences in *cos*. The recognition function of the small subunit of the enzyme appears to be mediated by an α helix-turn-α helix motif (8, 58). This subunit binds to three, possibly four, repeated sequences (59). The large subunit of terminase has nicking activity and binds to the region where the two nicking sites are located (R. R. Higgins and A. Becker, personal communication). The partial two-fold rotational symmetry of the nicking substrate suggests that the enzyme also has two-fold rotational symmetry. The presence in the large subunit of a putative leucine zipper domain in the vicinity of a region of basic amino acid residues lends support to this idea (A. Davidson and M. Gold, personal communication). Terminase binding is enhanced by the binding and bending of *cos* DNA by the host protein IHF (60–62). Other host proteins can substitute for IHF (63). The involvement of numerous sites in the DNA to which terminase and host factor(s) bind results in the formation of a complex structure, which, by analogy to other nucleoprotein complexes (e.g. the intasome), has been named termisome (9).

Cutting in vivo of λ DNA does not occur in the absence of procapsids (7, 64–67), and whether it happens before or after packaging initiation is still a controversial issue. The mechanism that impedes cutting in vivo in the absence of proheads remains a mystery (68). In vitro cutting is not coupled to packaging, and any topological form of DNA containing *cos* is a substrate (48, 49, 63, 69).

The actual mechanism of DNA translocation into the interior of the procapsid remains unknown; the most popular assumption views vectorial translotation driven by ATP hydrolysis by the terminase itself (6, 70). The main events that occur during DNA maturation and packaging in vivo are terminase binding to *cos*, prohead binding to the terminase-*cos* complex, initial *cos* cutting and melting of the joint, DNA packaging, prohead expansion and incorporation of gpD into the shell of the procapsid, terminal *cos* cutting, disengagement of the filled prohead from terminase and the rest of the concatemer, and stabilization of the encapsidated DNA.

# HEAD ASSEMBLY: AN OVERVIEW

## Prohead Structure

Because DNA maturation and DNA packaging are normally coupled in vivo, it seems pertinent to summarize the main features of prohead morphogenesis (Figure 2). The process requires four phage genes and at least two bacterial genes. The products of the two bacterial genes, *groEL* and *groES*, appear to assist in the proper folding of *gpB*, which forms the head-tail connector or doorway of the head. This structure is composed of 12 gpB molecules disposed in the form of a ring with a central orifice, which most likely provides an entrance for DNA during packaging and an exit during injection. gpNu3 plays a transient role during connector assembly and may also function as a disposable scaffold during the assembly of the prohead icosahedral shell. With the exception of one among the 12 corners of the icosahedron, the shell is a lattice built entirely of gpE molecules, 405 in all. The exceptional corner contains 10 molecules derived from a covalent association of a portion of gpC with a portion of gpE. These 10 hybrid molecules, called pX, appear to form the collar that holds the connector in place (7, 64–67).

## DNA Maturation, Packaging, and Head Completion

Packaging initiates when terminase binds at *cos* to form a structure known as complex I (Figure 2) (70, 71). The selection of a particular *cos* among those present on a concatemer appears to occur at random (50, 51, 72). In vitro, the binding to *cos* of at least one of the two host factors, IHF and THF, is essential for DNA cleavage (61–63, 70, 73–79). Complex I then binds a prohead to generate complex II, a reaction that is strongly enhanced by the product of the *FI* gene of λ (70, 80, 81). Chromosome entry into the prohead during packaging is polar, from the *Nu1* end to the *R* end (50, 72, 82). Thus, by its interaction with *cos,* terminase not only catalyzes the formation of ends but, in coupling proheads to DNA, establishes the origin and direction of packaging. Whether cutting of the initial *cos* precedes or follows packaging is still an unsettled issue. In vivo, however, it appears as though all *cos* cleavage is conditional on the packaging into the prohead of a minimum amount of DNA (see below) (83). Packaging is processive, so that on the average two to three chromosomes in series are packaged from a concatemer (50, 72, 46). At a certain time during the process of packaging, the procapsid expands, increasing in volume by about 75% (84–86). The expansion creates sites in the shell lattice into which gpD molecules enter in numbers equivalent to those of gpE; this provides structural stability to the head (83, 87–90). gpD may also stabilize the outer layers of DNA in the head (91). At this point, the terminal *cos* is cut and the DNA-filled head disengages from the packaging complex (92). Terminase remains, in many cases, bound to the left end of the next, downstream chromosome and, upon prohead capture, can begin a second round of packaging (50, 72, 93, 46). The assembly of the head is completed by the addition of two phage proteins, gpW and gpFII, which by binding to the head-tail connector prevent the exit of the DNA and provide the proper binding site for tail attachment (92, 94–97).

# STRUCTURE AND FUNCTION OF TERMINASE

## Purification and Properties of the Protomer and of the Small Subunit

Prior to the cloning of its genes, the enzyme had been purified 60,000-fold from crude extracts of phage-infected cells to about 50% homogeneity. Terminase holoenzyme is a heterooligomer, the subunits of which are the products of *Nu1* and *A,* the two leftmost genes of the $\lambda$ chromosome (Figure 2) (53–57). The molecular weights predicted from the DNA sequence are 20,444 for gpNu1 and 72,280 for gpA (12); predictions that were based, in part, on the molecular weight estimated from gel electrophoresis (55, 96). The subunit stoichiometry remains ambiguous. The molecular weight of protomers in solution suggests the composition $gpNu1_2gpA_1$. The tendency of gpNu1 to aggregate, however, even at relatively low concentrations (98), casts an uncertainty on the functional form of the enzyme. The actual stoichiometry of the DNA-terminase complex (complex I) or of the packaging apparatus (termisome), however, is likely to differ (see MODEL OF COMPLEX I below) (9).

Besides its endonucleolytic and packaging activities, the enzyme displays a DNA-dependent ATPase activity. The hydrolysis generates ADP and Pi and is supported by double-stranded DNA of many sources (56).

The small subunit of terminase has been purified starting from a gpNu1 overproducer strain. The purified gpNu1 has a DNA-independent ATPase activity. The sequence of a few of its N-terminal amino acids and the determination of its molecular weight (98) match the calculated values derived from the predicted sequence of the gene (12). The protein has the tendency to form aggregates with $M_r > 500,000$ (98, 99), a property that may be related to its mechanism of action (see MODEL OF COMPLEX I, below].

## Cloning of the Terminase Genes and Protein Overproduction

Only about one to two hundred molecules of enzyme are synthesized per bacterium in the course of a normal infection cycle (96). The low productivity is due to a very low efficiency of mRNA translation (57, 100–103). The two terminase genes have been cloned both separately and together in inducible expression vectors, and their ribosome-binding sites have been replaced by those of λ genes which are efficiently expressed (57, 102). Induction of plasmid-borne gene expression in cells harboring the plasmid with both terminase genes leads to accumulation of gpNu1 and gpA with a combined mass that may reach about 15–20% of the total cell protein (57). These plasmids have been used as the source of material for the preparation of the individual subunits and of the holoenzyme (52, 98; W. Parris, W. Yang, M. Gold, personal communication). The overproduced proteins, especially gpNu1, accumulate in expressing cells as insoluble particulates, and much of the research effort expended during their purification has been directed to devise procedures for their solubilization in active form (W. Parris, W. Yang, M. Gold, personal communication).

## Functional Domains

The enzyme's primary functions are the maturation and packaging of the DNA. To carry out these processes, the enzyme resorts to multiple activities, which include site-specific DNA binding (59, 70, 74, 104–106), site-specific nicking (48, 49, 57), prohead binding (70, 80, 107, 108), possible gpFI binding (80, 81, 109), cohesive end melting (52), possible DNA translocation (6, 11, 70), DNA sequence scanning (46, 93), ATP binding (52, 56, 70); and ATP hydrolysis (52, 56, 70, 98).

Genetic and biochemical experiments have shown that some of these activities are localized to functional domains along the polypeptide chains of the two subunits. The genetic experiments made use of φ21, a "lambdoid" phage with two genes analogous to λ genes *Nu1* and *A,* called *1* and *2* respectively, which code for corresponding terminase subunits. The terminases of λ and φ21 are specific for their own DNA and prohead. Viable φ21/λ hybrid phages that code for hybrid terminases (8) can be constructed by genetic crosses, however. The phages with hybrid terminases were analyzed for their ability in mixed infections to (*a*) package DNA of one or the other specificity, (*b*) complement λ and φ21 defective in each of the terminase genes, and (*c*) complement λ and φ21 defective in prohead synthesis. Further, the exact points of the cross-overs in the hybrid phages were determined by DNA sequencing of the φ21-λ junctions, thereby allowing the localization of the domain that determines DNA-binding specificity to within the first 91-amino terminal residues of gpNu1, part of the prohead-binding domain to within the 38 carboxy-terminal residues of gpA, and the domains responsible for the specificity of subunit assembly to the carboxy terminus of gpNu1 and the amino terminus of gpA (Figure 3) (107, 108, 110).

Within the domain of gpNu1 that determines DNA-binding specificity, there is a putative α2 helix-turn-α3 helix (HTH) motif (111), which com-

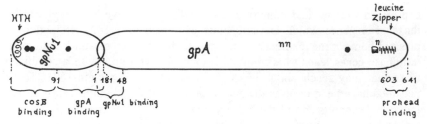

*Figure 3*    Genetic map of the λ terminase genes and the approximate locations of the functional domains and other features. The numbers refer to amino acid residues of gpNu1 and gpA. ●: location of putative ATP-binding domains. *Helix:* location of HTH. *Interdigitated lines:* location of putative leucine zipper. *n:* location of mutations that impair nucleolytic, but not packaging, activity.

prises amino acid residues 5 to 24 and which is likely to be responsible for the specificity of binding to *cos* DNA (Figure 3) (8, 58) (see below).

There is genetic and biochemical evidence that the endonucleolytic activity is a function of gpA. The genetic evidence involves the analysis of a large number of mutants in *Nu1* and *A*. These mutants were isolated by making use of the observation that *E. coli* cells deficient in recombination are killed by levels of terminase normally present in λ infected cells. Thus, whereas induction of the terminase genes cloned in a plasmid under the control of an inducible promoter resulted in death of the majority of the cells (78), most of the survivors consisted of cells harboring plasmids with mutations in genes *Nu1* or *A* (A. Davidson, P. Yau, H. Murialdo, M. Gold, personal communication). Three of these mutants produced terminase that lacked nucleolytic activity but retained packaging activity (A. Davidson, P. Yau, H. Murialdo, M. Gold, personal communication). The three mutations all mapped in gene *A* (Figure 3). These results suggest that DNA cutting is a function of gpA. This interpretation is strengthened by biochemical experiments showing that purified gpA had endonucleolytic activity, whereas gpNu1 does not (98; R. R. Higgins and A. Becker, personal communication).

gpA's specificity for DNA may be mediated by a putative basic region associated to a leucine zipper (bZIP). This domain spans amino acid residues 570 to 620, near the carboxy terminus of the polypeptide chain (Figure 3) (A. Davidson and M. Gold, personal communication). Leucine zippers are α-helices in which the first (*a*) and fourth (*d*) positions within the heptad repeat are occupied by large apolar residues and in which one of these positions is almost invariably occupied by Leu (112, 113). In the bZIPs of most proteins, if the first *a* position of the series is closer to the basic region than the first *d* position of the series, then the Leu repeat is at position *d*. gpA differs from other proteins containing leucine-zipper domains in that the leucine repeat occurs in the *a* instead of the *d* position. The position of the basic region relative to the residues that interact in the formation of the coiled coil, however, is the same (A. Davidson and M. Gold, personal communication). Since $\phi21/\lambda$ hybrid phages with the putative bZIP domain of gpA package $\phi21$ DNA (110), this DNA-binding domain does not distinguish between $\lambda$ or $\phi21$ cos.

The locations of the ATP-interactive sites on both subunits (Figure 3) were predicted by comparisons of sequence homology (99, 114). The presence of these sites in gpNu1 (99, 98) and gpA (R. R. Higgins and A. Becker, personal communication) has been verified experimentally.

## STRUCTURE OF COS

The site of action of terminase is called cos (42, 50, 51) and includes all the sequences necessary for terminase to bind specifically, cleave, and start packaging. It comprises about 200 base pairs and its structure is summarized in Figure 4A. cos consists of several subunits. The site nicked by the enzyme and recognized by the gpA subunit is called cosN (105, 106). The rest of cos contains a series of protein-binding sites and spacers and is called cosB (60, 104–106, 115–117). cosB has elements on both sides of cosN. These sides are designated here cosBL and cosBR, where the L and the R signify to the left and to the right, respectively, of cosN.

cosN contains, in addition to the 12 base pairs that form the cohesive ends, a few neighboring base pairs (105) that play a role in terminase recognition and DNA ejection (S.-Y. Xu and M. Feiss, personal communication). The sequence has two-fold hyphenated rotational symmetry (Figure 4A) (15, 118), and the two pseudo-symmetry elements are referred to as cosNL and cosNR (52). The phosphodiester bonds hydrolyzed by terminase are symmetrically placed within this dyad.

cosBR comprises a region of about 150 base pairs between cosN and Nu1 (Figure 4). Within this region are three segments called R3, R2, and R1. These are 16 base pairs long and are identical at 12 positions (74, 116) (Figure 4B). The Rs bind gpNu1 in vitro (59). In addition, cosBR contains several consensus sequences for IHF binding (called I in Figure 4A) (74). In vitro, however, IHF binds with high affinity only to site I1 and with only moderate affinity to site I2 (61, 62). THF also binds to cosBR and can stimulate the activities of terminase (63, 79).

cosBL is located to the left of cosN and contains one R site, called R4, which shares only 8 of the 12 base pairs common among the other three R sites (Figure 4B). R4 does not bind gpNu1 in vitro (59) but does bind holoenzyme, at least when ATP is present and when cosN is occupied by terminase (R. R. Higgins and A. Becker, personal communication).

The structural organization of $\phi21$ cos is quite similar to that of $\lambda$. $\phi21$

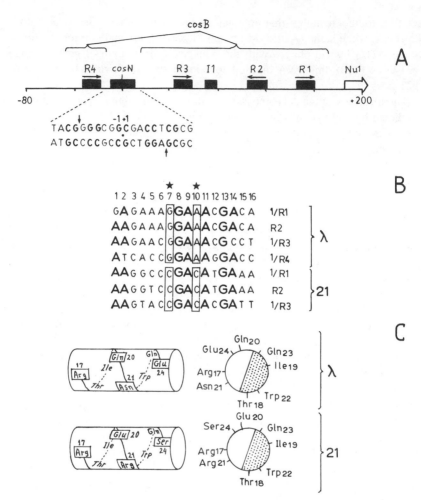

*Figure 4*   (A) Map of *cos*. *Filled blocks* represent binding sites for proteins. Only one IHF binding site is shown (see text). *Arrows* on top of the blocks representing *R*s indicate their relative orientation. The *open arrow* indicates the start of gene *Nu1*. The sequence of *cosN* is shown in the lower expansion with its axis of pseudo-two-fold symmetry indicated by a *dot* and the *vertical arrows* showing the sites of normal nicking. The nucleotide numbering is centered at the axis of symmetry. (B) Nucleotide sequence of the *R*s. Numbering is arbitrary. Because of their orientations, some of the sequences correspond to that of the 1-strand *(R)*, and some to that of the r-strand (1/*R*). Boldface type indicates nucleotides present in six or seven of the *R*s. (C) Schematic axial and end projection representations of the α3 helices of gpNu1 and gp1. Residue numbering is as deduced from the nucleotide sequences of the genes. (From Ref. 9)

DNA contains a *cosN* with a nucleotide sequence identical to that of λ and a *cosBR*, with three *R* sites and IHF-binding consensus sequences in the same relative positions and orientations as those of λ (119, 120). The base sequence of the *R* sites *(R3, R2, R1)* of φ21 *cos* is similar but not identical to that of λ's (Figure 4*B*), and the differences may underlie the DNA recognition specificities of the two terminases. The sequence of *R4* in φ21 is unknown.

# THE FUNCTIONS OF TERMINASE

## *Formation of the Initiation Complex*

COS BINDING   Terminase binds to *cos* to form a binary complex (71). The reaction can also take place in vitro and its product, called complex I or initiation complex, can be separated from the reactants in biologically active form by sedimentation through a density gradient (70).

Terminase is known to have two different site specificities for DNA: one for *cosN* and another for an *R* segment of *cosB*. The specificity of terminase for *cosN* is independent of *cosB* binding, that is, specific nicking is seen in substrates from which all four *R* sites have been deleted (52; R. R. Higgins and A. Becker, personal communication). Further, the enzyme protects specific bases from DNase I cleavage within and around *cosN* (R. R. Higgins and A. Becker, personal communication). For reasons outlined above, nicking is most certainly a function of the gpA subunit. The specificity of gpA for *cosN* is error prone and the *cosN* recognition sequence is somewhat degenerate (52; R. R. Higgins and A. Becker, personal communication; I. S. Xu and M. Feiss, personal communication). In vitro and in vivo, many mutations in the interval between positions $-7$ to $+8$ (see Figure 4*A*) have little or no effect on terminase function. The exceptions are a G:C→T:A transition at position $-7$, which reduces cleavage moderately, and a G:C→C:G transversion at position $-5$, which eliminates cleavage almost completely. Certain combinations of the more innocuous mutations have, however, serious effects. Hence, it has been suggested that at least part of the recognition is based on structural features of *cosN* (S.-Y. Xu and M. Feiss, personal communication), in a way similar to that proposed for the recognition of the *trp* operator by its repressor (121). Interestingly, mutations in symmetrical positions do not have similar effects. In particular, whereas the transversion at position $-5$ drastically reduces cleavage, the symmetrical transversion at position $+5$ has no detectable effect (S.-Y. Xu and M. Feiss, personal communication). The seemingly poor nucleotide specificity of terminase for *cosN* raises the important issue of how, during packaging, the terminal *cos* at the end of a chromosome is recognized, since it is known that this process does not require the recognition of the *R*s of *cosB* (see below: *Cutting of the Terminal* cos).

λ terminase can also initiate packaging of bacterial DNA at low frequency. The choice of the bacterial DNA packaged seems to be for the most part random (121a,b). This nonspecific DNA packaging may be related to the ability of terminase to package with low-efficiency cosmids missing *cosBR* (see below). The fact that temperature-sensitive mutations in *A* lower the frequency of transducing particles in the lysate suggests that gpA plays a preponderant role in the recognition and capture of the bacterial DNA (121a,b).

Terminase binds to *cosB* in a phage-specific manner (59, 105, 106). Since the nucleotide sequence of the *R* sites of *cosB* of λ and φ21 differ (120), the small subunit of terminase that binds to the *R* sites (59) must mediate this discrimination. Based on a comparison of the nucleotide sequence of the known *R*s, Becker & Murialdo (9) have suggested that at least some of the invariant positions define the target for gpNu1 (gp1) binding, and that the two positions that are identical among the *R*s of a particular phage, but that differ between the two phages (star bases), provide the basis of discrimination (Figure 4*B*). Experimental evidence that at least one of the invariant positions is important in recognition comes from the analysis of a λ mutant called *cos*154, a transition at position 8 of *R3*. The mutation impairs *cos* function, partially in wild-type cells and completely in IHF mutants (74; S.-Y. Xu and M. Feiss, personal communication) and abolishes gpNu1 binding in vitro (59).

Fittingly, a comparison of the amino acid sequences of the putative HTHs in the DNA-binding domains of gpNu1 and gp1 suggests that the four identical residues in the α3 helices of both gpNu1 and gp1 lie in the relatively hydrophobic face of the helices, presumably confronting the interior of the proteins (Figure 4*C*), and that of the four residues on the hydrophilic face, one is identical (Arg17) and three differ. It has been speculated that Arg17 touches one of the bases in the invariant positions and that one or more of the other residues discriminate between λ and φ21 *R*s by bonding with the "star" bases (Figure 4*B,C*) (9).

The role of the four *R* sites has been studied with the help of cosmids. Because of the presence of *cos*, cosmids are packaged upon superinfection with λ. The packageability in vivo and in vitro of *cos*154 and a series of deletion mutants of *cos* has shown that *R1* is not needed for in vivo packaging (74, 104, 116). Since *cos*154 is IHF-dependent for growth, it seems likely that cosmids with *cos* lacking *R1* will not be packaged in vivo in IHF⁻ cells. The ability to bind terminase by the various mutants showed that, whereas *R1* is dispensable, *R2* and *R3* are essential for efficient binding.

Cosmids lacking *R3, R2,* and *R1* are packaged in vivo with an efficiency of about 10% that of the wild-type cosmid (104, 116). Therefore terminase can bind, cleave, and initiate packaging at reduced efficiency in the absence of *cosB*. Interestingly, cosmids missing *R2* and *R1* but with an intact *R3* are packaged much less efficiently than those lacking all three R sites, as though *R3* when present alone were inhibitory. It is possible that binding to *R3* alone results in the formation of a sterile enzyme:DNA complex, which can only be corrected by the binding of a second terminase or gpNu1 molecule to *R2*.

Terminase binds cooperatively to DNA containing both *R2* and *R3* in vitro (106). The cooperativity in binding works better between *R3* and *R2* than between *R2* and *R1*. This effect is also observed in DNAse footprinting studies with gpNu1 alone (60), thus suggesting that this subunit is the one responsible for the effect observed with the holoenzyme. $\lambda$ point mutants in one, two, or three R sites grow poorly. The yield of phage is reduced even further if the host cells are IHF$^-$. The degrees of impairment indicate that the order of importance for the sites is *R3>R2>R1* (D. Cue and M. Feiss, personal communicatiaon). Despite its inability to bind gpNu1 in vitro, the *R4* site in *cosBL* binds the holoenzyme (R. R. Higgins and A. Becker, personal communication) and is essential for packaging in vivo (D. Cue and M. Feiss, personal communicataon). It is apparent, therefore, that for optimal function all four *R* sites are needed and that *R1*, although not absolutely essential, does stimulate binding (60).

To accommodate the dual interaction of a terminase protomer with *cos*, i.e. gpA at *cosNR*, and gpNu1 at, say, *R3*, the DNA between those two sites would have to bend or loop (see Figure 5). Structurally, the situation appears analogous to that of the site-specific recombination system of $\lambda$, in which two autonomous DNA-binding domains of one integrase monomer link two separate sites in *att* DNA, a core-type binding site (which, like *cosN*, is nicked specifically) and an arm-binding site, respectively (9, 122, 123).

In a biochemical pathway, two or more sequential reactions—nicking and packaging initiation, for example—may be coupled on a multifunctional protein. It has been argued that DNA bending or looping occurs when the distance between DNA targets and the distance between the domains of the multifunctional protein that catalyzes the reactions do not coincide (9). Alternatively, the bends may provide a structural arrangement that facilitates the movement of DNA during recombination and initiation of packaging. Up to now, there is no direct evidence that phage terminases (sometimes called pacases) bend or loop DNA. Experimental results on DNA packaging in phage Mu (124) and in phage P1 (125), however, can be best interpreted by assuming the formation of loops (9, 125).

Analysis of nicking specificity in vitro on a series of substrates (see below) has provided evidence that ATP serves as an allosteric ligand in complex I formation. It markedly enhances specific contacts of terminase at *R4*, rectifies the specificity of nicking of both strands, and stimulates nicking activity (52; R. R. Higgins and A. Becker, personal communication).

Quasi-equilibrium and kinetic studies suggest that the number of terminase molecules involved in complex I formation depends on the type of substrate.

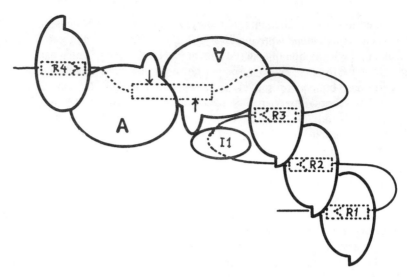

*Figure 5*   Schematic model representing the initiation complex, complex I, or termisome. *Boxes* represent terminase binding sites. Shown are two gpA subunits (A), four gpNu1 subunits (occupying the *R* sites), and one IHF molecule. The lengths of the various segments of the line that represents the DNA are drawn to scale. (Modified from Ref. 9)

Thus, whereas complex I formation with immature DNA requires three to four terminase molecules, only two are needed with mature DNA (70).

ROLE OF IHF    IHF is a small, site-specific, DNA-binding protein (126). Binding introduces a very acute bend to DNA (62, 127–129). $\lambda$ *cos* has several putative IHF-binding sites, but the analysis of small deletions and insertions in these putative IHF-binding sites (130), and their affinity for IHF (60–62), strongly suggest that under normal physiological conditions only *I1* is used (Figure 4*A*) (9). The burst size of $\lambda$ in IHF$^-$ cells is three- to four-fold less than the burst in wild-type cells (77, 130). The $\lambda$ mutant *cos*154, a GC to AT transition in position 8 of *R1* (Figure 4*B*), is IHF-dependent for growth (74), as are phages carrying the same transition in position 8 of *R3* or of *R2* and the mutants $R2^-R1^-$ and $R3^-R1^-$, which carry the transition in two *Rs*. The double and triple mutant phages $R3^-R2^-$ and $R3^-R2^-R1^-$, respectively, are absolutely defective (D. Cue and M. Feiss, personal communication). Other IHF-dependent mutants of $\lambda$ include some 2 or 3 base-pair deletions or insertions in the spacer between *cosN* and *R3* and a series of insertions ranging from 5 to 18 base-pairs in the spacer between *R2* and *R1* (D. Cue and M. Feiss, personal communication).

   In vitro, IHF introduces a very sharp bend in the DNA when bound to *I1* (62). It stimulates the binding of gpNu1 to the *R* sites (60), either by direct

IHF-gpNu1 interaction or, more likely, by the ability of IHF to bring into close proximity separate protein-binding sites due to its DNA-bending action. Sequence-induced bending has also been detected to the left of *cosN* and immediately adjacent to the right side of *I1* (131). Therefore it is likely that the role of IHF binding is to increase the degree of bending to facilitate the interaction between gpNu1 subunits momentarily bound to *R*s.

The cooperativity of gpNu1 binding suggests that interaction between gpNu1 molecules stabilizes their association with the *R*s. IHF would facilitate this process better than other cellular factors, so that when gpNu1 binding to one or more *R*s is weakened by mutation in the sites, its presence becomes essential.

IHF-independent variants of many of the IHF-dependent λ mutants mentioned above have been isolated (76, 77; D. Cue and M. Feiss, personal communication). In addition, these latter authors have isolated pseudorevertants of the absolute defective mutant λ*R3⁻R2⁻R1⁻*. All the pseudorevertants characterized so far carry amino acids substitutions in *Nu1*. These suppressor mutations are not allele sensitive, i.e. any of them suppresses the IHF-dependent phenotype of any IHF-dependent λ mutant, whether the crippling mutation is a point mutation in one of the *R*s, a small deletion or insertion in the *cosN-R3* space, or an insertion in the *R2-R1* spacer (D. Cue and M. Feiss, personal communication). All the suppressor mutations are outside the HTH motif (amino acid positions 5 to 24) believed to mediate the binding of gpNu1 to the *R*s. Although it is possible that the amino acid changes enhance DNA binding through an alteration in protein conformation, it is also possible that the changes enhance the interaction between gpNu1 monomers, thereby stabilizing the oligomeric state to the point where interacting gpNu1 monomers can sustain DNA bending in the absence of IHF. This suggests that under certain conditions of weakened protein-DNA interaction, binding to *R* sites is enhanced by either IHF-assisted DNA bending or stronger interprotein interactions.

MODEL OF COMPLEX I    A schematic model for the structure of complex I (Figure 5) has been presented (9). The proposal takes into consideration the following experimental observations: (*a*) the very sharp bend (about 140°) that IHF confers to DNA upon binding to *I1*, (*b*) the cooperative binding of gpNu1 to *R3, R2,* and *R1*, and (*c*) the recognition of *cosN* as a function of gpA. The interpretation of these observations is summarized below. The DNA bending approximates in space *R3* and *R2* and aligns them in the same orientation. This would facilitate the interaction between gpNu1 subunits bound to these sites and would provide an explanation for the cooperativity of their binding. The two-fold rotational symmetry of *cosN* requires that the two molecules of gpA bind in a way that is also related by rotational symmetry. The apparent necessity for DNA bending between *cosN* and *R3* to accommodate the dual interaction of a terminase protomer, of gpA to *cosNR* and of gpNu1 to *R3*, would position *R4* and *R3* along a two-fold rotational axis, neatly providing the appropriate configuration for the binding to the equally related gpNu1 molecules of the two protomers bound to *cosN*.

In any nucleoprotein complex in which distant DNA sites are brought together by bound proteins, the length of the spacers is critical (132, 133). Therefore, of particular interest are the 2 or 3 base-pair deletions or insertions in the *cosN-R3* spacer that render the mutants IHF-dependent for growth (130). A change in distance of 2 or 3 base pairs alters the relative orientation of *cosN* with respect to *R3* by 70° or 105°, respectively. It seems likely that IHF stimulates growth not by altering the conformation of DNA in the *cosN-R3* region directly, but rather indirectly, by increasing the efficiency of complex I formation through its capacity to bring *R3* and *R2* together.

As mentioned above, the pseudo order of the reaction for mature DNA as substrate for complex I formation is smaller than that obtained with immature DNA (70). An interpretation of these results, taking into consideration the above model for the structure of complex I, would be that complex I formed with mature DNA would be missing the terminase protomer that binds to *R4* and *cosNL*.

## Endonucleolytic Activity

IN VIVO    In λ infection, the endonucleolytic activity of the enzyme is coupled to packaging. DNA maturation does not take place in the absence of proheads (i.e. when wild-type cells are infected with λ mutants in any of the four prohead genes, when *groEL* or *groES* mutants of *E. coli* are infected by wild-type phage) or in the absence of the products of genes *D* or *FI* (33–36, 73, 134–136). These results suggest that the association of a prohead with complex I activates the nucleolytic activity of terminase, possibly by altering its conformation. This hypothesis also explains why gpFI is essential for packaging initiation, since it functions by enhancing, in an unknown way, the rate of prohead binding to complex I (80, 81).

More difficult to understand is the failure to mature DNA in cells infected with mutants in *D*. gpD, a histone-like protein (91), does not seem to act until about 82% of the λ DNA molecule is inside the head, at a time when the prohead expands, creating sites in the shell for gpD incorporation (83, 88, 92). Prohead expansion is triggered, at least in vitro, by packaging, provided the fragment used as DNA substrate is longer than 45% the wild-type length (cited in 117). The addition of gpD to the expanded head appears to stabilize the head and to allow the packaging of the remaining 18% of the chromosome. This has been deduced from the fact that λ *D* mutants with deletions larger than 18% of the λ chromosomes are viable, producing a burst of phage with normal, expanded heads (83). It appears that in the absence of gpD, packaging of wild-type genomes would cease at about 82% entry, failing to reach the terminal *cos* and eventually dissociating from the complex. Therefore, since *D* mutants fail to cleave their DNA (even at the initial *cos*) and yet about 82% of the λ genome is packaged before any necessity of gpD, the DNA must be packaged from a continuum (not an end), thereby invaginating into the prohead in the form of a loop. Cleavage of the initial *cos* would then take place at any moment after prohead expansion and stabilization by gpD binding. If this were indeed the case, the generation of ends would be essential only for DNA injection and/or for terminal events of packaging.

The looping model of packaging is not supported by other experimental results. For instance, cells infected with λ*D⁻* produce expanded (but empty) proheads, as if packaging had been attempted and aborted (83, 137, 138). A similar phenomenon takes place with a certain class of *E* mutants (139). Yet, lysates of cells infected with *cosN* mutants are devoid of expanded proheads, suggesting that cutting of the initial *cos* is a prerequisite for packaging initiation (S.-Y. Xu and M. Feiss, personal communication). The explanation often offered for the apparent paradox is that following *cos* cutting and partial packaging, the fragile complex that is formed in the absence of gpD collapses, releasing the DNA, the ends of which would then be rapidly ligated by ligase (83; S.-Y. Xu and M. Feiss, personal communication). Thus, according to this model, gpD would have no role in the cleavage of the initial *cos*.

Prophages that cannot excise from the integrated state in the bacterial chromosome, due to mutations that impair excision, are substrates for packaging. The process starts at the only available *cos* and proceeds in the normal direction, eventually crossing the λ–*E. coli* DNA junction. Packaging ceases once the head is full of DNA, but because of the absence of a terminal *cos,* the bacterial DNA remains hanging out, uncut (Figure 1). This DNA can be cut, upon lysis of the cells, by the addition of DNAse. Phage particles, called λdocL, are then assembled by interaction with free gpW, gpFII, and tails (43). This experiment shows that (*a*) cleavage of the initial *cos* is independent of terminal *cos* cleavage or even the presence of a terminal *cos* and (*b*) the terminal *cos,* or right-end, is not what stabilizes full heads, because λdocL heads are relatively stable, as stable as *FII⁻* or *W⁻* heads (43, 95, 137).

In some circumstances, terminase can work as a free nuclease in the absence of proheads, gpD, and gpFI. Plasmids have been constructed that contain terminase genes, which, upon derepression, induce the synthesis of levels of terminase comparable to those found in λ wild-type infected cells. The enzyme cleaves *cos* sites located in the bacterial chromosome and, if the bacterium is deficient in DNA damage repair (as in recombination-deficient strains of *E. coli*), the *cos* cleavage causes its death (78). Also, if the cells are recombination-proficient and if the plasmid contains *cos,* the plasmid will commit suicide by cutting itself (140). Derepression of cells carrying the suicidal plasmid and a λ prophage defective in a prohead gene, or *FI*, or *D*, leads to failure to mature both the plasmid and the phage DNA. This suggests the production by the phage of a transactive inhibitor of DNA maturation (68). A search for such a factor has, so far, been unsuccessful (H. Murialdo, unpublished). The DNAs of replicating phage and plasmid have different topological forms. It is possible that in vivo certain topological forms of DNA are refractory to *cos* cleavage when the system is uncoupled from packaging. Since phage superinfection transforms plasmid circles into concatemers (141, 142), it may be that the cleavage of this form of DNA is prohead- (and gpFI-) dependent.

IN VITRO     In vitro *cos* cutting is not coupled. The reaction is catalyzed by purified terminase and is augmented by host factors (48, 49, 63). Spermidine is essential and although hydrolysis of ATP is not required, ATP works as an allosteric effector in nicking (52, 56). The reaction can take place with substrates deleted for some of the *R* sites, and under these circumstances the enzyme makes mistakes, cutting at sites other than the normal ones. The normal nicking sites are $N_2$ in the l-strand and $N_1$ in the r-strand. Two aberrant sites are $N_y$ in the l-strand and $N_x$ in the r-strand. The latter two are located exactly 8 base-pairs to the left of the former ones (Figure 6) (52; R. R. Higgins and A. Becker, personal communication).

The purified gpA subunit has a relatively unspecific nicking activity on *cos* DNA and fails to nick at $N_1$ and $N_2$. On the other hand, holoenzyme nicks $N_1$, $N_x$, $N_2$, and $N_y$ (Figure 6) on a fragment of DNA containing *cosN* but devoid of all the four *R*s, in a reaction that is strictly dependent on ATP. This observation indicates that gpNu1 imparts nicking-site specificity independently of its R-site binding function (52; R. R. Higgins and A. Becker,

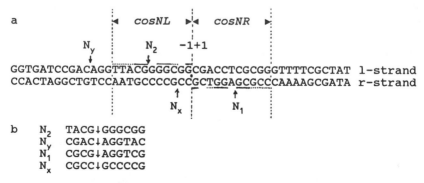

Figure 6     (*a*) Nomenclature of the *cosN* region. The *vertical arrows* represent the nicking sites. Overlined and underlined letters with the same absolute coordinates are identical (*continuous lines*) or of the same type (*dotted line*) (*b*) Nucleotide sequence surrounding the normal and aberrant nicking sites.

personal communication). The sequences that surround the $N_1$ and $N_2$ nicking sites are similar to those that surround $N_x$ and $N_y$ (Figure 6B). To nick efficiently and with high specificity (exclusively at $N_2$ and $N_1$), R3 must be present in cis and ATP must be provided as an allosteric effector. R3 and ATP play distinct but interrelated regulatory roles in cosN nicking (R. R. Higgins and A. Becker, personal communication).

If R3 is present, nicking at cosN is seen whether or not ATP is added to the reaction; nicking of the l-strand takes place at $N_2$ in either case, but nicking of the r-strand is incorrect (at $N_x$) in the absence of ATP. This observation suggests that an R3-anchored enzyme can undergo a conformational adjustment upon ATP binding that redirects the nicking active center for the r-strand 8 bases to the right.

If R3 is deleted, neither l- nor r-strand is nicked in the absence of ATP. In the presence of ATP, cosN nicking is restored at $N_2$ in the l-strand, but the enzyme cannot distinguish between $N_x$ and $N_1$ on the r-strand, and nicking at these two sites occurs with about equal frequency. These observations indicate that (a) the anchoring of terminase at R3 is important for enzymatic attack on cosN and for nicking, not only of the r-strand but also of the l-strand, across the symmetry axis, and (b) although the ATP-induced conformational change in terminase allows site-specific attack on cosN, only a R3-bound enzyme can be fully redirected to cut at $N_1$ rather than at $N_x$ in the r-strand. Under certain conditions with both R3 and R4 missing (but not one or the other), cutting of the l-strand can also slip 8 bases to the left ($N_y$), thus suggesting a barrier or buttress function for R4 that may be redundant if fixation at R3 is possible.

Finally, whereas the presence of R3 activates nicking in both strands of cosN, the presence of R4 is not sufficient to trigger activity; the addition of ATP is essential (R. R. Higgins and A. Becker, personal communication). The contrasting behavior of the two R sites is probably due to their sequence differences. It should be remembered that R4 shares only 8 base-pairs instead of 12 with the other Rs (Figure 3B) and that R4, as opposed to the other Rs, does not readily bind gpNu1 in vitro (59).

Together, the results suggest a model in which one terminase protomer bound to R3 via its gpNu1 subunit nicks $N_1$ at cosNR via its gpA subunit, while another protomer bound to R4 nicks the l-strand at $N_2$. Strong terminase footprinting in the R4-cosN-R3 interval supports this idea. The model in Figure 5 incorporates these notions and, to account for the influence of R3 binding on nicking in cosNL, the two protomers are shown communicating across the cosN symmetry axis via their gpA subunits. Some of the A mutants defective in nicking but with an unaltered packaging activity for mature DNA have been mapped by nucleotide sequencing to the region coding for the putative bZIP domain of gpA (Figure 3). In particular, one of these mutants changes an amino acid residue in the spacer between the basic region and the leucine zipper (A. Davidson and M. Gold, personal communication). Such a mutation may prevent the proper assembly of gpA dimers, as depicted in the model, and interfere with the communication between the two protomers.

The melting of the cohesive ends following nicking of the two strands requires ATP (or dATP) hydrolysis, an activity of the enzyme that could be described as site-specific helicase (52).

## Prohead Binding

In vitro, complex I interacts with a prohead to generate the ternary complex DNA:terminase:prohead called complex II. The reaction is pseudo-first order for proheads (70), and gpFI acts as a catalyst, enhancing the rate of the reaction and the final yield (80, 81). The requirements for complex II formation, starting from complex I and proheads, have not been investigated. Complex II can be isolated by velocity sedimentation in sucrose gradients, however, and upon addition of gpD, gpW, gpFII, and tails it is converted into biologically active phage (84, 70). The sedimentation properties of complex II suggest that most of the chromosome lies outside the capsid, as yet unpackaged (70, 80, 92).

As mentioned above, a region within the 38 carboxy-terminal amino acid residues of gpA determines the specificity of prohead binding (107). More recently, it has been found that deletion of the five carboxy-terminal amino acid residues of gpA blocks prohead binding and hence the ability of the enzyme to package. The mutation does not abolish nucleolytic activity, either in vitro or in vivo (J. Sippy and M. Feiss, personal communication).

It is presumed that proheads bind to terminase via their connector. Evidence for this putative terminase-connector interaction in phage λ is lacking, but there is some genetic evidence for this in phage T4 (143). Also, certain mutants in the gene that codes for the connector protein of phage P22 permit the packaging of an excess of DNA (S. Casjens, personal communication).

As mentioned above, in vitro experiments have shown that gpFI plays a catalytic role in complex II formation (80). These results are supported by the genetic analysis of pseudorevertant isolates of $FI^-$ of λ. These pseudorevertants, called fin (for FI independence), map in the terminase genes, and are of two classes. One class results in an increase in the level of gpA production upon infection. Apparently, the absence of gpFI can be compensated in part by this slight (about five-fold) increase in intracellular gpA concentration (109). A second class of mutants seem to alter the A gene, but the nature of the change is unknown (109).

## DNA Packaging

Proheads are metastable structures and, at some point during DNA packaging, they expand. The expansion is due to a rearrangement of the major capsid protein, creating sites into which gpD binds (144). gpD stabilizes the expanded capsid and this enables the encapsidation of enough additional DNA until the next *cos* is reached (83, 84, 92). The continued action of terminase is required for packaging after initial cleavage. If the in vitro staged reactions are carried out with a temperature-sensitive terminase derived from cells infected with *Ats* phage, the formation of phage, starting from already assembled, biologically active complex II, is temperature-sensitive (70). λ DNA packaging in vitro requires ATP hydrolysis (70), a property common to all double-stranded DNA phages (6).

gpD is incorporated into complex II to form complex III. With concatemeric DNA as a substrate for packaging, complex III retains the sedimentation properties of complex II; however, a substantial proportion of the phage precursors are DNAse resistant. The interpretation of these results is that the heads in complex III contain a full chromosome but that the rest of the concatemer protrudes. The addition of gpW converts complex III into fast-sedimenting full heads (92). Since gpW is not needed for the formation of full heads when mature DNA is used as a packaging substrate, it appears that gpW is needed for the disengagement of the packaged DNA from the rest of the concatemer (not necessarily the same as melting of the cohered ends). This is not the case in vivo, since cells infected with *W* mutants produce heads that, although unstable, contain fully mature DNA. Therefore, gpW is not needed for packaging or for DNA muturation, but rather for stabilization of the packaged DNA and for the formation of sites for the binding of gpFII, which, in turn, allows tail attachment (95, 97). The reason for the apparent necessity for gpW in vitro to achieve concatemer disengagement is unknown.

Over time, many mechanisms and sources of energy have been proposed to explain DNA packaging and the associated condensation. They will not be discussed here as they have been reviewed extensively by others; most recently by L. Black (6). Many of the proposed mechanisms apply to one or another specific bacteriophage. The only one that seems to be common to all the bacteriophages with double-stranded DNA, and the one that seems most widely accepted today, is the mechanism first proposed by Becker et al (70). It invokes sequential vectorial DNA translocation catalyzed by the terminase and driven by ATP hydrolysis (70). The mechanism would be similar to the movement of polymerases and of the EcoK restriction enzyme along the DNA (6).

Given the anatomical properties of connectors (65), it seemed obvious that translocation of the DNA would have to proceed through their central orifice. Hendrix has proposed an elegant model for DNA packaging. In it he assumed that the mismatch in rotational symmetry between the connector and the collar of the head allows free rotation of one with respect to the other. Therefore, if the terminase and connector adhere to each other, the passive rotation of the terminase-connector complex would release the torsional stress created by the rotation of the DNA (in relation to the terminase-connector machine) as it is translocated into the cavity of the head (145). In a variation of this model, the ATP-driven machine would be the surface of the connector that is in contact with the prohead collar. As the connector rotates, the DNA would be "screwed" toward the inside, much like the rotation of a nut along a threaded rod (145). The thread in the nut could be an integral part of the connector's orifice, or it could be provided by terminase. In any case, since terminase is essential during packaging (70) and since it scans for the terminal *cos* (93), it must remain attached and in contact with the DNA.

The model in Figure 5 shows complex I with two protomers of terminase related by a two-fold symmetry axis. If terminase were indeed the translocase, however, then only one of the two protomers shown should function during packaging. If both worked simultaneously, they would translocate in opposite orientations. One possibility is that both protomers travel during packaging and are available for $N_2$ and $N_1$ nicking of the terminal *cos*. Another is that the protomer that was sitting at *cosNL* is left behind and the one that was positioned at *cosNR* translocates. Upon reaching the terminal *cos* it waits for another free protomer to bind to the other half of *cosN*. Either way it seems that only one of the protomers has translocating activity. These considerations, and the observation that λ DNA molecules in which one of the strands contains a deletion or an insertion of up to 19 nucleotides are efficiently packaged in vitro (146, 147), suggest that the integrity of the double helix is not an absolute requirement; only one of the two strands may be used as handrail for packaging. This result suggests, in addition, that the diameter of the orifice in the connector is large enough to accommodate small loops of one of the two strands (146, 147).

There is still a controversy as to the packageability of the various topological configurations of λ DNA due to the apparently contradictory findings of in vivo and in vitro experiments. Infection of a lysogen carrying two prophages in tandem (so that the DNA between the two *cos* could be packaged) with a homoimmune phage and a heteroimmune helper phage produced, in addition to the burst of heteroimmune phages, a small yield of phages of the prophage type. The yield of phages of the superinfecting homoimmune type was tenfold lower than that of the prophage type (41, 42). Since the infecting homoimmune phage DNA is converted to covalently closed monomeric rings (17, 18), it has been argued that, in vivo, monomeric circles are not a substrate for packaging (41, 42). On the other hand, polymeric rings formed by recombination and monomeric rings containing two *cos,* due to a chromosomal duplication, are packaged very efficiently. Therefore it has been postulated that only DNA containing two *cos* can be packaged (37, 42). In vitro, monomeric rings are packaged well (69). The paradox between the sets of in vitro and in vivo data may only be apparent, as it has been found that during a normal infection, the intracellular concentration of the right end is only a fraction of that of the left end (H. Murialdo, P. Gajardo, and D. Tzamtzis, unpublished). These results suggest that each time *cos* is cut, the left end is protected by terminase and packaging, but the right end is left exposed to degradation by nucleases. Thus, opening of a monomeric ring would invariably result in a damaged chromosome, whereas opening of polymers would leave time for the packaging machinery to get to the terminal *cos* before the nuclease(s) (see also 5).

Also controversial is whether mature or concatemeric DNA is the preferential substrate for packaging. Rosenberg et al (148) have argued, based on the difference in the order of the packaging reactions as a function of DNA concentration, that the real substrates are *cos* sites and not ends, and that packaging in vitro of mature DNA molecules occurs because of their annealing.

## Cutting of the Terminal cos

Cutting of the terminal *cos* depends on packaging. This conclusion comes from the ability of λ terminase to cut a terminal *cos* site of φ21 specificity even though λ terminase cannot bind to a *cos* of φ21 specificity. The terminase molecule(s) that binds and cleaves the initial *cos* must also cleave the terminal *cos* and thus be brought into contact with the terminal *cos* by the packaging process (46, 93). In a trilysogen, three prophages are integrated in the bacterial chromosome in a head-to-tail tandem arrangement. Such a concatemer can be built so that only the left-most *cos* is of λ specificity and the other two are of φ21 specificity. Infection of the trilysogen by a heteroimmune λ helper phage yields, aside from the helper phage, phages with the DNA between the leftmost and center *cos*s and also with the DNA between the center and rightmost *cos*s. Since λ terminase cannot bind to the central and rightmost *cos*s, it must reach them by packaging. The results also imply that after cutting the central *cos*, terminase remains attached to and proceeds to package the distal chromosome in a typical processive fashion (46). These experiments also show that terminase scans for the terminal *cosN* (not for *cosB*), that is, it remains in contact with the DNA during packaging and reads its sequence (46, 93), an observation that is in agreement with the continuous necessity for this enzyme during packaging (70). It should be mentioned that in the experiment with the trilysogen, the rightmost *cos* is of φ21 origin on both sides of *cosN*, that is, all the *R*s and spacers are φ21-like. Since the sequence of *cosBL* (which includes *R4*) of φ21 is unknown, however, the role, if any, of sites in this portion of DNA in terminal *cos* scanning also remains unknown.

Helper-packaging experiments with a series of monolysogens in which the prophage contained duplications or deletions showed that packaging is optimal when the distance between initial and terminal *cos*s is within the range of 0.80 to 1.05 times the length of the wild-type DNA (44). Similar experiments have shown that the probability of terminal *cos* cleavage depends on the distance between the initial and the terminal *cos* (45). To explain this observation, it has been proposed that DNA packaging triggers prohead expansion, which in turn activates terminase to start scanning for the sequences of *cosN*. If the terminal *cos* were located before terminase became active, it would be ignored (45). Another possibility is that the velocity of DNA entry may be length dependent, decreasing as packaging proceeds. The ability of terminase to cleave the terminal *cos* might depend inversely on the packaging rate (10).

If each time a concatemer were cut at *cos* the right end was damaged, processivity of packaging would be of a tremendous advantage. For processivity to be of any advantage, however, the intracellular concentration of *cos* must be in excess of terminase, or complex II formation must be relatively inefficient.

## CONCLUSIONS

As stated at the beginning of this review, the reason(s) for linearization of the intracellular replicating DNA among the Myoviruses is not obvious. Yet, with some variations among them (see for instance 6), Myoviruses have evolved highly sophisticated processes for the cutting and packaging of their DNAs and exquisite mechanisms for regulating these events. One is left to wonder whether packaging of double-stranded circular DNA would be even more complicated and therefore eliminated by natural selection. Alternatively, such rings may not be more difficult to package than linear molecules, and the only reason that this mechanism is not found in nature may be a mere distraction of Chance. The process is a fascinating one, however, and many mysteries still remain. The most puzzling ones—such as the structure of the termisome, the nature of the coupling in vivo of cutting and packaging, and the mechanism of DNA translocation into the capsid—are likely to keep many of us busy for many years to come.

ACKNOWLEDGMENTS

I express my most sincere appreciation to Andrew Becker, Alan Davidson, and Marvin Gold for innumerable useful comments on the scientific content of the manuscript and for editing. Research in the author's laboratory has been funded by the Medical Research Council, Canada, and by the National Institutes of Health, USA.

## Literature Cited

1. Casjens, S. 1985. *Virus Structure and Assembly,* ed. S. Casjens, pp. 1–28. Boston: Jones & Barlett. 295 pp.
2. Keppel, F., Fayet, O., Georgopoulos, C. 1988. See Ref. 149, pp. 145–262.
3. Greenstein, M., Skalka, A. 1975. *J. Mol. Biol.* 97:543–59
4. Goldberg, E. 1983. *Bacteriophage T4,* pp. 32–39. Washington, DC: ASM. 410 pp.
5. Sternberg, N., Coulby, J. 1987. *J. Mol. Biol.* 194:453–68
6. Black, L. W. 1988. See Ref. 149, pp. 321–73
6a. Black, L. W. 1989. *Annu. Rev. Microbiol.* 43:267–92
7. Murialdo, H., Becker, A. 1978. *Microbiol. Rev.* 42:529–76
8. Feiss, M. 1986. *Trends Genet.* 2:100–4
9. Becker, A., Murialdo, H. 1990. *J. Bacteriol.* 172:2819–24
10. Feiss, M., Becker, A. 1983. See Ref. 150, pp. 305–30
11. Earnshaw, W. C., Casjens, S. R. 1980. *Cell* 21:319–31
12. Sanger, F., Coulson, A. R., Hong, G. F., Hill, D. F., Petersen, G. B. 1982. *J. Mol. Biol.* 162:729–73
13. Hershey, A. D., Burgi, E., Ingraham, L. 1963. *Proc. Natl. Acad. Sci. USA* 49:748–55
14. Hershey, A. D., Burgi, E. 1965. *Proc. Natl. Acad. Sci. USA* 53:325–28
15. Wu, R., Taylor, E. 1971. *J. Mol. Biol.* 57:491–511
16. Young, E. T., Sinsheimer, R. L. 1964. *J. Mol. Biol.* 10:562–64
17. Dove, W., Weigle, J. 1965. *J. Mol. Biol.* 12:620–29
18. Bode, V. C., Kaiser, A. D. 1965. *J. Mol. Biol.* 14:399–417
19. Lipton, A., Weissbach, A. 1966. *J. Mol. Biol.* 21:517–25
20. Ogawa, H., Tomizawa, J. I. 1967. *J. Mol. Biol.* 23:265–76
21. Dove, W. 1968. *Annu. Rev. Genet.* 2:305–40
22. Wang, J. C., Davidson, N. 1968. *Cold Spring Harbor Symp. Quant. Biol.* 33:409–15
23. Miller, G. M., Feiss, M. 1981. *Virology* 109:379–90
24. Gellert, M. 1967. *Proc. Natl. Acad. Sci. USA* 57:148–55
25. Kaiser, D., Masuda, T. 1973. *Proc. Natl. Acad. Sci. USA* 70:260–64
26. Kaiser, D., Syvanen, M., Masuda, T. 1974. *J. Supramol. Struct.* 2:318–28
27. Hohn, B., Hohn, T. 1974. *Proc. Natl. Acad. Sci. USA* 71:2372–76
28. Smith, M. G., Skalka, A. 1966. *J. Gen. Physiol.* 49(pt.2):127–42
29. Salzman, L. A., Weissbach, A. 1967. *J. Mol. Biol.* 28:53–70
30. Carter, B. J., Smith, M. G. 1968. *Biochim. Biophys. Acta* 166:735–37
31. Carter, B. J., Shaw, B.D., Smith, M. G. 1969. *Biochim. Biophys. Acta* 195:494–505
32. Carter, B. J., Smith, M. G. 1970. *J. Mol. Biol.* 50:713–18
33. MacKinlay, A. G., Kaiser, A. D. 1969. *J. Mol. Biol.* 39:679–83
34. Skalka, A., Poonian, M., Bartl, P. 1972. *J. Mol. Biol.* 64:541–50
35. Wake, R.G., Kaiser, A. D., Inman, R. B. 1972. *J. Mol. Biol.* 64:519–40
36. McClure, S. C. C., MacHattie, L., Gold, M. 1973. *Virology* 54:1–18
37. Enquist, L. W., Skalka, A. 1973. *J. Mol. Biol.* 75:185–212
38. Better, M., Friefelder, D. 1983. *Virology* 126:168–82
39. McClure, S. C. C., Gold, M. 1973. *Virology* 54:19–27
40. Mousset, S., Thomas, R. 1969. *Nature* 221:242–44
41. Szpirer, J., Brachet, P. 1970. *Mol. Gen. Genet.* 108:78–92
42. Feiss, M., Margulies, T. 1973. *Mol. Gen. Genet.* 127:285–95
43. Sternberg, N., Weisberg, R. 1975. *Nature* 256:97–103
44. Feiss, M., Fisher, R. A., Crayton, M. A., Egner, C. 1977. *Virology* 77:281–93
45. Feiss, M., Siegele, D. A. 1979. *Virology* 92:190–200
46. Feiss, M., Sippy, J., Miller, G. 1985. *J. Mol. Biol.* 186:759–71
47. Yarmolinsky, M. B. 1971. See Ref. 151, pp. 97–111
48. Wang, J. C., Kaiser, A. D. 1973. *Nature New Biol.* 241:16–17
49. Becker, A., Gold, M. 1978. *Proc. Natl. Acad. Sci. USA* 75:4199–4203
50. Emmons, S. W. 1974. *J. Mol. Biol.* 83:511–25
51. Feiss, M., Campbell, A. 1974. *J. Mol. Biol.* 83:527–40
52. Higgins, R. R., Lucko, H. J., Becker, A. 1988. *Cell* 54:765–75
53. Weisberg, R., Sternberg, N., Gallay, E. 1979. *Virology* 95:99–106
54. Sumner-Smith, M., Becker, A., Gold, M. 1981. *Virology* 111:642–46

55. Sumner-Smith, M., Becker, A. 1981. *Virology* 111:629–41
56. Gold, M., Becker, A. 1983. *J. Biol. Chem.* 258:14619–25
57. Chow, S., Daub, E., Murialdo, H. 1987. *Gene* 60:277–89
58. Kypr, J., Mrázek, J. 1986. *J. Mol. Biol.* 191:139–40
59. Shinder, G., Gold, M. 1988. *J. Virol.* 62:387–92
60. Shinder, G., Gold, M. 1989. *Nucleic Acids Res.* 17:2005–22
61. Xin, W., Feiss, M. 1988. *Nucleic Acids Res.* 16:2015–30
62. Kosturko, L. D., Daub, E., Murialdo, H. 1989. *Nucleic Acids Res.* 17:317–34
63. Gold, M., Parris, W. 1986. *Nucleic Acids Res.* 14:9797–9809
64. Georgopoulos, C., Tilly, K., Casjens, S. 1983. See Ref. 150, pp. 279–304
65. Bazinet, C., King, J. 1985. *Annu. Rev. Microbiol.* 39:109–29
66. Casjens, S., Hendrix, R. 1988. See Ref. 149, 1:15–91
67. Hohn, T., Katsura, I. 1977. *Curr. Top. Microbiol. Immunol.* 78:69–110
68. Murialdo, H., Fife, W. L. 1987. *Genetics* 115:3–10
69. Gold, M., Hawkins, D., Murialdo, H., Fife, W. L., Bradley, B. 1982. *Virology* 119:35–41
70. Becker, A., Marko, M., Gold, M. 1977. *Virology* 78:291–305
71. Hohn, B. 1975. *J. Mol. Biol.* 98:93–106
72. Feiss, M., Bublitz, A. 1975. *J. Mol. Biol.* 94:583–94
73. Becker, A., Murialdo, H., Gold, M. 1977. *Virology* 78:277–90
74. Bear, S. E., Court, D. L., Friedman, D. I. 1984. *J. Virol.* 52:966–72
75. Feiss, M., Frackman, S., Sippy, J. 1985. *J. Mol. Biol.* 183:239–49
76. Feiss, M., Fogarty, S., Christiansen, S. 1988. *Mol. Gen. Genet.* 212:142–48
77. Granston, A. E., Alessi, D. M., Eades, L. J., Friedman, D. I. 1988. *Mol. Gen. Genet.* 212:149–56
78. Murialdo, H. 1988. *Mol. Gen. Genet.* 213:42–49
79. Shinder, G., Parris, W., Gold, M. 1988. *Nucleic Acids Res.* 16:2765–85
80. Becker, A., Murialdo, H., Lucko, H., Morell, J. 1988. *J. Mol. Biol.* 199:597–607
81. Davidson, A., Gold, M. 1987. *Virology* 161:305–14
82. Padmanabhan, R., Wu, R., Bode, V. C. 1972. *J. Mol. Biol.* 69:201–7
83. Sternberg, N., Weisberg, R. 1977. *J. Mol. Biol.* 117:717–31
84. Kaiser, D., Syvanen, M., Masuda, T. 1975. *J. Mol. Biol.* 91:175–86

85. Hohn, B., Wurtz, M., Klein, B., Lustig, A., Hohn, T. 1974. *J. Supramol. Struct.* 2:302–17
86. Hohn, T., Wurtz, M., Hohn, B. 1976. *Philos. Trans. R. Soc. London Ser. B* 276:51–61
87. Hohn, T., Morimasa, T., Tsugita, A. 1976. *J. Mol. Biol.* 105:337–24
88. Hohn, T., Hohn, B. 1973. *J. Mol. Biol.* 79:649–62
89. Casjens, S. R., Hendrix, R. W. 1974. *J. Mol. Biol.* 88:535–45
90. Imber, R., Tsugita, A., Wurtz, M., Hohn, T. 1980. *J. Mol. Biol.* 139:277–95
91. Witkiewicz, H., Schweiger, M. 1982. *EMBO J.* 1:1559–64
92. Perucchetti, R., Parris, W., Becker, A., Gold, M. 1988. *Virology* 165:103–14
93. Feiss, M., Widner, W. 1982. *Proc. Natl. Acad. Sci. USA* 79:3498–3502
94. Casjens, S. 1971. See Ref. 151, pp. 725–32
95. Casjens, S., Hohn, T., Kaiser, A. D. 1972. *J. Mol. Biol.* 64:551–63
96. Murialdo, H., Siminovitch, L. 1972. *Virology* 48:785–823
97. Casjens, S. 1974. *J. Mol. Biol.* 90:1–23
98. Parris, W., Davidson, A., Keeler, C. L., Gold, M. 1988. *J. Biol. Chem.* 263:8413–19
99. Becker, A., Gold, M. 1988. *J. Mol. Biol.* 199:219–22
100. Ray, R. N., Pearson, M. L. 1974. *J. Mol. Biol.* 85:163–75
101. Ray, P. N., Pearson, M. L. 1975. *Nature* 253:647–50
102. Murialdo, H., Davidson, A., Chow, S., Gold, M. 1987. *Nucleic Acids Res.* 15:119–40
103. Sampson, L. L., Hendrix, R. W., Huang, W. M., Casjens, S. R. 1988. *Proc. Natl. Acad. Sci. USA* 85:5439–43
104. Miwa, T., Matsubara, K. 1983. *Gene* 24:199–206
105. Feiss, M., Kobayashi, I., Widner, W. 1983. *Proc. Natl. Acad. Sci. USA* 80:955–59
106. Feiss, M., Widner, W., Miller, G., Johnson, G., Christiansen, S. 1983. *Gene* 24:207–18
107. Frackman, S., Siegele, D. A., Feiss, M. 1984. *J. Mol. Biol.* 180:283–300
108. Wu, W. F., Christiansen, S., Feiss, M. 1988. *Genetics* 119:477–84
109. Murialdo, H., Fife, W. L., Becker, A., Feiss, M., Yochem, J. 1981. *J. Mol. Biol.* 145:375–404
110. Frackman, S., Siegele, D. A., Feiss, M. 1985. *J. Mol. Biol.* 183:225–38
111. Pabo, C. O., Sauer, R. T. 1984. *Annu. Rev. Biochem.* 53:293–321

112. Landschulz, W. H., Johnson, P. F., McKnight, S. L. 1988. *Science* 240:1759–64
113. Cohen, C., Parry, D. A. D. 1990. *Proteins Struct. Funct. Genet.* 7:1–15
114. Guo, P. X., Peterson, C., Anderson, D. 1987. *J. Mol. Biol.* 197:229–36
115. Feiss, M., Fisher, R. A., Siegele, D. A., Nichols, B. P., Donelson, J. E. 1979. *Virology* 92:56–67
116. Miwa, T., Mitsubara, K. 1982. *Gene* 20:267–79
117. Hohn, B. 1983. *Proc. Natl. Acad. Sci. USA* 80:7456–60
118. Weigel, P. H., Englund, P. T., Murray, K., Old, R. W. 1973. *Proc. Natl. Acad. Sci. USA* 70:1151–55
119. Nichols, B. P. 1978. PhD thesis. Univ. Iowa, Iowa City
120. Miller, G., Feiss, M. 1985. *J. Mol. Biol.* 183:246–49
121. Otwinowski, R. W., Schevitz, R. W., Zhang, R. G., Lawson, C. L., Joachimiak, A., et al. 1988. *Nature* 335:321–29
121a. Sternberg, N., Weisberg, R. 1977. *J. Mol. Biol.* 117:717–31
122b. Sternberg, N. 1986. *Gene* 50:69–85
122. Moitoso de Vargas, L., Pargellis, C. A., Hasan, N. M., Bushman, E. W., Landy, A. 1988. *Cell* 54:923–29
123. Moitoso de Vargas, L., Kim, S., Landy, A. 1989. *Science* 244:1457–61
124. George, M., Bukhari, A. I. 1981. *Nature* 292:175–76
125. Sternberg, N., Coulby, J. 1987. *J. Mol. Biol.* 194:469–79
126. Friedman, D. I. 1988. *Cell* 55:545–54
127. Prentki, P., Chandler, M., Galas, D. J. 1987. *EMBO J.* 6:2479–87
128. Stenzel, T. T., Patel, P., Bastia, D. 1987. *Cell* 49:709–17
129. Yang, C. C., Nash, H. A. 1989. *Cell* 57:869–80
130. Miller, G., Feiss, M. 1988. *Mol. Gen. Genet.* 212:157–65
131. Leo, A., Kosturko, L. D., Feiss, M. 1990. *Virology* 174:329–34
132. Hochschild, A., Prashne, M. 1986. *Cell* 44:681–87
133. Thompson, J. F., Snyder, U. K., Landy, A. 1988. *Proc. Natl. Acad. Sci. USA* 85:6323–27
134. Dove, W. F. 1966. *J. Mol. Biol.* 19:187–201
135. Boklage, C. E., Wong, E. C., Bode, V. C. 1973. *Genetics* 75:221–30
136. Georgopoulos, C. P., Hendrix, R. W., Casjens, S. R., Kaiser, A. D. 1973. *J. Mol. Biol.* 76:45–60
137. Murialdo, H., Siminovitch, L. 1972. *Virology* 48:824–35
138. Lickfeld, K. G., Menge, B., Hohn, B., Hohn, T. 1976. *J. Mol. Biol.* 103:299–318
139. Katsura, I. 1989. *J. Mol. Biol.* 205:397–405
140. Murialdo, H., Fife, W. L. 1984. *Gene* 30:183–94
141. Feiss, M., Siegele, D. A., Rudolph, C. F., Frackman, S. 1982. *Gene* 17:123–30
142. Biek, D. P., Cohen, S. N. 1986. *J. Bacteriol.* 167:594–603
143. Hsiao, C. L., Black, L. W. 1977. *Proc. Natl. Acad. Sci. USA* 74:3652–56
144. Wurtz, M., Kistler, J., Hohn, T. 1976. *J. Mol. Biol.* 101:39–56
145. Hendrix, R. W. 1978. *Proc. Natl. Acad. Sci. USA* 75:4779–83
146. Pearson, R. K., Fox, M. S. 1988. *Genetics* 118:5–12
147. Pearson, R. K., Fox, M. S. 1988. *Genetics* 118:13–19
148. Rosenberg, S. M., Stahl, M. M., Kobayashi, I., Stahl, F. W. 1985. *Gene* 38:165–75
149. Calendar, R., ed. 1988. *The Bacteriophages,* Vol. 2. New York: Plenum
150. Hendrix, R. W., Roberts, J. W., Stahl, F. W., Weisberg, R. A., eds. 1983. *Lambda II.* Cold Spring Harbor, NY: Cold Spring Harbor Lab. 694 pp.
151. Hershey, A. D., ed. 1971. *The Bacteriophage Lambda.* Cold Spring Harbor, NY: Cold Spring Harbor Lab. 792 pp.

*Annu. Rev. Biochem. 1991. 60:155–90*

# CELL ADHESION MOLECULES:
# Implications for a Molecular Histology

*Gerald M. Edelman and Kathryn L. Crossin*

The Rockefeller University, 1230 York Avenue, New York, NY 10021

KEY WORDS:   morphogenesis, development, cell surface modulation, substratum adhesion, cell junctions.

## CONTENTS

## PERSPECTIVES AND SUMMARY

More than a decade has passed since the first definitive demonstrations (1–3) of molecular activity mediating cell-cell adhesion. Since that time, interest in this subject has grown explosively, and at least 1250 articles directly con-

0066-4154/91/0701-0155$02.00

cerned with cell adhesion molecules (CAMs) have been published. The reasons for this interest are not difficult to find: cell-cell adhesion was essential in the evolution of metazoans, it is a key element in cell regulation and tissue physiology, it is fundamental in morphogenesis (4, 5), metamorphosis, and regeneration, and it is disrupted in a great variety of disease processes. All of these subjects have been discussed in a number of useful books and reviews (6–15); we extend them here as appropriate. But our main thrust in the present review is to build the case that an understanding of the properties of CAMs and their relationship to other morphoregulatory molecules provides one of the major bases for the development of molecular histology. By this term, we mean an understanding of the molecular regulatory and specificity properties of collections of cells that lead to the macroscopic form of defined tissues. This field lies between molecular embryology and tissue physiology, and its pursuit should provide deep insights into normal function as well as into the diagnosis and treatment of disease. We pick up this theme at the end of this review.

The bulk of our remarks are aimed at providing a picture of what is known as of August 1990 of the structure, molecular genetics, biochemistry, and role in tissue formation of CAMs. Give that the subject has been well reviewed in the general sense, we emphasize contributions made in the past five years and especially those having a direct bearing on the relationship between chemical structure and tissue function. We make no attempt to be exhaustive in our coverage.

In our opinion, three main new conclusions have emerged from the accumulated work on these molecules. The first is that CAMs function in a dynamic and highly regulated fashion to form and maintain tissue structure. The dynamics involve more than just specificity of binding, and are concerned with many levels of cellular control of function. The second is that CAM precursors in evolution have provided a motif for a remarkably diverse set of molecules with many functions, of which the Ig superfamily is a key example. The third is that a network of evolutionary, structural, and functional themes connects CAMs to two other extensive families of morphoregulatory molecules: the substrate adhesion molecules (SAMs) and cell junctional molecules (CJMs). It is the connection of data behind these three conclusions with work on cell specification and differentiation that suggests that the time is ripe for a three-dimensional (and indeed four-dimensional) analysis of tissues in molecular terms.

## CAM TYPES, STRUCTURE, AND BINDING

As with all other classificatory schemes, a certain arbitrariness enters into the grouping of CAM types. But given a tolerance for exceptions, it may be said that CAMs fall into three groups: molecules related to the Ig superfamily that generally have $Ca^{2+}$-independent binding (16), members of a $Ca^{2+}$-

**Table 1**  Cell adhesion molecules and related proteins[a,b]

## N-CAM Superfamily

| Neural | Immune | Miscellaneous | Insects |
|---|---|---|---|
| N-CAM[a] (20) | I-CAM[a] (242) | CEA[a] (81) | fasciclin II[a] (230) |
| Ng-CAM (291)/G4 (298) | I-CAM-2 (300) | NCA (304) | fasciclin III (232) |
| | CD4 (301) | C-CAM[a]/BGP-I (284) | amalgam (233) |
| L1 (292) (NILE)[a] (147) | LAR (142) | V-CAM (305) | neuroglian (220) |
| Nr-CAM (291) | PECAM-1 (302)/ endo CAM (303) | poliovirus receptor (249) | DLAR (307) |
| Contactin (293) | | | |
| F11 (294)/F3 (295) | | OB-CAM (306) | DPTP (307) |
| TAG-1[a] (82) | | PDGF R (144) | |
| MAG[a] (296, 299) | | MUC 18 (285) | |
| P$_0$[a] (297) | | DCC (286) | |

## Calcium-Dependent CAMs

L-CAM (21) [Uvomorulin (23), E-cadherin (44), cell CAM 120/80 (308), ARC-1 (309)]
N-cadherin (25)/A-CAM[a] (26)
P-cadherin[a] (24)
T-cadherin (45)
V-cadherin (310)

## Others

AMOG (311)
Fasciclin I (231)
cs A (227)
Lymphocyte homing receptors (312)
Neurofascin (313)
Chaoptin (85)
galactosyl transferase (314)
axonin-1 (315)

[a] Molecules with demonstrated adhesion activity.
[b] Multiple names applied to what are likely to be identical proteins or homologs of the same protein in different species are connected by slashes.

dependent family most generally called cadherins (reviewed in 14), and a third group containing molecules with no known structural relation to either of the other two families (Table 1).

## Structure

We give a basic description of one example from each of the first two families of CAMs to provide a ground for discussion of binding and expression of these molecules. The paradigm molecules are the two primary CAMs, N-CAM (neural cell adhesion molecule, the first to be isolated by strict criteria), and L-CAM (liver cell adhesion molecule or uvomorulin/E-cadherin). All known CAMs are large intrinsic cell-surface glycoproteins that are mobile in the plane of the membrane (17, 18). The complete primary structures of

**N-CAM**

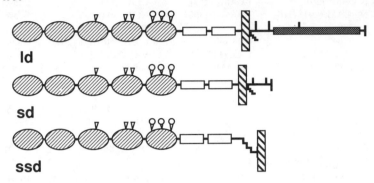

ld

sd

ssd

**Ng-CAM, L1**

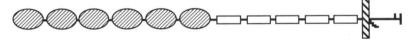

**Nr-CAM**

**Contactin, F11, F3, Tag-1**

**MAG**                                    $P_0$

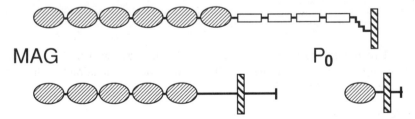

N-CAM and L-CAM have been deduced from sequence analyses of their respective cDNAs in the chicken (19–21) and subsequently in the mouse (22, 23). In both cases, there is a high degree of homology between corresponding CAMs in different species; molecules of the cadherin group exist that are structurally (and, in all likelihood, evolutionarily) related to L-CAM (24–26; reviewed in 14). Diagrams of the two main paradigm molecules are given in Figure 1, along with those of several related CAMs to indicate how various

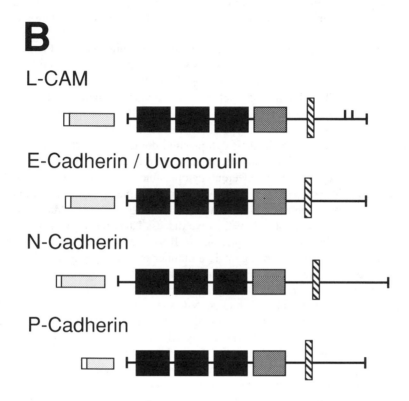

*Figure 1* (Part A on page 158) Structural features of cell adhesion molecules related to N-CAM (A) and L-CAM (B). A, the major forms of N-CAM differ in their mode of attachment to the membrane and the size of their cytoplasmic domains. Their immunoglobulin-like domains (ovals) and fibronectin-like repeats (open rectangles) are otherwise identical. Other closely related neural CAMs resemble N-CAM in that they contain both Ig-like segments and fibronectin-like repeats that are about 20–30% identical to those in N-CAM. Of these, Ng-CAM (291), resembles L1 (292), and Nr-CAM (291) and contactin (293) (or F11, 294) resemble F3 (295) and TAG-1 (82) more closely, than either resembles N-CAM. Other molecules such as myelin-associated gly-coprotein (MAG, 296) and $P_0$ (297) have immunoglobulin-like segments only. The large stairstep symbol denotes a phosphatidylinositol link to the membrane, whereas the small symbol denotes other lipids. Arrowheads on N-CAM indicate N-linked oligosaccharides and open circles polysialic acid. B, calcium-dependent CAMs are all 50% or greater in identity. They contain internal repeats (large shaded rectangles) of ~100 amino acids but do not resemble im-munoglobulin; the first three repeats (dark shading) are most closely related to one another. The number of apparent repeats may vary from three to five depending on the alignments and the criteria chosen for similarity. L-CAM (21) and E-cadherin/uvomorulin (23, 44) are thought to be comparable molecules in chicken and mouse. All molecules have typical signal sequences (thin open rectangles). N-cadherin (24) is slightly larger than the others, whereas P-cadherin (25) has a substantially shorter precursor sequence (stippled thin rectangles). In both A and B, the hatched vertical bar denotes the cell membrane and short vertical lines denote phosphoserine and phosphothreonine.

structural motifs are shared in each family. The main paradigms, N-CAM and L-CAM, are completely different in structure and appear to be evolutionarily unrelated.

N-CAM is expressed as a number of major polypeptide chains (Figure 1), each having five successive homologous domains, the first of which begins close to the N-terminus (20). Each of these five domains has a single intrachain disulfide bond, and each is homologous to domains of the immunoglobulin superfamily (19, 20). Beyond the fifth domain are repeated regions resembling fibronectin type III repeats. Major differences among the polypeptides occur either in their cytoplasmic domains or in their means of attachment to the cell membrane. The ld (large domain) and sd (small domain) polypeptides have different cytoplasmic domains; the ssd (small surface domain) polypeptide has no cytoplasmic segment, but is linked to the cell surface by a phosphatidylinositol intermediate (27–31). Many other minor spliced forms exist (32–39); as we discuss later, the various forms of N-CAM arise by alternative splicing of RNA transcribed from a single N-CAM gene. As shown in Figure 1, a number of other Ig-like CAMs have been discovered; they differ in the number of Ig motifs or fibronectin repeats but are generally similar in organization. Attached to the fifth immunoglobulin-like domain of chicken N-CAM at up to three sites (40) are oligosaccharides containing a carbohydrate that is unusual in vertebrates, $\alpha$-2,8-polysialic acid (41, 42). The embryonic (E) form of N-CAM contains about 30 g of this carbohydrate per 100 g of polypeptide, and the adult (A) form about one-third as much. "Embryonic" and "adult" refer to the statistical predominance of these forms at different ages: molecules having low amounts of polysialic acid can be seen in early embryos and molecules with high amounts can be seen in certain locations in adult brain, skin, and muscle. As we discuss below, polysialic acid does not participate directly in N-CAM binding, but the presence of higher amounts of the carbohydrate does diminish binding rates (43).

L-CAM (Figure 1), which is unrelated to N-CAM in sequence, has three and possibly four regions of internal homology of about 110 amino acids each (21, 23, 44). At least two other molecules, N-cadherin or A-CAM (24, 26) and P-cadherin (25), are structurally related to L-CAM. Several new members of this family have recently been reported (45–47). Aside from these molecules, which apparently constitute an evolutionarily related cadherin group, no homology to other proteins has yet been detected. Unlike N-CAM, there appear to be no disulfide bonds in L-CAM homology regions, and although oligosaccharides are attached at four sites, no unusual carbohydrates have been found. A characteristic property of L-CAM and of all other cadherins isolated to date is a dependence upon $Ca^{2+}$ for both their conformation and their binding; in the absence of this ion, the molecules are rapidly cleaved by proteases.

The three-dimensional structure of CAMs has been examined by electron microscopy of rotary-shadowed CAMs or CAM fragments. Electron microscopic analysis of N-CAM and L-CAM (48) suggests that both molecules have a hinged structure with their binding regions amino-terminal to the hinge. Earlier (49, 50) electron microscopic analyses are consistent with such a structure. A similar hinged structure has recently been reported for the extracellular region of I-CAM-1 (51). These hinged structures may facilitate *trans*-homophilic binding in the presence of cell-shape changes that would otherwise sterically hinder homophilic binding. The effect of $\alpha$-2,8-polysialic acid on N-CAM binding may result from its net negative charge, its large excluded volume, and, given its location near the hinge, alteration of the various hinge angles, which may be required to form multiple homophilic attachments between cells with flexible membranes. Such a structure has also been supported by molecular modelling (52). As discussed above, E forms may predominate when adhesion is being established (such as in the formation of new neural connections), and A forms may predominate when stabilization is required (as in the adult brain).

The existence of multiply spliced N-CAM forms (20, 33, 38, 39, 53, 54) with variations mainly in regions outside the binding regions (to be discussed below) may alter the ways in which these hinged molecules interact with each other. Whatever the case, different CAMs do not bind to each other; this is true even of related molecules such as N- and Ng-CAM or E-, P- and N-cadherins. A given cell has multiple different CAMs on its surface, and obviously these must rearrange in the plane of the membrane to form appropriate specific homophilic attachments. The mobility of N-CAM in the plane of the membrane (17, 18), for example, is consistent with this possibility; other CAMs are likely to be equally mobile, but may be modulated by particular cytoskeletal attachments, as discussed below. Thus, the total complement of different CAMs exhibited by a given cell may influence particular second messengers (55) and subsequent cell behavior.

## Binding

CAM binding is remarkable because it depends not only upon specific binding regions, but also upon various forms of cell-surface modulation (56). The occurrence of modulating mechanisms and cytoskeletal interactions indicates that the binding properties of homophilic CAMs are under direct and immediate control of the cells that they ligate. This is in contrast to certain SAMs, which bind as extracellular molecules to independent receptors (or integrins) on adhering cells; thus, their multimolecular interactions are more complex and their control or their effects are more indirect.

Modulation events (56) include changes in surface density (prevalence modulation), distribution on the cell surface (polarity modulation), and carbohydrate structure or posttranslational modifications (chemical modulation).

All of these forms of cell-surface modulation have been observed for various CAMs and would be expected to change the amount of cell binding or cell responses to local signals by changing the rate or strength of binding, the spacing between bound cells, or the cytoskeletal dynamics affecting signal processing and cell shape.

The binding mechanisms of several CAMs have now been analyzed by kinetic assays or by use of transfection techniques (see below) applied to cells lacking a given CAM. In general, CAM binding is homophilic (CAM on one cell to the same CAM on the opposing cell). This conclusion has been reached by performing assays on N-CAM and Ng-CAM linked to Covaspheres or other solid substrates, or in artificial lipid vesicles (43, 57, 58) and for numerous CAMs by transfection with appropriate cDNAs of cell lines that lack these CAMs. We discuss these experiments in the next section. In addition to mediating homophilic binding between neurons, Ng-CAM binds neurons to astrocytes by a second mechanism, which is presumably heterophilic inasmuch as astrocytes lack Ng-CAM (58–60). Kinetic and binding analyses indicate that the binding mechanisms of N-CAM (43, 57, 61–63), L-CAM (44, 61), and Ng-CAM (58, 59) are specific and independent of each other. N-CAM and other Ig-related homophilic CAMs mediate homophilic binding by mechanisms that are $Ca^{2+}$-independent. Calcium dependence is more complex when the molecules are involved in heterophilic interactions; the interaction of I-CAM with its ligand LFA-1, an integrin, is, for example, $Ca^{2+}$-dependent, while its interaction with rhinovirus is $Ca^{2+}$-independent (51). This leaves open the possibility that levels of extracellular calcium could affect the occurrence of certain molecular interactions or the choice of binding mechanisms of other Ig-like CAMs.

Physicochemical assays (43, 57, 58) that depend upon measurements of rates of aggregation for both N-CAM and Ng-CAM indicate that a twofold increase in surface density can result in a more than 30-fold increase in binding rates. Such nonlinear changes in CAM concentrations (prevalence modulation) could lead to rapid changes in cell interactions in vivo by changing the linkage of a cell collective bound by a given CAM or in forming borders between two such collectives linked by different CAMs (see below for discussion of CAMs in border formation and cell sorting). This is also the case for chemical modulation: although α-2,8-polysialic acid is not directly involved in binding, changes in the amount of this negatively charged sugar (chemical modulation) can result in three- to fourfold changes in binding rates (43). Consistent with these results, in solid phase binding assays (62), binding constants were reported to be approximately 20-fold higher for N-CAM isolated from adult rat brain than that from newborn brain.

Binding to other surface molecules may also be a means by which CAM function is modulated. For example, N-CAM, but not Ng-CAM, copurifies with a heparan sulfate proteoglycan (64) to which it binds in a specific and saturable manner (65) in a 25-kDa amino-terminal domain of N-CAM (66–68). This interaction may modulate N-CAM homophilic binding (68, 69), or may serve as a means of N-CAM binding to the extracellular matrix (70).

Recent studies have suggested that N-CAM enhances the effects of L1 binding (71), and that this cooperation is carbohydrate dependent (72). Such interactions among CAMs affecting their binding or signalling mechanisms, if verified, may help to explain the presence of multiple CAMs on a single cell. The occurrence of such interactions may perhaps be further revealed by transfection studies as discussed below.

Sequences in the cadherin family have been identified that are required for homophilic binding and for molecular specificity among the different cadherins. Using a combination of site-directed mutagenesis and constructs of chimeric cadherin molecules, it has been shown that the binding specificity of cadherins resides primarily in their amino-terminal 113 amino acids (14, 73). Within this sequence is a highly conserved sequence surrounding a tripeptide HAV. Moreover, synthetic peptides containing this sequence were shown to block certain biological functions mediated by E-cadherin and N-cadherin (74).

## Transfection as a Molecular Histological Tool

Another powerful means of exploring the effects of CAM binding in addition to its physicochemical specificity is to show that cells that ordinarily do not make a given CAM will aggregate specifically after being transfected with a cDNA for that CAM. Mouse L-cells, which do not ordinarily express any of these CAMs, were transfected with cDNAs for each of the N-CAM polypeptides (61) and for L-CAM (61, 75) and each of the cadherins (24, 44, 76). In each case, the appropriate CAM was synthesized. Similar experiments using various cell lines have now been performed for N-CAM (61, 77), L-CAM (44, 61, 75, 77), N-cadherin (24, 78), P-cadherin (76), MAG (79), $P_0$ (80), CEA (81), Tag-1 (82), PE-CAM (W. Muller, P. Newman, S. Albelda, C. Buck, in preparation), fasciclin I (83), fasciclin III (84), and chaoptin (85) with striking and clearly interpretable results demonstrating the cell-binding activity and the homophilic or heterophilic nature of such binding.

Binding assays on transfected cells show that the sd and ld chains of N-CAM and L-CAM each at their respective cell surfaces were capable of linking the cells in aggregates and, in the case of N-CAM expression, of binding neuronal vesicles containing N-CAM. In all cases, binding was inhibited by appropriately specific antibody fragments. L-CAM aggregation was particularly striking and provided strong evidence for a homophilic mechanism of binding inasmuch as untransfected L-cells would not bind with transfected cells expressing L-CAM (44, 61).

Transfection is particularly useful in investigating the role of cytoplasmic domains in CAM binding, as well as in exploring the consequences of cell binding on the expression and function of other morphoregulatory molecules, c.g. CJMs. A number of studies (86–92) have shown that cytoplasmic domains of CAMs interact directly or indirectly with proteins of the cytoskeleton. Deletion of as few as 30 amino-acid residues of the C-terminal portion of E-cadherin (88, 89, 93) or L-CAM (75) leads to loss of cell binding despite adequate surface expression of the modified CAM. Recently, it has been shown that E-cadherin and uvomorulin cytoskeletal domains interact with two components of 94 kDa and 102 kDa, named catenins, which appear to interact with the cytoskeleton (88, 89). In other studies, L-CAM was found to copurify with complexes containing fodrin, ankyrin, and the $Na^+,K^+$-ATPase (91). The ld form of N-CAM was shown to interact with brain spectrin (86). Clearly, differential cytoskeletal attachments may provide additional mechanisms for cell-surface modulation.

Besides its usefulness in binding studies, transfection has allowed exploration of how cell collectives sort out (4, 5) from each other in vitro to form boundaries and pseudo-tissues. Sorting into collectives can occur if cells with different amounts of the same CAM are mixed or if cells transfected with two CAMs of different homophilic specificities are mixed (76, 83, 94). This approach has been used to explore the functional consequences of structural differences in CAM cytoplasmic domains. A chimeric CAM was constructed by cDNA manipulation to contain the L-CAM extracellular domains linked to the N-CAM (sd) transmembrane and cytoplasmic domains (75). Cells transfected with the chimera expressed as much CAM on the surface of L-cells as the control L-CAM transfectants; they also bound as well in specific binding assays. Cell binding of the L-CAM transfectants was disturbed by cytochalasin D, an actin filament dissociating drug, whereas binding of the chimeric L-CAM/N-CAM transfectants was unaffected. Reciprocally, nocodazole (a drug that enhances microtubule disassembly) diminished cell binding of the chimeric transfectants with no effect on L-CAM transfectants. Even more striking was the fact that the L-CAM transfectants sorted readily from untransfected L-cells whereas the chimeric transfectants did not. This suggests that there is a definite coupling of different cytoplasmic domains to different cytoskeletal elements to affect higher-order functions such as cell migration and sorting.

The transfection technique has also been applied to reveal the expression of junctional molecules, such as CJMs, after transfecting with CAMs. S180 mouse sarcoma cells (spindle-shaped transformed cells) transfected with L-CAM (77) or with N-cadherin (78) changed their phenotype when the cells became confluent, becoming polygonal and resembling a pseudo-epithelial or epithelioid sheet. This sheet of polygonal cells was linked by adherens junctions and communicating gap junctions (77, 78) with cardiac-type connexin molecules that became phosphorylated in functional junctions (95). Dissociation of the sheet by anti-L-CAM Fab' fragments led to a great diminution in the number of recognizable junctions, loss of coupling, and decrease in the phosphorylation of connexin (77, 78, 95). In general, the cell collectives that formed in mixed cultures of L-CAM- and N-cadherin-transfected cells were linked by a single CAM, and there was no junctional coupling across L-CAM–N-cadherin boundaries (78). These findings suggest that CAM interaction may be required for adequate expression and modification of junction-associated molecules. N-cadherin and uvomorulin both appear to participate in formation of adherens junctions (26, 96–98). A recent study (99) indicates that CAM expression by transfection can induce polar expression of molecules such as the $Na^+$, $K^+$-ATPase, presumably by altering cytoskeletal structures. A similar conclusion was drawn from perturbation studies in the early mouse blastocyst: the disruption of cell interactions by antibodies to uvomorulin severely affected the distribution of the $Na^+,K^+$-ATPase and the appropriate polarization of cells of the blastocyst (100).

Clearly, CAM binding is not simply a matter of homophilic specificity, but depends on and is coupled to various intracellular events related to functions of the cytoskeleton. It is thus a cell-surface modulatory event (56).

## CHROMOSOMAL LOCALIZATION, GENE STRUCTURE, AND ALTERNATIVE RNA SPLICING

So far the known CAMs have each been found to be specified by single genes. N-CAM is specified by a single gene at the *n-cam* locus on chromosome 9 in the chicken (101) and on chromosome 11 in the human (102). The structurally different major polypeptides of N-CAM arise in development at different times and places by alternative splicing of mRNA (39, 53, 54). As shown by a detailed analysis of the chicken N-CAM gene (103), which spans over 80 kb, the three main types of chains arise by splicing of nucleotide sequences corresponding to different combinations of exons. Alternative splicing has now been observed for several Ig-like CAMs, including MAG (104), CEA (105), and neuroglian (106). In particular cases, this event can be tissue specific. For example, the ld chain of N-CAM is expressed only in the nervous system (54, 107) and is enriched in the postsynaptic densities of certain neurons (108), and other tissue-specific variants are seen in tissues such as skeletal and heart muscle (32, 33, 38, 39, 54). Soluble N-CAM forms also exist (35, 36, 38). Such splicing events would be expected to alter the position, mobility, and cytoskeletal interactions of a given CAM chain, and they may also alter the effects of CAM interactions on signaling in the cell cortex. As we mentioned above, more recent studies have revealed a profusion of possible splice sites in the extracellular domains.

cDNA analyses of L-CAM (21) and the cadherins (14, 23) have shown these proteins to be specified by different genes, having close structural relationships and homology. Genomic analysis of L-CAM (109) reveals a smaller (16 kb) genomic structure than N-CAM. No definitive evidence of RNA splicing events has emerged in this family, although a cadherin-like molecule with a PI-linked carboxyl terminal segment has been described (45). Uvomorulin is specified on chromosomes 8 in the mouse (110) and chromosome 16 in the human (111). The N-cadherin gene has been mapped to human chromosome 18, clearly distinct from the uvomorulin gene (112).

Analyses of upstream regulatory segments, and of promotor or enhancer activities, have been reported recently for mouse N-CAM (113), rat N-CAM (114), and human N-CAM (115). Tissue-specific activity of these regions in vitro appears to be controlled by both positive and negative regulatory elements. At least eight domains are present that interact with nuclear proteins. The upstream sequences contain consensus sites for known transcription factors including Spl and NF I factor binding sites and motifs resembling the DNA-binding domains of Antennepedia and Hox 1.3 (113). The upstream sequences of cytotactin (a substrate adhesion molecule, see below) have also been shown to include sequences resembling the binding domains of various homeotic genes (116). The details of the coordinate control of CAMs with homeotic genes in vertebrate species should provide exciting insights into the mechanisms of morphoregulation.

Most recently, it has been shown that transgenic mice produced with the rat insulin promoter or the neurofilament promoter fused to L-CAM genomic DNA resulted in L-CAM expression in certain regions where L-CAM is normally expressed as well as in the known target regions of the promoters (117). It appears likely that a *cis*-regulatory element in the L-CAM genomic sequence was responsible for this retargeting of expression. In this same study, expression of chicken L-CAM in the $\beta$ cells of the pancreas suppressed endogenous mouse L-CAM expression in the same cells.

## EXPRESSION AND SYNTHESIS

Inasmuch as CAMs are expressed in a place-dependent and tissue-specific manner (10, 11), one would expect their synthesis to be differentially responsive to various signals and *trans*-acting elements. Moreover, a number of posttranslational events can modify CAM activity. We briefly discuss the latter events and then take up the few cases where known agents stimulate or alter CAM expression.

*Biochemical Modifications*

We have already discussed the synthesis of N-CAM variants as a result of alternative RNA splicing. While L-CAM shows no such phenomenon, it is synthesized via a precursor that is posttranslationally cleaved to produce the membrane form of the molecule (118, 119).

N-CAM and L-CAM both exist in phosphorylated forms (118, 120). Phosphoserine and phosphothreonine were found in the COOH-terminal third of the N-CAM molecule. Two protein kinases (glycogen synthase kinase 3 and casein kinase I) that phosphorylate chicken N-CAM were isolated from avian and mammalian brain (121), but the sites of phosphorylation differed from those found in N-CAM molecules isolated from tissue. The L1 molecule (see Table 1) was phosphorylated at serine residues by two kinase activities that copurified with it, and one of these was also casein kinase I (122).

More extensive studies have been carried out on the $\alpha$-2,8-polysialic acid of N-CAM, probably prompted by its unusual structure and its indirect role in the modulation of binding. As we discussed above, changes in the amounts of this negatively charged sugar have dramatic effects on binding activity. Treatment of the E form by an endoneuraminidase from phage that cleaves the polysialic acid enhanced binding and led to in vivo abnormalities (123–125), but this is not the means by which the forms are expressed during development. In vivo, the change from the embryonic highly sialylated N-CAM forms to the adult less sialylated forms occurs on a definite schedule in the developing mouse brain (126), and is the result of turnover of E forms and de novo replacement by A forms (127), rather than changes in extracellular neuraminidase activity acting to remove the polysialic acid from the E forms. Moreover, the sialylation state of N-CAM has been shown to be controlled by a developmentally regulated Golgi sialyltransferase (128, 129).

A carbohydrate epitope, HNK-1, is found on some N-CAM molecules, L1, MAG, and various SAMs, including cytotactin and CTB proteoglycan (130–132). Interestingly, certain similar carbohydrate determinants are present in various invertebrate species (84, 133). Except in the case of human natural killer lymphocytes, on which HNK-1 was first detached (134), HNK-1 appears to be present only on the neural forms of the adhesion molecules listed above and not their non-neural counterparts (131). HNK-1 is associated with a sulfated glucuronic acid (135), but its function remains unknown. Sulfation was found in asparagine-linked carbohydrates of N-CAM, but not in sialic acid (120), and could be part of the HNK-1 epitope. It has been suggested that the HNK-1 epitope (136) and another type of carbohydrate unit (137, 138) are involved in cell-cell interactions. The HNK-1 determinant has subsequently been shown not to have a role in adhesion (132, 139).

In general, aside from the modulation of binding by $\alpha$-2,8-polysialic acid, most biochemical modifications of CAMs have yet to be assigned functional roles. A recent striking finding, however, is that C-CAM, a CAM found in liver, is identical to an ecto ATPase, which is the apparent rat homolog of BGP1, a member of the Ig superfamily (140, 141) (Table 1). The intracellular domains of LAR and PDGF-R (Table 1) include regions homologous to protein tyrosine phosphatase (142, 143) and tyrosine kinase (144), respectively. It will be fascinating to see whether other CAMs in the Ig family have enzymatic activities, and whether any such activities are associated with the Ig domains or with regions outside these domains.

## Signals Regulating CAM Synthesis In vitro and In vivo

Although a number of different growth factors, morphogens, tumor promoters, and hormones have been shown to influence synthesis of various CAMs, only one, thyroxine, has been definitively demonstrated to regulate CAM synthesis in tissues in vivo (145).

Nerve growth factor (NGF) stimulates increased production of Ng-CAM/ L1 in PC12 cells (146). This was not a surprise inasmuch as the molecules were shown (146, 147) to be closely related to if not identical to NILE, a glycoprotein first identified because its production was stimulated by NGF in PC12 cells (148). NGF has also been reported to affect N-CAM levels in these cells (149, 150). A study of the effects of intracellular cyclic AMP on the expression of N-CAM in PC12 cells suggested the ability of this compound to modulate N-CAM and L1 synthesis in these cells (151). NGF has also been shown to affect L1 expression in Schwann cells (152).

Several other agents affect N-CAM expression in vitro. The E form of N-CAM was up-regulated by treatment of embryonal carcinoma cell lines with retinoic acid (153). In contrast, rat neuroblastoma cell line B104 transfected with an N-myc oncogene showed a dramatic down-regulation of N-CAM expression (154). More recently, N-CAM expression in 3T3 cells was induced by transforming growth factor $\beta$, and it was tentatively concluded that the growth factor acted by a transcriptional mechanism (155).

Hormone treatment also affects CAM expression. Estradiol has been shown to enhance E-cadherin expression by rat ovarian granulosa cells in vitro (156). Progesterone increases the expression of an unidentified cadherin-like molecule in Xenopus oocytes (157). Steroid hormones affect expression of C-CAM in rat uterus in vivo (158). Thyroid hormones regulate N-CAM expression in adult skeletal muscle in vivo; thyroidectomy leads to increased expression on the muscle surface, which is reversible by thyroxine treatment (159). Perhaps the most dramatic finding concerning the stimulation of N-CAM expression in vivo is related to metamorphosis in Xenopus laevis. Prior to metamorphosis, the major CAM in Xenopus liver is L-CAM; after metamorphosis, a novel

N-CAM of 160 kDa replaces most of the L-CAM (145, 160). Treatment of stage 51 tadpoles with thyroxine converts the liver CAM expression from L-CAM to N-CAM within 48 hours, a period of time too short for the majority of liver cells to be replaced (145). Clearly, the hormone is acting directly or indirectly to regulate CAM expression.

A host of agents have been shown to affect the expression of I-CAM in various cell types. Activation of protein kinase C by phorbol esters increases I-CAM levels in human endothelial cells (161, 162). Exposure of tumor necrosis factor and gamma interferon increases endothelial I-CAM expression near the site where the compounds were injected (162–164). These inductions could be prevented by treatment with 3-deaza-adenosine (165).

## MORPHOREGULATORY ACTIVITY OF CAMs

A large body of data has accumulated on the specific place-dependent expression (10) of different CAMs during embryogenesis and regeneration. In addition, various in vitro and in vivo experiments have shown that interference with CAM binding or synthesis results in perturbation of morphology. It has also been found that interruption of tissue interactions by the perturbation of morphology alters the level and location of CAM expression. Much of this literature has been reviewed elsewhere (8, 11, 12), Here, we only present sufficient examples to make three points related to an understanding of molecular histology:

1. Topobiological (10, 11) interactions, i.e. interactions of cell surfaces with other cell surfaces or cell substrates in a place-dependent manner, result in changes in CAM expression.
2. There is evidence for a coordinated control of the expression of CAMs and SAMs to yield morphological change [we discuss cytotactin (tenascin) as a key example of a SAM involved in such coregulation].
3. Adhesion is not the only means of altering morphology. Evidence is accumulating for molecules the presence of which discourages cell migration ("repulsins"). In part, cytotactin has such a function.

### Place-Dependent Expression and Interactions

Early studies (49, 166, 167) indicated that some CAMs such as N-CAM and L-CAM were primary CAMs, i.e. they appear early in development in tissues of all germ layers but are also used in later histogenesis. An example of a secondary CAM, which appears somewhat later in development mainly in neural tissues, is Ng-CAM (59, 168, 169). Some primary CAMs can be detected either in oocytes (170) or as early as the two-cell stage under control of maternal message (160, 170). Later, the distribution of the primary CAMs changes. N-CAM and L-CAM, for example, are expressed on all cells at early stages, e.g. blastoderm or blastula of chick and frog embryos, respectively (8, 49, 160, 166, 167). Upon formation of mesenchyme to form the middle layers or mesoderm of the embryo, N-CAM is down-regulated (often during cell movement) only to be reexpressed when mesenchymal cells condense to form epithelia. After cell movement and gastrulation take place, a remarkable segregation of CAM expression occurs: N-CAM is seen on cells of the neural plate and reappears on condensing mesoderm, whereas L-CAM is seen alone or appears together with N-CAM in epithelia that will give rise to the skin (ectoderm) or gut (endoderm).

During embryogenesis, transformation from epithelia (polarized sheets of cells linked by CAMs, SAMs, and CJMs) to mesenchyme (collections of loosely associated cells interacting with SAMs of the extracellular matrix and generally having down-regulated CAMs at their surface) occurs. Separate epithelia and mesenchymes can interact with each other or with themselves, resulting in embryonic induction, a complex process of milieu-dependent differentiation (171) that leads to expression of genes for the morphoregulatory molecules (CAMs, SAMs, and CJMs) as well as of those for tissue- and cell-specific molecules.

As noted above, on early epithelia, one observes both N-CAM and L-CAM; subsequently, one or the other is lost. These phenomena are regular enough to allow formulation of a set of place-dependent rules (167). Other CAMs have also been discovered to follow place-dependent expression, a notable example being N-cadherin, which follows a distribution pattern roughly similar to that of N-CAM, beginning at a point just before neural plate formation (12). Often, one may observe in the same developing tissue (e.g. the kidney or the feather) multiple cycles of primary CAM expression in which CAMs are repeatedly up-regulated and down-regulated in this fashion at different stages of histogenesis. Later, secondary CAMs appear at specific sites. Some examples of these topobiological patterns are revealing. Aside from the formation of the neural tube, down-regulation of N-CAM (166) and N-cadherin (12) is seen during migration of neural crest cells, and Ng-CAM is expressed when neurites are extended from neurons, mediating fasciculation of neural tracts. In certain cases, where such tracts change direction, new secondary CAMs such as Ng-CAM and TAG-1 can be modulated in their expression (172). In muscle formation, N-CAM appears on the entire cell surface of myoblasts and then localizes at the neuromuscular end plate when functional synapses are formed (173, 174). In the formation of the gut and lung, N-CAM and L-CAM are seen in epithelia and N-CAM disappears (167). In kidney, a remarkable series of CAM cycles is observed: L-CAM is seen on Wolffian ducts (the inductor tissue) and after induction, N-CAM appears on mesonephric mesenchyme, while L-CAM appears on collecting tubules (175).

An example of place-dependent CAM cycles is seen in the formation of the feather, where N-CAM and L-CAM expression are tightly coordinated with border formation and cell differentiation to produce keratin, and also with cell death to yield the final structure (reviewed elsewhere, 11, 176, 177). Given the complexity of various tissues, these examples could be extended, but the main point is that CAM expression is modulated in a space- and time-dependent fashion to produce cell collectives and borders in a manner that is coordinated with cell differentiation. That such patterns depend on dynamic synthetic events has been shown by an analysis of N-CAM mRNA distribution using in situ hybridization (178, 179).

## Coregulation with SAM Expression

An important principle of molecular histology is that regulation of the expression of CAMs is coordinated with that of the other morphoregulatory molecules. We have already seen that CAM binding is necessary for the expression of certain junctions and their associated molecules in vitro. Given the close interdependency of cell adhesion and migration, it is perhaps not surprising that there is quite a refined place-dependent regulation of the appearance of CAMs and SAMs. A striking example is provided by cytotactin (also known as tenascin) (180, 181, reviewed in 182). Cytotactin is an extracellular matrix protein consisting in the chicken (180, 183) of 190-, 200-, and 220-kDa polypeptides. cDNA studies have shown that it consists of a series of EGF-like repeats, followed by fibronectin type III repeats, and has a fibrinogen-like region at its C-terminus (184–186). It is usually found as a hexabrachion (187), a structure made of six such chains linked by S-S bonds near their amino termini. It is synthesized by glia but not by neurons, by somites, smooth muscle cells, immature chondrocytes, and perichondrial cells, by myotendinous regions (183, 188), and by various other cells (132, 180, 189, 190). It binds a cytotactin-binding (CTB) proteoglycan, which is synthesized by central nervous system neurons and, in various forms, also by non-neural cells elsewhere in the body (131, 132). Cytotactin also binds fibronectin and it has been shown to bind to fibroblasts (132, 191). It also interacts with various other matrix components (192, 193). Cytotactin has been shown to affect cell movement in a fashion that is strikingly different from that of other extracellular matrix proteins such as fibronectin and laminin (194–196). Its most striking functional effect is that it causes cells in culture to round up (197); it also tends to inhibit cellular migration and neurite invasion (198). When mixed with fibronectin, which generally supports such migration, these effects are mitigated (194). There is evidence that cytotactin also modulates adhesive activity. It might be considered an amphitropic molecule, i.e. one that carries out either adhesive or repulsive functions at different sites or in combination with various other extracellular matrix proteins.

Cytotactin gained widespread attention when it was shown that it appears in a series of cephalocaudal waves of expression during development (189). This observation, together with its amphitropic functions, fit the concept of topobiology as well as did the various CAMs. These waves are reflections of the temporal growth gradients related to macromolecular synthesis and are an important part of morphogenesis. The promotor region of the cytotactin gene shows a very rich series of regulatory motifs, including DNA sequences similar to binding regions for the products of homeotic genes, consistent with tight spatiotemporal expression (116). In situ hybridization experiments confirm and extend these observations and reveal tissue sites where different alternatively spliced forms appear (199).

A specific example in which there is spatial and temporal coregulation of CAM and SAM expression is in the formation of the somites, periodic structures which arise in development in a craniocaudal sequence from the so-called segmented plate of mesoderm. After their initial appearance, somites undergo further patterning changes, e.g. to provide tissue for vertebrae, muscle, and pigment cells, and also to allow for periodic accumulations of migrating cells from the neural crest to form ganglia. An examination of somitic mesoderm for N-CAM and N-cadherin during the early stages of somite formation has revealed a dynamic pattern of expression (200) that is correlated with somite segmentation and rearrangement. Antibody fragments to each CAM, but particularly to N-cadherin, cause somites to disperse in organ culture, suggesting that somitic cells are linked by these CAMs.

After somite formation, which appears to be correlated with CAM activity, cytotactin appears in the basement membrane of the somites in a cephalocaudal sequence (189) and is more concentrated in the posterior portion of each somite. Somewhat later, a part of each somite becomes mesenchymal, forming the sclerotome. Neural crest cells that have down-regulated N-CAM (168, 201) enter the anterior region of each sclerotome; they will later express N-CAM and form dorsal root ganglia. At this time of entry or just before, both fibronectin and the cytotactin-binding proteoglycan are present throughout the somite. Cytotactin then appears (194) in just the anterior portion of the nascent sclerotome, and after this patterned appearance, CTB proteoglycan decreases in the anterior half and is seen mainly in the posterior half; fibronectin remains distributed throughout both halves.

The resultant pattern of alternating stripes of the interactive couple, cytotactin and CTB proteoglycan, is correlated with the position of the neural crest cells in the cytotactin-rich anterior half and with their absence in the CTB proteoglycan-rich posterior half. The concurrent finding that cytotactin alters neural crest cell shape and movement (194, 195) is consistent with the hypothesis that, in the anterior region of each somite, cytotactin causes cell-surface modulation of migrating neural crest cells. Clearly, this topo-biological example suggests a complex regulatory sequence of patterned expression of CAMs and SAMs, showing definite correlation with cell function and distribution and with the primary processes of development. Of course, additional molecules may be involved, including others that show restricted distributions within the sclerotome (202, 203).

This work has been challenged (204) with the claims that neural crest cells regulate expression of the cytotactin in the sclerotome and that the timing of expression of the molecule is different from that initially reported. A series of careful experiments (S.-S. Tan et al, in preparation) refute this criticism and demonstrate clearly by in situ hybridization the exact sequence of cytotactin appearance: first in the posterior half and then in the anterior half of the somite as it becomes the sclerotome. The new studies also show that somite cells synthesize cytotactin without the need for regulation by neural crest cells.

From a knowledge of their structure and location, it is clear that CAMs and SAMs cannot be controlled in exactly the same fashion. CAMs are generally homophilic and are controlled by the cells they ligate, whereas adhesion via SAMs involves interactions between cell-surface receptors or integrins (205, 206) and one or more extracellular SAMs. Indeed, extensive analysis of SAM molecules such as fibronectin (207) indicates a multiplicity of cellular binding sites as well as binding sites for other extracellular matrix molecules, which may number as many as 50 different kinds. The presence of a combinatorially variant network of different interactive SAMs (e.g. cytotactin, fibronectin, CTB proteoglycan) in different distributions is highly likely to be significant in the alteration of morphogenetic patterns. Equally important is the kind of coregulation with CAM expression we have just discussed.

## *"Repulsins"*

Another important principle in the molecular regulation of tissue formation is hinted at in some of the effects of cytotactin on cells. Adhesion may not be the only way to regulate cell location. Some molecules may act to discourage cells from extending their lamellipodia (or growth cones in the case of neurons) into a particular area. Cytotactin is an example and has complex amphitropic properties: it can mediate cell binding but also causes modulation and rounding of cells (132, 197) with retraction or collapse of their processes (198). It can thus function, in one of its amphitropic roles, as a "repulsin." Several other structurally unrelated molecules have been reported to collapse growth cones and alter cell attachment and migration in the nervous system (202, 208–210). Thus, a number of different molecular structures may act as repulsins and affect neural morphogenesis by altering neurite extension. A balance between adhesion and repulsion may regulate morphogenesis. With molecules as complex in their interactions as cytotactin, however, this activity is only one of many different activities and it can be altered by the composition of other molecules in the extracellular matrix.

## Perturbation of CAM Binding and Expression

The regulation and coregulation of CAMs and SAMs are dramatically shown by perturbation experiments. Blockade of CAM binding both in vitro and in vivo leads to alterations of morphology. Examples include interference with side-to-side interaction of axonal fibers of neurons by Fab' fragments to Ng-CAM (211, 212), perturbation of layer formation in the retina in organ culture by Fab' fragments of antibodies to N-CAM (213) or antibodies to N-cadherin (214), and alteration of orderly mapping of the retina to the optic tectum in vivo by anti-N-CAM fragments (215, 216). More recently, it has been shown that misexpression of N-CAM (217) or N-cadherin (218, 219) by microinjection of RNA into *Xenopus* embryos disrupts normal morphogenesis. Amphibian gastrulation is perturbed by the microinjection of exogenous cytotactin (tenascin) into the blastocoel (196). Similar perturbations occur in *Drosophila* (83, 220).

A striking example of perturbation that also reveals the role of coregulation is seen in experiments using antibodies to perturb layer formation in the developing cerebellum. In this neural tissue, layers of neurons and their neurites are formed as a result of migration of neurons called external granule cells from the exterior (pial) surface to the interior. This migration takes place on radial glial cells (221) through their interaction with postmitotic neurons in the so-called external granular layer. These neurons send out neurites into the subjacent molecular layer and finally translocate on the radial glia via a leading process to migrate and form the internal granule layer.

Granule cells express both N-CAM and Ng-CAM (169). Addition of Fab' fragments of anti-Ng-CAM to cerebellar tissue slices, in which movements of postmitotic cells marked with radioactive thymidine can be followed in culture, inhibited external granule cell migration into the molecular layer (211, 222). Fab' fragments of anti-cytotactin antibodies had no such effect (222), but instead caused a pileup in the molecular layer of those granule cells that had already entered that layer. Thus, a CAM of neuronal origin played a functional role in cell migration at an early stage (possibly by regulating fasciculation), while a SAM of glial origin played a different role at a later stage in the same morphogenetic process. Inasmuch as each of these molecules appears in a sequence prior to these events, there must be a precise coregulation of signals for their expression as well as coordination of their different binding functions in affecting migration. Furthermore, given the possible heterophilic interaction of Ng-CAM with glia, and of cytotactin with neurally produced CTB proteoglycan, at least four molecules must be involved.

Disruption of CAM binding obviously alters morphology. But consistent with the observation that CAMs are involved via a regulatory loop in controlling tissue structure, two other examples of perturbation actually show that disruption of signaling between two different tissues can affect CAM expression and thus morphogenesis. The first concerns regeneration of cut nerves to reinnervate muscles. As already mentioned, after neuromuscular synaptic contact with myotubes occurs, N-CAM is down-regulated and is barely expressed on the surface of muscle fibers. When the sciatic nerve is crushed or cut (223), N-CAM message (224) and protein (173) reappear in the affected muscle cells and cell surfaces, respectively. Moreover, N-CAM and Ng-CAM expression are up-regulated in the dorsal root ganglion of the affected segment and are down-regulated in the ventral horn of the segment on the affected side. Only after regeneration occurs, more than 200 days later, are the normal patterns restored. In addition, striking patterns of increased expression of N-CAM, Ng-CAM (223), and cytotactin (190) are observed in Schwann cells appearing at the cut or crush site. It is obvious that a long signal loop extending both distally and proximally from the cut site is altered in regeneration, which utilizes the same CAMs and SAMs used in embryogenesis in a different context of control.

Another example is provided by the embryogenesis of the feather, in which perturbations of explanted skin by anti-L-CAM Fab' fragments cause a marked change in the shape of underlying N-CAM-linked dermal condensations (225). The shape change, from blobs to stripes, in these dermal structures could not have been caused directly by the antibodies inasmuch as the condensations do not express L-CAM. The conclusion is that by loosening connectivity in epidermal cells (possibly by altering gap junctions), the anti-L-CAM indirectly affected signals being exchanged between the epidermis and the dermis. This affected primary processes and significant timing events. A computer model showed the self-consistency of such ideas connecting CAM linkage to signals and morphogenesis in the feather (225).

These examples, taken in the context of the transfection experiments and the effects of various signaling factors on CAM synthesis, suggest that maintaining tissue integrity via CAMs and SAMs is a dynamic regulatory process. The establishment and maintenance of anatomy are in large measure the result of regulatory biochemistry.

## EVOLUTIONARY RELATIONSHIPS

We have already touched on the fact that CAMs and SAMs share common protein structural motifs. Two additional fundamental questions concern the origins of the earliest precursors of CAMs in each taxonomic group and the relationships between CAMs and related molecules appearing later in evolution. We can say very little about the very earliest origins of CAMs; data are lacking and evolutionary divergence makes judgment difficult. It is clear, however, that other kinds of adhesion systems have evolved, for example, in *Dictyostelium discoideum* (226–228) and in sponges (229). These are not necessarily simpler because they occur in earlier precursors.

Following upon the work on CAMs pioneered in avian and mammalian systems, explorations of CAMs in insects, particularly *Drosophila*, have revealed molecules similar to N-CAM and L-CAM (220, 230; C. S. Goodman, unpublished observations) as well as several others that are different in structure (85). These include fasciclins I (231), II (230), and III (232), and neuroglian (220). Moreover, the same principles of CAM modulation have been confirmed by using mutational deletions in *Drosophila*. The interesting result is that deletion of one CAM gene had little effect on morphology. Deletion of either of two neurally important CAMs led to some distortion of neural commissure formation in *Drosophila* embryos (83, 220), although null mutants of neuroglian were lethal later at the early larvae stages (220). Only when a mutation was also introduced in a gene for Abelson tyrosine kinase along with a mutation for fasciclin I were major defects seen in commissural pathways (83). The significance of this result is not yet obvious. One possibility is that, because cells contain multiple CAMs, deletion of one leads to compensatory changes in others. Moreover, the control loops mediating such change may involve regulatory pathways for kinases. Obviously, given what we have said earlier about multiple CAM expression on a single cell, the combinatorial effects of switching CAMs on their expression and function deserve close attention and exploration.

Although much remains to be done in the comparative biochemistry of CAMs in descendants of very early precursor species, rather extraordinary insights into evolutionary opportunism have been provided by structural analysis of N-CAM and its relatives. The first complete cDNA sequence for a CAM was that for N-CAM (20), and it showed up to 23% homology to immunoglobulin V and C regions. This led to the hypothesis (16) that a precursor for the N-CAM gene led not only to the emergence of other Ig-related CAMs of the nervous system but also to the entire Ig superfamily. The finding of immunoglobulin homologies in *Drosophila* (220, 230, 233), a species without an adaptive immune system involving immunoglobulin molecules, strongly confirms the hypothesis, which has been reviewed in detail elsewhere (10, 11, 16). Moreover, N-CAM-like molecules that are down-regulated on neurons, before sprouting events that are induced by the stimulation that leads to habituation, have recently been described in *Aplysia califor-*

*nica* (234). This result confirms the proposed notion of modulation by signal loops and also extends the evolutionary family of N-CAM-related structures.

Several intracellular molecules have been described that are members of the immunoglobulin superfamily. These include a number of muscle proteins, some of which also exhibit kinase activity and many of which may bind myosin, including C-protein (235), titin (236), myosin light chain kinase, 86-kDa protein, and twitchin (237). A number of these proteins also contain motifs homologous to fibronectin type III repeats. Twitchin may be of particular evolutionary significance since it was discovered in an invertebrate species, *Caenorhabditis elegans*.

It is obvious that the gene for an N-CAM-like cell adhesion molecule precursor provided the basis for the evolution of a wide variety of molecular types with diverse functions, including antigen recognition and restriction, cell-surface binding of Igs and growth factors, and, of course, cell adhesion itself. So far, although cadherin-like molecules have been found in insects (C. S. Goodman, unpublished observations; 238), no such florid connection of cadherin structure with other molecular families has yet been found. It is worth noting the occurrence of CAMs unrelated to the two main groups (Table 1); these deserve closer scrutiny for their possible evolutionary relatives.

It is probably not an accident that members of each of the main CAM families are expressed together at some time in the life of embryonic cells and that a given cell usually expresses several CAMs. The analysis of this phenomenon across species lines and the possible divergence of detailed function of members of each family deserve greater attention in connection with molecular histology. In one study, it was found that N- and L-CAM appear across similar tissue borders in similar patterns in the development of *Xenopus laevis* and chicken (160). But the relative times of appearance clearly differed in the two species, suggesting the possibility that mutation in regulatory elements for a given CAM might lead to molecular heterochrony (alteration in timing of expression with respect to other molecules) and thus to significant morphologic change in evolution (10, 239).

The origin of the entire Ig superfamily from an early N-CAM-like gene precursor has deep implications for the understanding of the role of adhesion in processes that are not concerned with morphogenesis but rather with immune defense, inflammation, and repair. After the discovery of N-CAM, an exploration of the role of adhesion in lymphocyte helper function, killing function, and homing responses led to the isolation of several interesting molecules (reviewed in 240). The most germane to our present interests are I-CAM-1 (intercellular CAM-1) (241–245) and I-CAM-2 (246). I-CAMs have a key function in mediating adhesion necessary for helper and killer functions of lymphocytes along with other members of the Ig superfamily. Although they are most closely related to N-CAM and are found on a variety of different tissues, they do not bind homophilically. Instead, they bind to LFA-1, a receptor on certain lymphocytes that has the structure of an integrin (i.e. a receptor for fibronectin or other SAMs). Thus, the ligand for I-CAM resembles more closely those for SAMs. This opens a pathway for exploring whether morphologically significant interactions may also be mediated in this way. An interesting finding is that I-CAM-1 is a receptor for rhinovirus (246–248), and that the receptor for poliovirus is also a member of the immunoglobulin superfamily (249). The possibility that other viruses and microorganisms (250) exploit CAMs as receptors deserves further exploration.

Finally, structural studies and those showing functional dependencies of CAMs with the other morphoregulatory families, SAMs and CJMs, may be relevant to various functional and evolutionary relationships in these families. The presence of fibronectin-like domains in members of the N-CAM group indicates that precursors of CAMs and of SAMs underwent gene shuffling events to yield current molecular types. More recently, it has been shown (251) that desmoglein, a CJM and a key transmembrane molecule of the desmosome, is homologous to N-cadherin. Whether these varied relationships have a central evolutionary basis in early cell-cell interactions is not known. But the evidence supports a rich set of structural and functional dependencies amongst the three families of morphoregulatory molecules.

The results of this brief survey of CAMs and other molecules mentioned in regard to the evolutionary connections to other morphoregulatory molecules suggest that additional important evolutionary relationships will be revealed by further structural and genetic analysis. Obviously, the origins of molecular histology must be understood in terms of such relationships.

## CAMs IN DISEASE

Given the role of CAMs in immune defense and the results of perturbation experiments reviewed above, it is not surprising that the onset of disease would alter CAM expression, tissue pattern, and function. We have already mentioned trauma (nerve-muscle regeneration), but other processes involved in bone repair and degenerative disease also result in altered CAM expression and function. A variety of conditions, differing in both etiology and pathogenesis, involve CAMs. The importance for a molecular histology and the implications for pathology are obvious. A brief summary of some key findings is thus in order. A more complete description and further references may be found in (252). Given its widespread changes during the immune response (240), we will not consider I-CAM in disease.

## Neuromuscular and Neurological Disorders

CAM-expression patterns are changed in a variety of neuromuscular disorders. The changes appear to parallel those seen in regenerating muscles discussed above. In a variety of muscle diseases, where hematoxylin-eosin staining reveals regenerating fibers, N-CAM reactivity is observed (253–256). In diseased regenerating myofibers, N-CAM is present all along the muscle fibers and in the cytoplasm, thus resembling the distribution in experimentally induced regeneration.

In dysmyelinating mutants of mice, CAM expression patterns are disturbed in a fashion concordant with the known aberrations in Schwann cell-neuron interactions (257). For example, in the *trembler* mutant, in which Schwann cells are defective and no nodes of Ranvier are observed, N-CAM and Ng-CAM remain diffusely distributed along the nerve fiber; cytotactin synthesized by Schwann cells (which normally becomes localized to the region of the node), is intensely expressed in Schwann cell bodies. In contrast, in motor end-plate disease *(med/med)* mutants, N-CAM and Ng-CAM are localized in the widened nodes characteristic of this disease state. Cytotactin is expressed in the Schwann cells in *med/med*, but does not become localized to the nodal region. In the *quaking* mutant mouse, decreased levels of the ld form of N-CAM are observed (258).

Various alterations of N-CAM levels have been seen in degenerative disease, developmental defects, and toxic conditions. Increases in the levels of N-CAM in the cerebrospinal fluid of patients with multiple sclerosis have been observed to parallel their clinical improvement (259). Levels of N-CAM were reported to be elevated in the amniotic fluid of mothers carrying fetuses with neural tube defects (260). Since many such defects are likely to be due to mechanical aberrations rather than genetic defects, confirmation of these results would provide a new diagnostic component for prenatal testing. Another provocative finding relates to observations on the stimulation of Golgi sialyltransferases by lead (129, 261). Exposure to lead chloride markedly stimulated sialyltransferase activity from postnatal days 16 to 30 in rat. This time is coincident with the period when N-CAM normally becomes less sialylated. Thus exposure to lead at critical developmental periods would presumably lead to more highly sialylated, less adhesive, forms of N-CAM; this prevention of E-A conversion could have significant effects on neural development. E-A conversion itself has been found to be delayed in the mouse mutant *staggerer* (262) in conjunction with the connectivity changes associated with the mutation.

## Transformed Cells, Tumors, and Metastasis

The down-regulation of N-CAM in migrating cells and mesenchyme led to the hypothesis (263) that CAM expression might also be altered in transformed cells. To address this possibility, neural retinal cells were transformed with temperature-sensitive mutants of Rous sarcoma virus. These cells normally express N-CAM and aggregate by an N-CAM-dependent mechanism. Transformants behave similarly at the nonpermissive temperature for the virus. At the permissive temperature, however, cells fail to aggregate to the same extent and have decreased amounts of N-CAM mRNA and surface N-CAM. This change is reversed when cells are returned to the nonpermissive temperature. These early results suggested that CAM expression might be inversely correlated with the transformation state of certain cells and possibly with their metastatic potential.

Similar results have recently been reported for a calcium-dependent CAM, uvomorulin (264). MDCK cells, an epithelial line normally expressing high levels of uvomorulin, were transformed with either Harvey or Moloney sarcoma viruses. Transformed cells were selected that retained both their epithelial character and high levels of surface uvomorulin, or that had lost both of these characteristics. The ability of these cells to invade collagen gels or chick heart explants was measured and found to correlate with expression of uvomorulin, i.e. cells that retained uvomorulin expression were noninvasive. Conversely, untransformed MDCK cells could be made invasive by treatment with antibodies that block uvomorulin-mediated adhesion.

Although loss of CAM expression in cell lines can be correlated with decreased aggregation, increased motility, and increased invasiveness, this correlation appears not to apply to all CAMs, since accumulating observations indicate that different CAMs are expressed differently in a variety of tumors. Decreases in CAMs may only function in certain instances and, in certain invasive tumors, an increase in CAM at the cell surface might actually enhance metastasis and survival. N-CAM expression has been observed in Wilms tumor (265–269), small cell lung carcinomas (270–274), Ewing's sarcoma (275–277), and human neuroblastoma (270, 272, 277). With the exception of Ewing's sarcoma, N-CAM is present in all of these tumors in its embryonic form containing high levels of $\alpha$-2,8-linked polysialic acid, which, as discussed above, is less adhesive than the adult form (43). This observation may be relevant to the metastatic potential of these tumors. It may also be relevant that N-CAM is expressed in tumors that are derived from tissues such as kidney and lung, which transiently express N-CAM in their normal developmental program (49, 166, 167, 175).

The expression of calcium-dependent CAMs has also been examined in two tumor types in vivo (278, 279). Expression of uvomorulin (L-CAM or E-cadherin) has been shown to be inversely correlated with metastatic potential in sublines of the murine ovarian tumor line OV2944 (278). In lung carcinomas, both E- and P-cadherins are present although at varying levels and regions, and their expression parallels that of normal tissues. P-cadherin

is absent from squamous cell carcinomas (resembling upper layers of strati-
fied epithelia) and is expressed only at low levels in well-differentiated
adenocarcinomas (resembling simple epithelia). The cadherins were also
detected at similar levels in lymph node metastases of these tumors. The
involvement of cadherins in metastasis requires further study to reconcile the
in vivo and in vitro observations.

Several molecules identified in connection with tumor formation have been
found to be homologous to N-CAM (see Table 1). These include the
carcinoembryonic antigen (CEA) (280–283), BGP1 (284), MUC-18, a melanoma antigen (285), and a gene deleted in colon carcinoma (DCC) (286).

CEA is present in embryonic and transformed colonic epithelia in a different pattern from that of normal adult colon, and it has been definitively shown
to be a homophilic CAM by transfection techniques (81). The localization of
CEA in normal adult colonic epithelium is restricted to the apical cell surface.
In contrast, in embryonic or transformed epithelium, the protein is expressed
on the entire surface of the cell. The presence of MUC-18 in normal adult
tissues is limited to vascular smooth muscle (287). The presence of this
protein on the surface of melanoma cells is positively correlated with metastatic potential and poor prognosis. It has been postulated (286) that the
function of the DCC gene product is as a tumor suppressor, the loss of which
forms the final step in a multi-step cascade of carcinogenesis and metastasis
(288). It will be important to see whether such correlations with tumor
suppressors are general.

It is clear that the appearance of CAMs in tumors may be disturbed in terms
of amount and location and that these changes may have diagnostic potential
in terms of disturbance of molecular histology. It is equally clear, however,
that CAM levels are not necessarily prognostic of metastatic potential in any
obvious way and that the mechanism must be explored in each independent
tumor. Nonetheless, the role of CAMs in connection with and deviation from
the normal molecular histology is revealing and important in understanding
pathogenesis.

# PROSPECTS FOR A MOLECULAR HISTOLOGY

We feel that enough data have been accumulated on morphoregulatory molecules to suggest the feasibility of correlating the biochemical controls and
interactions of their functions with the three-dimensional structures of normal
tissues, both as a function of time and of cellular differentiation state. This
arena of study, molecular histology, obviously connects with molecular
embryology (10) on one hand and with adult physiology and pathology on the
other. We hope that at least the outline of some biochemical principles
applicable to molecular histology has emerged from this admittedly selective

review of CAM biochemical structures and function. We summarize them here, knowing that our list is partial and that future studies will modify some of the conclusions.

1. Morphoregulatory molecules—CAMs, SAMs, and CJMs—form different families with occasional evolutionary relationships and coordinated functions in tissues. All are involved in dynamic regulatory loops. CAM function and regulation are intimately tied to initial boundary formation in tissues, embryonic induction and migration, tissue stabilization, and regeneration. The known functions of SAMs are related to cell migration, stabilization of epithelia, and the development of hard tissues. The known functions of CJMs are formation of specialized cell connections, cell communication (as seen in gap junctions), and the sealing of the surfaces of epithelial sheets (see 6, 15 for extensive discussions by various authors). These structural and functional distinctions must be made clear in order to avoid confusion about the criteria for adhesion at different times and in different circumstances, particularly because cells may "adhere" by CAMs, SAMs, or CJMs in any combination.

2. Certain spatiotemporal sequences of expression of morphoregulatory molecules are characteristic for a given normal tissue in development, and certain topobiological patterns are characteristic of the normal adult state. Disruption of such patterns may be indicative of disease and may in some cases be diagnostic.

3. In vitro experiments employing transfection may allow determination of the dependencies of CJM function on CAM action, of the role of combinatorial expression of different CAMs on the same cell in border formation, and of the influence of CAM expression and SAM modulation on cell shape.

4. Analysis of *cis* and *trans* regulation of CAM expression should allow dissection of their place-dependent sequences of expression during tissue formation. The correlation with the expression of homeotic genes (289, 290) and specific historegulatory genes (10) is extremely important and should give further insight into differentiation patterns. At the same time, the effect of CAM binding on the biochemical and enzymatic functions (kinases, phosphatases, etc) of cells forming tissues should help close the loop in deciphering the alterations induced by linking cells together in particular tissue patterns.

5. A combination of biochemical, molecular biological, immunological, and microscopic techniques may make it possible to map the dynamic expression of such molecules (and particularly CAMs) stereologically onto known three-dimensional histologic structure. While this goal is still far off, its achievement would have great significance for understanding morphologic evolution, embryogenesis, and the relation of tissue structure to physiological function. Its practical achievement would enhance our capabilities of medical diagnosis to the same degree as biochemical serum analysis, dissociative cytology, and molecular biology have already done.

Perhaps the most well-developed instance of molecular histology is the immune system. This system consists of a morphology in solid tissue to which CAMs, SAMs, and CJMs specifically contribute: the lymphatics, lymph nodes, spleen, and thymus. At the same time, however, in functioning as a fluid tissue, the immune system exploits adhesive interactions and modifications of precursor genes for adhesion molecules in order to carry out the most sophisticated recognition functions that are not morphogenetic. Much remains to be worked out even in this well-explored case.

It is possible to envision a time when the regulation and function of tissue architecture can be understood in terms of a remarkable set of biochemical control loops. These would range from the action of morphoregulatory molecules up to the physiological regulation of organs and organ systems and back again. While this picture is just beginning to be filled in, it is already clear that the anatomy that underlies function is a product of regulatory biochemistry acting simultaneously over many parallel scales of tissue organization. CAMs are extremely important in the establishment and maintenance of all levels of that organization.

ACKNOWLEDGMENTS

We thank the following for communicating results prior to publication: W. W. Franke, R. Heimark, C. S. Goodman, E. Kandel, W. Muller, B. Ranscht, L. Reichardt, M. Takeichi.

*Literature Cited*

1. Brackenbury, R., Thiery, J.-P., Rutishauser, U., Edelman, G. M. 1977. *J. Biol. Chem.* 252:6835–40
2. Thiery, J.-P., Brackenbury, R., Rutishauser, U., Edelman, G. M. 1977. *J. Biol. Chem.* 252:6841–45
3. Takeichi, M. 1977. *J. Cell Biol.* 75:464–74
4. Holtfreter, J. 1948. *Symp. Soc. Exp. Biol.* 11:17
5. Holtfreter, J. 1948. *Ann. NY Acad. Sci.* 49:709–60
6. Edelman, G. M., Thiery, J.-P., eds. 1985. *The Cell in Contact—Adhesions and Junctions as Morphogenetic Agents.* New York: Wiley
7. Edelman, G. M. 1985. *Annu. Rev. Biochem.* 54:135–69
8. Edelman, G. M. 1986. *Annu. Rev. Cell. Biol.* 2:81–116
9. Öbrink, B. 1986. *Exp. Cell Res.* 163:1–21
10. Edelman, G. M. 1988. *Topobiology: An Introduction to Molecular Embryology.* New York: Basic
11. Edelman, G. M. 1988. *Biochemistry* 27:3533–43
12. Takeichi, M. 1988. *Development* 102:639–55
13. Jessell, T. M. 1988. *Neuron* 1:3–13
14. Takeichi, M. 1990. *Annu. Rev. Biochem.* 59:237–52
15. Edelman, G. M., Cunningham, B. A., Thiery, J.-P., eds. 1990. *Morphoregulatory Molecules.* New York: Wiley
16. Edelman, G. M. 1987. *Immunol. Rev.* 100:11–45
17. Gall, W. E., Edelman, G. M. 1981. *Science* 213:903–5
18. Pollerberg, G. E., Schachner, M., Davoust, J. 1986. *Nature* 324:462–65
19. Hemperly, J. J., Edelman, G. M., Cunningham, B. A. 1986. *Proc. Natl. Acad. Sci. USA* 83:9822–26
20. Cunningham, B. A., Hemperly, J. J., Murray, B. A., Prediger, E. A., Brackenbury, R., Edelman, G. M. 1987. *Science* 236:799–806
21. Gallin, W. J., Sorkin, B. C., Edelman,

G. M., Cunningham, B. A. 1987. *Proc. Natl. Acad. Sci. USA* 84:2808–12
22. Barthels, D., Santoni, M.-J., Wille, W., Ruppert, C., Chaix, J.-C., et al. 1987. *EMBO J.* 6:907–14
23. Ringwald, M., Schuh, R., Vestweber, D., Eistetter, H., Lohspeich, F., et al. 1987. *EMBO J.* 6:3647–53
24. Hatta, K., Nose, A., Nagafuchi, M., Takeichi, M. 1988. *J. Cell Biol.* 106:873–81
25. Nose, A., Nagafuchi, A., Takeichi, M. 1987. *EMBO J.* 6:3655–61
26. Volk, T., Geiger, B. 1984. *EMBO J.* 3:2249–60
27. Nybroe, O., Albrechtsen, M., Dahlin, J., Linnemann, D., Lyles, J. M., et al. 1985. *J. Cell Biol.* 101:2310–15
28. Hemperly, J. J., Murray, B. A., Edelman, G. M., Cunningham, B. A. 1986. *Proc. Natl. Acad. Sci. USA* 83:3037–41
29. He, H. T., Barbet, J., Chaix, J. C., Goridis, C. 1986. *EMBO J.* 5:2489–94
30. Sadoul, K., Meyer, A., Low, M., Schachner, M. 1986. *Neurosci. Lett.* 72:341–46
31. He, H.-T., Finne, J., Goridis, C. 1987. *J. Cell Biol.* 105:2489–500
32. Dickson, G., Gower, H. J., Barton, C. H., Prentice, H. M., Elsom, V. L., et al. 1987. *Cell* 50:1119–30
33. Prediger, E. A., Hoffman, S., Edelman, G. M., Cunningham, B. A. 1988. *Proc. Natl. Acad. Sci. USA* 85:9616–20
34. Small, S. J., Haines, S. L., Akeson, R. A. 1988. *Neuron* 1:1007–17
35. Nybroe, O., Linnemann, D., Bock, E. 1989. *J. Neurochem.* 53:1372–78
36. Dalseg, A. M., Linnemann, D., Bock, E. 1989. *Int. J. Dev. Neurosci.* 7:209–17
37. Pizzey, J. A., Rowett, L. H., Barton, C. H., Dickson, G., Walsh, F. S. 1989. *J. Cell Biol.* 109:3465–76
38. Thompson, J., Dickson, G., Moore, S. E., Gower, H. J., Putt, W., et al. 1989. *Genes Dev.* 3:348–57
39. Santoni, M. J., Barthels, D., Vopper, G., Boned, A., Goridis, C., Wille, W. 1989. *EMBO J.* 8:385–92
40. Crossin, K. L., Edelman, G. M., Cunningham, B. A. 1984. *J. Cell Biol.* 99:1848–55
41. Rothbard, J. B., Brackenbury, R., Cunningham, B. A., Edelman, G. M. 1982. *J. Biol. Chem.* 257:11064–69
42. Finne, J., Finne, U., Deagostini-Bazin, H., Goridis, C. 1983. *Biochem. Biophys. Res. Commun.* 112:482–87
43. Hoffman, S., Edelman, G. M. 1983. *Proc. Natl. Acad. Sci. USA* 80:5762–66
44. Nagafuchi, A., Shirayoshi, Y., Okaza-

ki, K., Yasuda, K., Takeichi, M. 1987. *Nature* 329:341–43
45. Ranscht, B., Dours, M. T. 1989. *Soc. Neurosci. Abstr.* 15:959 (Abstr.)
46. Reichardt, L. 1991. *Cold Spring Harbor Symp. Quant. Biol.* 55: In press
47. Takeichi, M., Inuzuka, H., Shimamura, K., Fujimori, T., Nagafuchi, A. 1991. *Cold Spring Harbor Symp. Quant. Biol.* 55: In press
48. Becker, J. W., Erickson, H. P., Hoffman, S., Cunningham, B. A., Edelman, G. M. 1989. *Proc. Natl. Acad. Sci. USA* 86:1088–92
49. Edelman, G. M., Gallin, W. J., Delouvée, A., Cunningham, B. A., Thiery, J.-P. 1983. *Proc. Natl. Acad. Sci. USA* 80:4384–88
50. Hall, A. K., Rutishauser, U. 1987. *J. Cell Biol.* 104:1579–86
51. Staunton, D. E., Dustin, M. L., Erickson, H. P., Springer, T. A. 1990. *Cell* 61:243–54
52. Santoni, M. J., Goridis, C., Fontecilla-Camps, J. C. 1988. *J. Neurosci. Res.* 20:304–10
53. Murray, B. A., Hemperly, J. J., Prediger, E. A., Edelman, G. M., Cunningham, B. A. 1986. *J. Cell Biol.* 102:189–93
54. Murray, B. A., Owens, G. C., Prediger, E. A., Crossin, K. L., Cunningham, B. A., Edelman, G. M. 1986. *J. Cell Biol.* 103:1431–39
55. Schuch, U., Lohse, M. J., Schachner, M. 1989. *Neuron* 3:13–20
56. Edelman, G. M. 1976. *Science* 192:218–26
57. Sadoul, R., Hirn, M., Deagostini-Bazin, H., Rougon, G., Goridis, C. 1983. *Nature* 304:347–49
58. Grumet, M., Edelman, G. M. 1988. *J. Cell Biol.* 106:487–503
59. Grumet, M., Edelman, G. M. 1984. *J. Cell Biol.* 98:1746–56
60. Grumet, M., Hoffman, S., Edelman, G. M. 1984. *Proc. Natl. Acad. Sci. USA* 81:267–71
61. Edelman, G. M., Murray, B. A., Mege, R.-M., Cunningham, B. A., Gallin, W. J. 1987. *Proc. Natl. Acad. Sci. USA* 84:8502–6
62. Moran, N., Bock, E. 1988. *FEBS Lett.* 242:121–24
63. Hall, A. K., Nelson, R., Rutishauser, U. 1990. *J. Cell Biol.* 110:817–24
64. Cole, G. J., Burg, M. 1989. *Exp. Cell Res.* 182:44–60
65. Nybroe, O., Moran, N., Bock, E. 1989. *J. Neurochem.* 52:1947–49
66. Cole, G. J., Loewy, A., Glaser, L. 1986. *Nature* 320:445–47

67. Cole, G. J., Loewy, A., Cross, N. V., Akeson, R., Glaser, L. 1986. *J. Cell Biol.* 103:1739–44
68. Cole, G. J., Glaser, L. 1986. *J. Cell Biol.* 102:403–12
69. Reyes, A. A., Akeson, R., Brezina, L., Cole, G. J. 1990. *Cell Regul.* 1:567–76
70. Probstmeier, R., Kühn, D., Schachner, M. 1989. *J. Neurochem.* 53:1794–801
71. Kadmon, G., Kowitz, A., Altevogt, P., Schachner, M. 1990. *J. Cell Biol.* 110:193–208
72. Kadmon, G., Kowitz, A., Altevogt, P., Schachner, M. 1990. *J. Cell Biol.* 110:209–18
73. Nose, A., Tsuji, K., Takeichi, M. 1990. *Cell* 61:147–55
74. Blaschuk, O. W., Pouliot, Y., Holland, P. C. 1990. *J. Mol. Biol.* 211:679–82
75. Jaffe, S. H., Friedlander, D. R., Matsuzaki, F., Crossin, K. L., Cunningham, B. A., Edelman, G. M. 1990. *Proc. Natl. Acad. Sci. USA* 87:3589–93
76. Nose, A., Nagafuchi, A., Takeichi, M. 1988. *Cell* 54:993–1001
77. Mege, R.-M., Matsuzaki, F., Gallin, W. J., Goldberg, J. I., Cunningham, B. A., Edelman, G. M. 1988. *Proc. Natl. Acad. Sci. USA* 85:7274–78
78. Matsuzaki, F., Mege, R.-M., Jaffe, S. H., Friedlander, D. R., Gallin, W. J., et al. 1990. *J. Cell Biol.* 110:1239–52
79. Johnson, P. W., Abramow-Newerly, W., Seilheimer, B., Sadoul, R., Tropak, M. B., et al. 1989. *Neuron* 3:377–85
80. Filbin, M. T., Walsh, F. S., Trapp, B. D., Pizzey, J. A., Tennekoon, G. I. 1990. *Nature* 344:871–72
81. Benchimol, S., Fuks, A., Jothy, S., Beauchemin, N., Shirota, K., Stanners, C. P. 1989. *Cell* 57:327–34
82. Furley, A. J., Morton, S. B., Manalo, D., Karagogeos, D., Dodd, J., Jessel, T. M. 1990. *Cell* 61:157–70
83. Elkins, T., Zinn, K., McAllister, L., Hoffman, F. M., Goodman, C. S. 1990. *Cell* 60:565–75
84. Snow, P. M., Bieber, A. J., Goodman, C. S. 1989. *Cell* 59:313–23
85. Krantz, D. E., Zipursky, S. L. 1990. *EMBO J.* 9:1969–77
86. Pollerberg, G. E., Burridge, K., Krebs, K. E., Goodman, S. R., Schachner, M. 1987. *Cell Tissue Res.* 250:227–36
87. Hirano, S., Nose, A., Hatta, K., Kawakami, A., Takeichi, M. 1987. *J. Cell Biol.* 105:2501–10
88. Ozawa, M., Baribault, H., Kemler, R. 1989. *EMBO J.* 8:1711–17
89. Nagafuchi, A., Takeichi, M. 1989. *Cell Regul.* 1:37–44
90. Trapp, B. D., Andrews, S. B., Wong, A., O'Connell, M., Griffin, J. W. 1989. *J. Neurocytol.* 18:47–60
91. Nelson, W. J., Shore, E. M., Wang, A. Z., Hammerton, R. W. 1990. *J. Cell Biol.* 110:349–57
92. Ozawa, M., Ringwald, M., Kemler, R. 1990. *Proc. Natl. Acad. Sci. USA* 87:4246–50
93. Nagafuchi, A., Takeichi, M. 1988. *EMBO J.* 7:3679–94
94. Friedlander, D. R., Mege, R.-M., Cunningham, B. A., Edelman, G. M. 1989. *Proc. Natl. Acad. Sci. USA* 86:7043–47
95. Musil, S. M., Cunningham, B. A., Edelman, G. M., Goodenough, D. A. 1990. *J. Cell Biol.* 111:2077–88
96. Boller, K., Vestweber, D., Kemler, R. 1985. *J. Cell Biol.* 100:327–32
97. Volk, T., Geiger, B. 1986. *J. Cell Biol.* 103:1441–50
98. Volk, T., Geiger, B. 1986. *J. Cell Biol.* 103:1451–64
99. McNeil, H., Ozawa, M., Kemler, R., Nelson, W. J. 1990. *Cell* 62:309–16
100. Watson, A. J., Damsky, C. H., Kidder, G. M. 1990. *Dev. Biol.* 141:104–14
101. D'Eustachio, P., Owens, G. C., Edelman, G. M., Cunningham, B. A. 1985. *Proc. Natl. Acad. Sci. USA* 82:7631–35
102. Nguyen, C., Mattei, M. G., Goridis, C., Mattei, J. F., Jordan, B. R. 1985. *Cytogenet. Cell Genet.* 40:713
103. Owens, G. C., Edelman, G. M., Cunningham, B. A. 1987. *Proc. Natl. Acad. Sci. USA* 84:294–98
104. Lai, C., Brow, M. A., Nave, K.-A., Noronha, A. B., Quarles, R. H., et al. 1987. *Proc. Natl. Acad. Sci. USA* 84:4337–41
105. Barnett, T. R., Kretschmer, A., Austen, D. A., Goebel, S. J., Hart, J. T., et al. 1989. *J. Cell Biol.* 108:267–76
106. Hortsch, M., Bieber, A. J., Patel, N. H., Goodman, C. S. 1990. *Neuron* 4:697–709
107. Pollerberg, E. G., Sadoul, R., Goridis, C., Schachner, M. 1985. *J. Cell Biol.* 101:1921–29
108. Persohn, E., Pollerberg, G. E., Schachner, M. 1989. *J. Comp. Neurol.* 288:92–100
109. Sorkin, B. C., Hemperly, J. J., Edelman, G. M., Cunningham, B. A. 1988. *Proc. Natl. Acad. Sci. USA* 85:7617–21
110. Eistetter, H. R., Adolph, S., Ringwald, M., Simon-Chazottes, D., Schuh, R., et al. 1988. *Proc. Natl. Acad. Sci. USA* 85:3489–93
111. Mansouri, A., Spurr, N., Goodfellow, P. N., Kemler, R. 1988. *Differentiation* 38:67–71

112. Walsh, F. S., Barton, C. H., Putt, W., Moore, S. E., Kelsell, D., et al. 1990. *J. Neurochem.* 55:805–12
113. Hirsch, M.-R., Gaugler, L., Deagostini-Bazin, H., Bally-Cuif, L., Goridis, C. 1990. *Mol. Cell. Biol.* 10:1959–68
114. Chen, A. S., Reyes, A., Akeson, R. 1990. *Mol. Cell. Biol.* 10:3314–24
115. Barton, C. H., Mann, D. A., Walsh, F. S. 1990. *Biochem. J.* 268:161–68
116. Jones, F. S., Crossin, K. L., Cunningham, B. A., Edelman, G. M. 1990. *Proc. Natl. Acad. Sci. USA* 87:6497–501
117. Begemann, M., Tan, S.-S., Cunningham, B. A., Edelman, G. M. 1990. *Proc. Natl. Acad. Sci. USA* 87:9042–46
118. Cunningham, B. A., Leutzinger, Y., Gallin, W. J., Sorkin, B. C., Edelman, G. M. 1984. *Proc. Natl. Acad. Sci. USA* 81:5787–91
119. Peyrieras, N., Hyafil, F., Louvard, D., Ploegh, H. L., Jacob, F. 1983. *Proc. Natl. Acad. Sci. USA* 80:6274–77
120. Sorkin, B. C., Hoffman, S., Edelman, G. M., Cunningham, B. A. 1984. *Science* 225:1476–78
121. Mackie, K., Sorkin, B. C., Nairn, A. C., Greengard, P., Edelman, G. M., Cunningham, B. A. 1989. *J. Neurosci.* 6:1883–96
122. Sadoul, R., Kirchhoff, F., Schachner, M. 1989. *J. Neurochem.* 53:1471–78
123. Vimr, E. R., McCoy, R. D., Vollger, H. F., Wilkison, N. C., Troy, F. A. 1984. *Proc. Natl. Acad. Sci. USA* 81:1971–75
124. Finne, J., Makela, P. H. 1985. *J. Biol. Chem.* 260:1265–70
125. Rutishauser, U., Watanabe, M., Silver, J., Troy, F. A., Vimr, E. R. 1985. *J. Cell Biol.* 101:1842–49
126. Chuong, C.-M., Edelman, G. M. 1984. *J. Neurosci.* 4:2354–68
127. Friedlander, D. R., Brackenbury, R., Edelman, G. M. 1985. *J. Cell Biol.* 101:412–19
128. Breen, K. C., Kelly, P. G., Regan, C. M. 1987. *J. Neurochem.* 48:1486–93
129. Breen, K. C., Regan, C. M. 1988. *Development* 104:147–54
130. Kruse, J., Mailhammer, R., Wernecke, H., Faissner, A., Sommer, I., et al. 1984. *Nature* 311:153–55
131. Hoffman, S., Edelman, G. M. 1987. *Proc. Natl. Acad. Sci. USA* 84:2523–27
132. Hoffman, S., Crossin, K. L., Edelman, G. M. 1988. *J. Cell Biol.* 106:519–32
133. Dennis, R. D., Antonicek, H., Weigandt, H., Schachner, M. 1988. *J. Neurochem.* 51:1490–96
134. Abo, T., Balch, C. M. 1981. *J. Immunol.* 127:1024–29
135. Chou, K. H., Ilyas, A. A., Evans, J. E., Quarles, R. H., Jungalwala, F. B. 1985. *Biochem. Biophys. Res. Commun.* 128:383–88
136. Künemund, V., Jungalwala, F. B., Fischer, G., Chou, D. K., Keilhauer, G., Schachner, M. 1988. *J. Cell Biol.* 106:213–23
137. Niedieck, B., Lohler, J. 1987. *Acta Neuropathol.* 75:173–84
138. Loveless, W., Bellairs, R., Thorpe, S. J., Page, M., Feizi, T. 1990. *Development* 108:97–106
139. Cole, G. J., Schachner, M., Fliesler, S. J. 1988. *Neurosci. Lett.* 93:170–75
140. Lin, S., Guidotti, G. 1989. *J. Biol. Chem.* 264:14408–14
141. Aurivillius, M., Hansen, O. C., Lazrek, M. B. S., Bock, E., Öbrink, B. 1990. *FEBS Lett.* 264:267–69
142. Streuli, M., Krueger, N. X., Hall, L. R., Schlossman, S. F., Saito, H. 1988. *J. Exp. Med.* 168:1553–62
143. Charbonneau, H., Tonks, N. K., Walsh, K. A., Fischer, E. H. 1988. *Proc. Natl. Acad. Sci. USA* 85:7182–86
144. Yarden, Y., Escobedo, J. A., Kuang, W. J., Yang-Feng, T. L., Daniel, T. O., et al. 1986. *Nature* 323:226–32
145. Levi, G., Broders, F., Dunon, D., Edelman, G. M., Thiery, J.-P. 1990. *Development* 109:681–92
146. Friedlander, D. R., Grumet, M., Edelman, G. M. 1986. *J. Cell Biol.* 102:413–19
147. Bock, E., Richter-Landsberg, C., Faissner, A., Schachner, M. 1985. *EMBO J.* 4:2765–68
148. McGuire, J. C., Greene, L. A., Furano, A. V. 1978. *Cell* 15:357–65
149. Prentice, H. M., Moore, S. E., Dickson, J. G., Doherty, P., Walsh, F. S. 1987. *EMBO J.* 6:1859–63
150. Doherty, P., Mann, D. A., Walsh, F. S. 1988. *J. Cell Biol.* 107:333–40
151. Mann, D. A., Doherty, P., Walsh, F. S. 1989. *J. Neurochem.* 53:1581–88
152. Seilheimer, B., Schachner, M. 1987. *EMBO J.* 6:1611–16
153. Husmann, M., Gorgen, I., Weisgerber, C., Bitter-Suermann, D. 1989. *Dev. Biol.* 136:194–200
154. Akeson, R., Bernards, R. 1990. *Mol. Cell. Biol.* 10:2012–16
155. Roubin, R., Deagostini-Basin, H., Hirsch, M.-R., Goridis, C. 1990. *J. Cell Biol.* 111:673–84
156. Blaschuk, O. W., Farookhi, R. 1989. *Dev. Biol.* 136:564–67
157. Choi, Y. S., Sehgal, R., McCrea, P.,

Gumbiner, B. 1990. *J. Cell Biol.* 110:1575–82

158. Svalander, P. C., Odin, P., Nilsson, B. O., Öbrink, B. 1990. *J. Reprod. Fertil.* 88:213–21

159. Thompson, J., Moore, S. E., Walsh, F. S. 1987. *FEBS Lett.* 219:135–38

160. Levi, G., Crossin, K. L., Edelman, G. M. 1987. *J. Cell Biol.* 105:2359–72

161. Lane, T. A., Lamkin, G. E., Wancewicz, E. 1989. *Biochem. Biophys. Res. Commun.* 161:945–52

162. Griffiths, C. E., Esmann, J., Fisher, G. J., Voorhees, J. J., Nickoloff, B. J. 1990. *Br. J. Dermatol.* 122:333–42

163. Munro, J. M., Pober, J. S., Cotran, R. S. 1989. *Am. J. Pathol.* 135:121–33

164. Barker, J. N., Allen, M. H., MacDonald, D. M. 1989. *J. Invest. Dermatol.* 93:439–42

165. Jurgensen, C. H., Huber, B. E., Zimmerman, T. P., Wolberg, G. 1990. *J. Immunol.* 144:653–61

166. Thiery, J.-P., Duband, J.-L., Rutishauser, U., Edelman, G. M. 1982. *Proc. Natl. Acad. Sci. USA* 79:6737–41

167. Crossin, K. L., Chuong, C.-M., Edelman, G. M. 1985. *Proc. Natl. Acad. Sci. USA* 82:6942–46

168. Thiery, J.-P., Delouvée, A., Grumet, M., Edelman, G. M. 1985. *J. Cell Biol.* 100:442–56

169. Daniloff, J., Chuong, C.-M., Levi, G., Edelman, G. M. 1986. *J. Neurosci.* 6:739–58

170. Kintner, C. R., Melton, D. A. 1987. *Development* 99:311–25

171. Gurdon, J. B. 1987. *Development* 99:285–306

172. Dodd, J., Jessell, T. M. 1988. *Science* 242:692–99

173. Rieger, F., Grumet, M., Edelman, G. M. 1985. *J. Cell Biol.* 101:285–93

174. Covault, J., Sanes, J. R. 1985. *Proc. Natl. Acad. Sci. USA* 82:4544–48

175. Thiery, J.-P., Delouvée, A., Gallin, W. J., Cunningham, B. A., Edelman, G. M. 1984. *Dev. Biol.* 102:61–78

176. Edelman, G. M., Gallin, W. J. 1987. *Am. Zool.* 27:645–56

177. Edelman, G. M., Gallin, W. J. 1987. In *Banbury Report 26—Developmental Toxicology: Mechanisms and Risk*, pp. 109–21. New York: Cold Spring Harbor Lab.

178. Prieto, A. L., Crossin, K. L., Cunningham, B. A., Edelman, G. M. 1989. *Proc. Natl. Acad. Sci. USA* 86:9579–83

179. Goldowitz, D., Barthels, D., Lorenzon, N., Jungblut, A., Wille, W. 1990. *Dev. Brain Res.* 52:151–60

180. Grumet, M., Hoffman, S., Crossin, K. L., Edelman, G. M. 1985. *Proc. Natl. Acad. Sci. USA* 82:8075–79

181. Chiquet-Ehrismann, R., Mackie, E. J., Pearson, C. A., Sakakura, T. 1986. *Cell* 47:131–39

182. Erickson, H. P., Bourdon, M. A. 1989. *Annu. Rev. Cell Biol.* 5:71–92

183. Chiquet, M., Fambrough, D. M. 1984. *J. Cell Biol.* 98:1937–46

184. Jones, F. S., Burgoon, M. P., Hoffman, S., Crossin, K. L., Cunningham, B. A., Edelman, G. M. 1988. *Proc. Natl. Acad. Sci. USA* 85:2186–90

185. Jones, F. S., Hoffman, S., Cunningham, B. A., Edelman, G. M. 1989. *Proc. Natl. Acad. Sci. USA* 86:1905–9

186. Spring, J., Beck, K., Chiquet-Ehrismann, R. 1989. *Cell* 59:325–34

187. Erickson, H. P., Iglesias, J. L. 1984. *Nature* 311:267–69

188. Chiquet, M., Fambrough, D. M. 1984. *J. Cell Biol.* 98:1926–36

189. Crossin, K. L., Hoffman, S., Grumet, M., Thiery, J.-P., Edelman, G. M. 1986. *J. Cell Biol.* 102:1917–30

190. Daniloff, J. K., Crossin, K. L., Pinçon-Raymond, M., Murawsky, M., Rieger, F., Edelman, G. M. 1989. *J. Cell Biol.* 108:625–35

191. Chiquet-Ehrismann, R. 1990. *FASEB J.* 4:2598–604

192. Faissner, A., Kruse, J., Kuhn, K., Schachner, M. 1990. *J. Neurochem.* 54:1004–15

193. Probstmeier, R., Kuhn, K., Schachner, M. 1990. *J. Neurochem.* 54:1016–19

194. Tan, S.-S., Crossin, K. L., Hoffman, S., Edelman, G. M. 1987. *Proc. Natl. Acad. Sci. USA* 84:7977–81

195. Mackie, E. J., Tucker, R. P., Halfter, W., Chiquet-Ehrismann, R., Epperlein, H. H. 1988. *Development* 102:237–50

196. Riou, J. F., Shi, D. L., Chiquet, M., Boucaut, J. C. 1990. *Dev. Biol.* 137:305–17

197. Friedlander, D. R., Hoffman, S., Edelman, G. M. 1988. *J. Cell Biol.* 107:2329–40

198. Crossin, K. L., Prieto, A. L., Hoffman, S., Jones, F. S., Friedlander, D. R. 1990. *Exp. Neurol.* 109:6–18

199. Prieto, A. L., Jones, F. S., Cunningham, B. A., Crossin, K. L., Edelman, G. M. 1990. *J. Cell Biol.* 111:685–98

200. Duband, J.-L., Dufour, S., Hatta, K., Takeichi, M., Edelman, G. M., Thiery, J.-P. 1987. *J. Cell Biol.* 104:1361–74

201. Thiery, J. P., Tucker, G. C., Aoyama, H. 1985. In *Molecular Bases of Neural Development*, ed. G. M. Edelman, W.

E. Gall, W. M. Cowan, pp. 181–211. New York: Wiley

202. Davies, J. A., Cook, G. M. W., Stern, C. D., Keynes, R. J. 1990. *Neuron* 2:11–20

203. Ranscht, B., Bronner-Fraser, M. 1991. *Development*. In press

204. Stern, C. D., Norris, W. E., Bronner-Fraser, M., Carlson, G. J., Faissner, A., et al. 1989. *Development* 107:309–19

205. Hynes, R. O. 1987. *Cell* 48:549–54

206. Ruoslahti, E., Pierschbacher, M. D. 1987. *Science* 238:491–97

207. Yamada, K. M. 1983. *Annu. Rev. Biochem.* 52:761–99

208. Caroni, P., Schwab, M. E. 1988. *J. Cell Biol.* 106:1281–88

209. Cox, E. C., Muller, B., Bonhoeffer, F. 1990. *Neuron* 2:31–37

210. Raper, J. A., Kapfhammer, J. P. 1990. *Neuron* 2:21–29

211. Fischer, G., Künemund, V., Schachner, M. 1986. *J. Neurosci.* 6:605–12

212. Hoffman, S., Friedlander, D. R., Chuong, C.-M., Grumet, M., Edelman, G. M. 1986. *J. Cell Biol.* 103:145–58

213. Buskirk, D. R., Thiery, J.-P., Rutishauser, U., Edelman, G. M. 1980. *Nature* 285:488–89

214. Matsunaga, M., Hatta, K., Takeichi, M. 1988. *Neuron* 4:289–95

215. Fraser, S. E., Murray, B. A., Chuong, C.-M., Edelman, G. M. 1984. *Proc. Natl. Acad. Sci. USA* 81:4222–26

216. Fraser, S. E., Carhart, M. S., Murray, B. A., Chuong, C.-M., Edelman, G. M. 1988. *Dev. Biol.* 129:217–30

217. Kintner, C. 1988. *Neuron* 1:545–55

218. Detrick, R. J., Dickey, D., Kintner, C. R. 1990. *Neuron* 4:493–506

219. Fujimori, T., Miyatani, S., Takeichi, M. 1990. *Development* 110:97–104

220. Bieber, A. J., Snow, P. M., Hortsch, M., Patel, N. H., Jacobs, J. R., et al. 1989. *Cell* 59:447–60

221. Rakic, P. 1971. *J. Comp. Neurol.* 141:283–312

222. Chuong, C.-M., Crossin, K. L., Edelman, G. M. 1987. *J. Cell Biol.* 104:331–42

223. Daniloff, J. K., Levi, G., Grumet, M., Rieger, F., Edelman, G. M. 1986. *J. Cell Biol.* 103:929–45

224. Covault, J., Merlie, J. P., Goridis, C., Sanes, J. R. 1986. *J. Cell Biol.* 102:731–39

225. Gallin, W. J., Chuong, C.-M., Finkel, L. H., Edelman, G. M. 1986. *Proc. Natl. Acad. Sci. USA* 83:8235–39

226. Siu, C. H., Lam, T. Y., Choi, A. H. C. 1985. *J. Biol. Chem.* 260:16030–36

227. Noegel, A., Gerisch, G., Stadler, J., Westphal, M. 1986. *EMBO J.* 5:1473–76

228. Siu, C.-H. 1990. *BioEssays* 12:357–62

229. Wilson, H. V. 1907. *J. Exp. Zool.* 5:245–58

230. Harrelson, A. L., Goodman, C. S. 1988. *Science* 242:700–7

231. Zinn, K., McAllister, L., Goodman, C. S. 1988. *Cell* 53:577–87

232. Patel, N. H., Snow, P. M., Goodman, C. S. 1987. *Cell* 48:975–88

233. Seeger, M. A., Haffley, L., Kaufman, T. C. 1988. *Cell* 55:589–600

234. Kandel, E. 1991. *Cold Spring Harbor Symp. Quant. Biol.* 55: In press

235. Einheber, S., Fischman, D. A. 1990. *Proc. Natl. Acad. Sci. USA* 87:2157–61

236. Labeit, S., Barlow, D. P., Gautel, M., Gibson, T., Holt, J., et al. 1990. *Nature* 345:273–76

237. Benian, G. M., Kiff, J. E., Neckelmann, N., Moerman, D. G., Waterson, R. H. 1989. *Nature* 342:45–50

238. Klambt, C., Muller, S., Lutzelschwab, R., Rossa, R., Totzke, F., Schmidt, O. 1989. *Dev. Biol.* 133:425–36

239. Edelman, G. M. 1986. *Chem. Scr.* 26:363–75

240. Springer, T. A. 1990. *Nature* 346:425–34

241. Simmons, D., Makgoba, M. W., Seed, B. 1988. *Nature* 331:624–27

242. Staunton, D. E., Marlin, S. D., Stratowa, C., Dustin, M. L., Springer, T. A. 1988. *Cell* 52:925–33

243. Ballantyne, C. M., O'Brien, W. E., Beaudet, A. L. 1989. *Nucleic Acids Res.* 17:5853

244. Siu, G., Hedrick, S. M., Brian, A. A. 1989. *J. Immunol.* 143:3813–20

245. Horley, K. J., Carpenito, C., Baker, B., Takei, F. 1989. *EMBO J.* 8:2889–96

246. Staunton, D. E., Merluzzi, V. J., Rothlein, R., Barton, R., Marlin, S. D., Springer, T. A. 1989. *Cell* 56:849–53

247. Greve, J. M., Davis, G., Meyer, A. M., Forte, C. P., Yost, S. C., et al. 1989. *Cell* 56:839–47

248. Tomassini, J. E., Graham, D., DeWitt, C. M., Lineberger, D. W., Rodkey, J. A., Colonno, R. J. 1989. *Proc. Natl. Acad. Sci. USA* 86:4907–11

249. Mendelsohn, C. L., Wimmer, E., Racaniello, V. R. 1989. *Cell* 56:855–65

250. Berendt, A. R., Simmons, D. L., Tansey, J., Newbold, C. I., Marsh, K. 1989. *Nature* 341:57–59

251. Koch, P. J., Walsh, M. J., Schmelz, M., Goldschmidt, M. D., Zimbelmann, R., Franke, W. W. 1990. *Eur. J. Cell Biol.* 53:1–12

252. Crossin, K. L. 1991. *Ann. NY Acad. Sci.* 615:172–86
253. Walsh, F. S., Moore, S. E. 1985. *Neurosci. Lett.* 59:73–78
254. Cashman, N. R., Covault, J., Wollman, R. L., Sanes, J. R. 1987. *Ann. Neurol.* 21:481–88
255. Walsh, F. S., Moore, S. E., Lake, B. D. 1987. *J. Neurol. Neurosurg. Psychiatry* 50:439–42
256. Walsh, F. S., Moore, S. E., Dickson, J. G. 1988. *J. Neurol. Neurosurg. Psychiatry* 51:136–38
257. Rieger, F., Daniloff, J. K., Pinçon-Raymond, M., Crossin, K. L., Grumet, M., Edelman, G. M. 1986. *J. Cell Biol.* 103:379–91
258. Bhat, S., Silberberg, D. H. 1988. *Brain Res.* 452:373–77
259. Massaro, A. R., Albrechtsen, M., Bock, E. 1987. *Ital. J. Neurol. Sci. Suppl.* 6:85–88
260. Ibsen, S., Berezin, V., Norgaard-Petersen, B., Bock, E. 1983. *J. Neurochem.* 41:363–66
261. Cookman, G. R., King, W., Regan, C. M. 1987. *J. Neurochem.* 49:399–403
262. Edelman, G. M., Chuong, C.-M. 1982. *Proc. Natl. Acad. Sci. USA* 79:7036–40
263. Brackenbury, R., Greenberg, M. E., Edelman, G. M. 1984. *J. Cell Biol.* 99:1944–54
264. Behrens, J., Mareel, M. M., VanRoy, F. M., Birchmeier, W. 1989. *J. Cell Biol.* 108:2435–47
265. Bitter-Suermann, D., Roth, J. 1987. *Immunol. Res.* 6:225–37
266. Roth, J., Zuber, C., Wagner, P., Taatjes, D. J., Weisgerber, C., et al. 1988. *Proc. Natl. Acad. Sci. USA* 85:2999–3003
267. Roth, J., Brada, D., Blaha, I., Ghielmini, C., Bitter-Suermann, D., et al. 1988. *Virchows Arch.* 56:95–102
268. Roth, J., Blaha, I., Bitter-Suermann, D., Heitz, P. U. 1988. *Am. J. Pathol.* 133:596–608
269. Zuber, C., Roth, J. 1990. *Eur. J. Cell Biol.* 51:313–21
270. Patel, K., Moore, S. E., Dickson, G., Rossell, R. J., Beverley, P. C., et al. 1989. *Int. J. Cancer* 44:573–78
271. Kibbelaar, R. E., Moolenaar, C. E., Michalides, R. J., Bitter-Suermann, D., Addis, B. J., Mooi, W. J. 1989. *J. Pathol.* 159:23–28
272. Moolenaar, C. E. C. K., Muller, E. J., Schol, D. J., Figdor, C. G., Bock, E., Bitter-Suermann, D., Michalides, R. J. 1990. *Cancer Res.* 50:1102–106
273. Wilson, B. S., Petrella, E., Lowe, S.

R., Lien, K., Mackensen, D. G., et al. 1990. *Cancer Res.* 50:3124–30
274. Aletsee-Ufrecht, M. C., Langley, K., Rotsch, M., Havemann, K., Gratzl, M. 1990. *FEBS Lett.* 267:295–300
275. Lipinski, M., Braham, K., Philip, I., Wiels, J., Philip, T., et al. 1986. *J. Cell. Biochem.* 31:289–96
276. Lipinski, M., Braham, K., Philip, I., Wiels, J., Philip, T., et al. 1987. *Cancer Res.* 47:183–87
277. Lipinski, M., Hirsch, M. R., Deagostini-Bazin, H., Yamada, O., Tursz, T., Goridis, C. 1987. *Int. J. Cancer* 40:81–88
278. Hashimoto, M., Niwa, O., Nitta, Y., Takeichi, M., Yokoro, K. 1989. *Jpn. J. Cancer Res.* 80:459–64
279. Shimoyama, Y., Hirohashi, S., Hirano, S., Noguchi, M., Shimosato, Y., et al. 1989. *Cancer Res.* 49:2128–33
280. Zimmermann, W., Ortlieb, R., Friedrich, R., vonKleist, S. 1987. *Proc. Natl. Acad. Sci. USA* 84:2960–64
281. Oikawa, S., Nakazato, H., Kozaki, G. 1987. *Biochem. Biophys. Res. Commun.* 142:511–18
282. Kamarck, M. E., Elting, J. J., Hart, J. T., Goebel, S. J., Rae, P. M. M., et al. 1987. *Proc. Natl. Acad. Sci. USA* 84:5350–54
283. Beauchemin, N., Benchimol, S., Cournoyer, D., Fuks, A., Stanners, C. P. 1987. *Mol. Cell. Biol.* 7:3221–30
284. Hinoda, Y., Neumaier, M., Hefta, S. A., Drzeniek, Z., Wagener, C., et al. 1988. *Proc. Natl. Acad. Sci. USA* 85:6959–63
285. Lehmann, J. M., Riethmüller, G., Johnson, J. P. 1989. *Proc. Natl. Acad. Sci. USA* 86:9891–95
286. Fearon, E. R., Cho, K. R., Nigro, J. M., Kern, S. E., Simons, J. W., et al. 1990. *Science* 247:49–56
287. Lehmann, J. M., Holzmann, B., Breitbart, E. W., Schmiegelow, P., Riethmüller, G., Johnson, J. P. 1987. *Cancer Res.* 47:841–45
288. Stanbridge, E. J. 1990. *Science* 247:12–13
289. DeRobertis, E. M., Oliver, G., Wright, C. V. E. 1990. *Sci. Am.* 263:46–52
290. Kessel, M., Gruss, P. 1990. *Science* 249:374–79
291. Grumet, M., Burgoon, M. P., Mauro, V., Edelman, G. M., Cunningham, B. A. 1989. *Soc. Neurosci. Abstr.* 15:568a Abstr.
292. Moos, M., Tacke, R., Scherer, H., Teplow, D., Fruh, K., Schachner, M. 1988. *Nature* 334:701–3
293. Ranscht, B. 1988. *J. Cell Biol.* 107:1561–73

294. Brummendorf, T., Wolff, J. M., Frank, R., Rathjen, F. G. 1989. *Neuron* 2:1351–61
295. Gennarini, G., Cibelli, G., Rougon, G., Mattei, M.-G., Goridis, C. 1989. *J. Cell Biol.* 109:775–88
296. Arquint, M., Roder, S., Chia, L. S., Down, J., Wilkinson, D., et al. 1987. *Proc. Natl. Acad. Sci. USA* 84:600–4
297. Lemke, G., Axel, R. 1985. *Cell* 40: 501–8
298. Rathjen, F. G., Schachner, M. 1984. *EMBO J.* 3:1–10
299. Salzer, J. G., Holmes, W. P., Colman, D. R. 1987. *J. Cell Biol.* 104:957–65
300. Staunton, D. E., Dustin, M. L., Springer, T. A. 1989. *Nature* 339:61–64
301. Maddon, P. J., Littman, D. R., Godfrey, M., Maddon, D. E., Chess, L., Axel, R. 1985. *Cell* 42:93
302. Newman, P. J., Berndt, M. C., Gorski, J., White, G. C. II, Lyman, S., et al. 1990. *Science* 247:1219–22
303. Albelda, S. M., Oliver, P. D., Romer, L. H., Buck, C. A. 1990. *J. Cell Biol.* 110:1227–37
304. Grunert, F., Abuharfeil, N., Schwartz, K., vonKleist, S. 1985. *Int. J. Cancer* 36:357–62

305. Osborn, L., Hession, C., Tizard, R., Vassallo, C., Luhowskyj, S., et al. 1989. *Cell* 59:1203–11
306. Schofield, P. R., McFarland, K. C., Hayflick, J. S., Wilcox, J. N., Cho, T. M., et al. 1989. *EMBO J.* 8:489–95
307. Streuli, M., Krueger, N. X., Tsai, A. Y. M., Saito, H. 1989. *Proc. Natl. Acad. Sci. USA* 86:8698–702
308. Damsky, C. H., Richa, J., Solter, D., Knudsen, K., Buck, C. A. 1983. *Cell* 34:455–66
309. Behrens, J., Birchmeier, W., Goodman, S. L., Imhof, B. A. 1985. *J. Cell. Biol.* 101:1307–15
310. Heimark, R. L., Degner, M., Schwartz, S. M. 1990. *J. Cell Biol.* 110:1745–56
311. Gloor, S., Antonicek, H., Sweadner, K. J., Pagliusi, S., Frank, R., et al. 1990. *J. Cell Biol.* 110:165–74
312. Stoolman, L. M. 1989. *Cell* 56:907–10
313. Rathjen, F. G., Wolff, J. M., Chang, S., Bonhoeffer, F., Raper, J. A. 1987. *Cell* 51:841–49
314. Bayna, E. M., Shaper, J. H., Shur, B. D. 1988. *Cell* 53:145–57
315. Ruegg, M. A., Stoeckli, E. T., Lanz, R. B., Streit, P., Sonderegger, P. 1989. *J. Cell Biol.* 109:2363–78

*Annu. Rev. Biochem. 1991. 60:191–227*

# RIBOSOMAL RNA AND TRANSLATION

*Harry F. Noller*

Sinsheimer Laboratories, University of California, Santa Cruz, California 95064

KEY WORDS:   ribosomes, protein synthesis, transfer RNA, ribozymes.

## CONTENTS

## PERSPECTIVES AND SUMMARY

There is extensive biochemical, genetic, and phylogenetic evidence for the functional role of ribosomal RNA (rRNA). This includes the participation of

0066-4154/91/0701-0191$02.00

rRNA in mRNA selection, tRNA binding (in A, P, and E sites), ribosomal subunit association, proofreading, factor binding, antibiotic interactions, nonsense and frame-shift suppression, termination, and the peptidyl transferase function. So far, only mRNA selection has been firmly established as an rRNA-based mechanism. However, rRNA has been placed at the sites of all functional interactions that have so far been investigated, either by affinity labelling or by chemical footprinting. Many of the implicated bases have been shown to be important for ribosome function from mutations that confer deleterious phenotypes, by specific biochemical inactivation experiments, and from their high phylogenetic conservation.

A general picture of the structural arrangement of rRNA in the ribosome is emerging, and in the case of 16S rRNA, there are well-constrained models for its folding. The locations of functionally implicated sites have been identified, and are in good agreement with previous information obtained from electron microscopy and other approaches. mRNA is selected by interaction with the 3'-terminus of 16S rRNA in the platform region of the small subunit. tRNA interacts with 16S rRNA (either directly or indirectly) as well as with mRNA A- and P-site codons in the cleft of the small subunit via its anticodon stem-loop. In the P site, it interacts with several unpaired bases in the central and 3'-domains that surround the cleft. In the A site, it interacts with nucleotides near the base of the penultimate stem, but also (probably indirectly) with bases in the structurally remote 530 loop. The role of the latter effect is not understood, but may have a relationship to the interactions between the elongation factors and the ribosome.

In the large subunit, tRNA interacts primarily via its CCA end with sites in domain V of 23S rRNA, particularly around the universally conserved central loop. This region of the RNA may be involved in the peptidyl transferase function, a possibility that is supported by the unusual stability of this activity to protein extraction and proteolysis procedures. The peptidyl transferase region has been located between the L1 ridge and the central protuberance of the large subunit, opposite the cleft of the small subunit. Elongation factors EF-Tu and EF-G interact with the large subunit on the opposite side, around the base of the L7/L12 stalk, opposite the 530 loop region of the small subunit. Both factors interact with the universally conserved alpha-sarcin loop of 23S rRNA, and EF-G interacts in addition with the 1067 region, near the binding sites for proteins L7/L12 and L11. A general idea of the locations of tRNA and EF-Tu relative to the two ribosomal subunits is suggested in Figure 1. Zimmermann and coworkers (26a) have proposed a similar arrangement.

Subunit interaction appears to involve elements of both 16S and 23S rRNA, although it is not yet clear whether this occurs via RNA-RNA or RNA-protein interactions. Elements of the central and 3' minor domains of 16S rRNA, located in the platform region of the small subunit, have been identified with

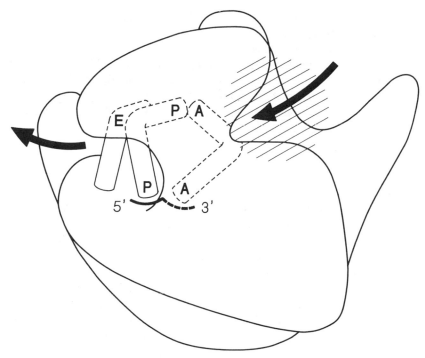

*Figure 1*   Plausible locations for ribosome-bound tRNAs in the A/A, P/P, and E states (27). The shaded area at right shows the approximate site of interaction of EF-Tu. The polarity of a fragment of mRNA containing the A- and P-site codons is shown. The arrows indicate the likely path of a tRNA as it transits the ribosome.

this function. There is increasing evidence that the apparently distinct functions of subunit association, tRNA binding, and translocation may be closely interrelated.

Translocation now appears to involve two distinct steps. In the first step, the acceptor ends of the A- and P-site tRNAs move with respect to the large subunit, to their 23S rRNA P and E sites, respectively. This step occurs spontaneously upon peptide bond formation and is independent of factors or GTP. It results in hybrid states of binding for the two tRNAs, the A/P and P/E states, in which the tRNAs are in different "sites" on the two subunits, as defined by the classical model. This allows the nascent polypeptide chain to remain in a fixed position on the ribosome during elongation. In the second step, the tRNAs and associated mRNA move with respect to the small subunit, dependent on EF-G and GTP. Six different states of tRNA binding have been identified: A/A, A/P, P/P, P/E, E, and A/T (the first state of tRNA selection, in which tRNA interaction with the large subunit is mediated by EF-Tu). This mechanism raises the possibility that translocation may involve relative movement of the large and small subunits, explaining the two-subunit architecture of all ribosomes.

Ribosomal proteins are important for their effects on folding and "fine tuning" of the conformation of rRNA during ribosome assembly. There is still no conclusive evidence that the r-proteins themselves are directly involved in any translational function. Certain protein mutations are known to affect the conformation of functionally implicated regions of rRNA, suggesting that their effects on function may be indirect.

## INTRODUCTION

Protein synthesis is a very complicated process, and the structural complexity of the translational machinery is no doubt a reflection of this. Attempts to understand the mechanism of translation have progressed through several discernible phases, beginning with the Adaptor Hypothesis (1). The realization that tRNA was the adaptor, recognizing triplet codons in mRNA via simple base-pairing rules, characterized this first stage. The essence of the translational mechanism could be symbolized by tRNA—at one end, the anticodon; at the other, the aminoacyl or peptidyl group. Although this amounted to a characteristically RNA-based mechanism (apart from charging each tRNA with its cognate amino acid, catalysis of peptide bond formation, and other "details"), the ribosome (and rRNA) seemed to play only a passive role. This was reinforced in the next stage, when the use of purified in vitro systems led to the discovery of several essential protein factors. During the factor stage, crucial events such as initiation, translocation, and peptide bond formation tended to be ascribed to enzymatic properties of the protein factors. It was already known that initiation could take place in the absence of initiation factors under certain conditions. However, the discovery of factor-independent translocation (2, 3) and peptide bond formation (4) clearly demonstrated that the fundamental, underlying mechanism resides in the ribosome itself.

With this realization came the expectation that the ribosomal functions would be carried out by the various ribosomal proteins (r-proteins), and that the ribosome could be viewed as a "multienzyme complex." The role of ribosomal RNA was to provide a structural scaffold for the positioning of the proteins. This view was based on the precedent that catalytic and other functional roles were traditionally carried out by proteins, and was reinforced by the discovery that mutations in the gene for protein S12 conferred strep-tomycin resistance (5) or dependence (6). In vitro reconstitution methods led to the assignment of various ribosomal functions to specific proteins, although the omission of individual proteins almost always failed to show all-or-nothing effects, and in many cases resulted in only modest decreases in ribosome function (7). In no case could any activity be demonstrated for a purified r-protein, or indeed for any RNA-free mixture of r-proteins.

Meanwhile, evidence for a functional role for ribosomal RNA began to appear, as discussed in the following pages. Although the evidence from biochemical, genetic, and phylogenetic approaches continued to mount, the notion that rRNA could be an important determinant of ribosomal function did not gain wide acceptance until the discovery of self-splicing introns and catalytic RNA (8, 9). It is the purpose of this review to take stock of the rRNA paradigm, to review the relevant biochemical, genetic, and evolutionary evidence, and to draw attention to some of the important unanswered questions. The reader is referred to a number of other review articles relevant to various aspects of this topic (10–26, 26a, 27, 27a, 28–30).

# EVOLUTIONARY ARGUMENTS

## Origins of Translation

It has been argued that the design of the translation apparatus is imprinted with and reflects its evolution, and to understand one is to understand the other (11, 30a). The sheer complexity of even the simplest bacterial translational systems, which require well over a hundred different species of macromolecules, makes it clear that protein synthesis could not have evolved in a single step. The crucial role of translation, linking genotype with phenotype, only compounds the difficulty of imagining how it could have evolved. A striking characteristic of the translational apparatus is that it is rich in RNA. Crick suggested that this is because the RNA components were part of the primitive machinery for protein synthesis, and wondered whether the original ribosomes were made entirely of RNA (31). The "chicken-or-the-egg" question of whether proteins or nucleic acids (or both) arose prior to the evolution of translation was addressed by Orgel, who pointed out a key problem: the unlikelihood of "reverse translation" of a polypeptide sequence into a polynucleotide sequence (32). Thus, if any proteins existed prior to the evolution of translation, they are no longer with us.

A way out of the RNA-protein paradox has been provided by the discovery of catalytic RNA (8, 9). If RNA molecules can have enzymatic activity, one can imagine the existence of an RNA-based genetic system, in which genotype and phenotype are embodied in the same molecule. It can then be postulated that within such an RNA-based system, a primitive translational apparatus evolved. But this somewhat plausible sequence of events implies evolutionary selection for a relatively complex system, even in its simplest conceptual form. That is, the RNA-based system could not have "known" in advance that it was preparing for a more powerful protein-based system during the time that the precursors to tRNA, mRNA, and rRNA were evolving. More likely, the translational apparatus emerged from the modification of some pre-existing functional RNA molecules; the question is, what was their function?

## Phylogenetic Conservation

Conservation of primary structure, first noted in comparison of sequences of RNase $T_1$ oligonucleotides from 16S-like rRNAs from different organisms (33), was extended with the comparison of newly available full-length sequences of both 16S and 23S rRNA across a wide phylogenetic spectrum (12, 16). These studies revealed several sequences on the order of ten to twenty nucleotides long that were essentially invariant across all lines of descent, an unprecedented level of phylogenetic conservation. This degree of sequence conservation is expected for molecules that carry out some crucial function, as opposed to a purely structural role. The availability of the conserved sequences on the surface of the ribosome (33, 34) is consistent with their participation in ribosomal function, as discussed below.

Other regions of rRNA show increasing levels of sequence variability, which made it possible to deduce their secondary structures (11, 35–38). Convincing evidence for pairing of specific sequences came from analysis of Watson-Crick covariance. All of the helices, and most of the individual base pairs in the rRNA secondary structures, are now supported by this type of evidence, as well as a wide variety of experimental data. Even more striking than the primary structure conservation is the similarity between the various secondary structures. The implication is that there is a very high degree of conservation of the three-dimensional structure of rRNA. This recalls the universal conservation of tRNA structure, and the fact that tRNAs and ribosomes from very highly divergent organisms are intercompatible. At this writing, there are insufficient complete r-protein sequences to make a fair comparison, although so far it would appear that the degree of r-protein sequence conservation is lower than that observed for rRNA (39, 40). An important goal for the near future will be to deduce the primary structures of the complete set of r-proteins for one or more representatives each of the eukaryotic and archaebacterial kingdoms, as has been done for *Escherichia coli* among the eubacteria. It will be then an interesting theoretical challenge to derive methods for comparing the relative evolutionary rates of RNA vs protein sequences.

## BIOCHEMICAL EVIDENCE

### Activity of Protein-Depleted RNA

Some support for the importance of rRNA comes from the observation that elimination of certain r-proteins in vitro (7) or in vivo (41) fails to inactivate ribosomes, and that there is, to my knowledge, no substantiated report of ribosomal activity by RNA-free preparations of r-proteins. By far the most convincing evidence for a functional role for rRNA would be in vitro demonstration of ribosomal function in protein-free preparations of rRNA. Some degree of success has been reported, although certain results have yet to be confirmed.

Most clearly established is the base pairing between the purine-rich initiation region of mRNA and the 3'-terminal CCUCC sequence of 16S rRNA, which takes place during mRNA selection in bacteria, the Shine-Dalgarno interaction (42). This has been demonstrated directly by characterization of in vitro initiation complexes subjected to deproteinization (43). The in vitro results are supported by ample genetic evidence, including studies where mutations in both the mRNA (44, 45) and 16S rRNA (46, 47) sequences confer the predicted results. Apart from the Shine-Dalgarno interaction, there is so far no solid evidence for the interaction of mRNA with rRNA. It is even possible that the main mRNA attachment points during translation are its interactions with the tRNA anticodons. This would help to rationalize some of the extraordinary frame-shifting events that have recently been observed (47a).

At least one antibiotic, thiostrepton, has been shown to bind directly to rRNA. Thiostrepton inhibits protein synthesis by interfering with EF-G-related functions. It binds to a fragment comprising residues 1052–1110 of 23S rRNA, albeit with lower affinity than to the intact 50S ribosomal subunit (48). Thiostrepton and protein L11, which also binds to this region of 23S rRNA, show mutual cooperativity of binding to this RNA fragment, implying that both ligands interact preferentially with the same specific conformer of the RNA. Methylation of the 2'-hydroxyl of A1067 within this fragment provides resistance to the drug by abolishing its binding (49). Although it can be argued that antibiotic binding is a ribosomal "activity" only in the broadest sense, it is nevertheless significant in this context, since these relatively small molecules often exert profound effects on crucial ribosomal functions such as translocation, codon recognition, and peptidyl transferase. Evidence that most, if not all, ribosome-directed antibiotics interact with rRNA is steadily accumulating.

Association of protein-free 16S and 23S rRNAs has been reported (50). However, it remains to be shown whether the interaction bears any relationship to the biologically relevant association of 30S and 50S subunits. Since subunit association protects specific bases in 16S and 23S rRNA from chemical attack (discussed below), protection of these same bases by association of naked RNA would provide support for the physiological relevance of these observations.

Peptidyl transferase is the sole catalytic activity that has unambiguously been shown to be a property of the ribosome itself (4, 51). The simplest assay for peptidyl transferase is the "fragment reaction" (52), in which an N-blocked aminoacyl-CCA terminal fragment of tRNA such as CACCA (Ac-Phe) reacts with puromycin (an aminoacyl-tRNA analogue) in the presence of ethanol or methanol to yield Ac-Phe-puromycin in a peptide-bond-forming reaction. This reaction requires only the 50S ribosomal subunit, eliminating the need for 30S subunits, mRNA, factors, GTP, or intact tRNA. Its sensitivity to antibiotics such as chloramphenicol, carbomycin, and anisomycin (53), and the stereochemical requirements of the substrates provide ample evidence that the fragment reaction is indeed catalyzed by peptidyl transferase.

The simple substrate requirements for the fragment reaction make it one of the least demanding and most specific assays for a single ribosome function. Peptidyl transferase activity has been demonstrated in severely protein-depleted 50S subunits (54, 55). Recently, some progress has been made toward the preparation of protein-free RNA that retains activity (56). *E. coli* ribosomes and 50S subunits retain high levels of peptidyl transferase activity after treatment with 1% SDS and 10 mg/ml protease K at 37°C, followed by extraction with chloroform. Exposure to EDTA or phenol, however, abolishes activity. If the reaction is truly RNA-catalyzed, the latter observation could be explained by collapse of some crucial RNA structural feature upon removal of protein. On the possibility that such a structure might be more robust in ribosomes from a thermophilic bacterium, the extraction experiments were carried out on *Thermus aquaticus* ribosomes. Again, the activity was resistant to SDS, protease K, and chloroform treatments, but in this case it was also stable to multiple extractions with phenol (although it was again sensitive to EDTA treatment). Using $^{35}$S-methionine to label r-proteins in vivo, it was estimated that more than 95% of the protein had been removed from preparations that retained 20–40% of their peptidyl transferase activity. Although these studies provide hope for a direct demonstration of a protein-free ribosomal function, it could as well be argued that the activity depends on an unusually extraction-resistant r-protein moiety. Clear proof that peptidyl transferase is in fact rRNA will require strategies for eliminating all of the associated protein.

## Affinity Labeling and Cross-Linking

The power of affinity labeling (actually, a type of site-directed cross-linking) is that it can provide unambiguous information about the proximity of a functional ligand to specific structural features on the ribosome. This approach is most convincing when it can be shown that the ligand retains its functional capacity in its cross-linked state. The latter condition has been met in two classic studies of tRNA affinity labeling, in which the two functional extremities of tRNA were photo-cross-linked to the ribosome, at specific sites in 16S and 23S rRNA, respectively (57, 58).

Certain tRNAs can be cross-linked directly to ribosomes by virtue of the intrinsic photoreactivity of their 5-alkoxyuridine residues at position 34 (59, 60). Irradiation of tRNA-ribosome complexes at 310–325 nm provides cross-linked products with single-hit kinetics at up to 70% yield. Cross-linking is dependent on the cognate mRNA codon, with the tRNA bound to the P site. Reactivity of the cross-linked tRNA with puromycin provides clear evidence that the tRNA is in fact cross-linked to the P site in a state closely resembling the physiologically correct interaction. The tRNA is cross-linked exclusively to 16S rRNA, at C1400, by 5,6-cyclobutane dimer formation between the two pyrimidines.

This study carries with it a number of important implications. First, the high yield and specificity of the cross-linked product argue that the anticodon is held in a fixed position with respect to C1400 of 16S rRNA, and that the two bases come within 3–4 Å of each other, i.e. at Van der Waals contact distance. This contact occurs between the wobble base of the anticodon, one of the two most important functional sites in tRNA, and C1400, which is located in the middle of one of the most highly conserved sequences in 16S rRNA. Finally, the cross-linked tRNA has been located in the cleft of the 30S subunit by both immuno-electron microscopy (IEM) (61) and by direct electron microscopy, using computerized image averaging (62). This is perhaps the most convincing of a large body of evidence localizing the ribosomal decoding site in the cleft of the 30S subunit.

Peptide bond formation involves the 3' CCA end of tRNA, and so the ribosomal environment of the acceptor end of A- and P-site tRNA identifies the peptidyl transferase region of the ribosome, an integral part of the 50S subunit. Several types of reactive group have been attached to the aminoacyl/peptidyl moiety of tRNA in efforts to identify the molecular components of peptidyl transferase. In some cases, proteins were labeled; however, in at least the case of protein L27, one of the most efficiently cross-linked proteins (63), it could be shown that protein-depleted 50S subunits lacking L27 retained full peptidyl transferase activity (54).

In other studies, the major cross-linked component was 23S rRNA (64–66). A complete analysis of the reaction using 3-(4'-benzoylphenyl)propionyl-Phe-tRNA has been carried out (67). Here, the benzophenone moiety, attached to the $\alpha$-amino group of phenylalanine, has potential photochemical reactivity toward both protein and RNA. When bound to the ribosome in a mRNA-dependent manner, this derivative reacts exclusively with 23S rRNA, with a yield of greater than 70%. The cross-linked tRNA retains its puromycin reactivity, showing that it occupies the ribosomal P site in a physiologically correct state. Under these conditions, the cross-linked residues are A2451 and C2452 in the universally conserved central loop of domain V of 23S rRNA (58). In the presence of excess deacylated tRNA, peptidyl-tRNA analogues bind in a puromycin-unreactive state, originally thought to correspond to A-site binding of peptidyl-tRNA prior to translocation (68). More recent evidence, discussed below, indicates that the peptidyl-tRNA analogue N-acetyl-Phe-tRNA is in a "hybrid" binding state, the A/P state, in these complexes, in which the 3'-end of the tRNA occupies the P site of the large subunit, but in a puromycin-unreactive manner (69). When the benzophenone-derivatized Phe-tRNA is bound under these conditions, it reacts instead with U2584 and U2585, on the opposite side of the central loop of domain V (70). These findings indicate that there is a distinctly different orientation of the peptidyl moiety in the A/P state, compared with the puromycin-reactive P/P state. They also provide evidence that the two opposite sides of the central loop must be folded into proximity with one another in three dimensions. Both kinds of complex give weaker cross-linking to bases 2503–2056, also in the central loop. The location of certain antibiotic resistance mutations and footprints in these same regions confirms the connection between the central loop of domain V and the peptidyl transferase function (see below).

The distance between the photoreactive position of the benzophenone derivative and the 3'-end of tRNA is 8–12 Å (67). In studies by Zimmermann and coworkers, incorporation of photoreactive base analogues directly into tRNA allowed cross-links to regions of the ribosome that are in even closer proximity to bases in the 3'-terminus of tRNA (70a). Introduction of 2-azido- or 8-azido-adenosine at positions 73 or 76 of yeast tRNA[Phe] had little effect on charging the tRNA with phenylalanine, and the resulting Ac-Phe-(azido-A)-tRNAs were bound to E. coli ribosomes and photo-cross-linked to 50S subunits with yields of 6 to 13% (71). The covalently cross-linked Ac-Phe derivatives were fully reactive toward puromycin, again showing that correct P-site binding was maintained. The nature of the cross-linked ribosomal components depended on the position of the azido-A derivative in the tRNA, and in the case of 8-azido-A73-tRNA, on whether an Ac-Phe moiety was present. Whereas the 2-azido-A73 derivatives reacted only with protein L27, both the 2-azido-A76 and 8-azido-A73 derivatives reacted with both proteins (mainly L27, with minor cross-linking to L2, L15, and L16) and 23S rRNA. In the case of 2-azido-A76-tRNA, the site of cross-linking was tentatively assigned to G1945, which lies in a conserved region of domain IV. The nearby bases A1916, A1918, and U1926 were found to be protected by binding tRNA to the P site, and the protection was found to be independent of whether the tRNA contained an intact 3' CA (72). More recently, it has been found that the latter bases are protected by 30S-50S subunit association (73).

Thus, the tRNA-dependent protection can be reinterpreted to indicate stimulation of subunit association by P-site tRNA. The cross-linking of derivatized A76 to this region thus places the CCA terminus of tRNA close to a region of 23S rRNA that interacts (directly or indirectly) with the 30S subunit. This is part of a growing body of evidence that seems to indicate that the apparently separate functions of tRNA binding, subunit interaction, and translocation are in fact closely interrelated. By virtue of the benzophenone cross-links, it also places this region of domain IV in close proximity to the central loop of domain V, a conclusion that is supported by an RNA-RNA cross-link between the 1780 region of domain IV and the 2585 region of domain V (74).

Affinity-labeling experiments have also placed mRNA close to the 1400 and 3'-terminal regions of 16S rRNA, results that are in accord with the proximities of the anticodon and Shine-Dalgarno sequences, respectively, to these same sites. In experiments using nonderivatized poly A mRNA bound in the presence of tRNA$^{Lys}$, the poly A was cross-linked by UV irradiation to a site within positions 1394–1399 of 16S rRNA (75). Cross-linking to this same site was observed in the absence of tRNA$^{Lys}$, but with a fivefold lower efficiency, providing evidence for binding of the poly A to the mRNA-binding site. In another study, defined mRNA sequences from 24 to 34 nucleotides long, substituted with 4-azidophenacyl-derivatized 4-thiouridine, were photo-cross-linked to ribosomes (76). In addition to tRNA-dependent cross-linking to protein S7, the mRNA analogues were found linked to sites within the 3'-terminal 30 nucleotides of 16S rRNA. The latter result is not unexpected, since the mRNA contained a strong Shine-Dalgarno sequence. In all cases, the photoreactive base was located a few nucleotides to the 5'-side of the cognate codons.

## Protection from Chemical and Enzymatic Probes

If functional ligands interact directly with rRNA during their association with the ribosome, they will limit or prevent access to the sites of interaction. This provides an approach to identification of ligand interaction sites, using chemical probes that attack the bases, ribose, or phosphate groups in the RNA, or single- and double-strand-specific nucleases. Ligand binding may also induce conformational changes in the ribosome, and if this results in increased or decreased access to regions of the RNA, these changes can also be detected by the same probes. This raises the main shortcoming of this approach, that one cannot easily distinguish whether protection is caused by direct ligand-RNA contact, or by ligand-induced conformational change. Of course, both are interesting from the point of view of the functional role of ribosomal RNA.

In the case of certain small chemical probes, the site of attack can be exquisitely specific, permitting localization to individual atoms in the RNA. Among the more commonly used probes (and their main sites of attack) are kethoxal (N1 and N2 of G), dimethyl sulfate (N1 of A, N3 of C, N7 of G), CMCT (N3 of U, N1 of G), hydroxyl radical generated by $Fe^{2+}$ (C4' of ribose), and ethylnitrosourea (phosphate oxygen) (77–79). These probes are especially useful for studies on ribosomal RNA, since they have generally mild effects on proteins, and can be used at moderate concentrations under conditions where ribosomes are known to be functionally active in vitro. The precise targets of modification as well as the relative extent of modification can readily be assessed by primer extension (77, 80). This is based on the fact that reverse transcriptase either pauses or stops at modified nucleotides, or in the cases of DMS modification of G at N7, hydroxyl radical attack at ribose, or ethylnitrosourea attack at phosphate, reverse transcriptase is terminated at the point of chain scission. Use of the primer extension method also permits very low levels of modification (minimizing any structural damage to the ribosome or ligands), since less than one percent derivatization at a given position can easily be detected.

## Binding of tRNA to the A, P, and E Sites

Earlier studies showed that kethoxal modification of a few guanines in the 16S rRNA abolished the ability of 30S subunits to bind tRNA (81). Reconstitution showed that loss of activity was indeed due to RNA modification. Loss of activity could be prevented by binding tRNA to the ribosomes prior to the reaction. More recently, the approach described above has been used to identify the bases in rRNA that are protected when tRNA is bound to the ribosome. Binding was directed specifically to the A, P, and E sites using established criteria. Characteristic sets of bases were protected by tRNA in 16S rRNA for A- and P-site binding (82, 83; Figure 2), and in 23S rRNA for A-, P-, and E-site binding (72; Figure 3). Nearly all of the protected bases are universally conserved, supporting the interpretation that their protection by tRNA reflects their functional importance.

In 16S rRNA, A-site bound tRNA protects two clusters of bases, located in the 530 loop and in the 1400/1500 region, respectively (82, 83). These are the two most highly conserved regions in the 16S-like rRNAs (12, 16). Most strongly protected are G530, A1492, and A1493. The latter two bases, along with the less strongly protected A1408 and G1494, lie in close proximity to C1400, the site of cross-linking to the P-site anticodon, and provide further evidence that the high sequence conservation in this region is due to its importance for the codon recognition function. A-site binding also causes enhanced reactivity of A892 and G1405, indicating tRNA-induced conformational changes. Interestingly, A892 is protected by tetracycline (84), an antibiotic that is believed to inhibit A-site binding. Perhaps the action of tetracycline is to interfere with the A-site tRNA-dependent conformational change involving A892. There is evidence for two other functionally related conformational changes involving the 900 region; one, involving A908 (85), appears to be modulated by ribosomal proteins, and the other, involving A909 (84), is one of the "class III" sites, and is affected by tRNA binding, subunit association, and drug binding (see below). This region of 16S rRNA appears to have significant conformational mobility, and is sensitive to the interaction of the ribosome with a number of important functional ligands, including streptomycin (84).

P-site tRNA protects a distinctly different set of bases in 16S rRNA (82, 83). In contrast to the A-site protections, they are distributed widely throughout the RNA. The most strongly protected bases (A693, A794, C795, G926, and G1401) appear to be localized near one another in three dimensions, however, based on models for the folding of 16S rRNA in the 30S subunit (86–88). Protection of G1401, and weaker protection of C1399 and C1400, are in good agreement with the P-site cross-link between the anticodon and C1400. The generality of these results is supported by the fact that several different tRNA species protect the same set of bases when bound to the ribosome under P-site conditions (83). These data provide the first strong evidence that the A and P sites are in fact distinct, specific sites on the ribosome structure; this is further supported by corresponding data for 23S rRNA (below).

All of the bases in 16S rRNA that are protected by tRNA are also protected by a 15-nucleotide tRNA anticodon stem-loop (82, 83). This indicates that, insofar as 16S rRNA is concerned, interactions between tRNA and the 30S subunit involve only the anticodon stem-loop region of tRNA. This result raises an interesting question concerning the protection of bases in the 530 loop region, most notably the strong protection of the highly kethoxal-reactive G530 by A-site tRNA. By all available structural evidence the 530 loop is located on the opposite side of the 30S subunit from the decoding site, at a distance of 70–100 Å (86–88). Since the maximum dimension of the stem-loop fragment is no more than 25 Å (89), protection of bases in the 530 loop by tRNA appears to be indirect, most likely the result of an allosterically induced conformational change. Genetic studies, described below, provide further evidence for the importance of the 530 loop in A-site function.

In 23S rRNA, almost all of the tRNA-protected bases are found in domain V (72). Distinct, characteristic sets of protected bases correspond to A-, P-, and E-site binding. The only overlap occurs between A- and P-site bases, where A2439 and A2451 are protected in both cases. Removal of the aminoacyl moiety abolishes protection of these two bases, indicating that it is the presence of the acyl group, in either the A or P site, that is responsible for their protection. Another base, A2602, is protected by A-site tRNA and

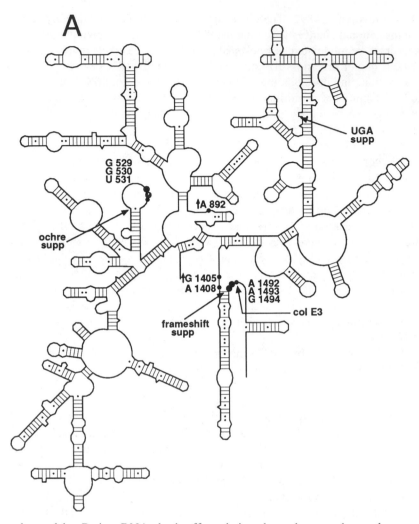

enhanced by P-site tRNA, both effects being dependent on the acyl group. Clearly, there is sufficient flexibility around the 3'-terminal acyl linkage to allow the acyl group to exert its effects on these bases from either the A or P sites. This should not be surprising, since the respective 3'-hydroxyl groups of the two tRNAs must approach each other within a few angstroms when a peptide bond is formed.

There is extensive evidence that the interactions with 23S rRNA corresponding to the other protected bases involve mainly, if not exclusively, the conserved CCA terminus of tRNA. Progressive removal of the 3'-terminal A and penultimate C progressively eliminates protection of most of the remain-

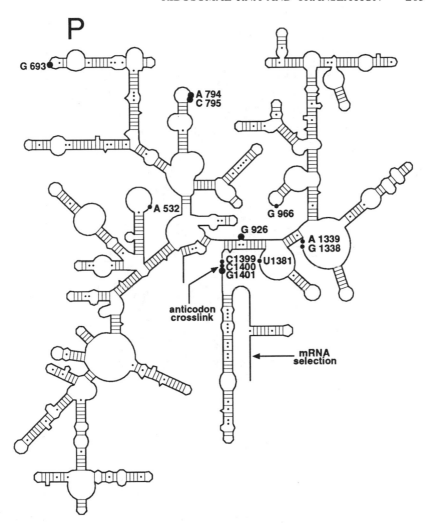

*Figure 2*   (see also p. 204) Bases in 16S rRNA protected by A- and P-site tRNA (82, 83). Also shown are sites of mutations conferring nonsense and frame-shift suppression (155–157), the site of cleavage by colicin E3 (128, 129), and the site of base pairing with mRNA initiation regions (42).

ing bases (72). The group of bases in domain IV that remain protected by tRNA lacking its 3'-terminal CA are also protected by 30S subunits (73). Thus, it seems likely that these effects are attributable to stimulation of 30S-50S interaction by tRNA binding. In a complementary study, 3'-terminal oligonucleotides containing the CCA-aminoacyl end of tRNA were bound to the 50S P site in the presence of sparsomycin and methanol (90). These oligonucleotide fragments give 23S rRNA footprints that are almost identical

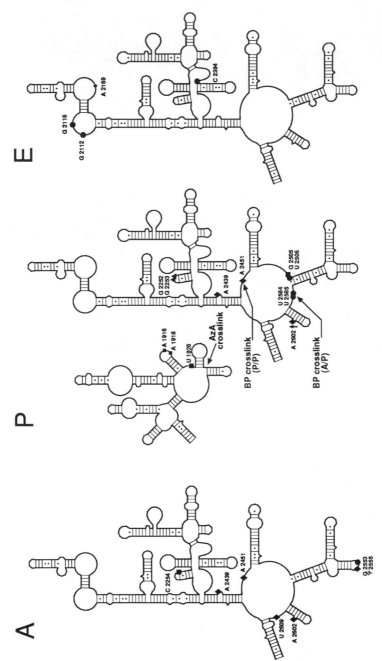

*Figure 3* Bases in 23S rRNA protected by A-, P- and E-site tRNA (72). Symbols show whether the protections require the acyl (aminoacyl or peptidyl) moiety (◆), the 3'-terminal A (▲), or the 3'-terminal CA (●) of tRNA. Also shown are the sites of cross-linking by azidoadenosine-containing tRNA in the P/P state (AzA) (71), and benzophenone-derivatized tRNA in the A/P or P/P states (BP) (58, 70).

to those of intact tRNA. It can be concluded that interaction of tRNA with rRNA takes place mainly, if not exclusively, through its anticodon and CCA extremities.

Aminoacyl tRNA can be bound stably as a ternary tRNA·EF-Tu·GTP complex, if GTP hydrolysis is prevented by addition of kirromycin or substitution of nonhydrolyzable GTP analogues. Protection of bases in such complexes differs from the standard A-site pattern. Whereas a normal A-site footprint is observed on 16S rRNA, none of the 23S A-site bases are protected by the ternary complex. Instead, the Ef-Tu dependent footprint in the $\alpha$-sarcin loop, around position 2660 (discussed below), is the only 23S rRNA protection found (72, 91). This result shows that the aminoacyl end of the incoming aminoacyl tRNA is shielded from interaction with the peptidyl transferase site while it is in a ternary complex, and can only participate in peptide bond formation following hydrolysis of GTP and release of EF-Tu·GDP. This is almost surely the physical manifestation of a proofreading mechanism. It has been proposed that the inherent accuracy of codon recognition is amplified by a branched kinetic scheme involving two separate steps in which noncognate aminoacyl tRNAs may be rejected (92, 93). The independence of these two steps is ensured by an intervening irreversible step (GTP hydrolysis), such that the accuracy of the overall recognition process is the product of the two events. The footprinting results suggest that in the first step of aminoacyl tRNA binding, interaction with the small subunit (and by implication, with mRNA) is normal, but interaction with the large subunit is mediated by EF-Tu. The proofreading step must follow release of the EF-Tu·GDP complex, and may correspond to a transient state following the first step, but preceding the docking of the aminoacyl end of tRNA with the 23S A site.

## Intermediate States in Translocation: Hybrid Sites

The tRNA footprints on 16S and 23S rRNA can be used as a direct assay for occupancy of the A, P, and E sites on the two ribosomal subunits. The state of occupancy of the tRNA binding sites has been examined during each step of a round of elongation in vitro (69). This study shows that the CCA end of peptidyl tRNA moves spontaneously from the P site to the E site of 23S rRNA upon peptide bond formation, while remaining in the P site with respect to 16S rRNA. This results in a hybrid state of binding, called the P/E state. Similarly, the CCA end of aminoacyl tRNA shifts from the A site to the P site on 23S rRNA, while remaining in the 16S rRNA A state. Thus, upon peptidyl transfer, the two tRNAs shift from A/A and P/P to A/P and P/E states, respectively. This movement occurs independently of factors or GTP. In the EF-G and GTP-dependent step, both tRNAs and mRNA move relative to 16S rRNA. Thus, translocation of tRNA occurs in two steps (Figure 3); in the first step, the tRNAs move relative to 23S rRNA, resulting in hybrid binding

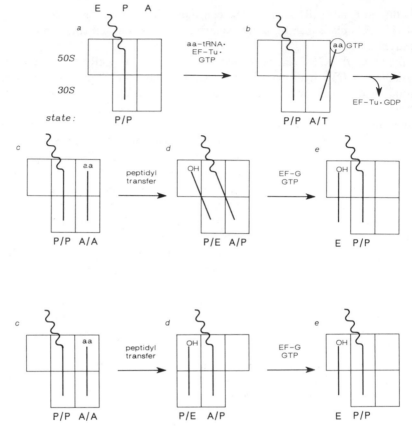

*Figure 4*  Hybrid site model for the movement of tRNA during translation (69). The binding states (P/P, A/T, etc) are shown at the bottom.

states, and the second step returns them to the nonhybrid state. Instead of two or three states of binding for tRNA, there are at least six: A/T, A/A, A/P, P/P, P/E, and E (Figure 4). Use of fluorescent-labeled tRNAs has provided physical evidence for this mechanism (94). Independent movement of the two ends of tRNA allows one end to be fixed while the other moves relative to the ribosome. Another important implication is that the nascent polypeptide chain remains in the same ribosomal location during the elongation cycle. Finally, the hybrid site model suggests that translocation may involve relative movement of 30S and 50S subunits, which would account for the fact that ribosomes are always made up of the two subunits. Evidence supporting this possibility has been discussed (95).

## Interactions of Antibiotics with Ribosomal RNA

Using the same chemical footprinting methods, many of the best characterized ribosome-directed antibiotics have been shown to protect specific bases in rRNA (84, 96). This, in itself, argues for the functional importance of rRNA, since antibiotics seem to act by directly interfering with specific steps in the translational process. Their mode of action recalls that of enzyme inhibitors, which have usually been found to bind to catalytic or regulatory sites of enzymes. In nearly every case so far, the bases protected by antibiotics occur in regions of rRNA that have independently been implicated in functions that are affected by the corresponding drugs. Moreover, there is a striking correlation between the sites of protection and the positions of mutations or posttranscriptional modifications that confer resistance to the same antibiotic.

Neomycin and related aminoglycoside antibiotics (paromomycin, gentamycin, and kanamycin) induce translational errors and inhibit translocation. These antibiotics protect A1408 and G1494 from chemical probes, and cause enhanced reactivity of C525 (84). A1408 and G1494 are two of the bases that are weakly protected by A-site tRNA, and are adjacent to the two strongly protected A-site bases, A1492 and G1493. Another aminoglycoside with a similar mode of action, hygromycin, protects G1494 but causes enhancement of A1408. Mutations conferring resistance to paromomycin occur at positions 1409 and 1491 (97, 98), and to hygromycin at position 1495 (97), all directly adjacent to the A-site tRNA and antibiotic footprints. A simple explanation for the mode of action of these antibiotics is that they bind very close to the site of codon recognition in the small subunit A site, increasing the nonspecific affinity for A-site tRNA, either by direct contact with tRNA or by distortion of the ribosomal decoding site. This increased affinity not only interferes with rejection of noncognate tRNA, but impedes translocation between the subunit A and P sites.

Edeine is an inhibitor of P-site tRNA binding. This antibiotic protects G693, A794, C795, and G926 in 16S rRNA (84); all four of these bases are also strongly protected by P-site tRNA (82, 83), and are believed to line the cleft of the 30S subunit (86). Thus, edeine appears to exert its effects by direct blockage of the 30S P site.

Chloramphenicol and carbomycin are well-known inhibitors of peptidyl transferase. These drugs protect A2058, A2059, A2062, A2451, and G2505 in the central loop of domain V of 23S rRNA (96). Chloramphenicol-resistance mutations have been identified at positions 2057, 2447, 2451, 2452, 2503, and 2504, adjacent to the sites of drug protection. As discussed above, benzophenone-derivatized Phe-tRNA has been cross-linked to positions 2451, 2452, and 2503–2506. Finally, protection of A2451 by the aminoacyl or peptidyl moiety of A- or P-site tRNA, and protection of G2505 and U2506 by P-site tRNA extend the correlation. Thus, four independent lines of evidence place the central loop of domain V in close proximity to the peptidyl transferase site. Whether nucleotides in this region are directly involved in this catalytic function remains to be seen.

EF-G-dependent hydrolysis of GTP is inhibited by thiostrepton but stimulated by micrococcin. Both drugs protect bases in the 1067–1098 region of 23S rRNA from chemical and enzymatic probes (98a). However, they have opposite effects on the reactivity of A1067; thiostrepton protects N1 from dimethyl sulfate attack, while micrococcin renders it more reactive, a striking parallel to their respective physiological effects.

In chemical protection studies, it is not often possible to distinguish whether ligand-RNA contact and or ligand-induced conformational changes are responsible for the observed effects. In at least one case, the 16S rRNA class III sites (84), there is a strong case for the latter explanation. These bases are protected by binding of tRNA to isolated 30S subunits (82), are also protected by 50S subunits in the absence of tRNA (73), and are protected by certain antibiotics in the absence of tRNA or 50S subunits (84). Since it is known that tRNA, 50S subunits, and these antibiotics can all bind simultaneously to 30S subunits, and moreover cooperatively stimulate each others' binding, it is unlikely that any of these ligands interacts directly with class III bases. The simplest explanation is that they are protected by a conformational change in the 30S subunit, such that binding of any of the three ligands stabilizes the protected conformation. Three of the bases, A790, A791, and A1394, are protected by edeine, while the other three, A909, A1413, and G1487, are protected by streptomycin, suggesting that two distinct conformational events occur, one of them P-site-related and the other A-site-related. Neomycin-class antibiotics, tRNA, or 50S subunits protect all six bases. These findings point to a connection between the processes of tRNA binding and subunit association and may be a clue to their respective mechanisms.

## Elongation Factors and Ribosomal RNA

Early studies showed that the two GTP-dependent elongation factors EF-Tu and EF-G bind to overlapping sites on the ribosome (99, 100). Chemical footprinting studies show that EF-Tu and EF-G protect overlapping sets of nucleotides in the universally conserved $\alpha$-sarcin loop at position 2660 of 23S rRNA (91). This result correlates with the observed loss of elongation factor function upon cleavage of the RNA following G2661 by $\alpha$-sarcin (101, 102). Both factors protect G2655, A2660, and G2661 [A2660 is the base removed by ricin, another cytotoxin that is known to inactivate ribosomes (103)]; A2665 is protected by EF-Tu, but not EF-G. EF-G (but not EF-Tu) protects

A1067 (91), a base whose corresponding ribose is methylated in thiostrepton-producing organisms (49), and which is located in the region of 23S RNA to which thiostrepton binds (48), and where EF-G has been cross-linked (104). This is also a region of interaction with r-proteins L11 and L7/L12, all of which have been linked to EF-G-related functions of the ribosomes.

## Initiation Factors and Ribosomal RNA

Initiation factor IF-3 promotes dissociation of ribosomes (by binding preferentially to free 30S subunits), and interaction of the initiator tRNA anticodon with start codons (105–108). IF-3 interacts with the central domain and with the 3'-terminal helix region of 16S rRNA. Chemical and enzymatic footprinting experiments and cross-linking have localized its effects to the 700 and 790 loops, the 840 stem, and the 1500 linker region (109–112). Proton NMR spectroscopy has shown that it interacts with the 3'-terminal stem, destabilizing the universally conserved G-U and adjacent A-U pairs (113). Mutation of G790 to A results in decreased binding of IF-3 (114).

The locations of these IF-3-related effects correlate well with what is known about their functions. These sites have all been located in (700 and 790 loops and 3'-helix) or near (840 stem) the platform of the 30S subunit, surrounding the site of codon-anticodon interaction. The effects in the 700 and 790 regions are closely adjacent to P-site tRNA-protected bases, consistent with the fact that IF-3 promotes recognition of initiation codons by initiator tRNA in the P site. Methylation of the 3'-terminal hairpin loop confers sensitivity to kasugamycin, an antibiotic that blocks translational initiation (115). Finally, the 700 and 790 loops and 3'-terminal helix have all been implicated in subunit association (116–118), in keeping with the subunit dissociation role of IF-3.

A model for the mode of action of IF-3 that accounts for these data and reconciles both of its functional roles is the following. IF-3 binds to the periphery of the 30S P site, loosening its nonspecific affinity for tRNA. This could occur through reorientation of the four or five helical elements that interact with the anticodon stem-loop of P-site tRNA (83, 86). The result would be to shift the burden of the P-site binding energy to codon-anticodon interaction. Normally, P-site binding is relatively strong, and is even independent of codon-anticodon interaction at high $Mg^{2+}$ concentrations, indicating a strong nonspecific affinity for tRNA. In contrast, A-site binding is weaker, and depends absolutely on mRNA. Thus, one role of IF-3 may be to transform the 30S P site into a quasi-A site, in which accurate codon recognition can occur. Since several 50S subunit contacts appear to be located at the periphery of the 30S P site, binding of IF-3 accomplishes a second functional task, subunit dissociation (or, anti-association).

IF-1 is known to stimulate the binding and activities of the other two factors, but its own functional role has remained vague. A possible insight into the function of IF-1 comes from chemical footprinting studies. Quite unexpectedly, IF-1 protects G529, G530, A1492, and A1493, giving a footprint virtually identical to that of A-site tRNA (112). The only differences are that the reactivities of A909 and A1408 are enhanced, rather than weakly protected. This result suggests that IF-1 acts as a surrogate A-site tRNA during initiation. The purpose of such a function is not immediately obvious, other than simply to block the A site. Perhaps occupying the A site helps to optimize other 30S functions, such as selection of the proper reading frame, and correct docking of the 50S subunit with the 30S initiation complex.

Among the three initiation factors, only IF-2 has so far failed to yield a detectable "footprint" on rRNA (119). However, its effect on the rRNA footprint of fMet-tRNA is revealing. IF-2 is a G-protein that appears to interact with both 30S and 50S subunits, as well as with fMet-tRNA. It is important for discrimination of formylated vs nonformylated tRNA, and activates the ribosome-dependent GTPase upon subunit association (105–108). When fMet-tRNA is bound with IF-2 in the presence of a nonhydrolyzable GTP analogue, a normal 16S rRNA P-site footprint is observed, but no protection of 23S rRNA can be detected (119). This result is a striking parallel to the results found with EF-Tu for A-site binding, and suggests that selection of initiator tRNA may be subject to a closely analogous kinetic proofreading mechanism, in the first step of which IF-2 shields its acceptor end from the peptidyl transferase P site. Although it constitutes only negative evidence, the absence of rRNA protection by IF-2 suggests that it may interact with the ribosome mainly by protein-protein contacts.

## Subunit Association

There is extensive evidence that rRNA is involved in subunit association. Treatment of 30S subunits with the RNA-specific reagent kethoxal abolishes their ability to associate with 50S subunits (117). Several conserved bases and a number of phosphates in 16S rRNA are protected by subunit association (73, 116, 118, 120, 121). Many of the bases appear to be important for the association process, in that their kethoxal modification interferes with this function in damage/selection experiments (117). Oligonucleotides complementary to the 50S-protected 790 loop sequences in 16S rRNA interfere with subunit association (122), and mutation of bases in this loop causes loss of the association function (123). Two dimethyl A residues at positions 1518 and 1519 are inaccessible to antibodies (124), consistent with the protection of G1517 from kethoxal by 50S subunits (118).

Similarly, bases in 23S rRNA have been identified that are protected by 30S subunits (125). It is not yet clear, however, which 16S sites go with which 23S sites, nor is it obvious that the interactions are in fact RNA-RNA, as opposed to RNA-protein, in nature. One report of a psoralen-mediated 16S-23S rRNA cross-link suggests interaction with a site near one end of 16S rRNA with a site around 40% of the distance from one end of 23S rRNA (126). This would be consistent with interactions of the 3'-terminal region of 16S rRNA with the domain IV region of 23S rRNA, to correlate the result with known protected sites. In view of the increasing implication of subunit association in several fundamental aspects of the translation process, identification of 16S-23S rRNA contacts (if, indeed, any exist) is clearly a problem that warrants further attention.

There are some subtleties of interpretation involving subunit association experiments. First, there is the omnipresent danger that subunits may become structurally damaged during the low $Mg^{2+}$ conditions of their isolation, and so reactivity differences between subunits and ribosomes may also include spurious effects. Second, if "native" 70S are compared with subunits (rather than reassociated 70S) there may be differences due to 70S-bound ligands, such as tRNA, factors, etc. Apart from these problems, there are at least two classes of conformational changes in the 30S subunit that need to be recognized. One is the aforementioned class III phenomenon, in which six bases in 16S rRNA appear to be conformationally protected upon subunit association (84). The other is the well-studied active-inactive transition, a reversible inactivation of the 30S subunit caused by depletion of $Mg^{2+}$ or monovalent cations (127). If specific steps are not taken to activate 30S subunits, they will typically exist as the inactive conformer as a result of low $Mg^{2+}$ isolation procedures. Thus, many of the subunit association results described in the literature may well include the effects of structural damage or conformational changes.

## RNA-Directed Biochemical Inactivation

Another important line of evidence for the functional role of rRNA is that various types of biochemical modification of the RNA, often very minor, result in loss of ribosomal activity. Although it can never be excluded that inactivation is the result of structural damage, or interference with r-proteins, many of these experiments bear close similarity to classical enzyme inactivation studies, in which the targets of modification were found to be components of catalytic sites or substrate-binding sites. In a broad sense, rRNA mutants, discussed in the following section, also represent a type of biochemical modification.

One of the first studies of this kind, and in retrospect, one of the most dramatic, was the inactivation of ribosomes by colicin E3 (128, 129). This is a highly specific nuclease that cleaves a single phosphodiester bond between A1493 and G1494 of 16S rRNA in 70S ribosomes. The site of cleavage is a universally conserved sequence associated with the codon recognition function, by several lines of evidence, including the facts that the A-site tRNA anticodon stem-loop protects A1492, A1493, and G1494 from chemical attack (82, 83), that the aminoglycoside antibiotics protect G1494 (84), and that a U to C change at position 1495 confers hygromycin resistance (97). Accordingly, it is found that attack by colicin E3 is blocked by A-site tRNA, and by the A-site-related antibiotics streptomycin, tetracycline, and gentamycin (130). That scission of a single bond in a macromolecular structure as vast as the ribosome results in complete inactivation argues strongly that the site of attack is crucially and directly involved in translational function.

The counterparts to colicin E3 in the large subunit are $\alpha$-sarcin and ricin. Again, these cytotoxins cause complete loss of translational activity by cleavage of a single bond. In the case of $\alpha$-sarcin, it is cleavage of the phosphodiester bond following G2661 of 23S rRNA (102), and for ricin, cleavage of the glycosyl linkage of A2660, releasing free adenine from its ribose (103). As discussed above, loss of activity is correlated with inactivation of functions that depend on the elongation factors EF-Tu and EF-G in prokaryotes, or EF-1 and EF-2 in eukaryotes (101). EF-Tu and EF-G give overlapping footprints at the sites of cleavage by the cytotoxin (91), suggesting that their interaction with the $\alpha$-sarcin loop in 23S rRNA is in some way crucial for their function. Protection of this loop from $\alpha$-sarcin cleavage by bound elongation factors provides further evidence in support of this (131). The actions of both factors require GTP hydrolysis, and in fact the "G domain" of EF-Tu, a proteolytic fragment containing its GTP-binding site, gives a similar footprint (132). These findings raise the possibility that the $\alpha$-sarcin loop may actually be involved in the ribosome-dependent GTPase function. A third GTPase event is promoted by IF-2; no such footprint has been detected for IF-2-containing complexes, suggesting that its GTP-dependent function is somewhat different from that of the elongation factors.

Peptidyl transferase has been inactivated by digestion of ribosomes with ribonuclease T1 (133). Activity in the fragment reaction was lost in parallel with loss of binding of the Ac-Leu-oligonucleotide to the 50S P site. Binding of a nonacetylated aminoacyl-oligonucleotide to the A site was unaffected. Erythromycin protected ribosomes against inactivation.

Another approach exploits the use of DNA oligomers that are complementary to putative functional sites in rRNA. Thus, oligonucleotides complementary to the 790 loop inhibit subunit association and IF-3 binding (122); oligonucleotides complementary to the 810 region of 23S rRNA inhibit tRNA binding (134); an oligomer complementary to the Shine-Dalgarno sequence has been used as an indicator for the ability of ribosomes to bind natural mRNA (135). The oligonucleotide-binding approach has the advantage of being highly site-specific, although it is important to demonstrate directly that binding indeed takes place at the intended site, and that no significant binding occurs elsewhere. One drawback to this approach is that the oligonucleotides themselves are large and, by virtue of their base pairing with rRNA, cause obligatory conformational changes upon binding. This can result in disruption of higher-order structure in the rRNA, thereby making the interpretation of any functional effects equivocal. Indeed, chemical and enzymatic modification studies suggest that very few sequences in rRNA of significant length are actually free to base pair; unpaired hairpin loop and linker sequences usually behave as if they are involved in tertiary or quaternary interactions in active ribosomes. Another potential hazard is that endogenous ribonuclease H may cleave the rRNA once the DNA oligonucleotide binds, and so loss of function may in some cases be caused by cleavage events, although this can also provide important information.

Depletion of monovalent cations or magnesium ions causes a reversible conformational change in 30S subunits (136). Loss of activity in nonenzymatic binding of tRNA, dihydrostreptomycin binding, and association with 50S subunits characterize this change in conformation. All of the activities are coordinately restored by heating the particles in the presence of replenished cations. Since ribosomes that are inactive in nonenzymatic tRNA binding become active in factor-promoted binding, it was suggested that the conformational change may have physiological relevance (137). In studies using $N$-ethylmaleimide modification, the subunits were irreversibly inactivated when exposed to the reagent in the inactive state, probably due to modification of protein S18, and possibly S21 (137). These findings suggested that the active-inactive transition could be related to some translational event that normally involves these proteins.

Chemical and enzymatic probing of 16S rRNA in the active and inactive forms of the 30S subunit revealed interesting differences in the accessibility of the RNA (138). The transition does not appear to be simply a loosening of the structure due to ion depletion, since both increases and decreases in reactivity are observed upon inactivation, giving the appearance of a reciprocal interconversion between two differently structured states. Changes in reactivity are almost exclusively confined to the decoding region centered around positions 1400 and 1500, but significant effects are also detected at U723 and G791, which are believed to be proximal to the decoding region in three dimensions. The Shine-Dalgarno region is less accessible in the inactive form (135). The most dramatic changes in reactivity involve the universally conserved G926, C1395, A1398, and G1401. Interestingly, the three purines show reciprocal behavior at the N1 vs N7 positions. G926 loses its reactivity at N1, becoming highly reactive at N7; in contrast, A1398 and G1401 become reactive at N1 but lose their hyper-reactivity at N7. These changes in reactivity show that the higher-order structure of 16S rRNA is altered by the active-inactive transition, and raises an alternative explanation for the inactivation, that a crucial functional site (or sites) in the RNA is disrupted. Indeed, the proteins whose chemical reactivities are altered by the transition (S1, S18, and S21) are all believed to be located in close proximity to the conformationally altered region of 16S rRNA. Loss of nonenzymatic tRNA binding can be rationalized by these results. Under these conditions, tRNA is most likely bound to the 30S P site. Both N1 of G926 and N7 of G1401 are strongly protected by P-site binding (82, 83), consistent with the interpretation that the unavailability of these positions is, at least in part, responsible for loss of tRNA binding. The causes of loss of dihydrostreptomycin and 50S interactions are less obvious, but changes in the reactivities of A907 and of nucleotides in the penultimate stem suggest that these effects may also be related to changes in the conformation of 16S rRNA.

## GENETIC EVIDENCE

### Detection of Ribosomal RNA Phenotypes

The first ribosomal mutations to be characterized were streptomycin resistance and dependence mutations in ribosomal protein S12 (5, 6). These findings tended to reinforce the ribosomal protein paradigm, even though it was known that ribosomal proteins are encoded by single-copy genes, whereas there are many copies of the rRNA genes in *E. coli*, the organism of choice for these studies. It is now clear that the phenotypes of organisms bearing mutations in rRNA often depend crucially on the relative amounts of mutant and wild-type ribosomes in the cell (139, 140). Indeed, in organisms or organelles bearing single-copy rRNA operons, drug resistance (as well as other) mutations in rRNA are readily detected.

Most genetic studies still use *E. coli* as the experimental organism. One reason for this is that the vast majority of the biochemical and structural studies have been done on *E. coli* ribosomes, providing the most complete context for interpretation of genetic results. Another reason is the power of the available methods for manipulating *E. coli* using both recombinant and classical genetics. Since there are seven rRNA operons in *E. coli*, there is a danger of mutant phenotypes being masked by the presence of wild-type copies. Use of multicopy plasmids containing a cloned rRNA operon results in cellular ribosome populations in which as much as 65% of the ribosomes contain plasmid-encoded rRNA (139, 140). This often allows expression of the mutant phenotype over that of the wild-type, in which cells bearing plasmid-carrying mutant rRNA genes exhibit impaired growth rates or resistance to antibiotics. However, some mutations give rise to recessive phenotypes that would escape detection by these criteria (141). This emphasizes the subtlety of interpretation of rRNA phenotypes. Since many published studies have relied on differences in growth rate, it should be kept in mind that some of the mutations exhibiting little or no effect may in fact cause significant impairment of ribosome function that is, for some reason, recessive in the context of a significant proportion of wild-type ribosomes. Another difficulty arises in attempting to identify the precise functional impairment in the mutant ribosomes, due to the fact that in vitro assays must be carried out on a mixed population of ribosomes.

One solution to these problems is the use of in vitro expression. Mutant rRNA genes are transcribed using T7 RNA polymerase and reconstituted in vitro with r-proteins (142). This approach has provided extensive information concerning the functional effects of mutations in the decoding region (143). The drawbacks involve the usual problems of in vitro vs in vivo characterization. Thus, mutation of certain universally conserved bases has little or no observed effect on any of the in vitro assays, prima facie evidence for the inadequacy of the assays. In at least one case, the same set of mutations has been characterized independently by in vitro and in vivo methods. These mutations involve substitution of the conserved C1400. In vitro, the most pronounced effects were seen for the A-site tRNA-binding assay, where G, A, and U substitutions gave 60, 80, and 130% activity, respectively, compared with C (143). In contrast, in vivo activity using "specialized ribosomes" (see below) indicated that the G substitution was virtually nonfunctional; however, the effects of the mutations were again (in order of decreasing severity) $G > A > U$ (144). Similarly, a C to A substitution at position 1402 gave 60–90% activity in in vitro assays, but the in vivo assay again showed no activity for the same mutation. This shows that in vitro assays run the risk of underestimating the physiological consequences of rRNA mutations.

Another approach utilizes selectable markers in the cloned rRNA genes (141, 145). For example, C1192 to U in 16S rRNA confers spectinomycin resistance and A2058 to G in 23S rRNA erythromycin resistance (146, 147). By selecting on antibiotic-containing media, cells are forced to use ribosomes containing rRNA from the cloned genes. Recessive phenotypes are easily detected using this approach (141). The presence of the marker mutations is also useful for accurately measuring the relative amounts and cellular distribution of cloned vs genomic rRNA, using primer extension (140).

An elegant in vivo approach is the use of "specialized ribosomes," in which the Shine-Dalgarno sequence in the cloned 16S rRNA gene is engineered to recognize predominantly a single mRNA whose translational product acts as a reporter for the functional activity of the cloned mRNA (148). As mentioned above, this system has proven its usefulness in characterization of site-directed mutations in the decoding region of 16S rRNA (144).

Because of the essential role of ribosomes, mutations affecting central functions may be lethal. If these mutants are able to compete with wild-type ribosomes for limiting components of the translational machinery, they may have dominant phenotypes. Although the failure to obtain stable transformants containing certain mutations is often taken to indicate a dominant lethal phenotype, it is important to demonstrate this directly. This has been done by expression of the mutant rRNA from a lambda $P_L$ promoter, in conjunction with the temperature-sensitive cI857 repressor (139, 149). One caution regarding this method is that levels of expression of rRNA genes from this phage promoter are significantly lower than observed with bacterial rRNA promoters, so that the lower proportion of mutant ribosomes may in some cases result in misleadingly mild phenotypes.

## Antibiotic Resistance

The number of antibiotic resistance mutations in rRNA is now so great that space limitations do not permit adequate discussion of this topic. For further details, the reader should refer to several recent reviews (22, 23, 29, 30, 150–152). Antibiotic resistance can be conferred by several types of genetic (or phylogenetic) changes involving rRNA, including base methylation, absence of base methylation or ribose methylation (reviewed in 151), point mutation (reviewed in 22, 23, 30, 150), and deletion (153, 154). As mentioned above, the sites of antibiotic resistance mutations are more often than not correlated with both the sites that are protected by the drug from chemical modification and sites that have been implicated in ribosomal function, usually the function targeted by the drug itself.

## Translational Suppressors

The first mutant of this type to be characterized was a spontaneous chromosomal mutation in E. coli 16S rRNA that was found to suppress UGA nonsense mutations (155). It suppresses UGA mutations at only two of four positions tested in the trpA gene, and is therefore context-dependent. The alteration in 16S rRNA was found to be deletion of the bulged nucleotide C1054. It has been proposed that this region of 16S rRNA is involved in reading the UGA codon by base pairing. This model must be reconciled with current models for the folding of 16S rRNA, which place this region rather far from the decoding site. Alternatively, the mutation may interfere with proper functioning of the release factor RF-2. This intriguing mutant clearly deserves further study.

Another nonsense suppressor mutation in rRNA is the ochre suppressor MSU1 from yeast mitochondria (156). Ochre suppression is conferred by a G to A mutation in 15S rRNA at a position corresponding to G517 of E. coli 16S rRNA. This universally conserved base is located in the top of the stem below the 530 loop, a structural feature implicated in the A-site (decoding) function of the 30S subunit, as discussed above. The effect of the mutation is to change a potential G-C pair to an A-C mismatch; it should be noted that there is no phylogenetic evidence to support the existence of this base pair, and so the mutated G may in fact be involved in tertiary or quaternary interactions. Although it might be imagined that a general disruption of the decoding process could give rise to ochre suppression, the MSU1 mutation was ineffective in attempts to suppress various amber, missense, and frame-shift mutations. Thus, like the *E. coli* C1054 mutation, the MSU1 suppressor appears to be quite codon-specific. As for position 1054, the location of G517 is remote from the decoding site according to our current understanding of the folding of 16S rRNA in the 30S subunit, and so the mechanism of suppression is likely to be indirect.

A third example of an rRNA translational suppressor is a frame-shift suppressor isolated from yeast mitochondria. This has been shown to be conferred by a C to G mutation at a position corresponding to nucleotide C1409 in *E. coli* 16S rRNA (157). The effect on the secondary structure is to disrupt the 1409-1491 base pair, the terminal base pair of the penultimate stem, which lies directly adjacent to A1408, A1492, and A1493, bases that are believed to be at the site of interaction of the anticodon stem-loop of A-site tRNA with the 30S subunit, as described above. Here, it seems more likely that the mechanism of suppression will turn out to be direct, resulting from perturbation of the structure of the decoding site itself. Although there is extensive phylogenetic evidence for the existence of this base pair in bacterial and organelle 16S rRNAs, it is mispaired in the small subunit rRNAs of several eukaryotic species (16). If this structural feature is important for maintenance of the translational reading frame, it raises the question of how these eukaryotic ribosomes accommodate its disruption.

## Conditional Mutants

Both temperature-sensitive (ts) and cold-sensitive (cs) mutants of rRNA have been isolated (141, 158, 159). Several recessive ts mutants in 16S rRNA have been characterized in 16S rRNA, and have been found to be caused by disruption of base pairing (141). More unexpected is the isolation of a dominant cold-sensitive mutation in 16S rRNA that is caused by conversion of a G-C base pair to G-U (158). This mutant is suppressed by mutation of the G-U pair to A-U, but also by other mutations outside the pairing region. There are indications that cold sensitivity may be caused by formation of alternate conformations of 16S rRNA.

Even more puzzling is a class of ts mutants affecting 16S rRNA (as determined by expression of the selectable $Spc^R$ marker) which map upstream of the 5'-terminus of mature 16S rRNA (159). These mutations, which occur at positions $-13$, $-30$, and $-59$, lie well outside all known transcriptional and processing signals, and have no detectable effect on maturation of the 5'-end of 16S rRNA. The relative distribution of plasmid and chromosomally derived 16S rRNA in vivo is similar between 30S subunits, 70S ribosomes, and polysomes at the permissive and restrictive temperatures. Several second-site suppressors were found; all but one (at position 21 of mature 16S rRNA) also map to the upstream region. These mutations emphasize our ignorance concerning the role of this conserved upstream region.

## Site-Directed Mutations

Site-directed mutagenesis is best used when models are sufficiently well developed that critical tests of their correctness can be devised. Most researchers would admit that most aspects of the mechanism of action of the ribosome are still poorly understood, and that unbridled site-directed mutagenesis at this stage of our knowledge raises the danger of creating more mysteries than it solves. Nevertheless, some testable models are emerging, in both the structural and functional realms, and it is worth examining how they are standing up to this type of scrutiny.

Site-directed mutagenesis has been used by two groups to provide a particularly convincing demonstration of the Shine-Dalgarno mechanism. In one study, a C to U change at position 1538 was made, converting the CCUCC sequence to CCUUC (46). The mutation was lethal when expressed at high copy number, but when expressed conditionally at lower levels from the lambda $P_L$ promoter caused a dramatically altered pattern of protein synthesis. In instances where the proteins with altered levels were identified, it was found that their increased or decreased synthesis correlated with the ability of their mRNAs to pair with the mutant 16S rRNA. In another study, the CCUCC sequence was changed drastically to GGAGG or CACAC, and used in vivo to translate human growth hormone mRNA containing a CCUCC or GUGUG sequence upstream from its start codon (47). Translation of the target mRNA depended on complementarity between the corresponding 16S rRNA and mRNA sequences. The resulting "specialized ribosomes," in conjunction with a $Spc^R$ mutation, can be used to overproduce specific proteins in vivo, as well as serving as a powerful system for assaying the effects of rRNA mutations in vivo, as mentioned above.

The importance of bases that have been implicated in function by biochemical experiments can be tested by directed mutagenesis. The universally conserved G530, which is protected by A-site tRNA, is a particularly clear example (139). Base substitution mutations at position 530 could not be cloned using the rrnB promoter, suggesting that they are lethal. This was confirmed by using conditional expression from the lambda $P_L$ promoter. Mutant 16S rRNA was incorporated into 30S subunits and 70S ribosomes, but only at low levels into polysomes, consistent with the proposed involvement of G530 in A-site function. A proposed tertiary interaction (160) between positions 524–526 in this loop and 505–507 in the adjacent bulge loop has been tested by directed mutagenesis (161). A change in any of these bases interferes with the function of the 16S rRNA; however, if compensating pairs of changes are made, preserving the Watson-Crick relationships, function is maintained, providing direct evidence for the proposed tertiary interaction. Interestingly, certain mutations, in which non-Watson-Crick (e.g. wobble) pairings are created, confer low levels of resistance to streptomycin. This is all the more interesting in light of the fact that protein S12, mutations in which also confer streptomycin resistance, appears to stabilize the 505–507/524–526 interaction during in vitro assembly (162).

Another example is the directed mutagenesis of bases in the central loop of domain V of 23S rRNA, in the region associated with the peptidyl transferase function. Again, using a heat-inducible promoter, it was shown that changing A2060 to C or A2450 to C are dominant lethal events (149). The mutant rRNAs were incorporated into 50S subunits, but strongly excluded from 70S ribosomes.

A provocative result was obtained with an $\alpha$-sarcin loop mutation, in which a G to C substitution was made at position 2661 of 23S rRNA (163). Here, the distribution of mutant RNA among 50S, 70S, and polysomes was similar to that of wild-type rRNA, in both DH1 and HB101 hosts. However, the growth rate in HB101, which carries an *rpsL* 20 mutation in r-protein S12, was drastically impaired. Testing of other *rpsL* strains showed that the impairment of growth depends on the presence of a restrictive (i.e. hyperaccurate) *rpsL* phenotype, and is more drastic with more restrictive strains. These results provide evidence for intersubunit communication between the $\alpha$-sarcin loop in the 50S subunit and protein S12 in the 30S subunit. The effects of protein S12 on the 530 loop, described above, then draw a link between the universally conserved 530 and $\alpha$-sarcin loops, both of which have been implicated in functions related to binding of aminoacyl tRNA. In vitro studies on the C2661 mutant ribosomes show a decreased elongation rate and a tighter binding of EF-Tu. Since both EF-Tu and EF-G interact with the $\alpha$-sarcin loop, it has been suggested that this part of 23S rRNA might be part of the ribosomal GTPase (91). The observed stabilization of binding of EF-Tu to the C2661 ribosomes could be explained by blocking of ribosomal GTPase, since GTP cleavage is required for release of EF-Tu.

Other site-directed mutations have provided direct evidence for the role of the 790 loop in subunit association and IF-3 binding (114, 123), and the importance of bases in the 1400 region for tRNA binding (143, 144). Many other rRNA mutants have been constructed, although in most instances their specific effects on the translation process are not yet clear. It is abundantly evident that the major challenge here lies in characterization of rRNA mutants, not in constructing them.

## FUNCTIONAL ROLE OF RIBOSOMAL PROTEINS

The lack of all-or-nothing effects when ribosomal proteins are omitted, either biochemically (7) or genetically (41), is consistent with the idea that function is determined fundamentally by the rRNA. However, several lines of evidence support the importance of r-proteins in translation. First, omission of individual proteins usually impairs ribosome activity. Second, chemical modification of r-proteins causes partial loss of ribosomal activity in many cases. Third, many mutations in r-protein genes affect ribosomal function. How can this be reconciled with the notion that rRNA is the functional molecule? The generality that emerges from these results is that perturbation of the proteins tends to have only partial effects on function. This can be interpreted in a number of ways. If protein assembly is important for folding of the RNA into a compact particle, omission of one or more proteins could result in incompletely folded particles, or a mixed population of particles, some of which manage to fold correctly and function. On a more subtle level, certain proteins may be important for "fine tuning" of the conformational details of the RNA, for example in functional sites. It remains an important possibility that the proteins are themselves directly involved in certain aspects of ribosomal function, but this would be expected to give all-or-nothing effects. Yet another possibility is that the protein and RNA cofunction. Catalytic sites or binding sites for functional ligands could be composed of both protein and RNA; proteins could modulate the movement of the RNA, or could transmit conformational effects between remote sites in the RNA. Our present knowledge is incomplete, but there is already evidence to support several of these possibilities.

### Protein Mutants and Ribosomal RNA

A long-standing question is whether mutations in r-proteins might perturb the conformation of rRNA. Early genetic studies on ribosomal proteins showed that streptomycin-dependent strains of *E. coli*, in which mutations in protein S12 confer the Sm$^D$ phenotype, could be suppressed by mutations in protein S4 or S5. When segregated from the Sm$^D$ background, the S4 mutations had an abnormally high translational error rate and were referred to as *ram* [ribosomal ambiguity] mutants (164). Suppression was attributed to compensation for the abnormally low error frequency of the Sm$^D$ mutant by the *ram* mutation. Evidence for the influence of mutations in proteins S4 and S12 on 16S rRNA has been obtained by chemical probing of the RNA in mutant ribosomes (85): in S4 ram mutants, the reactivities of A8 and A16 are enhanced, while in Sm$^D$ mutants of S12, the reactivities of A908, A909, A1413, and G1487 are decreased. There is also an apparent correlation between the miscoding potential of mutant ribosomes and the dimethyl sulfate reactivity of A908. The data are consistent with a simple model in which the reactivity of A908 reflects the position of an equilibrium between two conformational states of the 30S subunit, which is in turn influenced by proteins S4 and S12. Recently, a mutation in 16S rRNA has been found that suppresses an Sm$^D$ mutation (165), because of a C to U change at position 1469 of 16S rRNA. As for the S4 and S5 mutations, the U1469 mutation confers a *ram* phenotype when present in a wild-type background.

Another example of suppression of a mutation in a ribosomal protein by an rRNA mutation concerns the large subunit. A Gly to Asp mutation at position 84 of protein L24 gives a temperature-sensitive phenotype (166). Among the spontaneous revertants found were mutants in which the ts phenotype is suppressed by a C to U substitution at position 33 of 23S rRNA (167), which lies within the L24 binding site (168). These studies show that genetic approaches can also be used effectively to dissect ribosomal protein-rRNA interactions.

## Assembly Experiments

Results from in vitro assembly experiments have an interesting bearing on the possibility that certain r-proteins are involved in the fine tuning of RNA functional sites. One cluster of proteins that has traditionally been associated with ribosomal function is S6, S11, S18, and S21, which surround the cleft of the small subunit. Assembly of these proteins affects the reactivities of bases in parts of the 16S rRNA that are also located in the cleft region (169, 170). Bases in the vicinity of tRNA-protected sites around positions 690, 790, and 926 are affected, resulting in a pattern of base reactivity resembling that of the folded, active subunit. An example is the reactivity of G926, a base that is protected from kethoxal attack by P-site tRNA (82, 83). In naked or partly assembled 16S rRNA, the bulged G926 is only weakly reactive, as are the other guanines immediately adjacent to it, as if there are several coexisting structures, each with a different guanine bulged. The effect of assembly (and proteins S11 and S21 contribute to this effect) is to simultaneously decrease the reactivities of G925, G927, and G928, while rendering G926 hyperreactive. Assembly of the proteins selectively stabilizes one of the four apparently thermodynamically equivalent conformers, presumably converting this part of the structure into its active form. It has been shown that protein S21 is essential for initiation of translation on natural mRNA (171), probably because S21 is required for the availability of the Shine-Dalgarno CCUCC sequence in 16S rRNA for base pairing (135). This is a particularly clear case of a protein-dependent conformational change in rRNA that is required for ribosomal function. Similar effects occur in the A-site-related 530 loop, where proteins S4 and S12 (themselves heavily implicated in A-site-related functions) produce conformational effects that appear to convert the loop into its active form (170). There is a curious aspect to these findings: Why is it left to protein assembly (and in fact, to rather late-assembling proteins in some cases) to render these conformations? It is difficult to see why these steps could not be effected by primary binding proteins or by the naked RNA itself. One possibility is that these reactivities are the manifestations of intricate three-dimensional folding of the RNA, and this folding depends on earlier steps of folding and assembly. Another possibility is that these amount to regulatory steps, postponing the emergence of the active conformations of local sites until the entire particle is near its final structure.

## Literature Cited

1. Crick, F. H. C. 1960. In *The Nucleic Acids*, ed. E. Chargaff, J. N. Davidson, Vol. 3, pp. 400–1. New York: Academic
1a. Hoagland, M. B. 1960. In *Nucleic Acids*, ed. E. Chargaff, J. N. Davidson, 3:400–1. New York: Academic
2. Pestka, S. 1974. *Methods Enzymol.* 30:462–70
3. Gavrilova, P., Spirin, A. S. 1974. *Methods Enzymol.* 30:452–62
4. Monro, R. E., Maden, B. E. H., Traut, R. R. 1967. *Symp. Fed. Eur. Biochem. Soc.*, ed. D. Shugar, pp. 179–203
5. Ozaki, M., Mizushima, S., Nomura, M. 1969. *Nature* 222:333–39
6. Birge, E. A., Kurland, C. G. 1969. *Science* 166:1292–84
7. Nomura, M., Mizushima, S., Ozaki, M., Traub, P., Lowry, C. V. 1969. *Cold Spring Harbor Symp. Quant. Biol.* 34:49–61
8. Cech, T. R., Zaug, A. J., Grabowski, P. J. 1981. *Cell* 27:487–96
9. Guerrier-Takada, C., Gardiner, K., Marsh, T., Pace, N., Altman, S. 1983. *Cell* 35:849–57
10. Chambliss, G., Craven, G. R., Davies, J., Davis, K., Kahan, L., Nomura, M., eds. 1980. *Ribosomes. Structure, Function and Genetics*. Baltimore: Univ. Park Press
11. Noller, H. F., Woese, C. R. 1981. *Science* 212:403–11
12. Woese, C. R., Gutell, R. R., Gupta, R., Noller, H. F. 1983. *Microbiol. Rev.* 47:621–69
13. Thompson, J. F., Hearst, J. E. 1983. *Cell* 33:19–24
14. Brimacombe, R., Maly, P., Zwieb, C. 1983. *Prog. Nucleic Acid Res. Mol. Biol.* 28:1–48
15. Noller, H. F. 1984. *Annu. Rev. Biochem.* 53:119–62
16. Gutell, R. R., Weiser, B., Woese, C. R., Noller, H. F. 1985. *Prog. Nucleic Acids Res. Mol. Biol.* 32:155–216

17. Hardesty, B., Kramer, G., eds. 1986. *Structure, Function and Genetics of Ribosomes*. New York: Springer-Verlag
18. Noller, H. F., Asire, M., Barta, A., Douthwaite, S., Goldstein, T., et al. 1986. See Ref. 17, pp. 143–63
19. Nomura, M. 1987. *Cold Spring Harbor Symp. Quant. Biol.* 52:653–63
20. Noller, H. F., Stern, S., Moazed, D., Powers, T., Svensson, P., Changchien, L.-M. 1987. *Cold Spring Harbor Symp. Quant. Biol.* 52:695–708
21. Moore, P. B. 1988. *Nature* 331:223–27
22. De Stasio, E. A., Göringer, H. U., Tapprich, W. E., Dahlberg, A. E. 1988. *Genetics of Translation*, ed. M. F. Tuite, M. Picard, M. Bolotin-Fukuhara, pp. 17–41. Berlin: Springer-Verlag
23. Morgan, E. A., Gregory, S. T., Sigmund, C. D., Borden, A. 1988. See Ref. 22, pp. 43–53
24. Raué, H. A., Klootwijk, J., Musters, W. 1988. *Prog. Biophys. Mol. Biol.* 51:77–129
25. Dahlberg, A. E. 1989. *Cell* 57:525–29
26. Hill, W. E., Dahlberg, A. E., Garrett, R. A., Moore, P. B., Schlessinger, D., Warner, J. R. 1990. *The Ribosome. Structure, Function and Evolution*. Washington, DC: Am. Soc. Microbiol.
26a. Zimmermann, R. A., Thomas, C. L., Wower, J. 1990. See Ref. 26, pp. 331–47
27. Noller, H. F., Moazed, D., Stern, S., Powers, T., Allen, P. N., et al. 1990. See Ref. 26, pp. 73–92
27a. Noller, H. F. 1990. *Harvey Lect.* 84:97–117
28. Brimacombe, R., Greuer, B., Mitchell, P., Osswald, M., Rinke-Appel, J., et al. 1990. See Ref. 26, pp. 93–106
29. Tapprich, W. E., Göringer, H. U., de Stasio, E., Prescott, C., Dahlberg, A. E. 1990. See Ref. 26, pp. 236–42
30. Cundliffe, E. 1990. See Ref. 26, pp. 479–90

30a. Woese, C. R. 1980. See Ref. 10, pp. 357–76
31. Crick, F. H. C. 1968. J. Mol. Biol. 38:367–79
32. Orgel, L. E. 1968. J. Mol. Biol. 38:381–93
33. Woese, C. R., Fox, G. E., Zablen, L. B., Uchida, T., Bonen, L., et al. 1975. Nature 254:83–86
34. Noller, H. F. 1974. Biochemistry 13: 4694–4703
35. Woese, C. R., Magrum, L. J., Gupta, R., Siegel, R. B., Stahl, D. A., et al. 1980. Nucleic Acids Res. 8:2275–93
36. Noller, H. F., Kop, J., Wheaton, V., Brosius, J., Gutell, R. R., et al. 1981. Nucleic Acids Res. 9:6167–89
37. Stiegler, P., Carbon, P., Zuker, M., Ebel, J.-P. 1981. Nucleic Acids Res. 9:2153–72
38. Zwieb, C., Glotz, C., Brimacombe, R. 1981. Nucleic Acids Res. 9:3621–40
39. Wittmann-Liebold, B. 1986. See Ref. 17, pp. 326–61
40. Wittmann-Liebold, B., Köpke, A. K. E., Arndt, E., Krömer, W., Hatakeyama, T., Wittmann, H.-G. 1990. See Ref. 26, pp. 598–616
41. Dabbs, E. R. 1986. See Ref. 17, pp. 733–48
42. Shine, J., Dalgarno, L. 1974. Proc. Natl. Acad. Sci. USA 71:1342–46
43. Steitz, J. A., Jakes, K. 1975. Proc. Natl. Acad. Sci. USA 72:4734–38
44. Dunn, J. J., Buzash-Pollert, E., Studier, F. W. 1978. Proc. Natl. Acad. Sci. USA 75:2741–45
45. Taniguchi, T., Weissmann, C. 1978. J. Mol. Biol. 118:533–65
46. Jacob, W. F., Santer, M., Dahlberg, A. E. 1987. Proc. Natl. Acad. Sci. USA 84:4757–61
47. Hui, A., de Boer, H. A. 1987. Proc. Natl. Acad. Sci. USA 84:4762–66
47a. Weiss, R., Dunn, D., Atkins, J., Gesteland, R. 1990. See Ref. 26, pp. 534–40
48. Cundliffe, E. 1986. See Ref. 17, pp. 586–604
49. Thompson, J., Schmidt, F., Cundliffe, E. 1982. J. Biol. Chem. 257:7915–17
50. Burma, D. P., Nag, B., Tewari, D. S. 1983. Proc. Natl. Acad. Sci. USA 80:4875–78
51. Staehelin, T., Maglot, D., Monro, R. E. 1969. Cold Spring Harbor Symp. Quant. Biol. 34:39–48
52. Monro, R. E. 1971. Methods Enzymol. 20:472–81
53. Monro, R. E., Vazquez, D. 1967. J. Mol. Biol. 28:161–65
54. Moore, V. G., Atchison, R. E., Thom-
as, G., Moran, M., Noller, H. F. 1975. Proc. Natl. Acad. Sci. USA 72:844–48
55. Hampl, H., Schulze, H., Nierhaus, K. H. 1981. J. Biol. Chem. 256:2284–88
56. Noller, H. F., Ares, V., Zagórska, L. Unpublished.
57. Prince, J. B., Taylor, B. H., Thurlow, D. H., Ofengand, J., Zimmermann, R. A. 1982. Proc. Natl. Acad. Sci. USA 79:5450–54
58. Barta, A., Steiner, G., Brosius, J., Noller, H. F., Kuechler, E. 1984. Proc. Natl. Acad. Sci. USA 81:3607–11
59. Ofengand, J. 1980. See Ref. 10, pp. 497–529
60. Ofengand, J., Ciesiolka, J., Denman, R., Nurse, K. 1986. See Ref. 17, pp. 473–94
61. Gornicki, P., Nurse, K., Hellmann, W., Boublik, M., Ofengand, J. 1984. J. Biol. Chem. 259:10493–98
62. Wagenknecht, T., Frank, J., Boublik, M., Nurse, K., Ofengand, J. 1988. J. Mol. Biol. 203:753–60
63. Oen, H., Pellegrini, M., Eilat, D., Cantor, C. R. 1973. Proc. Natl. Acad. Sci. USA 70:2799–803
64. Girshovich, A. S., Bochkareva, E. S., Kramarov, V. M., Ovchinnikov, Yu A. 1974. FEBS Lett. 45:213–17
65. Barta, A., Kuechler, E., Branlant, C., SriWidada, J., Krol, A., Ebel, J.-P. 1975. FEBS Lett. 56:170–74
66. Breitmeyer, J. B., Noller, H. F. 1976. J. Mol. Biol. 101:297–306
67. Barta, A., Kuechler, E., Steiner, G. 1990. See Ref. 26, pp. 358–65
68. de Groot, N., Panet, A., Lapidot, Y. 1971. Eur. J. Biochem. 23:523–27
69. Moazed, D., Noller, H. F. 1989. Nature 342:142–48
70. Steiner, G., Kuechler, E., Barta, A. 1988. EMBO J. 7:3949–55
70a. Wower, J., Hixson, S. S., Zimmermann, R. A. 1988. Biochemistry 27: 8114–21
71. Wower, J., Hixson, S. S., Zimmermann, R. A. 1989. Proc. Natl. Acad. Sci. USA 86:5232–36
72. Moazed, D., Noller, H. F. 1989. Cell 57:585–97
73. Moazed, D., McWhirter, J., Noller, H. F. Unpublished.
74. Stiege, W., Atmadja, J., Zobawa, M., Brimacombe, R. 1986. J. Mol. Biol. 191:135–38
75. Stiege, W., Stade, K., Schüler, D., Brimacombe, R. 1988. Nucleic Acids Res. 16:2369–88
76. Stade, K., Rinke-Appel, J., Brima-

combe, R. 1989. *Nucleic Acids Res.* 17:9889–908
77. Moazed, D., Stern, S., Noller, H. F. 1986. *J. Mol. Biol.* 187:399–416
78. Latham, J. A., Cech, T. R. 1989. *Science* 245:276–82
79. Vlassov, V. V., Giegé, R., Ebel, J.-P. 1981. *Eur. J. Biochem.* 119:51–59
80. Inoue, T., Cech, T. R. 1985. *Proc. Natl. Acad. Sci. USA* 82:648–52
81. Noller, H. F., Chaires, J. B. 1972. *Proc. Natl. Acad. Sci. USA* 69:3115–18
82. Moazed, D., Noller, H. F. 1986. *Cell* 47:985–94
83. Moazed, D., Noller, H. F. 1990. *J. Mol. Biol.* 211:135–45
84. Moazed, D., Noller, H. F. 1987. *Nature* 327:389–94
85. Allen, P. N., Noller, H. F. 1989. *J. Mol. Biol.* 208:457–68
86. Stern, S., Weiser, B., Noller, H. F. 1988. *J. Mol. Biol.* 204:447–81
87. Brimacombe, R., Atmadja, J., Stiege, W., Schüler, D. 1988. *J. Mol. Biol.* 199:115–36
88. Oakes, M., Kahan, L., Lake, J. A. 1990. *J. Mol. Biol.* 211:907–18
89. Kim, S. H. 1976. *Prog. Nucleic Acid Res. Mol. Biol.* 17:181–216
90. Moazed, D., Noller, H. F. 1990. *Proc. Natl. Acad. Sci. USA.* Submitted
91. Moazed, D., Robertson, J. M., Noller, H. F. 1988. *Nature* 334:362–64
92. Hopfield, J. J. 1974. *Proc. Natl. Acad. Sci. USA* 71:4135–39
93. Ninio, J. 1974. *J. Mol. Biol.* 84:297–313
94. Hardesty, B., Odom, O. W., Czworkowski, J. 1990. See Ref. 26, pp. 366–72
95. Spirin, A. S. 1969. *Cold Spring Harbor Symp. Quant. Biol.* 34:197–207
96. Moazed, D., Noller, H. F. 1987. *Biochimie* 69:879–84
97. Spangler, E. A., Blackburn, E. H. 1985. *J. Biol. Chem.* 260:6334–40
98. Li, M., Tzagoloff, A., Underbrink-Lyon, K., Martin, N. C. 1982. *J. Biol. Chem.* 257:5921–28
98a. Egebjerg, J., Douthewaite, S., Garrett, R. A. 1989. *EMBO J.* 8:607–11
99. Cabrer, B., Vazquez, D., Modolell, J. 1972. *Proc. Natl. Acad. Sci. USA* 69:733–36
100. Richman, N., Bodley, J. W. 1972. *Proc. Natl. Acad. Sci. USA* 69:688–89
101. Hausner, T.-P., Atmadja, J., Nierhaus, K. H. 1987. *Biochimie* 69:911–23
102. Endo, Y., Wool, I. G. 1982. *J. Biol. Chem.* 257:9054–60
103. Endo, Y., Mitsui, K., Motizuki, M.,

Tsurugi, K. 1987. *J. Biol. Chem.* 262:5908–12
104. Sköld, S. 1983. *Nucleic Acids Res.* 11:4923–32
105. Gualerzi, C. O., Pon, C. L. 1990. *Biochemistry* 29:5881–89
106. Hartz, D., McPheeters, D. S., Gold, L. 1989. *Genes Dev.* 3:1899–912
107. Hartz, D., McPheeters, D. S., Gold, L. 1990. See Ref. 26, pp. 275–80
108. Gualerzi, C. O., La Teana, A., Spurio, R., Canonaco, M. A., Severini, M., Pon, C. L. 1990. See Ref. 26, pp. 281–91
109. Wickstrom, E. 1983. *Nucleic Acids Res.* 11:2035–52
110. Muralikrishna, P., Wickstrom, E. 1989. *Biochemistry* 28:7505–10
111. Ehresmann, C., Moine, H., Mougel, M., Dondon, J., Grunberg-Manago, M., et al. 1986. *Nucleic Acids Res.* 14:4803–21
112. Moazed, D. Unpublished.
113. Wickstrom, E., Heus, H. A., Haasnot, C. A. G., von Knippenberg, P. H. 1986. *Biochemistry* 25:2770–77
114. Tapprich, W. E., Goss, D. J., Dahlberg, A. E. 1989. *Proc. Natl. Acad. Sci. USA* 86:4927–31
115. Helser, T. L., Davies, J., Dahlberg, J. E. 1972. *Nature New Biol.* 235:6–8
116. Chapman, N. M., Noller, H. F. 1977. *J. Mol. Biol.* 109:131–49
117. Herr, W., Chapman, N. M., Noller, H. F. 1979. *J. Mol. Biol.* 130:433–49
118. Brow, D. A., Noller, H. F. 1983. *J. Mol. Biol.* 163:27–46
119. Moazed, D., Gualerzi, C. O., Noller, H. F. Unpublished.
120. Baudin, F., Mougel, M., Romby, P., Eyermann, F., Ebel, J.-P., et al. 1989. *Biochemistry* 28:5847–55
121. Vassilenko, S. K., Carbon, P., Ebel, J.-P., Ehresmann, C. 1981. *J. Mol. Biol.* 152:699–721
122. Tapprich, W. E., Hill, W. E. 1986. *Proc. Natl. Acad. Sci. USA* 83:556–60
123. Santer, M., Bennett-Guerrero, E., Byahatti, S., Czarnecki, S., O'Connell, D., et al. 1990. *Proc. Natl. Acad. Sci. USA* 87:3700–4
124. Politz, S. M., Glitz, D. G. 1977. *Proc. Natl. Acad. Sci. USA* 74:1468–72
125. Herr, W., Noller, H. F. 1979. *J. Mol. Biol.* 130:421–32
126. Cantor, C. R. 1980. See Ref. 10, pp. 23–49
127. Zamir, A., Miskin, R., Vogel, Z., Elson, D. 1974. *Methods Enzymol.* 30:406–26
128. Bowman, C. M., Dahlberg, J. E., Ikemura, T., Konisky, J., Nomura, M.

1971. *Proc. Natl. Acad. Sci. USA* 68:964–68
129. Senior, B. W., Holland, I. B. 1971. *Proc. Natl. Acad. Sci. USA* 68:959–63
130. Nomura, M., Sidikaro, J., Jakes, K., Zinder, N. 1974. In *Ribosomes*, ed. M. Nomura, A. Tissières, P. Lengyel, pp. 805–14. New York: Cold Spring Harbor Lab.
131. Leffers, H., Egebjerg, J., Anderson, A., Christensen, T., Garrett, R. A. 1988. *J. Mol. Biol.* 204:507–22
132. Robertson, J. M., Parmeggiani, A., Noller, H. F. Unpublished.
133. Černá, J., Rychlík, I., Jonák, J. 1973. *Eur. J. Biochem.* 34:551–56
134. Marconi, R. T., Hill, W. E. 1989. *Biochemistry* 28:893–99
135. Backendorf, C., Ravensbergen, C. J. C., Van der Plas, J., Van Boom, J. H., Veeneman, G., Van Duin, J. 1981. *Nucleic Acids Res.* 9:1425–44
136. Zamir, A., Miskin, R., Elson, D. 1969. *FEBS Lett.* 3:85–88
137. Ginzberg, I., Miskin, R., Zamir, A. 1973. *J. Mol. Biol.* 79:481–94
138. Moazed, D., Van Stolk, B. J., Douthwaite, S., Noller, H. F. 1986. *J. Mol. Biol.* 191:483–93
139. Powers, T., Noller, H. F. 1990. *Proc. Natl. Acad. Sci. USA* 87:1042–46
140. Sigmund, C. D., Ettayebi, M., Borden, A., Morgan, E. A. 1988. *Methods Enzymol.* 164:673–90
141. Triman, K., Becker, E., Dammel, C., Katz, J., Mori, H., et al. 1989. *J. Mol. Biol.* 209:645–53
142. Krzyzosiak, W., Denman, R., Nurse, K., Hellmann, W., Boublik, M., et al. 1987. *Biochemistry* 26:2553–64
143. Denman, R., Weitzmann, C., Cunningham, P. R., Nègre, D., Nurse, K., et al. 1989. *Biochemistry* 28:1002–11
144. Hui, A. S., Eaton, D. H., de Boer, H. A. 1988. *EMBO J.* 7:4838–88
145. Sigmund, C. D., Ettayebi, M., Prasad, S. M., Flatow, B. M., Morgan, E. A. 1985. In *Sequence Specificity in Transcription and Translation*, ed. R. Calendar, L. Gold, pp. 409–17. New York: Liss
146. Sigmund, C. D., Morgan, E. A. 1982. *Proc. Natl. Acad. Sci. USA* 79:5602–6
147. Sigmund, C. D., Ettayebi, M., Morgan, E. A. 1984. *Nucleic Acids Res.* 12:4653–63
148. Hui, A., Jhurani, P., de Boer, H. 1987. *Methods Enzymol.* 153:432–52
149. Vester, B., Garrett, R. A. 1988. *EMBO J.* 7:3577–87
150. Hummel, H., Böck, A. 1989. *Handb. Exp. Pharmacol.* 91:193–225
151. Cundliffe, E. 1989. *Annu. Rev. Microbiol.* 43:207–33
152. Cunningham, P. R., Weitzmann, C. J., Nègre, D., Sinning, J. G., Frick, V., et al. 1990. See Ref. 26, pp. 243–52
153. Douthwaite, S., Prince, J. B., Noller, H. F. 1985. *Proc. Natl. Acad. Sci. USA* 82:8330–34
154. Douthwaite, S., Powers, T., Lee, J. Y., Noller, H. F. 1989. *J. Mol. Biol.* 209: 655–65
155. Murgola, E. J., Hijazi, K. A., Göringer, H. U., Dahlberg, A. E. 1988. *Proc. Natl. Acad. Sci. USA* 85:4162–65
156. Shen, Z., Fox, T. D. 1989. *Nucleic Acids Res.* 17:4535–39
157. Weiss-Brummer, B., Hüttenhofer, A. 1989. *Mol. Gen. Genet.* 217:362–69
158. Dammel, C. S., Triman, K., Noller, H. F. Unpublished.
159. Mori, H., Dammel, C., Becker, E., Triman, K., Noller, H. F. 1990. *Biochim. Biophys. Acta* 1050:323–27
160. Woese, C. R., Gutell, R. R. 1989. *Proc. Natl. Acad. Sci. USA* 86:3119–22
161. Powers, T., Noller, H. F. 1991. In preparation
162. Stern, S., Powers, T., Changchien, L.-M., Noller, H. F. 1988. *J. Mol. Biol.* 201:683–95
163. Tapprich, W. E., Dahlberg, A. E. 1990. *EMBO J.* 9:2649–55
164. Rosset, R., Gorini, L. 1969. *J. Mol. Biol.* 39:95–112
165. Allen, P. N., Noller, H. F. 1991. In preparation
166. Cabezon, T., Herzog, A., Petre, J., Yaguchi, M., Bollen, A. 1977. *J. Mol. Biol.* 116:361–74
167. Nishi, K., Schnier, J. 1986. *EMBO J.* 5:1373–76
168. Sloof, P., Hunter, J. B., Garrett, R. A., Branlant, C. 1978. *Nucleic Acids Res.* 5:3503–13
169. Svensson, P., Changchien, L.-M., Craven, G. R., Noller, H. F. 1988. *J. Mol. Biol.* 200:301–8
170. Stern, S., Powers, T., Changchien, L.-M., Noller, H. F. 1988. *J. Mol. Biol.* 201:683–95
171. Van Duin, J., Wijnands, R. 1981. *Eur. J. Biochem.* 118:615–19

*Annu. Rev. Biochem. 1991. 60:229–55*

# ATRIAL NATRIURETIC FACTOR AND RELATED PEPTIDE HORMONES

*Anthony Rosenzweig*

Cardiac Unit, Massachusetts General Hospital, Boston, Massachusetts 02114 and Department of Genetics, Harvard Medical School, Boston Massachusetts 02115

*Christine E. Seidman*

Cardiology Division, Brigham and Women's Hospital, Boston, Massachusetts 02115

KEY WORDS:  ANF, BNP, CNP, cardiac hormones.

## CONTENTS

229

0066-4154/91/0701-0229$02.00

## PERSPECTIVES AND SUMMARY

Over the past decade, a considerable amount of research has led to the identification and characterization of a family of peptide hormones that can participate in salt and water homeostasis. Members of this family of natriuretic hormones share many common features, including tissue distribution of gene expression, biosynthetic pathways, and pharmacologic effects in target organs. The prototype of these natriuretic hormones is Atrial Natriuretic Factor (ANF), which has also been called atrial natriuretic peptide, atriopeptin, and more recently A-type natriuretic peptide. Like all members of this family, ANF is produced primarily in the heart and specialized areas of the central nervous system. It is synthesized as a precursor molecule that, upon appropriate stimuli for release, is cleaved to yield a small peptide that produces vasodilation and natriuresis. The natriuretic effect results from direct inhibition of sodium absorption in the collecting duct, increased glomerular filtration, and inhibition of aldosterone production and secretion.

Since the initial identification of ANF in 1981, there has been an explosion of information on the molecular biology, biochemistry, and pharmacology of this hormone. The genes encoding ANF from many diverse species have been cloned and characterized. The biosynthetic steps required for processing the precursor molecule into a mature hormone have been identified, and the structural features of the resultant peptide that are critical for function have been defined. Hormone interactions with ANF receptors have been elucidated, and a signal-transduction mechanism has recently been proposed.

These studies have not only characterized a novel class of hormones with important physiologic functions, but have also provided insights and posed questions regarding physiologic and pathologic functions of the heart. This manuscript reviews the advances made in our understanding of cardiac peptide hormones and focuses on studies of ANF. Many groups have demonstrated that the gene encoding this peptide hormone is expressed at very high levels in atrial myocytes. Ventricular ANF levels are normally low, but increase dramatically with cardiovascular diseases. While these patterns of expression have been well characterized, the molecular signals responsible for tissue-specific expression, developmental regulation, or induction of ventricular expression in response to pathology have only recently begun to be defined. Atrial granules store large quantities of ANF precursor molecules, which must be cleaved to liberate the circulating hormone. Neither the cell responsible for this maturation nor the enzymes involved have been conclusively identified. The predominant signal responsible for ANF release is stretch; however, the signals that transduce mechanical events into intracellular stimuli that release ANF are unknown. Elucidation of these aspects of the molecular biology and biochemistry of ANF and related peptides should increase our understanding of the function of these hormones and also further our knowledge of cardiac physiology.

# CLASSIFICATION OF CARDIAC PEPTIDE HORMONES

## Type A Natriuretic Peptides

The discovery of ANF was the culmination of many years of morphologic and biochemical analyses. In 1956 Kisch (1) demonstrated membrane-bound granules in the atria, but not the ventricles of guinea pigs. These structures appeared to be similar to secretory granules found in endocrine tissues. The presence of atrial granules was confirmed subsequently in a variety of animal species including man (2, 3); however the content of these remained unclear until de Bold et al (4) demonstrated that a granule-enriched extract of rat atrial tissue provoked a prompt, though short-lived, natriuresis in living rats. They postulated the existence of a natriuretic agent contained in the granules, which they named atrial natriuretic factor (ANF) (5, 6). Several groups subsequently isolated peptides of varying length (7–11) from rat atrial homogenates that showed natriuretic and smooth muscle-relaxant (12) activity.

All ANF peptides identified shared a common amino acid sequence, which differed only in length. The precise relationship of these peptides was most clearly elucidated by molecular studies that identified and characterized the ANF gene and encoded mRNA. A single ANF gene has been identified in all mammalian species examined to date. In man this gene is located on chromosome 1, band p36, and is one of a syntenic group that is present on chromosome 4 of the mouse (13). There is significant homology in the organization and nucleotide sequences of ANF genes from many species. In each the gene consists of three exons separated by two introns. Transcription of the ANF gene yields an mRNA species that encodes a 152-amino-acid precursor, preproANF (14–18). PreproANF contains a 24-amino-acid hydrophobic leader sequence, which is probably important for the translocation of the nascent peptide precursor from the ribosome into the rough endoplasmic reticulum (19, 20). PreproANF is rapidly converted to a 126-amino-acid peptide, proANF, the predominant storage form of ANF (21), and major constituent of atrial granules. Upon appropriate signals for hormone release, proANF is cleaved into an amino-terminal fragment, ANF1–99 (22), and the biologically active hormone, amino acid residues 99–126 (11, 23, 24).

The amino acid sequence of the precursor molecule proANF is strikingly homologous among many species. Differences are clustered at the extreme carboxyl terminus of the molecule, where a terminal Arg-Arg sequence that is present in rats (14, 16, 18, 24–26), mice (27), cows (28), and rabbits (29) is absent in humans (15, 27, 30–32) and dogs (29). These dibasic residues have not been detected in purified rat atrial secretory granules (33, 34), suggesting that they are posttranslationally removed. Carboxypeptidase E (CPE) or enkephalin convertase has been implicated as the enzyme responsible for this cleavage, because it can be cofractionated with ANF-containing secretory granules in the rat (35). Phosphorylation is the other major posttranslational modification of the prohormone, and several serine residues (36–38) have been shown to be substrates for phosphorylation within the proANF. However, the physiologic significance of this modification is uncertain, as both phosphorylated and nonphosphorylated forms of ANF are present in atrial granules (39).

The amino acid sequences of the biologically active hormone ANF99–126 are identical in all mammalian species except at residue 110, which is methionine in humans (15, 27, 30–32), dogs (29), and cows (28), but isoleucine in rats (14, 16, 18, 24–26), mice (27), and rabbits (29). Studies on ANF have delineated many molecular and biochemical features of this hormone that are fundamental to understanding of structure-function relationships that are important for the physiologic effects of all natriuretic peptide hormones. A 17-member ring that contains an intrachain disulfide bond between cysteine residues 105 and 121 is required for biologic activity (40). This structure is conserved in all members of the natriuretic peptide hormone family (Figure 1).

Several other compounds that are closely related to ANF have been independently identified by several laboratories. Isolation of these new peptides has been facilitated by both molecular and biochemical features that are shared by members of the natriuretic peptide hormone family. As can be seen in Figure 1, there is a significant amount of structural and amino acid sequence homology between A, B, and C types of natriuretic hormones. Because this homology also extends to the nucleic acid level, ANF cDNAs and genes have been useful probes for identifying the nucleic acids encoding related peptide hormones. Further, these peptides all produce similar pharmacologic effects of natriuresis and muscle relaxation. Assays for novel peptides with muscle relaxant activity have led to the identification of several new members of this family. The classification shown in Figure 1 has recently been suggested by Sudoh et al (41). Hormones are grouped into the A, B, or C type based on their genetic or amino acid homology with the prototypes, ANF (A type), Brain Natriuretic Peptide (BNP; B type), or C type Natriuretic Peptide.

Urodilatin designates a peptide identified in human urine (42), with an amino acid sequence identical with that of ANF, although extended at the amino terminus by four residues. This addition appears responsible for reduc-

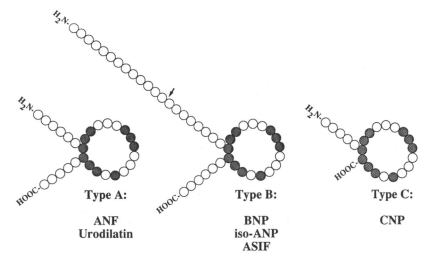

*Figure 1* Members of the Natriuretic Peptide Family. The structural homology between A-, B-, and C-type hormones is schematically shown. Amino acids within the 17-member ring that are identical in all natriuretic peptides are shaded. An arrow by the B-type peptides identifies the human and porcine prohormone cleavage site that yields BNP (32 amino acids) versus rat iso-ANP (45 amino acids).

ing the rate of degradation by the brush border of renal cells (43), and initially suggested that urodilatin was a novel hormone. However, recent data by Ritter et al (44) have demonstrated that ANF prohormone is synthesized and secreted by primary cultures of fetal and adult renal cells. These data imply that urodilatin may simply represent atypical processing of proANF by renal tissues. Since sodium reabsorption by the collecting duct is inhibited by ANF, production of the hormone by renal cells classifies it as a paracrine substance.

## Type B Natriuretic Peptides

Other compounds that have been isolated recently are closely related but clearly distinct from ANF (Figure 1). BNP and iso-ANP can be classified as Type B natriuretic hormones. BNP is a small peptide, which like ANF contains a 17-member ring that is formed by a disulfide bond (45, 46). Within this ring structure, the amino sequence of BNP differs from that of ANF at only seven residues. Greater divergence from ANF sequences is exhibited by the residues in the amino acid and carboxyl extensions of the molecule. BNP was initially identified in porcine brain based on its muscle relaxant activity, but is more abundant in cardiac atria than in the central nervous system (47). Subsequent molecular analyses have identified the comparable hormone in humans (46), rats (48), and dogs (49). The amino acid sequences of these BNPs are less conserved across species than are those of the ANFs.

Within the 17-member ring structure of the ANF molecule, all except one amino acid residue are invariant between different species, whereas only 59% of these residues are identical in BNP molecules from different species. Greater homology exists between BNP and ANF molecules in a species than between BNP molecules from different species.

The biosynthetic pathway of BNP is strikingly similar to that of ANF, and transcription yields a precursor molecule containing a hydrophobic leader segment that is cleaved to produce a prohormone. Subsequent production of the active hormone is species dependent. In humans and pig a 32-amino-acid hormone is the circulating form of the hormone (48). A functional but not circulating, 26-amino-acid species has also been detected in brain tissues of the pig (45). In rodents, mature BNP contains 45 amino acids. This peptide was independently identified by several laboratories and is also called iso-ANP (50, 51). The amino acid sequence of iso-ANP was initially believed to be sufficiently different from that of porcine BNP to suggest that iso-ANP represented a novel hormone. However, subsequent nucleic acid sequence analyses of rat cDNA (48) and genomic clones (52) have demonstrated the relatedness of this hormone to human and porcine BNP, as well as confirmed that processing of prohormone (Figure 1, arrow) accounts for the differences in the mature hormone sizes. The nucleic acid sequence of rat BNP gene is particularly divergent from other species in exon II, which encodes amino acid residues that are involved in prohormone processing. In both ANF and BNP, a Pro-Arg sequence is important for appropriate cleavage of pro-hormone to yield the mature peptides. In the rodent BNP gene, this cleavage sequence is located 13 residues closer to the amino terminus than in other species. Processing therefore produces a 45-amino-acid peptide, the circulating form of BNP found in the rat. Whether processing differences between species have any effect on the pharmacologic function of these molecules is unknown.

A 35-amino-acid bovine peptide, aldosterone secretion inhibitory factor (ASIF), appears to represent a third example of species-specific processing of BNP compounds (53). This molecule contains sequences identical with that found in porcine BNP, but contains three additional amino acids residues. It has been identified in chromaffin granules of bovine adrenal cells, a location that suggests that an intra-adrenal paracrine mechanism may be important for regulating aldosterone secretion.

## Type C Natriuretic Peptide

The newest member of the natriuretic hormone family is a 22-amino-acid peptide called Type C natriuretic peptide (CNP) (41). In contrast to other family members, this peptide terminates with the second cysteine residue that is critical for a 17-member ring structure. Preliminary data of a cDNA encoding porcine CNP demonstrates that a termination codon directly follows that for cysteine. Hence, the absence of the carboxyl-terminal residues in CNP does not appear to reflect posttranslational processing, but rather results from a unique gene transcript.

## MOLECULAR BIOLOGY OF ANF GENE EXPRESSION

*Overview*

In recent years there has been an explosive growth in our understanding of eukaryotic gene regulation (reviewed in 54). These insights have arisen largely through development of new technologies, many of which have recently been applied to understanding ANF gene expression. This undertaking is of interest for several reasons beyond the intrinsic merit of dissecting the specific mechanisms involved in ANF expression. First, regulation of ANF transcription appears to be an important mechanism for modulating hormone production, as peptide levels parallel mRNA abundance. In the adult, mRNA encoding ANF is found at the highest level in the atria, where it constitutes approximately 1–3% of the total mRNA (32). Ventricular expression, which is approximately 100–fold lower than that in the atria, nevertheless is next in abundance of ANF message (55). Gene transcription has also been documented in a variety of extra-cardiac tissues including hypothalamus (56), pituitary (57), aortic arch (58), lung (57), and adrenal medulla (59). However extra-cardiac transcription is considerably less than even that of the adult ventricle. While extra-cardiac ANF may have important local effects, the major site of gene transcription is the heart, and ANF is a reasonable model to study the elements controlling a cardiac-specific gene.

Second, expression is strictly modulated during development. Atrial expression continues to rise during fetal and postnatal development, while ventricular expression, high in utero, falls quickly after birth (60). ANF expression is therefore a useful tool for analyzing the signals involved in early cardiac development.

Third, ventricular expression can be induced by a variety of pathologic conditions, most notably cardiac hypertrophy. This induction has been documented both in whole animal (61–64) and cellular (65) models of hypertrophy. It is associated with myosin heavy chain isoform switching (62) and proto-oncogene expression, leading some to suggest that it may represent regression or dedifferentiation (62, 64) to a fetal pattern of gene expression. ANF may therefore provide a window on the molecular mechanisms responsible for altered gene expression and ultimately the altered phenotype of cardiac hypertrophy.

It is also likely that many of the features important for the regulated expression of the ANF gene will be applicable to the B type or C type natriuretic hormone genes. As more of the members of this family have been identified from different tissues and species, it has become obvious that each of these hormones is expressed within certain tissues responsible for salt and water homeostasis. Cardiac and central nervous system expression has been demonstrated for all natriuretic peptide family members (41, 45, 56, 57, 66) except bovine ASIF, which has only recently been identified in the adrenal gland. Both ANF (59) and BNP-like peptides (53) have been detected in chromaffin cells of the adrenal gland, and pathologies that alter cardiovascular hemodynamics similarly affect ventricular expression of ANF and BNP. However, it may be too simplistic to assume that these hormones are always produced in parallel. Recent data demonstrate that the induction of BNP by the ventricle in response to cardiac hypertrophy is greater than the induction of ANF (67). Hence the precise mechanisms that regulate these hormones may be different, and may indicate subtle differences in their physiologic effects.

## Factors that Regulate ANF Gene Transcription

A variety of stimuli can modulate ANF levels. Posttranslational effects such as peptide processing are likely to be important in regulating overall peptide levels, but are distinct from the direct transcriptional modulation involved in gene regulation. Water or sodium deprivation in intact animals leads to a reversible decrease in ANF mRNA of approximately 30–70% (67, 68, 169). Correspondingly, administration of the mineralocorticoid desoxycorticosterone acetate (DOCA) to adrenalectomized animals results in an increase of ANF mRNA by 70% as early as 12 hours after administration, which may contribute to the renal "escape" from chronic mineralocorticoid exposure (70). It is difficult to know what the direct mediators of these in vivo effects are. Alterations in atrial stretch associated with volume shifts may be important in these models, especially since cardiocytes appear not to have mineralocorticoid receptors (71).

In contrast, glucocorticoids and thyroid hormone increase the level of ANF mRNA both in intact animals (61, 72, 73) and in cultured cardiocytes in vitro (74–76), suggesting these are due to direct effects on gene transcription or effects on the stability of mRNA. Other data support a primary role of glucocorticoids on gene transcription. Expression of reporter genes is increased by glucocorticoids when these are transcribed under the direction of ANF rather than other regulatory sequences (77). To the extent that these effects are primary, one would expect this increased gene expression to be mediated by appropriate hormone receptors that interact with specific DNA sequences, as has been demonstrated in other systems (78). A binding site homologous to the glucocorticoid receptor–binding consensus sequence has been identified in the second intron of the human gene (27). The mouse gene lacks this binding site (27). The rat gene contains multiple fragments with partial homology to this consensus sequence (77). Argentin and colleagues have recently reported that purified glucocorticoid receptors bind an upstream element required for hormone responsiveness in rat gene (79).

Calcium plays a critical role in the electro-mechanical functioning of the heart, and recently has been shown to affect cardiac hormonal function also. Gardner and colleagues have demonstrated that extracellular calcium increases the amount of mRNA for ANF by two- to three-fold in primary cardiocyte cultures (80). Calcium antagonists decreased ANF transcripts. Modest increases of ANF gene expression were also detected after treatment with either the calcium ionophore A23187, which facilitates intracellular calcium entry, or the phorbol ester 12-O-tetradecanoyl phorbol 13-acetate (TPA), which activates protein kinase C. However, the latter results did not reach statistical significance unless the two agents were used in concert.

The effects of calcium on gene transcription are particularly interesting because of the stimulatory role of this ion on secretory granules. In cells with regulated secretory pathways, secretory granules fuse with the cell membrane in response to a cytoplasmic messenger, usually calcium (81). If pharmacologic agents or myocyte stretch increase intracellular calcium, both ANF release and gene transcription would be enhanced. Sorbera & Morad (82) have recently shown that ANF suppresses both the sodium and calcium currents in primary cultures of cardiocytes. This provocative finding suggests that ANF secretion may then provide a negative feedback mechanism for calcium-mediated ANF synthesis and release.

The recently identified peptide vasoconstrictor endothelin has also been shown to increase both ANF secretion (83) and mRNA (84) in cardiocytes. Shubeita and colleagues have recently reported that endothelin-1 induces a pattern of early gene expression in cardiocytes that includes not only ANF expression but also expression of the proto-oncogene, *c-fos,* and the immediate early response gene, *egr-1* (85). As seen in cardiac hypertrophy (62), the induction of ANF expression by endothelin may provide another model for the study of early cardiac gene regulation.

## Cis-Acting Sequences

Eukaryotic gene expression appears to be controlled to a large extent by the interaction of specific DNA sequences, often but not necessarily located adjacent to the gene, termed cis-acting regulatory elements. These sequences interact with proteins called trans-acting factors (54). Analyses of these elements in many other eukaryotic genes have provided significant insights into the molecular mechanisms mediating the physiologic and pathophysiologic responses of these genes. Such an analysis of ANF gene regulation is a first step toward dissecting the precise mechanisms mediating its expression.

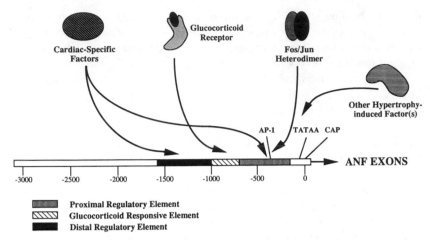

*Figure 2* Regulation of ANF gene expression. A 3000-base-pair region (enumerated as in Ref. 77) located 5' of the rat ANF cap site is schematically shown. Cis-acting sequences and related trans-acting factors that have been identified as important for modulating ANF gene expression are indicated and detailed in the text.

The structural organization of the ANF gene has been highly conserved through evolution (27). The gene is composed of three exons separated by two introns. The proximal flanking sequence upstream of the transcription initiation site is schematically shown in Figure 2. This region contains elements characteristic of eukaryotic promoters including a TATAA box approximately 30 bp upstream of the transcription initiation or CAP site (77). This sequence is essential for transcription of many eukaryotic genes and has recently been shown to bind the trans-acting factor TFIID in other systems (86). By convention, base position is given with respect to the initiation site so that the TATAA site approximately 30 bp upstream or 5' of this would be given as $-30$ bp.

To identify the other sequences important for ANF transcription in the rat gene, Seidman et al (77) attached portions of the 5' flanking sequence to the prokaryotic reporter gene, chloramphenicol acetyltransferase *(CAT)*. These constructs were then transfected into atrial or ventricular cardiocytes. Since *CAT* is not found in mammalian tissues, its presence provides an easily assayed indicator of whether the ANF sequences attached can direct transcription (87–89). A construct with 2.4 kb of 5' upstream sequence contained sufficient information to direct high-level, atrial-specific transcription in primary cardiocytes (77). Constructs containing less than 1 kb of upstream sequence produced less *CAT* activity, indicating the presence of one or multiple cis-acting regulatory element(s) between $-2.4$ and $-1.0$ kb. More recently these results have been extended to analyze sequences more proximal to the initiation site. Internal deletions of constructs that contained at least 2.4 kb of 5' sequences were studied, and important regulatory sequences between $-137$ and $-693$ bp were identified (89). Deletion of these proximal regulatory sequences reduced ANF gene transcription by 20- to 30-fold. These studies indicate that regulation of the ANF gene may involve sequences that are proximal as well as distal to the promoter apparatus (77, 89; Figure 2). In addition, Argentin et al have found two regions that are important to regulating ANF expression (79). These elements, located at positions $-136$ to $-700$ bp and $-1000$ to $-1600$ bp, further delineate the boundaries of the proximal and distal regulatory elements. These studies have also identified a region that is necessary for glucocorticoid responsiveness that resides between $-700$ and $-1000$ bp.

In the human gene, LaPointe et al (90) found that 409 bp of upstream sequence was sufficient to direct tissue-specific expression in primary cardiocyte cultures. Deletion analysis further demonstrated that this human cis-acting element corresponded in location to positions $-389$ to $-191$ in the rat gene. The human gene element appears to be able to direct transcription in both atrial and ventricular cells but not in several nonmyocardial cell types. Wu and colleagues found that the corresponding region of the rat gene was unable to direct atrial transcription in primary culture of rat neonatal cardiocytes (91), suggesting that unlike those sequences that encode the hormone, ANF gene regulatory regions have evolved considerably between these species. Within the regulatory region of the human ANF gene, an 18-bp binding site was identified by DNaseI footprinting and methylation interference experiments (91). This small region is homologous to the sequences in the rat ANF gene, at positions $-1220$ to $-1203$ pb, which is encompassed in the distal regulatory region identified by others (77; Figure 2). It is therefore possible that some similar cis-acting elements have a different structural organization between rat and human.

The importance of those sequences identified in the human ANF gene has been confirmed by Field. A segment of human ANF gene containing 423 bp of regulatory sequences was sufficient to target expression of the SV-40 large T antigen in a transgenic mouse model (92). This study further delineated that these sequences promoted atrial-specific expression, as expression of the oncogene by transgenic animals produced marked hyperplasia only of the right atrium, while the left atrium remained relatively normal in size.

To further elucidate the regulation of rat ANF expression, Seidman et al (93) created transgenic mice carrying a construct containing 2.4 kb of 5' flanking sequence in front of the bacterial reporter gene *CAT*. The tissue and developmental regulation of the ANF transgene could then be followed by assaying *CAT* activity. These in vivo studies support the earlier findings in cardiocytes: 2.4 kb of upstream sequence is sufficient to direct high-level, atrial-specific, and developmentally appropriate transcription. Of interest, the developmental expression pattern of the rat transgene was quite different from that of the endogenous murine ANF gene. While murine ventricular ANF mRNA content remained constant from gestational day 15 through adulthood, transgene expression as reflected by *CAT* activity fell dramatically between one and seven days after birth. This decline parallels that seen in native rat ventricular ANF expression (60). The dominance of the rat pattern suggests developmental differences between mouse and rat are due to the cis-acting regulatory sequences rather than trans-acting factors. These studies also demonstrated a low level of hypothalamic *CAT* activity, corresponding to the low level of mRNA for ANF detected in rat hypothalamus (56, 66), suggesting that these 2.4 kb contain cis-acting elements important for transcription in brain as well as the heart.

## Cardiac Trans-Acting Factors

Cis-acting regulatory elements generally work by binding proteins called trans-acting factors to modulate transcription. The cis-acting element identified by LaPointe et al in the human gene (90) appears to interact with a cardiac-specific nuclear protein. The cis-acting element can direct transcription in both atrial and ventricular cells but not in several nonmyocardial cell types. These data further suggest the existence of other protein(s) to account for the difference between atrial and ventricular expression. Argentin et al (79) recently reported that a cardiac-specific nuclear protein binds the rat proximal and distal (Figure 2), and the human cis-acting control element. The relative abundance of this protein appears to correlate with the level of ANF expression in cardiac tissues during development. This protein may play a role in differential expression of the ANF gene between atria and ventricles.

The molecular basis for ventricular ANF gene expression is poorly understood. Prior to birth, ventricular myocytes produce both ANF mRNAs and hormone (60). ANF gene expression declines rapidly at birth, but recurs in response to cardiac hypertrophy (62). Whether this phenomenon represents a compensatory response to the hemodynamic changes that accompany cardiac disease or simply a part of a phenotypic response to pathology is unknown.

Cardiac hypertrophy is a particularly important stimulus for ventricular induction of ANF. Understanding the mechanisms by which this induction occurs may provide insights into the molecular basis of this important pathologic process. There has been speculation that the regression to an early pattern of gene expression seen in hypertrophy may involve proto-oncogenes (reviewed in 94). Izumo et al have shown that *c-fos* and *c-myc* increase early and dramatically in cardiac hypertrophy due to aortic constriction (62). Other investigators have also found increased *c-fos* expression in in vivo and in vitro models of hypertrophy (63, 68, 95–98). This increase precedes the increase in

ANF expression (99). The product of the *c-fos* proto-oncogene appears to function as a transcription factor (99). It does not bind DNA alone, but forms a heterodimer with the product of the *c-jun* proto-oncogene (99–101). Of interest, *c-jun* expression has recently been shown to increase dramatically in an in vitro model of hypertrophy (96). The fos/jun heterodimer can then stimulate transcription by binding to a specific site, called an AP-1 site (99, 102). Rosenzweig et al have recently identified a region within the proximal cis-acting regulatory element of the rat ANF gene with significant homology to the consensus AP-1 sequence (103; Figure 2). Because of the exquisite specificity of these protein-DNA interactions, it cannot be assumed to actually bind the fos/jun heterodimer. However, recent experiments demonstrate that in vitro translated fos/jun heterodimer does in fact specifically bind this site in gel-retardation assays (103). Since expression of these proto-oncogenes increases dramatically in hypertrophy before increases in ANF expression, interaction of these known transcriptional activators with this site provides a likely molecular mechanism underlying increased ANF expression in hypertrophy. This argument is strengthened by the recent observation by Chien et al that a cis-acting element located at position −648 of the rat ANF gene responds to α-adrenergic pharmacologic agents that produce some of the features of hypertrophy (104). This hypertrophy-responsive element maps to the same location of the AP-1 site, within the proximal regulatory element.

Further characterization of the genetic elements that are responsible for atrial-specific expression of the ANF gene and ventricular induction in response to pathology should increase our knowledge of the molecular basis for cardiac gene expression. These data should also provide new insights into the role of this hormone in myocardial function. Is ANF one of a group of cardiac genes that are coordinately regulated in response to development and pathology? It was recently discovered that the early pattern of gene expression critical to development of skeletal muscle can be controlled to a large extent by a single master gene, *MyoD*, which appears to encode a trans-acting transcriptional factor (105). Transfection of a fibroblast cell line with the cDNA for this gene alone is sufficient to cause the fibroblast to metamorphose into a primitive muscle cell. It is reasonable to think cardiocyte development may involve similar master control genes. Since ANF is expressed as part of the early pattern of cardiac gene expression, it may be useful in seeking the cardiac equivalent of *MyoD*.

If ANF is one of several cardiac genes expressed in a coordinated pattern, identification of this group should provide insights into the interactive role of these gene products in cardiovascular function. Delineating the molecular basis for regulated expression of the ANF gene should provide new tools to answer these questions.

Substantial progress has recently been made in understanding the control of

ANF gene expression. There appear to be multiple elements involved in the constitutive expression, hormone responsiveness, and pathologic induction of ANF. We hope that further elucidation of the details of these processes will continue to provide insights into the physiologic and pathophysiologic molecular biology of the heart.

## PEPTIDE PROCESSING AND RELEASE

After ANF messenger RNA is translated, the nascent peptide undergoes a complex process of posttranslational modifications, which ultimately results in the secreted hormone. This process has been extensively reviewed recently (106, 107) and is summarized here.

Similar to other peptide hormones, ANF is initially translated as a pre-prohormone. The 24 amino-terminal residues likely constitute a hydrophobic leader sequence (108) involved in peptide transport and packaging. This segment is cleaved to yield the prohormone pro-ANF, which is the principle storage form. Of interest, neonatal atria and ventricles secrete ANF via different pathways (109). Atrial cells store ANF at high concentration in secretory granules, which provided one of the first clues to the endocrine function of the heart (1, 2, 4). This stored hormone is then available for regulated, bolus release in response to appropriate secretagogues. Ventricular cells are virtually devoid of granules, and appear to secrete ANF con-stitutively (109). Because ventricular myocytes do not appear to store signifi-cant quantities of the hormone, gene transcription and translation in these cells result rapidly in hormone synthesis and release. ANF provides the first example of tissue-specific alternative secretory pathways involving a single peptide hormone (109). Further understanding of these processes may shed light on the developmental differences between atrial and ventricular cells.

### ANF Processing

Three major types of posttranslational modification have been implicated in the processing of ANF. The first is formation of an intrachain disulfide bond between cysteines 105 and 121 to form the 17-member ring structure (Figure 1) that appears essential for bioactivity of ANF (110). Second, ANF can act as a substrate for phosphorylation, and both the phosphorylated and un-phosphorylated forms are found in atrial secretory granules (111). The physi-ological significance and precise site of phosphorylation remain unclear. Third, processing involves three possible cleavage events. As mentioned above, the leader segment of preproANF is probably cleaved during transla-tion to yield proANF. In several species, although nucleotide sequences predict two arginines immediately after the tyrosine residue at position 126, purified proANF lacks these Arg residues (33). There are two possible explanations for this. Either these residues are never translated, or they are rapidly cleaved to generate ANF1–126. Thibault et al (34) have recently identified multiple different molecules immunoreactive with antibodies directed against different parts of proANF. One of these molecules appears to be intact ANF1–128. The authors speculate that this is cleaved in situ by carboxypeptidase E, which has been detected in atrial granules (112).

The most important posttranslational modification of proANF is cleavage to produce ANF99–126, the principle, bioactive, and circulating form of the hormone (24). While a dibasic Arg101-Arg102 site exists within the molecule, cleavage of prohormone occurs exclusively at the more proximal monobasic site, Arg98-Ser99. Despite the wealth of data regarding other aspects of ANF biosynthesis, the cell type, intracellular site, and enzyme responsible for prohormone cleavage remain uncertain.

Whole animal studies and isolated perfused hearts have demonstrated release of ANF99–126 (22, 113). In contrast, studies with primary neonatal cardiocyte cultures released only the larger proANF species (114). This dichotomy suggested that cleavage might occur in nonmyocyte cells. Recently, neonatal cardiocytes cultured in the presence of dexamethasone have been demonstrated to cleave ANF1–126 to release ANF99–126 (115, 116). Dexamethasone was found to stimulate ANF processing and secretion in a dose-dependent, reversible manner (117) in both atrial and ventricular cells. Appropriate processing of proANF by cardiocytes supplemented with steroids suggests that prohormone cleavage is a myocyte function and that the enzyme responsible for this is induced by dexamethasone treatment.

Several proteolytic enzymes have been suggested as candidates for processing proANF to ANF. Both kallikrein (118) and thrombin (22) can catalyze this reaction in vitro, but their physiologic relevance is uncertain. Three serine proteases have been purified from atrial tissue, two of which appropriately yield ANF99–126. One of these, called atrioactivase (119), is present in the microsomal fractions of bovine atria, while the other has been detected in bovine secretory granules (120). Cathepsin B, a lysosomal protease, has recently been localized in rat atrial secretory granules and could also play a role in catalyzing this reaction (121). If the cleavage enzyme is located in atrial granules, a mechanism must exist for preventing cleavage until appropriate stimuli for hormone release are present, as granules contain predominantly proANF. Whether one of these enzymes is physiologically important for prohormone cleavage, as well as how this important processing step is regulated, remains an area of active research.

## ANF Release

The primary stimulus for ANF secretion appears to be atrial distention or stretch (Figure 3). Katsube and colleagues found that increases in right atrial

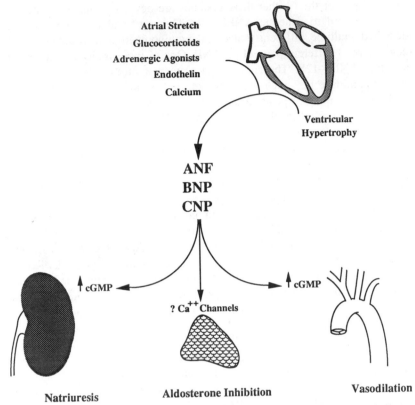

*Figure 3*   Mechanisms of ANF release and resulting physiologic effects. Factors that act as secretagogues for ANF are detailed in the text and may also promote the release of BNP (67) and CNP. These hormones cause natriuresis and vasodilation by receptor-coupled guanylate cyclase activation (150). The mechanism(s) by which ANF inhibits aldosterone secretion may involve inhibition of adrenal calcium channels (177).

tension due either to vasoconstriction or water immersion in anesthetized rats produced a significant increase in ANF release (122). Left atrial pressure as reflected by left ventricular end diastolic pressure did not correlate with ANF release, suggesting that the right atrium produces most of the ANF released in response to volume expansion. In humans and dogs, ANF secretion is increased by left atrial hypertension caused by mitral valve obstruction (123, 124). Canine studies have also demonstrated that this release was not neurally mediated, since neither bilateral cervical vagotomy nor beta blockade prevented the response, suggesting it is a direct effect of stretch of the atrial wall. Discrimination between the effects of atrial volume and pressure is particularly difficult, because in most models used these parameters tend to change in parallel. Mancini and colleagues (125) studied cardiac tamponade in an open-chest dog model to address this issue. They demonstrated that cardiac tamponade that increased right atrial pressure without affecting atrial volume did not induce ANF release. Furthermore, intra-atrial pressures did not correlate with ANF levels. These data suggest that atrial volume is a more potent stimulus for ANF release than is pressure.

In vitro models have confirmed these interpretations. Schiebinger & Linden (126) measured ANF secretion while the resting tension of explanted rat atria was adjusted. Threefold increases in resting tension resulted in a 35% increase in immunoreactive ANF. Moreover, these results were not due to adrenergic or cholinergic influences from endogenous nerve endings, since they were unaffected by pharmacologic blockade of these systems. Taken together, these data suggest that atrial distention, especially right atrial, is the primary stimulus for ANF secretion and the mechanism of the hormone's responsiveness to changes in intravascular volume status. The intracellular signals that transduce atrial stretch into ANF release have not yet been established.

A variety of humoral factors have been implicated as secretagogues of cardiac sources of ANF (Figure 3). These include adrenergic agonists (127), glucocorticoids (117), endothelin (83), acetylcholine (128), and calcium (117). Experiments testing the effects of these agents in intact animals are often difficult to interpret because the agents themselves invoke a complex array of hemodynamic responses, which can affect ANF release. For this reason, their direct effects on cardiocyte secretion may be best studied in models employing perfused explants or primary cardiocyte cultures. The latter system is complicated by the fact that cultured cardiocytes do not fully process ANF unless glucocorticoids are included in the medium. Therefore, many but not all studies addressing ANF release in cardiocyte cultures have employed dexamethasone treatment so that the prohormone is cleaved to ANF99–126 as in vivo. This treatment may, however, influence the results in nonphysiologic ways. Furthermore, most but not all of these studies have used neonatal rather than adult cardiocytes. These differences may be important when attempting to understand occasionally conflicting results.

Norepinephrine, an adrenergic agonist with both $\alpha$ and $\beta$ properties, has been shown to stimulate ANF secretion from isolated, perfused rat hearts (127). This effect appeared mediated via $\alpha$-adrenergic receptors, since it is induced by phenylephrine but not isoproterenol and is blocked by phentolamine. A selective $\alpha_2$ agonist, BHT-920, had no effect, suggesting involvement of $\alpha_1$ receptors. These results have been confirmed and extended using cardiocytes cultured in glucocorticoid-containing, serum-free media (129). In this system phenylephrine-stimulated ANF secretion was not blocked by the $\beta$-adrenergic antagonist propanolol, but was decreased in a dose-dependent manner by the $\alpha_1$-adrenergic receptor antagonist, prazosin. Activation of

$\alpha_1$-adrenergic receptors may not fully account for all signalling of ANF release, because the $\beta$ agonist isoproterenol actually inhibited the phenylephrine-stimulated ANF secretion by as much as 50%. The effect of $\beta$-adrenergic stimulation on basal ANF secretion was not tested.

The diverse effects of $\alpha$- and $\beta$-adrenergic agonists are mediated through different second-messenger systems (reviewed in 130). $\beta$ agonists lead to activation of adenylate cyclase, increased cAMP levels, and subsequent activation of cAMP-dependent protein kinase A. Stimulation of $\alpha$-adrenergic receptors induces hydrolysis of membrane-associated phosphotidylinositol, 4,5-bisphosphate to yield inositol phosphate (IP$_3$) and diacyl glycerol (DAG). IP$_3$ stimulates intracellular calcium release; DAG activates protein kinase C. A large variety of pharmacologic agents that directly influence these two systems are available. The effect of many on ANF secretion has been tested, and in general the results are consistent with predictions based on the different effects of $\alpha$ and $\beta$ agonists. The phorbol ester TPA, which directly activates protein kinase C, stimulates ANF release in both perfused heart preparations (128) and cardiocytes cultured with (129) and without (131) glucocorticoids. In contrast, forskalin, which activates adenylate cyclase and indirectly protein kinase A, inhibited ANF release from noncontracting cultured atrial myocytes (132). Membrane-permeable analogs of cAMP had a similar effect (129, 132).

The role of intracellular calcium in stimulating ANF release is particularly inportant because of its role as a second messenger, independent of the phosphoinositol system. Calcium fluxes are critical for both the electrical and mechanical functions of the heart. The calcium-channel agonist, BAY K8644, has been demonstrated to increase ANF secretion both in cultured cardiocytes (131) and isolated, perfused hearts (133). This effect was blocked by the calcium channel antagonist, nifedipine, in both studies. Similarly, secretion in isolated heart preparations increased in response to the calcium ionophore A 23187 (128), an effect that was synergistic in combination with the phorbol ester, TPA. Similarly, Sei & Glembotski (134) found in neonatal atrial cultures that both phenylephrine- and endothelin-induced stimulation of ANF secretion was dependent on intracellular calcium. Stimulation of ANF secretion could be maintained in extracellular calcium concentrations as low as 2 $\mu$M; however, reduction to 2 nM decreased phenylephrine or endothelin-induced secretion. Treatment of cultures with BAPTA/acetoxymethyl ester, an intracellular calcium chelator, or ionomycin abolished this stimulation.

Interpretation of these experiments is complicated by the chronotropic and inotropic effects of these agents, which may contribute to ANF secretion. To discriminate between these effects, Iida & Page (135) utilized primary cardiocyte cultures that had been paralyzed with tetrodotoxin. In this system, extracellular calcium did not affect ANF secretion as it did with spontaneously contracting myocytes. Moreover, treatment of cardiocytes with ryanodine to inhibit sarcoplasmic reticulum release of intracellular calcium did not affect ANF release under these conditions. Furthermore, the same investigators have reported that forskolin reduced ANF secretion in nonbeating cardiocyte cultures (132).

Recently, Uusimaa and colleagues (136) have been able to use fluorescent calcium indicators to observe intracellular calcium changes during a variety of pharmacologic manipulations on myocardial cells cultured on microcarriers. In these studies, although KCl-induced depolarization and BAY K8644 produced significant increases in intracellular calcium, they had little effect on ANF release. In contrast, even under these conditions, TPA was able to stimulate ANF secretion, confirming the importance of the protein kinase C system. This work is in agreement with that of Greenwald and colleagues (137), who found that although stimulation of ANF secretion by osmotic stretch is associated with increased intracellular calcium, such stimulation is actually increased in calcium-free medium or when intracellular calcium stores are chelated. These studies suggest that ANF release does not require calcium and may be inhibited by it.

Methological differences, particularly with regard to the species and age of myocytes examined, may explain some of the differences in results. Nevertheless, although some issues remain unresolved, there appears to be growing consensus on certain points. First, the primary and physiological stimulus for ANF secretion is stretch of the atrial cardiocyte. Neurohumoral secretagogues appear to exert their effects largely through the protein kinase C system. Furthermore, activators of protein kinase A can actually inhibit release under a variety of conditions. Future investigations will be necessary to elucidate the precise mechanism of stretch-induced secretion and the role that calcium plays in mediating these effects.

Molecular studies have identified low levels of ANF mRNA species from a variety of tissues, including lung, central nervous system, adrenal, and renal tissues. While stretch or distention is the critical stimulus for ANF release from the heart, humeral factors may be particularly important for ANF release from noncardiac sources. Unfortunately, both the quantities of ANF produced and localization of hormone expression to specialized cells within noncardiac tissues have hindered definition of the stimuli that are important for release from these tissues.

# BIOCHEMISTRY OF PHARMACOLOGIC ACTIONS

## ANF Receptors

As with other peptide hormones, ANF exerts its effects by binding specific membrane-bound receptors. Two types of ANF-specific receptors have been identified in target tissues: two different guanylate cyclase-linked receptors that appear to mediate most of ANF's biological effects, and a clearance receptor. Initial characterization of ANF receptors involved photoaffinity labeling and affinity cross-linking studies (138–141). Functional evidence of multiple subpopulations was provided by examination of binding of a variety of truncated ANF analogs to bovine aortic smooth muscle cells (142). These studies demonstrated that biologically active ANF peptides bound specifically to cells and stimulated cyclic GMP (cGMP) accumulation (Figure 3). However, biologically inactive ANF analogs also bound specifically to membrane receptors without stimulating cGMP accumulation. Moreover, binding of these inactive analogs did not antagonize the cGMP response to biologically active ANF species. This suggested the existence of at least two subpopulations of specific ANF receptors. Most tissues express ANF clearance or silent receptors (142), which do not mediate known physiologic effects of the hormone. In vivo experiments, however, suggest that these receptors may play an important role in regulating plasma ANF levels by affecting storage and clearance (143).

Proof of distinct ANF receptors was provided by the purification (144–148) and subsequent cloning, sequencing, and expression (149–152) of three distinct receptors: two guanylate cyclase–linked receptors (120–140 kDa) present as single-chain polypeptides, and one clearance receptor (60–70 kDa) that forms homodimers. The high-molecular-weight ANF receptor copurifies with a functional guanylate cyclase activity (147, 148), and amino acid sequences deduced from cDNAs encoding these molecules suggested the existence of both an extracellular ligand-binding domain and an intracellular guanylate cyclase catalytic domain (150). The bifunctional nature of this receptor was conclusively demonstrated by expression of an ANF receptor cDNA in cultured mammalian cells, which conferred both ANF-binding and guanylate cyclase activity (150). The same molecule also contains a cytoplasmic domain with structural homology to protein kinase, which appears to function as a regulatory element (153). After deletion of this kinase-like domain, guanylate cyclase activity becomes independent of ANF binding. It appears that the kinase-like domain may repress guanylate cyclase activity until ANF binds the extracellular domain (153).

The guanylate cyclase–coupled ANF receptor initially isolated has been used to identify additional members of the guanylate cyclase/natriuretic peptide receptor family in man (151) and rat (152). In rat, these receptors have been termed GC-A and GC-B in order of their discovery (152). A third ANF receptor (180 kDa) has been identified in rat adrenal cells that has partial nucleotide sequence homology to the GC-A receptor (154), and may represent another member of this family. In man, the comparable receptors are referred to as ANP-A and -B; ANP-C denotes the human clearance receptor (151). In both species, the intracellular regions are quite similar (73–74% amino acid identity), but the extracellular regions diverge considerably (43–44% identity). Consequently, the receptors have different specificities for natriuretic peptide binding. ANF is more potent than BNP in stimulating the guanylate cyclase activity of GC-A; the reverse is true of GC-B (152). ANP-A responds to both ANF and BNP; ANP-B responds preferentially to BNP (151). To date, a receptor with highest affinity for CNP has not been identified.

Several lines of evidence suggest that our understanding of ANF receptors and how they mediate ANF's biological effects is incomplete (Figure 3). First, although the guanylate cyclase–associated receptor has been identified in and isolated from the adrenal cortex (139, 140, 146, 148), the inhibition of aldosterone production by ANF is not mediated through cGMP (155). It is not clear whether these effects are mediated by an undiscovered receptor or as unknown activity of the previously described receptors. The recent identification of aldosterone secretion inhibitory factor (53) in bovine adrenal cells suggests that paracrine mechanisms may be important for aldosterone inhibition. Second, ANF binding inhibits adenylate cyclase through inhibitory G-proteins (156), but the precise receptor unit involved and the biological significance of this interaction remain a mystery. Third, another receptor type that exists as a 70 kDa monomer has been described in rat thoracic aortic cultured smooth muscle cells (157). The abundance of these binding sites in the face of low cGMP stimulation led the investigators involved to postulate these receptors may be linked to an alternative signal transduction system (157). It may be, however, that these receptors represent another form of the clearance receptor.

## Physiologic Effects of ANF

It is ironic that the renal effects of ANF (so important to its discovery that they provided its name) remain controversial both in terms of their precise mechanism and their physiological importance. It is clear that acute administration of exogenous ANF results in natriuresis and diuresis. These effects are cGMP mediated (158, 159; Figure 3). ANF increases the glomerular filtration rate (160, 161), but this rate does not always correlate with the natriuretic and diuretic effects (162), suggesting a tubular effect is involved. Microperfusion studies do not detect a direct effect of ANF on proximal tubule transport (163), suggesting a distal segment may be involved. In addition, ANF inhibits aldosterone synthesis and secretion, which will also promote natriuresis (see below).

The importance of ANF in the physiological maintenance of body fluid and sodium homeostasis is still debated (for recent discussion see 164, 165). One strong argument against a major role for ANF is that although ANF levels usually correlate with natriuresis, they do not always do so. As noted above, a mitral valve obstruction model in the dog results in left atrial hypertension and increased circulating levels of ANF (124). This increase occurs even if the hearts are denervated. However, neither the increase in urine flow nor sodium excretion occur in the denervated model, suggesting that ANF does not account for the effects elicited during left atrial distention. Furthermore, chronic infusions of ANF have produced no or unimpressive changes in renal sodium excretion (166, 167). The counterargument would be that this homeostasis is controlled by multiple interacting factors of which ANF is but one and that ANF may be necessary but not sufficient to induce diuresis or natriuresis (164).

Experimental approaches to these questions may now be achieved using molecular biologic techniques to produce animals lacking or overproducing ANF. Homologous recombination provides a method for genetically engineering an animal without endogenous ANF. While this technology has been successfully applied to the study of several other genes (168), animals deficient in ANF have not yet been produced. An alternative approach to delineating the function of a particular peptide is to study the effects of its overproduction. Steinhelper and colleagues have recently reported the generation of transgenic mice with elevated plasma ANF concentrations (169). In these animals, hepatic production of an ANF transgene was directed by either the transthyretin or albumin promoters and resulted in immunoreactive ANF increases of 2–10-fold. Essentially normal fluid and electrolyte homeostasis was maintained, which argues against a central role of ANF in water and sodium homeostasis. However, the mice with the highest ANF levels were significantly hypotensive. Perhaps the pharmacologic renal effects that led to ANF's discovery are less important physiologically than, for example, its hemodynamic effect.

ANF is a potent vasorelaxant in large arteries (12); the relaxation appears to be endothelium-independent and mediated through cGMP (171). However, the acute hypotensive response to ANF appears to be due to decreased cardiac output resulting from reduced preload and consequently stroke volume (167, 172). Rather than an effect on venous capacitance, this preload reduction may be mediated by reduced net capillary absorption (173). Thus ANF alters the extracellular fluid distribution to favor interstitium over plasma. In sheep, after five days of ANF infusion, these initial effects were superseded by a decrease in total peripheral resistance accounting for the continued hypotensive effect (167). In transgenic mice overproducing ANF, the mechanism of hypotension has not yet been determined (169). These studies of artificially increased ANF suggest it could play a major role in the maintenance of normal blood pressure.

Hemodynamic homeostasis is a complex, imperfectly understood process. The renin-angiotension-aldosterone pathway plays a major role in this process, and ANF interacts with both renin and aldosterone. ANF inhibits renin release from juxtaglomerular cells (174, 175). This inhibition correlates with increases in cGMP, is potentiated by a cGMP-phosphodiesterase inhibitor, and is undermined by the cGMP-inhibitor, methylene blue (174), all of which suggest it occurs by a cGMP-mediated pathway consistent with the guanylate cyclase–associated receptor. ANF also inhibits aldosterone synthesis and release (155, 176). In contrast to renin, however, alterations in cGMP do not account for this inhibition (155). This effect of ANF must therefore be mediated either through an as yet undiscovered receptor population or through the previously described receptors utilizing a different signal transduction pathway. Barrett and colleagues (177) recently showed that ANF differentially affects two populations of voltage-dependent calcium channels that may be involved in sustaining aldosterone secretion. It seems possible that the demonstrated inhibition of one of these channels could contribute to inhibition of aldosterone secretion (Figure 3). The effects of ANF on adrenal cells is also demonstrated by the transgenic mice generated by Steinhelper et al (169). Overproduction of ANF actually causes marked hyperplasia of the zona glomerulosa and approximately twofold increased aldosterone levels. While it is possible that chronic hypotension could induce a secondary hyperaldosteronism even in the presence of increased ANF, preliminary results suggest plasma renin activity is normal (178, 179). If the renin level is in fact normal, it would appear the increased aldosterone levels must be mediated through a novel mechanism. Further studies on these mice should provide invaluable information for elucidating these complex, interacting endocrine functions.

ACKNOWLEDGMENTS

This work was supported in part by the American Heart Association (CES) and National Institutes of Health grants HL02228 (AR), HL41474, and HL19259 (CES).

*Literature Cited*

1. Kisch, B. 1956. *Exp. Med. Surg.* 114:99–112
2. Palade, G. E. 1961. *Anat. Rec.* 139:262
3. Jamieson, J. D., Palade, G. E. 1964. *J. Cell Biol.* 23:151–72
4. de Bold, A. J., Borenstein, H. B., Veress, A. T., Sonnenberg, H. 1981. *Life Sci.* 28:89–94
5. de Bold, A. J. 1982. *Can. J. Physiol. Pharmacol.* 60:324–30
6. de Bold, A. J. 1985. *Science* 230:767–70
7. Currie, M. G., Geller, D. M., Cole, B. R., Siegel, N. R., Fox, K. F., et al. 1984. *Science* 223:67–69
8. Atlas, S. A., Kleinert, H. D., Camargo, M. J., Januszewicz, A., Sealey, J. E., et al. 1984. *Nature* 309:717–19
9. de Bold, A. J., Flynn, T. G. 1983. *Life Sci.* 33:297–302
10. Seidah, N. G., Lazure, C., Chretien, M., Thibault, G., Garcia, R., et al. 1984. *Proc. Natl. Acad. Sci. USA* 81: 2640–44
11. Flynn, T. G., de Bold, M. L., de Bold, A. J. 1983. *Biochem. Biophys. Res. Commun.* 117:859–65
12. Currie, M. G., Geller, D. M., Cole, B. R., Boylan, J. G., YuSheng, W., et al. 1983. *Science* 221:71–73
13. Yang-Feng, T. L., Floyd-Smith, G., Nemer, M., Drouin, J., Francke, U. 1985. *Am. J. Hum. Genet.* 37:1117–28
14. Maki, M., Takayanagi, R., Misono, K. S., Pandey, K. N., Tibbets, C., et al. 1984. *Nature* 309:722–24

15. Oikawa, S., Imai, M., Ueno, A., Tanaka, S., Noguchi, T., et al. 1984. *Nature* 309:724–26
16. Seidman, C. E., Duby, A. D., Choi, E., Graham, R. M., Haber, E., et al. 1984. *Science* 225:324–26
17. Nakayama, K., Ohkubo, H., Hirose, T., Inayama, S., Nakanishi, S. 1984. *Nature* 310:699–701
18. Yamanaka, M., Greenberg, B., Johnson, L., Seilhamer, J., Brewer, M., et al. 1984. *Nature* 309:719–22
19. Blobel, G., Dobberstein, B. 1975. *J. Cell Biol.* 67:852–62
20. Perlman, D., Halvorson, H. O. 1983. *J. Mol. Biol.* 167:391–409
21. Miyata, A., Kangawa, K., Toshimori, T., Hatoh, T., Matsuo, H. 1985. *Biochem. Biophys. Res. Commun.* 129: 248–55
22. Michener, M. L., Gierse, J. K., Seetharam, R., Fok, K. F., Olins, P. O., et al. 1987. *Mol. Pharmacol.* 30:552–57
23. Schwartz, D., Geller, D. M., Manning, P. T., Siegel, N. R., Fok, K. F., et al. 1985. *Science* 229:397–400
24. Thibault, G., Lazure, C., Schiffrin, E. L., Gutkowska, J., Chartier, L., et al. 1985. *Biochem. Biophys. Res. Commun.* 130:981–86
25. Argentin, S., Nemer, M., Drouin, J., Scott, G. K., Kennedy, B. P., et al. 1985. *J. Biol. Chem.* 260:4568–71
26. Kangawa, K., Tawaragi, T., Oikawa, S., Mizuno, A., Sakuragawa, Y., et al. 1984. *Nature* 312:152–55
27. Seidman, C. E., Bloch, K. D., Klein, K. A., Smith, J. A., Seidman, J. G. 1984. *Science* 226:1206–9
28. Vlasuk, G. P., Miller, J., Bencen, G. H., Lewicki, J. A. 1986. *Biochem. Biophys. Res. Commun.* 136:396–403
29. Oikawa, S., Imai, M., Inuzuka, C., Tawaragi, T., Nakazato, H., et al. 1985. *Biochem. Biophys. Res. Commun.* 132: 892–99
30. Greenberg, B. D., Bencen, G. H., Seilhamer, J. J., Lewicki, J. A., Fiddes, J. C. 1984. *Nature* 312:656–58
31. Nemer, M., Chamberland, M., Sirois, D., Argentin, S., Drouin, J., et al. 1984. *Nature* 312:654–56
32. Seidman, C. E., Bloch, K. D., Zisfein, J. B., Smith, J. A., Haber, E., et al. 1985. *Hypertension* 7:31–34
33. Thibault, G., Garcia, R., Gutkowska, J., Bilodeau, J., Lazure, C., et al. 1987. *Biochem. J.* 241:265–72
34. Thibault, G., Lazure, C., Chretien, M., Cantin, M. 1989. *J. Biol. Chem.* 264: 18796–802
35. Lynch, D. R., Venable, J. C., Snyder,

S. H. 1988. *Endocrinology* 122:2683–91
36. Rittenhouse, J., Moberly, L., O'Donnell, M. E., Owen, N. E., Marcus, F. 1986. *J. Biol. Chem.* 261:7607–10
37. Bloch, K. D., Jones, S. W., Preibisch, G., Seipke, G., Seidman, C. E., et al. 1987. *J. Biol. Chem.* 262:9956–61
38. Ruskoaho, H., Toth, M., Ganten, D., Unger, T., Lang, R. E. 1986. *Biochem. Biophys. Res. Commun.* 139:266–74
39. Wildey, G. M., Fischman, A. J., Fallon, J. T., Homcy, C. J., Graham, R. M. 1988. *Circulation* 78:II–426
40. Schiller, P. W., Bellini, R., Dionne, G., Mazick, L. A., Garcia, R., et al. 1986. *Biochem. Biophys. Res. Commun.* 138: 880–86
41. Sudoh, T., Minamino, N., Kangawa, K., Matsuo, H. 1990. *Biochem. Biophys. Res. Commun.* 168:863–70
42. Feller, S. M., Schulz-Knappe, P., Forssmann, W.-G. 1988. *Circulation* 78:II–429
43. Gagelmann, M., Hock, D., Forssmann, W.-G. 1988. *FEBS Lett.* 233:249–54
44. Ritter, D., Needleman, P., Greenwald, J. E. 1991. *J. Clin. Invest.* 87:In press
45. Sudoh, T., Kangawa, K., Minamino, N., Matsuo, H. 1988. *Nature* 332:78–81
46. Sudoh, T., Maekawa, K., Kojima, M., Minamino, N., Kangawa, K., et al. 1989. *Biochem. Biophys. Res. Commun.* 159:1427–34
47. Saito, Y., Nakao, K., Itoh, H., Yamada, T., Mukoyama, M., et al. 1989. *Biochem. Biophys. Res. Commun.* 158: 360–68
48. Kojima, M., Minamino, N., Kangawa, K., Matsuo, H. 1989. *Biochem. Biophys. Res. Commun.* 159:1420–26
49. Seilhamer, J. J., Arfsten, A., Miller, J. A., Lundquist, P., Scarborough, R. M., et al. 1989. *Biochem. Biophys. Res. Commun.* 165:650–58
50. Flynn, T. G., Brar, A., Tremblay, L., Sarda, I., Lyons, C., et al. 1989. *Biochem. Biophys. Res. Commun.* 161: 830–37
51. Kambayashi, Y., Nakao, K., Itoh, H., Hosoda, K., Saito, Y., et al. 1989. *Biochem. Biophys. Res. Commun.* 163: 233–40
52. Roy, R. N., Flynn, T. G. 1990. *Biochem. Biophys. Res. Commun.* 171: 416–23
53. Nguyen, T. T., Lazure, C., Babinski, K., Chretien, M., Ong, H., et al. 1989. *Endocrinology* 124:1591–93
54. Johnson, P. F., McKnight, S. L. 1989. *Annu. Rev. Biochem.* 58:799–839
55. Lewicki, J. A., Greenberg, B., Yama-

naka, M., Vlasuk, G., Brewer, M., et al. 1986. *Fed. Proc.* 45:2086–90
56. Gardner, D. G., Vlasuk, G. P., Baxter, J. D., Fiddes, J. C., Lewicki, J. A. 1987. *Proc. Natl. Acad. Sci. USA* 84: 2175–79
57. Gardner, D. G., Deschepper, C. F., Ganong, W. F., Hane, S., Fiddes, J., et al. 1986. *Proc. Natl. Acad. Sci. USA* 83:6697–701
58. Gardner, D. G., Deschepper, C. F., Baxter, J. D. 1987. *Hypertension* 9: 103–6
59. Morel, G., Chabot, J.-G., Gossard, F., Heisler, S. 1989. *Endocrinology* 124: 1703–10
60. Zeller, R., Bloch, K. D., Williams, B. S., Arceci, R. J., Seidman, C. E. 1987. *Genes Dev.* 1:693–93
61. Day, H. L., Schwartz, D., Wiegand, R. C., Stockman, T., Brunnert, S. R., et al. 1987. *Hypertension* 9:485–91
62. Izumo, S., Nadal-Ginard, B., Mahdavi, V. 1988. *Proc. Natl. Acad. Sci. USA* 85:339–43
63. Komuro, I., Kurabayashi, M., Takaku, F., Yazaki, Y. 1988. *Circ. Res.* 62: 1075–79
64. Lee, R. T., Bloch, K. D., Pfeffer, J. M., Pfeffer, M. A., Neer, E. J., Seidman, C. E. 1988. *J. Clin. Invest.* 81: 431–34
65. Knowlton, K. U., Ross, R. S., Chien, K. R. 1990. *Clin. Res.* 38:471A (Abstr.)
66. Standaert, D. G., Needleman, P., Day, M. L., Weigand, R., Krause, J. E. 1988. *Mol. Brain Res.* 4:7–13
67. Mukoyama, M., Nakao, K., Saito, Y., Ogawa, Y., Hosoda, K., et al. 1990. *New Engl. J. Med.* 323:757–58
68. Takayanagi, R., Tanaka, I., Maki, M., Inagami, T. 1985. *Life Sci.* 36:1843–48
69. Zisfein, J. B., Matsueda, B. R., Fallon, J. T., Bloch, K. D., Seidman, C. E., et al. 1986. *J. Mol. Cell. Cardiol.* 18:917–29
70. Ballermann, B. J., Bloch, K. D., Seidman, J. G., Brenner, B. M. 1986. *J. Clin. Invest.* 78:840–43
71. Funer, J. W., Duval, D., Meyer, P. 1973. *Endocrinology* 93:1300–8
72. Gardner, D. G., Hane, S., Trachewsky, D., Schenk, D., Baxter, J. D. 1986. *Biochem. Biophys. Res. Commun.* 139: 1047–54
73. Ladenson, P. W., Bloch, K. D., Seidman, J. G. 1988. *Endocrinology* 123: 652–57
74. Matsubara, H., Hirata, Y., Yoshimi, H., Takata, S., Takagi, Y., et al. 1987. *Biochem. Biophys. Res. Commun.* 148: 1030–38
75. Nemer, M., Argentin, S., Lavigne, J. P., Chamberland, M., Drouin, J. 1987. *J. Cell. Biochem.* Suppl. 11A:121 (Abstr.)
76. Argentin, S., Drouin, J., Nemer, M. 1987. *Biochem. Biophys. Res. Commun.* 146:1336–41
77. Seidman, C. E., Wong, D. W., Jarcho, J. A., Bloch, K. D., Seidman, J. G. 1988. *Proc. Natl. Acad. Sci. USA* 85: 4104–8
78. Chandler, V. L., Maler, B. A., Yamamoto, K. R. 1983. *Cell* 33:489–99
79. Argentin, S., Ardati, A., Lihrman, I., Su, Y. L., Drouin, J., et al. 1990. *Endocrinol. Soc. 72nd Annu. Meet.: Program Abstr.* 72:1415 (Abstr.)
80. LaPointe, M. C., Deschepper, C. F., Wu, J., Gardner, D. G. 1990. *Hypertension* 15:20–28
81. Kelly, R. B. 1985. *Science* 230:25–32
82. Sorbera, L. A., Morad, M. 1990. *Science* 247:969–76
83. Mantymaa, P., Leppaluoto, J., Ruskoaho, H. 1990. *Endocrinology* 126: 587–95
84. Fukuda, Y., Hirata, Y., Shigeru, T., Takatsugu, K., Oikawa, S., et al. 1989. *Biochem. Biophys. Res. Commun.* 164: 1431–36
85. Shubeita, H. E., Harris, A. N., McDonough, P. M., Knowlton, K. U., Glembotski, C., et al. 1990. *Circulation* Suppl. 82:III–687. (Abstr.)
86. Buratowski, S., Hahn, S., Guarente, L., Sharp, P. A. 1989. *Cell* 56:549–61
87. Gorman, C. M., Moffat, L. F., Howard, B. H. 1982. *Mol. Cell. Biol.* 2:1044–51
88. Prost, E., Moore, D. D. 1986. *Gene* 45:107–11
89. Rosenzweig, A., Selvaraj, S., Seidman, J. G., Seidman, C. E. 1989. *Circulation* 80:II–110 (Abstr.)
90. LaPointe, M. C., Wu, J., Greenberg, B., Gardner, D. G. 1988. *J. Biol. Chem.* 263:9075–78
91. Wu, J., LaPointe, M. C., West, B. L., Gardner, D. G. 1989. *J. Biol. Chem.* 264:6472–79
92. Field, L. J. 1988. *Science* 239:1029–33
93. Seidman, C. E., Schmidt, E., Seidman, J. G. 1991. *Can. J. Physiol. Pharmacol.* 68:In press
94. Simpson, P. C. 1988. *Annu. Rev. Physiol.* 51:189–202
95. Barka, T., Noen, H. V. D., Shaw, P. A. 1987. *Oncogene* 1:439–43
96. Iwaki, K., Sukhatme, V. P., Shubeita, H. E., Chien, K. R. 1990. *J. Biol. Chem.* 265:13809–17
97. Komuro, I., Kaida, T., Shibazaki, Y.,

Kurabayashi, M., Katoh, Y. et al. 1990. *J. Biol. Chem.* 265:3595–98
98. Yazaki, Y., Tsuchimochi, H., Kurabayashi, M., Komuro, I. 1989. *J. Mol. Cell. Cardiol.* 21(5):91–101
99. Chiu, R., Boyle, W. J., Meek, J., Smeal, T., Hunter, T., et al. 1988. *Cell* 54:541–52
100. Halazonetis, T. D., Georgopoulos, K., Greenberg, M. E., Leder, P. 1988. *Cell* 55:917–24
101. Rauscher, F. J., Voulalas, P. J., Franza, B. R., Curran, T. 1988. *Genes Dev.* 2:1687–99
102. Sonnenberg, J. L., Rauscher, F. J., Morgan, J. I., Curran, T. 1989. *Science* 246:1622–25
103. Rosenzweig, A., Halazonetis, T. D., Seidman, J. G., Seidman, C. E. 1990. *Clin. Res.* 38:286A (Abstr.)
104. Knowlton, K. U., Ross, R. S., Chien, K. R. 1990. *Clin. Res.* 38:471A (Abstr.)
105. Davis, R. L., Weintraub, H., Lasar, A. B. 1987. *Cell* 51:987–1000
106. Wildey, G. M., Misono, K. S., Graham, R. M. 1991. In *The Heart and Cardiovascular System: Scientific Foundations,* ed. H. A. Fozzard, E. Haber, R. B. Jennings, A. M. Katz, H. E. Morgan. New York: Raven. In press
107. Needleman, P., Blaine, E. H., Greenwald, J. E., Michener, M. L., Sapier, C. B., et al. 1989. *Annu. Rev. Pharmacol. Toxicol.* 29:23–54
108. Flynn, T. G., Davies, P. L., Kennedy, B. P., de Bold, M. L., de Bold, A. J. 1985. *Science* 228:323–25
109. Bloch, K. D., Seidman, J. G., Naftilan, J. D., Fallon, J. T., Seidman, C. E. 1986. *Cell* 47:695–702
110. Misono, K. S., Fukumi, H., Grammer, R. T., Inagami, T. 1984. *Biochem. Biophys. Res. Commun.* 119:524–29
111. Wildey, G. M., Fischman, A. J., Fallon, J. T., Homcy, C. J., Graham, R. M. 1988. *Circulation* 78:II-426 (Abstr.)
112. Lynch, D. R., Venable, J. C., Snyder, S. H. 1988. *Endocrinology* 122:2683–91
113. Shields, P. P., Glembotski, C. C. 1987. *Biochem. Biophys. Res. Commun.* 146:547–53
114. Bloch, K. D., Scott, J. A., Zisfein, J. B., Fallon, J. T., Margolies, M. N., et al. 1985. *Science* 230:1168–71
115. Shields, P. P., Glembotski, C. C. 1988. *J. Biol. Chem.* 263:8091–98
116. Glembotski, C. C., Dixon, J. E., Gibson, T. R. 1988. *J. Biol. Chem.* 263:16073–81
117. Shields, P. P., Dixon, J. E., Glembot-

ski, C. C. 1988. *J. Biol. Chem.* 263:12619–28
118. Currie, M. G., Geller, D. M., Chao, J., Margolius, H. S., Needleman, P. 1984. *Biochem. Biophys. Res. Commun.* 120:461–66
119. Imada, T., Takayanagi, R., Inagami, T. 1988. *J. Biol. Chem.* 263:9515–19
120. Wypij, D. M., Harris, R. B. 1988. *J. Biol. Chem.* 263:7079–86
121. Watanabe, T., Watanabe, M., Ishii, Y., Matsuba, H., Kimura, S., et al. 1989. *J. Histochem. Cytochem.* 37:347–51
122. Katsube, N., Schwartz, D., Needleman, P. 1985. *Biochem. Biophys. Res. Commun.* 133:937–44
123. Ledsome, J. R., Wilson, N., Courneya, C. A., Rankin, A. J. 1985. *Can. J. Physiol. Pharmacol.* 63:739–42
124. Goetz, K. L., Wang, B. C., Geer, P. G., Leadley, R. J., Reinhardt, H. W. 1986. *Am. J. Physiol.* 250:R946–50
125. Mancini, G. B. J., McGillem, M. J., Bates, E. R., Weder, A. B., Deboe, S. F., et al. 1987. *Circulation* 76:884–90
126. Schiebinger, R. J., Linden, J. 1986. *Circ. Res.* 59:105–9
127. Currie, M. G., Newman, W. H. 1986. *Biochem. Biophys. Res. Commun.* 137:94–100
128. Ruskoaho, H., Toth, M., Lang, R. E. 1985. *Biochem. Biophys. Res. Commun.* 133:581–88
129. Shields, P. P., Glembotski, C. C. 1989. *J. Biol. Chem.* 264:9322–28
130. Darnell, J., Lodish, H., Baltimore, D., eds. 1986. *Molecular Cell Biology,* pp. 676–93. New York: Sci. Am. Books. 1192 pp.
131. Matsubara, H., Hirata, Y., Yoshimi, H., Takata, S., Takagi, Y., et al. 1988. *Am. J. Physiol.* 155:H405–9
132. Iida, H., Page, E. 1988. *Biochem. Biophys. Res. Commun.* 157:330–36
133. Saito, Y., Nakao, K., Morii, N., Sugawara, A., Shiono, S., et al. 1986. *Biochem. Biophys. Res. Commun.* 138:1170–76
134. Sei, C. A., Glembotski, C. C. 1990. *J. Biol. Chem.* 265:7166–72
135. Iida, H., Page, E. 1989. *Am. J. Physiol.* 256:C608–13
136. Uusimaa, P. A., Ruskoaho, H., Leppäluoto, J., Hassinen, I. E. 1991. *Mol. Cell Endocrinol.* In press
137. Greenwald, J. E., Apkon, M., Hruska, K. A., Needleman, P. 1989. *J. Clin. Invest.* 83:1061–65
138. Vandlen, R. L., Arcuri, K. E., Napier, M. A. 1985. *J. Biol. Chem.* 260:10889–92

139. Hirose, S., Akiyama, F., Shinjo, M., Ohno, H., Murakami, K. 1985. *Biochem. Biophys. Res. Commun.* 130: 574–79

140. Meloche, S., Ong, H., Cantin, M., De-Lean, A. 1986. *J. Biol. Chem.* 261: 1525–28

141. Leitman, D. C., Andresen, J. W., Kuno, T., Kamisaki, Y., Chang, J.-K., et al. 1986. *J. Biol. Chem.* 261:11650–55

142. Scarborough, R. M., Schenk, D. B., McEnroe, G. A., Arfsten, A., Kang, L.-L., et al. 1986. *J. Biol. Chem.* 261:12960–64

143. Maack, T., Suzuki, M., Almeida, F. A., Nussenzveig, D., Scarborough, R. M., et al. 1987. *Science* 238:675–78

144. Shimonaka, M., Saheki, T., Hagiwara, H., Ishido, M., Nogi, A., et al. 1987. *J. Biol. Chem.* 262:5510–14

145. Schenk, D. B., Phelps, M. N., Porter, J. G., Fuller, F., Cordel, B., et al. 1987. *Proc. Natl. Acad. Sci. USA* 84:1521–25

146. Takayanagi, R., Inagami, T., Snajdar, R. M., Imada, T., Tamura, M., et al. 1987. *J. Biol. Chem.* 262:12104–13

147. Kuno, T., Andresen, J. W., Kamisaki, Y., Waldman, S. A., Chang, L. Y., et al. 1986. *J. Biol. Chem.* 261:5817–23

148. Paul, A. K., Marala, R. B., Jaiswal, R. K., Sharma, R. K. 1987. *Science* 235: 1224–26

149. Fuller, F., Porter, J. G., Arfsten, A. E., Miller, J., Schilling, J. W., et al. 1988. *J. Biol. Chem.* 263:9395–401

150. Chinkers, M., Garbers, D. L., Chang, M.-S., Lowe, D. G., Chin, H., et al. 1989. *Nature* 338:78–83

151. Chang, M.-S., Lowe, D. G., Lewis, M., Hellmiss, R., Chen, E., et al. 1989. *Nature* 341:68–72

152. Schulz, S., Singh, S., Bellet, R. A., Singh, G., Tubb, D. J., et al. 1989. *Cell* 58:1155–62

153. Chinkers, M., Garbers, D. L. 1989. *Science* 245:1392–94

154. Sharma, R. K., Chalberg, S. C., Marala, R. B. 1991. *Can. J. Physiol. Pharmacol.* 68:In press

155. Vari, R. C., Freeman, R. H., Davis, J. O., Villarreal, D., Verburg, K. M. 1986. *Am. J. Physiol.* 251:R48–52

156. Anand-Srivastava, M. B., Srivastava, A. K., Cantin, M. 1987. *J. Biol. Chem.* 262:4931–34

157. Pandey, K. N., Pavlou, S. N., Inagami, T. 1988. *J. Biol. Chem.* 263:13406–13

158. Appel, R. G., Wang, J., Simonson, M. S., Dunn, M. J. 1986. *Am. J. Physiol.* 251:F1036–42

159. Ardaillou, N., Nivez, M-P., Ardaillou, R. 1985. *FEBS Lett.* 189:8–12

160. Burnett, J. C., Granger, J. P., Opgenorth, T. J. 1984. *Am. J. Physiol.* 247:F863–66

161. Bourgoignie, J. J., Gavellas, G., Hwang, K. H. 1986. *Am. J. Physiol.* 251:F1049–54

162. Mendez, R. E., Dunn, B. R., Troy, J. L., Breener, B. M. 1986. *Circ. Res.* 59:605–11

163. Baum, M., Toto, R. D. 1986. *Am. J. Physiol.* 250:F66–69

164. Blaine, E. H. 1990. *Hypertension* 15:2–8

165. Goetz, K. L. 1990. *Hypertension* 15:9–19

166. Granger, J. P., Opgenorth, T. J., Salazar, J., Romero, J. C., Burnett, J. C. 1986. *Hypertension* 8II112–16

167. Parkes, D. G., Coghlan, J. P., McDougall, J. G., Scoggins, B. A. 1988. *Am. J. Physiol.* 254:H811–15

168. Zijlstra, M., Bix, M., Simister, N. E., Loring, J. M., Raulet, D. H., et al. 1990. *Nature* 344:742–46

169. Steinhelper, M. E., Cochrance, K. L., Field, L. J. 1990. *Hypertension* 16:301–7

170. Deleted in proof

171. Winquist, R. J., Faison, E. P., Waldman, S. A., Schwartz, K., et al. 1984. *Proc. Natl. Acad. Sci. USA* 81:7661–64

172. Lappe, R. W., Smits, J. F. M., Todt, J. A., Debets, J. J. M., Wendt, R. L. 1985. *Circ. Res.* 56:606–12

173. Trippodo, N. C., Barbee, R. W. 1987. *Am. J. Physiol.* 252:R915–20

174. Kurtz, A., Della Bruna, R., Pfeilschifter, J., Taugner, R., Bauer, C. 1986. *Proc. Natl. Acad. Sci. USA* 83:4769–73

175. Scheuer, D. A., Thrasher, T. N., Quillen, E. W., Metzler, C. H., Ramsay, D. J. 1987. *Am. J. Physiol.* 251:R423–27

176. Atarashi, K., Mulrow, P. J., Franco-Saenz, R., Snajdar, R., Rapp, J. 1984. *Science* 224:992–94

177. McCarthy, R. T., Isales, C. M., Bollag, W. B., Rasmussen, H., Barrett, P. Q. 1990. *Am. J. Physiol.* 258:F473–78

178. Steinhelper, M. E., Cochrance, K. L., Field, L. J. 1991. *Am. J. Physiol.* In press

179. Steinhelper, M. E., Cochrance, K. L., Veress, A., Sonnenberg, H., Field, L. J. 1990. In *Mouse Molecular Genetics*, p. 45. Cold Spring Harbor, NY: Cold Spring Harbor Lab. (Abstr.)

Annu. Rev. Biochem. 1991. 60:257–80

# LYSOSOMAL STORAGE DISEASES

## Elizabeth F. Neufeld

Department of Biological Chemistry and Brain Research Institute, School of Medicine, and Molecular Biology Institute, University of California, Los Angeles, California 90024

KEY WORDS: lysosomal enzymes, mutation analysis, replacement therapy, sequence homologies.

## CONTENTS

0066-4154/91/0701-0257$02.00

# INTRODUCTION

The concept of "lysosomal storage disorders" was introduced by Hers in 1965 (1) to explain how genetically determined absence of a seemingly unimportant enzyme, $\alpha$-glucosidase, could lead to the fatal condition known as Pompe disease. The undegraded substrate would gradually accumulate within lysosomes, causing progressive increase in the size and number of these organelles; the cellular pathology would eventually lead to malfunction of the affected organ. This concept quickly led to the discovery of additional lysosomal storage disorders, which at this time number three dozen (Table 1). The majority are caused by deficiency of a single lysosomal enzyme; others, for which I-cell disease is the paradigm, result from pleiotropic loss of several lysosomal enzymes because of an underlying defect in some common protein. Although the total number of known lysosomal storage diseases has not changed much since the subject was last reviewed in this series (2), research activity in the field has been intense in the intervening years. The 1989 edition of *The Metabolic Basis of Inherited Disease* (3) devotes 300 pages, documented by more than 3500 references, to the topic. The reader is directed in Table 1 to chapters in this compendium and other reviews for detailed accounts of the clinical, pathological, biochemical, genetic, and molecular aspects of individual disorders or classes of related disorders (4–20). The current review concentrates on recent developments in the area of lysosomal enzyme deficiency diseases, and focuses on general issues such as processing of normal and mutant enzymes, the structure of normal and mutant genes, correlation of genotype with phenotype, and development of therapeutic strategies. The interesting areas of transport of small molecules out of lysosomes and genetic defects thereof (20, 20a) are not in the purview of this chapter.

# PROCESSING OF LYSOSOMAL ENZYMES

## Normal Enzymes

PATHWAYS FOR SOLUBLE AND MEMBRANE-BOUND ENZYMES    The pathway of synthesis, transport, and processing of soluble lysosomal enzymes and related proteins is well known and well reviewed (21, 22). Briefly, the early events—insertion into the endoplasmic reticulum, removal of the signal peptide, and $N$-glycosylation—are the same as for secretory proteins. A reaction specific to soluble proteins destined to lysosomes is acquisition of the mannose 6-phosphate marker for targeting to the organelles (23). The mannose 6-phosphate structure mediates binding to one of two mannose 6-phosphate receptors [one of which, the calcium-independent receptor, is a bifunctional protein, which also serves as the insulin-like growth factor II receptor (reviewed in 24)]. Association with the receptor occurs in a late Golgi compartment, and dissociation in a prelysosomal acidic compartment. As the system is not fully efficient, a fraction of newly made enzyme is secreted; this can be enhanced in cell culture by amines that raise the pH of acidic compartments, thereby preventing the regeneration of free mannose 6-phosphate receptors. The calcium-independent mannose 6-phosphate receptor also participates in endocytosis, bringing in extracellular mannose 6-phosphate–containing enzymes to a prelysosomal compartment through coated pits and vesicles.

Since the function of the mannose 6-phosphate marker is to bind a soluble protein to a membrane receptor, membrane-bound proteins should not require mannose phosphorylation in order to be targeted to lysosomes. It is therefore not surprising that glucocerebrosidase, an enzyme associated with the lysosomal membrane though not a transmembrane protein, does not acquire phosphorylated mannose residues (25). Acid phosphatase has an unusual biosynthetic pathway in that it is synthesized as a transmembrane protein (26, 27), transported to lysosomes via the cell surface (28), and cleaved to a soluble form by the sequential action of a cytoplasmic thiol protease and a lysosomal aspartyl protease (29). Internalization depends on the presence of a tyrosine residue in the cytoplasmic tail (29a); the mannose 6-phosphate pathway is not involved.

PROCESSING OF CARBOHYDRATE CHAINS    Synthesis of the mannose 6-phosphate structure requires two reactions: transfer of phospho-$N$-acetylglucosamine from UDPGlcNAc to give a phosphodiester, followed by removal of the GlcNAc to uncover the mannose 6-phosphate monoester (21, 22). The first reaction itself was recently found to occur in two stages: transfer of PGlcNAc to a mannose residue on the $\alpha$-1,6 branch of a high-mannose oligosaccharide and addition of a second PGlcNAc to the $\alpha$-1,3 branch, thought to occur in a pre-Golgi compartment and the cis-Golgi, respectively (30, 31). The diesterase reaction takes place in the medial Golgi. Subsequently, some of the high-mannose oligosaccharides, even those bearing phosphate groups, may be further modified to hybrid structures, and those that are not phosphorylated may be modified to the complex type.

The major conceptual question in the synthesis of the mannose 6-phosphate recognition marker—how the PGlcNAc transferase recognizes its substrate among the many glycoproteins that pass through the endoplasmic reticulum—is closer to resolution (32). As lysosomal enzymes do not share linear amino acid sequences, the determinant for PGlcNAc transferase recognition had been suspected to reside in the three-dimensional structure. Kornfeld and coworkers approached the problem by engineering chimeric molecules from parts of two homologous proteins, cathepsin D (lysosomal) and glycosylated

**Table 1** Summary of lysosomal storage disorders

| Disorder | Primary deficiency [secondary deficiency] | Substrate | Major review |
|---|---|---|---|
| Disorders of sphingolipid degradation | | | |
| Fabry disease | α-galactosidase | Gal-Gal-Glu-ceramide | 4 |
| Farber disease | ceramidase | ceramide | 5 |
| Gaucher disease | glucocerebrosidase | glucosylceramide | 6, 7 |
| G_M1 gangliosidosis | β-galactosidase | G_M1 ganglioside, galactosyl oligosaccharides, keratan sulfate | 8 |
| G_M2 gangliosidoses | | | |
| Tay-Sachs disease | β-hexosaminidase, α-subunit [hexosaminidase A] | G_M2 ganglioside | 9 |
| Sandhoff disease | β-hexosaminidase, β-subunit [hexosaminidases A and B] | G_M2 ganglioside, oligosaccharides | 9 |
| Activator deficiency | G_M2 activator | G_M2 ganglioside | 9 |
| Krabbe disease | galactosylceramidase | galactosylceramide, galactosylsphingosine | 10 |
| Metachromatic leukodystrophy | | | |
| enzyme-deficient form | arylsulfatase A | galactosylsulfatide | 11 |
| activator-deficient form | sulfatide activator/saposin | galactosylsulfatide | |
| Mucolipidosis IV | primary defect unknown [ganglioside sialidase] | | 8 |
| Multiple sulfatase deficiency | primary defect unknown [deficiency of all sulfatases] | [sulfatase substrates] | 11 |
| Niemann-Pick disease | sphingomyelinase | sphingomyelin | 12 |
| Schindler disease | α-N-acetylgalactosaminidase | α-galNAc glycolipids, glycoproteins | 4 |
| Disorders of glycoprotein degradation | | | |
| Aspartylglycosaminuria | aspartylglycosaminidase | N-linked oligosaccharides | 13 |
| Fucosidosis | α-L-fucosidase | α-L-Fuc oligosaccharides | 13 |
| Galactosialidosis | protective protein/cathepsin [β-galactosidase and sialidase] | ? | 8 |
| α-Mannosidosis | α-mannosidase | [substrates of β-galactosidase and sialidase] α-Man oligosaccharides | 13 |

| Disorder | Defective enzyme/protein | Accumulated substrate | Ref. |
|---|---|---|---|
| β-Mannosidosis | β-mannosidase | β-Man oligosaccharides | 13 |
| Sialidosis | sialidase | sialyl oligosaccharides | 13 |
| **Disorders of glycosaminoglycan degradation** | | | |
| Hunter syndrome | iduronate sulfatase | dermatan sulfate, heparan sulfate | 14, 15 |
| Hurler and Scheie syndromes | α-L-iduronidase | dermatan sulfate, heparan sulfate | 14, 15 |
| Maroteaux-Lamy syndrome | GalNAc 4-sulfatase/arylsulfatase B | dermatan sulfate | 14, 15 |
| Morquio syndrome | | | |
| A subtype | Gal 6-sulfatase | keratan sulfate, chondroitin 6-sulfate | 14, 15 |
| B subtype | β-galactosidase | keratan sulfate | 14, 15 |
| Sanfilippo syndrome | | | |
| A-subtype | heparan N-sulfatase | heparan sulfate | |
| B-subtype | α-N-acetylglucosaminidase | heparan sulfate | |
| C-subtype | AcetylCoA: glucosamine N-acetyltransferase | heparan sulfate | |
| D-subtype | GlcNAc 6-sulfatase | heparan sulfate | |
| Sly syndrome | β-glucuronidase | dermatan sulfate, heparan sulfate, chondroitin 4,6 sulfates | 14–16 |
| **Other single enzyme deficiency disorders** | | | |
| Pompe disease (glycogenosis II) | α-glucosidase | glycogen | 17 |
| Wolman disease | acid lipase | cholesteryl esters, triglycerides | 18 |
| **Disorders of lysosomal enzyme biosynthesis** | | | |
| I-cell disease and pseudoHurler polydystrophy | 6-phospho-N-acetylglucosamine transferase | nascent hydrolases | 19 |
| | [mislocalization of many lysosomal enzymes] | [substrates of mislocalized enzymes] | |
| **Disorders of lysosomal membrane transport** | | | |
| Cystinosis | cystine transport | cystine | 20, 20a |
| Sialic storage and Salla disease | sialic acid transport | sialic acid | 20, 20a |

pepsinogen (secretory), of which the three-dimensional structure is known. The simultaneous presence of two noncontiguous portions of cathepsin D was found to be the minimal requirement for recognition by PGlcNAc transferase in an expression system, though additional sites could improve the effectiveness of the chimeric protein as substrate. The current model is that PGlcNAc transferase recognizes a number of topological features on the surface of polypeptides destined for lysosomes, some combination of which is sufficient to produce an identifying signal (32).

Though they do not acquire mannose 6-phosphate, the two membrane-associated enzymes studied do require $N$-glycosylation, as shown by lack of active glucocerebrosidase (33) or acid phosphatase (34) when cells are grown in tunicamycin. In the latter case, the newly made polypeptide was found to aggregate (34). It is likely that glycosylation is necessary for correct folding of the enzyme.

Once in lysosomes, both soluble and membrane-bound enzymes lose some of their carbohydrate by the action of lysosomal exoglycosidases. The deglycosylation may be extensive; for instance, a high-mannose chain that is phosphorylated in newly made $\beta$-subunit of $\beta$-hexosaminidase (35) is reduced to a single $N$-acetylglucosamine stub in enzyme purified from placenta (36). Trimming of carbohydrate chains does not occur in fibroblasts from patients with deficiency of lysosomal sialidase or $\beta$-galactosidase (37–40), and such fibroblasts have been used to show that in the course of their biosynthesis lysosomal enzymes pass through the compartments where galactosylation and sialylation occur (37, 38). The trimming of terminal carbohydrate residues is not necessary for the function of the enzymes.

PROTEOLYTIC PROCESSING    In addition to removal of the signal peptide in the endoplasmic reticulum, the newly synthesized protein may be subject to additional limited proteolysis during transport or more often after it has reached lysosomes. This process, often called maturation, is generally deduced from changes in migration in polyacrylamide gels under reducing conditions (41, 42). Edman degradation of amino termini of processing intermediates and products can provide precise identification of cleavage sites, as originally applied to cathepsin D (43) and more recently to $\beta$-hexosaminidase (39, 44–47). The reduction in size had been interpreted as loss of a segment of the polypeptide, but reexamination of the structure and biosynthesis of the two subunits of $\beta$-hexosaminidase showed that the precursors are cleaved internally and that the fragments remain attached by disulfide bonds (47, 48).

Much of this proteolytic processing, like the trimming of oligosaccharides described above, is a consequence of the degradative lysosomal environment and is not required for acquisition of activity; most lysosomal enzymes found in cell secretions are in precursor form and catalytically active. Proteases, however, are an exception; they are inactive in precursor form and require proteolytic cleavage for activation, as has been shown for cathepsin D (49) and for protective protein/cathepsin, a recently discovered serine protease (50).

A different kind of proteolytic processing has been discovered in the study of a group of small nonenzymatic lysosomal proteins, called "saposins," that are required for the activity of certain glycolipid hydrolases. A combination of protein sequencing (51), cloning (52–55), and biosynthetic studies (56) has shown that four saposins, of similar sequence but distinct function (although some overlap exists), are carved out of a polyprotein precursor (reviewed in 57). Whether the proteolytic cleavage occurs in lysosomes or in a pre-lysosomal compartment is not yet known.

## Mutant Enzymes

Mutant enzymes, if made at all, often are not processed to the mature lysosomal form, as observed in biosynthetic labeling experiments (41, 58). Excluding I-cell disease and pseudoHurler polydystrophy, in which defective processing of lysosomal enzymes is a consequence of a primary deficiency of phospho-$N$-acetylglucosamine transferase (19), the processing abnormality usually indicates that the mutant enzyme (or more precisely, the immunologically cross-reactive protein) either failed to reach lysosomes or was unstable in those acidic organelles. In the latter case, the enzyme may be stabilized within lysosomes if the cells are grown in the presence of protease inhibitors such as leupeptin (59, 60). Alternatively, the enzyme may be secreted and recovered from the medium if the cells are incubated in the presence of 10 mM $NH_4^+$ (59, 61, 62). But it is not unusual that the mutant enzyme is not secreted even in the presence of $NH_4^+$ and that it remains in the cell in the form of an early precursor without mannose 6-phosphate, and in the case of $\beta$-hexosaminidase, without formation of heterodimers. It has been suggested that like many other improperly folded proteins (63), such mutant lysosomal enzymes are trapped and degraded within the endoplasmic reticulum (e.g. 64).

# MOLECULAR ANALYSIS OF NORMAL GENES

The cloning and characterization of nearly 20 complementary DNAs (cDNAs) and a half dozen genes encoding lysosomal enzymes have been reported to date. While often undertaken as a prelude to analysis of mutations underlying storage diseases, the study of the normal genes and transcripts has revealed several interesting common features.

## Sequence Homologies

There appear to be no short or long stretches of DNA or amino acid sequence that are specific to lysosomal enzymes. On the other hand, there is strong sequence similarity between lysosomal and nonlysosomal enzymes catalyzing closely related reactions; the similarity may extend to enzymes from lower eukaryotes and even prokaryotes. Lysosomal enzymes therefore belong to families linked by functional and structural relationships and presumably by a common evolutionary origin. As summarized in Table 2, there are families of

**Table 2**  Enzyme families based on homology

| Enzyme and source[a] | Ref. |
|---|---|
| $\beta$-Hexosaminidase $\alpha$-subunit, human | 65–67 |
| $\beta$-Hexosaminidase $\beta$-subunit, human | 66, 68, 69 |
| $\beta$-Hexosaminidase $\beta$-subunit, mouse | 70 |
| $\beta$-N-Acetylglucosaminidase A, *Dictyostelium discoideum* | 71 |
| $\beta$-Glucuronidase, human | 72, 73 |
| $\beta$-Glucuronidase, rodent | 74–76 |
| $\beta$-Glucuronidase, *Escherichia coli* | compared in Ref. 75 |
| $\alpha$-Galactosidase, human | 77, 78 |
| $\alpha$-N-Acetylgalactosaminidase ($\alpha$-galactosidase B), human | 79, 80 |
| $\alpha$-Galactosidase, yeast | compared in Refs. 79, 80 |
| $\alpha$-Galactosidase, *E. coli* | compared in Ref. 80 |
| $\alpha$-Glucosidase, human | 80a, 80b |
| Intestinal brush border sucrase/isomaltase, rabbit | compared in Refs. 80a,b |
| Acid phosphatase, human | 26, 81 |
| Prostatic acid phophatase, human, secretory | compared in Ref. 82 |
| Arylsulfatase A (galactosylsulfatide sulfatase), human | 83, 84 |
| Arylsulfatase B (GalNAc 4-sulfatase), human | 85, 86 |
| N-Acetylglucosamine 6-sulfatase, human | 87 |
| Iduronate sulfatase, human | 88 |
| Steroid sulfatase, human, microsomal | 89, 90 |
| Arylsulfatase, sea urchin | compared in Refs. 85, 88 |
| Protective protein/cathepsin, human | 91 |
| Protective protein/cathepsin, mouse | 50 |
| Carboxypeptidase Y, yeast | compared in Ref. 91 |
| KEX 1 protease, yeast | compared in Ref. 91 |
| Cathepsin D, human, lysosomal | 92 |
| Renin, human, secretory | compared in Ref. 92 |
| Pepsinogen, human, secretory | compared in Ref. 92 |

[a] Mammalian enzymes are lysosomal unless indicated otherwise.

$\beta$-hexosaminidases, $\beta$-glucuronidases, $\alpha$-galactosidases, $\alpha$-glucosidases, phosphatases, sulfatases, serine proteases, and aspartyl proteases (50, 65–80, 80a, 80b, 81–92). Intramolecular homology of the four saposins within the polyprotein precursor has been mentioned above.

## Promoters

Lysosomal enzymes and lysosomal enzyme deficiency can be found in every type of cell except for the organelle-less mature erythrocyte; it is the tissue specifity of the substrate rather than of the enzyme that determines the tissues affected in lysosomal enzyme deficiency diseases. It is therefore not surprising that genes encoding some human lysosomal enzymes [the $\beta$-subunit of $\beta$-hexosaminidase (69), $\alpha$-glucosidase (80b), acid phosphatase (81), and arylsulfatase A (84)] have regions upstream of the coding sequence that are common to promoters of housekeeping genes: i.e. regions with no TATA box, and with a very high GC content with potential SP1-binding sites (93). On the other hand, the genes for $\alpha$-galactosidase (94), murine $\beta$-glucuronidase (76), and the $\alpha$-subunit of $\beta$-hexosaminidase (67) do have a TATA box within the GC-rich region.

The gene encoding glucocerebrosidase does not conform at all to the housekeeping model; its promoter has a TATA box, no SP1 binding sites, and causes differential expression of a reporter gene in different cell types, similar to the level of expression of endogenous glucocerebrosidase in the same cells (95, 96). However, the level of differential expression should not be extrapolated directly to normal tissues, because transformed and clearly abnormal lines had been used as representative cell types.

## Alternate Transcripts

Evidence for transcripts that may be derived from alternate splicing has been found in the course of cloning cDNA encoding $\beta$-glucuronidase (72), sphingomyelinase (97), $\beta$-galactosidase (98, 99), sulfatide activator (54), and $\alpha$-N-acetylgalactosaminidase (100). When expressed in transfected cells, the alternate cDNA specified a protein product that did not have the catalytic activity of the corresponding hydrolase (72, 98) and was not targeted to lysosomes (98). The physiological function of these alternate transcripts, if any, is not known.

# MOLECULAR ANALYSIS OF MUTATIONS
## Heterogeneity of Allelic Mutations

Multiplicity of mutant alleles had long been suspected from the clinical heterogeneity associated with any one lysosomal enzyme deficiency and from differences in the properties of residual enzyme. Later, heterogeneity was

**Table 3**  Common mutations in genes encoding lysosomal enzymes

| Enzyme | Mutation | Disease (when homozygous) |
|---|---|---|
| Arylsulfatase A | splice site, intron 2 | metachromatic leukodystrophy, infantile |
| | $^{426}$Pro→Leu | metachromatic leukodystrophy, adult |
| | polyadenylation site[a] | pseudodeficiency, asymptomatic |
| $\alpha$-L-Fucosidase | $^{442}$Gln→Stop | fucosidosis, infantile |
| Glucocerebrosidase | $^{370}$Asn→Ser[b] | Gaucher, non-neuronopathic |
| | $^{444}$Leu→Pro | Gaucher, neuronopathic |
| $\beta$-Hexosaminidase | | |
| $\alpha$-subunit | insertion, exon 11[b] | Tay-Sachs, infantile |
| $\alpha$-subunit | splice site, intron 12[b] | Tay-Sachs, infantile |
| $\alpha$-subunit | deletion, exon 1[c] | Tay-Sachs, infantile |
| $\alpha$-subunit | $^{269}$Gly→Ser[d] | $G_{M2}$ gangliosidosis, adult |
| $\beta$-subunit | deletion | Sandhoff disease, infantile |

[a] Does not cause disease
[b] common among Ashkenazi Jews
[c] common among French-Canadians
[d] among Ashkenazi Jews, adult disease is caused by compound heterozygosity with infantile mutation.

inferred from differences in synthesis and processing of the protein, and when probes became available, from differences in Southern or Northern blots. But it has only been in the past few years, as technology for characterizing mutations has become simpler, that extensive heterogeneity has become evident. Though one or a few mutations, often clustered in particular ethnic groups, may account for a significant fraction of patients or carriers (Table 3), a very large number of rare mutations account for the remainder. Thus, patients are likely to be compound heterozygotes (i.e. carry two different mutant alleles); as is seen below, homozygosity can no longer be assumed even for well-defined populations. Mutations observed in specific disorders are presented in some detail, as they illustrate a number of genetic principles.

## Specific Disorders

$G_{M2}$ GANGLIOSIDOSES    Three genes are required for the lysosomal degradation of $G_{M2}$ ganglioside, an important consitutent of neuronal membranes— the *HEX A* and *HEX B* genes encoding the $\alpha$- and $\beta$-subunits of $\beta$-hexosaminidase, respectively, and the gene encoding the $G_{M2}$ activator, a lipid-binding protein that presents the $G_{M2}$ substrate to the enzyme (9). Failure to degrade the glycolipid because of mutation in any of the three genes gives rise to disorders known collectively as the $G_{M2}$ gangliosidoses.

Mutations in the *HEX A* gene cause infantile Tay-Sachs disease as well as disorders of later onset and slower course. Historically, Tay-Sachs disease was found with greatest frequency in the Ashkenazi Jewish and French-Canadian populations, until carrier detection programs greatly reduced its incidence in these groups. The first of the mutations to be characterized was the French-Canadian, a deletion of exon 1 and flanking sequences due to recombination at *Alu* repetitive sequences (101, 102). Then surprisingly, not one, but two Tay-Sachs mutations were found among Ashkenazi Jews: a base substitution in the conserved donor splice site of intron 12 (103–105) that prevents normal maturation of heterogeneous nuclear RNA (106), and a four-nucleotide insertion in exon 11 that causes premature termination and unstable messenger RNA (mRNA) (107). Analysis of DNA in carriers of enzyme deficiency has shown that the exon 11 insertion accounts for about 70% and the splice site mutation for less than 20% of the Tay-Sachs alleles in the Jewish population (107–109). The insertion also represents a significant proportion of Tay-Sachs alleles in the general population of the United States (109). The less common $^{269}$Gly→Ser substitution, which accounts for 3–4% of mutant *HEX A* alleles (108, 109), gives rise to chronic/adult-onset $G_{M2}$ gangliosidosis when in compound heterozygosity with one of the infantile alleles (110, 111) or in very rare cases of homozygosity (112). The $^{269}$Gly→Ser allele encodes an $\alpha$-subunit that does not properly dimerize with the $\beta$-subunit and yields only a very low level of functional A ($\alpha\beta$) isoenzyme (113).

A rare enzymatic variant of Tay-Sachs disease, called "B1," was long known for the diagnostic confusion that it caused, as the A ($\alpha\beta$) isoenzyme has the substrate specificity of the B ($\beta\beta$)—i.e. it hydrolyzes neutral $\beta$-hexosaminides but not $G_{M2}$ ganglioside or a synthetic sulfated substrate. The B1 phenotype can be produced by one of two mutations that change arginine at position 178 of the $\alpha$-subunit to either histidine (114) or cysteine (115). Since kinetic studies had shown that the $\alpha$- and $\beta$-subunits of the heterodimeric isoenzyme bind acidic and neutral substrates, respectively (116), the apparent change in substrate specificity of B1 can be interpreted as loss of the $\alpha$-subunit-binding site as a consequence of the amino acid substitution at position 178. Evidence for such an interpretation comes from the catalytically inactive $\beta\beta$ dimer formed when $\beta$-subunit, mutagenized at the homologous arginine, is expressed in Cos1 cells (117). These results do not necessarily imply participation of $^{178}$Arg in catalysis, as the effect of the mutation might be indirect.

The two B1 mutations occur at a CpG dinucleotide, as do several other amino acid substitutions [$^{170}$Arg→Gln (118), $^{482}$Glu→Lys (119), $^{499}$Arg→His (62), $^{504}$Arg→His (62)] and a nucleotide deletion in codon 504 causing premature termination (64). CpG dinucleotides are known to be mutagenic hotspots because they are frequent sites of cytosine methylation, and upon spontaneous deamination of methyl cytosine to thymidine may go unrepaired (120, 120a). Mutations of the *HEX A* gene also occur elsewhere— e.g. $^{420}$Trp→Cys (121). In yet another non-CpG mutation, a change of one nucleotide at the 3' end of exon 5 does not change the sense of the codon but interferes with processing of transcripts (122). With two exceptions that cause a juvenile form of $G_{M2}$ gangliosidosis (arginine to histidine substitutions in codons 499 and 504), all these rare mutations result in infantile forms of Tay-Sachs disease.

A common mutation in the *HEX B* gene is a deletion of exons 1–5 by recombination at flanking *Alu* sequences; this deletion accounts for one quarter of the known alleles in Sandhoff disease (123). The deletion mutation does not seem to be enriched in any one ethnic group. Two interesting mutations that create new splice sites in the gene give rise to milder conditions such as juvenile Sandhoff disease or asymptomatic states, depending on the level of residual activity. One is G→A transition at a CpG site in intron 12 (124, 125), the other a duplication of a sequence straddling intron 13 and exon 14 (125). Although the upstream splice sites created by the mutation are used preferentially, the normal splice sites may be used to a very small extent, permitting the formation of some normal β-subunit (125).

The cDNA encoding the $G_{M2}$ activator protein (genetically unrelated to the polyprotein saposin activators) has been cloned, but mutation analysis has not yet been reported (126).

GAUCHER DISEASE    Gaucher disease is caused by deficiency of glucocere-brosidase (also called glucosylceramidase or acid β-glucosidase) and the resulting accumulation of glucosylceramide; it is the most common of the lysosomal storage diseases, with the highest frequency encountered in the Ashkenazi Jewish population (6, 7). In this ethnic group, the disease is of the Type 1 variety, affecting primarily the reticuloendothelial system; the much rarer but more widely distributed Types 2 and 3 (infantile and juvenile neuronopathic forms, respectively) affect the nervous system as well (6, 7). The major mutation in the non-neuronopathic form, $^{370}$Asn→Ser, accounts for three quarters of the Gaucher alleles among Ashkenazi Jewish patients (127–129) and is found also in non-Jewish patients with Type 1 Gaucher disease (130). This allele encodes an enzyme with reduced specific activity and other abnormal properties, as seen in studies of patients' enzyme and in expression of mutagenized cDNA in transfected Sf9 cells (131). The $^{370}$Asn→Ser allele may be found in Type 1 patients in homozygous form or in compound heterozygous form; it is not seen in patients with neuronopathic disease, showing that the residual activity is sufficient to protect against the most severe manifestations of the disease. The $^{370}$Asn→Ser mutation in Ashkenazi Jews is linked to a PvuII polymorphism in intron 6 (132).

The second most common mutation is $^{444}$Leu→Pro, which is found in all three disease types, but causes neuronopathic disease in homozygous form (133, 134). The altered amino acid is in the catalytic domain of glucocerobrosidase, which had been identified previously by reaction with the substrate analog, conduritol B epoxide (135).

A unique feature of glucocerebrosidase genetics is the existence of a pseudogene that is in close proximity to the functional gene and that lacks *Alu* sequences in four introns, has a deletion in exon 9, and has a number of nucleotide changes scattered throughout (136). Several Gaucher alleles have been found to differ from the normal allele in as many as four codons, which are similar to their counterpart in the pseudogene. These complex alleles have been identified as chimeric molecules, part functional gene and part pseudogene, that could have been formed by gene conversion or unequal crossing over (137–140). The [444]Leu→Pro mutation is also present in the pseudogene, and some of the alleles carrying this mutation were subsequently shown to be of the complex type. Homozygosity for [444]Leu→Pro alone appears to cause a somewhat milder neurodegenerative disease than a combination with a complex allele (138).

A number of additional rare mutations, all amino acid substitutions, have been described and are reviewed in Ref. 7. In contrast to the profusion of CpG mutations in the *HEX A* gene, only 2 of 13 point mutations in the glucocerebrosidase gene are at CpG sites.

One instance of Gaucher disease with deficiency of an activator protein (variously known as coglucosidase, sulfatide activator 2, or saposin C) rather than of glucocerebrosidase has been reported (141). Further studies of this disorder would be of great interest, as they would clarify the physiological roles of two activator proteins (saposins A and C) that can stimulate glucocerebrosidase activity in vitro (57).

METACHROMATIC LEUKODYSTROPHY    Metachromatic leukodystrophy is caused by failure to degrade cerebroside sulfate (sulfatide), a component of myelin. Except for an unusually high frequency in the small isolate of Habbanite Jews in Israel, it is panethnic. The storage disease may be caused by deficiency of the enzyme, arylsulfatase A (cerebroside sulfatase), or of the sulfatide activator protein (11).

Two mutant alleles of the arylsulfatase A gene have been found to predominate among patients with metachromatic leukodystrophy (142). An allele common among adult-onset patients carries a [426]Pro→Leu substitution that permits some residual activity. A different allele common among patients with the late infantile form of the disease has been found to carry two mutations: a [193]Trp→Cys substitution and a nucleotide substitution in the donor splice site of intron 2; transfection experiments showed that the amino acid substitution was innocuous and that loss of activity was caused by the splice site mutation (142). Some patients with the juvenile form of the disease have the adult and late infantile mutations in compound heterozygosity (142). By analogy with findings in other disorders, the remainder of mutations causing metachromatic leukodystrophy are likely to be heterogeneous.

Biochemical diagnosis and heterozygote detection for metachromatic leukodystrophy are complicated by the common occurrence of a "pseudodeficiency" allele that specifies a product with very low enzymatic activity but does not cause disease (143, 144). The pseudodeficient enzyme has only one of the two normal glycosylation sites, but mutagenesis and transfection experiments showed that this change was of no consequence (145). The cause of the enzyme deficiency proved to be a base change in the consensus polyadenylation site; the large transcripts made by use of alternate polyadenylation sites are relatively unstable and give rise to only small amounts of polyadenylated mRNA (145).

Metachromatic leukodystrophy may also be caused by deficiency of saposin B, variously known as sulfatide activator protein or sphingolipid activator 1 (11, 57). Although saposin B stimulates the hydrolysis in vitro of several glycolipids by their respective hydrolases, its physiological function must reside in the degradation of cerebroside sulfate, the only glycolipid to accumulate in the deficiency state. Two mutations affecting this protein have been identified. One is a substitution of isoleucine for threonine, obliterating a glycosylation site and possibly exposing the saposin to proteolytic degradation at a nearby arginine residue (146, 146a). The other is a 33-nucleotide insertion into mRNA between nucleotides 777 and 778, counting from the initiation codon of the saposin polyprotein precursor (147). Some normal individuals also have the nine downstream nucleotides of the insertion, probably due to alternate splicing (54, 146a, 147).

FABRY DISEASE    Deficiency of $\alpha$-galactosidase causes accumulation of glycosphingolipids with terminal $\alpha$-galactosyl residues, primarily in capillary walls (4). The disease is one of two lysosomal storage disorders to have a locus on the X-chromosome (Hunter syndrome is the other). A number of mutations have been characterized, including amino acid substitution [$^{356}$Arg→Trp (148), $^{301}$Arg→Gln (149)], premature termination (149), and major rearrangements to which the $\alpha$-galactosidase gene appears especially prone (148, 150, 151). Although the gene is rich in *Alu* repetitive elements, only one of the rearrangements involves recombination at *Alu* sites; the remainder, mostly deletions but also a duplication, have been attributed to slipped mispairing of short direct repeats (150).

GALACTOSIALIDOSIS    The combined deficiency of $\beta$-galactosidase and sialidase, known as galactosialidosis, is a distinct clinical and biochemical entity caused by deficiency of yet another lysosomal protein, which is required for protecting the activity of the two glycosidases (152). Stabilization of $\beta$-galactosidase and activation of sialidase is achieved by association of the three proteins into a multimeric complex (153–155). As the sequence of the cDNA encoding this "protective protein" was found to be similar to that of the yeast serine proteases, carboxypeptidase Y and KEX 1, the protective protein was predicted to be a protease (91), and was found to react with diisopropylfluorophosphate (50). Examination of its substrate specificity suggested that the "protective protein" was a lysosomal carboxypeptidase [carboxypeptidase L (156)] and was identical to cathepsin A (A. d'Azzo, personal communication). However, its function in protecting the glycosidases in the complex is distinct from its function as a cathepsin (A. d'Azzo, personal communication). In another development in this intriguing story, a newly characterized deamidase has been found to be closely related to the protective protein/cathepsin (157).

Analysis of mutations in galactosialidosis and of the consequences thereof should clarify the role of this potentially multifunctional protein.

OTHER DISORDERS    Characterization of mutations has begun for many other lysosomal storage disorders. One fifth of fucosidosis patients have been found to carry a mutant allele of the gene encoding $\alpha$-L-fucosidase, in which a nucleotide substitution creates a stop at codon 442, 120 nucleotides upstream of the normal termination site (158, 159; K. A. Kretz, personal communication). A$^{163}$Cys→Ser substitution in the cDNA encoding aspartylglycosaminidase has been found among Finnish patients with aspartylglycosaminuria (159a). Heterogeneity, identified at the level of Northern or Southern analysis, has been reported for $\alpha$-glucosidase deficiency, Pompe disease (160, 161), as well as for iduronate sulfatase deficiency (Hunter syndrome) (66); in the latter case, complete deletion of the gene has been observed in some patients (66). Precise analysis of mutations are anticipated in the near future for these disorders, as well as for the many other lysosomal storage disorders for which cloned probes are or soon will be available.

## Genotype-Phenotype Correlations

Clinical heterogeneity is characteristic of lysosomal storage disorders, most of which display a spectrum ranging from very severe to relatively benign. It seems intuitively reasonable to assume that the more profound the enzyme deficiency, the more severe the disorder, and conversely, the higher the residual activity of the mutant enzyme, the greater the protection against disease. The level of activity needed to protect against the most severe manifestations of a disease may be a small fraction of the normal level and perhaps difficult to measure by standard enzyme assays. Conzelmann &

Sandhoff (162) have provided a formal kinetic treatment of this concept. Their model proposes a "critical threshold" of activity, above which the enzyme is capable of keeping up with substrate influx and below which it cannot keep up and accumulation of undegraded substrate occurs. The critical threshold and the rate of accumulation can differ between cells (e.g. between subsets of neurons), or between the same cells at different times in development, depending on the rate of influx of the substrate. Thus, a very small difference in residual activity can markedly affect the age of onset of the disease, the rate of progression, and the symptomatology. Experimental verification of this model showed a good inverse correlation of residual activity of $\beta$-hexosaminidase A and arylsulfatase A with severity of $G_{M2}$ gangliosidosis and metachromatic leukodystrophy, respectively (P. Leinekugel, S. Michel, E. Conzelmann, and K. Sandhoff, unpublished results).

Because of implications for prognosis and counseling of carriers, a major objective of mutation analysis has been to correlate the nature of the mutation with the clinical manifestations of the disease. The multiplicity of mutant alleles discussed above means that many or perhaps most patients have a combination of two different alleles. Alleles that cause total loss of enzyme activity can be expected to be very heterogeneous, as there are innumerable ways to prevent the formation of functional enzyme, and to cause severe disease in any combination. On the other hand, mutations that permit some function are expected to be limited in number, as the residual phenotype is itself selective, and to cause a milder disease. There appears to be some dose dependence for the milder alleles: homozygotes for the [370]Asn→Ser glucocerebrosidase allele (129, 163), for the [269]Gly→Ser *HEX A* allele (112, 163a), and for the [426]Pro→Leu arylsulfatase A allele (142) are likely to have a somewhat milder form of Type 1 Gaucher disease, adult $G_{M2}$ gangliosidosis, and late-onset metachromatic leukodystrophy, respectively, than individuals who carry these same alleles in compound heterozygosity with alleles for the severe form of the disease.

However, current correlations of genotype and clinical presentation are imperfect; for instance, juvenile as well as adult-onset metachromatic leukodystrophy may result from homozygosity of [426]Pro→Leu (142), and the homozygosity for the [444]Leu→Pro mutation, generally associated with the neuronopathic form of Gaucher disease, has been found among Japanese patients with the non-neuronopathic form (164). Such discrepancies emphasize that the consequences of specific mutations can be modulated by unknown variables, some of which, such as infections or use of lysosomotropic drugs, might be environmental. Caution should be exercised in predicting clinical outcome from genotype, particularly with novel combinations of alleles.

# PROSPECTS FOR THERAPY

The discovery of lysosomal storage disease was accompanied by the suggestion that this class of disorders could be treated by administration of exogenous enzymes, which would find their way to lysosomes by the process of endocytosis (1). Hopes were raised by the ease with which storage could be reduced in cell culture (2). Numerous experiments were conducted in the 1960s and 1970s, in which purified enzymes, or plasma and leukocytes as sources of enzymes, were administered to patients with lysosomal storage diseases, particularly mucopolysaccharidoses (14). The results were generally disappointing, in part because only minute amounts of enzyme were administered and in part because the need for signals for receptor-mediated endocytosis was not yet appreciated. Both problems were addressed in a recent study of enzyme replacement for patients with Type 1 Gaucher disease (165). Glucocerebrosidase, which normally carries complex oligosaccharide chains, was modified after purification to expose terminal mannose residues; these allow the enzyme to bind to the mannose receptors present on the surface of reticulo-endothelial cells (166), which are specifically affected in Type 1 Gaucher disease. The modified enzyme was administered in substantial quantity over a two-year period, with significant clinical benefit. This promising report suggests that enzyme replacement, carried out on an appropriate scale and with due consideration to receptor-mediated endocytosis in target cells, remains a viable concept for therapy of lysosomal storage disease.

Because of the failure of early attempts at enzyme replacement, other alternatives have been pursued, including bone marrow transplantation, even though lysosomal storage disorders are not specific to blood cells. The underlying hypothesis is that cells derived from hematopoietic progenitors of the donor (circulating leukocytes and tissue macrophages) can donate lysosomal enzyme to the deficient cells in all tissues of the host, either through secretion or direct cell-cell interaction. Both biochemical and clinical benefit has been observed in some patients, mainly with Hurler or Maroteaux-Lamy syndrome, who received allogeneic bone marrow transplantation (167, 168). Better controlled experiments than would be possible with human patients have been performed in animals with lysosomal storage disease. Bone marrow transplantation was shown to have significant positive effects in dogs with $\alpha$-L-iduronidase deficiency, including decrease in glycosaminoglycan storage in all organs examined and a much slower progression of disease symptoms (169, 170). Clinical amelioration was also seen in dogs with fucosidosis, especially if the transplantation was performed at a very young age (171, 172), and in a cat with arylsulfatase B deficiency (173). Beneficial effects on neuronal pathology have been found in the twitcher mouse, which has a galactocerobrosidase deficiency analogous to Krabbe disease in the human; the blood-brain barrier is believed to have been crossed by donor-derived macrophages (174). On the other hand, bone marrow transplantation had no effect on the progression of neurological disease in dogs with $G_{M1}$ gangliosidosis (175).

But even at best, the benefits of bone marrow transplantation are modest, while the risks and cost of the procedure are very great. However, its limited success has encouraged the prospect of gene replacement through transfection of hematopoietic progenitor cells. Retroviral vectors containing human glucocerebrosidase cDNA have been used to express glucocerebrosidase in culture fibroblasts (176, 177) and hematopoietic cells (178) from Gaucher patients. Mouse hematopoietic progenitor cells transfected with human glucocerebrosidase cDNA in a retroviral vector expressed the human enzyme both in long-term culture (179) and after transplantation into mice (180). These results seem particularly promising for eventual gene replacement in Gaucher disease Type 1, which affects bone marrow–derived cells of the monocyte/macrophage lineage.

A retroviral vector carrying a rat cDNA encoding $\beta$-glucuronidase was used to correct the deficiency in both human and canine fibroblasts deficient in that enzyme, as well as in retinal epithelial cells and hematopoietic cells derived from the deficient dog (181). Finally, insertion of the human $\beta$-glucuronidase gene into the germ line of a mouse model for complete $\beta$-glucuronidase defiency resulted in the expression of human $\beta$-glucuronidase in all tissues of a transgenic animal, and a completely normal phenotype even though the animal was a homozygote for the murine deficiency disease (182). This study illustrates the value of mouse models with total enzyme deficiency, which could serve as background for the insertion of mutant human genes to create better models of human disease for testing novel therapies.

## CONCLUDING REMARKS

Over the past five years, molecular studies have expanded our understanding of lysosomal enzymes and their genes both in the normal state and in lysosomal storage diseases. The limiting step in this progress has been the initial purification of the enzymes to homogeneity, in order to obtain antibodies as well as partial sequence for cloning cDNA. Some lysosomal enzymes, mainly those that are membrane-bound, have resisted attempts at purification and may require different cloning strategies.

Numerous disease-producing mutations in genes encoding lysosomal enzymes have been characterized, including deletions, insertions, splice site and polyadenylation site alterations, and missense and nonsense mutations. Illegitimate recombination at *Alu* repetitive sequences and other sites, recombination of gene and pseudogene sequences (for glucocerebrosidase), and base changes at CpG dinucleotides appear to be frequent mechanisms of mutagenesis. Heterogeneity of mutant alleles appears to be the rule for every lysosomal storage disease studied to date, although one or a small number of alleles may be widely distributed in particular populations, due to founder effect or selection. Patients are therefore likely to carry two different mutant alleles, the combination of which must be considered in order to understand the manifestations of the disease.

A true understanding of the effect of mutations on the transport and function of lysosomal enzymes has been hampered by lack of information about their three-dimensional structure. As a result of their residence in the degradative organelles, lysosomal enzymes purified from tissues tend to be nicked and partially deglycosylated, and thus too heterogeneous for crystallization. The problem will no doubt be remedied in the next few years by the development of overexpressing cell lines that can secrete large quantities of lysosomal enzymes in intact form and in quantities sufficient for structural studies.

The ability to produce large quantities of lysosomal enzymes by recombinant DNA technology should also encourage new attempts at enzyme replacement. The recombinant enzymes can be modified chemically or enzymatically to enhance their receptor-mediated entry into target cells. This approach can be expected to proceed in parallel with the development of gene replacement via retroviral vectors, for which certain lysosomal storage disorders are considered to be good candidates. The existing animal models of lysosomal storage disorders, and perhaps new mouse models created by gene disruption, should be invaluable for such therapeutic endeavors.

ACKNOWLEDGMENTS

I thank my many colleagues who generously provided unpublished manuscripts; Drs. Gideon Bach, Alessandra d'Azzo, Leonard Rome, Kurt von Figura, and members of my laboratory for critical reading of this review; and the National Institutes of Health (NS 22376 and DK 38857) for support.

*Literature Cited*

1. Hers, H. G. 1965. *Gastroenterology* 48:625–33
2. Neufeld, E. F., Lim, T. W., Shapiro, L. J. 1975. *Annu. Rev. Biochem.* 44:357–76
3. Scriver, C. R., Beaudet, A. L., Sly, W. S., Valle, D., eds. 1989. *The Metabolic Basis of Inherited Disease*. New York: McGraw-Hill. 3006 pp. 6th ed.
4. Desnick, R. J., Bishop, D. F. 1989. See Ref. 3, pp. 1751–96
5. Moser, H. W., Moser, A. B., Chen, W. W., Schram, A. W. 1989. See Ref. 3, pp. 1645–54
6. Barranger, J. A., Ginns, E. I. 1989. See Ref. 3, pp. 1677–98
7. Grabowski, G. A., Gatt, S., Horowitz, M. 1990. *CRC Crit. Rev. Biochem.* In press
8. O'Brien, J. S. 1989. See Ref. 3, pp. 1797–806
9. Sandhoff, K., Conzelmann, E., Neufeld, E. F., Kaback, M. M., Suzuki, K. 1989. See Ref. 3, pp. 1807–39
10. Suzuki, K., Suzuki, Y. 1989. See Ref. 3, pp. 1699–720
11. Kolodny, E. H. 1989. See Ref. 3, pp. 1721–50

12. Spence, M. W., Callahan, J. W. 1989. See Ref. 3, pp. 1655–76
13. Beaudet, A. L., Thomas, G. H. 1989. See Ref. 3, pp. 1603–21
14. Neufeld, E. F., Muenzer, J. 1989. See Ref. 3, pp. 1565–87
15. Hopwood, J. J., Morris, C. P. 1990. Mol. Biol. Med. 7:381–404
16. Paigen, K. 1989. Prog. Nucleic Acid Res. Mol. Biol. 37:155–205
17. Hers, H. G., van Hoof, F., de Barsy, T. 1989. See Ref. 3, pp. 425–52
18. Schmitz, G., Assmann, G. 1989. See Ref. 3, pp. 1623–44
19. Nolan, C. M., Sly, W. S. 1989. See Ref. 3, pp. 1589–601
20. Gahl, W. A., Renlund, M., Thoene, J. G. 1989. See Ref. 3, pp. 2619–47
20a. Gahl, W. A. 1989. Annu. Rev. Nutr. 9:39–61
21. von Figura, K., Hasilik, A. 1986. Annu. Rev. Biochem. 55:167–93
22. Kornfeld, S., Mellman, I. 1989. Annu. Rev. Cell Biol. 5:483–525
23. Kaplan, A., Achord, D. T., Sly, W. S. 1977. Proc. Natl. Acad. Sci. USA 74:2026–30
24. Dahms, N. M., Lobel, P., Kornfeld, S. 1989. J. Biol. Chem. 264:12115–18
25. Aerts, J. M. F. G., Schram, A. W., Strijland, A., van Weely, S., Jonsson, L. M. V., et al. 1988. Biochim. Biophys. Acta 964:303–8
26. Pohlmann, R., Krentler, C., Schmidt, B., Schröder, W., Lorkowski, G., et al. 1988. EMBO J. 7:2343–50
27. Waheed, A., Gottschalk, S., Hille, A., Krentler, C., Pohlmann, R., et al. 1988. EMBO J. 7:2351–58
28. Braun, M., Waheed, A., von Figura, K. 1989. EMBO J. 8:3633–40
29. Gottschalk, S., Waheed, A., Schmidt, B., Laidler, P., von Figura, K. 1989. EMBO J. 8:3215–19
29a. Peters, C., Braun, M., Weber, B., Wendland, M., Schmidt, B., et al. 1990. EMBO J. 9:3497–506
30. Lazzarino, D. A., Gabel, C. A. 1988. J. Biol. Chem. 263:10118–26
31. Lazzarino, D. A., Gabel, C. A. 1989. J. Biol. Chem. 264:5015–123
32. Baranski, T. J., Faust, P. L., Kornfeld, S. 1990. Cell 63:281–91
33. Grace, M. E., Grabowski, G. A. 1990. Biochem. Biophys. Res. Commun. 168:771–77
34. Gottschalk, S., Waheed, A., von Figura, K. 1989. Biol. Chem. Hoppe-Seyler 370:75–80
35. Sonderfeld-Fresko, S., Proia, R. L. 1989. J. Biol. Chem. 264:7692–97
36. O'Dowd, B. F., Cumming, D. A.,

Gravel, R. A., Mahuran, D. 1988. Biochemistry 27:5216–26
37. Vladutiu, G. D. 1983. Biochim. Biophys. Acta 760:363–70
38. Fedde, K. N., Sly, W. S. 1985. Biochem. Biophys. Res. Commun. 133:614–20
39. Little, L. E., Lau, M. M. H., Quon, D. V. K., Fowler, A. V., Neufeld, E. F. 1988. J. Biol. Chem. 263:4288–92
40. van Weely, S., Aerts, J. M. F. G., van Leeuwen, M. B., Heikoop, J. C., Donker-Koopman, W. E., et al. 1990. Eur. J. Biochem. 191:669–77
41. Hasilik, A., Neufeld, E. F. 1980. J. Biol. Chem. 255:4937–45
42. Hasilik, A., von Figura, K. 1984. Lysosomes in Biology and Pathology, ed. J. T. Dingle, R. T. Dean, W. S. Sly, 7:3–16. Amsterdam: Elsevier. 479 pp.
43. Erickson, A. H., Conner, G. E., Blobel, G. 1981. J. Biol. Chem. 256:11224–31
44. Quon, D. V. K., Proia, R. L., Fowler, A. V., Bleibaum, J., Neufeld, E. F. 1989. J. Biol. Chem. 264:3380–84
45. Mahuran, D. J., Neote, K., Klavins, M. H., Leung, A., Gravel, R. A. 1988. J. Biol. Chem. 263:4612–18
46. Stirling, J., Leung, A., Gravel, R. A., Mahuran, D. 1988. FEBS Lett. 231:47–50
47. Mahuran, D. J. 1990. J. Biol. Chem. 265:6794–99
48. Hubbes, M., Callahan, J., Gravel, R., Mahuran, D. 1989. FEBS Lett. 249: 316–20
49. Hasilik, A., von Figura, K., Conzelmann, E., Nehrkorn, H., Sandhoff, K. 1982. Eur. J. Biochem. 125:317–21
50. Galjart, N. J., Gillemans, N., Meijer, D., d'Azzo, A. 1990. J. Biol. Chem. 265:4678–84
51. Furst, W., Machleidt, W., Sandhoff, K. 1988. Biol. Chem. Hoppe-Seyler 369:317–28
52. O'Brien, J. S., Kretz, K. A., Dewji, N., Wenger, D. A., Esch, F., Fluharty, A. L. 1988. Science 241:1098–101
53. Rorman, E. G., Grabowski, G. A. 1989. Genomics 5:486–92
54. Nakano, T., Sandhoff, K., Stumper, J., Christomanou, H., Suzuki, K. 1989. J. Biochem. 105:152–54
55. Reiner, O., Dagan, O., Horowitz, M. 1989. J. Mol. Neurosci. 1:225–33
56. Fujibayashi, S., Wenger, D. A. 1986. J. Biol. Chem. 261:15339–43
57. O'Brien, J. S., Kishimoto, Y. 1990. FASEB J. In press
58. Tager, J. M., Jonsson, L. V. M., Aerts, J. M. F. G., Elferink, R. P. J. O.,

Schram, A. W., et al. 1984. *Biochem. Soc. Trans.* 12:902–5

59. von Figura, K., Steckel, F., Hasilik, A. 1983. *Proc. Natl. Acad. Sci. USA* 80:6066–70

60. Jonsson, L. M. V., Murray, G. J., Sorrell, S. H., Strijland, A., Aerts, J. F. G. M., et al. 1987. *Eur. J. Biochem.* 164:171–79

61. Lemansky, P., Bishop, D. F., Desnick, R. J., Hasilik, A., von Figura, K. 1987. *J. Biol. Chem.* 262:2062–65

62. Paw, B. H., Moskowitz, S. M., Uhrhammer, N., Wright, N., Kaback, M. M., Neufeld, E. F. 1990. *J. Biol. Chem.* 265:9452–57

63. Lodish, H. F. 1988. *J. Biol. Chem.* 263:2107–10

64. Lau, M. M. H., Neufeld, E. F. 1989. *J. Biol. Chem.* 264:21376–80

65. Myerowitz, R., Piekarz, R., Neufeld, E. F., Shows, T. B., Suzuki, K. 1985. *Proc. Natl. Acad. Sci. USA* 82:7830–34

66. Korneluk, R. G., Mahuran, D. J., Neote, K., Klavins, M. H., O'Dowd, B. F., et al. 1986. *J. Biol. Chem.* 261:8407–13

67. Proia, R. L., Soravia, E. 1987. *J. Biol. Chem.* 262:5677–81

68. Proia, R. L. 1988. *Proc. Natl. Acad. Sci. USA* 85:1883–87

69. Neote, K., Bapat, B., Dumbrille-Ross, A., Troxel, C., Schuster, S. M., et al. 1988. *Genomics* 3:279–86

70. Bapat, B., Ethier, M., Neote, K., Mahuran, D., Gravel, R. A. 1988. *FEBS Lett.* 237:191–95

71. Graham, T. R., Zassenhaus, H. P., Kaplan, A. 1988. *J. Biol. Chem.* 263:16823–29

72. Oshima, A., Kyle, J. W., Miller, R. D., Hoffmann, J. W., Powell, P. P., et al. 1987. *Proc. Natl. Acad. Sci. USA* 84:685–89

73. Miller, R. D., Hoffmann, J. W., Powell, P. P., Kyle, J. W., Shipley, J. M., et al. 1990. *Genomics* 7:280–83

74. Nishimura, Y., Rosenfeld, M. G., Kreibich, G., Gubler, U., Sabatini, D. D., et al. 1986. *Proc. Natl. Acad. Sci. USA* 83:7292–96

75. Powell, P. P., Kyle, J. W., Miller, R. D., Pantano, J., Grubb, J. H., Sly, W. S. 1988. *Biochem. J.* 250:547–55

76. d'Amore, M. A., Gallagher, P. M., Korfhagen, T. R., Ganschow, R. E. 1988. *Biochemistry* 27:7131–40

77. Bishop, D. F., Calhoun, D. H., Bernstein, H. S., Hantzopoulos, P., Quinn, M., Desnick, R. J. 1986. *Proc. Natl. Acad. Sci. USA* 83:4859–63

78. Kornreich, R., Desnick, R. J., Bishop,

D. F. 1989. *Nucleic Acids Res.* 17:3301–2

79. Tsuji, S., Yamauchi, T., Hiraiwa, M., Isobe, T., Okuyama, T., et al. 1989. *Biochem. Biophys. Res. Commun.* 163:1498–504

80. Wang, A. M., Bishop, D. F., Desnick, R. J. 1990. *J. Biol. Chem.* In press

80a. Hoefsloot, L. H., Hoogeveen-Westerfeld, M., Kroos, M. A., van Beeumen, J., Reuser, A. J. J., Oostra, B. A. 1988. *EMBO J* 7:1697–704

80b. Martiniuk, F., Mehler, M., Tzall, S., Meredith, G., Hirschhorn, R. 1990. *DNA Cell Biol.* 9:85–94

81. Geier, C., von Figura, K., Pohlmann, R. 1989. *Eur. J. Biochem.* 183:611–16

82. Peters, C., Geier, C., Pohlmann, R., Waheed, A., von Figura, K., et al. 1989. *Biol. Chem. Hoppe-Seyler* 370:177–81

83. Stein, C., Gieselmann, V., Kreysing, J., Schmidt, B., Pohlmann, R., et al. 1989. *J. Biol. Chem.* 264:1252–59

84. Kreysing, J., von Figura, K., Gieselmann, V. 1990. *Eur. J. Biochem.* 191:627–31

85. Peters, C., Schmidt, B., Rommerskirch, W., Rupp, K., Zühlsdorf, M., et al. 1990. *J. Biol. Chem.* 265:3374–81

86. Schuchman, E. H., Jackson, C. E., Desnick, R. J. 1990. *Genomics* 6:149–58

87. Robertson, D. A., Freeman, C., Nelson, P. V., Morris, C. P., Hopwood, J. J. 1988. *Biochem. Biophys. Res. Commun.* 157:218–24

88. Wilson, P. J., Morris, C. P., Anson, D. S., Occhiodoro, T., Bielicki, J., et al. 1990. *Proc. Natl. Acad. Sci. USA* 87:8531–35

89. Yen, P. H., Allen, E., Marsh, B., Mohandas, T., Wang, N., et al. 1987. *Cell* 49:443–54

90. Stein, C., Hille, A., Seidel, J., Rijnbout, S., Waheed, A., et al. 1989. *J. Biol. Chem.* 264:13865–72

91. Galjart, N. J., Gillemans, N., Harris, A., van der Horst, G. T. J., Verheijen, F. W., et al. 1988. *Cell* 54:755–64

92. Faust, P. L., Kornfeld, S., Chirgwin, J. M. 1985. *Proc. Natl. Acad. Aci. USA* 82:4910–14

93. Dynan, W. S. 1986. *Trends Genet.* 2:196–97

94. Bishop, D. F., Kornreich, R., Desnick, R. J. 1988. *Proc. Natl. Acad. Sci. USA* 85:3903–7

95. Reiner, O., Widgerson, M., Horowitz, M. 1988. *DNA* 7:107–16

96. Reiner, O., Horowitz, M. 1988. *Gene* 73:469–78

97. Quintern, L. E., Schuchman, E. H.,

Levran, O., Suchi, M., Ferlinz, K., et al. 1989. *EMBO J.* 8:2469–73

98. Morreau, H., Galjart, N. J., Gillemans, N., Willemsen, R., van der Horst, G. T. J., d'Azzo, A. 1989. *J. Biol. Chem.* 264:20655–63

99. Yamamoto, Y., Hake, C. A., Martin, B. M., Kretz, K. A., Ahern-Rindell, A. J., et al. 1990. *DNA Cell Biol.* 9:119–27

100. Yamauchi, T., Hiraiwa, M., Kobayashi, H., Uda, Y., Miyatake, T., Tsuji, S. 1990. *Biochem. Biophys. Res. Commun.* 170:231–37

101. Myerowitz, R., Hogikyan, N. D. 1986. *Science* 232:1646–48

102. Myerowitz, R., Hogikyan, N. D. 1987. *J. Biol. Chem.* 262:15396–99

103. Myerowitz, R. 1988. *Proc. Natl. Acad. Sci. USA* 85:3955–59

104. Arpaia, E., Dumbrille-Ross, A., Maler, T., Neote, K., Tropak, M., et al. 1988. *Nature* 333:85–86

105. Ohno, K., Suzuki, K. 1988. *Biochem. Biophys. Res. Commun.* 153:463–69

106. Ohno, K., Suzuki, K. 1988. *J. Biol. Chem.* 263:18563–67

107. Myerowitz, R., Costigan, F. C. 1988. *J. Biol. Chem.* 263:18587–89

108. Triggs-Raine, B. L., Feigenbaum, A. S. J., Natowicz, M., Skomorowski, M. A., Schuster, S. M., et al. 1990. *New Engl. J. Med.* 323:6–12

109. Paw, B. H., Tieu, P. T., Kaback, M. M., Lim, J., Neufeld, E. F. 1990. *Am. J. Hum. Genet.* 47:698–705

110. Navon, R., Proia, R. L. 1989. *Science* 243:1471–74

111. Paw, B. H., Kaback, M. M., Neufeld, E. F. 1989. *Proc. Natl. Acad. Sci. USA* 86:2413–17

112. Navon, R., Kolodny, E. H., Mitsumoto, H., Thomas, G. H., Proia, R. L. 1990. *Am. J. Hum. Genet.* 46:817–21

113. d'Azzo, A., Proia, R. L., Kolodny, E. H., Kaback, M. M., Neufeld, E. F. 1984. *J. Biol. Chem.* 259:11070–74

114. Ohno, K., Suzuki, K. 1988. *J. Neurochem.* 50:316–18

115. Tanaka, A., Ohno, K., Sandhoff, K., Maire, I., Kolodny, E. H., et al. 1990. *Am. J. Hum. Genet.* 46:329–39

116. Kytzia, H. J., Sandhoff, K. 1985. *J. Biol. Chem.* 260:7568–72

117. Brown, C. A., Neote, K., Leung, A., Gravel, R. A., Mahuran, D. J. 1988. *J. Biol. Chem.* 264:21705–10

118. Nakano, T., Nanba, E., Tanaka, A., Ohno, K., Suzuki, Y., Suzuki, K. 1990. *Ann. Neurol.* 27:465–73

119. Nakano, T., Muscillo, M., Ohno, K., Hoffman, A. J., Suzuki, K. 1988. *J. Neurochem.* 51:984–87

120. Coulondre, C., Miller, J. H., Farabaugh, P. J., Gilbert, W. 1978. *Nature* 274:775–80

120a. Barker, D., Schafer, M., White, R. 1984. *Cell* 36:131–38

121. Tanaka, A., Punnett, H. H., Suzuki, K. 1990. *Am. J. Hum. Genet.* 47:567–74

122. Akli, S., Chelly, J., Mezard, C., Gandy, S., Kahn, A., Poenaru, L. 1990. *J. Biol. Chem.* 265:7324–30

123. Neote, K., McInnes, B., Mahuran, D. J., Gravel, R. A. 1990. *J. Clin. Invest.* 86:1524–31

124. Nakano, T., Suzuki, K. 1989. *J. Biol. Chem.* 264:5155–58

125. Dlott, B., d'Azzo, A., Quon, D. V. K., Neufeld, E. F. 1990. *J. Biol. Chem.* 265:17921–27

126. Schröder, M., Klima, H., Nakano, T., Kwon, H., Quintern, L. E., et al. 1989. *FEBS Lett.* 251:197–200

127. Tsuji, S., Martin, B. M., Barranger, J. A., Stubblefield, B. K., LaMarca, M. E., Ginns, E. I. 1988. *Proc. Natl. Acad. Sci. USA* 85:2349–52

128. Zimran, A., Sorge, J., Gross, E., Kubitz, M., West, C., Beutler, E. 1989. *Lancet* 2:349–52

129. Theophilus, B., Latham, T., Grabowski, G. A., Smith, F. I. 1989. *Am. J. Hum. Genet.* 45:212–25

130. Firon, N., Eyal, N., Kolodny, E. H., Horowitz, M. 1990. *Am. J. Hum. Genet.* 46:527–32

131. Grace, M. E., Graves, P. N., Smith, F. I., Grabowski, G. A. 1990. *J. Biol. Chem.* 265:6827–35

132. Zimran, A., Gelbart, T., Beutler, E. 1990. *Am. J. Hum. Genet.* 46:902–5

133. Tsuji, S., Choudary, P. V., Martin, B. M., Stubblefield, B. K., Mayor, J. A., et al. 1987. *New Engl. J. Med.* 315:570–75

134. Wigderson, M., Firon, N., Horowitz, Z., Wilder, S., Frishberg, Y., et al. 1989. *Am. J. Hum. Genet.* 44:365–77

135. Dinur, T., Osiecki, K. M., Legler, G., Gatt, S., Desnick, R. J., Grabowski, G. A. 1986. *Proc. Natl. Acad. Sci. USA* 83:1660–64

136. Horowitz, M., Wilder, S., Horowitz, Z., Reiner, O., Gelbart, T., Beutler, E. 1989. *Genomics* 4:87–96

137. Zimran, A., Sorge, J., Gross, E., Kubitz, M., West, C., Beutler, E. 1990. *J. Clin. Invest.* 85:219–22

138. Latham, T., Grabowski, G. A., Theophilus, B. D. M., Smith, F. I. 1990. *Am. J. Hum. Genet.* 47:79–86

139. Hong, C. M., Ohashi, T., Yu, X. J., Weiler, S., Barranger, J. A. 1990. *DNA Cell Biol.* 9:233–41

140. Eyal, N., Wilder, S., Horowitz, M. 1990. *Gene.* In press
141. Christomanou, H., Aignesberg, A., Linke, R. P. 1986. *Biol. Chem. Hoppe-Seyler* 367:879–90
142. Polten, A., Fluharty, A. L., Fluharty, C. B., Koppler, J., von Figura, K., Gieselmann, V. 1991. *New Engl. J. Med.* In press
143. Chang, P. L., Davidson, R. G. 1983. *Proc. Natl. Acad. Sci. USA* 80:7323–27
144. Hertz, B., Bach, G. 1984. *Hum. Genet.* 66:147–50
145. Gieselmann, V., Polten, A., Kreysing, J., von Figura, K. 1989. *Proc. Natl. Acad. Sci. USA* 86:9436–40
146. Kretz, K. A., Carson, G. S., Morimoto, S., Kishimoto, Y., Fluharty, A. L., O'Brien, J. S. 1990. *Proc. Natl. Acad. Sci. USA* 87:2541–44
146a. Rafi, M. A., Zhang, X. L., DeGala, G., Wenger, D. A. 1990. *Biochem. Biophys. Res. Commun.* 166:1017–23
147. Zhang, X. L., Rafi, M. A., DeGala, G., Wenger, D. A. 1990. *Proc. Natl. Acad. Sci. USA* 87:1426–30
148. Bernstein, H. S., Bishop, D. F., Astrin, K. H., Kornreich, R., Eng, C. M., et al. 1989. *J. Clin. Invest.* 83:1390–99
149. Sakuraba, H., Oshima, A., Fukuhara, Y., Shimmoto, M., Nagao, Y., et al. 1990. *Am. J. Hum. Genet.* 47:784–89
150. Kornreich, R., Bishop, D. F., Desnick, R. J. 1990. *J. Biol. Chem.* 235:9319–26
151. Fukuhara, Y., Sakuraba, H., Oshima, A., Shimmoto, M., Nagao, Y., et al. 1990. *Biochem. Biophys. Res. Commun.* 170:296–300
152. d'Azzo, A., Hoogeveen, A., Reuser, A. J. J., Robinson, D., Galjaard, H. 1982. *Proc. Natl. Acad. Sci. USA* 79:4535–39
153. Hoogeveen, A., Verheijen, F. W., Galjaard, H. 1983. *J. Biol. Chem.* 258:12143–46
154. Verheijen, F. W., Palmeri, S., Hoogeveen, A. T., Galjaard, H. 1985. *Eur. J. Biochem.* 149:315–21
155. Potier, M., Michaud, L., Tranchemontage, J., Thauvette, L. 1990. *Biochem. J.* 267:197–202
156. Tranchemontage, J., Michaud, L., Potier, M. 1990. *Biochem. Biophys. Res. Commun.* 168:22–29
157. Jackman, H. L., Tan, F., Tamei, H., Beurling-Harbury, C., Li, X. Y., et al. 1990. *J. Biol. Chem.* 265:11265–72
158. Willems, P. J., Darby, J. K., DiCioccio, R. A., Nakashima, P., Eng, C., et al. 1988. *Am. J. Hum. Genet.* 43:756–63
159. Kretz, K. A., Darby, J. K., Willems, P.

J., O'Brien, J. S. 1989. *J. Mol. Neurosci.* 1:177–80
159a. Ikonen, E., Baumann, M., Grön, K., Syvänen, A. C., Enomaa, N., et al. 1991. *EMBO J.* In press
160. Martiniuk, F., Mehler, M., Pellicer, A., Tzall, S., LaBadie, G., et al. 1986. *Proc. Natl. Acad. Sci. USA* 83:9641–44
161. Martiniuk, F., Mehler, M., Tzall, S., Meredith, G., Hirschhorn, R. 1990. *Am. J. Hum. Genet.* 47:73–78
162. Conzelmann, E., Sandhoff, K. 1983/84. *Dev. Neurosci.* 6:58–71
163. Kolodny, E. H., Firon, N., Eyal, N., Horowitz, M. 1990. *Am. J. Med. Genet.* 36:467–72
163a. Proia, R. L., Kolodny, E. H., Navon, R. 1990. *Am. J. Hum. Genet.* 47:881–82
164. Masuno, M., Tomatsu, S., Sukegawa, K., Orii, T. 1990. *Hum. Genet.* 84:203–6
165. Barton, N. W., Furbish, F. S., Murray, G. J., Garfield, M., Brady, R. O. 1990. *Proc. Natl. Acad. Sci. USA* 87:1913–16
166. Stahl, P. D., Wileman, T. E., Diment, S., Shepherd, V. L. 1984. *Biol. Cell* 51:215–18
167. Krivit, W., Paul, N. W., eds. 1986. *Bone Marrow Transplantation for Treatment of Lysosomal Storage Diseases.* New York: Liss. 202 pp.
168. Krivit, W., Whitley, C. B., Chang, P. N., Shapiro, E., Belani, K. G., et al. 1990. *Bone Marrow Transplantation in Children,* ed. F. L. Johnson, C. Pochedly, pp. 261–87. New York: Raven. 529 pp.
169. Shull, R. M., Hastings, N. E., Selcer, R. R., Jones, J. J., Smith, J. R., et al. 1987. *J. Clin. Invest.* 79:435–43
170. Breider, M. A., Shull, R. M., Constantopoulos, G. 1989. *Am. J. Pathol.* 134:677–92
171. Taylor, R. M., Stewart, G. J., Farrow, B. R. H. 1989. *Transpl. Proc.* 21:3818–19
172. Taylor, R. M., Stewart, G. J., Farrow, B. R. H., Bryne, J., Healy, P. J. 1989. *Transpl. Proc.* 21:3074–75
173. Gasper, P. W., Thrall, M. A., Wenger, D. A., Macy, D. W., Ham, L., et al. 1984. *Nature* 312:467–69
174. Hoogerbrugge, P. M., Suzuki, K., Suzuki, K., Poorthuis, B. J. H. M., Kobayashi, T., et al. 1988. *Science* 239:1035–38
175. O'Brien, J. S., Storb, R., Raff, R. F., Harding, J., Appelbaum, F., et al. 1990. *Clin. Genet.* 38:274–80

176. Choudary, P. V., Tsuji, S., Martin, B. M., Guild, B. C., Mulligan, R. C., et al. 1986. *Cold Spring Harbor Symp. Quant. Biol.* 86:1047–52

177. Sorge, J., Kuhl, W., West, C., Beutler, E. 1987. *Proc. Natl. Acad. Sci. USA* 84:906–9

178. Fink, J. K., Correll, P. H., Perry, L. K., Brady, R. O., Karlsson, S. 1990. *Proc. Natl. Acad. Sci. USA* 87:2334–38

179. Nolta, J. A., Sender, L. S., Barranger, J. A., Kohn, D. B. 1990. *Blood* 75:787–97

180. Correll, P. H., Fink, J. K., Brady, R. O., Perry, L. K., Karlsson, S. 1989. *Proc. Natl. Acad. Sci. USA* 86:8912–16

181. Wolfe, J. W., Schuchman, E. H., Stramm, L. E., Concaugh, E. A., Haskins, M. E., et al. 1990. *Proc. Natl. Acad. Sci. USA* 87:2877–81

182. Kyle, J. W., Birkenmeier, E. H., Gwynn, B., Vogler, C., Hoppe, P. C., et al. 1990. *Proc. Natl. Acad. Sci. USA* 87:3914–18

*Annu. Rev. Biochem. 1991. 60:281–319*

# PRIMARY RESPONSE GENES INDUCED BY GROWTH FACTORS AND TUMOR PROMOTERS

*Harvey R. Herschman*

Department of Biological Chemistry, and Laboratory of Biomedical and Environmental Sciences, UCLA School of Medicine, Los Angeles, California 90024

KEY WORDS:   immediate early genes, early response genes, competence genes, signal transduction, transcription.

## CONTENTS

0066-4154/91/0701-0281$02.00

## SUMMARY AND PERSPECTIVES

All nucleated eukaryotic cells exhibit ligand-induced, transcription-dependent changes in phenotype. Even the single-celled yeasts respond to polypeptide mating factors by alterations in cellular proliferation and mating capability that are dependent on receptor-mediated transcriptional activation. In multi-cellular organisms, individual cells constantly respond to a complex environment in which gene expression and consequent cellular phenotype are mediated by ligands such as growth factors, peptide hormones, neurotransmitters, antigens, morphogens, and other hydrophilic agents that bind to cell-surface, transmembrane receptors. Researchers studying areas as seemingly diverse as mitogenesis, endocrinology, immunology, neurobiology, and developmental biology are all confronted with a similar set of questions: How is the information that a ligand-receptor interaction has occurred at the cell surface transmitted across the membrane? What are the intracellular mechanisms that relay, from the membrane to the nucleus, the information that a cell-surface, ligand-receptor interaction has occurred? What genes are influenced as a result of this new information, and how is their expression selectively modulated? What are the roles of these new gene products, expressed in response to ligand-receptor interaction, in effecting the subsequent alteration in phenotype? One major approach to these questions has been to identify those genes whose expression is initially induced by ligands, then work "upstream" to the signalling mechanisms that lead to their expression and "downstream" to their roles in cellular responses.

The first experiment reporting the cloning of cDNAs for mitogen-inducible messages, using differential hybridization to filter-replicas of cDNA libraries, demonstrated the feasibility of this approach and caught the interest of biologists studying the molecular basis of the mitogenic response (1). The subsequent identification of the cellular homologues of two oncogenes, v-fos and v-myc, as mitogen-responsive genes likely to play a role in the growth response (2–4), demonstrated a satisfying logical relationship between normal and abnormal growth control. This relationship fired the imagination and enthusiasm of molecular oncologists, and greatly increased interest in identifying additional genes that shared the characteristics of rapid, transient induction by mitogens.

Differential screening of cDNA libraries has been used to identify genes induced by a variety of agents. These include, but are not restricted to, appropriate "target" cells treated with platelet-derived growth factor (1), serum (5–7), tetradecanoyl phorbol acetate (8, 9), epidermal growth factor (10), nerve growth factor (11–13), interferon (14, 15), insulin-like growth factor (16), tumor necrosis factor (17, 18), ultraviolet light (19), and retinoic acid (20). Similar procedures have been used to identify elevated message levels in cytotoxic T cells (21), cells transformed by oncogenic viruses (22), differentiated muscle cells (23), and cells induced by serum starvation to exit the cell cycle and enter growth arrest (24). A substantial number of cDNAs for "primary response genes" (genes whose induction can occur with no intervening protein synthesis, but requires only the modification of pre-existing transcriptional modulators; Ref. 25), induced by a variety of ligands in a wide range of biological contexts, have been identified in the seven years between the preparation of this review and the observation that mitogens induce rapid and transient *c-fos* expression. The proteins encoded by many of the primary response genes induced in response to growth factors and tumor promoters have been deduced (Table 1); they include transcriptional modulators, structural proteins, cytokines, and proteins of as yet unknown function.

Until recently, most hypotheses to explain both (*a*) ligand-specific responses in a common cell type, and (*b*) distinct responses to a common ligand among different cell types, invoked highly restricted patterns of unique gene expression. Surprisingly, however, many of the same primary response genes are induced in a variety of biological contexts as diverse as mitogenic responses and neuronal responses to neurotransmitters and sensory stimuli. Indeed, it has now become clear that a variety of ligand-induced phenotypic changes, which are either ligand-specific in the context of a single cell type or cell-type specific in response to a common ligand, are accompanied by initial transcriptional responses that include many common genes. At first the common induction of the same genes in distinct biological responses was perplexing. However, a substantial body of work now suggests that biological diversity in response to alternative inductive signals can be understood, at least in principle, to occur as a consequence of unique combinatorial utilization of a common pool of genes, rather than as the result of the induction of transcripts unique to a particular ligand or cell type. The variability of biological response resulting from the employment of genes whose products are used in many contexts occurs as a consequence of (*a*) differential quantitative induction of primary response genes by distinct ligands, (*b*) differing kinetics of induction of the primary response genes and their products, (*c*) formation of changing patterns of heterodimeric transcription factors, with alternative transcriptional capacities, during the course of ligand-stimulated responses, (*d*) cell-type specific restriction of the expression of subsets of primary response genes, (*e*) ligand- and cell-specific posttranslational modification of primary response gene products and consequent alterations in the biological properties of these proteins, and (*f*) differential production of autocrine and paracrine factors that modulate initial inductive responses.

The basic outlines of several pathways leading to ligand-induced transient *c-fos* expression are now clear: ligand-receptor interaction, second messenger

**Table 1** Primary response genes cloned from cells treated with growth factors or tumor promoters.[a]

| Name | mRNA size (kb) | Open reading frame (amino acids) | Cell localization | Gene structure | Related genes |
|---|---|---|---|---|---|
| c-fos | 2.0 | 380 | nuclear | 4 kb; 4 exons, 3 introns | fosB, fra-1, fra-2 |
| fosB | 5.1 | 338 | nuclear | not known | c-fos, fra-1, fra-2 |
| fra-1 | 1.8 | 275 | nuclear | not known | c-fos, fosB, fra-2 |
| fra-2 | 6.0 | 323 (chick) 326 (human) | nuclear | 4 exons, 3 introns | c-fos, fosB, fra-1 |
| c-jun | 2.7, 3.2 | 334 | nuclear | 3.5 kb; no introns | jun-B, jun-D |
| junB | 2.1 | 344 (mouse) 347 (human) | nuclear | 2.3 kb; no introns | c-jun, jun-D |
| egr-1, krox24, zif-268, NGFI-A, d2, TIS8, 225, CEF-5 | 3.3–3.7 | 508 (rat) 533 (mouse) 543 (human) | nuclear | 4 kb; 2 exons, 1 intron | egr-2 |
| krox20, egr-2 | 3.2 | 470 (mouse) 426 (human) | nuclear | 5 kb; 3 exons, 2 introns (human) 4.3 kb; 2 exons, 1 intron (mouse) | egr-1 |
| nurk77, N10, NGFI-B, TIS1 | 2.5–2.7 | 601 (mouse) 563 (rat) | nuclear | 8 kb; 7 exons, 6 introns | steroid receptors |
| TIS11, TTP, Nup475 | 2.3 | 319 | nuclear | not known | cMG1/TIS11b, TIS11d |
| cMG1, TIS11b | 3.0 | 334 | nuclear (by analogy) | not known | TIS11, TIS11d |

| Gene | mRNA size (kb) | Amino acids | Localization | Gene structure | Related genes |
|---|---|---|---|---|---|
| TIS11d | | 357 | nuclear (by analogy) | not known | TIS11, cMGI/TIS11b |
| c25, rIRF-1 | 2.1 | 328 | nuclear | not known | see text |
| SRF (serum response factor) | 2.9, 4.5 | 508 | nuclear | not known | yeast MCM1 |
| c-myc | 2.3 | 439 | nuclear | 7 kb; 3 exons, 2 introns | L-myc, N-myc |
| actin[b] | *[b] | *[c] | cytoskeletal | *[b] | *[b] |
| fibronectin[b] | *[b] | *[c] | extracellular | *[b] | *[b] |
| fibronectin receptor[b] | *[b] | *[c] | transmembrane | *[b] | *[b] |
| tropomyosin | 1.7 | 282 | cytoskeletal | not known (mouse) | not known |
| p27 | 1.2 | 201 | cytoskeletal | not known | not known |
| mtf | 2.4 | 294 | transmembrane | not known | human tissue factor |
| glucose transporter | 2.8 | 492 | transmembrane | not known | not known |
| KC, N51 | 1.0 | 96 | secreted | not known | MGSA, gro, IP-10, 9E3/CEF4, HPBP, PF4 |
| JE | 1.0 | 148 | secreted | 2 kb; 3 exons, 2 introns | LD78/AT464, MIP1, act-2/AT744, H400, TCA3, RANTES |
| TIS10 | 4.0 | 600 | not known | not known | prostaglandin synthase |
| cyr61 | 2.0 | 379 | cytoplasmic, secreted | not known | CEF-10, IGF-binding protein |
| PC4, TIS7 | 1.9 | 449 | cytoplasmic | not known | interferon γ see text |
| TIS21 | 2.7 | 158 | not known | not known | not known |
| Pip92 | 1.5 | 221 | cytoplasmic | not known | not known |

[a] The isolation of the cDNAs, the characteristics of the protein products, and the regulation of the expression of these genes are discussed in the text.
[b] The particular genes of these potential gene families were not characterized.

elevation, activation of an appropriate protein kinase, subsequent phosphorylation and activation of pre-existing transcription factors, initiation of *c-fos* transcription, transcriptional repression of the *c-fos* gene by the induced FOS protein, and message degradation. Moreover, the specific DNA sequence recognized by the FOS/JUN heterodimer has been identified, and requirements for FOS expression in both the mitogenic response and the expression of a specific secondary response gene, stromolysin/transin, have been demonstrated. Similar studies on regulation and function are under way for many of the primary response genes listed in Table 1.

Many primary response genes remain uncharacterized. The mechanisms by which nearly all primary response genes are regulated and the causal roles of their gene products in ligand-induced cellular responses await investigation. Reasonable models both for ligand- and cell-specific transcriptional activation of secondary response genes and for subsequent changes in cellular physiology, whether it be proliferation, differentiation, or storage of a memory, are now emerging. However, the cascades of gene expression comprising specific physiological responses need to be elucidated: How do distinct combinatorial assortments of transcription factors regulate downstream genes, and how do autocrine and paracrine factors influence these responses?

## IDENTIFICATION OF THE FIRST MITOGEN-INDUCIBLE GENES

The mitogen-induced proliferative response of murine 3T3 cells is central in the history of the identification of genes responsive to extracellular signals. Resting 3T3 cells, when stimulated to re-enter the cell cycle by mitogens, undergo extensive alterations in transport of ions and small molecules, increased rates of overall protein and RNA synthesis, and elevated levels of activity in the rate-limiting enzymes for several anabolic pathways (reviewed in Ref. 26). Pardee and his colleagues (27) demonstrated that serum-induced division of 3T3 cells was accompanied by rapidly increased synthesis of specific cytoplasmic and nuclear proteins. Several groups subsequently identified both nuclear and cytoplasmic proteins whose synthesis was increased in response to mitogenic stimulation of quiescent 3T3 cells by serum or platelet-derived growth factor (PDGF; 28–30). When Hendrickson & Scher (31) demonstrated that increased levels of translatable mRNAs could be induced in 3T3 cells by mitogens such as PDGF or epidermal growth factor (EGF), the groundwork was set for the identification of mitogen-inducible genes.

The serum- and growth factor–induced mitogenic response of quiescent cells was one of the first ligand-induced phenotypic responses to be studied with recombinant DNA cloning technology. Cochran et al (1) were the first to use cDNA cloning techniques to identify transcripts whose levels are increased in cells in response to a mitogenic signal. They prepared a cDNA library from polyA+ mRNA of BALB/c 3T3 cells treated for four hours with partially purified PDGF. Individual bacterial colonies were picked, plated in ordered arrays, and screened by differential colony hybridization, to identify colonies that gave positive signals with $^{32}$P-labelled cDNA probe prepared from mitogen-treated cells but not with probe prepared from quiescent cells. The *JE* and *KC* partial cDNAs were identified in this study (Table 1). While the accumulation of the *JE* message continued for the four hours of these experiments, the *KC* mRNA level peaked at one hour, then fell. Moreover cycloheximide inhibition did not block induction of these messages; indeed, the PDGF-induced accumulation of *KC* mRNA was dramatically "superinduced" in the presence of cycloheximide.

Cochran et al (2) also identified a third cDNA, *pBC-JB*, related to the cellular *c-fos* gene. They demonstrated that the level of *c-fos* mRNA was, like that of *KC*, induced rapidly and transiently in quiescent 3T3 cells treated with PDGF. Moreover, PDGF-stimulated elevation of *c-fos* mRNA was also superinduced by cycloheximide. Several other groups quickly confirmed the rapid and transient induction of *c-fos* expression in mitogen-stimulated 3T3 cells (32, 33). Greenberg & Ziff (3), using a different approach, characterized the transcriptional responses of known cellular proto-oncogenes to mitogenic stimulation of 3T3 cells. Transcription of the *c-fos* gene was rapidly and transiently elevated in response to both PDGF and the tumor promoter tetradecanoyl phorbol acetate (TPA). After the identification of the *JE, KC,* and *JB* genes as mitogen-inducible sequences in 3T3 cells, Kelly et al (4) demonstrated that PDGF, TPA, and fibroblast growth factor (FGF) could also induce *c-myc* message. Ziff & Greenberg (3), in their study of mitogen-inducible transcription of proto-oncogenes, also demonstrated a serum-stimulated increase in *c-myc* transcription and message accumulation.

## THE *FOS* AND *JUN* GENE FAMILIES

The observations that serum and growth factors could induce the rapid, transient, and protein synthesis–independent expression of *c-fos* mRNA were seminal in studies of early gene responses to mitogens. Soon after these observations were made, however, rapid and transient induction of *c-fos* expression was shown to accompany nerve growth factor (NGF)-induced differentiation of PC12 cells (34, 35). PC12 cells, derived from a rat pheochromocytoma (36), cease dividing when challenged with NGF and differentiate morphologically, biochemically, and electrophysiologically into cells that resemble sympathetic neurons (reviewed in Ref. 37). Moreover, *c-fos* induction accompanies a myriad of additional ligand-induced cellular responses including, but not limited to, vitamin $D_3$–driven differentiation of macrophages, thyrotropin action on thyroid cells, activation of postmitotic neutrophils by granulocyte-macrophage colony stimulating factor (GM-CSF), and depolarization of postmitotic central and peripheral neurons (38).

FOS protein is found in the nucleus (39). Although speculation existed that FOS might be involved in transcriptional regulation (40), conclusive evidence was sparse until 1987, when Distel et al (41) demonstrated that antisera to FOS could block gel-shift activity of adipocyte nuclear extracts for the "fat-specific element" FSE2, a regulatory element of genes transcriptionally activated during adipocyte differentiation. The FSE2 sequence is similar to the consensus sequence TGACTCA for the DNA-binding site of AP-1 transcription factor (42, 43); FOS (and/or FOS-related proteins) is a component of AP-1 transcription factor activity.

FOS is often isolated as an immunoprecipitable complex with a protein termed p39 (44). p39 was subsequently identified as the product of the *c-jun* proto-oncogene. *c-jun* is the cellular homologue of *v-jun*, the oncogene of avian sarcoma virus ASV17 (45). Sequence analysis (46) and "domain-swap" (47) experiments demonstrated that the N-terminal portion of the JUN/p39 protein could replace the DNA-binding domain of the yeast GCN4 transcription factor (47, 48). Sequence similarities between the GCN4 DNA-binding site and the AP-1 consensus DNA-binding site led to experiments that identified JUN/p39 as another protein present in the AP-1 binding activity purified by DNA-affinity chromatography (49–51); AP-1 transcription factor preparations include JUN as well as FOS. Like *c-fos*, rapid, transient, protein synthesis–independent expression of *c-jun* is induced by serum and other mitogens (52–55) in 3T3 cells and by NGF in PC12 cells (56); *c-jun* is also a growth factor–induced primary response gene.

FOS and JUN form a heterodimer (57) held together by a "leucine zipper" of the type first identified by Landschulz et al (58) for the C/EBP transcription factor. While JUN can form homodimers, FOS cannot (57). Hetero- and homodimer formation, DNA binding to the AP-1 consensus element by FOS-JUN and JUN-JUN dimers, and transcriptional activation by FOS and JUN proteins have been reviewed extensively (48, 59). Two laboratories (60, 61) have recently isolated a *Drosophila c-jun* homologue. The DJUN and human JUN proteins share 84% similarity over their DNA-binding and leucine zipper domains. Moreover, DJUN can, in cooperation with FOS, transactivate in mammalian cells a reporter gene driven by an AP-1 promoter. The cloning of *djun* should allow the full power of *Drosophila* genetics to be focused on the role of the *jun* gene in signal transduction.

*c-fos* and *c-jun* are the first genes isolated from what we now know to be two families of transcriptional regulators. In the course of screening a cDNA library prepared from 3T3 cells treated with serum plus cycloheximide, Ryder et al (62) isolated a cDNA, *jun-B,* that coded for a transiently expressed, cycloheximide superinducible message related in sequence to *c-jun.* JUN-B can form heterodimers, which bind to AP-1 sequences, with both JUN and FOS proteins. However, JUN-B/JUN heterodimers cannot transcriptionally activate single AP-1 sequences (63, 64); JUN-B inhibits JUN transactivation of the AP-1 enhancer element. A third member of the JUN family, JUN-D, is also able to form AP-1-binding homodimers and heterodimers with either FOS or the other JUN proteins (65). It is not yet known whether JUN-D heterodimers are transcriptional activators or whether, like JUN-B, JUN-D can block transcriptional activation by JUN. Unlike the *c-jun* and *jun-B* genes, the *jun-D* gene is usually constitutively expressed, and is not inducible by serum or growth factors (66, 67).

cDNAs for three FOS-related proteins, *fra-1* (68, 69), *fra-2* (69, 70), and *fos-B* (71), have been cloned (Table 1), using either degenerate oligonucleotide probes, low-stringency screening with cDNA probes, or serologic screens of expression libraries with anti-FOS antibodies. All three FOS-related proteins can form JUN heterodimers that bind to the AP-1 enhancer element (68–72). Whether JUN/FRA heterodimers act as transcriptional activators or repressors has not yet been reported. Like *c-fos,* the *fra-1, fra-2,* and *fos-B* genes can be transiently induced by serum, growth factors, and TPA. However, the accumulation of message for *fra-1* and *fra-2* following mitogen stimulation is substantially slower than that observed for *c-fos* mRNA (68, 70, 71). As a result of the different kinetics of expression of the various members of the FOS and JUN gene families following stimulation, the composition of induced complexes binding to the AP-1 enhancer changes over time (73).

Macgregor et al (74) used a cloning strategy based on leucine-zipper interactions to isolate additional cDNAs for proteins that could form heterodimers with JUN. Surprisingly, they cloned a molecule that had previously been identified: a cyclic AMP response element binding protein, or CREB (75, 76). The cyclic AMP response element, or CRE, is a cis-acting enhancer element that is responsible for cyclic AMP regulation of gene expression (77). The JUN/CREB heterodimer binds to CRE sequences rather than AP-1 sequences (74). It is, therefore, possible that elevated levels of JUN induced in response to ligand stimulation may regulate genes that have CRE sequences in their regulatory regions, as well as genes with AP-1 sequences. Since CREB is but one of a number of related mammalian transcription factors recently cloned (78, 79), substantial possibilities for alternative regulatory responses exist if JUN proteins can also form heterodimers with some or all of these CRE-binding proteins. Habener (79) has recently reviewed the literature on CRE-binding proteins.

## PRIMARY RESPONSE GENES FROM CELLS TREATED WITH GROWTH FACTORS OR TUMOR PROMOTERS

The successful isolation of cDNAs for PDGF-inducible genes by differential screening (1), and the observation that c-fos is inducible by mitogens (2, 3), encouraged investigators to look, by differential screening, for additional ligand-inducible genes. Several groups have used serum stimulation of either BALB/c or Swiss 3T3 cells (5–7), because of the extensive use of these cells in the study of cellular proliferation. Alternatively, several laboratories (11–13) have used the NGF-induced differentiation of PC12 cells as a model system in which to identify primary response genes potentially associated with neuronal differentiation.

TPA is a potent mitogen for 3T3 cells (80, 81). It is also the most intensively studied of the tumor promoters. Most biological activities of TPA are thought to occur as a result of its ability to activate protein kinase C (PKC). We selected TPA for our own analysis of mitogen-stimulated primary response genes for three reasons: (a) TPA is thought to stimulate only a single second messenger pathway. Perhaps we would identify primary response genes induced by TPA, but not by mitogens that work via other pathways. (b) We might also identify primary response genes that play a role in tumor promotion. (c) We previously isolated several mutant 3T3 cell lines that are unable to mount a mitogenic response to TPA, but still retain TPA-activatable PKC (82–84). Since these TPA-nonproliferative mutants can still respond to other mitogens (85), cloning TPA-inducible primary response genes might subsequently identify genes unable to be induced in the TPA-nonproliferative variants. Such genes would be strong candidates to play causal roles in the TPA-induced mitogenic response. Using differential hybridization, we identified cDNAs for TPA-Inducible-Sequences, or TIS genes (8).

One of the most critical variables in identifying the first genes induced in response to ligand binding, based on the rapid and transient induction of the KC (1) and c-fos genes (32, 33), would appear to be the time following ligand addition at which mRNA is isolated for cDNA library construction. The observation that cycloheximide could prolong and superinduce accumulation of KC (1) and c-fos (33) mRNA led many workers in this area to search for genes whose expression can be induced by mitogens in the presence of cycloheximide. Yamamoto & Alberts (25) termed genes whose induction can be stimulated in the absence of protein synthesis "primary response genes." These genes have also been called "immediate-early" (5), "competence" (1), and "early growth response" (6) genes. We prefer the term "primary response genes," since (a) this terminology is based on a mechanistic definition (25), and (b) those genes whose induction requires intervening protein synthesis are often termed "secondary response genes." At the level of sensitivity afforded by differential hybridization of duplicate filter replicas, there are likely to be between 100 and 200 primary response genes (7). The names used in the headings below correspond, where appropriate, to the first characterization of each cDNA from serum (5–7), NGF (11, 13), and TPA (8) primary response cDNA libraries.

## 1. Nuclear Proteins

egr-1/NGFI-A/TIS8    Sukhatme et al (6) were the first to describe isolation, from 3T3 cells, of a cDNA for a mitogen-inducible transcription factor not previously identified as the cellular homologue of a retroviral oncogene. The *egr-1* gene product is a 533-amino-acid protein containing three potential zinc fingers (6, 86). Following mitogen induction, *egr-1* mRNA is expressed with rapid, transient kinetics very similar to those of *c-fos*. cDNAs for the *egr-1* gene were also identified and sequenced, from serum-induced 3T3 cDNA libraries, as *krox24* (87) and *zif-268* (88). The *NGFI-A* cDNA cloned from NGF/cycloheximide-treated PC12 cells encodes a protein of 508 amino acids, which is the rat homologue of *egr-1* (11). Milbrandt (11) was the first to report the sequence of this cDNA, and identify its predicted protein as a potential transcription factor. He did not, however, examine other ligands for their ability to induce *NGFI-A* expression in PC12 cells, nor did he study *NGFI-A* induction in other cell types. The sequence of a 1.1-kb partial cDNA of the TPA-inducible *TIS8* gene was identical with that of *egr-1* (89). *TIS8* expression can be induced in PC12 cells by EGF, TPA, or $K^+$-induced depolarization, as well as by NGF (90). Thus, as for *c-fos* and *c-jun*, induction of the *egr-1/NGFI-A/TIS8* gene is not restricted to the mitogenic response (11, 90). Cho et al (12) also cloned a partial cDNA, *d-2*, for this gene from NGF-treated PC12 cells. A cDNA called gene *225* (91) for the human *egr-1* homologue was cloned and sequenced from PHA/TPA-stimulated T cells. *egr-1/225* message levels are transiently induced in normal human lymphocytes, but are chronically elevated in HTLV-I- and HTLV-II-infected cells (91). A partial cDNA for this gene, termed *CEF-5*, was also cloned by differential screening of a cDNA library prepared from chicken fibroblasts infected with a temperature-sensitive retrovirus (22). The murine *egr-1/zif-268* (92, 93) gene and the rat *NGFI-A* gene (94) are 4 kb in length and contain two exons and one intron. *egr-1* and *c-fos* are coregulated in several tissues, including cartilage and bone (95). The *egr-1* gene product is a nuclear protein subject to phosphorylation (96–98). The DNA consensus–binding site for the protein encoded by the *egr-1/zif-268/krox24* gene has recently been identified as GCG(G/T)GGGCG (96, 97, 99), a sequence present twice in the immediate upstream 5'-region of the *egr-1/zif-268/krox24/TIS8* gene itself. The sequence is also present in other primary response genes, including *c-myc*, *krox20/egr-2*, and *NGFI-B/N10*. Cotransfection experiments with an expression plasmid encoding the *krox24/egr-1* protein and a reporter gene containing four copies of the sequence GCGGGGCG demonstrated that the *krox24/egr-1* gene product can act as a sequence-specific transactivating factor (97).

KROX20    The serum-inducible gene *krox20* was identified by screening a collection of serum-induced primary response cDNAs with a conserved zinc-finger domain from the *Drosophila kruppel* gene (100). The human *krox20* homologue, *egr-2*, was identified by low-stringency hybridization with *egr-1* cDNA (101). The proteins encoded by the murine *egr-1/krox24* and *krox20/egr-2* genes share strong sequence similarity in the zinc finger binding domain, and in several regions outside this domain. The murine *krox20* gene is also small, covering approximately 4 kb. However, the gene has two introns, and can be alternatively spliced to produce two messages that code for proteins with distinct amino-terminal ends (102). In contrast, the human *egr-2* gene, like the murine *egr-1* gene, has only a single intron, and only a single transcript is detectable (103). As might be expected from their strong sequence similarity in the zinc-finger regions, *krox20* protein binds to the same DNA sequence as *krox24/egr-1* protein (104). In the same transactivation system used to characterize *egr-1/krox24*, *krox20/egr-2* protein also acts as a sequence-specific transactivating factor (104). Despite their similarity in sequence and enhancer-binding specificity, the *krox24/egr-1* and *krox20* transcription factors have distinct developmental patterns and tissue distributions (87). A yeast gene, *MIG1*, whose protein product has zinc-finger domains and DNA-binding characteristics very similar to the *egr-1/krox24* and *krox20/egr-2* proteins, has recently been described (105).

nur77/NGFI-B/TIS1    Sequence of the *nur77* cDNA, another rapid, transient, and cycloheximide-superinducible message isolated from a library of serum/cycloheximide-treated 3T3 cells, showed it to be a member of the superfamily of ligand-binding transcription factors that includes the steroid and thyroid hormone receptors (106). Sequence similarity between the 601-amino-acid *nur77* protein and other members of this family of ligand-induced transcription factors is particularly striking in a 66-amino-acid region, containing two zinc fingers, responsible for DNA binding. Sequence conservation also occurs in a region of the molecule presumed to be responsible for ligand binding. The gene for this serum-inducible murine message has been cloned and named *N10* (107). *NGFI-B*, a cDNA cloned from NGF-treated PC12 cells, encodes a 563-amino-acid protein, which is the rat homologue of *nur77* (108). Milbrandt's report (108) was the first to describe the sequence of this cDNA, and predict that its encoded protein is a transcription factor related to the steroid hormone receptor superfamily. The sequence of a cDNA clone for the TPA-inducible *TIS1* gene (109) was identical to that reported for *nur77*

(106). Like *TIS8/egr-1/NGFI-A*, the *NGFI-B/nur77/TIS1* gene can be induced by EGF, TPA, and depolarization, as well as by NGF, in PC12 cells (90). The transcription units of both the rat *NGFI-B* and murine *N10/nur77/TIS1* genes are 8 kb in length; both are split into seven exons (107, 110). The structure of the gene, while similar to that of other members of the superfamily, has enough distinct characteristics to suggest that it belongs to a new subfamily.

TIS11/TTP/Nup475[1]    Like *c-fos*, *TIS11* is rapidly and transiently induced by serum, TPA, EGF, and FGF in 3T3 cells (8), and by NGF, EGF, TPA, and depolarization in PC12 cells (90). Analysis of the sequence of the TIS11 open reading frame did not suggest similarity to any known protein (149). However, a repeat of the sequence YKTELC, as well as a repeated motif, $CX_8CX_5CX_3H$, beginning at each of the cysteines in the YKTELC sequence, is present near the center of the TIS11 protein sequence.

The tris-tetraprolin *(TTP)* cDNA, cloned from insulin-treated 3T3 cells (149a),[1] and the *Nup475* cDNA, cloned from serum-treated 3T3 cells (149b), encode identical proteins that share a 102-amino-acid sequence identical to a region of the original *TIS11* cDNA. *TTP/Nup475* differs in the amino- and carboxyl-terminal portions from the *TIS11* sequence. Using polymerase chain reaction, we have shown that it is likely the *TIS11* cDNA is a chimeric cDNA created during the cloning process. Sequencing of the cloned product of the PCR reaction demonstrated that the 5' sequence of the message encoded by the *TIS11* gene is identical to that of the *TTP/Nup475* gene, and that the presence of one C residue shifts the reading frame of the 3' portion of the *TIS11* cDNA to that of *TTP/Nup475*. The *TIS11* gene encodes the same message and protein as the *TTP/Nup475* gene (Q. Ma, H. Herschman, submitted for publication). The TIS11/TTP/Nup475 protein product has several serine/proline-rich regions, similar to many transcription factors. Using antisera to recombinant Nup475 protein, DuBois et al (149b) have localized the protein to the nucleus by immunofluorescence. It is likely that the *TIS11/TTP/Nup475* gene product is another rapidly inducible transcriptional regulator.

cMG1/TIS11b    Differential screening of a library prepared from the rat RIE-1 epithelial cell line following exposure to EGF plus cycloheximide identified a gene, *cMG1*, related to *TIS11* (10). *TIS11* and *cMG1* proteins share a 67-amino-acid region that is 72% identical. The YKTELC repeat and the

---

[1]The DuBois et al publication (149b) appeared as this review was being edited. Their cell localization data for TIS11/TTP/Nup475 protein required shifting the discussion of this gene from "proteins of unknown function" to "nuclear proteins" at a late date in preparation of the manuscript. Consequently, the reference numbering for discussion of this gene and the two related genes, *cMG1/TIS11b* and *TIS11d*, are not in consecutive order.

$CX_8CX_5CS_3H$ sequences are conserved in cMG1 and TIS11. Within the 67-amino-acid region are 11 basic amino acids. Additional similarities in the sequences of the two proteins also exist outside the central 67-amino-acid region (149, 10). Gomperts et al (10) speculate that *TIS11* and *cMG1* may be members of a new family of transcription factors. We used a degenerate oligonucleotide for the YCTELC sequence and cloned a *TIS11/TTP/Nup475*-related cDNA from 3T3 cells, which we named *TIS11b*. *TIS11b* is the murine homologue of *cMG1*. TPA induces *TIS11b* only poorly in 3T3 cells, and superinduction by cycloheximide does not occur (B. Varnum, Q. Ma, B. Fletcher, T. Chi, H. Herschman, in preparation).

TIS11d    We used degenerate oligonucleotides to the conserved sequences of *TIS11* and *cMG1/TIS11b* to amplify DNA, by polymerase chain reaction, from murine genomic DNA. Cloning and sequencing the amplification products identified products from *TIS11* and *TIS11b*, and two additional genomic sequences. One of these amplified genomic sequences identified a third related clone from a cDNA library prepared from BALB/MK cells, a murine epithelial cell line. The sequence of this cDNA, which we call *TIS11d*, encodes a 367-amino-acid protein that is 94% identical to *TIS11b* over the conserved 67-amino-acid domain (B. Varnum, Q. Ma, B. Fletcher, T. Chi, H. Herschman, in preparation). *TIS11*, *cMG1/TIS11b*, and *TIS11d* are clearly members of a gene family. Arguing by analogy to the immunofluorescence data for Nup475 (149b), it seems likely that cMG1/TIS11b and TIS11d are also found in the nucleus.

SRF    The *c-fos* gene, and several other primary response genes, contain a common cis-acting enhancer element necessary for serum induction. This element is termed the "serum response element," or SRE (111). A trans-acting factor, "serum response factor" or SRF, binds to the SRE and regulates serum-induced expression (112, 113). Surprisingly, the gene encoding SRF protein is a primary response gene, rapidly and transiently induced by serum (114). However, the basal level of SRF is higher than that of other primary response genes, serum induction of SRF is somewhat more modest (approximately fivefold) than that observed for the other serum-inducible transcription factors, and cycloheximide-dependent superinduction of SRF mRNA is substantially less than that observed for these genes.

c25    Prolactin is a potent mitogen for the rat Nb2 T-cell lymphoma cell line (115). When a cDNA library prepared from prolactin/cycloheximide-treated Nb2 cells was screened by differential hybridization, the transiently expressed, highly inducible *c25* clone was isolated (116). Sequencing of the *c25* cDNA showed it to be the rat homologue of the transcription factor known as interferon-regulatory factor 1 (IRF-1), a protein previously shown to be a potent transcriptional regulator of both virally induced interferon and of interferon-induced genes (117). The *c25/IRF-1*-encoded transcription factor apparently plays a role in both proliferative and antiproliferative cellular responses. It will be of great interest to see if *c25* is expressed in the wide context in which many of the serum/NGF/TPA-inducible transcription factors such as *c-fos, c-jun, egr-1/NGFI-A/TIS8,* and *nur77/NGFI-B/TIS1* can be expressed.

## 2. Cytoskeletal and Extracellular Matrix Proteins

Growth factors play key roles in embryogenesis and wound healing. It seems likely that genes whose products are involved in processes of cell movement might also be induced by these agents. One might predict, however, that such genes would be induced as part of a secondary response, inaugurated by primary response transcription factors. While this is often the case, Ryseck et al (118) have identified genes for actin, fibronectin, fibronectin receptor, and tropomyosin among their collection of serum-inducible primary response gene cDNAs. It should be noted that actin was one of the first genes shown to be induced by serum (27). In contrast to mRNAs for serum-inducible transcription factors, message levels for these four genes remain elevated for a substantial period after serum addition. The cDNA for *p27,* a serum-inducible primary response gene (119) whose protein product was previously identified immunochemically as an actin-associated protein (120), has also recently been cloned (121).

## 3. Transmembrane Proteins

MTF    The predicted protein for the serum-inducible *482* gene (122) shows strong similarity to human "tissue factor," a membrane protein that initiates blood clotting following injury. This gene has been renamed mouse tissue factor, or *mtf.* The serum-induced expression of tissue factor in human fibroblasts is also independent of protein synthesis (123).

GLUCOSE TRANSPORTER    Elevated glucose transport is one of the first observable physiological responses to growth factors in cultured fibroblasts (124). Serum and purified growth factors such as PDGF, FGF, and EGF, as well as TPA, rapidly and transiently stimulate the mRNA level of the transporter in BALB/c 3T3 cells (125, 126). Modest cycloheximide superinduction can also be demonstrated.

## 4. Cytokines

KC    A surprisingly long time passed between the original isolation (1) of partial cDNA clones for the PDGF/serum-inducible genes *JE* and *KC* and the identification of proteins encoded by these genes. The *KC* gene codes for a secreted protein with an open reading frame of 96 amino acids that includes a presumptive signal sequence of 20 residues (127). *N51*, a serum-induced primary response gene, is identical to *KC* (128). The sequence of the human cDNA homologue of the *KC/N51* gene (127) is identical to the sequence of the cDNA of Melanoma Growth Stimulating Activity (MGSA), a protein secreted by human melanoma cells (129). *MGSA* is, in turn, the human homologue of the *gro* gene, a constitutively expressed message in transformed Chinese Hamster fibroblasts (130). The *KC/N51, MGSA*/human *gro*, and hamster *gro* genes are apparently all the same gene, whose cDNAs have been isolated either as mitogen-inducible mRNAs *(KC/N51)* or messages constitutively expressed in transformed cell lines *(MGSA/gro)*. Sequence similarities suggest that the gamma-interferon-induced gene *IP-10* (131), the *v-src*-induced chicken gene known as *9E3* (132) or *CEF4* (133), the human platelet basic protein *HPBP* (134), and platelet factor 4 (135) are members of the same gene family of secreted proteins.

JE    The protein encoded by the *JE* gene is 148 amino acids in length, and contains a possible leader sequence of 29 amino acids (136, 137). The *JE* gene has a three-exon, two-intron structure that is found within two kb of contiguous sequence (136). The human *JE* homologue has been cloned from both serum-stimulated fibroblasts (138) and a human monocyte cDNA library (139). *JE* also appears to be a member of a family of genes, thought to include: *LD78/AT464*, cDNAs cloned from PHA-treated human lymphocytes (140) and PHA/TPA-induced T cells (141); *MIP1*, an LPS-inducible macrophage protein (142); *act-2*, a PHA/TPA-inducible T cell gene distinct from *AT464*, but the same as *AT744* (143, 142); *H400* (144) and *TCA-3* (145), two conA-inducible T-cell genes; and *RANTES*, a human T-cell c-DNA (146).

While the sequence similarities within the members of the *KC* and *JE* families are in some instances clear and compelling, in other cases assignment of family membership based on sequence similarity is not so clear. The gene for the mitogen-stimulated lymphocyte cDNA *3-10c* (147) has been assigned to both the *JE* family (138) and the *KC* family (127). Moreover, sequence similarities between the *JE* protein and the *KC/gro/MGSA* family of proteins have been pointed out by Kawahara & Deuel (137). The structural relationships among these proteins has recently been reviewed (148). It is clear that there exists a group of small inducible genes that code for secreted cytokines expressed not only in response to mitogenic stimulation, but also in a number of other physiological contexts. While the roles of these proteins are not clear at present, they are likely to be important in the intercellular communication that follows a variety of cellular activation processes.

## 5. Proteins of Unknown Function

TIS10    *TIS10* is rapidly induced in 3T3 cells by EGF, FGF, and TPA (8). In contrast to the other *TIS* genes, *TIS10* expression cannot be induced by any extracellular ligand in PC12 cells (90); *TIS10* expression is "restricted" in this latter cell line. The *TIS10* gene encodes an mRNA of approximately four kilobases. Recent sequencing studies (B. Fletcher, D. Kujubu, R. Lim, B. Varnum, H. Herschman, unpublished observations) have demonstrated that the *TIS10* protein product has extensive sequence similarity to the murine prostaglandin endoperoxide synthase gene (148a). Since mitogen induction also causes substantial elevation of prostaglandins, it will be of great interest to characterize the potential catalytic functions of the *TIS10* gene product.

cyr61    The predicted *cyr61* open reading frame encodes a 379-amino-acid protein that includes 38 cysteines (150). The central region of the predicted protein has no cysteines and is highly acidic, with 13 aspartic or glutamic acid residues between amino acids 159 and 188. The Cyr61 protein sequence also contains a hydrophobic N-terminal sequence, with a consensus serine-threonine cleavage site at residues 24 and 25. The Cyr61 protein has many similarities, including the presence of a secretory signal, a 10% cysteine content, and a central acidic region, to low-molecular-weight IGF-binding proteins (151). However, antisera directed against a fusion protein detected labelled Cyr61 intracellular protein following serum stimulation, but could not detect Cyr61 protein chased to the culture medium (150). O'Brien et al (150) speculate that the protein might be only inefficiently secreted, or sequestered at either the cell surface or the extracellular matrix. The Cyr61 protein is similar in sequence to the protein of *CEF-10,* a gene cloned by differential hybridization as a *v-src*-inducible cDNA in chicken embryo fibroblasts (22). However, *cyr61* is induced by serum and TPA in 3T3 cells (150), whereas TPA transiently represses *CEF-10* mRNA levels in chick embryo fibroblasts (22).

PC4/TIS7    The open reading frame predicted by the NGF-inducible *PC4* cDNA from PC12 cells encodes a 449-amino-acid protein, which Tirone & Shooter (13) suggest is related to rat gamma interferon. The TPA-inducible *TIS7* cDNA is the murine homologue of *PC4;* the predicted protein sequences of *TIS7* and *PC4* are greater than 90% identical (152). *TIS7* mRNA accumulation can also be induced by EGF and FGF in 3T3 cells (8), and by NGF, EGF, TPA, and depolarization in PC12 cells (90). We have prepared antisera to a *TIS7* fusion protein, and demonstrated TPA-inducible incorporation of $^{35}$S-cysteine into *TIS7* protein (B. Varnum, R. Koski, and H. Herschman;

unpublished observations). Although the PC4 protein is reported to have sequence homology to interferon gamma (13), we are unable to demonstrate secretion of *TIS7* protein from 3T3 cells into the medium. It should be noted that each of the primary response genes cloned from PC12 cells *(PC4/TIS7, NGFI-A/TIS8,* and *NGFI-B/TIS1)* is inducible by TPA, EGF, and depolarizing conditions in PC12 cells (90), as well as by NGF (90, 13, 11, 108); none is uniquely associated with NGF-induced differentiation of PC12 cells

TIS21    *TIS21* is induced in 3T3 cells by all mitogens tested (8) and in PC12 cells by EGF, NGF, FGF, TPA, carbachol, and depolarization (90). The 2.7-kb *TIS21* message encodes a protein of 158 amino acids (B. Fletcher, R. Lim, B. Varnum, R. Koski, D. Kujubu, H. Herschman, in preparation). No significant sequence similarities exist between *TIS21* protein and any protein whose sequence is in the current data banks. The predicted *TIS21* amino acid sequence does not contain any motifs that suggest potential function. Twelve percent of *TIS21* protein is arginine or lysine. A short, 40–50 nucleotide 5' untranslated region in the *TIS21* message is followed by the open reading frame and a 3' untranslated region greater than two kb. Ninety nucleotides 3' from the termination codon there is a 117-nucleotide sequence that is 85% A and U, with a surprising run of 18 consecutive uridines. The *TIS21* message is encoded in a 4-kb genomic sequence containing two exons and a single 1.4-kb intron. The first kilobase of 5' flanking sequence contains several consensus Sp1 sites, two possible CRE elements, a potential SRE element, and a region of Z-DNA composed of 29 CA repeats.

Pip92    The *Pip92* gene is rapidly and transiently induced in 3T3 cells stimulated with serum, PDGF, FGF, or TPA and in PC12 cells by NGF, EGF, or potassium depolarization. Sequence of the *Pip92* cDNA identified an open reading frame of 221 amino acids. The protein is proline rich, highly hydrophilic, and is not related to any sequence present in the current data bases. Antibodies generated to recombinant Pip92 protein have been used to show that the protein, as well as the mRNA, has a very short half-life. The Pip92 protein is found in the cytoplasm (152a).

## 6. Primary Response Genes from a "Late" Library

Hirschhorn et al (153) prepared a cDNA library from quiescent hamster fibroblasts treated for six hours with serum. No cycloheximide was present during the induction. cDNAs from this library should include both secondary response genes and primary response genes that have relatively long-lived messages. Among the primary response cDNAs present in this collection are clones for vimentin (154), the ADP/ATP translocator (155), and the heavy chain of calpactin I (156). A fourth protein, named calcyclin, has strong sequence similarity to the calcium-binding S100 protein (157). Two NGF-induced clones, isolated from PC12 cells, also encode mRNAs for proteins with sequences similar to that of the S100 protein (158).

# REGULATION OF PRIMARY RESPONSE GENE EXPRESSION

Induction of primary response genes is characterized by a rapid, protein-synthesis-independent increase in transcription. In some cases, mRNA accumulation is only transient, as a result of both a subsequent inhibition of transcription and rapid degradation of accumulated message. Frequent re-isolation of many of the primary response genes following growth factor or TPA exposure (see Table 1) suggests either that these ligands all induce the commonly responsive genes by signal transduction pathways that converge or, alternatively, that the regulatory regions of these genes all contain cis-acting response elements for alternative ligand-activated signal transduction pathways. As might be expected, regulation is better characterized for c-fos than for other primary response genes. However, the regulatory regions for other primary response genes have recently been cloned, and there will, undoubtedly, soon be an avalanche of data concerning the regulation of expression of other primary response genes.

## 1. c-fos

Transient transfection experiments employing deletions of the c-fos promoter localized serum-inducibility to the sequence, termed the "serum response element" or SRE, located between nucleotides −322 and −276 (111). Nuclear extracts contain a protein, the scrum response factor or SRF, which binds to the c-fos SRE (112). SRF was purified (113, 159, 160) and cloned (114); it is a 508-amino-acid protein that has no DNA-binding motifs previously described. Actin genes, which were among the first genes shown to be serum-inducible in quiescent 3T3 cells (27), contain a conserved promoter element, termed the CArG element [CC(A/T)$_6$GG], which is related to the SRE (161). SRF and the CArG-binding protein can both bind to either the c-fos SRE or the CArG element (162). Biochemical studies suggest that SRF and CArG-BP are very similar, if not identical. Thus, induction of c-fos and actin genes by serum is mediated by similar cis-acting elements and trans-acting factors.

Yeast cells contain a transcription system related to SRF/SRE. Yeast mating factors bind to cell-surface transmembrane receptors and alter cell cycle progress and mating characteristics. The sequence CCTAATTAGG, a CArG element, is present in several yeast genes whose transcription is induced by mating factors, via a protein known alternatively as pheromone/

receptor transcription factor (PTRF, 163), general regulator of mating type (GRM, 164), or maintenance of minichromosome 1 (MCM1, 165). MCM1 can bind to the *c-fos* SRE, and the *c-fos* SRE can act as a yeast upstream activating sequence in vivo (166, 167). Not surprisingly, substantial amino acid conservation exists between the DNA-binding domains of SRF and MCM1 (114, 165). Thus, substantial structural and functional similarities exist between transcription factors that relay signals from ligand-occupied cell-surface receptors in eukaryotic organisms as distant as yeast and man.

Serum can induce transcription of *c-fos* in the presence of cycloheximide, suggesting that ligand-induced posttranslational modification of SRF is sufficient to activate its transcriptional function. SRF contains potential casein kinase II and protein kinase A phosphorylation sites (114). EGF treatment of cells induces phosphorylation of SRF in vivo (168); phosphatase treatment destroys the ability of SRF to bind to the SRE. While recombinant bacterial SRF can bind to the SRE, its DNA-binding ability is greatly enhanced by phosphorylation, using HeLa cell nuclear extracts. The active kinase in these extracts is casein kinase II (169). These results suggest that phosphorylation of SRF plays a role in either constitutive and/or ligand-regulated activity of SRF at the *c-fos* SRE. In vivo footprinting indicates that SRF is bound to the *c-fos* SRE both before and after growth factor stimulation (170).

A second protein, p62, forms a complex with SRF only when SRF is bound to DNA. Sequences outside the CC(A/T)$_6$GG of the *c-fos* SRE are required for this ternary interaction; these sequences are not required for binding of SRF alone. These same flanking sequences are also required for effective *c-fos* transcriptional activation in vivo, suggesting that the ternary SRE/SRF/p62 complex is important in serum-induced transcriptional activation of *c-fos* (171, 172).

The signal transduction pathways for *c-fos* induction in response to a number of ligands appear to converge at the SRE of the *c-fos* promoter. Mutations within the SRE can block serum, TPA, and much (but not all) PDGF induction of *c-fos* in 3T3 cells (173, 174). Mutations in the SRE also abolish NGF, EGF, TPA, and FGF induction of *c-fos* in PC12 cells (175). The SRE must, therefore, mediate *c-fos* induction by NGF, EGF, EGF, and TPA, as well as by serum. It is not clear whether all these inductions are mediated through a common SRF molecule, or whether alternative SRF-like proteins are involved. Although TPA induces *c-fos* via the SRE, down-regulation of protein kinase C (PKC) does not block serum or PDGF induction of *c-fos* mRNA accumulation (174). The SRE, therefore, mediates both PKC-dependent and PKC-independent pathways of *c-fos* induction.

Interestingly, NGF-induced *c-fos* expression in PC12 cells occurs only in G1; cells in S or G2 do not express elevated *c-fos* message following NGF exposure (176). Since several ligands (NGF, EGF, FGF, TPA) appear to induce *c-fos* via the common cis-acting SRE, it will be of considerable interest to determine (*a*) whether this cell cycle restriction of *c-fos* induction is unique to NGF in PC12 cells, and (*b*) whether other primary response genes show a similar cell-cycle specificity in their response to NGF.

A second, somewhat less potent, serum and NGF-responsive element, SRE-2, has been reported in the *c-fos* promoter. A sequence-specific binding activity for this element, distinct from SRF, is present in nuclear extracts (177). Hayes et al (178) have described yet another regulatory element of the *c-fos* gene, located in the region $-346$, which is partially responsible for induction by PDGF or conditioned medium from *v-sis* transformed cells, but does not respond to EGF, TPA, or insulin. Gel-shift experiments identified a DNA-binding activity, induced following administration of PDGF or conditioned medium, for this region.

Mutation of the SRE does not prevent *c-fos* induction by agents that elevate cAMP in 3T3 cells (174), suggesting that protein kinase A (PKA)–mediated *c-fos* induction does not occur via the SRE. Similarly, calcium-mediated *c-fos* induction by depolarization of PC12 cells is not blocked by mutations in the SRE (175). The *c-fos* promoter contains a cyclic AMP response element, or CRE, at position $-60$. This CRE can confer cAMP responsiveness on heterologous promoters and can bind CREB. It seems likely that *c-fos* induction by cAMP is mediated by PKA phosphorylation of CREB, since *c-fos* induction by cyclic AMP is greatly reduced in a PC12 mutant deficient in PKA (179). Mutational analysis has shown that the cis-acting element responsible for calcium induction of *c-fos* in PC12 cells, the calcium response element or CaRE, is identical with the CRE (180). Moreover, depolarization of PC12 cells leads to rapid phosphorylation of CREB in a cAMP-independent fashion. The data suggest that cAMP-mediated *c-fos* induction occurs via PKA-dependent phosphorylation of CREB and transactivation at the CRE, while calcium-stimulated *c-fos* induction occurs via activation of a different protein kinase (perhaps $Ca^{2+}$/calmodulin dependent protein kinase), phosphorylation of CREB, and transactivation at the same cis-acting CaRE/CRE element in the *c-fos* promoter (180). The proposed pathways for ligand activation of *c-fos* transcription are illustrated in Figure 1.

Lamb et al (181) have recently demonstrated the presence of an intragenic regulatory element, located at the end of exon-1 of the *c-fos* gene. The *fos* intragenic regulatory element, or FIRE, appears to be a negative regulatory element whose repression must be relieved for induced *fos* message expression. Robbins et al (182) suggest that the retinoblastoma gene product can also exert a negative regulatory influence on the expression of the *c-fos* gene, and have mapped the cis-acting element to a region between nucleotides $-102$ and $-71$ of the *c-fos* promoter.

Ligand-stimulated *c-fos* transcription is transient, and is rapidly repressed.

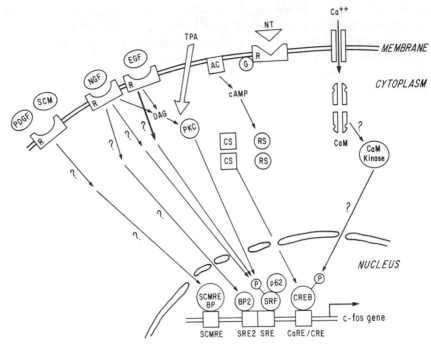

*Figure 1* Proposed pathways for induction of *c-fos* transcription by extracellular ligands. Abbreviations: CaM, calmodulin; CaM kinase, calmodulin-dependent protein kinase; NT, neurotransmitter; R, receptor; G, G-protein complex; AC, adenylate cyclase; RS, regulatory unit of cAMP-dependent protein kinase; CS, catalytic subunit of cAMP dependent protein kinase; PKC, protein kinase C; TPA, tetradecanoyl phorbol acetate; DAG, diacyl glycerol; EGF, epidermal growth factor; NGF, nerve growth factor; PDGF, platelet-derived growth factor; SCM, factor from medium conditioned by *v-sis*-transformed cells; CaRE/CRE; the calcium and cAMP response element (180); SRE, the serum response element (111); SRE-2, a second serum- and NGF-responsive element of the *c-fos* promoter (177); SCMRE, a cis-acting element of the *c-fos* promoter responsible for induction in response to PDGF and/or the mitogenic factor from medium conditioned by *v-sis*-transformed cells (178); CREB, the CRE-binding protein (79); SRF, the serum response factor (112); p62, a protein that forms a ternary complex with SRF and SRE (171); BP2, a factor defined by gel-shift experiments that binds to SRE-2 (177); SCMRE-BP, a factor defined by gel-shift experiments that binds to SCMRE (178). Question marks indicate either proposed intermediates suggested as candidates to mediate *c-fos* induction (e.g. CaM and CaM kinase) or signal transduction intermediates and reactions that have not yet been identified.

Mutational analysis with transfected reporter genes containing the *c-fos* regulatory region demonstrated that the SRE also mediates repression of *c-fos* (183–185). Transient cotransfection experiments suggest that the FOS protein itself participates in *c-fos* repression. Both FOS and FRA-1 can repress expression of *egr-1/krox24* and *egr-2/krox20,* as well as *c-fos* expression

(186). JUN has no effect on repression by FOS. The C-terminal 27 amino acids of FOS are required for, and can confer, repression of c-fos transcription; the DNA-binding domain of the FOS protein is not necessary (186). Since FOS cannot interact directly with the SRE, the mechanism of transcriptional repression by FOS remains unexplained. Clearly, however, FOS can serve in the dual roles of activator and repressor of transcription in different contexts, via distinct cis-acting elements.

mRNAs from primary response genes are often subject to rapid degradation. Many of these messages, including c-fos mRNA, have in their 3' untranslated regions repeated copies of the $AU_3A$ element reported to confer destabilization on the mRNA of GM-CSF and other cytokines (187). c-fos mRNA degradation requires ongoing protein synthesis. The first protein-synthesis-dependent alteration that occurs in c-fos mRNA is the loss of the polyA tail, followed by rapid degradation of the remaining message (188). The 3' AU-rich region of the c-fos message apparently directs removal of the poly(A) tail, producing a labile message. A cytoplasmic protein that binds to RNA molecules with reiterated copies of the $AU_3A$ sequence has recently been isolated (189). A second destabilizing element exists within the protein-coding region of c-fos mRNA (190). Degradation via the sequence present in the c-fos open reading frame is independent of continued transcription, while degradation via the 3' AU-rich element requires continued RNA synthesis.

## 2. c-jun

Although both c-fos and c-jun are rapidly and transiently induced in 3T3 cells by serum and in PC12 cells by NGF, these two genes are not always coregulated. Growth factors (NGF and EGF) induce both c-fos and c-jun in PC12 cells. In contrast, depolarization and nicotine can induce c-fos in PC12 cells, but not c-jun (191). c-jun contains an AP-1 element which may be involved in an autoregulatory induction loop (192, 193); TPA appears to induce c-jun expression at an AP-1 site, rather than through an SRE, as occurs for c-fos. Serum and TPA induction of endogenous c-jun in Balb/c 3T3 cells, presumably mediated by the AP-1 binding site, can be blocked by forskolin (194). These data suggest that a PKA-mediated event, presumably phosphorylation of CREB, can block serum- and TPA-induced c-jun expression in 3T3 cells. In contrast, transfection experiments in 3T3 cells (195) suggest that (a) CREB can bind to the AP-1 site of c-jun, (b) serum- and TPA-induced expression from an exogenous c-jun promoter is blocked if CREB is present, and (c) CREB repression can be alleviated if CREB is phosphorylated by PKA. At present it is not clear why PKA activation has opposite effects of induced expression from endogenous (194) and exogenous (195) c-jun promoters.

## 3. JE

*JE* and *KC/N51* were isolated as PDGF and/or serum-inducible cDNAs (1). PDGF and serum induce *JE* and *KC/N51* by a pathway, shared by polyI:polyC, distinct from the PDGF/serum pathway responsible for *c-fos* induction (196). No SRE or IRS (interferon response sequence) is present in the first 500 nucleotides of the *JE* promoter (136). It will be of great interest to determine the regulatory element(s) and signal transduction pathway that confer PDGF- and serum-inducible characteristics to *JE,* since this induction does not occur via a recognizable SRE.

## 4. egr-1/zif-268/krox24/NGFI-A/TIS8

PKC down-regulation by prior treatment with TPA reduces, but does not eliminate, *TIS8/egr-1/NGFI-A* induction by EGF or FGF in 3T3 cells (197, 198), or by NGF, FGF, or EGF in PC12 cells (J. Altin, D. Kujubu, S. Raffioni, D. Eveleth, H. Herschman, R. Bradshaw, submitted for publication). Moreover, when cells are exposed to both maximally inducing levels of TPA and either additional EGF or FGF, *TIS8/egr-1/NGFI-A* induction is synergistically elevated (199). Multiple, independent, additive pathways for the induction of this gene must exist. When both PKC and PKA pathways are eliminated from PC12 cells, however, by down-regulating PKC in a PKA-deficient PC12 variant, NGF induction of *NGFI-A/egr-1/TIS8* transcription is substantially attenuated (200). It will be of great interest to repeat this experiment with other inducers. The mouse and rat *NGFI-A/egr-1* genes have multiple SREs, four Sp1-binding sites, two potential AP-1 elements, and two potential CaRE/CRE sequences in the first 500 nucleotides of the promoter (92–94, 201). *NGFI-A/egr-1/TIS8* is induced in PC12 cells by carbachol, $K^+$ depolarization, EGF and FGF, in addition to NGF (90). Sheng et al (180) suggest that *c-fos* induction by both depolarization and forskolin occurs through the common CaRE/CRE in PC12 cells, as a consequence of phosphorylation of CREB in response to distinct kinases.

## 5. egr-2/krox20

Although the *krox24/zif-268/egr-1* and *krox20/egr-2* protein products have similar DNA-binding domains and recognize identical transcriptional enhancer motifs (96, 97, 99, 104), the regulatory regions of these genes are quite distinct. *krox20/egr-2* has two SRE sequences, three potential AP-1 sites, two Sp1 sites, and two possible CRE/CaRE elements 5' to its mRNA initiation site (102). Thus, several regulatory elements are found in common in the promoters of these two genes, but the regulatory elements are arranged differently. Transfection experiments with promoter constructs demonstrated the functionality of the most 3' SRE (103).

## 6. N10/nur77/NGFI-B/TIS1

The *N10/nur77/NGFI-B/TIS1* gene is inducible in 3T3 cells by serum, TPA, EGF, and FGF (8). However, it does not have a true SRE/CArG sequence (107, 110). Two candidate AP-1 sequences, as well as three possible Sp1 sequences, are present in the first 400 nucleotides 5' of the transcription start site of the murine gene (107). In contrast to the *egr-1/NGFI-A/TIS8* gene, expression of the *NGFI-B/nur77/N10/TIS1* gene is much more inducible by $K^+$ depolarization than by growth factors (191). However, no obvious CaRE/CRE element is present in *NGFI-B/N10* (107, 110). Deletion and mutation studies will be required both to identify enhancer elements conferring serum and growth factor responsiveness on this gene, and to determine whether depolarization-induced expression of the *NGFI-A* and *NGFI-B* genes occur via pathways different (*a*) from the CaRE/CRE-mediated mechanism observed with *c-fos*, and (*b*) from one another.

## CELL-SPECIFIC RESTRICTION OF PRIMARY RESPONSE GENES

Many primary response genes are induced in a wide variety of contexts. *c-fos*, *egr-1/NGFI-A/TIS8*, *nur77/NGFI-B/TIS1*, and *PC4/TIS7* are induced by mitogens, NGF, depolarization, and neurotransmitters (38), to cite only a few such contexts. We found, in characterizing *TIS* gene expression, that the *TIS10* gene could be induced by a variety of mitogens in 3T3 cells, but could not be induced by any ligand in PC12 cells; *TIS10* expression is "restricted" in PC12 cells (90). Although the *egr-1/krox24/NGFI-A/TIS8* and *egr-2/krox20* proteins are similar in structure and function, and their promoters contain many of the same regulatory motifs, *egr-2/krox20* also cannot be expressed in PC12 cells (101).

We subsequently turned to cells of a very different lineage, and demonstrated that murine and human myeloid cells, while they could express *TIS10* and the remaining *TIS* genes, could not express *TIS1/nur77/NGFI-B* in response to either TPA or GM-CSF (202); *TIS1* expression is restricted in these myeloid cells. We have also demonstrated cell-type-specific restriction of *c-jun* and *egr-1/NGFI-A/TIS8* (203). Similarly, several of the cytokine-like genes cloned from mitogen-stimulated T cells cannot be induced in cultured fibroblasts (9, 204). Cell-type-specific restriction, analyzed by in situ hybridization of TIS probes to cells treated with TPA plus cycloheximide, also occurs in cells present in bone marrow and in primary cell cultures of brain (A. Arenander, J. de Vellis, H. Herschman, and P. Koeffler, unpublished). At present the molecular basis for the cell-type restriction of these primary response genes is, in general, not known. Possibilities include the presence of a trans-acting repressor in restricted cells (205), the absence of a required transcription factor (206, 207), the presence of a transcription factor inhibitor (208), cell-specific methylation of the gene (209), and cis-acting changes in chromatin structure (210). The *egr1/TIS8* gene cannot be induced in WEHI-231 cells, a murine B cell line which possesses a stable immature B cell phenotype (210a). Lack of *egr1* expression in these cells is correlated with hypermethylation of the *egr1* gene, in comparison to B cells able to express *egr1* message (210b). It is likely that cell-specific restriction of primary response gene expression is a developmental process that plays a fundamental role in determining the permissible repertoire of biological responses, following ligand binding, for distinct differentiated cells.

## SECONDARY RESPONSE GENES

A number of cDNAs have been isolated from libraries prepared following stimulation with mitogen, NGF, or TPA alone; i.e. without cycloheximide. Often their message levels begin to accumulate later than those of the majority of the primary response genes, and remain elevated for longer periods. Although some of the sequences cloned in this way can be induced in the presence of cycloheximide, and are therefore primary response genes, the message levels for other genes cloned from such libraries cannot be elevated if cycloheximide is included with ligand. By definition (25), such messages are the products of secondary response genes. Included in the category of mitogen-induced secondary response genes are ornithine decarboxylase, asparagine synthetase, proliferin/MRP, MEP/cathepsin-L, phorbin/TIMP/16C8, C122/osteopontin, and transin/stromolysin. The induction characteristics of these genes and the nature of their encoded proteins are described in the recent review by Lau & Nathans (211). Because mitogen-stimulated accumulation of these messages requires prior protein synthesis, it seems likely that the products of some of the primary response genes are involved in their induction, either directly as transcriptional modulators or indirectly as components of autocrine loops or paracrine responses.

## DO PRIMARY RESPONSE GENES PLAY CAUSAL ROLES IN MODULATING SECONDARY RESPONSE GENES AND CELLULAR PHYSIOLOGY?

Experimental evidence that ligand-induced primary response gene products play causal roles in subsequent cellular responses has been reported only for FOS. Both antisense *c-fos* oligonucleotides (212, 213) and microinjection of FOS-specific antibodies (214) can block serum/PDGF-induced DNA synthesis and proliferation of 3T3 cells, suggesting that *c-fos* induction is requisite for the mitogenic response of these agents.

Cells stably transfected with a dexamethasone-inducible *c-fos* antisense vector were used to examine the role of FOS protein in the expression of transin/stromolysin, a secondary response gene (215, 216). Production of antisense *c-fos* completely blocked transin induction by PDGF, but had no effect on EGF-induced transin expression. However, the antisense *c-fos* transcript prevented DNA synthesis in response to both PDGF and EGF. These data suggest that, while both PDGF and EGF induce DNA synthesis by FOS-dependent pathways, the PDGF pathway for transin expression requires *c-fos* induction, but the EGF pathway for transin expression is *c-fos* independent. Double-stranded oligonucleotides identical to the AP-1 site present in the transin promoter, when transfected into cells, blocked both PDGF and EGF induction of transin, suggesting that EGF induction of transin may require expression of another factor(s), e.g. one or more FRA or JUN proteins, acting at the AP-1 site.

TGFβ inhibits growth factor–induced transin expression (217). Inhibition is mediated through a cis-acting sequence, the TGFβ inhibitory element (TIE), of the transin gene. The TIE is conserved in several genes whose expression is inhibited by TGFβ (218). Although the TIE sequence (GnnTTGGtGa) and the AP-1 sequence (TGAg/cTCA) are not similar, the TIE binds a protein complex, induced in TGFβ-treated 3T3 cells, that contains FOS. Moreover, AP-1 sequences can compete, in gel-shift experiments, for binding to the TIE by the TGFβ-inducible, FOS-containing protein complex isolated from nuclei of TGFβ-treated cells (218). While these data may require re-examination of the conclusion that growth factor induction of transin transcription occurs via the AP-1 element of the transin promoter (217), it seems clear that FOS protein is necessary, in different contexts, for both induction (216) and inhibition (218) of transin gene expression. No SRE elements are present in the transin promoter; TGFβ-induced repression of transin expression by FOS protein clearly occurs by a mechanism distinct from the SRE-mediated repression of *c-fos* and *egr-1* expression by FOS protein (186).

# COMMON PRIMARY RESPONSE GENES AND CELL- AND LIGAND-SPECIFIC RESPONSES

When the search for primary response genes in various biological responses began, the assumption was that cell-type specific and/or ligand-specific genes would be identified. It seemed almost certain that proliferation-specific primary response genes would be identified. On the other hand, it seemed likely that ligand-specific primary response genes would also exist for a response as distinct as the NGF-directed differentiation of neuronal precursor cells. It was, therefore, initially perplexing that *c-fos* could be induced in so many different contexts. The overlapping induction of additional primary response genes (90), emphasized by the redundant cloning of many of these genes (Table 1), at first simply added to the confusion. If the same battery of genes is induced by so many ligands, in such different biological contexts, how can we account for the ligand-specific and/or cell-specific responses that follow?

As we learn more about the primary response genes themselves and the proteins they encode, the bases for ligand- and cell-specific responses are beginning to emerge. Several cytokine genes (e.g. *KC, JE)* are preferentially induced by agents such as PDGF, whose normal biological role may be concerned with wound healing. Cytokine production may be particularly important in this and similar contexts, and these primary response genes may have regulatory elements that make them more responsive to appropriate ligands.

Although many of the same primary response genes are induced by alternative ligands in a common cell, the quantitative levels of their gene products often differ in response to distinct inducing agents. Recall, for example, that *c-fos* and *c-jun* expression are regulated differently by growth factors and depolarizing agents in PC12 cells (191); both are extensively induced by growth factors, while *c-jun* is relatively refractory to induction by depolarization. The genes encoding the transcription factors *NGFI-A/egr-1/TIS8* and *NGFI-B/zif-268/TIS1* are also differentially regulated by growth factors and depolarization in PC12 cells (191). As we learn more about the enhancer motifs present in the various primary response genes and the nature of the second-messenger activity induced by alternative ligands, the basis for these differential transcriptional responses will be clarified. Cell-specific restriction of primary response gene expression (90, 101, 202–204), is also likely to play a major role in determining the molecular basis for distinct responses to ligands.

The multiple, interacting members of the FOS and JUN transcription factor families illustrate the possibilities for subtle combinatorial regulation possible with heterodimeric proteins. At least 15 known heterodimeric complexes of these molecules have been observed (48). Moreover, we now know that JUN can also form heterodimers with at least one member of the CREB family (74), raising the possibility of an extraordinary number of alternative heterodimers within the leucine-zipper family of transcriptional modulators. The functional diversity of heterodimers of this family has already been demonstrated; JUN/JUN-B heterodimer acts as a transcriptional repressor in a context where JUN/JUN and JUN/FOS heterodimers act as transcriptional activators (63, 64).

Members of the FOS and JUN families are subject to regulated posttranslational modification. FOS protein is phosphorylated in a ligand-dependent manner; both depolarization and NGF can induce *c-fos* in PC12 cells, but the FOS of NGF-treated cells is much more extensively phosphorylated than the FOS induced following depolarization (219, 220). Since phosphorylation can potentially affect the dimerization, DNA-binding, and transactivation properties of transcription factors, their ligand-specific and cell-specific posttranslational modification creates another level of diversity.

The FOS-related genes are also induced with differing kinetics following ligand stimulation (68, 70, 71). The potential for shifting functional capabilities in this family of transcriptional regulators is illustrated in the changing composition of the AP-1 complex induced in brain following depolarization (73). Similar alterations in structure and function due to ligand-specific and/or cell-specific posttranslational modifications, and distinct kinetics of expression, are likely to occur for the other transcription factors that are products of primary response genes (Table 1).

## PRIMARY RESPONSE GENE EXPRESSION IN THE NERVOUS SYSTEM

The expression of *c-fos* and other primary response genes has received substantial attention from the neurobiology community. Indeed, the literature on expression of these genes in the nervous system rivals that in mitogenesis: a tribute to intense interest in both molecular neurobiology and the role of primary response genes in mediating responses to extracellular signals. The literature on primary response gene expression in the nervous system has recently been well reviewed by experts in this area (221–224). While I do not provide an exhaustive review here, a primarily biochemical presentation of the major observations and current issues is appropriate.

The observation that *c-fos* could be induced by depolarization (219) and nicotinic agonists (225) in PC12 cells first suggested that FOS might mediate neuronal responses to external stimuli. *TIS1/NGFI-A/nur77, TIS7/PC4, TIS8/ NGFI-B/egr-1, TIS11*, and *TIS21* can also be induced by depolarization (90) and cholinergic agonists (198a) in PC12 cells, suggesting that these mitogen-inducible primary response genes might also play a role in neuronal responses. These results have prompted a number of investigations into the expression of primary response genes in central and peripheral neural tissue, in response to various stimuli.

Metrazole, a seizure-inducing agent, induces rapid and transient accumulation of *c-fos* message in brain. Accumulation of *c-fos* message occurs after convulsions, and is apparently a consequence of seizure, since anticonvulsants prevent metrazole-induced accumulation of *c-fos* mRNA (226).

Seizure-induced *c-fos* expression can also be blocked by antagonists of the N-methyl-D-aspartate (NMDA) receptor, one of several pharmacologically distinguishable glutamate receptors (227). Both NMDA and kainic acid, an agonist for another of the glutamate receptors, induce *c-fos* in brain, suggesting activation of glutamate receptors is a major contributor to seizure-induced *c-fos* expression (228). Studies with cultured cerebellar granular cells demonstrate that glutamate can directly induce *c-fos*, through NMDA receptors (229).

Regional expression of *c-fos* mRNA following seizure induction was demonstrated both by in situ hybridization and by immunohistochemistry; dramatic expression occurs in the neurons of the piriform cortex, dentate gyrus, and hippocampus after metrazole administration (226, 230). Focal lesions (231), electroconvulsive shock (232), and picrotoxin (228, 233) all induce both convulsions and *c-fos* expression. *zif-268/NGFI-A/TIS8, c-jun,jun-B,* and *NGFI-B/N10/nur77/TIS1* are also induced in brain by convulsants (86, 233, 234). While metrazole-induced *c-fos* mRNA peaks at one hour in brain and disappears by two hours, induced AP-1-binding activity remains elevated well beyond six hours (73). Metrazole also induces FOS-related antigens; however, their kinetics of appearance and disappearance are substantially slower than those of *c-fos*. Although FOS protein can initially be found in the induced AP-1-binding activity, the protein composition of the AP-1-binding complex induced in brain following metrazole administration changes with time (73). It is likely that the functional characteristics of the AP-1-binding activity expressed following seizure are also changing.

Other global treatments, including stress, caffeine, morphine, vitamin B6, estrogen, heat shock, ischemia, and activation of adrenergic receptors also induce *c-fos* in brain (for reviews, see Refs. 222 and 224). Both heat shock (235) and focal brain injury (236, 237) induce *c-fos* expression in non-neuronal brain cells: glia, ependyma, and pia. Rolipram, a cyclic AMP–dependent phosphodiesterase inhibitor, also induces *c-fos* expression in glial cells in brain (238). *c-fos* and the *TIS* genes can be induced in cultured neocortical astrocytes both by adrenergic and cholinergic agonists, as well as by TPA and growth factors such as FGF and EGF (199, 239, 240). Distinct *TIS* gene induction patterns, reminiscent of the differences observed for induction by growth factors and depolarization in PC12 cells (191, 198a), occur in cultured neocortical astrocytes in response to alternative ligands (199, 239, 240). It will be of great interest to determine whether the *c-fos* induction mediated by adrenergic receptors in brain (241) occurs in glia, neurons, or both cell types.

Morgan & Curran (222, 226) have suggested *c-fos* immunohistochemistry as an alternative to classical methods, such as 2-deoxyglucose (2-DG) uptake, to map neuronal connections in the nervous system. Sensory stimulation can induce trans-synaptic *c-fos* expression in the spinal cord (242). Electrical stimulation in the brain can also cause trans-synaptic *c-fos* induction (243). A direct comparison of FOS and 2-DG mapping has been reported (244). The use of *c-fos* as a metabolic marker to trace neuronal pathways has recently been reviewed (245), and guidelines and pitfalls have been discussed.

Several paradigms have been employed to investigate the relationship between localized stimuli, induction of brain primary response gene expression, and "behavioral" responses. "Kindling" is a protein-synthesis dependent process in which repeated small increments of regional electrical or pharmacologic stimulation, below convulsive levels, are administered. Kindling results in a permanent change such that, following what would be a subconvulsive challenge in a naive animal, convulsions occur in the kindled individual. The kindling stimulus can induce local *c-fos* expression (231). However, *c-fos* induction following stimulation does not differ in kindled and naive animals, despite their difference in convulsive response, and the basal *c-fos* level is not altered in kindled animals (222). These data suggest that *c-fos* induction following kindling stimuli may not be causal in establishing the kindled phenotype. The issue is controversial, however, and the hypothesis that induction of primary response genes participates in establishing the kindled phenotype is supported by other arguments (224).

Long-term potentiation (LTP), like kindling, is an electrophysiological paradigm in which local electrical stimulation causes long-term alteration in neuronal excitability, in this case in the hippocampus. LTP does not involve seizure induction. Although the stimuli used to induce LTP also induce *c-fos* and *NGFI-A/zif-268/TIS8* (246), and both phenomena are mediated by NMDA receptors (227), induction of LTP and *c-fos* can be dissociated; LTP can be induced under conditions where *c-fos* induction does not occur (247, 248). The correlation between *NGFI-A/zif-268/TIS8* induction and LTP is somewhat stronger (227, 246). However, activation of a neuronal inhibitory input can block LTP induction but not *NGFI-A/zif-268/TIS8* induction, suggesting that induction of this gene may be necessary, but is not sufficient, for establishment of LTP (227). Recently Otani et al (249) reported that actinomycin D had no effect on establishment of LTP or on its maintenance for three hours, suggesting that induction of primary response genes is not involved in this stimulus-response model, at least for the time period studied. However, although general mRNA synthesis decreased to 5% of control values in actinomycin D–treated animals, *c-fos* and/or *NGFI-A/zif-268/TIS8* induction under these conditions was not examined.

Circadian rhythms are controlled in mammals by a "pacemaker" in the suprachiasmatic nucleus (SCN), located in the hypothalamus. The endogenous 24-hour periodicity of the pacemaker can be "entrained" by altering activity in the cells of the pacemaker in response to artificial light-dark cycles. Exposure of rats and hamsters to light pulses during the dark cycle, when light can shift pacemaker activity, caused extensive c-fos (250–253) and *NGFI-A* (253) induction in the SCN. In contrast, the pineal gland experiences a rapid, transient induction of c-fos soon after the onset of darkness (254). These systems provide interesting new models to investigate the relationship between stimulus, primary response gene expression, and behavior.

The implicit assumption in all the neuronal stimulus-response paradigms described above is that the transcription factors induced as a primary response to the stimulus serve as intermediates that regulate expression of secondary response genes, leading to an altered phenotype. The best evidence to date to support this hypothesis comes from studies of seizure-induced expression of proenkephalin (PENK). During pentylenetetrazole-induced seizure, PENK message accumulates in the hippocampus (255, 256). However, PENK message first appears in the hippocampus one hour after pentylenetetrazole administration, while c-fos, c-jun, jun-B, and *egr-1/NGFI-A/TIS8* messages have peaked at this point (256). The *PENK* gene has an AP-1 site in its regulatory region (257, 258). Transient transfection experiments demonstrate that JUN and FOS synergistically transactivate the PENK promoter (256). Thus AP-1 proteins induced in the hippocampus in response to neuronal stimulation may mediate subsequent proenkephalin gene expression.

The existing studies of c-fos induction in brain by various stimuli and subsequent biological behavior, whether it be kindling, LTP, circadian rhythm entrainment, or proenkephalin expression, all suffer from the same criticism; while in some cases good correlations between c-fos induction and a subsequent phenotype can be shown, the experiments do not provide evidence for cause and effect. Until one can do experiments that inactivate FOS protein or other primary response gene products in brain during the response to stimuli, as has been done for cellular proliferation (Refs. 212–214; see above), it will be difficult to determine cause-and-effect relationships between induction of primary response gene expression in brain and behavior. One such relationship has, however, recently been established. The learned gill and siphon withdrawal reflex in the invertebrate sea slug *Aplysia* requires protein synthesis for long-term storage (259). An *Aplysia* cell culture model has been established in which a single application of serotonin to a sensory neuron causes short-term facilitation, while repeated serotonin administration can establish transcription- and translation-dependent long-term facilitation (260). Serotonin elicits cAMP accumulation, and cAMP injections can cause long-lasting, protein-synthesis-dependent increases in synaptic strength, suggesting that a PKA/CREB-mediated transcriptional response may contribute to the long-term facilitation (261, 262). Injection of CRE element sequences into the nucleus of the afferent neuron blocks the long-term facilitation induced by serotonin, but not the short-term response (263). These data suggest that primary-response genes—induced in response to serotonin-stimulated activation of PKA, phosphorylation of CREB, and transactivation at a CRE—play a causal role in the consolidation of the facilitation engram.

ACKNOWLEDGMENTS

I thank Drs. R. Bravo, P. Charnay, T. Curran, M. Greenberg, M. Karin, L. Lau, C. Stiles, V. Sukhatme, and J. Wagner for allowing me to use information prior to publication, and the members of my laboratory and Drs. Alaric Arenander, Judith Gasson, and Oliver Hankinson for particularly helpful comments on the text. Thanks also to Raymond Basconcillo and Sharon Solomon for their help in literature searching and preparation of the manuscript. HRH is the recipient of NIH grant GM24797 and DOE contract DE FC03 87ER 60615.

*Literature Cited*

1. Cochran, B. H., Reffel, A. C., Stiles, C. D. 1983. *Cell* 33:939–47
2. Cochran, B. H., Zullo, J., Verma, I. M., Stiles, C. D. 1984. *Science* 226:1080–82
3. Greenberg, M. E., Ziff, E. B. 1984. *Nature* 311:433–38
4. Kelly, K., Cochran, B. H., Stiles, C. D., Leder, P. 1983. *Cell* 35:603–10
5. Lau, L. F., Nathans, D. 1985. *EMBO J.* 4:3145–51
6. Sukhatme, V. P., Sreedharan, K., Toback, F. G., Taub, R., Hoover, R. G., Tsai-Morris, C.-H. 1987. *Oncogene Res.* 1:343–55
7. Almendral, J. M., Sommer, D., Macdonald-Bravo, H., Burckhardt, J., Perera, J., Bravo, R. 1988. *Mol. Cell. Biol.* 8:2140–48
8. Lim, R. W., Varnum, B. C., Herschman, H. R. 1987. *Oncogene* 1:263–70
9. Irving, S. G., June, C. H., Zipfel, P. F., Siebenlist, U., Kelly, K. 1989. *Mol. Cell. Biol.* 9:1034–40
10. Gomperts, M., Pascall, J. C., Brown, K. D. 1990. *Oncogene* 5:1081–83
11. Milbrandt, J. 1987. *Science* 238:797–99
12. Cho, K., Skarnes, W. C., Minsk, B. M., Palmieri, S., Jackson-Grusby, L., Wagner, J. A. 1988. *Mol. Cell. Biol.* 9:135–43
13. Tirone, F., Shooter, E. M. 1989. *Proc. Natl. Acad. Sci. USA* 86:2088–92
14. Friedman, R. L., Manly, S. P., McMahon, M., Kerr, I. M., Stark, G. R. 1984. *Cell* 38:745–55
15. Luster, A. D., Unkeless, J. C., Ravetch, J. V. 1985. *Nature* 315:672–76

16. Zumstein, P., Stiles, C. D. 1987. *J. Biol. Chem.* 262:11252–60
17. Dixit, V. M., Marks, R. M., Sarma, V., Prochownik, E. V. 1989. *J. Biol. Chem.* 264:16905–9
18. Dixit, V. M., Green, S., Sarma, V., Holzman, L. B., Wolf, F. W., et al. 1990. *J. Biol. Chem.* 265:2973–78
19. Fornace, A. J. Jr., Alamo, I. Jr., Hollander, M. C. 1988. *Proc. Natl. Acad. Sci. USA* 85:8800
20. LaRosa, G. J., Gudas, L. J. 1988. *Proc. Natl. Acad. Sci. USA* 85:329–33
21. Lobe, C. G., Havele, C., Bleackley, R. C. 1986. *Proc. Natl. Acad. Sci. USA* 83:1448–52
22. Simmons, D. L., Levy, D. B., Yannoni, Y., Erikson, R. L. 1989. *Proc. Natl. Acad. Sci. USA* 86:1178–82
23. Hastings, K. E. M., Emerson, C. P. Jr. 1982. *Proc. Natl. Acad. Sci. USA* 79:1553–57
24. Schneider, C., King, R. M., Philipson, L. 1988. *Cell* 54:787–93
25. Yamamoto, K. R., Alberts, B. M. 1976. *Annu. Rev. Biochem.* 45:721–46
26. Rollins, B. J., Stiles, C. D. 1989. *Adv. Cancer Res.* 5:1–32
27. Riddle, V. G. H., Dubrow, R., Pardee, A. B. 1979. *Proc. Natl. Acad. Sci. USA* 76:1298–302
28. Thomas, George, Thomas, Gary, Luther, H. 1981. *Proc. Natl. Acad. Sci. USA* 78:5712–16
29. Pledger, W. J., Hart, C. A., Locatell, K. L., Scher, C. D. 1981. *Proc. Natl. Acad. Sci. USA* 78:4358–62

30. Olashaw, N. E., Pledger, W. J. 1983. *Nature* 306:272–74
31. Hendrickson, S. L., Scher, C. D. 1983. *Mol. Cell. Biol.* 3:1478–87
32. Kruijer, W., Cooper, J. A., Hunter, T., Verma, I. M. 1984. *Nature* 312:711–16
33. Müller, R., Bravo, R., Burckhardt, J., Curran, T. 1984. *Nature* 312:716–20
34. Greenberg, M. E., Greene, L. A., Ziff, E. B. 1985. *J. Biol. Chem.* 260:14101–10
35. Kruijer, W., Schubert, D., Verma, I. 1985. *Proc. Natl. Acad. Sci. USA* 82:7330–34
36. Greene, L. A., Tischler, A. S. 1976. *Proc. Natl. Acad. Sci. USA* 73:2424–28
37. Greene, L. A., Tischler, A. S. 1982. *Adv. Cell. Neurobiol.* 3:373–414
38. Herschman, H. R. 1989. *Trends Biochem. Sci.* 14:455–58
39. Curran, T., Miller, A. D., Zokas, L., Verma, I. M. 1984. *Cell* 36:259–68
40. Setoyama, C., Frunzio, R., Liau, G., Mudryj, M., de Crombrugghe, B. 1986. *Proc. Natl. Acad. Sci. USA* 83:3213–17
41. Distel, R. J., Ro, H.-S., Rosen, B. S., Groves, D. L., Spiegelman, B. M. 1987. *Cell* 49:835–44
42. Lee, W., Haslinger, A., Karin, M., Tjian, R. 1987. *Nature* 325:368–72
43. Piette, J., Yaniv, M. 1987. *EMBO J.* 6:1331–37
44. Curran, T., Van Beveren, C., Ling, N., Verma, I. M. 1985. *Mol. Cell. Biol.* 5:167–72
45. Maki, Y., Bos, T. J., Davis, C., Starbuck, M., Vogt, P. K. 1987. *Proc. Natl. Acad. Sci. USA* 84:2848–52
46. Vogt, P. K., Bos, T. J., Doolittle, R. F. 1987. *Proc. Natl. Acad. Sci. USA* 84:3316–19
47. Struhl, K. 1988. *Nature* 332:649–50
48. Vogt, P. K., Bos, T. J. 1990. *Adv. Cancer Res.* 55:1–35
49. Bohmann, D., Bos, T. J., Admon, A., Nishimura, T., Vogt, P. K., Tjian, R. 1987. *Science* 238:1386–92
50. Rauscher, F. J. III, Cohen, D. R., Curran, T., Bos, T. J., Vogt, P. K., et al. 1988. *Science* 240:1010–16
51. Sassone-Corsi, P., Lamph, W. W., Kamps, M., Verma, I. M. 1988. *Cell* 54:553–60
52. Ryseck, R.-P., Hirai, S. I., Bravo, R. 1988. *Nature* 334:535–37
53. Quantin, B., Breathnach, R. 1988. *Nature* 334:538–39
54. Ryder, K., Nathans, D. 1988. *Proc. Natl. Acad. Sci. USA* 85:8464–67
55. Lamph, W. W., Wamsley, P., Sassone-Corsi, P., Verma, I. M. 1988. *Nature* 334:629–31
56. Wu, B. Y., Fodor, E. J., Edwards, R. H., Rutter, W. J. 1989. *J. Biol. Chem.* 264:9000–3
57. Halazonetis, T. D., Georgopoulos, G., Greenberg, M. E., Leder, P. 1988. *Cell.* 55:917–24
58. Landschulz, W. H., Johnson, P. F., McKnight, S. L. 1988. *Science* 240:1759–64
59. Busch, S. J., Sassone-Corsi, P. 1990. *Trends Genet.* 6:36–40
60. Perkins, K. K., Admon, A., Paten, N., Tjian, R. 1990. *Genes Dev.* 5:822–34
61. Zhang, K., Chaillet, J. R., Perkins, L. A., Halazonetis, T. D., Perrimon, N. 1990. *Proc. Natl. Acad. Sci. USA* 87:6281–85
62. Ryder, K., Lau, L. F., Nathans, D. 1988. *Proc. Natl. Acad. Sci. USA* 85:1487–91
63. Chiu, R., Angel, P., Karin, M. 1989. *Cell* 59:979–86
64. Schütte, J., Viallet, J., Nau, M., Segal, S., Fedorko, J., Minna, J. 1989. *Cell* 59:987–97
65. Nakabeppu, Y., Ryder, K., Nathans, D. 1988. *Cell* 55:907–15
66. Ryder, K., Lanahan, A., Perez-Albuerne, E., Nathans, D. 1989. *Proc. Natl. Acad. Sci. USA* 86:1500–3
67. Hirai, S.-I., Ryseck, R.-P., Mechta, F., Bravo, R., Yaniv, M. 1989. *EMBO J.* 8:1433–39
68. Cohen, D. R., Curran, T. 1988. *Mol. Cell. Biol.* 8:2063–69
69. Matsui, M., Tokuhara, M., Konuma, Y., Nomura, N., Ishizaki, R. 1990. *Oncogene* 5:249–55
70. Nishina, H., Sato, H., Suzuki, T., Sato, M., Iba, H. 1990. *Proc. Natl. Acad. Sci. USA* 87:3619–23
71. Zerial, M., Toschi, L., Ryseck, R.-P., Schuermann, M., Müller, R., Bravo, R. 1989. *EMBO J.* 8:805–13
72. Cohen, D. R., Ferreira, C. P., Gentz, R., Franza, B. R. Jr., Curran, T. 1989. *Genes Dev.* 3:173–84
73. Sonnenberg, J. L., Macgregor-Leon, P. F., Curran, T., Morgan, J. I. 1989. *Neuron* 3:359–65
74. Macgregor, P. F., Abate, C., Curran, T. 1990. *Oncogene* 5:451–58
75. Maekawa, T., Sakura, H., Kanei-Ishii, C., Sudo, T., Yoshimura, T., et al. 1989. *EMBO J.* 8:2023–28
76. Hoeffler, J. P., Meyer, T. E., Yungdae, Y., Jameson, J. L., Habener, J. F. 1988. *Science* 242:1430–33
77. Montminy, M. R., Sevarino, K. A., Wagner, J. A., Mandel, G., Goodman, R. H. 1986. *Proc. Natl. Acad. Sci. USA* 83:6682–86

78. Hai, T., Liu, F., Coukos, W. J., Green, M. R. 1989. *Genes Dev.* 3:2083–90
79. Habener, J. F. 1990. *Mol. Endocrinol.* 4:1087–94
80. Sivak, A. 1972. *J. Cell. Physiol.* 80:167–74
81. Herschman, H. R., Passovoy, D. S., Pruss, R. M., Aharonov, A. 1978. *J. Supramol. Struct.* 8:41–46
82. Butler-Gralla, E., Herschman, H. R. 1981. *J. Cell. Physiol.* 107:59–68
83. Blumberg, P. M., Butler-Gralla, E., Herschman, H. R. 1981. *Biochem. Biophys. Res. Commun.* 102:818–23
84. Bishop, R., Martinez, R., Weber, M. J., Blackshear, P. J., Beatty, S., et al. 1985. *Mol. Cell. Biol.* 5:2231–37
85. Herschman, H. R. 1988. *Growth Factors, Tumor Promoters, and Cancer Genes*, Vol. 58, pp. 295–310. New York: Liss
86. Sukhatme, V. P., Cao, X.-M., Chang, L. C., Tsai-Morris, C.-H., Stamenkovich, D., et al. 1988. *Cell* 53:37–43
87. Lemaire, P., Revelant, O., Bravo, R., Charnay, P. 1988. *Proc. Natl. Acad. Sci. USA* 85:4691–95
88. Christy, B. A., Lau, L. F., Nathans, D. 1988. *Proc. Natl. Acad. Sci. USA* 85:7857–61
89. Varnum, B. C., Lim, R. W., Sukhatme, V. P., Herschman, H. R. 1989. *Oncogene* 4:119–20
90. Kujubu, D. A., Lim, R. W., Varnum, B. C., Herschman, H. R. 1987. *Oncogene* 1:257–62
91. Wright, J. J., Gunter, K. C., Mitsuya, H., Irving, S. G., Kelly, K., Siebenlist, U. 1990. *Science* 248:588–91
92. Tsai-Morris, C. H., Cao, X., Sukhatme, V. P. 1988. *Nucleic Acids Res.* 16:8835–45
93. Janssen-Timmen, U., Lemaire, P., Mattéi, M.-G., Revelant, O., Charnay, P. 1989. *Gene* 80:325–36
94. Changelian, P. S., Feng, P., King, T. C., Milbrandt, J. 1989. *Proc. Natl. Acad. Sci. USA* 86:377–81
95. McMahon, A. P., Champion, J. E., McMahon, J. A., Sukhatme, V. P. 1990. *Development* 108:281–87
96. Cao, X., Koski, R. A., Gashler, A., McKiernan, M., Morris, C. F., et al. 1990. *Mol. Cell. Biol.* 10:1931–39
97. Lemaire, P., Vesque, C., Schmitt, J., Stunnenberg, H., Frank, R., Charnay, P. 1990. *Mol. Cell. Biol.* 10:3456–67
98. Waters, C. M., Hancock, D. C., Evan, G. I. 1990. *Oncogene* 5:669–74
99. Christy, B., Nathans, D. 1989. *Proc. Natl. Acad. Sci. USA* 86:8737–41
100. Chavrier, P., Zerial, M., Lemaire, P., Almendral, J., Bravo, R., Charnay, P. 1988. *EMBO J.* 7:29–35
101. Joseph, L. J., Le Beau, M. M., Jamieson, G. A., Acharya, S., Shows, T. B., et al. 1988. *Proc. Natl. Acad. Sci. USA* 85:7164–68
102. Chavrier, P., Janssen-Timmen, U., Mattéi, M.-G., Zerial, M., Bravo, R., Charnay, P. 1989. *Mol. Cell. Biol.* 9:787–97
103. Rangnekar, V. M., Aplin, A. C., Sukhatme, V. P. 1990. *Nucleic Acids Res.* 18:2749–57
104. Chavrier, P., Vesque, C., Galliot, B., Vigneron, M., Dollé, P., et al. 1990. *EMBO J.* 9:1209–18
105. Nehlin, J. O., Ronne, H. 1990. *EMBO J.* 9:2891–98
106. Hazel, T. G., Nathans, D., Lau, L. F. 1988. *Proc. Natl. Acad. Sci. USA* 85:8444–48
107. Ryseck, R.-P., Macdonald-Bravo, H., Mattéi, M.-G., Ruppert, S., Bravo, R. 1989. *EMBO J.* 8:3327–35
108. Milbrandt, J. 1988. *Neuron* 1:183–88
109. Varnum, B. C., Lim, R. W., Herschman, H. R. 1989. *Oncogene* 4:1263–65
110. Watson, M. A., Milbrandt, J. 1989. *Mol. Cell. Biol.* 9:4213–19
111. Treisman, R. 1985. *Cell* 42:889–902
112. Treisman, R. 1986. *Cell* 46:567–74
113. Treisman, R. 1987. *EMBO J.* 6:2711–17
114. Norman, C., Runswick, M., Pollock, R., Treisman, R. 1988. *Cell* 55:989–1003
115. Gout, P. W., Beer, C. T., Noble, R. L. 1980. *Cancer Res.* 40:2433–36
116. Yu-Lee, L.-Y., Hrachovy, J. A., Stevens, A. M., Schwarz, L. A. 1990. *Mol. Cell. Biol.* 10:3087–94
117. Miyamoto, M., Fujita, T., Kimura, Y., Maruyama, M., Harada, H., et al. 1988. *Cell* 54:903–13
118. Ryseck, R.-P., Macdonald-Bravo, H., Zerial, M., Bravo, R. 1989. *Exp. Cell Res.* 180:537–45
119. Santarén, J. F., Bravo, R. 1986. *EMBO J.* 5:877–82
120. Santarén, J. F., Blüthman, H., Macdonald-Bravo, H., Bravo, R. 1987. *Exp. Cell Res.* 173:341–48
121. Almendral, J. M., Santarén, J. F., Perera, J., Zerial, M., Bravo, R. 1989. *Exp. Cell Res.* 181:518–30
122. Hartzell, S., Ryder, K., Lanahan, A., Lau, L. F., Nathans, D. 1989. *Mol. Cell. Biol.* 9:2567–73
123. Bloem, L. J., Chen, L., Konigsberg, W. H., Bach, R. 1989. *J. Cell. Physiol.* 139:418–23

124. Kletzien, R., Purdue, J. F. 1974. *J. Biol. Chem.* 249:3366–74
125. Birnbaum, M. J., Haspel, H. C., Rosen, O. M. 1986. *Proc. Natl. Acad. Sci. USA* 83:5784–88
126. Hiraki, Y., Rosen, O. M., Birnbaum, M. J. 1988. *J. Biol. Chem.* 263:13655–62
127. Oquendo, P., Alberta, J., Wen, D., Graycar, J. L., Derynck, R., Stiles, C. D. 1989. *J. Biol. Chem.* 264:4133–37
128. Ryseck, R.-P., Macdonald-Bravo, H., Mattéi, M.-G., Bravo, R. 1989. *Exp. Cell Res.* 180:266–75
129. Richmond, A., Balentien, E., Thomas, H. G., Flaggs, G., Barton, D. E., et al. 1988. *EMBO J.* 7:2025–33
130. Anisowicz, A., Bardwell, L., Sager, R. 1987. *Proc. Natl. Acad. Sci. USA* 84:7188–92
131. Luster, A. D., Ravetch, J. V. 1987. *Mol. Cell. Biol.* 7:3723–31
132. Sugano, S., Stoeckle, M. Y., Hanafusa, H. 1987. *Cell* 49:321–28
133. Bedard, P.-A., Alcorta, D., Simmons, D. L., Luk, K.-C., Erikson, R. L. 1987. *Proc. Natl. Acad. Sci. USA* 84:6715–19
134. Deuel, T. F., Keim, P. S., Farmer, M., Heinrikson, R. L. 1977. *Proc. Natl. Acad. Sci. USA* 74:2256–58
135. Holt, J. C., Harris, M. E., Holt, A. M., Lange, E., Henschen, A., Nieurarowski, S. 1986. *Biochemistry* 25:1988–96
136. Rollins, B. J., Morrison, E. D., Stiles, C. D. 1988. *Proc. Natl. Acad. Sci. USA* 85:3738–42
137. Kawahara, R. S., Deuel, T. F. 1989. *J. Biol. Chem.* 264:679–82
138. Rollins, B. J., Stier, P., Ernst, T., Wong, G. G. 1989. *Mol. Cell. Biol.* 9:4687–95
139. Yoshimura, T., Yuhki, N., Moore, S. K., Appella, E., Lerman, M. I., Leonard, E. J. 1989. *FEBS Lett.* 244:487–93
140. Obaru, K., Fukuda, M., Maeda, S., Shimada, K. 1986. *J. Biochem.* 99:885–94
141. Zipfel, P. F., Balke, J., Irving, S. G., Kelly, K., Siebenlist, U. 1989. *J. Immunol.* 142:1582–90
142. Davatelis, G., Tekamp-Olson, P., Wolpe, S. D., Hermsen, K., Luedke, C., et al. 1988. *J. Exp. Med.* 167:1939–44
143. Lipes, M. A., Napolitano, M., Jeang, K.-T., Chang, N. T., Leonard, W. J. 1988. *Proc. Natl. Acad. Sci. USA* 85:9704–8
144. Brown, K. D., Zurawski, S. M., Mosmann, T. R., Zurawski, G. 1989. *J. Immunol.* 142:679–87
145. Burd, P. R., Freeman, G. J., Wilson, S. D., Berman, M., DeKruyff, R., et al. 1987. *J. Immunol.* 139:3126–31
146. Schall, T. J., Jongstra, J., Dyer, B. J., Jorgensen, J., Clayberger, C., et al. 1988. *J. Immunol.* 141:1018–25
147. Schmid, J., Weissmann, C. 1987. *J. Immunol.* 139:250–56
148. Stoeckle, M. Y., Barker, K. A. 1990. *New Biol.* 2:313–23
148a. DeWitt, D. L., El-Harith, E. A., Kraemer, S. A., Andrews, M. J., Yao, E. F., et al. 1990. *J. Biol. Chem.* 265:5192–98
149. Varnum, B. C., Lim, R. W., Sukhatme, V. P., Herschman, H. R. 1989. *Oncogene* 4:119–20
149a. Lai, W. S., Stumpo, D. J., Blackshear, P. J. 1990. *J. Biol. Chem.* 265:16556–63
149b. DuBois, R. N., McLane, M. W., Ryder, K., Lau, L. F., Nathans, D. 1990. *J. Biol. Chem.* 265:19186–91
150. O'Brien, T. P., Yang, G. P., Sanders, L., Lau, L. F. 1990. *Mol. Cell. Biol.* 10:3569–77
151. Brown, A. L., Chiariotti, L., Orlowski, C.C., Mehlman, T., Burgess, W. H., et al. 1989. *J. Biol. Chem.* 264:5148–54
152. Varnum, B. C., Lim, R. W., Herschman, H. R. 1989. *Oncogene* 4:1263–65
152a. Charles, C. H., Simske, J. S., O'Brien, T. P., Lau, L. F. 1990. *Mol. Cell. Biol.* 10(12): In press
153. Hirschhorn, R. R., Aller, P., Yuan, Z.-A., Gibson, C. W., Baserga, R. 1984. *Proc. Natl. Acad. Sci. USA* 81:6004–8
154. Ide, T., Ninomiya-Tsuji, J., Ferrari, S., Philiponis, V., Baserga, R. 1986. *Biochemistry* 25:7041–46
155. Battini, R., Ferrari, S., Kaczmarek, L., Calabretta, B., Chen, S.-T., Baserga, R. 1987. *J. Biol. Chem.* 262:4355–59
156. Keutzer, J. C., Hirschhorn, R. R. 1990. *Exp. Cell Res.* 188:153–59
157. Calabretta, B., Venturelli, D., Kaczmarek, L., Narni, F., Talpaz, M., et al. 1986. *Proc. Natl. Acad. Sci. USA* 83:1495–98
158. Masiakowski, P., Shooter, E. M. 1988. *Proc. Natl. Acad. Sci. USA* 85:1277–81
159. Prywes, R., Roeder, R. G. 1987. *Mol. Cell. Biol.* 7:3482–89
160. Schroter, H., Shaw, P. E., Nordheim, A. 1987. *Nucleic Acids Res.* 15:10145–58
161. Minty, A., Kedes, L. 1986. *Mol. Cell. Biol.* 6:2125–36
162. Boxer, L. M., Prywes, R., Roeder, R. G., Kedes, L. 1989. *Mol. Cell. Biol.* 9:515–22

163. Bender, A., Sprague, G. F. Jr. 1987. *Cell* 50:681–91
164. Keleher, C. A., Goutte, C., Johnson, A. D. 1988. *Cell* 53:927–36
165. Passmore, S., Maine, G. T., Elble, R., Christ, C., Tye, B.-K. 1988. *J. Mol. Biol.* 204:593–606
166. Hayes, T. E., Sengupta, P., Cochran, B. H. 1988. *Genes Dev.* 2:1713–22
167. Passmore, S., Elble, R., Tye, B.-K. 1989. *Genes Dev.* 3:921–35
168. Prywes, R., Dutta, A., Cromlish, J. A., Roeder, R. G. 1988. *Proc. Natl. Acad. Sci. USA* 85:7206–10
169. Manak, J. R., de Bisschop, N., Kris, R. M., Prywes, R. 1990. *Genes Dev.* 4: 955–67
170. Herrera, R. E., Shaw, P. E., Nordheim, A. 1989. *Nature* 340:68–70
171. Shaw, P. E., Schröter, H., Nordheim, A. 1989. *Cell* 56:563–72
172. Schröter, H., Mueller, C. G. F., Meese, K., Nordheim, A. 1990. *EMBO J.* 9:1123–30
173. Greenberg, M. E., Siegfried, Z., Ziff, E. B. 1987. *Mol. Cell. Biol.* 7:1217–25
174. Gilman, M. Z. 1988. *Genes Dev.* 2:394–402
175. Sheng, M., Dougan, S. T., McFadden, G., Greenberg, M. E. 1988. *Mol. Cell. Biol.* 8:2787–96
176. Rudkin, B. B., Lazarovici, P., Levi, B.-Z., Abe, Y., Fujita, K., Guroff, G. 1989. *EMBO J.* 8:3319–25
177. Visvader, J., Sassone-Corsi, P., Verma, I. M. 1988. *Proc. Natl. Acad. Sci. USA* 85:9474–78
178. Hayes, T. E., Kitchen, A. M., Cochran, B. H. 1987. *Proc. Natl. Acad. Sci. USA* 84:1272–76
179. Sassone-Corsi, P., Visvader, J., Ferland, L., Mellon, P. L., Verma, I. M. 1988. *Genes Dev.* 2:1529–38
180. Sheng, M., McFadden, G., Greenberg, M. E. 1990. *Neuron* 4:571–82
181. Lamb, N. J. C., Fernandez, A., Tourkine, N., Jeanteur, P., Blanchard, J.-M. 1990. *Cell* 61:485–96
182. Robbins, P. D., Horowitz, J. M., Mulligan, R. C. 1990. *Nature* 346:668–71
183. Rivera, V. M., Sheng, M., Greenberg, M. E. 1990. *Genes Dev.* 4:255–68
184. Konig, H., Ponta, H., Rahmsdorf, U., Buscher, M., Schonthal, A., et al. 1989. *EMBO J.* 8:2559–66
185. Lucibello, F. C., Lowag, C., Neuberg, M., Müller, R. 1989. *Cell* 59:999–1007
186. Gius, D., Cao, X.-M., Rauscher, F. J. III, Cohen, D. R., Curran, T., Sukhatme, V. P. 1990. *Mol. Cell. Biol.* 10:4243–55

187. Shaw, G., Kamen, R. 1986. *Cell* 46: 659–67
188. Wilson, T., Treisman, R. 1988. *Nature* 336:396–99
189. Malter, J. S. 1989. *Science* 246:664–66
190. Shyu, A.-B., Greenberg, M. E., Belasco, J. G. 1989. *Genes Dev.* 3:60–72
191. Bartel, D. P., Sheng, M., Lau, L. F., Greenberg, M. E. 1989. *Genes Dev.* 3:304–13
192. Hattori, K., Angel, P., LeBeau, M. M., Karin, M. 1988. *Proc. Natl. Acad. Sci. USA* 85:9148–52
193. Angel, P., Hattori, K., Smeal, T., Karin, M. 1988. *Cell* 55:875–85
194. Mechta, F., Piette, J., Hirai, S.-I., Yaniv, M. 1989. *New Biol.* 1:297–304
195. Lamph, W. W., Dwarki, V. J., Ofir, R., Montminy, M., Verma, I. M. 1990. *Proc. Natl. Acad. Sci. USA* 87:4320–24
196. Hall, D. J., Jones, S. D., Kaplan, D. R., Whitman, M., Rollins, B. J., Stiles, C. D. 1989. *Mol. Cell. Biol.* 9:1705–13
197. Jamieson, G. A. Jr., Mayforth, R. D., Villereal, M. L., Sukhatme, V. P. 1989. *J. Cell. Physiol.* 139:262–68
198. Lim, R. W., Varnum, B. C., O'Brien, T. G., Herschman, H. R. 1989. *Mol. Cell. Biol.* 9:1790–93
198a. Altin, J. G., Kujubu, D., Raffionio, S., Eveleth, D., Herschman, H. R., Bradshaw, R. A. 1991. *J. Biol. Chem.* In press
199. Arenander, A. T., Lim, R. W., Varnum, B. C., Cole, R., de Vellis, J., Herschman, H. R. 1989. *J. Neurosci. Res.* 23:257–65
200. Damon, D. H., D'Amore, P. A., Wagner, J. A. 1990. *J. Cell Biol.* 110:1333–39
201. Christy, B., Nathans, D. 1989. *Mol. Cell. Biol.* 9:4889–95
202. Varnum, B. C., Lim, R. W., Kujubu, D. A., Luner, S. J., Kaufman, S. E., et al. 1989. *Mol. Cell. Biol.* 9:3580–83
203. Herschman, H. R. 1991. *Advances in Regulation of Cell Growth*, Vol. 2, ed. J. Mond, A. Weiss, J. Cambier. New York: Raven. In press
204. Gunter, K. C., Irving, S. G., Zipfel, F., Siebenlist, U., Kelly, K. 1989. *J. Immunol.* 142:3286–91
205. Boshart, M., Weih, F., Schmidt, A., Fournier, R. E. K., Schütz, G. 1990. *Cell* 61:905–16
206. Bergman, Y., Strich, B., Sharir, H., Ber, R., Laskov, R. 1990. *EMBO J.* 9:849–55
207. McCormick, A., Wu, D., Castrillo, J.-L., Dana, S., Strobl, J., et al. 1988. *Cell* 55:379–89

208. Baeuerle, P. A., Baltimore, D. 1988. *Science* 242:540–46
209. Cedar, H. 1988. *Cell* 53:3–4
210. Ip, Y. T., Fournier, R. E. K., Chalkley, R. 1990. *Mol. Cell. Biol.* 10:3782–87
210a. Seyfert, V. L., Sukhatme, V. P., Monroe, J. G. 1989. *Mol. Cell. Biol.* 9:2083–88
210b. Seyfert, V. L., McMahon, B., Glenn, W. D., Yellen, A. J., Sukhatme, V. P., et al. 1990. *Science* 250:797–800
211. Lau, L. F., Nathans, D. 1991. *Molecular Aspects of Cellular Regulation: Hormonal Regulation of Transcription,* Vol. 6. Amsterdam: Elsevier. In press
212. Holt, J. T., Gopal, T. V., Moulton, A. D., Nienhuis, A. W. 1986. *Proc. Natl. Acad. Sci. USA* 83:4794–98
213. Nishikura, K., Murray, J. M. 1987. *Mol. Cell. Biol.* 7:639–49
214. Riabowol, K. T., Vosatka, R. J., Ziff, E. B., Lamb, N. J., Feramisco, J. R. 1988. *Mol. Cell. Biol.* 8:1670–76
215. Matrisian, L. M., Leroy, P., Ruhlman, C., Gesnel, M.-C., Breathnach, R. 1986. *Mol. Cell. Biol.* 6:1679–86
216. Kerr, L. D., Holt, J. T., Matrisian, L. M. 1988. *Science* 242:1424–27
217. Kerr, L. D., Olashaw, N. E., Matrisian, L. M. 1988. *J. Biol. Chem.* 263:16999–7005
218. Kerr, L. D., Miller, D. B., Matrisian, L. M. 1990. *Cell* 61:267–78
219. Morgan, J. I., Curran, T. 1986. *Nature* 322:552–55
220. Curran, T., Morgan, J. I. 1986. *Proc. Natl. Acad. Sci. USA* 83:8521–24
221. Morgan, J. I., Curran, T. 1989. *Genes and Signal Transduction in Multistage Carcinogenesis,* ed. N. H. Colburn, pp. 377–90. New York: Dekker
222. Morgan, J. I., Curran, T. 1991. *Annu. Rev. Neurosci.* 14:421–51
223. Sheng, M., Greenberg, M. E. 1990. *Neuron* 4:477–85
224. Dragunow, M., Currie, R. W., Faull, R. L. M., Robertson, H. A., Jansen, K. 1989. *Neurosci. Biobehav. Rev.* 13:301–13
225. Greenberg, M. E., Ziff, E. B., Greene, L. A. 1986. *Science* 234:80–83
226. Morgan, J. I., Cohen, D. R., Hempstead, J. L., Curran, T. 1987. *Science* 237:192–97
227. Wisden, W., Errington, M. L., Williams, S., Dunnett, S. B., Waters, C., et al. 1990. *Neuron* 4:603–14
228. Sonnenberg, J. L., Mitchelmore, C., Macgregor-Leon, P. F., Hempstead, J., Morgan, J. I., Curran, T. 1989. *J. Neurosci. Res.* 24:72–80
229. Szekely, A. M., Barbaccia, M. L.,

Alho, H., Costa, E. 1989. *Mol. Pharmacol.* 35:401–8
230. Dragunow, M., Robertson, H. A. 1988. *Brain Res.* 440:252–60
231. Dragunow, M., Robertson, H. A. 1987. *Nature* 329:441–42
232. Daval, J.-L., Nakajima, T., Gleiter, C. H., Post, R. M., Marangos, P. J. 1989. *J. Neurochem.* 52:1954–57
233. Saffen, D. W., Cole, A. J., Worley, P. F., Christy, B. A., Ryder, K., Baraban, J. M. 1988. *Proc. Natl. Acad. Sci. USA* 85:7795–99
234. Watson, M. A., Milbrandt, J. 1989. *Mol. Cell. Biol.* 9:4213–19
235. Dragunow, M., Currie, R. W., Robertson, H. A., Faull, R. L. M. 1989. *Exp. Neurol.* 106:105–9
236. Dragunow, M., Robertson, H. A. 1988. *Brain Res.* 455:295–99
237. Dragunow, M., Goulding, M., Faull, R. L. M., Ralph, R., Mee, E., Frith, R. 1990. *Exp. Neurol.* 107:236–48
238. Dragunow, M., Faull, R. L. M. 1989. *Brain Res.* 501:382–88
239. Arenander, A. T., Lim, R. W., Varnum, B. C., Cole, R., de Vellis, J., Herschman, H. R. 1989. *J. Neurosci. Res.* 23:247–56
240. Arenander, A. T., de Vellis, J., Herschman, H. R. 1989. *J. Neurosci. Res.* 24:107–14
241. Gubits, R. M., Smith, T. M., Fairhurst, J. L., Yu, H. 1989. *Mol. Brain Res.* 6:39–45
242. Hunt, S. P., Pini, A., Evan, G. 1987. *Nature* 328:632–34
243. Sagar, S. M., Sharp, F. R., Curran, T. 1988. *Science* 240:1328–31
244. Sharp, F. R., Gonzalez, M. F., Sharp, J. W., Sagar, S. M. 1989. *J. Comp. Neurol.* 284:621–36
245. Dragunow, M., Faull, R. 1989. *J. Neurosci. Methods* 29:261–65
246. Cole, A. J., Saffen, D. W., Baraban, J. M., Worley, P. F. 1989. *Nature* 340:474–76
247. Douglas, R. M., Dragunow, M., Robertson, H. A. 1988. *Mol. Brain Res.* 4:259–62
248. Dragunow, M., Abraham, W. C., Goulding, M., Mason, S. E., Robertson, H. A., Faull, R. L. M. 1989. *Neuroscience Lett.* 101:274–80
249. Otani, S., Marshall, C. J., Tate, W. P., Goddard, G. V., Abraham, W. C. 1989. *Neuroscience* 28:519–26
250. Rea, M. A. 1989. *Brain Res.* 23:577–81
251. Aronin, N., Sagar, S. M., Sharp, F. R., Schwartz, W. J. 1990. *Proc. Natl. Acad. Sci. USA* 87:5959–62
252. Kornhauser, J. M., Nelson, D. E.,

Mayo, K. E., Takahashi, J. S. 1990. *Neuron* 5:127–34
253. Rusak, B., Robertson, H. A., Wisden, W., Hunt, S. P. 1990. *Science* 248:1237–40
254. Carter, D. A. 1990. *Biochem. Biophys. Res. Commun.* 166:589–94
255. White, J. D., Gall, C. M. 1987. *Mol. Brain Res.* 3:21–29
256. Sonnenberg, J. L., Rauscher, F. J. III, Morgan, J. I. 1989. *Science* 246:1622–25
257. Hyman, S. E., Comb, M., Lin, Y.-S., Pearlberg, J., Green, M. R., Goodman, H. M. 1988. *Mol. Cell. Biol.* 8:4225–33

258. Comb, M., Mermod, N., Hyman, S. E., Pearlberg, J., Ross, M. E., Goodman, H. M. 1988. *EMBO J.* 7:3793–805
259. Castellucci, V. F., Blumenfeld, H., Goelet, P., Kandel, E. R. 1989. *J. Neurobiol.* 20:1–9
260. Montarolo, P. G., Goelet, P., Castellucci, V. F., Morgan, J., Kandel, E. R., et al. 1986. *Science* 234:1249–54
261. Schacher, S., Castellucci, V. F., Kandel, E. R. 1988. *Science* 240:1667–69
262. Scholz, K. P., Byrne, J. H. 1988. *Science* 240:1664–66
263. Dash, P. K., Hochner, B., Kandel, E. R. 1990. *Nature* 345:718–21

*Annu. Rev. Biochem. 1991. 60:321–47*

# MOLECULAR CHAPERONES

## R. John Ellis

Department of Biological Sciences, University of Warwick, Coventry, United Kingdom

## Saskia M. van der Vies

Molecular Biology Division, Central Research and Development Department, E. I. Du Pont de Nemours and Company, Experimental Station, Wilmington, Delaware

KEY WORDS:    assisted self-assembly, chaperonins, heat-shock proteins, protein folding, protein assembly.

## CONTENTS

## PERSPECTIVES

A continuing challenge for biochemists is to unravel those cellular processes that use the one-dimensional information in the genetic material to produce the three-dimensional structures that give proteins their biological properties. The detailed mechanism of polypeptide chain synthesis is well established, but it remains to "crack the second half of the genetic code," which ensures that these chains attain their functional conformations (1, 1a). The current paradigm for protein biogenesis presented in textbooks of biochemistry (e.g.

321

0066-4154/91/0701-0321$02.00

Ref. 2) derives from the classic work of Anfinsen (3) on the in vitro renaturation of ribonuclease, and is based on the hypothesis of self-assembly. This hypothesis postulates that each polypeptide chain interacts with itself as it is synthesized to assume a folded conformation of lower free energy. This spontaneous process requires neither further input of energy nor any steric information extrinsic to the polypeptide itself. The conformation assumed by each polypeptide possesses the functional activity unique to that polypeptide. Such activity often includes the ability to bind to macromolecules, including other polypeptides, in a highly specific manner, so that complex oligomeric structures can also self-assemble from their component subunits. Such complex structures often mediate their cellular functions in ways that require their subunits to transiently dissociate and reassociate. Many proteins are also unfolded to an approximately linear conformation in order to traverse membranes en route to a wide array of subcellular compartments or to the extracellular medium; such proteins refold into their functional conformations once membrane transport is complete.

The hypothesis of protein self-assembly postulates that in all these situations the interactions that occur within and between polypeptides are both necessary and sufficient to produce the functional conformation. According to this hypothesis it is thus sufficient to specify the sequences of nucleotides in the genes encoding the amino acid sequences of polypeptides in order to specify the information sufficient for the biogenesis of these polypeptides in their functional forms. Self-assembly is not presented as the universal mechanism for the genesis of biological structures, since exceptions, such as the requirement of scaffolding proteins for the assembly of some bacteriophages, are well-established. Nevertheless, most textbooks give the impression that self-assembly as described above is the predominant process for the biogenesis of protein-containing structures.

The concept of molecular chaperones (4–10) qualifies this self-assembly hypothesis by suggesting that in many cases interactions within and between polypeptides and other molecules need to be controlled to reduce the probability of formation of incorrect structures, i.e. structures that lack the functions required by their biological context. This control is exerted by pre-existing proteins acting as chaperones to inhibit incorrect molecular interactions. It is argued that in any given assembly process there is a certain probability that incorrect interactions will produce nonfunctional structures. Where this probability is small, self-assembly needs no assistance, but where it is high, molecular chaperones are essential to produce sufficient correct structures for cellular needs.

Molecular chaperones are currently defined as a family of unrelated classes of protein that mediate the correct assembly of other polypeptides, but are not themselves components of the final functional structures (6). Known molecular chaperones do not convey steric information for either polypeptide folding or oligomerization, so the principle of self-assembly is not violated in this sense. Rather the principle is qualified by the need in many cases for assistance by molecular chaperones; these function by binding to specific structural features that are exposed only in the early stages of assembly, and so inhibit unproductive assembly pathways that would otherwise act as kinetic dead-end traps and produce incorrect structures. The term chaperone is appropriate for this family of proteins because the role of the human chaperone is to prevent incorrect interactions between people, not to provide steric information for these interactions.

The general concept of molecular chaperones developed over several years as a result of studies at Warwick University on the biogenesis of the chloroplast enzyme that fixes carbon dioxide in photosynthesis—ribulose bisphosphate carboxylase-oxygenase, or rubisco (8–10). It became clear by 1987 that this concept applies generally to a wide variety of processes involving proteins within all types of cell—plant, animal, and bacterial. The growing number and variety of proteins that can be regarded as molecular chaperones suggest both that there is a widespread cellular need for chaperone function, and that self-assembly in its strictest formulation may not be the predominant process for the biogenesis of macromolecular structures. The consequences of this conclusion for our understanding of the evolution of these structures have yet to be explored. The concept of molecular chaperones may be useful in medicine, since chaperone diseases may exist in which correct folding and/or oligomerization of particular proteins may fail due to changes in their chaperone. It is also possible that some human viruses require host chaperones for their replication. Some, but not all, molecular chaperones are stress proteins, so it is possible that all stress proteins act as molecular chaperones. The molecular chaperone concept may also be useful in biotechnology, where problems often arise in using heterologous systems to produce valuable proteins in the required active form.

In this article we first review the concept of molecular chaperones, and then discuss a few selected examples and speculate about directions for future research. The literature on molecular chaperones is growing rapidly as this novel concept is stimulating research in many laboratories, so that exhaustive coverage is no longer possible in one article. Several reviews have been published (6–9), as well as the first issue of a new journal, *Seminars in Cell Biology,* which is entirely devoted to all aspects of molecular chaperones (10).

## THE MOLECULAR CHAPERONE CONCEPT

Molecular chaperones are defined as a family of unrelated classes of protein that mediate the correct assembly of other polypeptides but that are not components of the functional assembled structures (6). The imprecise word mediate is used deliberately in this definition because of our ignorance about the molecular details of chaperone action. The proposed function of chaperone proteins is to assist polypeptides to self-assemble by inhibiting alternative assembly pathways that produce nonfunctional structures. The term polypeptide chain—binding proteins has been proposed for essentially the same group of proteins in a stimulating review by Rothman (11). The history of the development of the molecular chaperone concept from work on the assembly of rubisco has been described (9), so only the key points are presented here.

## The Principle of Protein Self-Assembly

The term self-assembly is used in the literature with different meanings (12). In the context of molecular chaperones we use the term self-assembly to mean that the information that is both necessary and sufficient in principle to specify the three-dimensional structure, and hence the function, of each polypeptide, resides solely within the amino acid sequence of that polypeptide; thus a newly synthesized polypeptide chain should be able to attain its functional conformation in the intracellular environment with no assistance from other molecules. We are not referring to the more general idea that complex structures can reform after dissociation in vitro into their component parts with no further input of information; this clearly does not hold true for structures as complex as membranes, organelles, and some viruses, or even for proteins whose assembly involves irreversible processing steps. It is also important to appreciate that the term assembly in the chaperone context is used in a wide sense. It embraces both the folding of polypeptide chains immediately after synthesis, and any oligomerization into larger structures with proteins or other macromolecules that may occur, as well as changes in the degree of folding or oligomerization that take place as part of the way in which these structures function subsequently. Such changes may occur when proteins are transported across membranes and are reactivated during recovery from stresses such as heat shock.

Anfinsen and his colleagues found that denatured purified ribonuclease refolds spontaneously in the absence of other proteins into an active enzyme (3). Many other protein-containing structures have been successfully renatured in vitro to their functional conformations from their purified and separated components (13–16), including structures as complex as the tobacco mosaic virus (17), and the subunits of bacterial ribosomes (18). These classic demonstrations are commonly cited in textbooks to support the notion that self-assembly is the dominant process for the biogenesis of functional protein-containing structures. However, it has been pointed out by Creighton (19) that all the evidence for self-assembly comes from in vitro studies with isolated proteins so that the potential involvement of other proteins has been either ignored or discounted. It is instructive to quote his words (19) on this point: "These generalizations about protein folding in vivo are extrapolations from in vitro studies and assume that folding is strictly a self-assembly process. There is no need to invoke special factors to explain folding in vivo, since intact proteins can fold by themselves in vitro, but their existence in the biosynthetic apparatus cannot by ruled out." The fact that the denaturation of many proteins is not completely reversible in vitro, especially at high protein concentrations where incorrect interactions may predominate, has not until recently raised serious concern about the general validity of the self-assembly hypothesis to describe the in vivo situation. One consequence of the assumption that self-assembly operates widely in vivo is that much effort is currently being expended to determine the rules by which the primary structure of a polypeptide determines its functional conformation, an effort that has yet to bear fruit (1, 1a).

The molecular chaperone concept postulates that self-assembly as defined above is not the predominant process of protein assembly in vivo because increasingly, examples are being discovered where proteins will not assemble correctly unless assisted by other proteins. It is this latter group of proteins that we term molecular chaperones, a suggestion made first at a meeting in Copenhagen in 1987 (20) and subsequently published in *Nature* (4). The chaperone function is required because many cellular processes that involve protein assembly carry an inherent risk of malfunction owing to the number, variety, and flexibility of the weak interactions that hold proteins in their functional conformations. The cell thus continually faces the problem that incorrect interactions will generate nonfunctional structures. It is the role of molecular chaperones to combat this problem.

## The Cellular Problem of Interactive Surfaces

A number of fundamental cellular processes involve the transient exposure of interactive protein surfaces to the intracellular environment. The term interactive surfaces refers to any regions of intramolecular or intermolecular contact between parts of the protein-containing structure that are significant in maintaining that structure. All these processes may have to combat the problem of incorrect interactions. Examples of such processes are as follows:

1. Protein synthesis: the aminoterminal region of each polypeptide is made before the carboxyterminal region. If the normal fate of the aminoterminal region is to interact with the carboxyterminal region of the same chain, what happens to the aminoterminal region before the carboxyterminal region is made—might it undergo incorrect interactions with itself or other cellular components? The chance of incorrect interactions occurring will be greater if the rate of synthesis is much slower than the rate of folding, so chaperones may be required during synthesis only for proteins that fold rapidly. The rate

of in vitro renaturation often exceeds that of in vivo assembly by several orders of magnitude, so it is not surprising that incorrect interactions are commonly observed in vitro (1). Thus the time required to refold denatured proteins in vitro varies from milliseconds to hours (13, 21), but the average protein in yeast is synthesized in about two minutes (22). Chaperones belonging to the chaperonin group (Table 1) bind to nascent polypeptides synthesized by cell-free extracts of *Escherichia coli* (23), while chaperones of the heat shock protein 70 (hsp 70) class bind transiently to many newly synthesized polypeptides during their synthesis in vivo (24). It has been suggested that the efficiency of folding of some proteins is increased by controlled rates of translation in vivo (1, 25).

2. Protein transport: proteins that enter organelles such as the endoplasmic reticulum, mitochondrion, plastid, and the bacterial periplasm traverse the membrane in an unfolded or partially unfolded state (26). Such proteins are often synthesized by cytosolic ribosomes in the form of precursors. Such precursors have to be prevented from folding into a translocation-incompetent conformation before transport can proceed (27), and once inside the organelle the polypeptide refolds into its final functional conformation. Thus mechanisms to mediate folding and refolding are required on both sides of such membranes, and much current work on molecular chaperones concerns these processes in both bacteria (23, 26, 28, 29) and eukaryotic cells (30–32).

3. Protein function: in several examples, the normal functioning of oligomeric complexes involves changes in subunit-subunit interactions, so that regions previously involved in subunit contacts are transiently exposed to the intracellular environment. Such processes include DNA replication (33), the recycling of clathrin cages (34), and the assembly of microtubules.

4. Organelle biogenesis: many of the protein complexes found inside organelles such as plastids and mitochondria consist of subunits synthesized inside the organelle bound to other subunits synthesized by cytosolic ribosomes; the enzyme rubisco is a good example of such a complex. Thus in some cases modulation of the binding propensities of the subunits may be required before they can all be located together in the same organelle. It was the discovery of a chloroplast protein that binds noncovalently to newly synthesized subunits of rubisco, and thereby prevents incorrect interactions, that eventually led to the molecular chaperone concept (reviewed in 8, 9).

5. Stress responses: environmental stresses such as excessive heat often cause the denaturation of proteins and the formation of aggregates. To protect against such stresses, cells accumulate proteins that prevent the production of such aggregates by inhibiting the incorrect interactions that cause them and/or unscrambling the aggregates to allow correct reassembly (31, 35). Many molecular chaperones are stress proteins that are abundant even in the absence of stress; thus the stress response can be viewed as an amplification of the basic chaperone function (24, 35, 36).

The self-assembly hypothesis in its strictest formulation implies that all the interactions between protein surfaces exposed in such processes as those listed above are totally correct, i.e. that they are both necessary and sufficient to produce the functional conformation. Since incorrect interactions have been observed both in vivo and in vitro (8, 14, 16), it seems more plausible to suggest that during a given assembly process there is a certain probability that incorrect interactions will lead to the formation of nonfunctional structures. Where this probability is low, molecular chaperones may not be required, but where it is high, they may be essential to produce enough functional structures to meet cellular needs. There is evidence that molecular chaperones are involved in all these processes as indicated above, and it was the accumulation of this evidence that spawned the general concept of molecular chaperones (4, 6, 7, 9). How do chaperones mediate their essential functions?

## Assisted Self-Assembly: a Model for Chaperone Action

The molecular chaperone concept does not necessarily imply that chaperones convey steric information specifying assembly. Present knowledge on the formation of nucleosomes (37) and rubisco (38) shows that the chaperones involved in their assembly do not convey steric information, because correct structures can be formed in the absence of chaperones under certain, albeit unphysiological, conditions. Instead chaperones function by recognizing structural feature(s) of the interactive surface(s), which are accessible only during stages in the assembly process, or which appear as a result of stress of a mutation in an already assembled structure; chaperones have not been observed to bind to assembled functional proteins or to all unfolded proteins during in vitro renaturation (28). The chaperone binds to this feature noncovalently to form a stable complex in which incorrect assembly pathways are inhibited. This binding is reversed under, as yet, undefined circumstances, which permit correct interactions to predominate. In some cases, but not all, this reversal requires the hydrolysis of ATP, probably to release the ligand, and the presence of another protein to mediate the release step. The best-studied systems supporting this model are the renaturation of bacterial rubisco (8, 38, 39) and the assembly of lambda phage heads (40).

This model proposed for the action of molecular chaperones suggests that their action is rather subtle and is best described as assisting self-assembly. Thus the principle of protein self-assembly is not violated by the molecular chaperone concept, rather it is qualified. On this basis we distinguish two types of self-assembly:

1. Strict self-assembly: no factors other than the primary structure are required for the polypeptide to have a high probability of assembling correctly within the intracellular environment.

2. Assisted self-assembly: the appropriate molecular chaperone is required in addition to the primary structure to allow correct assembly to predominate over incorrect assembly; such chaperones convey no steric information over and above that in the primary structure of the ligand.

It is conceivable that some chaperones do convey steric information for other polypeptides to assemble correctly, i.e. that there are cases of directed assembly. No examples of such chaperones have yet been reported, and it is rather hard to imagine how evidence for them could be obtained until our knowledge of the structural aspects of chaperone action is much better developed.

It is possible to represent this model for chaperone action graphically by plotting the free energy for protein assembly within the cell against two possible outcomes—either the protein assembles correctly to give a functional structure or it misassembles to give a nonfunctional structure. For some proteins the formation of the correct structure may be favored both kinetically and thermodynamically, and in such cases chaperones are not required (Figure 1a). In other cases misassembly may be kinetically favored, so that chaperones are required either to inhibit the misassembly pathway by presenting an energy barrier or to lower the activation energy of the correct pathway (Figure 1b). The molecular details of chaperone action are totally obscure at the present time, but studies have commenced in several crystallographic laboratories to determine the three-dimensional structures of chaperone-ligand complexes, and reports have already appeared describing the three-dimensional structures of two chaperones (a fragment of hsp70 and the PapD protein) in the absence of ligand (41, 42). In vitro assays for chaperone function in the renaturation of rubisco (38, 39) and beta-lactamase (43) have been developed using purified components, and their further exploitation should provide much valuable information.

The basic postulate of the chaperone hypothesis is that chaperones function by preventing incorrect interactions, and thereby allow correct interactions encoded within the primary structure to predominate. Gething & Sambrook (44) have made the intriguing speculation that some chaperones might also maintain parts of a polypeptide chain in conformations that can eventually be converted into a wider variety of shapes than would otherwise be possible. Such hypothetical chaperones would thus expand the repertoire of interactions within and between polypeptides to produce structures that would not be possible in their absence.

## Origin of Molecular Chaperone

The history of the term molecular chaperone has been described repeatedly (4–10), but is rarely referred to by other authors. Some other authors also err

(a)                                                          (b)

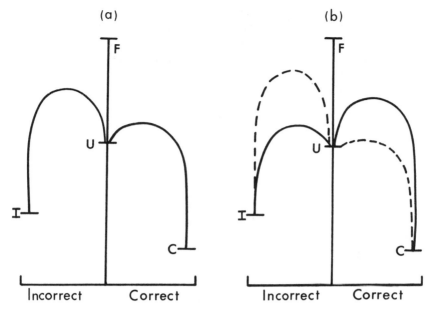

*Figure 1* Graphical representation of the energetic effects of chaperone action. The vertical axis (F) represents the free energy of protein assembly from the unfolded state (U) within the cellular environment, while the horizontal axis represents two possible results of assembly—either an inactive misassembled structure (I) or a functional correctly assembled structure (C). In example (*a*) the formation of the correct structure is favored both kinetically and thermodynamically; this is a case of strict self-assembly where chaperones are not required. In example (*b*) the formation of an incorrect structure is favored kinetically; the addition of a chaperone, indicated by the dotted lines, could either block the misassembly pathway, or lower the activation energy of the correct pathway. Example (*b*) is a case of assisted self-assembly. This method of representation was suggested by R. B. Freedman. Reprinted with permission from Ellis (10).

in restricting the term to proteins that mediate the folding of polypeptides but not other aspects of protein assembly and vice-versa, while others confuse the word chaperone with the word chaperonin, which describes just one class of the larger family of molecular chaperones. For these reasons we summarize the history of the term here, as well as justify its present biochemical usage.

The term molecular chaperone was first used in the scientific literature by Laskey et al (45) as a succinct way of describing the function of nucleoplasmin. Nucleoplasmin is an acidic soluble nuclear protein that mediates the in vitro assembly of nucleosomes from separated histones and DNA. When DNA is mixed with histone monomers at physiological ionic strength, self-assembly fails in a spectacular way—an instant precipitate forms. However, if the histones are first mixed with a molar excess of nucleoplasmin before addition of DNA, soluble nucleosome cores form; this assembly process does not require ATP (37, 46). Nucleoplasmin binds to histones and thereby reduces their strong positive charge, but it does not bind to DNA or to assembled nucleosomes. Nucleoplasmin is required only for assembly, and is not a component of the functional nucleosome. Separate experiments show that the steric information for nucleosome assembly resides in the histones and not in nucleoplasmin, because nucleosomes can be formed in the absence of nucleoplasmin under appropriate but unphysiological conditions. The role of the nucleoplasmin is thus not to provide steric information for nucleosome assembly but to reduce the positive charges of the histone monomers, so that correct interactions with DNA are not swamped by incorrect interactions. In the words of Laskey et al (45), "We suggest that the role of the protein we have purified is that of a molecular chaperone which prevents incorrect ionic interactions between histones and DNA." Nucleoplasmin is thus the archetypal molecular chaperone; our contribution has been to point out that the term can be usefully extended to describe a much larger range of unrelated proteins that function to prevent incorrect interactions in a variety of cellular processes.

The word chaperone is normally used to describe a particular and largely outdated form of social behavior by human beings. The *Oxford English Dictionary* (2nd edition) defines chaperone, or more correctly chaperon, as a "person, usually a married or elderly woman who, for the sake of propriety, accompanies a young unmarried lady in public as guide and protector." Thus the traditional role of the human chaperone, if described in biochemical terms, is to prevent improper interactions between potentially complementary surfaces. Moreover, human chaperones do not possess the steric information by which people interact nor are they present during married life. Molecular chaperone is thus a very appropriate term to describe proteins which prevent incorrect interactions between parts of other molecules, but which do not impart steric information or form part of the final functional structures.

## EXAMPLES OF MOLECULAR CHAPERONES

Table 1 lists some of the proteins that meet the definition of molecular chaperones; this list is not exhaustive since potential candidates are appearing rapidly, especially with respect to protein transport. It is also a matter for debate where to draw the line between molecules whose sole function is to act as chaperones and those with additional functions (see below). The family of molecular chaperones is defined functionally, whereas the classes within this family are defined on the basis that members within a class are sequence-related to one another but not to members of any other class. Three recent examples are of especial interest because these particular chaperones are

Table 1  The molecular chaperone family

| Class | Members | Alternative names | Refs. |
|---|---|---|---|
| 1. Nucleoplasmins | Nucleoplasmin | — | 37, 45, 52 |
|  | Protein XLNO-38 | — | 51, 53 |
|  | Protein Ch-NO38 | — | 55 |
|  | Nucleoplasmin S | — | 56 |
| 2. Chaperonins | Chaperonin 60 | Bacterial: | |
|  |  | groEL (*E. coli*) | 5, 8, 23, 26, 28, 29, 39, 40, 43, 57–66, 72–82 |
|  |  | 65-kDa common antigen (mycobacteria) | 62–64 |
|  |  | Plastid: | |
|  |  | rubisco subunit-binding protein | 5, 8, 9, 65, 67, 83–90 |
|  |  | Mitochondrial: | |
|  |  | hsp 60 | 32, 68, 92–98 |
|  |  | mitonin | |
|  |  | HuCha60 | 102 |
|  |  | Eukaryotic cytosol: | |
|  |  | TCP-1 | 9, 59 |
|  | Chaperorin 10 | Bacterial: | |
|  |  | groES | 40, 71 |
|  |  | Mitochondrial | 70 |
| 3. Heat shock proteins 70 | Hsp68, 72, 73; DnaK; clathrin uncoating ATPase; BiP; grp75, 78, 80; hsc70, KAR2; SSA1–4; SSB1; SSC1; SSD1 | | 35, 42, 44, 97, 105–109 |
| 4. Heat shock proteins 90 | Hsp83, 87; HtpG | | 35, 110–114 |
| 5. Signal recognition particle | | SRP | 50 |
| 6. Subtilisin prosequence | | | 47a |
| 7. *Alpha*-lytic factor prosequence | | | 47b |
| 8. Ubiquitinated ribosomal proteins | | | 48 |
| 9. Trigger factor | | | 28 |
| 10. Sec B protein | | | 28, 29 |
| 11. Pap D protein | | | 41 |

attached covalently to the molecules whose assembly they mediate, i.e. the pro-sequence of pro-subtilisin (47a) and *alpha*-lytic protease (47b), and the ubiquitin sequence present at the aminoterminus of two ribosomal precursor proteins in yeast and other eukaryotes (48). In all three cases the extra sequences improve the correct assembly of the functional structures, but are then removed. The advantage of these cotranslational chaperones is presumably that they can bind their ligands without having to seek them out amid the crowd of other molecules.

It can be argued that the presequences present on many proteins that are transported across membranes should also be regarded as cotranslational chaperones, since there is evidence that they function to prevent the folding of the remainder of the polypeptide into a transport-incompetent conformation (49). However, in several cases it has been found that the chaperone function of these presequences is not completely effective without the assistance of separate chaperones, e.g. the signal recognition particle in the case of cotranslational protein transport (50), and the trigger factor, the SecB protein, and the bacterial chaperonin 60 protein (groEL) in the case of posttranslational protein transport (23, 28, 29). These observations raise the possibility that the correct assembly of some proteins requires the assistance of more than one chaperone, some of which may be transient sequences within the protein itself, and that in some cases these chaperones interact with one another to achieve the ultimate goal of promoting efficient correct assembly. Evidence for the involvement of more than one chaperone in the transport and assembly of imported mitochondrial proteins is beginning to emerge (see later).

Two enzymes involved in protein assembly, i.e. protein disulfide isomerase and peptidyl-prolyl-*cis-trans*-isomerase, are not regarded as molecular chaperones, since the evidence suggests that, for example, disulfide bridges stabilize the native state but do not determine the spatial arrangement of the polypeptide backbone (1a, 13).

We predict that the list of molecular chaperones will grow steadily longer, not only because the recent recognition of the chaperone function is stimulating much research, but also because this function is apparently required by many cellular processes. Owing to space limitations, only the first four classes in Table 1 are now discussed.

## The Nucleoplasmins

Processes such as the replication of DNA, the transcription and processing of RNA, and the assembly of nucleosomes and ribosomes involve strong ionic interactions between positively charged proteins and negatively charged nucleic acids. In eukaryotic organisms, additional processes transport these proteins into the nucleus, and transport ribonucleoprotein particles and ribosomes out of the nucleus. Recent evidence has suggested that a group of nucleoplasmin-like proteins exist that may perform a variety of chaperone functions during these processes (51); this evidence, however, rests largely on guilt-by-association, since direct demonstrations of the functions of any of these nucleoplasmin-like proteins are lacking.

The role of nucleoplasmin in the assembly of chromatin in vitro and in vivo has been reviewed recently (37), and is not discussed here. Immunogold localization studies show that nucleoplasmin is present not only throughout the nucleoplasm, but also along the whole length of lampbrush chromosome loops and with the nucleoli. Double-labeling experiments with an anti-nucleoplasmin antibody and an anti-ribonucleoprotein antiserum show that both antigens are present on the same ribonucleoprotein particles on these loops, suggesting that nucleoplasmin may be involved in the assembly of these particles (52).

Since the studies that led to the description of nucleoplasmin as the first molecular chaperone (45), further work has revealed the existence of several other nuclear proteins that may fulfill a number of different but related chaperoning functions (51). One such protein (XLNO-38) shows striking sequence similarity to nucleoplasmin, but is located in the granular component of the nucleolus (53). The first 124 amino acids of this protein show 61% identity with the aminoterminal region of nucleoplasmin (53). This protein is not a component of mature ribosomes, but is associated with both large and small preribosomal subunit particles. Thus protein XLNO-38 may carry out chaperoning functions specific to the nucleolus, such as the assembly of basic proteins on ribosomal RNA precursors. It will be recalled that the self-assembly of ribosomes in vitro from purified RNA and ribosomal proteins proceeds with difficulty (54), so it would be worth repeating these studies in the presence of purified protein XLNO-38. The homologous nucleolar protein from chicken (termed protein Ch-NO38) has been shown to shuttle between the nucleus and cytoplasm (55), suggesting that this protein may also escort newly synthesized ribosomal proteins into the nucleus or be required to maintain preribosomal particles in a conformation compatible with export from nucleus to cytoplasm.

A protein closely related to nucleoplasmin has been purified from a *Xenopus* kidney cell line (56). This protein is called nucleoplasmin S (S for somatic), and cross-reacts with monoclonal antibodies used to identify protein XLNO-38. However, sequence information will be required to determine the relationship of nucleoplasmin S to nucleoplasmin and protein XLNO-38.

## The Chaperonins

The chaperonins were originally defined (5) as a class of sequence-related molecular chaperones found in all bacteria examined (including the eubacteria, archaebacteria, cyanobacteria, and rickettsiae), in all mitochondria examined (including those from yeast, *Tetrahymena, Drosophila, Xenopus, Zea mais* leaves, and human cells), and in all plastids examined (including chloroplasts, chromoplasts, and etioplasts). The degree of sequence relatedness within the chaperonin group is high; for example the yeast mitochondrial chaperonin 60 protein shares 45% identical residues with the wheat plastid chaperonin 60, and 54% identical residues with the *E. coli* chaperonin 60, so it is probable that these proteins are true evolutionary homologues. The proteins from bacteria and plastids now known as chaperonins have been studied over many years in different laboratories under a variety of names (Table 1). The distribution of the chaperonins, coupled with their high degree of sequence identity, is consistent with an ancient origin for this group, and with their appearance in mitochondria and plastids via endosymbiosis (57, 58). Recently, sequence comparisons have revealed a lesser similarity between the chaperonins and the protein TCP-1, encoded in the *t*-complex of mouse chromosome 17 (9, 59). This protein of previously unknown function occurs in the cytosolic compartment of animal cells (V. Lewis, K. Willison, unpublished data), and a monoclonal antibody against it detects a protein with a 60-kilodalton subunit in the cytosol of *Pisum sativum* leaves (G. M. F. Watson, N. H. Mann, R. J. Ellis, unpublished data). Presumably the TCP-1 protein performs the same chaperone functions in the cytosolic compartment of eukaryotic cells that its relatives are known to perform inside bacteria, mitochondria, and plastids, and this possibility warrants investigation.

The chaperonins are all abundant constitutive proteins that increase in amount after stresses such as heat shock (10), bacterial infection of macrophages (60), and an increase in the cellular content of unfolded proteins (61). This stress response makes the chaperonins potent antigens in a variety of infections (62), and their presence in all organisms raises the potential danger of autoimmune disease (63, 64). The chaperonins are sometimes referred to as the heat shock 60 group of proteins, but we suggest that this is a poor name because it emphasizes just one specialized aspect of the broader roles of these proteins.

There are two types of chaperonin that are sequence-related to each other. The larger type is called chaperonin 60 (cpn 60), since its subunit $M_r$ is about 60,000, while the smaller type is called chaperonin 10 (cpn 10) with a subunit $M_r$ of about 10,000. Comparison of sequences of chaperonin 10 from four bacterial species with those of chaperonin 60 from eight species (including plastid and mitochondrial representatives) reveals an amino acid match in 58 out of the total 103 residues of chaperonin 10; a match is scored whenever an amino acyl residue occurs in the same position in at least one chaperonin homologue and in at least one chaperonin 60 homologue (65). The region of the chaperonin 60 sequence that shows this similarity to the entire chaperonin 10 sequence extends from residue 101 of the chaperonin 60 sequence; the remainder of the chaperonin 60 sequence shows no detectable similarity to the chaperonin 10 sequence so the evolutionary relationship of the two chaperonins remains unclear. Chaperonin 60 purified from bacteria (66), plastids (67), or mitochondria (68) is an oligomeric protein with a distinctive structure of 14 subunits, arranged in two rings of seven subunits each (the "double-donut"). Chaperonin 60 from *Escherichia coli* (originally termed groEL) and mitochondria is a homo-oligomer, but the plastid chaperonin 60 contains equal amounts of two highly related polypeptides termed *alpha* and *beta* (65, 69). The chaperonin 10 type was until recently reported only from bacteria, where it occurs as an oligomer of seven identical subunits arranged in a single ring (40). However, a functional and structural equivalent to chaperonin 10 has been partially purified from the mitochondrial fraction of beef and rat liver (70). Evidence is accumulating to support the view that these two chaperonins interact in a functional manner to mediate polypeptide assembly inside bacteria and mitochondria. It is therefore likely that a chaperonin 10–like protein occurs inside plastids and perhaps also in the cytosol of eukaryotic cells.

The bacterial and plastid chaperonin 60 oligomers respond to added MgATP by hydrolyzing it and dissociating reversibly to smaller forms; bacterial and mitochondrial chaperonin 10 bind to chaperonin 60 in the presence of MgATP and suppress the ATPase activity of the latter (38, 70, 71). These responses to ATP are involved in the mechanism of action of the chaperonins; information about this mechanism is becoming available for the bacterial and mitochondrial chaperonins.

THE BACTERIAL CHAPERONINS    The best-studied example is the chaperonin 60 of *E. coli*, whose function was first identified by its requirement for the assembly of bacteriophages lambda, T4, and T5 (40). The first step in the assembly of bacteriophage lambda is the formation of an oligomeric structure of 12 identical phage-encoded polypeptides. The correct formation of this preconnector, the basic structure on which head phage proteins assemble, involves the intermediate formation of a 1:1 noncovalent complex between a phage protein monomer and the chaperonin 60 oligomer (72, 73). In the absence of chaperonin 60 function, the head proteins of phage T4 form large insoluble aggregates that associate with the cell membrane (74). The function of chaperonin 10 is also required for correct phage head formation, since it is involved in the release of the phage protein from chaperonin 60, but neither chaperonin forms part of the assembled phage.

The two chaperonins have been implicated in a number of processes in uninfected bacterial cells, and are required at all temperatures (40). For example, genetic evidence suggests a role in DNA replication (75, 76), cell division (77), and protein secretion (78, 79). Increased levels of the two chaperonins in *E. coli* will suppress a number of temperature-sensitive mutations that cause defects in enzymes of intermediary metabolism and protein secretion (80). Such proteins include Sec Y, which is an integral membrane protein, suggesting that chaperonins may be involved in the assembly of membrane proteins as well as soluble proteins. High levels of the two chaperonins have also been shown to enhance the assembly of the foreign oligomeric protein rubisco into a functional enzyme when the latter is expressed in *E. coli* from a cloned cyanobacterial gene (81). It seems that by increasing the amount of the two chaperonins in the *E. coli* cell, either by heat shock, overexpression from a multicopy plasmid, or the presence of unfolded protein (61), the folding capacity can be enhanced, allowing the cell to cope with an increased demand for folding activity. This conclusion should be borne in mind by biotechnologists experiencing difficulty in persuading bacterial cells to produce large amounts of foreign proteins in the required active form (82).

In most cases it has been found that processes mediated by chaperonin 60 in vivo also require the function of chaperonin 10, an observation consistent with the ability of the two chaperonins to form a complex with each other (40). However, in vitro studies have shown that chaperonin 60 alone is capable of maintaining precursor polypeptides in conformations compatible with subsequent transport across membranes (28). The binding of the precursor molecule to the chaperonin 60 oligomer may either retard folding of the precursor or fix its folded state, thereby gaining time for the precursor to interact correctly with the transport machinery. In this model the precursor is released from chaperonin 60 by some component of the transport machinery rather than by chaperonin 10.

Recently an in vitro system employing purified components has been developed that demonstrates directly the requirement of the two bacterial chaperonins for protein assembly (8, 38, 39). Rubisco purified from *Rhodospirillum rubrum* is a dimeric protein whose active sites involve both subunits so that enzymic activity requires correctly assembled dimers. Treatment with high concentrations of guanidine HCl or urea denatures the dimer to inactive monomers containing little secondary structure. When the inactive protein solution is diluted 200-fold into buffer lacking denaturant at 25°C, less than 2% of the original rubisco activity appears after incubation for 2 h and less than 10% after 24 h; at high protein concentrations insoluble aggregates form in a few minutes. However, if the temperature of incubation is reduced below 25°C, the recovery of active enzyme increases, until all the activity is recovered at 10°C and below on incubation for 24 h. If the buffer at 25°C contains a molar excess of *E. coli* chaperonin 60 oligomer to rubisco monomer, a stable 1:1 complex forms between the rubisco monomer and the oligomeric form of chaperonin 60. No other cellular factors are required for this binary complex to form, and it is stable for at least 16 h at 25°C. Addition of MgATP, potassium ions, and chaperonin 10 then results in the discharge of this complex, and the recovery of up to 80% of the original rubisco activity in the form of dimers. The number and order of steps in this latter process are not known, so it is not clear if the chaperonins are required in the dimerization step as well as in the folding step. What is clear, however, is that the formation of a binary complex between chaperonin 60 and some folding intermediate of rubisco is an obligatory step in the chaperonin-dependent refolding pathway. This folding intermediate is also observed in the spontaneous folding of rubisco at 15°C and can be trapped by chaperonin 60 into a stable complex, thus inhibiting the spontaneous folding process. A similar observation has been made in the refolding of pre*beta*-lactamase; however, in this system retardation of refolding is reversed by ATP hydrolysis in the absence of chaperonin 10 (43). Recent work has extended the findings made with the rubisco refolding assay to citrate synthase from pig heart mitochondria, the reactivation of which after dilution from guanidine HCl also requires the presence of the two bacterial chaperonins and MgATP (103).

Chloroplast and mitochondrial chaperonin 60 will partially replace bacterial chaperonin 60 in the rubisco refolding assay (39), while the mitochondrial chaperonin 10 will partially replace the bacterial chaperonin 10 (70). Addition of the chaperonins at temperatures below 25°C stimulates the spontaneous renaturation rate by tenfold. While it is likely that the chaperonins act catalytically in vivo, this has not been demonstrated so far in this in vitro system because the tendency of unfolded rubisco subunits to aggregate unless mopped up immediately by chaperonin 60 is so great.

These in vitro observations confirm the proposal (6) that chaperones convey no steric information for protein assembly, but rather act kinetically to assist self-assembly. The development of such in vitro reconstitution systems will pave the way for a more detailed understanding of the steps in chaperonin action. It should be possible to use optical techniques to monitor changes in the environment of suitable chromophores present in the renaturing polypeptide to supplement information about the nature of the chaperonin-ligand interaction that will eventually come from structural studies of crystals of these complexes.

The general picture that is emerging for the function of the bacterial chaperonins has been summarized by Georgopoulos & Ang (40) as follows. Chaperonin 60 binds to unfolded forms of many polypeptides during processes such as protein synthesis and protein transport. This binding maintains the polypeptides in a state that (a) prevents them from misfolding, (b) assists them in arriving at their correct intra- and inter-molecular folding pattern, (c) allows them to be more readily transported, and (d) permits proteases to degrade them. For some of these processes chaperonin 10 and MgATP are also required. Whether the chaperonins also bind to the interactive surfaces of folded subunits that have transiently dissociated from a larger complex as part of the function of that complex remains to be established, but such binding might account for the suppression of mutations in proteins of the bacterial DNA replication complex by increases in the cellular content of the chaperonins (75, 76).

THE PLASTID CHAPERONINS    The plastid chaperonin 60 was discovered in the author's (RJE) laboratory during studies on the biogenesis of the chloroplast enzyme rubisco, which catalyzes the first step in the photosynthetic pathway of carbon dioxide fixation (8, 9, 83). The chloroplast rubisco enzyme consists of eight large subunits that are synthesized inside the chloroplast and eight small subunits that are imported into the chloroplast after synthesis as higher-molecular-weight precursors in the cytosol. From the observation made by Roger Barraclough in 1980, that rubisco large subunits newly synthesized by isolated intact chloroplasts associate with a large oligomeric chloroplast protein prior to their assembly into the rubisco holoenzyme, it was hypothesized that the binding of the large subunits to this chloroplast protein may be an obligatory step in the assembly of the rubisco enzyme, thus suggesting that the latter is not a spontaneous process (83). The binding of newly synthesized rubisco large subunits to the oligomeric chloroplast protein resembles that of denatured *Rhodospirillum rubrum* rubisco subunits to the bacterial chaperonin 60 in its stoichiometry and stability. It was later shown that imported rubisco small subunits also bind noncovalently to the same chloroplast protein (84, 85). Since then it has been found from cDNA sequence determinations that this chloroplast protein is closely related to both the groEL protein of *E. coli* and the heat shock protein 60 of mitochondria, a finding that led to the new name of the chaperonins for this group (5).

Compelling evidence that the plastid chaperonin 60 is obligatorily involved in the assembly of plant rubisco has proved elusive; the best evidence in favor of this view is the inhibition by plastid chaperonin 60 antibodies of the appearance of in vitro-synthesized rubisco large subunits in an oligomeric form comigrating with pre-existing rubisco holoenzyme in chloroplast extracts (86, 87).

The plastid chaperonin 60 contains equal amounts of two polypeptides called *alpha* and *beta,* which show about 50% sequence identity (65, 88). The arrangement and distribution of these two polypeptides within the chaperonin 60 oligomer is not known. Each polypeptide is encoded by a small number of nuclear genes, whose transcription is slightly enhanced by heat shock. Etioplasts also contain plastid chaperonin 60, and the amount of this protein increases when etiolated shoots of *Pisum sativum* are exposed to light (89).

It is likely that plastid chaperonin 60 fulfills a multifunctional role within this organelle, analogous to bacterial chaperonin 60. A wide range of polypeptides are found associated with chaperonin 60 after transport as precursors into isolated intact chloroplasts; these include foreign polypeptides such as chloramphenicol acetyltransferase, as well as glutamine synthetase and the *beta* subunit of the thylakoid ATP synthase. However, not all imported polypeptides associate with the chaperonin 60 in a detectable fashion, e.g. ferredoxin and superoxide dismutase, indicating some specificity of binding (90). It is possible that the presequence of these imported polypeptides is either required for or enhances the binding of the polypeptide to chaperonin 60; mature rubisco small subunit, for example, does not bind to chaperonin 60. It has been reported that both the presequence and regions within the mature *E. coli* outer membrane protein A (omp A) participate in the recognition of the *E. coli* chaperonin 60 in vitro (28). Release of the imported chloroplast polypeptides by the plastid chaperonin 60 has been shown to require the hydrolysis of ATP (90). It is highly probable that plastids contain a form of chaperonin 10 (70), but no purification has yet been reported.

The range of plastid processes requiring chaperonin 60 has yet to be explored, but by analogy with the bacterial situation may include plastid DNA replication, plastid division, and the assembly of oligomeric complexes such as the photosystems of the thylakoid membranes. The fact that some photosystem polypeptides are transported into the thylakoid lumen at the non-appressed regions, where they are transported for some distance before combining with other polypeptides anchored in appressed regions, suggests another potential intracellular site for chaperone action (91); by analogy with the endoplasmic reticulum however, any chaperones functioning in the thylakoid lumen may be members of the hsp 70 class rather than the chaperonin class.

Another obvious goal for future research is the development of an in vitro system for the renaturation of the chloroplast rubisco enzyme from denatured subunits, based on the success that has been achieved with the simpler bacterial rubisco (8, 38, 39). Crop plant rubisco is a potential target for improvement by genetic engineering in view of its major influence on crop productivity, but all attempts to produce active mutant forms of such rubiscos in *E. coli* have so far been unsuccessful, apparently because of the failure of the rubisco subunits to be released from the bacterial chaperonin 60 (8). The development of an in vitro reconstitution system for crop plant rubisco using the plastid chaperonins would overcome this block.

THE MITOCHONDRIAL CHAPERONINS The most recently identified chaperonin 60 is the one located in the matrix of mitochondria from plant, animal, and protistan cells (68, 92, 93). The gene encoding the mitochondrial chaperonin 60 is part of the nuclear genome in yeast, and as in *E. coli*, the protein is essential for cell viability at all temperatures. The concentration of this protein increases twofold when *Tetrahymena* cells are exposed to high temperatures (94).

The proposed requirement for chaperones in the refolding and oligomerization of polypeptides that are transported across membranes (5, 7, 68) was elegantly confirmed with the aid of a yeast strain carrying a mutation in the mitochondrial chaperonin 60 gene. In this temperature-sensitive mutant *(mif4)*, human ornithine transcarbamylase fails to form an enzymatically active trimer in the mitochondrial matrix, but transport and processing of the precursor polypeptide is unaffected (95). Similarly, there are defects in the assembly of the mitochondrial ATPase complex in this mutant; the imported *beta* subunit fails to assemble into the ATPase complex, while two other subunits normally destined for the intermembrane space stay in the matrix because of a failure of the mechanism that re-exports them from the matrix (95). Analogous to the observations made with the chloroplast chaperonin 60, the mitochondrial chaperonin 60 is capable of binding both to polypeptides newly synthesized inside mitochondria (93) and to a number of unrelated proteins that are imported into isolated mitochondria (96). Whereas imported polypeptides bound to partially purified chaperonin 60 are sensitive to protease and hence regarded as unfolded, the addition of ATP reduces their protease sensitivity without leading to release from the chaperonin; this observation suggests that folding takes place on the surface of the chaperonin 60 oligomer (96). Subsequent release of this folded polypeptide presumably requires another factor, the most likely candidate being chaperonin 10. The report that a protein partially purified from liver mitochondrial fractions can substitute for the *E. coli* chaperonin 10 function in the in vitro renaturation of bacterial rubisco strongly suggests the existence of a chaperonin 10 homologue in mitochondria (70).

These initial findings have sparked more detailed studies on the involvement of chaperonin 60 and other chaperones in protein transport and assembly in mitochondria, some of which are leading to a reassessment of earlier conclusions. Thus it used to be thought that ATP is required to unfold precursor proteins prior to their transport across the mitochondrial membranes, as well as to provide the energy for transport across contact sites. Recent findings suggest instead that ATP is required only to prevent the formation of incorrectly folded structures in the cytosol in conjunction with the cytosolic heat-shock 70 proteins (hsp 70), and to release imported polypeptides from binding to chaperonin 60 and the mitochondrial hsp 70 chaperone after import (32). The new model postulates that it is the binding to the matrix hsp 70 chaperone of the polypeptide that is traversing the contact site that provides the driving force for transport. This binding also facilitates the unfolding of the precursor polypeptide on the cytosolic side of the contact site and provides both the energy and the vectorial aspects of the transport process—in other words this binding pulls the polypeptide into the mitochondrion. The hsp 70 chaperone then mediates the transfer of the imported polypeptide to the matrix chaperonin 60 for correct assembly to be assisted (97). There is some evidence to support such a model for animal mitochondria; many newly synthesised mitochondrial proteins made by HeLa cells are transiently attached to either hsp 70 or the chaperonin 60, as judged by immunoprecipitation of extracts of pulse-labeled cells; this binding is released by incubation with ATP. Some evidence was also obtained for the existence of an ATP-dissociable complex of chaperonin 60 and hsp 70 (L. A. Mizzen, A. N. Kabiling, W. J. Welch, submitted for publication).

If all imported mitochondrial proteins need the aid of the chaperonins in order to assemble, it is predicted that imported chaperonin polypeptides would also require the assistance of the already-present chaperonin 60 oligomer. This appears to be the case; when synthesis of the wild-type chaperonin 60 is induced in the yeast *mif4* mutant under conditions where the mitochondrial chaperonin function is absent, the imported chaperonin 60 polypeptides fail to form the characteristic double-donut structure (97a). In other words, functional preexisting chaperonin 60 is required inside the mitochondrion in order that new chaperonin 60 can be assembled from imported subunits. This requirement can be viewed as one aspect of the principle that mitochondria do not arise by self-assembly but only by the division and growth of existing mitochondria. Chaperonin 60 from *E. coli,* on the other hand, is capable of self-assembly after denaturation with urea in vitro (97b). This self-assembly of bacterial chaperonin 60 requires MgATP and is stimulated by chaperonin 10. These observations are interpreted to mean that the chaperonin 60 oligomers that are formed initially then self-chaperone the assembly of the remaining monomers.

A more speculative function for mitochondrial chaperonin 60 has recently been proposed (98, 115). Mutant lines of Chinese hamster ovary cells were selected for their resistance to anti-mitotic drugs such as podophyllotoxin, and a number of these were found to contain electrophoretic variants of either the mitochondrial chaperonin 60 (called P1 by Gupta) or the constitutive cytosolic hsp 70 protein (called P2). Podophyllotoxin is believed to act by inhibiting the formation of tubulin dimers and so preventing the assembly of microtubules. Gupta suggests that hsp 70 facilitates the transfer of newly synthesized tubulin polypeptides into the mitochondrial matrix where the chaperonin 60 assists their assembly into dimers. These dimers are then exported without unfolding into the cytosol for use in microtubule assembly. The processes of microtubule assembly and disassembly are in principle excellent candidates for assistance by molecular chaperones, since they involve the transient exposure of interactive protein surfaces, while the in vitro assembly of tubulin dimers, as distinct from that of microtubules, has never been reported. However, the presumed cytosolic chaperonin 60 homologue, protein TCP-1, would seem a more plausible candidate for such a chaperone than the homologue located inside the mitochondrion, and there is no evidence that tubulin polypeptides are imported into mitochondria. The model also possesses two features for which there are no precedents; these are the export of a mitochondrial protein from the matrix into the cytosol and the passage of a dimeric protein across the mitochondrial membrane without disassembly. However, this model has several testable features and these should be explored, since progress in science often comes from pursuing the unexpected.

## Heat-Shock Proteins 70 and 90

The molecular chaperone concept can accommodate the functions of stress proteins if it is assumed that the primary effect of stress is to cause the appearance of interactive surfaces that are recognized by chaperones. Thus heat shock causes proteins to denature and form incorrect aggregates, while heat-shock proteins inhibit these processes by binding to the interactive surfaces exposed by high temperature. The stress response can thus be viewed as an amplification of a preexisting function that all cells require for their operation under nonstress conditions.

Recent studies of the hsp 70 and 90 classes of stress protein have been reviewed and support the notion that these proteins act as molecular chaperones (35, 104). Members of the hsp 70 class occur in both prokaryotes (Dna K protein in *E. coli*) and in several compartments of eukaryotic cells, e.g. several members in the cytosol, including the clathrin-uncoating ATPase, the BiP protein in the endoplasmic reticulum, and a member in the mitochondrial matrix. There is evidence that these proteins can interact with polypeptides during a variety of assembly processes in such a way as to prevent the formation of nonfunctional structures. For example, the BiP protein stabilizes nascent unfolded polypeptides as they emerge into the lumen of the endoplasmic reticulum during cotranslational protein transport (44), while the cytosolic hsp 70 proteins encoded by the *SSA1* and *SSA2* genes of yeast maintain precursor polypeptides destined to enter the endoplasmic reticulum or mitochondrion in a translocation-competent conformation (105–107). The hsp 70 member in the mitochondrial matrix is required for protein translocation and the correct folding of proteins imported into the matrix (97).

Pelham suggested in a seminal review (108) that heat-shock proteins not only mediate assembly processes but also promote the disassembly of proteins that have been damaged as a result of stress. Cycles of binding of a heat-shock protein to interactive surfaces exposed by the stress were suggested to be followed by ATP-mediated release, which triggers conformational changes in the denatured protein that favor correct reassembly. No report has yet appeared that any chaperone can mediate the reappearance of the original biological activity from insoluble protein aggregates in vitro, but a recent remarkable finding shows that the Dna K protein from *E. coli* causes the reappearance of RNA polymerase activity when incubated with heat-denatured inactive polymerase and MgATP (109).

Evidence is emerging that suggests that members of the hsp 90 class act as molecular chaperones in the mechanism of signal transduction by steroid receptors. In the absence of steroid hormones, the steroid receptor is bound to hsp 90 and is unable to activate transcription of steroid-controlled genes. Addition of steroid hormone displaces the bound hsp 90 and produces a receptor capable of activating transcription (110–113). However, this effect is not just one of steric hindrance of the DNA-binding site by hsp 90 since genetic studies in yeast suggest that receptors can be activated by hormone only if they are bound to hsp 90 initially. In other words, hsp 90 facilitates the subsequent response of the receptor to the hormone by binding to the receptor and allowing it to maintain an activatable conformation (114).

## IMPLICATIONS AND SPECULATIONS

The growing evidence for the involvement of several distinct classes of molecular chaperone in a variety of fundamental cellular processes is forcing a re-examination of some basic concepts. Self-assembly in the strict sense may be much more limited than the current paradigm implies, and it may have to be replaced by assisted self-assembly as the predominant process for the biogenesis of protein-containing structures. It follows that the evolution of these structures must involve not only the amino acyl sequences of the structural components but also those of their chaperones. The involvement of chaperones in the assembly of many proteins must be taken into account by those endeavoring to "crack the second half of the genetic code."

There are also implications of the molecular chaperone concept for medicine and biotechnology. Besides the fact that those molecular chaperones that are stress proteins are also dominant immunogens in human bacterial infections and are involved in certain autoimmune diseases (62–64), there is the possibility that chaperone diseases exist in which the correct assembly of a particular protein fails because of changes in its sequence that affect recognition by the chaperone, or because of changes in the sequence of the chaperone itself; the mitochondrial myopathies are obvious candidates here. Diseases with such a cause may in the future become treatable by overexpression of the appropriate chaperone from inserted genes, analogous to the suppression by this means of certain mutations affecting protein assembly in *E. coli* (80). The fact that some bacteriophages require the chaperonins of *E. coli* in order to replicate raises the possibility that some viruses of plant and animal cells also require the assistance of host chaperones for their replication. This possibility should be explored to see if it is possible to mutate these chaperones to the disadvantage of viral replication without harming the host cell.

There is much commercial interest in expressing genes for useful proteins in heterologous systems, which offer advantages such as cost, purity, and the readiness with which structural alterations in the proteins can be made. However, production of recombinant proteins in bacteria is limited in many cases by the failure of the protein to assemble correctly into an active conformation; a common problem is that the protein forms insoluble inclusion bodies (99, 100). The molecular chaperone concept suggests that researchers encountering such problems should re-examine the biogenesis of the protein of interest by the cells in which it occurs naturally to determine whether chaperones are involved. Such re-examination should include short pulse-label experiments, since chaperones are typically involved early in biogenesis, and the use of nondenaturing analytical techniques that allow the detection of complexes linked by noncovalent bonds. If chaperones are found to be involved, consideration should be given to the desirability of co-expressing the genes for this chaperone in the same heterologous cells that are making the protein of interest in the hope that correct assembly will be favored. It may also be possible to develop defined in vitro systems containing chaperones that will produce active proteins from denatured inclusion bodies, analogous to the one used with such success for bacterial rubisco (8, 38, 39).

There are two key questions to be addressed by future research on molecular chaperones. Firstly, what is the structural basis by which a given chaperone recognizes and binds to some feature(s) present in a wide variety of unrelated proteins but which is accessible only in the early stages of assembly? Studies on the binding of short synthetic peptides to chaperones are already shedding some light on this problem (101), but the final answer will come from studies of the crystal structures of chaperones containing bound polypeptide ligands. Secondly, how do bound chaperones exert their function? It is our current view that chaperones work by inhibiting incorrect assembly pathways, but the answer to this question will come from the development of in vitro assembly systems using pure components whose conformations can be readily measured.

ACKNOWLEDGMENTS

We thank the following for generous access to unpublished data: A. A. Gatenby, R. S. Gupta, S. M. Hemmingsen, A. L. Horwich, R. Jaenicke, V.

Lewis, N. H. Mann, W. Neupert, N. Pfanner, G. M. F. Watson, and W. J. Welch; and the Science and Engineering Research Council for financial support.

*Literature Cited*

1. Jaenicke, R. 1988. *Protein Structure and Protein Engineering,* pp. 16–36. Mossbach Colloq. 39. Berlin/Heidelberg: Springer-Verlag
1a. Jaenicke, R. 1991. *Biochemistry.* In press
2. Watson, J. D., Hopkins, N. H., Roberts, J. W., Steitz, J. A., Weiner, A. M. 1987. *Molecular Biology of the Gene,* 1:159. Menlo Park, Calif: Benjamin/Cummings. 744 pp. 4th ed.
3. Anfinsen, C. B. 1973. *Science* 181:223–30
4. Ellis, R. J. 1987. *Nature* 328:378–79
5. Hemmingsen, S. M., Woolford, C., van der Vies, S. M., Tilly, K., Dennis, D. T., et al. 1988. *Nature* 333:330–34
6. Ellis, R. J., Hemmingsen, S. M. 1989. *Trends Biochem. Sci.* 14:339–42
7. Ellis, R. J., van der Vies, S. M., Hemmingsen, S. M. 1989. *Biochem. Soc. Symp.* 55:145–53
8. Gatenby, A. A., Ellis, R. J. 1990. *Annu. Rev. Cell. Biol.* 6:125–49
9. Ellis, R. J. 1990. *Science* 250:954–59
10. Ellis, R. J., ed. 1990. *Seminars in Cell Biology,* Vol. 1. Philadelphia: Saunders Scientific. 72 pp.
11. Rothman, J. E. 1989. *Cell* 59:591–601
12. Miller, A. 1984. *Developmental Control in Animals and Plants,* ed. C. F. Graham, P. F. Wareing, pp. 373–95. Oxford: Blackwell Scientific. 519 pp. 2nd ed.
13. Fischer, G., Schmidt, F. X. 1990. *Biochemistry* 29:2205–12
14. Jaenicke, R., Rudolph, R. 1989. *Protein Structure: A Practical Approach,* ed. T. E. Creighton, pp. 191–223. Ithaca, NY: ILR Press
15. Creighton, T. E. 1990. *Biochem. J.* 270:1–16
16. Jaenicke, R. 1987. *Prog. Biophys. Mol. Biol.* 49:117–237
17. Klug, A. 1979. *Harvey Lect.* 74:141–72
18. Nomura, M. 1973. *Science* 179:864–73
19. Creighton, T. E. 1984. *The Proteins,* p. 312. New York: Freeman. 515 pp.
20. Ellis, R. J., van der Vies, S. M., Hemmingsen, S. M. 1987. *Plant Molecular Biology, Life Sci., NATO-ASI Ser. A,*

140:33–40, ed. D. von Wettstein, N-H. Chua. New York: Plenum. 697 pp.
21. Jaenicke, R., Rudolph, R. 1986. *Methods Enzymol.* 131:218–50
22. Petersen, N. S., McLaughlin, C. S. 1973. *J. Mol. Biol.* 81:33–45
23. Bochkareva, E. S., Lissin, N. M., Girshovich, A. S. 1988. *Nature* 336:254–57
24. Beckman, R. P., Mizzen, L. A., Welch, W. J. 1990. *Science* 248:850–54
25. Purvis, I. J., Bettany, A. J. E., Santiago, T. C., Coggins, J. R., Duncan, K., et al. 1987. *J. Mol. Biol.* 193:413–17
26. Meyer, D. I. 1988. *Trends Biochem. Sci.* 13:471–74
27. Rothman, J. E., Kornberg, R. D. 1986. *Nature* 322:209–10
28. Lecker, S., Lill, R., Ziegelhofer, T., Georgopoulos, C., Bassford, P. J., et al. 1989. *EMBO J.* 8:2703–9
29. Weiss, J. B., Ray, P. H., Bassford, P. J. 1988. *Proc. Natl. Acad. Sci. USA* 85:8979–82
30. Hartl, F-U., Neupert, W. 1990. *Science* 247:930–38
31. Pelham, H. R. B. 1989. *EMBO J.* 8:3171–76
32. Pfanner, N., Rassow, J., Guiard, B., Sollner, T., Hartl, F-U., Neupert, W. 1990. *J. Biol. Chem.* 265:16324–29
33. Alberts, B. M. 1987. *Philos. Trans. R. Soc. London, Ser. B* 317:395–420
34. Rothman, J. E., Schmid, S. L. 1986. *Cell* 46:5–9
35. Morimoto, R. T., Tissieres, A., Georgopoulos, C., eds. 1990. *Stress Proteins in Biology and Medicine,* Monogr. 19. New York: Cold Spring Harbor Lab. 450 pp.
36. Pelham, H. R. B. 1990. See Ref. 35, pp. 287–99
37. Dilworth, S. M., Dingwall, C. 1988. *BioEssays* 9:44–49
38. Viitanen, P. V., Lubben, T. H., Reed, J., Goloubinoff, P., O'Keefe, D. P., et al. 1990. *Biochemistry* 29:5665–71
39. Goloubinoff, P., Christeller, J. T., Gatenby, A. A., Lorimer, G. H. 1989. *Nature* 342:884–89

40. Georgopoulos, C., Ang, D. 1990. See Ref. 10, pp. 19–25
41. Holmgren, A., Branden, C-I. 1989. *Nature* 342:248–51
42. Flaherty, K. M., DeLuca-Flaherty, C., McKay, D. B. 1990. *Nature* 346:623–28
43. Laminet, A. A., Ziegelhoffer, T., Georgopoulos, C., Pluckthun, A. 1990. *EMBO J.* 9:2315–19
44. Gething, M-J., Sambrook, J. 1990. See Ref. 10, pp. 65–72
45. Laskey, R. A., Honda, B. M., Mills, A. D., Finch, J. T. 1978. *Nature* 275:416–20
46. Blow, J. J., Dilworth, S. M., Dingwall, C., Mills, A. D., Laskey, R. A. 1987. *Philos. Trans. R. Soc. London, Ser. B* 317:483–94
47a. Zhu, X., Ohta, Y., Jordan, F., Inouye, M. 1989. *Nature* 339:483–84
47b. Silen, J. L., Agard, D. A. 1989. *Nature* 341:462–64
48. Finley, D., Bartel, B., Varsharsky, A. 1989. *Nature* 338:394–401
49. Verner, K., Schatz, G. 1988. *Science* 241:347–58
50. Walter, P., Lingappa, V. R. 1986. *Annu. Rev. Cell Biol.* 2:499–516
51. Dingwall, C., Laskey, R. A. 1990. See Ref. 10, pp. 11–17
52. Moreau, N., Angelier, N., Bonnanfant-Jais, M. L., Gonnon, P., Kubisz, P. 1986. *J. Cell Biol.* 103:683–90
53. Schmidt-Zachmann, M. S., Hugle-Darr, B., Franke, W. W. 1987. *EMBO J.* 6:1881–90
54. Nomura, M., Held, W. A. 1974. *Ribosomes*, ed. M. Nomura, A. Tissieres, P. Lengyel. Cold Spring Harbor, NY: Cold Spring Harbor Lab.
55. Borer, A. A., Lehner, C. F., Eppenberger, H. M., Nigg, E. A. 1989. *Cell* 56:379–90
56. Cotten, M., Chalkey, P. 1987. *EMBO J.* 6:3945–54
57. Picketts, D. J., Mayanil, C. S. K., Gupta, R. S. 1989. *J. Biol. Chem.* 264:12001–8
58. Gupta, R. S., Picketts, D. J., Ahmad, S. 1989. *Biochem. Biophys. Res. Commun.* 163:780–87
59. Gupta, R. S. 1990. *Biochem. Int.* 20:833–41
60. Buchmeier, N. A., Heffron, F. 1990. *Science* 248:730–32
61. Parsell, D. A., Sauer, R. T. 1989. *Genes Dev.* 3:1226–32
62. Young, D. B., Lathigra, R., Hendrix, R. W., Sweetser, D., Young, R. A. 1988. *Proc. Natl. Acad. Sci. USA* 85:4267–70
63. Young, D. B. 1990. See Ref. 10, pp. 27–35
64. Young, D. B., Mehlert, A., Smith, D. F. 1990. See Ref. 35, pp. 131–65
65. Martel, R., Cloney, L. P., Pelcher, L. E., Hemmingsen, S. M. 1990. *Gene* 94:181–87
66. Hendrix, R. W. 1979. *J. Mol. Biol.* 129:375–92
67. Pushkin, A. V., Tsuprun, V. L., Solovjera, N. A., Shubin, V. V., Evstigneera, Z. G., Kretovitch, W. L. 1982. *Biochim. Biophys. Acta* 704:379–84
68. McMullin, T. W., Hallberg, R. L. 1988. *Mol. Cell. Biol.* 8:371–80
69. Hemmingsen, S. M., Ellis, R. J. 1986. *Plant Physiol.* 80:269–76
70. Lubben, T. H., Gatenby, A. A., Donaldson, G. K., Lorimer, G. H., Viitanen, P. V. 1990. *Proc. Natl. Acad. Sci. USA* 87:7683–87
71. Chandrasekhar, G. N., Tilly, K., Woolford, C., Hendrix, R. W., Georgopoulos, C. 1986. *J. Biol. Chem.* 261:12414–19
72. Murialdo, H. 1979. *Virology* 96:341–67
73. Kochan, J., Murialdo, H. 1983. *Virology* 131:100–15
74. Takano, T., Kakefuda, T. 1972. *Nature New Biol.* 239:34–37
75. Fayet, O., Louarn, J-M., Georgopoulos, C. 1986. *Mol. Gen. Genet.* 202:435–45
76. Jenkins, A. J., March, J. B., Oliver, I. R., Masters, M. 1986. *Mol. Gen. Genet.* 202:446–54
77. Miki, T., Orita, T., Furuno, M., Horiuchi, T. 1988. *J. Mol. Biol.* 201:327–38
78. Phillips, G. J., Silhavy, T. J., 1990. *Nature* 344:882–84
79. Kusukawa, N., Yura, T., Ueguchi, C., Akiyama, Y., Ito, K. 1989. *EMBO J.* 8:3517–21
80. Van Dijk, T. K., Gatenby, A. A., LaRossa, R. A. 1989. *Nature* 342:451–53
81. Goloubinoff, P., Gatenby, A. A., Lorimer, G. H. 1989. *Nature* 337:44–47
82. Gatenby, A. A., Viitanen, P. V., Lorimer, G. H. 1990. *Trends Biotechnol.* 8:354–58
83. Barraclough, B. R., Ellis, R. J. 1980. *Biochim. Biophys. Acta* 608:19–31
84. Ellis, R. J., van der Vies, S. M. 1988. *Photosynth. Res.* 16:101–15
85. Gatenby, A. A., Lubben, T. H., Ahlqvist, P., Keegstra, K. 1988. *EMBO J.* 7:1307–14
86. Cannon, S., Wang, P., Roy, H. 1986. *J. Cell Biol.* 103:1327–35
87. Roy, H. 1989. *Plant Cell* 1:1035–42

88. Hemmingsen, S. M. 1990. See Ref. 10, pp. 47–54
89. Lennox, C. R., Ellis, R. J. 1986. *Biochem. Soc. Trans.* 14:9–11
90. Lubben, T. H., Donaldson, G. K., Viitanen, P. V., Gatenby, A. A. 1989. *Plant Cell* 1:1223–30
91. Anderson, J. M., Andersson, B. 1988. *Trends Biochem. Sci.* 13:351–55
92. Horwich, A. L., Neupert, W., Hartl, F-U. 1990. *Trends Biotechnol.* 8:126–31
93. Hallberg, R. L. 1990. See Ref. 10, pp. 37–45
94. McMullin, T. W., Hallberg, R. L. 1987. *Mol. Cell Biol.* 7:4414–23
95. Cheng, M. Y., Hartl, F-U., Martin, J., Pollock, R. A., Kalousek, F., et al. 1989. *Nature* 337:620–25
96. Ostermann, J., Horwich, A. L., Neupert, W., Hartl, F-U. 1989. *Nature* 341:125–30
97. Kang, P-J., Ostermann, J., Shilling, J., Neupert, W., Craig, E. A., et al. 1990. *Nature* 348:137–43
97a. Cheng, M. Y., Hartl, F-U., Horwich, A. L. 1990. *Nature* 348:455–58
97b. Lissin, N. M., Venyaminov, S. Yu., Girshovich, A. S. 1990. *Nature* 348:339–42
98. Gupta, R. S. 1990. *Trends Biochem. Sci.* 15:415–18
99. Mitraki, A., King, J. 1989. *Bio/Technol.* 7:690–97
100. Schein, C. H. 1989. *Bio/Technol.* 7:1141–49
101. Flynn, G. C., Chappell, T. G., Roth-man, J. E. 1989. *Science* 245:385–90
102. Jindal, S., Dudani, A. K., Singh, B., Harley, C. B., Gupta, R. S. 1989. *Mol. Cell. Biol.* 9:2279–83
103. Buchner, J., Schmidt, M., Fuchs, M., Jaenicke, R., Rudolph, R., et al. 1991. *Biochemistry.* In press
104. Ellis, R. J. 1990. *Nature* 346:710
105. Deshaies, R. J., Koch, B. D., Werner-Washburne, M., Craig, E. A., Schekman, R. 1988. *Nature* 332:800–5
106. Chirico, W. J., Waters, M. G., Blobel, G. 1988. *Nature* 332:805–9
107. Murakami, H., Pain, D., Blobel, G. 1988. *J. Cell Biol.* 107:2051–57
108. Pelham, H. R. B. 1986. *Cell* 46:959–61
109. Skowyra, D., Georgopoulos, C., Zylicz, M. 1990. *Cell* 62:939–44
110. Groyer, A., Schweizer-Groyer, G., Cadepond, F., Mariller, M., Baulieu, E-E. 1987. *Nature* 328:624–26
111. Sanchez, E. R., Meshinchi, S., Tienrungroj, W., Schlesinger, M. J., Toft, D. O., Pratt, W. B. 1987 *J. Biol. Chem.* 262:6986–91
112. Willmann, T., Beato, M. 1986. *Nature* 324:688–91
113. Klein-Hitpass, L., Tsai, S. Y., Weigel, N. L., Allen, G. F., Riley, D., et al. 1990. *Cell* 60:247–57
114. Picard, D., Khurseed, B., Garabedian, M. J., Fortin, M. G., Lindquist, S., Yamamoto, K. R. 1990. *Nature* 348:166–68
115. Gupta, R. S. 1990. *Biochem. Cell Biol.* 68:1352–63

*Annu. Rev. Biochem. 1991. 60:349-400*

# STRUCTURE AND FUNCTION OF SIGNAL-TRANSDUCING GTP-BINDING PROTEINS

*Yoshito Kaziro,[1,2] Hiroshi Itoh,[3] Tohru Kozasa,[4] Masato Nakafuku,[2] and Takaya Satoh[2]*

Institute of Medical Science, University of Tokyo, Shirokanedai, Minato-ku, Tokyo 108, Japan

KEY WORDS:    elongation factor, Ras proteins, G proteins, GTPase cycle, conformational switch.

## CONTENTS

[1]To whom correspondence should be addressed.

[2]Present address: DNAX Research Institute of Molecular & Cellular Biology, 901 California Avenue, Palo Alto, California 94304-1104

[3]Present address: Department of Pharmacology, The University of Texas, SouthWestern Medical Center, Dallas, Texas 75235-9041

[4]Present address: Department of Pharmacology, The University of Illinois at Chicago, Chicago, Illinois 60612

349

0066-4154/91/0701-0349$02.00

# I. PERSPECTIVES AND SUMMARY

Extracellular signals are recognized by receptors on the surface of the cytoplasmic membrane and transmitted through transducers to amplifiers, which synthesize catalytically many molecules of second messengers that reach final targets through many additional steps. Transducers, which regulate the opening and closing of the signalling pathways, are GTP-binding proteins, which we discuss in detail in this review article.

Signal-transducing GTP-binding proteins, or GTPases, are classified largely into two groups. One is the high-molecular-weight or heterotrimeric GTP-binding proteins (G proteins), consisting of three subunits, $\alpha$, $\beta$, and $\gamma$. Another group is the low-molecular-weight, monomeric GTP-binding proteins (often abbreviated as "small Gs"). The two groups, both growing in size, contain a number of subgroups, each of which in turn contains several members. GTP-binding proteins are widely distributed in various tissues and among organisms, and the structures of individual proteins are highly conserved among distant organisms.

The basic mechanism of the reactions catalyzed by GTP-binding proteins is essentially analogous to that proposed for translational factors (1). The GTP-bound form is an active conformation that turns on the transmission of signals, and the hydrolysis of bound GTP to bound GDP is required to shift the conformation to an inactive form, i.e. to turn off the signal transduction.

Interconversion of the GTP-bound form (E·GTP) and the GDP-bound form (E·GDP) is regulated by two mechanisms. The conversion of E·GDP to E·GTP by nucleotide exchange is stimulated by an exchange-promoting protein (EP), which increases the off-rate ($K_{\text{diss.GDP}}$) of GDP from E·GDP. On the other hand, the conversion of E·GTP to E·GDP is accelerated by a protein that stimulates the intrinsic GTPase activity. A GTPase-activating protein (GAP) functions by increasing the first-order rate constant of GTP hydrolysis ($k_{\text{cat.GTP}}$), often by several orders of magnitude.

The above regulatory mechanism, of course, varies with individual GTP-binding proteins. In the classical example of *E. coli* elongation factors, the exchange reaction, EF-Tu·GDP $\rightleftharpoons$ EF-Tu·GTP is stimulated by EF-Ts, and the hydrolysis of GTP bound to EF-Tu by ribosomes (1). In the case of signal-transducing GTP-binding proteins, heterotrimeric G proteins are activated through the ligand·receptor-induced stimulation of the exchange reaction. However, the hydrolysis of bound GTP appears to be determined by the intrinsic (built-in) GTPase activity of G proteins ($\sim 3$–$5$ min$^{-1}$). On the other hand, in *ras* p21 proteins, where the rates of GDP release and GTP hydrolysis are exceedingly small ($<0.01$ min$^{-1}$), both EP and GAP are required for the GTPase cycle to proceed. We discuss later the detailed reaction mechanisms of individual GTP-binding proteins as well as the regulation of the GTPase cycle by signals coming from outside the cells.

Finally, we emphasize that no matter how complex the diverse reactions appear, the basic principle of the function of GTP-binding proteins as molecular switches is universal. All GTP-binding proteins utilize the same cycle of reactions, i.e. activation by GTP binding and relaxation by GTP hydrolysis. They undergo sequential conformational alterations depending on the phosphate potential of their ligand, and their abilities to interact with their cognate macromolecules are altered qualitatively. This cycle is irreversible and unidirectional by virtue of GTP hydrolysis.

The most important problem still remaining is the identification of the downstream targets of GTP-binding proteins. Except for a few cases, the assignment of functions for each individual GTP-binding protein is still under way. For example, in spite of much effort, the immediate upstream and downstream molecules for mammalian Ras proteins are still unknown. The precise role of individual GTP-binding proteins in switching the signalling for cellular growth and differentiation as well as the response to hormones, neurotransmitters, and sensory stimuli remains for further investigation.

## II. INTRODUCTION

The superfamily of GTP-binding proteins consists of several families including (*a*) translational factors, (*b*) heterotrimeric GTP-binding proteins involved in transmembrane signalling processes (abbreviated as G proteins), (*c*) proto-oncogenic *ras* proteins (Ras proteins), (*d*) other low-molecular-weight GTP-binding proteins, including the products of *rab, rap, rho, rac, smg21, smg25, YPT, SEC4, ARF* genes (abbreviated as "small Gs"), and (*e*) tubulins. There are also other metabolic enzymes that interact with GTP (such as succinate thiokinase, and phosphoenolpyruvate carboxykinase), but they are not discussed here.

In this review article, we limit our discussions mainly to the proteins that belong to Groups (*b*) and (*c*), i.e. G proteins and Ras proteins. We describe mainly progress achieved since the previous reviews by Gilman (2) and Barbacid (3) in the *Annual Review of Biochemistry*. This review is not comprehensive; rather, we emphasize our special interest, which is how GTP-binding proteins function as a "molecular switch" that regulates the opening and closing of signal transmission. For a broader view, we recommend the more recent reviews by Neer & Clapham (4), Freissmuth et al (5), Ross (6), Gibbs & Marshall (7), Kaziro (8), Hall (9), and Bourne et al (10). Several monographs have also been published recently; they include those edited by Bosch, Kraal, and Parmeggiani (11), Iyenger and Birnbaumer (12), Houslay and Milligan (13), and Moss and Vaughan (14).

All signal-transducing GTP-binding proteins bind and hydrolyze GTP, properties that are crucial to their function as a molecular switch for diverse cellular functions. Each GTP-binding protein, however, has its own dissociation constant for guanine nucleotides, $K_d$ value for GTP and GDP, and rate constant for GTP hydrolysis ($k_{cat.GTP}$).

GTP-binding proteins undergo two alternate conformations (and reactivities) depending on the phosphate potentials of the ligand (Figure 1A). The GTP-bound form is an active conformation; in this form the protein can recognize and interact with its target molecules. On hydrolysis of the bound GTP to GDP and inorganic phosphate, the conformation as well as the reactivity of the protein is shifted to the GDP-bound form (an inactive form). The conversion of the GDP-bound form to the GTP-bound form is achieved by the exchange of the bound GDP with an external GTP. This is the step in which the energy from outside is fed into this system in the form of GTP. On the other hand, in the relaxation process, the bound GTP is hydrolyzed to GDP and inorganic phosphate. Depending on the kinetic parameters of each GTP-binding protein, factors that stimulate the exchange reaction (EP or exchange-promoting proteins), or factors that increase the rate of hydrolysis of bound GTP (GAP or GTPase-activating proteins) are required.

The above basic mechanism, originally proposed from studies on translational factors (see the next section), has been found in a variety of systems to be dependent on either GTP or ATP. It has been known that replication, recombination, and repair of DNA are dependent on ATP hydrolysis, and a number of DNA-dependent ATPases are involved in these processes (15). These ATPases function in a manner analogous to translation GTPases; i.e. in the presence of ATP, the proteins are bound to DNA, inducing structural changes that are required for the sequential events in the processes of DNA replication, recombination, and repair.

Furthermore, we can find an analogous mechanism in protein phosphorylation and dephosphorylation reactions. Proteins can be activated or inactivated by a covalent phosphorylation. In this case, a protein can undergo, again, two alternate conformational changes, i.e. between phosphorylated and nonphos-

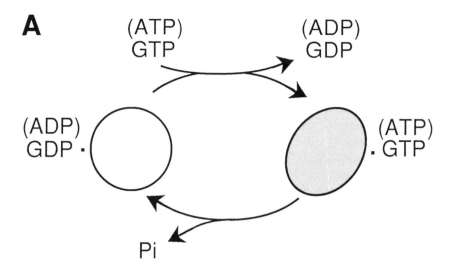

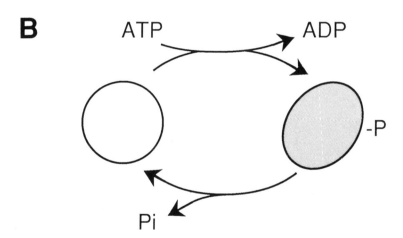

*Figure 1* General reaction mechanism of GTP- (or ATP-) binding proteins (*A*) and protein kinases (*B*). For details, see text.

phorylated forms (Figure 1*B*). As described below, intracellular signals are mostly transmitted via protein-protein interactions. The interactions are modulated by conformational change induced by ligands covalently or noncovalently associated with the proteins. Recently, it was found that in growth regulation, many protein kinases as well as protein phosphatases play an essential role in response to growth factor stimuli.

## III. GENERAL CHARACTERISTICS

In this section, we first outline the classical GTP-utilizing reaction, i.e. the reaction catalyzed by the translational elongation factors Tu and G. Then, we compare this reaction with newly developed systems.

### Translational Factors

Several GTP-binding proteins are known to be involved in protein biosynthesis. These include initiation factor 2 (IF-2), elongation factor Tu (EF-Tu), and elongation factor G (EF-G) for prokaryotes, and initiation factor 2 (eIF-2), elongation factor 1α (EF-1α), and elongation factor 2 (EF-2) for eukaryotes. Below we discuss some general characteristics of the polypeptide chain elongation cycle (1). Figure 2 summarizes the reaction mechanism of

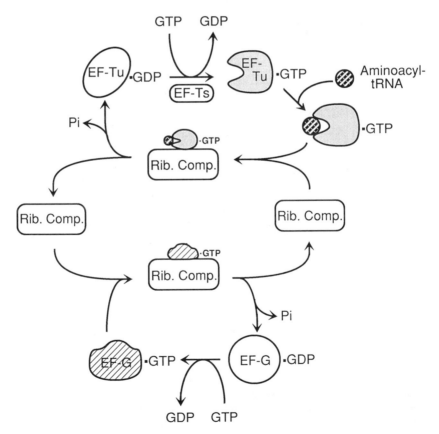

*Figure 2*   Role of EF-Tu and EF-G in polypeptide chain elongation [modified from Kaziro (1)].

two GTP-requiring processes in polypeptide chain elongation. The first reaction, a GTP-dependent binding of aminoacyl-tRNA to the ribosomal A site, is promoted by EF-Tu, a 43-kDa protein, one of the most abundant proteins in *Escherichia coli*. The second reaction, a GTP-dependent translocation of the peptidyl-tRNA·mRNA complex from the A to the P sites on ribosomes, is promoted by EF-G.

EF-Tu has one tightly bound GDP ($K_d = 10^{-9}$M) per mole of protein and is converted to EF-Tu·GTP through displacement of bound GDP with external GTP. This process is promoted by EF-Ts. EF-Tu·GTP is an active form in which EF-Tu can interact with aminoacyl-tRNA and subsequently with ribosomes to form a ribosomal intermediate. Then, the GTP moiety of the complex is hydrolyzed to GDP and inorganic phosphate. The most remarkable feature of this reaction is that the reactivity as well as the conformation of the EF-Tu molecule is reversibly and qualitatively modulated through interaction with GTP or GDP, i.e. by the "phosphate-energy level" of the ligand. EF-Tu·GTP is an active form and EF-Tu·GDP is an inactive form, and the reaction is driven by the conformational change induced by the "high-energy" phosphate. An essentially analogous mechanism has been found for the EF-G-promoted translocation reaction. EF-G·GTP can interact with ribosomes to form a stable complex, whereas EF-G·GDP has no affinity to ribosomes. Interaction of EF-G·GTP with ribosomes promotes the translocation of the peptidyl-tRNA from the A to the P sites of ribosomes and the movement of mRNA by three-nucleotides distance with the noncomitant release of deacylated tRNA from ribosomes. The hydrolysis of GTP is not required up to this step, but is needed later for the release of EF-G from ribosomes.

In the polypeptide chain elongation cycle, EF-Tu and EF-G interact sequentially with ribosomes. Since the binding of EF-Tu and EF-G to ribosomes is mutually exclusive, EF-Tu has to be released from ribosomes before EF-G binds ribosomes and vice versa. The release of EF-Tu and EF-G from ribosomes requires GTP hydrolysis. Consequently, a stoichiometric relationship holds between the number of amino acids incorporated into the polypeptide chain and the number of GTPs hydrolyzed. This is in contrast to G proteins and Ras proteins which, as discussed below, continue to transmit signals as long as they remain in the GTP-bound form. This property is important to the amplification of signals. The hydrolysis of GTP is apparently required for the shut-off of signal transduction. In the absence of GTP hydrolysis, the proteins remain persistently in an active conformation, and the signals become overamplified.

The nature of the conformational change between the GTP- and GDP-bound forms is extensively studied in the case of EF-Tu [for earlier works, see (1)]. Several biochemical and biophysical studies have revealed that a local conformational change restricted to the neighborhood of the active site interacting with aminoacyl-tRNA and ribosomes (near Cys-81) is induced by the ligand change from GDP to GTP (or vice versa) at the nucleotide interaction site, [G site which includes the NKXD sequence; Asn(135)-Lys-Cys-Asp(138)]. X-ray analysis, which has been carried out by several groups for almost 15 years, has demonstrated the core structure of the GTP-binding pocket (16, 17). However, the key structural difference between EF-Tu·GTP and EF-Tu·GDP has not been obtained. As is discussed below, the studies on the 3D structure of Ras protein have been successful in solving the differential conformation between Ras·GDP and Ras·GTP (18, 19).

On binding of the aminoacyl-tRNA·EF-Tu·GTP complex to ribosomes, the GTP-binding site of EF-Tu is converted to the catalytic site for GTP hydrolysis to hydrolyze bound GTP. It would be interesting to see what kind of structural change takes place in EF-Tu·GTP after interaction with ribosomes and how its latent GTPase function becomes activated.

The exchange of GDP bound to EF-Tu with external GTP was thought to proceed by the following reactions:

$$
\begin{array}{l}
\text{EF-Tu} \cdot \text{GDP} \quad + \text{EF-Ts} \rightleftharpoons \text{EF-Tu} \cdot \text{EF-Ts} \quad + \text{GDP} \\
\text{EF-Tu} \cdot \text{EF-Ts} \quad + \quad \text{GTP} \quad \rightleftharpoons \text{EF-Tu} \cdot \text{GTP} \quad + \text{EF-Ts}
\end{array}
$$

$$
\text{sum:} \quad \text{EF-Tu} \cdot \text{GDP} \quad + \text{GTP} \overset{\text{EF-Ts}}{\rightleftharpoons} \text{EF-Tu} \cdot \text{GTP} \quad + \text{GDP}
$$

However, this mechanism has been questioned, since in higher eukaryotes, EF-1$\beta\gamma$ (counterpart of EF-Ts) always occurs as a complex with EF-1$\alpha$, i.e. as an EF-1$\alpha\beta\gamma$ complex. Likewise, in an extreme thermophile, *Thermus thermophilus HB8,* there is a tight complex of EF-Tu·EF-Ts, which does not dissociate in the presence of an excess GTP or GDP. The proposed revised mechanism for GDP exchange is as follows:

$$
\begin{array}{l}
\text{EF-Tu} \cdot \text{GDP} \quad\quad\quad + \text{GTP} \cdot \text{EF-Tu} \cdot \text{EF-Ts} \quad \rightleftharpoons \quad\quad \text{EF-Tu} \cdot \text{GTP} \quad\quad + \text{GDP} \cdot \text{EF-Tu} \cdot \text{EF-} \\
\text{GDP} \cdot \text{EF-Tu} \cdot \text{EF-Ts} + \quad\quad\quad \text{GTP} \quad\quad\quad \rightleftharpoons \quad \text{GTP} \cdot \text{EF-Tu} \cdot \text{EF-Ts} + \quad\quad\quad\quad \text{GDP}
\end{array}
$$

$$
\text{sum:} \ \text{EF-Tu} \cdot \text{GDP} \quad + \quad\quad\quad \text{GTP} \quad\quad \overset{\text{EF-Tu·EF-Ts}}{\rightleftharpoons} \ \text{EF-Tu} \cdot \text{GTP} \quad + \quad\quad\quad \text{GDP}
$$

In support of this revised mechanism, a rapid exchange of the EF-Tu·GTP moiety of GTP·EF-Tu·EF-Ts with EF-Tu·GDP is observed. Also, the dissociation constant for GDP from GDP·EF-Tu·EF-Ts was shown to be about 600-fold greater than that from EF-Tu·GDP (1). This means that the binding of GDP to EF-Tu is weakened by association of EF-Ts. On the other hand, the affinity for GTP of EF-Tu is not affected by interaction with EF-Ts.

## Heterotrimeric G Proteins

Heterotrimeric G proteins, which are involved in a variety of hormonal and sensory signal transductions, consist of three subunits: $\alpha$ (39–52 kDa), $\beta$

(35–36 kDa), and $\gamma$ (7–10 kDa). $\beta$ and $\gamma$ subunits are tightly associated and can be dissociated only by treatment with denaturants.

The GTPase cycle of heterotrimeric G proteins, which is derived from many lines of evidence [reviewed by Gilman, (2)], is represented in Figure 3. Activation of the receptors on the cytoplasmic membrane by extracellular ligands induces the release of GDP, tightly associated with the $\alpha$-subunit of G protein (G$\alpha$), to exchange with GTP. This leads to the dissociation of the receptor-associated G$\alpha\beta\gamma$ complex into G$\alpha$·GTP and G$\beta\gamma$. G$\alpha$·GTP then activates an amplifier or an effector (including an ion channel) to generate catalytically many molecules of the second messenger (including cations). After hydrolysis of bound GDP, G$\alpha$·GDP is released from an amplifier or an effector. Now G$\beta\gamma$, which has a high affinity to G$\alpha$·GDP, can associate to form the GDP·G$\alpha\beta\gamma$ complex, which interacts with the membrane receptors. The modification of G$\gamma$ subunit through isoprenylation (20–22) facilitates the association of G$\alpha\beta\gamma$ with a membrane receptor.

In the above reaction cycle, the ligand-activated receptor plays a role as an EP, facilitating the exchange of G$\alpha$-bound GDP. Therefore, the activity of EP is regulated by extracellular signals. On the other hand, the rate of GTP hydrolysis appears to be determined by an intrinsic "built-in" timer of the G$\alpha$-GTPase activity. The intrinsic $k_{\text{cat.GTP}}$ values of G proteins are about 2–4 min$^{-1}$, much higher than the intrinsic $k_{\text{cat.GTP}}$ value of EF-Tu or Ras proteins. The extent of amplification of G protein–mediated signals appears to be predetermined by the intrinsic GTPase activity of each G protein $\alpha$-subunit.

As is discussed in the following sections, the presence of at least nine distinct genes for G$\alpha$ [G$_s$1$\alpha$, G$_s$2$\alpha$ (or G$_{\text{olf}}\alpha$), G$_i$1$\alpha$, G$_i$2$\alpha$, G$_i$3$\alpha$, G$_o\alpha$, G$_t$1$\alpha$, G$_t$2$\alpha$, and G$_x\alpha$] in mammalian cells is known. Since the G$_s$1$\alpha$ gene and G$_o\alpha$ gene produce 4 and 2, respectively, different species of cDNAs by alternative splicing, the total number of cDNA species for mammalian G$\alpha$ is 13. More recently, four additional cDNAs for G$\alpha$ have been isolated by the polymerase chain reaction technique (23).

## Ras Proteins

*ras* genes were found first as the oncogenes of rat sarcoma virus and later identified as cellular genes responsible for certain human tumors (3). The high degree of conservation of Ras protein structures among eukaryotes points to their essential role in cellular functions. The fact that Ras proteins bind and hydrolyze GTP, as well as their association with cytoplasmic membranes through modification by lipids, suggests that Ras proteins, like G proteins, serve as transducers in transmembrane signalling systems.

Several differences, however, are noted in properties of Ras proteins and G proteins. Ras proteins are monomeric, while G proteins are trimeric. The intrinsic GTPase activity of Ras proteins ($< 0.01$ min$^{-1}$) is much lower than that of G proteins (3–5 min$^{-1}$). Therefore, besides the EP, which is required

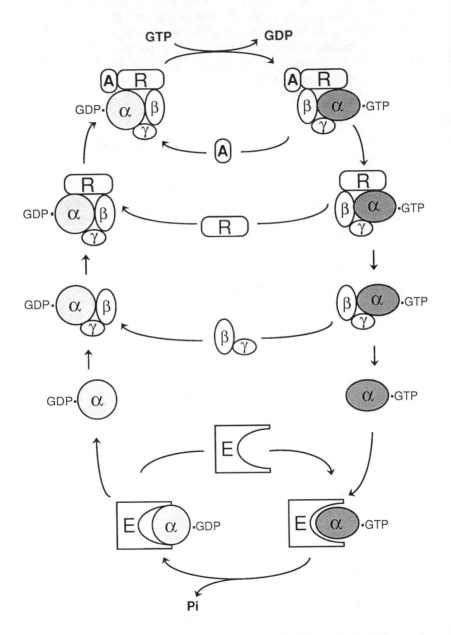

*Figure 3* Function of G proteins in signal transduction. Abbreviations: A, agonist; R, receptor; and E, effector.

for the promotion of the exchange reaction, stimulation by GAP of GTP hydrolysis is essential for Ras protein to continue its GTPase cycle. With purified recombinant Ras proteins, the rate of exchange of bound GDP as well as the hydrolysis of bound GTP is very slow. The EP and GAP for mammalian Ras and the corresponding proteins for yeast Ras are described in detail in Sections VII and VIII.

The function of Ras proteins appears to depend on cell type. In fibroblast cells, Ras·GTP stimulates proliferations, whereas it stimulates the differentiation of rat pheochromocytoma (PC) 12 cells to neurite cells (24, 25). More interestingly, Ras proteins in yeast *Saccharomyces cerevisiae,* Ras1 and Ras2, stimulate adenylate cyclase, as does $G_s\alpha$ in mammalian cells (7). On the other hand, the product of *ras1,* a homolog of *ras* in fission yeast *Schizosaccharomyces pombe,* does not activate adenylate cyclase, but participates in the mating factor signal transduction (26, 27). In *Saccharomyces cerevisiae,* a G protein termed GP1$\alpha$ (product of *GPA1*) participates in the mating factor signal transduction as a signal transducer (see Section V). The function of Ras proteins in mammalian and yeast cells is discussed in detail in Sections VII and VIII.

## *Structural Similarities Between Various GTP-Binding Proteins*

The search for similarities of primary structure between various GTP-binding proteins has revealed the presence of several regions of high degree of homology (see Figures 4 and 5).

(a) P region. This region, with the consensus sequence motif, G-X-X-X-X-G-K-S/T, is common not only in GTP-binding proteins but also in many nucleotide triphosphate-utilizing enzymes. This region is probably involved directly in the hydrolytic process. Mutation of Gly-12 of Ha-*ras* p21 protein to Val or other amino acids results in deficient GTPase activity and increased transforming activity (3). X-ray analysis of EF-Tu (16, 17) and Ras (18, 19, 28), proteins has indicated that this region is close to the $\beta$ and $\alpha$ phosphoryl groups of bound GDP. Also, the amino group of the Lys-16 forms hydrogen bonds with the $\beta$ and/or $\gamma$ phosphate of GTP and with the main-chain oxygens of residues 10 and 11 in the loop (18). The hydroxy oxygen of Ser-17 in Ras protein interacts with $Mg^{2+}$, which binds phosphates of guanine nucleotide.

(b) E region. This region, covering amino acid residues 32–42 of Ras, is called an effector (E) region, the site interacting with the putative effector molecule (3). The mutation in this region of *ras* p21 protein leads to the loss of transforming activity without affecting the GTP-binding property. The sequence in this region is not strongly conserved between different families but only within the same family.

(c) G' region. This region is involved also in the GTPase function in the case of Ras proteins. The consensus sequence motif, D-X-X-G-Q, is highly

(A)

```
                                    G X X X X X G KS/T
 EF-Tu:   13 -  28:  N V G T I G H V D H G K T T L T
 Gsα:     42 -  57:  R L L L L G A G E S G K S T I V
 Gi1α:    35 -  50:  K L L L L G A G E S G K S T I V
 Gi2α:    35 -  50:  K L L L L G A G E S G K S T I V
 Gi3α:    35 -  50:  K L L L L G A G E S G K S T I V
 Goα:     35 -  50:  K L L L L G A G E S G K S T I V
 Gxα:     35 -  50:  K L L L L G T S N S G K S T I V
 GP1α:    43 -  58:  K L L L L G A G E S G K S T V L
 GP2α:   125 - 140:  K V L L L G A G E S G K S T V L
 Ha-ras:   5 -  20:  K L V V V G A G G V G K S A L T
```

(B)

```
                              D X X G Q
 EF-Tu:   80 -  97:  D C P G H A D Y V K N M I T G A A Q
 Gsα:    223 - 240:  D V G G Q R D E R R K W I Q C F N D
 Gi1α:   200 - 217:  D V G G Q R S E R K K W I H C F E G
 Gi2α:   201 - 218:  D V G G Q R S E R K K W I H C F E G
 Gi3α:   200 - 217:  D V G G Q R S E R K K W I H C F E G
 Goα:    201 - 218:  D V G G Q R S E R K K W I H C F E D
 Gxα:    201 - 218:  D V G G Q R S E R K K W I H C F E G
 GP1α:   319 - 336:  D A G G Q R S E R K K W I H C F E G
 GP2α:   296 - 313:  D V G G Q R S E R K K W I H C F D N
 Ha-ras:  57 -  74:  D T A G Q E E Y S A M R D Q Y M R T
```

(C)

```
                              N K X D
 EF-Tu:  127 - 144:  V P Y I I V F L N K C D M
 Gsα:    285 - 296:  I S - V I L F L N K Q D L
 Gi1α:   262 - 273:  T S - I I L F L N K K D L
 Gi2α:   263 - 274:  T S - I I L F L N K K D L
 Gi3α:   262 - 273:  T S - I I L F L N K K D L
 Goα:    263 - 274:  T S - I I L F L N K K D L
 Gxα:    263 - 274:  T S - L I L F L N K K D L
 GP1α:   381 - 392:  T P - F I L F L N K I D L
 GP2α:   358 - 369:  T S - V V L F L N K I D L
 Ha-ras: 109 - 120:  V P - M V L V G N K C D L
```

*Figure 4*  Conserved sequences of GTP-binding proteins. *E. coli* EF-Tu and human c-Ha-*ras* are aligned with G protein α-subunits. Sequences of P (*A*), G' (*B*), and G (*C*) sites are shown.

conserved in most GTP-binding proteins. The Asp-57 of Ras protein as well as the corresponding Asp residue in other GTP-binding proteins is probably coordinated to $Mg^{2+}$ (19, 29, 30). The amide hydrogen of Gly-60 in Ras protein can form a hydrogen bond with the γ-phosphate of GTP. This Gly-60 appears to be involved in conformational change between the GDP- and the GTP-bound states. The mutation of Gln-61 to leucine strongly impairs GTPase activity and potentiates the transforming activity. In recent reports,

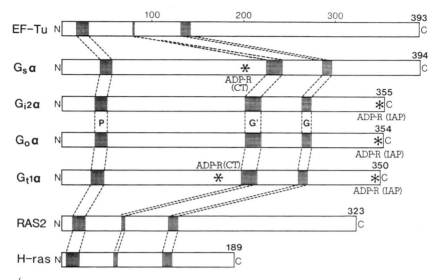

*Figure 5*   Schematic representation of structures of EF-Tu, G$_s\alpha$ (G$_s$1$\alpha$), G$_i$2$\alpha$, G$_o\alpha$ (G$_o\alpha$-1), G$_t$1$\alpha$, *RAS2*, and Ha-*ras* p21. The positions of P, G', and G sites are shown. *, the sites modified by CT and IAP. Abbreviations: ADP-R, ADP-ribose; CT, cholera toxin; and IAP, islet-activating protein (pertussis toxin).

Bourne and his collaborators (31, 32) have demonstrated that the mutation of G protein in this region also gives an oncogenic phenotype in certain cell-types (see Section IV). The sequence in EF-Tu at the corresponding region is D-C-P-G-H; this region had been assigned as the site where the conformational change induced by the ligand change from GDP to GTP takes place (1). Supporting this assignment, comparison of the 3D structures of Ras·GTP and Ras·GDP revealed the most remarkable structural change in this region: in various G protein $\alpha$-subunits, the sequence succeeding D-X-X-G-Q is highly conserved.

(d) G region. This region interacts directly with the guanine ring of GTP. The consensus motif, N-K-X-D, is conserved in all GTP-binding proteins. Earlier studies with elongation factor EF-Tu indicated that the modification of the N-K-C-D sequence at Cys-137 destroyed the interaction with GTP (1). Later, the four residues N-K-C-D were found by X-ray crystal analysis to be situated close to the guanine ring (16). In Ras protein, the carboxylate group of the invariant Asp-119 interacts with the exocyclic amino group and endocyclic nitrogen N1 of the guanine ring (29).

(e) G'' region. This is another region that is closely situated to guanine ring of GTP in the 3D structure of Ras protein. The conserved sequence motif of this region in Ras is E-T-S-A-K. In G proteins, H-(F/M)-T-C-A-(T/V)-D-T may correspond to this region.

**Table 1** Molecular cloning of $G_s$, $G_i$, $G_o$, $G_x$, and $G_t\alpha$ subunit genes and cDNAs from mammalian cells

| DNA library | $G_{s1}\alpha$ | $G_{s2}\alpha$ | $G_{i1}\alpha$ | $G_{i2}\alpha$ | $G_{i3}\alpha$ | $G_o\alpha$ | $G_x\alpha$ | $G_{t1}\alpha$ | $G_{t2}\alpha$ | Ref. |
|---|---|---|---|---|---|---|---|---|---|---|
| Genomic library | | | | | | | | | | |
| Human | $G_s\alpha$ | | | | | | | | | 33 |
| Human | | | $G_{i1}\alpha$ | $G_{i2}\alpha$ | $G_{i3}\alpha$ | | | | | 34 |
| Human | | | | $G_{i2}\alpha$ | | | | | | 35 |
| Human | | | | | | $G_o\alpha$ | | | | 36a |
| Human | | | | | | $G_o\alpha$ | | | | 37 |
| Human | | | | | | | $G_x\alpha$ | | | 38 |
| Mouse | | | | | | | | $G_{t1}\alpha$ | | 39 |
| cDNA library | | | | | | | | | | |
| Human brain | $G_s\alpha$ | | $\alpha_{i-1}$ | | | | | | | 40, 41 |
| Human brain | | | | | | $G_o\alpha$ | | | | 36a |
| Human T cells | | | | $\alpha_{i2}$ | $\alpha_{i3}$ | | | | | 42 |
| Human T cells | | | | | $\alpha_{i-3}$ | | | | | 43 |
| Human monocytes (U-937) | | | $G_i\alpha$ | | | | | | | 44 |
| Human granulocytes (HL-60) | | | | | | $G_x\alpha$ | | | | 45 |
| Human liver | $G_s\alpha$ | | | | | | | | | 46 |
| Human liver | | | | $\alpha_{i-3}$ | | | | | | 47 |
| Human liver | | | | $\alpha_{i-3}$ | | | | | | 48 |
| Human retina | | | | | | | $G_z\alpha$ | | | 49 |
| Bovine brain | $G_s\alpha$ | | | | | | | | | 50 |
| Bovine adrenal gland | $G_s\alpha$ | | | | | | | | | 51 |
| Bovine cerebral cortex | $G_s\alpha$ | | $G_i\alpha$ | | | | | | | 52, 53 |
| Bovine pituitary gland | | | $\alpha_i$ | $\alpha_h$ | | | | | | 54 |
| Bovine retina | | | | | | $G_o\alpha$ | | | | 55 |
| Bovine cerebellum | | | | | | $G_{39}$ | | | | 56 |
| Bovine retina | | | | | | $G_o\alpha$ | | | | 57 |
| Bovine retina | | | | | | | | $T\alpha$ | | 58 |
| Bovine retina | | | | | | | | $T\alpha$ | | 59 |
| Bovine retina | | | | | | | | $Gt\alpha$ | | 60 |
| Bovine retina | | | | | | | | | $T\alpha$ | 61 |
| Hamster insulinoma | | | | | | $\alpha_{o1}$ $\alpha_{o2}$ | | | | 62 |
| Hamster lung fibroblasts (CCL39) | $G_s\alpha$ | | | | | | | | | 63 |
| Rat brain | | | | | | $G_x\alpha$ | | | | 36b |
| Rat glioma cells (C6) | $G_s\alpha$ | | | $G_{i2}\alpha$ | $G_{i3}\alpha$ | $G_o\alpha-1$ | | | | 34, 64 |
| Rat PC12 cells | | | | | | $G_o\alpha-2$ | | | | 37 |
| Rat olfactory epithelium | $G\alpha_s$ | $G_{olf}$ | $G\alpha_{il}$ | $G\alpha_{i2}$ | $G\alpha_{i3}$ | $G\alpha_o$ | | | | 65, 66 |
| Mouse macrophages (PU-5) | $\alpha_s$ | | | $\alpha_i$ | | | | | | 67 |
| Mouse lymphoma cells (S49) | $G_s\alpha$ | | | | | | | | | 68 |
| Mouse lymphoma cells (S49) | $G_s\alpha$ | | | | | | | | | 69 |
| Mouse brain | | | | | | $G_oA\alpha$ $G_oB\alpha$ | | | | 70 |

# IV. MAMMALIAN G PROTEINS

Recent developments in molecular cloning have revealed the existence of multiple G protein $\alpha$-subunits. At least nine genes code for G protein $\alpha$-subunit in mammalian cells, some of which give rise to more than one cDNA by alternative splicing. Table 1 lists the cDNA clones that have been isolated so far. Some molecular and biochemical properties of various $G\alpha s$ are presented in Table 2.

## Molecular Entities and Gene Organizations of G Proteins

$G_s\alpha$    $G_s$ is involved in hormonal stimulation of adenylate cyclase and opening of $Ca^{2+}$ channels. Modification by cholera toxin of $G_s\alpha$ leads to ADP-ribosylation of Arg-201 and to persistent activation of adenylate cyclase. Since there is a specific $G_s\alpha$ expressed only in olfactory cells ($G_{olf}$), we refer to the classical $G_s\alpha$ that is expressed in all kinds of tissues as $G_s1\alpha$, and to the one expressed exclusively in olfactory cells as $G_s2\alpha$. Human $G_s1\alpha$ gene contains 13 exons and 12 introns and spans about 20 kb (33) (Figure 6). Exons 3 and 12 encode the unique sequences of $G_s1\alpha$ at N- and C-terminal regions, respectively. Bray et al (40) isolated four different $G_s1\alpha$ cDNAs (termed $G_s1\alpha-1$ to $-4$) from the human brain and characterized their partial structure. $G_s1\alpha-1$ and $G_s1\alpha-3$ differ in that $G_s1\alpha-3$ lacks a stretch of 45 nucleotides or 15 amino acids coded by exon 3. $G_s1\alpha-2$ and $G_s1\alpha-4$ have three additional nucleotides (CAG) to $G_s1\alpha-1$ and $G_s1\alpha-3$, respectively, 3' to the above 45 nucleotides. Comparison of the four types of human $G_s1\alpha$ cDNAs with the sequence of human $G_s1\alpha$ gene suggests that four types of $G_s1\alpha$ mRNA are generated from a single $G_s1\alpha$ gene by alternate use of exon 3 and/or of two 3' alternative splice sites of intron 3. $G_s\alpha$ was purified as a mixture of 52-kDa and 45-kDa proteins on SDS-polyacrylamide gels. Two $G_s1\alpha$ cDNAs, which correspond to $G_s1\alpha-1$ and $G_s1\alpha-4$, generate a 52-kDa and a 45-kDa protein, respectively, on SDS-polyacrylamide gel when they are expressed in COS-m6 cells or in *E. coli* (51, 71, 72). The proportions of 52-kDa and 45-kDa $G_s1\alpha$ protein vary among tissues and cells, but the functional difference of these two types of $G_s1\alpha$ proteins has not yet been clarified.

$G_s2\alpha$ ($G_{olf}\alpha$) is expressed exclusively in olfactory sensory neurons and is therefore thought to be involved specifically in odorant signal transduction (66). Rat $G_s2\alpha$ cDNA codes for 381 amino acid residues with the calculated molecular weight of 44,322. $G_s2\alpha$ has 88% homology with $G_s1\alpha$ in amino acid sequence. $G_s2\alpha$ retains a sequence that corresponds to exon 12 of $G_s1\alpha$ but lacks a sequence corresponding to exon 3. $G_s2\alpha$ stimulates adenylate cyclase when expressed in S49 cyc$^-$ kin$^-$ cells that lack $G_s1\alpha$ (66).

**Table 2**  Properties of G protein $\alpha$-subunits

| Species | No. of amino acids | Size[a] (kDa) | $M_r$ | Expression | Substrate for: | Function |
|---|---|---|---|---|---|---|
| Mammalian | | | | | | |
| hGs$\alpha$-1 | 394 | 52 | 45,664 | | | |
| hGs$\alpha$-2 | 395 | 52 | 45,769 | All tissues | CTX | { ACase (+) |
| hGs$\alpha$-3 | 379 | 45 | 44,189 | | | Ca$^{2+}$ channel (+) |
| hGs$\alpha$-4 | 380 | 45 | 44,294 | | | |
| rGs2$\alpha$ | 381 | 45 | 44,322 | Olfactory neurons | CTX | ACase (+) |
| hGi1$\alpha$ | 354 | 41 | 40,345 | Brain | PTX | { ACase (−) |
| hGi2$\alpha$ | 355 | 40 | 40,479 | All tissues | PTX | PLC (+) |
| hGi3$\alpha$ | 354 | 41 | 40,522 | All tissues | PTX | K$^+$ channel (+) } |
| hGo$\alpha$-1 | 354 | 39 | 40,053 | Brain | PTX | Ca$^{2+}$ channel (−) |
| hGo$\alpha$-2 | 354 | 39 | 40,075 | | | |
| bGt1$\alpha$ | 350 | 39 | 39,971 | Retina (rod) | CTX, PTX | { Cyclic GMP PDEase (+) |
| bGt2$\alpha$ | 354 | 40 | 40,143 | Retina (cone) | CTX, PTX | |
| hGx$\alpha$ (Gz$\alpha$) | 355 | 41 | 40,920 | Brain | – | PLC (+)? |
| Drosophila (D. melanogaster) | | | | | | |
| DGs$\alpha$L | 385 | 51 | 45,003 | Brain | CTX | { ACase (+)? |
| DGs$\alpha$S | 382 | 48 | 44,704 | | CTX | |

| | | | | | | |
|---|---|---|---|---|---|---|
| DGiα (DGα1) | 355 | 40,612 | 41 | Embryo, pupae | (−) | ? |
| DGoα−1 | 354 | 40,430 | 40 | Brain | PTX | ? |
| DGoα−2 | 354 | 40,491 | 40 | Brain | PTX | ? |
| **Yeast** | | | | | | |
| *S. cerevisiae* | | | | | | |
| GP1α | 472 | 54,075 | 54 | Haploid cells | ? | ? (Mating) |
| GP2α | 449 | 50,516 | 50 | also diploid cells | ? | ? (cAMP pathway) |
| *S. pombe* | | | | | | |
| GP1α | 407 | 46,254 | 46 | Inducible by nitrogen starvation | ? | ? (Mating) |
| **Dictyostelium** *(D. discoideum)* | | | | | | |
| Gα1 | 356 | 40,621 | 41 | Vegetative cells | ? | ? |
| Gα2 | 357 | 41,323 | 41 | Aggregated cells Inducible by cAMP pulse | ? | PLC (+)? |
| **Plant** *(A. thaliana)* | | | | | | |
| GPα1 | 383 | 44,582 | 45 | Leaves and roots | ? | ? |

Abbreviations: h, human; r, rat; b, bovine; CTX, cholera toxin; PTX, pertussis toxin; ACase, adenylate cyclase; PLC, phospholipase C; PDEase, phosphodiesterase; (+), activation; (−), inhibition.
[a] Size refers to the position of migration on the SDS polyacrylamide gel

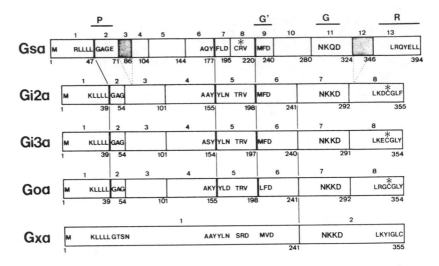

*Figure 6*  Organization of human genes for Gα subtypes.

$G_i\alpha$    Studies of cDNA cloning of $G_i\alpha$ have revealed the existence of three kinds of distinct $G_i\alpha$ cDNAs, i.e. $G_i1\alpha$, $G_i2\alpha$, and $G_i3\alpha$ (34, 64, 65). Their amino acid sequences show more than 85% identity with each other. The three types of $G_i\alpha$ cDNAs are encoded by the distinct genes. So far, the entire organization of human $G_i2\alpha$ and $G_i3\alpha$ genes and the partial structure of human $G_i1\alpha$ gene have been determined (34, 35). The coding region of human $G_i2\alpha$ and $G_i3\alpha$ is encoded by eight exons, and their exon-intron organization is completely identical (Figure 6). The same splice sites are also conserved in the partial sequence (exons 1 to 3) of human $G_i1\alpha$ gene. From Southern blot analysis, it appears that each of the three $G_i\alpha$ genes occurs as a single copy per haploid human genome.

Two kinds of $G_i\alpha$ protein, 40-kDa and 41-kDa, have been purified from porcine brains as substrates for ADP-ribosylation with pertussis toxin. Partial amino acid sequences of 40-kDa and 41-kDa proteins have been found to be identical to those of $G_i2\alpha$ and $G_i1\alpha$, respectively (73). $G_i\alpha$ purified from human erythrocyte (48) and 41-kDa pertussis toxin substrate from bovine spleen (74) showed partial amino acid sequence identity to $G_i3\alpha$.

In Northern blot analysis, $G_i2\alpha$ and $G_i3\alpha$ mRNAs are expressed in almost all kinds of tissues and cells. However, the relative abundance of these two mRNAs varies among different tissues (42, 43, 47, 65, 75). The expression of $G_i1\alpha$ mRNA seems to be more specific to tissue and cell types. $G_i1\alpha$ mRNA is predominantly expressed in brain (41, 75). Among several neuronal cell lines, $G_i1\alpha$ mRNA is detected in pheochromocytoma PC12 cells and neuroblastoma SK-N-SH cells, but not in glioma C6 cells and neuroblastoma-glioma hybridoma NG108-15 cells (76).

$G_i\alpha$s are thought to couple not only to adenylate cyclase but also to phospholipase C or $K^+$ channel, but the functional specificity of each $G_i\alpha$ is not yet known. The region encoded by exons 3 and 4 is the most heterogeneous among the three $G_i\alpha$s; however, the function of this region has not yet been determined.

$G_o\alpha$    Recently two types of $G_o\alpha$ cDNAs ($G_o\alpha-1$ and $G_o\alpha-2$) were isolated from mouse (70), hamster (62), and rat (37). Amino acid sequences of $G_o\alpha-1$ and $G_o\alpha-2$ were identical except in the C-terminal region. The human $G_o\alpha$ gene spans more than 100 kb and contains 11 exons including one at the 3' noncoding region (37). It was found that exons 7 and 8 coding for amino acid residues 242 to 354 of $G_o\alpha$ protein were duplicated (referred to as exons 7A, 7B, 8A, and 8B). The amino acid sequence identity of exons 7A and 7B, and exons 8A and 8B, is 88% and 68%, respectively. It was found that exons 7A and 8A, and 7B and 8B, code for $G_o\alpha-1$ and $G_o\alpha-2$, respectively. Thus, two different $G_o\alpha$ mRNAs may be generated by alternative splicing of a single $G_o\alpha$ gene. The positions of the splice junctions of human $G_o\alpha$ gene are identical to those of the human $G_i2\alpha$ and $G_i3\alpha$ gene in the coding region (Figure 6).

$G_o\alpha$-like protein distinct from the major species of $G_o\alpha$ was identified from bovine brain (77) and rat myometrium (78). Katada and his collaborators purified four types of $G_o\alpha$ protein from rat brain. The four types of $G_o\alpha$ protein reacted equally to anti-$G_o\alpha$ antibody raised against synthetic peptide with the amino acid sequence of $G_o\alpha$ of Glu-94 to Cys-108 (79a). One of them may correspond to $G_o\alpha-2$. In Northern blot analysis, $G_o\alpha$ mRNA is expressed mainly in neuronal tissues (75, 79b).

$G_t\alpha$    Two types of $G_t\alpha$ cDNA, $G_t1\alpha$ and $G_t2\alpha$, were isolated. Their amino acid sequences show 80% identity. $G_t1\alpha$ and $G_t2\alpha$ mRNAs are expressed in rods and cones, respectively (80). The exon-intron organization of mouse $G_t1\alpha$ gene (39) is identical to that of human $G_i2\alpha$, $G_i3\alpha$, and $G_o\alpha$ genes. In $G_i\alpha$, $G_o\alpha$, and $G_t\alpha$, the cysteine residue fourth from the C-terminus is ADP-ribosylated by pertussis toxin.

$G_x\alpha$    $G_x\alpha$ (36b), also referred to as $G_z\alpha$ (49), is a new member of the $G\alpha$ family, and is expressed mainly in neuronal cells. In amino acid sequence, $G_x\alpha$ has the highest identity with $G_i\alpha$ (60%) of all $G\alpha$s. The sequence of $G_x\alpha$ in the P region is different from other $G\alpha$s in that the consensus A-G-E is replaced by T-S-N (see Figure 4), and the rate of GTP hydrolysis (about 0.05 $min^{-1}$) is about 200 times slower than that of other $G\alpha$ subunits (81). The fourth amino acid residue from the C-terminus of $G_x\alpha$ is not cysteine but isoleucine. $G_x\alpha$ is ADP-ribosylated by neither pertussis toxin nor cholera toxin. $G_x\alpha$ may be involved in the signal transduction pathway, which is

**Table 3**  Conservation of G protein α-subunit sequences among different mammalian species

| Species[a] | Amino acid sequences | Nucleotide sequences |
|---|---|---|
| rGsα vs hGsα | 393/394 (99.7%) | 1128/1182 (95.4%) |
| bGi1α vs hGi1α | 354/354 (100 %) | 998/1062 (94.0%) |
| rGi2α vs hGi2α | 350/355 (98.6%) | 985/1065 (92.4%) |
| rGi3α vs hGi3α | 349/354 (98.6%) | 981/1062 (92.4%) |
| rGoα vs bGoα | 348/354 (98.3%) | 992/1062 (93.4%) |
| rGxα vs hGxα | 349/355 (98.3%) | 977/1065 (91.7%) |

[a] h, human; r, rat; b, bovine

insensitive to pertussis toxin, possibly in the activation of phospholipase C (82) or the closing of $K^+$ channel (83). The gene organization of $G_x\alpha$ is also quite different from other Gαs (Figure 6). The human $G_x\alpha$ gene contained only two exons for its coding region. However, an additional exon codes for 170 bp far upstream (more than 20 kb) of the 5'-noncoding region (38).

## Overall Structure of G Protein α-Subunits

The availability of complete amino acid sequences of nine G protein α-subunits allows comparison of their primary structures as shown in Figure 7. The overall structure homology is striking. Table 3 shows, in addition to remarkable homologies of the overall structures, a strong conservation of the amino acid and the nucleotide sequences in each group of G protein α-subunit. The amino acid sequence of $G_s\alpha$ is strongly conserved between human and rat; only 1 out of 394 amino acid residues is different. The sequence of $G_i1\alpha$ is completely identical between bovine and human. For $G_i2\alpha$, $G_i3\alpha$, $G_o\alpha$, and $G_x\alpha$, more than 98% identity of amino acid sequences is maintained among different mammalian species. An evolutionary tree of G protein α-subunits can be drawn on the basis of the differences among the predicted amino acid sequences obtained from various mammalian sources (Figure 8). There are three subfamilies of G protein α-subunits, $G_s$, $G_i$, and $G_t$. It is remarkable that the homologies among the three $G_i\alpha$ species are higher than those of the $G_s\alpha$ and $G_t\alpha$ subfamilies.

## Structure-Function Relationship of G Protein α-Subunits

G protein α-subunits have multiple domains, which are involved in GTP binding, GTP hydrolysis, conformation change by different guanine nucleotides, βγ-subunits association, receptor recognition, and effector interaction and regulation (2). The strong conservation of the amino acid sequence of each G protein α-subunit among distant mammalian species may reflect evolutionary pressure to maintain the multiple and specific physiological functions of each G protein gene product.

CONSERVED REGIONS OF G PROTEIN IN α-SUBUNITS    Figure 5 shows a schematic representation of the structure of *Escherichia coli* EF-Tu, G protein α-subunits (Gα), yeast Ras2 protein, and Ha-Ras p21 protein. G protein α-subunits contain several regions of highly conserved amino acid sequences. Small portions of these conserved sequences are homologous to regions of sequences that contribute to the guanine nucleotide-binding domain of EF-Tu and Ras proteins (see Figure 4 and Section III). These regions presumably form a core structure of GTP binding, GTP hydrolysis, and conformation change by different guanine nucleotides. These regions are designated P, E, G', G, and G'' (see Section III). The amino acid replacements at the P, G', and G regions of G proteins have been carried out by several groups in analogy to the transforming mutation of Ras proteins (see Table 4).

The replacement of Gly-49 by Val in $G_s\alpha$ did in fact reduce the rate of $k_{cat}$ of GTP hydrolysis (84, 85). As shown in Figure 4A, three amino acid residues, T-S-N, in the P site of $G_x\alpha$ are different from the residues, A-G-E, in other α-subunits. The differences are presumed to be responsible for the remarkably lower GTPase activity of $G_x\alpha$ compared with other α-subunits (81). Casey et al reported that a mutant of recombinant $G_s\alpha$, whose sequence in this region was changed to that present in $G_x\alpha$ by site-directed mutagenesis, showed a 50-fold reduction in $k_{cat.GTP}$ compared with its wild-type $G_s\alpha$. These results indicate the similarity of structure and function in GTP-binding domains between Ras proteins and G protein α-subunits.

As described above, a conformational change of E·GTP and E·GDP takes place in the G' region of GTP-binding proteins. The conformational change of G protein α-subunits has been studied by sensitivity to tryptic cleavage. In the case of normal α-subunits, GTP analog protects the tryptic cleavage at a conserved site, probably corresponding to Arg-231 of $G_s\alpha$ (located downstream of G' region). H21a mutant of $G_s\alpha$ can bind and hydrolyze GTP, as well as interact with β-adrenergic receptor, but the GTP-bound H21a protein is not protected from trypsin (86). H21a mutation of $G_s\alpha$ replaced Gly-226, corresponding to Gly-60 of Ras protein, by Ala. This substitution of H21a may prevent a critical conformational change.

In Ras proteins, replacement of Gln-61 by Leu shows the reduction of GTP hydrolysis and leads to malignant transformation (3). Recent work has revealed oncogenic mutations of $G_s\alpha$ in growth hormone–secreting pituitary tumors (31, 32, 87). One of the mutation sites is Gln-227 of $G_s\alpha$, which is equivalent to Gln-61 of Ras proteins. On the other hand, site-directed mutagenesis studies (84, 85) demonstrated that the Gln→Leu mutation at position 227 of $G_s\alpha$ reduced the intrinsic GTPase activity, and induced the constitutional activation of the $G_s\alpha$.

CHOLERA TOXIN ADP-RIBOSYLATION SITE    Cholera toxin catalyzes the ADP-ribosylation of $G_s\alpha$ and $G_t\alpha$, and the modification reduces GTPase

*Multiple sequence alignment of G protein α subunits (hGs1α, rGs2α, hGi1α, hGi2α, hGi3α, hGo1α, hGxα, bGt1α, bGt2α).*

Block 1 (end positions): 48, 50, 42, 41, 41, 41, 41, 37, 41

Block 2 (end positions): 98, 85, 78, 78, 78, 78, 76, 72, 76

Block 3 (end positions): 148, 135, 128, 126, 126, 128, 121, —, 125

Block 4 (end positions): 198, 185, 178, 176, 175, 178, 171, —, 175

Block 5 (end positions): 248, 235, 225, 226, 225, 226, —, 221

| Label   | Block 1 | Block 2 | Block 3 | Block 4 | Block 5 |
|---------|---------|---------|---------|---------|---------|
| hGs1α:  | 1 –     | 49 –    | 99 –    | 149 –   | 199 –   |
| rGs2α:  | 1 –     | 51 –    | 86 –    | 136 –   | 186 –   |
| hGi1α:  | 1 –     | 43 –    | 77 –    | 126 –   | 176 –   |
| hGi2α:  | 1 –     | 42 –    | 77 –    | 127 –   | 177 –   |
| hGi3α:  | 1 –     | 42 –    | 77 –    | 128 –   | 176 –   |
| hGo1α:  | 1 –     | 42 –    | 77 –    | 127 –   | 177 –   |
| hGxα:   | 1 –     | 42 –    | 73 –    | 122 –   | 177 –   |
| bGt1α:  | 1 –     | 38 –    | —       | 122 –   | 172 –   |
| bGt2α:  | 1 –     | 42 –    | 77 –    | 126 –   |         |

```
hGs1α: 249 - 298: A S S S Y N M V I R E D N Q T N R L Q E A L N L F K S I W N N R W L R T I S V I L F L N K Q D L L A
rGs2α: 236 - 285: A C S S Y N M V I R E D N N T N R L R E S L D L F E S I W N N R W L R T I S I I L F L N K Q D M L A
hGi1α: 226 - 275: A L S D Y D L V L A E D E E M N R M H E S M K L F D S I C N N K W F T D T S I I L F L N K K D L F E
hGi2α: 227 - 276: A L S A Y D L V L A E D E E M N R M H E S M K L F D S I C N N K W F T D T S I I L F L N K K D L F E
hGi3α: 226 - 275: A L S D Y D L V L A E D E E M N R M H E S M K L F D S I C N N K W F T E T S I I L F L N K K D L F E
hGoα1: 227 - 276: A L S G Y D Q V L H E D E T T N R M H E S L M L F D S I C N N K F F I D T S I I L F L N K K D L F G
hGoα2: 242 - 276:                               N R M H E S L K L F D S I C N N K W F T D T S I I L F L N K K D I F E
 hGxα: 227 - 276: E L S A Y D L K L Y E D N Q T S R M A E S L R L F D S I C N N N W F I N T S L I L F L N K K D L L A
bGt1α: 222 - 271: A L S A Y D M V L V E D D E V N R M H E S L H L F N S I C N H R Y F A T T S I V L F L N K K D V F S
bGt2α: 226 - 275: A L S A Y D M V L V E D D E V N R M H E S L H L F N S I C N H K F F A A T S I V L F L N K K D L F E

hGs1α: 299 - 348: E K V L A G K S K I E D Y F P E F A R Y T T P E D A T P E P G E D P R V T R A K Y F I R D E F L R I
rGs2α: 286 - 335: E K V L A G K S K I E D Y F P E Y A N Y T V P E D A T P D A G E D P K V T R A K F F I R D L F L R I
hGi1α: 276 - 310: E K I K - - K S P L T I C Y P E Y A G S N T Y E E - - - - - - - - - - - - - A A A Y I Q C Q F E D L
hGi2α: 277 - 311: E K I T - - H S P L T I C F P E Y T G A N K Y D E - - - - - - - - - - - - - A A S Y I Q S K F E D L
hGi3α: 276 - 310: E K I K - - R S P L T I C Y P E Y T G S N T Y E E - - - - - - - - - - - - - A A A Y I Q C Q F E D L
hGoα1: 277 - 310: E K I K - - K S P L T I C F P E Y T G P N T Y E D - - - - - - - - - - - - - A A A Y I Q A Q F - E S
hGoα2: 277 - 310: E K I K - - K S P L T I C F P E Y T G P S A F T E - - - - - - - - - - - - - A V A Y I Q A Q Y - E S
 hGxα: 277 - 311: E K I R - - R I P L T I C F P E Y K G Q N T Y E E - - - - - - - - - - - - - A A V Y I Q R Q F E D L
bGt1α: 272 - 306: E K I K - - K A H L S I C F P D Y N G P N T Y E D - - - - - - - - - - - - - A G N Y I K V Q F L E L
bGt2α: 276 - 310: E K I K - - K V H L S I C F P E Y D G N N S Y E D - - - - - - - - - - - - - A G N Y I K S Q F L D L

hGs1α: 349 - 394: S T A S G D G R H Y C Y P H F T C A V D T E N I R R V F N D C R D I I Q R M H L R Q Y E L L
rGs2α: 336 - 381: S T A T G D G K H Y C Y P H F T C A V D T E N I R R V F N D C R D I I Q R M H L K Q Y E L L
hGi1α: 311 - 354: N K R K D T K E I Y - - T H F T C A T D T K N V Q F V F D A V T D V I I K N N L K D C G L F
hGi2α: 312 - 355: N K R K D T K E I Y - - T H F T C A T D T K N V Q F V F D A V T D V I I K N N L K D C G L F
hGi3α: 311 - 354: N R R K D T K E I Y - - T H F T C A T D T K N V Q F V F D A V T D V I I K N N L K E C G L Y
hGoα1: 311 - 354: K N R S P N K E I Y - - C H M T C A T D T N N I Q V V F D A V T D I I I A N N L R G C G L Y
hGoα2: 311 - 354: K N K S A H K E I Y - - S H V T C A T D T N N I Q F V F D A V T D V I I A K N L R G C G L Y
 hGxα: 312 - 355: N R N K E T K E I Y - - S H F T C A T D T S N I Q F V F D A V T D V I I Q N N L K Y I G L C
bGt1α: 307 - 350: N M R R D V K E I Y - - S H M T C A T D T Q N V K F V F D A V T D I I I K E N L K D C G L F
bGt2α: 311 - 354: N M R K D V K E I Y - - S H M T C A T D T Q N V K F V F D A V T D I I I K E N L K D C G L F
```

*Figure 7* Deduced amino acid sequences of human Gs1α, Gi1α, Gi2α, Gi3α, Goα-1, Goα-2, and Gxα, rat Gs2α, and bovine Gt1α, and Gt2α. Sources for sequences are as follows: hGs1α, Kozasa et al (33); rGs2α, Jones & Reed (66); hGi1α, Bray et al (41) and Itoh et al (34); hGi2α, Beals et al (42), Didsbury et al (44), and Itoh et al (34); hGi3α, Beals et al (42), Itoh et al (34), and Kim et al (43); hGoα-1 and hGoα-2, Tsukamoto et al (37); hGxα, Fong et al (49), and Matsuoka et al (38); bGt1α, Tanabe et al (58), Medynski et al (59), and Yatsunami & Khorana (60); and bGt2α, Lochrie et al. (61).

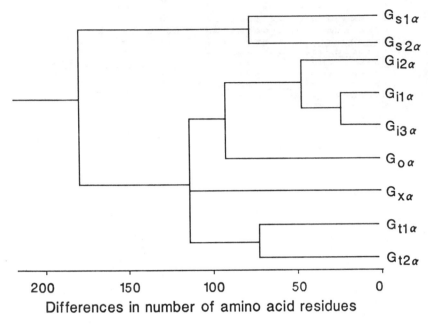

*Figure 8* Structural relationships among different mammalian Gα species [modified from Itoh et al (34)].

activity. Arg-174 is the site of modification in $G_t\alpha$ (88). The corresponding amino acid residue in $G_s\alpha$ is Arg-201. Freissmuth & Gilman (89) demonstrated that the replacement of this Arg by Ala, Glu, and Lys reduced $k_{cat.GTP}$, and these mutant proteins activated adenylate cyclase constitutively in the presence of GTP. Mutations of $G_s\alpha$ in pituitary tumors inhibit GTP hydrolysis and cause constitutive elevation of adenylate cyclase activity (31, 32). One mutation replaces an Arg-201 residue. All known G protein α-subunits contain an Arg residue at the position corresponding to Arg-201. In adrenal cortex tumors, mutations have been found in $G_i2\alpha$ that replace the conserved Arg-179 by Cys or His. The sequence around Arg-201 of $G_s\alpha$ is very much homologous not only to that of $G_t\alpha$, which is ADP-ribosylated by cholera toxin, but also to those of $G_i\alpha$ and $G_o\alpha$, which are not ADP-ribosylated by the toxin under normal conditions. More recently, it has been revealed that the activated receptor can support the ADP-ribosylation of pertussis toxin substrates by cholera toxin in HL60 cells (90, 91). These results suggest that the conserved Arg residue among all G protein α-subunits may be unmasked by interaction with an activated receptor, and may play an important role in the regulation of GTP hydrolysis.

**Table 4**  Genetic and biochemical studies of G protein $\alpha$-subunits

| Type | Implication | Ref. |
|---|---|---|
| **Mutation** | | |
| UNC ($\alpha_s$, R389P) | Uncoupling of $G_s$ with receptor | 68, 69 |
| I121a ($\alpha_s$, G226A) | Prevents GTP-dependent conformational change | 86 |
| Phosphoryl-binding region | ~5-fold reduction in $k_{cat.GTP}$ hydrolysis | 84, 85, 121 |
| ($\alpha_s$, G49V) | | |
| ($\alpha_s$, AGE48-50TSN) | 50-fold reduction in $k_{cat.GTP}$ | 81 |
| Conserved Gln ($\alpha_s$, Q227L) | >100-fold reduction in $k_{cat.GTP}$. Constitutively activates adenylate cyclase | 84, 85 |
| Conserved Arg ($\alpha_s$, R201A/Q/K)[a] | >100-fold reduction in $k_{cat.GTP}$ | 89 |
| Pertussis toxin modification site ($\alpha_s$, QYE390-392DCG) | Indirect enhancement of rate of GDP dissociation | 89 |
| Effector interaction ($\alpha_s$, LR,282,283,FT)[a] | Reduction of ability to stimulate adenylate cyclase | 105 |
| **Oncogenic mutations** | | |
| gsp ($\alpha_s$, Q227R/H/L) | Found in GH-secreting pituitary tumors and thyroid tumors | 31, 32, 87 |
| gsp ($\alpha_s$, R201C/H) | Found in GH-secreting pituitary tumors | 31, 32 |
| gip2 ($\alpha_{i2}$, R179C/H) | Found in ovarian sex cord stromal tumors and adrenal cortical tumors | 32 |
| **Chimera** | | |
| $\alpha_{i2}$ (1–212)/$\alpha_s$(235–394) | Receptor and effector interaction sites are in the carboxy-terminal 40% of $\alpha_s$ | 98 |
| $\alpha_s$ (1–356)/$\alpha_{i2}$(320–355) | Constitutively activates adenylate cyclase. Dominant negative effects on receptor-stimulated activation of phospholipase $A_2$. | 99, 122 |
| $\alpha_{i2}$(1–54)/$\alpha_s$(62–356)/$\alpha_{i2}$(320–355) | Dominant negative effects on receptor-stimulated activation of phospholipase $A_2$. | 122 |
| $\alpha_{i2}$(1–54)/$\alpha_s$(62–394) | Constitutively activates adenylate cyclase | 123 |

**Table 4**  (*continued*)

| Type | Implication | Ref. |
|---|---|---|
| **N-Terminal cleavage** | | |
| By trypsin ($\alpha_o$, K21) | Involved in the association with $\beta\gamma$-subunits | 101 |
| By V8 protease ($\alpha_{t1}$, E21) | Involved in the association with $\beta\gamma$-subunits | 100 |
| **Internal cleavage** | | |
| By trypsin ($\alpha_{t1}$, R204) ($\alpha_o$, R209) | GTP analog prevents cleavage at these sites | 116 |
| **ADP-ribosylation** | | |
| By pertussis toxin ($\alpha_{i1}$, $\alpha_{i3}$, $\alpha_o$, C351) ($\alpha_{i2}$, C352) ($\alpha_{t1}$, C347) | Impairs the interaction with receptor | 92, 93 |
| By cholera toxin ($\alpha_{t1}$, R174), ($\alpha_s$, R201) | Reduces GTPase activity and decreases affinity for $\beta\gamma$-subunits | 88, 124 |
| **Myristoylation** | | |
| ($\alpha_i$, $\alpha_o$, $\alpha_t$, $\alpha_x$, G2) | Enhances interaction with $\beta\gamma$-subunits and/or with membranes | 102, 104, 251 |
| **Synthetic peptide** | | |
| C-terminal regions ($\alpha_{t1}$, 311–328), ($\alpha_{t1}$, 340–350), ($\alpha_s$, 379–394) | Inhibits receptor–G protein interaction | 96, 97 |

ᵃ The exact number of these residues depends on the splice variants of $\alpha_s$. In this table, we use the number of $\alpha_s$ long form, which consists of 394 amino acid residues.

RECEPTOR INTERACTION SITE    Several lines of evidence strongly suggest that the predicted $\alpha$-helix at the carboxy terminus of G$\alpha$ subunits is a contact site for receptors. The pertussis toxin–catalyzed ADP-ribosylation of a cysteine located four amino acid residues from the C-terminus prevents the interaction of G proteins with receptors (92, 93). UNC mutation replaces Arg-389 of G$_s\alpha$ by Pro, at the position six amino acid residues from the C-terminal end (68, 69). The internal amino acid sequence of arrestin of retinal rod cells resembles the sequence of the C-terminus of G$_t\alpha$ (94). Arrestin competes with G$_t\alpha$ for binding to the phosphorylated photorhodopsin (95). The synthetic peptides corresponding to the C-terminus of G protein $\alpha$-subunits can inhibit the interaction of G proteins with receptors (96, 97). Analysis of chimeric $\alpha$-subunit indicated that the C-terminal 40% of G$_s\alpha$ is involved in the specificity of interaction with the receptor (98). Although the replacement of the sequence near the C-terminal region of G$_s\alpha$ by that of G$_i\alpha$ does not alter the rate of GDP dissociation and GTP hydrolysis, mutation of G$_s\alpha$ enhances the activation of adenylate cyclase in the presence of GTP after reconstitution of the $G_s\alpha$ into $cyc^-$ cell membranes (89). A chimeric $\alpha$-protein, in which the C-terminal 38 residues of $G_s\alpha$ are replaced by the C-terminal 36 residues of $G_i2\alpha$, has been expressed in mammalian cells (99). In the cells expressing this $G_s\alpha/G_i2\alpha$ chimera, adenylate cyclase appears to be constitutively activated by the receptor even in the absence of agonist. The lag phase of the cAMP synthesis by GTP analog in the absence of agonist was not observed in either study. The lag phase is due to a slow dissociation of GDP from $\alpha$-subunit. The activated receptor can abolish the lag phase by accelerating the rate of GDP dissociation. These data indicate that the replacement of the C-terminus of $G_s\alpha$ results in the indirect enhancement of the rate of GDP dissociation. The detailed mechanism by which the activated receptor can accelerate the rate of GDP dissociation from $\alpha$-subunit is still unclear.

INTERACTION WITH $\beta\gamma$-SUBUNITS    Proteolytic cleavage of the amino-terminal regions of $\alpha$-subunits decreases their ability to interact with $\beta\gamma$ (100, 101). Recent studies indicate that the most $\alpha$-subunits, except $G_s\alpha$, are myristoylated at the amino terminus when their cDNAs are expressed in cultured cells (102, 251). The myristoylation of the $\alpha$-subunits may play an important role in the enhancement of the interaction of $\alpha$ with $\beta\gamma$-subunits, because recombinant $G_o\alpha$ expressed in *E. coli*, which is not myristoylated, has a low affinity for $\beta\gamma$-subunits compared with native $G_o\alpha$ (103). Myristoylation of the recombinant protein by coexpressing of the $\alpha$-subunits and N-myristoyl transferase in *E. coli* restores the affinity of the recombinant $\alpha$-subunit for $\beta\gamma$-subunits to that of the native protein (104).

EFFECTOR CONTACT SITE    Analysis of the adenylate cyclase activity in the transfected cells expressing the $G_i2\alpha$ (1–212)/$G_s\alpha$ (235–394) chimera protein (98) and the $G_s\alpha$ (1–356)/$G_i2\alpha$ (320–355) chimera protein (99) suggested that an adenylate cyclase activation site may be located in the residues 235–356 of $G_s\alpha$. More recently, Itoh & Gilman have reduced an adenylate cyclase–stimulating activity of $G_s\alpha$ by site-directed mutagenesis (105). The amino acid residues of L282, and R283, which are specific for $G_s\alpha$, were replaced by F and T, respectively, corresponding to the $G_i\alpha$ residues. The mutant protein has been expressed and purified from *E. coli*. The kinetic properties of GTP binding and hydrolysis of the mutant $G_s\alpha$ were almost the same as that of wild-type $G_s\alpha$. Neither the affinity for the $\beta\gamma$-subunits, the interaction with $\beta$-adrenergic receptor, nor GTP-induced conformational change was altered by the mutation. However, the adenylate cyclase–stimulating activity of the mutant $G_s\alpha$ was less than 10% of the activity of wild-type $G_s\alpha$. The mutation is located just prior to the consensus sequence, N-K-X-D, of the GTP-binding proteins. This region corresponds the R6 loop in the schematic model of the G$\alpha$s proposed by Holbrook & Kim (106), which was based on the crystal structure of the Ras protein. These data strongly suggest that the loop of $G_s\alpha$ is responsible for the interaction with the adenylate cyclase.

THREE-DIMENSIONAL STRUCTURE MODEL OF G PROTEIN $\alpha$-SUB-UNITS    Although an X-ray crystal analysis of G protein $\alpha$-subunits has not yet been reported, some models of the tertiary structures of G protein $\alpha$-subunits have been proposed on the basis of the crystal structure of EF-Tu and Ras proteins (106–109). These models should be useful in interpreting experimental data and testing hypotheses of G protein function.

### Structure of βγ-Subunits of G Protein

The $\beta$-subunit of G proteins other than $G_t$ is a mixture of 35-kDa and 36-kDa proteins. On the other hand, $G_t$ has only a 36-kDa $\beta$-subunit. Two $\beta$-subunit cDNAs, $\beta1$ and $\beta2$, have been isolated (110–112). $\beta1$ and $\beta2$ cDNAs encode 36-kDa and 35-kDa subunits, respectively, and their amino acid sequences have 90% identity (113, 114). For $\gamma$-subunit, cDNA for $G_t\gamma$- and $\gamma$-subunit from brain are known (115, 117, 118). Although $\beta\gamma$-subunits of G protein from nonretinal sources are functionally interchangeable (119), there is a marked difference between the coupling of $G_t\beta\gamma$ and $\beta\gamma$ of G protein from nonretinal sources with $G_s\alpha$ (120).

## V. G PROTEINS IN YEAST SACCHAROMYCES CEREVISIAE

As described in the previous section, one of the major characteristic features of the G protein gene family is its strong conservation of structure among different mammalian species. Thus, G$\alpha$-homologous cDNAs and genes have been isolated from many sources by the cross-hybridization technique. From yeast Saccharomyces cerevisiae, two genes, GPA1 and GPA2, were initially isolated by cross-hybridization studies with rat G$\alpha$ cDNAs (125, 126). Later, two genes independently isolated by genetic analyses, SCG1 and CDC70, were found to be identical with GPA1 (see below).

### Comparison of Structures of Yeast and Mammalian Gα Proteins

GPA1 and GPA2 encode proteins of 472 and 449 amino acid residues, respectively, with calculated molecular weights of 54,075 and 50,516. The deduced amino acid sequences of these proteins, termed GP1$\alpha$ and GP2$\alpha$, are compared with those of mammalian G$\alpha$ proteins in Figure 9. GP1$\alpha$ and GP2$\alpha$ are about 100 amino acid residues larger than mammalian G$\alpha$s. The additional stretches, 110 amino acids (residues 126–235) in GP1$\alpha$ and 83 amino acids (residues 37–119) in GP2$\alpha$, are inserted near the N-terminal portion at the different sites. The two extra sequences had no homology with each other or with any other proteins reported.

Disregarding the unique stretches, the overall structural similarities of yeast GP1α and GP2α with rat Giα and Gsα are remarkable. All of the domains of Gα proteins discussed in the previous sections are highly conserved from humans to yeast. As shown in Figure 4, in the region responsible for GTP hydrolysis (P site), the amino acid sequences of yeast GP1α, GP2α, rat Gi2α, and Gsα (residues 43–58 of GP1α) are identical at 12 out of 16 positions, with conservative substitutions at the other 4 positions. In the guanine ring–binding site (G site, residues 382–397 in GP1α), 9 out of 16 amino acids are identical with the other 7 conservative substitutions. The conservation is highest in the G' site (residues 319–335 of GP1α); 14 contiguous amino acids are all identical in yeast GP1α, GP2α, and rat Gi2α. The calculated overall homology between GP1α and GP2α is 53% in nucleotide sequence and 60% in amino acid sequence. Both proteins also share a similar extent of homologies with rat Gi2α. These homologies are lower than those between Gi2α and Goα (85% in amino acid sequence), but comparable with those between Gi2α and Gsα (60%).

The cysteine residue located fourth (in GP1α) or fifth (in GP2α) from the C-terminal end, which corresponds to the site of ADP-ribosylation by pertussis toxin in mammalian Gα subunits, is replaced by isoleucine in GP1α or serine in GP2α. Therefore, yeast Gα proteins are probably refractory to modification by pertussis toxin. On the other hand, the cholera toxin–catalyzed ADP-ribosylation site, i.e. the sequence surrounding Arg-201 of rat Gsα, is conserved in both GP1α (around Arg-297) and GP2α (around Arg-273).

## The Role of G Proteins in Yeast Signal Transduction

GENE EXPRESSION OF G PROTEIN GENES IN YEAST CELLS    The first clue for solving the function of *GPA1* in yeast cells was obtained by the analysis of its expression. Northern blot analysis indicated that *GPA1* was expressed only in haploid cells (127), whereas *GPA2* was expressed in both haploid and diploid cells (125). Later, it was found that the level of *GPA1* transcript was increased severalfold in response to mating factors as in the case of other haploid-specific genes (128). Miller et al (129) found a 20-bp consensus sequence in the 5'-upstream regions of the reported haploid-specific genes. Mutational analysis of this consensus sequence revealed that the sequence is responsible for cell-type-specific gene expression. In *GPA1*, a sequence similar to the haploid-specific consensus motif was found about 50 bp upstream of the translation start site (126). The apparent difference of the expression pattern of the *GPA1* and *GPA2* genes strongly suggests differential functions for these two genes. Subsequent studies have shown that *GPA1* is

scGP1α:289 - 314:
scGP2α:245 - 291:
spGP1α:190 - 235:
ddGα1:152 - 197:
ddGα2:154 - 199:
atGPα1:162 - 211:
rGi2α:151 - 198:
rGsα:173 - 218:

scGP1α:315 - 362:
scGP2α:292 - 339:
spGP1α:236 - 283:
ddGα1:198 - 245:
ddGα2:200 - 247:
atGPα1:212 - 261:
rGi2α:197 - 244:
rGsα:219 - 266:

scGP1α:363 - 408:
scGP2α:340 - 385:
spGP1α:284 - 329:
ddGα1:246 - 291:
ddGα2:248 - 292:
atGPα1:262 - 309:
rGi2α:245 - 290:
rGsα:267 - 314:

scGP1α:409 - 437:
scGP2α:386 - 413:
spGP1α:330 - 373:
ddGα1:292 - 320:
ddGα2:293 - 321:
atGPα1:310 - 347:
rGi2α:291 - 320:
rGsα:315 - 359:

scGP1α:438 - 472:
scGP2α:414 - 449:
spGP1α:374 - 407:
ddGα1:321 - 356:
ddGα2:322 - 357:
atGPα1:348 - 383:
rGi2α:321 - 355:

```
 scGP1α:  1 -  40: M G - - C T V S T Q T I G D E S D P F L Q N K R A N D V I E Q S L Q L E K Q R                     D K N
 scGP2α:  1 - 122: M G - L C A S S E K N G S T P D T Q T A S A G S D N - - V G K A K V P P K Q E - ( 8 3 a a ) - N D K
 spGP1α:  1 -  73: M G - - C M S S K Y A D T S G G E V I Q K K L S D T Q T S N S S T T G S Q N A - ( 3 3 a a ) - G G N
  ddGα1:  1 -  35: M G N I C G K P E L G S P - - - - E E I K A N Q H - - - I N S L L K Q A R S K                     L E G
  ddGα2:  1 -  30: M G - I C A S S M E G - - - - - - E K T N T D I N L S - I E - - - K E R K K K                     - H N
 atGPα1:  1 -  37: M G L L C S R S R H H T - - - - - E D T D E N T Q A A E I E R R I E Q E A K A                     E K H
  rGi2α:  1 -  32: M G - - C T V S A - - - - - - - - - E D K A A A E R S K M I D K N L R E D G E K                   A A R
   rGsα:  1 -  39: M G - - C L G N S K T E D Q R N - E E K A Q R E A N K K I E K Q L Q K D K Q V                     Y R A

 scGP1α: 41 -  75: E I K L L L L G A G E S G K S T V L K Q L K L L H Q G G F S H Q - - - - - - - - - - - - - - - E R L
 scGP2α:123 - 157: E L K V L L L G A G E S G K S T V L Q Q L K I L H Q N G F S E Q - - - - - - - - - - - - - - - E I K
 spGP1α: 74 - 108: D I K V L L L G A G D S G K T T I M K Q M R L L Y S P G F S Q V - - - - - - - - - - - - - - - V R K
  ddGα1: 36 -  70: E I K L L L L G A G E S G K S T I A K Q M K I I H L N G F N D E - - - - - - - - - - - - - - - E K S
  ddGα2: 31 -  65: E V K L L L L G A G E S G K S T I S K Q M K I I H Q S G Y S N E - - - - - - - - - - - - - - - E R K
 atGPα1: 38 -  72: I R K L L L L G A G E S G K S T I F K Q I K L L F Q T G F D E G - - - - - - - - - - - - - - - E L K
  rGi2α: 33 -  67: E V K L L L L G A G E S G K S T I V K Q M K I I H E D G Y S E E - - - - - - - - - - - - - - - E C R
   rGsα: 40 -  89: T H R L L L L G A G E S G K S T I V K Q M R I L H V N G F N G E G G E E D P Q A A R S N S D G E K A

 scGP1α: 76 - 125: Q Y A Q V I W A D A I Q S M K I L I I Q A R K L G I Q L D C D D P I N N K D L F A C K R I L L K A K
 scGP2α:158 - 207: E Y I P L I Y Q N L L E I G R N L I Q A R T R F N V N L E P E C E L T Q Q D L S R T M S Y E M P N N
 spGP1α:109 - 158: Q Y R V M I F E N I I S S L C L L L E A M D N S N V S L L P E N E K Y R A V I L R K H T S Q P N E P
  ddGα1: 71 - 119: S Y K T I I Y N N T V G S M R V L V N A A E E L K I G I S E N N K E A A S R I S N D L G D H F N G -
  ddGα2: 66 - 115: E F K P I I T R N I L D N M R V L L D G M G R L G M T I D P S N S D A A V M I K E L T S L Q A S I V
 atGPα1: 73 - 122: S Y V P V I H A N V Y Q T I K L L H D G T K E F A Q N E T D S A K Y M L S S E S I A I G E K L S E I
  rGi2α: 68 - 117: Q Y R A V V Y S N T I Q S I M A I V K A M G N L Q I D F A D P Q R A D D A R Q L F A L S C A A E E Q
   rGsα: 90 - 139: T K V Q D I K N N L K E A I E T I V A A M S N L V P P V E L A N P E N Q F R V D Y I L S V M N V P N

 scGP1α:126 - 268: A L D Y I N A - ( 1 0 3 a a ) - L I H E D I A K A I K Q L W N N D K G - I K Q C F A R S N E F - Q L E
 scGP2α:208 - 244: Y T G - - - -                   Q F P E D I A G V I S T L W A L P S T Q D L V N G P N A S K F - Y L M
 spGP1α:159 - 189: - - - - - - -                   - F S P E I Y E A V H A L T L D T K - - L R T V Q S C G T N L - S L L
  ddGα1:120 - 151: - - - - - - -                   V L T A E L A Q D I K A L W A D P G - - I Q N T F Q R S S E F - Q L N
  ddGα2:116 - 153: T D C W G - -                   E L N E D Q G K K I K A L W T D P G - - V K Q A M R R A N E F S T L P
 atGPα1:123 - 161: G G R L D Y P                   R L T K D I A E G I E T L W K D P A - - I Q E T C A R G N E L - Q V P
  rGi2α:118 - 150: G - - - - - -                   M L P E D L S G V I R R L W A D H G - - V Q A C F G R S R E Y - Q L N
   rGsα:140 - 172: F - - - - - -                   D F P P E F Y E H A K A L W E D E G - - V R A C Y E R S N E Y - Q L I
```

*Figure 9* Comparison of the predicted amino acid sequences of G protein $\alpha$-subunits from different organisms. Identical or conservative amino acid residues are enclosed with solid lines. Aligned sequences are: scGP1$\alpha$ and scGP2$\alpha$, *Saccharomyces cerevisiae* GP1$\alpha$ and GP2$\alpha$ [Nakafuku et al (126) and (125)]; spGP1$\alpha$, *Schizosaccharomyces pombe* GP1$\alpha$ [Obara et al (174)]; ddG$\alpha$1 and ddG$\alpha$2, *Dictyostelium discoideum* G$\alpha$1 and G$\alpha$2 [Pupillo et al (167)]; atGP$\alpha$1, *Arabidopsis thaliana* GP$\alpha$1 [Ma et al (173)]; rGi2$\alpha$ and rGs$\alpha$, rat Gi2$\alpha$ and rat Gs1$\alpha$ [Itoh et al (64)].

directly involved in mating factor signal transduction. On the other hand, genetic and biochemical analyses have suggested that *GPA2* may participate in regulation of the intracellular levels of cAMP.

*GPA1*-INVOLVED MATING SIGNAL TRANSDUCTION    Two haploid cells, **a** and $\alpha$, can mate to yield **a**/$\alpha$ diploids through a multistep process. This mating event is initiated by pheromones secreted by each haploid cell type: $\alpha$ cells secrete $\alpha$-factor, which is composed of 13 amino acid residues and acts on **a** cells; whereas **a** cells produce **a**-factor, which is composed of 12 amino acids and acts on $\alpha$ cells. Mating factors induce cell-cycle arrest at the late G1 phase prior to the onset of DNA replication, and concomitantly alter the expression of a set of genes. Both processes are required to initiate the mating program of haploid cells (for a review, see Ref. 130). A number of genes involved in this signal transduction pathway have been identified. Among them, *STE2* and *STE3* code for the specific receptors for $\alpha$-factor and **a**-factor, respectively (131–133). The products of the *STE2* and *STE3* genes possess seven putative transmembrane domains similar to those of the $\beta$-adrenergic receptor (134) and rhodopsin (135, 136), suggesting that they may interact with G protein. Subsequent genetic studies have shown that the *GPA1* gene is involved in the mating factor receptor–mediated signal transduction.

Independent of the gene cloning of *GPA1*, an allelic gene of *GPA1* named *SCG1* was isolated as a multicopy suppressor of the *sst2* mutation, which confers a pheromone-supersensitive phenotype on haploid cells (137). A cell-cycle–defective temperature-sensitive mutation called *cdc70*, which allows conjugation of the cells lacking a mating factor receptor (receptor-bypass mutation) (128), and a newly isolated mating pheromone-resistant mutation, *dac1* (138), were also mapped in the *GPA1* locus. These findings again support the idea that *GPA1* functions in mating factor signal transduction. Comparison of the sequences of *GPA1*, *SCG1*, and *CDC70* showed that the three genes are the same alleles, although nucleotide and amino acid sequences of *SCG1* differ at five positions from *GPA1* and *CDC70* (the latter two have an identical sequence), presumably due to a strain polymorphism.

IDENTIFICATION OF $\beta$- AND $\gamma$-SUBUNITS IN YEAST    More recently, it has been demonstrated that the *STE4* and *STE18* genes of *S. cerevisiae* encode proteins highly homologous to mammalian G protein $\beta$- and $\gamma$-subunits (139). DNA sequence analysis has shown that *STE4*-encoded protein is composed of 432 amino acids sharing 49% homology with two kinds of mammalian G$\beta$, G$\beta$1(110, 140, 141) and G$\beta$2 (111, 112). A unique repetitive structure of G$\beta$, which is composed of seven homologous units of 40–50 amino acid residues, was well conserved in Ste4 protein. The deduced sequence of Ste18 also exhibited partial similarities with G$\gamma$-subunits. Ste18 protein of 110 amino acid residues shares 35% and 40% homologies with bovine $G_t\gamma$ (115, 142) and a $G\gamma$ for brain $G_i\alpha$ and $G_o\alpha$ (117), respectively. Although these similarities are not remarkable, they are significant in that two bovine $G\gamma$-subunits also show the same extent of sequence divergence (117). Genetic study clearly shows that the Ste4 and Ste18 proteins are the components of the mating factor signaling system. Both genes were expressed only in haploid cells, and the defective mutations of these genes resulted in the loss of mating abilities. Products of $GPA1(\alpha)$, $STE4(\beta)$, and $STE18(\gamma)$ may constitute a heterotrimeric G protein, which is directly coupled with $STE2(\alpha)$ and $STE3$(a) mating factor receptors.

POSTTRANSLATIONAL MODIFICATION OF YEAST G PROTEIN $\gamma$-SUB-
UNIT    At the C-terminus of Ste18 proteins, a Cys-A-A-X sequence (A represents aliphatic amino acid and X is the last amino acid) common to all the $G\gamma$-subunits is found. In Ras p21, this motif signals posttranslational modification of the C-terminus, which is required for membrane association and biological activity (143). Recent studies have demonstrated that the conserved cysteine in the motif is the site of polyisoprenylation of both Ras (144, 252) and $G\gamma$ proteins (20–22). It was also shown that a mutational change of the cysteine (Cys-107) to serine resulted in the loss of function of Ste18 (145). Furthermore, a yeast *dpr1/ram1* mutation, which was originally isolated as a defective mutation in posttranslational processing of yeast Ras proteins, was found to affect the membrane association and biological activity of Ste18 proteins. These results indicate that G protein $\gamma$-subunits and Ras proteins may share a set of the same modification process.

IMPLICATIONS FROM STUDIES OF THE YEAST MATING PATHWAY    The observations described above demonstrate that the transmembrane-signalling system, similar to the mammalian system, is conserved in yeast. The next interest is to identify the effector molecule(s) operating in the mating signalling, which still remains unknown.

An important conclusion from the above studies is that the apparent mode of action of the yeast mating factor signalling system is quite different from that of the mammalian system. In mammalian systems, biochemical studies demonstrated that $G\alpha s$ play a major role in interacting with the downstream effector molecules (2). Although independent roles of $\beta\gamma$-subunits in interacting with effectors have been assumed in some systems (146, 147, 253), precise mechanisms remain unclear. In contrast, in the yeast mating pathway, a series of genetic data suggests that the $\alpha$-subunit (*GPA1*) plays a negative role in signal transduction, whereas the $\beta\gamma$-subunits (*STE4/STE18*) function as positive activators of the downstream signalling pathway (148). The following observations can explain this: disruption of the *GPA1* gene resulted in the constitutive activation of mating factor signalling pathway (127, 137). This was evidenced by the fact that $gpa1^-$ mutations bypassed the requirement of factor receptors for conjugation as well as for induction of expression of pheromone-inducible genes (128, 149). In contrast to this situation, loss of functions of either *STE4* or *STE18* caused sterile phenotypes (150–152), which mean defects in mating responses. Moreover, $ste4^-$ and $ste18^-$ mutations quenched the activation of the pathway caused by $gpa1^-$ mutations (139, 149, 153), whereas overexpression (154) or an "activated" mutation (153) of *STE4* appeared to elicit signals constitutively irrespective of the presence or absence of *GPA1* gene products. These results indicate that the $\beta\gamma$-complex but not the $\alpha$-subunit is responsible for transducing signals necessary for mating.

Recent studies have hinted at more complex situations. Using genetic approaches, several kinds of "activated" mutations of the *GPA1* gene have been characterized. A $GPA1^{Val-50}$ mutation, which has a substitution of Gly-50 with valine, was introduced by site-directed mutagenesis based on the analogy with the Val-12 mutation of Ras (155). This mutation of Ras decreases GTPase activity and increases transformation activity (156, 157). The other mutations, $GPA1^{Lys-355}$ and $GPA1^{Lys-364}$, were selected from a pool of mutants based on phenotypic changes (158). The alignment of primary structures shows that the mutation corresponds to Val-49 mutation of Gs$\alpha$ proteins. According to the model described above, constitutive activation of GP1$\alpha$ would cause phenotypes supersensitive to the mating factors. This has turned out to be the case in short-term responses; i.e. growth arrest and gene inductions of cells carrying these *GPA1* mutations were elicited by a 100-fold lower concentration of mating factors than that required for wild-type cells. More interestingly, however, these mutations also enhanced recovery from factor-induced growth arrest, and after long-term incubation with factors, mutant cells finally showed phenotypes of factor-resistant growth (155, 158). One possible explanation is that independent of growth arrest and gene-induction pathways driven by the $\beta\gamma$-subunit, GP1$\alpha$ can turn on another signalling pathway which leads to a recovery from mating factor responses. All of the evidence described above has relied on genetic studies. Further biochemical studies are necessary to elucidate the precise molecular mechanisms of G protein function in mating factor signal transduction.

CYCLIC AMP PATHWAY AND *GPA2*    In addition to the mating factor signalling system, *S. cerevisiae* has another signal transduction pathway, which operates in the early G1 phase of the cell cycle. This is mediated by nutrients such as glucose, which serves as an extracellular signal for the activation of adenylate cyclase, and cAMP plays a crucial role in cell cycle progression at this stage (159). It is well known that GTP-binding proteins encoded by *RAS1* and *RAS2*, which are yeast counterparts of mammalian Ras, participate in the control of the activity of adenylate cyclase (discussed in detail in Section VIII). In contrast to yeast adenylate cyclase, mammalian adenylate cyclase activity is regulated by two G proteins, $G_s$ and $G_i$. A recent study reports that a yeast G protein *GPA2*, in addition to Ras proteins, is involved in the regulation of cAMP.

Yeast cells cultured under starvation conditions transiently accumulate cAMP in response to glucose (160). Introduction of YEp*GPA2* (a multicopy plasmid carrying the *GPA2* gene) in wild-type cells was found to enhance glucose-induced cAMP accumulation remarkably (125). In addition, YEp*GPA2* suppressed the growth defect by a temperature-sensitive mutation of the *RAS2* gene [*ras2-101*(ts)] (125). In *ras2-101*(ts) cells, mutant RAS2 proteins would not support the activation of adenylate cyclase at nonpermissive temperature, and therefore glucose could not induce cAMP formation. Introduction of YEp*GPA2* restored the cAMP response in the mutant cells at high temperature. These results suggest that *GPA2*, in addition to Ras proteins, is involved in the regulation of cAMP levels in *S. cerevisiae*.

Since *GPA2* could not restore *gpa1⁻* phenotypes (I. Miyajima, K. Matsumoto, unpublished observation), *STE4*- and *STE18*-encoded β- and γ-subunits and $DG_o\alpha$, cannot interact with *GPA2*. This implies that an additional set of genes that code for β- and γ-subunits interacting with *GPA2* must be present in yeast cells.

# VI. G PROTEINS FROM OTHER ORGANISMS

## *G Protein of* Drosophila melanogaster

G protein homologous genes in *D. melanogaster* were isolated by homology with mammalian G protein genes. For α-subunit, genes homologous to mammalian $G_s\alpha$, $G_i\alpha$, and $G_o\alpha$ (designated $DG_s\alpha$, $DG_i\alpha$, and $DG_o\alpha$ respectively) have been characterized (161–163, 164a, 164b). β-subunit homologous cDNA was also isolated (165). These genes have 70–80% identity in amino acid sequence with their mammalian counterparts.

The $DG_s\alpha$ gene is composed of nine exons, and its structure is similar to that of human $G_s\alpha$ (166). However, $DG_s\alpha$ lacks the exon that corresponds to exon 3 of human $G_s\alpha$. Two types of $DG_s\alpha$ mRNA are generated by alternative use of two 3' splice sites of intron 7, which are nine nucleotides apart. The two types of $DG_s\alpha$ thus generated are composed of 385 and 382 amino acids. In mammalian $G_s\alpha$, no variation in amino acid sequence is reported in this region (the C-terminal end of exon 11 of human Gsα).

The $DG_o\alpha$ gene is split into seven exons in the coding region (163). Its splice junctions are identical to those of human $G_o\alpha$. However, $DG_o\alpha$ does not have an intron corresponding to intron 2 of human $G_o\alpha$; exons 2 and 3 of human $G_o\alpha$ are fused in $DG_o\alpha$. Two kinds of $DG_o\alpha$ proteins, which differ only in their N-terminal 40 amino acids, are generated by alternate use of two kinds of exon 1.

The $DG_i\alpha$ gene, composed of five exons, also has a structure similar to that of human $G_i\alpha$ gene (162). The amino acid residue at the fourth position from the C-terminus is isoleucine rather than cysteine. Therefore, $DG_i\alpha$ is probably insensitive to pertussis toxin. By Northern blot analysis, $DG_s\alpha$ and $DG_o\alpha$ are expressed mainly in nervous systems. $DG_i\alpha$ is highly expressed in embryo and pupae.

## G Proteins from Dictyostelium discoideum

Recently, the existence of two G proteins, $G\alpha1$ and $G\alpha2$, was demonstrated in the amoeba *Dictyostelium discoideum* (167). $G\alpha1$ encodes a protein of 356 amino acid residues, whereas $G\alpha2$ codes for a protein of 357 residues. The calculated molecular weights of $G\alpha1$ and $G\alpha2$ are 40,621 and 41,323, respectively. The primary sequences of $G\alpha1$ and $G\alpha2$ proteins show clear overall homologies with other $G\alpha$s; $G\alpha1$ and $G\alpha2$ share 45% identity with each other and show the same degree of identity to mammalian $G_i\alpha$, $G_o\alpha$, and $G_t\alpha$ subtypes. Identities with $G_s\alpha$ and yeast GPA1 proteins are slightly lower, about 35% (see Figure 9). Neither $G\alpha1$ nor $G\alpha2$ had the cysteine residue at the ADP-ribosylation site with pertussis toxin. Since biochemical analysis has identified a potential candidate for the pertussis toxin substrate in *D. discoideum* (168), there may be at least one additional $G\alpha$ protein in this organism.

*Dictyostelium* has been utilized as an experimental organism for understanding the mechanism of cell development and differentiation (169). Cyclic AMP, which was originally identified as acrasin in this organism, is a chemotactic substance that elicits a variety of responses including the activation of adenylate cyclase and phospholipase C. Previous biochemical and genetic analyses have implicated the involvement of G proteins in this signalling pathway (170). The receptor for cAMP has a secondary structure similar to that of G protein-coupled receptors (171). At present it is suggested that $G\alpha2$ is coupled to the chemotactic receptor to activate phospholipase C (170). The $G\alpha2$ gene was mapped in the *FrigidA* locus of the *D. discoideum* genome, the defect of which causes an aggregation-minus phenotype (172). The function of $G\alpha1$ appears to be different from that of $G\alpha2$. Ten- to 20-fold overexpression of $G\alpha1$ resulted in large multinuclear cells, which aggregated and differentiated only very poorly (172). Further studies on $G\alpha1$ and $G\alpha2$ in *Dictyostelium* are required to understand their roles in the regulation of developmental processes.

## Plant G Protein

A plant G protein $\alpha$-subunit gene, *GPA1*, was isolated from *Arabidopsis thaliana* (173). *GPA1* encodes a protein of 383 amino acid residues with a calculated molecular weight of 44,582. The size of the predicted protein was larger than those of mammalian Gi $\alpha$-subunits, rather close to that of mammalian Gs$\alpha$ due to the presence of several short insertions. One of these inserts was found at a unique position upstream of G' site. Other short stretches were located near the sites of unique insertions present in mammalian Gs$\alpha$ and yeast *GPA1*. The overall structure of *Arabidopsis GPA1* shares 36% identity and 73% homology with mammalian Gi $\alpha$-subunits (Figure 9). Comparison of the organization of *GPA1*, composed of 14 exons and 13 introns, with those of mammalian G$\alpha$ genes, indicated that several exon-intron junctions are conserved. The third intron of *GPA1* was at the same position as the first intron of the human Gs$\alpha$ and Gi$\alpha$ genes. The positions of the fifth and sixth introns of *GPA1* corresponded to those of the fourth intron of human Gs$\alpha$ and the third introns of human Gi$\alpha$ genes, respectively. These observations again emphasize a strong structural conservation of G proteins through evolution.

## G Protein from Schizosaccharomyces pombe

Recently a G protein–homologous gene has been isolated from *Schizosaccharomyces pombe* (174). The gene, designated *gpa1*, encodes a protein (GP1$\alpha$) of 407 amino acid residues with a calculated molecular weight of 46,254 (see Figure 9). Preliminary experiments have shown that the expression of *gpa1* is induced under nitrogen-starvation conditions. In *S. pombe*, nitrogen starvation leads to the conjugation of haploid cells. Genetic analyses suggest the involvement of the *gpa1* gene in mating factor signal transduction (174).

# VII. MAMMALIAN RAS PROTEINS

## Regulation of Ras Activity by a Bound Guanine Nucleotide

The mammalian *ras* gene family (Ha-, Ki-, and N-*ras*) contains some of the most well-characterized proto-oncogenes (3). The genes encode a 21-kDa protein (p21), and studies using recombinant p21 purified from *E. coli* have revealed that p21 binds GDP/GTP and hydrolyzes the bound GTP to GDP and Pi (Figure 10). The GTPase activity of Ras p21 is greatly enhanced by a cytosolic protein named GAP (175–177). Some constitutively active mutant Ras p21s with transforming activity have decreased GTPase, some have an increased GDP/GTP exchange rate, and some have both. Analysis of the functional domain by reverse genetics shows the importance of bound GDP/GTP in the biological function of p21. Therefore, *ras* gene products are thought to be signal-transducing GTP-binding proteins regulating long-term cellular responses, namely proliferation and differentiation. Comparison of the biological activities of GDP-bound p21 and GTP-bound p21 shows that the former is inactive and the latter is active (175, 178). It has also been shown that constitutively active Ras p21, for example the Val-12 mutant, and viral Ras p21, are present more in the GTP-bound form than the wild-type Ras p21 in *S. cerevisiae*, PC12 rat pheochromocytoma cells, and NIH 3T3 mouse cells (179–181).

Analyses of three-dimensional structures have shown that there are two regions of conformational differences between GDP-complexed and GTP-complexed forms of Ras p21 (19, 182). The first region spans residues 30–38, corresponding to the effector recognition domain. The second consists of residues 60–76. A monoclonal antibody Y13-259 binds to the second region inhibiting the Ras-mediated signal transduction (3). Both regions are essential for the interaction of Ras p21 with GAP (183–185).

Measurements of Ras p21–bound GDP/GTP in living cells are very useful in understanding the signal transduction network upstream of Ras. Since the ratio of the amount of GTP-bound Ras p21 to the total amount of Ras p21 in the cells reflects the activity of Ras p21, we can quantitate the Ras-activity in response to the upstream signal by measuring the change in the population of GTP-bound p21 even though the effector of Ras has not been identified. Comparison of activities of Ras p21 in growth factor–treated and untreated cells has shown that platelet-derived growth factor (PDGF), epidermal growth

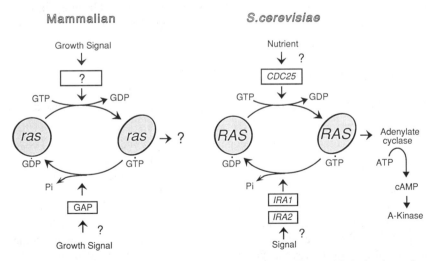

*Figure 10*  Regulation of Ras proteins in mammalian and yeast (*S. cerevisiae*) signal transduction systems. For details, see text.

factor (EGF), and colony stimulating factor-1 (CSF-1) activate Ras p21 (by increasing the GTP-bound population) within a few minutes after the addition of ligands, but basic fibroblast growth factor (bFGF), insulin, and bombesin have no effect (Table 5) (186–188). The growth factor receptors linked to Ras p21 have intrinsic tyrosine kinase activities, and tyrosine-phosphorylation of unknown target molecules are thought to be important for the transduction of growth signals. It remains unclear what mediates the signals from growth factor receptors to Ras p21. Accumulation of GTP-bound Ras p21 is also observed in transformants, with oncogenes encoding nonreceptor-type (*src*, *abl*) or growth factor receptor–derived (*erbB*-2) tyrosine kinases (Table 5) (187, 188). These results demonstrate that Ras functions downstream of certain kinds of tyrosine kinases, and that unregulated activation of these kinases causes the transformation, at least in part, through constitutive activation of Ras.

Microinjection of anti–Ras p21 antibody Y13-259 into NIH 3T3 cells inhibits the DNA synthesis induced by PDGF and EGF (189). Transformation by *src* or *fms* tyrosine kinase was also inhibited by this antibody (190). These results strongly suggest that Ras plays an essential role in proliferation or transformation by tyrosine kinases, and closely coincide with the results of the analyses of Ras p21–bound guanine nucleotide. The involvement of Ras p21 in the signal transduction pathways from PDGF or EGF has also been suggested in previous works with normal and *ras*-transformed cell lines (3).

Rapid activation of Ras p21 was also observed after the treatment of T lymphocyte with phytohemagglutinin (PHA) or an anti-CD3 antibody. In this

**Table 5**  Tyrosine phosphorylation of GAP and the increase of p21·GTP by growth factor stimuli and oncogene activation

| Tyrosine kinase | Phosphorylation of GAP | Increase of p21·GTP |
|---|---|---|
| Growth factor receptor | | |
| PDGF-R | + | + |
| EGF-R | + | + |
| FGF-R | − | − |
| IGF-1-R | − | − |
| CSF-1-R (c-*fms*) | + | + |
| | | |
| Oncogene product | | |
| *src* | + | + |
| *fps* | + | ? |
| *abl* | + | + |
| *erb*B-2 | + | + |

The data are taken from Refs. 186–188, 209, 210.

case, protein kinase C is involved between the T cell receptor and Ras p21, and reduction of GAP activity was observed after the stimulation of protein kinase C (191).

Regulatory mechanisms of the activity of Ras p21 have not been clarified so far. Two types of mechanism may activate Ras p21. One is the enhancement of GDP/GTP exchange. In the case of heterotrimeric G proteins, the agonist-associated receptor stimulates the release of GDP from the $\alpha$-subunit, and the $\alpha$-subunit converts to the active GTP-bound conformation. Similarly, it was proposed that when the agonists bind, some growth factor receptors may stimulate the exchange of GDP/GTP bound to Ras p21. EGF and PDGF were found to stimulate the exchange of p21-bound guanine nucleotide in crude membrane systems (192); however, exact kinetics or the interactions between the receptor and Ras p21 have not been analyzed. It is also possible that some other components may be involved between growth factor receptors and Ras p21. Recently, several groups have purified a protein that stimulates the release of Ras p21–bound GDP/GTP (193–195).

The other possibility for the activation of Ras p21 is inhibition of GTPase. The GTPase activity of Ras p21 is greatly enhanced by GAP in the cells. If the GAP activity is reduced by upstream growth signals, active Ras p21 complexed with GTP may accumulate. Therefore, GAP may be a probable candidate as the regulator of Ras p21.

## GAP

Trahey & McCormick (175) have found a protein that enhances the GTPase of normal Ras p21, but not of constitutively activated mutants in the crude extract of *Xenopus* oocytes. They named this protein GAP (GTPase activating protein), and GAP activity was detected in many kinds of tissues and cultured cell lines (175, 196). Experiments with recombinant Ras p21 purified from *E. coli* showed approximately 10-fold reduced GTPase activity of a transforming mutant Ras p21 (3). However, in the cells, the half-life of the GTP-bound conformation of mutated Ras p21 (Val-12) is much longer than that of the normal Ras p21 because of the effect of GAP (175).

GAP was purified as a 125-kDa protein from the bovine brain (196), and the primary structure of GAP was determined by cDNA cloning (197, 198). Bovine GAP is composed of 1044 amino acids and has SH2 (*src*-homology 2) and SH3 domains. The C-terminal one-third of GAP is sufficient to stimulate the GTPase of Ras p21 (199). The gene encoding GAP was mapped on human chromosome 5, near bands q13–q15 (200).

Although GTPase of transforming Ras p21s, for example, Val-12, Thr-59, and Leu-61, is not stimulated by GAP, these mutants can compete with the normal Ras p21 to interact with GAP (184, 197). In contrast, mutated Ras p21s with an amino acid substitution in the effector-binding domains, for example at Thr-35, Ile-36, and Asp-38, do not affect the interaction between normal Ras p21 and GAP, and GAP cannot enhance the GTPase of these mutants (183, 184). Rap1A protein (201) [also called Krev-1 protein (202) or Smg p21 (203)], a suppressor of Ras-induced transformation, has an amino acid sequence identical to that of Ras p21 at the effector region. It has been reported that GAP-stimulation of GTPase of Ras p21 was inhibited by Rap1A protein in the reconstitution system with purified components (204, 205). The GTPase activity of Rap1A protein is not stimulated by GAP, and a GTPase-stimulating protein that is specific for Rap1A protein has also been reported (206). Mutations at the epitope for the antibody Y13-259 also make Ras p21 unable to interact with GAP (185).

The effect of GAP is inhibited by some kinds of phospholipids, for example, phosphatidic acid, phosphatidylinositol phosphates, and arachidonic acid, that are known to be produced in response to growth factor stimulation (207). Since certain phospholipid-related materials have mitogenic activities that are dependent on Ras (208), it is possible that the phospholipids may regulate Ras activity via GAP.

Biochemical links between GAP and various tyrosine kinases have been reported recently. The first is the tyrosine phosphorylation of GAP, and the second is the physical association of GAP with PDGF receptor. GAP is rapidly phosphorylated on tyrosine residues by PDGF and EGF, but insulin and basic FGF have no effect (Table 5) (209, 210). Most tyrosine-phosphorylated GAP is located in the cell membrane, in contrast to the unphosphorylated GAP, which is predominantly in the cytosol (209). GAP is also phosphorylated on tyrosine in transformants by cytoplasmic and receptor-like oncogene tyrosine kinases, for example, the products of *src, fps, fms,* and *abl* (Table 5) (210). Two proteins, p62 and p190, were also phosphorylated on tyrosine and coprecipitated with GAP, but these two proteins have not yet been precisely characterized (210). It is possible that the tyrosine phosphorylation may regulate the function of GAP. Phosphorylated GAP itself may have decreased stimulating activity of GTPase of Ras p21, or the phosphorylation of GAP may change the affinity for other signal transduction components that regulate the function and localization of GAP. However, up to now, no evidence has supported these ideas.

PDGF receptor has been shown to be associated with phospholipase C-γ, phosphatidylinositol 3-kinase, and *raf* proto-oncogene product, when the ligand binds to the receptor. The complex is called STP (Signal Transfer Particle) (211). The associated components are also direct substrates for the receptor-tyrosine kinase, and the enzymatic activities of the components are increased by phosphorylation. GAP was recently shown to be associated in the PDGF-STP (212, 213). A deletion of the kinase insert domain results in inability of the receptor to bind phosphatidylinositol 3-kinase and GAP, but not phospholipase C-γ and Raf protein, suggesting that different types of growth signals may be transmitted by these components.

GAP is the only well-characterized protein that directly and specifically interacts with *ras* p21, but the physiological roles of GAP are not yet clarified (176, 177). It has been proposed that GAP may be a possible target molecule of Ras p21. Several findings support this: (*a*) Ras p21 binds to GAP in a GTP-dependent manner (197); (*b*) GAP acts on the effector domain of Ras p21, which is thought to be the essential region for the interaction of Ras p21 with an effector (183, 184); (*c*) transforming mutants of Ras p21 can bind GAP (197). Inhibition of Ras action in *Xenopus* oocytes by a cytosol-localized mutant Ras p21 and the reduction of this effect by GAP suggest that GAP may transduce Ras-mediated signals (214). A recent study of atrial cells showing that the Ras p21·GAP complex prevents the coupling of muscarinic receptors to K$^+$ channel (215) also agrees with the above idea. On the other hand, direct association of GAP to the PDGF receptor and tyrosine-phosphorylation of GAP by various tyrosine kinases, including nonreceptor and receptor-derived oncogene products, strongly suggest that GAP is an upstream regulator of Ras p21 (177). The fact that these kinases increase the population of the activated (GTP-bound) form of p21 also supports this idea. Inhibition of GAP by phospholipids may reflect another situation: that the phospholipids are the mediators of growth signal that reduce the activity of GAP, and then activate Ras p21.

It was predicted that GAP-like activity specific for yeast Ras may be present in *S. cerevisiae* (179), and actually two genes, *IRA1* (216) and *IRA2* (217), have been found to encode proteins structurally related to GAP. These proteins have been shown to act as GAP for Ras proteins in *S. cerevisiae* by genetic and biochemical experiments (217, 218; and see Section VIII).

## VIII. RAS PROTEINS IN YEAST

### S. cerevisiae *Ras and Cyclic AMP Pathway*

Yeast Ras1 and Ras2 proteins play a crucial role in regulation at the G1 phase of the cell cycle. Ras proteins exert their effect through the control of adenylate cyclase, and modulate the activity of cAMP-dependent protein kinases (A-kinases). A-kinases display pleiotropic control through phosphorylation of proteins, and promote progression of the cell cycle (159; see Figure 10). The cell cycle control and Ras/cAMP pathway in yeast have been recently reviewed by Gibbs & Marshall (7) and Broach & Deschenes (219). We focus our discussion on the function of Ras proteins in yeast and try to compare it with that of mammalian Ras p21 protein.

Although in mammalian cells, Ras does not regulate the activity of adenylate cyclase, two observations indicate a strong conservation of the basic biological properties between mammalian Ras and yeast Ras proteins. The first is that mammalian *ras* genes can complement the function of yeast *RAS* (220, 221). The second is that a truncated yeast *RAS1* carrying an oncogenic Val mutation (*RAS1*$^{Val-9}$) can transform mammalian cells (220).

## Downstream Effector of Yeast Ras

In mammalian cells, the direct downstream effector of Ras p21 protein is still unknown. In contrast, in *S. cerevisiae*, adenylate cyclase appears to be an effector of Ras protein. Purified yeast Ras proteins can activate membrane-bound adenylate cyclase in a cell-free system (222–224). The activation depends on the nature of the Ras-bound nucleotide; GTP or its analog, Gpp(NH)p, potentiated Ras proteins, but GDP or GDP$\beta$S was inactive (223). This indicates that the Ras·GTP complex can specifically interact with yeast adenylate cyclase. Genetic evidence has also supported this idea. Mutational analyses of the mammalian *ras* gene have identified a domain that is indispensable for biological activity of Ras. This domain, encompassing residues 32–40 of Ras p21 proteins, is proposed as the "effector domain," which is the interaction site for a putative effector molecule (225). Sigal et al (226a) have demonstrated that the corresponding region of yeast Ras proteins is essential to stimulate adenylate cyclase. This observation indicates that the common domain is responsible for activating effector molecules in mammalian and yeast Ras.

Recently, it has been found that a protein named p70 (CAP) is copurified with yeast adenylate cyclase (226b, 227). The RAS-responsive adenylate cyclase may be a complex of the *CYR1* gene product and the Cap protein. The molecular cloning and genetic characterization of the gene (*CAP*) encoding the Cap protein have also been reported (227). Defective mutations of CAP showed pleiotropic phenotypes other than those due to the defect of adenylate cyclase. An allele of the *CAP* gene, named *SRV2*, was identified as an extragenic suppressor of the activated *RAS2*$^{Val-19}$ mutation (228). The precise role of the Cap protein in the interaction between Ras proteins and adenylate cyclase remains unknown.

Although the regulation of the cAMP pathway is the major function of *RAS* in *S. cerevisiae*, some genetic evidence indicates that Ras proteins may have other functions. *S. cerevisiae* cells have three distinct genes, *TPK1*, *TPK2*, and *TPK3*, encoding catalytic subunits of A-kinase (229). Upon disruption of all three *TPK* genes, cells either become nonviable or germinate and grow slowly (229). This slow growth of A-kinase-deficient mutants appears to be in sharp contrast to the absolute requirement of *RAS* for growth. Genetic studies showed that *ras1/ras2* double disruptants could not germinate under the same conditions (230). The discrepancy may be explained by assuming an alternative function of *RAS* in *S. cerevisiae*.

## Regulation of Ras Proteins in Yeast

Genetic studies have so far identified two classes of regulatory molecules for yeast *RAS*. One is the *CDC25* gene, which is considered to be an upstream activator of yeast Ras proteins. The other class contains negative regulators of *RAS*, named *IRA1* and *IRA2*.

CDC25    The *CDC25* gene was originally identified as a temperature-sensitive (ts) mutation (*cdc25*) which showed growth defects at the G1 phase of the cell cycle (231). Subsequent genetic experiments indicate that the *CDC25* gene product is involved in the regulation of *RAS*/cAMP pathway. First, the addition of cAMP corrected the growth defect of *cdc25* mutation at the restrictive temperature (232). Then, mutations identified as suppressors of *ras2⁻* were capable of suppressing ts-growth of *cdc25* (233–235). Finally, $RAS2^{Val-19}$ could bypass the requirement of the *CDC25* gene for growth (233, 236). These results suggest that the Cdc25 protein is a component of the cAMP pathway, probably acting upstream of *RAS*. The ability of activated $RAS2^{Val-19}$ but not normal *RAS2* to suppress the growth defects of *cdc25* suggests that *CDC25* may function as an activator of Ras proteins. By analogy with other GTP-binding proteins, it is assumed that the Cdc25 protein may promote a nucleotide-exchange reaction of Ras proteins. Biochemical studies have provided evidence consistent with this hypothesis. Membrane fractions prepared from *cdc25* strains showed impaired adenylate cyclase activities compared with the wild-type strains, and the addition of GTP or Gpp(NH)p restored the defect (233, 236, 237). However, some reports claim that membranes from *cdc25* mutants show reduced adenylate cyclase activity in response to added guanine nucleotides (238, 239).

The *CDC25* gene has been cloned, and its DNA sequence has revealed the presence of an open reading frame coding for a protein of 1578 amino acids with a predicted molecular size of 178 kDa (233, 240). More recently, a gene showing a strong sequence homology to the *CDC25* gene was isolated as a multicopy suppressor of the *cdc25* mutation. The gene, named *SCD25*, contained a truncated open reading frame encoding a protein 45% homologous to the *CDC25* C-terminal domain (241). Recombinant Scd25 protein dramatically promoted the in vitro nucleotide-exchange reaction of yeast and mammalian Ras proteins (242). This effect of the Scd25-truncated protein was specific to Ras·GDP. The dissociation rate of Ras·GDP was significantly increased in the presence of partially purified Scd25 proteins, but the release of GTP from Ras·GTP was not affected. The similar activity was not detected by the full-length Scd25 or Cdc25 proteins, possibly because the N-terminal portion of these proteins may have some negative regulatory functions. This may be a plausible explanation, because the activation step of Ras proteins would be regulated by upstream signals under physiological conditions.

*IRA1* AND *IRA2*    In *S. cerevisiae,* genetic analyses identified components negatively regulating the activity of *RAS*. Two genes, *IRA1* and *IRA2*, encode large proteins of 2938 and 3079 amino acid residues, respectively, with a high degree of overall structural similarity (216, 217). Disruption of either the *IRA1* or *IRA2* gene resulted in an increased sensitivity to heat shock and nitrogen starvation, as well as sporulation deficiency (217). These phenotypes are analogous to those of the activated *RAS2* mutation, $RAS2^{Val-19}$. Further genetic experiments showed that *IRA1* and *IRA2* gene products function as negative regulators of *RAS*, antagonistic to *CDC25*. This was demonstrated by biochemical analysis of the guanine-nucleotides bound to Ras proteins (218). In the wild-type strains, overexpressed Ras1 or Ras2 protein existed mostly in the GDP-bound form, whereas in the *ira1-* and *ira2*-disrupted cells, the amount of the GTP-bound form was significantly increased (218).

As described in Section VII, in mammalian cells, the cytoplasmic factor GAP stimulates the GTPase activity of Ras p21 proteins. It was found that GAP can act on the normal p21 but not on $p21^{Val-12}$ or $p21^{Thr-59}$ (176). In agreement with this finding, the corresponding mutants of yeast *RAS2*, $RAS2^{Val-19}$ and $RAS2^{Thr-66}$, showed a high percentage of GTP-bound form in both wild-type and *ira1/ira2* mutant cells (218). This suggests that the *IRA* may function in a manner similar to mammalian GAP. In fact, a significant structural homology was found between *IRA* and GAP (216, 217). The catalytic domains of Ira1 and Ira2 proteins shared 45% homology with the C-terminal part of bovine GAP. It has been reported that the C-terminal portion of GAP is sufficient to stimulate GTPase of Ras p21 proteins (199). Furthermore, bovine GAP could effectively suppress *ira⁻* phenotypes, by reducing the increased level of Ras·GTP in *ira⁻* cells (218, 243a). Recently, Tanaka et al (243b) have found that recombinant Ira protein has GAP activity for yeast RAS proteins.

A human gene named *NF1*, which is involved in the pathogenesis of the neurofibromatosis type1 disease, was found to code for a protein homologous to GAP and Ira (244a, b). Remarkably, a homologous region between yeast Ira and human NF1 was larger than that between human GAP and human NF1. The sequence alignment of Ira1 and NF1 identified a contiguous homologous stretch of 1500 amino acids. This suggests that NF1 proteins may be human counterparts of yeast Ira proteins. In fact, the presence of the GAP activity in recombinant NF1 proteins had recently been demonstrated (244c, d, e).

PHOSPHORYLATION OF RAS PROTEIN    Both Ras1 and Ras2 proteins are phosphorylated in yeast cells (245, 246). Resnick & Racker (247) have shown that purified Ras proteins can be phosphorylated by A-kinase and the phosphorylated Ras protein is twofold less active in stimulating adenylate cyclase. Since A-kinase is downstream of *RAS* in the cAMP pathway, such phosphorylation may be the mechanism of feedback control of the pathway.

## *RAS Proteins in* Schizosaccharomyces pombe

Only a single *ras* gene, *ras1*, has been identified in *Schizosaccharomyces pombe* (26, 248). *S. pombe ras1* gene was neither essential for growth nor involved in the control of adenylate cyclase (27, 249a). Instead, *ras1* mutant cells showed a sterile phenotype in haploid cells. Disruption of *ras1* did not impair secretion of a diffusible pheromone-like factor from h⁻ cells, but *ras1⁻*/h⁺ cells were not able to respond to the factor (249b). These results suggest the involvement of *S. pombe ras1* in the mating factor signal transduction. Disruption of *ras1* also resulted in other apparently divergent phenotypes. Cells lacking *ras1* had some defects in morphology, and showed round and small cell shapes. In addition, the sporulation efficiency of *ras1⁻* diploid cells was severely decreased.

Investigations of *ras1*-mediated signalling pathways in *S. pombe* have identified several components. A gene coding for a protein kinase with unknown function was isolated as a multicopy suppressor of the *ras1* mutation (254). Recently, *S. pombe ste6* gene was found to encode a protein homologous to *S. cerevisiae CDC25*. Phenotypes of *ste6* mutants were analogous to those of *ras1⁻* cells, and an activated allele of *ras1*, *ras1*$^{Val-17}$, was able to suppress the mating defect of *ste6* cells (250). The downstream target of *ras1* and the GAP/IRA counterpart in *S. pombe* remain to be identified.

ACKNOWLEDGMENTS

We thank our collaborators, whose names are cited in the references, for their valuable contributions and stimulating discussions throughout this work. The work, which has been carried out at the Institute of Medical Science, University of Tokyo, was supported by the Grant-in-Aid from the Ministry of Education, Science, and Culture, Japan. Thanks are due to Dr. J. Allan Waitz, for hosting us at DNAX Research Institute since April 1990, and also for critical reading of the manuscript. We thank Robin Rodgers, Masahide Kikkawa, Gary Burget, and Dovie Wylie for their help in preparing the manuscript.

## Literature Cited

1. Kaziro, Y. 1978. *Biochim. Biophys. Acta* 505:95–127
2. Gilman, A. G. 1987. *Annu. Rev. Biochem.* 56:615–49
3. Barbacid, M. 1987. *Annu. Rev. Biochem.* 56:779–827
4. Neer, E. J., Clapham, D. E. 1988. *Nature* 333:129–34
5. Freissmuth, M., Casey, P. J., Gilman, A. G. 1989. *FASEB J.* 3:2125–31
6. Ross, E. M. 1989. *Neuron* 3:141–52
7. Gibbs, J. B., Marshall, M. S. 1989. *Microbiol. Rev.* 53:171–85
8. Kaziro, Y. 1990. See Ref. 14, pp. 189–206
9. Hall, A. 1990. *Science* 249:635–40
10. Bourne, H. R., Sanders, D. A., McCormick, F. 1990. *Nature* 348:125–32; 1991. *Nature* 349:117–27
11. Bosch, L., Kraal, B., Parmeggiani, A., eds. 1988. *The Guanine Nucleotide Binding Proteins, Common Structural and Functional Properties.* New York: Plenum. 409 pp.
12. Iyengar, R., Birnbaumer, L., eds. 1990. *G Proteins.* New York: Academic. 651 pp.
13. Houslay, M. D., Milligan, G., eds. 1990. *G-proteins as Mediators of Cellular Signaling Processes.* New York: Wiley. 232 pp.
14. Moss, J., Vaughan, M., eds. 1990. *ADP-Ribosylating Toxins and G Proteins, Insight into Signal Transduction.* Washington, DC: Am. Soc. Microbiol. 567 pp.
15. Kornberg, A. 1980. *DNA Replication.* San Francisco: Freeman. (See also 1982 suppl.)
16. Jurnak, F. 1985. *Science* 230:32–36
17. LaCour, T. F. M., Nyborg, J., Thirup, S., Clark, B. F. C. 1985. *EMBO J.* 4:2385–88
18. Pai, E. F., Kabsch, W., Krengel, U., Holmes, K. C., John, J., et al. 1989. *Nature* 341:209–14
19. Milburn, M. V., Tong, L., deVos, A. M., Brunger, A., Yamaizumi, Z., et al. 1990. *Science* 247:939–45
20. Yamane, H. K., Fransworth, C. C., Xie, H., Howald, W., Fung, B. K.-K., et al. 1990. *Proc. Natl. Acad. Sci. USA* 87:5868–72
21. Mumby, S. M., Casey, P. J., Gilman, A. G., Gutowski, S., Sternweis, P. C. 1990. *Proc. Natl. Acad. Sci. USA* 87:5873–77
22. Fukada, Y., Takao, T., Ohguro, H., Yoshizawa, T., Akino, T., et al. 1990. *Nature* 346:658–60
23. Strathmann, M., Wilkie, T. M., Simon, M. I. 1989. *Proc. Natl. Acad. Sci. USA* 86:7407–9
24. Noda, M., Ko, M., Ogura, A., Liu, D., Amano, T., et al. 1985. *Nature* 318:73–75
25. Bar-Sagi, D., Ferumisco, J. R. 1985. *Cell* 42:841–48
26. Fukui, Y., Kaziro, Y. 1985. *EMBO J.* 4:687–91
27. Fukui, Y., Kozasa, T., Kaziro, Y., Takeda, T., Yamamoto, M. 1986. *Cell* 44:329–36
28. deVos, A. M., Tong, L., Milburn, M. V., Matias, P. M., Jancarik, J., et al. 1988. *Science* 239:888–93
29. Pai, E. F., Krengel, U., Petsko, G. A., Googy, R. S., Kabsch, W., et al. 1990. *EMBO J.* 9:2351–59
30. Jurnak, F., Heffron, S., Bergmann, E. 1990. *Cell* 60:525–28
31. Landis, C. A., Masters, S. B., Spada, A., Pace, A. M., Bourne, H. R., et al. 1989. *Nature* 340:692–96
32. Lyons, J., Landis, C. A., Harsh, G., Vallar, L., Grunewald, K., et al. 1990. *Science* 249:655–59
33. Kozasa, T., Itoh, H., Tsukamoto, T., Kaziro, Y. 1988. *Proc. Natl. Acad. Sci. USA* 85:2081–85
34. Itoh, H., Toyama, R., Kozasa, T., Tsukamoto, T., Matsuoka, M., Kaziro, Y. 1988. *J. Biol. Chem.* 263:6656–64
35. Weinstein, L. S., Spiegel, A. M., Carter, A. D. 1988. *FEBS Lett.* 232:333–40
36a. Lavu, S., Clark, J., Swarup, R., Matsushima, K., Paturu, K., et al. 1988. *Biochem. Biophys. Res. Commun.* 150:811–15
36b. Matsuoka, M., Itoh, H., Kozasa, T., Kaziro, Y. 1988. *Proc. Natl. Acad. Sci. USA* 85:5384–88
37. Tsukamoto, T., Toyama, R., Itoh, H., Kozasa, T., Matsuoka, M., Kaziro, Y. 1991. *Proc. Natl. Acad. Sci. USA.* In press
38. Matsuoka, M., Itoh, H., Kaziro, Y. 1990. *J. Biol. Chem.* 265:13215–20
39. Raport, C. J., Dere, B., Hurley, J. B. 1989. *J. Biol. Chem.* 264:7122–28
40. Bray, P., Carter, A., Simons, C., Guo, V., Puckett, C., et al. 1986. *Proc. Natl. Acad. Sci. USA* 83:8893–97
41. Bray, P., Carter, A., Guo, V., Puckett, C., Kamholz, J., et al. 1987. *Proc. Natl. Acad. Sci. USA* 84:5115–19
42. Beals, C. R., Wilson, C. B., Perlmutter, R. M. 1987. *Proc. Natl. Acad. Sci. USA* 84:7886–90
43. Kim, S., Ang, S. L., Bloch, D. B., Bloch, K. D., Kawahara, Y., et al. 1988. *Proc. Natl. Acad. Sci. USA* 85:4153–57
44. Didsbury, J. R., Ho, Y., Snyderman, R. 1987. *FEBS. Lett.* 211:160–64

45. Didsbury, J. R., Snyderman, R. 1987. *FEBS Lett.* 219:259–63
46. Mattera, R., Codina, J., Crozat, A., Kidd, V., Woo, S. L. C., et al. 1986. *FEBS Lett.* 206:36–41
47. Suki, W. N., Abramowitz, J., Mattera, R., Codina, J., Birnbaumer, L., 1987. *FEBS Lett.* 220:187–92
48. Codina, J., Olate, J., Abramowitz, J., Mattera, R., Cook, R. G., Birnbaumer, L. 1988. *J. Biol. Chem.* 263:6746–50
49. Fong, H. K. W., Yoshimoto, K. K., Eversole-Cire, P., Simon, M. I. 1988. *Proc. Natl. Acad. Sci. USA* 85:3066–70
50. Harris, B. A., Robishaw, J. D., Mumby, S. M., Gilman, A. G. 1985. *Science* 229:1274–77
51. Robishaw, J. D., Russell, D. W., Harris, B. A., Smigel, M. D., Gilman, A. G. 1986. *Proc. Natl. Acad. Sci. USA* 83:1251–55
52. Nukada, T., Tanabe, T., Takahashi, H., Noda, M., Hirose, T., et al. 1986. *FEBS Lett.* 195:220–24
53. Nukada, T., Tanabe, T., Takahashi, H., Noda, M., Haga, K., et al. 1986. *FEBS Lett.* 197:305–10
54. Michel, T., Winslow, J. W., Smith, J. A., Seidman, J. G., Neer, E. J. 1986. *Proc. Natl. Acad. Sci. USA* 83:7663–67
55. Van Meurs, K. P., Angus, C. W., Lavu, S., Kung, H.-F., Czarnecki, S. K., et al. 1987. *Proc. Natl. Acad. Sci. USA* 84:3107–11
56. Ovchinnikov, Y. A., Slepak, V. Z., Pronin, A. N., Shlensky, A. B., Levina, N. B., et al. 1987. *FEBS Lett.* 226:91–95
57. Price, S. R., Murtagh, J. J., Tsuchiya, M., Serventi, I. M., Van Meurs, K. P., et al. 1990. *Biochemistry* 29:5069–76
58. Tanabe, T., Nukada, T., Nishikawa, Y., Sugimoto, K., Suzuki, H., et al. 1985. *Nature* 315:242–45
59. Medynski, D. C., Sullivan, K., Smith, D., Van Dop, C., Chang, F.-H., et al. 1985. *Proc. Natl. Acad. Sci. USA* 82:4311–15
60. Yatsunami, K., Khorana, H. G. 1985. *Proc. Natl. Acad. Sci. USA* 82:4316–20
61. Lochrie, M. A., Hurley, J. B., Simon, M. I. 1985. *Science* 228:96–99
62. Hsu, W. H., Rudolph, U., Sanford, J., Bertrand, P., Olate, J., et al. 1990. *J. Biol. Chem.* 265:11220–26
63. Merchen, L., Moras, V., Tocque, B., Mayaux, J.-F. 1990. *Nucleic Acids Res.* 18:662
64. Itoh, H., Kozasa, T., Nagata, S., Nakamura, S., Katada, T., et al. 1986. *Proc. Natl. Acad. Sci. USA* 83:3776–80
65. Jones, D. T., Reed, R. R. 1987. *J. Biol. Chem.* 262:14241–49
66. Jones, D. T., Reed, R. R. 1989. *Science* 244:790–95
67. Sullivan, K. A., Liao, Y.-C., Alborzi, A., Beiderman, B., Chang, F.-H., et al. 1986. *Proc. Natl. Acad. Sci. USA* 83:6687–91
68. Sullivan, K. A., Miller, R. T., Masters, S. B., Beiderman, B., Heideman, W., et al. 1987. *Nature* 330:758–60
69. Rall, T., Harris, B. A. 1987. *FEBS Lett.* 224:365–71
70. Strathmann, M., Wilkie, T. M., Simon, M. I. 1990. *Proc. Natl. Acad. Sci. USA* 87:6477–81
71. Robishaw, J. D., Smigel, M. D., Gilman, A. G. 1986. *J. Biol. Chem.* 261:9587–90
72. Graziano, M. P., Casey, P. J., Gilman, A. G. 1987. *J. Biol. Chem.* 262:11375–81
73. Itoh, H., Katada, T., Ui, M., Kawasaki, H., Suzuki, K., et al. 1988. *FEBS Lett.* 230:85–89
74. Morishita, R., Asano, T., Kato, K., Itoh, H., Kaziro, Y. 1989. *Biochem. Biophys. Res. Commun.* 161:1280–85
75. Brann, M. R., Collins, R. M., Spiegel, A. 1988. *FEBS Lett.* 222:191–98
76. Garibay, J. L. R., Kozasa, T., Itoh, H., Tsukamoto, T., Matsuoka, M., Kaziro, Y. 1991. *Biochim. Biophys. Acta.* Submitted
77. Goldsmith, P., Backlund, P. S., Rossiter, K., Carter, A., Milligan, G., et al. 1988. *Biochemistry* 27:7085–90
78. Milligan, G., Tanfin, Z., Goureau, O., Unson, C., Harbon, S. 1989. *FEBS Lett.* 244:411–16
79a. Inanobe, A., Shibasaki, H., Takahashi, K., Kobayashi, I., Tomita, U., et al. 1990. *FEBS Lett.* 263:369–72
79b. Price, S. R., Tsai, S.-C., Adamik, R., Angus, C. W., Serventi, I. M., et al. 1989. *Biochemistry* 28:3803–7
80. Lerea, C. L., Somers, D. E., Hurley, J. B., Klock, I. B., Bunt-Milam, A. H. 1986. *Science* 234:77–80
81. Casey, P. J., Fong, H. K. W., Simon, M. I., Gilman, A. G. 1990. *J. Biol. Chem.* 265:2383–90
82. Masters, S. B., Martin, M. W., Harden, T. K., Brown, J. H. 1985. *Biochem. J.* 227:933–37
83. Nakajima, Y., Nakajima, S., Inoue, M. 1988. *Proc. Natl. Acad. Sci. USA* 85:3643–47
84. Graziano, M. P., Gilman, A. G. 1989. *J. Biol. Chem.* 264:15475–82
85. Masters, S. B., Miller, R. T., Chi, M.-H., Chang, F.-H., Beiderman, B., et al. 1989. *J. Biol. Chem.* 264:15467–74
86. Miller, R. T., Masters, S. B., Sullivan, K. A., Beiderman, B., Bourne, H. R. 1988. *Nature* 334:712–15

87. Clementi, E., Malgaretti, N., Meldolesi, J., Taramelli, R. 1990. *Oncogene* 5:1059–61
88. Van Dop, C., Tsubokawa, M., Bourne, H. R., Ramachandran, J. 1984. *J. Biol. Chem.* 259:696–98
89. Freissmuth, M., Gilman, A. G. 1989. *J. Biol. Chem.* 264:21907–14
90. Iiri, T., Tohkin, M., Morishima, N., Ohoka, Y., Ui, M., et al. 1989. *J. Biol. Chem.* 264:21394–400
91. Gierschik, P., Sidiropoulos, D., Jakobs, K. H. 1989. *J. Biol. Chem.* 264:21470–73
92. Ui, M. 1984. *Trends Pharmacol. Sci.* 5:277–79
93. West, R. E. Jr., Moss, J., Vaughan, M., Liu, T., Liu, T.-Y. 1985. *J. Biol. Chem.* 260:14428–30
94. Shinohara, T., Dietzschold, B., Craft, C. M., Wistow, G., Early, J. J., et al. 1987. *Proc. Natl. Acad. Sci. USA* 84:6975–79
95. Wilden, U., Hall, S. W., Kuhn, H. 1986. *Proc. Natl. Acad. Sci. USA* 83:1174–78
96. Hamm, H. E., Deretic, D., Arendt, A., Hargrave, P. A., Koenig, B., et al. 1988. *Science* 241:832–35
97. Palm, D., Munch, G., Malek, D., Dees, C., Hekman, M. 1990. *FEBS Lett.* 261:294–98
98. Masters, S. B., Sullivan, K. A., Miller, R. T., Beiderman, B., Lopez, N. G., et al. 1988. *Science* 241:448–51
99. Woon, C. W., Soparkar, S., Heasley, L., Johnson, G. L. 1989. *J. Biol. Chem.* 264:5687–93
100. Navon, S. E., Fung, B.-K. 1987. *J. Biol. Chem.* 262:15746–51
101. Neer, E. J., Pulsifer, L., Wolf, L. G. 1988. *J. Biol. Chem.* 263:8996–9000
102. Jones, T. L. Z., Simonds, W. F., Merendino, J. J. Jr., Brann, M. R., Spiegel, A. M. 1990. *Proc. Natl. Acad. Sci. USA* 87:568–72
103. Linder, M. E., Ewand, D. A., Miller, R. J., Gilman, A. G. 1990. *J. Biol. Chem.* 265:8243–51
104. Linder, M. E., Pang, I. H., Buroino, R. J., Gordon, J. I., Sternweis, P. C., Gilman, A. G. 1991. *J. Biol. Chem.* 266: In press
105. Itoh, H., Gilman, A. 1991. Submitted
106. Holbrook, S., Kim, S.-H. 1989. *Proc. Natl. Acad. Sci. USA* 86:1751–55
107. Masters, S. B., Stroud, R. M., Bourne, H. R. 1986. *Protein Eng.* 1:47–54
108. Deretic, D., Hamm, H. E. 1987. *J. Biol. Chem.* 262:10839–47
109. Hingorani, V. N., Ho, Y.-K. 1987. *FEBS Lett.* 220:15–22
110. Sugimoto, K., Nukada, T., Tanabe, T., Takahasi, H., Noda, M., et al. 1985. *FEBS Lett.* 191:235–40
111. Fong, H. K. W., Amatruda, T. T., Birren, B. W., Simon, M. I. 1987. *Proc. Natl. Acad. Sci. USA* 84:3792–96
112. Gao, B., Gilman, A. G., Robishaw, J. D. 1987. *Proc. Natl. Acad. Sci. USA* 84:6122–25
113. Amatruda, T. T., Gautam, N. H., Fong, H. K. W., Northup, J. K., Simon, M. I. 1988. *J. Biol. Chem.* 263:5008–11
114. Gao, B., Mumby, S. M., Gilman, A. G. 1987. *J. Biol. Chem.* 262:17254–57
115. Hurley, J. B., Fong, H. K. W., Teplow, D. B., Dreyer, W. J., Simon, M. I. 1984. *Proc. Natl. Acad. Sci. USA* 81:6948–52
116. Hurley, J. B., Simon, M. I., Teplow, D. B., Robishaw, J. D., Gilman, A. G. 1984. *Science* 226:860–62
117. Gautam, N., Baetscher, M., Aebersold, R., Simon, M. I. 1989. *Science* 244:971–74
118. Robishaw, J. D., Kalman, V. K., Moomaw, C. R., Slaughter, C. A. 1989. *J. Biol. Chem.* 264:15758–61
119. Katada, T., Bokoch, G. M., Northup, J. K., Ui, M., Gilman, A. G. 1984. *J. Biol. Chem.* 259:3568–77
120. Casey, P. J., Graziano, M. P., Gilman, A. G. 1989. *Biochemistry* 28:611–16
121. Woon, C. W., Heasley, L., Osawa, S., Johnson, G. L. 1989. *Biochemistry* 28:4547–51
122. Gupta, S. K., Diez, E., Heasley, L. E., Osawa, S., Johnson, G. L. 1990. *Science* 249:662–66
123. Osawa, S., Heasley, L. E., Dhanasekaran, N., Gupta, S. K., Woon, C. W. 1990. *Mol. Cell. Biol.* 10.2931–40
124. Kahn, R., Gilman, A. G. 1984. *J. Biol. Chem.* 259:6235–40
125. Nakafuku, M., Obara, T., Kaibuchi, K., Miyajima, I., Miyajima, A., et al. 1988. *Proc. Natl. Acad. Sci. USA* 85:1374–78
126. Nakafuku, M., Itoh, H., Nakamura, S., Kaziro, Y. 1987. *Proc. Natl. Acad. Sci. USA* 84:2140–44
127. Miyajima, I., Nakafuku, M., Nakayama, N., Brenner, C., Miyajima, A., et al. 1987. *Cell* 50:1011–19
128. Jahng, K.-Y., Ferguson, J., Reed, S. I. 1988. *Mol. Cell. Biol.* 8:2484–93
129. Miller, A. M., MacKay, V. L., Nasmyth, K. A. 1985. *Nature* 314:598–603
130. Herskowitz, I. 1989. *Nature* 342:749–57
131. Burkholder, A. C., Hartwell, L. H. 1985. *Nucleic Acids Res.* 13:8463–75
132. Nakayama, N., Miyajima, A., Arai, K. 1985. *EMBO J.* 4:2643–48
133. Hagen, D. C., McCaffrey, G., Sprague,

G. F. Jr. 1986. *Proc. Natl. Acad. Sci. USA* 82:1418–22
134. Dixon, R. A. F., Kobilka, B. K., Strader, D. J., Benovic, J. L., Dohlman, H. G., et al. 1986. *Nature* 321:75–79
135. Ovchinnikov, Y. A. 1982. *FEBS Lett.* 148:179–91
136. Nathans, J., Hogness, D. S. 1983. *Cell* 34:807–14
137. Dietzel, C., Kurjan, J. 1987. *Cell* 50:1001–10
138. Fujimura, H. 1989. *Mol. Cell. Biol.* 9:152–58
139. Whiteway, M., Hougan, L., Dignard, D., Thomas, D. Y., Bell, L., et al. 1989. *Cell* 56:467–77
140. Fong, H. K. W., Hurley, J. B., Hopkins, R. S., Miake-Lye, R., Johnson, M. S., et al. 1986. *Proc. Natl. Acad. Sci. USA* 83:2162–66
141. Codina, J., Stengel, D., Woo, S. L. C., Birnbaumer, L. 1986. *FEBS Lett.* 207:187–92
142. Yatsunami, K., Pandya, B. V., Oprian, D. D., Khorana, H. G. 1985. *Proc. Natl. Acad. Sci. USA* 82:1936–40
143. Willumsen, B. M., Christensen, A., Hubbert, N. L., Papageorge, A. G., Lowy, D. R. 1984. *Nature* 310:583–86
144. Hancock, J. F., Magee, A. I., Childs, J. E., Marshall, C. J. 1989. *Cell* 57:1167–77
145. Finegold, A. A., Schafer, W. R., Rine, J., Whiteway, M., Tamanoi, F. 1990. *Science* 249:165–69
146. Katada, T., Oinuma, M., Ui, M. 1986. *J. Biol. Chem.* 261:5215–21
147. Jelsema, C. L., Axelrod, J. 1987. *Proc. Natl. Acad. Sci. USA* 84:3623–27
148. Matsumoto, K., Miyajima, I., Nomoto, S., Nakayama, N., Arai, K. 1990. *Developmental Biology*, pp. 289–306. Wiley-Liss
149. Nakayama, N., Kaziro, Y., Arai, K., Matsumoto, K. 1988. *Mol. Cell. Biol.* 8:3777–83
150. MacKay, V., Manney, T. R. 1974. *Genetics* 76:255–71
151. MacKay, V., Manney, T. R. 1974. *Genetics* 76:273–88
152. Hartwell, L. H. 1980. *J. Cell Biol.* 85:811–22
153. Blinder, D., Bouvier, S., Jenness, D. D. 1989. *Cell* 56:479–86
154. Nomoto, S., Nakayama, N., Arai, K., Matsumoto, K. 1990. *EMBO J.* 9:691–96
155. Miyajima, I., Arai, K., Matsumoto, K. 1989. *Mol. Cell. Biol.* 9:2289–97
156. Gibbs, J. B., Sigal, I. S., Poe, M., Scolnick, E. M. 1984. *Proc. Natl. Acad. Sci. USA* 81:5704–8
157. Seeburg, P. H., Colby, W. W., Capon,

D. J., Goeddel, D. V., Levinson, A. D. 1984. *Nature* 312:71–75
158. Stone, D. E., Reed, I. S. 1990. *Mol. Cell. Biol.* 10:4439–46
159. Matsumoto, K., Uno, I., Ishikawa, T. 1985. *Yeast* 1:15–24
160. Eraso, P., Gancedo, M. 1985. *FEBS Lett.* 191:51–54
161. Quan, F., Wolfgang, W. J., Forte, M. A. 1989. *Proc. Natl. Acad. Sci. USA* 86:4321–25
162. Provost, N. M., Somers, D. E., Hurley, J. B. 1988. *J. Biol. Chem.* 263:12070–76
163. Yoon, J. S., Shortridge, R. D., Bloomquist, B. T., Schneuwly, S., Perdew, M. H., Pak, W. L. 1989. *J. Biol. Chem.* 264:18536–43
164a. de Sousa, S. M., Hoveland, L. L., Yarfitz, S., Hurley, J. B. 1989. *J. Biol. Chem.* 264:18544–51
164b. Thambi, N. C., Quan, F., Wolfgang, W. J., Spiegel, A., Forte, M. 1989. *J. Biol. Chem.* 264:18552–60
165. Yarfitz, S., Provost, N. M., Hurley, J. B. 1988. *Proc. Natl. Acad. Sci. USA* 85:7134–38
166. Quan, F., Forte, M. A. 1990. *Mol. Cell. Biol.* 10:910–17
167. Pupillo, M., Kumagai, A., Pitt, G. S., Firtel, R. A., Devreotes, P. N. 1989. *Proc. Natl. Acad. Sci. USA* 86:4892–96
168. Snaar-Jagalska, B. E., De Wit, R. J. W., Van Haastert, P. J. M. 1988. *FEBS Lett.* 232:148–52
169. Janssens, P. M. W., Van Haastert, P. J. M. 1987. *Microbiol. Rev.* 51:396–418
170. Firtel, R. A., Van Haastert, P. J. M., Kimmel, A. R., Deveotes, P. N. 1989. *Cell* 58:235–39
171. Klein, P., Sun, T. J., Saxe, C. L., Kimmel, A. R., Devreotes, P. N. 1988. *Science* 241:1467–72
172. Kumagai, A., Pupillo, M., Gundersen, R., Miake-Lye, R., Devreotes, P. N., et al. 1989. *Cell* 57:265–75
173. Ma, H., Yanofsky, M. F., Meyerowitz, E. M. 1990. *Proc. Natl. Acad. Sci. USA* 87:3821–25
174. Obara, T., Nakafuku, M., Yamamoto, M., Kaziro, Y. 1991. *Proc. Natl. Acad. Sci. USA.* Submitted
175. Trahey, M., McCormick, F. 1987. *Science* 238:542–45
176. McCormick, F. 1989. *Cell* 56:5–8
177. Hall, A. 1990. *Cell* 61:921–23
178. Satoh, T., Nakamura, S., Kaziro, Y. 1987. *Mol. Cell. Biol.* 7:4553–56
179. Gibbs, J. B., Schaber, M. D., Marshall, M. S., Scolnick, E. M., Sigal, I. S. 1987. *J. Biol. Chem.* 262:10426–29
180. Satoh, T., Endo, M., Nakamura, S.,

Kaziro, Y. 1988. *FEBS Lett.* 236:185–89

181. Hoshino, M., Kawakita, M., Hattori, S. 1988. *Mol. Cell. Biol.* 8:4169–73

182. Schlichting, I., Almo, S. C., Rapp, G., Wilson, K., Petratos, K., et al. 1990. *Nature* 345:309–15

183. Cales, C., Hancock, J. F., Marshall, C. J., Hall, A. 1988. *Nature* 332:548–51

184. Adari, H., Lowy, D. R., Willumsen, B. M., Der, C. J., McCormick, F. 1988. *Science* 240:518–21

185. Srivastava, S. V., Donato, A. D., Lacal, J. C. 1989. *Mol. Cell. Biol.* 9:1779–83

186. Satoh, T., Endo, M., Nakafuku, M., Nakamura, S., Kaziro, Y. 1990. *Proc. Natl. Acad. Sci. USA* 87:5993–97

187. Satoh, T., Endo, M., Nakafuku, M., Akiyama, T., Yamamoto, T., Kaziro, Y. 1990. *Proc. Natl. Acad. Sci. USA* 87:7926–29

188. Gibbs, J. B., Marshall, M. S., Scolnick, E. M., Dixon, R. A. F., Vogel, U. S. 1990. *J. Biol. Chem.* 265:20437–42

189. Mulcahy, L. S., Smith, M. R., Stacey, D. W. 1985. *Nature* 313:241–43

190. Smith, M. R., DeGudicibus, S. J., Stacey, D. W. 1986. *Nature* 320:540–43

191. Downward, J., Graves, J. D., Warne, P. H., Rayter, S., Cantrell, D. A. 1990. *Nature* 346:719–23

192. Kamata, T., Feramisco, J. R. 1984. *Nature* 310:147–50

193. West, M., Kung, H., Kamata, T. 1990. *FEBS Lett.* 259:245–48

194. Wolfman, A., Macara, I. G. 1990. *Science* 248:67–69

195. Downward, J., Riehl, E., Wu, L., Weinberg, R. A. 1990. *Proc. Natl. Acad. Sci. USA* 87:5998–6002

196. Gibbs, J. B., Schaber, M. D., Allard, W. J., Sigal, I. S., Scolnick, E. M. 1988. *Proc. Natl. Acad. Sci. USA* 85:5026–30

197. Vogel, U. S., Dixon, R. A. F., Schaber, M. D., Diehl, R. E., Marshall, M. S., et al. 1988. *Nature* 335:90–93

198. Trahey, M., Wong, G., Halenbeck, R., Rubinfeld, B., Martin, G. A., et al. 1988. *Science* 242:1697–700

199. Marshall, M. S., Hill, W. S., Ng, A. S., Vogel, U. S., Schaber, M. D., et al. 1989. *EMBO J.* 8:1105–10

200. Lemons, R. S., Espinosa, R., Rebentisch, M., McCormick, F., Ladner, M., et al. 1990. *Genomics* 6:383–85

201. Pizon, V., Chardin, P., Lerosey, I., Olofsson, B., Tavitian, A. 1988. *Oncogene* 3:201–4

202. Kitayama, H., Sugimoto, Y., Matsuzaki, T., Ikawa, Y., Noda, M. 1989. *Cell* 56:77–84

203. Kawata, M., Matsui, Y., Kondo, J., Hishida, T., Teranishi, Y., et al. 1988. *J. Biol. Chem.* 263:18965–71

204. Hata, Y., Kikuchi, A., Sasaki, T., Schaber, M. D., Gibbs, J. B., et al. 1990. *J. Biol. Chem.* 265:7104–7

205. Frech, M., Jhon, J., Pizon, V., Chardin, P., Tavitian, A., et al. 1990. *Science* 249:169–71

206. Kikuchi, A., Sasaki, T., Araki, S., Hata, Y., Takai, Y. 1989. *J. Biol. Chem.* 264:9133–36

207. Tsai, M.-H., Yu, C.-L., Wei, F.-S., Stacey, D. W. 1989. *Science* 243:522–26

208. Yu, C.-L., Tsai, M.-H., Stacey, D. W. 1988. *Cell* 52:63–71

209. Molloy, C. J., Bottaro, D. P., Fleming, T. P., Marshall, M. S., Gibbs, J. B., et al. 1989. *Nature* 342:711–14

210. Ellis, C., Moran, M., McCormick, F., Pawson, T. 1990. *Nature* 343:377–81

211. Ullrich, A., Schlessinger, J. 1990. *Cell* 61:203–12

212. Kazlauskas, A., Ellis, C., Pawson, T., Cooper, J. A. 1990. *Science* 247:1578–81

213. Kaplan, D. R., Morrison, D. K., Wong, G., McCormick, F., Williams, L. T. 1990. *Cell* 61:125–33

214. Gibbs, J. B., Schaber, M. D., Schofield, T. L., Scolnick, E. M., Sigal, I. S. 1989. *Proc. Natl. Acad. Sci. USA* 86:6630–34

215. Yatani, A., Okabe, K., Polakis, P., Halenbeck, R., McCormick, F., Brown, A. M. 1990. *Cell* 61:769–76

216. Tanaka, K., Matsumoto, K., Toh-e, A. 1989. *Mol. Cell. Biol.* 9:757–68

217. Tanaka, K., Nakafuku, M., Tamanoi, F., Kaziro, Y., Matsumoto, K., et al. 1990. *Mol. Cell. Biol.* 10:4303–13

218. Tanaka, K., Nakafuku, M., Satoh, T., Marshall, M. S., Gibbs, J. B., et al. 1990. *Cell* 60:803–7

219. Broach, J. R., Deschenes, R. J. 1990. *Adv. Cancer Res.* 54:79–139

220. DeFeo-Jones, D., Tatchell, K., Robinson, L. C., Sigal, I. S., Vass, W. C., et al. 1985. *Science* 228:179–84

221. Kataoka, T., Powers, S., Cameron, S., Fasano, O., Goldfarb, M., et al. 1985. *Cell* 40:19–26

222. Broek, D., Samiy, N., Fasano, O., Fujiyama, A., Tamanoi, F., et al. 1985. *Cell* 41:763–69

223. Field, J., Broek, D., Kataoka, T., Wigler, M. 1987. *Mol. Cell. Biol.* 7:2128–33

224. Gibbs, J. B., Marsico-Ahern, J. D., Scolnick, E. M., Sigal, I. S. 1988. *Biochem. J.* 252:289–92

225. Willumsen, B. M., Papageorge, A. G.,

Kung, H., Bekesi, E., Robins, T., et al. 1986. *Mol. Cell. Biol.* 6:2646–54

226a. Sigal, I. S., Gibbs, J. B., D'alonzo, J. S., Scolnick, E. M. 1986. *Proc. Natl. Acad. Sci. USA* 83:4725–29

226b. Field, J., Nikawa, I., Broek, D., Mac-Donald, B., Rodgers, L., et al. 1988. *Mol. Cell. Biol.* 8:2159–65

227. Field, J., Vojtek, A., Ballester, R., Bolger, G., Colicelli, J., et al. 1990. *Cell* 61:319–27

228. Fedor-Chaiken, M., Deschenes, R. J., Broach, J. R. 1990. *Cell* 61:329–40

229. Toda, T., Cameron, S., Sass, P., Zoller, M., Wigler, M. 1987. *Cell* 50:277–88

230. Kataoka, T., Powers, S., McGill, C., Fasano, O., Strathern, J., et al. 1984. *Cell* 37:437–46

231. Hartwell, L. H. 1973. *Exp. Cell Res.* 76:111–17

232. Martegani, E., Baroni, M., Wanoni, M. 1986. *Exp. Cell Res.* 162:544–48

233. Broek, D., Toda, T., Michaeli, T., Levin, L., Birchmeier, C., et al. 1987. *Cell* 48:789–800

234. Cannon, J. F., Gibbs, J. B., Tatchell, K. 1986. *Genetics* 113:247–64

235. Lisziewicz, J., Godany, A., Forster, H. H., Kuntzel, H. 1987. *J. Biol. Chem.* 262:2549–53

236. Robinson, L. C., Gibbs, J. B., Marshall, M. S., Sigal, I. S., Tatchell, K. 1987. *Science* 235:1218–21

237. Marshall, M. S., Gibbs, J. B., Scolnick, E. M., Sigal, I. S. 1987. *Mol. Cell. Biol.* 7:2309–15

238. Daniel, J., Becker, J. M., Enari, E., Levitzki, A. 1987. *Mol. Cell. Biol.* 7:3857–61

239. Petitjean, A., Hilger, F., Tatchell, K. 1990. *Genetics* 124:797–806

240. Camonis, J. H., Kalekine, M., Gondre, B., Garreau, H., Boy-Marcotte, E., et al. 1986. *EMBO J.* 5:375–80

241. Boy-Marcotte, E., Damak, F., Camonis, J., Garreau, H., Jacquet, M. 1989. *Gene* 77:21–30

242. Créchet, J. B., Poullet, P., Mistou, M. Y., Parmeggiani, A., Camonis, J., et al. 1990. *Science* 248:866–68

243a. Ballester, R., Michaeli, T., Ferguson, K., Xu, H.-P., McCormick, F., Wigler, M. 1989. *Cell* 59:681–86

243b. Tanaka, K., Lin, B. K., Wood, D. R., Tamanoi, F. 1991. *Proc. Natl. Acad. Sci. USA* 88:468–72

244a. Xu, G., O'Connell, P., Viskochil, D., Cawthon, R., Robertson, M., et al. 1990. *Cell* 62:599–608

244b. Buchberg, A. M., Cleveland, L. S., Jenkins, N. A., Copeland, N. G. 1990. *Nature* 347:291–94

244c. Xu, G., Lin, B., Tanaka, K., Dunn, D., Wood, D., et al. 1990. *Cell* 63:835–41

244d. Martin, G. A., Viskochil, D., Bollag, G., Mc Cabe, P. C., Crosier, W. J., et al. 1990. *Cell* 63:843–49

244e. Ballester, R., Marchuk, D., Bogskio, M., Saulino, A., Letcher, R., et al. 1990. *Cell* 63:851–59

245. Sreenath, T. L. V., Breviario, D., Ahmed, N., Dhar, R. 1988. *Biochem. Biophys. Res. Commun.* 157:1182–89

246. Corbitz, A. R., Yim, E. H., Brown, W. R., Perou, C. M., Tamanoi, F. 1989. *Proc. Natl. Acad. Sci. USA* 86:858–62

247. Resnick, R. J., Racker, E. 1988. *Proc. Natl. Acad. Sci. USA* 85:2474–78

248. Nadin-Davis, S. A., Yang, R. C. A., Narang, S. A., Nasim, A. 1986. *J. Mol. Evol.* 23:41–52

249a. Nadin-Davis, S. A., Nasim, A., Beach, D. 1986. *EMBO J.* 5:2963–72

249b. Fukui, Y., Kaziro, Y., Yamamoto, M. 1986. *EMBO J.* 5:1991–93

250. Hughes, D. A., Fukui, Y., Yamamoto, M. 1990. *Nature* 344:355–57

251. Mumby, S. M., Heukeroth, R. O., Gordon, J. I., Gilman, A. G. 1990. *Proc. Natl. Acad. Sci. USA* 87:728–32

252. Casey, P. J., Solski, P. A., Der, C. J., Buss, J. 1989. *Proc. Natl. Acad. Sci. USA* 86:8323–27

253. Logothetis, D. E., Kurachi, Y., Galper, J., Neer, E. J., Clapham, D. E. 1987. *Nature* 325:321–26

254. Nadin-Davis, S. A., Nasim, A. 1988. *EMBO J.* 7:985–97

*Annu. Rev. Biochem. 1991. 60:401–441*

# SIGNAL TRANSDUCTION PATHWAYS INVOLVING PROTEIN PHOSPHORYLATION IN PROKARYOTES

*Robert B. Bourret, Katherine A. Borkovich, and Melvin I. Simon*

Division of Biology, California Institute of Technology, Pasadena, California 91125

KEY WORDS:   two-component regulatory systems, bacterial chemotaxis, regulation of gene expression, phosphohistidine, phosphoaspartate.

## CONTENTS

401

0066-4154/91/0701-0401$02.00

## INTRODUCTION

Bacteria are capable of sensing a wide variety of environmental signals, including changes in chemical concentrations, the presence of a host organism, or variation in physical parameters such as temperature, osmolarity, viscosity, or light. One response to changing conditions is to move to a more "favorable" locale; changes in locomotive behavior can be observed less than one second after a change in chemical composition of the medium. Another possible course of action is to adapt the cell to the new environment, either by changing enzyme activity or by altering expression of specific genes or groups of genes. The bacterium may use the modified enzymes or new gene products to adjust to its surroundings temporarily, or to establish a new long-term state (e.g. the sporulation response to starvation).

Genes important for controlling responses to environmental signals can be identified by analysis of mutant bacteria deficient in the response. Comparison of DNA sequences from many such genes reveals two sets of proteins that share amino acid sequence similarities. These form a widespread family of regulatory systems, originally called "two-component regulatory systems," in which pairs of proteins transduce environmental signals into metabolic changes (1, 2; reviewed in 3–9). Members of the family are defined by the presence of conserved amino acid sequence motifs spanning ~250 amino acids near the carboxy-terminus (C-terminus) of the first protein and ~120 amino acids near the amino-terminus (N-terminus) of the second protein (1, 3). Examples of the diverse processes controlled by two-component regulatory systems include chemotaxis, nitrogen assimilation, and outer membrane protein expression in *Escherichia coli;* virulence in *Bordetella pertussis, Salmonella typhimurium,* and *Agrobacterium tumefaciens;* $N_2$ fixation and dicarboxylic acid transport in *Rhizobium* species; and sporulation in *Bacillus subtilis.*

The evidence now suggests that in addition to amino acid sequence similarities, members of the two-component family share a common mechanism of action involving transient protein phosphorylation. One component is a protein kinase that autophosphorylates by catalyzing transfer of the $\gamma$-phosphate of ATP to a histidine residue. The phosphoryl group is then transferred to an aspartate residue of another component that has been called the regulator, modifying the regulator's activity. Finally, a specific phosphatase activity may reset the system by restoring the regulator to its unphosphorylated state. In each case, regulator-phosphate levels are controlled in response to environmental conditions.

A role for phosphorylation was first found in the nitrogen regulation system (10). This paradigm now appears adaptable to many signal transduction tasks, including chemotaxis, where the response time is measured in milliseconds, and sporulation, where responses occur over periods of hours. Underlying this flexibility is a series of protein domains that apparently have been modified and rearranged in evolution to accommodate the regulatory properties of a wide variety of systems. Although we retain the term "two-component regulatory systems" because of its historical significance, it should not be taken as a literal description. The different transducing systems often have more than two components and can be quite complex. We use the italicized terms *"kinase"* and *"regulator"* when specifically referring to the two components that bear conserved sequence motifs, although (*a*) these proteins frequently contain functional elements in addition to those implied by the names kinase and regulator, and (*b*) kinase activity has not yet been demonstrated in all of the so-called *kinase* proteins. Finally, we generically term any activity that stimulates *regulator* dephosphorylation a "phosphatase," although the actual mechanism(s) involved is unknown.

Discoveries in the bacterial chemotaxis system have played a major role in shaping our understanding of the two-component family. This article first explores newer insights into the mechanisms underlying chemotaxis, derived from work published since two comprehensive reviews in 1987 (11, 12). Some of this material has been covered in more recent reviews (13–21). Other two-component regulatory systems are then compared, with an emphasis on signal processing strategies.

## BACTERIAL CHEMOTAXIS: A PARADIGM FOR SIGNAL TRANSDUCTION

*E. coli* swims by rotating its flagella (22). Normal behavior in a homogeneous environment is smooth swimming interrupted by brief periods of tumbling, which serve to reorient the cell (23). The direction of flagellar rotation is correlated with swimming behavior. In general, counterclockwise (CCW) rotation leads to smooth swimming whereas clockwise (CW) rotations leads to tumbling (24). Pausing of the flagellar motor, apparently the consequence of aborted reversal attempts, has recently been investigated and may also play a role in generating tumbles (25–28). The state of flagellar rotation is characterized by the "bias" of the flagellar motor, which is defined as the fraction of time spent rotating CCW.

*E. coli* moves towards or away from a wide variety of compounds, which are empirically classified as "attractants" or "repellants," respectively (29). The chemosensory mechanism treats repellants as the opposite of attractants (30). As a bacterium swims through a spatial chemical gradient, it measures changes over time in local chemoeffector concentration (31), implying that the bacterium has a memory for chemoeffector concentrations experienced in the recent past. Increasing attractant or decreasing repellant concentrations (positive stimuli) reduce tumbling frequency, thereby extending travel in a favorable direction (23, 30, 31). Conversely, sharply increasing repellant or decreasing attractant concentrations (negative stimuli) increase tumbling frequency, reorienting the bacterium for travel in a potentially more favorable direction. Thus chemotaxis is achieved by biasing a three-dimensional random walk. Although the response to negative chemotactic stimuli can be easily demonstrated experimentally, the response threshold is higher for negative than for positive stimuli (19, 23, 32).

Changes in chemoeffector concentration elicit a brief (~200 msec) "latent" phase of no behavioral change while the signal is processed (33). This is followed by a rapid "excitation" phase, characterized by an extreme (very high or very low) tumble frequency. Finally, behavior gradually returns to the basal state ("adaptation") if there are no further changes in chemoeffector concentration (34, 35). Recovery time is related to the magnitude of the stimulus.

## Components of the Chemotactic Signal Transduction Pathway

An overview of the information flow between signal transduction components of the *E. coli* pathway, deduced primarily from genetic and physiological experiments, is given in Figure 1 and summarized below. The first step is detection of external environmental changes and transmission of this information across the membrane. Once inside the cell, the signal proceeds along both the excitation and adaptation pathways.

SIGNAL DETECTION COMPONENTS    In order for the cell to respond to a chemoeffector, there must be an appropriate pore in the outer membrane to allow the compound to reach receptors located in the periplasm or inner

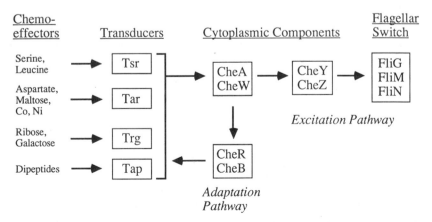

*Figure 1*    Scheme of information flow in the chemotaxis signal transduction system of *E. coli*.

membrane (36). The transducer proteins Tsr, Tar, Trg, and Tap each relay information about a specific subset of chemoeffectors across the inner membrane. For example, Tsr recognizes the attractant serine and the repellant leucine, Tar recognizes the attractants aspartate and maltose and the repellants $Co^{2+}$ and $Ni^{2+}$, and Trg recognizes the attractants ribose and galactose. The transducers have a common transmembrane topology and homologous cytoplasmic domains (37–39). Typically, a transducer periplasmic domain initiates a signal by either binding ligands directly (e.g. serine, aspartate) or interacting with a periplasmic binding protein, which can bind a specific compound (e.g. maltose, ribose, galactose). This periplasmic detection event is communicated to intracellular components via the transducer cytoplasmic domain. Some chemoeffectors are detected independently of the transducers; these cases have been reviewed (11, 12, 18) and are not discussed in this article.

EXCITATION COMPONENTS    Integration of signals from the transducers and transmission to the switch controlling flagellar rotation involves the cytoplasmic proteins CheA, CheW, CheY, and CheZ. In strains with mutant transducers that result in constantly tumbling cells, the presence of *cheA, cheW,* or *cheY* mutations results in smooth swimming, indicating that these functions are required for transmission of a "tumble signal" from transducer to switch (40). Mutations in these three genes also cause smooth swimming in strains with wild-type transducers (41). Mutations in the *cheZ* gene result in bacteria that tumble (41), suggesting a role for CheZ in transmission of a smooth signal or destruction of a tumble signal.

Genetic analysis suggests that both CheY and CheZ physically interact with the switch proteins (Figure 1) (42, 43). Furthermore, overproduction of CheY in either a wild-type or Δ*cheAWYZ* strain causes tumbling behavior (44). CheA and CheW are therefore presumed to function between the transducers and CheY/CheZ in the intracellular signalling pathway. Examination of amino acid sequences reveals that CheA is a *kinase* and CheY is a *regulator* in the two-component family, consistent with their relative order in the scheme for chemotaxis.

The flagellar switch is composed of the FliG, FliM, and FliN proteins. Mutations in these proteins can lead to nonflagellate, nonmotile, or nonchemotactic (either CW- or CCW-biased) phenotypes (45).

ADAPTATION COMPONENTS    A great deal of evidence indicates that changes in transducer methylation levels correlate with the ability of the organism to adapt to different chemoeffector concentrations (46). Negative chemotactic stimuli (decreased attractant concentration) rapidly decrease the methylation level whereas positive stimuli (increased attractant) slowly increase it. The altered methylation level turns off the excitation signal sent by the transducer, even though the altered conditions that triggered the signal persist. This results in behavioral adaptation to the new environment and permits response to subsequent changes in chemoeffector concentration. CheR is a methyltransferase that transfers the methyl group from S-adenosylmethionine to the γ-carboxyl group of specific glutamate residues in the transducer cytoplasmic domains (11, 13, 47). CheB, a member of the *regulator* family, is a methylesterase that hydrolyzes the glutamyl methylesters to liberate methanol (48). CheB and CheR are specific to the adaptation pathway; mutants lacking either of these functions are nonchemotactic but remain capable of generating an excitation response, i.e. switching the direction of flagellar rotation in response to a change in chemoeffector concentration (49–51).

This background provides the context for the following discussion of recent work, which has filled in details for many of the steps outlined in Figure 1.

## Structure and Function of Transducers

The transducers perform a number of functions. They interact with the ligand, mediate transmembrane signal transmission, initiate excitation signals, and are modulated and desensitized by adaptation signals. It has been possible to map these functions to discrete structural domains of the proteins (Figure 2).

LIGAND SPECIFICITY DETERMINANTS    Sequences of the transducer genes suggested that the transducers are transmembrane proteins, with the first 30–40 amino acids from the N-terminus forming a segment that spans the membrane (TM1) (37, 38). The next ~150 amino acids correspond to a periplasmic domain that has been assigned the function of binding ligands directly and interacting with specific periplasmic ligand-binding proteins (52).

Mutations have been isolated in *tar* and *tsr* that result in transducers with specifically lowered affinity for aspartate or serine respectively, but not other Tar- and Tsr-mediated chemoeffectors such as maltose, $Ni^{2+}$, or leucine (53–57). The mutations alter Arg64, Arg69, Arg73, and Thr154 in Tar, and Arg64 and Thr156 (corresponding to Tar Thr154) in Tsr. These four residues are conserved in each of the five aspartate and serine transducers of *Enterobacter aerogenes, E. coli,* and *S. typhimurium* for which sequences are available, but not in Tap or Trg (13, 58). The latter two transducers signal by interacting with periplasmic binding proteins rather than by directly binding a small molecule chemoeffector. These data together suggest that the conserved arginine and threonine side chains may interact directly with features common to amino acid ligands, i.e. the α-amino or α-carboxyl groups (54–56, 59). The potential ligand-binding role of another conserved motif (Glu-Leu-Ile, corresponding to amino acids 136–138 of Tar and 138–140 of Tsr), which is missing in Tap and Trg (58), has not been investigated.

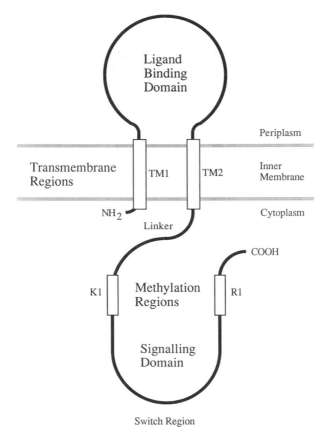

*Figure 2*   Structural and functional organization of chemotaxis transducer proteins.

Information is also available regarding transducer interaction with periplasmic binding proteins. Galactose and ribose inhibit one another's response via the Trg protein (60), suggesting a common interaction site for galactose-binding protein and ribose-binding protein. However, a number of mutant Trg proteins have been identified that exhibit a significant response to ribose and a greatly reduced response to galactose, or vice versa (61; R. Yaghmai, G. L. Hazelbauer, personal communication). The ribose and galactose responses are thus separable to some degree at the transducer level; presumably the binding sites on Trg for galactose- and ribose-binding proteins overlap but are nonidentical. Results concerning the interactions of maltose-binding protein and Tar are more difficult to understand. Tar appears to process aspartate and maltose information independently, i.e. the presence of one stimulus does not block response to the other (62). Maltose-binding protein mutants exist that support maltose transport but not chemotaxis. Maltose chemotaxis is restored in some of these mutants by the presence of mutant Tar proteins altered at Arg69 or Arg73 (63). This is somewhat surprising, because these same residues of Tar are also implicated in aspartate signalling (see above). Thus, the mechanism of signal processing by transducers is more complicated than the simple extreme possibilities of complete independence or commonality for multiple inputs.

TRANSMEMBRANE REGIONS    The transducers contain two stretches of largely hydrophobic amino acids, TM1 and TM2, which span the inner membrane (Figure 2). The sequence of TM1 is fairly conserved among different transducers, whereas the TM2 sequence is nonconserved (11, 13). Introduction of a positively charged residue into the middle of TM1 by site-specific mutagenesis of *tar* results in loss of function (64). Most suppressor mutations restoring Tar function alter amino acids in the portion of the cytoplasmic domain (linker) between TM2 and the first methylation region, suggesting a possible role for TM1 in transmitting information between the periplasmic and cytoplasmic domains of Tar (64).

Biochemical and biophysical studies are also consistent with a role for the transmembrane segments in transmembrane signal transduction. Wild-type *S. typhimurium* Tar does not contain any cysteine residues. Introduction of single cysteine residues by site-specific mutagenesis at seven different positions in the periplasmic or cytoplasmic domains or flanking each transmembrane segment has little effect on Tar function (65). The cysteine-substituted proteins have been used in a variety of biochemical experiments e.g. oxidative crosslinking of cysteine residues or reaction with sulfhydryl-modifying reagents. The extent of crosslinking observed when wild-type (no cysteine) and monocysteine Tar molecules are mixed together in different proportions suggests that the transducer exists as a dimer (66). The size of the Tar oligomer in either detergent or mixed micelles is not altered by aspartate, apparently ruling out change in oligomeric size as a possible signalling mechanism (66). Aspartate alters the rates of disulfide crosslinking between two monomers within a Tar oligomer, either detergent solubilized or in the native membrane (65). The sign and magnitude of change varies with the location of the cysteine residue, but an effect is observed for six of the seven cysteine substitutions. These results suggest that aspartate binding alters the conformation of Tar on both sides of the membrane.

The cysteine-substituted transducers have also been used to examine the spatial arrangement of Tar (67). The relative proximity of various residues in the Tar dimer was determined by measuring oxidative crosslinking rates for all possible combinations of singly substituted molecules. Accessibility differences between sulfhydryl groups in Tar were revealed by differences in reaction rates with various reagents. Taken together, these results suggest a model in which the two TM1 segments are in contact in the center of the dimer, the TM1 and TM2 segments of either monomer are close together, and the TM2 segments of each monomer are farthest apart. Packing of transducer transmembrane segments has also been investigated using a set of Trg proteins marked with cysteine residues at each of seven different positions within TM1 (21). The physical accessibility of cysteines, reflected by the extent of labelling with [$^3$H]$N$-ethylmaleimide, and the proximity of two cysteines within a dimer, assessed by the extent of oxidative crosslinking between the two TM1 segments, varies in a manner consistent with the model of two adjacent TM1 segments.

METHYLATION REGIONS    Details of transducer methylation chemistry have been reviewed (11, 13), and are only briefly mentioned here. The transducers contain four to six methylation sites located in two regions of the cytoplasmic domain termed K1 and R1 (Figure 2). Some methylation sites are synthesized in the transducer as glutamine residues that are later deamidated by CheB to form glutamate. The sites in K1 are spaced seven amino acids apart and thus would lie on the same face of a putative $\alpha$-helix. The different sites vary in reactivity, but are all capable of methylation or demethylation in response to stimuli.

It is not known how methylation influences transducer signalling. Glutamines and methylated glutamates are functionally similar, although not identical, with respect to transducer signalling, and quite different from glutamate (68). Thus charge differences between the modified (methylated or amidated) and unmodified states are one possible basis for a mechanism of signal control by methylation.

SIGNALLING REGIONS    The regions of the transducers responsible for various aspects of signalling function can be deduced to some extent by isolation and characterization of a variety of mutants with signalling defects. The central portion of the cytoplasmic domain plays an important role. Missense mutations resulting in defects in aspartate taxis cluster in the region of *tar* corresponding to the cytoplasmic domain between the methylation regions, thus defining the "signalling domain," whereas nonsense mutations are scattered throughout *tar* (40).

Dominant mutations, resulting in transducers "locked" into CW (tumble) or CCW (smooth) signalling modes, alter cytoplasmic domain residues highly conserved among the various transducers (40, 69). Many alterations that result in a strong bias for flagellar rotation in the CW or CCW direction cluster in the center of the signalling domain (69). Some amino acid substitutions that apparently give rise to defects in switching from one direction to the other are also found here. Thus, this portion of the protein may be primarily responsible for switching between CW and CCW signalling states (Figure 2).

The anatomy of the transducer suggests a crude picture of how it may function. The transducer is imagined to exist in at least two different conformational states (see 62, 70 for discussion). In the absence of ligand, the conformation of the cytoplasmic portion of the transducer would generate a "tumble" swimming signal. In the presence of an attractant that interacts with the periplasmic portion of the transducer, a different specific conformational state could be induced or stabilized. This change would be transmitted via the transmembrane segments to the cytoplasmic portion of the transducer and result in a "smooth" signal. The adaptation system would increase transducer methylation to counteract the smooth signal. The methylation state of the transducer would affect the lifetime of the altered conformation or its ability to generate a signal.

## Phosphate-Transfer Reactions Catalyzed by Chemotaxis Proteins

The biochemical basis of the interactions between chemotaxis proteins outlined in Figure 1 was first revealed by the demonstration in vitro of a phosphate-transfer cascade which has CheA as its central component (71, 72). Subsequently, a function for the transducer and the CheW protein in regulating these phosphate-transfer reactions was also uncovered (73), allowing successful reconstitution of an excitation response in vitro.

CHEA IS AN AUTOPHOSPHORYLATING KINASE    In the presence of $Mg^{2+}$, CheA autophosphorylates using the $\gamma$-phosphate of ATP to yield CheA-phosphate and ADP (71). CheA-phosphate is relatively stable, with a stoichiometry of one mole of phosphate per CheA monomer (71, 72). The reaction is accelerated by addition of $K^+$ ions; $Na^+$ has no effect (72). Measurements of initial rates of CheA-phosphate formation as a function of ATP concentration yield an apparent $K_M$ of approximately 0.29 mM ATP (74; K. A. Borkovich, M. I. Simon, unpublished observations). ADP functions as a competitive inhibitor of ATP in the autophosphorylation reaction with a $K_i$ of approximately 70 $\mu$M ADP (K. A. Borkovich, M. I. Simon, unpublished observations). Since both the $K_M$ for ATP and the $K_i$ for ADP are below the normal cellular concentrations of these two nucleotides (75), the rate of CheA autophosphorylation in vivo should be responsive to the energy charge of the cell (75a). Furthermore, the observation that a normal tumbling frequency is not observed at intracellular ATP concentrations below 0.2 mM (76) is consistent with the argument that attainment of a certain level of CheA-phosphate is critical to generation of tumbles in vivo.

The phosphate linkage in CheA-phosphate is stable to alkali but not to acid and has properties most characteristic of phosphohistidine (74, 77). Enzymatic digestion of $^{32}$P-labelled CheA, sequencing of radioactive peptides, and the identification of phosphohistidine by thin-layer chromatography indicated that His48 is the site of autophosphorylation (78). Subsequent $^{31}$P-NMR studies are consistent with phosphorylation at the N-3 position of a histidine imidazole ring (G. Lucat, J. Stock, cited in 8). The site of phosphorylation and the importance of phosphorylation in chemotaxis were confirmed by site-specific mutagenesis of *cheA*. Three different substitutions at position 48 each eliminate CheA autophosphorylation in vitro (78, 79). Furthermore, among 14 different amino acids tested in position 48, only histidine results in a CheA protein that supports chemotaxis (78).

CheA PHOSPHORYLATES CheY AND CheB    Phosphorylated CheA can rapidly transfer its phosphate to the CheY protein (72, 74), with a stoichiometry of approximately one phosphate per CheY monomer (72). CheY-phosphate spontaneously autodephosphorylates in a $Mg^{2+}$-dependent manner with a half-life in the range of 6–15 sec (72, 80, 81). CheZ greatly accelerates the dephosphorylation of CheY-phosphate, but has a relatively small effect in increasing the rate of CheA-phosphate dephosphorylation (72, 80).

CheA can also phosphorylate CheB (72, 82, 83). CheB-phosphate, like CheY-phosphate, is unstable and rapidly autodephosphorylates (72, 82, 83). In contrast to CheY-phosphate, dephosphorylation of CheB-phosphate is not accelerated by CheZ under the conditions used (72). A maximum incorporation of 0.1 to 0.2 moles phosphate per mole CheB monomer has been observed; this substoichiometric labelling probably results from the rapid rate at which CheB-phosphate autodephosphorylates (72).

DOMAIN STRUCTURE OF CheA    CheA occupies a central position in the information-processing pathway of bacterial chemotaxis. It integrates sensory information provided by the four transducers in a process that requires the participation of CheW (Figure 1). It is capable of autophosphorylation and the transfer of phosphate to CheY and CheB. Genetic studies probing the structure of CheA have yielded evidence for the division of the protein into at least two different functional units, corresponding to the N- and C-terminal regions.

Mutant CheA proteins generated by random mutagenesis exhibit one of four different in vitro phosphorylation phenotypes: (*a*) nonphosphorylated, (*b*) partially phosphorylated, (*c*) autophosphorylated but defective in transfer to CheY and CheB, and (*d*) normally autophosphorylated and dephosphorylated by CheY and CheB (84). The distribution of the mutations suggests that the N-terminus of CheA is involved in transfer of phosphate to

CheY and CheB, whereas the middle and C-terminal portions of the protein are essential for the autophosphorylation activity. The ability of an isolated N-terminal fragment of CheA-phosphate to transfer phosphate but not autophosphorylate (78) is consistent with this scheme, as are the results of *cheA* intragenic suppression and complementation studies (85, 86). Since sedimentation equilibrium experiments indicate that purified CheA is a dimer in solution (86a), intragenic complementation could be due to subunit interactions. For example, phosphorylation of His48 in the N-terminus of one CheA monomer could result from the kinase activity residing in the C-terminus of the other.

MODULATION OF THE PHOSPHORYLATION REACTIONS BY TRANSDUCER AND CHEW   The finding that CheA phosphorylates CheY and CheZ dephosphorylates CheY, together with the knowledge that *cheA* and *cheY* mutants are smooth swimming whereas *cheZ* mutants tumble, supports the hypothesis that CheY-phosphate is the tumble regulator (72). These reactions cannot be the entire basis for an excitation signal, however, since excitation in vivo is regulated by ligand-responsive transducers and the CheW protein (Figure 1). The system has been further reconstituted by the addition of Tar transducer-containing membranes and the CheW protein. This results in an approximately 300-fold stimulation in the level of CheY-phosphate within 5 sec (73). The activation is observed in reactions containing the wild-type Tar protein or a tumble mutant Tar protein (the product of a dominant mutation in *tar* that causes constant tumbling), but not a smooth mutant Tar (the product of a dominant mutation in *tar* that causes constant smooth swimming). The attractant aspartate inhibits activation of CheY-phosphate levels using wild-type transducer but not tumble mutant Tar. Titration of reactions containing wild-type Tar with increasing concentrations of aspartate indicates that aspartate inhibits activation in a dose-dependent manner, with a $K_D$ of 3 $\mu$M aspartate. This value is in the range that influences chemotaxis in vivo (87), and is approximately the same as the $K_D$ for aspartate binding to Tar in vitro (87, 88). The rapidity of this reaction and the magnitude of the enhancement are features consistent with the tumble response in vivo.

Increased levels of CheY-phosphate observed in reconstitution with transducer and CheW are due to activation of CheA kinase activity. The direct activation of CheA autophosphorylation can be demonstrated in reactions containing CheW and a wild-type or tumble mutant transducer (89). CheA-phosphate formation is maximal within seconds. The phosphate group on activated-CheA-phosphate (formed in the presence of transducer and CheW) is labile and can be rapidly chased using either ATP or ADP, whereas that of nonactivated CheA-phosphate is relatively stable and can be exchanged using ADP, but not ATP. Furthermore, the properties of activated-CheA-phosphate depend on the state of the transducer. Thus, addition of the attractant ligand aspartate immediately renders activated-CheA-phosphate refractory to exchange using either ATP or ADP. CheA-phosphate formed under such conditions can transfer its phosphate to CheY, but is incapable of new rounds of activated phosphorylation.

INHIBITORY ACTION OF TRANSDUCER BOUND TO ATTRACTANT    Decreased levels of CheY-phosphate could result from the inactivation of CheA. Evidence for a transducer-mediated inhibitory or "smooth" signal, which operates at the level of CheA autophosphorylation, has also been observed (89). Attractant-bound transducer or smooth mutant transducer inhibits CheA autophosphorylation and thus CheY-phosphate formation in reactions containing a wild-type or tumble mutant transducer and CheW. Inhibitory activity has been demonstrated for wild-type Tar bound to aspartate, wild-type Tsr liganded to serine, and a smooth mutant Tar transducer.

The above studies were performed in the absence of the adaptation response, with transducers extracted from cells lacking both CheR and CheB. Therefore, the activities observed most closely correspond to those essential for the excitation response. In this simplified system, CheA is postulated to exist in three forms (Figure 3). In the absence of transducer and CheW, a nonactivated or "closed" form predominates; this form acts as a pool of recruitable CheA. The closed form is converted to an activated or "open" form in the presence of CheW and a tumble mutant or a wild-type transducer in the absence of attractant. This form of CheA is directly involved in generating the tumble signal by efficiently transferring phosphate to CheY. Finally, there is a "sequestered" form of CheA, which is found in reactions containing smooth mutant or attractant bound wild-type transducer and CheW. The sequestered form appears to be responsible for generating the smooth signal and may act by inhibiting CheA kinase activity. The level of open CheA, and hence tumbling behavior, depends on the ligand-bound status of the transducers. For example, if the transducers are not bound with attractant, the open form of CheA will predominate, leading to increased levels of CheY-phosphate, and a tumble response. Upon binding of attractant ligand to transducers, CheA would be converted to the sequestered form and the concentration of free CheA available for activation by unliganded transducers would be depleted. This would lead to a decrease in CheY-phosphate levels, as net hydrolysis stimulated by CheZ exceeds new synthesis by CheA-phosphate, and the cell would resume smooth-swimming behavior. These simple behavioral responses are not observed directly in wild-type cells, because the presence of feedback (e.g. methylation) modifies the excitation response. The predicted behavior is approached, however, in cells lacking the methylation system.

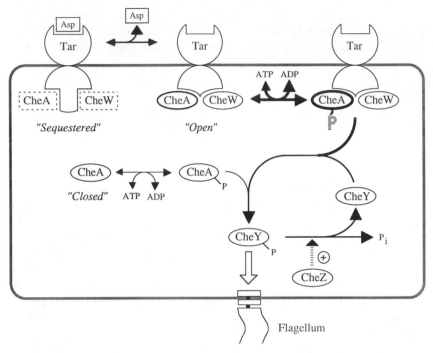

*Figure 3*   The three functional states of CheA in the excitation pathway of chemotaxis signal transduction. The physical relationship of transducer and CheW to CheA in the open and sequestered states is uncertain; the closed state of CheA by definition excludes transducer and CheW. The dashed arrow with a plus sign signifies stimulation of the indicated reaction; the open arrow indicates the final target of the regulatory pathway.

## Genetic and Biochemical Evidence for Interactions between Components of the Excitation Pathway

The results obtained from both genetic analysis and reconstitution of the phosphorylation reactions in vitro are summarized in the model for an excitation response diagrammed in Figure 3. This model implies the existence of specific interactions between cytoplasmic Che proteins and the transmembrane transducers. Recent studies involving the transducers, CheA, CheW, CheY, and CheZ provide more direct evidence that associations formed between these proteins are important for normal chemotaxis.

EVIDENCE FOR INTERACTIONS BETWEEN CHEA, CHEW, AND THE TRANS-
DUCERS    Evidence to support the notion that specific interactions between CheA, CheW, and transducer are required for signalling comes from a number of sources:

   1. Overexpression of either CheA, CheW, or the transducers results in the

same phenotype as that observed for their deletion; i.e. smooth-swimming (90–92). This implies that there is a fairly constrained concentration range for these proteins over which normal chemotaxis is observed. The inhibitory action of excess CheW requires the presence of relatively low levels of transducers, whereas the inhibitory effect of excess transducer does not require CheW (91). However, overexpression of both CheW and transducer restores chemotaxis (91). These results suggest formation of a complex between CheW and the transducer that is important for signalling.

2. A strain expressing the *cheA, cheW, cheY, trg,* and *tar* gene products can change its swimming bias in response to either aspartate or $Ni^{2+}$ (93). Strains containing CheY but deleted for all of the transducers plus CheA or CheW are smooth swimming (bias equal to 1.0) (94). Expression of CheA, CheW, and CheY in the absence of transducers causes the bias to drop. When Tar is expressed in addition to CheA, CheW, and CheY, the cells can respond to aspartate. Thus, reconstitution of an attractant response in vivo requires the presence of the CheA, CheW, and CheY proteins as well as the transducer.

3. Equilibrium analytical sedimentation experiments show that purified CheW exists as a monomer, whereas CheA is a homodimer in solution (86a). The equilibrium column chromatography method of Hummel & Dreyer was used to study the interaction between CheA and CheW. This analysis showed formation of a complex between the two proteins with a dissociation constant of 17 $\mu$M for CheW binding to CheA, which is in the concentration range of these proteins in vivo (5 $\mu$M CheW and 2.5 $\mu$M CheA; 86a). The data also indicate that there are two CheW binding sites per CheA dimer, and that the sites are independent and noncooperative.

4. Antibodies directed against CheA precipitate a mixture of $CheA_L$,[1] $CheA_S$,[1] and CheW from crude extracts; anti-CheW antibody has a similar effect (D. F. McNally, P. Matsumura, submitted for publication). Coprecipitation of these proteins with either antibody suggests that they form a complex. When present in such a complex, $CheA_L$ exhibits a 70-fold lower $K_M$ for ATP binding compared to CheA alone (5 $\mu$M vs 340 $\mu$M ATP).

5. Using an affinity column consisting of immobilized CheY, a complex containing a 1:1:1 stoichiometric mixture of $CheA_L$, $CheA_S$, and CheW can be isolated (D. F. McNally, P. Matsumura, submitted for publication). Furthermore, a complex between $CheA_S$ and CheW can be demonstrated in the absence of $CheA_L$, although no biological role for $CheA_S$ has been established.

6. A collection of *cheW* and *tsr* mutations that show patterns of allele-specific suppression has been isolated (85). The *tsr* mutations cluster in the

---

[1]The *cheA* locus encodes two forms of the CheA protein, $CheA_L$ (long form) and $CheA_S$ (short form) (86, 221). Only $CheA_L$ contains the site of phosphorylation.

portion of the gene encoding the cytoplasmic signalling domain of the transducer (Figure 2), and without exception affect residues conserved among the
various transducers. The *cheW* mutations are distributed throughout the gene,
but with a regular spacing; this periodicity may correspond to mutation of the
same position in a repeating structural unit. These results suggest that the
C-terminus of transducers interacts with the CheW protein to augment tumble
generation; formation of this complex may activate autophosphorylation of
CheA.

7. Mutant alleles of *tsr* have also been used for the isolation of second-site
suppressors in the *cheA* locus (85). The *cheA* suppressor mutations cluster in
the region of the gene corresponding to the C-terminus of the protein. Most of
the suppressed *tsr* alleles alter the signalling region of Tsr. These results
suggest that the C-terminal domain of CheA may interact with the signalling
region of Tsr.

ANTAGONISTIC ACTIVITIES OF CHEY AND CHEZ    CCW (smooth-
swimming) is the default rotational bias for the flagellar motor in a strain
deleted for all the *che* genes and the transducers (a "gutted" strain; 93).
Adding back CheY alone is sufficient to induce tumbling in a gutted strain
(93). Both the concentration of CheY needed to sustain tumbling, and
identification of the other gene products that modulate the effect of CheY in
vivo have been determined. Increasing levels of CheY protein correlate with
an increase in the percentage of tumbling cells (93, 95). However, the
concentration of CheY required for half-maximal levels of tumbling, 30 $\mu$M,
is much higher than the concentration of CheY protein in wild-type cells (8
$\mu$M; 95), suggesting that only a subpopulation of CheY expressed in a gutted
strain is capable of mounting a tumble response in the absence of the other
chemotaxis proteins. This phenomenon can be explained in terms of the
model shown in Figure 3 by arguing either (*a*) that a subset of the unphosphorylated CheY molecules are in the same conformation as the phosphorylated form and can activate CW rotation, or (*b*) that CheY is inefficiently phosphorylated by other "crosstalking" *kinases* in gutted cells,
causing a requirement for higher levels of total protein to allow accumulation of a concentration of the phosphorylated form sufficient for tumble
activity.

Deletion of CheZ in an otherwise wild-type background results in a tumble
phenotype (41). Expression of CheZ in cells producing large amounts of
CheY results in smooth swimming (93, 95), even when the only source of
CheZ is from one chromosomal copy of the gene. Thus, CheZ has an activity
antagonistic to that of CheY, even at apparently substoichiometric levels of
CheZ protein, presumably due to its role as a catalyst activating dephosphorylation of CheY-phosphate. Further evidence of physical interaction

between CheY and CheZ is provided by retention of CheZ on a CheY affinity column (D. F. McNally, P. Matsumura, submitted for publication).

## CheY Structure and Function

The results summarized thus far demonstrate that specific protein/protein interactions are essential for information transfer through the chemotaxis signalling pathway. Detailed understanding of the molecular mechanisms involved in this process requires higher resolution than that obtainable by genetic and biochemical approaches alone. Thus many laboratories have been attempting to determine the structures of the various chemotaxis proteins by X-ray crystallography.

CheY STRUCTURE    The structure of CheY from both *S. typhimurium* (96) and *E. coli* (K. Volz, P. Matsumura, submitted for publication) has been determined. These proteins, which differ by only three of 128 amino acids (97, 98), have essentially the same structures. CheY consists of a compact single domain formed from a core of five parallel $\beta$-strands surrounded by five $\alpha$-helices (Figure 4). The $\beta$-strands and $\alpha$-helices alternate in the CheY primary structure. Three aspartate residues (Asp12, Asp13, Asp57) cluster together at the C-terminal end of two adjacent $\beta$-strands to form an "acid pocket" (96). These three aspartate residues are highly conserved among *regulator* proteins of the two-component family (8), suggesting an important functional role. Another highly conserved residue, Lys109, is positioned on

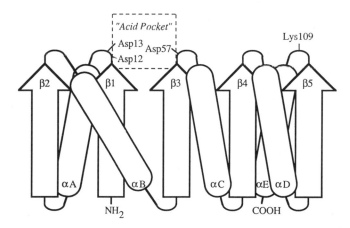

*Figure 4*    Organization of $\alpha$-helices and $\beta$-strands in the CheY structure. The general topology of CheY is illustrated, rather than precise structural relationships. Adapted from Figure 5 of Ref. 8.

the same face of the molecule. Comparison of *regulator* protein primary sequences with the CheY crystal structure suggests structural roles for about 25 conserved residues in forming the protein's hydrophobic core (9, 96).

CHEY FUNCTIONAL ELEMENTS   The functional importance of numerous amino acids in CheY has been tested by site-specific mutagenesis. Substitutions at the conserved residues Asp12, Asp13, Asp57, and Lys109 result in loss of chemotaxis (81, 99, 100). In contrast, mutants bearing substitutions at any of 10 different nonconserved aspartate and serine residues in CheY remain chemotactic (96, 99).

The stability of CheY-phosphate following various chemical treatments indicated that CheY is phosphorylated on an acidic (aspartate or glutamate) residue (82). The sequence conservation and crystal structure data immediately pointed to the acid pocket as a strong candidate for the site of phosphorylation (82, 96). The normal site of CheY phosphorylation was in fact recently demonstrated to be Asp57 (100). Proteolytic peptides derived from CheY-phosphate reduced with [$^3$H]sodium borohydride were compared with those from nonphosphorylated CheY by liquid secondary ion mass spectrometry. Fragments differing in mass by the 14 Da predicted from reduction of a phosphorylated carboxyl group were radioactive as expected and contained amino acid 57. In addition, homoserine was found in place of Asp57 when relevant peptides were sequenced by tandem mass spectrometry. The identification of Asp57 as the site of phosphorylation is further supported by an analysis of the biochemical properties of mutant CheY proteins bearing single and double substitutions at Asp12, Asp13, and Asp57 in all possible combinations (99). The phosphorylation properties of all the mutant proteins are modified, but those bearing substitutions at position 57 are most aberrant. Furthermore, a mutant CheY protein bearing the nonphosphorylatable amino acid asparagine in place of Asp57 does not support chemotaxis (99, 100).

A divalent metal ion is required for phosphotransfer from CheA to CheY and for autodephosphorylation of CheY-phosphate (80, 81, 100). These roles can be fulfilled by $Co^{2+}$, $Mg^{2+}$, $Mn^{2+}$, or $Zn^{2+}$, or to a lesser extent $Cd^{2+}$ (81). $Ca^{2+}$ supports slow transfer but not dephosphorylation. Fluorescence studies of the single tryptophan in CheY (Trp58), located immediately adjacent to the acid pocket, have been informative (81). Scatchard analysis of data derived from quenching of tryptophan fluorescence by metal ions is consistent with a single metal-binding site in CheY. Metal ion quenching of tryptophan fluorescence is greatly reduced in mutant CheY proteins bearing substitutions at Asp13 or Asp57 but not in mutants altered at Lys109, suggesting that the acid pocket is the metal-binding site.

TUMBLE GENERATION BY CHEY    Phosphorylation of CheY could lead to tumbling in several ways. The phosphoryl group could be directly involved in the mechanism of tumble generation by phosphotransfer from CheY to the flagellar switch proteins. Alternatively, the phosphoryl group could have a more indirect role. It may be required to generate or stabilize a conformational change in CheY that allows interaction with the switch resulting in tumbling. The phenotype of a CheY mutant bearing lysine in place of Asp13 (termed "CheY13DK") suggests the latter explanation is correct (99). Bacteria with the *cheY13DK* mutation tumble, yet the mutant CheY protein is not detectably phosphorylated by CheA in vitro. Furthermore, bacteria with CheY13DK tumble even in the absence of CheA. Thus, the altered CheY protein apparently causes tumbling in the absence of phosphorylation, presumably by mimicking the structure of wild-type CheY-phosphate.

The crystal structure of CheY will permit a more in depth investigation of the nature of interactions between CheY and CheA, CheY and CheZ, or CheY and the flagellar switch. There has been some progress with regard to CheY-switch interactions. The FliG protein is thought to be part of the switch that controls the bias of flagellar rotation. Mutations in *cheY* suppress *fliG* mutations in an allele-specific manner, consistent with a direct physical interaction between the CheY and FliG proteins. The *cheY* mutations result in amino acid substitutions that appear to cluster on the three-dimensional structure of CheY, with most of the altered residues located at the N-terminal ends of $\alpha$D and $\alpha$E, and on the loop between $\beta$5 and $\alpha$E (Figure 4) (101). These positions potentially define a face of CheY that may interact with FliG.

ACETATE EFFECT    Adding acetate to cells lacking all known chemotaxis genes except *cheY* greatly increases CW rotation (tumbling) (102). This effect is antagonized by CheZ, and not observed when CheY Asp57 is replaced by glutamate (H. C. Berg, F. E. Dailey, A. J. Wolfe, personal communication), suggesting that acetate-induced tumbling involves CheY-phosphate formation. Studies with a variety of metabolic inhibitors and mutants suggest that acetate metabolism influences chemotactic signal transduction at the level of CheY, but the exact mechanism is unknown.

Acetate can be metabolized to acetyl-CoA via either acetyladenylate or acetylphosphate intermediates. Acetate kinase, the enzyme that generates acetylphosphate, has also been shown to influence other metabolic pathways. Acetate kinase can use ATP to phosphorylate correctly Enzyme I of the phosphotransferase system, an enzyme that normally autophosphorylates with phosphoenolpyruvate (103). Acetate kinase can also activate the *pho* regulon in the absence of the PhoR and PhoM *kinases* (104). The mechanism of this activation is unknown; phosphorylation of the *regulator* PhoB by acetate kinase has not been observed in vitro. Acetate kinase may possess a previously unappreciated role in bacterial metabolism and signal transduction.

## Adaptation via Transducer Methylation

The foregoing discussion of chemotaxis has focussed on molecular details of the excitation pathway. Adaptation via transducer methylation is also a crucial part of the chemotaxis response. In the absence of stimuli, absolute methylation levels remain steady, but methyl groups constantly turn over (105). Transient changes in demethylation rate (106, 107) in response to stimuli result in changes in transducer methylation levels that modulate the excitation signal.

REGULATION OF CHEB FUNCTION    The methylesterase activity of CheB is located in the nonconserved C-terminal portion of the protein (108). CheA phosphorylates the N-terminal portion of CheB, increasing in vitro methylesterase activity more than 10-fold (79, 83). The increase in transducer demethylation rate observed in vivo following negative chemotactic stimuli requires CheA, CheW, and intact CheB, but not CheY or CheZ (79, 83, 90, 109, 110). Furthermore, point mutations in *cheA* or *cheB* that block phosphotransfer from CheA to CheB also eliminate increased demethylation (79). Thus, negative stimuli apparently cause transducer-mediated phosphorylation of CheB via CheW and CheA, and increased methylesterase activity leads to the increased demethylation rate. The instability of CheB-phosphate (72) accounts for the transient nature of the increased demethylation rate.

Decreased methylesterase activity following positive chemotactic stimuli does not require CheA, CheY, CheZ, the N-terminal portion of CheB, or CheB phosphorylation (79, 83, 90, 109, 110). Data regarding the necessity of CheW are contradictory; inhibition of demethylation in four different strains that should express little or no CheW ranged from apparently normal to nonexistent (79, 83, 90). A simple way to account for these results is to propose that two separate processes inhibit demethylation and the relative contribution of each varied in the different strains under the test conditions (R. C. Stewart, personal communication). One pathway requires CheW and could involve interaction with the C-terminal portion of CheB or with the transducers (79). The other putative inhibitory process does not require CheW and could simply result from transducer conformational changes altering access of CheB to methylation sites (see next section).

HOMOLOGOUS AND HETEROLOGOUS ADAPTATION RESPONSES    Changes in methylation levels during adaptation occur primarily on the transducer that was stimulated (homologous adaptation). Attractants and repellants influence in vivo methylation levels of the specific transducers through which they act

(111, 112), presumably by stabilizing conformational changes that affect the accessibility of methylation sites to CheB and CheR (113–115). Reducing the number of methylation sites on a transducer, e.g. Trg, by site-specific mutagenesis increases the time required for adaptation by cells containing a normal complement of transducers when exposed to Trg-mediated stimuli, emphasizing the importance of homologous adaptation (116).

There is also a modest heterologous methylation effect manifested by the alteration of methylation levels on the nonstimulated transducers (117). The heterologous effect is due to changes in CheB methylesterase activity, which acts on all transducers to the extent that substrate is available (there is no evidence that CheR methyltransferase activity is regulated). The heterologous effect was illustrated in an experiment in which the methylation sites in Trg were eliminated by site-specific mutagenesis (118). Bacteria carrying this mutant form of Trg as their sole transducer respond to Trg-mediated attractant stimuli, but do not adapt and therefore are nonchemotactic. The presence of additional transducers (Tsr, Tar, Tap) restores chemotaxis to Trg-mediated stimuli. This phenomenon can be explained by increased methylation of the unstimulated transducers (118), presumably causing them to send tumble signals that balance any smooth signals sent by the stimulated, unmethylated transducers. Thus excitation and adaptation signals were transmitted by different transducers.

BEHAVIOR OF *CHEB CHER* MUTANTS    *cheRB* bacteria lacking both methyltransferase and methylesterase activities are defective in chemotaxis (119–121). Bacteria containing transducers that lack all methylation sites are also nonchemotactic (118). The essential role of methylation-dependent adaptation in transducer-mediated chemotaxis has been challenged by behavior of *cheRB* strains that resembles residual chemotaxis in certain circumstances (122). However, satisfactory explanations exist for the three observations of apparent chemotaxis by *cheRB* bacteria: (*a*) *cheRB* bacteria migrate on swarm plates at a significant fraction of the *che*$^+$ rate (49, 51, 123) because they tumble enough to get around obstructions in the agar (124). (*b*) Partial recovery of prestimulus behavior by *cheRB* cells following small stimuli (51, 120) could reflect a transient burst of CheY-phosphate formation or destruction preceding the establishment of new steady-state levels (51). (*c*) The slight accumulation of *cheRB* bacteria in capillary tubes (123, 125) could result from random diffusion of cells to the capillary, which are then influenced by the steep gradient at the mouth of the capillary tube (125).

## Future Directions

SOME REMAINING QUESTIONS    Many questions remain as to the mechanism by which bacteria sense and respond to chemotactic stimuli. Any model of chemotaxis must account for the following signal processing characteristics: (*a*) *Dynamic range*—Bacteria respond to changes in chemoeffector concentration in a range spanning about five orders of magnitude (126). (*b*) *Amplification*—Bacteria are sensitive to chemoeffector concentration changes that alter the ligand-binding status of less than 1% of the cell's receptors (119). (*c*) *Integration of multiple inputs*—Bacteria presented with multiple or conflicting stimuli process this information to arrive at an average best response (34, 35). (*d*) *Speed of response*—The time from chemoeffector stimulus through detection and internal processing to flagellar response is a fraction of a second (33). (*e*) *Memory*—The decision to continue swimming in a particular direction or to alter course is based on a comparison of the past and present environments, and implies continuous updating and interaction between the excitation and adaptation mechanisms (119). A model can be assembled that qualitatively explains most of these features, but a quantitative accounting of the actual in vivo physiology is not yet possible. Accurate kinetic data are necessary to make predictions testing the validity of models based on reactions reconstituted in vitro.

Regulated formation of CheY-phosphate is a plausible means of generating a tumble signal, although enhancement of this reaction by a repellant compound in vitro has not been reported. The response to positive stimuli can be explained partially by inhibition of CheA autophosphorylation activity (89). However, prevention of CheY-phosphate formation appears insufficient to explain the observed signal amplification; stimulation of about one receptor in 600 can cause a change in bias of 10% (119). Amplification could occur by a variety of possible mechanisms: e.g. it could result from activation of CheZ. The observation that *cheZ⁻* bacteria take longer than wild-type bacteria to transmit excitation signals (33) appears to be consistent with this proposal.

The interactions between transducer, CheW, and CheA also remain to be clarified. CheA and CheW may form a ternary complex with the transducer. Alternatively, CheA-CheW complexes could be activated by the transducer and become diffusible catalysts. Finally, it is also possible that CheW shuttles between the transducer and CheA. Any consideration of CheA activation mechanisms must also take into account chemosensory input that bypasses the transducers, e.g. responses to oxygen, membrane potential, or carbohydrates transported by the phosphotransferase system (11, 12, 18).

The beginning and the end of the signalling pathway require further study. The mechanism of transmembrane signal transduction by the transducers is unknown. Similarly, events in the tumble generation process downstream of CheY-phosphate formation are uncharacterized. Furthermore, the degree to which other two-component systems "crosstalk" with and contribute to chemotaxis has not been clarified.

Methylation-dependent adaptation poses a number of questions. It is not known how transducer signalling is modulated by methylation, nor exactly how "memory" works. The interactions of the excitation and adaptation pathways are critical but not entirely identified. One possibility for coordinate regulation of the two pathways is that the relative rates of phosphotransfer to CheB or CheY may vary depending on the extent of CheA activation. These interactions can be reconstituted in vitro.

Determination of the structures of chemotaxis proteins will aid in answering many questions. Analysis of active site determinants and conformational changes induced by phosphorylation will be possible. Associations between components may be mapped and then verified by structure determination of cocrystals.

CHEMOTAXIS IN OTHER BACTERIA    In general, motile bacteria move to a favorable environment using a biased random walk (127). We have described the sensory transduction mechanism responsible for this behavior in *E. coli* and *S. typhimurium*, where it is best understood. This mechanism may be in large part common to all chemotactic bacteria. For example, signal transduction components involved in controlling motility of the nonflagellated gliding bacteria *Myxococcus xanthus* share significant sequence similarities with those of *E. coli*, in spite of an entirely different mechanism of locomotion (128–130). Furthermore, similar transducer proteins appear to be present in many (but not all) chemotactic species (21, 127). However, two observations suggest that exceptions to the putative common mechanism must exist: (*a*) The properties of chemotaxis-associated protein methylation are different in *B. subtilis* (131) and *Halobacterium halobium* (132) than in *E. coli*. (*b*) Membrane potential appears to play a primary role in chemotaxis by *Spirochaeta aurantia* (133) but not by *E. coli*. Further investigation in species other than *E. coli* may reveal alternative designs for sensory transduction pathways.

# OTHER BACTERIAL TWO-COMPONENT REGULATORY SYSTEMS

Examination of two-component regulatory systems besides chemotaxis provides a useful perspective. The two best-studied systems are those controlling nitrogen assimilation and outer membrane protein expression. These are reviewed briefly before surveying properties common to many systems.

## Nitrogen Assimilation

Many species of bacteria regulate both the amount and enzymatic activity of glutamine synthetase in response to nitrogen availability (reviewed in 134–136). We focus on the roles of the *kinase* NtrB ($NR_{II}$) and the *regulator* NtrC

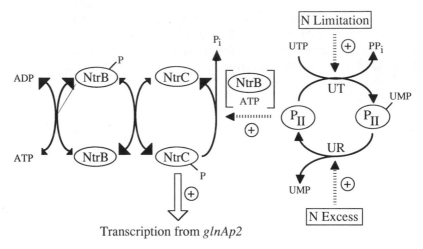

*Figure 5* Regulation of glutamine synthetase transcription by the Ntr signal transduction system. Dashed arrows with plus signs signify stimulation of the indicated reactions; the open arrow indicates the final target of the regulatory pathway.

(NR$_I$) in controlling expression of *glnA*, the gene for glutamine synthetase (Figure 5). *glnA* is transcribed from a major promoter *glnAp2* and a minor promoter *glnAp1* (137). Phosphorylation of NtrC stimulates transcription from *glnAp2* (10).

NtrB autophosphorylates on a histidine residue (138, 139). Isolation and sequencing of $^{32}$P-labelled proteolytic peptides from *E. coli* NtrB-phosphate indicates that the site of phosphorylation is His139 (222). The phosphate is transferred from NtrB to an aspartate residue in the N-terminal portion of NtrC (138, 139). The site of phosphorylation in *S. typhimurium* NtrC has been identified as Asp54 using the same mass spectrometry method used for determining the site of CheY phosphorylation (D. A. Sanders, D. E. Koshland Jr., personal communication). NtrC-phosphate is labile, with a half-life of ~4 min (138, 139). Hydrolysis of NtrC-phosphate is accelerated by the combined presence of NtrB, ATP, and another protein, P$_{II}$ (10, 138).

P$_{II}$ provides the link between the phosphorylation reactions and the nitrogen sensing apparatus. Nitrogen availability influences the ratio of glutamine and 2-ketoglutarate in the cell, which in turn influences the activity of UT/UR, the uridylyltransferase/uridylyl-removing enzyme (Figure 5). UT/UR alters the activity of P$_{II}$ by uridylylation or deuridylylation. Only the unmodified form of P$_{II}$ stimulates NtrC dephosphorylation (138). Under conditions of limiting nitrogen, 2-ketoglutarate levels increase and stimulate uridylylation of P$_{II}$, which permits NtrC-phosphate levels to increase, resulting in increased transcription from *glnAp2*. Under conditions of excess nitrogen, $P_{II}$ is in the unmodified form and prevents NtrC-phosphate buildup, so *glnAp2* activation ceases.

Rapid activation and deactivation of *glnAp2* transcription in response to changes in nitrogen availability requires NtrB (137). Steady-state levels of glutamine synthetase under nitrogen-limiting conditions, however, are essentially normal in the absence of NtrB (140, 141), even though NtrC-phosphate is required for *glnAp2* activation. Complete lowering of glutamine synthetase levels under nitrogen excess conditions requires NtrB (140, 141). These results can be explained by the hypothesis that NtrC is phosphorylated by *kinases* other than NtrB ("crosstalk"), and an important function of NtrB under conditions of nitrogen excess is to hydrolyze NtrC-phosphate via the NtrB-$P_{II}$-ATP route.

The mechanism of *glnAp2* transcriptional activation is now understood in general terms. NtrC and the $\sigma^{54}$-subunit of RNA polymerase are absolutely required (141–144). NtrC binds upstream of *glnAp2* (143, 145, 146), but does not influence RNA polymerase binding (134, 145–147). The NtrC-binding sites remain fully functional for transcriptional activation when moved various distances to either side of the promoter, so long as they are not much closer than in the wild-type arrangement (145, 148, 149). Furthermore, activation of transcription is almost as great when the NtrC-binding sites and *glnAp2* are located on two different interwound DNA molecules as when the binding sites and *glnAp2* are on the same piece of DNA (150). Activation is lost when the binding sites and promoter are on unlinked DNA molecules. Activation of *glnAp2* also occurs in the absence of NtrC-binding sites, although a higher concentration of NtrC is needed than when binding sites are present (134, 145, 148). Together these results suggest that NtrC does not activate transcription by transmitting a signal to RNA polymerase through the DNA molecule; rather the NtrC-binding sites serve to increase the local concentration of NtrC and thus increase the chance of protein/protein interaction between NtrC and RNA polymerase. NtrC and RNA polymerase have been observed by electron microscopy in apparent physical contact under conditions of open complex formation (151).

NtrC promotes conversion of RNA polymerase-*glnAp2* complexes from the closed to the open form and then leaves the transcription complex (146, 152, 153). ATP is needed for open complex formation (152). This requirement is observed under conditions where ATP is neither incorporated into the transcript nor used as a substrate for NtrC phosphorylation, and apparently reflects hydrolysis of ATP.

Sequence similarities with various other proteins suggest NtrC has three functional domains (3, 7, 9): (*a*) The phosphorylated N-terminal regulatory domain is shared with other *regulators* of the two-component family. (*b*) A central domain that may interact with RNA polymerase is shared with other transcriptional regulators (e.g. NifA). (c) A helix-turn-helix DNA-binding motif is present in the C-terminal domain. Mutations resulting in phosphorylation-independent activation of NtrC have been found in the central domain (152, 154). The central domain also contains a putative nucleotide-binding site (155), consistent with the ATPase activity inferred for open complex formation. Finally, mutations in the helix-turn-helix region affect DNA binding of NtrC (152, 156).

## Outer Membrane Protein Expression

One response to changes in osmolarity involves modification of the outer membrane by the introduction of new porins. Levels of the *E. coli* outer membrane porins OmpC and OmpF are reciprocally regulated in response to osmolarity by the *kinase* EnvZ and the *regulator* OmpR (Figure 6) (reviewed in 8, 157–159). OmpF is preferentially expressed at low osmolarity whereas OmpC predominates at high osmolarity.

EnvZ is located in the inner membrane and has a topology similar to that of the chemotaxis transducers, i.e. a periplasmic domain, a cytoplasmic domain, and two transmembrane regions (160). The mechanism of osmolarity sensing by EnvZ is unknown, but appears to involve the periplasmic domain since a cytoplasmic fragment supports OmpC and OmpF expression but not proper osmoregulation (161). Both intact EnvZ and soluble C-terminal fragments of EnvZ autophosphorylate and transfer phosphate to OmpR (161–164; S.

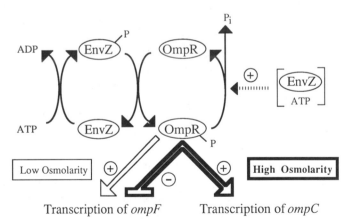

*Figure 6*  Regulation of outer membrane protein transcription by the Omp signal transduction system. Although EnvZ senses osmolarity changes, it is not known which specific step(s) in the phosphorylation pathway are influenced by osmolarity. The dashed arrow with a plus sign signifies stimulation of the indicated reaction; open arrows indicate the final targets of the regulatory pathway. OmpR-phosphate stimulates *ompF* transcription at low osmolarity. At high osmolarity, OmpR-phosphate represses *ompF* transcription and stimulates *ompC* transcription.

Tokishita et al, cited in 159; A. Rampersaud, cited in 165). Hydrolysis of OmpR-phosphate is greatly accelerated by the joint presence of EnvZ and ATP (or a nonhydrolyzable analog of ATP) (163, 166, 167).

OmpR is essential for *ompC* and *ompF* expression (168, 169) and has both positive and negative effects on transcription (Figure 6) (170). EnvZ transfers phosphate to a residue in the N-terminal portion of OmpR (171). The C-terminal portion of OmpR is capable of DNA binding but not transcriptional activation (171, 172). Phosphorylation of OmpR stimulates both binding (173) and activation (162, 174) of the *ompC* and *ompF* promoters in vitro. Bacteria bearing dominant mutations that either reduce EnvZ phosphatase activity or intrinsically stabilize the OmpR phosphoryl group both have the same $OmpF^-$ $OmpC^+$ phenotype, suggesting that high levels of OmpR-phosphate result in *ompF* repression and *ompC* expression in vivo (167). Similarly, increasing levels of OmpR-phosphate first activate *ompF* transcription in vitro, which then declines as *ompC* transcription increases (174). It is not known whether quantitative or qualitative differences in OmpR-phosphate are responsible for the differential regulation of *ompC* and *ompF*. OmpR is capable of stimulating expression from a variety of artificial promoters and has been proposed to function by facilitating binding of RNA polymerase to weak promoters (175). OmpR binding sites must be positioned stereospecifically with respect to the *ompC* promoter for OmpR to activate transcription (176, 177), consistent with this mechanism. OmpR can also repress expression from a strong promoter (175).

OmpR-phosphate has a half-life of ~1.5 hr, which is much longer than that of the Che or Ntr *regulators* (166). The relative stability of OmpR-phosphate has been exploited to observe OmpR phosphorylation in vivo (165). This provides important support for a large body of work based on in vitro phosphorylation, both in osmosensing and in other systems. High osmolarity increases the level of OmpR-phosphate, in accordance with expectations based on in vitro results. Surprisingly, cells either expressing an EnvZ missense mutant protein that does not autophosphorylate in vitro or containing an *envZ* nonsense mutation also form OmpR-phosphate in an apparently osmolarity dependent manner. Although it is conceivable the mutant EnvZ proteins retain some phosphorylation activity in vivo, this result strongly suggests OmpR can be phosphorylated correctly in vivo by a *kinase* other than EnvZ.

## Additional Two-Component Regulatory Systems

Bacterial two-component regulatory systems in addition to Che, Ntr, and Omp have begun to be characterized (Table 1). The *regulators* can be grouped into three obvious subclasses based on extensive sequence similarities within

**Table 1** Two-component regulatory systems of bacteria[a]

| Function | Kinase(s) Protein | Kinase(s) Phosphorylation site[b] | Regulator(s) Protein | Regulator(s) Phosphorylation site[b] | Species[c] | Refs.[d] |
|---|---|---|---|---|---|---|
| NtrC regulator subclass[e] | | | | | | |
| Nitrogen assimilation | NtrB | His139 | NtrC | Asp54[f] | Bsp, Ec, Kp, Rm | [10, 138, 139, 222] |
| Dicarboxylic acid transport | DctB | — | DctD | — | Rl, Rm | 178, 179 |
| Phosphoglycerate transport | PgtB | — | PgtA | — | St | 180 |
| OmpR regulator subclass | | | | | | |
| Outer membrane protein expression | EnvZ | Histidine? | OmpR | Aspartate? | Ec, St | [161–167] |
| Phosphate assimilation | PhoR,[g] | Histidine? | PhoB | Aspartate? | Bs, Ec, Kp, Pa, Sd | [181, 181a], 182–184 |
| | PhoM | Histidine | PhoMORF2 | Aspartate? | | |
| Agrobacterium virulence | VirA[g] | Histidine | VirG | Asp52 | At | [185–187], 188–192 |
| Anaerobic repression, | ArcB, | — | ArcA | — | Ec | 193–195 |
| Conjugation? | CpxA | — | | | | |
| Salmonella virulence | PhoQ | — | PhoP | — | St | 196 |
| FixJ regulator subclass | | | | | | |
| N$_2$ fixation | FixL | — | FixJ | — | Rm | 197 |
| Nodulation | NodV | — | NodW | — | Bj | 198 |
| Nitrate reductase | NarX | — | NarL | — | Ec | 199 |
| Sugar phosphate transport | UhpB | — | UhpA | — | Ec | 200 |
| Capsule synthesis | RcsC | — | RcsB | — | Ec | 201 |

| | | Histidine? | | Aspartate? | | |
|---|---|---|---|---|---|---|
| Degradative enzymes | DegS | Histidine? | DegU | Aspartate? | Bs | [201a], 202 |
| Competence | ComP | — | ComA | — | Bs | 203 |
| Bordetella virulence | BvgS | — | BvgA | — | Bp | 204–206 |
| Unique regulator subclass | | | | | | |
| Chemotaxis | CheA | His48 | CheB, CheY | —, Asp57 | Ea, Ec, St | [71–74, 78–80, 82, 83, 89, 99, 100], 58 |
| Motility/development | FrzE | — | FrzE, FrzG | Aspartate?, — | Mx | [206a], 128–130 |
| Sporulation | KinA, KinB | Unknown, — | SpoOA, SpoOF | Unknown, Unknown | Bs | [207, 208], 209, 210 |

[a] Only systems in which the genes for both components had been sequenced and mutated as of November 1990 are listed. There are additional putative members of the two-component family (see 8) for which an open reading frame has been identified in the absence of any genetic data, or only one component has been satisfactorily characterized.

[b] Phosphorylated amino acid listed where known. A "?" signifies chemical properties that are consistent with the indicated phosphoamino acid, but conclusive identification has not been made. "Unknown" indicates that phosphorylation of the protein has been demonstrated but there is no information regarding the site. A "—" indicates phosphorylation has not been reported.

[c] Only species for which sequence data are available are listed. Related species may also possess such systems. At, *Agrobacterium tumefaciens* (four different Ti plasmids); Bj, *Bradyrhizobium japonicum*; Bp, *Bordetella pertussis*; Bs, *Bacillus subtilis*; Bsp, *Bradyrhizobium* species; Ea, *Enterobacter aerogenes*; Ec, *Escherichia coli*; Kp, *Klebsiella pneumoniae*; Mx, *Myxococcus xanthus*; Pa, *Pseudomonas aeruginosa*; Rl, *Rhizobium leguminosarum*; Rm, *Rhizobium meliloti*; Sd, *Shigella dysenteriae*; St, *Salmonella typhimurium*.

[d] This list of recent references is not comprehensive, but is intended to provide entry into the primary literature; see Ref. 8 for additional references. References for phosphorylation are in brackets.

[e] Subclasses are based on sequence similarities in the *regulator* C-terminal regions. In the "unique" subclass, CheY, FrzE, and SpoOF do not extend beyond the N-terminal *regulator* domain, whereas SpoOA possesses a C-terminal region unlike any other currently known *regulator*. CheB and FrzG are similar over their entire lengths.

[f] D. A. Sanders, D. E. Koshland Jr., personal communication.

[g] Phosphorylation reactions for PhoR and VirA utilized truncated or fusion proteins retaining the C-terminal portion of the protein.

their nonconserved C-terminal portions (Table 1; 210a), perhaps reflecting different mechanisms of transcriptional regulation (7, 8).

Protein phosphorylation has so far been demonstrated in vitro for eight of the 19 two-component regulatory systems recognized by amino acid sequence similarity and genetic analysis (Table 1). The reaction characteristics are remarkably similar. *Kinase* autophosphorylation specifically uses the γ-phosphoryl group of ATP; *kinase* labelling has not been observed with $[\gamma^{32}P]GTP$ (71, 161, 185) or $[\alpha^{32}P]ATP$ (71, 138, 139, 161, 163, 181, 181a, 185, 186, 201a). The known sites of phosphorylation are a *kinase* histidine residue and a *regulator* aspartate residue (Table 1). Together these observations strongly suggest a common phosphorylation reaction mechanism. This is further emphasized by the finding that various *kinase* proteins are capable of correctly phosphorylating noncognate *regulator* proteins in vitro, although to a much lesser extent than cognate *regulators*. For example, CheA can phosphorylate NtrC (211), OmpR (166), and SpoOA (208), NtrB can phosphorylate SpoOA (138) and CheY (211), EnvZ can phosphorylate NtrC (166), and PhoM can phosphorylate PhoB (181a).

## PROPERTIES OF INFORMATION PROCESSING SYSTEMS

### Role of Conserved Domains

The original model of signalling in the two-component systems proposed communication between the conserved *kinase* and *regulator* domains, with the nonconserved domains fulfilling signal detection and response functions unique to each system (1, 3). Autophosphorylation and phosphotransfer determinants generally appear to map to the conserved *kinase* domain: (*a*) The most likely candidate for the site of autophosphorylation is a conserved histidine located in the conserved *kinase* domain (8), because NtrB is phosphorylated at this histidine (222) and mutation of the conserved histidine in EnvZ or VirA results in loss of both autophosphorylation and biological activity (164, 165, 185). (*b*) Deleted versions of EnvZ, PhoM, PhoR, and VirA lacking their N-terminal nonconserved portions autophosphorylate and transfer phosphate to *regulators* (161, 163, 164, 181, 181a, 185, 186), suggesting that the autophosphorylation and phosphotransfer functions of these *kinases* could reside in the conserved domain.

CheA, DegS, FrzE, KinA, and NtrB are cytoplasmic *kinases* (8) and thus are not subject to the topological constraints imposed by a periplasmic location for the N-terminal nonconserved region in transmembrane *kinase* proteins. The location of phosphorylation functions within NtrB fits the above model; however, CheA is different. The nonconserved N-terminal domain of CheA contains the autophosphorylation site and phosphotransfer determinants (78).

It is now well established that the conserved *regulator* domain serves as a phosphoacceptor. CheY and SpoOF, which consist exclusively of this domain (8), are phosphorylated (72, 74, 207). Truncated or fusion proteins and proteolytic fragments containing only the conserved N-terminal domains of CheB (79, 83), NtrC (138), and OmpR (171) are also phosphorylated. A highly conserved aspartate has been identified as the site of phosphorylation in the *regulators* CheY, VirG, and NtrC (Table 1). Mutation of the conserved aspartate in CheB results in loss of phosphorylation and biological activity (79). Thus all available data is consistent with phosphorylation on a conserved N-terminal aspartate. Altered activity following phosphorylation in vitro has been demonstrated for CheB, NtrC, and OmpR, as described previously.

## Role of Phosphorylation

Detailed understanding of how phosphorylation leads to *regulator* activation is currently lacking. Mutations that result in increased activity have also been isolated in the genes for a variety of *regulator* proteins, including CheY (99), DegU (202, 212), NtrC (152, 154), OmpR (170, 213), PhoQ (196), and SpoOA (208, 209). Mutations have been identified that activate the *regulator* by making it a better substrate for phosphorylation (208), by stabilizing the phosphoryl group (167), or by triggering function in the absence of phosphorylation (99, 152, 154). The latter class suggests that the role of phosphorylation in activation is indirect, e.g. stabilizes a conformational change.

If *regulators* have common structural and functional features leading to activation, then mutations in conserved positions might be active in any *regulator*, whereas those in nonconserved residues might be *regulator*-specific. This hypothesis predicts that replacing the conserved aspartate corresponding to Asp13 of CheY with lysine generally may result in *regulator* activation, because this substitution activates CheY (99). The wild-type *regulators* FrzG (130) and FlbD (214) have lysine instead of aspartate at the corresponding positions; however, it is not known whether FrzG and FlbD are constitutively active. Thus, the hypothesis remains to be rigorously tested.

## Functional Modules and Circuit Design

Two-component regulatory systems each have several functional elements in common: (*a*) a detector to sense environmental signals and in some manner influence *regulator* phosphorylation, (*b*) an autophosphorylating protein kinase, (*c*) a regulator with phosphorylation-sensitive activity, (*d*) a target of *regulator* action, and (*e*) a specific phosphatase activity that dephosphorylates the *regulator*. These functional elements are modular in the sense that they apparently can be arranged to suit the requirements of each signalling system (Table 2; 215) and functional chimeras can be assembled (216). The number of each type of element used varies in different systems, and multiple functional elements may be present on a single protein molecule. Considering the common phosphotransfer chemistry, the diversity among the analyzed systems is striking. No two systems have been found to be similarly organized. Examples of different circuit designs and their characteristic properties are described in the following sections.

DETECTOR/KINASE RELATIONSHIPS    The detector proteins for the Ntr and Che systems are physically distinct from the respective *kinase* proteins. This arrangement permits branched regulatory connections, either from the detector to downstream elements or from upstream elements to the kinase, forming distributive or integrative circuits respectively. The Ntr detector is UT/UR, which modifies $P_{II}$ to regulate both glutamine synthetase synthesis (via NtrC dephosphorylation) and glutamine synthetase activity (via adenylylation). Here separation of the detector from the phosphorylation reactions allows coordination of two different responses to one stimulus. In the chemotaxis system, CheW functionally couples the transducers to the CheA *kinase*. CheA responds to input from many chemoreceptors, enabling integration of multiple inputs and chemotaxis to many chemoeffectors.

In contrast to the cytoplasmically located CheA and NtrB, many of the known *kinases* are transmembrane proteins with the conserved C-terminal domain located in the cytoplasm (8). By analogy with the chemotaxis transducers, the *kinases* have been proposed to fuse periplasmic detector and cytoplasmic kinase elements into a single molecule (1, 3). This topology would seem ideal for detecting environmental signals and communicating with a cytoplasmic *regulator*. The available evidence supports a C-terminal kinase location, as discussed previously. Detector function in *kinases* is poorly characterized, however, and the specific stimulus detected remains unidentified in many systems. The first clear demonstration of a detector function for the periplasmic portion of a *kinase* has only recently been reported. Phenolic compounds (e.g. acetosyringone) released by wounded plant tissue are believed to be a major signal for virulence induction in *A. tumefaciens,* yet deletion of the periplasmic domain of VirA does not impair acetosyringone induction (188–190). Certain sugars (e.g. galactose) significantly enhance acetosyringone induction, apparently by binding to the periplasmic ChvE protein, which in turn interacts with the periplasmic portion of VirA (189, 190, 216a). This chain of events is strongly reminiscent of sugar-binding proteins interacting with the Trg and Tar chemotaxis transducers in *E. coli.* Further support for a N-terminal detector location comes from the observation that a truncated EnvZ has lost osmoregulation (161). The

**Table 2**  Organization of functional elements in selected two-component regulatory systems

| Element | Chemotaxis | Nitrogen assimilation | Outer membrane protein expression | A. tumefaciens virulence |
|---|---|---|---|---|
| Detector(s) | Periplasmic binding proteins, Transducers | UR/UT | EnvZ | ChvE, VirA |
| Kinase | CheA | NtrB | EnvZ cytoplasmic domain | VirA cytoplasmic domain |
| Regulator(s) | CheB, CheY | NtrC | OmpR | VirG |
| Target(s) | Methylesterase, Flagellar switch | *ntr* promoters, RNA polymerase | *ompC/ompF* promoters, RNA polymerase | *vir* promoters, RNA polymerase |
| Phosphatase(s) | CheB autophosphatase, CheY autophosphatase, CheZ | NtrC autophosphatase, NtrB-$P_{II}$-ATP | EnvZ cytoplasmic domain-ATP | None identified[a] |

[a]No effect on VirG-phosphate observed with three different C-terminal fragments of VirA, in either the presence or absence of ATP (187).

osmolarity signal is unidentified, however, and other explanations for loss of function are possible.

The mechanism by which the detector influences *regulator* phosphorylation has been established only for the Che and Ntr systems. The strategies employed for altering the balance between protein phosphorylation and dephosphorylation appear to be different. The chemotaxis transducers influence CheA kinase activity (73, 89), whereas the Ntr detector UT/UR influences dephosphorylation via $P_{II}$ (10, 138). This scheme could allow the Ntr system to maintain environmental control in spite of a high level of NtrC phosphorylation generated by crosstalking *kinases*. In the chemotaxis system, which is designed to operate on a more rapid time scale, any "noise" from crosstalk may be erased by dephosphorylation through the constitutive activity of CheZ before it has a chance to build up and obscure the true signal generated by the regulated CheA kinase.

As an enzymatic reaction, *regulator* phosphorylation is presumably one step in the sensory transduction pathway responsible for signal amplification or "gain."

KINASE/REGULATOR RELATIONSHIPS     Most two-component systems have one *kinase* acting on one *regulator* (e.g. NtrB → NtrC or EnvZ → OmpR). Different numbers of these elements may be linked together, however, allowing one *kinase* to initiate multiple responses (distributive circuit) or multiple *kinases* to initiate a common response (integrative circuit). CheA regulates CheB and CheY to control both adaptation and excitation responses. The sporulation system apparently involves two *kinases,* KinA and KinB, and two *regulators,* SpoOA and SpoOF (207, 209, 210). PhoR normally regulates PhoB, but in the absence of PhoR, a different *kinase* (PhoM) can act on PhoB (181a, 217). The *kinases* ArcB and CpxA apparently detect different signals to control two entirely different responses (218). Both *kinases* act through the same *regulator,* however, suggesting ArcA may be modified differently by ArcB and CpxA.

The *kinases* ArcB (193), BvgS (204), FrzE (129), RscC (201), and VirA (8) contain *regulator* domains in their C-terminal portions, i.e. kinase and regulator elements appear to be fused in the same protein. However, additional *regulators* have been identified for each of these *kinases* (Table 1). The function of the "extra" fused *regulator* domains is unknown but apparently essential, because mutations deleting or disrupting this region of ArcB, BvgS, or VirA result in inactive proteins (188, 193, 204).

REGULATOR/TARGET INTERACTION     The conserved *regulator* domain can interact with protein targets in an intra- or intermolecular fashion. CheY, consisting in its entirety of this domain, acts on the flagellar switch proteins. In contrast, the regulatory N-terminal domain in CheB interacts with the

C-terminal methylesterase domain, which in turn covalently modifies the transducers. Deletion of the conserved N-terminal domain activates CheB (108) and DctD (179), but does not activate OmpR (172). The most common target of *regulator* action is a promoter, with both DNA and protein (RNA polymerase) interactions involved. The unique C-terminal portion of the *regulators* DctD (179), NtrC (152, 156), OmpR (171, 172), and VirG (192) appears to be responsible for DNA binding, and in the cases of DctD (179) and NtrC (152, 154), interactions with RNA polymerase as well.

PHOSPHATASE ACTIVITIES    *Regulator*-phosphate stabilities vary widely. The phosphorylated forms of CheB (72), CheY (72, 80), NtrC (138, 139), and PhoB (181) are very labile, with half-lives on the order of seconds to minutes, whereas phosphorylated FrzE (206a), OmpR (166), SpoOA (E. Ninfa, J. Stock, & P. Youngman, cited in 8), SpoOF (A. Ninfa & I. Smith, cited in 8), and VirG (187) have half-lives of $\sim 1$ hour. The intrinsic half-life of an acyl phosphate is on the order of hours (219), suggesting the local environment within some *regulator* proteins destabilizes the phosphate group and promotes autodephosphorylation. Disruption of this structure by removal of $Mg^{2+}$ or protein denaturation increases the half-lives of CheY-phosphate and NtrC-phosphate to hours, as expected (80, 82, 139).

Other proteins can accelerate *regulator* dephosphorylation (Table 2). ATP is required for stimulation of dephosphorylation by NtrB-$P_{II}$ (10, 138) or EnvZ (163, 166, 167). A comparison of the $K_M$s for ATP in the phosphorylation and dephosphorylation reactions has not been made; the reported experiments used 0.1–1 mM ATP for both reactions. In the only other system with an identified trans-acting phosphatase, the possible effect of CheZ, CheA, and ATP together on CheY dephosphorylation has not been investigated. VirG dephosphorylation is not affected by VirA in either the presence or absence of ATP (187).

Phosphatase activities destroy the signal generated by outdated sensory data and thus permit responses on time scales shorter than the intrinsic chemical lifetime of aspartyl phosphates. Regulation of phosphatase activity could have a dramatic impact on the dynamic properties of a signal transduction system based on protein phosphorylation. However, $P_{II}$ is currently the only phosphatase element known to be regulated in a two-component system. Further research may uncover additional regulatory circuits that affect phosphatase activities.

The timing requirements of different two-component regulatory systems vary greatly. It is important to integrate environmental information over an appropriate period of time before implementing a response. The stability of various *regulator*-phosphate species may correlate with these needs. Chemotaxis depends on rapid environmental sensing and response. CheB and CheY are appropriately the least stable of any known *regulator*-phosphates, with in vitro half-lives on the order of seconds. A range of time constants is evident for transcriptional responses. The phosphorylated forms of the *regulators* for nitrogen and phosphate assimilation are stable for minutes in vitro, allowing reaction to changing conditions in less than one cell generation. In contrast, OmpR-phosphate, which has an in vitro half-life greater than one hour, would be expected to persist for many cell generations in the absence of a phosphatase. This property fits nicely with the fact that porin synthesis patterns apparently switch rapidly following an osmolarity change, but altering porin composition in the outer membrane requires about two generations to dilute out old porin molecules (220). When osmolarity decreases, the EnvZ phosphatase could be transiently activated to hydrolyze preexisting OmpR-phosphate and thus permit a rapid shift to a new pattern of porin synthesis. Finally, long-term responses like sporulation or symbiosis may require a high threshold and integration of information over a long time period in order to avoid premature implementation; this could explain why FrzE, SpoOA, SpoOF, and VirG lack autophosphatase activity and why other phosphatases have not yet been identified.

## CONCLUDING REMARKS

Several themes emerge from the comparison of bacterial signal transduction systems. First, signalling is probably accomplished via conformational changes stabilized by transient covalent modifications. Second, a variety of functional domains or elements can be organized in an almost limitless number of permutations to yield a circuit having the desired characteristics with regard to rapidity and duration of response, signal amplification, etc. By varying the number of each type of functional element used, networks with integrative and/or distributive properties can be assembled. The challenge that these systems pose in the next few years will be to clarify the nature of the relationships between circuit elements in vivo and the precise roles of these elements in the adaptation of microorganisms to their environment.

ACKNOWLEDGMENTS

We thank R. L. Bennett, H. C. Berg, F. W. Dahlquist, F. E. Dailey, J. J. Falke, J. A. Gegner, G. L. Hazelbauer, D. E. Koshland Jr., S. Kustu, J. Liu, P. Matsumura, D. F. McNally, A. J. Ninfa, J. S. Parkinson, D. A. Sanders, T. J. Silhavy, R. C. Stewart, K. Volz, R. M. Weis, A. J. Wolfe, R. Yaghmai, and P. Youngman for communication of unpublished results and/or helpful discussions; and L. Alex, A. Glasgow, and C. O'Day for comments on the manuscript. This work was supported by National Research Service Award Fellowships AI07798 (to R.B.B.) and GM11223 (to K.A.B.) and Grant AI19296 from the National Institutes of Health (to M.I.S.).

## Literature Cited

1. Nixon, B. T., Ronson, C. W., Ausubel, F. M. 1986. *Proc. Natl. Acad. Sci. USA* 83:7850–54
2. Winans, S. C., Ebert, P. R., Stachel, S. E., Gordon, M. P., Nester, E. W. 1986. *Proc. Natl. Acad. Sci. USA* 83:8278–82
3. Ronson, C. W., Nixon, B. T., Ausubel, F. M. 1987. *Cell* 49:579–81
4. Miller, J. F., Mekalanos, J. J., Falkow, S. 1989. *Science* 243:916–22
5. Bourret, R. B., Hess, J. F., Borkovich, K. A., Pakula, A. A., Simon, M. I. 1989. *J. Biol. Chem.* 264:7085–88
6. Gross, R., Arico, B., Rappuoli, R. 1989. *Mol. Microbiol.* 3:1661–67
7. Albright, L. M., Huala, E., Ausubel, F. M. 1989. *Annu. Rev. Genet.* 23:311–36
8. Stock, J. B., Ninfa, A. J., Stock, A. M. 1989. *Microbiol. Rev.* 53:450–90
9. Stock, J. B., Stock, A. M., Mottonen, J. M. 1990. *Nature* 334:395–400
10. Ninfa, A. J., Magasanik, B. 1986. *Proc. Natl. Acad. Sci. USA* 83:5909–13
11. Stewart, R. C., Dahlquist, F. W. 1987. *Chem. Rev.* 87:997–1025
12. Macnab, R. M. 1987. In *Escherichia coli and Salmonella typhimurium: Cellular and Molecular Biology*, ed. F. C. Neidhardt, pp. 732–59. Washington, DC: Am. Soc. Microbiol.
13. Hazelbauer, G. L. 1988. *Can. J. Microbiol.* 34:466–74
14. Koshland, D. E. Jr. 1988. *Biochemistry* 27:5829–34
15. Parkinson, J. S. 1988. *Cell* 53:1–2
16. Berg, H. C. 1988. *Cold Spring Harbor Symp. Quant. Biol.* 53:1–9
17. Simon, M. I., Borkovich, K. A., Bourret, R. B., Hess, J. F. 1989. *Biochemie* 71:1013–19
18. Taylor, B. L., Lengeler, J. W. 1990. In *Membrane Transport and Information Storage,* ed. R. Aloia, pp. 69–90. New York: Liss
19. Berg, H. C. 1990. *Cold Spring Harbor Symp. Quant. Biol.* 55:539–45
20. Eisenbach, M. 1991. *Mod. Cell. Biol.* In press
21. Hazelbauer, G. L., Yaghmai, R., Burrows, G. G., Baumgartner, J. W., Dutton, D. P., et al. 1990. In *Biology of the Chemotactic Response,* ed. J. P. Armitage, J. Lackie, pp. 107–34. Cambridge: Cambridge Univ. Press
22. Silverman, M., Simon, M. 1974. *Nature* 249:73–74
23. Berg, H. C., Brown, D. A. 1972. *Nature* 239:500–4
24. Larsen, S. H., Reader, R. W., Kort, E. N., Tso, W.-W., Adler, J. 1974. *Nature* 249:74–77
25. Lapidus, I. R., Welch, M., Eisenbach, M. 1988. *J. Bacteriol.* 170:3627–32
26. Eisenbach, M., Wolf, A., Welch, M., Caplan, S. R., Lapidus, I. R., et al. 1990. *J. Mol. Biol.* 211:551–63
27. Eisenbach, M. 1990. *Mol. Microbiol.* 4:161–67
28. Kudo, S., Magariyama, Y., Aizawa, S.-I. 1990. *Nature* 346:677–80
29. Adler, J. 1966. *Science* 153:708–16
30. Tsang, N., Macnab, R., Koshland, D. E. Jr. 1973. *Science* 181:60–63
31. Macnab, R. M., Koshland, D. E. Jr. 1972. *Proc. Natl. Acad. Sci. USA* 69:2509–12
32. Block, S. M., Segall, J. E., Berg, H. C. 1983. *J. Bacteriol.* 154:312–23
33. Segall, J. E., Manson, M. D., Berg, H. C. 1982. *Nature* 296:855–57
34. Spudich, J. L., Koshland, D. E. Jr. 1975. *Proc. Natl. Acad. Sci. USA* 72:710–13
35. Berg, H. C., Tedesco, P. M. 1975. *Proc. Natl. Acad. Sci. USA* 72:3235–39
36. Ingham, C., Buechner, M., Adler, J. 1990. *J. Bacteriol.* 172:3577–83
37. Krikos, A., Mutoh, N., Boyd, A., Simon, M. I. 1983. *Cell* 33:615–22
38. Russo, A. F., Koshland, D. E. Jr. 1983. *Science* 220:1016–20
39. Manoil, C., Beckwith, J. 1986. *Science* 233:1403–8
40. Mutoh, N., Oosawa, K., Simon, M. I. 1986. *J. Bacteriol.* 167:992–98
41. Parkinson, J. S. 1978. *J. Bacteriol.* 135:45–53
42. Parkinson, J. S., Parker, S. R., Talbert, P. B., Houts, S. E. 1983. *J. Bacteriol.* 155:265–74
43. Yamaguchi, S., Aizawa, S.-I., Kihara, M., Isomura, M., Jones, C. J., et al. 1986. *J. Bacteriol.* 168:1172–79
44. Clegg, D. O., Koshland, D. E. Jr. 1984. *Proc. Natl. Acad. Sci. USA* 81:5056–60
45. Yamaguchi, S., Fujita, H., Ishihara, A., Aizawa, S.-I., Macnab, R. M. 1986. *J. Bacteriol.* 166:187–93
46. Springer, M. S., Goy, M. F., Adler, J. 1979. *Nature* 280:279–84
47. Springer, W. R., Koshland, D. E. Jr. 1977. *Proc. Natl. Acad. Sci. USA* 74:533–37
48. Stock, J. B., Koshland, D. E. Jr. 1978. *Proc. Natl. Acad. Sci. USA* 75:3659–63
49. Stock, J. B., Maderis, A. M., Koshland, D. E. Jr. 1981. *Cell* 27:37–44
50. Block, S. M., Segall, J. E., Berg, H. C. 1982. *Cell* 31:215–26
51. Stock, J., Kersulis, G., Koshland, D. E. Jr. 1985. *Cell* 42:683–90

52. Krikos, A., Conley, M. P., Boyd, A., Berg, H. C., Simon, M. I. 1985. *Proc. Natl. Acad. Sci. USA* 82:1326–30
53. Hedblom, M. L., Adler, J. 1980. *J. Bacteriol.* 144:1048–60
54. Wolff, C., Parkinson, J. S. 1988. *J. Bacteriol.* 170:4509–15
55. Lee, L., Mizuno, T., Imae, Y. 1988. *J. Bacteriol.* 170:4769–74
56. Lee, L., Imae, Y. 1990. *J. Bacteriol.* 172:377–82
57. Mowbray, S. L., Koshland, D. E. Jr. 1990. *J. Biol. Chem.* 265:15638–43
58. Dahl, M. K., Boos, W., Manson, M. D. 1989. *J. Bacteriol.* 171:2361–71
59. Ames, P., Chen, J., Wolff, C., Parkinson, J. S. 1988. *Cold Spring Harbor Symp. Quant. Biol.* 53:59–65
60. Strange, P. G., Koshland, D. E. Jr. 1976. *Proc. Natl. Acad. Sci. USA* 73:762–66
61. Park, C., Hazelbauer, G. L. 1986. *J. Bacteriol.* 167:101–9
62. Mowbray, S. L., Koshland, D. E. Jr. 1987. *Cell* 50:171–80
63. Kossmann, M., Wolff, C., Manson, M. D. 1988. *J. Bacteriol.* 170:4516–21
64. Oosawa, K., Simon, M. 1986. *Proc. Natl. Acad. Sci. USA* 83:6930–34
65. Falke, J. J., Koshland, D. E. Jr. 1987. *Science* 237:1596–600
66. Milligan, D. L., Koshland, D. E. Jr. 1988. *J. Biol. Chem.* 263:6268–75
67. Falke, J. J., Dernburg, A. F., Sternberg, D. A., Zalkin, N., Milligan, D. L., et al. 1988. *J. Biol. Chem.* 263:14850–58
68. Park, C., Dutton, D. P., Hazelbauer, G. L. 1990. *J. Bacteriol.* 172:7179–87
69. Ames, P., Parkinson, J. S. 1988. *Cell* 55:817–26
70. Asakura, S., Honda, H. 1984. *J. Mol. Biol.* 176:349–67
71. Hess, J. F., Oosawa, K., Matsumura, P., Simon, M. I. 1987. *Proc. Natl. Acad. Sci. USA* 84:7609–13
72. Hess, J. F., Oosawa, K., Kaplan, N., Simon, M. I. 1988. *Cell* 53:79–87
73. Borkovich, K. A., Kaplan, N., Hess, J. F., Simon, M. I. 1989. *Proc. Natl. Acad. Sci. USA* 86:1208–12
74. Wylie, D., Stock, A., Wong, C.-Y., Stock, J. 1988. *Biochem. Biophys. Res. Commun.* 151:891–96
75. Bochner, B. R., Ames, B. N. 1982. *J. Biol. Chem.* 257:9759–69
75a. Atkinson, D. E. 1977. *Cellular Energy Metabolism and its Regulation.* New York: Academic. 293 pp.
76. Shioi, J., Galloway, R. J., Niwano, M., Chinnock, R. E., Taylor, B. L. 1982. *J. Biol. Chem.* 257:7969–75
77. Hultquist, D. E., Moyer, R. W., Boyer, P. D. 1966. *Biochemistry* 5:322–31
78. Hess, J. F., Bourret, R. B., Simon, M. I. 1988. *Nature* 336:139–43
79. Stewart, R. C., Roth, A. F., Dahlquist, F. W. 1990. *J. Bacteriol.* 172:3388–99
80. Hess, J. F., Bourret, R. B., Oosawa, K., Matsumura, P., Simon, M. I. 1988. *Cold Spring Harbor Symp. Quant. Biol.* 53:41–48
81. Lukat, G. S., Stock, A. M., Stock, J. B. 1990. *Biochemistry* 29:5436–42
82. Stock, A. M., Wylie, D. C., Mottonen, J. M., Lupas, A. N., Ninfa, E. G., et al. 1988. *Cold Spring Harbor Symp. Quant. Biol.* 53:49–57
83. Lupas, A., Stock, J. 1989. *J. Biol. Chem.* 264:17337–42
84. Oosawa, K., Hess, J. F., Simon, M. I. 1988. *Cell* 53:89–96
85. Liu, J. 1990. *Molecular genetics of the chemotactic signalling pathway in* Escherichia coli. PhD thesis. Univ. Utah
86. Smith, R. A., Parkinson, J. S. 1980. *Proc. Natl. Acad. Sci. USA* 77:5370–74
86a. Gegner, J. A., Dahlquist, F. W. 1991. *Proc. Natl. Acad. Sci. USA* 88:750–54
87. Clarke, S., Koshland, D. E. Jr. 1979. *J. Biol. Chem.* 254:9695–702
88. Bogonez, E., Koshland, D. E. Jr. 1985. *Proc. Natl. Acad. Sci. USA* 82:4891–95
89. Borkovich, K. A., Simon, M. I. 1990. *Cell* 63:1339–48
90. Stewart, R. C., Russell, C. B., Roth, A. F., Dahlquist, F. W. 1988. *Cold Spring Harbor Symp. Quant. Biol.* 53:27–40
91. Liu, J., Parkinson, J. S. 1989. *Proc. Natl. Acad. Sci. USA* 86:8703–7
92. Sanders, D. A., Mendez, B., Koshland, D. E. Jr. 1989. *J. Bacteriol.* 171:6271–78
93. Wolfe, A. J., Conley, M. P., Kramer, T. J., Berg, H. C. 1987. *J. Bacteriol.* 169:1878–85
94. Conley, M. P., Wolfe, A. J., Blair, D. F., Berg, H. C. 1989. *J. Bacteriol.* 171:5190–93
95. Kuo, S. C., Koshland, D. E. Jr. 1987. *J. Bacteriol.* 169:1307–14
96. Stock, A. M., Mottonen, J. M., Stock, J. B., Schutt, C. E. 1989. *Nature* 337:745–49
97. Matsumura, P., Rydel, J. J., Linzmeier, R., Vacante, D. 1984. *J. Bacteriol.* 160:36–41
98. Stock, A., Koshland, D. E. Jr., Stock, J. 1985. *Proc. Natl. Acad. Sci. USA* 82:7989–93
99. Bourret, R. B., Hess, J. F., Simon, M. I. 1990. *Proc. Natl. Acad. Sci. USA* 87:41–45
100. Sanders, D. A., Gillece-Castro, B. L., Stock, A. M., Burlingame, A. L., Koshland, D. E. Jr. 1989. *J. Biol. Chem.* 264:21770–78

101. Matsumura, P., Roman, S., Volz, K., McNally, D. 1990. See Ref. 21, pp. 135–54
102. Wolfe, A. J., Conley, M. P., Berg, H. C. 1988. *Proc. Natl. Acad. Sci. USA* 85:6711–15
103. Fox, D. K., Meadow, N. D., Roseman, S. 1986. *J. Biol. Chem.* 261:13498–503
104. Lee, T.-Y., Makino, K., Shinagawa, H., Nakata, A. 1990. *J. Bacteriol.* 172:2245–49
105. Kort, E. N., Goy, M. F., Larsen, S. H., Adler, J. 1975. *Proc. Natl. Acad. Sci. USA* 72:3939–43
106. Toews, M. L., Goy, M. F., Springer, M. S., Adler, J. 1979. *Proc. Natl. Acad. Sci. USA* 76:5544–48
107. Kehry, M. R., Doak, T. G., Dahlquist, F. W. 1984. *J. Biol. Chem.* 259:11828–35
108. Simms, S. A., Keane, M. G., Stock, J. 1985. *J. Biol. Chem.* 260:10161–68
109. Springer, M. S., Zanolari, B. 1984. *Proc. Natl. Acad. Sci. USA* 81:5061–65
110. Stewart, R. C., Dahlquist, F. W. 1988. *J. Bacteriol.* 170:5728–38
111. Springer, M. S., Goy, M. F., Adler, J. 1977. *Proc. Natl. Acad. Sci. USA* 74:3312–16
112. Silverman, M., Simon, M. 1977. *Proc. Natl. Acad. Sci. USA* 74:3317–21
113. Kleene, S. J., Hobson, A. C., Adler, J. 1979. *Proc. Natl. Acad. Sci. USA* 76:6309–13
114. Bogonez, E., Koshland, D. E. Jr. 1985. *Proc. Natl. Acad. Sci. USA* 82:4891–95
115. Borczuk, A., Staub, A., Stock, J. 1986. *Biochem. Biophys. Res. Commun.* 141:918–23
116. Nowlin, D. M., Bollinger, J., Hazelbauer, G. L. 1988. *Proteins* 3:102–12
117. Sanders, D. A., Koshland, D. E. Jr. 1988. *Proc. Natl. Acad. Sci. USA* 85:8425–29
118. Hazelbauer, G. L., Park, C., Nowlin, D. M. 1989. *Proc. Natl. Acad. Sci. USA* 86:1448–52
119. Segall, J. E., Block, S. M., Berg, H. C. 1986. *Proc. Natl. Acad. Sci. USA* 83:8987–91
120. Weis, R. M., Koshland, D. E. Jr. 1988. *Proc. Natl. Acad. Sci. USA* 85:83–87
121. Berg, H. C., Turner, L. 1990. *Biophys. J.* 58:919–30
122. Stock, J., Stock, A. 1987. *Trends Biochem. Sci.* 12:371–75
123. Stock, J., Borczuk, A., Chiou, F., Burchenal, J. E. B. 1985. *Proc. Natl. Acad. Sci. USA* 82:8364–68
124. Wolfe, A. J., Berg, H. C. 1989. *Proc. Natl. Acad. Sci. USA* 86:6973–77
125. Weis, R. M., Chasalow, S., Koshland, D. E. Jr. 1990. *J. Biol. Chem.* 265:6817–26
126. Adler, J. 1969. *Science* 166:1588–97
127. Armitage, J. P., Sockett, R. E. 1987. *Ann. NY Acad. Sci.* 510:9–15
128. McBride, M. J., Weinberg, R. A., Zusman, D. R. 1989. *Proc. Natl. Acad. Sci. USA* 86:424–28
129. McCleary, W. R., Zusman, D. R. 1990. *Proc. Natl. Acad. Sci. USA* 87:5898–902
130. McCleary, W. R., McBride, M. J., Zusman, D. R. 1990. *J. Bacteriol.* 172:4877–87
131. Thoelke, M. S., Casper, J. M., Ordal, G. W. 1990. *J. Biol. Chem.* 265:1928–32
132. Alam, M., Lebert, M., Oesterhelt, D., Hazelbauer, G. L. 1989. *EMBO J.* 8:631–39
133. Goulbourne, E. A. Jr., Greenberg, E. P. 1983. *J. Bacteriol.* 153:916–20
134. Keener, J., Wong, P., Popham, D., Wallis, J., Kustu, S. 1986. In *RNA Polymerase and the Regulation of Transcription*, ed. W. S. Reznikoff, C. A. Gross, R. R. Burgess, M. T. Record Jr., J. E. Dahlberg, M. P. Wickens, pp. 159–75. New York: Elsevier
135. Reitzer, L. J., Magasanik, B. 1987. See Ref. 12, pp. 302–20
136. Magasanik, B. 1989. *Biochemie* 71:1005–12
137. Reitzer, L. J., Magasanik, B. 1985. *Proc. Natl. Acad. Sci. USA* 82:1979–83
138. Keener, J., Kustu, S. 1988. *Proc. Natl. Acad. Sci. USA* 85:4976–80
139. Weiss, V., Magasanik, B. 1988. *Proc. Natl. Acad. Sci. USA* 85:8919–23
140. Chen, Y.-M., Backman, K., Magasanik, B. 1982. *J. Bacteriol.* 150:214–20
141. Bueno, R., Pahel, G., Magasanik, B. 1985. *J. Bacteriol.* 164:816–22
142. Backman, K. C., Chen, Y.-M., Ueno-Nishio, S., Magasanik, B. 1983. *J. Bacteriol.* 154:516–19
143. Hirschman, J., Wong, P.-K., Sei, K., Keener, J., Kustu, S. 1985. *Proc. Natl. Acad. Sci. USA* 82:7525–29
144. Hunt, T. P., Magasanik, B. 1985. *Proc. Natl. Acad. Sci. USA* 82:8453–57
145. Ninfa, A. J., Reitzer, L. J., Magasanik, B. 1987. *Cell* 50:1039–46
146. Sasse-Dwight, S., Gralla, J. D. 1988. *Proc. Natl. Acad. Sci. USA* 85:8934–38
147. Kustu, S., Sei, K., Keener, J. 1986. In *Regulation of Gene Expression*, ed. I. R. Booth, C. F. Higgins, pp. 139–54. Cambridge, England: Cambridge Univ. Press
148. Reitzer, L. J., Magasanik, B. 1986. *Cell* 45:785–92

149. Reitzer, L. J., Movsas, B., Magasanik, B. 1989. *J. Bacteriol.* 171:5512–22
150. Wedel, A., Weiss, D. S., Popham, D., Droge, P., Kustu, S. 1990. *Science* 248:486–90
151. Su, W., Porter, S., Kustu, S., Echols, H. 1990. *Proc. Natl. Acad. Sci. USA* 87:5504–8
152. Popham, D., Szeto, D., Keener, J., Kustu, S. 1989. *Science* 243:629–35
153. Ninfa, A. J., Brodsky, E., Magasanik, B. 1989. In *DNA-Protein Interactions in Transcription*, ed. J. Gralla, pp. 43–52. New York: Liss
154. Weglenski, P., Ninfa, A. J., Ueno-Nishio, S., Magasanik, B. 1989. *J. Bacteriol.* 171:4479–85
155. Ronson, C. W., Astwood, P. M., Nixon, B. T., Ausubel, F. M. 1987. *Nucleic Acids Res.* 15:7921–34
156. Contreras, A., Drummond, M. 1988. *Nucleic Acids Res.* 16:4025–39
157. Forst, S., Inouye, M. 1988. *Annu. Rev. Cell Biol.* 4:21–42
158. Igo, M. M., Slauch, J. M., Silhavy, T. J. 1990. *New Biol.* 2:5–9
159. Mizuno, T., Mizushima, S. 1990. *Mol. Microbiol.* 4:1077–82
160. Forst, S., Comeau, D., Norioka, S., Inouye, M. 1987. *J. Biol. Chem.* 262:16433–38
161. Igo, M. M., Silhavy, T. J. 1988. *J. Bacteriol.* 170:5971–73
162. Igo, M. M., Ninfa, A. J., Silhavy, T. J. 1989. *Genes Dev.* 3:598–605
163. Aiba, H., Mizuno, T., Mizushima, S. 1989. *J. Biol. Chem.* 264:8563–67
164. Forst, S., Delgado, J., Inouye, M. 1989. *Proc. Natl. Acad. Sci. USA* 86:6052–56
165. Forst, S., Delgado, J., Rampersaud, A., Inouye, M. 1990. *J. Bacteriol.* 172:3473–77
166. Igo, M. M., Ninfa, A. J., Stock, J. B., Silhavy, T. J. 1989. *Genes Dev.* 3:1725–34
167. Aiba, H., Nakasai, F., Mizushima, S., Mizuno, T. 1989. *J. Biol. Chem.* 264:14090–94
168. Hall, M. N., Silhavy, T. J. 1981. *J. Mol. Biol.* 151:1–15
169. Mizuno, T., Mizushima, S. 1987. *J. Biochem.* 101:387–96
170. Slauch, J. M., Silhavy, T. J. 1989. *J. Mol. Biol.* 210:281–92
171. Kato, M., Aiba, H., Tate, S., Nishimura, Y., Mizuno, T. 1989. *FEBS Lett.* 249:168–72
172. Tsung, K., Brissette, R. E., Inouye, M. 1989. *J. Biol. Chem.* 264:10104–9
173. Aiba, H., Nakasai, F., Mizushima, S., Mizuno, T. 1989. *J. Biochem.* 106:5–7
174. Aiba, H., Mizuno, T. 1990. *FEBS Lett.* 261:19–22
175. Tsung, K., Brissette, R., Inouye, M. 1990. *Proc. Natl. Acad. Sci. USA* 87:5940–44
176. Maeda, S., Ozawa, Y., Mizuno, T., Mizushima, S. 1988. *J. Mol. Biol.* 202:433–41
177. Maeda, S., Mizuno, T. 1988. *J. Biol. Chem.* 263:14629–33
178. Jiang, J., Gu, B., Albright, L. M., Nixon, B. T. 1989. *J. Bacteriol.* 171:5244–53
179. Ledebur, H., Gu, B., Sodja, J., Nixon, B. T. 1990. *J. Bacteriol.* 172:3888–97
180. Jiang, S.-Q., Yu, G.-Q., Li, Z.-G., Hong, J.-S. 1988. *J. Bacteriol.* 170:4304–8
181. Makino, K., Shinagawa, H., Amemura, M., Kawamoto, T., Yamada, M., et al. 1989. *J. Mol. Biol.* 210:551–59
181a. Amemura, M., Makino, K., Shinagawa, H., Nakata, A. 1990. *J. Bacteriol.* 172:6300–7
182. Seki, T., Yoshikawa, H., Takahashi, H., Saito, H. 1988. *J. Bacteriol.* 170:5935–38
183. Lee, T.-Y., Makino, K., Shinagawa, H., Amemura, M., Nakata, A. 1989. *J. Bacteriol.* 171:6593–99
184. Anba, J., Bidaud, M., Vasil, M. L., Lazdunski, A. 1990. *J. Bacteriol.* 172:4685–89
185. Jin, S., Roitsch, T., Ankenbauer, R. G., Gordon, M. P., Nester, E. W. 1990. *J. Bacteriol.* 172:525–30
186. Huang, Y., Morel, P., Powell, B., Kado, C. I. 1990. *J. Bacteriol.* 172:1142–44
187. Jin, S., Prusti, R. K., Roitsch, T., Ankenbauer, R. G., Nester, E. W. 1990. *J. Bacteriol.* 172:4945–50
188. Melchers, L. S., Regensburg-Tuink, T. J. G., Bourret, R. B., Sedee, N. J. A., Schilperoort, R. A., et al. 1989. *EMBO J.* 8:1919–25
189. Shimoda, N., Toyoda-Yamamoto, A., Nagamine, J., Usami, S., Katayama, M., et al. 1990. *Proc. Natl. Acad. Sci. USA* 87:6684–88
190. Cangelosi, G. A., Ankenbauer, R. G., Nester, E. W. 1990. *Proc. Natl. Acad. Sci. USA* 87:6708–12
191. Pazour, G. J., Das, A. 1990. *J. Bacteriol.* 172:1241–49
192. Roitsch, T., Wang, H., Jin, S., Nester, E. W. 1990. *J. Bacteriol.* 172:6054–60
193. Iiuchi, S., Matsuda, Z., Fujiwara, T., Lin, E. C. C. 1990. *Mol. Microbiol.* 4:715–27
194. Rainwater, S., Silverman, P. M. 1990. *J. Bacteriol.* 172:2456–61

195. Iuchi, S., Chepuri, V., Fu, H.-A., Gennis, R. B., Lin, E. C. C. 1990. *J. Bacteriol.* 172:6020–25
196. Miller, S. I., Mekalanos, J. J. 1990. *J. Bacteriol.* 172:2485–90
197. de Philip, P., Batut, J., Boistard, P. 1990. *J. Bacteriol.* 172:4255–62
198. Gottfert, M., Grob, P., Hennecke, H. 1990. *Proc. Natl. Acad. Sci. USA* 87:2680–84
199. Egan, S. M., Stewart, V. 1990. *J. Bacteriol.* 172:5020–29
200. Weston, L. A., Kadner, R. J. 1988. *J. Bacteriol.* 170:3375–83
201. Stout, V., Gottesman, S. 1990. *J. Bacteriol.* 172:659–69
201a. Mukai, K., Kawata, M., Tanaka, T. 1990. *J. Biol. Chem.* 265:20000–6
202. Msadek, T., Kunst, F., Henner, D., Klier, A., Rapoport, G., et al. 1990. *J. Bacteriol.* 172:824–34
203. Weinrauch, Y., Penchev, R., Dubnau, E., Smith, I., Dubnau, D. 1990. *Genes Dev.* 4:860–72
204. Arico, B., Miller, J. F., Roy, C., Stibitz, S., Monack, D., et al. 1989. *Proc. Natl. Acad. Sci. USA* 86:6671–75
205. Roy, C. R., Miller, J. F., Falkow, S. 1990. *Proc. Natl. Acad. Sci. USA* 87:3763–67
206. Scarlato, V., Prugnola, A., Arico, B., Rappuoli, R. 1990. *Proc. Natl. Acad. Sci. USA* 87:6753–57
206a. McCleary, W. R., Zusman, D. R. 1990. *J. Bacteriol.* 172:6661–68
207. Perego, M., Cole, S. P., Burbulys, D., Trach, K., Hoch, J. A. 1989. *J. Bacteriol.* 171:6187–96
208. Olmedo, G., Ninfa, E. G., Stock, J., Youngman, P. 1990. *J. Mol. Biol.* 215:359–72
209. Spiegelman, G., Van Hoy, B., Perego, M., Day, J., Trach, K., et al. 1990. *J. Bacteriol.* 172:5011–19
210. Antoniewski, C., Savelli, B., Stragier, P. 1990. *J. Bacteriol.* 172:86–93
210a. Henikoff, S., Wallace, J. C., Brown, J. P. 1990. *Methods Enzymol.* 183:111–32
211. Ninfa, A. J., Ninfa, E. G., Lupas, A. N., Stock, A., Magasanik, B., et al. 1988. *Proc. Natl. Acad. Sci. USA* 85:5492–96
212. Henner, D. J., Yang, M., Ferrari, E. 1988. *J. Bacteriol.* 170:5102–9
213. Nara, F., Matsuyama, S., Mizuno, T., Mizushima, S. 1986. *Mol. Gen. Genet.* 202:194–99
214. Ramakrishnan, G., Newton, A. 1990. *Proc. Natl. Acad. Sci. USA* 87:2369–73
215. Kofoid, E. C., Parkinson, J. S. 1988. *Proc. Natl. Acad. Sci. USA* 85:4981–85
216. Utsumi, R., Brissette, R. E., Rampersaud, A., Forst, S. A., Oosawa, K., et al. 1989. *Science* 245:1246–49
216a. Ankenbauer, R. G., Nester, E. W. 1990. *J. Bacteriol.* 172:6442–46
217. Wanner, B. L. 1987. See Ref. 12, pp. 1326–33
218. Iuchi, S., Furlong, D., Lin, E. C. C. 1989. *J. Bacteriol.* 171:2889–93
219. Koshland, D. E. Jr. 1952. *J. Am. Chem. Soc.* 74:2286–92
220. Van Alphen, W., Lugtenberg, B. 1977. *J. Bacteriol.* 131:623–30
221. Kofoid, E. C., Parkinson, J. S. 1991. *J. Bacteriol.* 173:In press
222. Ninfa, A. J., Bennett, R. L. 1991. *J. Biol. Chem.* 266:In press

*Annu. Rev. Biochem. 1991. 60:443–75*

# PROTEOGLYCANS: STRUCTURES AND INTERACTIONS[1]

## Lena Kjellén and Ulf Lindahl

Department of Veterinary Medical Chemistry, Swedish University of Agricultural Sciences, The Biomedical Center, Box 575, S-751 23 Uppsala, Sweden

KEY WORDS:   glycosaminoglycan, sulfation, L-iduronic acid, glycosaminoglycan-protein interactions, extracellular matrix.

## CONTENTS

## PERSPECTIVES AND SUMMARY

Proteoglycans (PGs) are macromolecules composed of glycosaminoglycan (GAG) chains covalently bound to a protein core. The GAG components consist of hexosamine [D-glucosamine (GlcN) or D-galactosamine (GalN)] and either hexuronic acid [D-glucuronic acid (GlcA) or L-iduronic acid (IdoA)] or galactose units (in keratan sulfate) that are arranged in alternating, unbranched sequence, and carry sulfate substituents in various positions. The common GAGs [in addition to hyaluronan (HA), which does not occur as a PG but in free form] include the galactosaminoglycans, chondroitin sulfate

443

(CS) and dermatan sulfate (DS), and the glucosaminoglycans, heparan sulfate (HS), heparin, and keratan sulfate (KS). DS, HS, and heparin contain both GlcA and IdoA units, whereas CS has GlcA as the only hexuronic acid constituent. Owing to the variability in sulfate substitution, all GAGs, and particularly the GlcA/IdoA copolymeric species, display considerable sequence heterogeneity both within and between different GAG chains. For previous reviews on PG structure and/or biosynthesis see (1–7); references to older literature are found in (8, 9). Information regarding HA is found in (9a).

The number of GAG chain substituents (generally on serine residues) on a protein core may vary from one to > 100. The presence of GAG as well as oligosaccharide substituents has hampered the sequence analysis of these proteins, which has therefore depended heavily on the application of recombinant DNA techniques. A large number of core protein structures have been identified, and some "families" may be discerned based on sequence information. However, many of the core proteins identified show no distinctive features, and it is not possible at present to predict, solely on the basis of sequence data, whether a particular protein will carry a GAG chain and thus be classified as a PG. In fact, some proteins [e.g. a lymphocyte "homing" receptor (10)] are referred to as "parttime PGs" (although "glycanated variant" would be more correct) because of their inconsistent substitution with GAG chains. In addition to the GAG-substituted region, the protein core may contain other functional domains, designed for anchoring the PG to the cell surface (via a transmembrane peptide sequence or a phosphoinositol linkage) or to macromolecules of the extracellular matrix, or for interaction with other proteins. Continued analysis of primary structures for core proteins will undoubtedly unravel additional domain structures, which may be expected to provide further clues to the functional properties of the PGs.

PGs may occur intracellularly (usually in secretory granules), at the cell surface, or in the extracellular matrix. Their biological roles are highly diversified, ranging from relatively simple mechanical support functions, to more intricate, as yet rather poorly understood, effects on various cellular processes such as cell adhesion, motility, and proliferation, further to differentiation and tissue morphogenesis. Most of these effects depend on binding of proteins to the GAG chains, the association varying from charge interactions of relatively low affinity and specificity to highly specific, high-affinity binding involving a particular saccharide region of constant structure.

[1]Abbreviations used: CS, chondroitin sulfate; DS, dermatan sulfate; GAG, glycosaminoglycan; GalNAc, 2-deoxy-2-acetamido-D-galactose ($N$-acetyl-D-galactosamine); GlcA, D-glucuronic acid; GlcNAc, 2-deoxy-2-acetamido-D-glucose ($N$-acetyl-D-glucosamine); HA, hyaluronan; HexA, unspecified hexuronic acid; HS, heparan sulfate; IdoA, L-iduronic acid; KS, keratan sulfate; $-NSO_3$, N-sulfate group; $-OSO_3$, O-sulfate, ester sulfate group (the locations of O-sulfate groups are indicated in parentheses); PG, proteoglycan; TM, thrombomodulin.

Certain functions, such as the anticoagulant and antiproliferative activities of heparin/HS, are expressed by the free GAG chain. However, most biological activities attributed to PGs depend to some extent on the presence of the protein core, which may contribute in various ways. The protein core may simply provide a scaffold for the appropriate immobilization and spacing of the attached GAG chains, as required, for instance, to mask receptors or other sites by steric exclusion (see e.g. 11, 12). Attached to the cell surface the protein core assumes an anchoring function, which may be essential for the appropriate positioning of a GAG-bound ligand [e.g. an enzyme or enzyme inhibitor (see 13, 14)]. More complex relations apply when core proteins mediate cellular responses (for instance, reorganization of the actin cytoskeleton) to interactions involving GAG chain substituents (see 15, 16), or actively participate in a particular process in concert with the GAG component [see the antithrombin-dependent anticoagulant activity of thrombomodulin (17)].

The abundant experimental evidence implicating PGs in various cellular processes is mostly indirect or circumstantial, e.g. a concurrent change in PG structure, the transient appearance of a particular PG species, or induced effects of administered exogenous GAGs. The functional diversity of PGs may be expressed even in relation to an isolated phenomenon such as cell proliferation, which may be variously modulated by heparin-related GAGs with widely different structures. Attempts to explain in molecular terms the mechanisms behind such effects will benefit from the development of improved technology regarding, for instance, sequence analysis of GAGs, tailoring of specific GAG structures by organic synthesis, and selective genetic manipulation of PG biosynthesis. Conversely, it will be important to elucidate the mechanisms of unperturbed PG biosynthesis and its regulation, both as regards the initiation of GAG chains and the generation of specific GAG sequences. It is somewhat disconcerting that we still do not understand how a particular protein is selected for promotion to PG status.

## CLASSIFICATION AND NOMENCLATURE

PGs are widely distributed in animal tissues and appear to be synthesized by virtually all types of cells. A simple mode of classification takes note of the basic structure of the glycan backbone, which may be composed of 1. $[\text{HexA-GalN}]_n$, 2. $[\text{HexA-GlcN}]_n$, or 3. $[\text{Gal-GlcN}]_n$ disaccharide units. While the type-3 disaccharide unit occurs in KS only, types 1 and 2 may be further subdivided, type 1 into CS and DS, type 2 into HS and heparin. Unfortunately, this definition of subspecies is complicated by extensive microheterogeneity of the glycan structures (see section below on structure of

GAG chains). A further obstacle is posed by hybrid PGs in which different types of GAG chains are bound to the same core protein.

The topographical distribution of PGs in relation to the cell provides a different means of classification. We thus differentiate between PGs that occur intracellularly (generally in secretory granules), associated with the cell surface, or in the extracellular matrix. The ultimate classification of PGs, however, will rely on the structures of core proteins. This area has only recently become readily amenable to exploration, thanks to the dramatically improved possibilities for structural analysis of glycosylated proteins offered by recombinant DNA techniques. Comparison of the core proteins sequenced so far fails to show any common structural feature that clearly distinguishes the PG core proteins from any other proteins. A classification of PG core proteins analogous to that applied to the collagens thus does not appear feasible. Instead, many of the PGs are being increasingly referred to by trivial names ("aggrecan," "decorin," "syndecan," "versican," etc) that relate to structural or functional characteristics of the respective macromolecules. For practical purposes, some PGs may be grouped into "families" based on the amino-acid sequence homologies established so far, immunological cross-reactivities, and/or topographical/functional properties (see section below on core proteins). Table 1 lists PGs primarily with regard to topographical distribution, and includes designation of "families" and additional structural/functional characteristics. The table is not meant to be comprehensive, nor is it entirely consistent (e.g. regarding the allocation of PGs to the various topographical subgroups). The table does, however, provide an overview and may be useful as a source of reference and comparison.

## STRUCTURE OF PROTEOGLYCANS

### Carbohydrate Components

The carbohydrate constituents unique to PGs are sulfated GAG chains composed of alternating hexosamine (GlcN or GalN) and hexuronic acid (GlcA or IdoA) or galactose units (Figure 1). The characteristic structural features of these polymers are discussed below along with some recent findings regarding the glycopeptide "linkage regions" that join the polysaccharide chains to the core proteins [for references to older literature see reviews (1, 8, 9)]. In addition to the GAG chains, most core proteins carry N- and/or O-linked oligosaccharides, which appear similar in type to those present in other proteins (see e.g. 18, 19). These components are not considered further in this review.

GLYCOSAMINOGLYCAN CHAINS     The basic inherent structural feature of GAG chains, i.e. the alternation of two types of monosaccharide units, would

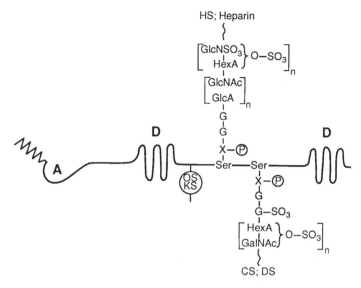

*Figure 1*  Conjectured model of a PG ("imagican"), which combines the structural features typical of PGs into one imaginary macromolecule. The GAG-substituted portion of the core protein is shown to carry both CS/DS and HS chains linked to serine units by galactosyl-galactosyl-xylosyl (GGX) trisaccharide sequences. Potential phosphate (P) and sulfate substitution sites on the GGX sequence are indicated. OS, oligosaccharides; KS, keratan sulfate. The anchoring (A) and other (D) domains of the core protein are described in the text. The terminal acyl substituent symbolizes the phosphatidyl inositol group that anchors certain PGs to plasma membranes, but may also indicate direct fatty acylation of the core protein (19a). For additional information see the text and Table 1.

seem to provide for the occurrence of regular polymer sequences. However, analyses of GAG preparations often reveal marked heterogeneity within as well as between the individual polysaccharide chains. This feature is readily explained in terms of the mechanisms of GAG biosynthesis. The initial polymerization reactions will generate products composed of strictly repeating disaccharide units. The subsequent modification process, which involves sulfate substitution at various positions and may include C5 epimerization of GlcA to IdoA units, is incomplete and will yield polymer sequences at various stages of modification.

The biosynthesis of heparin/HS (20–23) is initiated by formation of a polysaccharide chain with the structure $[GlcA \xrightarrow{\beta1,4} GlcNAc \xrightarrow{\alpha1,4}]_n$. This polymer is N-deacetylated/N-sulfated and subsequently undergoes C5 epimerization of GlcA to IdoA units, 2-O-sulfation of IdoA, and 6-O-sulfation of GlcN units. Additional, more rare, O-sulfate substituents are incorporated at C3 of GlcN units (24) and at C2 (or C3) of GlcA units (25,

26). The final products are highly diverse; in all, the various monosaccharide structures and substituents that have been identified account for four distinct HexA and six GlcN units, allowing for some 17 (or possibly 18) different → HexA → GlcN → and 10 (possibly 12) → GlcN → HexA → disaccharide sequences (Figure 2). The preferential codistribution of certain disaccharide units results in the generation of block structures, most commonly seen in HS (Figure 1), where extended regions composed of alternating, nonsulfated, GlcNAc and GlcA units are interspersed between N-sulfated, more extensively modified, sequences (1, 21, 27–32). However, isolated GlcNAc units surrounded by N-sulfated GlcN residues (typical for heparin) and, conversely, $GlcNSO_3$ units interrupting extended N-acetylated block structures, or alternating with N-acetylated disaccharide units (typical for HS), are common features and particularly contribute to the structural variability that is a hallmark of the heparin/HS family. That such irregular sequences may be of functional importance is demonstrated by the antithrombin-binding pentasaccharide region that is essential for the blood anticoagulant activity of heparin/ HS (Figure 3). This sequence is composed of three GlcN units, one of which is preferentially N-acetylated, one GlcA unit, and one IdoA unit, with O-sulfate groups in various positions (see further discussion of PG functions).

The galactosaminoglycans (CS and DS) are generated principally by similar modifications of an initial polymerization product with the structure [GlcA $\xrightarrow{\beta1,3}$ GalNAc $\xrightarrow{\beta1,4}$ ]$_n$ (1, 9). However, since the GalNAc residues remain acetylated, the structural diversity is less pronounced than for the heparin-related species. Nevertheless, owing to the presence of GlcA as well as IdoA units, and the variable location of (O-)sulfate groups, as many as nine different HexA-GalNAc disaccharide units have been identified (33).

KSs are generated by O-sulfate substitution of a basic polylactosamine structure, [Gal $\xrightarrow{\beta1,4}$ GlcNAc $\xrightarrow{\beta1,3}$ ]$_n$. The sulfate groups may be located at C6 of either or both of the Gal and GlcNAc units, and thus may give rise to mono- as well as disulfated disaccharide units [see (34) for references to previous work]. Studies on corneal KS (35) indicated a nonrandom distribution of such units, a number of monosulfated disaccharide units being inserted between the linkage region to protein and a sequence of disulfated units towards the nonreducing terminus of the chain. Recently, some preparations of skeletal KS were found to contain a small proportion of fucose residues along the glycan chain, apparently in $\beta1,3$ linkage to GlcNAc units (36). Moreover, "capping" sialic acid units at the nonreducing termini have been reported (18, 36).

A further note on nomenclature may be appropriate at this stage. Problems of designation are not restricted to the PG level (see above), but may apply even to single GAG chains. Consider, for instance, the distinction between heparin and HS. HS, first described by Jorpes & Gardell (37), was initially

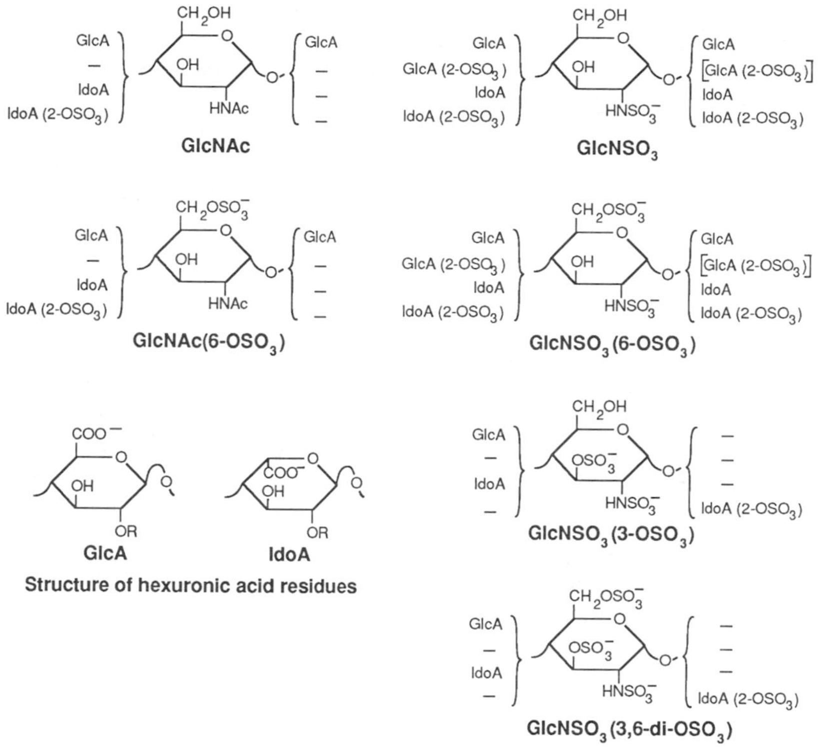

GlcA
−
IdoA
IdoA (2-OSO₃)  }  CH₂OH  OH  HNAc  GlcNAc  {  GlcA
−
−
−

GlcA
GlcA (2-OSO₃)
IdoA
IdoA (2-OSO₃)  }  CH₂OH  OH  HNSO₃⁻  GlcNSO₃  {  GlcA
[GlcA (2-OSO₃)]
IdoA
IdoA (2-OSO₃)

GlcA
−
IdoA
IdoA (2-OSO₃)  }  CH₂OSO₃⁻  OH  HNAc  GlcNAc(6-OSO₃)  {  GlcA
−
−
−

GlcA
GlcA (2-OSO₃)
IdoA
IdoA (2-OSO₃)  }  CH₂OSO₃⁻  OH  HNSO₃⁻  GlcNSO₃(6-OSO₃)  {  GlcA
[GlcA (2-OSO₃)]
IdoA
IdoA (2-OSO₃)

COO⁻  OH  OR  GlcA      COO⁻  OH  OR  IdoA
**Structure of hexuronic acid residues**

GlcA
−
IdoA
−  }  CH₂OH  OSO₃⁻  HNSO₃⁻  GlcNSO₃(3-OSO₃)  {  −
−
−
IdoA (2-OSO₃)

GlcA
−
IdoA
−  }  CH₂OSO₃⁻  OSO₃⁻  HNSO₃⁻  GlcNSO₃(3,6-di-OSO₃)  {  −
−
−
IdoA (2-OSO₃)

*Figure 2*  Scheme of disaccharide sequences identified in heparin and heparan sulfate. The six variously substituted internal GlcN residues are combined with the HexA units indicated at C4 and C1 to give 17 (possibly 18) different → HexA → GlcN → sequences, and 10 (possibly 12) → GlcN → HexA → sequences, respectively. The structures of the HexA units are indicated in the lower left corner (R = –H or –SO₃). Gaps indicated by (–) denote combinations that either do not exist (e.g. → GlcNAc → IdoA →) or occur very infrequently. Plausible sequences, still to be demonstrated, are indicated by [ ] around the HexA concerned.

believed to be fully N-acetylated, devoid of anticoagulant activity, and clearly distinct from the N-sulfated, biologically active heparin (38). We now know that both heparin and HS contain N-acetylated as well as N-sulfated GlcN units and, furthermore, that both types of polysaccharide may have or lack anticoagulant activity. In general, heparin contains more N- and O-sulfate groups and a higher proportion of IdoA units than HS, which instead contains more GlcNAc and GlcA residues. However, the information available reveals no qualitative differences between the two species. It appears questionable whether a distinction based solely on quantitative data [for instance, designating all species in which > 80% of the GlcN units are N-sulfated as "heparin" (39)] will invariably tell a low-sulfated "heparin" from a high-sulfated "HS."

We propose that for the time being the term "heparin" be reserved for the glucosaminoglycan component of serglycin PGs synthesized by mast cells; all other structurally related GAGs should be named "heparan sulfate." Similarly, there is considerable confusion as to the distinction between CS and DS. By definition, "chondroitin sulfate" is devoid of IdoA units; hence any galactosaminoglycan with a measurable proportion of this component should be termed "dermatan sulfate." In practice, this distinction, generally based on differential degradation by the two bacterial lyases, chondroitinase ABC and AC (1), is not always clear-cut, and chondroitinase-degradable material in general thus is often referred to as "chondroitin sulfate."

LINKAGE REGIONS    With the exception of HA, which appears to be assembled as a free polysaccharide chain (40), all GAGs are synthesized as PGs, i.e. in covalent linkage to protein. The linkage region is basically the same in PGs that carry CS/DS or heparin/HS chains, and consists of a $\rightarrow$ GlcA $\xrightarrow{\beta 1,3}$ Gal $\xrightarrow{\beta 1,3}$ Gal $\xrightarrow{\beta 1,4}$ Xyl $\xrightarrow{\beta}$ O-serine bridge between the GAG chain proper and the polypeptide core [Figure 1; see (9) for references]. More recent work showed that a variable proportion of the xylose units in linkage regions from CS- (41) as well as HS- (42) PGs are phosphorylated at C2. In contrast, the substitution of galactose residues in the linkage region with O-sulfate groups, at either C4 or C6, appears to be restricted to CSPGs and was not found in heparin (43–45).

Further intriguing development in this area was reported by Sundblad et al (46, 47), who found that digestion with peptide:N-glycosidase F (PNGaseF) released CS- and HS-like components from as yet unidentified proteins present in certain mammalian cells. Such presumably asparagine-linked GAGs were synthesized by bovine pulmonary arterial endothelial cells and by human erythroleukemia cells, but not by a large number of other cell types examined, including normal or transformed fibroblasts, various carcinoma cells, or Chinese hamster ovary cells. The structural and functional characterization of this novel type of PG will be of great interest.

Corneal KS is N-linked to asparagine units in protein via mannose and GlcNAc residues, whereas skeletal KS is O-linked to serine/threonine units via a GalNAc residue (see 48, 49). Recent structural refinement locating a 6-O-sulfated GlcNAc residue next to the GalNAc linkage unit was described recently (50).

## Core Proteins

Table 1 lists a number of PGs along with some of their basic structural and functional characteristics. While information regarding the structure of many core proteins (see Table 1 for references) is still lacking or incomplete, an attempt to sort the PGs into families seems justified, even if the basis for such classification will not be entirely consistent.

The families are: (a) Large-molecular extracellular CSPGs that are capable of specifically interacting with HA. For two of these PGs, the cartilage PG, aggrecan, and the fibroblast PG, versican, the amino-acid sequences of the entire core proteins are known (see further information below). A number of large extracellular matrix PGs, purified from different connective tissues, apparently belong to this group (Table 1), although their detailed structural relationship to aggrecan or versican remains to be elucidated. (b) PGs with small homologous core proteins substituted with one or two GAG chains (CS, DS, or KS). This family includes decorin, biglycan, and fibromodulin, which have all been sequenced at the cDNA and/or protein level. A fourth member may be PG 100, the newly discovered CSPG synthesized by osteosarcoma cells. (c) HSPGs of extracellular matrix/ basement membrane, related to the large-molecular HSPG synthesized by the EHS tumor. The evidence for relationship between the various members of this group is tentative, and is mainly based on immunological cross-reactivity of similar-sized core proteins. More detailed comparison of the various protein species will show if they are encoded for by more than one gene. (d) Membrane-intercalated cell-surface PGs related to the CS/HS hybrid PG, syndecan. At least one more member of this family, a fibroblast HSPG with a 48-kDa core protein, has been identified. (e) Serglycins, a group of largely intracellular PGs, containing core proteins with extended sequences of alternating serine and glycine units, heavily substituted with CS and/or heparin chains.

Many PGs require further detailed analysis with regard to core protein structure before they become amenable to classification, even at the crude level outlined above. It is already obvious, however, that numerous PG species will fall outside this scheme, since their core proteins bear no resemblance to those of any of the model PGs; instead, they may be closely similar to conventional proteins that are devoid of GAG chains. Notable in this context are the so-called "parttime PGs," including receptors for transferrin (51; however, see 67) and "homing" lymphocytes (10), as well as the invariant chain of the type II histocompatibility antigens (52), and collagen type IX (53), which may be synthesized either with or without GAG substituents on the protein moieties.

As illustrated for the imagican model in Figure 1, the core proteins of most PGs can be subdivided into various domains. All core proteins, by definition, contain a GAG-substitution domain. In addition, many PGs are anchored to the cell surface or to macromolecules in the extracellular matrix through specific domains in the core proteins. Some core proteins contain yet other domains with specific interaction properties.

GLYCOSAMINOGLYCAN-SUBSTITUTION DOMAINS    GAG chains, whether of the CS/DS or heparin/HS type, are bound to serine residues in the core

**Table 1** Properties and functions of proteoglycans

| Source/designation | Core protein size (kDa)[a] | GAG chains | Proposed functions/characteristics | Refs. |
|---|---|---|---|---|
| **Extracellular** | | | | |
| Cartilage/aggrecan | 208–221[b] | >100 CS, 20–30 KS | Aggregated by HA. | 54–63 |
| Fibroblasts/versican | 265[b] | 12–15 CS | Provide mechanical support; fixed negative charge. | 64–67 |
| Embryonic mesenchyme/PG-M | 550 | CS | | 68–70 |
| Aorta | | | | 71–74 |
| Aortic endothelial cells | | | | 75 |
| Aortic smooth muscle cells | >100–550 | CS/DS | Regulate cell migration and cell aggregation? | 76 |
| Embryonic skeletal muscle | | | | 77 |
| Developing bone | | | | 78 |
| Glial cells | | | | 79 |
| Nucleus pulposus | | | | 80 |
| Sclera | | | | 81 |
| Tendon | | | | 82 |
| Brain | 75–420 | CS | Role in development? | 83–85 |
| Neurons/CAT-301 | 580 | CS | Stabilize synaptic structure? | 86, 87 |
| Endothelial cells | >200 | DS | Role in endothelial regeneration? | 88 |
| Reicherts membrane | 160 | 13–22 CS | Stabilize basement membranes? | 89, 89a, 255, 269 |
| Connective tissues/decorin | 36[b] | 1 CS/DS | Modulate collagen fibrillogenesis; regulate cell growth. | 90–102 |
| Connective tissues/biglycan | 38[b] | 2 CS/DS | | 94, 97, 102–104 |
| Connective tissues/fibromodulin | 41[b] | 1 KS | Modulate collagen fibrillogenesis. | 105–108 |
| Osteosarcoma cells, fibroblasts/PG 100 | 106 | 1 CS | | 108a |
| Cartilage, vitreous humor/Collagen type IX | 68[b] | 1 CS | Binds to collagen fibrils. | 53, 109, 110 |

Family a (brackets spanning Fibroblasts/versican through Tendon rows)

Family b (brackets spanning Connective tissues/biglycan through Osteosarcoma cells rows)

| Source | MW | GAG | Function | Ref. |
|---|---|---|---|---|
| Cornea, other tissues | 25, 37–51 | KS | Organize collagen fibrils, required for corneal transparency. | 111–116 |
| Hepatocytes | 28 | 1–2 CS | Urinary trypsin inhibitor | 116a |
| EHS-tumor | 400–450 | 2–3 HS | | 117–122 |
| Fibroblasts | 250–400 | HS | | 67, 123–125 |
| Colon carcinoma cells | 240–400 | 10–15 HS | Modulate assembly of basement membranes. | 19a, 126, 127 |
| Basement membranes | 150–400 | HS | Provide filtration barrier. | 128–131 |
| Endothelial cells | 350 | HS | | 132–134 |
| Mammary epithelial cells | ? | HS | | 135 |

Family c

Cell surface

Family d

| | | | | |
|---|---|---|---|---|
| Mammary epithelial cell/syndecan | 31[b] | 1–2 CS/ 1–2 HS | Role in morphogenesis; link cytoskeleton to extracellular matrix | 16, 136–141 |
| Fibroblasts/fibroglycan | 48(90) | HS | | 142, 143 |
| Liver | 35, 44, 49, 77 | HS | | 144–146a |
| Brain | 55 | CS/HS | | 147–149 |
| Fibroblasts | 125, 35 | HS | | 67, 142, 150, 151 |
| Fibroblasts | 90, 52 | CS/DS | | 67, 152, 153 |
| Neurons | 280 | CS | | 154–156 |
| Marine ray electric organ/TAP-1 | 180 | ≈20 CS | Mediate cell adhesion. Role in morphogenesis? | 157 |
| Melanoma cells | 250 | 3–12 CS | Modulate assembly of matrix. | 158–160 |
| Hepatocyte cell line | ? | HS | | 161 |
| Ovarian granulosa cells | 80 | 2–4 HS | Phosphatidyl-inositol linked PGs. | 162 |
| Schwann cells | ? | HS, CS? | | 163 |
| Central nervous system | 50, 59 | HS | | 85 |
| Fibroblasts | 64 | HS | | 163a |
| Fibroblasts/betaglycan | 100–120 | CS/HS | Provide reservoir for TGF-β | 164, 165 |

**Table 1** (*Continued*)

| Source/designation | Core protein size (kDa)[a] | GAG chains | Proposed functions/characteristics | Refs. |
|---|---|---|---|---|
| Endothelial cells/thrombomodulin | 58–60[b] | 1 CS | Regulate blood coagulation. | 17, 166–169 |
| Lymphocytes/CD44, homing receptor | 37[b] | CS | Mediate cell adhesion. | 10, 170–174 |
| Fibroblasts/transferrin receptor | 2 × 90[b] | 4–6 HS | Mediate uptake of transferrin. | 29, 67 |
| B-cells, antigen presenting cells/invariant chain PG | 31[b] | CS | Role in antigen presentation. | 52, 175, 176 |
| **Intracellular** | | | Part-time PGs. | |
| Chromaffin granules/Chromogranin A PG | 48[b] | 1 CS | | 177, 178 |
| Family e — Yolk sac carcinoma cells[c] / Mast cells / Basophilic leukemia cells / Platelets / NK-cells / Macrophage-like cells / Eosinophils — serglycin | 17–19[b] | CS, DS, HS, heparin | Store and modulate activity of granular proteases. Prevent blood coagulation (heparin). | 179–195 |
| Chromaffin granules | 14 | 1–2 CS/DS | | 177 |

[a] If not otherwise indicated, $M_r$ was estimated from the migration of core proteins in SDS-PAGE after removal of GAG chains with endoglycosidases.

[b] $M_r$ calculated from amino acid/cDNA sequence data.

[c] In these cells, serglycin is primarily located extracellularly.

proteins (9). The attachment regions for KS are not considered further in this review [see (34) for information]. Chain elongation is initiated by xylosylation of selected serine residues, catalyzed by a xylosyltransferase using UDP-xylose as a sugar donor (9). Early efforts to define a recognition sequence of amino acids around the target serine units, using conventional sequencing techniques, were essentially inconclusive. Subsequent analysis of a large number of core proteins, using recombinant-DNA sequencing methods, revealed novel patterns, which led to the proposal of a "recognition consensus amino acid sequence" for CS/DS substitution (196). Serine units amenable to xylosylation thus would occur in S-G-X-G tetrapeptide sequences (where X is variable), usually preceded by a few acidic residues. Synthetic peptides containing such sequences are good substrates for xylosyltransferase in vitro (196, 197). "Consensus" xylosylation applies in the biosynthesis of the epithelial cell CS/HS hybrid PG syndecan (only one of the GAG attachment sites) (136), a related HSPG from fibroblasts (143), the CS/DSPG decorin (90), and the CS-substituted invariant chain of the type II histocompatibility antigens (52). However, a more detailed survey shows numerous exceptions to this rule [see also (30)]. In the serglycins, which contain up to 24 consecutive, heavily GAG-substituted, Ser-Gly repeats (180, 183, 187, 188, 198), obviously only one of these repeats will be flanked by acidic amino acid residues. Moreover, variability has been demonstrated also for the actual tetrapeptide acceptor sequence, which is S-G-E/D-X$^2$ (where X is variable) in versican (64) and syndecan (GAG attachment site different from that indicated above) (136), and (by inference, since GAG substitution at the implicated sites has yet to be actually demonstrated), S-G-E-Y$^2$ in the large-molecular basement-membrane HSPG (122). Biglycan (103) and aggrecan (54, 55, 60) contain GAG-substituted Ser-Gly sequences that are mostly single and associated with acidic and hydrophobic amino acids, without defining any consensus sequence. Not even the requirement for a glycine unit in juxtaposition to the substituted serine is absolute, the glycine being replaced by an alanine residue in collagen type IX [which carries a single CS chain (110)]. Taken together, these findings do not support the notion of a universal consensus sequence for GAG attachment, nor is there any apparent relationship between the structures of GAG-substitution domains and the type of GAG substituent (CS/DS vs heparin/HS; see further discussion regarding PG biosynthesis).

OTHER DOMAINS    In addition to GAG-substitution domains, core proteins of many PGs contain specific regions that are devoid of GAG chains, yet have

___

$^2$While these structures do not conform to any consensus recognition sequence, it is noted that the potential CS-substitution sites in versican consistently occur in a E/D-G-_S_-G-E/D structure (the substituted serine is underlined), whereas the HS chains in the HSPG would substitute D-_S_-G-E-Y sequences.

important functional roles. A major proportion of such regions serve to anchor the PG either to the cell surface or to other macromolecules of the extracellular matrix. Additional regions appear to be designed for other kinds of interactions, the functional purposes of which remain largely obscure.

The interaction between the large CSPG of cartilage, aggrecan, and HA is mediated by a globular domain (G1) at the N-terminal portion of the protein core, which binds to a decasaccharide sequence of the HA molecule (199). A short, extended peptide stretch separates the HA-binding G1 domain of the core protein from a second globular domain (G2), which does not bind HA (62, 200) despite its homology with G1 (54).

The binding to HA is stabilized by a link protein, which shows major homology with the G1 domain (201–203). Since several PG molecules can bind to the same HA chain, large aggregates that become entrapped in the collagen network anchor aggrecan in the cartilage matrix. Other connective tissue PGs with large-sized core proteins are related to aggrecan and share its ability to interact with HA (see Table 1). The core protein of the fibroblast PG versican (64), which is distinct from that of aggrecan, contains an N-terminal HA-binding region homologous to the corresponding region in aggrecan, but lacks the G2 globular domain. A HA-binding domain, similar to that found in aggrecan and versican, is present also in the homing receptor (10), which mediates binding of lymphocytes to the high endothelium of venules in lymphoid organs (171, 172).

Many PGs have the ability to interact with extracellular matrix proteins through their GAG chains. However, specific matrix interactions mediated by core proteins have also been demonstrated. The core protein of the matrix-associated HSPG, synthesized by human lung fibroblasts, thus binds with high affinity to a defined domain of the fibronectin molecule (125). This PG may be identical, or at least closely related, to the large basement membrane HSPG (124), which also interacts with fibronectin (125).

The core protein of the connective-tissue PG decorin (also denoted PG 40, DSPG II, PG-S2) binds fibronectin as well as collagen (100, 204, 205). The latter interaction has been shown to influence collagen fibrillogenesis (91, 106). Binding to collagen is also shown by fibromodulin (106), but not by biglycan (DSPG I, PG-S1), in spite of its homology with decorin (103). Presumably, these relationships may help to define the actual collagen-binding domains of the core proteins. The fourth member of the decorin family, PG 100, showed immunological cross-reactivity with the core protein of decorin, yet similar to biglycan, did not bind collagen (108a). Interestingly, both decorin and biglycan have the ability to bind TGF-$\beta$ (93), possibly through interactions between the leucin-rich domains of the core proteins and the growth factor.

The PGs considered so far are typical components of the extracellular matrix. Other species occur at the cell surface, to which they may be attached in a number of different ways. One group, represented by syndecan, is anchored to the plasma membrane through a hydrophobic core protein stretch that is intercalated in the lipid bilayer. The 33-kDa core protein of this PG contains a cytoplasmic, a transmembrane, and a N-terminal extracellular domain (136, 137, 137a). The transmembrane and cytoplasmic domains are highly homologous to the corresponding regions of the 48-kDa core protein of a fibroblast HSPG (143), whereas the extracellular domains of the two PGs appear to be unrelated (137). Some core proteins are bound to the cell surface through linkage to phosphatidyl-inositol, as first described for a HSPG produced by a rat hepatoma cell line (161; see also 85, 162, 163, 163a). Yet another type of interaction, apparently designed to enable internalization of PGs, is mediated by cell-surface receptors that specifically recognize certain core proteins [see (206, 207) for uptake of decorin]. Finally, PGs may also be retained at the cell surface through interactions not involving the protein component but rather the GAG chains (146).

The sequence information accumulated for core proteins has revealed the occurrence of various structural motifs that have been associated with certain functional characteristics of other proteins, yet are of unclear functional relevance. The cartilage PG, aggrecan (56, 208), as well as the related fibroblast PG, versican (209), thus both contain C-terminal core-protein domains that are homologous to a lectin isolated from chicken liver (210). Indeed, this region of aggrecan, which is highly conserved among species (55, 56, 208), is capable of binding galactose and fucose (211). Other regions with interesting functional correlates include domains resembling complement regulatory proteins (212), found in aggrecan and versican (64), and the EGF-like repeats (213) in versican (64), a variant of aggrecan (213a), and thrombomodulin (168, 169, 214). The EGF-like repeats in laminin actually express growth factor activity, promoting cell division (215). The presence of EGF-like, lectin-like, and complement regulatory protein–like regions in the versican core protein relates it to LEC-CAMs, a group of cell adhesion molecules (216). It has been speculated (64) that versican may act as a link between HA, in the extracellular matrix, and the cell membrane, the N-terminal G1 domain interacting with HA and the C-terminal CAM-like region with the cell surface. The core protein of the large basement membrane HSPG, which has been partially sequenced (122), shows homologies with laminin (217–219) and with N-CAM, a neuronal cell adhesion molecule (220) which belongs to the immunoglobulin superfamily (221). It is possible that the reported binding of this core protein to fibronectin [see above and (125)], as well as its self-aggregation (119), are mediated by the N-CAM like sequences. Finally, indications of extensive sequence homology between the core protein of a HSPG synthesized by a nerve cell line and the amyloid $\beta$

protein precursor implicated in Alzheimer's disease (222, 223; however, see 223a) and, further, between the latter protein and the serine proteinase inhibitor, protease nexin-II (224), point to novel intriguing structural and functional relationships involving PG core proteins.

## FUNCTIONS OF PROTEOGLYCANS

The biological roles of PGs are highly diversified (Table 1), ranging from relatively straightforward mechanical functions, essential for maintaining the structural integrity of various connective tissues, to effects on more dynamic processes such as cell adhesion and motility, to complex and, on the whole, poorly understood influences on cell differentiation and morphogenesis. A comprehensive treatise of these functions falls outside the scope of this review. Instead, we focus on some general mechanistic aspects, with emphasis on the respective roles of the core protein and GAG components. We then consider a few selected cellular processes, where the functional roles of PGs seem amenable to analysis at the molecular level.

### Mechanisms of Action

By and large, the biological functions ascribed to PGs depend on the concerted contributions of the GAG and the core protein components, which may cooperate at various levels of mechanistic complexity. However, important biological effects may be exerted by single GAG chains (see Table 2 and below). Conversely, parttime PGs have been described, in which the core proteins account for the only biological activities known to be associated with the conjugate, and which in fact may occur in two forms, with and without GAG chains.

INTERACTIONS OF GLYCOSAMINOGLYCAN CHAINS     Most of the functional properties of the GAG and protein moieties of PGs depend on interactions between these components and other macromolecules. The interaction properties of PG core proteins have been discussed above. A peculiar feature of GAG chains, expressed in particular by markedly copolymeric species with GlcA- and IdoA-containing disaccharide units arranged in alternating sequence, is the ability to self-associate into aggregates (1, 312). However, the large majority of the ligands identified are proteins, some of which are listed in Table 2 along with the proposed functional roles of the corresponding interactions.

Binding of proteins to GAGs is usually electrostatic in nature, although other types of interaction may occur. Binding may be tight and highly specific (lock-and-key binding), as in the case of antithrombin, which binds with high affinity to a particular pentasaccharide sequence in heparin/HS (see below and

**Table 2**  Heparin-binding proteins—functional role of interaction[a]

| Protein | Proposed functional role of interaction | Refs. |
|---|---|---|
| **Enzymes** | | |
| Mast cell proteinases | Provide intracellular storage of enzymes in inactive form; modulate enzyme activity | 190, 225–227 |
| Lipoprotein lipase | Anchor enzyme at cell surface; regulate enzyme expression | 228–238 |
| Coagulation enzymes | Facilitate enzyme inactivation by antithrombin | 239–244 |
| Elastase | Promote enzyme inhibition | 245 |
| Extracellular superoxide dismutase | Anchor enzyme at cell surface or in extracellular matrix | 246–249 |
| **Serine protease inhibitors** | | |
| Antithrombin | Promote inactivation of coagulation enzymes at vascular endothelium | 250–260 |
| Heparin cofactor II | Promote inactivation of thrombin | 261–264 |
| Protease nexins | Inactivate serine proteases in the extracellular matrix | 265–266 |
| **Extracellular matrix proteins** | | |
| Vitronectin | Modulate coagulation and complement processes | 283–284 |
| Fibronectin | | 15, 101, 150, 276–282 |
| Laminin | Mediate/modulate cell-substrate interations | 285–287 |
| Thrombospondin | | 288–292 |
| Collagens | | 293, 294 |
| **Growth factors** | | |
| Fibroblast growth factors | | 295–301 |
| Schwann cell growth factor | Sequester (protect) growth factors in the extracellular matrix | 302 |
| Retinal survival factor | | 303 |
| Hemopoietic growth factors | | 303a, b |
| **Other proteins** | | |
| Apolipoproteins | Mediate lipoprotein uptake by cells; role in atherosclerosis | 267–275 |
| Platelet factor 4 | Neutralize biological activity of heparin/HS | 304, 305 |
| N-CAM | Promote nerve cell adhesion | 306–308 |
| Viral coat protein | Mediate binding of virus (HIV; Herpes) to cells | 309 |
| Transcription factors? | Modulate gene expression | 310, 311 |

[a] The ligand-binding GAG in vivo is likely to be HS (or in some cases, a galactosaminoglycan) rather than heparin; see the text. For recent reviews on GAG-protein interactions see Cardin & Weintraub (313), Jackson et al (7), and Lane (311a).

Figure 3), but is often of lower affinity and less specific (cooperative electrostatic binding) such that a number of different sequences may interact with each ligand. Consensus sequences for GAG binding, containing clusters of basic amino acids, have been tentatively postulated for a number of proteins (7, 313). The term "heparin-binding" in Table 2 refers to the use of commercial heparin as the active component of affinity matrices used to isolate many of the proteins, but also reflects the fact that heparin shows the strongest binding of the various GAGs tested, or most readily induces the particular biological effects believed to involve the proteins (see references listed in Table 2). A caveat is warranted here: heparin is the strongest negatively charged polyelectrolyte found in mammalian tissues, and its ability to interact with positively charged regions in various proteins does not necessarily imply a biological function. In particular, effects seen only at high concentrations of heparin should be evaluated with caution.

Under physiological conditions in vivo, it is likely that HS or in some cases DS (see e.g. heparin cofactor II) rather than heparin (which is generally confined to the interior of mast cells), will provide the binding sites for the proteins. The superior binding capacity of these GAGs as compared to that of CS (of similar charge density) has been ascribed to the presence of IdoA units, which are believed to confer conformational flexibility to the polymers (314–318, 316a).

Biological activities expressed by single GAG chains (or oligosaccharides) can generally be attributed to specific carbohydrate structures. Examples of such activities are the CSPG-aggregating effect of HA (319), the blood anticoagulant activities of heparin (via antithrombin) and DS (via heparin cofactor II) (see Table 2 for references), and the postulated involvement of nuclear HS oligosaccharides in the regulation of cell proliferation (161, 320). The interaction between heparin and antithrombin is based on the occurrence in the GAG chain of the pentasaccharide sequence shown in Figure 3. The distinguishing structural feature of this region is the 3-O-sulfate group, which occurs on the internal GlcN unit and is essential for high-affinity binding of the protease inhibitor. Such binding requires, in addition, the appropriate sequence of monosaccharide units, with one GlcA and one IdoA unit, both N-sulfate groups, and, finally, the 6-O-sulfate group on the GlcN residue at the nonreducing end of the the pentasaccharide sequence. Recent evidence indicates that the interaction between DS and heparin cofactor II is also strongly favored by a specific binding region in the GAG molecule (264), here identified as a hexasaccharide sequence with the unusual structure, $[-IdoA(2-OSO_3)-GalNAc(4-OSO_3)-]_3$ (Figure 3). Protein ligands mediating the antiproliferative effect of nuclear HS have not been identified, but should be sought for in view of recent findings, which potentially link nuclear localization of GAGs to regulation of transcription factor activity and gene expression (321). Again, the unusual structure of the nuclear HS, with a high proportion of sulfated GlcA units, implies, albeit indirectly, that highly specific interactions may be involved (161, 320).

Further studies on biological effector functions of GAGs are likely to reveal

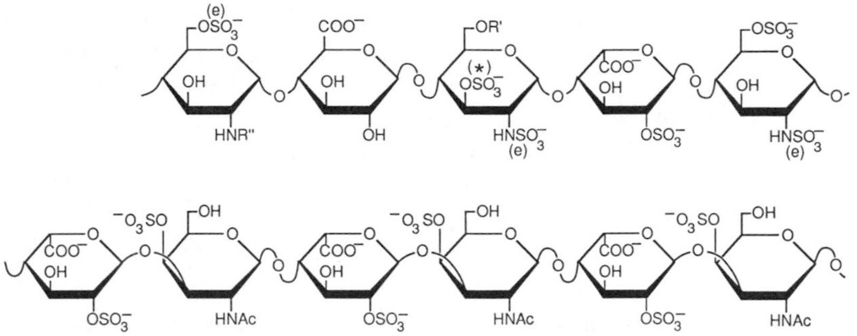

*Figure 3*  The binding regions for antithrombin in heparin [upper structure (253)] and for heparin cofactor II in DS [lower structure (264)]. Structural variants in the antithrombin-binding region are indicated by R' (-H or -SO$_3^-$) and R'' (-SO$_3^-$ or -COCH$_3$). The 3-O-sulfate group marked by an asterisk is a marker group for the antithrombin-binding region, and is essential for the high-affinity binding of heparin to antithrombin. The three sulfate groups marked (e) are also critically important to this interaction.

additional interaction systems involving specific recognition/binding sequences in the GAG chains. Identification of such regions will be facilitated by the introduction of improved or novel methods for sequence analysis of GAGs (see e.g. 322, 323), as well as by the availability of monoclonal antibodies with specificity toward various GAG structures (324–327). Another area of importance to the elucidation of structure/function relationships is the manufacture of defined GAG structures by organic synthesis (328).

CONCERTED EFFECTS OF GLYCOSAMINOGLYCAN AND PROTEIN COMPONENTS    The simplest mode of functional cooperation between the components of a PG features the core protein primarily as a scaffold for the appropriate immobilization and spacing of GAG chains. This property is prerequisite to numerous functions ascribed to PGs, for example the inhibitory (masking) effect of certain CSPGs on cell adhesion (4, 5, 6, 70). Serglycin, produced by rat yolk sac tumor cells, contains closely spaced CS chains (180) and inhibits the attachment of these cells to fibronectin (11). It has been proposed that binding of the CS-chains to the GAG-binding site of fibronectin will sterically hinder the interaction of the cell-attachment site of the protein with its cell-surface receptor (12). The effect is seen with the intact PG but not with free CS chains.

If the core is attached to the cell or to an extracellular macromolecule it assumes an anchoring function, which may be required for locating GAG chains at specific sites. Examples in this regard are the HSPGs, which locate antithrombin (13) and protease nexin (14, 266) to the surfaces of vascular endothelial cells and fibroblasts, respectively. Anchoring of PGs to the cell surface may be a prerequisite for their ability to influence the adhesion of cells to the extracellular matrix [e.g. through binding of HS-chains to fibronectin, which will expose the cell-binding site of this protein and thus promote the interaction with its integrin receptor (15, 329)], or to other cells (see e.g. the effect of HSPG on the N-CAM "adheron" of nerve cells, Table 2).

Core proteins display a more complex mode of action when they mediate a cellular response to an interaction involving a GAG chain substituent. Interaction of the GAG chains of cell-surface HSPGs with extracellular matrix components (such as fibronectin, various collagens, thrombospondin which, in fact, may cross-link the PGs) thus induces a reorganization of the actin cytoskeleton, resulting in the formation of focal cell adhesions, which appears to be mediated through the integral membrane protein core(s) (15, 16, 330, 330a). Such interactions may serve to modulate the formation and elimination of focal cell adhesions. Other cellular processes that depend on the mediating roles of HSPG core proteins include the capture and internalization of lipoprotein lipase (331) and thrombospondin (289, 290).

Further complexity as regards functional integration between the protein and GAG components of a PG is displayed by thrombomodulin (TM), an integral membrane 90 kDa protein on vascular endothelial cells, with important regulatory functions in blood coagulation. TM binds thrombin and acts as a cofactor in the thrombin-catalyzed activation of Protein C, which thereby assumes anticoagulant properties (332). TM is a PG which carries a [presumably single (167)] CS (or DS) chain (166, 167, 333, 334). Due to the CS component, the TM-bound thrombin has lost its ability to cleave fibrinogen and activate Factor V [probably due to a sterical hindering effect of the CS chain (17)] but has become more susceptible to inactivation by antithrombin, as compared to free thrombin (166, 333, 335–337; see TM model in Figure 4). All the biological activities of TM, except the ability to promote the activation of Protein C, are lost upon enzymatic removal of the CS chain. Interestingly, the antithrombin-dependent anticoagulant activity of TM (expressed only in the presence of the CS component) appears to result from a TM-induced increase in sensitivity of thrombin toward antithrombin (167, 334) (information included in Figure 4).

## Regulation of Cellular Processes

An abundance of experimental observations implicate PGs in various cellular processes. The evidence is often indirect or circumstantial, such as the often transient appearance of a particular PG or change in PG structure during morphogenesis/differentiation (see e.g. 16, 338–343). Nevertheless, this kind of information may indicate highly specialized roles for PGs, a striking example being the selective expression of a specific CSPG on mammalian neurons, apparently regulated by neuronal activity in early life (86, 87, 344).

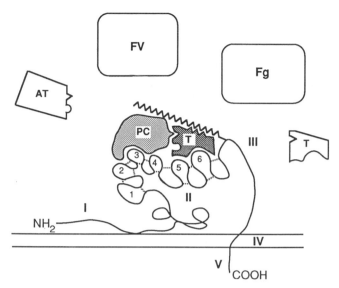

*Figure 4*   Model of rabbit thrombomodulin (TM) proteoglycan. The TM molecule is composed of (I) an N-terminal domain, (II) a domain comprising six EGF-like repeats, (III) a domain with potential glycosylation sites, (IV) a transmembrane hydrophobic region, and (V) a cytoplasmic C-terminal domain. The EGF-like regions numbered 2–6 provide the binding sites for thrombin (T) and Protein C (PC) required for efficient activation of the latter component. The CS chain, attached to domain III, does not influence the activation of Protein C by thrombin, but will potentiate the inhibition of thrombin by antithrombin (AT), and prevent the cleavage of fibrinogen (Fg) and the activation of Factor V (FV) by thrombin. For additional information see the text. The authors are grateful to M.-C. Bourin for help in preparing this figure [modified from (17)].

Conversely, initial observations regarding the effects of administering exogenous GAGs provided clues to the roles for PGs in diverse processes such as angiogenesis (345; see also refs. on fibroblast growth factors in Table 2), nerve cell adhesion (346; see also refs. on N-CAM in Table 2), and metastasis (347–351). It is now believed that angiogenesis requires basic fibroblast growth factor, which is kept sequestered, bound to HSPG of the extracellular matrix; nerve cell adhesion depends on a change in conformation of the adheron, N-CAM, induced by interaction with the GAG chains of cell-surface HSPG; whereas metastasis involves degradation of HSPG in the extracellular matrix, catalyzed by tumor cell–derived endoglucuronidase.

Arbitrarily selected as these examples may be, they do illustrate the functional diversity of PGs. Other observations confirm this impression, but also appear to some extent contradictory. We may consider, for instance, the structural properties of GAGs implicated in the regulation of cell proliferation. One line of evidence derives from studies on HSPGs synthesized by rat hepatoma cells. Following internalization, a portion of the HSPG is processed such that free chains, enriched in -GlcA(2-OSO$_3$)-GlcNSO$_3$(6-OSO$_3$)-disaccharide sequences, appear in the nucleus where they prevent cell division (161, 320, 352, 353). On the other hand, investigations on the antiproliferative effect of heparin on vascular smooth muscle cells suggested that the activity increased with the chain length and degree of sulfation of the saccharide molecule, but showed no requirement for 2-O-sulfated GlcA units (354, 355). A proposed role for the GlcN 3-O-sulfate group occurring as a marker component for the antithrombin-binding region in heparin (356) appeared untenable, in view of the subsequent finding that heparin fractions with high or low affinity for antithrombin showed equal antiproliferative potencies (354, 355). A third line of evidence implicates HS synthesized by the smooth muscle target cells as a strongly antiproliferative agent, the most active species being generated by postconfluent cells (357). Preliminary structural comparison of active and less active HS fractions indicated that the antiproliferative activity was associated with the presence of N-acetylated rather than N-sulfated GlcN units (358, 359). Taken together, these findings suggest that cell proliferation may be influenced by various heparin/HS-related GAGs with apparently widely differing structures, hence presumably by different mechanisms (see 311, 359a). Work with vascular endothelial cells showed, in line with this contention, that HS and heparin under certain conditions exert opposing modulatory effects on cell growth (360).

Current research in the area will undoubtedly reveal additional cellular processes that are modulated by PGs, and explain in molecular terms the mechanisms behind these effects. This development will be facilitated by the application of novel strategies, such as the comparison, with regard to specific functions, of wild-type and genetically manipulated cells, modified with regard to the expression of specified PGs. Genes for core proteins may be introduced by transfection into recipient cells (92, 93), leading to novel expression or overexpression of PG. Alternatively, genes coding for enzymes involved in PG biosynthesis (e.g. glycosyltransferases, sulfotransferases) may be deleted by chemical mutagenesis, followed by selection of the appropriate defective mutants (279, 288, 289, 294, 309, 361, 362).

## BIOSYNTHESIS OF PROTEOGLYCANS—CURRENT PROBLEMS

The biosynthesis of PGs has been the subject of previous reviews (9, 20, 22, 23) and is only considered here to the extent required for calling attention to some current enigmas. After formation of the protein core, most of the glycosylation and subsequent reactions take place in the Golgi (363, 364). The posttranslational modifications are initiated by xylosylation of serine residues in the core, and proceed by stepwise addition of monosaccharide units from the appropriate UDP-sugars to the nonreducing termini of the growing chains. To attain their final shape the chains are then carried through the series of modification reactions outlined in a previous section. A number of major questions regarding various aspects of regulation remain unresolved.

Which mechanism determines whether a protein, during transit through the Golgi system, will be converted into a PG? The above discussion regarding recognition sites for xylosylation of serine residues concludes that there is no single consensus sequence for chain initiation and, conversely, that proteins often contain potential GAG initiation sites that escape xylosylation (see 30). A recent report on the effects of site-directed mutagenesis on the expression of decorin in transfected cells raises new aspects of this problem (365). Since either one of the two glycine residues in the Ser-Gly-Ile-Gly sequence of the decorin core protein could be replaced by alanine units, without any appreciable decrease in GAG substitution, the authors argued that the conformation, rather than the primary structure of the substitution site, is important for xylosyltransferase recognition. Thus, it may be rewarding to study the conformation around substituted serine units in different core proteins. The serine residue in the xylose acceptor sequence of decorin is present in a $\beta$-turn (365), and it is conceivable that only serine units within $\beta$-structures are recognized by the enzyme. Silk cloth fibroin, a well-known example of a protein forming $\beta$-pleated sheets, is an excellent xylose acceptor in the xylosyltransferase reaction (366).

Which mechanism determines whether a GAG chain, once initiated through xylosylation of a serine residue in the core protein, will become a glucosaminoglycan (HS) or a galactosaminoglycan (CS/DS)? Characterization of animal cell mutants defective in xylosyltransferase (361), or in the galactosyltransferase that adds the first galactose residue to the serine-bound xylose unit (367), indicated that the same enzymes initiate both types of GAG chains. The committing step thus would be the incorporation of the first hexosamine, either GlcNAc or GalNAc, unit, immediately adjacent to the polysaccharide-protein linkage region. Interestingly, the GalNAc transferase that catalyzes this particular reaction in CS biosynthesis appears to differ from that involved in the formation of the more peripheral portions of the GAG chain (368). There may be a similarly unique GlcNAc transferase that catalyzes the incorporation of the first GlcNAc residue in heparin/HS chains. How are these particular enzymes regulated? What is the significance in this context of the sulfated galactose residues that seem to occur exclusively in the linkage region of galactosaminoglycans (Figure 1)? If, as seems likely, there are no distinct recognition signals for the initiation of glucosaminoglycan vs galactosaminoglycan chains close to the target serine units, the selection may be under control of more remote core protein determinants. Such a mechanism would explain how the same cell (e.g. a fibroblast) may synthesize different CS- and HSPGs with distinct core proteins (67). In serglycins, which may carry either CS or heparin chains on the same serine residues (181, 188), differentiating signals may be generated through posttranslational modification of the protein core.

Which mechanism(s) determine(s) the overall extent of polymer modification? Why does HS remain at a low level of modification (high contents of GlcA and GlcNAc; low of IdoA, $GlcNSO_3$ and O-sulfate) rather than develop into a heparin (with reversed proportions of the various constituents)? This question is of obvious importance in relation to the functional properties of the GAGs; see e.g. the previous discussion regarding structure/function relationships for species with antiproliferative activity. Interestingly, diabetes affects the extent of polymer modification, HS derived from liver (369), and glomerular basement membrane (370) of diabetic rats having lower sulfate contents than control material. The mechanism of regulation is unknown but relates to the level of N-deacetylase activity, which was found to be decreased in afflicted hepatocytes (370a). Since deacetylation is prerequisite to N-sulfation, and N-sulfate groups are required for substrate recognition by the GlcA C5-epimerase as well as by the O-sulfotransferases, the N-deacetylase has a key regulatory role (20, 22, 23). Purified N-sulfotransferase (371, 371a) generated N-deacetylase activity together with an as yet unidentified protein cofactor, in accord with the functional interdependence of the two enzyme activities (371a). Effects similar to those seen in diabetes were observed after treatment of cells with n-butyrate (372).

Finally, which mechanisms control the generation of specific GAG sequences that are required for functionally important interactions with other macromolecules? The number of well-defined ligand-binding regions is so far limited but is likely to increase. The importance of regulated polymer modification in connection with the biosynthesis of such regions is clearly demonstrated in studies on the biosynthesis of the antithrombin-binding region in heparin. The incorporation of the unique GlcN 3-O-sulfate group, essential for the biological activity (Figure 3), occurs through a specific sulfotransferase reaction that concludes the modification process (24). The substrate recognized by the 3-O-sulfotransferase apparently corresponds to the entire pentasaccharide sequence shown in Figure 3 (except the target 3-O-sulfate group), and thus contains all structural components (except the 3-O-sulfate group) required for binding to antithrombin (24, 373). On the other hand, some of the potential 3-O-sulfation sites appear inaccessible to the enzyme (374), probably due to the presence of inhibitory O-sulfate groups in the vicinity (N. Razi, U. Lindahl, unpublished data). These findings, together with the highly nonrandom distribution of antithrombin-binding regions between the GAG chains of a heparin PG (375), suggest the additional involvement of as yet poorly understood selection mechanisms.

## Literature Cited

1. Fransson, L.-Å. 1985. *The Polysaccharides,* Vol. III, ed. G. D. Aspinall, pp. 337–415. New York: Academic
2. Hassell, J. R., Kimura, J. H., Hascall, V. C. 1986. *Annu. Rev. Biochem.* 55:539–67
3. Poole, A. R. 1986. *Biochem. J.* 236:1–14
4. Ruoslahti, E. 1988. *Annu. Rev. Cell Biol.* 4:229–55
5. Ruoslahti, E. 1989. *J. Biol. Chem.* 264:13369–72
6. Gallagher, J. T. 1989. *Curr. Opin. Cell Biol.* 1:1201–18
7. Jackson, R. L., Busch, S. J., Cardin, A. D. 1991. *Physiol. Rev.* In press
8. Lindahl, U., Höök, M. 1978. *Annu. Rev. Biochem.* 47:385–417
9. Rodén, L. 1980. *The Biochemistry of Glycoproteins and Proteoglycans,* ed. W. J. Lennarz, pp. 267–371. New York: Plenum
9a. Evered, D., Whelan, J., eds. 1989. *The Biology of Hyaluronan.* Chichester: Wiley
10. Jalkanen, S., Jalkanen, M., Bargatze, R., Tammi, M., Butcher, E. C. 1988. *J. Immunol.* 141:1615–23
11. Brennan, M. J., Oldberg, Å., Hayman, E. G., Ruoslahti, E. 1983. *Cancer Res.* 43:4302–7
12. Hautanen, A., Gailit, J., Mann, D. M., Ruoslahti, E. 1989. *J. Biol. Chem.* 264:1437–42
13. Marcum, J. A., Atha, D. H., Fritze, L. M. S., Nawroth, P., Stern, D., Rosenberg, R. D. 1986. *J. Biol. Chem.* 261:7507–17
14. Wagner, S. L., Lau, A. L., Cunningham, D. D. 1989. *J. Biol. Chem.* 264:611–15
15. Woods, A., Couchman, J. R. 1988. *Collagen Rel. Res.* 8:155–82
16. Bernfield, M., Sanderson, R. D. 1990. *Philos. Trans. R. Soc. London, Ser B* 327:171–86
17. Bourin, M.-C., Lindahl, U. 1990. *Biochem. J.* 270:419–25
18. Lohmander, L. S., De Luca, S., Nilsson, B., Hascall, V. C., Caputo, C. B., et al. 1980. *J. Biol. Chem.* 255:6084–91
19. Nilsson, B., DeLuca, S., Lohmander, S., Hascall, V. C. 1982. *J. Biol. Chem.* 257:10920–27

19a. Iozzo, R. V., Kovalszky, I., Hacobian, N., Schick, P. K., Ellingson, J. S., Dodge, G. R. 1990. *J. Biol. Chem.* 265:19980–89
20. Lindahl, U., Feingold, D. S., Rodén, L. 1986. *Trends Biochem. Sci.* 11:221–25
21. Gallagher, J. T., Lyon, M., Steward, W. P. 1986. *Biochem. J.* 236:313–25
22. Lindahl, U., Kjellén, L. 1987. *Biology of Proteoglycans,* pp. 59–104. Orlando: Academic
23. Lindahl, U. 1989. *Heparin: Chemical and Biological Properties, Clinical Applications,* ed. D. A. Lane, U. Lindahl, pp. 159–89. London: Edward Arnold
24. Kusche, M., Bäckström, G., Riesenfeld, J., Petitou, M., Choay, J., Lindahl, U. 1988. *J. Biol. Chem.* 263: 15474–84
25. Bienkowski, M. J., Conrad, H. E. 1985. *J. Biol. Chem.* 260:356–65
26. Kusche, M., Lindahl, U. 1990. *J. Biol. Chem.* 265:15403–9
27. Cifonelli, J. A. 1968. *Carbohydr. Res.* 8:233–42
28. Höök, M., Lindahl, U., Iverius, P.-H. 1974. *Biochem. J.* 137:33–43
29. Fransson, L.-Å. 1987. *Trends Biochem. Sci.* 12:406–11
30. Fransson, L.-Å. 1989. See Ref. 23, pp. 115–33
31. Gallagher, J. T., Lyon, M. 1989. See Ref. 23, pp. 136–58
32. Turnbull, J. E., Gallagher, J. T. 1990. *Biochem. J.* 265:715–24
33. Seldin, D. C., Seno, N., Austen, K. F., Stevens, R. L. 1984. *Anal. Biochem.* 141:291–300
34. Greiling, H., Scott, J. E., eds. 1989. *Keratan Sulphate.* London: The Biochem. Soc.
35. Oeben, M., Keller, R., Stuhlsatz, H. W., Greiling, H. 1987. *Biochem. J.* 248:85–93
36. Nieduszynski, I. A., Huckerby, T. N., Dickenson, J. M., Brown, G. M., Guihua, T., Bayliss, M. T. 1990. *Biochem. Soc. Trans.* 18:792–93
37. Jorpes, E. J., Gardell, S. 1948. *J. Biol. Chem.* 176:267
38. Rodén, L. 1989. See Ref. 23, pp. 1–23
39. Gallagher, J. T., Walker, A. 1985. *Biochem. J.* 230:665–74

## 468    KJELLÉN & LINDAHL

40. Prehm, P. 1983. *Biochem. J.* 211:181–89
41. Oegema, T. R. Jr., Kraft, E. L., Jourdian, G. W., Van Valen, T. R. 1984. *J. Biol. Chem.* 259:1720–26
42. Fransson, L.-Å., Silverberg, I., Carlstedt, I. 1985. *J. Biol. Chem.* 260:14722–26
43. Sugahara, K., Yamashina, I., De Waard, P., Van Halbeek, H., Vliegenthart, J. F. G. 1988. *J. Biol. Chem.* 263:10168–74
44. Sugahara, K., Masuda, M., Yamada, S., Harada, T., Yoshida, K., Yamashina, I. 1989. *Proc. 10th Int. Symp. Glycoconjugates*, pp. 322–24. Jerusalem
45. Sugahara, K., Ohi, Y., Harada, T., Yamashina, I., De Waard, P., Vliegenthart, J. F. G. 1990. *25th Int. Carbohydrate Symp., Yokohama* (Abstr.)
46. Sundblad, G., Holojda, S., Roux, L., Varki, A., Freeze, H. H. 1988. *J. Biol. Chem.* 263:8890–96
47. Sundblad, G., Kajiji, S., Quaranta, V., Freeze, H. H., Varki, A. 1988. *J. Biol. Chem.* 263:8897–903
48. Stuhlsatz, H. W., Keller, R., Becker, G., Oeben, M., Lennartz, L., et al. 1989. See Ref. 34, pp. 1–11
49. Hascall, V. C., Midura, R. J. 1989. See Ref. 34, pp. 66–75
50. Dickenson, J. M., Huckerby, T. N., Nieduszynski, I. A. 1990. *Biochem. J.* 269:55–59
51. Cöster, L., Carlstedt, I., Kendall, S., Malmström, A., Schmidtchen, A., Fransson, L.-Å. 1987. *Trends Biochem. Sci.* 12:406–11
52. Sant, A. J., Cullen, S. E., Giacoletto, K. S., Schwartz, B. D. 1985. *J. Exp. Med.* 162:1916–34
53. Yada, T., Suzuki, S., Kobayashi, K., Kobayashi, M., Hoshino, T., et al. 1990. *J. Biol. Chem.* 265:6992–99
54. Doege, K., Sasaki, M., Horigan, E., Hassell, J. R., Yamada, Y. 1987. *J. Biol. Chem.* 262:17757–67
55. Oldberg, Å., Antonsson, P., Heinegård, D. 1987. *Biochem. J.* 243:255–59
56. Sai, S., Tanaka, T., Kosher, R. A., Tanzer, M. L. 1986. *Proc. Natl. Acad. Sci. USA* 83:5081–85
57. Antonsson, P., Heinegård, D., Oldberg, Å. 1989. *J. Biol. Chem.* 264:16170–73
58. Heinegård, D., Oldberg, Å. 1989. *FASEB J.* 3:2042–51
59. Heinegård, D., Hedbom, E., Antonsson, P., Oldberg, Å. 1990. *Biochem. Soc. Trans.* 18:209–12
60. Doege, K., Sasaki, M., Kimura, T., Yamada, Y. 1991. *J. Biol. Chem.* 265:894–902
61. Dudhia, J., Fosang, A. J., Hardingham,

T. E. 1990. *Biochem. Soc. Trans.* 18:198–200
62. Mörgelin, M., Paulsson, M., Hardingham, T. E., Heinegård, D., Engel, J. 1988. *Biochem. J.* 253:175–85
63. Perris, R., Johansson, S. 1990. *Dev. Biol.* 137:1–12
64. Zimmerman, D. R., Ruoslahti, E. 1989. *EMBO J.* 8:2975–81
65. Johansson, S., Hedman, K., Kjellén, L., Christner, J., Vaheri, A., Höök, M. 1985. *Biochem. J.* 232:161–68
66. Carlstedt, I., Cöster, L., Malmström, A., Fransson, L-Å. 1983. *J. Biol. Chem.* 258:11629–35
67. Schmidtchen, A., Carlstedt, I., Malmström, A., Fransson, L-Å. 1990. *Biochem. J.* 265:289–300
68. Yamagata, M., Yamada, K. M., Yoneda, M., Suzuki, S., Kimata, K. 1986. *J. Biol. Chem.* 261:13526–35
69. Kimata, K., Oike, Y., Tani, K., Shinomura, T., Yamagata, M., et al. 1986. *J. Biol. Chem.* 261:13517–25
70. Yamagata, M., Suzuki, S., Akiyama, S. K., Yamada, K. M., Kimata, K. 1989. *J. Biol. Chem.* 264:8012–18
71. Gardell, S., Baker, J., Caterson, B., Heinegård, D., Rodén, L. 1980. *Biochem. Biophys. Res. Commun.* 95:1823–31
72. Oegema, T. R. Jr., Hascall, V. C., Eisenstein, R. 1979. *J. Biol. Chem.* 254:1312–18
73. Heinegård, D., Björne-Persson, A., Cöster, L., Franzén, A., Gardell, S., et al. 1985. *Biochem. J.* 230:181–94
74. Mörgelin, M., Paulsson, M., Malmström, A., Heinegård, D. 1989. *J. Biol. Chem.* 264:12080–90
75. Morita, H., Takeuchi, T., Suzuki, S., Maeda, K., Yamada, K., et al. 1990. *Biochem. J.* 265:61–68
76. Chang, Y., Yanagishita, M., Hascall, V. C., Wight, T. N. 1983. *J. Biol. Chem.* 258:5679–88
77. Carrino, D. A., Caplan, A. I. 1989. *Conn. Tissue Res.* 19:35–50
78. Fisher, L. W., Termine, J. D., Dejter, S. W. Jr., Whitson, W., Yanagishita, M., et al. 1983. *J. Biol. Chem.* 258:6588–94
79. Norling, B., Glimelius, B., Wesrermark, B., Wasteson, Å. 1978. *Biochem. Biophys. Res. Commun.* 84:914–21
80. Oegema, T. R. Jr., Bradford, D. S., Cooper, K. M. 1979. *J. Biol. Chem.* 254:10579–81
81. Cöster, L., Fransson, L-Å. 1981. *Biochem. J.* 193:143–53
82. Vogel, K. G., Heinegård, D. 1985. *J. Biol. Chem.* 260:9298–306
83. Oohira, A., Matsui, F., Matsuda, M.,

Takida, Y., Kuboki, Y. 1988. *J. Biol. Chem.* 263:10240–46

84. Gowda, D. C., Margolis, R. U., Margolis, R. K. 1989. *Biochemistry* 28:4468–74

85. Herndon, M. E., Lander, A. S. 1990. *Neuron* 4:949–61

86. Zaremba, S., Guimaraes, A., Kalb, R. G., Hockfield, S. 1989. *Neuron* 2:1207–19

87. Kalb, R. G., Hockfield, S. 1990. *Neuroscience* 34:391–401

88. Kinsella, M. G., Wight, T. N. 1988. *J. Biol. Chem.* 263:19222–31

89. McCarthy, K. J., Accavitti, M. A., Couchman, J. R. 1989. *J. Cell Biol.* 109:3187–98

89a. McCarthy, K. J., Couchman, J. R. 1990. *J. Histochem. Cytochem.* 38:1479–86

90. Krusius, T., Ruoslahti, E. 1986. *Proc. Natl. Acad. Sci. USA* 83:7683–87

91. Vogel, K. G., Paulsson, M., Heinegård, D. 1984. *Biochem. J.* 223:587–97

92. Yamaguchi, Y., Ruoslahti, E. 1988. *Nature* 336:244–46

93. Yamaguchi, Y., Mann, D. M., Ruoslahti, E. 1990. *Nature* 346:281–84

94. Rosenberg, L. C., Choi, H. U., Tang, L-H., Johnson, T. L., Pal, S., et al. 1985. *J. Biol. Chem.* 260:6304–13

95. Sampaio, L. O., Bayliss, M. T., Hardingham, T. E., Muir, H. 1988. *Biochem. J.* 254:757–64

96. Kokenyesi, R., Woessner, J. F. 1989. *Biochem. J.* 260:413–19

97. Vogel, K. G., Fisher, L. W. 1986. *J. Biol. Chem.* 261:11334–40

98. Day, A. A., McQuillan, C. I., Termine, J. D., Young, M. R. 1987. *Biochem. J.* 248:801–5

99. Chopra, R., Pearson, C., Pringle, G., Fackkre, D., Scott, P. 1985. *Biochem. J.* 232:277–79

100. Schmidt, G., Robenek, H., Harrach, H., Glössl, J., Nolte, V., et al. 1987. *J. Cell Biol.* 104:1683–91

101. Lewandowska, K., Choi, H. U., Rosenberg, L. C., Zardi, L., Culp, L. A. 1987. *J. Cell Biol.* 105:1443–54

102. Roughley, P. J., White, R. J. 1989. *Biochem. J.* 262:823–27

103. Fisher, L., Termine, J., Young, M. 1989. *J. Biol. Chem.* 264:4571–76

104. Neame, P. J., Choi, H. U., Rosenberg, L. C. 1989. *J. Biol. Chem.* 264:8653–61

105. Oldberg, Å., Antonsson, P., Lindblom, K., Heinegård, D. 1989. *EMBO J.* 8:2601–4

106. Hedbom, E., Heinegård, D. 1989. *J. Biol. Chem.* 264:6898–905

107. Heinegård, D., Larsson, T., Sommarin, Y., Franzén, A., Paulsson, M., Hedbom, E. 1986. *J. Biol. Chem.* 261:13866–72

108. Plaas, A. H. K., Ison, A. L., Ackland, J. 1989. *J. Biol. Chem.* 264:14474–54

108a. Schwarz, K., Breuer, B., Kresse, H. 1990. *J. Biol. Chem.* 265:22023–28

109. McCormick, D., Van Der Rest, M., Goodship, J., Lozano, G., Ninomiya, Y., Olsen, B. R. 1987. *Proc. Natl. Acad. Sci. USA* 84:4044–48

110. Huber, S., Winterhalter, K. H., Vaughan, L. 1988. *J. Biol. Chem.* 263:752–56

111. Hassell, J. R., Cintron, C., Kublin, C., Newsome, D. A. 1983. *Arch. Biochem. Biophys.* 222:362–69

112. Nilsson, B., Nakazawa, K., Hassell, J. R., Newsome, D. A., Hascall, V. C. 1983. *J. Biol. Chem.* 258:6056–63

113. Funderburgh, J. L., Caterson, B., Conrad, G. W. 1987. *J. Biol. Chem.* 262:11634–40

114. Midura, R. J., Toledo, O. M. S., Yanagishita, M., Hascall, V. C. 1989. *J. Biol. Chem.* 264:1414–22

115. Midura, R. J., Hascall, V. C. 1989. *J. Biol. Chem.* 264:1423–30

116. Funderburgh, J. L., Conrad, G. W. 1990. *J. Biol. Chem.* 265:8297–303

116a. Sjöberg, E. M., Fries, E. 1990. *Biochem. J.* 272:113–18

117. Ledbetter, S. R., Tyree, B., Hassell, J. R., Horigan, E. A. 1985. *J. Biol. Chem.* 260:8106–13

118. Kato, M., Koike, Y., Ito, Y., Suzuki, S., Kimata, K. 1987. *J. Biol. Chem.* 262:7180–88

119. Yurchenko, P. D., Cheng, Y-S., Ruben, G. C. 1987. *J. Biol. Chem.* 262:17668–76

120. Kato, M., Koike, Y., Suzuki, S., Kimata, K. 1988. *J. Cell Biol.* 106:2203–10

121. Paulsson, M., Yurchenko, P. D., Ruben, G. C., Engel, J., Timpl, R. 1987. *J. Mol. Biol.* 197:297–313

122. Noonan, D. M., Horigan, E. A., Ledbetter, S. R., Vogeli, G., Sasaki, M., et al. 1988. *J. Biol. Chem.* 263:16379–87

123. Heremans, A., Cassiman, J-J., Van den Berghe, H., David, G. 1988. *J. Biol. Chem.* 263:4731–39

124. Heremans, A., Van Der Schueren, B., De Cock, B., Paulsson, M., Cassiman, J-J., et al. 1989. *J. Cell Biol.* 109:3199–211

125. Heremans, A., De Cock, B., Cassiman, J-J., Van den Berghe, H., David, G. 1990. *J. Biol. Chem.* 265:8716–24

126. Iozzo, R. V. 1984. *J. Cell. Biol.* 99:403–17

127. Iozzo, R. V., Hassell, J. R. 1989. *Arch. Biochem. Biophys.* 269:239–49

128. Andersson, M. J., Fambrough, D. M. 1983. *J. Cell Biol.* 97:1396–411
129. Klein, D. J., Brown, D. M., Oegema, T. R., Brenchley, P. E., Andersson, J. C., et al. 1988. *J. Cell Biol.* 106:963–70
130. Couchman, J. R., Ljubimov, A. V. 1989. *Matrix* 9:311–21
131. Pejler, G., Bäckström, G., Lindahl, U., Paulsson, M., Dziadek, M., et al. 1987. *J. Biol. Chem.* 262:5036–43
132. Kinsella, M. G., Wight, T. N. 1988. *Biochemistry* 27:2136–44
133. Saku, T., Furthmayer, H. 1989. *J. Biol. Chem.* 264:3514–23
134. Lindblom, A., Carlstedt, I., Fransson, L-Å. 1989. *Biochem. J.* 261:145–53
135. Jalkanen, M., Rapraeger, A., Bernfield, M. 1988. *J. Cell Biol.* 106:953–62
136. Saunders, S., Jalkanen, M., O'Farrell, S., Bernfield, M. 1989. *J. Cell Biol.* 108:1547–56
137. Mali, M., Jaakola, P., Arvilommi, A-M., Jalkanen, M. 1990. *J. Biol. Chem.* 265:6884–89
137a. Kiefer, M. C., Stephans, J. C., Crawford, K., Okino, K., Barr, P. J. 1990. *Proc. Natl. Acad. Sci. USA* 87:6985–89
138. Rapraeger, A., Jalkanen, M., Endo, E., Koda, J., Bernfield, M. 1985. *J. Biol. Chem.* 265:6884–89
139. Rapraeger, A., Jalkanen, M., Bernfield, M. 1986. *J. Cell Biol.* 103:2683–96
140. Rapraeger, A., Jalkanen, M., Bernfield, M. 1987. See Ref. 376, pp. 129–54
141. Sanderson, R. D., Lalor, P., Bernfield, M. 1989. *Cell Regul.* 1:27–35
142. Lories, V., Cassiman, J-J., Van den Berghe, H., David, G. 1989. *J. Biol. Chem.* 264:7009–16
143. Marynen, P., Zhang, J., Cassiman, J-J., Van Den Berghe, H., David, G. 1989. *J. Biol. Chem.* 264:7017–24
144. Kjellén, L., Oldberg, Å., Höök, M. 1980. *J. Biol. Chem.* 255:10407–13
145. Kjellén, L., Pettersson, I., Höök, M. 1981. *Proc. Natl. Acad. Sci. USA* 78:5371–75
146. Höök, M., Kjellén, L., Johansson, S., Robinson, J. 1984. *Annu. Rev. Biochem.* 53:847–69
146a. Lyon, M., Gallagher, J. T. 1990. *Biochem. J.* In press
147. Ripellino, J. A., Margolis, R. U. 1989. *J. Neurochem.* 52:807–12
148. Klinger, M. M., Margolis, R. U., Margolis, R. K. 1985. *J. Biol. Chem.* 260:4082–90
149. Margolis, R. K., Margolis, R. U. 1989. *Neurobiology of Glycoconjugates*, pp. 85–126. New York: Plenum
150. Woods, A., Couchman, J. R., Höök, M. 1985. *J. Biol. Chem.* 260:10872–79
151. Cöster, L., Carlstedt, I., Kendall, S.,

Malmström, A., Schmidtchen, A., Fransson, L-Å. 1986. *J. Biol. Chem.* 261:12079–88
152. David, G., Lories, V., Heremans, A., Van Der Schueren, B., Cassiman, J-J., Van Den Berghe, H. 1989. *J. Cell Biol.* 108:1165–75
153. Hedman, K., Christner, J., Julkunen, I., Vaheri, A. 1983. *J. Cell Biol.* 97:1288–93
154. Hoffman, S., Edelman, G. M. 1987. *Proc. Natl. Acad. Sci. USA* 84:2523–27
155. Tan, S-S., Crossin, K. L., Hoffman, S., Edelman, G. M. 1987. *Proc. Natl. Acad. Sci. USA* 84:7977–81
156. Hoffman, S., Crossin, K. L., Edelman, G. M. 1988. *J. Cell Biol.* 106:519–32
157. Carlson, S. S., Wight, T. N. 1987. *J. Cell Biol.* 105:3075–86
158. Garrigues, H. J., Lark, M. W., Lara, S., Hellström, I., Hellström, K. E., Wight, T. N. 1986. *J. Cell Biol.* 103:1699–710
159. Spiro, R. C., Parsons, W. G., Perry, S. K., Caulfield, J. P., Hein, A., et al. 1986. *J. Biol. Chem.* 261:5121–29
160. Harper, J. R., Reisfeld, R. A. 1987. See Ref. 140, pp. 345–64
161. Ishihara, M., Fedarko, N. S., Conrad, H. E. 1987. *J. Biol. Chem.* 262:4708–16
162. Yanagishita, M., McQuillan, D. J. 1989. *J. Biol. Chem.* 264:17551–58
163. Carey, D. J., Evans, D. M. 1989. *J. Cell Biol.* 108:1891–97
163a. David, G., Lories, V., Decock, B., Marynen, P., Cassiman, J-J., Van den Berghe, H. 1990. *J. Cell Biol.* 111:3165–76
164. Cheifetz, S., Andres, J. L., Massagué, J. 1988. *J. Biol. Chem.* 263:16984–91
165. Andres, J. L., Stanley, K., Cheifetz, S., Massagué, J. 1989. *J. Biol. Chem.* 109:3137–45
166. Bourin, M.-C., Öhlin, A.-K., Lane, D., Stenflo, J., Lindahl, U. 1988. *J. Biol. Chem.* 263:8044–52
167. Bourin, M.-C., Lundgren-Åkerlund, E., Lindahl, U. 1990. *J. Biol. Chem.* 265:15424–31
168. Suzuki, K., Kusumoto, H., Deyashiki, Y., Nishioka, J., Maruyama, I., et al. 1987. *EMBO J.* 6:1891–97
169. Dittman, W. A., Kumada, T., Sadler, J. E., Majerus, P. W. 1988. *J. Biol. Chem.* 263:15815–22
170. Carter, W. G., Wayner, E. A. 1988. *J. Biol. Chem.* 263:4193–201
171. Goldstein, L. A., Zhou, D. F. H., Picker, L. J., Minty, C. N., Bargatze, R. F., et al. 1989. *Cell* 56:1063–72
172. Stamenkovic, I., Amiot, M., Pesando, J. M., Seed, B. 1989. *Cell* 56:1057–62
173. Idzerda, R. L., Carter, W. G., Notten-

burg, C., Wayner, E. A., Gallatin, W. M., St. John, T. 1989. *Proc. Natl. Acad. Sci. USA* 86:4659–63
174. St. John, T., Meyer, J., Idzerda, R., Gallatin, W. M. 1990. *Cell* 60:45–52
175. Sivak, L. E., Harris, M. R., Kindle, C., Cullen, S. E. 1987. *J. Immunol.* 138:1319–21
176. Miller, J., Hatch, J. A., Simonis, S., Cullen, S. E. 1988. *Proc. Natl. Acad. Sci. USA* 85:1359–63
177. Gowda, D. C., Hogue-Angeletti, R., Margolis, R. K., Margolis, R. U. 1990. *Arch. Biochem. Biophys.* 281:219–24
178. Wohlfarter, T., Fischer-Colbrie, R., Hogue-Angeletti, R., Eiden, L. E., Winkler, H. 1988. *FEBS Lett.* 231:67–70
179. Bourdon, M. A., Oldberg, Å., Pierschbacher, M., Ruoslahti, E. 1985. *Proc. Natl. Acad. Sci. USA* 82:1321–25
180. Bourdon, M. A., Shiga, M., Ruoslahti, E. 1986. *J. Biol. Chem.* 261:12534–37
181. Tantravahi, R. V., Stevens, R. L., Austen, K. F., Weis, J. H. 1986. *Proc. Natl. Acad. Sci. USA* 83:9207–10
182. Avraham, S., Stevens, R. L., Gartner, M. C., Austen, K. F., Lalley, P. A., Weis, J. H. 1988. *J. Biol. Chem.* 263:7292–96
183. Stevens, R. L., Avraham, S., Gartner, M. C., Bruns, G. A. P., Austen, K. F., Weis, J. H. 1988. *J. Biol. Chem.* 263:7287–91
184. Périn, J-P., Bonnet, F., Maillet, P., Jollés, P. 1988. *Biochem. J.* 255:1007–13
185. Allie, P. M., Périn, J-P., Maillet, P., Bonnet, F., Rosa, J-P., Jollés, P. 1988. *FEBS Lett.* 236:123–26
186. Rothenberg, M. E., Pomerantz, J. T.., Owen, W. F. Jr., Avraham, S., Soberman, R. J., et al. 1988. *J. Biol. Chem.* 263:13901–8
187. Avraham, S., Austen, K. F., Nicodemus, C. F., Gartner, M. C., Stevens, R. L. 1989. *J. Biol. Chem.* 264:16719–26
188. Kjellén, L., Pettersson, I., Lillhager, P., Steen, M-L., Pettersson, U., et al. 1989. *Biochem. J.* 263:105–13
189. McQuillan, D. J., Yanagishita, M., Hascall, V. C., Bickel, M. 1989. *J. Biol. Chem.* 264:13245–51
190. Stevens, R. L., Austen, K. F. 1989. *Immunol. Today* 10:381–86
191. Avraham, S., Stevens, R. L., Nicodemus, C. F., Gartner, M. C., Austen, K. F., Weis, J. H. 1989. *Proc. Natl. Acad. Sci. USA* 86:3763–67
192. Nicodemus, C. F., Avraham, S., Austen, K. F., Purdy, S., Jablonski, J., Stevens, R. L. 1990. *J. Biol. Chem.* 265:5889–96
193. Angerth, T., Huang, R., Aveskogh, M.,

Pettersson, I., Kjellén, L., Hellman, L. 1990. *Gene* 93:235–40
194. Lohmander, L. S., Arnljots, K., Yanagishita, M. 1990. *J. Biol. Chem.* 265:5802–8
195. Kolset, S. O., Gallagher, J. T. 1990. *Biochim. Biophys. Acta.* 1032:191–211
196. Bourdon, M. A., Krusius, T., Campbell, S., Schwartz, N. B., Ruoslahti, E. 1987. *Proc. Natl. Acad. Sci. USA* 84: 3194–98
197. Rodén, L., Koerner, T., Olson, C., Schwartz, N. B. 1985. *Fed. Proc.* 44:373–80
198. Robinson, H. C., Horner, A. A., Höök, M., Ögren, S., Lindahl, U. 1978. *J. Biol. Chem.* 253:6687–93
199. Hardingham, T. E., Beardmore-Gray, M., Dunham, D. G., Ratcliffe, A. 1986. *Ciba Found. Symp.* 124:30–39
200. Fosang, A. J., Hardingham, T. E. 1989. *Biochem. J.* 261:801–9
201. Neame, P., Christner, J., Baker, J. 1986. *J. Biol. Chem.* 261:8653–61
202. Deák, F., Kiss, I., Sparks, K. J., Argraves, W. S., Hampikian, G., Goetinck, P. 1986. *Proc. Natl. Acad. Sci. USA* 83:3766–70
203. Dudhia, J., Hardingham, T. E. 1989. *J. Mol. Biol.* 206:749–53
204. Scott, J. E., Orford, C. R. 1981. *Biochem. J.* 197:213–16
205. Scott, J. E. 1988. *Biochem. J.* 252:313–23
206. Hausser, H., Hoppe, W., Rauch, U., Kresse, H. 1989. *Biochem. J.* 263:137–42
207. Schmidt, G., Hausser, H., Kresse, H. 1990. *Biochem. J.* 266:591–95
208. Doege, K., Fernandez, P., Hassell, J. R., Sasaki, M., Yamada, Y. 1986. *J. Biol. Chem.* 261:8108–11
209. Krusius, T., Gehlsen, K. R., Ruoslahti, E. 1987. *J. Biol. Chem.* 262:13120–25
210. Sikder, S. K., Kabat, E. A., Steer, C. J., Ashwell, G. 1983. *J. Biol. Chem.* 258:12520–25
211. Halberg, D. F., Proulx, G., Doege, K., Yamada, Y., Drickamer, K. 1988. *J. Biol. Chem.* 263:9486–90
212. Müller-Eberhard, H. J. 1988. *Annu. Rev. Biochem.* 57:321–47
213. Appella, E., Weber, I. T., Blasi, F. 1988. *FEBS Lett.* 231:1–4
213a. Baldwin, C. T., Reginato, A. M., Prockop, D. J. 1989. *J. Biol. Chem.* 264:15747–50
214. Jackman, R. W., Beeler, D. L., Fritze, L., Soff, G., Rosenberg, R. D. 1987. *Proc. Natl. Acad. Sci. USA* 84:6425–29
215. Panayotou, G., End, P., Aumailley, M., Timpl, R., Engel, J. 1989. *Cell* 56:93–101

216. Coombe, D. R., Rider, C. C. 1989. *Immunol. Today* 10:289–91
217. Sasaki, M., Kato, S., Kohno, K., Martin, G. R., Yamada, Y. 1987. *Proc. Natl. Acad. Sci USA* 84:935–39
218. Sasaki, M., Yamada, Y. 1987. *J. Biol. Chem.* 262:17111–17
219. Sasaki, M., Kleinman, H. K., Huber, H., Deutzman, R., Yamada, Y. 1988. *J. Biol. Chem.* 263:16536–44
220. Edelman, G. M. 1988. *Biochemistry* 27:3533–43
221. Williams, A. F., Barclay, A. N. 1988. *Annu. Rev. Immunol.* 6:381–405
222. Schubert, D., Schroeder, R., LaCorbiere, M., Saitoh, T., Cole, G. 1988. *Science* 241:223–26
223. Snow, A. D., Mar, H., Nochlin, D., Kimata, K., Kato, M., et al. 1988. *Am. J. Pathol.* 133:456–63
223a. Gowda, D. C., Margolis, R. K., Frangione, B., Ghiso, J., Larrondo-Lillo, M., Margolis, R. U. 1989. *Science* 244:826–28
224. Van Nostrand, W. E., Wagner, S. L., Suzuki, M., Choi, B. H., Farrow, J. S., et al. 1989. *Nature* 341:546–49
225. Yurt, R., Austen, K. F. 1977. *J. Exp. Med.* 146:1405–19
226. Serafin, W. E., Dayton, E. T., Gravallese, P. M., Austen, K. F., Stevens, R. L. 1987. *J. Immunol.* 139:3771–76
227. Le Trong, H., Neurath, H., Woodbury, R. G. 1987. *Proc. Natl. Acad. Sci. USA* 84:364–67
228. Olivecrona, T., Bengtsson, G., Marklund, S.-E., Lindahl, U., Höök, M. 1977. *Fed. Proc.* 36:60–65
229. Cheng, C.-F., Oosta, G. M., Bensadoun, A., Rosenberg, R. D. 1981. *J. Biol. Chem.* 256:12893–98
230. Shimada, K., Gill, P. J., Silbert, J. E., Douglas, W. H. J., Fanburg, B. L. 1981. *J. Clin. Invest.* 68:995–1002
231. Bengtsson, G., Olivecrona, T., Höök, M., Riesenfeld, J., Lindahl, U. 1980. *Biochem. J.* 189:625–33
232. Bengtsson-Olivecrona, G., Olivecrona, T. 1985. *Biochem. J.* 226:409–13
233. Busch, S. J., Martin, G. A., Barnhart, R. L., Jackson, R. L. 1989. *J. Biol. Chem.* 264:9527–32
234. Wion, K. L., Kirchgessner, T. G., Lusis, A. J., Schotz, M. C., Lawn, R. M. 1987. *Science* 235:1638–41
235. Persson, B., Bengtsson-Olivecrona, G., Enerbäck, S., Olivecrona, T., Jörnvall, H. 1989. *Eur. J. Biochem.* 179:39–45
236. Cisar, L. A., Hoogewerf, A. J., Cupp, M., Rapport, C. A., Bensadoun, A. 1989. *J. Biol. Chem.* 264:1767–74
237. Cupp, M., Bensadoun, A., Melford, K. 1987. *J. Biol. Chem.* 262:6383–88
238. Olivecrona, T., Bengtsson-Olivecrona, G. 1989. See Ref. 23, pp. 335–62
239. Holmer, E., Lindahl, U., Bäckström, G., Thunberg, L., Sandberg, H., et al. 1980. *Thromb. Res.* 18:861–69
240. Griffith, M. J. 1982. *J. Biol. Chem.* 257:7360–65
241. Holmer, E., Kurachi, K., Söderström, G. 1981. *Biochem. J.* 193:395–400
242. Oosta, G. M., Gardner, W. T., Beeler, D. L., Rosenberg, R. D. 1981. *Proc. Natl. Acad. Sci. USA* 78:829–33
243. Lane, D. A., Denton, J., Flynn, A. M., Thunberg, L., Lindahl, U. 1984. *Biochem. J.* 218:725–32
244. Danielsson, Å., Raub, E., Lindahl, U., Björk, I. 1986. *J. Biol. Chem.* 261: 15467–73
245. Redini, F., Tixier, J.-M., Petitou, M., Choay, J., Robert, L., Hornebeck, W. 1988. *Biochem. J.* 252:515–19
246. Hjalmarsson, K., Marklund, S. L., Engström, A., Edlund, T. 1987. *Proc. Natl. Acad. Sci. USA* 84:6340–44
247. Karlsson, K., Lindahl, U., Marklund, S. L. 1988. *Biochem. J.* 256:29–33
248. Karlsson, K., Marklund, S. L. 1989. *Lab. Invest.* 60:659–66
249. Adachi, T., Marklund, S. L. 1989. *J. Biol. Chem.* 264:8537–41
250. Björk, I., Olson, S. T., Shore, J. D. 1989. See Ref. 23, pp. 229–57
251. Lindahl, U., Bäckström, G., Thunberg, L., Leder, I. G. 1980. *Proc. Natl. Acad. Sci. USA* 77:6551–55
252. Thunberg, L., Bäckström, G., Lindahl, U. 1982. *Carbohydr. Res.* 100:393–410
253. Lindahl, U., Thunberg, L., Bäckström, G., Riesenfeld, J., Nordling, K., Björk, I. 1984. *J. Biol. Chem.* 259:12368–76
254. Casu, B., Oreste, P., Torri, G., Zoppetti, G., Choay, J., et al. 1981. *Biochem. J.* 197:599–609
255. Iozzo, R. V., Clark, C. C. 1986. *J. Biol. Chem.* 261:6658–69
256. Atha, D. H., Stephens, A. W., Rimon, A., Rosenberg, R. D. 1984. *Biochemistry* 23:5801–12
257. Atha, D. H., Lormeau, J.-C., Petitou, M., Rosenberg, R. D., Choay, J. 1987. *Biochemistry* 26:6454–61
258. Marcum, J. A., Rosenberg, R. D. 1989. See Ref. 23, pp. 275–94
259. Craig, P. A., Olson, S. T., Shore, J. D. 1989. *J. Biol. Chem.* 264:5452–61
260. Petitou, M., Duchaussoy, P., Lederman, I., Choay, J., Jacquinet, J.-C., et al. 1987. *Carbohydr. Res.* 167:67–75
261. Tollefsen, D. M., Peacock, M. E., Monafo, W. J. 1986. *J. Biol. Chem.* 261:8854–58
262. Tollefsen, D. M. 1989. See Ref. 23, pp. 257–73

PROTEOGLYCANS 473

263. Ragg, H., Ulshöfer, T., Gerewitz, J. 1990. *J. Biol. Chem.* 265:5211–18
264. Maimone, M. M., Tollefsen, D. M. 1990. *J. Biol. Chem.* 265:18263–71
265. Farrell, D. H., Cunningham, D. D. 1986. *Proc. Natl. Acad. Sci. USA* 83:6858–62
266. Farrell, D. H., Cunningham, D. D. 1987. *Biochem. J.* 245:543–50
267. Camejo, G., Lalaguna, F., Lopez, F., Starosta, R. 1980. *Atherosclerosis* 35:307–20
268. Srinivasan, S. R., Vijayagopal P., Dalferes, E. R. Jr., Abbate, B., Radhakrishnamurthy, B., Berenson G. S. 1984. *Biochim. Biophys. Acta* 793:157–68
269. Couchman, J. R., Woods, A., Höök, M., Christner, J. E. 1985. *J. Biol. Chem.* 260:13755–62
270. Camejo, G., Olofsson, S.-O., Lopez, F., Carlsson, P., Bondjers, G. 1988. *Arteriosclerosis* 8:368–77
271. Cardin, A. D., Randall, C. J., Hirose, N., Jackson, R. L. 1987. *Biochemistry* 26:5513–18
272. Cardin, A. D., Jackson, R. L., Sparrow, D. A., Sparrow, J. T. 1989. *Ann. NY Acad. Sci.* 556:186–93
273. Weisgraber, K. H., Rall, S. C. Jr. 1987. *J. Biol. Chem.* 262:11097–103
274. Alves, C. S., Mourao, P. A. S. 1988. *Atherosclerosis* 73:113–24
275. Steele, R. H., Wagner, W. D., Rowe, H. A., Edwards, I. J. 1987. *Atherosclerosis* 65:51–62
276. Yamada, K. M., Kennedy, D. W., Kimata, K., Pratt, R. M. 1980. *J. Biol. Chem.* 255:6055–63
277. Khan, M. Y., Jaikaria, N. S., Frenz, D. A., Villanueva, G., Newman, S. A. 1988. *J. Biol. Chem.* 263:11314–18
278. Woods, A., Höök, M., Kjellén, L., Smith, C. G., Rees, D. A. 1984. *J. Cell Biol.* 99:1743–53
279. LeBaron, R. G., Esko, J. D., Woods, A., Johansson, S., Höök, M. 1988. *J. Cell Biol.* 106:945–52
280. Saunders, S., Bernfield, M. 1988. *J. Cell Biol.* 106:423–30
281. Ogamo, A., Nagai, A., Nagasawa, K. 1985. *Biochim. Biophys. Acta* 841:30–41
282. Benecky, M. J., Kolvenbach, C. G., Amrani, D. L., Mosesson, M. W. 1988. *Biochemistry* 27:7565–71
283. Preissner, K. T., Müller-Berghaus, G. 1987. *J. Biol. Chem.* 262:12247–53
284. Lane, D. A., Flynn, A. M., Pejler, G., Lindahl, U., Choay, J., Preissner, K. 1987. *J. Biol. Chem.* 262:16343–48
285. Martin, G. R., Timpl, R. 1987. *Annu. Rev. Cell Biol.* 3:57–85
286. Skubitz, A. P. N., McCarthy, J. B., Charonis, A. S., Furcht, L. T. 1988. *J. Biol. Chem.* 263:4861–68
287. Kouzi-Koliakos, K., Koliakos, G. G., Tsilibary, E. C., Furcht, L. T., Charonis, A. S. 1989. *J. Biol. Chem.* 264:17971–78
288. Kaesberg, P. R., Ershler, W. B., Esko, J. D., Mosher, D. F. 1989. *J. Clin. Invest.* 83:994–1001
289. Murphy-Ullrich, J. E., Westrick, L. G., Esko, J. D., Mosher, D. F. 1988. *J. Biol. Chem.* 263:6400–6
290. Murphy-Ullrich, J. E., Mosher, D. F. 1987. *J. Cell Biol.* 105:1603–11
291. Majack, R. A., Goodman, L. V., Dixit, V. M. 1988. *J. Cell Biol.* 106:415–22
292. Sun, X., Mosher, D. F., Rapraeger, A. 1989. *J. Biol. Chem.* 264:2885–89
293. Koda, J. E., Rapraeger, A., Bernfield, M. 1985. *J. Biol. Chem.* 260:8157–62
294. LeBaron, R. G., Höök, A., Esko, J. D., Gay, S., Höök, M. 1989. *J. Biol. Chem.* 264:7950–56
295. Burgess, W. H., Maciag, T. 1989. *Annu. Rev. Biochem.* 58:575–606
296. Folkman, J., Klagsbrun, M. 1987. *Science* 235:442–47
297. Folkman, J., Klagsbrun, M., Sasse, J., Wadzinski, M., Ingber, D., Vlodavsky, I. 1988. *Am. J. Pathol.* 130:393–400
298. Baird, A., Schubert, D., Ling, N., Guillemin, R. 1988. *Proc. Natl. Acad. Sci. USA* 85:2324–28
299. Bashkin, P., Doctrow, S., Klagsbrun, M., Svahn, C.-M., Folkman, J., Vlodavsky, I. 1989. *Biochemistry* 28:1737–43
300. Saksela, O., Moscatelli, D., Sommer, A., Rifkin, D. B. 1988. *J. Cell Biol.* 107:743–51
301. Sudhalter, J., Folkman, J., Svahn, C.-M., Bergendal, K., D'Amore, P. A. 1989. *J. Biol. Chem.* 264:6892–97
302. Ratner, N., Hong, D., Lieberman, M. A., Bunge, R. P., Glaser, L. 1988. *Proc. Natl. Acad. Sci. USA* 85:6992–96
303. Berman, P., Gray, P., Chen, E., Keyser, K., Ehrlich, D., et al. 1987. *Cell* 51:135–42
303a. Gordon, M. Y., Riley, G. P., Watts, S. M., Greaves, M. F. 1987. *Nature* 326:403–5
303b. Roberts, R., Gallagher, J., Spooncer, E., Allen, T. D., Bloomfield, F., Dexter, T. M. 1988. *Nature* 332:376–78
304. Lane, D. A., Denton, J., Flynn, A. M., Thunberg, L., Lindahl, U. 1984. *Biochem. J.* 218:725–32
305. Cowan, S. W., Bakshi, E. N., Machin, K. J., Isaacs, N. W. 1986. *Biochem. J.* 234:485–88
306. Cole, G. J., Glaser, L. 1986. *J. Cell Biol.* 102:403–12

307. Cole, G. J., Loewy, A., Glaser, L. 1986. *Nature* 320:445–47
308. Cole, G. J., Burg, M. 1989. *Exp. Cell Res.* 182:44–60
309. WuDunn, D., Spear, P. G. 1989. *J. Virol.* 63:52–58
310. Wright, T. C. Jr., Pukac, L. A., Castellot, J. J. Jr., Karnovsky, M. J., Levine, R. A., et al. 1989. *Proc. Natl. Acad. Sci. USA* 86:3199–203
311. Pukac, L. A., Castellot, J. J. Jr., Wright, T. C. Jr., Caleb, B. L., Karnovsky, M. J. 1990. *Cell Regul.* 1:435–43
311a. Lane, D. A. 1989. See Ref. 23, pp. 363–92
312. Fransson, L.-Å., Carlstedt, I., Cöster, L., Malmström, A. 1983. *J. Biol. Chem.* 258:14342–45
313. Cardin, A. D., Weintraub, H. J. R. 1989. *Arteriosclerosis* 9:21–32
314. Casu, B., Choay, J., Ferro, D. R., Gatti, G., Jacquinet, J.-C., et al. 1986. *Nature* 322:215–16
315. Sanderson, P. N., Huckerby, T. N., Nieduszynski, I. A. 1987. *Biochem. J.* 243:175–78
316. Casu, B. 1989. See Ref. 23, pp. 25–49
316a. Casu, B., Petitou, M., Provasoli, M., Sinay, P. 1988. *Trends Biochem. Sci.* 13:221–25
317. Nieduszynski, I. A. 1989. See Ref. 23, pp. 51–63
318. Ferro, D. R., Provasoli, A., Ragazzi, M., Casu, B., Torri, G., et al. 1990. *Carbohydr. Res.* 195:157–67
319. Mason, R. M., Crossman, M. V., Sweeney, C. 1989. See Ref. 9a, pp. 107–20
320. Fedarko, N. S., Conrad, H. E. 1986. *J. Cell Biol.* 102:587–99
321. Busch, S. J., Sassone-Corsi, P. 1990. *Trends Genet.* 6:36–40
322. Linhardt, R. J., Turnbull, J. E., Wang, H. M., Loganathan, D., Gallagher, J. T. 1990. *Biochemistry* 29:2611–17
323. Fransson, L.-Å., Havsmark, B., Silverberg, I. 1990. *Biochem. J.* 269:381–88
324. Caterson, B., Calabro, T., Hampton, A. 1987. See Ref. 376, pp. 1–26
325. Sorrell, J. M., Mahmoodian, F., Schafer, I. A., Davis, B., Caterson, B. 1990. *J. Histochem. Cytochem.* 38:393–402
326. Pejler, G., Lindahl, U., Larm, O., Scholander, E., Sandgren, E., Lundblad, A. 1988. *J. Biol. Chem.* 263: 5197–201
327. Feizi, T. 1989. See Ref. 34, pp. 21–29
328. Petitou, M. 1989. See Ref. 23, pp. 65–79
329. Johansson, S., Höök, M. 1984. *J. Cell Biol.* 98:810–17
330. Couchman, J. R., Austria, R., Woods, A., Hughes, R. C. 1988. *J. Cell. Physiol.* 136:226–36
330a. Murphy-Ullrich, J. E., Höök, M. 1989. *J. Cell Biol.* 109:1309–19
331. Saxena, U., Klein, M. G., Goldberg, I. J. 1990. *J. Biol. Chem.* 265:12880–86
332. Esmon, C. T. 1987. *Science* 235:1348–52
333. Bourin, M.-C., Boffa, M.-C., Björk, I., Lindahl, U. 1986. *Proc. Natl. Acad. Sci. USA* 83:5924–28
334. Preissner, K. T., Koyama, T., Müller, D., Tschopp, J., Müller-Berghaus, G. 1990. *J. Biol. Chem.* 265:4915–22
335. Bourin, M.-C. 1989. *Thromb. Res.* 54:27–39
336. Hofsteenge, J., Taguchi, H., Stone, S. R. 1986. *Biochem. J.* 237:243–51
337. Preissner, K. T., Delvos, U., Müller-Berghaus, G. 1987. *Biochemistry* 26: 2521–28
338. Kolset, S. O., Seljelid, R., Lindahl, U. 1984. *Biochem. J.* 219:793–99
339. Vainio, S., Lehtonen, E., Jalkanen, M., Bernfield, M., Saxén, L. 1989. *Dev. Biol.* 134:382–91
340. Vainio, S., Jalkanen, M., Thesleff, I. 1989. *J. Cell Biol.* 108:1945–54
341. Sanderson, R. D., Lalor, P., Bernfield, M. 1989. *Cell Regul.* 1:27–35
342. Thesleff, I., Jalkanen, M., Vainio, S., Bernfield, M. 1988. *Dev. Biol.* 129: 565–72
343. Carrino, D. A., Oron, U., Pechak, D. G., Caplan, A. I. 1988. *Development* 103:641–56
344. Kalb, R. G., Hockfield, S. 1988. *J. Neurosci.* 8:2350–60
345. Folkman, J., Langer, R., Linhardt, R. J., Haudenschild, C., Taylor, S. 1983. *Science* 221:719–25
346. Cole, G., Schubert, J. D., Glaser, L. 1985. *J. Cell. Biol.* 100:1192–99
347. Nakajima, M., Irimura, T., Di Ferrante, D., Di Ferrante, N., Nicolson, G. L. 1983. *Science* 220:611–13
348. Nakajima, M., Irimura, T., Di Ferrante, N., Nicolson, G. L. 1984. *J. Biol. Chem.* 259:2283–90
349. Nakajima, M., Irimura, T., Nicolson, G. L. 1988. *J. Cell Biochem.* 36:157–67
350. Parish, C. R., Coombe, D. R., Jakobsen, K. B., Bennett, F. A., Underwood, P. A. 1987. *Int. J. Cancer* 40:511–18
351. Bar-Ner, M., Eldor, A., Wasserman, L., Matzner, Y., Cohen, I. R., et al. 1987. *Blood* 70:551–57
352. Ishihara, M., Conrad, H. E. 1989. *J. Cell. Physiol.* 138:467–76
353. Fedarko, N. S., Ishihara, M., Conrad, H. E. 1989. *J. Cell. Physiol.* 139:287–94

354. Wright, T. C. Jr., Castellot, J. J. Jr., Diamond, J. R., Karnovsky, M. J. 1989. See Ref. 23, pp. 295–316
355. Wright, T. C. Jr., Castellot, J. J. Jr., Petitou, M., Lormeau, J.-C., Choay, J., Karnovsky, M. J. 1989. *J. Biol. Chem.* 264:1534–42
356. Castellot, J. J. Jr., Choay, J., Lormeau, J.-C., Petitou, M., Sache, E., Karnovsky, M. J. 1986. *J. Cell Biol.* 102:1979–84
357. Fritze, L. M. S., Reilly, C. F., Rosenberg, R. D. 1985. *J. Cell Biol.* 100:1041–49
358. Schmidt, A., Buddecke, E. 1990. *Eur. J. Cell Biol.* 52:229–35
359. Schmidt, A., Buddecke, E. 1990. *Exp. Cell Res.* 189:269–75
359a. Castellot, J. J. Jr., Pukac, L. A., Caleb, B. L., Wright, T. C. Jr., Karnovsky, M. J. 1989. *J. Cell Biol.* 109:3147–55
360. Imamura, T., Mitsui, Y. 1987. *Exp. Cell Res.* 172:92–100
361. Esko, J. D., Stewart, T. E., Taylor, W. H. 1985. *Proc. Natl. Acad. Sci. USA* 82:3197–201
362. Esko, J. D., Rostand, K. S., Weinke, J. L. 1988. *Science* 241:1092–96
363. Lohmander, L. S., Hascall, V. C., Yanagishita, M., Kuettner, K. E., Kimura, J. H. 1986. *Arch. Biochem. Biophys.* 250:211–27
364. Nuwayhid, N., Glaser, J. H., Johnson, J. C., Conrad, H. E., Hauser, S. C., Hirschberg, C. B. 1986. *J. Biol. Chem.* 261:12936–41
365. Mann, D. M., Yamaguchi, Y., Bour-don, M. A., Ruoslahti, E. 1990. *J. Biol. Chem.* 265:5317–23
366. Campbell, P., Jacobsson, I., Benzing-Purdie, L., Rodén, L., Fessler, J. H. 1984. *Anal. Biochem.* 137:505–16
367. Esko, J. D., Weinke, J. L., Taylor, W. H., Ekborg, G., Rodén, L., et al. 1987. *J. Biol. Chem.* 262:12189–95
368. Rohrmann, K., Niemann, R., Buddecke, E. 1985. *Eur. J. Biochem.* 148:463–69
369. Kjellén, L., Bielefeld, D., Höök, M. 1983. *Diabetes* 32:337–42
370. Cohen, M. P., Klepser, H., Wu, V-Y. 1988. *Diabetes* 37:1324–27
370a. Unger, E., Pettersson, I., Eriksson, U. J., Lindahl, U., Kjellén, L. 1991. *J. Biol. Chem.* In press
371. Brandan, E., Hirschberg, C. B. 1988. *J. Biol. Chem.* 263:2417–22
371a. Pettersson, I., Kusche, M., Unger, E., Wlad, H., Nylund, L., et al. 1991. *J. Biol. Chem.* In press
372. Jacobsson, K.-G., Riesenfeld, J., Lindahl, U. 1985. *J. Biol. Chem.* 260:12154–59
373. Kusche, M., Oscarsson, L.-G., Reynertson, R., Rodén, L., Lindahl, U. 1990. *J. Biol. Chem.* In press
374. Kusche, M., Torri, G., Casu, B., Lindahl, U. 1990. *J. Biol. Chem.* 265:7292–300
375. Jacobsson, K.-G., Lindahl, U., Horner, A. A. 1986. *Biochem. J.* 240:625–32
376. Wight, T. N., Mecham, R. P., eds. 1987. *Biology of Extracellular Matrix: Biology of Proteoglycans.* New York: Academic

Annu. Rev. Biochem. 1991. 60:477–511

# FIDELITY MECHANISMS IN DNA REPLICATION

*Harrison Echols*

Department of Molecular and Cell Biology, University of California, Berkeley, California 94720

*Myron F. Goodman*

Department of Biological Science, University of Southern California, Los Angeles, California 90089-1340

KEY WORDS: DNA polymerases, mutation mechanisms, DNA damage, editing by DNA polymerases.

## CONTENTS

0066-4154/91/0701-0477$02.00

## PERSPECTIVES AND SUMMARY

The survival of an organism requires precise duplication of the genome. Error frequencies are typically only $10^{-9}$ to $10^{-10}$ per base replicated. The biochemical problem posed by this demand for faithful duplication is exceptionally complex because the correct (Watson-Crick) base pairs do not exhibit large structural or energetic differences from some incorrect base pairs (e.g. G·T vs A·T). For the well-defined prokaryotic replication systems, the requirement for a low mutation rate is achieved by the sequential operation of three fidelity mechanisms: (a) selection of the correct deoxynucleoside triphosphate (dNTP) substrate in the polymerization reaction (base selection); (b) exonucleolytic removal of an incorrectly inserted deoxynucleoside monophosphate (dNMP) from the end of the growing chain (editing); (c) postreplicative excision of an incorrectly inserted deoxynucleoside monophosphate after the polymerase has extended the DNA chain (mismatch repair). Each of these mechanisms exhibits an impressive selectivity for excluding noncanonical base pairs; the sequential application of all three fidelity operations yields precisely duplicated DNA. For eukaryotic replication systems, only base selection has so far been clearly defined as a general fidelity mechanism. However, most cellular and viral eukaryotic polymerases exhibit exonucleolytic editing activities, and there has been excellent recent evidence for mismatch correction pathways. Thus the prokaryotic solutions described here are likely to be general.

In this review we consider in detail only the two coupled fidelity mechanisms that involve the direct participation of the DNA polymerase—base selection and editing. Excellent recent reviews have appeared on mismatch repair (1–3). We focus on the fundamental biochemical question: how does a DNA polymerase distinguish a correct from an incorrect base pair with such astounding precision? Recent structural and kinetic data indicate that polymerases achieve selectivity by mechanisms that differ in detail. However, from the available evidence, we believe that all polymerases follow general fidelity principles for base selection and editing.

For base selection, the most likely discrimination principle is an exquisite demand for the precise geometry of the Watson-Crick base pair. The most remarkable structural property of the A·T and G·C base pairs is their complete geometric equivalence as subunits of the double helix; all other non-Watson-Crick base pairs known to occur (e.g. the G·T "wobble" pair) have altered geometry. We presume that the active site of the polymerase exerts transition state selectivity for the identical geometry (C1' distances and bond angles) of the Watson-Crick base pairs. As a consequence, the correct dNTP is used $10^4$ or $10^5$ times as efficiently as an incorrect dNTP. This demand for Watson-Crick geometry at the transition state might be manifested either in more rapid dissociation of incorrect dNTP, or slower phosphodiester bond formation for misoriented bases, or most likely both. The active site of a polymerase probably evolved to catalyze phosphodiester bond formation with either A, T, G, or C in the dNTP site, with frequency dependent on pairing energetics between substrate and template. Thus a geometric selection mechanism could be achieved by fine-tuning the allowed template-substrate configurations.

For exonucleolytic editing, the most likely discrimination mechanism is the "melting capacity" of the mispaired DNA; a misaligned base at the 3' end will be more often in a single-strand configuration. This mechanism is an easy evolutionary transition from a free single-strand exonuclease. However, the degree of discrimination against a mispair is likely to be relatively small, especially with respect to the more thermally sensitive A·T correct pairs. Editing becomes an efficient process, with discrimination of $10^2$ or more, because of kinetic amplification mechanisms. In the most clearly defined mechanism, the next base addition reaction following a misinsertion is slow compared to base addition following a correct insertion. Thus the exonuclease has longer to act on the mispair, amplifying the editing discrimination.

Certain DNA lesions inhibit DNA replication. The damage site is presumably sufficiently distorted that the geometric recognition mechanism does not allow effective base insertion. In addition, the editing exonuclease will preferentially remove even a "correctly" inserted base because the resultant distorted base pair looks like a mismatch. The rescue of such a stalled polymerase requires a complex cellular response. In the last section of this review, we consider briefly the induced "SOS response" to replication-blocking lesions. The SOS response provides for a highly mutagenic pathway of "translesion" DNA replication, dependent on induced synthesis of the mutation proteins UmuC and UmuD, followed by regulated processing of UmuD to its active form, UmuD'. An interesting point is that mutations are introduced by an induced and highly regulated pathway; SOS mutagenesis is not a passive response to DNA damage. The existence of an induced mutation pathway might seem puzzling in view of our opening statement that the mutation rate must be kept low. However, mutation is the source of genetic variation, and a regulated loss of replication fidelity might serve to accelerate evolutionary adaptation in an endangered population. Thus the fidelity of DNA replication may be subject to environmental regulation.

# BASE INSERTION SELECTIVITY BY DNA POLYMERASES

## The Base Insertion Pathway

The role of a DNA polymerase is to catalyze the formation of phosphodiester bonds between primer 3'-termini and dNTP substrates complementary to the template base. In the enzyme active site, hydrogen bonds can form between substrate and template bases in accordance with Watson-Crick rules to guide this process. However, non-Watson-Crick base pairs may also form; the task of the base insertion pathway is to discriminate against these noncanonical base pairs.

There are several possible control points in the nucleotide insertion pathway that are likely to be used to discriminate between Watson-Crick and non-Watson-Crick base pairs. Discrimination might initially occur in the dNTP substrate-binding stage of the reaction. The length of time that a substrate remains bound on a polymerase-DNA complex may be determined by nonspecific nucleotide binding at the polymerase triphosphate binding site, hydrogen bonding between substrate and template bases, and nearest neighbor base-stacking interactions. Since base-pairing (H-bonding and base-stacking) interactions stabilize correct base pairs preferentially over incorrect base pairs, dissociation from the polymerase-DNA complex might occur more rapidly for a mispaired substrate (4–6). A second stage of discrimination could involve an enzyme conformational change strongly favoring correct over incorrect substrates (7–9). Finally, the rate of phosphodiester bond formation with release of pyrophosphate could occur more rapidly for correct dNTPs (7–9). Different steps in the insertion pathway could be used to a greater or lesser extent by individual polymerases to optimize insertion fidelity.

The recent introduction of enzyme kinetic approaches has sharpened the analysis of replication fidelity. As described below, pre-steady-state kinetic analysis with pol I of *Escherichia coli* has provided quantitative estimates of the discrimination for each step in the nucleotide insertion pathway (7–9). Steady-state kinetic methods have provided a simple and rapid approach to measurements of overall insertion fidelity (10–16).

## Structural Aspects to Base Selection by DNA Polymerases

In their original papers on DNA structure, Watson & Crick pointed out that errors in DNA replication were expected from the properties of base pairs (17, 18). These authors pointed out that A·C and G·T transition mispairs could occur as Watson-Crick structures consisting of two and three H-bonds respectively, provided that either of the mispaired bases was present as a disfavored imino (A or C) or enol (G or T) tautomer. In addition to tautomeric base shifts, disfavored tautomer models for transversions require rotating the base relative to sugar from an *anti* to *syn* conformation (19). Two key points are implied by tautomer shift schemes: geometrical constraints imposed by a polymerase active site strongly favor Watson-Crick over non-Watson-Crick structures; and two H-bonds are better than one for mismatch stabilization (17–20).

During the past several years, the structure of aberrant base pairs in duplex DNA oligonucleotides has been analyzed by X-ray crystallography (21–26) and nuclear magnetic resonance (NMR) (27–34). Experiments have been performed using natural bases (21–23, 25–30, 32, 33), base analogues (24, 34–37), and abasic (apurinic/apyrimidinic) lesions (38–40). These studies have revealed that base mispairs are generally stabilized by more than a single H-bond; in none of the studies, however, were disfavored tautomeric structures observed. Instead, the source of additional H-bonding between bases came from solvent-base interactions leading to either protonated or ionized base mispairs, or from wobble structures. A protonated A·C mispair (33) was observed to be in a wobble conformation (21, 32, 33); ionized Watson-Crick pairs in equilibrium with neutral wobble structures were found for the case of fluorouracil or bromorouracil paired with guanine (36, 37); G·T pairs were detected as wobble structures (21, 31); and G·A pairs were observed in an *anti-anti* glycosidic configuration (29, 31), as G·A(*syn*) (23), or as G(*syn*)·A (25). For DNA containing an abasic lesion, A located opposite the lesion was found to stack within the double helix with no discernible DNA distortion (38, 39); G and T revealed both intra- and extrahelical components; C appeared to be exclusively extrahelical (40).

When attempting to relate structural data on base mispairs to their formation by polymerases, it is important to keep in mind that the most stable mispaired structures derived from X-ray crystallography and NMR or by an analysis of melting temperatures are not necessarily the misinsertions made most efficiently by the enzyme. For the base analogue 2-aminopurine (2AP), DNA containing 2AP·A wobble pairs (35) was found to be more stable (41) than that containing 2AP·C protonated Watson-Crick pairs (34). Yet polymerase forms 2AP·C pairs much more efficiently than 2AP·A (42). The same discrepancy appears to be true for the base xanthine (X); insertion of C opposite X occurs much more efficiently than insertion of G opposite X, even though polymers containing X·G are more stable than those with X·C (43). The key point may be that polymerases strongly prefer to insert nucleotides in the configuration closest to a canonical Watson-Crick structure, even when the structure may be aberrant. Yet, in product DNA, a well-stacked wobble pair may result in a more stable structure than a poorly stacked base pair approximating Watson-Crick geometry more closely.

From energy considerations, the geometrical and electrostatic properties of the polymerase active site are likely to exert a strong influence on nucleotide insertion specificities. Based on measurements of melting temperature differences between matched and mismatched base pairs in aqueous solution, free energy differences are estimated to be in the range of 0.2 to 0.4 kcal/mol for terminal pairs (44) and 1 to 3 kcal/mol for internal pairs (45). Assuming that the energetic properties of a terminal mismatch approximate those of an incoming dNTP, the calculated discrimination between right and wrong base pairs would be only in the tenfold range (based on the calculated differential equilibrium dissociation constants for correct vs incorrect base pairs). DNA polymerase insertion accuracies have been measured in the range of $10^3$ to $10^5$

in vitro, requiring $\Delta\Delta G$ values to be much larger than those measured in water.

Petruska et al (44) have pointed out that the free energy differences between right and wrong base pairs in aqueous solution are small because of "entropy-enthalpy" compensation; $\Delta\Delta S°$ is proportional to $\Delta\Delta H°$ in aqueous medium. Thus, relatively large $\Delta\Delta H°$ values are nearly cancelled by large $T\Delta\Delta S°$ values, resulting in small $\Delta\Delta G°$ values ($\Delta\Delta G° = \Delta\Delta H° - T\Delta\Delta S°$). Entropy-enthalpy compensation in aqueous solution is observed not only for DNA molecules (44), but also for proteins (46, 47), protein-DNA complexes (48), and drug-DNA complexes (49). A possible means for a polymerase to exploit large enthalpy differences between right and wrong base pairs is to suppress differences in entropy (44). The geometric properties of the polymerase active site might provide this entropy suppression. Further amplification of enthalpy differences between right and wrong base pairs might be achieved by partial exclusion of water from the polymerase active site (50). A combination of enthalpy amplification and entropy suppression could markedly amplify the free energy differential between correct and incorrect base pairs.

## Polymerase Insertion Errors

Insertion of mismatched bases by DNA polymerases occur infrequently. Typically, error frequencies in vitro occur at the $10^{-3}$ to $10^{-5}$ level, although substantial variation has been observed dependent on the DNA polymerase, method of assay, and specific site investigated. The most common mispairs generally involve G pairing with T, with observed frequencies in a range of $10^{-2}$ to $10^{-4}$, and the least common mispairs involve pyr·pyr mispairs at frequencies of about $10^{-4}$ to $10^{-5}$ or less (13, 16, 45, 51–56). The wide variations in error range for a given mismatch can be attributed to several factors. First, DNA polymerases differ in their base misinsertion specificities and intrinsic fidelity. Second, there are specific template-primer sites where base insertion errors are found to occur at anomalously high and low frequencies (hot and cold spots, respectively).

In an effort to focus on how DNA polymerase chooses the correct base, we have emphasized single base insertion errors in this review. It is important to note, however, that misalignment of the template-primer constitutes a major source of mutation in vivo and in vitro (see Kunkel & Bebenek (57) for an excellent review of recent work). Misalignment errors can yield either frameshift mutations (58–61) or substantial rearrangements (62). Kunkel and collaborators have recently provided an important new insight about misalignment errors—that transient misalignment can yield base substitution mutations (59, 63, 64) (Figure 1). Because misalignment errors are dependent on DNA sequence, a mutational hot spot may be generated (58, 65). The majority of hot and cold mutational sites, however, probably cannot be explained solely in terms of misalignment errors (52, 66, 67).

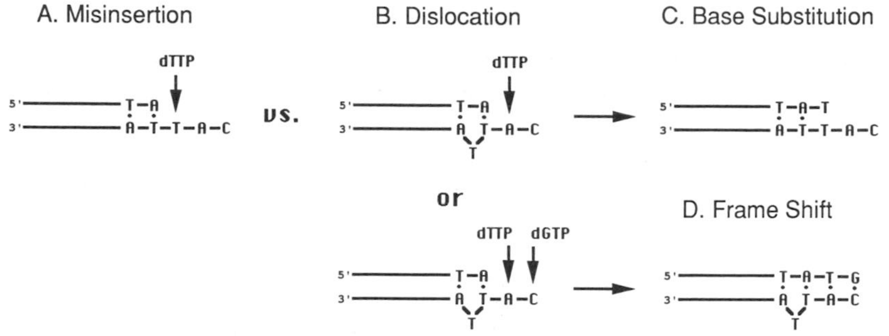

A. Misinsertion          B. Dislocation          C. Base Substitution

dTTP                     dTTP

5'————T–A        vs.    5'————T–A               5'————T–A–T
3'————A–T–T–A–C         3'————A T–A–C           3'————A–T–T–A–C
                                   T

or                       D. Frame Shift

dTTP dGTP

5'————T–A               5'————T–A–T–G
3'————A T–A–C           3'————A T–A–C
         T                       T

Figure 1   Mutagenic misalignment of the primer-template for DNA replication. Base substitution mutations are typically caused by mispairing of an incorrect dNTP opposite a template base (A). A transient misalignment ("dislocation") of the primer-template (B), however, can yield base substitution (C) or frame-shift mutations (D) even though the base insertion event involves a Watson-Crick pair.

From data obtained recently with *E. coli* strains defective in mismatch correction and editing, a rough estimate is possible of the frequency and type of base insertion error produced by pol III holoenzyme in vivo. As described in more detail below, the *mutD5 (dnaQ)* mutation confers a severe defect on the editing exonuclease ($\epsilon$ subunit) of pol III holoenzyme (68–70). Recent work has indicated strongly that *E. coli* carrying the *mutD5* mutation are also phenotypically defective in mismatch repair, probably because the mismatch repair pathway is saturated at high mutation rates (65, 71, 72). Because *mutD5* can provide an increase in mutation rate up to $10^5$ (65, 73, 74), the maximum base insertion error frequency by DNA pol III holoenzyme is probably in the $10^{-5}$ range ($10^{-5}$ out of $10^{-10}$ overall); this number is within an order of magnitude of that observed by the $\alpha$ polymerase subunit of pol III holoenzyme in vitro (51).

Of the base insertion mutations found in the *mutD5* strain, transition mutations predominate (90%) (65). These mutations would arise from G·T and A·C wobble pairings. Transversion mutations that would arise from. pyr·pyr mispairs are very rare. As described below, the class of transversion mutations arising from G·A mispairs would be relatively high, but these errors are prevented by a special editing system, mutT. Thus, G·T, A·C, and G·A mispairings are probably the most frequent allowed by the polymerase site of pol III in vivo. These misinsertion errors are also probably the most frequent in vitro for the $\alpha$-subunit of pol III holoenzyme (51).

## Fidelity Assays and Major Conclusions

Four main assays are presently used to analyze the fidelity of replication by purified polymerases in vitro or the process of genome replication in vivo.

MISINCORPORATION AND PROOFREADING IN VITRO MEASURED BY A
"STANDARD" POLYMERASE ASSAY    In the first misincorporation demonstration in 1962, Trautner et al showed with *E. coli* DNA polymerase I that G was misinserted opposite bromouracil on d(ABrU) but not opposite T on d(AT) (75). Subsequently, Hall & Lehman used a similar assay to show that a mutator mutant of T4 polymerase misincorporated T opposite C more frequently than wild type (76). The incorporation assay was generalized to include a measurement of deoxynucleotide turnover, the conversion of newly inserted nucleotides into acid-soluble dNMPs by polymerase-associated exonuclease (77, 78). In the "standard" polymerase assay, measurements are generally carried out with both matched and mismatched dNTPs present in the reaction simultaneously, so that the two can compete directly for insertion into DNA. The misinsertion ratio, $f_{ins}$, is defined as the ratio of wrong to right substrates incorporated into DNA, normalized to equimolar substrate concentrations. Fidelity is defined to be the reciprocal of the misinsertion ratio, $F = 1/f_{ins}$. The misinsertion ratio can be determined by tagging right and wrong dNTPs with different radioactive labels and measuring their relative incorporation into DNA and turnover as dNMPs, where nucleotide insertion is given by the sum of incorporation plus turnover (77–80); alternatively, competition for incorporation can be measured between an unlabeled and labeled substrate (10). The insertion efficiencies for mismatched nucleotides are generally between three to five orders of magnitude less than for correctly matched nucleotides. Thus, to detect incorporation of mismatched bases, it is frequently necessary to provide a substantial pool bias by increasing the relative concentration of mismatched to matched dNTP.

The measurement of insertion of radioactive deoxynucleotide substrates into DNA and subsequent proofreading of newly inserted substrates by exonuclease continues to serve as a standard "workhorse" assay to address a wide variety of important questions concerning fidelity of DNA synthesis. Fidelity measurements have been made for a large number of polymerases by using the "standard" assay to estimate average fidelity values and by applying more recent assays to determine fidelity at individual template sites (for previous reviews, see Refs. 45, 57, 81). Similar studies have been carried out to determine base mispairing properties of nucleotide analogues and the effects of mutagens and carcinogens on polymerase fidelity (11, 12, 15, 42, 75, 80, 82–102).

REVERSION, TRANSFECTION, AND SHUTTLE VECTOR SYSTEMS USING REPORTER GENES    An understanding of replication fidelity requires data on the mutation process in vivo. The importance of the DNA polymerase in providing fidelity was initially established by forward mutational and genetic reversion studies using a dispensable phage T4 gene (e.g. *rII* locus) as a marker (103–107). An assay measuring spontaneous or mutagen-induced reversion of nonsense mutations in the *lacI* gene of *E. coli* was introduced by J. Miller and coworkers to determine the mutational spectrum associated with genomic duplication in vivo (108, 109). Recently, a series of *lacZ* mutations were defined that allow rapid detection by reversion of all base substitution mutations (109a, 109b).

Another general protocol that has been used extensively involves expression of a "reporter gene" present on a plasmid or phage that has been replicated in vitro or in vivo. Mutant phenotypes are revealed by plaque or colony color or other morphologic properties, and the specific type of mutation is identified by direct sequencing of the reporter gene. *E. coli lacZ* and *lacI* genes are examples of commonly used reporter genes (110–112). The use of DNA sequencing methodology allows a determination of the complete spectrum of mutational alterations in a defined genetic region. Most importantly, replication errors other than base substitutions are detected.

Current applications of reporter-gene techniques use: (*a*) a variety of DNA polymerases in vitro to copy a template-primer containing a reporter sequence in which mutations are readily detected (see e.g. Refs. 113, 114); (*b*) a reporter gene present on a plasmid or phage for replication in vivo in bacteria with different genetic backgrounds with or without mutagens (61, 65, 112, 115–117); (*c*) shuttle vector systems in which DNA is replicated in animal cells in vivo and screened for mutations in *E. coli* (110, 118–122); (*d*) an animal cell chromosomal reporter gene (e.g. *aprt*) from which mutations are analyzed by cloning and sequencing in order to determine the mutational spectrum of the chromosomal replication system (123, 124).

Reversion of defined single-site bacteriophage markers was the initial assay for in vitro replication and in vivo assay (125–127); current versions measure forward mutation spectra, which are sensitive to virtually all base substitutions and frameshifts. Analysis of DNA sequence surrounding mutational hot spots led to models for base substitution and frameshift mutations involving short-range primer-template misalignment (59, 63, 128) and long-range DNA strand displacements (61–63, 113, 129). Ripley has written an excellent recent review on this topic (130).

Transfection assays of polymerase fidelity in vitro include contributions from both nucleotide insertion and editing steps. The insertion component of base substitution and frameshift mutational spectra can be measured using polymerases lacking in proofreading. However, these assays require that two kinetically slow steps take place prior to transfection: base misinsertion; and continued polymerization beyond the error, extending the mismatched primer terminus (44, 55, 131). Primer strands that are only elongated partially because of inefficient mismatch extension may be subject to error correction or strand-selective elimination following transfection. To avoid overestimating polymerase fidelities, careful controls are required to correct for strand loss (45, 132). In two recent experiments, genetic transfection and steady-state kinetic fidelity assays (described below) were compared at identical template-primer locations and found to agree (56, 64).

USE OF GEL ASSAYS: STEADY-STATE KINETIC STUDIES    Fidelity assays have made liberal use of DNA sequencing technology, especially polyacrylamide gel electrophoresis to resolve radioactively labelled primer molecules extended by polymerase, followed by autoradiography to visualize the extended primers. In a series of studies by Strauss and colleagues, DNA polymerases from a variety of sources were used to copy DNA containing different template lesions (98, 101, 102). Sites of termination of in vitro DNA synthesis can be defined precisely by comparing replication stops with DNA sequence using gel electrophoresis. An analysis of replication with DNA containing abasic sites yielded the conclusion that dATP is inserted preferentially at abasic lesions (15, 97–100). Using a similar technique, but excluding one of the dNTPs from the reaction mix, Beattie and coworkers were able to identify base misinsertions opposite template sites complementary to a single missing substrate (87, 133, 134). Toorchen & Topal used DNA sequencing gels to show that *E. coli* pol I incorporates $O^6$-methyl-dGTP opposite T and C template residues, with a greater than 20-fold preference for T (85). B. Singer and colleagues have used sequencing gels to investigate the kinetics of forming base mispairs for a variety of alkylated base analogues present either as dNTP substrate or incorporated at a specific template site (94–96).

Recently, Boosalis et al derived a method to convert a radioactive signal from a gel band into steady-state nucleotide insertion velocities at individual sites on natural or synthetic primer-template DNA (16). This gel kinetic assay has been used to analyze nucleotide insertion fidelity (15, 16, 51–56, 135). The assay has also been applied to study the efficiency of extending mismatched compared to correctly matched primer 3'-termini (44, 55, 131, 135). The gel assay provides a rapid method to measure fidelity at many template sites. Although fidelity can be estimated by competition between matched and mismatched dNTPs (16), a much more convenient approach is to measure kinetics of insertion for right and wrong dNTPs in separate reactions. The ratio $V_{max}/K_m$ measures the efficiency of nucleotide insertion by polymerase (136). At equimolar concentrations of right (r) and wrong (w) dNTP substrates, the ratio of insertion efficiencies for wrong versus right base pairs is equal to the misinsertion efficiency ($f_{ins}$):

$$f_{ins} = \frac{\dfrac{V_{maxw}}{K_{mw}}}{\dfrac{V_{maxr}}{K_{mr}}} = \frac{K_{mr}}{K_{mw}} \frac{V_{maxw}}{V_{maxr}} = f_k \cdot f_v \qquad \qquad 1.$$

The ratio of $K_m$ values for right and wrong substrates, $f_k = K_{mr}/K_{mw}$, is generally much less than 1. Much higher concentrations of an incorrectly paired dNTP are required to achieve half maximum velocity ($V_{max}/2$). The ratio, $f_v = V_{maxw}/V_{maxr}$, is generally less than 1, since $V_{max}$ is typically lower for incorrect than for correct nucleotides. If dNTP concentrations were not equimolar, then substrate concentrations would enter into the expression for relative insertion efficiency solely as the ratio [dWTP]/[dRTP], describing the effects of mass action under pool-biased conditions (136).

PRE-STEADY-STATE KINETICS    The recent introduction of pre-steady-state kinetic assays has allowed detailed insights into the reaction pathway for a DNA polymerase, and into mechanisms for base insertion and editing fidelity. In a series of papers, Benkovic, Johnson, and coworkers have outlined a model for the series of steps occurring during the replication of DNA by *E. coli* pol I Klenow fragment (7–9, 137–139). By a combination of rapid single-turnover and isotope-trapping experiments, the kinetic constants were measured for a single base insertion step. These results were combined with data on elemental effects of $\alpha$-thio dNTP (137, 138) to define the series of steps involved in polymerization and the points at which the polymerase can discriminate between correct and incorrect base pairs. The kinetic scheme is shown in Figure 2.

Important aspects of this model for fidelity include the conformational changes before and after the chemical step, and the occurrence of different rate-limiting steps for insertion of correct and incorrect nucleotides (step B for the correct nucleotide and step C for the incorrect nucleotide). As indicated in the diagram, the polymerase can discriminate between nucleotides at three basic stages. Stage 1 defines events before and during chemical bond formation. Stage 1 consists of steps A, B and C—triphosphate binding, an initial conformational change, and the formation of the chemical bond. Among these three steps, the marked discrimination between correct and incorrect dNTP is in step C, the formation of the chemical bond. There is approximately a 6000-fold difference between the polymerization rate for right versus wrong. Stage 2 is the pause time before pyrophosphate release. Stage 2 discrimination occurs after the chemical bond is formed, but before the pyrophosphate has been released from the enzyme. For the incorrect nucleotide, the rate of step D, the second conformation change, is postulated to be slower than the rate of exonuclease action. As a result, most of the incorrect product is excised by the editing exonuclease. Stage 3 is the pause time before subsequent polymerization. Stage 3 discrimination occurs after the pyrophosphate has been released but before the next correct nucleotide has been added. In a manner similar to stage 2, the competition between additional polymerization (step F) and

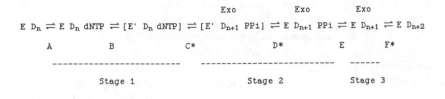

*Figure 2*   Kinetic pathway for DNA polymerase I (adapted from Refs. 137, 138). The individual steps and fidelity checkpoints are discussed in the text.

exonuclease leads to an additional preferential excision of the incorrect product.

## Error Avoidance in Base Insertion—General Conclusions

In general terms, we can divide base insertion specificity into two components. The most clearly defined is the free energy of base pairing between template base and incoming dNTP. This interaction provides for preferential occupancy of the active site by the correct dNTP. The energetic advantage for correct over incorrect dNTP will be very substantial for those potential mispairs that are incapable of effective hydrogen bonding. As noted above, however, differences in base-pairing free energies measured in aqueous solution cannot account for the extraordinary selectivity exhibited by DNA polymerases. We suggest that the second specificity component is a demand for the equivalent geometry (C1' distances and bond angles) of the Watson-Crick base pairs (23, 51, 140). This geometric mechanism can provide for the most spectacular feature of base selection—the ability of DNA polymerases to discriminate $10^3$- to $10^5$-fold against mispairs such as G·T that form efficient hydrogen bonds and can fit into standard B-type DNA. Geometric selection might operate by a direct requirement for the canonical C1' bond distances and angles and/or by the manifestation of this geometry in side-group configuration.

The geometric selection principle predicts that the most frequent misinsertions will involve base pairs that are closest to Watson-Crick geometry. For the polymerase subunit of pol III holoenzyme, this prediction appears to be fulfilled (51). The most frequent (though rare) mispairs are G·T, G·A, and C·A. These are base pairings that have been observed as constituents of B-DNA by X-ray crystallography and NMR studies with oligonucleotides (Figure 3). Each of these miscreant base pairs, however, has substantially different bond angles from those of the structurally identical Watson-Crick pairs. Thus, an efficient geometric selection process by the DNA polymerase will discriminate effectively against acceptance of even these incorrect base pairs that approximate Watson-Crick geometry most closely and fit into B-DNA. Other base pairs that deviate more from Watson-Crick geometry would be discriminated against more effectively.

For other DNA polymerases, the principle of geometric selection appears to work reasonably well with one major exception—A·A is a relatively frequent error for eukaryotic polymerase $\alpha$ (52). One possible explanation is based on the previous proposal that A insertion is a default mechanism at a "noninformational" lesion such as an abasic site (97–99). The enzyme must insert something or replication fails. The rare insertion of A opposite A might be an aberrant application of a default function carried by certain DNA polymerases. DNA polymerase III of *E. coli* depends on SOS-induced proteins to bypass replication-blocking DNA lesions, and so this enzyme might not need a self-contained default function.

As discussed later, the concept of geometric recognition is also useful in considering the response of a DNA polymerase to a damaged base in the template. Replication is inhibited by DNA lesions that distort the normal template configuration even if capacity for base pair recognition is retained (e.g. a cyclobutane pyrimidine dimer). Presumably the polymerase rejects even the Watson-Crick pair in this instance because of noncanonical geometry.

In the base insertion reaction by a DNA polymerase, there are two obvious check points for the correct geometric alignment: the initial binding of dNTP, and the chemical step of phosphodiester bond formation. The detailed kinetic analysis of pol I has introduced a third potential check point—a kinetically slow step after binding, but preceding the chemical step (8, 138). A very recent study with phage T7 DNA polymerase has revealed a similar stage in the polymerization reaction (141). This "extra" step, attributed to a conformation change of the enzyme, might involve postbinding selection for the correct geometry by an induced-fit mechanism. Alternatively, the slow step might be the time required for alignment of the bound triphosphate into the appropriate

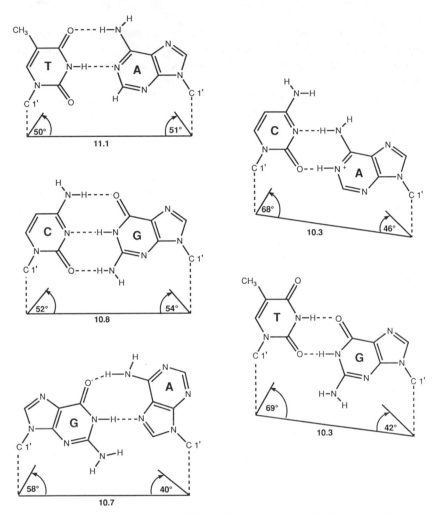

*Figure 3*  Geometric characteristics of Watson-Crick and of mismatched base pairs. The figure is based on X-ray crystallography of duplex B-DNA oligonucleotides (reproduced from Ref. 253). The striking geometric identity of the Watson-Crick pairs is not matched by the A·C and G·T wobble pairs or by the G(*anti*)·A(*syn*) pair.

transition state geometry. For pol I, the major selectivity for base insertion fidelity occurs at the chemical step (138). In contrast, based on measurements from pre-steady-state kinetics, there is a very large contribution to fidelity from the dNTP binding stage for the phage T7 polymerase (141). As inferred from large $K_m$ differences in steady-state kinetics, binding differences between correct and incorrect dNTP may be major contributors to the fidelity of eukaryotic polymerase $\alpha$ (16), *E. coli* pol III (51), and T4 polymerase (11, 142).

From the available data, there are clearly differences in the extent to which different polymerases use different check points. However, we believe that the extraordinary base insertion fidelity of DNA polymerases in general derives from the sequential application of each check point to provide an exquisite sensitivity to Watson-Crick geometry at the transition state for phosphodiester bond formation.

# EDITING BY DNA POLYMERASES

## *Exonucleolytic Editing by Prokaryotic Polymerases*

The concept of exonucleolytic editing was introduced by Brutlag & Kornberg, who showed that the 3'-exonuclease of DNA pol I removed incorrectly paired terminal nucleotides more rapidly than correctly paired nucleotides (143). These experiments also defined the "melting capacity" of the 3' terminus as the likely mechanism for the editing exonuclease. The nuclease worked best with single-strand DNA, and the activity with double-strand DNA increased markedly with temperature (143).

A role for editing in vivo was defined clearly by experiments with wild-type and mutant T4 polymerases by Bessman and coworkers. Certain "mutator" T4 polymerases were found to have a lower exonuclease to polymerase activity in vitro (77, 80, 144), correlated with a higher mutation rate in vivo (103, 104, 144). "Antimutator" polymerases exhibited a higher exonuclease to polymerase activity (77, 80, 144), correlated with a lower mutation rate in vivo (106, 107, 144). Experiments by Nossal and colleagues with mutant T4 polymerases established a role for base insertion specificity by the polymerase and a functional separation of editing and polymerization (78, 79, 145–147).

A critical role for exonucleolytic editing in replication fidelity has also been established by biochemical studies with wild-type and mutant DNA pol III of *E. coli*—the enzyme responsible for genome duplication (148, 149). The strongest mutator type mutations of *E. coli* have been located in the *dnaQ* gene (originally termed *mutD*) (70, 73, 74, 150). The mutant pol III from a strong mutator *mutD5* was found to have a severely reduced 3'→5' exonuclease activity (68, 69). In addition, a temperature-sensitive mutator mutation exhibited thermally sensitive exonuclease activity in vitro (68). In contrast to pol I of *E. coli* and T4 polymerase, the polymerase and exonuclease activities of pol III holoenzyme reside on different subunits. The *dnaQ* gene encodes the $\epsilon$ exonuclease subunit (70); the *dnaE* gene encodes the $\alpha$ polymerase subunit (151).

In principle, the *dnaQ* mutants allow an assessment of the importance of editing in vivo. However, the $10^4$- to $10^5$- fold increase in mutation rate observed for these mutator defects is not all attributable to an editing failure. Under the maximal mutator condition in rich medium, the mismatch repair system becomes ineffective (65, 71). Apparently, the mismatch repair capacity becomes saturated at high mutation rates (65, 72). The situation is further complicated by the fact that the editing mutants are not completely defective in exonuclease in vitro (68, 69). The best current guess for the contribution of exonucleolytic editing in vivo is probably in the $10^2$ to $10^3$ range (65). This number also correlates with estimates in vitro for pol III (13). A similar estimate can be inferred for the editing contribution for T4 polymerase (152–154).

One important property of exonucleolytic editing is the substantial cost of the process in terms of excision of correctly inserted nucleotides. This aspect can be measured in vitro during DNA synthesis by the "turnover" of correct dNTP into dNMP by base insertion, followed by exonucleolytic removal. For *E. coli* pol III and T4 polymerase, about 5–15% of the correctly inserted bases are removed (11, 13). These data indicate that the process of editing could not be improved substantially for these enzymes without severe problems in the rate of replication and replenishment of dNTP substrates. Indeed, antimutator mutants of T4 phage exhibit a decrease in phage production compared to wild-type T4 (154a).

## Editing in Eukaryotes

At least five distinct DNA polymerases are present in eukaryotic cells: $\alpha$, $\beta$, $\gamma$, $\delta$, and $\epsilon$ (155, 156). Recent data using SV40 as a model system have focused special attention on DNA polymerases $\delta$ and $\alpha$; pol $\delta$ and pol $\alpha$ are believed to carry out leading- and lagging-strand replication, respectively (157). Pol $\beta$ is thought to be involved in DNA repair, either short-patch repair of damaged templates or postreplication excision of single base mismatches (158, 159). Pol $\epsilon$ (formerly pol $\delta_2$) has been implicated in DNA repair of UV lesions (160). DNA polymerase $\gamma$ is specific for mitochondrial replication (161, 162).

For bacterial and bacteriophage replication, the biological role of exonucleolytic editing is clear because mutations in proofreading exonucleases have been correlated with mutator and antimutator phenotypes (68, 69, 77, 163); there is, as yet, no direct evidence in eukaryotes that proofreading occurs in vivo. However, highly purified forms of pol $\delta$ (164) and pol $\epsilon$ (165, 166) have been shown to contain associated 3'→5' exonuclease activity. Exonuclease activity is typically found for animal virus polymerases (156) and mitochondrial pol $\gamma$ (167, 168). Pol $\beta$ appears to lack an exonuclease (169).

Cellular pol $\alpha$ is believed to be required for lagging-strand synthesis based on its association with RNA primase activity and properties in the SV40 replication reaction (170, 171). Whether or not pol $\alpha$ has an associated 3' exonuclease activity remains an open question. Several groups have demonstrated comigration of exonuclease and polymerase $\alpha$ activities during purification (172–175). However, in the absence of detailed subunit analysis and reconstitution experiments of the type used for *E. coli* pol III (148, 149), the possibility remains that the presence of associated exonuclease may be adventitious. In the case of *Drosophila* pol $\alpha$, a "cryptic" exonuclease activity was observed following dissociation of the polymerase-primase holoenzyme complex (176). Teresa Wang and colleagues have cloned the pol $\alpha$ gene from a human KB cell line (177). Although the highly purified form of this enzyme has no detectable exonuclease activity, the pol $\alpha$ sequence may contain a region with similarity to the 3' exonuclease domain for prokaryotic cellular and phage polymerases (178). As an alternative to editing by pol $\alpha$, Perrino & Loeb have made the interesting suggestion that pol $\delta$ might edit both leading and lagging strands if replication is carried out by an asymmetric dimer of pol $\alpha$ and pol $\delta$ (179).

## Kinetic Amplification of Editing: Inefficient Extension from a Mismatched 3'-Terminus

An extremely important component of editing is the competition between chain extension from a misinserted base and exonucleolytic excision. This point was initially inferred from the increase in base misincorporation in vitro in the presence of high concentrations of the correct following dNTP (11, 126). Very rapid chain elongation could lower editing efficiency. More recently, DNA polymerases have been found to be capable of a direct kinetic contribution to editing discrimination by their reluctance to extend a mismatched 3'-terminus (44, 55, 131, 135, 138). As noted earlier, work with pol I has clearly established the contribution to exonucleolytic editing of this kinetic lag in mismatch extension (138).

The kinetics of mismatch extension can be studied most easily with polymerases lacking exonuclease proofreading. Eukaryotic $\alpha$ polymerase (44, 55, 131) and viral reverse transcriptases (54, 131, 135) appear to catalyze extension of G·T and T·G mismatches most easily, with efficiencies in the range of $10^{-1}$ to $10^{-3}$, compared to extension of a correct pair. On the other end of the scale, mismatches involving purines in all combinations appear to be extended least efficiently, in the range of $10^{-4}$ to $10^{-5}$ compared to the correct pair (55, 131, 135). If the efficiency for a polymerase to extend a mismatch is compared to the efficiency to form the same mismatch, the ratio of misextension/misinsertion for eukaryotic pol $\alpha$ fluctuates around 1. However, for avian myeloblastosis reverse transcriptase (AMV RT), this ratio is much greater than 1 for almost all mispairs, and is approximately 100 for extension of G(primer)·T mismatches (131). Indeed, AMV RT extends G·T mispairs at a remarkably high rate, roughly 10% the rate of extending G·C base pairs. Because reverse transcriptases lack an editing exonuclease, an ability to extend mismatches is presumably necessary for effective, if error-prone, genome duplication.

Assays that measure nucleotide misinsertion efficiencies are also used to measure mismatch extension efficiencies (16, 52, 131, 138). However, although misinsertion and mismatch extension are both nucleotide insertion reactions, there are important distinctions between the two. For measurements of base insertion fidelity, DNA polymerase binds to a correctly matched primer 3' terminus, at which wrong and right dNTP substrates compete for insertion. For the extension reaction, the enzyme binds to either a matched or mismatched primer terminus. Direct competition between two different primer termini for binding to polymerase means that polymerase-DNA dissociation constants for matched and mismatched termini appear as explicit terms in the expression for mismatch extension efficiency, $f_{ext}$ (131). The enzyme-DNA binding constants cancel from the ratio describing insertion fidelity (cf. Equation 1) because the primer terminus is the same for right and wrong dNTP substrates.

Why are mispairs extended poorly compared to correct pairs? One possibility is that polymerases may have an inherent difficulty binding to mismatched termini. An alternative possibility is that polymerases bound to mismatched termini have an intrinsic difficulty adding a next nucleotide. To distinguish between these possibilities, it is necessary to compare polymerase-DNA dissociation constants for the two types of primer termini. Although there is currently a paucity of binding data, there have been recent measurements on binding of all naturally occurring matched and mismatched template-primer termini using AMV RT (131; S. Creighton, M. M. Huang, N. Arnheim, M. F. Goodman, unpublished work). The results suggest that the AMV enzyme binds equally well to correct and incorrect primer termini. A similar conclusion has been derived for pol I of *E. coli* (138). Thus for a polymerase bound to a mismatched terminus, there appears to be a significant kinetic barrier to DNA chain extension.

## Error Avoidance in Editing: General Conclusions

In principle, there are two plausible mechanisms for specificity in exonucleolytic editing: geometric recognition (as in the base insertion model); and melting capacity of the terminus (143, 179a). In a pure melting mechanism, the exonuclease is intrinsically a single-strand nuclease that preferentially removes misinserted bases because the mismatch at the 3' end is more often in a single-strand configuration. The general features of the melting mechanism are solidly supported by data for pol I and pol III of *E.*

*coli* and T4 DNA polymerase. Specificity in editing also depends on events in the polymerase active site, however. The exonucleases are outstanding editors because the addition of the next base after a mismatch is slow, giving the exonuclease more time to work. The constraint on base addition from a mismatch presumably reflects the operation of the geometric recognition mechanism governing base insertion. Thus, the exceptional specificity of the editing exonuclease derives from the combined geometric demands of the polymerase site and the single-strand recognition by the exonuclease.

Recent work has established an important structural principle about polymerase and exonuclease function—the two activities reside in separate domains or subunits of the DNA polymerase enzyme. For pol III of *E. coli*, the exonuclease is a distinct subunit ($\epsilon$) from the polymerase ($\alpha$) subunit (70, 163, 180). For pol I, the exonuclease domain is separated by some 30 Å from the polymerase domain (181–183). Based on these studies with pol I and pol III, the cooperation between polymerase and exonuclease is likely to involve communication between spatially separate active sites.

Structural studies with pol I have revealed that the exonuclease domain is indeed a single-strand nuclease; cocrystals have been obtained with single-strand DNA in the active site (184). Remarkably, the pol I exonuclease site must function solely as a single-strand exonuclease because there is no room for double-strand DNA. Thus pol I represents an unexpected type of editing specificity because the exonuclease site makes no direct selection for a mispaired base.

To explain editing by pol I, Steitz, Joyce, and their colleagues have proposed a "melt and slide" model (183). A rapid equilibrium is proposed between two forms of the DNA substrate—duplex DNA in the polymerase site and single-strand DNA in the exonuclease site (Figure 4). For a mismatched 3' terminus, the equilibrium might be expected to favor single-strand DNA in the exonuclease site. This pathway seems unlikely to provide much specificity, however, because four bases must be unwound in the melt step to accommodate the structural relationship between the two sites. The major determinants of editing specificity for pol I appear to be kinetic constraints delaying chain elongation from a misinserted base. The kinetic analysis of the pol I reaction has indicated that nearly all of the editing specificity resides in two steps that allow the exonuclease more time to work: delayed pyrophosphate release after a misinsertion; and delayed base addition to a mismatched 3' terminus (138).

The functional analysis of the exonuclease of pol III has revealed a somewhat different picture. For pol III, the editing subunit can be isolated free from the polymerase subunit and studied as a separate enzyme (163). The $\epsilon$-subunit exerts direct selection for a mispaired base on its own (163, 184a). This intrinsic selectivity appears to be based on a "classical" melting mech-

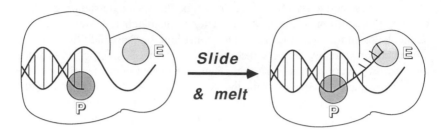

*Figure 4*   Exonucleolytic editing by DNA polymerase I. The model is derived from X-ray crystallography of the enzyme (adapted from Ref. 183). Duplex primer-template DNA occupies the polymerase site (P) for the base insertion reaction; for the editing reaction, the duplex DNA unwinds and single-strand primer DNA is transferred to the exonuclease site (E).

anism. The $\epsilon$ exonuclease prefers single-strand DNA as a substrate; with a double-strand DNA, the exonuclease activity is strongly temperature dependent and is greater at an A·T terminus than a G·C (184a). Thus, for pol III, there is probably a more equal division of labor between intrinsic exonuclease discrimination and the kinetic amplification provided by delayed chain elongation from the mismatched base. However, the editing function operates by the same fundamental mechanism as pol I—single-strand exonuclease.

To summarize our current picture, exonucleolytic editing proceeds by a single-strand exonuclease designed to accept preferentially the misinserted base by a melting mechanism. The very efficient selectivity of editing depends markedly on the ability of the polymerase site to resist further base addition after a misinsertion, providing a kinetic amplification to the selectivity of editing.

## Possible Additional Editing Mechanisms: the MutT Pathway

Are there editing mechanisms that operate during replication in addition to the $3' \rightarrow 5'$ editing exonuclease? Recent experiments strongly suggest the existence of at least one special mechanism—the MutT function. The *MutT* gene was initially characterized by mutator mutations with a highly specific phenotype: an extremely high mutation rate for a single class of transversion mutations, A·T→C·G (185). Recent studies have indicated that the MutT protein acts during DNA replication to prevent formation of G:A mispairs (115, 186); thus loss of MutT activity will yield abundant A·T→C·G transversions.

The initial biochemical studies defined the G·A correction activity in crude extracts in which several possible mechanisms might be involved (115). More recent experiments have demonstrated that misinsertion of G opposite A was prevented by MutT in concert with purified pol III holoenzyme (186). Moreover, the G·A mispairs were not efficiently removed by the editing exonuclease of pol III in the absence of MutT (186). Another study has revealed an intriguing property for the purified MutT protein, a dGTPase activity with a very high $K_m$ for the substrate (187). One possible explanation of the existing data is that MutT normally cooperates with pol III to excise dGMP misinserted opposite A or eliminate the substrate dGTP mispaired with A. That is, MutT provides a specific editing activity to prevent G·A mispairs. However, MutT might work in other ways. For example, MutT might act without pol III to attack preferentially a configuration of dGTP especially prone to make G·A mispairs (e.g. *Gsyn*).

The existence of a special mechanism to prevent G·A mispairs is a surprising aspect of replication fidelity in view of exonucleolytic and mismatch correction. However, G·A mispairs are poorly corrected by the methyl-directed mismatch system in vitro and in vivo (1, 188). Moreover, G·A mispairs appear to be ineffectively edited by the editing exonuclease of pol III. In base misinsertion assays, the polymerase subunit of pol III inserts G opposite A as a relatively frequent error (51, 186). Thus a special correction system is needed to avoid the error catastrophe defined by the MutT mutator phenotype. The inefficiency of the normal error correction pathways for G·A mispairs might derive from the close approximation of this structure to the Watson-Crick pair (23, 51) (Figure 3); alternatively, the formation of *syn-anti* base pairs might provoke special problems (188) (Figure 3).

There is also a special postreplicative mismatch repair reaction (MutY) that executes the complementary correction to MutT—excision of A from a G·A pair (189, 190). This correction mechanism is not a component of the strand-specific methyl-directed mismatch repair system. The reaction presumably improves fidelity in vivo because the misinsertion of G opposite template A is prevented by MutT; thus, the remaining G·A mismatches mainly arise from A misinserted opposite template G. The complementary functions of MutT and MutY provide a fascinating example of the sophistication and importance of error-correction pathways.

# MUTATIONS GENERATED BY DAMAGE TO DNA

## Response of DNA Polymerase to DNA Lesions

DNA polymerase must have extraordinary accuracy to accomplish the precise duplication of the genome. In addition, a variety of repair mechanisms are needed to remove DNA lesions ahead of the replication fork and to correct mistakes left behind by the replication machinery. Mechanisms must also exist to deal with the potentially catastrophic encounter of a DNA polymerase with a replication-blocking lesion. In this section, we consider replication errors that arise as a consequence of damage to DNA, with an emphasis on the bacterial SOS response induced by a replication-blocking lesion.

Mutagenic lesions in DNA fall into two general (but overlapping) classes: those that inhibit DNA replication (replication-blocking); and those that allow replication to proceed with diminished fidelity (error-promoting). The introduction of replication-blocking lesions into the DNA of *E. coli* induces the SOS response; as a consequence DNA replication is restored, and the lesions are converted functionally into error-promoting sites. Error-promoting lesions typically increase mispairing of bases during DNA replication (e.g. $O^6$-methyl guanine and 2-aminopurine). Replication-blocking lesions typically introduce substantial distortion into the DNA (e.g. pyrimidine dimer).

ERROR-PROMOTING MUTAGENIC EVENTS    A common class of error-promoting mutagens consists of analogues of normal nucleotides that exhibit a propensity for mispairing. For example, thymine analogues such as 5-bromouracil and 5-fluorouracil form normal Watson-Crick base pairs with A, but also mispair with G more often than the normal T (36, 37, 86, 88). The adenine analogue 2-aminopurine pairs with T but can mispair with C (34, 42, 80, 191). These mispairs occur with the analogue present either as triphosphate substrates for insertion by polymerase or as template bases, stimulating both $A \cdot T \rightarrow G \cdot C$ and $G \cdot C \rightarrow A \cdot T$ transitions (192).

A variety of chemicals modify normal DNA bases and alter base-pairing specificities. $N$-methyl-$N'$-nitro-$N$-nitrosoguanidine (MNNG) and $N$-methyl-$N$-nitrosourea alkylate normal DNA bases and are toxic and mutagenic to cells. When G is alkylated at the 6 position on the purine ring, the $G \cdot T$ mispair is stimulated strongly (85, 94). In a similar manner, $O^4$-MeT exhibits an increased probability to form mispairs with G (96). There exist well-characterized constitutive and inducible enzyme systems to rid the cells of alkylation damage (193).

REPLICATION-BLOCKING MUTAGENIC EVENTS    As the term implies, certain DNA lesions effectively inhibit DNA replication. The damage site is presumably sufficiently distorted that the geometric recognition mechanism does not allow effective base insertion. In the extreme, the template base may be completely absent as in depurination events. In addition, at a highly distorted site even a correctly inserted base will look like a mismatch to the editing exonuclease. For certain cases, such as cyclobutane pyrimidine dimers, replication probably does not stop because the damaged base is incapable of base pairing, but because the distorted base pair is unacceptable to the enzyme (194). The capacity of pyrimidine dimers and abasic sites to block DNA replication in vitro has been firmly established (98, 195). Other experiments with *E. coli* have established the same point in vivo; an especially clear demonstration has been provided by recent work with single-strand M13 viral DNA containing single *cys-syn* cyclobutane dimers or abasic sites (196, 197). Some in vitro experiments with random pyrimidine dimers appear to indicate a limited capacity of pol III to bypass the DNA lesion (198); however, all experiments in vivo and in vitro concur that pyrimidine dimers are effective in blocking elongation of DNA chains in the absence of intervention by the SOS response.

## The SOS Response of Escherichia coli

The reaction of *E. coli* to a replication-blocking lesion is an induced multi-gene response termed SOS (199–202). The induced genes code for a set of proteins that are biochemically disparate, but physiologically related to provide the needed salvational response. Many of the induced proteins have obvious rescue functions—excision repair, recombination, and inhibition of cell division (allowing time for replication of a complete chromosome before division). Of less obvious benefit, the induced UmuC and UmuD proteins specify a replication pathway that markedly elevates the mutation rate (201, 203–205). Most of the mutations appear to be targeted to the sites of DNA damage (206–208); in addition, there are untargeted mutagenic events in undamaged DNA (209).

Associated with the SOS response is a recovery in the capacity to replicate DNA (210, 211). This renewed DNA replication is probably partitioned between two pathways: a minor pathway of "translesion" replication that is responsible for targeted mutations (140, 212); and a major "replication-restart" pathway that provides for an error-free bypass of the DNA lesions (202, 210, 211) (Figure 5). As noted below, both pathways require the RecA

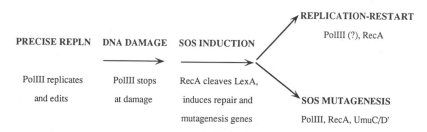

SOS REPLICATION

PRECISE REPLN    DNA DAMAGE    SOS INDUCTION

REPLICATION-RESTART
PolIII (?), RecA

PolIII replicates    PolIII stops    RecA cleaves LexA,
and edits            at damage       induces repair and        SOS MUTAGENESIS
                                     mutagenesis genes          PolIII, RecA, UmuC/D'

*Figure 5*   The SOS response and the recovery of DNA replication. DNA polymerase III stops at a structurally distorting lesion in the DNA. Associated with the induced SOS response, replication resumes by two general pathways: the mutagenic "translesion" pathway; and the error-free "replication-restart" pathway that somehow bypasses the DNA lesion.

protein, but translesion replication depends in addition on UmuC and UmuD.

The DNA polymerase used for the replicative rescue pathways of SOS is probably pol III, the enzyme responsible for normal chromosomal replication, but this point has not been completely established. Genetic and physiological experiments with pol III mutants have argued strongly for an essential role for pol III in SOS-induced mutagenesis (213, 214). However, pol II activity is markedly induced by SOS (215), and pol II polymerase is encoded by the SOS-induced *dinA* gene (216, 217). Although perhaps not active in translesion replication (based on the defined role for pol III), pol II appears likely to have an important function in the SOS response, perhaps in DNA repair events. Pol I has also been implicated in the SOS response by the observation of an altered form of the enzyme termed pol I* (218). Pol I* has lowered fidelity, as judged by in vitro assays (218). UV-induced mutagenesis, however, appears to be normal in *polA⁻* mutants (219), including those in a recent deletion study (220). Because the pathways of replicative repair are multiple and complex, both pol I* and pol II may have special functions.

## Multiple Roles of RecA in the SOS Response

The RecA protein is the central mediator of the SOS response, providing for several salvational functions that are distinct from its classical role in homologous genetic recombination. RecA has been extensively characterized as a recombination protein. RecA forms a nucleoprotein filament on single-strand DNA in the presence of a mononucleotide cofactor (ATP, dATP, or ATP-γ-S). In an ATP-consuming reaction, RecA then searches out homology in duplex DNA and catalyzes the branch migration reaction that generates heteroduplex DNA (221). All of the SOS activities depend on the ability of RecA to bind to single-strand or duplex DNA, but the functional diversity is astounding.

RecA has two regulatory roles in the SOS response and probably three distinct "direct" biochemical activities. The regulatory aspects are clearly defined at a molecular level: cleavage of the LexA repressor for SOS genes to turn on the induced response; and cleavage of the UmuD protein to activate the mutagenic pathway. The direct activities are not understood in any detail. RecA appears to facilitate both the translesion replication and the replication-restart pathways that bypass the DNA lesions. In addition, RecA is required for another SOS-induced mutagenic pathway responsible for large duplications (222); this special recombination function is not considered further here. We summarize briefly the regulatory and replicative activities of RecA; these aspects of the SOS response are described in more detail in other reviews (201, 202, 223).

REGULATORY ROLES OF RECA    The first SOS function of RecA to be identified is proteolytic cleavage of λ cI protein, allowing phage λ to escape a potentially doomed cell (224, 225). The critical role for regulating induction of bacterial proteins is cleavage of the LexA repressor of SOS genes (223, 226). The cleavage reaction as initially defined requires single-strand DNA (polynucleotide cofactor) and a mononucleotide cofactor (223, 224). More recent experiments have indicated that association of RecA with damaged duplex DNA might constitute a second pathway to SOS induction. Purified RecA associates much more efficiently with UV-irradiated duplex DNA than with non-irradiated duplex DNA (227, 228). The product of this reaction is a nucleoprotein filament covering a long stretch of DNA (228). Presumably the DNA lesion serves as a "micro single-stranded" region that provides a nucleation site for RecA to initiate cooperative binding on duplex DNA (227, 228). RecA also mediates cleavage of LexA after association with duplex DNA (227, 229).

Based on the mechanisms defined in vitro, RecA is probably activated for its regulatory role in vivo by association with a polynucleotide cofactor, either single-strand or damaged duplex DNA. The cleavage reaction also requires an unidentified mononucleotide cofactor; this molecule might be ATP, dATP, or a tight-binding cofactor analogous to ATP-γ-S, which works best in vitro. Full SOS induction probably requires a blocked replication fork as well as damaged DNA (230, 231). The replication block might serve to generate downstream single-strand DNA; alternatively (or additionally), a highly effective mononucleotide cofactor activity might be produced as a consequence of an "idling" reaction by the polymerase (e.g. higher concentration of dATP or a tight-binding cofactor).

In addition to RecA, SOS mutagenesis requires the UmuC and UmuD proteins, synthesis of which is induced by cleavage of LexA (201). Recent work has revealed another regulatory role of RecA in the SOS response— cleavage of UmuD to a processed product, UmuD', which is probably the active form for mutagenesis. Sequence homology between LexA and UmuD near the LexA cleavage site had been noted (232). The purification of UmuD allowed a direct test; RecA carries out cleavage of UmuD (233). UmuD is also cleaved in vivo (234); cleavage occurs at the site expected from homology with LexA (235). The biological importance of this cleavage reaction has been demonstrated by the engineered synthesis of UmuD' in bacterial cells. Production of UmuD' in vivo is sufficient for SOS mutagenesis in bacteria unable to cleave UmuD because of a mutation in RecA (236). Thus, RecA-mediated cleavage of UmuD to UmuD' is required for mutation by the UmuCD pathway.

Although RecA mediates cleavage of λ cI, LexA, and UmuD, RecA is not itself a protease. Each of the three proteins exhibits a self-cleavage reaction at alkaline pH (233, 237, 238). The single scissile peptide bond in each protein probably undergoes hydrolysis by a mechanism similar to that of serine proteases (238). Thus all three proteins carry a homologous "self-destruct" domain, for which the reactivity is greatly accelerated by RecA.

REPLICATIVE ROLES OF RECA    As noted above, the SOS-response is triggered by a replication-blocking lesion. The recovery of the capacity to replicate depends on RecA, whereas normal chromosomal duplication does not. There appear to be at least two replication recovery pathways active at the lesion site, operationally termed "translesion replication" and "replication restart." The translesion pathway is responsible for targeted mutagenesis, presumably through replication across the lesion site. This mutagenic pathway requires UmuC and UmuD' (201, 202, 236). The replication-restart pathway is the major route of replication recovery, as judged by measurements of DNA synthesis with radioactive precursors (210, 211). This pathway was initially termed induced replisome reactivation (IRR) (210); we have suggested replication-restart, however, as a more direct operational definition (202, 235). The term "pathway" is used loosely because more than one biochemical mechanism may be involved. Replication restart is normally independent of UmuC and UmuD', which are required for SOS mutagenesis; therefore this pathway presumably involves an error-free bypass of the DNA lesion.

Both translesion replication and replication-restart appear to require the direct participation of RecA protein. This point is complicated by the regulatory activities of RecA protein. A variety of recent genetic approaches, however, have argued strongly that RecA is a direct participant because the regulatory activities can be bypassed by the genetic inactivation of LexA and the production of engineered UmuD' protein (236, 239, 240).

The biochemical mechanisms involved in translesion replication and replication-restart have not been established. There are enough biochemical inferences however, to define a plausible mechanism for the mutagenic pathway (Figure 6). UmuC and UmuD' have been purified and shown to form a tight complex (235). The direct participation of RecA can be most easily understood by an action at the lesion site (140, 227–229). Therefore, translesion replication seems likely to depend on a damage-localized nucleoprotein complex involving RecA, UmuC-UmuD', and pol III—a "mutasome" (235). The mutasome presumably provides for a reduced fidelity mode of DNA replication, in which base insertion can occur opposite and beyond the lesion, often producing a mutation.

The presence of RecA at the lesion site might help in two ways: alter the distorted template conformation to one more palatable for pol III; and inhibit

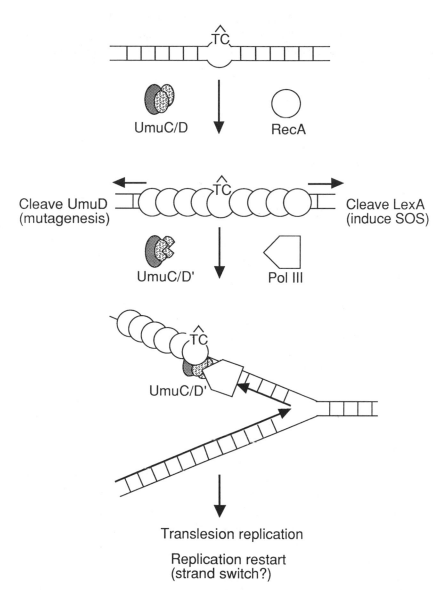

*Figure 6*   Possible mechanisms involved in translesion replication and restart. A multiprotein complex is assembled at the DNA lesion, including UmuC, UmuD', RecA, and DNA polymerase III holoenzyme. This nucleoprotein aggregate provides for replication across the lesion with the introduction of mutations. A possible alternative pathway for error-free replicative bypass is DNA synthesis with the daughter strand as template in a three-stranded complex generated by RecA (reproduced from Ref. 235).

the editing exonuclease (14, 227, 229). UmuC-UmuD' might facilitate DNA chain elongation by keeping the polymerase engaged in bypass attempts (either preventing dissociation or favoring reassociation) (229, 241–243). The assembly of the mutasome might begin with a RecA nucleoprotein filament at the site of the lesion (Figure 6). UmuD will be cleaved to UmuD' either at this site or another lesion site, and the UmuC-UmuD' complex will be available to facilitate a productive encounter of pol III with the site of DNA damage. UmuC-UmuD' might be brought to the mutasome by association with RecA (240, 244), or by interaction with pol III (or both). The UmuC protein is highly basic, a property likely to facilitate association of the UmuC-UmuD' complex with DNA. An alternative pathway for assembly of the same nucleoprotein structure would be created by RecA binding to single-strand DNA generated downstream from the lesion after the polymerase stalls (225, 231) (bottom of Figure 6).

The mechanism of the replication-restart reaction is more difficult to anticipate at present. There may well be more than one route to replicative bypass. One possible mechanism involving the known biochemistry of RecA is evident from considering the nucleoprotein assembly at the bottom of Figure 6. The RecA-coated single-strand DNA downstream from the blocked polymerase will presumably associate with the replicated duplex from the complementary strand because this interaction is equivalent to the initial step in formation of heteroduplex DNA in general recombination. In the resultant three-stranded structure, pol III might bypass the lesion by copying the undamaged daughter strand instead of the blocked parental strand, switching back after clearing the lesion (140, 235). UmuC-UmuD' would not normally be required for this reaction. However, there are two mutant RecA proteins for which replication restart depends on UmuC-UmuD' (211, 244a). These observations indicate that translesion replication and replication-restart are related pathways. One set of experiments that argue for a replicative bypass mechanism of this type has been carried out with phage λ (245) (λ uses the *E. coli* pol III replication system). After infection with λ DNA carrying several pyrimidine dimers, efficient semiconservative replication was observed without evidence of multiple recombination exchanges. The proposed replication-restart pathway would augment and might sometimes initiate the previously defined recombinational repair pathway (246).

This proposed role of RecA in replication-restart is different from the classical model of Rupp & Howard-Flanders (247). These authors found that newly synthesized DNA in cells containing pyrimidine dimers was mainly of low molecular weight, as judged by sucrose gradient centrifugation under alkaline conditions. These experiments indicated that DNA synthesis might reinitiate downstream from the dimer, leaving a "daughter-strand gap" that was repaired by RecA-mediated recombination (246, 247). The proposed mechanism makes sense in terms of the available data, but does not readily explain the later finding that recovery of most DNA synthesis at a blocking lesion appears to depend on RecA (not just repair) (210, 211). The most likely conclusion at present seems to be that two replication-restart mechanisms can operate at a blocked replication fork: a downstream priming mechanism needed for the Rupp & Howard-Flanders model; and a RecA-mediated replicative bypass. More experiments in vivo and in vitro are required, however, to define clearly the nature of the replication-restart pathway.

## Some Biological Comments

SOS mutagenesis is remarkable as a biological as well as a biochemical phenomenon. The mutation pathway depends on two regulatory events, SOS-induced synthesis of UmuC and UmuD and processing of UmuD to its active form, UmuD'. Thus the introduction of mutations is not a passive consequence of the encounter of DNA polymerase III with a damaged template base, but a highly regulated and biochemically complex pathway. From a biological perspective, the enhanced frequency of point mutations associated with SOS induction can be considered from two (not necessarily exclusive) points of view. The UmuCD mutagenic pathway is an important component of an induced cellular survival mechanism, and mutations are a "side-effect"; or, the UmuCD pathway exists as a population survival mechanism that enhances genetic variation, increasing the chances for a fortunate variant adapted to survive the environmental stress (140, 248, 249). The UmuCD pathway contributes essentially all of the point mutations and also contributes to survival, but to a small extent compared to the error-free mechanisms of excision repair, recombinational repair, and replication-restart. The genetic variation hypothesis suggests that SOS mutagenesis might be expected to include other types of mutations, such as duplications and rearrangements, which might have profound phenotypic effects. The recent finding of an SOS-inducible pathway for large tandem duplications demonstrates that SOS produces a broad mutagenic response (222). In this sense, SOS mutagenesis might be considered an example of "inducible evolution" (140, 249).

In other recent experiments of notable evolutionary impact, clear evidence has been presented for mechanisms that accelerate the appearance of adaptively favorable mutations under conditions of extreme starvation (250–252). This work has led to the fascinating suggestion that the favorable mutations might somehow be directed to the appropriate gene (250). More recently, Hall has noted the appearance of additional nonselected mutations in the bacteria carrying the salvational mutation (252); this work indicates that a sophisticated combination of an induced increase in the general mutation rate and selection is involved. Whatever the detailed mechanisms, the starvation mutagenesis experiments clearly define an inducible mutational response. How the fidelity of DNA replication is regulated promises to be a fascinating topic for the next few years.

ACKNOWLEDGMENTS

We thank Stephen Benkovic, Steven Creighton, Thomas Kunkel, Marie Petit, John Petruska, and Marcie Rosenberg for reading the manuscript and offering valuable suggestions. We also thank Richard Eisner for editorial assistance and Olga Kennard, Catherine Joyce, and Thomas Steitz for figures. Research and writing support for the authors has been provided in part by National Institutes of Health Grants GM21422, GM42554, and CA41655.

*Literature Cited*

1. Modrich, P. 1989. *J. Biol. Chem.* 264:6597–600
2. Radman, M. 1988. *Genetic Recombination,* ed. R. Kucherlapati, G. R. Smith, pp. 169–92. Washington, DC: Am. Soc. Microbiol.
3. Meselson, M. 1988. *Recombination of the Genetic Material,* ed. K. B. Low, pp. 91–113. New York: Academic
4. Hopfield, J. J. 1974. *Proc. Natl. Acad. Sci. USA* 71:4135–39
5. Galas, D. J., Branscomb, E. W. 1978. *J. Mol. Biol.* 124:653–87
6. Goodman, M. F., Branscomb, E. W. 1986. *Accuracy in Molecular Processes: Its Control and Relevance to Living Systems,* ed. T. B. L. Kirkwood, R. F. Rosenberger, D. J. Galas, pp. 191–232. New York: Chapman & Hall
7. Bryant, F. R., Johnson, K. A., Benkovic, S. J. 1983. *Biochemistry* 22:3537–46
8. Kuchta, R. D., Mizzrahi, V., Benkovic, P. A., Johnson, K. A., Benkovic, S. J. 1987. *Biochemistry* 26:8410–17
9. Mizrahi, V., Henrie, R. N., Marlier, J. F., Johnson, K. A., Benkovic, S. J. 1987. *Biochemistry* 24:4010–18
10. Travaglini, E. C., Mildvan, A. S., Loeb, L. A. 1975. *J. Biol. Chem.* 250:8647–56
11. Clayton, L. K., Goodman, M. F., Branscomb, E. W., Galas, D. J. 1979. *J. Biol. Chem.* 254:1902–12
12. Watanabe, S. M., Goodman, M. F. 1982. *Proc. Natl. Acad. Sci. USA* 79:6429–33
13. Fersht, A. R., Knill-Jones, J. W., Tsui, W.-C. 1982. *J. Mol. Biol.* 156:37–51
14. Fersht, A. R., Knill-Jones, J. W. 1983. *J. Mol. Biol.* 165:669–82
15. Randall, S. K., Eritja, R., Kaplan, B. E., Petruska, J., Goodman, M. F. 1987. *J. Biol. Chem.* 262:6864–70
16. Boosalis, M. S., Petruska, J., Goodman, M. F. 1987. *J. Biol. Chem.* 262:14689–96
17. Watson, J. D., Crick, F. H. C. 1953. *Nature* 171:964–67
18. Watson, J. D., Crick, F. H. C. 1953. *Cold Spring Harbor Symp. Quant. Biol.* 18:123–31
19. Topal, M. D., Fresco, J. R. 1976. *Nature* 263:285–89
20. Frese, E. 1959. *J. Mol. Biol.* 1:87–105
21. Hunter, W. N., Brown, T., Anand, N. N., Kennard, O. 1986. *Nature* 320:552–55
22. Hunter, W. N., Kneale, G., Brown, T., Rabinovich, D., Kennard, O. 1986. *J. Mol. Biol.* 190:605–18
23. Brown, T., Hunter, W. N., Kneale, G., Kennard, O. 1986. *Proc. Natl. Acad. Sci. USA* 83:2402–6
24. Brown, T., Kneale, G., Hunter, W. N., Kennard, O. 1986. *Nucleic Acids Res.* 14:1801–909
25. Brown, T., Leonard, G. A., Booth, E. D., Chambers, J. 1989. *J. Mol. Biol.* 207:445–57
26. Prive, G. G., Heinemann, U., Chandrasegaran, S., Kan, L.-S., Kopka, M. L., et al. 1987. *Science* 238:498–504
27. Patel, D. J., Kozlowski, S. A., Marky, L. A., Broka, C., Rice, J. A., et al. 1982. *Biochemistry* 21:428–36
28. Patel, D. J., Kozlowski, S. A., Marky, L. A., Rice, J. A., Broka, C., et al. 1982. *Biochemistry* 21:437–44
29. Kan, L.-S., Chandrasegaran, S., Pulford, S. M., Miller, P. S. 1983. *Proc. Natl. Acad. Sci. USA* 79:6429–33
30. Patel, D. J., Ikuta, S., Kozlowski, S., Itakura, K. 1983. *Proc. Natl. Acad. Sci. USA* 80:2184–88
31. Patel, D. J., Kozlowski, S. A., Ikuta,

S., Itakura, K. 1984. *Fed. Proc., Fed. Am. Soc. Exp. Biol.* 43:2663–70
32. Patel, D. J., Kozlowski, S. A., Ikuta, S., Itakura, K. 1984. *Biochemistry* 23:3218–26
33. Sowers, L. C., Fazakerley, G. V., Kim, H., Dalton, L., Goodman, M. F. 1986. *Biochemistry* 25:3983–88
34. Sowers, L. C., Fazakerley, G. V., Eritja, R., Kaplan, B. E., Goodman, M. F. 1986. *Proc. Natl. Acad. Sci. USA* 83:5434–38
35. Fazakerley, G. V., Sowers, L. C., Eritja, R., Kaplan, B. E., Goodman, M. F. 1987. *Biochemistry* 26:5641–46
36. Sowers, L. C., Eritja, R., Kaplan, B., Goodman, M. F., Fazakerley, G. V. 1988. *J. Biol. Chem.* 263:14794–801
37. Sowers, L. C., Goodman, M. F., Eritja, R., Kaplan, B., Fazakerley, G. V. 1989. *J. Mol. Biol.* 205:437–47
38. Cuniasse, P., Sowers, L. C., Eritja, R., Kaplan, B., Goodman, M. F., et al. 1987. *Nucleic Acids Res.* 15:8003–22
39. Kalnik, M. W., Chang, C. N., Grollman, A. P., Patel, D. J. 1988. *Biochemistry* 27:924–31
40. Cuniasse, P., Fazakerley, G. V., Guschlbauer, W., Kaplan, B. E., Sowers, L. C. 1990. *J. Mol. Biol.* 213:303–14
41. Eritja, R., Kaplan, B. E., Mhaskar, D., Sowers, L. C., Petruska, J., et al. 1986. *Nucleic Acids Res.* 14:5869–84
42. Mhaskar, D. N., Goodman, M. F. 1984. *J. Mol. Chem.* 259:11713–17
43. Eritja, R., Horowitz, D. M., Walker, P. A., Ziehler-Martin, J. P., Boosalis, M. S., et al. 1986. *Nucleic Acids Res.* 14:8135–53
44. Petruska, J., Goodman, M. F., Boosalis, M. S., Sowers, L. C., Cheong, C., et al. 1988. *Proc. Natl. Acad. Sci. USA* 85:6252–56
45. Loeb, L. A., Kunkel, T. A. 1982. *Annu. Rev. Biochem.* 52:429–57
46. Lumry, R., Rajenders, S. 1970. *Biopolymers* 9:1125–227
47. Creighton, T. E. 1984. *Proteins, Structure and Molecular Properties.* New York: Freeman
48. Ha, J.-H., Spolar, R. S., Record, T. Jr. 1989. *J. Mol. Biol.* 209:801–16
49. Breslauer, K. J., Remeta, D. P., Chou, W.-Y., Ferrante, R., Curry, J., et al. 1987. *Proc. Natl. Acad. Sci. USA* 84:8922–26
50. Petruska, J., Sowers, L. C., Goodman, M. F. 1986. *Proc. Natl. Acad. Sci. USA* 83:1559–62
51. Sloane, D. L., Goodman, M. F., Echols, H. 1988. *Nucleic Acids Res.* 16:6465–75
52. Mendelman, L. V., Boosalis, M. S.,

Petruska, J., Goodman, M. F. 1989. *J. Biol. Chem.* 264:14415–23
53. Ricchetti, M., Buc, H. 1990. *EMBO J.* 9:1583–93
54. Preston, B. D., Poiesz, B. J., Loeb, L. A. 1988. *Science* 242:1168–71
55. Perrino, F. W., Loeb, L. A. 1989. *J. Biol. Chem.* 264:2898–905
56. Bebenek, K., Joyce, C. M., Fitzgerald, M. P., Kunkel, T. A. 1990. *J. Biol. Chem.* 265:13878–87
57. Kunkel, T. A., Bebenek, K. 1988. *Biochim. Biophys. Acta* 951:1–15
58. Streisinger, G., Okada, Y., Emrich, J., Newton, J., Tsugita, A., et al. 1966. *Cold Spring Harbor Symp. Quant. Biol.* 31:77–84
59. Kunkel, T. A. 1986. *J. Biol. Chem.* 261:13581–87
60. Fowler, R. G., Schaaper, R. M., Glickman, B. W. 1986. *J. Bacteriol.* 167:130–37
61. deBoer, J. G., Ripley, L. S. 1984. *Proc. Natl. Acad. Sci. USA* 81:5528–31
62. Papanicolaou, C., Ripley, L. S. 1989. *J. Mol. Biol.* 207:335–53
63. Kunkel, T. A., Soni, A. 1988. *J. Biol. Chem.* 263:14784–89
64. Boosalis, M. S., Mosbaugh, D. W., Hamatake, R., Sugino, A., Kunkel, T. A., et al. 1989. *J. Biol. Chem.* 264:11360–66
65. Schaaper, R. M. 1988. *Proc. Natl. Acad. Sci. USA* 85:8126–30
66. Ronen, A., Rahat, A. 1976. *Mutat. Res.* 34:21–34
67. Leong, P.-M., Hsia, H. C., Miller, J. H. 1986. *J. Bacteriol.* 168:412–16
68. Echols, H., Lu, C., Burgers, P. M. J. 1983. *Proc. Natl. Acad. Sci. USA* 80:2189–93
69. DiFrancesco, R., Bhatnagar, S. K., Brown, A., Bessman, M. J. 1984. *J. Biol. Chem.* 259:5567–73
70. Scheuermann, R., Tam, S., Burgers, P. M. J., Lu, C., Echols, H. 1983. *Proc. Natl. Acad. Sci. USA* 80:7085–89
71. Schaaper, R. M. 1989. *Genetics* 121:205–12
72. Schaaper, R. M., Radman, M. 1989. *EMBO J.* 8:3511–16
73. Degnen, G. E., Cox, E. C. 1974. *J. Bacteriol.* 117:477–87
74. Cox, E. C., Horner, D. L. 1982. *Genetics* 100:7–18
75. Trautner, T. A., Swartz, M. N., Kornberg, A. 1962. *Proc. Natl. Acad. Sci. USA* 48:449–55
76. Hall, Z. W., Lehman, I. R. 1968. *J. Mol. Biol.* 36:321–33
77. Muzyczka, N., Poland, R. L., Bessman, M. J. 1972. *J. Biol. Chem.* 247:7116–22

78. Hershfield, M. S., Nossal, N. G. 1972. *J. Biol. Chem.* 247:3393–404
79. Hershfield, M. S. 1973. *J. Biol. Chem.* 248:1417–23
80. Bessman, M. J., Muzyczka, N., Goodman, M. F., Schnaar, R. L. 1974. *J. Mol. Biol.* 88:409–21
81. Fry, M., Loeb, L. A. 1986. *Animal Cell DNA Polymerases.* Boca Raton, Fla: CRC Press
82. Engel, J. D., von Hippel, P. H. 1978. *J. Biol. Chem.* 253:927–34
83. Engel, J. D., von Hippel, P. H. 1978. *J. Biol. Chem.* 253:935–39
84. Topal, M. D., Hutchison, C. A. III, Baker, M. S. 1982. *Nature* 298:863–65
85. Toorchen, D., Topal, M. D. 1983. *Carcinogenesis* 4:1591–97
86. Lasken, R. S., Goodman, M. F. 1984. *J. Biol. Chem.* 259:11491–95
87. Hillebrand, G. G., Beattie, K. L. 1985. *J. Biol. Chem.* 260:3116–25
88. Lasken, R. S., Goodman, M. F. 1985. *Proc. Natl. Acad. Sci. USA* 82:1301–5
89. Singer, B., Kusmierek, J. T., Fraenkel-Conrat, H. 1983. *Proc. Natl. Acad. Sci. USA* 80:969–72
90. Singer, B., Chavez, F., Spengler, S. J. 1986. *Biochemistry* 25:1201–5
91. Preston, B. D., Singer, B., Loeb, L. A. 1986. *Proc. Natl. Acad. Sci. USA* 83:8501–5
92. Huff, A. C., Topal, M. D. 1987. *J. Biol. Chem.* 262:12843–50
93. Singer, B., Spengler, S. J., Fraenkel-Conrat, H., Kusmierek, J. T. 1986. *Proc. Natl. Acad. Sci. USA* 83:28–32
94. Singer, B., Chavez, F., Goodman, M. F., Essigmann, J. M., Dosanjh, M. K. 1989. *Proc. Natl. Acad. Sci. USA* 86:8271–74
95. Singer, B., Chavez, F., Spengler, S. J., Kusmierek, J. T., Mendelman, L., et al. 1989. *Biochemistry* 28:1478–83
96. Dosanjh, M. K., Essigmann, J. M., Goodman, M. F., Singer, B. 1990. *Biochemistry* 29:4698–703
97. Boiteux, S., Laval, J. 1982. *Biochemistry* 21:6746–51
98. Sagher, D., Strauss, B. 1983. *Biochemistry* 22:4518–26
99. Schaaper, R. M., Kunkel, T. A., Loeb, L. A. 1983. *Proc. Natl. Acad. Sci. USA* 80:487–91
100. Takeshita, M., Chang, C.-N., Johnson, F., Will, S., Grollman, A. P. 1987. *J. Biol. Chem.* 262:10171–81
101. Moore, P. D., Strauss, B. 1979. *Nature* 278:664–66
102. Moore, P. D., Rabkin, S. D., Osborn, A. L., King, C. M., Strauss, B. S. 1982. *Proc. Natl. Acad. Sci. USA* 79:7166–70
103. Speyer, J. F. 1965. *Biochem. Biophys. Res. Commun.* 21:6–8
104. Speyer, J. F., Karam, J. D., Lenny, A. B. 1966. *Cold Spring Harbor Symp. Quant. Biol.* 31:693–97
105. Freese, E. B., Freese, E. F. 1967. *Proc. Natl. Acad. Sci. USA* 57:650–57
106. Drake, J. W., Allen, E. F. 1968. *Cold Spring Harbor Symp. Quant. Biol.* 33:339–44
107. Drake, J. W., Allen, E. F., Forsberg, S. A., Preparata, R., Greening, E. O. 1969. *Nature* 221:1128–31
108. Coulondre, C., Miller, J. H. 1977. *J. Mol. Biol.* 117:577–606
109. Coulondre, C., Miller, J. H., Farabaugh, P. J., Gilbert, W. 1979. *Nature* 274:775–80
109a. Cupples, C. G., Miller, J. H. 1989. *Proc. Natl. Acad. Sci. USA* 86:5345–49
109b. Cupples, C. G., Cabrera, M., Cruz, C., Miller, J. H. 1990. *Genetics* 125:275–80
110. Calos, M. P., Lebkowski, J. S., Botchan, M. B. 1983. *Proc. Natl. Acad. Sci. USA* 80:3015–19
111. Kunkel, T. A. 1984. *Proc. Natl. Acad. Sci. USA* 81:1494–98
112. Schaaper, R. M., Dunn, R. 1987. *Proc. Natl. Acad. Sci. USA* 84:6220–24
113. Kunkel, T. A. 1985. *J. Biol. Chem.* 260:5787–96
114. Conrad, M., Topal, M. D. 1986. *J. Biol. Chem.* 261:16226–32
115. Schaaper, R. M., Dunn, R. L. 1987. *J. Biol. Chem.* 262:16267–70
116. Wood, R. D., Skopek, T. R., Hutchinson, F. 1984. *J. Mol. Biol.* 173:273–91
117. Glickman, B. W., Schaaper, R. M., Haseltine, W. A., Dunn, R. L., Brash, D. E. 1986. *Proc. Natl. Acad. Sci. USA* 83:6945–49
118. Ashman, C. R., Davidson, R. L. 1984. *Mol. Cell. Biol.* 4:2266–72
119. Proti-Sabljic, M. N., Tuteja, N., Munson, P. J., Hauser, J., Kraemer, K. H., et al. 1986. *Mol. Cell. Biol.* 6:3349–56
120. Glazer, P. M., Sarkar, S. N., Summers, W. C. 1986. *Proc. Natl. Acad. Sci. USA* 83:1041–44
121. Drinkwater, N. R., Klinedinst, D. K. 1986. *Proc. Natl. Acad. Sci. USA* 83:3402–6
122. DeMarini, D. M., Brockman, H. E., de Serres, F. J., Evans, H. H., Stankowski, L. F. Jr., et al. 1989. *Mutat. Res.* 220:11–29
123. Phear, G., Nalbantoglu, J., Meuth, M. 1987. *Proc. Natl. Acad. Sci. USA* 84:4450–54
124. Tindall, K. R., Stankowski, L. F. Jr. 1989. *Mutat. Res.* 220:241–53

125. Weymouth, L. A., Loeb, L. A. 1978. *Proc. Natl. Acad. Sci. USA* 75:1924–28
126. Fersht, A. R. 1979. *Proc. Natl. Acad. Sci. USA* 76:4946–50
127. Hibner, U., Alberts, B. M. 1980. *Nature* 285:300–5
128. Kunkel, T. A. 1985. *J. Biol. Chem.* 260:12866–74
129. Ripley, L. S., Papanicolaou, C. 1988. *DNA Replication and Mutagenesis*, ed. R. E. Moses, W. C. Summers, pp. 227–36. Washington, DC: Am. Soc. Microbiol.
130. Ripley, L. S. 1990. *Annu. Rev. Genet.* 24:189–213
131. Mendelman, L. V., Petruska, J., Goodman, M. F. 1990. *J. Biol. Chem.* 265:2338–46
132. Kunkel, T. A., Alexander, P. S. 1986. *J. Biol. Chem.* 261:160–66
133. Hillebrand, G. G., McCluskey, A. H., Abbott, K. A., Revich, G. G., Beattie, K. L. 1984. *Nucleic Acids Res.* 12:3155–71
134. Hillebrand, G. G., Beattie, K. L. 1984. *Nucleic Acids Res.* 12:3173–83
135. Perrino, F. W., Preston, B. D., Sandell, L. L., Loeb, L. A. 1989. *Proc. Natl. Acad. Sci. USA* 86:8343–47
136. Fersht, A. 1985. *Enzyme Structure and Mechanism*, p. 361. New York: Freeman
137. Mizrahi, V., Henrie, R. N., Marlier, J. F., Johnson, K. A., Benkovic, S. J. 1985. *Biochemistry* 24:4010–18
138. Kuchta, R. D., Benkovic, P., Benkovic, S. J. 1988. *Biochemistry* 27:6716–25
139. Mizrahi, V., Benkovic, S. J. 1988. *Adv. Enzymol.* 61:437–57
140. Echols, H. 1982. *Biochimie* 64:571–75
141. Wong, I., Patel, S., Johnson, K. A. 1990. *Biochemistry.* In press
142. Goodman, M. F. 1988. *Mutat. Res.* 200:11–20
143. Brutlag, D., Kornberg, A. 1972. *J. Biol. Chem.* 247:241–48
144. Reha-Krantz, L. J., Bessman, M. J. 1977. *J. Mol. Biol.* 116:99–113
145. Nossal, N. G. 1969. *J. Biol. Chem.* 244:218–22
146. Nossal, N. G., Hershfield, M. S. 1971. *J. Biol. Chem.* 246:5414–26
147. Gillen, F. D., Nossal, N. G. 1976. *J. Biol. Chem.* 251:5219–24
148. Kornberg, A. 1988. *J. Biol. Chem.* 263:1–4
149. McHenry, C. S. 1988. *Annu. Rev. Biochem.* 57:519–50
150. Maruyama, M., Horiuchi, T., Maki, H., Sekiguchi, M. 1983. *J. Mol. Biol.* 167:757–71
151. Welch, M. W., McHenry, C. S. 1982. *J. Bacteriol.* 152:351–56
152. Sinha, N. K., Haimes, M. D. 1981. *J. Biol. Chem.* 256:10671–83
153. Sinha, N. K., Goodman, M. F. 1983. *Bacteriophage T4*, ed. C. K. Mathews, E. M. Kutter, G. Mosig, P. B. Berget, pp. 131–37. Washington, DC: Am. Soc. Microbiol.
154. Kunkel, T. A., Loeb, L. A., Goodman, M. F. 1984. *J. Biol. Chem.* 259:1539–45
154a. Allen, E. F., Albrecht, I., Drake, J. W. 1970. *Genetics* 65:187–200
155. Bambara, R. A., Jessee, C. B. 1990. *Biochim. Biophys. Acta.* In press
156. Wang, T. S.-F. 1991. *Annu. Rev. Biochem.* 60:513–52
157. Kelly, T. 1988. *J. Biol. Chem.* 263:17889–92
158. Hammond, R. A., McClung, J. K., Miller, M. R. 1990. *Biochemistry* 29:286–91
159. Wiebauer, K., Jiricny, J. 1990. *Proc. Natl. Acad. Sci. USA* 87:5842–45
160. Nishida, C., Reinhard, P., Linn, S. 1988. *J. Biol. Chem.* 263:501–10
161. Bolden, A., Noy, G. P., Weissbach, A. 1977. *J. Biol. Chem.* 252:3351–56
162. Hubscher, U., Kuenzle, C. C., Spardari, S. 1977. *Eur. J. Biochem.* 81:249–58
163. Scheuermann, R. H., Echols, H. 1984. *Proc. Natl. Acad. Sci. USA* 81:7747–51
164. Byrnes, J. J., Downey, K. M., Black, V. L., So, A. G. 1976. *Biochemistry* 15:2817–23
165. Crute, J. J., Wahl, A. F., Bambara, R. A. 1986. *Biochemistry* 25:26–36
166. Syväoja, J., Linn, S. 1989. *J. Biol. Chem.* 264:2489–97
167. Insdorf, N. F., Bogenhagen, D. F. 1989 *J. Biol. Chem.* 264:21498–503
168. Kunkel, T. A., Mosbaugh, D. W. 1989. *Biochemistry* 28:988–95
169. Abbotts, J., SenGupta, D. N., Zmudzka, B. Z., Widen, S. G., Notario, V., et al. 1988. *Biochemistry* 27:901–9
170. Lehman, I. R., Kaguni, L. S. 1989. *J. Biol. Chem.* 264:4265–68
171. Challberg, M. D., Kelly, T. J. 1989. *Annu. Rev. Biochem.* 58:671–718
172. Chen, Y.-C., Bohn, E. W., Planck, S. R., Wilson, S. H. 1979. *J. Biol. Chem.* 254:11678–87
173. Kaguni, L. S., Rossignol, J.-M., Conway, R. C., Lehman, I. R. 1983. *Proc. Natl. Acad. Sci. USA* 80:2221–25
174. Ottiger, H. P., Hubscher, U. 1984. *Proc. Natl. Acad. Sci. USA* 81:3993–97
175. Skarnes, W., Bonin, P., Baril, E. 1986. *J. Biol. Chem.* 261:6629–36
176. Cotterill, S. M., Reyland, M. E., Loeb, L. A., Lehman, I. R. 1987. *Proc. Natl. Acad. Sci. USA* 84:5635–39

177. Wong, S. W., Wahl, A. F., Yuan, P.-M., Ari, N., Pearson, B. E., et al. 1990. *EMBO J.* 7:37–47
178. Bernad, A., Blanco, L., Lazaro, J. M., Martin, G., Salas, M. 1989. *Cell* 59:219–28
179. Perrino, F. W., Loeb, L. A. 1990. *Biochemistry* 29:5226–31
179a. Bessman, M. J., Reha-Krantz, L. J. 1977. *J. Mol. Biol.* 116:115–23
180. Maki, H., Kornberg, A. 1985. *J. Biol. Chem.* 260:12987–92
181. Ollis, D. L., Brick, P., Hamlin, R., Xuong, N. G., Steitz, T. A. 1985. *Nature* 313:762–66
182. Derbyshire, V., Freemont, P. S., Sanderson, M. R., Beese, L., Friedman, J. M., et al. 1988. *Science* 240:199–201
183. Joyce, C. M., Friedman, J. M., Beese, L., Freemont, P. S., Steitz, T. A. 1988. See Ref. 129, pp. 220–26
184. Steitz, T. A., Beese, L., Freemont, P. S., Friedman, J. M., Sanderson, M. R. 1987. *Cold Spring Harbor Symp. Quant. Biol.* 52:465–72
184a. Brenowitz, S., Kwack, S., Goodman, M. F., O'Donnell, M., Echols, H. 1991. *J. Biol. Chem.* In press
185. Cox, E. C. 1973. *Genetics* 73 (Suppl.):67–80
186. Akiyama, M., Maki, H., Sekiguchi, M., Horiuchi, T. 1989. *Proc. Natl. Acad. Sci. USA* 86:3949–52
187. Bhatnagar, S. K., Bessman, M. J. 1988. *J. Biol. Chem.* 263:8953–57
188. Schaaper, R. M., Bond, B. I., Fowler, R. G. 1989. *Mol. Gen. Genet.* 219:256–62
189. Au, K. G., Cabrera, M., Miller, J. H., Modrich, P. 1988. *Proc. Natl. Acad. Sci. USA* 85:9163–66
190. Lu, A.-L., Chang, D.-Y. 1988. *Cell* 54:805–12
191. Watanabe, S. M., Goodman, M. F. 1981. *Proc. Natl. Acad. Sci. USA* 78:2864–68
192. Freese, E. 1959. *Proc. Natl. Acad. Sci. USA* 45:622–33
193. Lindahl, T., Sedgwick, B., Sekiguchi, M., Nakabeppu, Y. 1988. *Annu. Rev. Biochem.* 57:133–57
194. Lawrence, C. W., Banerjee, S. K., Borden, A., LeClerc, J. E. 1990. *Mol. Gen. Genet.* 222:166–68
195. Moore, P. D., Bose, K. K., Rabkin, S. D., Strauss, B. S. 1981. *Proc. Natl. Acad. Sci. USA* 78:110–14
196. Banerjee, S. K., Christensen, R. B., Lawrence, C. W., LeClerc, J. E. 1988. *Proc. Natl. Acad. Sci. USA* 85:8141–45
197. Lawrence, C. W., Borden, A., Banerjee, S. K., LeClerc, J. E. 1990. *Nucleic Acids Res.* 18:2153–57
198. Livneh, Z. 1986. *Proc. Natl. Acad. Sci. USA* 83:4599–603
199. Radman, M. 1975. *Molecular Mechanisms for Repair of DNA, Part A*, ed. P. Hanawalt, R. B. Setlow, pp. 355–67. New York: Plenum
200. Witkin, E. M. 1976. *Bacteriol. Rev.* 40:869–907
201. Walker, G. C. 1984. *Microbiol. Rev.* 48:60–93
202. Echols, H., Goodman, M. F. 1990. *Mutat. Res.* 236:301–11
203. Kato, T., Shinoura, Y. 1977. *Mol. Gen. Genet.* 156:121–31
204. Steinborn, G. 1978. *Mol. Gen. Genet.* 165:87–93
205. Bagg, A., Kenyon, C. J., Walker, G. C. 1981. *Proc. Natl. Acad. Sci. USA* 78:5749–53
206. Miller, J. H. 1982. *Cell* 31:5–7
207. Wood, R. D., Hutchinson, F. 1984. *J. Mol. Biol.* 173:293–305
208. Miller, J. H. 1985. *J. Mol. Biol.* 182:45–68
209. Caillet-Fauquet, P., Maenhaut-Michel, G., Radman, M. 1984. *EMBO J.* 3:707–12
210. Khidhir, M. A., Casaregola, S., Holland, I. B. 1985. *Mol. Gen. Genet.* 199:133–40
211. Witkin, E. M., Roegner-Maniscalco, V., Sweasy, J. B., McCall, J. O. 1987. *Proc. Natl. Acad. Sci. USA* 84:6805–9
212. Villani, G., Boiteux, S., Radman, M. 1978. *Proc. Natl. Acad. Sci. USA* 75:3037–41
213. Bridges, B. A., Mottershead, R. P., Sedgwick, S. G. 1976. *Mol. Gen. Genet.* 144:53–58
214. Hagensee, M. E., Timme, T., Bryan, S. K., Moses, R. E. 1987. *Proc. Natl. Acad. Sci. USA* 84:4195–99
215. Bonner, C. A., Randall, S. K., Rayssiguier, C., Radman, M., Eritja, R., et al. 1988. *J. Biol. Chem.* 263:18946–52
216. Bonner, C. A., Hays, S., McEntee, K., Goodman, M. F. 1990. *Proc. Natl. Acad. Sci. USA* 87:7663–67
217. Iwasaki, H., Nakata, A., Walker, G. C., Shinagawa, H. 1990. *J. Bacteriol.* 172:6268–73
218. Lackey, D., Krauss, S. W., Linn, S. 1982. *Proc. Natl. Acad. Sci. USA* 79:330–34
219. Witkin, E. M. 1970. *Nature New Biol.* 229:81–82
220. Bates, H., Randall, S. K., Rayssiguier, C., Bridges, B. A., Goodman, M. F., et al. 1989. *J. Bacteriol.* 171:2480–84
221. Cox, M. M., Lehman, I. R. 1987. *Annu. Rev. Biochem.* 56:229–62

222. Dimpfl, J., Echols, H. 1989. *Genetics* 123:255–60
223. Little, J. W., Mount, D. W. 1982. *Cell* 29:11–22
224. Craig, N. L., Roberts, J. W. 1980. *Nature* 283:26–30
225. Roberts, J. W., Devoret, R. 1983. *Lambda II*, ed. R. Hendrix, J. Roberts, F. Stahl, R. Weisberg, pp. 123–44. Cold Spring Harbor, NY: Cold Spring Harbor Lab.
226. Little, J. W., Mount, D. W., Yanisch-Perron, C. R. 1981. *Proc. Natl. Acad. Sci. USA* 78:4199–203
227. Lu, C., Scheuermann, R. H., Echols, H. 1986. *Proc. Natl. Acad. Sci. USA* 83:619–23
228. Rosenberg, M., Echols, H. 1990. *J. Biol. Chem.* 265:20641–45
229. Lu, C., Echols, H. 1987. *J. Mol. Biol.* 196:497–504
230. Ennis, D. G., Fisher, B., Edmiston, S., Mount, D. W. 1985. *Proc. Natl. Acad. Sci. USA* 82:3325–29
231. Sassanfar, M., Roberts, J. W. 1990. *J. Mol. Biol.* 212:79–96
232. Perry, K. L., Elledge, S. J., Mitchell, B. B., Marsh, L., Walker, G. C. 1985. *Proc. Natl. Acad. Sci. USA* 82:4331–35
233. Burckhardt, S. E., Woodgate, R., Scheuermann, R. H., Echols, H. 1988. *Proc. Natl. Acad. Sci. USA* 85:1811–15
234. Shinagawa, H., Iwasaki, H., Kato, T., Nakata, A. 1988. *Proc. Natl. Acad. Sci. USA* 85:1806–10
235. Woodgate, R., Rajagopalan, M., Lu, C., Echols, H. 1989. *Proc. Natl. Acad. Sci. USA* 86:7301–5
236. Nohmi, T., Battista, J. R., Dodson, L. A., Walker, G. C. 1988. *Proc. Natl. Acad. Sci. USA* 85:1816–20

237. Little, J. W. 1984. *Proc. Natl. Acad. Sci. USA* 81:1375–79
238. Slilaty, S. N., Little, J. W. 1987. *Proc. Natl. Acad. Sci. USA* 84:3987–91
239. Dutriex, M., Moreau, P. L., Bailone, A., Galibert, F., Battista, J. R., et al. 1989. *J. Bacteriol.* 171:2415–23
240. Sweasy, J. B., Witkin, E. M., Sinha, N., Roegner-Maniscalco, V. 1990. *J. Bacteriol.* 172:3030–36
241. Bridges, B. A., Woodgate, R. 1985. *Mutat. Res.* 150:133–39
242. Hevroni, D., Livneh, Z. 1988. *Proc. Natl. Acad. Sci. USA* 85:5046–50
243. Battista, J. R., Nohmi, T., Dodson, L. A., Walker, G. W. 1988. *UCLA Symp. Mol. Cell. Biol.: Mechanisms and Consequences of DNA Damage Processing*, ed. E. C. Friedberg, P. C. Hanawalt, 83:455–59. New York: Liss
244. Freitag, N., McEntee, K. 1989. *Proc. Natl. Acad. Sci. USA* 86:8363–67
244a. Sweasy, J. B., Witkin, E. M. 1991. *Biochimie.* In press
245. Defais, M., Lesca, C., Monsarrat, B., Hanawalt, P. 1989. *J. Bacteriol.* 171:4938–44
246. Rupp, W. D., Wilde, C. E., Reno, D. L., Howard-Flanders, P. 1971. *J. Mol. Biol.* 61:25–44
247. Rupp, W. D., Howard-Flanders, P. 1968. *J. Mol. Biol.* 31:291–301
248. Radman, M. 1980. *Photochem. Photobiol.* 32:823–30
249. Echols, H. 1981. *Cell* 25:1–2
250. Cairns, J., Overbaugh, J., Miller, S. 1988. *Nature* 335:142–45
251. Hall, B. G. 1988. *Genetics* 120:887–97
252. Hall, B. G. 1990. *Genetics* 126:5–16
253. Kennard, O. 1987. *Nucleic Acids Mol. Biol.* 1.25–52

Annu. Rev. Biochem. 1991. 60:513–52

# EUKARYOTIC DNA POLYMERASES

*Teresa S.-F. Wang*

Laboratory of Experimental Oncology, Department of Pathology, Stanford University School of Medicine, Stanford, California 94305

KEY WORDS:   DNA replication proteins, DNA replication, cell growth control, cell cycle, gene expression.

## CONTENTS

0066-4154/91/0701-0513$02.00

## PERSPECTIVES

Understanding the molecular mechanisms regulating eukaryotic cell growth and division is a major endeavor in biological science. To achieve this goal it is essential to comprehend two related but separate biological problems. First, what molecular mechanisms determine whether cells should be maintained in quiescence, allowed to terminally differentiate, or actively proliferate? Second, in actively proliferating cells, what mechanisms regulate the progression through stages of the cell cycle? DNA replication is essential not only for duplication of the genome but also for maintenance of genome integrity during DNA repair. Understanding the mechanisms that ensure that DNA replication does not occur in nonproliferating cells, and those that ensure that each chromosome is replicated only once during the S phase of the cell cycle, are major goals of biological science. A prerequisite to understanding the molecular mechanisms of DNA replication is a comprehension of the proteins involved in this process. Among the essential proteins for DNA replication are the DNA polymerases.

Progress in studying DNA polymerases from eukaryotic cells has been problematic due to the low abundance of DNA polymerases in cells, proteolysis during purification, and the lack of amenable genetics. Much progress has been made in the past few years, predominantly by the development of a cell-free reconstituted Simian virus (SV40) DNA replication system (1, 2), and the isolation and identification of genes and cDNAs of DNA polymerases from *Saccharomyces cerevisiae* (hereafter, budding yeast) and from mammalian cells (3–8). The cell-free reconstitution of SV40 viral DNA replication provides a system for identifying polymerases that are essential for SV40 viral DNA replication and most likely function similarly in host mammalian cells. Three types of eukaryotic DNA-polymerizing enzymes have been described to date. One type of enzyme, named DNA polymerase, polymerizes DNA utilizing DNA as template; a second type, named reverse transcriptase, synthesizes DNA utilizing RNA as template; and a third type, named terminal transferase, polymerizes on the 3'OH-terminus of DNA independent of template. Five eukaryotic cellular DNA template–dependent DNA polymerases exist, designated as DNA polymerase $\alpha$, $\beta$, $\gamma$, $\delta$, and $\epsilon$ and encoded by distinct genes. Several recent reviews have discussed the roles of polymerases involved in cell-free replication of viral DNA (1, 2, 9, 10). Other reviews have described polymerase protein structure and enzymology in budding yeast and other higher eukaryotic cells (9, 11–13). This review focuses on the five cellular eukaryotic DNA polymerases. I discuss their properties, their associated activities and effectors, their genetic loci, their genetic structures, the conservation of their genetic and protein sequences, their functional domains, and the regulation of their gene expression during cell growth and through the cell cycle. The functional role of each DNA polymerase in replication and repair synthesis is also discussed.

## NOMENCLATURE

The recent identification of genes from budding yeast and mammalian cells has defined five distinct species of cellular DNA polymerases (3–8, 14, 15). DNA polymerase $\alpha$, a relatively high-molecular-weight enzyme, was the first eukaryotic DNA polymerase identified (16). A low-molecular-weight nuclear DNA polymerase found only in vertebrates and not in budding yeast nor other lower eukaryotes is designated as DNA polymerase $\beta$ (17, 18). The polymerase required for mitochondrial DNA replication but encoded in the nucleus is named DNA polymerase $\gamma$ (19–24). This polymerase is the product of the *MIP1* gene of budding yeast (15). The catalytic subunits for the three budding yeast DNA polymerases are encoded by the genes *POL1*, *POL2*, and *POL3* for DNA polymerases I, II, and III, respectively. The previous Roman numeral designation of the three budding yeast DNA polymerases, PolI, PolII, and PolIII, has now been revised to polymerases $\alpha$, $\epsilon$, and $\delta$, respectively (25). In mammalian cells, two DNA polymerases that contain tightly associated 3'-5' exonuclease activities but respond differently to proliferating cell nuclear antigen (PCNA) for processive DNA synthesis have been described (26–35). These two forms of polymerase $\delta$ are distinct based on their dependency on PCNA for DNA synthetic processivity, and have therefore been described as PCNA-dependent and PCNA-independent polymerase $\delta$ (33, 34). The latter was also described as DNA polymerase $\delta$II (27, 36, 37). The PCNA-dependent polymerase $\delta$ has now formally been designated as DNA polymerase $\delta$, while the polymerase $\delta$II or PCNA-independent polymerase $\delta$ (homolog of yeast polymerase II) has now been named DNA polymerase $\epsilon$ (25, 33). Much confusion was created in the literature in the past few years by naming exonuclease-associated DNA polymerases "polymerase $\delta$." Two nuclease-associated DNA polymerase activities, one from bone marrow (38) and another 170-kDa exonuclease-associated polymerase purified from human placenta, were both described as polymerase $\delta$ (39). However, their dependency on PCNA for DNA synthetic processivity is still unknown, and this criterion cannot be used to classify these two polymerases as either polymerase $\delta$ or polymerase $\epsilon$. Another polymerase activity from calf thymus with a readily dissociable nuclease and an associated primase activity was described as DNA polymerase $\delta$I (27, 36). It has been suggested that this represents an enzyme fraction containing polymerase $\alpha$ and a 3'-5' exonuclease (27, 40).

## PROPERTIES

### Protein Structures and Enzymatic Properties

PROTEIN STRUCTURE    The protein structure of DNA polymerase $\alpha$ was a controversial issue for decades. Improvement of the conventional biochemical purification procedure and development of the imunnoaffinity purification method have facilitated the resolution of the structure of this enzyme. The subunit components of DNA polymerase $\alpha$ isolated from a wide range of phylogenetic species either by immunoaffinity purification protocol or by conventional chromatographic methods contain a remarkably similar set of four polypeptides (12, 41–56). Polymerase $\alpha$ is composed of (*a*) a cluster of related high-molecular-weight polypeptides predominantly of 165–180 kDa containing the catalytic function that are derivatives of a single primary polypeptide (44, 45); (*b*) a polypeptide of about 70 kDa with unknown catalytic function; and (*c*) two polypeptides of 55–60 kDa and 48–49 kDa associated with primase activity (reviewed in 57, 43, 46–48, 50, 58–70). Thus an appropriate designation for this four-subunit enzyme complex is DNA polymerase $\alpha$/primase (11, 12). The isolation of polymerase $\alpha$ full-length complementary DNA (cDNA) from human cells and the gene from budding yeast defines the catalytic polypeptide as a 165-kDa protein (4, 5). The often observed cluster of polypeptides with a molecular mass higher than the deduced primary sequence results from posttranslational modification, while the often observed lower-molecular-mass polypeptides ranging from 120 to 165 kDa result from proteolysis of the posttranslationally modified protein at specific labile site(s) within the N-terminus (71).

DNA polymerase $\beta$ isolated from all vertebrate species is a 40-kDa protein (18, 72–78). Isolation of the DNA polymerase $\beta$ gene from rodent and human cells confirms the structure of this enzyme as a 335-amino-acid protein with a molecular size of 39 kDa (14, 79, 80). A functional homolog has not yet been found in budding yeast.

The protein structure of DNA polymerase $\gamma$ has also been controversial due to its low abundance and proteolytic degradation during purification. Purification of DNA polymerase $\gamma$ from chick embryo suggested that the enzyme is a homotetramer of 47-kDa subunits (81, 82). This enzyme purified from *Drosophila melanogaster* is composed of a 125-kDa catalytic polypeptide and an additional subunit of 35 kDa in a 1 : 1 ratio (23). Polymerase $\gamma$ purified from *Xenopus laevis* contains a single polypeptide of 140 kDa reported to be associated with catalytic function (24). The confirmation that *MIP1* from budding yeast is the gene encoding the yeast mitochondrial DNA replication enzyme, polymerase $\gamma$, clearly defines the structure of this polymerase as a protein of 143.5 kDa (15).

Identification of the budding yeast *POL3 (CDC2)* gene encoding DNA polymerase δ defines the molecular mass of the catalytic subunit of this polymerase as 125 kDa (6). Biochemical purification of polymerase δ activity from calf thymus and HeLa cells demonstrates an enzyme composed of a 125-kDa catalytic polypeptide and an associated 48-kDa subunit of unknown function (29, 33).

Polymerase ε purified from HeLa cells was reported to be a protein of 215 kDa (32, 33). DNA polymerase ε (polymerase II) from budding yeast was described as a group of proteins ranging in molecular size from 132 to 200 kDa containing polymerase activity (83, 84). The recently isolated gene for polymerase ε from budding yeast encodes a protein of 255 kDa that is by far the largest DNA polymerase yet described (8). It is assumed that the previously observed lower-molecular-mass proteins from either mammalian cells or yeast were proteolytically degraded products of the 255-kDa polypeptide. Two polypeptides of 80 and 34 kDa from yeast polymerase ε enzyme preparations have been provisionally assigned as subunit components of this enzyme (A. Sugino, personal communication). In contrast, polymerase ε purified from HeLa cells does not contain these subunits (32, 33).

ENZYMATIC PROPERTIES    The general enzymatic properties such as primer-template preference, divalent cation preferences, salt sensitivity, and inhibitors of these five distinct polymerases have been described by others (13, 18, 29, 83, 85). Properties that can be used to distinguish these DNA polymerases are summarized in Table 1 and are not discussed here in detail.

Two notable enzymatic properties of DNA polymerases are processivity and fidelity. Processivity of a polymerase is the extent of DNA chain elongation by DNA polymerase per interaction with the 3'OH-end of the DNA primer. Processivity of DNA polymerases α and β has been studied extensively. For DNA polymerase α, depending on the assay conditions, purity of the enzyme, integrity of the catalytic polypeptide, and the subunit components of the enzyme fraction used, a large variation in the number of nucleotides polymerized per binding event has been reported (86–88). Compared to the other four eukaryotic DNA polymerases, polymerase α is generally regarded as moderately processive (Table 1). Despite the dependency of processivity on reaction conditions and polymerase α integrity, processivity has been used as a parameter to identify a variety of effectors and accessory proteins for polymerase α, which have been reported to enhance its processivity (89–92). With the exception of an accessory protein RF-A, the in vivo significance of these identified effectors is not yet clear.

DNA polymerase β is the smallest DNA polymerase in eukaryotic cells, but it has several unique and distinct enzymatic properties. It is insensitive to aphidicolin, sulfhydryl reagents, salt, and butylphenyl-dGTP (18, 93), but is sensitive to ddNTPs (18, 86). Processivity of polymerase β depends on the

**Table 1**  Eukaryotic DNA polymerases: protein structure, cellular location, and properties

| | DNA polymerase | | | | |
|---|---|---|---|---|---|
| | $\alpha$ | $\beta$ | $\gamma$ | $\delta$ | $\epsilon$ |
| **Protein structure** | | | | | |
| catalytic polypeptide (kDa) | 165[a] | 40 | 140 | 125 | 255 |
| associated subunits (kDa) | 70 58 48 | none | unknown | 48 | unknown |
| **Yeast gene** | POL1 (CDC17) | none | MIP1 | POL3 (CDC2) | POL2 |
| **Cellular location** | nuclear | nuclear | mito | nuclear | nuclear |
| **Associated enzyme activities** | | | | | |
| 3'-5' exonuclease | none[b] | none | yes | yes | yes |
| primase | yes | none | none | none | none |
| **Properties** | | | | | |
| processivity | | | | | |
| inherent | med | low | high | low | high |
| with PCNA | — | — | — | high | — |
| fidelity | high | low | high | high | high |
| **Inhibitors** | | | | | |
| aphidicolin | yes | no | no | yes | yes |
| N-ethylmaleimide | yes | no | yes | yes | unknown |
| butylphenyl-dGTP | yes | no | no | no | no |
| dideoxynucleoside 5'-triphosphate (ddNTP) | no | yes | med | no | no |
| carbonyldiphosphonate | no | unknown | unknown | yes | med |

[a] Molecular mass deduced from primary protein sequence.
[b] Cryptic exonuclease in the isolated catalytic polypeptide was reported in *Drosophila melanogaster* embryo (102).

metal ion in the reaction; it is five times higher in reactions with $Mn^{2+}$ than in reactions with $Mg^{2+}$ as activator (94).

Polymerase $\gamma$ is aphidicolin resistant but N-ethylmaleimide sensitive and is moderately inhibited by dideoxythymidine triphosphate (ddTTP) (24). Polymerase $\gamma$ has been reported to be highly processive in DNA synthesis (18).

Polymerases $\delta$ and $\epsilon$ are distinguished from the other DNA polymerases by their readily detectable proofreading 3'-5' exonuclease activity. Both are aphidicolin sensitive (29, 32, 33, 36), and both are rather insensitive to the inhibition of butylphenyl-dGTP. Furthermore, while polymerase $\delta$ is stimulated by 10% dimethyl sulfoxide (DMSO), polymerase $\epsilon$ is inhibited by this reagent (33). These two DNA polymerases are immunologically and structurally distinct from polymerase $\alpha$, but have some immunological and structural similarity with each other (34). Interestingly, these two polymerases are strikingly distinct in their processivity. Polymerase $\delta$ has low processivity with long single-stranded templates, but, in the presence of PCNA, polymerase $\delta$ can perform highly processive DNA synthesis on this template. Thus, PCNA has been designated as the auxiliary protein of DNA polymerase $\delta$ (95). In contrast to polymerase $\delta$, DNA polymerase $\epsilon$ has an inherently high processivity on this template in the absence of PCNA in reactions with low concentrations of $Mg^{2+}$ (30, 32, 33, 84).

DNA synthetic fidelity and mutation specificity of polymerases $\alpha$, $\beta$, and $\gamma$ have been studied extensively (reviewed in 18, 96–99). A detailed discussion of fidelity studies and the mutation specificity of each polymerase is beyond the scope of this review. A summary of these studies is presented in Table 1. Several interesting findings are noteworthy. Despite the lack of a measurable proofreading 3'-5' exonuclease in polymerase $\alpha$/primase complex, the misincorporation rate of this enzyme is low. Two factors are suggested to influence the fidelity of polymerase $\alpha$. First, the more intact form of the four-subunit complex of DNA polymerase $\alpha$/primase has higher DNA synthetic accuracy than aged and proteolyzed enzyme (100, 101). A second factor is the presence of a proofreading exonuclease activity. A cryptic 3'-5' exonuclease activity in the large catalytic subunit of polymerase $\alpha$ from $D.$ $melanogaster$ embryo has been described when this subunit is separated from the 70-kDa subunit (102). This unmasked cryptic exonuclease in the catalytic polypeptide enhances the fidelity of DNA synthesis 100-fold. Furthermore, this separated polymerase $\alpha$ catalytic polypeptide with the cryptic exonuclease has altered sensitivity to inhibitors. Like polymerases $\delta$ and $\epsilon$, it becomes less sensitive to butylphenyl-dGTP, whereas this analog is a potent inhibitor of the intact polymerase $\alpha$/primase complex (103). Moreover, a polymerase $\alpha$ enzyme preparation from human lymphocyte containing a loosely associated proofreading exonuclease was reported to have higher DNA synthetic fidelity (104). Consistent with the notion that proofreading exonuclease enhances polymerase DNA synthetic fidelity, the fidelity of calf thymus DNA polymerase $\epsilon$ (previously described as polymerase $\delta$II) has an accuracy 10- and 500-fold greater than that of polymerases $\alpha$ and $\beta$, respectively (37).

## Catalytic Mechanisms

Using steady-state kinetics analysis and supplemented by direct physical measurement, the catalytic mechanisms of DNA polymerases $\alpha$ and $\beta$ were thoroughly investigated (reviewed in 105 and 45, 86, 94, 105–111). Extensive studies of DNA polymerase $\alpha$ and substrate interaction were performed with the biochemically purified but proteolyzed form of the enzyme (112). Most of these polymerase $\alpha$ and substrate interaction mechanisms were later proven to be qualitatively identical when reassessed with the immunopurified polymerase $\alpha$/primase complex (44). In a separate study, analysis with immunoaffinity-purified calf thymus DNA polymerase $\alpha$ demonstrated qualitatively higher affinity for deoxynucleoside 5'-triphosphates (dNTPs) and primer-template (52).

These studies suggest that the intact form of DNA polymerase $\alpha$ might yield quantitatively more efficient enzyme for the assessment of catalytic mechanism. The interaction of polymerase $\alpha$ and substrates was found to obey a rigidly ordered sequential *ter* mechanism. The order of the sequential *ter* mechanism is first, interaction with template (single-stranded DNA), second, interaction with primer stem, and third, interaction with dNTP. Specification of the addition of dNTP to the enzyme is determined by the template sequence. A minimum length of 8 nucleotides of primer is required for effective binding with polymerase $\alpha$. The terminal 3–5 nucleotides of the primer must be template-complementary. A single mispaired terminal nucleotide prevents correct primer binding and the subsequent step of dNTP addition to the enzyme (106–110). This finding was later confirmed by assessment of rate of extension from mispaired primers by calf thymus polymerase $\alpha$ (113, 114). The rate of extension from mispairs is $10^3$ to $10^6$ times slower than from a paired primer terminus (114). Hydrolysis of the mispaired primer residue by either the $\epsilon$ subunit of *Escherichia coli* polymerase III or the proofreading 3' exonuclease associated in polymerase $\delta$ permits subsequent extension and increases the fidelity of DNA polymerase $\alpha$ (115, 116). 3'-Terminal H or OH, but not $PO_4$, can interact with polymerase $\alpha$. The primer binding occurs through the coordinated participation of four $Mg^{2+}$-primer binding subsites, which may serve as a shuttle to facilitate not only phosphodiester bond formation but also polymerase translocation. The catalytically active polymerase $\alpha$ appears to have two positively cooperative single-stranded DNA-binding sites. Thus, it was proposed that polymerase $\alpha$ is a conformationally active enzyme that responds to signals from template sequence that are transduced via template-binding site(s) interaction (109, 110).

Unlike polymerase $\alpha$, DNA polymerase $\beta$ has a weak affinity for single-stranded DNA that is not modulated by base sequence and not enhanced by the presence of potentially base-paired 3'-OH termini. Polymerase $\beta$ has no affinity for intact duplex DNA, but has a high affinity for nicked duplex DNA with nicked termini either bearing 3'-OH or 3'-$PO_4$ residues. Polymerase $\beta$ can perform a limited strand displacement synthesis at the 3'-OH termini of the nicked DNA (111). In corroboration of these earlier findings, polymerase

$\beta$ from HeLa cells was also demonstrated to perform limited strand displacement synthesis on 3'-OH termini in DNA produced by AP endonuclease at the 5' side of apurinic sites. Polymerase $\beta$ is able to insert a single complementary nucleotide into the single nucleotide gap created (117). Polymerase $\beta$ purified from human cells has less stringent requirements for primer interaction than polymerase $\alpha$. Polymerase $\beta$ is reactive with primers that contain 1–3 mispaired terminal residues. Polymerase $\beta$ purified from mouse cells was reported to carry out a two substrate–two product reaction that follows ordered bi-bi kinetics: interacting with primer template first, followed by dNTP (118). The mechanism of substrate interaction of human polymerase $\beta$ is also rigidly ordered, and sequential by binding to DNA first followed by dNTP. However, binding to primer-template appears to occur by a concerted mechanism. The primary signal for catalytic productive interaction of polymerase $\beta$ with DNA is a base-paired primer moiety that must be adjacent to a short length of potentially single-stranded template. The minimum length of this template and processivity are influenced by the divalent cation. With $Mg^{2+}$, minimum template length is $>5$ nucleotides and polymerization is distributive, whereas, with $Mn^{2+}$, the minimum required template is a single nucleotide and the polymerization is more processive by inserting 4–6 residues per cycle of primer interaction (94, 111). Consistent with the notion that higher processivity is responsible for higher fidelity, the error rate at hot spots for polymerase $\beta$ was also found to be lower in reactions with $Mn^{2+}$ as cation (119).

## ASSOCIATED ACTIVITIES AND EFFECTORS

### Primase

All DNA polymerases require a 3'-hydroxyl terminus of a preexisting primer for reaction (120, 121). DNA polymerase $\alpha$ is the only eukaryotic polymerase with a tightly associated primase. Primase activity resides in a heterodimer of 55–60 kDa and 48–49 kDa polypeptides that have been purified from various biological systems as the associated activity and subunit component of polymerase $\alpha$. The purification, enzymatic properties, and reaction mechanisms of primase reported from numerous laboratories have been reviewed (57). Some recent developments that were not included in the previous review are discussed here.

The genes for both primase subunits from budding yeast are now cloned and designated as *PRI1* and *PRI2*. By gene disruption, both genes were found to be essential (69, 122, 123). By affinity labeling, the ATP- or GTP-binding site(s) of primase from calf thymus and yeast were found to be localized exclusively in the 48-kDa subunit (67, 124). A monoclonal antibody against the 60-kDa subunit of chick primase has been produced. Immunofluorescence demonstrates the exclusive nuclear localization of primase in granular structures bound to the nuclear matrix. Furthermore, primase antigen was reduced in quiescent cells and induced when quiescent cells were activated to proliferate by serum addition. These results suggest that the expression and organization of primase in the granular structures on the nuclear matrix are regulated in correlation with cell proliferation (125). The cDNA of the 49-kDa subunit of mouse primase was cloned and characterized (126). Sequence comparison with yeast primase 49-kDa subunit shows a high conservation of amino acid sequences in the N-terminal halves of the polypeptides and a potential metal-binding domain (Zn finger) in this region. The C-terminal region diverges from yeast primase. During serum induction of quiescent cells to proliferate, expression of the mouse 49-kDa primase gene is positively induced ~10-fold at the steady state transcript level, but in actively proliferating cells the gene is constitutively expressed throughout the cell cycle (127). This finding is in agreement with the immunofluorescence data reported for chick primase (125). The cDNA for the 59-kDa large subunit of mouse primase has also been cloned (B. Y. Tseng, personal communication).

## 3'-5' Exonuclease

None of the five cellular DNA polymerases has an associated endonuclease or 5'-3' exonuclease. Polymerases $\gamma$, $\delta$, and $\epsilon$ do have an associated proofreading 3'-5' exonuclease. The proofreading 3'-5' exonuclease and polymerase $\gamma$ activity have been suggested to reside in the same polypeptide. From X. laevis, the 3'-5' exonuclease activity and polymerase $\gamma$ activity copurified in a constant ratio in the final three steps of purification. The two enzymatic activities also cosediment in a glycerol gradient under partially denaturing conditions (128). In chick embryo, the 3'-5' exonuclease was found to coprecipitate with DNA polymerase $\gamma$ during immunoprecipitation with anti-DNA polymerase $\gamma$ antibody (129). Like the classic proofreading nuclease of E. coli polymerase I and T4 polymerase (130, 131), the 3'-5' exonuclease of polymerase $\gamma$ hydrolyzes mismatched termini with significantly greater efficiency than base-paired termini (128, 129). The 3'-5' exonuclease activities in polymerases $\delta$ and $\epsilon$ also behave as typical proofreading nucleases with the preference for hydrolyzing mismatched over base-paired termini (32, 83, 84, 132). As with the proofreading 3'-5' exonuclease in prokaryotes, budding yeast polymerase $\delta$ also has a demonstrable turnover of dTTP to dTMP by the successive actions of polymerase and exonuclease (132). Based on the inhibitor studies and the constant ratio of nuclease to polymerase activity in the final steps of purification, it is suggested that the 3'-5' exonuclease activity of both polymerases $\delta$ and $\epsilon$ reside in the same polypeptide as the polymerase activity (26, 38, 83, 84). As described above, a cryptic 3'-5' exonuclease was also reported in the catalytic polypeptide of *Drosophila* polymerase $\alpha$ when isolated from the other subunits (102). An 3'-5' exonuclease activity has been described in recombinant yeast polymerase $\alpha$ 180-kDa subunit recently (132a). This 3'-5' exonuclease activity is not a conventional proofreading nuclease, however. The yeast 180-kDa subunit, as well as the 180-kDa/86-kDa complex and the four-subunit polymerase $\alpha$/primase complex, was capable of removing 3' terminal nucleotide of $poly(dT)_{600}$-$^{32}P$-$dNMP_{0.4-0.5}$. The physiological significance of this yeast 180-kDa associated exonuclease is not yet known. The catalytic polypeptide of human DNA polymerase $\alpha$ has been overproduced from recombinant baculovirus-infected insect cells as a single polypeptide of 180 kDa. This single-polypeptide polymerase $\alpha$ does not have a detectable proofreading 3'-5' exonuclease activity when assayed with either mismatched or matched base-paired termini (W. Copeland and T. Wang, manuscript submitted). It is interesting that the cryptic proofreading exonuclease of DNA polymerase $\alpha$ appears to be unique for the embryo of *Drosophila*.

## Proliferating Cell Nuclear Antigen (PCNA)

An auxiliary protein for DNA polymerase $\delta$ that enhances its processivity was found to be immunologically, structurally, and functionally identical with proliferating cell nuclear antigen, PCNA (29, 95, 133, 133a, 134). PCNA is a 36-kDa nuclear protein reactive with sera from a subset of patients with the autoimmune disease systemic lupus erythematosus (SLE), and this protein was only found in actively proliferating cells (135, 136). PCNA was also found to be essential for in vitro SV40 DNA replication (133, 137). Reviews of the role of PCNA in viral and cellular DNA replication and cell growth have been published recently (2, 13, 138). The gene and cDNA for PCNA have been isolated and characterized from rat, human, budding yeast, and *Drosophila*. Protein sequence comparison indicates that PCNA is a highly conserved protein among these eukaryotes (139–143). Budding yeast PCNA gene *(POL30),* like the budding yeast polymerase $\delta$, is periodically expressed at the level of steady-state mRNA in starve-feed cultures. During meiosis, the mRNA level also increases prior to initiation of premeiotic DNA synthesis (141). In contrast to budding yeast, the protein level of PCNA in mammalian cells was reported to be constitutively expressed throughout the cell cycle, although PCNA mRNA and protein levels were dramatically increased during activation of quiescent cells to proliferate (140, 144).

PCNA does not appear to have a stringent species specificity for its interaction with polymerase $\delta$ and other replication factors. Calf PCNA can stimulate yeast polymerase $\delta$ (145); calf thymus PCNA and human RF-C (a replication factor, see discussion below) are able to stimulate budding yeast polymerase $\delta$ DNA synthesis on singly primed M13 single-stranded DNA in an ATP-dependent manner (146). *Drosophila* PCNA could substitute for human PCNA, although with reduced efficiency, in stimulating SV40 in vitro DNA replication (143). PCNA has also been shown to affect the DNA synthetic activity of polymerase $\epsilon$. Under certain salt conditions in the presence of RF-C, calf thymus PCNA can enhance budding yeast polymerase $\epsilon$ DNA synthetic activity on singly primed single-stranded DNA and can enhance its synthetic processivity in homopolymer reactions (84, 146). In SV40 origin-dependent DNA replication, the absence of PCNA yields short and lagging-strand-specific products, which are predominantly derived from regions around the SV40 origin (137). PCNA also stimulates the ATPase activity of RF-C and is an essential component of the primer recognition complex, PCNA/RF-C/ATP (92).

Evidence from product analysis of the SV40 in vitro replication reactions suggests that PCNA is the functional equivalent of T4 phage gene 45 protein or *E. coli* DNA polymerase III $\beta$ subunit (147). The functional analogy between PCNA and T4 phage gene 45 protein is also reflected in the primary amino acid sequences of these proteins. These two proteins have a 31–50% primary sequence similarity within a limited region, but have no significant similarity to *dnaN* gene product, which encodes the $\beta$ subunit of *E. coli* DNA polymerase III holoenzyme (147).

In sum, several lines of evidence support the conclusion that PCNA does not directly affect the DNA synthetic activity of DNA polymerase $\alpha$. For polymerase $\delta$, synergistic interaction of PCNA with RF-C/ATP and RF-A is required for the optimal DNA synthetic activity on primed single-stranded DNA or in reconstituted SV40 origin containing DNA reaction. Depending on the reaction condition and the presence of RF-C and ATP, the conjecture is that PCNA might also have an effect on in vitro DNA synthetic activity of polymerase $\epsilon$ (84, 146).

## Replication Factor A (RF-A)

RF-A (also named replication protein A, RP-A, and human single-stranded DNA-binding protein, HSSB) is a multisubunit single-stranded DNA-binding protein. RF-A contains three polypeptides of 70, 34, and 11 kDa and functions as an auxiliary protein for both polymerases $\alpha$ and $\delta$. RF-A is required for the initiation and elongation stages of in vitro SV40 DNA replication (148–153). Biochemical and immunological evidence indicates that the single-stranded DNA-binding activity resides exclusively in the 70-kDa subunit (149, 151, 154). RF-A alone stimulates DNA polymerase $\alpha$ activity on either primed M13 DNA or oligo dT–primed polydA, resulting in short DNA products; PCNA had no effect on the size of the product. In contrast to the effect on polymerase $\alpha$, RF-A alone has no specific effect on the reactivity of polymerase $\delta$ even in the presence of PCNA. However, polymerase $\delta$ can be synergistically stimulated by PCNA, RF-A, and RF-C (another replication factor, see discussion below), yielding long DNA products (152–154).

Immunological evidence has suggested that the 70-kDa subunit may be required for DNA unwinding and polymerase $\delta$ stimulation, whereas both 70- and 32–34-kDa subunits have been implicated to be involved in the stimulation of polymerase $\alpha$ DNA synthetic activity (154). The gene for the 70-kDa subunit has been cloned from budding yeast and found to be essential (155). The deduced protein sequence reveals significant amino acid similarity of this yeast 70-kDa RF-A subunit with its human homolog (155). The cDNA for the 34-kDa subunit of RF-A from human cells was also cloned and overexpressed (156). The 34-kDa protein is a phosphoprotein, and phosphorylation of this subunit is cell cycle regulated with higher phosphorylation occurring at late $G_1$ and early S phase (157). Yeast RF-A or other single-stranded DNA-binding proteins, including $E.\ coli$ single-stranded binding protein (SSB) and adenovirus and herpes simplex virus DNA-binding proteins, are able to replace human RF-A in the unwinding step in the initial stage of SV40 in vitro DNA replication. However, yeast RF-A and these other SSBs are unable to substitute for human RF-A in the SV40 in vitro DNA replication reaction (1, 2, 153, 158). Thus, like polymerase $\alpha$, RF-A demonstrates stringent species specificity in SV40 in vitro DNA replication. It also appears to be involved in genetic recombination (155).

## Replication Factor C (RF-C)

RF-C (also named activator I, AI) is a multisubunit protein complex that has primer/template binding and DNA-dependent ATPase activity. The intrinsic DNA-dependent ATPase activity of RF-C is stimulated by PCNA (147), and addition of RF-A further stimulates the DNA-dependent ATPase activity (159). Purified RF-C is composed of subunits of 140, 41, and 37 kDa. The 140-kDa subunit binds primer/template DNA, whereas the 41-kDa subunit binds ATP (92). In SV40 in vitro DNA replication reactions, the absence of RF-C causes accumulation of early replicative intermediates and yields DNA products that hybridize only to the lagging-strand template (92, 152, 160–163). This observation suggests that RF-C has a profound effect on leading-strand DNA synthesis. RF-C appears to be the functional equivalent of the T4 phage gene 44/62 complex (reviewed in 164 and 165). Similar to the T4 phage DNA replication system, processive DNA synthesis by polymerase $\delta$ on primed M13 single-stranded DNA requires RF-A, RF-C, and PCNA and hydrolysis of ATP or dATP (92, 147, 159, 162, 163, 166). Furthermore, replication complexes from (a) T4 phage containing gene products 43, 44/62 and 45, or T7 polymerase/thioredoxin complex, (b) $E.\ coli$ DNA polymerase III holoenzyme, and (c) other DNA polymerases can substitute for the leading-strand polymerase complex from human cells in SV40 in vitro DNA replication reaction (162, 168).

Extensive product analysis of reactions with RF-C in various combinations with PCNA, ATP, and RF-A, in conjunction with isolation of the replication protein complex by gel filtration, and analysis of the kinetics of protein-complex formation, support a model of protein-protein interaction hierarchy at the replication fork. A model has been proposed that binding of RF-C to the lagging-strand polymerase complex at the replication origin occurs first, followed by binding of PCNA in an ATP-dependent manner for the formation of a "primer recognition complex," which in cooperation with RF-A plays a crucial role in initiating the leading-strand synthesis from the nascent DNA fragment synthesized by the lagging-strand polymerase complex (92, 162, 163, 166). A proposed model showing the effect of RF-A, RF-C, and PCNA on the lagging-strand and leading-strand switching is illustrated in Figure 3 and discussed further in the section of this review on Physiological Roles.

## T Antigen

In SV40 viral DNA replication, it is reasonable to assume that there are protein-protein interactions between T antigen and cellular replication proteins during initiation of the SV40 DNA replication. An earlier report described the association of DNA polymerase $\alpha$ activity from crude cell extracts with T antigen, but purified polymerase $\alpha$/primase complex failed to demonstrate significant interaction with T antigen (169). These observations suggest that the interaction between T antigen and polymerase $\alpha$/primase complex might be due to a third yet unidentified cellular protein in the crude extract. A subset of T antigen monoclonal antibodies was shown to prevent the binding of T antigen and polymerase $\alpha$/primase from crude cell lysate, and interestingly these antibodies also block T antigen interaction with murine p53 protein (170, 171). Recently, this issue has been investigated further with the highly purified four-subunit polymerase $\alpha$/primase complex and T antigen. It has been demonstrated that T antigen binds directly and specifically to the catalytic subunit of human DNA polymerase $\alpha$. The amino-terminal 83 amino acids of T antigen are both necessary and sufficient for the binding. Furthermore, polymerase $\alpha$ from human tissue has a 10-fold higher affinity for T antigen than does polymerase $\alpha$ from calf, and conversely human polymerase $\alpha$ that binds the SV40 T antigen with high affinity does not bind polyoma T antigen (172; I. Dorneiter, E. Fanning, personal communication). This is consistent with an earlier report of species specificity of polymerase $\alpha$/primase in SV40 in vitro DNA replication reaction (173). Thus, T antigen can be classified as another effector for DNA polymerase $\alpha$ in the viral SV40 replication system, perhaps by directing the polymerase to the origin of replication, where T antigen is bound.

Several other effectors for polymerase $\alpha$ have been reported. Among those, two proteins named C1 and C2 can increase the affinity of DNA polymerase $\alpha$ to primer stem 20–30-fold. Interestingly, the effect of these two proteins on polymerase $\alpha$ also demonstrates species specificity (174, 175). Another protein, alpha accessory factor (AAF), is highly specific for polymerase $\alpha$ with no effect on polymerases $\beta$, $\delta$, or $\gamma$. AAF increases the affinity of polymerase $\alpha$/primase for DNA templates, enhances its processivity, and allows this enzyme to traverse double-stranded DNA region (176, 177).

# GENETIC STRUCTURE AND PREDICTED FUNCTIONAL DOMAINS

## Chromosome Localization

The catalytic polypeptide of human DNA polymerase $\alpha$ has been mapped to the short arm of the X-chromosome at Xp21.3 to Xp22.1 by expression, and later confirmed by genomic Southern hybridization with the cDNA (4, 178). The gene for mouse polymerase $\alpha$ catalytic polypeptide was recently localized on the mouse X-chromosome in region C-D (179). The genes of the two mouse primase p49 and p58 subunits were localized to mouse chromosomes 10 and 1, respectively (179). The genes coding for all four subunits of budding yeast DNA polymerase $\alpha$ were mapped. The catalytic polypeptide maps on yeast chromosome XIV, the 74–86 kDa subunit on chromosome II, and the two primase 58- and 48-kDa subunits on chromosomes XI and IX, respectively (180). The chromosome localization of budding yeast polymerase $\delta$ is mapped to the left arm of chromosome IV (6), whereas the chromosome localization of polymerase $\epsilon$ is not yet known. The yeast *MIP1* gene encoding mitochondrial DNA polymerase $\gamma$ was mapped to the right arm of yeast chromosome XV (15). The genes for human and mouse DNA polymerase $\beta$ were both mapped to chromosome 8 of human and mouse, respectively (181–183). Further fine mapping of human and mouse DNA polymerase $\beta$ gene has localized the gene near localization of the tissue plasminogen activator on the proximal short arm of chromosome 8 (184).

## Conservation of Primary Sequences

Sequence comparison of cDNAs and genes of various prokaryotic and eukaryotic DNA polymerases has revealed an evolutionary conservation of three groups of DNA polymerases on the basis of their amino acid sequence relatedness: (*a*) *E. coli* DNA polymerase I and phage T7 DNA polymerase (185); (*b*) DNA polymerase $\beta$ and terminal transferase (80, 186); and (*c*) $\alpha$-like DNA polymerases, including protein-primed replicative DNA polymerases (4, 187).

Comparison of deduced amino acid sequences of DNA polymerase β and other proteins showed a sequence similarity with human terminal deoxynucleotidyltransferase (80, 186). Sequence alignments revealed conservation between residues 42–215 of polymerase β and residues 195–366 of terminal transferase. Nearly 30% of these 174 amino acid residues of polymerase β were identical with those of the 172-residue sequence of terminal transferase. Forty residues (23%) represented conservative substitution. Thus, the sum of identical residues, the favored substitutions, and the conservative substitutions in amino acid pairs was 124 out of 174 residues (71.3%). The overall predicted secondary structures of the two enzymes were also closely similar in their homologous regions (80). However, the C-terminal region (amino acid residues 216–335) of DNA polymerase β has no significant similarity with that of terminal transferase. Interestingly, the similar region is coded exclusively by the third and the later exons.

The isolation of the catalytic polypeptide of human DNA polymerase α cDNA and analysis of its sequence revealed striking sequence similarity with replicative DNA polymerases from both prokaryotes and eukaryotes (4, 187). Furthermore, by utilizing both immunologic and molecular genetic approaches, the differential evolutionary conservation of four unique epitopes on the catalytic polypeptide and the conservation of genomic DNA sequences have been demonstrated among organisms as phylogenetically disparate as primate and unicellular fungi (188). These findings have facilitated the identification of many DNA polymerase genes and cDNAs and have led to the recognition of an α-like family of DNA polymerase genes. From amino acid residues 609 to 1085 of human DNA polymerase α, six regions exist that are similar to regions of the three budding yeast DNA polymerases α, δ, and ε, a potential budding yeast polymerase gene, REV3, product that is involved in mutagenesis responses, DNA polymerases from the herpes simplex virus (HSV) family, cytomegalovirus (HCV), Epstein-Barr virus (EBV), and vaccinia virus polymerase (4–6, 8, 45, 187, 189–193). Five regions of similarity exist to adenovirus DNA polymerase and bacteriophage T4 DNA polymerase (194, 195). Three regions of similarity exist to DNA polymerases from Bacillus subtilis phage φ29 and E. coli phage PRD1 (196, 197), as well as to two potential DNA polymerases, yeast linear plasmid pGKL1 and maize S1 mitochondrial DNA (198–200).

The regions described above are designated I to VI according to their extent of similarity, with region I being the most conserved and region VI, the least similar. The significance of these six regions is further underscored by their relative linear spatial arrangements within each polymerase polypeptide (Figure 1). The order of these six regions in each polymerase polypeptide is IV-II-VI-III-I-V, while the distances between each of these conserved regions are variable. While regions I, II, and III appear to be the basic core regions of this α-like DNA polymerase family, yeast MIP1 gene encoding the yeast polymerase $\gamma$ revealed the presence of regions I, II, and VI but not regions III, IV, and V. Furthermore, the linear spatial order of the three regions, I, II, and VI, is different from that in the other $\alpha$-like family DNA polymerases (15). Although the sequence of yeast polymerase $\epsilon$ demonstrates the presence of these six conserved regions, and the linear spatial arrangements of these regions are similar to those of the other $\alpha$-like DNA polymerases, the similarity among these six regions is relatively weak. Region I, containing sequence -YGTDTS-, which is the most highly conserved sequence among the $\alpha$-like family DNA polymerases, is poorly conserved in yeast polymerase $\epsilon$ (8). Both yeast DNA polymerases $\delta$ and $\epsilon$ also showed distinct sequence similarity to viral polymerases (6, 8). This suggests that these two 3'-5' exonuclease-containing DNA polymerases of yeast might belong to a sub-family of $\alpha$-like polymerases. Recently the *dinA* gene of *E. coli* was found to encode the *E. coli* DNA polymerase II and, interestingly, the sequence of this gene reveals the presence of regions IV, II, III, I, and V sequences of the $\alpha$-like family polymerase (201). The presence of these conserved regions in DNA polymerases from human to bacteriophage suggests that this group of DNA polymerases may all have been derived from a common primordial gene.

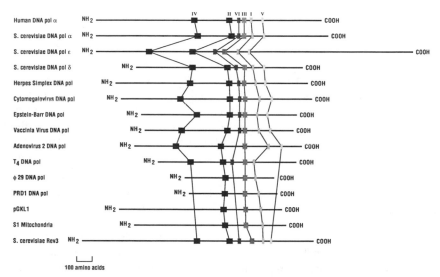

*Figure 1*   Relative linear spatial arangement of the conserved regions of $\alpha$-like DNA polymerases. Each DNA polymerase polypeptide is represented by a straight line with NH$_2$ and COOH denoting the amino- and carboxyl-terminus, respectively. The solid boxes represent the consensus sequences of each region. Similar regions of each DNA polymerase polypeptide are aligned by verticle lines. A portion of this figure is reprinted, with permission, from references (4) and (187).

## Functional Domains

Conservation of sequences suggests that these regions are required for enzymatic function. The isolation of genes and cDNAs of DNA polymerases from budding yeast and mammalian cells has made it possible to begin to understand the functional domains of these DNA polymerases.

DNA POLYMERASE $\beta$    The recombinant DNA polymerase $\beta$ from rat and human cells has been overexpressed in *E. coli* (202, 203). The expression of the recombinant DNA polymerase $\beta$ allows the enzyme to be evaluated directly for its structure-function relationships. Two approaches were taken by two independent laboratories. Based on the amino acid sequence conservation between DNA polymerase $\beta$ and terminal transferase and the locations of putative secondary and tertiary structures, five amino acid residues from 179 to 184 of polymerase $\beta$ were chosen for structure-function analysis (204). Three amino acid residues -F-R-R- (from amino acid residue 181 to 183) were replaced by site-directed mutagenesis. The results suggest that Arg182 might be involved in template binding, whereas Arg183 might play a role in primer recognition (204).

Another laboratory has approached the structure-function problem by characterization of domain structures of DNA polymerase $\beta$. By controlled proteolysis, polymerase $\beta$ was found to be organized in two relatively protease-resistant segments, of 8 kDa and 31 kDa, linked by a short protease-sensitive region (205, 206). Neither segment has active DNA polymerase activity, but the N-terminal segment of 80-amino-acid residues has single-stranded DNA– or template-binding activity. Further investigation demonstrates that the 31-kDa segment does have DNA polymerase activity but with different primer elongation properties than the intact enzyme (206). Thus, it is proposed that the C-terminal domain could be the catalytic domain of intact polymerase $\beta$, but that the 8-kDa domain might play a role in modulating the DNA elongation process. Pyridoxal 5'-phosphate modification followed by tryptic digestion has demonstrated that Lys71 may be at or near the binding site for dNTP (207). The different results for polymerase $\beta$ structure-function relationships derived from these two approaches can only be resolved physically by X-ray crystallographic analysis in the future.

DNA POLYMERASE $\alpha$    The available genetics of budding yeast provide a powerful tool to analyze the functional domains of DNA polymerase $\alpha$. Moreover, the recent accomplishment of functional expression of the recombinant human DNA polymerase $\alpha$ will allow the biochemical confirmation of these genetic data from yeast in a higher eukaryote, and the study of structure-function relationships of this large and complex polymerase molecule at the biochemical level.

*Protein-protein interaction domain*    The finding of a tightly associated DNA primase in DNA polymerase $\alpha$ purified from a variety of eukaryotic organisms suggests that a potential DNA primase interaction domain is present on the polymerase $\alpha$ molecule. Five temperature-sensitive budding yeast polymerase $\alpha$ mutants *(poll-17, poll-1, hpr3, cdc17-1, cdc17-2)* were characterized (208–210). Characterization of growth rate demonstrated *cdc17-1* and *poll-17* are quick-stop mutants at nonpermissive temperature, which is typical for mutants defective in the catalytic activity of such an essential enzyme (208, 210). Analysis of *poll-1, hpr3,* and *cdc17-2* has demonstrated that these strains are slow-stop mutants, which complete one round of cell division and DNA replication at nonpermissive temperature before growth arrest (210). DNA polymerase $\alpha$ partially purified from *poll-1* demonstrated that the mutation has led to a conformational change at nonpermissive temperature affecting the stability of the DNA primase–DNA polymerase I complex. Sequence analysis of *poll-1* and *hpr3* identified a point mutation that confers the temperature-sensitive phenotype and maps to a single amino acid substitution of glycine at residue 493 to arginine and glutamic acid, respectively (210). This region of yeast polymerase $\alpha$ exhibits near perfect homology with human DNA polymerase $\alpha$ (187) and is designated as region P for a domain that is responsible for DNA primase/DNA polymerase interaction (180, 210). Another slow-stop phenotype mutant, *cdc17-2,* has a point mutation in the conserved region IV at amino acid residue 637 by substitution of a glycine to aspartic acid (210). This region is also conserved in several DNA polymerases, including yeast DNA polymerase $\delta$ (180). Future isolation of new alleles, characterization of other yeast temperature-sensitive mutants, and identification of extragenic suppressors should be helpful in further defining the domain(s) required for protein-protein interaction.

*DNA interaction domain*    Sequence comparison of human DNA polymerase $\alpha$ and yeast DNA polymerase $\alpha$ reveals a cysteine-rich region towards the carboxyl terminus in both DNA polymerases (4, 5, 187). From amino acid numbers 1244 to 1391 of human DNA polymerase $\alpha$ and amino acid residues 1340 to 1390 of yeast polymerase $\alpha$, there is a region potentially able to form a DNA-binding motif (Zn-finger).

*Deoxynucleotide interaction domains*    The presence of the three consensus sequences, regions I, II, and III, in the DNA polymerases from bacteriophage to mammals suggests that these sequences serve essential functions in polymerase catalysis (4, 187). Sequence comparison of human DNA polymerase $\alpha$ and the herpes simplex virus DNA polymerase (HSV pol) reveals homology in all six conserved regions (4), with 87.5% in region I, 60% in region II, 47% in region III, 26% in region IV, 57% in region V, and

10.5% in region VI. Identified herpes simplex virus mutants that confer altered sensitivity to aphidicolin (Aph) or antiviral drugs like acyclovir (ACV), bromovinyldeoxyuridine (BVdU), gancyclovir (DHPG), vidarabine (araA), and phosphonoacetic acid (PAA) contain mutations in their DNA polymerase gene (192, 193, 211–213). These antiviral drugs are either analogs of dNTPs or pyrophosphate, or competitive inhibitors, such as Aph. Sequence analysis of many of these mutants from three strains of herpes simplex virus has demonstrated that most of the mutations are single amino acid substitutions within consensus regions II and III (192, 212, 213). Thus, these regions are predicted to be sites that are directly involved in dNTP binding or pyrophosphate hydrolysis (211). In addition to regions II and III, two other mutations have also been identified. One, in region V, has a single amino acid change of Asn961 to lysine, which confers altered sensitivity to Aph and DHPG; the corresponding residue at this site in human polymerase $\alpha$ is aspartate (4, 211). Another mutation, $PAA^rC$, localized between regions VI and III, confers resistance to phosphonoacetic acid and has a single amino acid substitution of Pro797 to threonine; interestingly, the corresponding residue in human DNA polymerase $\alpha$ is also a proline (4, 187, 211). Site-specific mutagenesis of adenovirus polymerase in region I has yielded non-functional enzyme (J. Engler, personal communication). This result and the failure to identify any herpes simplex virus mutants in region I suggests that region I may be extremely critical for the catalytic function of DNA polymerase $\alpha$ and the $\alpha$-like polymerases. The studies to date of herpes simplex DNA polymerase mutants with altered sensitivity to antiviral agents have not led to the identification of any region on the herpes polymerase molecule as the sole binding site for dNTPs (211). The interaction of dNTPs with DNA polymerase $\alpha$ as well as with other $\alpha$-like DNA polymerases might require several separate regions on the polymerase molecule. This question can be resolved in the future by combined effort of biochemical analysis of the recombinant polymerase $\alpha$ mutants, isolation of suppressor mutants, and physical studies.

*Proposed 3'-5' exonuclease regions*    The 3'-5' exonuclease active site of *E. coli* DNA polymerase I was defined by X-ray crystallographic and site-specific mutagenesis (214–217). It was proposed that the 3'-5' exonuclease active site of *E. coli* polymerase I might be conserved in the $\alpha$-like polymerases (218). Sequence alignment of *E. coli* polymerase I 3'-5' exonuclease active sites and $\epsilon$ subunit of *E. coli* DNA polymerase III, which contains the exonuclease activity, demonstrates the conservation of amino acids in three regions among the $\alpha$-like DNA polymerases (Figure 2). A direct test of these three proposed nuclease domains was performed by mutagenesis of the *Bacillus* phage $\phi$29 DNA polymerase in regions 1 and 2 by mutating amino acid Asp12, Glu14, or Asp66 to alanine. Mutation at these three amino acid residues resulted in a drastic reduction in the 3'-5' exonuclease activity of $\phi$29 DNA polymerase (218). Thus, these amino acids may be involved in exonuclease activity. The amino acids in human and yeast DNA polymerase $\alpha$ corresponding to $\phi$29 polymerase Asp12 are alanine and proline, respectively (Figure 2). The presence in human and yeast polymerase $\alpha$ of alanine and proline at this amino acid position in place of an aspartic acid argues for the absence of 3'-5' exonuclease in these two polymerases; however, yeast DNA polymerases $\delta$ and $\epsilon$ with known proofreading 3'-5' exonucleases both have alanine and asparagine, respectively, at this position (6, 8). The amino acid residue Asp66 of $\phi$29 polymerase in region 2 is conserved in yeast polymerases $\delta$ and $\epsilon$ but not in yeast or human polymerase $\alpha$ (Figure 2).

It should be noted that the primary sequences of human and yeast polymerase $\alpha$ are highly conserved and aligned throughout the entire protein sequences (187). In order to accommodate the conservation of amino acids in these proposed 3'-5' exonuclease domains, segments of amino acid sequence of human polymerase $\alpha$ starting from residues 606, 635, and 701 are chosen to conform to the three exonuclease domains of E. coli polymerase I, whereas segments of sequence of budding yeast polymerase $\alpha$ starting from residues 433, 474, and 715 are chosen (218). It is evident that the three exonuclease regions starting from residues 606, 635, and 701 of human polymerase $\alpha$ do

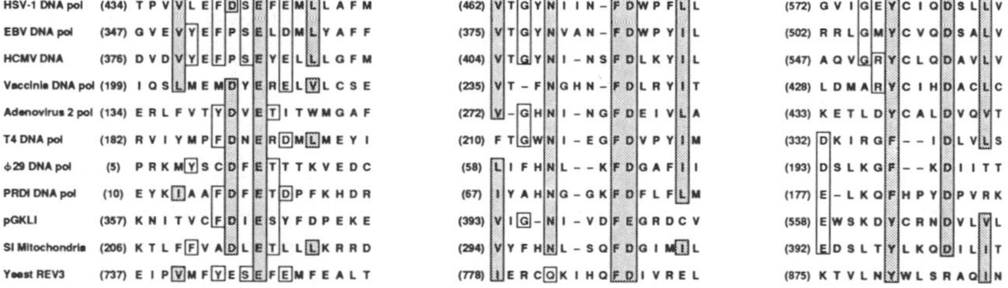

*Figure 2*  Proposed 3'-5' Exonuclease Regions. The three amino acid sequences of E. coli 3'-5' exonuclease domains and E. coli polymerase III holoenzyme $\epsilon$ subunit are aligned with $\alpha$-like DNA polymerases. Each region is represented with NH$_2$ terminus at the left and COOH at the right with the starting NH$_2$-terminal amino acid of each region numbered. Identical residues are boxed in shade and conserved substitutions are boxed. A portion of this figure is reprinted, with permission, from reference (218).

not align with yeast polymerase $\alpha$ exonuclease regions starting from residues 433, 474, and 715 (4, 5, 187). Moreover, in the original proposed exonuclease region 3 of budding yeast polymerase $\alpha$, a threonine residue at position 723 was missing, resulting in the aspartic acid at 726 being able to conform as the key consensus amino acid residue to *E. coli* polymerase I (218). Thus, in this review, the region of budding yeast polymerase $\alpha$ from amino acid residue 708 is chosen to conform to the amino acid alignment in the exonuclease region 3. Therefore, these proposed 3'-5' exonuclease regions presented in Figure 2 are a modification of those proposed in (218). Future X-ray crystallographic and extensive mutation analyses of the $\alpha$-like family of DNA polymerases are necessary to confirm the structural and functional relationships of these proposed 3'-5' exonuclease regions.

## GENE EXPRESSION

A comprehension of the control of the expression of DNA polymerase genes is a prerequisite to understanding how eukaryotic cell growth and division are regulated. Isolation of the genes of three DNA polymerases from budding yeast and the two major mammalian DNA polymerases, $\alpha$ and $\beta$, makes it possible to investigate these questions. The expression of these polymerase genes has been studied in transformed cells as compared to normal cells, in different cell types, in cellular proliferation and differentiation, and during the cell cycle. These studies have paved the way to an understanding of how the expression of these genes are regulated and whether these genes are the ultimate targets of the signal transduction cascade induced by environmental and extracellular growth signals.

### In Transformed Cells and Specific Cell Types

Early immunocytochemical detection with monoclonal antibodies against DNA polymerase $\alpha$ demonstrated the presence of polymerase $\alpha$ antigen in all cultured transformed cells. In cultured normal diploid fibroblasts, only a small fraction of the cells are antigen positive (219). Double immunoenzymatic staining with two types of monoclonal antibodies against polymerase $\alpha$ and lymphocyte membrane antigens has demonstrated a high percentage of polymerase $\alpha$ positive cells in malignant lymphomas, particularly those of peripheral T-cell type (220). The availability of human polymerase $\alpha$ cDNA permits a quantitative confirmation of these immunocytochemical observations at the molecular level. The level of polymerase $\alpha$ steady-state message from normal human cultured cell lines has been found to be 10–20-fold lower than those in transformed cells. Genomic Southern analysis indicates that the abundance of polymerase $\alpha$ message in transformed cells is not due to polymerase $\alpha$ gene amplification (4, 221). The level of DNA polymerase $\alpha$

enzymatic activity found in transformed cells is also greater than 10-fold higher than in normal proliferating cells (221). These data strongly suggest that the expression of polymerase $\alpha$ in transformed cells at the transcriptional and posttranscriptional levels is much higher than in normal actively cycling cells. The higher level of expression of polymerase $\alpha$ in transformed cells appears to be due to either higher transcription rate and/or greater stability of its message.

DNA polymerase $\beta$ is expressed in both proliferating and nonproliferating cells; however, a higher expression of the message has been detected in some transformed cells than in normal cultured fibroblast cells (78). Expression of this gene is cell type specific (222–224). In mouse tissue, it is most highly expressed in testis, with a lower level of expression in brain, and decreasing levels in thymus or spleen, kidney, heart, and liver, in that order. A study of polymerase $\beta$ message levels in testis during postnatal development and in isolated spermatogenic cells demonstrated that spermatocytes at early pachytene have the highest level of polymerase $\beta$ transcripts, implicating polymerase $\beta$ in repair synthesis during meiotic recombination (223).

## During Activation of Quiescent Cells to Proliferate

An important concept in the study of gene expression during cell growth is to distinguish the activation of quiescent cells to proliferate (transition from $G_0$ cells to actively dividing cells) from transition through the cell cycle. Several previous reports have described mammalian DNA polymerase $\alpha$ protein or enzymatic activity as being periodically expressed in the cell cycle with an increase at the beginning of S phase. This observation has been interpreted to mean that initiation of DNA replication during the cell cycle is temporally closely linked to the induction of this replicative DNA polymerase (225 and references cited in Table 1 of this reference). Conclusions based on studies performed in cells synchronized by metabolic block or by growth factor stimulation do not apply to gene expression during the cell cycle, but rather to gene expression during activation of $G_0$ cells into active proliferating cells. An early observation of the induction of polymerase $\alpha$ gene expression in normal human proliferative cells versus quiescent cells was demonstrated by immunocytochemical study using monoclonal antibodies against human DNA polymerase $\alpha$. The polymerase $\alpha$ antigen is not detectable in the closely apposed cells of microcolonies presumed to be contact-inhibited and nonproliferating (219). With the cDNA of human polymerase $\alpha$, this question has been extensively and quantitatively investigated at the molecular level. During activation of quiescent cells (in $G_0$ phase) to active proliferation ($G_1$/S phases), polymerase $\alpha$ steady-state transcript, rate of nascent protein synthesis, and enzymatic activity in vitro were positively and concordantly induced prior to the peak of DNA synthesis (221). Similar to the case in mammalian cells, in feed-starve budding yeast, expression of the steady-state transcripts of DNA polymerases $\alpha$ (POL1) and $\delta$ (POL3) and PCNA (POL30) were found to increase 10–100-fold in late $G_1$, peak in early S phase, and decrease to basal level in late S phase (141, 226).

A report described DNA polymerase $\beta$ steady-state transcript in contact-inhibited fibroblasts having the same level of mRNA as in growing cells. This study suggests polymerase $\beta$ transcript, unlike that of polymerase $\alpha$, is not growth responsive (227).

## In Terminal Differentiation

Differentiation is an essential process of cellular development and one of the principal differences between cancer and normal cells. In terminally differentiated cells, DNA replication ceases, while neoplastic cells do not differentiate and continue to replicate their DNA. The expression of two DNA polymerases, $\alpha$ and $\beta$, has been investigated in differentiated cells. It has been demonstrated that in DMSO-induced terminally differentiated human promyelocytic leukemia (HL60) cells, DNA polymerase $\alpha$ steady-state transcript and protein are down-regulated. Results from nuclear run-on experiments indicate that the down-regulation is due to reduction in transcription rate and not to transcriptional elongation block (A. Moore, T. Wang, personal communication). It will be interesting in the future to investigate the element(s) in the polymerase $\alpha$ gene responsive to terminal differentiation signals for the induction of the down-regulation. This finding provides the molecular evidence corresponding to an earlier immunocytochemical observation of rapid disappearance of chick polymerase $\alpha$ antigen in differentiated chick embryo lens epithelial cells (222). In contrast to DNA polymerase $\alpha$, DNA polymerase $\beta$ was detected in nuclei of both undifferentiated and differentiated neural cells in the chicken embryo (222). This observation was later confirmed at the molecular level. It was found that the mRNA level of mouse polymerase $\beta$ did not change in neuroblastoma cells (N18TG2) or rat glioma cells (C6Bu-1) after induction of differentiation by dibutyryl-cAMP (228).

## In the Cell Cycle

An in-depth study of human DNA polymerase $\alpha$ gene expression at the transcriptional, posttranscriptional, and posttranslational levels has been performed in actively cycling cells separated into progressive stages of the cell cycle by counter-flow elutriation or by mitotic shake-off (221). The levels of polymerase $\alpha$ steady-state transcript, translation rate, and enzymatic activity are constitutively expressed at all stages of the cell cycle. This study has demonstrated that regulation of the expression of the polymerase $\alpha$ gene is at the transcriptional level. It also indicates that the regulatory mechanisms for the expression of polymerase $\alpha$ during the activation of quiescent cells into the mitotic cycle (transition from $G_0$ to active cycling cells) are fundamentally different from those that regulate a cell during its progression through different stages of the cell cycle. This raises an interesting question: might post-translational modification of DNA polymerase $\alpha$ underlie the activation of this enzyme, or its interaction with DNA and other replicative proteins at the onset or the termination of DNA replication?

Human DNA polymerase $\alpha$ is phosphorylated in a cell cycle–dependent manner. DNA polymerase $\alpha$ catalytic polypeptide has been found to be phosphorylated throughout the cell cycle, and hyperphosphorylated in mitotic phase. The 70-kDa subunit is only phosphorylated in mitotic phase (228a). The cell cycle–dependent phosphorylation appears due to cell cycle–dependent kinase and not cell cycle–dependent phosphatase. Both in vitro and in vivo data show that one of the mitotic kinases for the catalytic polypeptide is the key mitotic regulatory kinase, p34[cdc2] kinase, which is the sole kinase responsible for phosphorylating the 70-kDa subunit in mitotic cells. A consensus substrate motif for this mitotic kinase is suggested as serine or threonine followed by proline (-S/T-P-) (229). Close examination of the catalytic polypeptide of human polymerase $\alpha$ reveals that 10 -S/T-P- motifs exist that are potential phosphorylation sites for p34[cdc2] kinase. Interestingly, none of these 10 p34[cdc2] kinase substrate motifs resides in the six conserved domains provisionally assigned as the substrate dNTP or dNTP-$Mg^{2+}$ interacting domains. Eight of these motifs are located toward the N-terminus within the region assigned for protein-protein interaction, and two of the motifs are located in the C-terminal region predicted to be the DNA interacting domain(s) (4, 187). Moreover, the hyperphosphorylated DNA polymerase $\alpha$ isolated from mitotic cells has lower affinity for single-stranded DNA than polymerase $\alpha$ from $G_1/S$ phase cells. These results imply that mitotic phosphorylation of polymerase $\alpha$ might influence the physical interaction of this enzyme with other replicative proteins and/or with DNA at the onset and termination of DNA replication. The cell cycle–dependent phosphorylation of DNA polymerase $\alpha$ by p34[cdc2] kinase provides biochemical evidence for the first time that a kinase that plays a key role in induction of mitosis phosphorylates a replicative enzyme, DNA polymerase $\alpha$, which plays an essential role in S phase. The two major biological events in the cell cycle, DNA replication and mitosis, might share at least one common regulator, p34[cdc2] kinase.

In contrast to the finding in mammalian cells, expression of the steady-state transcripts of DNA polymerases $\alpha$ (POL1) from elutriated budding yeast has shown periodicity with a dramatic increase in late $G_1$ and a peak around the $G_1/S$ boundry (226). Gene expression of budding yeast polymerases $\alpha$ *(POL1), $\delta$ (POL3),* and PCNA *(POL30)* in feed-starve cells demonstrated dramatic periodic expression of these three genes at the transcriptional level

(141, 226). Thus far, gene expression of budding yeast polymerases δ or ε at different stages of the cell cycle has not been studied. Studies of several DNA replication proteins/enzymes from budding yeast indicate that the steady-state transcripts of these genes are all expressed periodically (180, 226). Therefore, the prediction is that budding yeast polymerases ε and δ, like other yeast replicative proteins, will be expressed periodically during the cell cycle. These studies of DNA polymerase gene expression during cell growth indicated that the regulatory mechanisms for the expression of budding yeast replicative proteins appear to be fundamentally different from those regulating the expression of mammalian replicative proteins. Human cells synchronized by double thymidine block were used to investigate the steady-state transcript level of DNA polymerase β during the cell cycle. The result indicates that the abundance of polymerase β mRNA, like that of human polymerase α, does not vary at different stages of the cell cycle (227).

*Cis-Acting Elements in the Upstream Sequence*

Analysis of a genomic clone of human DNA polymerase α that spans the 5' end of this gene and includes 1.62 kb of upstream sequence has mapped the transcription start site at 46 ± 1 nucleotides upstream from the translation start site. The upstream sequence is highly GC rich, lacks a TATA sequence, but has a CCAAT sequence on the opposite strand. Transient transfections of deletion constructs with a reporter gene into cycling cells indicate that 248 bp of sequence upstream from the cap site modulates the transcription of the DNA polymerase α gene. Clustered within these 248 nucleotides are sequences similar to consensus sequences for Sp1-, Ap1-, Ap2-, and E2F-binding sites. These elements are bound by nuclear proteins as demonstrated by DNase I footprinting and gel mobility shift experiments. Serum activation of transiently transfected cells demonstrates that there are signals within the 1.62 kb upstream sequence that confer a late serum response on reporter gene expression. Transient transfection of deletion constructs indicates that elements within the 248 bp promoter region appear to be sufficient for the serum response of polymerase α expression during activation of quiescent cells to proliferate. However, no single sequence element within the 248 bp promoter region can be identified to confer the serum-inducible expression. Multiple sequence elements within the 248 bp might be required for the full extent of serum-inducible expression of DNA polymerase α (229a).

In budding yeast, the expression of DNA polymerases α *(POL1)* and δ *(POL3)*, PCNA *(POL30)*, and several other DNA replication–related genes such as primase large subunit *(PRI1)*, primase small subunit *(PRI2)*, thymidine kinase *(CDC8)*, ligase *(CDC9)*, and thymidylate synthetase *(CDC21)* suggests that expression of these genes may be regulated by common cis- or trans-acting elements. The upstream sequences of each of these genes including polymerase $\epsilon$ *(POL2)* contain at least one copy of 5'-ACGCGTA-3', which contains the sequence for MluI restriction site (141, 180; A. Sugino, personal communication). This 5'ACGCGT3' sequence or a similar hexanucleotide with a single base-pair mismatch in each gene is located within -100 to -200 position from the translation start site. It is suggested that these sequences are responsive to signal(s) for the synchronous periodic expression of these genes at the boundary of $G_1$ and S phases. Further investigation of the upstream sequences of these genes may enable definition of the physiological significance of this consensus sequence, as well as identification of other responsive elements.

Both human and mouse DNA polymerase $\beta$ upstream sequence have been characterized. The human polymerase $\beta$ upstream sequence contains no TATAA sequence or CAAT element. It does contain three GC boxes and a 10-nucleotide perfect palindrome 49 nucleotides upstream from the transcription start site. Sequence 114 base pairs 5' of the transcription start site is defined as the core sequence required for transcription (230). The palindrome sequence in the human polymerase $\beta$ upstream sequence has been extensively studied. The decanucleotide palindrome element has been found to be required for *ras*-mediated stimulation (231). It was reported that HTLV-I trans-activator protein Tax could down-regulate the expression of human DNA polymerase $\beta$ gene, but the decanucleotide palindrome is not required for the down-regulation (232). Besides being transactivated by p21$^{v-ras}$ and Tax, the promoter of human polymerase $\beta$ has also been reported to be transactivated by adenovirus E1A and E1B gene products, $N$-methyl-$N'$-nitrosoguanidine (MNNG) treatment of the cell (78, 230, 231). Surprisingly, the upstream sequence of mouse DNA polymerase $\beta$ is described quite differently from human polymerase $\beta$. Two cis-acting negative regulatory elements or silencers (NRE-I and NRE-II) are reported in the upstream sequence of the mouse polymerase $\beta$ gene (233). Deletion of the NRE-I results in 2–3-fold stimulation of reporter gene activity, and further deletion of NRE-II increases the transcriptional activity of polymerase $\beta$ promoter another 2–3-fold. These two silencers repress the function of not only polymerase $\beta$ promoter, but also several other heterologous promoter-enhancers (233, 234). Experiments have suggested that the cell type–specific expression of mouse polymerase $\beta$ is regulated by these silencer elements (228).

## PHYSIOLOGICAL ROLES

### Replicative DNA Polymerases

DNA polymerase $\alpha$ has long been recognized as a principal replicative DNA polymerase in eukaryotes. Several lines of evidence support this claim. The expression of DNA polymerase $\alpha$ both at the transcriptional and posttranscriptional levels is regulated by and positively correlated to activation of cell proliferation (221, 225). A temperature-sensitive mutant for DNA synthesis has been identified as a DNA polymerase $\alpha$ mutant (235). Monoclonal antibodies that specifically neutralize DNA polymerase $\alpha$ inhibit DNA synthesis when microinjected into nuclei or incubated with permeablized cells (236–238). Polymerase $\alpha$ is an essential component in reconstituted in vitro SV40 DNA replication reaction (1, 2). In the primary sequence of DNA polymerase $\alpha$ from human and budding yeast, six regions are conserved among a group of replicative DNA polymerases from both prokaryotes and eukaryotes (4, 187). Budding yeast *POL1* gene, which encodes the yeast polymerase $\alpha$, has been demonstrated as essential for yeast cell growth. Conditional lethal mutants were isolated, and clearly indicated the requirement of *POL1* gene for DNA synthesis during S phase (3, 5).

The recognition of DNA polymerase $\delta$ as another principal replicative DNA polymerase was initially based on indirect evidence from the requirement for PCNA, an auxiliary protein for polymerase $\delta$, in SV40 in vitro DNA replication reaction (133, 133a, 137). Moreover, other biochemical evidence has implicated another replicative DNA polymerase besides DNA polymerase $\alpha$ in DNA replication (239–241). The finding that the *CDC2* gene encodes the budding yeast polymerase $\delta$ provides genetic evidence that polymerase $\delta$ is essential for cell growth (6, 7). An enzyme fraction partially purified from HeLa cells required for SV40 DNA replication in vitro contains a PCNA-dependent DNA polymerase activity and has properties similar to those of DNA polymerase $\delta$ (242). Recently, a human cellular fraction essential for complete reconstitution of SV40 DNA replication was characterized. The key component purified from this cellular fraction has enzymological properties identical to those of DNA polymerase $\delta$. Purified calf thymus DNA polymerase $\delta$ can fully substitute for this purified component in SV40 in vitro DNA replication reaction (243). Thus, it is evident that DNA polymerase $\delta$ is a replicative DNA polymerase required in the reconstituted SV40 DNA replication reaction. DNA polymerase $\epsilon$ was originally identified as a DNA repair enzyme in permeablized mammalian cells (31). In budding yeast, like polymerases $\alpha$ and $\delta$, disruption of the polymerase $\epsilon$ gene results in dumbbell morphology, suggesting that polymerase $\epsilon$ might also be essential for yeast cell growth, and may be another replicative DNA polymerase (8).

## Roles in Replication

An important question is: What role does each of these replicative DNA polymerases play at the initiation of DNA replication and during the elongation of DNA chains? Based on the associated enzymatic activity and the processivity of polymerases $\alpha$ and $\delta$, DNA polymerase $\delta$ has been implicated as the leading-strand DNA synthesis polymerase, whereas DNA polymerase $\alpha$ is the lagging-strand polymerase (28, 137, 152). By use of the SV40 in vitro DNA replication assays, the interactions of DNA polymerase complexes and replication proteins, RF-A, RF-C, and PCNA, at the primer-template junction have been extensively investigated (92, 147, 159, 162, 163, 166–168). Hurwitz and his colleagues have examined the step of oligonucleotide primer synthesis, and demonstrated that the synthesis of small oligonucleotide primers by primase in the SV40 in vitro replication reaction requires the presence of polymerase $\alpha$, RF-A, and T antigen. Absence of polymerase $\alpha$ blocks the synthesis of the ribonucleotide primer in this system.

Furthermore, examining the elongation step by polymerase $\delta$ has also demonstrated that the interactions between T antigen, RF-A, and polymerase $\alpha$/primase complex on the SV40 Ori DNA are highly specific, whereas a variety of DNA polymerases can substitute for polymerase $\delta$ in this step (168). At a high level of polymerase $\alpha$ and a low primer concentration, even polymerase $\alpha$ can substitute for polymerase $\delta$ in the elongation step (167). With the purified replication factors and polymerases, Stillman and his colleagues have further demonstrated that RF-C plays a crucial role in the formation of a primer recognition complex and in modulating the switch from the nascent DNA synthesis by polymerase $\alpha$/primase complex to the leading-strand synthesis by polymerase $\delta$. The replication factors, PCNA and RF-C, are proposed as the functional equivalents of the T4 phage genes 45 and 44/62, respectively (147, 159, 166, 167). To prove this proposition, T4 DNA polymerases (gene 43 product), gene 44/62 product (DNA-dependent ATPase and primer-binding protein), and gene 45 product have been successfully demonstrated to be able to substitute for DNA polymerase $\delta$/RF-C/PCNA complex for SV40 in vitro DNA synthesis in reactions containing T antigen, topoisomerases I and II, RF-A, and polymerase $\alpha$/primase complex. In addition, phage T7 DNA polymerase coupled to *E. coli* thioredoxin can substitute for polymerase $\delta$ in this reaction (162). These results in conjunction with an extensive set of experiments assessing the replication products from reactions with different concentrations of replication proteins (92, 163), strongly suggest that initiation of DNA replication in vitro from the SV40 origin occurs by a sequentially ordered protein-protein interaction. Data from these studies support the notion that coordinated leading- and lagging-strand SV40 DNA synthesis in vitro requires the appropriate balance of leading- and lagging-strand DNA polymerases, and the replication factors, RF-A, RF-C, and PCNA (92, 163). In addition, only under conditions of optimally balanced protein factors, the in vitro reconstituted SV40 DNA replication reflects the synthetic mechanisms observed with crude cell extracts, and perhaps, the mechanisms occurring in vivo.

A model for DNA polymerase switching during initiation of the bidirectional DNA replication was proposed by Stillman and his colleagues (162, 163). The model presented in Figure 3 proposes: The Ori sequences are first unwound by the combined action of T antigen, topoisomerase, and RF-A. Polymerase $\alpha$/primase complex, by interacting with T antigen and/or RF-A, synthesizes the first short oligoribonucleotides by the primase at the origin. Polymerase $\alpha$ then uses this primer to synthesize the nascent DNA (the first Okazaki fragment). During the synthesis of the nascent DNA fragment, RF-C might interact with polymerase $\alpha$ to form a lagging-strand polymerase complex. RF-C in this lagging-strand polymerase complex, by interacting with PCNA and ATP, forms an active primer recognition complex. Once the primer recognition complex is formed, polymerase $\delta$ efficiently recognizes the 3'OH-end of the first nascent DNA fragment via protein-protein interactions with this complex, and polymerase $\alpha$/primases are released from the primer terminus, thus switching from a specialized lagging-strand polymerase complex to a specialized leading-strand complex. By optimally balanced but ordered interactions of the replication factors, polymerase $\alpha$/primase is specifically blocked from synthesizing on the leading strand. The leading-strand polymerase complex then translocates along the template to synthesize DNA continuously. This eukaryotic bidirectional initiation model involves two different DNA polymerases at the replication fork, whereas in the prokaryotic model, an asymmetric polymerase III functions in both leading- and lagging-strand synthesis. In the eukaryotic model, each DNA polymerase synthesizes one strand only. This model is based mainly on the reconstituted replication reactions with SV40 origin containing DNA in which T antigen appears to be the only viral encoded protein required. It is assumed that cellular origin–binding protein(s) and a DNA helicase are functional equivalents of the T antigen for cellular chromosome replication. Notwithstanding the requirement for different initiation protein(s) and helix-unwinding proteins in cellular replication, the SV40 replication system, like M13 and G4 in *E. coli,* probably provides an accurate reflection of host chromosome DNA replication.

It is interesting to note that data from these SV40 replication studies demonstrate that neither DNA polymerase $\alpha$/primase nor RF-A, the two proteins involved in the initiation process, can be substituted by functionally equivalent proteins or enzymes from other species (151, 162, 168). These findings are consistent with several other reports including the finding of the requirement for primate polymerase $\alpha$/primase for SV40 replication in vitro (173), the recent report that there is about a 10-fold higher binding between T antigen with polymerase $\alpha$/primase complex purified from human cells than from calf (172), and the inability of yeast RF-A to substitute for primate RF-A for SV40 in in vitro DNA replication (151). The stringent species-specific interactions between polymerase $\alpha$/primase complex, T antigen, and RF-A

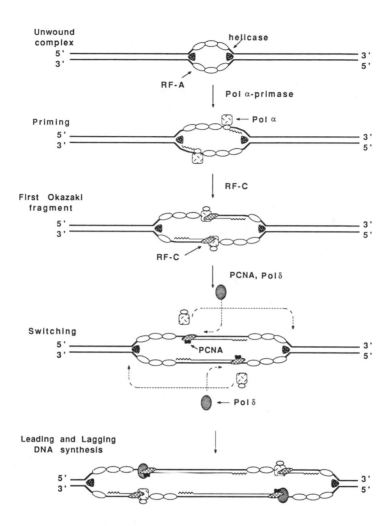

*Figure 3* Model for initiating leading- and lagging-strand synthesis with two DNA polymerases. The first line of the figure shows the unwound complex stage, followed by five subsequent stages: 1. primer RNA synthesis with primase (Priming), 2. first Okazaki fragment synthesis with polymerase $\alpha$, 3. switching of DNA polymerases at the 3'-end of this first Okazaki fragment, and finally, 4. leading- and 5. lagging-strand synthesis by polymerases $\delta$ and $\alpha$. Factors involved are DNA helicase (T antigen in the case of SV40), RF-A, polymerase $\alpha$/primase, RF-C, PCNA, and polymerase $\delta$. The factors are introduced according to the order of entry into the reaction. The roles of the topoisomerases are not indicated. Thick and thin lines are template and newly synthesized DNAs, respectively. Wavy lines are synthesized primer RNAs, and the dashed lines indicate the introduction of polymerase $\delta$ and the movement of polymerase $\alpha$/primase complex along the lagging-strand template without dissociation from the replication fork. Reprinted by permission from (162). Copyright 1990 Macmillan Magazines Limited.

appear not to be limited to the SV40 in vitro replication system. Recombinant human DNA polymerase $\alpha$ cannot complement several different alleles of temperature-sensitive polymerase $\alpha$ in budding yeast for growth at nonpermissive temperature, although recombinant human DNA polymerase $\alpha$ can be functionally expressed in these cells. Moreover, recombinant human polymerase $\alpha$ is also unable to complement diploid yeast when one copy of the yeast polymerase $\alpha$ gene is disrupted (W. Copeland, S. Francesconi, T. Wang, manuscript submitted). In contrast, polymerase $\delta$, RF-C, and PCNA as a complex can be replaced in vitro by prokaryotic homologs or eukaryotic homologs from different species (162, 167, 168). This is also consistent with the finding that calf PCNA and human RF-C are able to stimulate yeast polymerase $\delta$ DNA synthetic activity on primed single-stranded DNA (146). These results strongly suggest that there is a high stringency in the protein-protein interactions during the synthesis of the nascent DNA fragment at the origin of the replication fork.

The identification of budding yeast polymerase $\epsilon$ as a third essential DNA polymerase raises the question of whether polymerase $\epsilon$ also plays a role at the replication fork. To accommodate three DNA polymerases functioning at the replication fork, Sugino and his colleagues have proposed that polymerase $\alpha$ might function only as an initiation polymerase, whereas polymerases $\epsilon$ and $\delta$ function as the leading- and lagging-strand polymerases, respectively (8). This proposition is based solely on the biochemical properties of the three DNA polymerases such as in vitro processivity, association of primase activity with polymerase $\alpha$ but not polymerases $\delta$ and $\epsilon$, ability of polymerase $\alpha$ to interact with T antigen, and the observation of a mouse cell clone (tsFT20) that contains a thermolabile polymerase $\alpha$ in which the initiation event is stopped at nonpermissive temperature (244). Furthermore, under certain in vitro reaction conditions, RF-C and PCNA have an effect on the DNA synthetic capacity of polymerase $\epsilon$ (84, 146). A conditional mutant of budding yeast polymerase $\epsilon$ has been isolated recently; therefore, polymerase $\epsilon$ appears to be essential for cell growth during S phase (A. Sugino, personal communication).

Based on the biochemical evidence described above, Hurwitz and his colleagues have also proposed that polymerase $\alpha$/primase complex might be involved in the initiation of DNA chains for both leading and lagging strands. Polymerase $\delta$, with its auxiliary proteins, or polymerases $\delta$ and $\epsilon$, are involved in elongation of the chains initiated by polymerase $\alpha$/primase (168). In contrast to all these proposed models, a 21S enzyme complex from HeLa cells containing a 640-kDa multiprotein DNA polymerase $\alpha$/primase complex has been described that is able to support SV40 DNA replication in vitro. In the 21S replication protein complex, the only functional DNA polymerase identified has biochemical and immunological properties similar to those of polymerase $\alpha$ (245).

## Roles in DNA Repair Synthesis

DNA polymerases implicated in DNA repair synthesis are mostly based on results from differential inhibition by inhibitors such as aphidicolin, dideoxythymidine triphosphate (ddTTP), and butylphenyl-dGTP. This is an oversimplified approach to a very complex biological problem. In view of our current understanding of the response to inhibitors by eukaryotic polymerases, conclusions drawn from these earlier inhibitor studies of DNA repair synthesis enzymology need critical reevaluation. Despite these problems, these studies have provided some hints about the roles of polymerases in DNA repair synthesis. An updated and critical review of this topic is provided by Linn and his colleague (246), but is not discussed here.

A few notable conclusions are discussed here. DNA polymerase $\beta$ has been implicated in short-patch DNA repair synthesis. This notion is mainly based on the gap-filling property of DNA polymerase $\beta$ (18, 111, 117, 247), constitutive gene expression (78), and inhibitor studies (239, 248–253). By using antibody and inhibitors, the role of DNA polymerase $\beta$ in short-patch DNA repair has been further defined in the repair of G/T mispairs, which results from spontaneous hydrolytic deamination of 5'-methylcytosine (254). Polymerase $\beta$ is a constitutively expressed housekeeping polymerase in vertebrate cells. Wilson and his colleagues have shown induction of polymerase $\beta$ steady-state transcript after treatment with several monofunctional DNA-damaging agents such as MNNG (78). The transactivation of polymerase $\beta$ transcript is mediated through a decanucleotide palindrome sequence at $-49$ to $-40$ in the core promoter of polymerase $\beta$. There are nuclear protein(s) interacting with this element, and mutation of this decanucleotide element abolishes the nuclear binding as well as the response to MNNG (S. Wilson, personal communication). It should be noted that cells from invertebrates have no demonstrable polymerase $\beta$. An alternative repair mechanism must be utilized in short-patch repair by invertebrates. As discussed above, the roles of DNA polymerases $\alpha$, $\delta$, or $\epsilon$ in repair synthesis concluded from inhibitor studies are not definitive and should be further investigated.

Identification of polymerase $\epsilon$ as a factor in permeablized HeLa cells for UV-induced DNA repair synthesis strongly suggests that polymerase $\epsilon$ plays an essential role in UV-induced repair synthesis. In addition, in Xeroderma pigmentosum human diploid fibroblasts, UV-induced repair synthesis could be accomplished by polymerase $\epsilon$ and T4 endonuclease V (31). In this reaction, neither polymerase $\alpha$ nor polymerase $\beta$ can substitute for polymerase $\epsilon$ in the repair synthesis. This suggests that polymerase $\epsilon$, but not polymerases $\alpha$ or $\beta$, might be the limiting factor for UV-induced repair synthesis. Moreover, a direct comparison of purified polymerases $\delta$ and $\epsilon$ from HeLa cells has recently demonstrated that HeLa polymerase $\epsilon$, but not HeLa polymerase $\delta$, has UV-repair synthesis activity (246). These results strongly implicate polymerase $\epsilon$ as the major polymerase responsible for a majority of the UV-induced DNA repair syntheses in human fibroblasts. With the genes of polymerases $\alpha$, $\delta$, and $\epsilon$ from budding yeast clearly defined, the role of each of these three major polymerases in repair DNA synthesis induced by a specific damaging agent can be resolved by genetic approach. This is exemplified by a report of a temperature-sensitive mutant of budding yeast polymerase $\alpha$ (PolI-17) that has no demonstrable defective repair synthesis of X-ray-induced single-stranded breaks in nonpermissive temperature. This result provides unequivocal evidence that polymerase $\alpha$ is not the enzyme to repair X-ray-induced DNA damage (85).

## CONCLUSIONS AND PROSPECTS

The development of the cell-free SV40 DNA replication system and the isolation and characterization of genetic structures of DNA polymerases have greatly advanced our understanding of eukaryotic DNA polymerases and DNA replication. In retrospect, the present cell-free SV40 replication system resembles the bacteriophage replication systems that have served as models in identifying proteins and in understanding the basic mechanisms for *E. coli* chromosome replication. The accomplishments of this cell-free eukaryotic viral replication model system, and the identification of several polymerase genes and cDNAs, have now provided a starting point to investigate the functional roles of these polymerases in cellular chromosome replication and repair. To pursue this goal, it is necessary to elucidate the interactions of these replication proteins at the molecular level. Future structure-function studies of these eukaryotic DNA polymerases together with genetic studies in yeast will yield useful information for resolving this complex problem. The knowledge gained from understanding the mechanism of the interactions of each polymerase with other replicative or repair enzymes at the molecular level, and from understanding the molecular mechanisms regulating each polymerase gene expression during cell growth, will pave the way for us to comprehend the mechanisms of how chromosomes replicate, and how eukaryotic cell proliferation, cell differentiation, and the cell cycle are controlled.

Acknowledgment

I thank many colleagues for helpful discussions and generously providing me published and unpublished manuscripts. I am particularly grateful to Robert Lehman, Stuart Linn, and Bruce Stillman for discussions and critical reading of the manuscript. I thank Heinz-Peter Nasheuer, William C. Copeland, Amy Moore, Hyunsun Park, and Stefania Francesconi for reading the drafts of this manuscript, and Lynn Halpern and Alberto Martin for their help in organizing references. Research from the author's laboratory cited in this review is supported by a grant from the National Institutes of Health (CA14835).

## Literature Cited

1. Challberg, M. D., Kelly, T. J. 1989. *Annu. Rev. Biochem.* 58:671–717
2. Stillman, B. 1989. *Annu. Rev. Cell Biol.* 5:197–245
3. Johnson, L. M., Synder, M., Chang, L. M.-S., Davis, R. W., Campbell, J. L. 1985. *Cell* 43:369–77
4. Wong, S. W., Wahl, A. F., Yuan, P.-M., Arai, N., Pearson, B. E., et al. 1988. *EMBO J.* 7:37–47
5. Pizzagalli, A., Valsasnini, P., Plevani, P., Lucchini, G. 1988. *Proc. Natl. Acad. Sci. USA* 85:3772–76
6. Boulet, A., Simon, M., Faye, G., Bauer, G. A., Burgers, P. M. 1989. *EMBO J.* 8:1849–54
7. Sitney, K. C., Budd, M. E., Campbell, J. L. 1989. *Cell* 56:599–605
8. Morrison, A., Araki, H., Clark, A. B., Hamatake, R. K., Sugino, A. 1990. *Cell.* 62:1143–51
9. Campbell, J. L. 1986. *Annu. Rev. Biochem.* 55:733–71
10. Kelly, T. J. 1988. *J. Biol. Chem.* 263:17889–92
11. Kaguni, L. S., Lehman, I. R. 1988. *Biochim. Biophys. Acta* 950:87–101
12. Lehman, I. R., Kaguni, L. S. 1989. *J. Biol. Chem.* 264:4265–68
13. Burgers, P. M. J. 1989. *Progr. Nucleic Acid Res. Mol. Biol.* 37:235–80
14. Zmudzka, B. Z., SenGupta, D., Matsukage, A., Cobianti, F., Kumar, P., et al. 1986. *Proc. Natl. Acad. Sci. USA* 83:5106–10
15. Foury, F. 1989. *J. Biol. Chem.* 264:20552–60
16. Bollum, F. J. 1960. *J. Biol. Chem.* 235:2399–403
17. Weissbach, A., Baltimore, D., Bollum, F., Gallo, R., Korn, D. 1975. *Science* 90:401–2
18. Fry, M., Loeb, L. A. 1986. In *Animal Cell DNA Polymerases.* Florida: CRC
19. Bolden, A., Pedrali-Noy, G., Weissbach, A. 1977. *J. Biol. Chem.* 252:3351–56
20. Hubscher, U., Kuenzle, C. C., Silvio, S. 1979. *Proc. Natl. Acad. Sci. USA* 76:2316–20
21. Matsukage, A., Bohn, E. W., Wilson, S. H. 1975. *Biochemistry* 14:1006–20
22. Knopf, K.-W., Yamada, M., Weissbach, A. 1976. *Biochemistry* 15:4540–48
23. Wernette, C. M., Kaguni, L. S. 1986. *J. Biol. Chem.* 261:14764–70
24. Insdorf, N. F., Bogenhagen, D. F. 1989. *J. Biol. Chem.* 264:21491–97
25. Burgers, P. M. J., Bambara, R. A., Campbell, J. L., Chang, L. M. S.,

Downey, K. M., et al. 1990. *Eur. J. Biochem.* 191:617–618
26. Lee, M. Y. W. T., Tan, C-K., Downey, K. M., So, A. G. 1984. *Biochemistry* 23:1906–13
27. Crute, J. J., Wahl, A. F., Bambara, R. A. 1986. *Biochemistry* 25:26–36
28. So, A. G., Downey, K. M. 1988. *Biochemistry* 27:4591–95
29. Downey, K. M., Tan, C.-K., So, A. G., eds. 1990. *BioEssays* 12:231–36
30. Focher, F., Spadari, S., Ginelli, B., Hottiger, M., Gassmann, M., et al. 1988. *Nucleic Acids Res.* 16:6279–95
31. Nishida, C., Reinhard, P., Linn, S. 1988. *J. Biol. Chem.* 263:501–10
32. Syvaoja, J., Linn, S. 1989. *J. Biol. Chem.* 264:2489–97
33. Syvaoja, J., Suomensaari, S., Nishida, C., Goldsmith, J. S., Choi, G. S. J., et al. 1990. *Proc. Natl. Acad. Sci. USA* 87:6664–68
34. Wong, S. W., Syvaoja, J., Tan, C.-K., Downey, K. M., So, A. G., et al. 1989. *J. Biol. Chem.* 264:5924–28
35. Goulian, M., Herrmann, S. M., Sackett, J. W., Grimm, S. L. 1990. *J. Biol. Chem.* 265:16402–11
36. Wahl, A. F., Crute, J. J., Sabatino, R. D., Bodner, J. B., Marraccino, R. L., et al. 1986. *Biochemistry* 25:7821–27
37. Kunkel, T. A., Sabatino, R. D., Bambara, R. A. 1987. *Proc. Natl. Acad. Sci. USA* 84:4865–69
38. Goslin, L. A., Byrnes, J. J. 1982. *Biochemistry* 21:2513–18
39. Lee, M. Y. W. T., Toomey, N. L. 1987. *Biochemistry* 26:1076–85
40. Bambara, R. A., Myers, T. W., Sabatino, R. D. 1990. See Ref. 78, pp. 69–94
41. Mechali, M., Abadiedebat, J., de Recondo, A.-M. 1980. *J. Biol. Chem.* 255:2114–22
42. Kaguni, L. S., Rossignol, J.-M., Conaway, R. C., Lehman, I. R. 1983. *Proc. Natl. Acad. Sci. USA* 80:2221–25
43. Wang, T. S.-F., Hu, S.-Z., Korn, D. 1984. *J. Biol. Chem.* 259:1854–65
44. Wong, S. W., Paborsky, L. R., Fisher, P. A., Wang, T. S.-F., Korn, D. 1986. *J. Biol. Chem.* 261:7958–68
45. Wang, T. S.-F. 1990. See Ref. 78, pp. 147–98
46. Plevani, P., Foiani, M., Valsasnini, P., Badaracco, G., Cheriathundam, E., et al. 1985. *J. Biol. Chem.* 260:7102–7
47. Grosse, F., Krauss, G. 1985. *J. Biol. Chem.* 260:1881–88
48. Pausch, M. H., Peterson, B. C., Dumas, L. B. 1988. *Cancer Cells* 6:359–66

49. Wahl, A. F., Kowalski, S. P., Harwell, L. W., Lord, E. M., Bambara, R. A. 1984. *Biochemistry* 23:1894–99
50. Chang, L. M.-S., Rafter, E., Augl, C., Bollum, F. J. 1984. *J. Biol. Chem.* 259:14679–87
51. Masaki, S., Tanabe, K., Yoshida, S. 1984. *Nucleic Acids Res.* 12:4455–67
52. Nasheuer, H.-P., Grosse, F. 1987. *Biochemistry* 26:8458–66
53. Badaracco, G., Bianchi, M., Valsasnini, P., Magni, G., Plevani, P. 1985. *EMBO J.* 4:1313–17
54. Yamaguchi, M., Hendrickson, E. A., DePamphilis, M. L. 1985. *J. Biol. Chem.* 260:6254–63
55. Gronostajski, R. M., Field, J., Hurwitz, J. 1984. *J. Biol. Chem.* 259:9479–86
56. Goulian, M., Heard, C. J. 1989. *J. Biol. Chem.* 264:19407–15
57. Roth, Y.-F. 1987. *Eur. J. Biochem.* 165:473–81
58. Conaway, R. C., Lehman, I. R. 1982. *Proc. Natl. Acad. Sci. USA* 79:2523–27
59. Conaway, R. C., Lehman, I. R. 1982. *Proc. Natl. Acad. Sci. USA* 79:4585–88
60. Kaguni, L. S., Rossignol, J.-M., Conaway, R. C., Banks, G. R., Lehman, I. R. 1983. *J. Biol. Chem.* 288:9037–39
61. Shioda, M., Nelson, E. M., Bayne, M. L., Benbow, R. M. 1990. *Proc. Natl. Acad. Sci. USA* 79:7209–13
62. Riedel, H.-D., Konig, H., Stahl, H., Knippers, R. 1982. *Nucleic Acids Res.* 10:5621–35
63. Yagura, T., Kozu, T., Seno, T., Saneyoshi, M., Hiraga, S., et al. 1983. *J. Biol. Chem.* 258:13070–75
64. Yagura, T., Kozu, T., Seno, T., Tanaka, S. 1987. *Biochemistry* 26:7749–54
65. Tseng, B. Y., Ahlem, C. N. 1983. *J. Biol. Chem.* 258:9845–49
66. Yoshida, S., Suzuki, R., Masaki, S., Koiwai, O. 1983. *Biochim. Biophys. Acta* 741:348–57
67. Nasheuer, H.-P., Grosse, F. 1988. *J. Biol. Chem.* 263:8981–88
68. Hu, S. Z., Wang, T. S.-F., Korn, D. 1984. *J. Biol. Chem.* 259:2602–9
69. Plevani, P., Foiani, M., Francesconi, S., Mazza, C., Pizzagalli, A., et al. 1988. *Cancer Cells* 6:341–46
70. Singh, H., Dumas, L. B. 1984. *J. Biol. Chem.* 259:7936–40
71. Hsi, K.-L., Copeland, W. C., Wang, T. S-F. 1990. *Nucleic Acids Res.* 18:6231–37
72. Chang, L. M. S., Bollum, F. J. 1971. *J. Biol. Chem.* 246:5835–37
73. Chang, L. M. S., Bollum, F. J. 1972. *J. Mol. Biol.* 11:1264–72
74. Chang, L. M. S. 1973. *J. Biol. Chem.* 248:3789–95
75. Wang, T. S.-F., Sedwick, W. D., Korn, D. 1974. *J. Biol. Chem.* 249:841–50
76. Wang, T. S.-F., Eichler, D. C., Korn, D. 1977. *Biochemistry* 16:4927–34
77. Wilson, S., Abbotts, J., Widen, S. 1988. *Biochim. Biophys. Acta* 949:149–57
78. Strauss, P., Wilson, S. H., eds. 1990. *Molecular Biochemistry and Macromolecular Assemblies,* 1:199–233. New York: Telford
79. SenGupta, D. N., Zmudzka, B. Z., Kumar, P., Cobianchi, F., Skowronski, S., et al. 1986. *Biochem. Biophys. Res. Commun.* 136:341–47
80. Matsukage, A., Nishikawa, K., Ooi, T., Seto, Y., Yamaguchi, M. 1987. *J. Biol. Chem.* 262:8960–62
81. Yamaguchi, M., Matsukage, A., Takahashi, T. 1980. *J. Biol. Chem.* 255:7002–9
82. Yamaguchi, M., Tanabe, K., Takahashi, T., Matsukage, A. 1982. *J. Biol. Chem.* 57:4484–89
83. Budd, M. E., Sitney, K. C., Campbell, J. L. 1989. *J. Biol. Chem.* 264:6557–65
84. Hamatake, R. K., Hasegawa, H., Clark, A. B., Bebenek, K., Kunkel, T. A., et al. 1990. *J. Biol. Chem.* 265:4072–83
85. Budd, M. E., Wittrup, K. D., Bailey, J. E., Campbell, J. L. 1989. *Mol. Cell. Biol.* 9:365–76
86. Fisher, P. A., Wang, T. S.-F., Korn, D. 1979. *J. Biol. Chem.* 254:6128–37
87. Hohn, K.-T., Grosse, F. 1987. *Biochemistry* 26:2780–78
88. Sabatino, R. D., Myers, T. W., Bambara, R. A., Kwon-Shin, O., Marraccino, R. L., et al. 1988. *Biochemistry* 27:2998–3004
89. Cotterill, S., Chui, G., Lehman, I. R. 1987. *J. Biol. Chem.* 262:16105–8
90. Cotterill, S., Chui, G., Lehman, I. R. 1987. *J. Biol. Chem.* 262:16100–4
91. Goulian, M., Carton, C., DeGranpre, L., Heard, C., Olinger, B., et al. 1988. *Cancer Cells* 6:393–96
92. Tsurimoto, T., Stillman, B. 1991. *J. Biol. Chem.* 266:1950–60
93. Khan, N. N., Wright, G. E., Dudycz, L. W., Brown, N. C. 1985. *Nucleic Acids Res.* 13:6331–42
94. Wang, T. S.-F., Korn, D. 1982. *Biochemistry* 21:1597
95. Tan, C.-K., Castillo, C., So, A. G., Downey, K. M. 1986. *J. Biol. Chem.* 261:12310–16
96. Loeb, L. A., Kunkel, T. A. 1982. *Annu. Rev. Biochem.* 52:429–57
97. Loeb, L. A., Reyland, M. E., Eckstein, F., Lilley, D. M. J., eds. 1987. *Nucleic Acids and Molecular Biology,* pp. 157–73. Berlin/Heidelberg: Springer-Verlag

98. Kunkel, T. A., Bebenek, K. 1988. *Biochim. Biophys. Acta* 951:1–15
99. Kunkel, T. A. 1990. *Biochemistry* 29:8003–11
100. Brosius, S., Grosse, F., Krauss, G. 1983. *Nucleic Acids Res.* 11:193–202
101. Reyland, M. E., Loeb, L. A. 1987. *J. Biol. Chem.* 262:10824–30
102. Cotterill, S. M., Reyland, M. E., Loeb, L. A., Lehman, I. R. 1987. *Proc. Natl. Acad. Sci. USA* 84:5653–39
103. Reyland, M. E., Lehman, I. R., Loeb, L. A. 1988. *J. Biol. Chem.* 263:6518–24
104. Bialek, G., Nasheuer, H. P., Goetz, H., Grosse, F. 1989. *EMBO J.* 8:1833–39
105. Korn, D., Fisher, P. A., Wang, T. S.-F. 1983. In *New Approaches in Eukaryotic DNA Replication*, pp. 517–58. New York and London: Plenum
106. Fisher, P. A., Korn, D. 1979. *J. Biol. Chem.* 254:11033–39
107. Fisher, P. A., Korn, D. 1979. *J. Biol. Chem.* 254:11040–46
108. Fisher, P. A., Chen, J. T., Korn, D. 1981. *J. Biol. Chem.* 256:133–41
109. Fisher, P. A., Korn, D. 1981. *Biochemistry* 20:4560–69
110. Fisher, P. A., Korn, D. 1981. *Biochemistry* 20:4570–78
111. Wang, T. S.-F., Korn, D. 1980. *Biochemistry* 19:1782–90
112. Fisher, P. A., Korn, D. 1977. *J. Biol. Chem.* 252:6528–35
113. Reckmann, B., Grosse, F., Krauss, G. 1983. *Nucleic Acids Res.* 11:7251–60
114. Perrino, F. W., Loeb, L. A. 1989. *J. Biol. Chem.* 264:2898–905
115. Perrino, F. W., Loeb, L. A. 1989. *Proc. Natl. Acad. Sci. USA* 86:3085–88
116. Perrino, F. W., Loeb, L. A. 1990. *Biochemistry* 29:5226–31
117. Mosbaugh, D. W., Linn, S. 1984. *J. Biol. Chem.* 259:10247–51
118. Tanabe, K., Bohn, E. W., Wilson, S. H. 1979. *Biochemistry* 18:3401–6
119. Kunkel, T. A. 1985. *J. Biol. Chem.* 260:5787–96
120. Kornberg, A. 1980. In *DNA Replication*. San Francisco: Freeman
121. Kornberg, A. 1982 Supplement to *DNA Replication*. San Francisco: Freeman
122. Lucchini, G., Francesconi, S., Foiani, M., Badaracco, G., Plevani, P. 1987. *EMBO J.* 6:737–42
123. Foiani, M., Santocanale, C., Plevani, P., Lucchini, G. 1989. *Mol. Cell. Biol.* 9:3081–87
124. Foiani, M., Lindner, A. J., Hartmann, G. R., Lucchini, G., Plevani, P. 1989. *J. Biol. Chem.* 264:2189–94
125. Hirose, F., Yamamoto, S., Yamaguchi, M., Matsukage, A. 1988. *J. Biol. Chem.* 263:2925–33
126. Prussak, C. E., Almazan, M. T., Tseng, B. Y. 1989. *J. Biol. Chem.* 264:4957–63
127. Tseng, B. Y., Prussak, C. E., Almazan, M. T. 1989. *Mol. Cell. Biol.* 9:1940–45
128. Insdor, N. F., Bogenhagen, D. F. 1989. *J. Biol. Chem.* 264:21498–503
129. Matsukage, A., Nishimoto, Y. 1990. *J. Biochem.* 107:213–16
130. Brutlag, D., Kornberg, A. 1972. *J. Biol. Chem.* 247:241–48
131. Muzyczka, N., Poland, R. L., Bessman, M. J. 1972. *J. Biol. Chem.* 247:7116–22
132. Bauer, G. A., Heller, H-M., Burgers, P. M. 1988. *J. Biol. Chem.* 263:917–24
132a. Brooke, G. R., Singhal, R., Hinkle, D. C., Dumas, L. B. 1991. *J. Biol. Chem.* In press
133. Prelich, G., Kostura, M., Marshak, D. R., Mathews, M. B., Stillman, B. 1987. *Nature* 326:471–75
133a. Prelich, G., Tan, C., Kostura, M., Mathews, M. B., So, A. G., et al. 1987. *Nature* 326:517–20
134. Bravo, R., Frank, R., Blundell, P. A., Macdonald-Bravo, H. 1987. *Nature* 326:515–17
135. Miyachi, K., Fritzler, M. J., Tan, E. M. 1978. *J. Immunol.* 121:2228–34
136. Tan, C-K., Sullivan, K., Li, X., Tan, E. M., Downey, K. M., et al. 1987. *Nucleic Acids Res.* 15:9299–308
137. Prelich, G., Stillman, B. 1988. *Cell* 53:117–26
138. Mathews, M. B. 1989. *Growth Control During Cell Aging*, ed. H. R. Warner, E. Wang, pp. 89–116. Boca Raton: CRC
139. Matsumoto, K., Moriuchi, T., Koji, T., Nakane, P. K. 1987. *EMBO J.* 6:637–42
140. Almendral, B. M., Huebsch, D., Blundell, P. A., Macdonald-Bravo, H., Bravo, R. 1987. *Proc. Natl. Acad. Sci. USA* 84:1575–79
141. Bauer, G. A., Burgers, P. M. J. 1990. *Nucleic Acids Res.* 18:261–65
142. Yamaguchi, M., Nishida, Y., Moriuchi, T., Hirose, F., Hui, C.-C., et al. 1990. *Mol. Cell. Biol.* 10:872–79
143. Ng, L., Prelich, G., Anderson, C. W., Stillman, B., Fisher, P. A. 1990. *J. Biol. Chem.* 265:11948–54
144. Wold, M. S., Li, J. J., Weinberg, D. H., Virshup, D. M., Sherley, J. L., et al. 1988. *Cancer Cells* 6:133–41
145. Bauer, G. A., Burgers, P. M. J. 1988. *Proc. Natl. Acad. Sci. USA* 85:7506–10
146. Burgers, P. M. J. 1991. *Eukaryotic DNA Polymerases*, ed. J. A. Engler, T. S.-F. Wang. New York: Springer-Verlag. In press

147. Tsurimoto, T., Stillman, B. 1990. *Proc. Natl. Acad. Sci. USA* 87:1023–27
148. Wobbe, C. R., Weissbach, L. J., Borowiec, J. A., Dean, F. B., Mura Kami, Y., et al. 1987. *Proc. Natl. Acad. Sci. USA* 84:1834–38
149. Wold, M. S., Weinberg, D. H., Virshup, D. M., Li, J. J., Kelly, T. J. 1989. *J. Biol. Chem.* 264:2801–9
150. Fairman, M. P., Stillman, B. 1988. *EMBO J.* 7:1211–18
151. Brill, S. J., Stillman, B. 1989. *Nature* 342:92–95
152. Tsurimoto, T., Stillman, B. 1989. *EMBO J.* 8:3883–89
153. Kenny, M. K., Lee, S.-H., Hurwitz, J. 1989. *Proc. Natl. Acad. Sci. USA* 86:9757–61
154. Kenny, M. K., Schlegel, U., Furneaux, H., Hurwitz, J. 1990. *J. Biol. Chem.* 265:7693–700
155. Heyer, W.-D., Rao, M. R. S., Erdile, L. F., Kelly, T. J., Kolodner, R. D. 1990. *EMBO J.* 9:2321–29
156. Erdile, L. F., Wold, M. S., Kelly, T. J. 1990. *J. Biol. Chem.* 265:3177–82
157. Din, S., Brill, S. J., Fairman, M. P., Stillman, B. 1990. *Genes Dev.* 4:968–77
158. Wold, M. S., Kelly, T. 1988. *Proc. Natl. Acad. Sci. USA* 85:2523–27
159. Hurwitz, J., Dean, F. B., Kwong, A. D., Lee, S.-H. 1990. *J. Biol. Chem.* 265:18043–46
160. Tsurimoto, T., Stillman, B. 1989. *Mol. Cell. Biol.* 9:609–19
161. Tsurimoto, T., Fairman, M. P., Stillman, B. 1989. *Mol. Cell. Biol.* 9:3839–49
162. Tsurimoto, T., Melendy, T., Stillman, B. 1990. *Nature* 346:534–39
163. Tsurimoto, T., Stillman, B. 1991. *J. Biol. Chem.* 266:1961–68
164. Cha, T.-A., Alberts, B. M. 1988. *Cancer Cells* 6:1–10
165. Alberts, B. M. 1990. *Nature* 346:514–15
166. Lee, S.-H., Jurwitz, J. 1990. *Proc. Natl. Acad. Sci. USA* 87:5672–76
167. Ishimi, Y., Claude, A., Bullock, P., Hurwitz, J. 1988. *J. Biol. Chem.* 263:19723–33
168. Matsumoto, T., Eki, T., Hurwitz, J. 1990. *Proc. Natl. Acad. Sci. USA.* In press
169. Smale, S., Tjian, R. 1986. *Mol. Cell. Biol.* 6:4077–87
170. Gannon, J. V., Lane, D. P. 1987. *Nature* 329:456–58
171. Gough, G., Gannon, J. V., Lane, D. P. 1988. *Cancer Cells* 6:153–58
172. Dornreiter, I., Hoss, A., Arthur, A., Fanning, E. 1990. *EMBO J.* 9:3329–36

173. Murakami, Y., Wobbe, C. R., Weissbach, L., Dean, F. B., Hurwitz, J. 1986. *Proc. Natl. Acad. Sci. USA* 83:2869–73
174. Pritchard, C. G., DePamphilis, M. L. 1983. *J. Biol. Chem.* 258:9801–9
175. Pritchard, C. G., Weaver, D. T., Baril, E. F., DePamphilis, M. L. 1983. *J. Biol. Chem.* 258:9810–19
176. Goulian, M., Heard, C. J., Grimm, S. L. 1990. *J. Biol. Chem.* 265:13221–30
177. Goulian, M., Heard, C. J. 1990. *J. Biol. Chem.* 265:13231–39
178. Wang, T. S.-F., Pearson, B. E., Suomalainen, H. A., Mohandas, T., Shapiro, L. J., et al. 1985. *Proc. Natl. Acad. Sci. USA* 82:5270–74
179. Adler, D. A., Tseng, B. Y., Wang, T. S.-F., Disteche, C. M. 1991. *Genomics.* In press
180. Plevani, P., Foiani, M., Francesconi, S., Pizzagalli, A., Santocanale, C., et al. 1991. See Ref. 146
181. Matsukage, A., Yamaguchi, M., Utsumi, K. R., Hayashi, Y., Ueda, R., et al. 1986. *J. Cancer Res.* 77:330–33
182. McBride, O. W., Zmudzka, B. Z., Wilson, S. H. 1987. *Proc. Natl. Acad. Sci. USA* 84:503–7
183. Cannizzaro, L. A., Bollum, F. J., Huebner, K., Croce, C. M., Cheung, L. C., et al. 1988. *Cytogenet. Cell Genet.* 47:121–24
184. McBride, O. W., Kozak, C. A., Wilson, S. H. 1990. *Cytogenet. Cell Genet.* 53:108–11
185. Ollis, D. L., Kline, C., Steitz, T. A. 1985. *Nature* 313:818–19
186. Anderson, R. S., Lawrence, C. B., Wilson, S. H., Beattie, K. L. 1987. *Gene* 60:163–73
187. Wang, T. S.-F., Wong, S. W., Korn, D. 1989. *FASEB J.* 3:14–21
188. Miller, M. A., Korn, D., Wang, T. S.-F. 1988. *Nucleic Acids Res.* 16:7961–73
189. Morrison, A., Christensen, R. B., Alley, J., Beck, A. K., Bernstine, E. G., et al. 1989. *J. Bacteriol.* 171:5659–67
190. Earl, P. L., Jones, E. P., Moss, B. 1986. *Proc. Natl. Acad. Sci. USA* 83:3659–63
191. Gibbs, J. S., Chiou, H. C., Hall, J. D., Mount, D. W., Retondo, M. J., et al. 1985. *Proc. Natl. Acad. Sci. USA* 82:7969–73
192. Larder, B. A., Kemp, S. D., Darby, G. 1987. *EMBO J.* 6:169–75
193. Kouzardies, T., Banker, A. T., Satchwell, S. C., Weston, K., Tomlinson, P., et al. 1987. *J. Virol.* 61:125–33
194. Gingeras, T. R., Sciaky, D., Gelinas,

R. E., Bing-Dong, J., Parsons, B. L., et al. 1982. *J. Biol. Chem.* 257:13475–91
195. Spicer, E. K., Rush, J., Fung, C., Reha-Krantz, L. J., Karam, J. D., et al. 1988. *J. Biol. Chem.* 263:7478–86
196. Bernad, A., Zaballos, A., Salas, M., Blanco, L. 1987. *EMBO J.* 6:4219–25
197. Jung, G., Leavitt, M. C., Hsieh, J.-C., Ito, J. 1987. *Proc. Natl. Acad. Sci. USA* 84:8287–91
198. Paillard, M., Sederoff, R. R., Levings, C. S. 1985. *EMBO J.* 4:1125–28
199. Stark, M. J., Mileham, A. J., Romanos, M. A., Boyd, A. 1984. *Nucleic Acids Res.* 12:6011–30
200. Kikuchi, Y., Hirai, K., Gunge, N., Hishinuma, F. 1985. *EMBO J.* 4:1881–86
201. Bonner, C. A., Hays, S., McEntee, K., Goodman, M. F. 1990. *Proc. Natl. Acad. Sci. USA.* 87:7663–67
202. Abbotts, J., SenGupta, D. N., Zmudzka, B., Widen, S. G., Notario, V., Wilson, S. H. 1988. *Biochemistry* 27:901–9
203. Date, T., Yamaguchi, M., Hirose, F., Nishimoto, Y., Tanihara, K., et al. 1988. *Biochemistry* 27:2983–90
204. Date, T., Yamamoto, S., Tanihara, K., Nishimoto, Y., Liu, N., et al. 1990. *Biochemistry* 29:5027–34
205. Kumar, A., Widen, S. G., Williams, K. R., Kedar, P., Karpel, R. L., Wilson, S. H. 1990. *J. Biol. Chem.* 265:2124–31
206. Kumar, A., Abbotts, J., Karawya, E. M., Wilson, S. H. 1990. *Biochemistry* 29:7156–59
207. Basu, A., Kedar, P., Wilson, S. H., Modak, M. J. 1989. *Biochemistry* 28:6305–9
208. Budd, M., Campbell, J. L. 1987. *Proc. Natl. Acad. Sci. USA* 84:2838–42
209. Lucchini, G., Mazza, C., Scacheri, E., Plevani, P. 1988. *Mol. Gen. Genet.* 212:459–65
210. Lucchini, G., Falconi, M. M., Pizzagalli, A., Aguilera, A., Klein, H. L., Plevani, P. 1990. *Gene* 90:99–104
211. Gibbs, J. S., Chiou, H. C., Bastow, K. F., Cheng, Y-C., Coen, D. M. 1988. *Proc. Natl. Acad. Sci. USA* 85:6672–76
212. Coen, D. M., Fuman, P. A., Aschman, D., Schaffer, P. A. 1983. *Nucleic Acids Res.* 11:5287–97
213. Tsurumi, T., Maeno, K., Nishiyama, Y. 1987. *J. Virol.* 61:388–94
214. Ollis, D. L., Brick, P., Hamlin, R., Xuong, N. G., Steitz, T. A. 1985. *Nature* 313:762–66
215. Joyce, C. M., Steitz, T. A. 1987. *Science* 12:288–92
216. Freemont, P. S., Friedman, J. M., Beese, L. S., Sanderson, M. R., Steitz,

T. A. 1988. *Proc. Natl. Acad. Sci. USA* 85:8924–28
217. Derbyshire, V., Freemont, P. S., Sanderson, M. R., Beese, L., Friedman, J. M., et al. 1988. *Science* 240:199–201
218. Bernad, A., Blanco, L., Lazaro, J. M., Martin, G., Salas, M. 1989. *Cell* 59:219–28
219. Bensch, K. G., Tanaka, S., Hu, S.-Z., Wang, T. S.-F., Korn, D. 1982. *J. Biol. Chem.* 257:8391–96
220. Namikawa, R., Ueda, R., Suchi, T., Itoh, G., Ota, K., et al. 1987. *Am. J. Clin. Pathol.* 87:725–31
221. Wahl, A. F., Geis, A. M., Spain, B. H., Wong, S. W., Korn, D., et al. 1988. *Mol. Cell. Biol.* 8:5016–25
222. Matsukage, A., Kitani, H., Yamaguchi, M., Kusakabe, M., Morita, T., et al. 1986. *Dev. Biol.* 117:226–32
223. Hirose, F., Hotta, Y., Yamaguchi, M., Matsukage, A. 1989. *Exp. Cell. Res.* 181:169–80
224. Nowak, R., Siedlecki, J. A., Kaczmarek, L., Zmudzka, B. Z., Wilson, S. H. 1989. *Biochim. Biophys. Acta* 1008:203–7
225. Thommes, P., Reiter, T., Knippers, R. 1986. *Biochemistry* 25:1308–14
226. Johnston, L. H., White, J. H. M., Johnson, A. L., Lucchini, G., Plevani, P. 1987. *Nucleic Acids Res.* 15:5017–30
227. Zmudzka, B. Z., Fornace, A., Collins, J., Wilson, S. H. 1988. *Nucleic Acids Res.* 16:9587–96
228. Yamaguchi, M., Hayashi, Y., Ajiro, K., Matsukage, A. 1989. *J. Cell. Physiol.* 141:431–36
228a. Nasheuer, H.-P., Moore, A., Wahl, A. F., Wang, T. S.-F. 1991. *J. Biol. Chem.* In press
229. Moreno, S., Nurse, P. 1990. *Cell* 61:549–51
229a. Pearson, B. F., Nasheuer, H.-P., Wang, T. S.-F. 1991. *Mol. Cell Biol.* In press
230. Widen, S. G., Kedar, P., Wilson, S. H. 1988. *J. Biol. Chem.* 263:16992–98
231. Kedar, P. S., Lowy, D. R., Widen, S. G., Wilson, S. H. 1990. *Mol. Cell. Biol.* 10:3852–56
232. Jeang, K.-T., Widen, S. G., Semmes, O. J. IV, Wilson, S. H. 1990. *Science* 247:1082–84
233. Yamaguchi, M., Hayashi, Y., Matsukage, A. 1989. *J. Biol. Chem.* 105:79–83
234. Yamaguchi, M., Matsukage, A. 1989. *J. Biol. Chem.* 264:16887–91
235. Murakami, Y., Yasuda, H., Miyazawa, H., Hanaoka, F., Yamada, M. 1985. *Proc. Natl. Acad. Sci. USA* 83:2869–73
236. Tanaka, S., Hu, S.-Z., Wang, T. S.-F.,

552    WANG

Korn, D. 1982. *J. Biol. Chem.* 257:8386–90

237. Miller, M. R., Ulrich, R. G., Wang, T. S.-F., Korn, D. 1985. *J. Biol. Chem.* 260:134–38

238. Miller, M. R., Seighman, C., Ulrich, R. G. 1985. *Biochemistry* 24:7440–45

239. Dresler, S. L., Frattini, M. G. 1986. *Nucleic Acids Res.* 14:7093–102

240. Hammond, R. A., Byrnes, J. J., Miller, M. R. 1987. *Biochemistry* 26:6717–824

241. Decker, R. S., Yamaguchi, M., Possenti, R., Bradley, M. K., DePamphilis, M. L. 1987. *J. Biol. Chem.* 262:10863–72

242. Weinberg, D. H., Kelly, T. J. 1988. *Proc. Natl. Acad. Sci. USA* 86:9742–46

243. Melendy, T., Stillman, B. 1991. *J. Biol. Chem.* 266:1942–49

244. Eki, T., Murakami, Y., Enomoto, T., Hanaoka, F., Yamada, M-A. 1986. *J. Biol. Chem.* 261:8888–93

245. Malkas, L. H., Hickey, R. J., Li, C., Pedersen, N., Baril, E. F. 1990. *Biochemistry* 29:6362–74

246. Keeney, S., Linn, S. 1990. *Mutat. Res.* 236:239–52

247. Siedlecki, J. A., Szyszko, J., Pietrzykowska, I., Zmudzka, B. 1980. *Nucleic Acids Res.* 8:361–75

248. Dresler, S. L., Lieberman, M. W. 1983. *J. Biol. Chem.* 258:9990–94

249. Dresler, S. L., Kimbro, K. S. 1987. *Biochemistry* 26:2664–68

250. Dresler, S. L., Frattini, M. G. 1988. *Biochem. Pharmacol.* 37:1033–37

251. Miller, M. R., Chinault, D. N. 1982. *J. Biol. Chem.* 257:10204–9

252. Smith, C. A., Okumoto, D. S. 1984. *Biochemistry* 23:1383–90

253. DiGiuseppe, J. A., Dresler, S. L. 1989. *Biochemistry* 28:9515–20

254. Wiebauer, K., Jiricny, J. 1990. *Proc. Natl. Acad. Sci. USA* 87:5842–45

*Annu. Rev. Biochem. 1991. 60:553–75*

# SIGNAL TRANSDUCTION BY GUANYLYL CYCLASES

*Michael Chinkers*

Vollum Institute for Advanced Biomedical Research, Oregon Health Sciences University, Portland, Oregon 97201-3098

*David L. Garbers*

Howard Hughes Medical Institute and Department of Pharmacology, University of Texas Southwestern Medical Center, Dallas, Texas 75235-9050

KEY WORDS: cyclic GMP, receptors, natriuretic peptides, enterotoxins, endothelium-derived relaxing factor.

## CONTENTS

0066-4154/91/0701-0553$02.00

## PERSPECTIVES AND SUMMARY

After its discovery in 1963 (1), scientists were challenged to define the cellular function of guanosine 3',5'-monophosphate (cGMP). The gauntlet was taken up by many, but despite major advances, especially within the past few years, cGMP has continued to conceal many of its cellular responsibilities. Nevertheless, there is little doubt that cGMP is an important second messenger across the Animal Kingdom in almost every type of cell.

Guanylyl cyclase, responsible for the enzymatic synthesis of cGMP, was first described in 1969 (2–5) and shown to be distinct from adenylyl cyclase. Whereas adenylyl cyclase activity was found in only the particulate fractions of most tissue homogenates, guanylyl cyclase activity was found in both the soluble and particulate fractions, the relative distribution varying dependent on the tissue (6). Up until the mid-1970s, despite many reports showing that various agents could elevate cGMP levels in intact cells (7), no laboratories were able to reproducibly stimulate guanylyl cyclase activity in broken cell preparations. In the mid-to-late 1970s, sodium azide, nitric oxide, sodium nitroprusside, and various other agents were shown to activate the soluble form of guanylyl cyclase in broken cell preparations; these same agents also could cause marked elevations of cGMP in intact cells (8–10). Subsequently, Gerzer et al (11) reported that soluble guanylyl cyclase purified from bovine lung contained associated heme. Although not yet directly demonstrated, the subsequent model developed for regulation of the cytoplasmic form of guanylyl cyclase was that nitric oxide or a similar molecule binds to the associated heme, resulting in enzyme activation (12).

While work progressed on soluble enzyme regulation, research continued to falter on mechanisms of regulation of the particulate form. The most significant problem was the continued failure to activate the enzyme in broken cell preparations with potential regulatory molecules. This problem was first solved in the late 1970s, when small peptides obtained from *Escherichia coli*, known as heat-stable enterotoxins, were shown to activate a particulate form of guanylyl cyclase from intestinal mucosal cells (13–15). The intestinal enzyme, however, was not solubilized by nonionic detergents, unlike particulate guanylyl cyclases found in other tissues. In the early 1980s, peptides obtained from sea urchin eggs that activate spermatozoa (16–18), and a peptide obtained from the heart (atrial natriuretic peptide) that stimulates natriuresis, diuresis, and vasodilation (19, 20) were isolated and shown to cause elevations of cGMP in intact cells, and activation of a nonionic-detergent–soluble form of particulate guanylyl cyclase in broken cell preparations (21–25). Since these initial observations, much has been learned about the mechanisms of regulation of the particulate form of guanylyl cyclase, most importantly that particulate guanylyl cyclases can serve as cell surface receptors for a variety of extracellular peptide ligands.

This review concentrates on the mechanisms by which sea urchin egg peptides, natriuretic peptides, and bacterial heat-stable enterotoxins activate the particulate form of guanylyl cyclase, and by which nitric oxide and related molecules activate the soluble form of the enzyme. The guanylyl cyclase/cGMP system of retinal rod outer segments has been reviewed recently (26, 27), and we do not discuss it.

# ACTIVATORS OF GUANYLYL CYCLASES

## Sea Urchin Egg Peptides

Peptides produced and secreted by echinoderm eggs can activate and cause directional changes in sperm motility in a species-specific manner (28). One of the earliest biochemical events following binding of the peptides to their receptor is an elevation of intracellular cGMP, a response that appears to be explained by the activation of a plasma membrane form of guanylyl cyclase (25). Invertebrate spermatozoa contain little, if any, soluble guanylyl cyclase activity. Although the stimulation of motility can be mimicked by 8-bromo-cGMP, suggesting that the egg peptides may stimulate sperm motility by virtue of their effects on cGMP, motility stimulation has been observed at concentrations of peptide that do not detectably elevate cGMP, and by peptide analogues that do not measureably increase cGMP concentrations (29). Although the possibility of compartmentalization prevents ruling out cGMP as the second messenger, alternative possibilities are that multiple peptide receptors exist (30, 31) or that a single peptide receptor signals through multiple second messenger pathways.

In experiments designed to understand the mechanism by which the egg peptides activate the sperm plasma membrane form of guanylyl cyclase, it was discovered that one egg peptide, resact (CVTGAPGCVGGGRL-NH$_2$), could be covalently crosslinked to a cell surface protein that was subsequently identified as the membrane form of guanylyl cyclase by immunological and other criteria (32). Therefore, guanylyl cyclase appeared either to be the cell surface receptor for an egg peptide or to be closely associated with the receptor. A complementary DNA (cDNA) encoding the sea urchin guanylyl cyclase was subsequently cloned, and its structure is discussed in a later section.

The sea urchin sperm plasma membrane form of guanylyl cyclase also appears to be regulated by its state of phosphorylation (33–35). Under normal conditions the sperm guanylyl cyclase contains up to 17 mol phosphate/mol enzyme, all on serine residues, but after the treatment of sperm cells with resact the amount of phosphate associated with guanylyl cyclase falls to less than 2 mol/mol enzyme (36). Bentley et al (25) demonstrated that resact activates guanylyl cyclase before any loss of phosphate, and that the subsequent decrease in phosphate content is associated with a reduction in enzyme activity. Therefore, dephosphorylation of the sperm enzyme appears to act as a desensitization step. The state of enzyme phosphorylation also alters the kinetic behavior of the enzyme. The plasma membrane form of guanylyl cyclase has been distinguished from the soluble form based on velocity patterns as a function of the substrate, MeGTP (6, 37, 38); the plasma membrane form displays positive cooperative behavior, whereas the soluble form displays linear behavior as a function of MeGTP concentration. Ramarao & Garbers succeeded in purifying the phosphorylated form of the sea urchin sperm enzyme; it displayed positive cooperative behavior as a function of the MeGTP concentration (39). After dephosphorylation of the enzyme, however, linear kinetics were observed. Dephosphorylation of the purified enzyme also reduced its activity by approximately 80% (39), consistent with the hypothesis that the dephosphorylation observed in vivo after treatment with egg peptides is a mechanism for desensitization.

## Natriuretic Peptides

Several related peptides have now been described that cause vasodilation, diuresis, and natriuresis, and inhibit aldosterone synthesis; they have been designated atrial natriuretic peptide (ANP), brain natriuretic peptide (BNP), and C-type natriuretic peptide (CNP) (40–42). Another peptide that appears to be similar to, or the same as BNP, is named iso-atrial natriuretic peptide (43). Although these peptides are referred to as natriuretic peptides based on the original observations of deBold and colleagues (19), their physiological roles have not been fully elucidated. For example, ANP, the first peptide to be isolated from cardiac atrium, also increases testosterone production in Leydig cells (44), is synthesized in various tissues other than the heart (45), and has receptors in many different tissues including the brain (45–48).

The natriuretic peptides share a common core structure, including two cysteine residues that form a disulfide bond required for elevation of cGMP (49). Deletion of the three carboxyl-terminal amino acids of ANP also destroys its cGMP-elevating activity (50). ANP is highly conserved across various species (40), whereas the peptides designated as BNP show considerable diversity across species (Table 1). Like ANP, CNP is apparently highly conserved across species (50a, 50b). The diversity of BNP is intriguing since the individual natriuretic peptide receptors so far characterized are highly conserved between rat, mouse, and human (see later section); this implies that the BNP receptor has not yet been isolated.

Most of the biological effects of natriuretic peptides appear to be mediated by the activation of membrane-associated guanylyl cyclase activity (40, 45, 54, 55). cGMP analogues have been shown to mimic most of the effects of ANP, although the inhibition of aldosterone synthesis initially appeared to be an exception (reviewed in Refs. 40 and 45). Recent data have suggested,

**Table 1**  Variations in natriuretic peptide sequences across species

| Peptide | Species | Sequence |
|---|---|---|
| ANP[a] | rat, pig | S–L–R–R–S–S–C–F–G–G–R–I–D–R–I–G–A–Q–S–G–L–G–C–N–S–F–R–Y |
| | human, dog | S–L–R–R–S–S–C–F–G–G–R–M–D–R–I–G–A–Q–S–G–L–G–C–N–S–F–R–Y |
| BNP[b] | rat | N–S–K–M–A–H–S–S–S–C–F–G–Q–K–I–D–R–I–G–A–V–S–R–L–G–C–D–G–L–R–L–F |
| | human | S–P–K–M–V–Q–G–S–G–C–F–G–R–K–M–D–R–Q–S–S–S–S–G–L–G–C–K–V–L–R–R–H |
| | pig | S–P–K–T–M–R–D–S–G–C–F–G–R–L–D–R–I–G–S–L–S–G–L–G–C–N–V–L–R–R–Y |
| | dog | S–P–K–M–M–H–K–S–G–C–F–G–R–L–D–R–I–G–S–L–S–G–L–G–C–N–V–L–R–K–Y |

[a] Sequences from Ref. 40.
[b] Sequences from Refs. 51–53, 41.

however, that the inhibition of aldosterone synthesis may also be mediated by cGMP, through activation of a phosphodiesterase and the resulting decreases in cyclic AMP (55a). Two broad classes of natriuretic peptide receptors were initially distinguished by chemical crosslinking and receptor purification. The more abundant receptor subtype in most cells consists of a homodimer of a 65-kDa protein, which has been designated the $R_2$ or ANP clearance-receptor (ANP C-receptor) (56–61); this receptor was named the ANP C-receptor in conjunction with the hypothesis that its primary function is to clear excess ANP from the circulation. ANP C-receptors, which are often more abundant than the guanylyl cyclase-linked receptors, have been postulated to function as a buffer to prevent sudden and large changes in blood pressure when ANP is secreted from the atrium (57). The ANP C-receptor has a high affinity for ANP and for truncated analogues of ANP; the receptor also recognizes a large variety of different ANP-like structures including those that contain substituted D-amino acids (62, 63). The physiological role of the ANP C-receptor, however, remains controversial. Not only does one report (64) suggest that it is coupled to the activation of guanylyl cyclase, an observation not confirmed by others (65–67), but various data suggest that it signals through a guanine nucleotide–binding regulatory protein to activate phospholipase C and inhibit adenylyl cyclase (68–73). In the adenylyl cyclase experiments, inhibition of adenylyl cyclase by ANP or truncated analogues of ANP in isolated membranes was not observed unless GTPγS was added; the ANP effect could also be blocked by pertussis toxin (72).

The second type of ANP receptor is a monomeric 130-kDa protein that is coupled to the activation of guanylyl cyclase. This receptor, unlike the ANP C-receptor, has a low affinity for truncated analogues of ANP (62, 63); it specifically binds intact ANP with high affinity. Since activation of guanylyl cyclase appears to be a primary mechanism by which ANP exerts it physiological effects, and since most studies report no coupling between the ANP C-receptor and cGMP elevations, the 130-kDa receptor has been presumed to be the "biologically active" receptor in terms of the cGMP signaling pathway (65–67). The observation by several groups that this protein copurified with guanylyl cyclase activity suggested that, as in the case of the egg peptides, guanylyl cyclase might itself be a receptor for natriuretic peptides (58, 74, 75). One report also described the copurification of a 180-kDa protein with ANP-binding and guanylyl cyclase activities (76).

## Bacterial Heat-Stable Enterotoxins

Heat-stable enterotoxins are small peptides produced by *E. coli* and other bacteria that cause acute diarrhea (77–80). These peptides stimulate a particulate form of guanylyl cyclase found in small and large intestine resulting in marked elevations of cGMP (13–15), and subsequent effects on chloride and other ion transport (81–83). The enterotoxin peptides contain three disulfide bonds that are critical for activity; the pattern of disulfide linkages has been determined, and the importance of individual disulfide bonds has been ranked in order of importance for binding to intestinal receptors and production of diarrhea in suckling mice (84). The enterotoxin-stimulated guanylyl cyclase appears to differ from the plasma membrane forms of guanylyl cyclase described above in that it is relatively resistant to solubilization in nonionic detergents unless high concentrations of salt are present (85). It has been suggested that the intestinal enzyme is confined to the cytoskeleton and that the enterotoxin receptor and guanylyl cyclase are separate molecules (85, 86). However, even after detergent-solubilization of enterotoxin-binding proteins, guanylyl cyclase remaining in the insoluble fraction continues to be stimulated by the enterotoxin. Biochemical identification of the enterotoxin receptor has been difficult. The recent studies of Thompson & Giannella (87) demonstrated that, dependent on the crosslinking reagent used, different apparent enterotoxin-binding proteins were identified. With a crosslinking reagent (1-ethyl-3-(dimethylaminopropyl) carbodiimide) containing no spacer between adjacent carboxyl and amino groups, the major band radiolabeled had an apparent molecular weight of 125,000–130,000; this was suggested to be the protein most closely associated with the enterotoxin-binding site, possibly the receptor itself (87). Forte et al obtained similar results using disuccinimidyl suberate to identify an enterotoxin receptor on cultured cells (88). These results were intriguing in that the size of the crosslinked protein was similar to that of particulate guanylyl cyclases. In contrast, other studies, which obtained a number of different crosslinked proteins (89), suggested that the enterotoxin receptor was composed of multiple subunits, or that various different receptors exist. Based on equilibrium-binding studies, or guanylyl cyclase activation by the heat-stable enterotoxin, the receptor has been found only in the intestine of various mammals (15, 90); however, Forte and colleagues have found the receptor in epithelial cells of several different tissues in the North American opossum (91–93).

The normal function of the enterotoxin receptor is not yet known, but measurements of enterotoxin-stimulated guanylyl cyclase activity as a function of age in humans have shown that stimulatable guanylyl cyclase activity is much higher in the small intestine of younger than older children, with the highest activatable activity seen at one day of age (94). An as-yet-undiscovered physiological ligand may exist that could regulate ion transport in epithelial cells of the gut, or other tissues, by virtue of its interaction with the heat-stable-enterotoxin receptor.

## Endothelium-Derived Relaxing Factor

A labile factor produced by vascular endothelial cells that causes relaxation of smooth muscle has been named endothelium-derived relaxing factor (EDRF) (95). Substances interacting with the vascular endothelium that stimulate the release of EDRF include acetylcholine, serotonin, substance P, ATP, and ADP (96). EDRF is thought to be nitric oxide (NO) by some (12, 97–99) and a nitrosothiol derivative by others (100). There is evidence that additional relaxing factors may be produced by endothelial cells (reviewed in Ref. 96). EDRF directly activates the soluble form of guanylyl cyclase, and the resulting increases in cGMP levels correlate well with relaxation of blood vessels (96). Considerable data from various laboratories have supported cGMP as a mediator of smooth muscle relaxation (reviewed in Ref. 65). The cytoplasmic form of guanylyl cyclase is a heterodimeric protein that appears to contain heme as a prosthetic group (11, 101). Activation of the enzyme by NO is thought to be due to binding of NO to the heme prosthetic group, a well-characterized interaction in other systems (102); the mechanism by which such interaction may stimulate the soluble form of guanylyl cyclase, however, is unknown.

Nitric oxide synthase is responsible for the synthesis of NO and citrulline from arginine (103, 104). The enzyme is regulated by $Ca^{2+}$·calmodulin (105), and therefore in cells that contain the nitric oxide synthase, agents that elevate $Ca^{2+}$ may be expected to cause elevations of NO and subsequent elevations of cGMP. EDRF activity is known to be produced by many cells and tissues outside the cardiovascular system (106, 107), but its role as a messenger remains to be determined in most cases. Although NO or an NO-like molecule can diffuse from endothelial cells to smooth muscle cells to act as an intercellular messenger, it is also likely that NO or NO-like molecules act as second messengers within the cells in which they are synthesized.

# MOLECULAR CLONING OF GUANYLYL CYCLASES

## Membrane Guanylyl Cyclases

The sequences of cDNA clones encoding eight different membrane guanylyl cyclases have now been obtained (108–115). Although some of the sequences appear to encode the same protein from different species, at least four distinct proteins have been identified: one from sea urchin testis thought to mediate responses of spermatozoa to egg peptides (108, 113), two from various mammalian tissues (designated GC-A and GC-B) that appear to be natriuretic peptide receptors, (109–112, 115), and one from rat small intestinal mucosa (designated GC-C) that appears to be a receptor for bacterial heat-stable enterotoxins (114). Hydropathic analysis of the deduced amino acid sequences suggests that all of the membrane guanylyl cyclases contain a single transmembrane domain. The intracellular region, which represents about half of the protein, contains a protein kinase–like domain and a carboxyl region now known to be the cyclase catalytic domain (116, 117). The function of the protein kinase–like domain is unknown; protein kinase activity has not yet been detected. The extracellular domains of GC-A and GC-B show substantial sequence similarity to the ANP C-receptor, which contains an extracellular natriuretic-peptide-binding domain, a single transmembrane domain, and only 37 amino acids within the cytoplasm (118–120). GC-C, in contrast, shows virtually no similarity to GC-A or GC-B within the extracellular, putative ligand-binding domain (114). The topological model for membrane guanylyl cyclases is supported by the direct demonstration of cyclase catalytic activity within the carboxyl region (117) and by proteolytic fragmentation studies on the native enzyme (121). In addition, expression of a truncated recombinant ANP-clearance receptor, containing the sequences proposed to constitute the extracellular domain, results in secretion of a soluble protein having ANP-binding affinity indistinguishable from that of the native protein (122). A schematic diagram of the proposed topography of the membrane guanylyl cyclases is shown in Figure 1.

Studies of the genes encoding the membrane guanylyl cyclases have been undertaken, for the purposes of understanding the regulation of gene expression and the possible roles of defects in natriuretic peptide receptors in various disease states. The chromosomal locations of genes encoding the human GC-A, GC-B, and ANP C-receptor have been assigned as follows: the GC-A gene to 1q21-22, the GC-B gene to 9p12-p21, and the ANP C-receptor gene to 5p13-14 (123). The entire nucleotide sequence of the rat GC-A gene has been determined (124). It contains 22 exons spaced over a total of 17.5 kilobase pairs; analysis of the sequence of the 5' flanking region has shown the presence of Sp1-binding sites and no TATA box, but transcriptional regulation has yet to be studied.

## Soluble Guanylyl Cyclases

The sequences of both subunits of a soluble guanylyl cyclase from rat and bovine lung have been obtained (125–128). The carboxyl domains of both the 77-kDa $\alpha$-subunit and the 70-kDa $\beta$-subunit consist of amino acid sequences

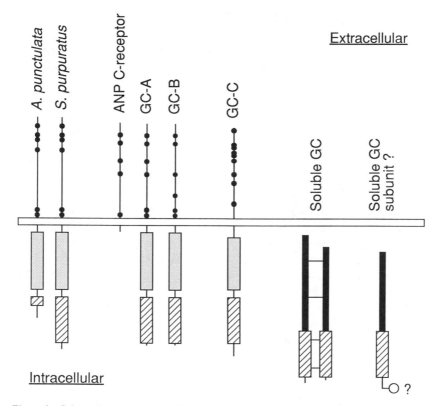

*Figure 1* Schematic representation of the topology of members of the guanylyl cyclase family. Two putative sea urchin egg peptide receptors are shown at the left (108, 113), followed by three natriuretic peptide receptors (the ANP C-receptor, GC-A, GC-B, Refs. 109–112, 118–120), an enterotoxin receptor, GC-C (114), the soluble guanylyl cyclase (125–128), and a novel guanylyl cyclase related to the β-subunit of the soluble enzyme (129). The guanylyl cyclase catalytic domains are indicated by hatched boxes, the protein kinase–like domains by gray boxes, and the amino-terminal regulatory domains of the soluble guanylyl cyclases by narrow black boxes. Locations of extracellular cysteine residues are indicated by black circles, and a putative isoprenylation site at the carboxyl terminus of the β-subunit homologue (129) is indicated by an open circle.

similar to those found in the carboxyl catalytic region of the membrane guanylyl cyclases, but neither contains a protein kinase–like domain. The carboxyl domains of the α- and β-subunits share approximately 45% amino acid sequence identity. The amino-terminal regions of the two subunits are also related to each other (20% identity), but not to other known proteins. These amino-terminal domains presumably play a regulatory role, although the specific sites responsible for subunit assembly, heme binding, and interaction with EDRF have yet to be identified. No homology of the soluble guanylyl cyclase (structure shown in Figure 1) to other heme-binding proteins has been observed (128).

Recently, Yuen et al used the polymerase chain reaction to identify a homologue of the $\beta$-subunit of soluble guanylyl cyclase (129). Steady-state levels of mRNA encoding the new protein were highest in kidney and liver, whereas Northern blot analysis of the previously described $\alpha$- and $\beta$-subunits showed highest mRNA levels in lung and brain (126, 128). The relatively kidney-specific protein is homologous across its entire sequence with the $\beta$-subunit of the soluble guanylyl cyclase from lung (34% amino acid sequence identity), including the amino-terminal regulatory domain, but it contains an additional 86 amino acids at the carboxyl terminus. The additional amino acid sequences do not resemble those of other known proteins, but contain a consensus sequence (CVVL) for isoprenylation at the carboxyl terminus; this sequence could play a key role in translocation of this guanylyl cyclase to the membrane. It remains to be determined whether the protein is isoprenylated and/or membrane-associated, and whether, like the soluble enzyme from lung, it contains an additional subunit. The structure of this protein is shown schematically in Figure 1.

Whether or not this novel protein homologous to the $\beta$-subunit of the soluble guanylyl cyclase is membrane-bound, its existence raises the possibility of new levels of regulation of cGMP production. If it is a subunit of a soluble guanylyl cyclase having a different tissue distribution than the previously described soluble enzyme, then mechanisms of regulation of soluble guanylyl cyclases may vary across tissues. If it is membrane-associated through isoprenylation, and is regulated by EDRF through the amino-terminal domain homologous to soluble guanylyl cyclases, then differential expression of soluble and membrane-associated EDRF-activated guanylyl cyclases could result in differential subcellular compartmentalization of cGMP produced in response to stimulation by EDRF. This could, in turn, lead to differential phosphorylation of proteins by cGMP-dependent protein kinase, or could favor interaction of cGMP with ion channels or phosphodiesterases.

## Comparison of Guanylyl Cyclases with Related Proteins

CATALYTIC DOMAINS OF GUANYLYL AND ADENYLYL CYCLASES Considerable homology exists between the catalytic domains of the various guanylyl cyclases and the two putative catalytic domains of a mammalian adenylyl cyclase (130); in addition, there is homology to the catalytic domain of an adenylyl cyclase from the nitrogen-fixing bacterium *Rhizobium meliloti* (131), and weak homology to the catalytic domains of adenylyl cyclases from the yeasts *Saccharomyces cerevisiae* (132) and *Schizosaccharo-*

*myces pombe* (133). Other bacterial adenylyl cyclases from *Escherichia coli* (134) and *Bordetella pertussis* (135) appear to be unrelated to each other and to the rest of the cyclase family at the amino acid sequence level; it is possible that a relationship will become more obvious when three-dimensional structures of adenylyl and guanylyl cyclases are determined. Alternatively, the different cyclase families may have arisen by convergent evolution. It has been argued (131) that the similarity between a prokaryotic adenylyl cyclase and eukaryotic adenylyl and guanylyl cyclases suggests a common descent of these enzymes, and that the strong inhibition of the *R. meliloti* enzyme by GTP suggests the existence of a primordial enzyme capable of catalyzing both cAMP and cGMP formation.

In Figure 2*a* we have aligned the deduced amino acid sequences of various members of the guanylyl cyclase family—the rat natriuretic peptide receptor, GC-A (109), the rat heat-stable enterotoxin receptor, GC-C (114), a sea urchin membrane guanylyl cyclase (113), the α- and β-subunits of the rat soluble guanylyl cyclase (126, 128), and the rat kidney homologue of the β-subunit of the soluble guanylyl cyclase (129). A consensus sequence, consisting of residues conserved among at least four of the six proteins, is indicated; invariant residues are marked with asterisks. In Figure 2*b* a similar alignment is shown, consisting of the two putative catalytic domains of bovine brain adenylyl cyclase (130), the *R. meliloti* adenylyl cyclase (131), and two yeast adenylyl cyclases (132, 133). Some caution should be exercised in interpreting the consensus sequences: The soluble guanylyl cyclase and the bovine adenylyl cyclase each contain two apparent catalytic domains, but it has not been shown that both actually possess catalytic function. If one of the catalytic domains of both the soluble guanylyl cyclase and the bovine adenylyl cyclase were excised from the alignments, a larger number of residues would be found within the consensus sequences, including a higher number of invariant residues. Therefore, if further studies show that only one of the apparent catalytic domains in the soluble guanylyl cyclase or bovine adenylyl cyclase has catalytic function, then residues not listed in the present consensus could prove to be important in catalysis. With this caveat in mind, we present in Figure 2*c* a consensus sequence for cyclases, derived from the adenylyl and guanylyl cyclase consensus sequences shown in Figure 2*a* and 2*b*. This sequence, particularly the invariant residues shown in boldface, may represent residues important in nucleotide binding and catalysis. Additional residues in the consensus sequences in Figure 2*a* and Figure 2*b* may be involved in specific recognition of the guanine and adenine moieties, respectively.

PROTEIN KINASE–LIKE DOMAIN OF MEMBRANE GUANYLYL CYCLASES AND PROTEIN KINASES    All known membrane guanylyl cyclases contain a

564     CHINKERS & GARBERS

a

```
                      1                                                    50
GC-A:                 ertqayleEK rKaeaLLyqi LPhsVAeQLk rGetVqAeaF dsVTIyFSDI
GC-C:                 ertqlykaEr dradhLnfml LPrlVvksLk ekgiVepely eeVTIyFSDI
S. purpuratus:        ertqelqkEK tKteqLLhrm LPpsiAsQLi kGiaVlpetF emVsIfFSDI
soluble β-subunit:    ltlraledEK kKtdtLLysv LPpsVAneLr hkrpVpAkry dnVTIlFSgI
soluble α-subunit:    hahqaleeEK kKtvdLLcsi fPseVAqQLw qGqiVqAkkF neVTmlFSDI
β-related GC:         vlsnhlaiEK kKtetLLyam LPehVAnQLk eGrkVaAgeF etcTIlFSDv
Consensus:            --------EK -K---LL--- LP--VA-QL- -G--V-A--F --VTI-FSDI
INVARIANT:                    *          *          *          *          *          **
```

```
                      51                                                   100
GC-A:                 VGFTAls... .aestPmQVV tlLNDLYTcF DavidnfD.. .VYKVETIGD
GC-C:                 VGFTtiCky. ...stPmeVV dmLNDiYksF Dqivdhh D.. .VYKVETIGD
S. purpuratus:        VGFTAls... .aastPiQVV nlLNDLYTlF DaiisnyD.. .VYKVETIGD
soluble β-subunit:    VGFnAfCskh asgegamkiV nlLNDLYTrF Dtltdsrknp fVYKVETvGD
soluble α-subunit:    VGFTAiCs.. ..qcsPlQVi tmLNaLYTrF DqqcgelD.. .VYKVETIGD
β-related GC:         VtFTniC... .aacePiQiv nmLNsmYskF DrltsvhD.. .VYKVETIGD
Consensus:            VGFTA-C--- -----P-QVV --LNDLYT-F D------D-- -VYKVETIGD
INVARIANT:            *  *                   **  *  *          ****** **
```

```
                      101                                                  150
GC-A:                 AYMVVSGLPv rngqlHArev ArmALallda vrsfrirHrp qEqlrlRIGI
GC-C:                 AYvVaSGLPm rngnrHAvdI skmALdilsf mgtfeleHlp GlpvwiRIGv
S. purpuratus:        AYMlVSGLPl rngdrHAgqI AstAhhlles vkgfivpHkp evflklRIGI
soluble β-subunit:    kYMtVSGLPe pcih.HArsI chlALdmmei aggvqv...d GEsvqitIGI
soluble α-subunit:    AYcVagGLhr es.dtHAvqI AlmALkmmel snevmspH.. GEpikmRIGl
β-related GC:         AYMVVgGvPv pv.esHAqrv AnfALgmris akevmnpv.t GEpiqiRvGI
Consensus:            AYMVVSGLP- -----HA--I A--AL----- -------H-- GE----RIGI
INVARIANT:            *  *             **   *                         *
```

```
                      151                                                  200
GC-A:                 HtGpVcAGVV GlKMPRYCLF GDTVNTASRM ESnGealKIH lSseTkavL.
GC-C:                 HsGpcaAGVV GiKMPRYCLF GDTVNTASRM EStGlplrIH mSssTiaiLr
S. purpuratus:        HsGscvAGVV GltMPRYCLF GDTVNTASRM ESnGlalrIH vSpwckqvL.
soluble β-subunit:    HtGeVvtGVi GqrMPRYCLF GnTVNltSRt EttGekgKIn vSeyTyrcLm
soluble α-subunit:    HsGsVfAGVV GvKMPRYCLF GnnVtlAnkf EScsvprKIn vSptTyrlLk
β-related GC:         HtGpVlAGVV GdKMPRYCLF GDTVNTASRM EShGlpsKvH lSptahraLk
Consensus:            H-G-V-AGVV G-KMPRYCLF GDTVNTASRM ES-G---KIH -S--T---L-
INVARIANT             *  *      **   *  ******* *   *          *          *    *
```

```
                      201                                                  240
GC-A:                 ..eefd.GFe lElRGdvemK GKGkvrTyWL lgergcstrg
GC-C:                 rt...dcqFl yEvRGetylK GrGtetTyWL tgmkdqeynl
S. purpuratus:        ..dklg.Gye lEdRGlvpmn GKGeihTfWL lgqdpsykit
soluble β-subunit:    spensdpqFh lEhRGpvsmK GKkepmqvWf lsrkntgtee
soluble α-subunit:    dc....pGFv ftpRsreelp pnfpsdipgi chfldayqhq
β-related GC:         ..nk...GFe ivrRGeievK GKGkmtTyfL iqnlnatede
Consensus:            -------GF- -E-RG----K GKG---T-WL ----------
INVARIANT:                      *
```

domain that is highly homologous to protein kinases. This domain is located between the plasma membrane and the guanylyl cyclase catalytic domain. In all cases, this region is more closely related to protein tyrosine kinases than to protein serine kinases. Detailed alignments of these protein kinase–like domains and known protein kinases have been presented elsewhere (108–110, 136). The vast majority of the residues that are highly conserved among the 11 subdomains of the catalytic domains of protein kinases (137) are also conserved in each of the membrane guanylyl cyclases (108–110, 136). The

b

```
              1                                                    50
Bovine domain 1: silFaDIvgf TgLasqctaq ElvklLnElf gkFdELaten hcr...rIKi
Bovine domain 2: ipnFnDfyie ldgnnmgv.. EclrlLnEii adFdELmdkd fYkdlekIKT
R. meliloti:      mFTDIydf TtisEgrspe EvvAmLsEyf dlFsEvvaah dgti...Iqf
S. cerevisiae:   amvFTDIkss TfLwElfpna mrtAikthnd imrrqLriyg gYe....vKT
S. pombe:        amvFTDIkns TlLwErhpia mrsAikthnt imrrqLratg gYe....vKT
Consensus        ---FTDI--- T-L-E----- E----L-E-- --F-EL---- -Y-----IKT
INVARIANT           *  *
              51                                                  100
Bovine domain 1: lGDcyycvsg ltqpktdhAh ccv....... .emgldmidt itsVaeatev
Bovine domain 2: iGstymaavg laptagtkAk kcisshLstl adfaIemfdv ldei.nyqsy
R. meliloti:     hGDsvfamwn apvadtrhAe hacrcaLave .....erlea fnsaqrasgl
S. cerevisiae:   eGDafmvafp tptsgltwcl sgqlklLdaq wpeeItsvqd gcqVtdrngn
S. pombe:        eGDafmvcfq tvpaallwcl svqlqlLsad wpneIvesvq grlVlgskne
Consensus        -GD------- --------A- ------L--- ----I----- ---V------
INVARIANT         *
              101                                                 150
Bovine domain 1: .d    LnmRv GlHtGrvlcG vlGlrkw.QY dvwsndVtlA nvmea..aGl
Bovine domain 2: nd    fvlRv GInvGpvVaG viGar.RpQY diwgntVNvA SRmds..tGv
R. meliloti:     pe    frtRf GIHtGtaVvG svGakeRlQY tamgdtVNvA SRlegmnkdy
S. cerevisiae:   iiyqgLsvRm GIHwGcpVpe ldlvtqRmdY lgpm..VNkA aRvqgvadGg
S. pombe:        vlyrgLsvRi GvnyGvtVse ldpitrRmdY ygpv..VNrt SRvvsvadGg
Consensus        -----L--R- GIH-G--V-G --G---R-QY ------VN-A SR------G-
INVARIANT         *    *   *             *             *  *
```

c

F-DI---T --(12-18)-- L --(6)-- F --(8-12)-- Y --(3-7)--

T-GD --(15-16)-- A --(7)-- L --(18-31)-- R-GIH-G --(1-2)--

V --(1-2)-- G--G --(5-6)-- Y --(4-6)-- VN-ASR --(4-6)-- G

*Figure 2*   Alignments of various members of the guanylyl and adenylyl cyclase family. (*a*) (See page 564) Alignment of guanylyl cyclases. A consensus sequence derived from amino acids conserved in at least four of the six sequences is shown below the alignment. Invariant residues are marked with asterisks. (*b*) Alignment of adenylyl cyclases. A consensus sequence and invariant residues are shown as in part *a* of the figure for guanylyl cyclases. (*c*) A consensus sequence for cyclases, consisting of residues present in both the guanylyl and adenylyl cyclase consensus sequences. Residues that are present in all of the guanylyl and adenylyl cyclases are shown in boldface.

existence of protein kinase activity intrinsic to the membrane guanylyl cyclases, however, remains to be demonstrated; despite considerable speculation, the function of the kinase-like regions remains an enigma. A particularly puzzling observation is that while 10 of the 11 protein kinase subdomains thought to be important in catalysis (137) are highly conserved in the membrane guanylyl cyclases, the subdomain containing the Gly-X-Gly-X-X-Gly sequence postulated to be important in nucleotide binding is poorly conserved in most of the membrane guanylyl cyclases, and completely absent from GC-C.

NATRIURETIC PEPTIDE-BINDING DOMAINS    The extracellular domains of the rat GC-A, rat GC-B, and the bovine ANP C-receptor are highly homologous to each other, showing 30–48% amino acid identity and 54–68% "similarity" in pairwise comparisons. Detailed alignments have been presented elsewhere (109–112). Across a region of 385 amino acids, 89 residues (23%) are identical among all three natriuretic peptide receptors. The roles of the conserved residues, and which residues interact with the natriuretic peptides, remain to be determined. It is clear, however, that the high degree of similarity in the extracellular domains of these receptors can be used to define a family of natriuretic-peptide-binding proteins. Schematic drawings of the natriuretic peptide-binding proteins are presented in Figure 1.

# FUNCTIONAL EXPRESSION OF RECOMBINANT GUANYLYL CYCLASES

## Sea Urchin Membrane Guanylyl Cyclases

Sea urchin guanylyl cyclases have been expressed in mammalian cells, as judged by production of immunoprecipitable proteins of the proper size (136), but the proteins have not displayed guanylyl cyclase or egg-peptide-binding activities. The lack of enzyme activity seen with the *Arbacia punctulata* cDNA now appears expected since the clone lacks most of the sequences encoding the catalytic domain (108, 116, 117). The cDNA cloned from *Strongylocentrotus purpuratus* testis, however, contains the entire predicted catalytic domain but enzyme activity has not yet been observed (D. S. Thorpe, D. L. Garbers, unpublished observations). Whether this is due to incorrect processing of the sea urchin enzyme, heat sensitivity of a protein designed to function at 20°C or less, or another defect is not known.

## Expression of GC-A

GC-A IS AN ANP RECEPTOR    Expression of rat or human GC-A in cultured cells has led to large increases in particulate guanylyl cyclase activity relative to that in control cells (109, 110). ANP also stimulated guanylyl cyclase activity of the recombinant protein (109, 110). GC-A-transfected cells bound severalfold more [125]I-labeled ANP than did control cells, and half-maximal inhibition of binding by unlabeled ANP occurred at approximately 3 nM. The truncated analogue of ANP, atriopeptin I, in contrast, required 300 nM concentrations for half-maximal competition against radiolabeled ANP (109, 110). These results agreed with previous data on receptor specificity obtained in various cell lines (45, 58–61). In further studies, [125]I-labeled ANP was crosslinked to cells transfected with the GC-A cDNA, and a single protein of apparent molecular weight of 130,000 was specifically labeled; this protein was not observed in control cells (109). The electrophoretic mobility of the labeled protein was consistent both with the predicted molecular weight of

GC-A and with the previously reported molecular size of the ANP receptor coupled to guanylyl cyclase activation in various cells (40, 45).

REGULATION OF GC-A ACTIVITY BY ADENINE NUCLEOTIDES    Kurose et al (138) first demonstrated that stimulation of particulate guanylyl cyclase activity by natriuretic peptides in membrane preparations from various tissues was potentiated by ATP and various ATP analogues. This observation was subsequently repeated by others (139–141). In some membrane preparations, ANP or ATP alone had little effect, but the combination of ANP and ATP produced a dramatic stimulation of guanylyl cyclase activity. In other preparations, ANP or ATP alone stimulated activity, possibly due to contamination of crude membrane preparations by ATP or endogenous natriuretic peptides. Alternatively, because there are multiple natriuretic peptide receptors and membrane guanylyl cyclases, different preparations may have contained different guanylyl cyclases; the question of which guanylyl cyclases were regulated by adenine nucleotides in the above studies was not clear, although GC-A was an obvious candidate. By using a recombinant baculovirus to express GC-A in insect cells lacking endogenous ANP receptors, stimulation of GC-A guanylyl cyclase activity by ANP in purified membrane preparations was shown to require adenine nucleotides (142). The rank order of potency of various adenine nucleotides was ATPγS > ATP > ADP, AMPPNP; AMP was without effect, and it is possible that ADP was active in membrane preparations only by virtue of being phosphorylated to form ATP. This nucleotide specificity was similar to that reported for the stimulation of guanylyl cyclase activity in membrane preparations from rat liver and lung (138, 140, 141).

The requirement of adenine nucleotides for activation of a guanylyl cyclase constitutes an intriguing parallel to the requirement of guanine nucleotides for activation of adenylyl cyclase. In both cases, nonhydrolyzable nucleotide analogues are functional (143), thus suggesting that ATP is not acting through phosphorylation. Unlike the adenylyl cyclase system, however, in which guanine nucleotides are known to act through binding to G proteins, the mechanism by which adenine nucleotides facilitate activation of guanylyl-cyclase activity by ANP is unknown. Are G cyclases regulated by A proteins as A cyclases are regulated by G proteins? Since the protein kinase–like domain of GC-A retains conserved motifs that may be involved in ATP binding, an attractive model would be that ATP binds to this domain of the cyclase, where it then serves as a regulatory molecule. Thus, GC-A would represent a receptor (extracellular domain), adenine-nucleotide-binding regulatory protein (protein kinase–like domain), and guanylyl cyclase (carboxyl domain); the presence of all three activities in one molecule would stand in contrast to the adenylyl cyclase system, in which receptor, nucleotide-binding, and cyclase activities reside in three separate proteins.

IDENTIFICATION OF CATALYTIC AND REGULATORY DOMAINS OF GC-A
Recombinant GC-A has been used to examine the roles of its two major
intracellular domains in signal transduction. Sequence analysis of two sea
urchin guanylyl cyclases (108, 113) and of a bovine adenylyl cyclase (130)
initially had resulted in some confusion regarding these roles: The observation
that the carboxyl domain homologous to soluble guanylyl cyclase was largely
missing from the *A. punctulata* clone (108) suggested that membrane
guanylyl cyclase activity resided in the kinase-like domain, since it was the
only region conserved among all membrane guanylyl cyclases. On the other
hand, a membrane guanylyl cyclase from the sea urchin *S. purpuratus* not
only had a large carboxyl domain homologous to soluble guanylyl cyclase,
but this region was more highly conserved between mammals and echi-
noderms than was the kinase-like domain (113). The adenylyl cyclase lacked
a kinase-like domain but had two domains homologous to the carboxyl
regions of guanylyl cyclases (130). These observations suggested that catalytic activity resided in the carboxyl domain, and that perhaps the *A. punctulata*
sequence in the carboxyl region was incomplete. Did guanylyl cyclase activity reside in the protein kinase–like domain, the carboxyl domain, or both?

To answer this question, the roles of the two major intracellular domains
were analyzed by deletion mutagenesis (116). Deletion of the carboxyl domain homologous to other cyclases resulted in a protein that bound ANP but
lacked guanylyl cyclase activity. Deletion of the kinase-like domain resulted
in a protein that bound ANP and was a constitutively active guanylyl cyclase.
Since guanylyl cyclase activity remained after deletion of the kinase-like
domain, it was concluded that the carboxyl region distal to the protein
kinase–like region of GC-A contained the guanylyl cyclase catalytic domain
(116). More direct evidence of this has been reported recently; in these
experiments, the carboxyl region of GC-A expressed in bacteria was found to
have guanylyl cyclase activity (117). Thus, guanylyl cyclase catalytic activity
was localized to the carboxyl sequences homologous to other cyclases. Because GC-A was constitutively active and no longer regulated by ANP after
deletion of the protein kinase–like domain, this domain appears to function as
a negative regulatory element, suppressing guanylyl cyclase activity in the
absence of ANP but not in its presence. Deletion of the kinase-like domain
from GC-A also resulted in loss of ATP regulation (116), consistent with the
possibility that this domain mediates activation by ATP.

## GC-B: A Second Natriuretic Peptide Receptor/Guanylyl Cyclase

As observed with GC-A, expression of rat or human GC-B in cultured
mammalian cells leads to dramatic increases in particulate guanylyl cyclase
activity, which can be stimulated by natriuretic peptides (111, 112). Cells

expressing GC-B specifically bind greater amounts of $^{125}$I-labeled ANP or $^{125}$I-labeled BNP than do control cells. However, the concentration response of natriuretic-peptide-stimulated cGMP formation in intact cells expressing GC-B is different from that observed in cells expressing GC-A. First, accumulation of cGMP in cells expressing GC-B is preferentially stimulated by BNP (111, 112), whereas GC-A is preferentially stimulated by ANP (112). Second, half-maximal stimulation of cGMP elevations in intact cells expressing rat GC-A occurred at concentrations of 3 nM and 25 nM ANP and BNP, respectively, whereas half-maximal cGMP elevations occurred at concentrations greater than 500 nM BNP in cells expressing GC-B (112). Since micromolar concentrations of BNP are likely to be out of the physiological concentration range (144, 145), the natural ligand for GC-B was suggested to be a natriuretic peptide other than ANP or BNP (112). Recently, Koller et al found that human CNP, while ineffective at stimulating cGMP formation in COS-7 cells transfected with human GC-A, was approximately tenfold more potent than porcine BNP in stimulating cGMP formation in COS-7 cells transfected with human GC-B (K. J. Koller, D. G. Lowe, G. L. Bennett, N. Minamino, K. Kangawa, H. Matsuo, D. V. Goeddel, personal communication). Human ANP and human BNP were unable to stimulate the human GC-B. These results suggested that CNP may be the natural ligand for human GC-B. In addition, the observation that porcine, but not human, BNP stimulates human GC-B suggests that caution must be used when interpreting studies in which cells are treated with BNP from a heterologous species.

The identification of multiple receptor/guanylyl cyclases, as well as multiple natriuretic peptides, indicates that the normal physiology of these peptides is more complex than was initially appreciated. The existence of at least three natriuretic peptides in conjunction with the isolation to date of only two guanylyl cyclase–linked natriuretic peptide receptors suggests that additional receptor/guanylyl cyclases exist. The physiological purpose of possessing at least three peptides with natriuretic properties remains unknown, although a family of such peptides and receptors may allow for more subtle regulation of cardiovascular function than a single peptide and receptor. Supporting this hypothesis is the observation that while the gross physiological effects of BNP and CNP are similar to those of ANP, the rank order of potency of the three peptides varies in different bioassays (42, 146), suggesting that they have unique physiological roles. In addition, the observation that CNP has been detected only in the central nervous system (50a, 50b), whereas ANP and BNP are found at highest levels in the heart, suggests that the normal function of CNP is quite different from that of ANP or BNP. Finally, as discussed above, these peptides are likely to have additional physiological roles unrelated to cardiovascular homeostasis. Complete elucidation of the members of this peptide/receptor family and their tissue-specific localization, and determining which peptides interact with which receptors, will be important from a therapeutic as well as a basic standpoint: The ability of ANP to induce vasodilation, natriuresis, and diuresis potently led to studies of clinical usefulness of the peptide in managing congestive heart disease and kidney failure (45). The clinical benefits of ANP have been limited, however, and its side effects, particularly hypotension, have been a severe problem (45). Systematic studies to localize GC-A and GC-B in the cardiovascular system have not yet been reported, but one recent study does indicate tissue-specific localization of these two receptors: Expression of GC-A and GC-B mRNA was examined in Rhesus monkey tissues by in situ hybridization (J. N. Wilcox, A. Augustine, D. V. Goeddel, D. G. Lowe, personal communication). It was demonstrated that mRNA for GC-A is present in kidney, adrenal cortex, heart, and cerebellum, while mRNA for GC-B was found in adrenal medulla, pituitary gland, and cerebellum. In conjunction with the primarily neural localization of CNP and the observation that CNP is the only known human ligand for GC-B, the finding that GC-B is localized primarily to neural and neural crest–derived tissues in man is consistent with its putative role as a CNP receptor. In conjunction with the development of receptor-specific agonists, a thorough understanding of receptor distribution may eventually allow targeting of such agonists to particular vascular beds for clinical purposes, avoiding the severe side effects of ANP.

## GC-C: A Receptor for Bacterial Heat-Stable Enterotoxins

To isolate DNA sequences encoding the particulate guanylyl cyclase activated by bacterial heat-stable enterotoxins (ST), the polymerase chain reaction was used to amplify cDNA from rat intestinal mucosa, with degenerate oligonucleotides encoding conserved guanylyl cyclase sequences as primers. The polymerase chain reaction product was then used to screen a rat intestinal mucosal cDNA library, resulting in the isolation of a cDNA clone related to GC-A and GC-B (see above) (114). The cDNA clone was designated GC-C. Expression of GC-C in COS-7 cells resulted in 10-fold higher levels of particulate guanylyl cyclase activity than found in control cells (114). Of even greater interest, cells expressing GC-C contained specific binding sites for $^{125}$I-labeled E. coli ST; no specific binding of the toxin was observed in control cells or in cells expressing GC-A or GC-B. ST also activated the guanylyl cyclase activity of GC-C, with half-maximal stimulation observed at concentrations of $10^{-7}$–$10^{-6}$ M. These concentrations were several orders of magnitude higher than were required to compete half-maximally for $^{125}$I-labeled ST-binding sites, but this concentration response was similar to that previously reported for ST stimulation of guanylyl cyclase activity in the intestine (82, 83). In control experiments, ANP and BNP failed to bind to or stimulate GC-C, and ST failed to stimulate guanylyl cyclase activity in control cells or in cells expressing GC-A or GC-B. Unlike the ST-stimulated guanylyl cyclase in the small intestine, GC-C expressed in COS-7 cells was soluble in nonionic detergent (114); this may indicate that the cultured cells lack the components that anchor GC-C to the cytoskeleton in the intestinal mucosa. Northern blot analysis detected GC-C mRNA only in intestine (114), consistent with the previously described tissue distribution of the ST receptor (15, 90).

These results demonstrated that an ST receptor is a plasma membrane form of guanylyl cyclase, and indicated further diversity in this rapidly growing family of receptors possessing intrinsic ligand-activated guanylyl cyclase activity. It seems likely that an as-yet-undiscovered endogenous ligand normally functions to regulate this receptor in the intestinal mucosa. As in the case of the natriuretic peptide receptors, there may be a diversity of ST receptors: Murad and coworkers have described what may be an ST receptor lacking guanylyl cyclase activity (85, 86), analogous to the ANP C-receptor that binds ANP but lacks enzyme activity, and Schulz et al reported the isolation from rat small intestinal mucosa of a cDNA clone that has not yet been fully characterized, but may be related to GC-C (114).

## Expression of Soluble Guanylyl Cyclase

Expression of the lower-molecular-weight $\beta$-subunit of the rat lung soluble guanylyl cyclase in cultured cells resulted in expression of immunoreactive protein, but no detectable guanylyl cyclase activity (126). Recently, cDNA clones encoding the larger ($\alpha$) subunit from lung were isolated (127, 128) and expressed in cultured cells (128) and, again, no enzyme activity was detected. When the subunits were coexpressed in transfected cells, however, not only was guanylyl cyclase activity measured, but the enzyme activity was responsive to sodium nitroprusside, a known activator of soluble guanylyl cyclase (128). These results suggested that, although both subunits appear to contain guanylyl cyclase catalytic domains based on sequence homology, assembly of the heterodimeric protein is required in order to obtain enzyme activity and nitroprusside stimulation. These results could occur for a variety of reasons, including: (a) Both subunits are required for the formation of an active catalytic domain. (b) The lack of the other subunit leads to folding problems. (c) The amino-terminal regions of each subunit are inhibitory unless bound to the opposite subunit. This model would be consistent with the observations that single catalytic domains derived from the rat GC-A or from the R. meliloti adenylyl cyclase are functional when expressed in bacteria (117, 131), and with the observation that the catalytic domain of GC-A is constitutively activated by removal of the protein kinase–like regulatory domain (116). Expression of the individual catalytic domains of the soluble guanylyl cyclase subunits in the absence of the amino-terminal domains has not yet been reported.

572 CHINKERS & GARBERS

## CONCLUSIONS

In the past several years, tremendous advances have been made in our understanding of signal transduction through activation of guanylyl cyclases. In particular, the discovery of a new family of cell surface receptors having intrinsic ligand-activated guanylyl cyclase activity has opened new horizons in cell biology. While the demonstration that receptors for egg peptides, atrial peptides, and heat-stable enterotoxins have guanylyl cyclase activity has clarified the mechanism by which these peptides elevate intracellular cGMP concentrations, a number of questions remain unanswered. In particular, the biochemical mechanisms by which cGMP exerts its biological effects after stimulation by the various peptides or EDRF remain unclear in most cases. In addition, the molecular cloning of the various guanylyl cyclases has raised new questions: How many members of the soluble and particulate guanylyl cyclase families exist, and what are their tissue distributions and functions? What are the particular physiological roles of the various natriuretic peptides and their receptors? What is the natural ligand for the heat-stable enterotoxin receptor, GC-C? What is the role of the protein kinase–like domain found in the membrane guanylyl cyclases—does it have protein kinase activity? What is the mechanism by which adenine nucleotides regulate guanylyl cyclase activity of cell surface receptors—and what is the biological significance of this mechanism? What is the mechanism of catalysis by the guanylyl cyclases—what sequences define a minimal catalytic domain, and which residues take part in the cyclase reaction? Over the next few years, as answers to these questions emerge, we will gain a better understanding of the biochemistry and physiology of the cGMP signal transduction system, as well as insights into potential therapeutic uses of this system.

*Literature Cited*

1. Ashman, D. F., Lipton, R., Melicow, M. M., Price, T. D. 1963. *Biochem. Biophys. Res. Commun.* 11:330–34
2. Hardman, J. G., Sutherland, E. W. 1969. *J. Biol. Chem.* 244:6363–70
3. Ishikawa, E., Ishikawa, S., Davis, J. W., Sutherland, E. W. 1969. *J. Biol. Chem.* 244:6371–76
4. White, A. A., Aurbach, G. D. 1969. *Biochim. Biophys. Acta* 191:686–97
5. Schultz, G., Böhme, E., Munske, K. 1969. *Life Sci.* 8:1323–32
6. Kimura, H., Murad, F. 1974. *J. Biol. Chem.* 249:6910–16
7. Goldberg, N. D., Haddox, M. K. 1977. *Annu. Rev. Biochem.* 46:823–96
8. Kimura, H., Mittal, C. K., Murad, F. 1975. *J. Biol. Chem.* 250:8016–22
9. DeRubertis, F. R., Craven, P. A. 1976. *Science* 193:897–99
10. Mittal, C. K., Murad, F. 1977. *J. Cyclic Nucleotide Res.* 3:381–91
11. Gerzer, R., Böhme, E., Hofmann, F., Schultz, G. 1981. *FEBS Lett.* 132:71–74
12. Ignarro, L. J. 1989. *FASEB J.* 3:31–36
13. Field, M., Graf, L. H., Laird, W. J., Smith, P. L. 1978. *Proc. Natl. Acad. Sci. USA* 75:2800–4
14. Hughes, J. M., Murad, F., Chang, B., Guerrant, R. L. 1978. *Nature* 271:755–56
15. Guerrant, R. L., Hughes, J. M., Chang, B., Robertson, D. C., Murad, F. 1980. *J. Infect. Dis.* 142:220–28
16. Hansbrough, J. R., Garbers, D. L. 1981. *J. Biol. Chem.* 256:1447–52

17. Suzuki, N., Nomura, K., Ohtake, H., Isaka, S. 1981. *Biochem. Biophys. Res. Commun.* 99:1238–44
18. Garbers, D. L., Watkins, H. D., Hansbrough, J. R., Smith, A., Misono, K. S. 1982. *J. Biol. Chem.* 257:2734–37
19. deBold, A. J., Bornstein, H. B., Veress, A. T., Sonnenberg, H. 1981. *Life Sci.* 28:89–94
20. Atlas, S. A. 1986. *Recent Prog. Horm. Res.* 42:207–42
21. Hamet, P., Tremblay, J., Pang, S. C., Garcia, R., Thibault, G., et al. 1984. *Biochem. Biophys. Res. Commun.* 123:515–27
22. Winquist, R. J., Faison, E. P., Waldman, S. A., Schwartz, K., Murad, F., Rapoport, R. M. 1984. *Proc. Natl. Acad. Sci. USA* 81:7661–64
23. Waldman, S. A., Rapoport, R. M., Murad, F. 1984. *J. Biol. Chem.* 259:14332–34
24. Tremblay, J., Gerzer, R., Pang, S. C., Cantin, M., Genest, J., et al. 1985. *FEBS Lett.* 194:210–14
25. Bentley, J. K., Tubb, D. J., Garbers, D. L. 1986. *J. Biol. Chem.* 261:14859–62
26. Yau, K.-W., Baylor, D. A. 1989. *Annu. Rev. Neurosci.* 12:289–327
27. Fain, G. L., Matthews, H. R. 1990. *Trends Neurosci.* 13:378–84
28. Garbers, D. L. 1989. *Annu. Rev. Biochem.* 58:719–42
29. Shimomura, H., Garbers, D. L. 1986. *Biochemistry* 25:3405–10
30. Dangott, L. J., Garbers, D. L. 1984. *J. Biol. Chem.* 259:13712–16
31. Dangott, L. J., Jordan, J. E., Bellet, R. A., Garbers, D. L. 1989. *Proc. Natl. Acad. Sci. USA* 86:2128–32
32. Shimomura, H., Dangott, L. J., Garbers, D. L. 1986. *J. Biol. Chem.* 261:15778–82
33. Ward, G. E., Vacquier, V. D. 1983. *Proc. Natl. Acad. Sci. USA* 80:5578–82
34. Suzuki, N., Shimomura, H., Radany, E. W., Ramarao, C. S., Ward, G. E., et al. 1984. *J. Biol. Chem.* 256:1447–52
35. Ward, G. E., Garbers, D. L., Vacquier, V. D. 1985. *Science* 227:768–70
36. Vacquier, V. D., Moy, G. W. 1986. *Biochem. Biophys. Res. Commun.* 137:1148–52
37. Garbers, D. L., Hardman, J. G., Rudolph, F. G. 1974. *Biochemistry* 13:4166–71
38. Chrisman, T. D., Garbers, D. L., Parks, M. A., Hardman, J. G. 1975. *J. Biol. Chem.* 250:374–81
39. Ramarao, C. S., Garbers, D. L. 1988. *J. Biol. Chem.* 263:1524–29
40. Inagami, T. 1989. *J. Biol. Chem.* 264:3043–46
41. Sudoh, T., Kangawa, K., Minamino, N., Matsuo, H. 1988. *Nature* 332:78–80
42. Sudoh, T., Minamino, N., Kangawa, K., Matsuo, H. 1990. *Biochem. Biophys. Res. Commun.* 168:863–70
43. Roy, R. N., Flynn, T. G. 1990. *Biochem. Biophys. Res. Commun.* 171:416–23
44. Pandey, K. N., Pavlou, S. N., Kovacs, W. J., Inagami, T. 1986. *Biochem. Biophys. Res. Commun.* 138:399–404
45. Brenner, B. M., Ballermann, B. J., Gunning, M. E., Zeidel, M. L. 1990. *Physiol. Rev.* 70:665–99
46. Bianchi, C., Gutkowska, J., Thibault, G., Garcia, R., Genest, J., et al. 1985. *Histochemistry* 82:441–50
47. Quirion, R., Dalpe, M., Dan, T. V. 1986. *Proc. Natl. Acad. Sci. USA* 83:174–80
48. Torda, T., Nazarali, A. J., Saavedra, J. M. 1989. *Biochem Biophys. Res. Commun.* 159:1032–38
49. Misono, K. S., Fukumi, H., Grammer, R. T., Inagami, T. 1984. *Biochem. Biophys. Res. Commun.* 119:524–29
50. Misono, K. S., Grammer, R. T., Rigby, J. W., Inagami, T. 1985. *Biochem. Biophys. Res. Commun.* 130:994–99
50a. Nakao, K. 1990. *Takeda Sci. Found. Symp. Biosci.*, p. 29 (Abstr.)
50b. Kangawa, K., Minamino, N., Matsuo, H. 1990. See Ref. 50a, p. 17 (Abstr.)
51. Kojima, M., Minamino, N., Kangawa, K., Matsuo, H. 1989. *Biochem. Biophys. Res. Commun.* 159:1420–26
52. Sudoh, T., Maekawa, K., Kojima, M., Minamino, N., Kangawa, K., et al. 1989. *Biochem. Biophys. Res. Commun.* 159:1427–34
53. Seilhamer, J. J., Arfsten, A., Miller, J. A., Lundquist, P., Scarborough, R. M., et al. 1989. *Biochem. Biophys. Res. Commun.* 165:650–58
54. Schulz, S., Chinkers, M., Garbers, D. L. 1989. *FASEB J.* 3:2026–35
55. Garbers, D. L. 1989. *J. Biol. Chem.* 264:9103–6
55a. MacFarland, R. T., Zelus, B. D., Beavo, J. A. 1991. *J. Biol. Chem.* 266:136–42
56. Leitman, D. C., Andresen, J. W., Kuno, T., Kamisaki, Y., Chang, J. K., et al. 1986. *J. Biol. Chem.* 261:11650–55
57. Maack, T., Suzuki, M., Almeida, F. A., Nussenzveig, D., Scarborough, R. M., et al. 1987. *Science* 238:675–78
58. Takayanagi, R., Inagami, T., Snajdar, R. M., Imada, T., Tamura, M., et al. 1987. *J. Biol. Chem.* 262:12104–13
59. Schenk, D. B., Johnson, L. K., Schwartz, K., Sista, H., Scarborough,

R. M., et al. 1985. *Biochem. Biophys. Res. Commun.* 127:33–42

60. Schenk, D. B., Phelps, M. N., Porter, J. G., Scarborough, R. M., McEnroe, G. A., et al. 1985. *J. Biol. Chem.* 260:14887–90

61. Shimonaka, M., Saheki, T., Hagiwara, H., Ishido, M., Nogi, A., et al. 1987. *J. Biol. Chem.* 262:5510–14

62. Scarborough, R. M., Schenk, D. B., McEnroe, G. A., Arfsten, A., Kang, L. L., et al. 1986. *J. Biol. Chem.* 261: 12960–64

63. Scarborough, R. M., McEnroe, G. A., Arfsten, A., Kang, L.-L., Schwartz, K., et al. 1988. *J. Biol. Chem.* 263:16818–22

64. Ishido, M., Fujita, T., Shimonaka, M., Saheki, T., Ohuchi, S., et al. 1989. *J. Biol. Chem.* 264:641–45

65. Murad, F. 1986. *J. Clin. Invest.* 78:1–5

66. Leitman, D. C., Murad, F. 1987. *Endocrinol. Metab. Clin. N. Am.* 16:79–105

67. Leitman, D. C., Andresen, J. W., Catalano, R. M., Waldman, S. A., Tuan, J. J., et al. 1988. *J. Biol. Chem.* 263: 3720–28

68. Anand-Srivastava, M. B., Cantin, M. 1986. *Biochem. Biophys. Res. Commun.* 138:427–36

69. Anand-Srivastava, M. B., Cantin, M., Genest, J. 1985. *Life Sci.* 36:1873–79

70. Anand-Srivastava, M. B., Franks, D. J., Cantin, M., Genest, J. 1984. *Biochem. Biophys. Res. Commun.* 121:855–62

71. Anand-Srivastava, M. B., Srivastava, A. K., Cantin, M. 1987. *J. Biol. Chem.* 262:4931–34

72. Anand-Srivastava, M. B., Sairam, M. R., Cantin, M. 1990. *J. Biol. Chem.* 265:8566–72

73. Hirata, M., Chang, C. H., Murad, F. 1989. *Biochim. Biophys. Acta* 1010: 346–51

74. Kuno, T., Andresen, J. W., Kamisaki, Y., Waldman, S. A., Chang, L. Y., et al. 1986. *J. Biol. Chem.* 261:5817–23

75. Meloche, S., McNicoll, N., Liu, B., Ong, H., DeLean, A. 1988. *Biochemistry* 27:8151–58

76. Paul, A. K., Marala, R. B., Jaiswal, R. K., Sharma, R. K. 1987. *Science* 235:1224–26

77. Giannella, R. A. 1981. *Annu. Rev. Microbiol.* 32:341–57

78. Chan, S. K., Giannella, R. A. 1981. *J. Biol. Chem.* 256:7744–46

79. Yoshimura, S., Ikemura, H., Watanabe, H., Aimoto, S., Shimonishi, Y., et al. 1985. *FEBS Lett.* 181:38–42

80. Kubota, H., Hidaka, Y., Ozaki, H., Ito, H., Hirayama, T., et al. 1989. *Biochem. Biophys. Res. Commun.* 161:229–35

81. Guandalini, S., Rao, M. C., Smith, P. L., Field, M. 1982. *Am. J. Physiol.* 243:G36–41

82. Guarino, A., Cohen, M., Thompson, M., Dharmsathaphorn, K., Giannella, R. 1987. *Am. J. Physiol.* 253:G775–80

83. Huott, P. A., Liu, W., McRoberts, J. A., Giannella, R. A., Dharmsathaphorn, K. 1988. *J. Clin. Invest.* 82:514–23

84. Gariepy, J., Judd, A. K., Schoolnik, G. K. 1987. *Proc. Natl. Acad. Sci. USA* 84:8907–11

85. Waldman, S. A., Kuno, T., Kamisaki, Y., Chang, L. Y., Gariepy, J., et al. 1986. *Infect. Immun.* 51:320–26

86. Kuno, T., Kamisaki, Y., Waldman, S. A., Gariepy, J., Schoolnik, G., et al. 1986. *J. Biol. Chem.* 261:1470–76

87. Thompson, M. R., Giannella, R. A. 1990. *J. Recept. Res.* 10:97–117

88. Forte, L. R., Krause, W. J., Freeman, R. H. 1988. *Am. J. Physiol.* 255:F1040–46

89. Ivens, K., Gazzano, H., O'Hanley, P., Waldman, S. A. 1990. *Infect. Immun.* 58:1817–20

90. Rao, M. C., Guandalini, S., Smith, P. L., Field, M. 1980. *Biochim. Biophys. Acta* 632:35–46

91. White, A. A., Krause, W. J., Turner, J. T., Forte, L. R. 1989. *Biochem. Biophys. Res. Commun.* 159:363–67

92. Forte, L. R., Krause, W. J., Freeman, R. H. 1989. *Am. J. Physiol.* 257:F874–81

93. Krause, W. J., Freeman, R. H., Forte, L. R. 1990. *Cell Tissue Res.* 260:387–94

94. Guarino, A., Cohen, M. B., Giannella, R. A. 1987. *Pediatr. Res.* 21:551–55

95. Furchgott, R. F., Zawadzki, J. V. 1980. *Nature* 288:373–76

96. Furchgott, R. F., Vanhoutte, P. M. 1989. *FASEB J.* 3:2007–18

97. Palmer, R. M. J., Ferrige, A. G., Moncada, S. 1987. *Nature* 327:524–26

98. Radomski, M. W., Palmer, R. M. J., Moncada, S. 1987. *Br. J. Pharmacol.* 92:181–87

99. Ignarro, L. J., Buga, G. M., Wood, K. S., Byrns, R. E., Chaudhuri, G. 1987. *Proc. Natl. Acad. Sci. USA* 84:265–69

100. Myers, P. R., Minor, R. L. Jr., Guerra, R. Jr., Bates, J. N., Harrison, D. 1990. *Nature* 345:161–63

101. Humbert, P., Niroomand, F., Fischer, G., Mayer, B., Koesling, D., et al. 1990. *Eur. J. Biochem.* 190:273–78

102. Edwards, S. L., Kraut, J., Poulos, T. L. 1988. *Biochemistry* 27:8074–81

103. Palmer, R. M. J., Ashton, A. S., Moncada, S. 1988. *Nature* 333:664–66
104. Schmidt, H. H. H. W., Nau, H., Wittfoht, W., Gerlach, J., Prescher, K.-E., et al. 1988. *Eur. J. Pharmacol.* 154:213–16
105. Bredt, D. S., Snyder, S. H. 1990. *Proc. Natl. Acad. Sci. USA* 87:682–85
106. Knowles, R. G., Palacios, M., Palmer, R. M. J., Moncada, S. 1989. *Proc. Natl. Acad. Sci. USA* 86:5159–62
107. Ishii, K., Gorsky, L. D., Forstermann, U., Murad, F. 1989. *J. Appl. Cardiol.* 4:505–12
108. Singh, S., Lowe, D. G., Thorpe, D. S., Rodriguez, H., Kuang, W.-J., et al. 1988. *Nature* 334:708–12
109. Chinkers, M., Garbers, D. L., Chang, M.-S., Lowe, D. G., Chin, H., et al. 1989. *Nature* 338:78–83
110. Lowe, D. G., Chang, M.-S., Hellmiss, R., Chen, E., Singh, S., et al. 1989. *EMBO J.* 8:1377–84
111. Chang, M.-S., Lowe, D. G., Lewis, M., Hellmiss, R., Chen, E., et al. 1989. *Nature* 341:68–72
112. Schulz, S., Singh, S., Bellet, R. A., Singh, G., Tubb, D. J., et al. 1989. *Cell* 58:1155–62
113. Thorpe, D. S., Garbers, D. L. 1989. *J. Biol. Chem.* 264:6545–49
114. Schulz, S., Green, C. K., Yuen, P. S. T., Garbers, D. L. 1990. *Cell* 63:941–48
115. Pandey, K. N., Singh, S. 1990. *J. Biol. Chem.* 265:12342–48
116. Chinkers, M., Garbers, D. L. 1989. *Science* 245:1392–94
117. Thorpe, D. S., Morkin, E. 1990. *J. Biol. Chem.* 265:14717–20
118. Fuller, F., Porter, J. G., Arfsten, A. E., Miller, J., Schilling, J. W., et al. 1988. *J. Biol. Chem.* 263:9395–401
119. Lowe, D. G., Camerato, T. R., Goeddel, D. V. 1990. *Nucleic Acids Res.* 18:3412
120. Porter, J. G., Arfsten, A., Fuller, F., Miller, J. A., Gregory, L. C., et al. 1990. *Biochem. Biophys. Res. Commun.* 171:796–803
121. Liu, B., Meloche, S., McNicoll, N., Lord, C., DeLean, A. 1989. *Biochemistry* 28:5599–605
122. Porter, J. G., Scarborough, R. M., Wang, Y., Schenk, D., McEnroe, G. A., et al. 1989. *J. Biol. Chem.* 264:14179–84
123. Lowe, D. G., Klisak, J., Sparks, R. S., Mohandas, T., Goeddel, D. V. 1990. *Genomics* 8:304–12
124. Yamaguchi, M., Rutledge, L. J., Garbers, D. L. 1990. *J. Biol. Chem.* 265:20414–20
125. Koesling, D., Herz, J., Gausepohl, H., Niroomand, F., Hinsch, K.-D., et al. 1988. *FEBS Lett.* 239:29–34
126. Nakane, M., Saheki, S., Kuno, T., Ishii, K., Murad, F. 1988. *Biochem. Biophys. Res. Commun.* 157:1139–47
127. Koesling, D., Harteneck, C., Humbert, P., Bosserhoff, A., Frank, R., et al. 1990. *FEBS Lett.* 266:128–32
128. Nakane, M., Arai, K., Sakeki, S., Kuno, T., Buechler, W., et al. 1990. *J. Biol. Chem.* 265:16841–45
129. Yuen, P. S. T., Potter, L. R., Garbers, D. L. 1990. *Biochemistry* 29:10872–78
130. Krupinski, J., Coussen, F., Bakalyar, H. A., Tang, W.-J., Feinstein, P. G., et al. 1989. *Science* 244:1558–64
131. Beuve, A., Boesten, B., Crasnier, M., Danchin, A., O'Gara, F. 1990. *J. Bacteriol.* 172:2614–21
132. Kataoka, T., Broek, D., Wigler, M. 1985. *Cell* 43:493–505
133. Yamawaki-Kataoka, Y., Tamaoki, T., Choe, H.-R., Tanaka, H., Kataoka, T. 1989. *Proc. Natl. Acad. Sci. USA* 86:5693–97
134. Aiba, H., Mori, K., Tanaka, M., Ooi, T., Roy, A., et al. 1984. *Nucleic Acids Res.* 12:9427–40
135. Glaser, P., Ladant, D., Sezer, O., Pichot, F., Ullmann, A., et al. 1988. *Mol. Microbiol.* 2:19–30
136. Garbers, D. L., Lowe, D. G., Dangott, L. J., Chinkers, M., Thorpe, D. S., et al. 1988. *Cold Spring Harbor Symp. Quant. Biol.* 53:993–1003
137. Hanks, S. K., Quinn, A. M., Hunter, T. 1988. *Science* 241:42–52
138. Kurose, H., Inagami, T., Ui, M. 1987. *FEBS Lett.* 219:375–79
139. Song, D.-L., Kohse, K. P., Murad, F. 1988. *FEBS Lett.* 232:125–29
140. Chang, C.-H., Kohse, K. P., Chang, B., Hirata, M., Jiang, B., et al. 1990. *Biochim. Biophys. Acta* 1052:159–65
141. Chang, C.-H., Jiang, B., Douglas, J. G. 1990. *Eur. J. Pharmacol.* In press
142. Chinkers, M., Singh, S., Garbers, D. L. 1991. *J. Biol. Chem.* 266:In press
143. Gilman, A. G. 1989. *J. Am. Med. Assoc.* 262:1819–25
144. Saito, Y., Nakao, K., Itoh, H., Yamada, T., Mukoyama, M., et al. 1989. *Biochem. Biophys. Res. Commun.* 158:360–68
145. Togashi, K., Hirata, Y., Ando, K., Takei, Y., Kawakami, M., et al. 1989. *FEBS Lett.* 250:235–37
146. Furuya, M., Takehisa, M., Minamitake, Y., Kitajima, Y., Hayashi, Y., et al. 1990. *Biochem. Biophys. Res. Commun.* 170:201–8

Annu. Rev. Biochem. 1991. 60:577–630

# THE BIOCHEMISTRY OF AIDS

*Yashwantrai N. Vaishnav and Flossie Wong-Staal*

Departments of Biology and Medicine, University of California, San Diego, La Jolla, California 92093

KEY WORDS:  HIV, virus replication, Tat and Rev, genetic heterogeneity, vaccine and antiviral therapy.

CONTENTS

0066-4154/91/0701-0577$02.00

# I. HISTORICAL PERSPECTIVES ON HUMAN RETROVIRUSES

Retroviruses are RNA-containing viruses that replicate through a DNA intermediate by virtue of a viral-coded RNA-dependent DNA polymerase, also called reverse transcriptase (1, 2). The family of retroviridae is divided into three subfamilies (3). Oncovirinae includes all the oncogenic retroviruses and many closely related non-oncogenic viruses. Lentivirinae includes the "slow" viruses such as visna-maedi virus, Caprine Arthritis Encephalitis Virus (CAEV), Equine Infectious Anemia Virus (EIAV), and the immunodeficiency viruses, including most notably Human Immunodeficiency Virus (HIV), the causative agent of Acquired Immunodeficiency Syndrome (AIDS) (4–6). The third subfamily, spumavirinae, consists of the "foamy" viruses that induce persistent infections without any clinical disease but cause vacuolization of cultured cells (3).

Recognition of a viral etiology of cancer dates back to the early 20th century when the transmission of chicken leukemia or sarcoma was accomplished through injection of filtered tumor extracts (7–9). The first mammalian retrovirus, murine leukemia virus, was isolated by Ludwik Gross in 1951 from an inbred mouse strain (10), which provided the first reproducible experimental system to study viral leukemogenesis. In the ensuing period, retroviruses were shown to cause lymphomas, leukemias, sarcomas, and various other cancers in species ranging from chicken, mice, and cattle to subhuman primates (2). The discovery of reverse transcriptase in 1970 by Temin & Mizutani (11), and independently, by Baltimore (12) substantiated the "provirus hypothesis" proposed by Temin in 1964 (13) and also established a genetic and biochemical basis for the classification of retroviruses.

With the role of retroviruses in carcinogenesis firmly established in animal models, massive efforts were then directed towards the search for human retroviruses in similar diseases (for reviews, see Refs. 14 and 15). Various approaches were taken towards the detection of retroviruses in normal or malignant human tissues: electron microscopy, detection of reverse transcriptase activity, study of virus-like RNA-protein complexes, evaluation of antigen or nucleic acid sequence relatedness to known animal retroviruses, rescue of defective viruses, and observation of reactivity of patients' sera with retroviral proteins. None of the above studies yielded unambiguous results. Parallel studies in the bovine leukemia virus model (16) provided valuable reminders that retroviral-induced diseases need not be associated with overt viremia, nor is the etiological agent necessarily related to previously known animal retroviruses and therefore detectable by probes derived from them. Appreciation of these possibilities, and of the importance of long-term culture of tumor cells in vitro, were critical for the isolation of human retroviruses.

The isolation of the first human retrovirus, human T-cell leukemia virus I (HTLV-I) (17) came on the heel of the discovery of interleukin 2 (IL-2), which enabled long-term cultivation of mature T-cells of normal and neoplastic origins (18). Both accomplishments took place in the laboratory of Gallo and associates. Although HTLV-I was first isolated from a patient diagnosed with mycosis fungoides, the virus had no etiological association with the disease. The establishment of a link between HTLV-I and malignancy rested on the recognition of a new disease entity termed "Adult T-cell Leukemia/ Lymphoma" (ATLL), which clusters in parts of southern Japan (19, 20) and was suspected to have a viral etiology. ATLL sera were found to be almost 100% positive for reactivity against HTLV-I (21, 22). A similar virus was subsequently isolated from ATLL patients and was designated as adult T-cell leukemia virus (ATLV) (23). Since HTLV-I and ATLV were practically identical viruses (24), the name HTLV-I was adopted as a consensus designation. The second human retrovirus, human T-cell leukemia virus type-II (HTLV-II), also was isolated by Gallo's team in 1982 from a patient with a T-cell variant of hairy cell leukemia (25).

In spite of HTLV-I and HTLV-II, human retroviruses would probably have remained relatively obscure in the United States and Europe had it not been for the sudden emergence of another retroviral disease that reached epidemic proportions. In 1981, five cases of *Pneumocystis carinii* pneumonia (PCP) occurring in previously healthy homosexual men were described in the United States (26). Outbreaks of this and other immunodeficiency-associated conditions such as Kaposi's Sarcoma, mucosal candidiasis, etc, soon followed (27–30). Strikingly, all the patients were either homosexual men or intravenous drug abusers and all had a decreased T-cell response to mitogens and antigens. The term Acquired Immunodeficiency Syndrome (AIDS) was coined to define clinically the various manifestations of this disease. By the end of 1982, the number of AIDS cases throughout the United States had exploded, and the disease was no longer restricted to homosexuals and drug abusers but also extended to hemophiliacs, blood transfusion recipients, Haitian immigrants, sex partners of risk-group members, and children born to mothers at-risk.

All the observations strongly suggested the etiological involvement of a transmissible agent spread through genital secretions and blood. A new retrovirus was an attractive idea since some animal retroviruses, e.g. feline leukemia virus (31), were known to induce immunodeficiency as well as leukemia/lymphoma. There were also some similarities between the previously discovered human retroviruses (HTLV-I and HTLV-II) and the putative AIDS agent with respect to host cell tropism and modes of transmission. In 1983, Barre-Sinoussi et al detected reverse transcriptase, cytopathic activity, and virus particles in phytohemaglutinin (PHA)- and IL-2-stimulated lymphocytes from a patient with lymphadenopathy, a frequent prodrome of AIDS (32). The virus was designated as LAV (lymphadenopathy-associated virus). Other similar cultures were subsequently reported (33). The causal association between LAV and AIDS remained equivocal, however, because no purified virus reagent was available to type the virus from the various cultures or the sera of different AIDS patients. A team headed by Gallo reported in 1984 the first long-term propagation of viruses from several AIDS patients in permanent CD4$^+$ T-cell lines (34, 35). The availability of continuous, high-titer producer cell lines allowed for the first time the development of highly purified and concentrated viral reagents necessary for the characterization of the virus and for the serological detection of exposed individuals (36, 37). Because of the tropism of this virus, Gallo et al designated it HTLV-III. Another team headed by Levy reported the isolation of similar viruses, which were designated as AIDS-related viruses (ARV) (38). Later on, HTLV-III, LAV, and ARV were found to be variants of the same AIDS virus (39). In 1986, an international committee recommended the name human immunodeficiency virus (HIV) for all isolates of the AIDS virus (40).

Since AIDS has become a global medical emergency, the scientific community has responded to the challenge by making progress at an unprecedented rate in the characterization of the viral agent. As a result, the life cycle of HIV is now understood more fully than that of any other retrovirus. Antiviral agents have been developed with some success, some of which may be applicable to other retroviral diseases of animals and humans. Massive efforts to develop an HIV vaccine are beginning to pay off. Furthermore, an unforeseen dividend from HIV research is the unraveling of the rich, intricate pathways of gene regulation utilized by the virus, which may well illuminate novel, fundamental cellular processes. The understanding of basic biology that we gain from the studies of HIV may be one major legacy of this epidemic to medical science.

## II. BIOLOGICAL PROPERTIES OF HIV

### A. Target Cell Tropism

One of the conspicuous features of HIV infection is a selective depletion of CD4-bearing (helper/inducer) T-lymphocytes, suggesting a selective tropism and cytopathic effect of HIV for this population. This suggestion was supported by in vitro demonstration that purified CD4$^+$ T-cell lines preferentially support the replication of HIV (41) and that syncytia formation and cell death are frequently associated with infection. Furthermore, it was shown that both infection and syncytia formation could be blocked by monoclonal antibodies directed against certain epitopes of the CD4 molecule (42–44). The HIV-1 envelope glycoprotein gp120 binds to CD4 as indicated by their coprecipitation by antibodies directed against either of the proteins (45). The definitive proof that CD4 is a receptor for HIV was provided by Maddon et al (46) when they showed that CD4⁻ cells (such as HeLa), which are ordinarily not targets for HIV infection, could be rendered infectable by transfection of a cloned CD4 gene (46). Interestingly, expression of human CD4 is not sufficient to confer infectivity to murine cells, suggesting that a second component present in human cells but not in murine cells is required for postbinding events in the infection process. Finally, soluble CD4 inhibits binding and infection of both HIV-1 and HIV-2 in vitro (47–49). In addition to helper T-cells and mono-cyte-macrophages (50–52), other cells such as Langerhans cells (53), follicular dendritic cells (54), glial cells (55, 56), and certain colon tumor cell lines (57, 58) are susceptible to HIV-1 infection. These cells express low levels of CD4 on their surface. Similarly, five hepatoma cell lines have been shown to be infected productively by HIV-1 despite the absence of detectable CD4 expression (59). Moreover, soluble CD4 and monoclonal antibodies against CD4 failed to block HIV-1 infection in these cell lines. However, the nature of an alternative receptor suggested by these studies remains to be determined.

## B. Syncytia Induction and Cytopathic Effect

Various mechanisms have been proposed to explain the cytopathic effects (CPE) of HIV on T-cells, particularly to accommodate the fact that a dramatic depletion of T-cells occurs even though relatively few cells are actually infected (60). One possible mechanism is through syncytia formation, which involves interaction of HIV-infected cells expressing viral glycoproteins (gp120 and gp40) on the cell surface and uninfected cells expressing CD4 (61–63). The syncytia are unstable in culture, do not proliferate, and usually die within 48 hours. In this way, a few infected cells may form the nuclei of many dying cells. However, syncytia formation is unlikely to be the major pathway for cell loss, as some cell lines that do not form syncytia are nevertheless susceptible to CPE and conversely, syncytia formation could be transient and not lead to significant cell death. Intracellular complexing of CD4 and gp120 or destruction of the permeability of the membrane owing to profuse virus budding have also been proposed to play a role in the single cell killing by HIV (64).

Like other cytopathic lentiviruses, HIV accumulates large amounts of unintegrated viral DNA upon infection (65). Although this feature has been suggested as an important factor in HIV-induced CPE (65), comparison with a mutant virus with greatly reduced cytopathicity did not reveal apparent differences in the level of unintegrated DNA (66).

Additional mechanisms of CD4 cell depletion can be invoked. For example, extracellular gp120 (shed by infected cells) may bind to CD4 expressed on the surface of uninfected T-helper cells and make the cell a target for antibody-dependent cell-mediated cytotoxicity (ADCC) (67). Other conceivable mechanisms include autoimmune phenomena directed against infected lymphocytes and HIV-induced terminal differentiation of infected cells.

Monocytes and macrophages have been shown to harbor HIV-1 in vivo and are likely to play an important role in the pathogenesis of the virus (50–52, 68). Monocytes/macrophages are relatively resistant to HIV CPE, probably due to the low level of CD4 expression. Therefore, monocytes may act as a reservoir of virus and may disseminate virus to various organs in the body. A noncytopathic, low-level infection of monocytes is also seen with other lentiviruses such as the visna virus of sheep (69). Also analogous to visna virus infection, monocytes appear to be the predominant cell type infected by HIV within the central nervous system of infected individuals (70–72). Infection of this target cell could be the basis of neurological abnormalities that are seen, to varying degrees, in at least 60% of AIDS patients. Lee et al (73) found that gp120 could inhibit the growth of neurons in the presence of neuroleukin, but not in the presence of nerve growth factor. It was postulated, based on the observed partial sequence homology between gp120 and neuroleukin, that gp120 could act as a competitive inhibitor of neuroleukin. AIDS-associated dementia could probably be explained by the observation that free gp120 even at low concentration could increase intracellular free calcium in rodent retinal ganglion cells and hippocampal neurons in culture and cause cell death within 24 hours (74). The specificity was evident as the effect could be abolished by antibodies against gp120. The mechanism of action is not clearly understood. The action of gp120 may be through a receptor, via a second messenger, or directly on $Ca^{2+}$ channels.

## C. Clinical Manifestations of HIV Infection

$CD4^+$ helper T-cells not only are quantitatively depleted but also are functionally abnormal in AIDS patients (75, 76). Other cells of the immune system are also affected, directly or indirectly, leading to the development of immunodeficiency which in turn allows opportunistic infections of a variety of agents. *Pneumocystis carinii* pneumonia (PCP) is the most common opportunistic infection in AIDS (77), and it appears to represent activation of latent infection in the majority of cases (78). Other opportunistic pathogens include *Candida albicans, Mycobacterium avium* complex, *Cryptococcus neoformans, Histoplasma capsulatum,* and papilloma viruses, all of which may cause skin lesions in HIV-infected persons. Reactivation of many other DNA viruses, including the herpesviruses, papovaviruses, and hepatitis virus, is also frequently seen. Another pathological consequence of HIV infection is the high incidence of Kaposi's Sarcoma (KS), particularly in homosexual males. A relatively benign form of KS has been previously observed in older Mediterranean men. KS in AIDS patients may involve skin and mucous membranes. Neurological manifestations could be a direct consequence of HIV infection in the brain, or due to opportunistic infections such as toxoplasmosis, cryptococcosis, JC papovavirus infection, and mycobacterial infections. For a more extensive account on the clinical aspects of HIV infection, please consult the review by Hirsch & Curran (79).

## III. REPLICATION CYCLE OF HIV

### A. Virus Binding, Fusion, and Entry

HIV is related to the animal lentiviruses on the basis of genetic, morphological, and pathological criteria. Electron microscopy has revealed a characteristic cylindrical core in HIV particles which is indistinguishable from that of animal lentiviruses (4) (Figure 1). The diameter of HIV virions is approximately 110 nm. Like other retroviruses, HIV has two copies of 35S single-stranded genomic RNA with the polarity of mRNA. The genomic RNA supposedly exists in the form of a ribonucleoprotein complex containing reverse transcriptase (p55/p61), endonuclease (p32), and retroviral RNA-binding Gag protein (p9). The viral core is composed of p24 as a major capsid antigen. The myristylated Gag protein (p17) forms the outer shell, and its amino-terminal end is inserted into the lipid membrane envelope derived from the infected cell membrane. The extracellular (gp120) and transmembrane (gp41) glycoproteins exist in the form of a noncovalent complex on the envelope.

The first step in the initiation of infection is the binding of a virus particle to a specific receptor on the surface of the target cell. In the case of retroviruses and other enveloped viruses, the interaction with the cell receptor is mediated by the envelope glycoprotein. As discussed earlier (Section IIA), the receptor for HIV is CD4, and the gp120 of HIV binds to CD4 on the target cell with an affinity constant of the order of $10^{-9}$ M (80). CD4 is not only the first retroviral receptor to be identified, but is also the most extensively studied. The only other retroviral receptor gene that has been identified, cloned, and sequenced is that of Molony murine leukemia virus (81). Mutational studies have implicated the amino-terminal half of gp120 in mediating the association with gp41 transmembrane protein in a head-to-head configuration and the carboxyl-terminal region in interaction with the CD4 receptor (82). The epitopes on gp120 recognized by murine monoclonal antibodies capable of inhibiting the interaction between gp120 and CD4 have been mapped to a relatively conserved region near the carboxy terminus spanning amino acids 397–439 (83). The deletion of 12 amino acids as well as a single amino acid substitution in this region resulted in a complete or substantial loss of binding

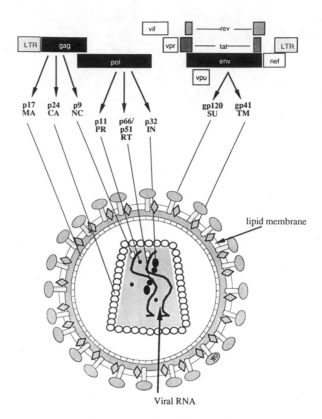

*Figure 1*    Structure of human immunodeficiency virus (HIV).

to CD4. Similarly, the binding site for gp120 has been mapped to a portion of the first amino-terminal variable domain of CD4 receptor (amino acid residues 16–84) that encompasses the region homologous to the second complementarity-determing region (CDR2) of immunoglobulin variable region (84–87). Additionally, amino acid residues of the second variable domain may contribute to gp120-binding (84, 86).

The envelope glycoprotein of most enveloped viruses exists as a heterodimer, often held by interchain disulfide bonds, with the extracellular component mediating the interaction with receptor and the transmembrane component inducing fusion with the host cell membrane. By analogy, the transmembrane protein gp41 of HIV has been thought to mediate fusion between the viral envelope and host cell membranes (82, 88). The transmembrane envelope glycoprotein of orthomyxoviruses and paramyxoviruses generally contains a stretch of hydrophobic amino acids at the amino terminus, which has been implicated in viral membrane fusion (89). HIV-1 gp41 also contains such a hydrophobic amino-terminal sequence. Linker insertion as well as point mutations in this region were found to abrogate cell fusion (82, 90). Single polar amino acid substitutions introduced throughout the stretch of 28 hydrophobic amino acids of gp41 dramatically reduced syncytia formation, suggesting a role for the entire hydrophobic terminus in cell fusion (91). Amino acid substitutions have also been introduced in the 15 amino-terminal residues of a simian immunodeficiency virus (SIVmac) gp32 transmembrane glycoprotein. It was found that mutations that increased the overall hydrophobicity of the gp32 amino terminus increased the ability of the viral envelope to induce syncytia formation, whereas introduction of polar or charged amino acids in the same region abolished the fusogenic function of the viral envelope (92).

The mode of entry of enveloped viruses can be broadly divided into pH-dependent and pH-independent processes (89). The former mode involves internalization of virus by receptor-mediated endocytosis into acidic compartments (endosomes), where low pH induces conformational changes in the transmembrane envelope glycoprotein, as a result of which the hydrophobic fusion domain is exposed, which in turn facilitates fusion of viral and endosomal membranes. As an example, this mechanism is used by influenza virus, an orthomyxovirus. The second mode, which is used by Sendai virus, a paramyxovirus, involves direct fusion of the virus envelope with the plasma membrane of the cell in a pH-independent manner. To distinguish these two modes, lysosomotropic agents such as weak bases (ammonium chloride, chloroquine, and amantadine) and carboxylic ionophores (monensin and nigericin), have been instrumental by virtue of their ability to neutralize the acidity of endosomes and thereby inhibit only the pH-dependent entry. In the case of HIV, the events that follow binding are less clear (for a review see Ref. 93). HIV infection was first reported to be inhibited by ammonium chloride, implying the pH-dependent mode of entry (46). However, it was later reported that ammonium chloride, chloroquine, amantadine, and monensin do not inhibit entry of HIV, suggesting that HIV does not require low pH for fusion (94, 95). Furthermore, electron microscopy has revealed direct fusion of the virus envelope with the plasma membrane (94). The fact that HIV-1-induced syncytia formation can occur in media maintained at neutral or mildly alkaline pH also supports the view that HIV enters susceptible cells via pH-independent membrane fusion. Mutant CD4 receptor with deletion of cytoplasmic domain was found to function as an effective receptor for HIV even though it was not internalized in response to phorbol esters (96). These results suggest that receptor-mediated endocytosis is not required for HIV entry. However, the possibility that the truncated CD4 molecules are still internalized in response to HIV binding cannot be excluded. Alternatively, internalization may still proceed independent of the cytoplasmic domain through an association between CD4 and another plasma membrane protein.

## B. Synthesis and Integration of Proviral DNA

After entry of a retroviral core into the cytoplasm of a susceptible cell, viral RNA is transcribed into double-stranded proviral DNA by the RNA/DNA-dependent DNA polymerase and ribonuclease H activities of reverse transcriptase. The proviral DNA, presumably in the form of a nucleoprotein complex, then migrates to the nucleus where it is integrated into the host cellular DNA by the endonuclease activity of the viral integrase, with possible participation of cellular proteins (for a review see Ref. 97). By analogy with animal retroviruses, HIV proviral DNA synthesis and integration are presumed to follow the same pathway. Retroviruses in general do not replicate in nondividing cells (98). Similarly, HIV has also been shown to require activated T-cells for productive infection in vitro (99, 100). HIV-1 infection of quiescent primary lymphocytes is stable but nonproductive until after mitogenic or antigenic stimulation (100–102). The restriction, the mechanism of which is unclear, appears to be at the level of complete proviral DNA synthesis. It has been shown that HIV-1 enters and initiates DNA synthesis in quiescent T-cells at levels comparable with those of activated T-cells. However, unlike that of activated T-cells, the viral RNA is not completely reverse transcribed in quiescent cells (103). The incompletely transcribed proviral DNA contains both minus and plus strands of one long terminal repeat (LTR) (3') with some further limited extension of the minus strand. This structure is labile with a half-life of about one day. The incomplete proviral DNA may account for the failure of some investigators to detect HIV-1 DNA in quiescent T-cells by Southern blot analysis, which was taken as an evidence that T-cell activation is required for virus entry (104). Another report suggested a defect at the level of proviral integration in the resting T-cells, which were found to contain a transcriptionally active extrachromosomal form of HIV-1 proviral DNA (105). The latter could persist for several weeks in resting T-cells and, upon T-cell activation, could integrate and act as a template for infectious virus. However, the requirement of cell proliferation for productive infection is not universal. For example, nondividing cultures of monocytes and macrophages can be infected by HIV. Furthermore, the PBJ strain of SIV, which is highly virulent in macaque monkeys, is capable of establishing productive infection in resting T-cells, while closely related SIV strains are not (P. Fultz, personal communication). A second level of restriction for virus replication is at the level of initiation of transcription, as some of the factors required for HIV promoter activity, such as NFκB, are activated upon mitogenic/antigenic stimulation. A detailed discussion of various factors interacting with HIV LTR is given in Section IIIC.

Most aspects of integration of HIV appear to be similar to other retroviral systems. For example, HIV-integrated DNA contains highly conserved dinucleotide sequences (5' TG—CA 3'), and there is a duplication of a short stretch of cellular DNA at the site of integration. Moreover, HIV viral termini contain short inverted repeats, which may be important for specificity of integrative recombination with respect to the viral DNA. Integration of HIV-1 DNA in vitro into heterologous DNA targets using extracts of cells infected with HIV-1 has been demonstrated (106, 107). As in the case of other retroviruses (97), the viral DNA active in integration was found to exist in the form of a nucleoprotein complex. Moreover, integration of linear HIV DNA occurred in cytoplasmic extracts that contained no detectable circular forms of viral DNA (106), consistent with other retroviral systems, in which linear DNA has been shown to be the immediate precursor for integration (97).

Whether proviral integration is necessary for HIV replication is yet to be established. Cells infected with HIV-1 carrying deletion in integrase gene were found to produce core and envelope antigens, but not infectious virus particles (105).

## C. Expression of Viral Genes

Once the provirus is integrated into the host DNA, it behaves as a resident cellular gene. For most retroviruses, the expression of integrated provirus follows the same general rules of eukaryotic gene expression and is governed exclusively by cellular proteins (97). The gene expression of human retroviruses (HIVs and HTLVs), however, is additionally controlled by viral regulatory proteins. Transcription is a complex process involving interplay between *cis*-acting regulatory sequences present in the viral LTR and *trans*-acting cellular transcription factors as well as viral transactivators (e.g. Tat of HIV and Tax of HTLV). The regulatory sequences present upstream of the transcription initiation site ($+1$) can be broadly catagorized into proximal (promoter) and distal (enhancer) elements (Figure 2). They are sequences typically utilized by cellular RNA polymerase II. The promoter element contains the TATA box, located between nucleotides $-22$ to $-27$, which is presumably recognized by transcription factor TFIID (108). The TATA sequence, the most highly conserved sequence in the RNA polymerase II–transcribed genes, determines the precise site of transcription initiation. The HIV-1 promoter also contains three GC-rich sequences, which bind to the ubiquitous transcription factor Sp1 (108, 109). Binding of Sp1 increases the rate of transcription initiation.

The enhancer is a distal region that stimulates transcription initiation in a distance- and orientation-independent manner. HIV-1 LTR contains such an element between nucleotides $-120$ and $-57$ (110). The enhancer also appears to govern the induction of HIV promoter activity in response to T-cell activation. It is recognized by EBP-1 in resting cells (111) and by NF$\kappa$B (112) and HIVEN 86A (113) in activated T-cells. There are two binding sites ($-109$, $-79$) for NF$\kappa$B in the form of imperfect repeats. Binding of purified

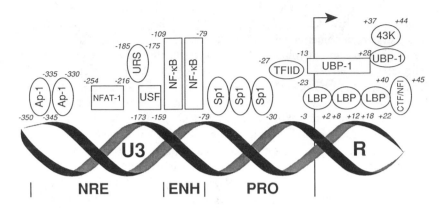

*Figure 2*   Organization of HIV-1 long terminal repeat (LTR). Binding sites of various factors are shown along with the coordinates of the recognition sequences. See text for details. NRE, negative regulatory element; ENH, enhancer; PRO, promoter. The approximate boundaries of NRE, ENH, and PRO are depicted by vertical lines. The arrow indicates the site of transcription initiation.

NFκB to HIV 1-LTR has been demonstrated in vitro (114). In resting T-cells, NFκB is present in the cytoplasm complexed with an inhibitory protein, IκB. Upon stimulation, NFκB is released from IκB and consequently translocated to the nucleus (for review see Ref. 115). The induction of HIV promoter activity in response to T-cell activation has been primarily attributed to the binding of NFκB (112, 114, 116). Activation of T-cells also induces a T-cell-specific protein HIVEN 86A (113). However, it is not clear whether the binding of NFκB and HIVEN 86A is simultaneous or mutually exclusive.

Recently, intragenic enhancers have also been reported in several retroviruses. A sequence with enhancer-like activity has been found in the *gag* region of avian leukemia-sarcoma viruses (117–119). Similarly, HIV-1 also contains an intragenic enhancer in the *gag-pol* region (between nucleotides 1711 and 6026) (120). However, the role of these intragenic enhancers in the initiation of transcription from the LTR is speculative at present.

The sequence located between nucleotides −410 and −157, designated as the negative regulatory element (NRE), has a negative effect on the rate of transcription in vivo (110, 121–124) and in vitro (125). Deletion of the NRE also increases the replication rate of the mutant virus 3–5-fold in CD4$^+$ cell lines (124). The NRE contains a number of recognition sequences for cellular transcription factors. The upstream binding factor (USF), originally identified as a cellular factor that binds to the adenovirus major late promoter (126), also binds to a consensus sequence located between nucleotides −173 and −159 of the HIV-1 LTR (W. Haseltine, personal communication). The sequence between −185 and −175 closely resembles the sequences (upstream repressor sequence, URS) that negatively regulate the expression of several yeast genes (127). A protein that is induced early in T-cell activation, designated as nuclear factor of activated T cells (NFAT-1), also binds to a site between nucleotides $-254$ and $-216$ in HIV-1 LTR (128). Two binding sites for transcription factor AP1 are located at nucleotides $-350$ to $-345$ and $-335$ to $-330$ (129). Deletion of AP1 and NFAT-1 sites has no measurable effect on HIV LTR-driven transcription in transient transfection assays. In contrast, deletion of the USF sequence leads to increased LTR-driven transcription as well as higher virus replication in $CD4^+$ T-cell lines (130). It would appear, therefore, that these sequences contain recognition sites for the cellular proteins, which could inhibit transcription initiation.

In addition to the upstream regulatory sequences, HIV promoter contains several protein-binding sites downstream of the RNA start site (Figure 2). A leader binding protein (LBP-1) binds to a region from $-17$ to $+27$. LBP-1 specifically recognizes three repeat elements within this region: $-3$ to $+2$, $+8$ to $+12$, and $+18$ to $+22$ (131). In vitro transcription reactions suggest a role for LBP-1 as an activator of initiation. CTF/NF-I, a factor previously shown to bind and activate HSV-1 thymidine kinase and human globin genes promoters, binds to a sequence from $+40$ to $+45$ (131). Another protein called untranslated binding protein (UBP-1) has been shown to recognize HIV LTR sequence from $-13$ to $+28$ nucleotides (111). Recently, a 43-kDa protein distinct from UBP-1 has been found to bind to nucleotides $+37$ to $+44$, a region that overlaps with the transactivation response element (TAR) (R. Gaynor, personal communication).

The precise role of these factors in the basal/induced transcription from HIV promoter remains to be established. Moreover, no information is available at present regarding the number and identity of factors that govern HIV transcription in monocytes, macrophages, and dendritic cells. It is not unreasonable to believe that cell type–specific factors play an important role in HIV gene expression.

Two viral transactivators, Tat and Rev, play extremely important roles in transcriptional and posttranscriptional regulation of HIV gene expression. The details of these viral transactivators are described in sections VA and VB.

## D. Assembly and Release of Mature Virus

The selective packaging of viral genomic RNA into particles during assembly is determined by interaction between cis-acting sequences present in the RNA and trans-acting gag-derived nucleocapsid (NC) protein. In all the retroviruses studied so far, the major packaging signal resides in the leader region between U5 and the gag gene initiation codon, although sequences located in the gag region appear to contribute towards packaging efficiency (132–137). A packaging signal of HIV-1 has also been described. It is located between the major splice donor and the *gag* initiation codon (138), a situation similar to type-C mammalian retroviruses. Deletion of a 19-base sequence within this region led to 98% reduction in the efficiency of viral RNA packaging. Since this sequence is located downstream of the major splice donor site, it is absent in all subgenomic spliced RNAs, and this could explain a selective encapsidation of genomic RNA. The *trans*-acting component required for viral RNA encapsidation is the RNA-binding nucleocapsid (NC) protein derived from Gag precursor. The NC protein of all retroviruses contain one or two copies of the invariant sequence Cys-X2-Cys-X4-His-X4-Cys, which is reminiscent of the "zinc finger" motif found in many DNA-binding proteins (139). Avian and murine type-C retroviruses with mutations in this sequence produce RNA-deficient, non-infectious virus particles, suggesting the role of this sequence in viral RNA packaging (140–142). Similarly, mutation or deletion of the zinc finger–like sequence of the p9 NC protein of HIV-1 was found to decrease the content of virion-associated genomic RNA to 2–20% of the wild type (143). Furthermore, no infectivity could be detected in the mutant particles, indicating either that these mutants have additional defect(s) for virus replication, or that the inefficient packaging of viral RNA resulted in subthreshold levels of infectious virus. Although it is not known whether retroviral NC proteins require $Zn^{2+}$ for their activity, peptides corresponding to zinc finger–like domains of these proteins do bind $Zn^{2+}$, and their structures have been determined by spectroscopic methods (144, 145).

Retroviral assembly is a unique process in which the products of *gag* and *pol* are incorporated into virions in the form of their polyprotein precursors during assembly and are proteolytically cleaved during or after budding to form the mature virus particles. One interesting implication of this process is that the individual components of the precursor proteins are active prior to proteolytic cleavage. The process of assembly of HIV particles appears to follow the same pathway as that of type-C retroviruses, with capsid assembly and budding occurring simultaneously. The Gag and Pol proteins are synthesized in the form of their respective precursors $Pr55^{gag}$, NH2-p17-(MA)-p24(CA)-p9(NC)-p7-COOH (146), and $Pr160^{gag-pol}$, NH2-p17(MA)-p24(CA)-p9(NC)-p7-p10(PR)-p66(RT)-p32(IN)-COOH (147, 148). The p17, p24, and p9 Gag domains as well as p10 protease domain of the Gag and Gag-Pol precursors play important roles in the assembly process. The interactions between the Gag proteins, the plasma membrane, and the virion RNA control the process. The aggregation of $Pr55^{gag}$ and $Pr160^{gag-pol}$ molecules under the plasma membrane mediated by $p17^{gag}$ protein initiates the complex process of assembly. In common with other retroviruses (97), the amino terminus of these precursors in HIV is posttranslationally modified by the addition of myristic acid (146–148), which provides a hydrophobic domain supposedly required for interaction with the membrane. As expected from similar studies carried out on other retroviruses (97), mutation of the amino-terminal glycine of HIV Gag precursor, which abolished myristylation, also resulted in complete disruption of virus assembly (149, 150). However, proteolytic processing of the capsid precursor in these mutants was unaffected, suggesting that the processing of the HIV *gag* gene products occurs before virus assembly. This suggestion is in contrast to results of similar mutational studies carried out on the type-C Mo-MuLv (151) and type-D Mason-Pfizer monkey virus (MPMV) (152).

As discussed earlier, the genomic RNA is brought to the site of assembly by the p9(NC) domain of Gag/Gag-Pol precursor. The p24 Gag, a major core protein, forms a protein shell surrounding the nucleocapsid and therefore determines the tubular shape of the core (153). The interaction between p24 domains of the Gag/Gag-Pol precursors is likely to play an important role in virion assembly. However, no information is available at present about the structure or mechanism of action of p24 in the assembly. The function of p7, a proline-rich carboxy-terminal product of Gag precursor, is not known.

The Gag and Gag-Pol precursors of HIV are cleaved during or after budding to produce individual proteins by the viral-coded protease, p10 (149, 154, 155). This process is responsible for morphological maturation of virions in which the core condenses to form an electron-dense cylindrical structure. Mutations in the HIV protease gene were shown to abolish proteolytic processing of capsid precursor Pr55$^{gag}$ (149, 154, 155). Electron microscopy revealed that assembly and budding steps were not affected in these mutants, but the virions produced resembled immature core particles, suggesting that the protease mediates a postbudding morphological maturation of the core structure (149, 155). Moreover, the mutant particles were non-infectious (154, 155).

As it is clear from the foregoing discussion that proteolytic processing of the polyprotein precursors takes place at a rather late stage of maturation, the question arises as to what prevents the action of the protease domain of Gag-Pol precursor until that stage. It is conceivable that if the enzyme acts only in *trans,* and is only slightly active in the precursor form, then no cleavage should occur until a very high local concentration of precursor is formed in the budding virion. Once the initial cleavage of precursor takes place, the liberated protease may exert its full activity and rapidly process the remaining precursors. An alternative explanation emerges from the deduced structure of avian leukosis-sarcoma virus (156) and HIV (157–159) proteases (for a review see Ref. 160). Both proteins were found to be dimers joined in such a way that the active site is formed by the dimer linkage region. If retroviral proteases are active only as dimers, which is very likely, then such dimers may not be able to form until the precursors have properly aligned themselves as in the late stages of budding. It may be noted that bacterially expressed HIV protease behaves as a dimer as determined by gel exclusion chromatography (161).

The envelope glycoproteins, synthesized initially as a precursor gp160, are incorporated into virions by a separate pathway. The precursor is cleaved intracellularly, and the products, gp120 and gp41, are inserted into the plasma membrane (162). Cleavage of gp160 is essential for viral infectivity, as mutation of the tryptic-like cleavage site to a chymotryptic site produces non-infectious particles, which can be converted to infectious particles by chymotrypsin treatment (163). The incorporation of envelope glycoproteins into the virions is probably mediated by the interaction between the p17 (MA) domain of Gag precursor and Env protein complex. Direct association between the Gag and Env proteins has also been described for Rous sarcoma virus (165) and murine leukemia virus (166). However, no information is presently available on this aspect in the case of HIV.

It appears then that only the Gag precursor, Pr55$^{gag}$, is required for assembly and release of HIV particles. Such a requirement is consistent with earlier studies on avian and murine retroviruses in which the expression of Gag precursor by itself was found to result in the formation of virus-like particles even in the absence of expression of viral envelope glycoproteins, reverse transcriptase, or viral RNA (167). The expression of *gag* gene alone of HIV in recombinant baculovirus-infected insect cells led to the synthesis, myristoylation, and targeting to the plasma membrane of Pr55$^{gag}$, resulting in the assembly and release of virus-like particles (168). Electron microscopy revealed immature particles, displaying the translucent center and a thick, peripheral electron-dense ring surrounded by an outer lipid membrane. Similarly, expression of the *gag* and *pol* genes of HIV from SV40 late replacement vector in transfected COS cells was shown to direct the assembly and release of virus particles resembling mature HIV particles (169). The particles also contained RNA, but the nature of the RNA was not investigated. It is possible that the RNA packaged was *gag-pol* mRNA containing packaging locus. Virus-like particles were also observed when *gag* and *pol* genes were expressed in a recombinant vaccinia virus system (170). Similarly, the cells coinfected with two recombinant vaccinia viruses, one carrying the HIV-1 *gag* and protease genes and the other the *env* gene, produced HIV-like mature particles, which also contained *gag* mRNA but not *env* mRNA (171). Thus, expression of the Gag precursor protein alone is sufficient for assembly and budding of retroviral particles in general.

Consistent with the role of multimeric Gag precursors in the assembly process, *gag* mutants with deletion or insertion in p24 domain as well as deletion in p17-p24 boundary were reported to behave as *trans*-dominant mutants, interfering with the assembly and/or release of wild-type HIV (172).

# IV. ORGANIZATION AND EXPRESSION OF HIV GENES

Molecular cloning (173) and sequencing (174–178) of HIV has revealed that the genomic structure of HIV is more complex than that of conventional type-C retroviruses, including HTLV-I and HTLV-II. In addition to typical retroviral genes, *gag*, *pol*, and *env*, HIV encodes at least eight other genes (Figure 4; see Section B). The HIV genome displays unprecedented economy in its coding potential as evident from the presence of nine overlapping genes in the 3′ half of the genome. For convention, we classify *gag*, *pol*, and *env* as structural genes, *tat*, *rev*, and *nef* as regulatory genes, and *vif*, *vpr*, *vpu*, *vpt*, and *tev/tnv* as accessory genes.

## A. *The Structural Genes*

The *gag* gene codes for various structural components of virus particles. It is translated from full-length viral mRNA to produce a polyprotein precursor Pr55$^{gag}$, NH2-p17(MA)-p24(CA)-p9(NC)-p7-COOH. The Gag precursor of HIV differs in some respects from that of other retroviruses. For example, there is no protein interposed between the amino terminus (p17) and the major capsid protein (p24) as there is for most other retroviruses. Also, two small proteins (p9 and p7) are derived from the carboxy terminus of the Gag precursor, unlike one in case of other retroviruses. A detailed scheme of proteolytic processing of the precursor has been proposed (146). Various domains of this precursor play pivotal roles in the assembly and release of virus particles as described earlier (Section IIID).

The *pol* gene of HIV codes for protease, reverse transcriptase, and integrase. In common with other retroviruses, it is translated from the genomic length viral mRNA to produce a polyprotein Pr160$^{gag-pol}$. The *gag* and *pol* genes overlap by 241 nucleotides and the *pol* gene is in $-1$ reading frame relative to *gag* reading frame. A ribosomal frameshift, as in the case of avian retroviruses, allows expression of the *pol* gene (179, 180). A frameshift event, which occurs at a low frequency of ~5%, is directed by a short homopolymeric sequence ("slippage sequence") at the *gag-pol* overlap region followed by a hairpin loop structure (for a review see Ref. 97). In the case of HIV, the slippage sequence alone appears to be sufficient to induce ribosomal frameshift (180). This unique strategy of gene expression ensures that the structural (Gag) proteins are made in large amounts and catalytic (Pol) proteins in relatively small amounts.

The precursor is cleaved by the viral protease to produce p10(PR), p66/p51(RT), and p32(IN). The reverse transcriptase polypeptide (p66) is partially cut again, apparently also by the viral protease (181, 182), to remove a 14-kDa polypeptide from the C-terminus. The N-terminal p51, resulting from this cleavage, forms a heterodimer with p66 (183, 184). The purpose of this additional cleavage is presently unclear. The fate and a possible function of the C-terminal 14-kDa protein are also not understood, though there is some evidence that it exists as a separate protein having RNase H activity (185).

The HIV reverse transcriptase, in common with other retroviral reverse transcriptases, has polymerase as well as RNase H activities. Mapping studies have shown that the N-terminal two-thirds of the molecule constitutes the polymerase domain, but no distinct domain could be assigned the RNase H function (186, 187). It appears that the optimum folding of the RNase H domain requires the entire protein sequence. Alternatively, discontinuous regions distributed throughout the molecule may directly participate in the formation of RNase H active site. Such a distribution is in contrast to the case in Molony-MuLV reverse transcriptase, in which two distinct regions carry out these two enzymatic activities (188). The RNase H activity plays a crucial role in proviral DNA synthesis. It carries out endo- as well as exo-nucleolytic hydrolysis of RNA from RNA:DNA hybrids with limit digestion products ranging in size from 5 to 20 nucleotides. Interestingly, a polypurine tract immediately upstream of the U3 region is left intact in order to create a primer for the synthesis of the plus strand of DNA (97).

Because of the lack of proofreading function, all reverse transcriptases (RTs) are notoriously error-prone (97). HIV RT seems to be significantly more error-prone than other RTs, with frequency of misincorporation ranging from $1:1700$ to $1:4000$ (189, 190), which appears to partly account for the high mutation rate of HIV in vivo.

The HIV-1 integrase has been expressed and purified from *Escherichia coli*. The purified protein has been shown to cleave adjacent to a conserved CA dinucleotide in a synthetic oligonucleotide substrate, resulting in the removal of two nucleotides from the 3' ends of both the U5 plus strand and the U3 minus strand (191). Thus, the purified protein simulates some of the steps in retroviral integration ascribed to integrase function.

The envelope glycoproteins of HIV are synthesized initially as an 88-kDa precursor inserted into the rough endoplasmic reticulum, where the addition of high-mannose N-linked carbohydrate chains as well as folding into an appropriate tertiary structure takes place (192, 193). The carbohydrate chains are terminally modified in the Golgi complex. The gp160 precursor is cleaved in a nonlysosomal acidic compartment by some cellular protease, and the mature envelope proteins (gp120 and gp41) are transported to the cell surface (162, 193). Only a small amount (5–15%) of gp160 is cleaved to produce gp120 and gp41, while most of the uncleaved gp160 is delivered to lysosomes, where it is degraded (162). It is not clear why most of the gp160 is routed to lysosomes, while gp120 and gp41 are transported to the cell surface. The possibility that such routing is a general feature of retroviral glycoprotein maturation is suggested by similar results obtained in case of Rous sarcoma virus (194–196). In any case, it explains the requirement of intracellular cleavage of gp160 for the production of infectious particles (163). In common with other retroviruses (197), the envelope glycoproteins of HIV also exist in oligomeric forms. It has been shown that the precursor gp160 forms stable homodimers that in turn assemble into higher-order structures, most likely representing tetramers composed of two dimers each. The molecule retains its oligomeric form after cleavage to gp120/gp41 (198). It appears that the extracellular domains of adjoining gp41 mediate oligomerization of gp160. Consistent with this, native gp41 in virions appears primarily to be a tetramer (199). Though noncovalently associated, these complexes were stable when boiled in the presence of a low concentration of SDS. Using various cross-linking reagents, another study has demonstrated tetrameric forms of gp160 and gp41 (200). The oligomeric form of gp160 is likely to be important for intracellular transport and stability. Moreover, it may increase the avidity of virus binding.

## B. The Regulatory Genes

The *tat* gene is encoded by two exons, one preceding the *env* gene coding for 76 amino acids and the other within the *env* gene coding for 12 amino acids (201). It is translated from various multiply spliced mRNAs (discussed in detail in Section VA). A predominant form of Tat in the infected cells is an 86-amino-acid long (16-kDa) protein derived from two-exon mRNAs (202, 203). Because of a stop codon immediately following the splice donor site of the first *tat*-coding exon, a minor form of 72-amino-acid (14 kDa) Tat is expressed from additional unspliced one-exon mRNAs. The Tat protein is a transactivator of LTR-directed gene expression (201, 204). It is absolutely essential for viral replication, as the Tat-defective mutants do not make virus unless Tat protein is provided in *trans* (205, 206). The N-terminal 72 amino acids of the Tat protein, encoded by the first exon, appear sufficient for full transactivation of HIV LTR-specific gene expression (204). Furthermore, mutational analysis has shown that the amino-terminal 58 amino acids of Tat are sufficient for transactivation, although with reduced activity (207, 208).

Three important functional domains have been identified in Tat (Figure 3). The acidic amino-terminal region has been proposed to have a periodic arrangement of acidic, polar, and hydrophobic residues consistent with an amphipathic α-helix (209), a feature reminiscent of activation domains of many transcription factors (210). Site-directed mutagenesis has shown that conservative changes in these acidic amino acids are well tolerated, whereas nonconservative changes markedly reduce Tat activity. Whether the acidic region of Tat folds into an α-helix and serves as an activation domain requires further study. A cluster of seven cysteine residues, highly conserved among divergent isolates of HIV-1, HIV-2, and SIV Tat proteins, constitutes the

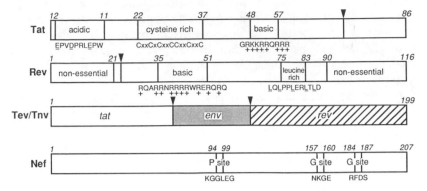

*Figure 3*  Schematic representation of the functional domains of HIV-1 regulatory proteins. Various numbers refer to the amino acid positions. Inverted triangles separate the regions contributed by adjacent exons. See text for details.

second domain (Figure 3). Mutation of all but one (residue 31) of these cysteine residues destroys activity (211–213). It has been proposed that Tat forms a metal-linked dimer with metal ions bridging cysteine-rich regions from each monomer (214). However, the existence of Tat dimers in vivo and their potential role in Tat activity are yet to be established.

A stretch of basic amino acids, from 49 to 57, constitutes the third domain and is required for nuclear localization (213, 215). Mutations within this region yield a cytoplasmic Tat protein, which is nonfunctional. The nuclear signal motif, GRKKR (amino acids 48 to 52), is sufficient to induce the nuclear localization of the normally cytoplasmic β-galactosidase gene product when introduced at the amino terminus (213). Interestingly, Tat has been shown to be predominantly localized to the nucleolus (213, 216). However, a sequence required for this subnuclear localization has not been identified. By analogy with the Rev protein, which is also a predominantly nucleolar protein, it is conceivable that the nucleolar targeting signal may be a bigger sequence extending on both the sides of the nuclear targeting signal.

Rev is the second regulatory protein of HIV-1 that is essential for viral replication (217, 218). It is a 19-kDa protein (116 amino acids) predominantly located in the nucleolus (219). Rev is expressed from two coding exons contained in a variety of multiply spliced mRNAs (211, 220). Although it has been shown to be phosphorylated at serine residues (221), phosphorylation is not essential for its transactivation function since serine substitution mutants that are not phosphorylated are fully active (222, 223). The amino-terminal 20 amino acids as well as the carboxy-terminal 25 amino acids are dispensable for Rev activity (222–224). Two distinct domains have been identified in Rev (Figure 3). A stretch of basic amino acids appears to be important, since deletions within it destroy the activity of Rev (223, 225). The motif NRRRRW confers nuclear localization (224, 226, 227), and it can also substitute for the nuclear localization domain of Tat (228). Removal of the amino-terminal portion of the basic region, which permits concentration of Rev in the nucleus but excludes it from the nucleolus, results in inactive proteins (224, 227, 228), indicating that nucleolar localization is crucial for Rev activity. Additionally, the basic sequence is responsible for interaction with the Rev RNA target sequence, RRE (229).

A leucine-rich sequence, LQLPPLERLTLD, constitutes the second important domain of Rev. Substitutions of the underlined leucine residues yield inactive Rev proteins that act as *trans*-dominant inhibitors of wild-type Rev (223, 230). The leucine-rich sequence has been thought to act as an "activation" domain. Thus, Rev appears to have a modular structure, as in the case of many eukaryotic transactivators (210), with a basic region acting as a "specificity" domain responsible for interaction with RRE-containing transcripts and an "activation" domain responsible for inducing nuclear export of such transcripts as a consequence of the binding event.

The *nef* gene is contained in a single open reading frame at the 3′ end of genome overlapping the *env* gene and the 3′ LTR. A number of multiply spliced mRNAs encode the Nef protein (Figure 4). Though the *nef* open reading frame is present in all HIV and SIV isolates, it frequently contains premature termination codons in virus strains that have been extensively passaged in tissue culture. Moreover, many lentiviruses such as visna virus, CAEV, and EIAV do not contain sequences similar to that of the *nef* gene. Nef is a 25–27-kDa cytoplasmic protein (231, 232). It is myristoylated at the penultimate glycine residue of the amino terminus, and is found in association with the inner plasma membrane presumably through this fatty acyl moiety (233). A free nonmyristoylated Nef has also been detected in the cytoplasm. Nef protein has also been found to be secreted from hamster kidney cells infected with recombinant vaccinia virus expressing *nef* gene (234).

Some features of Nef, such as its location at the inner face of plasma membrane and sequence similarity with guanine nucleotide–binding proteins (G proteins), led to the prediction that Nef is involved in signal transduction similar to the G proteins (235). The idea acquired momentum when it was reported that partially purified Nef protein expressed in *E. coli* indeed exhibited GTP-binding, and GTPase and autokinase activity (236). It was further supported by reports that Nef down-regulates HIV LTR-directed gene expression as well as viral replication (as discussed in section VC). However, many groups have since reported results that failed to support these claims. Recently, Nef protein from BH10 and LAV-1 strains of HIV-1 expressed in *E. coli* and purified to apparent homogeneity was found not to have GTP-binding or GTPase activity (237). Nevertheless, Nef protein does have sequence homol-

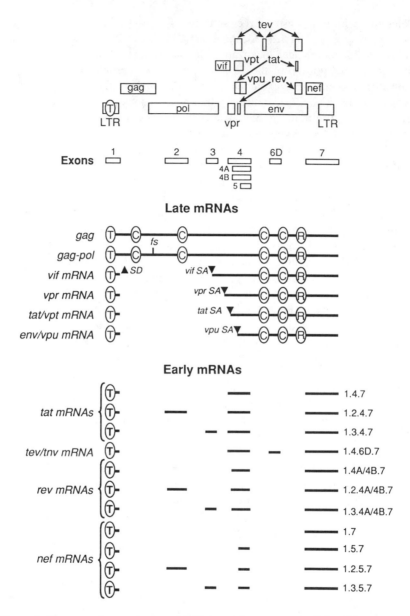

*Figure 4* The genomic organization of HIV-1 and the pattern of splicing. The exons are numbered essentially as described by Schwartz et al (220). T, *trans*-activation response (TAR) element; C, *cis*-acting repressor sequences (CRS); R, Rev response element (RRE). The major splice donor (SD) site is depicted by a triangle, and various splice acceptor (SA) sites are depicted by inverted triangles. fs represents the site of ribosomal frameshift during the translation of *gag-pol* mRNA. Combinatorial splicing of various exons in early mRNAs is shown on the right. See text for details.

ogy with GTP binding sites of G proteins (Figure 3). This includes a P site with a consensus sequence of GXXXXGK (KGGLEG in the case of Nef) and two G sites with consensus sequences NKXD and WRFD (NKGE and RFDS in the case of Nef). Nef mutants at the putative P site were found to be increasingly exported from the cell and had decreased phosphorylation. Deletion of the RFDS sequence resulted in a dramatic reduction in the half-life of Nef (234).

## C. The Accessory Genes

One common characteristic of the accessory genes is that many of them are defective in HIV strains that are extensively passaged in tissue culture, probably because they are dispensable and therefore are selected against during in vitro growth. However, they are likely to serve some important role during in vivo growth of virus and consequently are highly conserved in natural isolates.

A novel protein of 28 kDa designated as Tev (238) or Tnv (239) has recently been described in HIV-1-infected cells. It contains 72 amino acids from the first exon of Tat at its amino terminus, 38 amino acids from the *env* region in the middle, and 91 amino acids from the second exon of *rev* at its carboxy terminus (Figure 3). $p28^{tev}$ has both Tat and Rev activities, and its expression is temporally regulated as in the case of Tat and Rev. Its potential regulatory role, however, is yet to be established.

The *vif* gene encodes a 23-kDa cytoplasmic protein designated as virion infectivity factor (240, 241). The *vif* mRNA may additionally code for Vpu and one-exon Tat (Figure 4) and therefore act as a polycistronic mRNA. *Vif*-defective mutants are not compromised in the level of intracellular viral proteins and extracellular virus (242, 243). However, although the mutant viruses are morphologically and biochemically indistinguishable from wild-type, they exhibit a profound reduction in their efficiency to infect many $CD4^+$ cell lines (242, 243). The magnitude of reduction varies depending on the cell line. The *vif* mutation predominantly affects cell-free virus transmission, with little effect on the cell-to-cell mode of viral transmission. That the *vif* gene product exerts its effect in *trans* is evident by complementation of *vif* mutant viruses by cotransfection with *vif* expression plasmid (242). Considering the fact that Vif is not a virion-associated protein, it is difficult to explain its profound effect on virus infectivity. This protein may somehow modify some component of the virus particles in a way that makes them more infectious. The precise mechanism of action, however, remains to be established.

The *vpr* gene encodes a 15-kDa protein (96 amino acids) in most strains of HIV-1 (244), although the open reading frame is often truncated in viral strains extensively passaged in tissue culture. The *vpr* open reading frame is also present in HIV-2 isolates and in most but not all isolates of SIV. The Vpr protein contains numerous hydrophilic regions and several regions of clustered positive charges. It has recently been shown to be present in mature virus particles in multiple copies (245). Since Vpr is not known to be part of any larger precursor such as Gag or Gag-Pol, it must be incorporated independently into virions. Consistent with this possibility, Vpr was found to be incorporated in *trans* as shown by cotransfection of *vpr* expression plasmid and *vpr*⁻ provirus (E. Cohen and W. Haseltine, personal communication). Interestingly, Vpr is the first regulatory protein of any retrovirus found to be associated with virus particles. Other regulatory proteins, such as Tat, Rev, Nef, Vif, and Vpu, are not virion associated.

Vpr accelerates the replication and cytopathic effect of HIV-1 in CD4⁺ T-cells, with the most pronounced effect exerted early in infection (246, 247). The accelerating effect of Vpr on viral replication has been demonstrated not only in various T-cell lines but also in primary cultures of T-cells, monocytes, and macrophages. Vpr stimulates expression of reporter genes linked to HIV-1 LTR by about threefold (248); this increase is reflected by a proportionate increase in the steady state level of HIV LTR-directed RNA. Vpr-responsive *cis*-acting sequences are yet to be identified. Interestingly, Vpr and Tat appear to have a synergistic effect on HIV LTR-driven gene expression. In contrast to Tat, Vpr stimulates gene expression from a variety of viral and cellular promoters, such as LTRs from HTLV-1, Rous sarcoma virus, and the murine retrovirus (SL3-3) as well as SV40 early promoter and IL-2 promoter. The precise mode of action of Vpr is yet to be established. Some of the possibilities include stimulation of RNA initiation and stabilization of RNA. The presence of Vpr protein in virions suggests that it may function at some early steps in the virus replication cycle, such as the formation and integration of provirus or the initial rate of transcription from the provirus.

The *vpu* gene encodes a 15–20-kDa (81 amino acids) protein (249–251). It is expressed from a polycistronic mRNA that also encodes the envelope protein. The *vpu* reading frame lacks the initiation codon in many HIV-1 strains passaged extensively in tissue culture (E. Cohen and W. Haseltine; G. Pavlakis, personal communications). The *vpu* gene is unique to HIV-1 as it is absent in HIV-2 and in most SIV isolates. The Vpu protein has strong hydrophobic and hydrophilic regions at its amino- and carboxy-terminal regions, respectively. It is also phosphorylated at one or more serine residues (250). Although the *vpu* open reading frame is constant in length among different isolates, there is large variation in the apparent molecular weights of the Vpu proteins because of the anomalous electrophoretic mobilities on denaturing gels. For example, the apparent molecular weights determined on denaturing gels for ELI and IIIB Vpu are 15 and 20 kDa, respectively (252). This observation suggests that the protein is highly ordered, or small changes in charge distribution dramatically alter the electrophoretic mobility of the protein. Vpu is a cytoplasmic protein and is not found in the virions.

Vpu protein acts in a subtle way. While the total amount of viral protein made in infected cultures is the same for $vpu^+$ and $vpu^-$ viruses, the latter show a decrease in the amount of extracellular virus, which correlates with intracellular accumulation of viral proteins, and more rapid formation of syncytia and cytopathic effects (250, 252, 253). Additionally, T-cells infected with $vpu^-$ virus show numerous virus particles within intracytoplasmic vesicles (253). This analysis has suggested a role of Vpu in facilitating assembly and/or release of virus particles. However, the precise mode of action is not known.

The $vpx$ gene, present only in HIV-2 and SIV but not in HIV-1, encodes a 16-kDa virion-associated protein (254–256). Although it appears to be dispensable for viral replication in lymphocytic cell lines (257), $vpx^-$ mutants are less efficient in infecting primary lymphocytes (258). The $vpx$ gene and its product are yet to be characterized.

An open reading frame designated as $vpt$ has been detected in all HIV-1 isolates (259) as well as in closely related chimpanzee isolates (260), but not in HIV-2 and SIV isolates. This open reading frame does not contain an AUG initiation codon. However, a consensus $-1$ frameshift sequence exists near the 5' end that may permit expression of this open reading frame by ribosomes initiated at the $tat$ AUG codon. Consistent with this possibility, a 17-kDa Tat-T fusion protein was detected upon in vitro translation of RNA transcripts derived from this region (261). The fusion protein is recognized by antisera raised against the amino terminus of Tat and peptides derived from this predicted sequence at the carboxy terminus of T. Furthermore, elimination of the consensus frameshift sequence or the $tat$ AUG codon precludes synthesis of Vpt. This fusion protein has neither Tat nor Rev activity, and does not interfere in these activities. The expression and function of Vpt in natural infection have not been established. The protein has not been detected in HIV-1-infected cells, and antibodies to T-specific peptides have not been found in HIV-1-infected people (261). Nonetheless, the remarkable conservation of the $vpt$ reading frame and the consensus frameshift sequences among diverse HIV-1 isolates suggest an important role for $vpt$ in HIV infection.

Interestingly, an open reading frame has been detected on the plus strand of HIV-1 DNA and has been suggested to encode a protein (262). Recently, three RNAs of 1.6, 1.1, and 1.0 kb have been detected on Northern blot of poly(A)$^+$ RNA derived from HIV-1-infected H9 cells using HIV plus strand transcript as a probe (263). The expression of these RNAs was restricted to the early phase of infection. However, the significance of this observation awaits further studies.

## D. Splicing Pattern

HIV has evolved an unprecedented complexity in splicing, giving rise to a multitude of spliced mRNAs from a single primary transcript. The 3' half of the genome has a very intricate layout of multiple, often overlapping open reading frames (Figure 4). Most replication-competent retroviruses express two types of mRNAs: the genomic RNA, which also encodes the Gag and Gag-Pol precursors, and singly spliced subgenomic RNA, which encodes the Env precursor. In addition to these two species of mRNAs, HIV codes for a variety of multiply spliced mRNAs, some of which produce the regulatory and accessory proteins.

The organization of various exons and the pattern of splicing is depicted in Figure 4. Multiply spliced mRNA species have been cloned using polymerase chain reaction (PCR) amplification, and some of the clones analyzed functionally (220). Exon 1, derived from the 5' region of the genome (R, U5, and a small leader sequence) is common for all HIV RNAs. Similarly, exon 7 is present at the 3' end of most of the HIV RNAs. The regulatory proteins (Tat, Rev, and Nef) are each produced from at least three mRNAs differing in the presence or absence of one of the noncoding exons 2 and 3. For example, Tat is encoded by mRNAs generated by splicing of the central exon 4 (264) to exon 1 directly or via exon 2 or exon 3 (220). Two splice acceptor sites have recently been identified between the *tat* and the *rev* AUGs, which are used for *rev* expression (220). The corresponding exons are designated as 4A and 4B, and these splice sites are located 15 and 9 nucleotides upstream of the *rev* AUG, respectively. Thus, utilization of either splice acceptor 4A or 4B allows expression of *rev* but not *tat*. These two exons are spliced to exon 1 either directly or via exon 2 or exon 3, generating six different *rev* mRNAs (220). A splice site between the *rev* and *tat* AUGs has also been reported earlier by S1 nuclease analysis (211). Likewise, three *nef* mRNAs are generated by splicing of exon 5 (220, 264) to exon 1 directly or via exon 2 or exon 3 (220). Exons 2 and 3 are mutually exclusive within the multiply spliced mRNAs. There are conflicting reports as to whether exon 1 can be directly spliced onto exon 7 (220, 265). The presence or absence of small noncoding exons 2 and 3 plays no significant role in the expression of the multiply spliced mRNAs (220). Functional studies have shown that *tat* mRNAs, although polycistronic in nature, produce very low levels of Rev and Nef, while *rev* mRNAs produce high levels of both Rev and Nef proteins (220). It has also been shown that *nef* is the most abundant multiply spliced mRNA species in HIV-1-infected cells, independent of the strain of the virus and the cell type, and is seen in both chronic and short-term infections. Relatively, *tat* mRNA is expressed at extremely low levels, while *rev* mRNA is present at intermediate levels (265).

Another small exon located within the *env* gene, designated as 6D, has recently been identified. This exon is used to generate a hybrid protein Tev (238). Two other exons, 7A and 7B, have also been identified (Figure 4), with splice acceptors located 28 and 24 nucleotides upstream of exon 7, respectively (220). These exons are rarely used, however, and the presumptive mRNAs have not been functionally analyzed. Many additional multiply spliced mRNAs with different combinatorial association of various small exons, which could enormously expand and diversify the coding potential of HIV, have been detected using PCR. However, the functional utilization of such mRNAs and the significance of the protein products, many of which are expected to be variants, remain to be demonstrated. One common feature of all the multiply spliced mRNAs is the removal of RRE, and therefore nuclear export of these RNAs is independent of Rev. It is clear from the foregoing discussion that there is a great deal of structural and functional redundancy in HIV mRNAs, the significance of which is beyond our comprehension at present. The incorporation of additional splice signals and generation of many alternatively spliced mRNAs has been made possible by the acquisition of regulatory factors such as Rev, which ensures the utilization of appropriate quantities of unspliced/singly spliced structural mRNAs.

Proteins such as Vif, Vpr, Vpt, Vpu, and Env could be made from intermediate sized, singly or doubly spliced mRNAs. For example, a single splicing event between the splice donor site of exon 1 and splice acceptor site of exon 2 can generate *vif* mRNA. It is possible that Vpr as well as one-exon Tat can also be made from this RNA. *Vpr* mRNA is probably generated by splicing of exon 1 to the splice acceptor site of exon 3. Some of these intermediate size mRNAs are well characterized. Three different categories of *env/vpu* mRNAs utilizing exon 4A, 4B, or 5 have been demonstrated by PCR amplification and cloning (266). In each category the first exon was found to be spliced to the 3' end long exon, either directly or through exon 2 or exon 3, thereby generating a total of nine mRNA species. All the *env/vpu* mRNAs expressed both Env and Vpu proteins and therefore are functionally bicistronic. Translation of these mRNAs is consistent with the "scanning model" (267) where Env is produced by leaky scanning from mRNAs that contain *env* in the second or third reading frame. This is apparent as the AUG codons of the upstream open reading frames (*vpu* and/or *rev*) are not in a favorable context. Similarly, a group of three mRNAs coding for the one-exon Tat and Vpt have been shown to result from utilization of the splice acceptor site of exon 4 with or without exon 2 or 3 (266). This mRNA contains several downstream open reading frames (*rev, vpu,* and *env*). The *tat* reading frame, however, owing to the favorable context of its AUG codon, appears to block the expression of downstream open reading frames.

## V. NOVEL REGULATORY PATHWAYS IN HIV GENE EXPRESSION

### A. Transcriptional and Posttranscriptional Activation by Tat/TAR

Tat is absolutely essential for HIV transcription and consequently for viral replication. The cis-acting Tat-responsive element (TAR) was originally localized to nucleotides +1 to +80 within the viral LTR (268). Subsequent studies have further mapped its 3' boundary to +44 (269–271). The location of TAR within the transcribed region distinguishes it from cis-acting sequences of various cellular and viral transcriptional activators, which are almost always located upstream of the transcription initiation site. Because of its location, TAR is present at 5' end of all HIV RNAs. The activity of TAR is strictly position- and orientation-dependent. It functions only in its native orientation and when positioned immediately downstream of the site of transcription initiation (266, 270–274). Although it is not completely ruled out that TAR acts as DNA, much direct and indirect evidence strongly suggests that it acts in form of an RNA element. The TAR RNA assumes a stable stem-loop structure in vitro (Figure 5), as determined by nuclease mapping (273). Structural features of TAR that are important for Tat-mediated transactivation include the primary sequence in the loop (nucleotides 31 to 34), the 3-nucleotide bulge (nucleotides 23 to 25), and an intact stem (269–271, 275–277). Accordingly, nucleotide substitutions in the loop as well as the bulge dramatically reduce TAR activity, whereas mutations in the stem are tolerated as long as compensatory mutations are made to restore stem structure. Through use of a novel mutagenesis approach directed towards changing only the secondary structure of TAR, evidence has been provided in favor of TAR acting as RNA. The same study has also shown that the mutant transcripts that only transiently form a native TAR RNA structure, which is not maintained in the mature RNA, are efficiently transactivated by Tat, suggesting that TAR is recognized as a nascent RNA (278).

Purified Tat has been shown to interact with TAR RNA in vitro (279, 280). Use of mutant TAR demonstrated that the bulge was important for both interaction of Tat with TAR RNA as well as Tat-mediated transactivation. In contrast, mutations in the loop and the stem, which reduce Tat-mediated transactivation, had no effect on Tat binding (280). Therefore, direct binding of Tat to TAR appears to be important but not sufficient for transactivation. In conflict with this conclusion are recent studies that showed that Tat can activate transcription by direct binding to nascent RNA, and that the sole function of TAR may be to bring Tat to the vicinity of the promoter by virtue of providing a Tat-binding site. A Tat-Rev fusion protein has been reported to

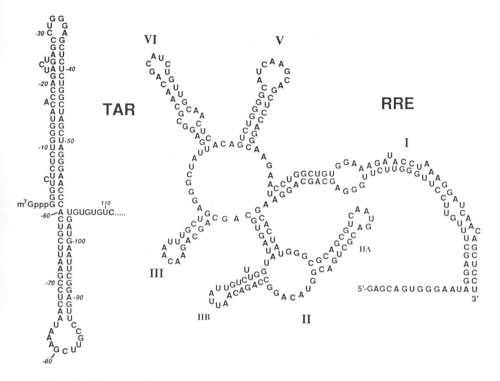

*Figure 5* The nucleotide sequence and predicted secondary structure of *trans*-activation response (TAR) element and Rev response element (RRE).

activate transcription from an HIV-1 promoter derivative, in which TAR is replaced by the Rev response element (RRE) (281). Similarly, Tat fused to the coat protein of bacteriophage MS2 can transactivate an HIV-1 LTR that contains a phage operator sequence in place of TAR (282). The use of fusion protein also identified a stretch of basic amino acids (from 49 to 57) as the RNA-binding domain and the N-terminal part comprising $\alpha$-helical and cysteine-rich regions as activation domains (282).

In the above reports, RNA-binding proteins were fused to Tat. Yet another study has shown that a Tat-Jun fusion protein, which contains the DNA-binding domain of v-Jun, could transactivate the reporter gene containing four copies of the AP-1/Jun-binding sites inserted in the TAR region (283). The insertion of AP-1 sites in TAR was shown to inactivate TAR function. Thus, the promoter-proximal positioning of Tat, either through RNA- or DNA-binding mechanisms, is essential for transactivation. The latter study (283) also showed that the deletion of NF$\kappa$B and Sp1 binding sites from the HIV promoter abrogates Tat-mediated transactivation, suggesting that once brought into the vicinity of the promoter, Tat may interact with transcription factor(s) bound at the NFκB/Sp1 region of the HIV promoter to stimulate transcription initiation and stabilize elongation complexes.

Cellular factors have been postulated to play a role in Tat-mediated transactivation. This suggestion stems from the observation that HIV-1 expression as well as HIV-1 LTR–directed heterologous gene expression are greatly reduced in rodent cells as compared to human cells. Through use of human–Chinese hamster ovary (CHO) hybrid cell clones, the presence of human chromosome 12 was found to be required for Tat-mediated transactivation (284). Several proteins present in nuclear extract of HeLa cells have also been shown to bind TAR RNA in vitro (285–287). A 68-kDa protein has been identified by UV cross-linking to TAR RNA, and the binding affinity of several TAR mutants for this protein was found to correlate with the ability of these mutants to support Tat-mediated transactivation (287). Similarly, a protein that interacts with Tat has been described. A cDNA coding for a 45-kDa nuclear protein, designated as Tat-binding protein-1 (TBP-1), has been cloned by screening a λ-gt11 expression library with Tat as a probe (288). The potential role of these Tat- or TAR-binding proteins in Tat-mediated transactivation needs further investigation. The fact that reporter constructs in which TAR is replaced by the recognition sequences of the RNA/DNA-binding proteins are efficiently transactivated by the Tat fusion proteins (281–283) indicates that transactivation by Tat can occur in the absence of cellular TAR DNA- and RNA-binding proteins.

Tat has also been proposed to act as an antiterminator of RNA elongation (274). When the HIV-1 LTR was placed on a replicating SV40 vector, the primary effect of Tat was reported to be permitting RNA polymerase II to continue to elongate past a strong termination site located approximately 60 nucleotides from the site of RNA initiation. Tat did not exert a pronounced effect on the rate of initiation in these experiments. A separate study (289) confirmed that Tat reduces polarity due to premature transcriptional termination in downstream sequences, but also showed that Tat stimulates the rate of transcription of the first 25 bp region by approximately 15-fold. It is difficult to reconcile the termination suppression model with the observations that removal of TAR does not increase the basal activity of HIV LTR, and that the TAR does not exert a negative effect on heterologous promoters when placed downstream of initiation site.

Most eukaryotic transactivators interact, directly or indirectly, through another protein, with their cognate DNA sequences in the vicinity of the promoter, and the resultant secondary interaction with the ubiquitous transcriptional factors (such as TF-IID) supposedly stimulates the rate of transcription. Tat may be considered as a unique transcription factor that accomplishes transcriptional stimulation by interacting with the nascent transcript. The following scenario could be envisioned. The basal activity of the HIV-1 LTR results in low-level initiation of viral RNA, and synthesis of a small amount of Tat protein. The newly made Tat protein binds to the TAR sequence at the 5' end of the RNA and interacts, directly or indirectly, with the transcription machinery to promote RNA initiation and/or efficiency of elongation of the initiated transcripts. Thus, TAR can be regarded as an "RNA enhancer" (290). The interaction of Tat with TAR RNA may recruit certain transcription factors and consequently facilitate assembly of the transcription initiation complexes at the promoter. However, since TAR acts as an RNA element, it shows orientation dependence due to the polarity of transcription and it functions only downstream of the site of initiation.

Another effect of this multifaceted protein has been reported at the post-transcriptional level. Tat dramatically induces the expression of viral proteins as well as heterologous proteins directed by HIV-1 LTR, attributable both to the low basal activity of the HIV-1 LTR in the absence of Tat and to a high induced level of protein synthesis in the presence of Tat. The extent to which increase in RNA concentration or increase in the efficiency of utilization of the transcribed RNA accounts for the dramatic increases in Tat-induced HIV-1 LTR-directed protein synthesis is controversial and appears to depend upon the experimental system used to measure the effect. Though all investigators concur that the concentration of all HIV-1 LTR–directed RNA rises in the presence of Tat, some investigators report that the increase in RNA is not enough to account for the total increase in the protein synthesis. For example, the Tat-induced 100-fold increase in gene expression could be divided into a 20-fold increase in mRNA synthesis and a 5-fold increase in the amount of protein synthesized per mRNA. A bimodal action of Tat has indeed been proposed (291). More direct evidence in favor of posttranscriptional activation by Tat comes from the study in which coinjection of purified Tat protein and in vitro synthesized TAR-containing CAT mRNA into the nuclei of *Xenopus* oocytes resulted in a substantial increase in CAT protein synthesis (292). Furthermore, this activation was not inhibited by transcriptional inhibitors, and it occurred only when the RNA was injected into the nucleus and not into the cytoplasm. It was proposed that in the presence of Tat, the TAR-containing RNA is either chemically modified or becomes associated in the nucleus with cellular factors and/or with Tat itself, resulting in enhanced translatability. The relevance of these observations to mammalian systems is not clear, however, especially in light of the observation that no transactivation is observed when purified Tat protein is added to mammalian cells in the presence of transcription inhibitors (293).

The observation that mutations that prevent nucleolar localization abrogate Tat activity raises the possibility that the nucleolus may be involved in Tat regulation of gene expression. It may be recalled that 5S RNA, made in the nucleoplasm, is transported into the nucleolus from which it is rapidly exported into the cytoplasm. The nucleolus also serves as the site of assembly and rapid transport of ribosomes from the nucleus to the cytoplasm. It is conceivable that Tat protein makes use of this cellular RNA transport process to achieve efficient export of TAR-containing RNA from the nucleus.

Until recently, Tat had been thought to be highly specific for the HIV LTR, unlike the HTLV transactivator Tax, which has been shown to transactivate a number of cellular gene promoters. It has now been reported, however, that Tat can transactivate a heterologous promoter, the JC virus late promoter (294). Transactivation of this promoter appears to be cell-type specific as it was observed only in glial cells, but not in T-cells or fibroblasts, indicating a possible involvement of cell-type specific factors. This activity of Tat may contribute to a neurological disease, progressive multifocal leukoencephalopathy (PML), by reactivation of latent JC virus.

## B. Regulation of Splicing/Transport of mRNA by Rev/RRE

HIV has a remarkable property of differentially regulating the expression of structural vs regulatory genes in a temporal manner. In the early phase, ~2 kb mRNAs encoding the regulatory proteins Tat, Rev, and Nef predominate. A marked shift takes place in the late phase, leading to the predominant expression of unspliced (~9 kb) and singly spliced (~4 kb) mRNAs encoding structural proteins including Gag, Pol, and Env precursors (295, 296) and many of the accessory proteins. The shift from early to late phase is mediated by the Rev protein (217, 218, 297, 298; for reviews see Refs. 295, 299). Transactivation of viral structural gene expression by Rev requires cis-acting RNA sequence designated as Rev response element (RRE), which is located in the env gene, and is removed by splicing to generate the small regulatory mRNAs (300–302). The RRE only functions in the sense orientation and within a transcription unit (300, 303). The RRE has been predicted to form a complex secondary structure (301) as shown in Figure 5. The central stem I is formed by complementary sequences at 5' and 3' termini of RRE. The intervening stem-loop structures are designated in a clockwise fashion stem-loops II, III, IV, and V, respectively. The stem-loop II, a hammerhead structure, consists of two smaller stem-loop structures, IIA and IIB. The minimal RRE required for biological activity is approximately 200 nucleotides as determined by deletion mutagenesis (304, 305). Within this structure, the stem-loop II appears to be the most important element for in vivo activity (304–306).

A specific and high-affinity interaction between purified Rev protein and RRE has been demonstrated in vitro (304–310). A single hairpin loop structure, stem-loop II, was found to be a primary determinant for Rev binding in vitro and Rev response in vivo (305, 306). Maintenance of secondary structure, rather than primary nucleotide sequence alone, appears to be necessary for Rev-RRE interaction (305). A 90-nucleotide subfragment of RRE at 5' end is sufficient for Rev binding (310). Interestingly, the same subfragment also binds specifically and predominantly to a seemingly ubiquitous mammalian nuclear factor designated as $NF_{RRE}$ (Nuclear Factor, RRE-binding). Furthermore, this ~56-kDa protein is the only detectable cellular factor that binds to the whole RRE (310a). The significance of this observation awaits further study. It may be noted, however, that the 90-nucleotide sequence is insufficient to confer Rev response in vivo, indicating that the RRE sequences 3' to this element are required for translating binding events into function (310a). The Rev protein has been suggested to bind RRE as a multimer on the basis of an intermediate gel shift of the labelled RRE in the presence of both the wild-type and a truncated Rev protein as compared to the extents of gel shift obtained with the two proteins separately (229). It is also consistent with the stoichiometric data obtained using a nitrocellulose filter-binding assay (307) as well as the observation of four distinct complexes detected with a smaller fragment of RRE as a probe in the gel shift assay (Y. Vaishnav and F. Wong-Staal, unpublished). Interestingly, a mutant Rev protein, previously shown to function as a *trans*-dominant inhibitor of Rev function (223), was found to bind RRE as a multimer to a similar extent as wild-type Rev (229). The results suggest the existence of cellular factor(s) that may interact with Rev or Rev/RRE complex, and this interaction may be important for transactivation. The *trans*-dominant Rev mutants probably lack the capacity to interact with these cellular factor(s) and therefore interfere with wild-type Rev by competing with them for binding to RRE.

Expression of unspliced and singly spliced RNA into protein is prevented by *cis*-acting repressive sequences (CRS) present in *gag, pol*, and *env* genes of the HIV genome (300). CRS elements repress expression when introduced into mRNA of heterologous genes such as CAT, but the repression is overcome by introducing HIV-1 RRE sequence in *cis* and supplying Rev in *trans*.

Rev is a posttranscriptional regulator of gene expression as it affects the fate of the completed RNA transcript. Rev was originally identified as an antirepressor of translation *(art)* because mutation in *rev* gene resulted in greatly diminished viral structural protein synthesis without affecting viral transcription (218). The suggestion that Rev might regulate the splicing process came from the observation that *rev⁻* mutant proviruses are defective in the expression of unspliced and singly spliced mRNAs, encoding structural proteins in the cytoplasm, and hence the term *trans*-regulator of splicing *(trs)* (217). Further studies have revealed, however, that the effect of Rev is primarily on the cytoplasmic mRNA and that the levels of spliced and unspliced viral RNA remained unchanged in the nucleus in the presence or absence of Rev (301, 302, 311, 312). Thus came the proposal that Rev induces structural gene expression by activating the sequence-specific nuclear export of incompletely spliced and unspliced RNA species. The role of splicing in Rev action was less favored mainly due to the observation that the cytoplasmic expression of a nonspliceable HIV-1 *env* gene sequence is also subject to Rev regulation (301). A subsequent study, however, showed that recognition of splice sites in addition to the presence of RRE is a must for Rev responsiveness (313). This study also showed that rapid splicing due to the presence of efficient splice sites precluded Rev responsiveness. The retention of incompletely spliced HIV RNAs within the nucleus, in the absence of Rev, appears to be due to the formation of splicing complexes that are slow in executing splicing because of inefficient splice sites. Rev probably promotes dissociation of the RRE-containing unspliced RNAs from these complexes and makes them available for export. The earlier observation that a nonspliceable HIV-1 *env* gene sequence, containing only one splice site, was also subject to Rev regulation (301), could be explained by the fact that a lone splice site, in the absence of the other, can still be recognized by splicing components. Retroviruses in general appear to contain inefficient splicing sites, which may account for the accumulation of an unusually large amount of unspliced viral RNAs in the nucleus (314, 315). This accumulation may reflect a requirement for transporting unspliced RNA to the cytoplasm. Mutations that enhance the rate of viral RNA splicing have been shown to result in defective replication (314, 316). Alternatively, Rev may act independently of spliceosome formation and establish an alternative transport pathway for RRE-containing mRNAs.

The role of the CRS elements is not clear. They may act as nuclear retention signals by virtue of binding to some cellular factor(s), and Rev-RRE interaction may counteract this nuclear retention effect. In some experiments, HIV-1 mRNAs made from recombinant plasmids accumulate in both the nucleus and the cytoplasm of the transfected cells in the absence of Rev, yet no envelope protein is made from these mRNAs even though they are polysome-associated (312, 317, 318). Evidently, under some conditions, the CRS elements can specify a block to the translation of cytoplasmic mRNA, which is overcome by Rev. This is somewhat similar to the observation made about posttranscriptional effect of Tat (as discussed in section VA).

## C. Transcriptional Down-Regulation by Rev

Rev appears to exhibit a dual regulatory role in determining the level of virus expression. At low levels, Rev stimulates virus expression by its posttranscriptional transactivation property, whereas at high level it appears to downregulate viral transcription. A *rev⁻* mutant of HIV-1, though replication-defective, expressed elevated levels of the 1.8 kb mRNA species but little or no higher-molecular-weight species (297). Further study has shown that Rev at relatively high levels exerts a negative effect on transcription in a dose-dependent manner (319). The *cis*-acting sequence responsive to Rev down-regulation appears to be located in the LTR upstream from −117. The inhibitory effect of Rev on transcription is not restricted only to HIV LTR; other promoters such as the immediate early promoter of cytomegalovirus (CMV), papovavirus JC promoter, and to a lesser extent the RSV LTR are also negatively affected. It is worthwhile to note that Rex, the HTLV protein functionally analogous to Rev, also down-regulates HTLV-LTR-directed gene expression when overexpressed (320). This observation strongly suggests an important role for this transcriptional inhibitory property of Rev in HIV pathogenesis.

## D. Is Nef a Negative Regulator?

Whether Nef protein negatively regulates virus expression or not is a controversial issue. Using isogenic strains of HIV-1 that differ only in their ability to produce Nef protein, it was found that *nef*-defective viruses replicate slightly faster than the wild-type viruses in CD4$^+$ T-cell lines (321–323). It was also reported that Nef down-regulates HIV-1 LTR-directed expression of heterologous genes (324, 325). Sequences within the LTR required for Nef activity were reported to be located either within the NRE or between the NRE and the site of RNA initiation.

In conflict with these studies, others have reported that isogenic strains of HIV-1 differing only in *nef* gene do not differ substantially in their ability to replicate in CD4$^+$ T-cell lines (326). Furthermore, it has also been shown that Nef does not inhibit transcription from HIV-1 LTR in a variety of cell types (327).

## E. Activation of HIV by Other Viruses

Although HIV is the etiological agent of AIDS, other cofactors have been implicated to play a role in the outcome of HIV infection, which ranges from asymptomatic carrier state to full-blown AIDS, as well as from latency to reactivation. Several viruses commonly detected in AIDS patients such as herpesviruses (328–330), papovaviruses (330), adenoviruses (331), HTLV-I (332), and human herpesvirus-6 (HHV-6) (333) are capable of transactivating the HIV-1 LTR. Many of these viruses infect and replicate in the same cell type as HIV-1. Interestingly, coinfection of the same cell by HIV and human cytomegalovirus (HCMV) has been detected in the brain tissue of AIDS patients (334). Some of these viruses also encode transactivator proteins, which have been shown to increase expression of the HIV LTR. These viruses include HTLV-I, Cytomegalovirus (328), Herpes Simplex Virus (329, 330), Epstein Barr Virus (335), HHV-6 (333), Adenovirus (336), Papovaviruses

(330), and Hepatitis B Virus (337). These viruses may contribute to AIDS progression by transcriptional activation or reactivation of HIV as well as through additional mechanisms (for a recent review, see Ref. 338). Of particular relevance as a cofactor may be HTLV, since these viruses follow the same routes of transmission as HIV (blood transfusion, sex) and target many of the same risk populations, e.g. intravenous drug users, hemophiliacs. In fact, epidemiological data indicate that patients doubly infected with HIV and HTLV have a much worse prognosis than those infected with HIV alone (339).

## VI. IMPACT OF TAT ON CELLULAR FUNCTIONS

Like most nuclear transactivators, Tat was expected to restrict its action to the cell in which it was produced. However, recent studies suggest that Tat may also function as an exogenous factor, based on observations that indicate that Tat is found extracellularly, and is taken up by other cells in a biologically active form. Purified Tat protein placed in the culture medium is taken up by the cells and is able to transactivate HIV LTR-directed gene expression (293, 340).

The second exon of Tat, which is completely dispensable for transactivation, contains a highly conserved tripeptide sequence Arg-Gly-Glu (RGD), which is a characteristic of many extracellular proteins that bind to cell adhesion molecules (341). RGD-containing peptides were found to compete with full-length Tat protein for binding to the cell surface (342). Since the one-exon Tat protein lacking the RGD sequence is also taken up by cells (340), it appears that the interaction of Tat with the cell surface through the RGD sequence is not involved in internalization. This interaction, however, may serve some other function such as cell-cell interaction through adhesion. A cell surface receptor for Tat has recently been identified as a novel integrin composed of the vitronectin receptor $\alpha_v$ subunit ($\alpha_v$) and a $\beta$ subunit that has not been identified previously (B. Vogel, S.-J. Lee, F. Wong-Staal, W. Craig, M. Pierschbacher, and E. Ruoslahti, unpublished observation). Interestingly, this interaction primarily requires the basic domain of Tat, while the RGD sequence possibly plays a secondary role. Since the basic domain is encoded by the first exon, it is possible that the interaction of Tat with this novel integrin is responsible for internalization. However, the biological significance of this interaction remains to be established.

Extracellular Tat protein has been detected in the medium of HIV-1-infected cells as well as cells transfected with *tat* DNA (343). The extracellular Tat is likely derived by secretion from the cell, since there was no correlation of the amount of Tat found in the medium with extent of cell lysis. Although Tat does not contain a classical signal peptide, a number of cytokines, such as IL-1$\alpha$, IL-1$\beta$, acidic and basic fibroblast growth factors, and platelet-derived endothelial growth factors, are secreted from cells in spite of the lack of a known signal peptide. Therefore, these cytokines, as well as Tat, may follow a secretory pathway yet to be elucidated.

A novel effect of Tat is its ability to affect cell proliferation as an exogenous factor. Two systems have been described. First, Tat inhibits antigen-induced proliferation of lymphocytes when added to the culture medium (344). This effect is specific as oxidized and mutant Tat proteins are inactive, and the inhibitory effect is abrogated by antibodies against Tat. The amount of exogenous Tat protein required to observe such an effect is comparable to that required for transactivation of a reporter gene by extracellular Tat. It was proposed that this may be one pathway whereby HIV can exert an effect on the immune system. The physiological significance of this observation, however, remains unclear, as it is unlikely that circulating Tat level reaches sufficiently high to exert a general regulatory effect.

A second system relates to the stimulatory effect of extracellular Tat protein on the growth of cell lines derived from Kaposi's Sarcoma tissues (343), and suggests that Tat may play a contributory role in the progression of KS. KS is an angiosarcoma of mesenchymal/endothelial origin frequently occurring in elderly men of African and Mediterranean origin as well as in immunosuppressed individuals. The incidence of Kaposi's Sarcoma, however, is increased a thousandfold in certain populations infected with HIV-1. Although HIV infection is a predisposing factor for the development of KS, HIV sequences are not found in the tumor cells, suggesting an indirect role of the virus (345–347). Similarly, transgenic mice carrying the *tat* gene, which develop KS-like lesions, express *tat* in the dermal tissues but not in the tumor cells (348). Several cell lines have been established from Kaposi's lesions of AIDS patients (349). These cell lines share many characteristics of the original tumor cells. For example, the cells are spindle-shaped and induce neoangiogenesis as well as KS-like lesions in nude mice (350). They also release a variety of cytokines with autocrine and paracrine growth effects (351). These cell lines were initially established using conditioned media from T-cells infected with human retroviruses (HTLV-I, HTLV-II, and HIV-1). These same conditioned media continue to enhance proliferation of the KS-derived cell lines in culture. The growth-promoting activity present in HIV-1-infected cell-conditioned medium turned out to be Tat protein itself as this activity is blocked by antibodies against Tat (343). Moreover, recombinant Tat protein was also found to have the same growth-promoting properties. The growth-promoting effect of Tat was detected at very low concentrations, which may be attained under physiological conditions. The optimal threshold of Tat for growth-promoting activity is much lower than that required for LTR transactivation, suggesting that the two phenomena may occur by different mechanisms. It is not clear whether Tat released from infected cell is taken up by the target cells and activates cellular genes required for growth promotion, or whether Tat acts directly as a growth factor and stimulates cell proliferation by signal transduction. Since Tat protein does not stimulate the growth of smooth muscle cells or endothelial cells, the putative progenitors of the KS cells, it is unlikely that Tat by itself induces Kaposi's Sarcoma. This idea is entirely consistent with epidemiological studies, which indicate that HIV-infection may increase the risk for KS, but is neither necessary nor sufficient for induction of the disease.

## VII. GENETIC HETEROGENEITY

Human and related simian immunodeficiency viruses (SIVs) exhibit a remarkable degree of plasticity in their genomes. Most of the analyses on genomic diversity so far have been carried out with HIV-1, but similar diversity is apparent of HIV-2 and various SIVs. One can study the divergence of these viruses at several levels. First, one can follow the complexity of the virus population in a single individual in the course of infection. This type of study gives some interesting insights into the various factors that may be at play in the generation of genetic variants, as well as the implications of such variation on disease progression and attempts at viral prevention and intervention. On a broader scale, one can assess the extent of variation among viruses obtained from different individuals. This comparison may be useful as a molecular epidemiological approach to determine virus transmission among potential donors and recipients. More importantly, in conjunction with immunological determinations, it also helps to circumscribe the regions that are important for vaccine development. Finally, the nature and degree of homology between the related human and simian immunodeficiency viruses will shed some light on the evolution and origin of this virus family.

### A. Genomic Complexity and Evolution in a Single Infection

The idea that each isolate of HIV-1 represents a single genetic entity was dispelled with the discovery of coexisting genotypes in in vitro cultured virus "strains" or uncultured tissue samples from infected individuals. Using restriction enzyme mapping of multiple virus clones from two patients, Saag et al (352) found that although in general clones obtained from the same patient are more related to each other, they are also distinguishable on the basis of their unique set of restriction enzyme sites. Goodenow et al (353) extended these studies, using the polymerase chain reaction (PCR) to amplify and sequence various regions of the HIV genome. They also compared such analyses on virus genomes that had been passaged in culture with those that were directly amplified from patient's peripheral blood mononuclear cells. Several conclusions can be drawn from these studies: (a) there is an inordinate degree of genetic complexity in HIV and that each "isolate" consists of a mixture or "swarm" of microvariants that are highly related but distinguishable. This finding is similar to that of other RNA viruses, giving rise to the concept of a "quasispecies" (354). (*b*) The predominant genotypes that are found in uncultured tissues are different from those present in cultured cells, suggesting that in vitro cultivation imposes a selective process (353). (*c*) There is a prevalence of defective genomes characterized by premature termination codons, insertions, deletions, and substitutions in several of the essential genes, which would prevent virus replication (353). Whether these defective genomes play a role in pathogenesis is an intriguing question.

Highly related HIV genomes obtained from the same patient can have very different biological properties with respect to replication efficiency, tropism, and susceptibility to virus neutralization (355, 356). For example, Koyanagi & Chen (355) found that virus isolates obtained from the brain and cerebral spinal fluid from the same patient could both replicate in peripheral blood lymphocytes, but only virus from the brain could efficiently infect macrophage/monocytes. Looney et al found that a single amino acid substitution in a critical region of the virus envelope could greatly affect the susceptibility of different cloned virus variants to neutralizing antibodies (356). These observations suggest that the genetic composition of an HIV quasispecies will also determine its predominant biological phenotype, and furthermore, such composition may change under selective pressures in the infected host. This concept of a rapidly evolving virus population also gains support from studies of sequential isolates. The coexistence of mixed genotypes makes it more difficult to assess divergence of the virus with time of infection. Thus, the earlier observation of differences in sequentially obtained isolates (357) could be due to selection from an existing population rather than change over time. However, in comparing the biological phenotypes of sequential isolates from patients, Asjo, Fenyo, and colleagues (358, 359) made the interesting observation that isolates obtained from asymptomatic patients in general display restricted growth in T-cell lines and reduced cytopathic effect (a slow/low phenotype), while viruses obtained from advanced AIDS patients infect T-cell lines readily, with a rapid rise in reverse transcriptase activity and production of large syncytia (a rapid/high phenotype). In some patients, a progression to the more highly cytopathic isolates occurs with progression to disease in the same host (360). In the one case study of a laboratory worker infected with an extensively tissue culture–passaged virus, HIV-IIIB, the emergence of variants that are resistant to neutralizing antisera raised against the parental virus was observed in the course of infection (W. Blattner, personal communication).

What may be some of the selective pressures for change? We hinted above at immunoselection as a force, similar to what has been observed in another lentivirus system, namely, equine infectious anemia virus (361). This possibility is also supported by in vitro studies, which demonstrated the generation of escape mutants when the virus is passaged in the presence of a selecting neutralizing serum (362). Similarly, resistant mutants can also be selected for in the presence of an antiviral drug in vivo, such as azidothymidine (AZT) (363). The emergence of more "virulent" forms of the virus, however, may represent removal of selective pressures rather than positive selection. A fast-replicating, highly cytopathic virus may be rapidly cleared by the immunosurveillance of a healthy infected host early after infection, and therefore a slow/low virus is selected for. As this low-level virus expression erodes the immune system, a more replication-efficient virus is allowed to dominate, which in turn causes more rapid deterioration of the immune system. Thus, one can envision a vicious cycle of ever-increasing viral virulence, and ever-decreasing immune competence of the host. This scenario also explains why the virus does not become more virulent with time in the course of the epidemic.

Selective pressure is only one consideration in retrovirus evolution. Another is the accumulation of mutations, predominantly as a function of the reverse transcription process, which is more error prone than DNA replication owing to the lack of a proofreading mechanism (97). The extent of mutation is therefore dependent on the error frequency of the reverse transcriptase and the number of replication cycles in a given time. HIV-1 undergoes more active replication than certain other retroviruses such as HTLV-1. In addition, HIV RT is found to have unusually poor fidelity, with an estimated error rate of 1 substitution in every 1700 to 4000 polymerized nucleotides (189, 190), which is an order of magnitude higher than that of the avian myeloblastosis virus or murine leukemia virus RT. Furthermore, these studies revealed that certain target sequences yielded a significantly higher error frequency, suggesting that there may be "hotspots" for mutation on the virus genome. In addition to mutational events, recombination among existing genotypes can also contribute to the generation of diversity.

## B. Divergence of HIV-1 in the Population

The degree of similarity among any two HIV isolates is in general less than that among viruses from the same person, and depends on the time they have diverged from a common progenitor. Therefore, it is expected that viruses derived from donor-recipient pairs would be highly related, and viruses from the same cohorts and same geographical areas more so than those from distant populations. Furthermore, the extent of divergence increases with the length of time the virus has been in a given population. For example, viruses from Africa are more divergent than viruses in the United States and Europe, supporting the notion that the virus originated in Africa. Attempts have been made to calculate the time of entry of HIV into man based on the genetic distance between isolates having a known time of divergence (364). Such calibrations, however, should be taken with caution since the necessary assumptions for constant mutation rate and replication rate are likely to be invalid.

One major implication of the extensive heterogeneity of HIV is its impact on vaccine development. This concern was fueled by the findings that the most divergent part of the genome was in fact the envelope gene (365, 366), which encodes critical products for immune recognition, and that the viral envelope protein indeed elicited predominantly a type-specific neutralization response (367). Amino acid variability can be as great as 15% over gp160, the envelope precursor protein, with discrete regions of conserved and hypervariable regions (260). As expected, there are functional constraints on the envelope protein that limit its variability. For example, all of the 18 cysteine residues in gp120 are conserved, presumably to maintain the tertiary structure conferred by the disulfide bonds. The region in gp120 responsible for binding to the cellular receptor, CD4, and that in gp41 for inducing cell fusion are also highly conserved. One might speculate that the hypervariable regions have no specific function other than to maintain the proper conformation of the envelope protein, and therefore mutations are more tolerated. Of interest, one of the hypervariable regions (V3) is recognized by both neutralizing antibodies (368) and cytotoxic T-lymphocytes (369), and immunologic selection may well be the mechanism for generating diversity at this site. The fact that the highly variable V3 sequence (as much as 50% variability among HIV-1 isolates) also appears to be the major immunodominant neutralization epitope(s) makes it both an attractive and a formidable target for vaccine development. The V3 sequence contains a loop formed by a disulfide bridge between two invariant cysteines at positions 303 and 338 (370). At the crown of the loop is a relatively conserved tripeptide G-P-G, which presumably allows the protein to adopt a $\beta$ bend structure. Recently, the V3 loop sequence was determined for a large number of HIV field isolates using PCR amplification and sequencing (370). Results from this study revealed that variation in this region is much less extensive than previously thought based on the few laboratory isolates. There is a greater than 80% conservation in 9 out of 14 amino acid positions in the central portion of the loop and a strong preference for particular oligopeptide sequences. Furthermore, the predicted structural motif ($\beta$ strand–type II $\beta$ turn–$\beta$ strand–$\alpha$-helix) of the consensus sequence is also prevalent. Of practical importance is the finding that antibodies that bind the consensus peptide sequence GPGRAF were able to neutralize 60% of random HIV isolates (371).

## C. Evolutionary Divergence of the Human and Simian Immunodeficiency Viruses

Two related but distinct immunodeficiency associated human lentiviruses have now been identified: HIV-1, the prototype AIDS virus prevalent in the

United States, Central Africa, Europe, and throughout most of the world, and HIV-2, which is largely confined to regions of West Africa and seems to exhibit a lesser virulence or penetrance than HIV-1. In addition, nonhuman primate lentiviruses have been obtained from African green monkeys (SIVagm), sooty mangabey monkeys (SIVsmm), baboons (SIVmnd), and captive rhesus monkeys (SIVmac). Most recently, a related virus was isolated from chimpanzees (SIVcpz) (372). Of these primates, the African green monkeys, sooty mangabey monkeys, and mandrill baboons appear to be infected in nature, while the others are more likely to be infected in captivity. Several parameters indicate that SIV has been indigenous to the three naturally infected monkey populations for some time: (a) Infection is prevalent. For example, 20–50% of green monkeys in Kenya, Ethiopia, and South Africa have antibodies to SIV proteins (373). (b) Lack of disease induction in either naturally infected or experimentally inoculated monkeys, suggesting a virus-host adaptation through coevolution. In contrast, the macaque monkeys are highly susceptible to disease induction by SIV. (c) More extensive variability. Using divergence in the *pol* gene as a measure, it is apparent that SIV has been in African green monkeys and sooty mangabey monkeys much longer than HIV-1 or HIV-2 has been in man. Up to 20% variation has been observed among SIVs from these monkeys, while the HIVs only vary as much as 7% in each group.

The question then arises as to whether some of these "older" SIVs could have been a precursor to other SIVs and to the HIVs. Nucleotide sequence analysis revealed four distinct groupings of primate lentiviruses, each about 55–60% related in their *pol* gene: HIV-1, SIVagm, SIVsmm, and SIVmnd being the prototypes of each group (Ref. 373, for review). Interestingly, SIVmac and HIV-2 both fall into the same group as SIVsmm, suggesting that macaques and humans were infected recently by a virus from sooty mangabey directly, or one closely related to it (374). Along the same argument, it is likely that the precursor to HIV-1 may be derived from a monkey virus, although the existence of such a virus has not been demonstrated. The virus obtained from captive chimpanzee was highly related to HIV-1 (372), but since chimpanzees are not commonly infected, it is possible that the chimpanzee was infected by a human virus. The evolutionary implications of these findings have to await additional isolations and genetic analyses.

## VIII. CURRENT AND FUTURE PROSPECTS FOR AN AIDS VACCINE AND ANTIVIRAL THERAPY

The optimism for an AIDS vaccine has waxed and waned in the years since HIV was first demonstrated to be the cause of AIDS. The extensive genetic heterogeneity of HIV, the early unsuccessful experiments in animals (chimpanzees and rhesus monkeys), and the lack of a clear correlation of antiviral immunity and disease progression in infected people were all foreboding observations. Recent progress, however, has injected a new sense of optimism for the prospect of designing a protective, broadly cross-reactive vaccine. As alluded to in Section VIIB, variability in the dominant neutralization and cytotoxic T-lymphocytes (CTL) epitope in gp120 was overestimated because of the choice of prototypes for comparison. It seems possible to induce a broad neutralization response (60% of random isolates) with a single peptide sequence (371). Furthermore, there is preliminary indication that serotype classes may exist in the human population, making it possible to use a cocktail approach to broaden the reactivity. In addition, other neutralization and T-cell epitopes have been mapped, and some of these are relatively conserved. More encouraging is the surge of positive results in monkeys where the vaccinated animals were shown to be protected either from infection or rapid disease induction by low-dose challenge viruses (375). The most effective immunogen in these studies seems to be whole-inactivated virus, but in one study, boosting with a V3-derived peptide in chimpanzees that were primed with inactivated HIV apparently played a key role in enhancing immunity (376), and in another study, recombinant gp160 conferred protection to two vaccinated chimpanzees (377). In view of these data, an AIDS vaccine for man is almost tangible. The difficult parts would be to assess safety and efficacy of candidate vaccines in man and to identify appropriate groups for clinical trials (378).

In contrast, the administration of antiviral drugs to AIDS patients would be less problematic, especially for those more advanced in disease. Nevertheless, to determine efficacy of a drug, the proper clinical trial with placebo controls is still mandatory. So far, the best known antiviral drugs for HIV are the family of 2'-3' dideoxynucleoside analogs, most notably 3'-azido-2',3'-dideoxythymidine (AZT) which has been proven for clinical use and has proven efficacy in prolonging patients' life span (379). Other compounds in this family, including 2',3'-dideoxycytidine (ddC) and 2',3'-dideoxyinosine (ddI) are also undergoing phase I/II clinical trials. Although these compounds have demonstrated antiviral activity in vitro and in vivo, there are also mitigating factors for their long-term use at high doses, such as toxicity of the drugs and emergence of drug-resistant mutants. For example, administration of AZT was found to be associated with bone marrow suppression, which warranted transfusions or dose reduction in one third of patients during the first six months of trial (380, 381). Furthermore, AZT-resistant mutants were isolated from patients who had been under therapy for more than six months (363). Such mutants are not cross-resistant to the other dideoxynucleoside analogs. It is hoped that using a combination regimen of more than one drug, one can lower the dosage of each drug, and thereby lower the toxicity and propensity for developing drug resistance.

The dideoxynucleoside analogs inhibit the reverse transcription process by acting as a chain terminator when incorporated into the viral DNA. There are many other steps in the HIV replication cycle that are amenable to interruption. Like the reverse transcriptase, the other retroviral enzymes protease and integrase are also potentially crucial targets. Synthetic peptides representing transition-state analogs of the protease substrate have been shown to be effective in inhibiting viral protease function as well as HIV replication in vitro, with little effect against cellular aspartic proteases (382–384). These results portend exciting new approaches for treatment of HIV infection.

In addition to targets that are common to all retroviruses, HIV also offers unique opportunities for intervention. Since the first identification of the CD4 molecule as a critical component of cellular receptor for HIV, different lines of approach have been taken to exploit this observation to the end of therapy. It was first shown that soluble recombinant CD4 (rCD4) acted as a competitor for HIV binding and inhibited virus replication in target T-cells (47–49, 80, 385). Attempts were also made to generate chimeric CD4-Ig molecules that make up the amino-terminal portion of CD4, which contains the gp120 binding site and the constant heavy chain domain of human (386) or murine (387) immunoglobulins. These "immunoadhesin" molecules retain antiviral activity and in some instances also mediate antibody-dependent cell cytotoxicity (ADCC). Further, they have the advantages of a longer plasma half-life, capacity to cross the placenta as well as to bind Fc receptors with high affinity. There is yet no evidence of clinical efficacy of either rCD4 or the immunoadhesin. At the preclinical level, soluble CD4 or monoclonal antibodies to gp120 have also been used as targeting devices that deliver toxin molecules to the infected cells to selectively destroy that cell population (388, 389).

A number of molecular approaches have been examined for their potential usefulness in virus inhibition. The following discussion is not meant to be exhaustive, but to indicate the trends of research in this area. As pointed out earlier, HIV replication is mediated by two essential transactivation pathways, both involving an RNA-binding protein (Tat and Rev) recognizing an RNA target sequence (TAR and RRE, respectively). In addition, cellular proteins participate in these pathways by interacting with the viral protein or RNA target. TAR-RNA decoys (390; J. Rappaport, personal communication) and antisense oligonucleotides directed at RRE (S. Daefler, M. Klotman, and F. Wong-Staal, unpublished), which would interfere with the natural RNA target binding to viral proteins, have been shown to inhibit transactivation and virus replication. Similarly, rev mutants with a negative dominant phenotype (223), which can still bind to RRE but may be unable to react with a cellular factor, can inhibit wild-type Rev function. Antisense oligonucleotides directed at the *tat/rev* coding sequences have also been shown to suppress virus expression greatly (391).

A somewhat novel technique to deplete viral-specific mRNA selectively is through the use of "ribozymes," catalytic RNA molecules with sequence-specific cleavage activity (392). Human cells expressing a hammerhead ribozyme targeted against the HIV *gag* gene transcripts were relatively resistant to HIV infection (393). The successful transition of these various genetic approaches to clinical application will depend critically on the technological advances in the frontier of gene therapy, which itself is a rapidly developing area.

In conclusion, there is a justifiable optimism for the prospects of successful treatment and prevention of HIV infection through the development of drugs and vaccine. The realization of these potentials may be too late, however, for those already inflicted with the disease and for those currently at risk of infection. In the meantime, public education and awareness are still the most powerful means to limit the spread of HIV and AIDS.

ACKNOWLEDGMENTS

We thank Elizabeth Swerda for excellent secretarial assistance, and various colleagues for sharing unpublished observations.

## Literature Cited

1. Fenner, F. 1975. *Intervirology* 6:1–12
2. Weiss, R., Teich, N., Varmus, H., Coffin, J., eds. 1984. *RNA Tumor Viruses*, Vols. 1, 2. Cold Spring Harbor: Cold Spring Harbor Lab. 1292 pp., 2344 pp. 2nd ed.
3. Teich, N. 1984. See Ref. 2, pp. 25–207
4. Gonda, M. A., Braun, M. J., Clements, J. E., Pyper, J. M., Wong-Staal, F., et al. 1986. *Proc. Natl. Acad. Sci. USA* 83:4007–11
5. Haase, A. T. 1986. *Nature* 322:131–36
6. Wong-Staal, F. 1990. *Virology*, ed. B. N. Fields, D. M. Knipe, R. M. Chanock, J. Melnick, B. Roizman, R. Slope, 2:1529–93. New York: Raven. 2nd ed.
7. Ellerman, V., Bang, O. 1908. *Zentralbl. Bakteriol.* 46:595–609
8. Ellerman, V., Bang, O. 1909. *Z. Hyg. Infektionskr.* 63:231–72
9. Rous, P. 1911. *J. Exp. Med.* 13:397–411
10. Gross, L. 1951. *Proc. Soc. Exp. Biol. Med.* 78:27–32
11. Temin, H. M., Mizutani, S. 1970. *Nature* 226:211–13
12. Baltimore, D. 1970. *Nature* 226:209–11
13. Temin, H. M. 1964. *Natl. Cancer Inst. Monogr.* 17:557–70
14. Weiss, R. 1989. See Ref. 2, pp. 1205–81
15. Dalgleish, A., Malkovsky, M. 1988. *Adv. Cancer Res.* 51:307–60
16. Burny, A., Bex, F., Chantrenne, H., Cleuter, Y., Dekegel, D., et al. 1978. *Adv. Cancer Res.* 28:251–59
17. Poiesz, B. J., Ruscetti, F. W., Gazdar, A. F., Bunn, P. A., Minna, J. D., et al. 1980. *Proc. Natl. Acad. Sci. USA* 77:7415–19
18. Morgan, D. A., Ruscetti, F. W., Gallo, R. C. 1976. *Science* 193:1007–8
19. Uchiyama, Y., Yodoi, J., Sagawa, K., Takatsuki, K., Uchino, H. 1977. *Blood* 50:481–92
20. Taijma, K., Tominaga, S., Kuroishi, T., Shimizu, H., Suchi, T. 1979. *Jpn. J. Clin. Oncol.* 9:495–504
21. Robert-Guroff, M., Fahey, A., Malda, M., Nakao, Y., Ito, Y., et al. 1982. *Virology* 122:297–305
22. Kalyanaraman, V., Sarngadharan, M., Nakao, Y., Ito, Y., Gallo, R. C. 1982. *Proc. Natl. Acad. Sci. USA* 79:1207–10
23. Hinuma, Y., Nagata, K., Hanaoka, M.

1981. *Proc. Natl. Acad. Sci. USA* 78:6476–80
24. Wong-Staal, F., Gallo, R. C. 1985. *Nature* 317:395–403
25. Kalyanaraman, V. S., Sarangadharan, M. G., Robert-Guroff, M., Miyoshi, I., Blayney, D., et al. 1982. *Science* 218: 571–73
26. Centers for Disease Control (CDC). 1981. *Morbid. Mortal. Wkly. Rep.* 30: 250–52
27. Centers for Disease Control (CDC). 1981. *Morbid. Mortal. Wkly. Rep.* 30: 305–8
28. Gottlieb, M. S., Schroff, R., Schanker, H. M., Weisman, J. D., Fan, P. T., et al. 1981. *New Engl. J. Med.* 305:425–31
29. Masur, H., Michelis, M. A., Greene, J. V., Onorato, F., Vande Stouwe, R. A., et al. 1981. *New Engl. J. Med.* 305: 1431–38
30. Siegal, F. P., Lopez, C., Hammer, G. S., Brown, A. E., Kornfeld, S. J., et al. 1981. *New Engl. J. Med.* 305:1439–44
31. Hardy, W. D. 1985. *Adv. Viral Oncol.* 5:1–37
32. Barre-Sinoussi, F., Chermann, J. C., Rey, F., Nugeyre, M. T., Chamaret, S., et al. 1983. *Science* 220:868–71
33. Montagnier, L., Dauguet, C., Axler, C., Chamaret, S., Gruest, J., et al. 1984. *Ann. Virol.* 135E:119–34
34. Popovic, M., Sarangadharan, M. G., Read, E., Gallo, R. C. 1984. *Science* 224:497–500
35. Gallo, R. C., Salahuddin, S. Z., Popovic, M., Shearer, G. M., Kaplan, M., et al. 1984. *Science* 224:500–3
36. Sarangadharan, M. G., Popovic, M., Bruch, L., Schupbach, J., Gallo, R. C. 1984. *Science* 224:506–8
37. Schupbach, J., Popovic, M., Gilden, R., Gonda, M. A., Sarangadharan, M. G., et al. 1984. *Science* 224:503–5
38. Levy, J. A., Hoffman, A. D., Kramer, S. M., Landis, J. A., Shimabukuro, J. M., et al. 1984. *Science* 225:840–42
39. Ratner, L., Gallo, R. C., Wong-Staal, F. 1985. *Nature* 313:636–37
40. Coffin, J. M., Haase, A., Levy, J. A., Montagnier, L., Oroszlan, S., et al. 1986. *Science* 232:697
41. Klatzmann, D., Barre-Sinoussi, F., Nugeyre, M. T., Dauguet, C., Vilmer, E., et al. 1984. *Science* 225:59–63
42. Dalgleish, A. G., Beverley, P. C. L., Clapham, P. R., Crawford, D. H., Greaves, M. F., et al. 1984. *Nature* 312:763–67
43. Klatzmann, D., Champagne, E., Chamaret, S., Gruest, J., Guetard, D., et al. 1984. *Nature* 312:767–68

44. McDougal, J. S., Mawle, A., Cort, S. P., Nicholson, J. K. A., Cross, G. D., et al. 1985. *J. Immunol.* 135:3151–62
45. McDougal, J. S., Kennedy, M. S., Sligh, J., Cort, S., Mawle, A., et al. 1986. *Science* 231:382–85
46. Maddon, P. J., Dalgleish, A. G., McDougal, J. S., Clapham, P. R., Weiss, R. A., et al. 1986. *Cell* 47:333–48
47. Smith, D. H., Byrn, R. A., Marsters, S. A., Gregory, T., Groopman, J. E., et al. 1987. *Science* 238:1704–7
48. Fisher, R. A., Bertonis, J. M., Meier, W., Johnson, V. A., Costopoulos, D. S., et al. 1988. *Nature* 331:76–78
49. Hussey, R. E., Richardson, N. E., Kowalski, M., Brown, N. R., Chang, H. C., et al. 1988. *Nature* 331:78–81
50. Gartner, S., Markovits, P., Markovitz, D. M., Kaplan, M. H., Gallo, R. C., et al. 1986. *Science* 233:215–19
51. Ho, D. D., Rota, T. R., Hirsch, M. S. 1986. *J. Clin. Invest.* 77:1712–15
52. Salahuddin, S. Z., Rose, R. M., Groopman, J. E., Markham, P. D., Gallo, R. C. 1986. *Blood* 68:281–84
53. Tschachler, E., Groh, V., Popovic, M., Mann, D. L., Konrad, K., et al. 1988. *J. Invest. Dermatol.* 88:233–37
54. Rappersberger, K., Gartner, S., Schenk, P., Stingl, G., Groh, V., et al. 1988. *Intervirology* 29:185–94
55. Cheng-Mayer, C., Rutka, J. T., Rosenbaum, M. L., McHugh, T., Stiles, D. P., et al. 1987. *Proc. Natl. Acad. Sci. USA* 84:3526–30
56. Dewhurst, S., Sakai, K., Bresser, J., Stevenson, M., Evinger-Hodges, M. J., et al. 1987. *J. Virol.* 61:3774–82
57. Adachi, A., Koenig, S., Gendelman, H. E., Daugherty, D., Gattoni-Celli, S., et al. 1987. *J. Virol.* 61:209–13
58. Nelson, J. A., Reynolds-Kohler, C., Margaretten, W., Wiley, C. A., Reese, C. E., et al. 1988. *Lancet* 1:259–62
59. Cao, Y., Friedman-Kien, A. E., Huang, Y., Li, X. L., Mirabile, M., et al. 1990. *J. Virol.* 64:2553–59
60. Fauci, A. S. 1988. *Science* 239:617–22
61. Lifson, J., Coutre, S., Huang, E., Engleman, E. 1986. *J. Exp. Med.* 164:2101–6
62. Lifson, J. D., Feinberg, M. B., Reyes, G. R., Rabin, L., Banapour, B., et al. 1986. *Nature* 323:725–28
63. Lifson, J. D., Reyes, G. R., McGrath, M. S., Stein, B. S., Engleman, E., et al. 1986. *Science* 232:1123–27
64. Hoxie, J. A., Alpers, J. D., Rackowski, J., Huebner, K., Haggarty, B. S., et al. 1986. *Science* 234:1123–27
65. Shaw, G. M., Hahn, B. H., Arya, S.

K., Groopman, J. E., Gallo, R. C., et al. 1984. *Science* 226:165–71

66. Fisher, A. G., Ratner, L., Mitsuya, M., Marselle, L. M., Harper, M. E., et al. 1986. *Science* 233:655–59

67. Lyerly, H. K., Matthews, T. J., Langlois, A. J., Bolognesi, D. P., Weinhold, K. J., et al. 1987. *Proc. Natl. Acad. Sci. USA* 84:4601–5

68. Nicholson, J. K. A., Cross, G. D., Callaway, C. S., McDougal, J. S. 1986. *J. Immunol.* 137:323–29

69. Gendelman, H. E., Narayan, O., Molineaux, S., Clements, J. E., Ghotbi, Z., et al. 1985. *Proc. Natl. Acad. Sci. USA* 82:7086–90

70. Koenig, S., Gendelman, H. E., Orenstein, J. M., Dalcanto, M. C., Pezeshkpour, G. H., et al. 1986. *Science* 233:1089–93

71. Wiley, C. A., Schrier, R. D., Nelson, J. A., Lampert, P. W., Oldstone, M. B. A. 1986. *Proc. Natl. Acad. Sci. USA* 83:7089–93

72. Gabuzda, D. H., Ho, D. D., De la Monte, S. M., Hirsh, M. S., Rota, T. R., et al. 1986. *Ann. Neurol.* 20:289–95

73. Lee, M. R., Ho, D. D., Gurney, M. E. 1987. *Science* 237:1047–51

74. Dreyer, E. B., Kaiser, P. K., Offermann, J. T., Lipton, S. A. 1990. *Science* 258:364–67

75. Lane, H. C., Depper, S. M., Greene, W. C., Whalen, G., Waldmann, J. A., et al. 1985. *New Engl. J. Med.* 313:79–84

76. Fauci, A. S. 1987. *Clin. Res.* 35:503–10

77. Murray, J. F., Garay, S. M., Hopewell, P. C., Mills, J., Snider, G. L., et al. 1987. *Annu. Rev. Resp. Dis.* 133:504–9

78. Mills, J. 1986. *Rev. Infect. Dis.* 8:1001–9

79. Hirsch, M. S., Curran, J. 1990. See Ref. 6, pp. 1545–70

80. Deen, K. C., McDougal, J. S., Inacker, R., Fulena-Wasserman, G., Arthos, J., et al. 1988. *Nature* 331:82–84

81. Albritton, L. M., Tseng, L., Scadden, D., Cunningham, J. M. 1989. *Cell* 57:659–66

82. Kowalski, M., Potz, J., Basiripour, L., Dorfman, T., Wei, C. G., et al. 1987. *Science* 237:1351–55

83. Lasky, L. A., Nakamura, G., Smith, D. H., Fennie, C., Shimasaki, C., et al. 1987. *Cell* 50:975–85

84. Clayton, L. K., Hussey, R. E., Steinbrich, R., Ramachandran, H., Husain, Y., et al. 1988. *Nature* 335:363–66

85. Jameson, B. A., Rao, P. E., Kong, L. I., Hahn, B. H., Shaw, G. M., et al. 1988. *Science* 240:1335–39

86. Landau, N. R., Warton, M., Littman, D. R. 1988. *Nature* 334:159–62

87. Peterson, A., Seed, B. 1988. *Cell* 54:65–72

88. Gallaher, W. R. 1987. *Cell* 50:327–28

89. White, J., Kielian, M., Helenius, A. 1983. *Q. Rev. Biophys.* 16:151–95

90. Felser, J. M., Klimkait, T., Silver, J. 1989. *Virology* 170:566–70

91. Freed, E. O., Myers, D. J., Risser, R. 1990. *Proc. Natl. Acad. Sci. USA* 87:4650–54

92. Bosch, M. L., Earl, P. L., Fargnoli, K., Picciafuoco, S., Giombini, F., et al. 1989. *Science* 244:694–97

93. Marsh, M., Dalgleish, A. 1987. *Immunol. Today* 8:369–71

94. Stein, B. S., Gowda, S. D., Lifson, J. D., Penhallow, R. C., Bensch, K. G., et al. 1987. *Cell* 49:659–68

95. McClure, M. O., Marsh, M., Weiss, R. A. 1988. *EMBO J.* 7:513–18

96. Bedinger, P., Moriarty, A., von Borstel, R. C. II, Donovan, N. J., Steimer, K. S., et al. 1988. *Nature* 334:162–64

97. Coffin, J. M. 1990. See Ref. 6, pp. 1437–500

98. Varmus, H. E., Swanstrom, R. 1984. See Ref. 2, pp. 369–512

99. McDougal, J. S., Mawle, A., Cort, S. P., Nicholson, J. K. A., Cross, G. D., et al. 1985. *J. Immunol.* 135:3151–62

100. Zagury, D., Bernard, J., Leonard, R., Chenier, R., Feldman, M., et al. 1986. *Science* 231:850–53

101. Folks, T., Kelly, J., Benn, S., Kinter, A., Justement, J., et al. 1986. *J. Immunol.* 136:4049–53

102. Zack, J. A., Cann, A. J., Lugo, J. P., Chen, I. S. Y. 1988. *Science* 240:1026–29

103. Zack, J. A., Arrigo, S. J., Weitsman, S. R., Go, A. S., Haislip, A., et al. 1990. *Cell* 61:213–22

104. Gowda, S. D., Stein, B. S., Mohaghagpour, N., Benike, C. J., Engleman, E. G. 1989. *J. Immunol.* 142:773–80

105. Stevenson, M., Stanwick, T. L., Dempsey, M. P., Lamonica, C. A. 1990. *EMBO J.* 9:1551–60

106. Farnet, C. M., Haseltine, W. A. 1990. *Proc. Natl. Acad. Sci. USA* 87:4164–68

107. Ellison, V., Abrams, H., Roe, T., Lifson, J., Brown, P. 1990. *J. Virol.* 64:2711–15

108. Garcia, J. A., Wu, F. K., Mitsuyasu, R., Gaynor, R. B. 1987. *EMBO J.* 63:3761–70

109. Jones, K. A., Kadonaga, J. T., Luciw, P. A., Tjian, R. 1986. *Science* 232:755–59

110. Rosen, C. A., Sodroski, J. G., Haseltine, W. A. 1985. *Cell* 41:813–23

111. Wu, F. K., Garcia, J. A., Harrich, D., Gaynor, R. B. 1988. *EMBO J.* 7:2117–29
112. Nabel, G., Baltimore, D. 1987. *Nature* 326:711–13
113. Franza, B. R. Jr., Josephs, S. F., Gilman, M. Z., Ryan, W., Clarkson, B. 1987. *Nature* 330:391–95
114. Kawasami, K., Schfidereit, C., Roeder, R. G. 1988. *Proc. Natl. Acad. Sci. USA* 85:4700–4
115. Lenardo, M. J., Baltimore, D. 1989. *Cell* 58:227–29
116. Dinter, H., Chiu, R., Imagawa, M., Karin, M., Jones, K. A. 1987. *EMBO J.* 6:4067–71
117. Arrigo, S., Yun, M., Beemon, K. 1987. *Mol. Cell Biol.* 7:388–92
118. Carlberg, K., Ryden, T. A., Beemon, K. 1988. *J. Virol.* 62:1617–24
119. Karnitz, L., Poon, D., Weil, A., Chalkley, R. 1989. *Mol. Cell. Biol.* 9:1929–39
120. Verdin, E., Becker, N., Bex, F., Droogmans, L., Burney, A. 1990. *Proc. Natl. Acad. Sci. USA* 87:4874–77
121. Siekevitz, M., Josephs, S. F., Dukovich, M., Peffer, N., Wong-Staal, F., et al. 1987. *Science* 238:1575–78
122. Horvat, R. T., Wood, C. 1989. *J. Immunol.* 143:2745–51
123. Bohan, C. A., Robinson, R. A., Luciw, P. A., Srinivasan, A. 1989. *Virology* 172:573–83
124. Lu, Y., Stenzel, M., Sodorski, J. G., Haseltine, W. A. 1989. *J. Virol.* 63: 4115–19
125. Okamoto, T., Benter, T., Josephs, S. F., Sadaie, M. R., Wong-Staal, F. 1990. *Virology* 177:606–14
126. Sawadogo, M., Roeder, R. G. 1985. *Cell* 43:165–75
127. Sumrada, R. A., Cooper, T. G. 1987. *Proc. Natl. Acad. Sci. USA* 84:3997–4001
128. Shaw, J. P., Ultz, P. J., Durand, D. B., Toole, J. J., Emmel, E. A., Crabtree, G. R. 1988. *Science* 241:202–5
129. Franza, B. R. Jr., Rauscher, F. J. III, Josephs, S. F., Curran, T. 1988. *Science* 239:1150–53
130. Lu, Y., Touzjian, N., Stenzel, M., Dorfman, T., Sodroski, J. G., et al. 1990. *J. Virol.* 64:5226–29
131. Jones, K. A., Luciw, P. A., Duchange, N. 1988. *Genes Dev.* 2:1101–14
132. Watanabe, S., Temin, H. M. 1982. *Proc. Natl. Acad. Sci. USA* 79:5986–90
133. Mann, R. S., Mulligan, R. C., Baltimore, D. 1983. *Cell* 32:871–79
134. Mann, R., Baltimore, D. 1985. *J. Virol.* 54:401–7
135. Katz, R. A., Teny, R. W., Skalka, A. M. 1986. *J. Virol.* 59:163–67
136. Bender, M. A., Palmer, T. D., Gelinas, R. E., Miller, A. D. 1987. *J. Virol.* 61:1639–46
137. Adam, M. A., Miller, A. D. 1988. *J. Virol.* 62:3802–6
138. Lever, A., Gottlinger, H., Haseltine, W., Sodroski, J. 1989. *J. Virol.* 63: 4085–87
139. Berg, J. M. 1986. *Science* 232:485–87
140. Meric, C., Spahr, P.-F. 1986. *J. Virol.* 60:450–59
141. Gorelick, R. J., Henderson, L. E., Hanser, J. P., Rein, A. 1988. *Proc. Natl. Acad. Sci. USA* 85:8420–24
142. Meric, C., Gouilloud, E., Spahr, P.-F. 1988. *J. Virol.* 62:3328–33
143. Gorelick, R. J., Nigida, S. M. Jr., Weiss, J. W. Jr., Arther, L. O., Henderson, L. E., et al. 1990. *J. Virol.* 64:3207–11
144. Green, L. M., Berg, J. M. 1989. *Proc. Natl. Acad. Sci. USA* 86:4047–52
145. South, T. L., Kim, B., Summers, M. F. 1989. *J. Am. Chem. Soc.* 111:395–96
146. Mervis, R. J., Ahmed, N., Lillehoj, E. P., Paum, M. G., Salazai, F. H. R., et al. 1988. *J. Virol.* 62:3993–4002
147. Veronese, F. D., Copeland, T. D., Oroszlan, S., Gallo, R. C., Sarngadharan, M. G. 1988. *J. Virol.* 62:795–801
148. Bathurst, I. C., Chester, N., Gibson, H. L., Dennis, A. F., Steimer, K. S., et al. 1989. *J. Virol.* 63:3176–79
149. Gottlinger, H. G., Sodroski, J. G., Haseltine, W. A. 1989. *Proc. Natl. Acad. Sci. USA* 86:5781–85
150. Pal, R., Reitz, M. S. Jr., Tschachler, E., Gallo, R. C., Sarngdharan, M. G. 1990. *AIDS Res. Hum. Retroviruses* 6:721–30
151. Rein, A., McClure, M. R., Rice, N. R., Luftig, R. B., Schultz, A. M. 1986. *Proc. Natl. Acad. Sci. USA* 83:7246–50
152. Rhee, S. S., Hunter, E. 1987. *J. Virol.* 61:1045–53
153. Gelderblom, H. R., Hausman, E. H. S., Ozel, M., Pauli, G., Koch, M. A. 1987. *Virology* 156:171–76
154. Kohl, N. E., Emini, E. A., Schleif, W. A., Davis, L. J., Heimbach, J. C., et al. 1988. *Proc. Natl. Acad. Sci. USA* 85:4686–90
155. Peng, C., Ho, B. K., Chang, T. W., Chang, N. T. 1989. *J. Virol.* 63:2550–56
156. Miller, M., Jaskolski, M., Rao, J. K. M., Leis, J., Wlodawer, A. 1989. *Nature* 337:576–79
157. Navia, M. A., Fitzgerald, P. M. D., McKeever, B. M., Leu, C.-T., Heim-

bach, J. C., et al. 1989. *Nature* 337: 615–20

158. Wlodawer, A., Miller, M., Jaskolski, M., Sathyanarayana, B. K., Baldwin, E., et al. 1989. *Science* 245:616–21

159. Weber, I. T., Miller, M., Jaskolski, M., Leis, J., Skalka, A. M., et al. 1989. *Science* 243:928–31

160. Skalka, A. M. 1989. *Cell* 56:911–13

161. Hansen, J., Billich, S., Schulze, T., Sukrow, S., Moelling, K. 1988. *EMBO J.* 7:1785–91

162. Willey, R. L., Bonifacino, J. S., Potts, B. J., Martin, M. A., Klausner, R. D. 1988. *Proc. Natl. Acad. Sci. USA* 85:9580–84

163. McCune, J. M., Rabin, L. B., Feinberg, M. B., Lieberman, M., Kosek, J. C., et al. 1988. *Cell* 53:66–67

164. Deleted in proof

165. Gebhardt, A., Bosch, J. V., Ziemieki, A., Friis, P. R. 1984. *J. Mol. Biol.* 174:297–317

166. Satake, M., Luftig, R. B. 1983. *Virology* 124:259–73

167. Perez, L. G., Davis, G. L., Hunter, E. 1987. *J. Virol.* 61:2981–88

168. Gheysen, D., Jacobs, E., deForesta, F., Thiriart, C., Francotte, M., et al. 1989. *Cell* 59:103–12

169. Smith, A. J., Cho, M.-I., Hammarskjold, M.-L., Rekosh, D. 1990. *J. Virol.* 64:2743–50

170. Karakostas, V., Nagashima, K., Gonda, M. A., Moss, B. 1989. *Proc. Natl. Acad. Sci. USA* 86:8964–67

171. Haffar, O., Garrigues, J., Travis, B., Moran, P., Zarling, J., et al. 1990. *J. Virol.* 64:2653–59

172. Trono, D., Feinberg, M. B., Baltimore, D. 1989. *Cell* 59:113–20

173. Hahn, B. H., Shaw, G. M., Arya, S. K., Popovic, M., Gallo, R. C., et al. 1984. *Nature* 312:166–69

174. Ratner, L., Haseltine, W., Patarca, R., Livak, K. J., Starcich, B., et al. 1985. *Nature* 313:277–84

175. Sanchez-Pescador, R., Power, M. D., Barr, P. J., Steimer, K. S., Stempien, M. M., et al. 1985. *Science* 227:485–92

176. Wain-Hobson, S., Sonigo, P., Danos, O., Cole, S., Alizon, M. 1985. *Cell* 40:9–17

177. Muesing, M. A., Smith, D. H., Cabridilla, C. D., Benton, C. V., Lasky, L. A., et al. 1985. *Nature* 313:450–58

178. Desai, S. M., Kalyanaraman, V. S., Casey, J. M., Srinivasan, A., Anderson, P. R., et al. 1986. *Proc. Natl. Acad. Sci. USA* 83:8380–84

179. Jacks, T., Power, M. D., Masiarz, F. R., Luciw, P. A., Barr, P. J., et al. 1988. *Nature* 331:280–83

180. Wilson, W., Braddock, M., Adams, S. E., Rathjen, P. D., Kingsman, S. M., Kingsman, A. J. 1988. *Cell* 55:1159–69

181. Farmerie, W. G., Loeb, D. D., Casavant, N. C., Hutchison, C. A. III, Edgell, M. H., Swanstrom, R. 1987. *Science* 236:305–8

182. Mous, J., Heimer, E. P., LeGrice, S. F. J. 1988. *J. Virol.* 62:1433–36

183. Lightfoote, M. M., Coligan, J. E., Folks, T. M., Fauci, A. S., Martin, M. A., et al. 1986. *J. Virol.* 60:771–75

184. DiMarzo, F., Veronese, F. D., Copeland, T. D., DeVico, A. L., Rahman, R., et al. 1986. *Science* 231:1289–91

185. Hansen, J., Schulze, T., Mellert, W., Moelling, K. 1988. *EMBO J.* 7:239–43

186. Prasad, V. R., Goff, S. P. 1989. *Proc. Natl. Acad. Sci. USA* 86:3104–8

187. Hizi, A., Barber, A., Hughes, S. H. 1989. *Virology* 170:326–29

188. Tanese, N., Goff, S. P. 1988. *Proc. Natl. Acad. Sci. USA* 85:1777–81

189. Preston, B. D., Poiesz, B. J., Loeb, L. A. 1988. *Science* 242:1168–71

190. Roberts, J. D., Bebenek, K., Kunkel, T. A. 1988. *Science* 242:1171–73

191. Sherman, P. A., Fyfe, J. A. 1990. *Proc. Natl. Acad. Sci. USA* 87:5119–23

192. Fennie, C., Lasky, L. A. 1989. *J. Virol.* 63:639–46

193. Kozarsky, K., Penman, M., Basiripour, L., Haseltine, W., Sodroski, J., et al. 1989. *J. AIDS* 2:163–69

194. Bosch, J. V., Schwartz, R. T. 1984. *Virology* 132:95–109

195. Wills, J. W., Srinivas, R. V., Hunter, E. 1984. *J. Cell Biol.* 99:2011–22

196. Davis, G. L., Hunter, E. 1987. *J. Cell Biol.* 105:1191–204

197. Einfeld, D., Hunter, E. 1988. *Proc. Natl. Acad. Sci. USA* 85:8688–92

198. Earl, P. L., Doms, R. W., Moss, B. 1990. *Proc. Natl. Acad. Sci. USA* 87: 648–52

199. Pinter, A., Honnen, W. J., Tilley, S. A., Bona, C., Zaghouani, H., et al. 1989. *J. Virol.* 63:2674–79

200. Schawaller, M., Smith, G. E., Skehel, J. J., Wiley, D. C. 1989. *Virology* 172:367–69

201. Arya, S. K., Guo, C., Josephs, S. F., Wong-Staal, F. 1985. *Science* 229:69–73

202. Aldovini, A., Debouck, C., Feinberg, M. B., Rosenberg, M., Arya, S. K., et al. 1986. *Proc. Natl. Acad. Sci. USA* 83:6672–76

203. Goh, W. C., Rosen, C., Sodroski, J., Ho, D. D., Haseltine, W. A. 1986. *J. Virol.* 59:181–84

204. Sodroski, J., Patarca, R., Rosen, C.,

Wong-Staal, F., Haseltine, W. 1985. *Science* 229:74–77

205. Fisher, A. G., Aldovini, A., Debouk, C., Gallo, R. C., Wong-Staal, F., et al. 1986. *Nature* 320:367–71

206. Dayton, A. I., Sodroski, J. G., Rosen, C. A., Goh, W. C., Haseltine, W. A. 1986. *Cell* 44:941–47

207. Siegel, L. J., Ratner, L., Josephs, S. F., Derse, D., Feinberg, M. B., et al. 1986. *Virology* 148:226–31

208. Wright, C. M., Felber, B. K., Paskalis, H., Pavlakis, G. N. 1986. *Science* 234:988–92

209. Rappaport, J., Lee, S.-J., Khalili, K., Wong-Staal, F. 1989. *New Biol.* 1:101–10

210. Ptashne, M. 1988. *Nature* 335:683–89

211. Sadaie, M. R., Rappaport, J., Benter, T., Josephs, S. F., Willis, R., et al. 1988. *Proc. Natl. Acad. Sci. USA* 85:9224–28

212. Garcia, J. A., Harrich, D., Soultanakis, E., Wu, F., Mituyasu, R., et al. 1989. *EMBO J.* 8:765–78

213. Ruben, S., Perkins, A., Purcell, R., Joung, K., Sia, R., et al. 1989. *J. Virol.* 63:1–8

214. Frankel, A. D., Bredt, D. S., Pabo, C. O. 1988. *Science* 240:70–73

215. Hauber, J., Malim, M. H., Cullen, B. R. 1989. *J. Virol.* 63:1181–87

216. Hauber, J., Perkins, A., Heimer, E. P., Cullen, B. R. 1987. *Proc. Natl. Acad. Sci. USA* 84:6364–68

217. Feinberg, M. B., Jarrett, R. F., Aldovini, A., Gallo, R. C., Wong-Staal, F. 1986. *Cell* 46:807–17

218. Sodroski, J., Goh, W. C., Rosen, C., Dayton, A., Terwilliger, E., et al. 1986. *Nature* 321:412–17

219. Cullen, B. R., Hauber, J., Campbell, K., Sodroski, J. G., Haseltine, W. A., et al. 1988. *J. Virol.* 62:2498–501

220. Schwartz, S., Felber, B. K., Benko, D. M., Fenyo, E.-M., Pavlakis, G. N. 1990. *J. Virol.* 64:2519–29

221. Hauber, J., Bouvier, M., Malim, M. H., Cullen, B. R. 1988. *J. Virol.* 62:4801–4

222. Cochrane, A. W., Golub, E., Volsky, D., Ruben, S., Rosen, C. A. 1989. *J. Virol.* 63:4438–40

223. Malim, M. H., Bohnlein, S., Hauber, J., Cullen, B. R. 1989. *Cell* 58:205–14

224. Venkatesh, L. K., Mohammed, S., Chinnadurai, G. 1990. *Virology* 176:39–47

225. Perkins, A., Cochrane, A., Ruben, A., Rosen, C. A. 1989. *J. AIDS* 2:256–63

226. Malim, M. H., Bohnlein, S., Fenrick, R., Le, S.-Y., Maizel, J. V., et al. 1989. *Proc. Natl. Acad. Sci. USA* 86:8222–26

227. Cochrane, A. W., Perkins, A., Rosen, C. A. 1990. *J. Virol.* 64:881–85

228. Subramanian, T., Kuppuswamy, M., Venkatesh, L., Srinivasan, A., Chinnadurai, G. 1990. *Virology* 176:178–83

229. Olsen, H. S., Cochrane, A. W., Dillon, P. J., Nalin, C. M., Rosen, C. A. 1990. *Genes Dev.* 4:1357–64

230. Mermer, B., Felber, B. K., Campbell, M., Pavlakis, G. N. 1990. *Nucleic Acids Res.* 18:2037–44

231. Allen, J. S., Coligan, J. E., Lee, T. H., McLane, M. F., Kanki, P., et al. 1985. *Science* 230:810–13

232. Franchini, G., Robert, G. M., Ghrayeb, J., Chang, N. T., Wong-Staal, F. 1986. *Virology* 155:593–99

233. Kan, N. C., Franchini, G., Wong-Staal, F., DuBois, G. C., Robey, W. G., et al. 1986. *Science* 231:1553–55

234. Guy, B., Acres, R. B., Kieny, M. P., Lecocq, J.-P. 1990. *J. AIDS* 3:797–809

235. Samuel, K. P., Seth, A., Konopka, A., Lautenberger, J. A., Papas, T. S. 1987. *FEBS Lett.* 218:81–86

236. Guy, B., Kieny, M. P., Riviere, Y., Le Peuch, C., Dott, K., et al. 1987. *Nature* 330:266–69

237. Kaminchik, J., Bashan, N., Pinchsi, D., Amit, B., Sarver, N., et al. 1990. *J. Virol.* 64:3447–54

238. Benko, D. M., Schwartz, S., Pavlakis, G. N., Felber, B. K. 1990. *J. Virol.* 64:2505–18

239. Salfeld, J., Gottlinger, H. G., Sia, R. A., Park, R. E., Sodroski, J., et al. 1990. *EMBO J.* 9:965–70

240. Lee, T. H., Coligan, J. E., Allan, J. S., McLane, M. F., Groopman, J. E., et al. 1986. *Science* 231:1546–49

241. Fisher, A. G., Ensoli, B., Ivanoff, L., Chamberlain, M., Petteway, S., et al. 1987. *Science* 237:888–93

242. Strebel, K., Daugherty, D., Clouse, K., Cohen, D., Folks, T., et al. 1987. *Nature* 328:728–30

243. Kan, N. C., Franchini, G., Wong-Staal, F., DuBois, G. C., Robey, W. G., et al. 1986. *Science* 231:1553–55

244. Wong-Staal, F., Chanda, P. K., Ghrayeb, J. 1987. *AIDS Res. Hum. Retroviruses* 3:33–39

245. Cohen, E. A., Dehni, G., Sodroski, J. G., Haseltine, W. A. 1990. *J. Virol.* 64:3097–99

246. Ogawa, K., Shibata, R., Kiyomasu, T., Higuchi, Y., Kishida, Y., et al. 1989. *J. Virol.* 63:4110–14

247. Cohen, E. A., Terwilliger, E. F., Jalinoos, Y., Proulx, J., Sodroski, J. G., et al. 1990. *J. AIDS* 3:11–18

248. Cohen, E. A., Lu, Y., Gottlinger, H., Dehni, G., Jalinoos, Y., et al. 1990. *J. AIDS* 3:601–8
249. Cohen, E. A., Terwilliger, E. F., Sodroski, J. G., Haseltine, W. A. 1988. *Nature* 344:532–34
250. Strebel, K., Klimkait, T., Martin, M. A. 1988. *Science* 241:1220–23
251. Matsuda, Z., Chou, M. J., Matsuda, M., Huang, J. H., Chen, Y. M., et al. 1988. *Proc. Natl. Acad. Sci. USA* 85:6968–72
252. Terwilliger, E. F., Cohen, E. A., Lu, Y. C., Sodroski, J. G., Haseltine, W. A. 1989. *Proc. Natl. Acad. Sci. USA* 86:5163–67
253. Klimkait, T., Strebel, K., Hoggan, M. D., Martin, M. A., Orenstein, J. M. 1990. *J. Virol.* 64:621–29
254. Franchini, G., Rusche, J. R., O'Keeffe, T. J., Wong-Staal, F. 1988. *AIDS Res. Hum. Retroviruses* 4:243–50
255. Henderson, L. E., Sowder, R. E., Copeland, T. D., Benveniste, R. E., Oroszlan, S. 1988. *Science* 241:199–201
256. Kappes, J. C., Morrow, C. D., Lee, S., Jameson, B. A., Kent, S. B., et al. 1988. *J. Virol.* 62:3501–5
257. Hu, W., Vander Heyden, N., Ratner, L. 1989. *Virology* 173:624–30
258. Guyader, M., Emerman, M., Montagnier, L., Peden, K. 1989. *EMBO J.* 8:1169–75
259. Sonigo, P., Alizon, M., Staskus, K., Klatzmann, D., Cole, S., et al. 1985. *Cell* 42:369–82
260. Myers, G., Rabson, A. B., Josephs, S. F., Smith, T. F., Wong-Staal, F. 1989. In *Human Retroviruses and AIDS*. Los Alamos, NM: Los Alamos Natl. Lab.
261. Cohen, E., Terwilliger, E. F., Lu, Y., Sodroski, J., Haseltine, W. A. 1989. *AIDS Res. Hum. Retroviruses* 6:56–57
262. Miller, R. H. 1988. *Science* 239:1420–22
263. Bukrinsky, M. I., Etkin, A. F. 1990. *AIDS Res. Hum. Retroviruses* 6:425–26
264. Arya, S. K., Gallo, R. C. 1986. *Proc. Natl. Acad. Sci. USA* 83:2209–13
265. Robert-Guroff, M., Popovic, M., Gartner, S., Markham, P., Gallo, R. C., et al. 1990. *J. Virol.* 64:3391–98
266. Schwartz, S., Felber, B. K., Fenyo, E.-M., Pavlakis, G. N. 1990. *J. Virol.* 64:5448–56
267. Kozak, M. 1989. *J. Cell Biol.* 108:229–41
268. Rosen, C. A., Sodroski, J. G., Haseltine, W. A. 1985. *Cell* 41:813–23
269. Hauber, J., Cullen, B. R. 1988. *J. Virol.* 62:673–79
270. Jacobovits, A., Smith, D. H., Jacobo-

vits, E. B., Capon, D. J. 1988. *Mol. Cell Biol.* 8:2555–61
271. Selby, M. J., Bain, E. S., Luciw, P. A., Peterlin, B. M. 1989. *Genes Dev.* 3:547–58
272. Peterlin, B. M., Luciw, P. A., Barr, P. J., Walker, M. D. 1986. *Proc. Natl. Acad. Sci. USA* 83:9734–38
273. Muesing, M. A., Smith, D. H., Capon, D. J. 1987. *Cell* 48:691–701
274. Kao, S.-Y., Calman, A., Luciw, P. A., Peterlin, B. M. 1987. *Nature* 330:489–93
275. Feng, S., Holland, E. C. 1988. *Nature* 334:165–67
276. Garcia, J. A., Harrich, D., Sultanakis, E., Wu, F., Mitsuyasu, R., et al. 1989. *EMBO J.* 8:765–78
277. Roy, S., Parkin, N. T., Rosen, C., Itovitch, J., Sonenberg, N. 1990. *J. Virol.* 64:1402–6
278. Berkhout, B., Silverman, R. H., Jeang, K.-T. 1989. *Cell* 59:273–82
279. Dingwall, C., Ernberg, I., Gait, M. J., Green, S. M., Heaphy, S., et al. 1989. *Proc. Natl. Acad. Sci. USA* 86:6925–29
280. Roy, S., Delling, U., Chen, C.-M., Rosen, C. A., Sonenberg, N. 1990. *Genes Dev.* 4:1365–73
281. Southgate, C., Zapp, M. L., Green, M. R. 1990. *Nature* 345:840–42
282. Selby, M. J., Peterlin, B. M. 1990. *Cell* 62:769–76
283. Berkhout, B., Gatignol, A., Rabson, A. B., Jeang, K.-T. 1990. *Cell* 62:757–67
284. Hart, C. E., Ou, C.-Y., Galphin, J. C., Moore, J., Bacheler, L. T., et al. 1989. *Science* 246:488–90
285. Gaynor, R., Soultanakis, E., Kuwabara, M., Garcia, J., Sigman, D. S. 1989. *Proc. Natl. Acad. Sci. USA* 86:4858–62
286. Gatignol, A., Kumar, A., Rabson, A., Jeang, K.-T. 1989. *Proc. Natl. Acad. Sci. USA* 86:7828–32
287. Marciniak, R. A., Garcia-Blanco, M. A., Sharp, P. A. 1990. *Proc. Natl. Acad. Sci. USA* 87:3624–28
288. Nelbock, P., Dillon, P. J., Perkins, A., Rosen, C. A. 1990. *Science* 248:1650–53
289. Laspia, M. F., Rice, A. P., Mathews, M. B. 1989. *Cell* 59:283–92
290. Sharp, P. A., Marciniak, R. A. 1989. *Cell* 59:229–30
291. Cullen, B. R. 1986. *Cell* 46:973–82
292. Braddock, M., Chambers, A., Wilson, W., Esnouf, M. P., Adams, S. E., et al. *Cell* 58:269–79
293. Gentz, R., Chen, C. H., Rosen, C. A. 1989. *Proc. Natl. Acad. Sci. USA* 86:821–29
294. Tada, H., Rappaport, J., Lashgari, M.,

Amini, S., Wong-Staal, F., et al. 1990. *Proc. Natl. Acad. Sci. USA* 87:3479–83
295. Wong-Staal, F., Sadaie, M. R. 1988. *The Control of Human Retrovirus Gene Expression*, ed. B. R. Franza, B. R. Cullen, F. Wong-Staal, pp. 1–10. Cold Spring Harbor Lab. NY: Cold Spring Harbor Press
296. Kim, S., Byrn, R., Groopman, J., Baltimore, D. 1989. *J. Virol.* 63:3708–13
297. Sadaie, M. R., Benter, T., Wong-Staal, F. 1988. *Science* 239:910–14
298. Terwilliger, E., Burghoff, R., Sia, R., Sodroski, J., Haseltine, W., et al. 1988. *J. Virol.* 62:655–58
299. Cullen, B. R., Greene, W. C. 1989. *Cell* 58:423–26
300. Rosen, C. A., Terwilliger, E., Dayton, A., Sodroski, J. G., Haseltine, W. A. 1988. *Proc. Natl. Acad. Sci. USA* 85:2071–75
301. Malim, M. H., Hauber, J., Le, S.-Y., Maizel, J. V., Cullen, B. R. 1989. *Nature* 338:254–57
302. Felber, B. K., Hadzopoulou-Cladaras, M., Cladaras, C., Copeland, T., Pavlakis, G. N. 1989. *Proc. Natl. Acad. Sci. USA* 86:1495–99
303. Hadzopoulou-Cladaras, M., Felber, B. K., Cladaras, C., Athanassopoulos, A., Tse, A., et al. 1989. *J. Virol.* 63:1265–74
304. Dayton, E., Powell, D., Dayton, A. 1989. *Science* 246:1625–29
305. Olsen, M., Melbock, P., Cochrane, A., Rosen, C. 1990. *Science* 247:845–48
306. Malim, M. H., Tiley, L. S., McCarn, D. F., Rusche, J. R., Hauber, J., et al. 1990. *Cell* 60:675–83
307. Daly, T., Cook, K., Gray, G., Maione, T., Rusche, J. 1989. *Nature* 342:816–19
308. Zapp, M., Green, M. R. 1989. *Nature* 342:714–16
309. Heaphy, S., Dingwall, C., Ernberg, I., Gait, M. J., Green, S. M., et al. 1990. *Cell* 60:685–93
310. Daefler, S., Klotman, M. D., Wong-Staal, F. 1990. *Proc. Natl. Acad. Sci. USA* 87:4571–75
310a. Vaishnav, Y. N., Vaishnav, M., Wong-Staal, F. 1991. *New Biol.* 3:142–50
311. Emerman, M., Vazeux, R., Peden, K. 1989. *Cell* 57:1155–65
312. Hammarskjold, M. L., Heimer, J., Hammarskjold, B., Sangwan, I., Albert, L., et al. 1989. *J. Virol.* 63:1959–66
313. Chang, D. D., Sharp, P. A. 1989. *Cell* 59:789–95
314. Arrigo, S., Beemon, K. 1988. *Mol. Cell. Biol.* 8:4858–67
315. Katz, R. A., Kotler, M., Skalka, A. M. 1988. *J. Virol.* 62:2686–95

316. Miller, C. K., Embretson, J. E., Temin, H. M. 1988. *J. Virol.* 62:1219–26
317. Knight, D. M., Flomerfelt, F. A., Ghrayeb, J. 1987. *Science* 236:837–40
318. Felber, B. K., Derse, D., Athanassopoulos, A., Campbell, M., Pavlakis, G. N. 1989. *New Biol.* 1:318–28
319. Sadaie, M. R., Benaissa, Z. N., Cullen, B. R., Wong-Staal, F. 1989. *DNA* 8:669–74
320. Rosenblatt, J. D., Cann, A. J., Slamon, D. J., Smalberg, I. S., Shah, N. P., et al. 1988. *Science* 240:916–19
321. Terwilliger, E., Sodroski, J. G., Rosen, C. A., Haseltine, W. A. 1986. *J. Virol.* 60:754–60
322. Luciw, P. A., Cheng-Mayer, C., Levy, J. A. 1987. *Proc. Natl. Acad. Sci. USA* 841:1934–38
323. Cheng-Mayer, C., Iannello, P., Shaw, K., Luciw, P. A., Levy, J. A. 1989. *Science* 246:1629–32
324. Ahmad, N., Venkatesan, S. 1988. *Science* 241:1481–85
325. Niederman, T. M., Thielan, B. J., Ratner, L. 1989. *Proc. Natl. Acad. Sci. USA* 86:1728–32
326. Kim, S., Ikeuchi, K., Byrn, R., Groopman, J., Baltimore, D. 1989. *Proc. Natl. Acad. Sci. USA* 86:9544–48
327. Hammes, S. R., Dixon, E. P., Malim, M. H., Cullen, B. R., Greene, W. C. 1989. *Proc. Natl. Acad. Sci. USA* 86:8549–53
328. Davis, M. G., Kenney, S. C., Kamine, J., Pagano, J. S., Huang, E. S. 1987. *Proc. Natl. Acad. Sci. USA* 84:8642–46
329. Mosca, J. D., Bednarik, D. P., Raj, N. B. K., Rosen, C. A., Sodroski, J. G., et al. 1987. *Nature* 325:67–70
330. Gendelman, H. E., Phelps, W., Feigenbaum, L., Ostrove, J. M., Adachi, A., et al. 1986. *Proc. Natl. Acad. Sci. USA* 83:9759–63
331. Nabel, G. J., Rice, S. A., Knipe, D. M., Baltimore, D. 1988. *Science* 239:1299–302
332. Siekevitz, M., Josephs, S. F., Dukovich, M., Peffer, N., Wong-Staal, F., et al. 1987. *Science* 238:1575–78
333. Lusso, P., Ensoli, B., Markham, P. D., Ablashi, D. V., Salahuddin, S. Z., et al. 1989. *Nature* 337:370–73
334. Nelson, J. A., Reynolds-Kohler, C., Oldstone, M. B. A., Wiley, C. A. 1988. *Virology* 165:386–90
335. Kenney, S., Kamine, J., Markovitz, D., Fenrrack, R., Pagano, F. 1988. *Proc. Natl. Acad. Sci. USA* 85:1652–56
336. Kliewer, S., Garcia, J., Pearson, L., Soultanakis, E., Dasgupta, A., et al. 1989. *J. Virol.* 63:4616–25
337. Twu, J. S., Rosen, C. A., Haseltine, W.

A., Robinson, W. S. 1989. *J. Virol.* 63:2857–60
338. Nelson, J. A., Ghazal, P., Wiley, C. A. 1990. *AIDS* 4:1–10
339. Bartholomew, C., Blattner, W., Cleghorn, F. 1987. *Lancet* 2:1469
340. Frankel, A. D., Pabo, C. O. 1988. *Cell* 55:1189–93
341. Ruoslahti, E., Pierschbacher, M. D. 1987. *Science* 238:491–97
342. Brake, D. A., DeBouck, C., Biesecker, G. 1990. *J. Cell Biol.* 111:1275–81
343. Ensoli, B., Barillari, G., Salahuddin, S. Z., Gallo, R. C., Wong-Staal, F. 1990. *Nature* 345:84–86
344. Viscidi, R. P., Mayur, K., Lederman, H. M., Frankel, A. D. 1989. *Science* 246:1606–8
345. Ensoli, B., Salahuddin, S. Z., Gallo, R. C. 1989. *Cancer Cells* 1:93–96
346. Roth, W. K., Werner, S., Werner, R., Remberger, K., Hofschneider, P. H. 1988. *Int. J. Cancer* 42:767–73
347. Delli-Bovi, P., Donti, E., Knowles, D. M. II, Friedman-Kien, A., Luciw, P. A., et al. 1986. *Cancer Res.* 46:6333–38
348. Vogel, J., Hinrichs, S. N., Reynolds, R. K., Luciw, P. A., Jay, G. 1988. *Nature* 335:606–11
349. Nakamura, S., Salahuddin, S. Z., Biberfeld, P., Ensoli, B., Markham, P. D., et al. 1988. *Science* 242:426–30
350. Salahuddin, S. Z., Nakamura, S., Biberfeld, P., Ensoli, B., Markham, P. D., et al. 1988. *Science* 242:430–33
351. Ensoli, B., Nakamura, S., Salahuddin, S. Z., Biberfeld, P., Larsson, L., et al. 1989. *Science* 243:223–26
352. Saag, M. S., Hahn, B. H., Gibbons, J., Li, Y., Parks, E. S., et al. 1988. *Nature* 334:440–44
353. Goodenow, M., Huet, T., Saurin, W., Kwok, S., Sninsky, J., et al. 1989. *J. AIDS* 2:344–52
354. Steinheuer, D. A., Holland, J. J. 1987. *Annu. Rev. Microbiol.* 41:409–17
355. Koyanagi, Y., Miles, S., Mitsuyasu, R. T., Merrill, J. E., Vinters, H. V., et al. 1987. *Science* 236:819–22
356. Looney, D. J., Fisher, A. G., Putney, S. D., Rusche, J. R., Redfield, R. R., et al. 1988. *Science* 241:357–59
357. Hahn, B. H., Shaw, G. M., Taylor, M. E., Redfield, R. R., Markham, P. D., et al. 1986. *Science* 232:1548–53
358. Asjo, B., Morfeldt-Manson, L., Albert, J., Biberfeld, G., Karlsson, A., et al. 1986. *Lancet* 2:660–62
359. Fenyo, E. M., Morfeldt-Manson, L., Chiodi, F., Lind, B., von Gegerfelt, A., et al. 1988. *J. Virol.* 62:4414–19
360. Cheng-Mayer, C., Seto, D., Tateno,

M., Levy, J. A. 1988. *Science* 240:80–82
361. Montelaro, R. C., Parekh, B., Orrego, A., Issel, C. J. 1984. *J. Biol. Chem.* 259:10539–44
362. Reitz, M. S. Jr., Wilson, C., Naugle, C., Gallo, R. C., Robert-Guroff, M. 1988. *Cell* 54:561–75
363. Larder, B. A., Darby, G., Richman, D. D. 1989. *Science* 243:1731–34
364. Smith, T. F., Srinivasan, A., Schochetman, G., Marcus, M., Myers, G. 1988. *Nature* 333:573–75
365. Starcich, B. R., Hahn, B. H., Shaw, G. M., McNeely, P. D., Modrow, S., et al. 1986. *Cell* 45:637–48
366. Hahn, B., Gonda, M., Shaw, G., Gallo, R. C., Wong-Staal, F. 1985. *Proc. Natl. Acad. Sci. USA* 82:4813–17
367. Nara, P. L., Robey, W. G., Pyle, S. W., Hatch, W. C., Dunlop, N. M., et al. 1988. *J. Virol.* 62:2622–28
368. Putney, S. D., Matthews, T. J., Robey, W. G., Lynn, D. L., Robert-Guroff, M., et al. 1986. *Science* 234:1392–95
369. Berzofsky, J. A., Bensussan, A., Cease, K. B., Bourge, J. F., Cheynier, R., et al. 1988. *Nature* 334:706–8
370. Palker, T. J., Clark, M. E., Langlois, A. J., Matthews, T. J., Weinhold, K. J., et al. 1988. *Proc. Natl. Acad. Sci. USA* 85:1932–36
371. LaRosa, G. J., Davide, J. P., Weinhold, K., Waterbury, J. A., et al. 1990. *Science* 249:932–35
372. Huet, T., Cheynier, R., Meyerhans, A., Roelants, G., Wain-Hobson, S. 1990. *Nature* 345:356–59
373. Desrosiers, R. C., Daniel, M. D., Li, Y. 1989. *AIDS Res. Hum. Retroviruses* 5:465–73
374. Hirsch, V. M., Olmsted, R. A., Murphey-Corb, M., Purcell, R. H., Johnson, P. R. 1989. *Nature* 339:389–92
375. Gardner, M. B. 1990. *AIDS Res. Hum. Retroviruses* 6:835–46
376. Girard, M. 1990. *Cancer Detect. Prev.* N3:411–13
377. Berman, P. W., Gregory, T. J., Riddle, L., Nakamura, G. R., Champe, M. A., et al. 1990. *Nature* 76:622–25
378. Koff, W., Fauci, A. 1989. *AIDS Res. Hum. Retroviruses* 3:S125–S129
379. Yarchoan, R., Berg, G., Brouwers, P., Fischl, M. A., Spitzer, A. R., et al. 1987. *Lancet* 1(8525):132–35
380. Fischl, M. A., Richman, D. D., Grieco, M. H., Gottlieb, M. S., Volberding, P. A., et al. 1987. *New Engl. J. Med.* 317:185–91
381. Richman, D. D. 1988. *Infect. Dis. Clin. North Am.* 2:397–407
382. Meek, T. D., Lambert, D. M., Dreyer,

G. B., Carr, T. J., Tomaszek, T. A. Jr.,
et al. 1990. *Nature* 343:90–92
383. McQuade, T. J., Tomasselli, A. G.,
Liu, L., Karacostas, V., Moss, B., et al.
1990. *Science* 247:454–56
384. Roberts, N. A., Martin, J. A., Kinchington, D., Broadhurst, A. V., Craig,
J. C., et al. 1990. *Science* 248:358–61
385. Traunecker, A., Luke, W., Karjalainen,
K. 1988. *Nature* 331:84–86
386. Capon, D. J., Chamow, S. M., Mordenti, J., Marsters, S. A., Gregory, T., et
al. 1989. *Nature* 337:525–31
387. Traunecker, A., Schneider, J., Kiefer,
H., Karjalainen, K. 1989. *Nature* 339:
68–70
388. Chaudhary, V. K., Mizukami, T., Fitz-

Gerald, D. J., Moss, B., Pastan, I., et
al. 1988. *Nature* 335:369–72
389. Berger, E. A., Clouse, K. A.,
Chaudhary, V. K., Chakrabarti, S.,
Fitzgerald, D. J., et al. 1989. *Proc.
Natl. Acad. Sci. USA* 86:9539–43
390. Graham, G. J., Maio, J. J. 1990. *Proc.
Natl. Acad. Sci. USA* 87:5817–21
391. Matsukura, M., Zon, G., Shinozuka,
K., Robert-Guroff, M., Shimada, T., et
al. 1989. *Proc. Natl. Acad. Sci. USA*
86:4244–48
392. Cech, T. R., Bass, B. L. 1986. *Annu.
Rev. Biochem.* 55:599–629
393. Sarver, N., Cantin, E. M., Chang, P.
S., Zaia, J. A., Ladne, P. A., et al.
1990. *Science* 247:1222–25

*Annu. Rev. Biochem. 1991. 60:631–52*

# ANTISENSE RNA[1]

## Yutaka Eguchi[2]

Laboratory of Molecular Biology, National Institute of Diabetes and Digestive and Kidney Diseases, National Institutes of Health, Bethesda, Maryland 20892

## Tateo Itoh

Department of Biology, Faculty of Science, Osaka University, Toyonaka, Osaka 560, Japan

## Jun-ichi Tomizawa

National Institute of Genetics, Mishima, Shizuoka-ken 411, Japan

KEY WORDS:  RNA-RNA interaction, ColE1, RNA I, RNA II, Rom.

---

## CONTENTS

[1]The US government has the right to retain a nonexclusive royalty-free license in and to any copyright covering this paper.

[2]Present Address: The Wistar Institute, 3601 Spruce Street, Philadelphia, Pennsylvania 19104

## PERSPECTIVES AND SUMMARY

RNA-RNA interactions have been shown to play key structural and enzymatic roles in several aspects of cellular processes, including transcription, translation, RNA processing, and regulation of DNA replication. Recently, it was found that a small RNA can bind to a complementary region of a target RNA and affect its function. An RNA that interferes with the activity of another RNA in this manner is defined as an *antisense RNA*.

Regulation by antisense RNA was first found during the study of replication of the *Escherichia coli* plasmid ColE1. Replication of this plasmid depends on formation of an RNA primer whose precursor RNA is functional only when it assumes a unique structure during its synthesis. Interaction of a small antisense RNA to the primer precursor inhibits formation of the unique structure, and consequently replication of the plasmid. Subsequently, many additional examples of regulation by antisense RNA have been observed, such as those for DNA replication and expression of gene functions. The ability of antisense RNA to inhibit the activity of a specific target mRNA has also led to increasing use of artificial antisense RNAs to study biological function in both prokaryotic and eukaryotic systems (reviewed in 1–3). It has also prompted consideration of antisense RNAs for medical purposes.

In this paper, we review regulation by antisense RNAs and examine the more general question of the mechanism of RNA-RNA interactions. An RNA-RNA interaction is biologically significant only when it occurs at a certain time in a biological process, because the effective interaction has to occur during transcription or because the transcript concerned has a certain limited lifetime. Therefore, the rate of RNA-RNA interaction is critically important for effective regulation by antisense RNA. In addition, most RNA transcripts exist in complex folded structures with one or more loops, and the difference in the structure has a profound effect on the kinetics of RNA-RNA interactions. Accordingly, in this article, the kinetics of interactions between folded RNA molecules are the major subject discussed. Most of the details on the subject are drawn from analysis of the interaction of the precursor RNA of ColE1 with its antisense RNA, as outlined above. Although the structures of RNAs and kinetics of their interaction are unique to each system, studies of ColE1 RNAs, together with those of complementary oligonucleotides, provide a useful conceptual framework for understanding the mechanisms of RNA-RNA interaction and its role in biological regulation.

## BASIC FEATURES OF RNA-RNA INTERACTION

### Binding Between Linear RNAs

Formation of a double-stranded RNA by binding of various complementary pairs of short oligoribonucleotides is bimolecular and second-order with the association rate constants mostly between $1 \times 10^6$ and $3 \times 10^6$ $M^{-1}$ $s^{-1}$. The initial phase of the binding process was studied by analyzing the effect of temperature on the association rate constants (for review, 4, 5). When the temperature increased, the rate constant of association between $A_8$ and $U_8$ (6), and that between two molecules of $A_nU_n$ ($n = 4$, 5, or 6) (7) decreased, giving a negative activation energy. If formation of the first base-pair were rate-determining, increasing temperature should be accompanied by an increasing association rate. On the other hand, if several base-pairs must form to commit the reactants to binding, the increasing temperature should result in a decreasing association rate, because base-pairing is less stable at a higher temperature. The activation energy thus obtained agrees with that calculated for involvement of two or three base-pairs in the rate-determining transition state (6). On the other hand, the association rate of $A_nGCU_n$ ($n = 2$, 3, or 4) increased by increasing temperature (8). Therefore, in this case the transition state is achieved by a diffusion-controlled reaction, probably formation of a nucleus of just one or two base-pairs. Thereafter, the pairing proceeds through the entire length of the complementary region of the components at the very fast rate of about $10^6$ bases per sec at around 20°C (7).

The stabilities of the complexes formed by short oligoribonucleotides depend strongly on the chain-length and base-composition of the components (9, 10). In addition, base sequence affects the stability of the complexes (11), indicating that nearest neighbor interactions must affect the contribution of each base-pair to the stability of the helix. The observation that the RNA double-helix is stabilized by the presence of terminal unpaired bases (dangling bases) (12–18) further suggests that stacking of nearest neighbors affects helix stability. Thermodynamic parameters for base-pairing and the effects of the nearest neighbors on the stability of the complex are summarized in Refs. 11, 15, 19–25. These well-characterized thermodynamic parameters allow us to predict the stability of a number of RNA duplexes and the secondary structures of RNA molecules.

## Binding Between Linear RNA and Stem-Loop RNA or Between Two Stem-Loop RNAs

Binding of a tRNA to a linear RNA containing a complementary triplet serves as a model for binding of linear RNA and stem-loop RNA. Complexes are formed by a tRNA and a triribonucleotide or a larger oligoribonucleotide containing a sequence that is complementary to the anticodon (26–30), under the condition where no complex has been detected by a pair of complementary triribonucleotides (31). Two tRNA molecules with complementary anticodons also form a complex (32–37). The association rate constants obtained for binding between linear RNA and the anticodon loop of tRNA or between two anticodon loops are similar to those for two linear complementary oligonucleotides (32–35, 38–41), even when using the isolated anticodon loop of a tRNA as one of the components (42) in either a linear-to-loop or loop-to-loop interaction. Because the association rates of these reactions increase by increasing temperature (34, 38), the binding is likely to begin with the diffusion-controlled nucleation of pairing of one or two base-pairs probably facilitated by stacking of bases in the loop.

Stability of the complex formed by an anticodon loop and a complementary linear RNA is much higher than expected from the interaction of two linear RNAs. Stability is dependent on the base sequence of the complementary region and increased by the presence of dangling base residues at both ends of the complementary region of the linear component (38, 43–46). Complexes formed by two anticodon loops were 100–10,000-fold more stable than those of the complexes formed by oligoribonucleotides with tRNA (32, 36, 37). The higher stability is probably due to the loop constraint and to stacking with dangling bases and with modified nucleotides (32). In the model proposed for the complex (32; Figure 1A), the three complementary central nucleotides (anticodons) in the seven-nucleotide loops form Watson-Crick base-pairs, and are sandwiched together with two unpaired bases located at the 3'-side of the paired bases between the stems of the components. The overall structure of the resulting complex is similar to a linear A-form double-stranded RNA with two stacked nucleotides looping out from a phosphodiester backbone.

## RNA Folding

Most RNA molecules probably do not exist as random coils, but form secondary structures through intramolecular interactions. Using free-energy parameters obtained for base-pairing and the effects of nearest neighbors and loops on the stability, a possible secondary structure of RNA can be predicted from the nucleotide sequence (19, 48; for recent review, 24). The structures thus predicted for small RNA species frequently match quite well with the biochemical or genetic properties of the molecules. However, agreement is more questionable for large RNA species. For example, the secondary structure of the wild-type ColE1 RNA II (555 nucleotides long) predicted by a computer program (20) is identical to that of a certain mutant RNA II with a single base change (Figure 2B). Nonetheless, the RNase sensitivity and the functional properties of these RNAs are quite different (49). A more plausible wild-type structure can be predicted by adding genetic and biological properties as constraints to the computer program (Figure 2A). The failure to predict RNA structures accurately by the energy parameters alone probably is due both to formation of tertiary structures and nonconventional structures and to inaccuracies in the values of the basic parameters themselves. The absence of an algorithm that incorporates tertiary interaction is an especially serious problem.

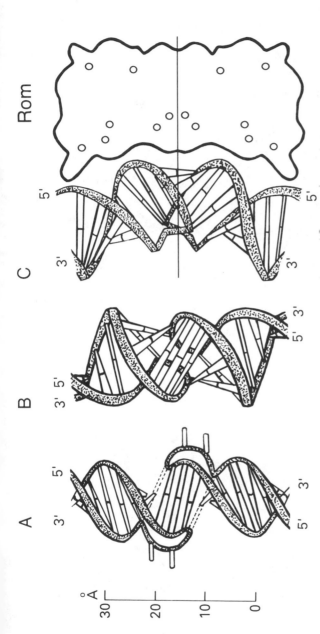

*Figure 1* (A) Model for complex of two anticodon stem-loops (32). Bases and phosphate backbones are shown by the rods and dotted areas, respectively. The three complementary central nucleotides (anticodon) in the seven-nucleotide loops pair with those of the other RNA and stack in a helical conformation together with the neighboring two bases on their 3′-side. The two bases on the 5′-side of the anticodon in the loop are stacked together and looped out from the phosphate backbone. The overall structure is similar to that of a linear A-form double-stranded RNA.

(B) Model for complex $C_{single}$ formed by two fully complementary stem-loop RNAs was constructed by a slight modification of that for a complex of two tRNAs (A). In this model, all seven nucleotides form base-pairs, and the phosphate backbone of each component kinks at the 5′-end of the loop. The overall structure is similar to that of a linear A-form double-stranded RNA with a little bend at the center of the complex. Each molecule in (A) and (B) has an axis of dyad symmetry perpendicular to the paper through the bond joining the pairing central bases.

(C) Model for complex $C^m_{single}$, which consists of a side view of the complex shown in (B) and schematic representation of Rom dimer (47). Open circles in the Rom molecule indicate the position of basic amino acid residues, which are localized on one face of the molecule. The axis of dyad symmetry of the complex (neglecting particular nucleotide sequences) is shown by a horizontal bar.

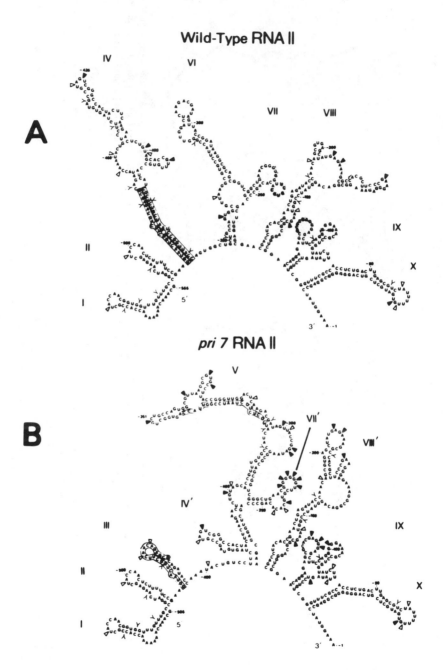

## Wild-Type RNA II

## pri 7 RNA II

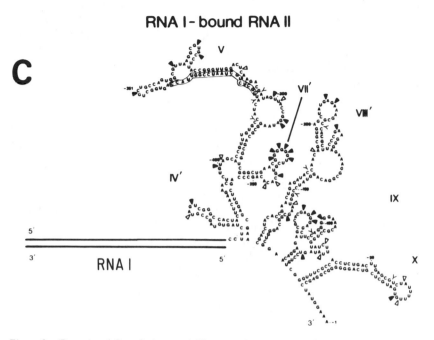

*Figure 2* (Parts *A* and *B* on facing page) The secondary structures of (*A*) the wild-type ColE1 RNA II, (*B*) *pri7* mutant RNA II (G to A change at 469 nucleotides upstream of the origin), and (*C*) the wild-type RNA II hybridized with RNA I with the lowest free energy of formation under certain constraints (48). Sites for preferential cleavage by RNase T1 (solid arrowhead), RNase A (open arrowhead), and RNase V1 (Y) are indicated. Roman numerals show the names of stem-loops. These figures are not intended to show the structures in detail, for which see references (48, 49).

Intramolecular interaction of a loop region with either another loop region or a linear sequence produces a tertiary structure. A simple structure formed by these interactions is called a pseudoknot. The biological significance of pseudoknotting has been reviewed (51).

# BIOLOGICAL REGULATION BY ANTISENSE RNA

## *Regulation of ColE1 DNA Replication*

Studies on the replication of plasmid ColE1 led to the discovery of biological regulation by antisense RNA. They also showed how folded RNA molecules interact with each other and how the interaction affects the control of ColE1 DNA replication (reviewed in 50, 52, 53). Here we describe the mode of ColE1 replication briefly, and then the process of binding of ColE1 RNAs in some detail. The sequence of reactions in ColE1 RNA primer formation is schematically shown in Figure 3 *(left)* (54). Replication of ColE1 is a multi-step process, which initiates with synthesis of an RNA primer for DNA synthesis, named RNA II, from a site 555 nucleotides upstream of the site (origin) where DNA replication initiates (55). Initially, the transcript is separated from the template as is normally the case in transcription (Figure 3, steps 1 and 2). However, as the RNA polymerase nears the origin, the transcription switches to a novel mode in which the transcript remains hybridized to the template DNA (Figure 3, steps 3 and 4). Then the hybrid is cleaved by RNase H (Figure 3, step 5), and the 3'-end of the cleaved RNA primes DNA synthesis by DNA polymerase I (Figure 3, steps 6 and 7).

What causes the transcript to remain hybridized to the template DNA? Computer analysis of the primer RNA, together with the results of genetic and biochemical analyses, shows that RNA II must fold into a unique structure to form the persistent hybrid with the template DNA (49; Figure 2A). An alteration of the nucleotide sequence of RNA II, even one as small as a single base change, can affect the tertiary structure (Figure 2B), and thereby reduce the ability of RNA II to function as a primer. The region of RNA II about 265 nucleotides upstream of the origin (in structure VII of Figure 2A), which contains six consecutive rG residues, probably interacts with the stretch of six consecutive dC residues in the template DNA strand about 20 nucleotides upstream of the replication origin to promote formation of the persistent hybrid (56). A change in the RNA sequence that prevents this interaction inhibits formation of the persistent hybrid (56).

The efficiency of primer formation is determined by a mechanism that regulates the formation of the persistent hybrid of RNA II to the template DNA. A key element of the mechanism is RNA I, a plasmid-specified RNA of about 108 nucleotides (57–59). This RNA is transcribed from the same region of the plasmid as the primer, but in the opposite direction (Figure 3 *(right)*, steps 1' and 2'). Binding of RNA I alters the structure of RNA II (Figure 2C) in a way that prevents RNA II from forming a persistent hybrid with the template DNA at the origin (49; Figure 3, steps 3' to 5'). However, to inhibit the formation of the persistent hybrid, RNA I must interact with the nascent RNA II transcript well before the RNA polymerase reaches the replication origin: the elongating RNA II must be between 100 and 360 nucleotides in length to be susceptible to inhibition by RNA I (60). Therefore, a knowledge of the kinetics of the RNA-RNA interaction is critically important for understanding the mechanism of regulation of ColE1 replication.

## Sequential Steps for Binding of ColE1 RNA I to RNA II

Binding of RNA I and RNA II was first detected by formation of a structure that is sensitive to RNase III that specifically cleaves double-stranded RNA

FORMATION OF PRIMER
AND ITS REMOVAL

INHIBITION OF PRIMER FORMATION
BY RNA I

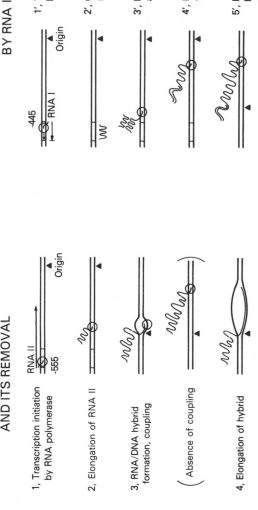

1, Transcription initiation
by RNA polymerase

2, Elongation of RNA II

3, RNA/DNA hybrid
formation, coupling

( Absence of coupling )

4, Elongation of hybrid

5, Cleavage by RNase H

6, Addition of dNMP by
DNA polymerase I

7, Elongation of DNA &
removal of primer

1', Transcription initiation
by RNA polymerase

2', Completion of RNA I
synthesis

3', Interaction of RNA I
and RNA II

4', RNA/RNA hybrid
formation

5', Inhibition of RNA/DNA
hybrid fromation,
uncoupling

*Figure 3* The processes of ColE1 primer formation and of inhibition of primer formation by RNA I are illustrated schematically (54). Straight lines represent double-strands of DNA; wavy lines, RNA transcripts; small circles, RNA polymerase. The DNA template forms an eye structure from the origin of DNA replication (filled triangle). One alternative of step 3 causes primer hybridization, and the other (shown in parentheses) does not. Arrows in the eye structures in steps 5 and 6 indicate cleavage of hybridized RNA by RNase H. The thick lines in the eye structures are newly synthesized DNA strands. Binding of RNA I to RNA II in step 3' inhibits formation of RNA/DNA hybrid and consequently primer formation.

# Process of Binding

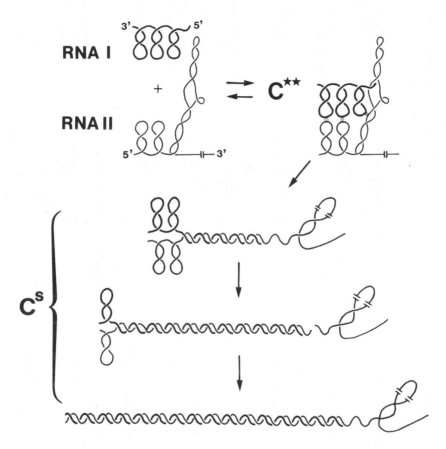

*Figure 4*    The stepwise process of binding of ColE1 RNA I to RNA II is illustrated schematically. RNA I and RNA II interact at the loops of their folded structures to form C**. This interaction facilitates pairing that starts at the 5'-end of RNA I. The pairing propagates progressively as the stem-loop structures unfold, and finally, RNA I hybridizes along the entire length to RNA II.

(54). Subsequently, binding of RNA I and RNA II was studied by quantitating the product by polyacrylamide gel electrophoresis (61). While RNA I has a unique structure consisting of three stem-loops and a short tail at its 5'-end (see Figure 4), an RNA II transcript changes both its folded structure and its propensity to bind RNA I during elongation (60, 62, 63). The use of a 241-nucleotide RNA II transcript in the binding studies as a representative of species whose binding to RNA I inhibits subsequent primer formation is considered to be reasonable (60).

Binding of RNA I to RNA II results in the formation of a complex, designated $C^s$ ($^s$ stands for "stable"), which has a distinct mobility in polyacrylamide gel electrophoresis. The rate constant of formation of $C^s$ between the wild-type RNA I and RNA II is about $1 \times 10^6$ $M^{-1}$ $s^{-1}$ at 25°C (60, 64). However, the rate constant is altered by even a single base change in any loop of the folded structures common to RNA I and RNA II. In particular, if the sequences at any corresponding pair of loops are not complementary, the binding rate is very low. These results suggest the importance of loop-to-loop interactions in the binding. The loop sequences in free RNA I can be cleaved with a low concentration (0.1 unit/ml) of RNase T1. If the two RNA components are mixed and then treated with the enzyme, the loop regions of RNA I become resistant to the enzyme quite rapidly (64). However, most complexes that are resistant to the enzyme at a low concentration are still sensitive to the enzyme at a high concentration (2000 units/ml). These complexes are converted very slowly to highly resistant forms: about half in 30 min (61). These results indicate the presence of a reversible loop-to-loop interaction followed by its slow conversion to a much more stable pairing. The complex thus formed by reversible loop-to-loop interactions is designated C**. The rate constant of association of the wild-type RNA I and RNA II to form C** is $3 \times 10^6$ $M^{-1}$ $s^{-1}$ (64). The conversion of the three loop structures to a more stable one is a sequential process that initiates at the loop and the tail region located near the 5'-end of RNA I, showing unidirectional progression of complete hybridization. The process of binding of the wild-type RNA I and RNA II, deduced from the properties of the binding described above, is schematically shown in Figure 4. On the polyacrylamide gel, the three types of elongating complexes have the same mobility as the completely hybridized complex. They are considered to be in the same class ($C^s$) in the kinetic analysis described below. When five or nine nucleotides are removed from the 5'-end of the wild-type RNA I, $C^s$ is not made with RNA II (61, 65). However, the removal does not prevent formation of C** (64). These results suggest that C** converts to $C^s$ through a process that requires pairing of the tail region of RNA I. The presence of a tail structure for at least one of the components and of a complementary single-stranded region in another component is generally required for stable binding of folded RNAs (60), even for an RNA with a single stem-loop (Y. Eguchi, J. Tomizawa, unpublished observation). The wild-type bacteria can eliminate five nucleotides from the 5'-end of RNA I. Loss of this cleavage activity by a bacterial mutation results in a decrease in plasmid copy number, indicating that the cleaved molecule is inactive in inhibition of in vivo ColE1 replication (66). Formation of the complete double-strand is a very slow process, because simultaneous melting of complementary stems is required for progression of hybridization. Extension of hybridization into the first stem-loops (top of the three $C^s$ diagrams in Figure 4) is sufficient to inhibit primer formation (61).

## Early Steps in Binding of RNA I to RNA II

The presence of a precursor of $C^{**}$ in the binding reaction is suggested by the following inhibition kinetics. RSF1030 is a close relative of ColE1. Its replication is regulated by RSF RNA I, which is very similar to ColE1 RNA I except for the loop III region (67, 68). RSF RNA I inhibits $C^s$ formation between ColE1 RNA I and RNA II (64, 69). The inhibition is presumably due to formation of a very unstable complex, named $C^*$ for the heterologous pair, whose rate constant of formation has been measured to be about $6 \times 10^6$ $M^{-1}$ $s^{-1}$ (64). If a corresponding complex is made between the homologous ColE1 pair, one would expect an even larger value for its association rate constant. The faster rate of formation of $C^*$ than of $C^{**}$, together with other evidence, suggests that $C^*$ is a precursor of $C^{**}$ (64). Formation of $C^*$ by a heterologous pair requires homology at both loop I and loop II and probably in the tail region as well. Because binding of RNA I and RNA II is likely to begin with an interaction at only one of these three regions, it is unlikely that $C^*$, which involves multiple regions of interaction, is the very first product of interaction. The very first product is likely to be an as-yet-undetected complex, named $C^x$, formed by a homologous pair by nucleation at one of the interacting loops. Its formation is probably the target of inhibition by formation of $C^*$ by the heterologous pair (64, 70). Thus, the binding of RNA I to RNA II consists of the following sequence of reactions, producing a series of progressively more stable intermediates leading to the final product (64):

$$\text{RNA I} + \text{RNA II} \rightleftarrows C^x \rightleftarrows C^* \rightleftarrows C^{**} \rightarrow C^s$$

## Binding of Two RNA Stem-Loops at Their Loops

The binding of two RNA stem-loops with fully complementary loop sequences was analyzed in detail using pairs of complementary single stem-loops bearing partial sequences of RNA I and RNA II and their derivatives (71, 72; Y. Eguchi, J. Tomizawa, in preparation). Each RNA has seven nucleotides in the loop and 5 or 20 base-pairs in the stem, so the structure may be similar to that of an anticodon loop. Binding of these stem-loops was examined by comparing the cleavage patterns by RNases. The complex formed by the binding of two single stem-loops, named complex $C_{single}$ (the subscript distinguishes it from complexes formed by RNA with multiple stem-loops), has a different cleavage pattern at the loops of the components from that of a duplex formed by annealing the components, and is much less stable than the duplex. Therefore, the complex is not an ordinary duplex but a hetero-dimeric molecule with interacting loops. The association rate constants for the interaction of two stem-loops with fully complementary sequences range from $1 \times 10^5$ to $3 \times 10^6 \, \text{M}^{-1} \, \text{s}^{-1}$, depending on the sequence. These values are similar to those obtained for the binding of the other types of RNAs described above. The observations of rapid and similar rates of binding of complementary RNAs with different structures suggest that their binding shares a common basic mechanism for the rate-determining steps, most likely the nucleation of base-pairing.

Replacement of a base at different positions in the loop of a component by a noncomplementary base reduces the stability of the resultant complex. The complexes formed by pairs in which only the central three or five bases are complementary are much less stable. These results suggest that all seven complementary nucleotides in the loops of the wild-type pair form base-pairs. The stability of the complex is highly dependent on the loop sequence as well, showing that base-stacking as well as base-pairing determines stability.

Analysis of the wild-type complex by native gel electrophoresis suggests that the complex is bending a little at the interacting region. Based on these observations, a structural model for the complex formed by fully complementary stem-loops has been proposed as illustrated in Figure 1B and 1C (72; Y. Eguchi, J. Tomizawa, in preparation). In the model, all seven nucleotides in the loops form base-pairs, and the phosphate backbone of each component kinks at the 5'-end of the loop. The overall structure is similar to a linear A-form double-stranded RNA with a small bend at the center of the complex. Bending assists in the pairing of all seven complementary nucleotides in the loop.

## Early Steps of Binding in the Presence of Rom

The presence of a certain region downstream from the origin of replication of ColE1 reduces the plasmid copy number (67, 73). The speculation that the region specifies a hypothetical repressor protein named Rop that inhibits synthesis of RNA II (74) was not supported experimentally (75, 76). Instead, the protein isolated affects the mode of binding of RNA I to RNA II and consequently, affects the inhibitory activity of RNA I (60, 70, 75). This protein of 63 amino acids was named Rom to emphasize its function as the "RNA-one modulator." The Rom protein acts at an early step in the binding process (60, 70). The following scheme has been proposed for formation of $C^s$ in the presence of Rom (70):

$$\text{RNA I} + \text{RNA II} \rightleftarrows C^x \rightleftarrows C^* \rightleftarrows C^{**} \searrow$$
$$+ \text{ Rom} \,\Updownarrow \qquad\qquad\qquad\qquad\; C^s$$
$$C^{m*} \rightleftarrows C^{m**} \nearrow$$

Binding of a single dimer molecule of Rom to C* forms $C^{m*}$, whose presence is shown by the effect of Rom on inhibition of binding by RSF RNA I. The complex is then converted to $C^{m**}$, whose RNase sensitivity differs from that of free RNA or C**. Rom does not convert C** to $C^{m**}$, but instead creates a second pathway for stable binding of ColE1 RNA I to RNA II by converting an unstable early intermediate to a more stable complex. The apparent rate constant of formation of $C^s$ from wild-type ColE1 RNA I and RNA II in the presence of Rom is $3 \times 10^6 \ M^{-1} \ s^{-1}$ and that of $C^{m**}$ is $5 \times 10^6 \ M^{-1} \ s^{-1}$. Rom does not always enhance the formation of $C^s$; it is actually inhibitory for certain lengths of RNA II (for example, 89-nucleotide long RNA II; see 60). Thus, the rate of conversion of $C^{m*}$ to $C^s$ is not always higher than the rate of C* to $C^s$.

## Stabilization by Rom of Complex Formed by RNA Stem-Loops

The molecular mechanism of the action of Rom has been studied in detail using the system of binding of two complementary single stem-loops (71, 72; Y. Eguchi, J. Tomizawa, in preparation). A single dimer molecule of Rom binds to the complex $C_{single}$, forming the complex $C^m_{single}$. Formation of $C^m_{single}$ was shown by resistance to cleavage by RNase $V_1$ at the loop region and on the 5'-side of the stem of the RNA components. $C^m_{single}$ is separable from $C_{single}$, single stem-loops, or Rom by gel filtration. While Rom binds neither to each component alone nor to A-form double-stranded RNA, it binds with a similar affinity to complexes formed by complementary stem-loops having various sequences. The affinity to $C_{single}$ decreases greatly when one or two noncomplementary bases are present at various positions of the loop sequence. Therefore, Rom recognizes the structure of a target rather than its exact nucleotide sequences. The binding of Rom decreases the rates of dissociation of the complexes about one hundred fold.

The gross structure of complex $C_{single}$, formed by pairing of all the bases in the loops, has a convex surface (Figure 1C). While all the phosphate groups of $C_{single}$ can be alkylated by ethylnitrosourea, binding of Rom suppresses alkylation of the groups on the convex surface (72). On the other hand, the Rom dimer has a concave surface on which are located several basic amino acid residues that are important for the function of Rom (77; see Figure 1C). These two surfaces fit to each other quite well, and their interaction must be responsible for formation of complex $C^m_{single}$ (Figure 1C).

Analysis of the interactions of pairs of single stem-loops to form $C_{single}$ and $C^m_{single}$ suggests a mechanism for complex formation between RNA I and RNA II. The apparent rate constant for formation of C** from RNA I and RNA II is about $2.5 \times 10^6 \ M^{-1} \ s^{-1}$, which is close to that for the formation of $C_{single}$. This suggests that the rate-determining step in formation of C** is interaction at either one of the loop regions, and this interaction induces interaction at other stem-loop regions to produce C** that is more stable than $C_{single}$. Because binding of only one dimer molecule of Rom is sufficient for the formation of $C^{m**}$, stabilization of a single pair of interacting stem-loops should be responsible for the direct effect of Rom on binding of RNA I and RNA II.

Complex C* is slightly more stable than $C_{single}$, yet C* is sensitive to RNase A while $C_{single}$ is resistant. The rate constant of formation of C* is also higher than that of $C_{single}$. It is possible that the rapid and weak interaction that produces C* does not involve simultaneously all the three loops or all the seven base-pairs of each loop.

## Regulation by Antisense RNA

A mutation in the RNA I region of the ColE1 plasmid alters both RNA I and RNA II. A comparison of the copy numbers of the mutant plasmids and the binding rates of their RNA I and RNA II shows an inverse correlation (61, 75). The characteristics of binding among homologous and heterologous pairs of RNA I and RNA II also explain very well the property of incompatibility, in which one member of a group of plasmids is excluded from a bacterial cell by another member. The inhibitory activity of ColE1 RNA I is suppressed by RSF RNA I and vice versa (69). This mutual suppression explains the increase in copy numbers of both ColE1 and RSF1030 by coexistence. These observations indicate that the binding of RNA I to RNA II is responsible for in vivo regulation of plasmid replication.

In several groups of ColE1-type plasmids, in vitro formation of the primer for DNA synthesis is inhibited by the group-specific RNA I (57). Regulation of copy number and incompatibility of these plasmids can be explained by the group-specific interaction of RNA I and RNA II. These RNA I species are very similar in size and structure; they have three similar-sized stem-loops and a similar short tail. A total of 12 stem-loops of four different groups of plasmids, ColE1, RSF1030, p15A, and CloDF13, consist of only four types of stem-loops with different nucleotide sequences in their loops (68). The specific regulatory system for each group of plasmids may have evolved by duplication and recombination of the DNA sequences for these stem-loops.

The efficiency of replication of ColE1 in vivo fits reasonably well with expectations based on the concentrations and rates of synthesis of RNA I, RNA II, and Rom, as well as the rate constant of formation of $C^s$ (78). Several theoretical models for binding of RNA I to RNA II and for replication of ColE1-type plasmid have been proposed (79–82).

Functional regulation of specific target RNAs by antisense RNA has been shown or proposed to operate in various prokaryotic systems besides the ColE1 system described above (reviewed in 83–85). In most of the systems listed in Table 1, antisense RNAs and their target RNAs are transcribed from

**Table 1**  Naturally occurring antisense RNAs

| | Antisense RNA | Target RNA | Function controlled | Level of control[a] | Type of mechanism | Selected refs. |
|---|---|---|---|---|---|---|
| **Plasmid** | | | | | | |
| ColE1 | RNA I | RNA II | Replication | Primer formation | II | |
| IncF | copA RNA | repA mRNA | Replication | mRNA stability | II | 86–91 |
| IncI | inc RNA | repZ mRNA | Replication | Translation | II | 92–96 |
| ColE2 | copRNA | rep mRNA | Replication | uncertain | II | 97, b |
| R1162 | ct RNA | rep1 mRNA | Replication | (Translation) | I | 98 |
| R6K | silencer | activator | Replication | uncertain | uncertain | 99 |
| pT181 | RNA I | repC mRNA | Replication | Transcription termination | II | 100–10? |
| IncF | finP RNA[c] | traJ mRNA | Conjugation | (Translation) | I | 103, 10■ |
| IncFII | sok RNA | hok mRNA | Host killing | (Translation) | II | 105 |
| **Phage** | | | | | | |
| lambda | aQ RNA | Q mRNA | Lysis/lysogeny | uncertain | uncertain | 106–1C |
| lambda | oop RNA | cII mRNA | Lysis/lysogeny | mRNA stability | I | 109 |
| P22 | sar RNA | ant mRNA | Lysis/lysogeny | (Translation) | I | 110 |
| P1, P7 | c4 repressor[d] | ant mRNA | Lysis/lysogeny | (Translation) | I | 111 |
| **Transposon** | | | | | | |
| IS10 | RNA-OUT | tnp mRNA | Transposition | Translation | I | 112–11◄ |
| **Bacterial** | | | | | | |
| E. coli | micF RNA[e] | ompF mRNA | OmpF synthesis | Translation | I | 115–11 |
| E. coli | tic RNA[f] | crp mRNA | Crp synthesis | Transcription termination | I | 118 |

[a] Speculated level of control is given in parentheses.
[b] H. Yasueda, S. Takechi, T. Itoh, in preparation
[c] The antisense control of conjugation by finP RNA requires the finO gene product, which might be RNA (119) o■ protein (120).
[d] The c4 repressor is an antisense RNA containing two short sequence elements that are complementary to tar■ sequences in the ant mRNA encompassing the ribosome-binding site involved in ant expression.
[e] The micF RNA is 70% complementary to the 5'-region of the ompF mRNA containing the ribosome-binding site a■ the initiation codon.
[f] The tic RNA is complementary to 10 out of the first 11 nucleotides of the crp mRNA.

the complementary strands of the same region of DNA. However, for the two *Escherichia coli* chromosomal genes, each antisense RNA is transcribed from a region away from the target gene (115–118). On the other hand, the *ant* mRNA of phage P1 (and P7) and its antisense RNA (*c4* repressor) are transcribed from the same promoter (111). The shorter antisense RNA probably interacts with a distal region of the target RNA.

Binding of an antisense RNA may disrupt the function of its target RNA either directly by blocking function of the region that is complementary to the antisense RNA (Type I), or indirectly by altering the structure of the target RNA (Type II). Inhibition of translation by blocking the ribosomal binding site and/or the initiation codon as seen for regulation of IS10 transposition (114) and *Escherichia coli* OmpF synthesis (116, 117) are examples of the Type I mechanism. Regulation by the Type II mechanism can take multiple forms. Regulation of ColE1 primer formation as described above results from the conformational change of RNA II that prevents it from making a structure necessary for primer formation. Binding of antisense RNA can also affect translation by altering the secondary structure of the target transcript in such a way that the ribosomes-binding site and/or the initiation codon is inaccessible to ribosome, as is proposed for regulation of replication of IncFII plasmid (121) and of IncIα (= IncI₁) plasmid ColIb-P9 (96), and host killing of the IncFII plasmid (105). In addition, binding of an antisense RNA could induce premature termination of transcription of the target RNA by forming a transcription terminator or terminator-like structure, as is the case for *repC* transcription of plasmid pT181 (102) or for *crp* transcription of *Escherichia coli* (118). Antisense RNA binding can also render a target RNA susceptible to digestion by RNase III or other RNases as seen for *rep* mRNA of IncFII plasmid R1 (91) and lambda *c*II mRNA (109).

All known antisense RNAs except the *tic* RNA are predicted to form very stable secondary structures. The *copA* RNA of IncFII plasmid R1 (122), *sar* RNA of bacteriophage P22 (123), and RNA-OUT of IS10 (124), in fact form stable stem-loop structures. Genetic analyses using mutants of IncFII plasmid (125, 126), ColE2 plasmid (S. Takechi, H. Yasueda, T. Itoh, in preparation), ColIb-P9 plasmid (96), bacteriophage P22 (cited in 85), bacteriophage P1 and P7 (111), or IS10 transposon (127), and phylogenetic comparisons of the *finP* RNAs of IncF plasmids (128), have suggested that the specificity of initial recognition between the antisense RNA and target RNA is determined by the nucleotide sequence of the loop portion of one major stem-loop structure of each antisense RNA.

The antisense RNAs named *cop* RNAs of closely related but mutually compatible plasmids of the IncF family (IncFI and IncFII) (86, 89, 90, 129) and those of the IncI and IncB families (92–94) share an identical sequence (5'-CGCCAA-3') in the loop portions of the major stem-loop structures. On the other hand, the *cop* RNAs of three closely related but mutually compatible plasmids of the ColE2 family share another identical sequence (5'-UUGGCG-3') in the loop portions of their major stem-loop structures (S. Hiraga, T. Itoh, unpublished). Because these shared sequences are complementary, the initial interaction of the antisense RNAs and the target RNAs of all these plasmids probably occurs between the identical combination of complementary nucleotide sequences, suggesting an evolutionary significance.

Binding between purified antisense RNA and the target RNA was demonstrated for the systems of IncFII plasmid R1 (*copA* RNA/*repA* mRNA) (130–132), of ColE2 plasmid (T. Sugiyama, T. Itoh, in preparation), of bacteriophage P22 (133), and of transposon IS10 (127). The second order rate constants of stable binding in these systems are $3 \times 10^5$ to $3 \times 10^6$ $M^{-1} s^{-1}$, comparable to that between RNA I and RNA II of ColE1. The effects of mutations in the loop portions of the antisense RNAs of R1, ColE2, and IS10 on the binding rates correlate very well with their in vivo phenotypes, such as increased copy numbers, reduced incompatibility, and reduced multicopy inhibition of IS10 transposition. By analogy with the ColE1 RNAs, binding of RNAs in systems other than IS10 is thought to be initiated by interaction between loops followed by stable binding that could start from the single-stranded tail extending from the stem-loop structure. However, for IS10 RNAs, pairing of RNA-OUT and *tnp* mRNA initiates between the free single-stranded 5'-end of *tnp* mRNA and the loop of RNA-OUT and then proceeds down one side of the RNA-OUT stem by strand displacement (127). In all cases examined, the final products of RNA-RNA interaction are complete duplex RNAs. However, it is not known whether formation of a complete hybrid is required for inhibition by an antisense RNA, except for ColE1 for which complete hybridization has been shown to be unnecessary (64).

So far no naturally occurring antisense RNA regulation has been proven in eukaryotic systems. However, there are numerous cases known in eukaryotic systems in which both strands of a DNA segment are transcribed, or (at least partially) complementary RNAs are detected (83, 134, 135). Some of these could be involved in antisense RNA regulation. For example, an RNA species found in neurons latently infected with herpes simplex virus is complementary to the mRNA for an immediate early viral protein required for the lytic process. Suppression of the lytic process by this RNA could be responsible for maintaining the latent state (136, 137). Small RNAs, called translational control (*tc*) RNAs, capable of specifically inhibiting in vitro translation for the chicken myosin heavy chain (MHC), are partially complementary to the 5'-end region and the 3'-part of MHC mRNA and might regulate expression of the mRNA (138). Synthesis of an RNA species complementary to the barley alpha-amylase mRNA is developmentally regulated. The antisense RNA might play a regulatory role in preventing expression of the alpha-amylase and thereby protecting the storage contents of the seeds (139). An RNA complementary to a part of the mRNA for the basic fibroblast growth factor in *Xenopus* oocyte directs the double-strand specific covalent modification (adenosine to inosine) (140) of the complementary region of the mRNA, and this modification might be the cause of the abrupt degradation of the mRNA during maturation of the oocyte (141).

## CONCLUSION

Nucleation by pairing of a very small number of nucleotides initiates binding of complementary oligonucleotides of various structures. The rate constant of association of the components is relatively independent of their nucleotide sequences and structures. On the other hand, the rate constant of dissociation depends very much on the nucleotide sequences and structures in the region of several nucleotides around the site of nucleation. Formation of a seven-nucleotide loop in the anticodon loop of tRNA or in the ColE1 RNA provides very favorable conditions for the binding through structural constraint by a particular base-stacking. The process of binding complex molecules consists of a sequence of reactions producing a series of progressively more stable intermediates, leading to the final stably bound product. Binding may express the regulatory function without completing the sequence.

Many functional RNA molecules form specific folded structures: for example, all kinds of tRNA molecules form L-shaped structures and all types of RNA I of ColE1 analogues are composed of three stem-loops. Interaction of folded structures has various advantages. If RNA molecules interact only in extended linear structures, the specificity of the interaction is determined solely by the nucleotide sequences. On the other hand, when one or both components of interaction is folded, the structure itself dictates an additional specificity of interaction and thus increases the functional and genetic versatility of the interaction. For example, a very small change in the primary sequence of folded structure can profoundly affect the rate and specificity of interaction of folded RNAs. In addition, formation of a group-specific structure allows a member to interact with a group-specific protein. This further increases the specificity and strength of the interaction. Furthermore, an analogous but genetically distinct regulatory system could evolve by genetically exchanging the stem-loop sequences. All these advantages of regulatory systems based on interactions between folded RNAs are clearly seen in regulation of the replication of plasmid ColE1 and its relatives.

RNA-RNA interaction through recognition of the complementarity in the nucleotide sequences is also found in systems other than those for antisense regulation. Recognition of the codon by the anticodon, determination of the sites of initiation of translation by rRNA, and selection of the site of splicing in RNA processing are just a few examples. These subjects were not discussed here as they have been reviewed previously (142–144).

ACKNOWLEDGMENT

We are thankful for valuable advice on the manuscript by Drs. Nobuo Shimamoto, Kiyoshi Mizobuchi, and Michael Brenner.

*Literature Cited*

1. Inouye, M. 1988. *Gene* 72:25–34
2. Toulme, J.-J., Helene, C. 1988. *Gene* 72:51–58
3. Weintraub, H. M. 1990. *Sci. Am.* 262:40–48
4. Pörschke, D. 1977. *Chemical Relaxation in Molecular Biology, Mol. Biol. Biochem, Biophys. Ser.*, ed. I. Pecht, R. Rigler, 24:191–218. Berlin, Heidelberg: Springer-Verlag
5. Cantor, C. R., Schimmel, P. R. 1980. *Biophysical Chemistry III*, pp. 1183–1264. New York: Freeman
6. Pörschke, D., Eigen, M. 1971. *J. Mol. Biol.* 62:361–81
7. Craig, M. E., Crothers, D. M., Doty, P. 1971. *J. Mol. Biol.* 62:383–401
8. Pörschke, D., Uhlenbeck, O. C., Martin, F. H. 1973. *Biopolymers* 12:1313–35

9. Martin, F. H., Uhlenbeck, O. C., Doty, P. 1971. *J. Mol. Biol.* 57:201–15
10. Uhlenbeck, O. C., Martin, F. H., Doty, P. 1971. *J. Mol. Biol.* 57:217–29
11. Borer, P. N., Dengler, B., Tinoco, I. Jr., Uhlenbeck, O. C. 1974. *J. Mol. Biol.* 86:843–53
12. Alkema, D. A., Bell, R. A., Hader, P. A., Neilson, T. 1981. *J. Am. Chem. Soc.* 103:2866–68
13. Petersheim, M., Turner, D. H. 1983. *Biochemistry* 22:256–63
14. Hickey, D. R., Turner, D. H. 1985. *Biochemistry* 24:3987–91
15. Freier, S. M., Kierzek, R., Caruthers, M. H., Neilson, T., Turner, D. H. 1986. *Biochemistry* 25:3209–13
16. Freier, S. M., Sugimoto, N., Sinclair, A., Alkema, D., Neilson, T., et al. 1986. *Biochemistry* 25:3214–19
17. Sugimoto, N., Kierzek, R., Turner, D. H. 1987. *Biochemistry* 26:4554–58
18. Sugimoto, N., Kierzek, R., Turner, D. H. 1987. *Biochemistry* 26:4559–62
19. Tinoco, I. Jr., Borer, P. N., Dengler, B., Levine, M. D., Uhlenbeck, O. C., et al. 1973. *Nature New Biol.* 246:40–41
20. Zuker, M., Stiegler, P. 1981. *Nucleic Acids Res.* 9:133–48
21. Papanicolaou, C., Gouy, M., Ninio, J. 1984. *Nucleic Acids Res.* 12:31–44
22. Jacobson, A. B., Good, L., Simonetti, J., Zuker, M. 1984. *Nucleic Acids Res.* 12:45–52
23. Freier, S. M., Kierzek, R., Jaeger, J. A., Sugimoto, N., Caruthers, M. H., et al. 1986. *Proc. Natl. Acad. Sci. USA* 83:9373–77
24. Turner, D. H., Sugimoto, N., Freier, S. M. 1988. *Annu. Rev. Biophys. Biophys. Chem.* 17:167–92
25. Jaeger, J. A., Turner, D. H., Zuker, M. 1989. *Proc. Natl. Acad. Sci. USA* 86: 7706–10
26. Hogenauer, G. 1970. *Eur. J. Biochem.* 12:527–32
27. Eisinger, J., Feuer, B., Yamane, T. 1970. *Proc. Natl. Acad. Sci. USA* 65:638–44
28. Eisinger, J., Feuer, B., Yamane, T. 1971. *Nature New Biol.* 231:126–28
29. Uhlenbeck, O. C., Baller, J., Doty, P. 1970. *Nature* 225:508–10
30. Uhlenbeck, O. C. 1972. *J. Mol. Biol.* 65:25–41
31. Jaskunas, S. R., Cantor, C. R., Tinoco, I. Jr. 1968. *Biochemistry* 7:3164–78
32. Grosjean, H., Söll, D. G., Crothers, D. M. 1976. *J. Mol. Biol.* 103:499–519
33. Labuda, D., Grosjean, H., Striker, G., Pörschke, D. 1982. *Biochim. Biophys. Acta* 698:230–36
34. Houssier, C., Grosjean, H. 1985. *J. Biomol. Struct. Dyn.* 3:387–408
35. Romby, P., Giege, R., Houssier, C., Grosjean, H. 1985. *J. Mol. Biol.* 184: 107–18
36. Eisinger, J. 1971. *Biochem. Biophys. Res. Commun.* 43:854–61
37. Eisinger, J., Gross, N. 1974. *J. Mol. Biol.* 88:165–74
38. Yoon, K., Turner, D. H., Tinoco, I. Jr. 1975. *J. Mol. Biol.* 99:507–18
39. Labuda, D., Pörschke, D. 1980. *Biochemistry* 19:3799–805
40. Labuda, D., Striker, G., Pörschke, D. 1984. *J. Mol. Biol.* 174:587–604
41. Labuda, D., Striker, G., Grosjean, H., Pörschke, D. 1985. *Nucleic Acids Res.* 13:3667–83
42. Bujalowski, W., Jung, M., McLaughlin, L. W., Pörschke, D. 1986. *Biochemistry* 25:6372–78
43. Eisinger, J., Spahr, P. F. 1973. *J. Mol. Biol.* 73:131–37
44. Pongs, O., Bald, R., Reinwald, E. 1973. *Eur. J. Biochem.* 32:117–25
45. Uhlenbeck, O. C., Chirikjian, J. G., Fresco, J. R. 1974. *J. Mol. Biol.* 89: 495–504
46. Freier, S. M., Tinoco, I. Jr. 1975. *Biochemistry* 14:3310–14
47. Banner, D. W., Kokkinidis, M., Tsernoglou, D. 1987. *J. Mol. Biol.* 196:657–75
48. Tinoco, I. Jr., Uhlenbeck, O. C., Levine, M. D. 1971. *Nature* 230:362–67
49. Masukata, H., Tomizawa, J. 1986. *Cell* 44:125–36
50. Tomizawa, J. 1987. *Molecular Biology of RNA: New Perspectives*, ed. M. Inouye, B. S. Dudock, pp. 249–59. New York: Academic
51. Pleij, C. W. A. 1990. *Trends Biochem. Sci.* 15:143–47
52. Cesareni, G., Banner, D. W. 1985. *Trends Biochem. Sci.* 10:303–6
53. Polisky, B. 1988. *Cell* 55:929–32
54. Tomizawa, J., Itoh, T. 1982. *Cell* 31: 575–83
55. Itoh, T., Tomizawa, J. 1980. *Proc. Natl. Acad. Sci. USA* 77:2450–54
56. Masukata, H., Tomizawa, J. 1990. *Cell* 62:331–38
57. Tomizawa, J., Itoh, T., Selzer, G., Som, T. 1981. *Proc. Natl. Acad. Sci. USA* 78:1421–25
58. Tomizawa, J., Itoh, T. 1981. *Proc. Natl. Acad. Sci. USA* 78:6096–100
59. Lacatena, R. M., Cesareni, G. 1981. *Nature* 294:623–26
60. Tomizawa, J. 1986. *Cell* 47:89–97
61. Tomizawa, J. 1984. *Cell* 38:861–70
62. Wong, E. M., Polisky, B. 1985. *Cell* 42:959–66
63. Polisky, B., Zhang, X.-Y., Fitzwater, T. 1990. *EMBO J.* 9:295–304

64. Tomizawa, J. 1990. *J. Mol. Biol.* 212: 638–94
65. Tamm, J., Polisky, B. 1985. *Proc. Natl. Acad. Sci. USA* 82:2257–61
66. Tomcsanyi, T., Apirion, D. 1985. *J. Mol. Biol.* 185:713–20
67. Som, T., Tomizawa, J. 1983. *Proc. Natl. Acad. Sci. USA* 80:3232–36
68. Selzer, G., Som, T., Itoh, T., Tomizawa, J. 1983. *Cell* 32:119–29
69. Tomizawa, J. 1985. *Cell* 40:527–35
70. Tomizawa, J. 1990. *J. Mol. Biol.* 212: 695–708
71. Eguchi, Y., Tomizawa, J. 1989. *Molecular Mechanisms in DNA Replication and Recombination, UCLA Symp. Molec. Cell Biol. New Ser.*, ed. C. C. Richardson, I. R. Lehman, 127:203–214. New York: Liss
72. Eguchi, Y., Tomizawa, J. 1990. *Cell* 60:199–209
73. Twigg, A. J., Sherratt, D. 1980. *Nature* 238:216–18
74. Cesareni, G., Muesing, M. A., Polisky, B. 1982. *Proc. Natl. Acad. Sci. USA* 79:6313–17
75. Tomizawa, J., Som, T. 1984. *Cell* 38: 871–78
76. Cesareni, G., Cornelissen, M., Lacatena, R. M., Castagnoli, L. 1984. *EMBO J.* 3:1365–69
77. Castagnoli, L., Scarpa, M., Kokkinidis, M., Banner, D. W., Tsernoglou, D., et al. 1989. *EMBO J.* 8:621–29
78. Brenner, M., Tomizawa, J. 1991. *Proc. Natl. Acad. Sci. USA* 88:405–9
79. Ataai, M. M., Shuler, M. L. 1986. *Plasmid* 16:204–12
80. Bremer, H., Lin-Chao, S. 1986. *J. Theor. Biol.* 123:453–70
81. Keasling, J. D., Palson, B. O. 1989. *J. Theor. Biol.* 141:447–61
82. Perelson, A. S., Brendel, V. 1989. *J. Mol. Biol.* 208:245–55
83. Green, P. J., Pines, O., Inouye, M. 1986. *Annu. Rev. Biochem.* 55:569–97
84. Simons, R. W. 1988. *Gene* 72:35–44
85. Simons, R. W., Kleckner, N. 1988. *Annu. Rev. Genet.* 22:567–600
86. Rosen, J., Ryder, T., Inokuchi, H., Ohtsubo, H., Ohtsubo, E. 1980. *Mol. Gen. Genet.* 179:527–37
87. Rosen, J., Ryder, T., Ohtsubo, H., Ohtsubo, E. 1981. *Nature* 290:794–97
88. Light, J., Molin, S. 1983. *EMBO J.* 2:93–98
89. Weber, P. C., Palchaudhuri, S. 1986. *J. Bacteriol.* 166:1106–12
90. Saadi, S., Maas, W. K., Hill, D. F., Bergquist, P. L. 1987. *J. Bacteriol.* 169:1936–46
91. Blomberg, P., Wagner, E. G. H., Nordstrom, K. 1990. *EMBO J.* 9:2331–40

92. Nikoletti, S., Bird, P., Praszkier, J., Pittard, J. 1988. *J. Bacteriol.* 170:1311–18
93. Praszkier, J., Bird, P., Nikoletti, S., Pittard, J. 1989. *J. Bacteriol.* 171:5056–64
94. Hama, C., Takizawa, T., Moriwaki, H., Urasaki, Y., Mizobuchi, K. 1990. *J. Bacteriol.* 172:1983–91
95. Shiba, K., Mizobuchi, K. 1990. *J. Bacteriol.* 172:1992–97
96. Asano, K., Kato, A., Moriwaki, H., Hama, C., Shiba, K., Mizobuchi, K. 1991. *J. Biol. Chem.* 266:3774–81
97. Yasueda, H., Horii, T., Itoh, T. 1989. *Mol. Gen. Genet.* 215:209–16
98. Kim, K., Meyer, R. J. 1986. *Nucleic Acids Res.* 14:8027–46
99. Patel, I., Bastia, D. 1987. *Cell* 51:455–62
100. Carlton, S., Projan, S. J., Highlander, S. K., Moghazeh, S. M., Novick, R. P. 1984. *EMBO J.* 3:2407–14
101. Kumer, C. C., Novick, R. P. 1985. *Proc. Natl. Acad. Sci. USA* 82:638–42
102. Novick, R. P., Iordanescu, S., Projan, S. J., Kornblum, J., Edelman, I. 1989. *Cell* 59:359–404
103. Mullineaux, P., Willetts, N. 1985. *Plasmids in Bacteria*, ed. D. R. Helinski, S. N., Cohen, D. B., Clewell, D. A. Jackson, A. Hollaender, pp. 605–14. New York: Plenum
104. Dempsey, W. B. 1987. *Mol. Gen. Genet.* 209:533–44
105. Gerdes, K., Helin, K., Christensen, O. W., Lobner-Olesen, A. 1988. *J. Mol. Biol.* 203:119–29
106. Stephenson, F. H. 1985. *Gene* 35:313–20
107. Hoopes, B. C., McClure, W. R. 1985. *Proc. Natl. Acad. Sci. USA* 82:3134–38
108. Ho, Y. S., Rosenberg, M. 1985. *J. Biol. Chem.* 260:11838–44
109. Krinke, L., Wulff, D. L. 1987. *Genes Dev.* 1:1005–13
110. Wu, T.-H., Liao, S.-M., McClure, W. R., Susskind, M. M. 1987. *Genes Dev.* 1:204–12
111. Citron, M., Schuster, H. 1990. *Cell* 62:591–98
112. Simons, R. W., Hoopes, B. C., McClure, W. R., Kleckner, N. 1983. *Cell* 34:673–82
113. Simons, R. W., Kleckner, N. 1983. *Cell* 34:683–91
114. Ma, C., Simons, R. W. 1990. *EMBO J.* 9:1267–74
115. Mizuno, T., Chou, M.-Y., Inouye, M. 1983. *Proc. Jpn. Acad.* 59:335–38
116. Aiba, H., Matsuyama, S., Mizuno, T., Mizushima, S. 1987. *J. Bacteriol.* 169: 3007–12
117. Andersen, J., Forst, S. A., Zhao, K., Inouye, M., Delihas, N. 1989. *J. Biol. Chem.* 264:17961–70

118. Okamoto, K., Freundlich, M. 1986. *Proc. Natl. Acad. Sci. USA* 83:5000–4
119. Yoshioka, Y., Ohtsubo, H., Ohtsubo, E. 1987. *J. Bacteriol.* 169:619–23
120. McIntire, S. A., Dempsey, W. B. 1987. *Nucleic Acids Res.* 15:2029–42
121. Womble, D. D., Dong, X., Rownd, R. H. 1987. See Ref. 50, pp. 225–47
122. Wagner, E. G. H., Nordstrom, K. 1986. *Nucleic Acids Res.* 14:2523–38
123. Liao, S.-M., Wu, T.-H., Chaing, C. H., Susskind, M. M., McClure, W. R. 1987. *Genes Dev.* 1:197–203
124. Case, C. C., Roles, S. M., Jensen, P. D., Lee, J., Kleckner, N., Simons, R. W. 1989. *EMBO J.* 8:4297–306
125. Brady, G., Frey, J., Danbara, H., Timmis, K. N. 1983. *J. Bacteriol.* 154:429–36
126. Givakov, M., Molin, S. 1984. *Mol. Gen. Genet.* 194:286–92
127. Kittle, J. D., Simons, R. W., Kleckner, N. 1989. *J. Mol. Biol.* 210:561–72
128. Finlay, B. B., Frost, L. S., Paranchych, W., Willetts, N. S. 1986. *J. Bacteriol.* 167:754–57
129. Couturier, M., Bex, F., Bergquist, P. L., Maas, W. K. 1988. *Microbiol. Rev.* 52:375–95
130. Persson, C., Wagner, E. G. H., Nordstrom, K. 1988. *EMBO J.* 7:3279–88
131. Persson, C., Wagner, E. G. H., Nordstrom, K. 1990. *EMBO J.* 9:3767–75

132. Persson, C., Wagner, E. G. H., Nordstrom, K. 1990. *EMBO J.* 9:3777–85
133. Liao, S.-M., McClure, W. R. 1989. *Molecular Biology of RNA*, ed. T. Cech, pp. 289–97. New York: Liss
134. Beck, C. F., Warren, A. J. 1988. *Microbiol. Rev.* 52:318–26
135. van der Krol, A. R., Mol, J. N. M., Stuitje, A. R. 1988. *BioTechniques* 6:958–76
136. Stevens, J. G., Wagner, E. K., Devi-Rao, G. B., Cook, M. L., Feldman, L. T. 1987. *Science* 235:1056–59
137. Croen, K. D., Ostrove, J. M., Dragovic, L. J., Smialek, J. E., Straus, S. E. 1988. *New Engl. J. Med.* 317:1427–32
138. Heywood, S. M. 1986. *Nucleic Acids Res.* 14:6771–72
139. Rogers, J. 1988. *Plant Mol. Biol.* 11:125–38
140. Bass, B. L., Weintraub, H. 1988. *Cell* 55:1089–98
141. Kimelman, D., Kirschner, M. W. 1989. *Cell* 59:687–96
142. Hardesty, B., Kramer, G., eds. 1986. *Structure, Function, and Genetics of Ribosomes.* New York: Springer
143. Cech, T. R., Bass, B. L. 1986. *Annu. Rev. Biochem.* 55:599–629
144. Padgett, R. A., Grabowski, P. J., Konarska, M. M., Seiler, S., Sharp, P. A. 1986. *Annu. Rev. Biochem.* 55:1119–50

*Annu. Rev. Biochem. 1991. 60:653–88*

# MODEL SYSTEMS FOR THE STUDY OF SEVEN-TRANSMEMBRANE-SEGMENT RECEPTORS

*Henrik G. Dohlman[1]*

Department of Biochemistry, Duke University Medical Center, Durham, North Carolina 27710

*Jeremy Thorner*

Division of Biochemistry and Molecular Biology, Department of Molecular and Cell Biology, University of California, Berkeley, California 94720

*Marc G. Caron*

Department of Cell Biology, Howard Hughes Medical Institute, Duke University Medical Center, Durham, North Carolina 27710

*Robert J. Lefkowitz[2]*

Department of Medicine, Howard Hughes Medical Institute, Duke University Medical Center, Durham, North Carolina 27710

KEY WORDS:  G protein, rhodopsin, desensitization, yeast, cloning.

## CONTENTS

[1]Present address: Division of Biochemistry and Molecular Biology, Department of Molecular and Cell Biology, University of California, Berkeley, California 94720
[2]Corresponding author.

0066-4154/91/0701-0653$02.00

# PERSPECTIVES AND SUMMARY

The past few years have been an exciting period in the field of receptor biology as remarkable similarities among various signal transduction systems have been revealed. Many such transmembrane signaling systems consist of three membrane-bound protein components: (*a*) a cell surface receptor; (*b*) an effector, such as an ion channel or the enzyme adenylyl cyclase; and (*c*) a guanine nucleotide-binding regulatory protein, or G protein, that is coupled to both the receptor and its effector. G protein–coupled receptors mediate the actions of extracellular signals as diverse as light, odorants, peptide hormones, and neurotransmitters, and they have been identified in organisms as evolutionarily divergent as yeast and man. Nearly all G protein–coupled receptors bear detectable sequence similarity with one another, and it is thought that all share a similar topological motif consisting of seven hydrophobic (and potentially $\alpha$-helical) segments that span the lipid bilayer (1).

The primary function of cell surface receptors is to discriminate the appropriate ligands from among multiple extracellular stimuli, and then to activate an effector system that produces an intracellular signal [e.g. the messenger cyclic AMP (cAMP)]. Hence, receptors not only transmit but also amplify and integrate extracellular signals, thereby controlling cellular processes. In this way, cells can communicate and coordinate their development and actions. Such intercellular communication has obvious utility in multicellular organisms, but also plays a role in the life cycle of unicellular eukaryotes.

Many G protein–coupled receptors of the seven-transmembrane-segment class (7-TMS) also share the ability to be desensitized, a process by which they become refractory to further stimulation after an initial response, despite the continued presence of a stimulus of constant intensity. Desensitization can result from a reduction in the number of cell surface receptors, as well as from an attenuation of receptor–G protein interactions. Indeed, following exposure to agonists, many receptors of the 7-TMS type are rapidly "uncoupled" and sequestered from the cell surface. Phosphorylation appears to be a critical event in regulating the uncoupling of receptors and their cognate G proteins (2–5).

The ligand-binding and G protein–coupling functions of two 7-TMS receptors, the $\beta_2$-adrenergic receptor ($\beta_2$-AR) and the visual pigment rhodopsin, are well established. An important objective, however, is to understand the extent to which the mechanisms elucidated for each of these receptors can be applied to other members of the 7-TMS receptor family.

## SCOPE

Our goal in this review is to draw attention to parallels among diverse types of signal transduction systems, and to illustrate how distinct experimental approaches have contributed to the understanding of a common problem in biology. This review emphasizes molecular mechanisms of signaling and desensitization at the level of the receptor, particularly those mechanisms that appear to be most conserved among the systems discussed: rhodopsin-mediated phototransduction; $\beta_2$-AR-mediated signal transduction in mammals; and intercellular signaling in the yeast *Saccharomyces cerevisiae*. The role of G proteins in signaling is summarized only briefly because several recent reviews of this subject are available (6–9).

## G PROTEIN SIGNALING

Early models for the control of cAMP production envisioned allosteric sites for hormone binding on adenylyl cyclase itself. It was later demonstrated that ligands that elevate cAMP, such as the catecholamines epinephrine and norepinephrine, bind to receptors that are distinct from the effector enzyme (10–12). Evidence for a third component of the $\beta_2$-AR-coupled adenylyl cyclase system emerged in the early 1970s from the work of Rodbell and colleagues (13, 14), who observed that GTP is required for hormonal activation of adenylyl cyclase in isolated plasma membranes. This in vitro observation led to the purification of an adenylyl cyclase–stimulatory, GTP-binding protein, $G_s$, and eventual cloning of its gene by Gilman and coworkers, and later to the discovery of an extended family of such regulatory G proteins (6–9).

All 7-TMS receptor-coupled G proteins consist of three tightly associated subunits, designated $\alpha$, $\beta$, and $\gamma$ (1:1:1) in order of decreasing mass. Following agonist binding to the receptor, a conformational change is transmitted to the G protein, which causes the $\alpha$-subunit to exchange a bound GDP for GTP and to disassociate from the $\beta\gamma$-subunits. The GTP-bound form of the $\alpha$-subunit is typically the effector-modulating moiety. Signal amplification results from the ability of a single receptor to activate many G protein molecules, and from the stimulation by $G\alpha$-GTP of several catalytic cycles of the effector.

The family of regulatory G proteins comprises a multiplicity of different $\alpha$-subunits (> 20 in man), which associate with a smaller pool of $\beta$- and $\gamma$-subunits (>4 each) (9). Thus, differences in the $\alpha$-subunits probably distinguish the various G protein oligomers, although the targeting or function of various $\alpha$-subunits might also depend on which $\beta\gamma$ they associate with (9). The $\alpha$-subunits also hydrolyze the bound GTP, and are often substrates for inhibitory modification by ADP-ribosylation catalyzed by the bacterial toxins of *Vibrio cholerae* and/or of *Bordetella pertussis* (15).

The adrenergic receptors interact with the endogenous catecholamines, epinephrine and norepinephrine, as well as with an enormous number of synthetic agonists and antagonists. Both of the natural catecholamines are neurotransmitters in sympathetic neural cells, while epinephrine acts as a hormone in peripheral tissues. Through the development of selective ligands and through molecular cloning of the receptors, it has become clear that distinct genes encode multiple subtypes of catecholamine receptors. Among the mammalian adrenergic receptors, agonist binding to the $\beta_1$-, $\beta_2$-, and $\beta_3$-receptor subtypes leads to activation of $G_s$ and, in turn, of adenylyl cyclase (16–19). In contrast, $\alpha_2$-ARs, of which there are at least three distinct sub-subtypes, mediate a decrease in adenylyl cyclase activity through the inhibitory G protein, $G_i$ (20, 21). The mechanism of $G_i$-mediated inhibition of adenylyl cyclase remains a matter of some controversy (6–9, 22). The three known $\alpha_1$-AR sub-subtypes are coupled predominantly to another G protein–regulated second messenger system involving hydrolysis of phosphatidylinositol by a phospholipase C (23–25).

An analogous pathway of signal transduction has been elucidated for the photoreceptor, rhodopsin, in the mammalian retina. Following capture of a single photon, the covalently bound chromophore of rhodopsin, all-*cis*-retinal, undergoes isomerization to the all-*trans* isomer. This structural change causes rhodopsin molecules to assume a number of conformational intermediates, one of which (metarhodopsin II) couples to the retinal G protein, transducin. The $\alpha$-subunit of rhodopsin-activated transducin blocks the activity of the inhibitory subunit of cyclic GMP (cGMP) phosphodiesterase, the effector enzyme (26–30). The consequent decrease in intracellular cGMP leads to closing of cGMP-gated cation-specific channels, and the resulting hyperpolarization of the plasma membrane is propagated to synaptic terminals on the cell (7, 30). Stimulation of rhodopsin also results in the activation of a phospholipase $A_2$ (31). Whereas activation of cGMP phosphodiesterase is mediated by the $\alpha$-subunit of transducin, the effect on phospholipase $A_2$ appears to be mediated by the $\beta\gamma$ moiety (31). As is discussed later, strong evidence also implicates $\beta\gamma$ as the effector-stimulating entity in the yeast pheromone response pathway.

Finally, receptors acting via G proteins have also been shown to modulate potassium (32, 33) and calcium channels (34–36).

## $\beta_2$-AR AND RHODOPSIN SIGNALING: SIMILARITIES AND DIFFERENCES

The process of phototransduction—the rhodopsin-mediated conversion of light energy into a neurochemical signal—is unique in several regards. It is highly specific (false signals are rare), sensitive (responding to a single photon), and extremely rapid (the cycle of stimulation and recovery takes less than a second). Moreover, rhodopsin is expressed only in the eye, and is compartmentalized at very high concentrations within a specialized organelle, the disks of retinal rod cells, which are tightly stacked within the rod outer segment. The abundance of rhodopsin in these cells has greatly facilitated its biochemical characterization (30).

The $\beta_2$-AR has long been a model system for studying how extracellular chemical signals influence intracellular events. In contrast to rhodopsin, $\beta$-receptors are expressed in very low numbers and in nearly all mammalian tissues, regulating processes as diverse as cellular metabolism, secretion, differentiation, and growth. Furthermore, the $\beta_2$-AR is coupled to the well-characterized cAMP second messenger system. Because this receptor-effector pathway is relatively well understood, it is a useful system for relating structural information with biological function (1).

Much of the recent progress in the study of adrenergic and other G protein–coupled receptors is a consequence of the early (and often heroic) efforts toward their purification, in some cases requiring enrichments of 100,000-fold. Typically, such purifications required the development of membrane solubilization procedures, unique affinity chromatography matrices, as well as specific radiolabeled ligands and affinity-labeling reagents (37). Each of the adrenergic receptors purified and characterized in this way (i.e. $\beta_1$-, $\beta_2$-, $\alpha_1$-, and $\alpha_2$-ARs) was found to consist of a single, highly hydrophobic, glycosylated polypeptide chain (37–41). Through reconstitution experiments, in which the pure receptor and G protein components were combined in lipid vesicles, it was established that both the ligand-binding and G protein–activating functions reside on a single receptor polypeptide (42, 43). Cloning and sequencing of the adrenergic receptor genes has revealed that they are structurally related to one another and to rhodopsin, suggesting that they have a common evolutionary origin (1, 16–25, 44, 45).

The gene and cDNA encoding the hamster $\beta_2$-AR were first cloned and sequenced in 1986 (16). By this time, the protein sequence of rhodopsin had already been known for several years (44, 45). A striking feature of the sequences of both rhodopsin and the $\beta_2$-AR is the presence of seven stretches of largely hydrophobic amino acids representing possible membrane-spanning domains. Using the hydropathy analysis method of Kyte & Doolittle (46), both receptors yield a profile that is remarkably similar to that of bacteriorhodopsin, a non–G protein–coupled light-driven proton pump from *Halobacterium halobium*. A detailed model of bacteriorhodopsin featuring seven $\alpha$-helices spanning the lipid bilayer had previously been determined by image reconstruction from electron microscopy (47, 48). Accordingly, a bacteriorhodopsin-like topography was proposed for the $\beta_2$-AR and for rhodopsin (16, 44, 45). For rhodopsin, this model is supported by proteolytic mapping and chemical modification studies (44, 45, 49–51). The $\alpha$-helical, transmembrane-spanning nature of the hydrophobic segments has been further characterized by a variety of physical methods, including spectroscopic and neutron diffraction studies, infrared absorption, circular dichroism, low-resolution electron microscopy, as well as neutron and X-ray scattering studies (reviewed in 30). Support for the 7-TMS topography of the $\beta_2$-AR comes from limited proteolysis experiments and from immunolocalization of antipeptide antibody-binding sites to the appropriate (external or cytoplasmic) face of the plasma membrane (52, 53). Both receptors have several sites for N-linked glycosylation near their N-termini (52, 54, 55), and multiple phosphorylation sites in putative cytoplasmic domains, especially the C-terminal segments (2–4). Neither protein has a cleaved signal sequence (16, 56). Clearly, comparisons with rhodopsin have been extremely useful in designing testable models of structure and function for the $\beta_2$-AR and other G protein–coupled receptors. As discussed below, the sequence similarities among these receptors reflect analogies not only in their topographical arrangement, but in their function and regulation as well.

## THE FAMILY OF G PROTEIN–COUPLED RECEPTORS

Improvements in cell culture and in pharmacological methods, and more recently in molecular cloning and gene expression techniques, have led to the identification and characterization of many new 7-TMS receptors, including new subtypes and sub-subtypes of previously identified receptors (Table 1). The $\alpha_1$- and $\alpha_2$-ARs, once thought to each consist of a single receptor species, are now known to each be encoded by at least three distinct genes (20, 21, 23–25). In addition to rhodopsin in rod cells, which mediates vision in dim light, three highly similar cone pigments mediating color vision have been cloned (57, 58). All appear to be homologous with other members of the family of G protein–coupled receptors (e.g. dopaminergic, muscarinic, serotonergic, tachykinin, etc; Table 1), and each appears to share the characteristic 7-TMS topography.

When comparing the 7-TMS receptors with one another, a discernable pattern of amino acid sequence conservation is observed. The transmembrane domains are often the most similar, whereas the N- and C-terminal regions and the cytoplasmic loop connecting transmembrane segments V and VI (loop V–VI; Figure 1) can be quite divergent (1). These observations have led to a great deal of speculation regarding the regions involved in membrane insertion, ligand binding, coupling to G proteins, and modification by regulatory kinases. For example, an early hypothesis was that each of the seven hydrophobic domains of the $\beta_2$-AR forms an $\alpha$-helix, and that these helices bundle together to form the ligand-binding pocket deep within the membrane (16). This model is analogous to that proposed for retinal binding to rhodopsin (44, 45). Interaction with cytoplasmic proteins, such as kinases and G proteins, was predicted to involve the hydrophilic loops connecting the transmembrane domains of the receptor. The challenge, however, has been to determine which features are preserved among 7-TMS receptors because of conservation of function, and which divergent features represent structural adaptations to new functions. A number of strategies have been used to test these ideas, including the use of recombinant DNA and gene expression techniques for the construction of substitution and deletion mutants, as well as of hybrid or chimeric receptors. Moreover, results from traditional biochemical approaches can be more easily interpreted, and more incisive experiments devised, now that the sequences of these receptors are known.

## DISTINCT DETERMINANTS FOR G PROTEIN COUPLING AND LIGAND BINDING

The construction of chimeric receptors, an approach whereby a domain of one protein is exchanged with an analogous (but perhaps functionally distinct) domain of another, is a useful starting point for elucidating structure and function relationships. In designing such chimeras, the aim is to correlate replacement of a particular structural domain with acquisition or alteration of a particular function, such as ligand binding properties or G protein coupling specificity. In contrast, in substitution or deletion mutations, a loss of func-

**Table 1**  7-TMS receptors for which the primary structure is known

| Receptor Subtype | Species | Ref. |
|---|---|---|
| Mammalian | | |
| $\beta_1$-adrenergic | Human | 18 |
| | Rat | 184 |
| $\beta_2$-adrenergic | Hamster | 16 |
| | Human | 185–188 |
| | Mouse | 189 |
| | Rat | 190, 191 |
| $\beta_3$-adrenergic | Human | 19 |
| $\alpha_{1B}$-adrenergic | Hamster | 23 |
| | Rat | 192 |
| $\alpha_{1C}$-adrenergic | Cow | 24 |
| $\alpha_{2A}$-adrenergic | Human | 20, 193 |
| | Rat | 193a |
| | Pig | 193b |
| $\alpha_{2B}$-adrenergic | Human | 21 |
| | Rat | 194 |
| $\alpha_{2C}$-adrenergic | Human | 25 |
| 5-HT1a-serotonergic | Human | 195, 196 |
| | Rat | 197 |
| 5-HT1c-serotonergic | Rat | 198 |
| 5-HT2-serotonergic | Rat | 199, 200 |
| M1-muscarinic | Pig | 201 |
| | Human | 202, 203 |
| | Rat | 204 |
| | Mouse | 205 |
| M2-muscarinic | Pig | 206, 207 |
| | Human | 202 |
| | Rat | 190, 204 |
| M3-muscarinic | Human | 202 |
| | Rat | 204 |
| M4-muscarinic | Human | 202 |
| | Rat | 208 |
| | Pig | 209 |
| M5-muscarinic | Human | 210 |
| | Rat | 210, 211 |
| $D_1$-dopaminergic | Human | 212–214 |
| | Rat | 214–215 |
| $D_2$-dopaminergic | Rat | 215–217 |
| | Human | 217–218 |
| | alternatively spliced | 219–226 |
| $D_3$-dopaminergic | Rat | 227 |
| Substance K | Cow | 228 |
| | Rat | 229 |
| | Human | 229a |
| Neuromedin K | Rat | 230 |

**Table 1**   (*Continued*)

| Receptor Subtype | Species | Ref. |
|---|---|---|
| Substance P | Rat | 231, 232 |
| F-Met-Leu-Phe | Human | 232a |
| Thyrotropin | Dog | 233, 234 |
| | Human | 235–236 |
| | Rat | 236a |
| Lutropin-choriogonadotropin | Rat | 237 |
| | Pig | 238 |
| Endothelin | Human | 238a |
| | Cow | 238b |
| Endothelin-ET$_B$ | Rat | 238c |
| Angiotensin (*mas*) | Human | 182, 183 |
| | Rat | 239 |
| Rhodopsin | Cow | 45, 44, 56 |
| | Human | 240 |
| | Mouse | 241 |
| Red opsin | Human | 57 |
| Green opsin | Human | 57 |
| Blue opsin | Human | 57 |
| Cannabinoid | Rat | 242 |
| Unknown-RDC1 | Dog | 243 |
| Unknown-RDC4 | Dog | 243 |
| Unknown-RDC7 | Dog | 243 |
| Unknown-RDC8 | Dog | 243 |
| Unknown-edg1 | Human | 244 |
| Unknown-RTA | Rat | 245 |
| Nonmammalian | | |
| Adrenergic ($\beta_1-$) | Turkey | 17 |
| Serotonergic | Fly | 245a |
| Muscarinic | Chicken | 246 |
| | Fly | 247, 248 |
| Opsin (*ninaE*) | Fly | 249, 250 |
| Opsin-Rh2 | Fly | 251 |
| Opsin-Rh3 | Fly | 252 |
| Opsin-Rh4 | Fly | 253, 254 |
| Rhodopsin | Fly | 255 |
| | Chicken | 256 |
| Octopamine | Fly | 257 |
| Mating factor (*STE2*) | Yeast | 60, 139, 140 |
| (*STE3*) | Yeast | 140, 141 |
| cAMP | Slime mold | 258 |
| Unknown-US27 | Viral | 177 |
| Unknown-US28 | Viral | 177 |
| Unknown-UL33 | Viral | 177 |

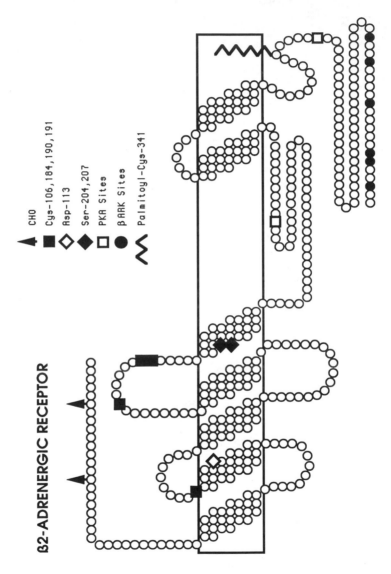

a

β2-ADRENERGIC RECEPTOR

CHO
Cys-106,184,190,191
Asp-113
Ser-204,207
PKA Sites
βARK Sites
Palmitoyl-Cys-341

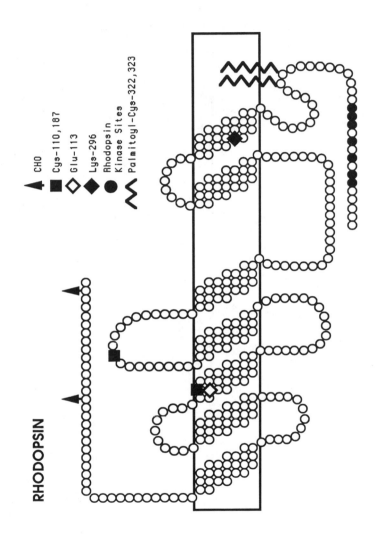

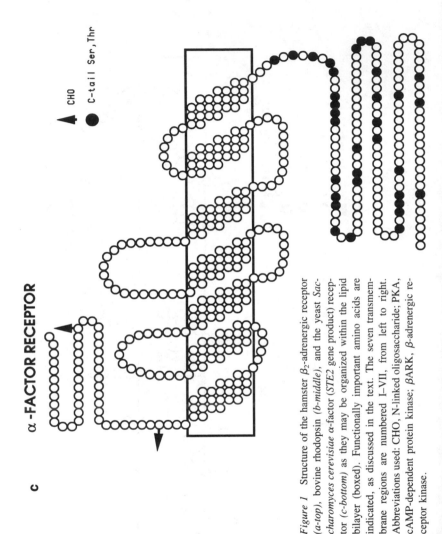

*Figure 1* Structure of the hamster $\beta_2$-adrenergic receptor (*a-top*), bovine rhodopsin (*b-middle*), and the yeast *Saccharomyces cerevisiae* α-factor (*STE2* gene product) receptor (*c-bottom*) as they may be organized within the lipid bilayer (boxed). Functionally important amino acids are indicated, as discussed in the text. The seven transmembrane regions are numbered I–VII, from left to right. Abbreviations used: CHO, N-linked oligosaccharide; PKA, cAMP-dependent protein kinase; βARK, β-adrenergic receptor kinase.

tion can be difficult to interpret in terms of receptor function, because such changes may simply interfere with folding or membrane insertion.

Chimeras of 7-TMS receptors have been constructed from the $\alpha_2$- and $\beta_2$-AR genes (59). These receptors were chosen for several reasons. First, their amino acid sequences have a high degree of similarity (70% in TMSs, 49% overall), making it likely that they would be sufficiently compatible so as to form stable and functional chimeras. Second, these receptors have similar (but readily distinguishable) ligand-binding characteristics; there are selective antagonists for each receptor, and the order of potency of agonists is different. Finally, the $\alpha_2$- and $\beta_2$-ARs are coupled to distinct G proteins that lead to the inhibition ($G_i$) or stimulation ($G_s$) of adenylyl cyclase, respectively.

It was observed that with increasing amounts of $\alpha_2$-AR sequence replacing the corresponding sequence of the $\beta_2$-AR, the order of potency of agonists was gradually changed such that a hybrid pattern of binding, neither $\alpha_2$- nor $\beta_2$-specific, was obtained. In contrast, the specificity of binding of antagonists appeared to be strongly correlated with the source of the seventh transmembrane domain. Replacement of TMS VII of the $\alpha_2$-AR with that from the $\beta_2$-AR resulted in a chimera with $\beta$-adrenergic antagonist binding specificity. Similarly, when hybrids are made between the $\alpha$-factor receptors of two closely related yeast species (*S. cerevisiae* and *S. kluyveri*), TMS VII seems to be an important determinant for recognition of the appropriate species-specific pheromone (60; L. Marsh, personal communication).

Distinct structural determinants for G protein coupling have also been identified using the chimera approach. A hybrid $\alpha_2$-AR in which TMS V, TMS VI, and the connecting loop were substituted with those from the $\beta_2$-AR, bound ligands with $\alpha_2$-AR specificity, yet stimulated adenylyl cyclase in the manner of the $\beta_2$-AR (59).

Two important conclusions can be drawn from these observations. First, the binding sites for agonists and antagonists have both distinct and shared determinants within these receptors, and major determinants of antagonist binding specificity lie in TMS VII. Second, determinants for G protein recognition lie in the domain encompassing TMS V and TMS VI and the connecting cytoplasmic loop.

The results implicating loop V-VI in G protein coupling have been corroborated using several other receptor hybrid constructions. A chimera in which loop V-VI of the $\beta_2$-AR was replaced with corresponding sequence from the $\alpha_1$-AR resulted in a receptor that bound ligands with $\beta_2$-AR specificity, but mediated agonist stimulation of phosphatidylinositol metabolism as efficiently as the native $\alpha_1$-AR (60a). Chimeras constructed from the M1- and M2-muscarinic cholinergic receptors similarly showed loop V-VI to have a dominant role in determining G protein coupling specificity (61).

An important point to consider when interpreting results from such chimeric receptor experiments is that, whereas these studies provide an indication of receptor domains that confer G protein coupling specificity, they do not necessarily identify structural domains that have functions common to both receptors. This caveat is highlighted by the recent work of Ross and colleagues (62). They observed that replacement of loop V-VI (or an N-terminal 11-amino-acid segment of loop V-VI) of the M1 muscarinic receptor with corresponding sequence from the turkey $\beta_1$-AR was sufficient to confer stimulation of adenylyl cyclase in response to a muscarinic agonist. However, these chimeric receptors were also comparable to the native muscarinic receptor in their ability to stimulate inositol phosphate release. Thus, substitution of the N-terminal portion of loop V-VI leads to promiscuous G protein coupling, and suggests that other domains must also participate in determining coupling specificity. Indeed, whereas replacement of loop III-IV by itself had no effect on native G protein coupling, simultaneous replacement of loops III-IV and V-VI enhanced the relative selectivity for $G_s$ and adenylyl cyclase stimulation over native coupling to the phospholipase C pathway (62).

The regions of the $\beta_2$-AR thought to comprise the G protein coupling sites have been further refined through the construction of chimeric receptors with smaller amounts of foreign sequence, as well as by site-directed and deletion mutagenesis experiments (63, 64). O'Dowd et al systematically replaced or deleted highly conserved residues near the cytoplasmic surface of the $\beta_2$-AR and evaluated their effects on agonist binding and adenylyl cyclase stimulation. The results of these experiments implicate cytoplasmic domains that are in juxtaposition to the plasma membrane, particularly near TMS V, TMS VI, and TMS VII, in determining G protein coupling and specificity (63, 64). One of the residues replaced was a conserved cysteine (Cys341) located near TMS VII in the C-terminal domain of the $\beta_2$-AR (Figure 1). This site was shown to be modified by palmitoylation in both the $B_2$-AR and rhodopsin, and replacement by mutagenesis in the $\beta_2$-AR results in a partial loss of G protein coupling activity, indicating a possible functional role for palmitoylation (65, 66).

Another approach taken by several groups has been to test synthetic peptides corresponding to the hydrophilic domains of G protein–coupled receptors for effects on receptor coupling in vitro. Several such peptides, derived from loops I-II and III-IV of the $\beta_1$-AR, and loops III-IV and V-VI and the C-terminal tail of rhodopsin, were shown to competitively inhibit G protein coupling (67, 68). Other intracellular and extracellular peptide segments of these receptors were without effect (67, 68). Strikingly, another $\beta_1$-AR peptide, corresponding to the C-terminal segment of loop V-VI, stimulated GTP binding and adenylyl cyclase activation of $G_s$ in the manner of the agonist-occupied receptor (67). Moreover, an amphipathic helix-forming peptide derived from wasp venom, mastoparan, was similarly found to promote activation of G proteins in a receptor-like manner (69). Despite the lack of apparent sequence similarity to mastoparan, the segments of loop V-VI of the $\beta_2$-AR nearest TMS V and TMS VI are also predicted to assume an amphipathic, cationic helical conformation (63). These observations suggest that there are several points of contact between the receptor and G protein, and they underscore the importance of conserved secondary structure (such as an amphipathic $\alpha$-helix), as well as of conserved amino acid sequence, in determining receptor function (70).

Recently, an intracellular neural growth cone protein, GAP-43, or a 24-amino-acid peptide corresponding to its N-terminus, was shown to promote GTP binding to a G protein, $G_o$ (71). The N-terminal domain of GAP-43 is palmitoylated and bears strong sequence similarity to the palmitoylated C-terminal domain of the $\beta_2$-AR, rhodopsin, and other 7-TMS receptors. These data confirm that small peptides resembling cytoplasmic domains of the receptors can regulate G protein function. These results also raise the possibility that an endogenous intracellular protein might regulate signaling at the level of the G protein in a manner similar to membrane-bound receptors.

## DETERMINANTS FOR LIGAND BINDING

The sequence conservation within the hydrophobic domains of 7-TMS receptors might indicate a functional role for these domains, other than to target and anchor the receptor to the membrane (1). An early hypothesis for rhodopsin was that these transmembrane domains assemble into an annular structure to form the retinal-binding pocket. Indeed, the hydrocarbon chain of retinal was shown to be covalently attached to Lys296 in the middle of TMS VII (72–75), and to lie close to the center of the transmembrane domains, parallel to the plane of the lipid bilayer (76).

The ligand-binding properties of the $\alpha_2$-$\beta_2$-AR chimeras described above indicate that sites in several of the TMSs contribute to forming the binding pocket (59). Additional experiments with chimeric and mutant receptors have provided new insights into the determinants of ligand-binding specificity. A series of $\beta_1$-$\beta_2$-AR chimeras were used to identify several TMSs, particularly TMS IV, as contributing to the binding selectivity of various agonists and antagonists for these closely related receptor subtypes (77, 78). These studies also showed that replacing particular transmembrane domains in some cases only affected the binding properties for either agonists or for antagonists, illustrating that there are distinct binding determinants that not only distinguish agonists from antagonists but also discriminate among ligands of each type (78).

In studies on the molecular biology of color-blindness, Nathans and colleagues observed several instances of apparent recombination between the red and green opsin genes in individuals with anomalous spectral sensitivity. These findings suggest that naturally occurring chimeric red-green pigments are being expressed in these individuals (58).

Affinity labeling experiments are also consistent with data obtained using chimeric receptors, indicating that no single transmembrane domain is the dominant contact site for ligand binding in all 7-TMS receptors. For example, the irreversible $\beta_2$-AR antagonist, [125]I-para-(bromoacetamido)-benzyl-1-iodocarazolol, was shown to bind covalently to a site within TMS II (79). In similar studies with rat M1-muscarinic receptor, the human $\alpha_{2A}$-AR and the avian $\beta_1$-AR, photoactivated antagonist probes reacted with sites in TMS III, IV, and VII, respectively (80–82). Additional sites of retinal binding to rhodopsin, in addition to the site of covalent attachment in TMS VII, were detected using a photoactivatable analog of retinal. Sites of labeling were detected in TMS III and TMS V, further implicating multiple membrane-spanning domains in forming the ligand-binding pocket of rhodopsin (83).

Site-directed mutagenesis is another approach that has been useful for the study of receptor-ligand interactions. The $\beta_2$-AR is particularly amenable to this type of analysis because of the availability of a wide variety of specific ligands, both agonists and antagonists, with which to monitor changes in ligand-binding properties. The endogenous agonists, epinephrine and norepinephrine, feature a catechol ring and a protonated amine group connected by a $\beta$-hydroxyethyl chain. Structure-activity correlations using synthetic adrenergic ligands have implicated the amine and $\beta$-hydroxyl substituents to be important for both agonist and antagonist binding, whereas the catechol moiety is characteristic of agonists (84). Accordingly, site-directed mutagenesis has been applied in attempts to identify the amino acid residues that accommodate binding of the amine, the $\beta$-hydroxyl, and the catechol substituents.

In an elegant example of this approach, two residues (Ser204, Ser207) in TMS V of the $\beta_2$-AR were shown to be required for normal binding and activation by catecholamine agonists (85). These two residues are conserved among all known 7-TMS receptors that bind catecholamines (the adrenergic and dopamine receptors). An amino acid substitution at either site markedly decreased the affinity and efficacy of agonists, suggesting that Ser204 and Ser207 might be involved in hydrogen bonding to the vicinyl hydroxyl groups of the catechol ring. Moreover, replacement of either Ser204 or Ser207 was mimicked by interaction of the native receptor with agonists lacking either the *meta*- or *para*-hydroxyl groups of catechol, respectively. Antagonist (non-catechol) binding was unaffected by these receptor mutations (85).

An acidic residue (Asp113) in TMS III of the $\beta_2$-AR is similarly conserved among all 7-TMS receptors that bind biogenic amines, and it was proposed that such an acidic residue could form an ion pair with the protonated amine moiety of the ligand (86). Indeed, replacement of Asp113 with asparagine resulted in an ~$10^5$-fold decrease in the potency of agonists in stimulating adenylyl cyclase. The affinity for antagonist binding was similarly diminished. Replacement of Asp113 with Glu resulted in a smaller (~$10^3$) decrease in agonist potency, suggesting that the presence of a negative charge could partially restore function (86). Analogous observations have been made for rhodopsin. In this case retinal is attached via a protonated Schiff base to the $\epsilon$ amino group of Lys296 (72–75). By itself, the presence of a positive charge in the hydrophobic transmembrane environment might be energetically unfavorable; however, a Glu residue that could serve as a stabilizing counterion has been identified by site-directed mutagenesis (87, 88). Of the acidic residues that lie in or near the TMSs of rhodopsin (i.e. Asp83, Glu113, Glu122, Glu134, Glu201), only substitution of Glu113 by a neutral residue (Gln) resulted in an appreciable alteration in the absorption wavelength of rhodopsin, whereas replacement with Asp resulted in a near normal absorption spectrum. In the Glu113-to-Gln mutant, the state of Schiff base protonation was rendered dependent on the pH of the medium, suggesting that another anionic side chain or solvent molecule can, at low pH, substitute for Glu113 (87, 88).

Unexpectedly, certain "classical" $\beta_2$-AR antagonists were found to have partial agonist activity in cells expressing the $\beta_2$-AR Asp113-to-Glu mutant (89). A similar observation has been made for the native $\beta_3$-AR. Several ligands that are antagonists toward the $\beta_1$-AR and $\beta_2$-AR act as partial agonists toward the $\beta_3$-AR, despite the fact that all three have the critical Asp residue in TMS III (19). These findings are consistent with results from the chimeric receptors indicating that there are overlapping determinants for agonist and antagonist binding, as well as with the demonstration that Asp113 is essential for binding to all $\beta_2$-AR ligands.

The blurring of the distinction between agonists and antagonists in certain receptor subtypes and receptor mutants is a reminder that biological activity is a function of both the structure of the ligand and the configuration of its binding site. This principle is well illustrated by examining the means by which visual receptors discriminate among external signals. Whereas individual hormone and neurotransmitter receptors can respond to several different ligands, rhodopsin and the three color pigments all contain an identical chromophore. Nonetheless, the absorption spectrum of retinal differs markedly in each receptor, as well as from that of free retinal in solution (30). Differences in the charge delocalization of retinal $\pi$-electrons by amino acids within the binding site of the different visual receptors have been proposed to be responsible for this so-called spectral tuning (90). Glu113 in rhodopsin is conserved among all of the mammalian opsins, so it is unlikely to play a role in determining the absorption maximum. However, the overall charge distribution within the predicted transmembrane segments of the four opsins does correlate with the direction of spectral shift; the net intramembrane charge being $+1$ (blue opsin, $\lambda_{max} \sim 430$ nm), 0 (rhodopsin, $\lambda_{max} \sim 500$ nm), or $-1$ (green opsin, $\lambda_{max} \sim 530$ nm, and red opsin, $\lambda_{max} \sim 560$ nm). Retinal in solution absorbs maximally at $\sim 440$ nm. Accordingly, Nathans (91) constructed a series of mutants of rhodopsin that were designed to mimic the distribution of charged residues within the transmembrane domains of the various color pigments. These mutants produced absorbance shifts far smaller than those seen in any of the cone pigments, suggesting that the characteristic absorption spectrum of each visual pigment is not due simply to a difference in the charge distribution in the retinal-binding pocket. Perhaps more subtle opsin-retinal interactions, involving multiple polar residues as well as single charged residues, contribute to spectral tuning.

Another feature of many 7-TMS receptors is the presence of several conserved cysteine residues. As discussed above, cysteines located in the C-terminal domain, just distal to TMS VII, in both the $\beta_2$-AR and rhodopsin are covalently modified by palmitoylation (65, 66; Figure 1). Through biochemical experiments with disulfide-reducing agents, at least two Cys-Cys disulfide bonds in the $\beta_2$-AR were shown to be required to preserve normal ligand-binding characteristics (92). Correspondingly, four cysteines (Cys106, 184, 190, 191; Figure 1) were identified by site-directed mutagenesis to be critical for normal ligand binding and proper cell surface expression of the $\beta_2$-AR (92–94). These cysteine residues are not, however, located within the hydrophobic transmembrane domains of the receptor where ligand binding is presumed to occur, but rather in the extracellular hydrophilic loops connecting the TMSs. Taken together, these findings indicate that disulfide bonds within the hydrophilic extracellular domains may be necessary to stabilize the correctly folded conformation of the $\beta_2$-AR.

A series of Cys-to-Ser substitution mutations of rhodopsin have been evaluated for effects on protein expression, retinal binding, and coupling to transducin (95). Replacement of Cys110 and Cys187 in the intradiscal (topographically extracellular) domains of rhodopsin resulted in abnormally low levels of expression, as well as in incomplete glycan processing and in an inability to bind 11-*cis*-retinal. These results indicate that, as in the $\beta_2$-AR, a disulfide bond in a hydrophilic loop is critical for directing and/or stabilizing interactions that form the retinal-binding pocket of rhodopsin. However, disulfide bonds are not universally required for the structure of 7-TMS receptors. The yeast $\alpha$-factor receptor has only two Cys residues (Cys59,

Cys252), and both can be replaced by site-directed mutagenesis without any deleterious effect on receptor function (D. Kaim, J. Reneke, J. Thorner, unpublished results).

The role of the transmembrane domains and the extracellular disulfide bonds in the $\beta_2$-AR and rhodopsin highlights the complexity of analyzing structure-function relationships for ligand binding. The same conclusion can be reached from attempts to identify sites of interaction between the receptors and G proteins. Clearly, while considerable progress has been made, much remains to be learned about how these proteins are assembled into a functional unit.

# NEGATIVE REGULATION OF RECEPTOR FUNCTION: DESENSITIZATION

A general property of signal-response systems in biology is that prolonged stimulation of a cell often results in reduced responsiveness to subsequent challenges by the same stimulus. This ability to adapt is mediated by processes that frequently occur at the level of the receptor itself, and have been extensively characterized for the $\beta_2$-AR and rhodopsin. In the visual system, rhodopsin signaling is rapidly inactivated ($< 0.5$ sec) following light stimulation, thereby preventing a brief flash of light from being perceived as continuous illumination (4, 30). Similarly, desensitization to drugs is a major factor that limits the efficacy and duration of action of many therapeutic agents, such as $\beta$-adrenergic agonists (96).

In principle, desensitization could occur by reducing expression of cell surface receptors, by attenuating the ligand-binding or G protein–coupling functions of receptors, or by modulating the function of downstream components in the signaling pathway. For the $\beta_2$-AR, desensitization can be broken down temporally into long- and short-term processes. Long-term desensitization, which will not be discussed in detail here, is characterized by a decrease in the total number of cell receptors over several hours, and by a requirement for de novo protein synthesis for recovery of receptor number. In contrast, short-term desensitization is characterized by rapid attenuation of signaling, within minutes of exposure to agonist, and a rapid recovery even in the absence of protein synthesis. To date, the study of $\beta_2$-AR desensitization has focused on the short-term attenuation of receptor-mediated $G_s$ coupling, and the consequent reduction in adenylyl cyclase activity, because the observed loss of G protein coupling appears sufficient to account for the attenuation of signaling. In this system, G protein functionality is not thought to be significantly altered (97).

Two distinct mechanisms underlie short-term desensitization. The first, termed homologous desensitization, leads to reduced responsiveness to the original stimulus. A second type, heterologous desensitization, leads to a decrease in responsiveness as the result of stimuli acting through other receptors. While homologous and heterologous desensitization are distinct processes, both are associated with receptor phosphorylation and with an impairment of G protein coupling. In contrast, receptor phosphorylation does not appear to play a role in long-term desensitization (98–100).

Heterologous desensitization of the $\beta_2$-AR involves phosphorylation of the receptor by the cAMP-dependent protein kinase (PKA) and follows any stimulus (e.g. $\beta$-adrenergic agonist, prostaglandin $E_2$) that raises cAMP levels in the cell. Thus, this type of desensitization also occurs following exposure of cells to agonist levels (nM, comparable to that of circulating epinephrine) that result in fractional occupancy of the receptor pool but maximal stimulation of cAMP production and activation of PKA (101–105). By reconstitution of purified $\beta_2$-AR with $G_s$, PKA-mediated phosphorylation of the receptor was shown to result in attenuated coupling to $G_s$ (104). Moreover, phosphorylation by PKA appears to be somewhat accelerated in vitro when the receptor is occupied by an agonist (104).

Two consensus sites for phosphorylation by PKA were identified in the deduced sequence of the $\beta_2$-AR (16; Figure 1). One site is in loop V-VI, and the other is in the C-terminal domain close to TMS VII. Substitution mutagenesis of the Ser residues at both consensus phosphorylation sites led to markedly diminished phosphorylation and impaired receptor desensitization in response to low concentrations of agonist (106). By replacing each phosphorylation site individually, it was found that the site in loop V-VI is sufficient for PKA-mediated desensitization of the $\beta_2$-AR (107). kin$^-$, a variant of S49 cells that lacks functional PKA, similarly lacks the ability to undergo $\beta_2$-AR desensitization in response to low concentrations of agonist (108).

The mechanism by which receptor phosphorylation leads to uncoupling is unknown. PKA phosphorylation sites in the $\beta_2$-AR are adjacent to cationic domains of the receptor that are implicated in G protein coupling. Thus phosphorylation could lead to receptor desensitization by adding a negatively charged phosphate group near sites that are required for G protein association.

Homologous desensitization of the $\beta_2$-AR occurs only at high levels of agonist ($\mu$M), which are likely to be reached in vivo only at neural synapses. Exposure of cells to a high concentration of agonist is accompanied by PKA-mediated phosphorylation of the $\beta_2$-AR, as well as by receptor phosphorylation catalyzed by a unique, non-cAMP-dependent, protein kinase, the $\beta$-adrenergic receptor kinase ($\beta$ARK) (109). The action of $\beta$ARK is unusual in that phosphorylation of the receptor is strictly dependent on agonist binding, and thus represents a rare example of a substrate-activated kinase (109). This feature may explain the receptor specificity of homologous desensitization.

Like PKA, crude $\beta$ARK preparations were shown to mediate agonist-dependent phosphorylation and uncoupling of the receptor in vitro. In contrast to PKA, however, phosphorylation of the $\beta_2$-AR by pure preparations of $\beta$ARK does not lead to a significant loss of $G_s$ coupling, suggesting that an essential cofactor for $\beta$ARK-mediated desensitization is lost during its purification (110). In an analogous manner, the enzyme rhodopsin kinase specifically phosphorylates light-bleached rhodopsin at serine and threonine residues clustered near the C-terminus of the protein (4, 45, 111, 112). When phosphorylated to high stoichiometry, rhodopsin has a reduced ability to couple to transducin (113); however, efficient uncoupling at low stoichiometry requires the binding of an additional protein, arrestin (also known as 48K protein or S antigen; 114–119). Arrestin is thought to block transducin activation by binding to the phosphorylated form of rhodopsin, resulting in competitive inhibition of rhodopsin-transducin signaling (118–121). It was later shown that high concentrations of retinal arrestin could also promote partial uncoupling of $\beta$ARK-phosphorylated $\beta_2$-AR and $G_s$ (110). Thus, a retinal arrestin cDNA was used to isolate a homologous cDNA encoding a $\beta_2$-AR-specific uncoupling factor, $\beta$arrestin (122). The $\beta$arrestin gene product was subsequently expressed, partially purified, and shown to block coupling of $\beta$ARK-phosphorylated $\beta_2$-AR and $G_s$ in vitro. Moreover, the deduced amino acid sequences of arrestin and $\beta$arrestin are 59% identical, and both have several stretches of sequence similarity to G protein $\alpha$-subunits, possibly indicating regions that are involved in receptor recognition (122).

The C-terminus of the $\beta_2$-AR, like that of rhodopsin, is rich in hydroxyl amino acids that might represent sites of phosphorylation by $\beta$ARK. This has been tested by peptide mapping experiments (52), and more recently by site-directed mutagenesis and expression of the $\beta_2$-AR (98–100, 106, 123). Replacement of all of the 11 Ser and Thr residues closest to the C-terminus results in marked attenuation of both receptor desensitization and agonist-stimulated receptor phosphorylation in cells (98, 106). Similar results were obtained for a mutant in which the C-terminus was simply truncated to remove these sites (98). Moreover, impaired desensitization of these C-terminal mutants of the receptor was observed only at high (i.e. $\mu$M) concentrations of agonist, whereas signaling and phosphorylation of the mutants exposed to low (i.e. nM) agonist concentrations was indistinguishable from the wild type (106). These findings are consistent with the scheme described above in which $\beta$ARK phosphorylates the C-terminal domain of the $\beta_2$-AR only at high agonist concentrations, whereas low agonist concentrations are sufficient for PKA-dependent phosphorylation and desensitization.

Little is known about the specificity of $\beta$ARK in vivo for substrates other than the $\beta_2$-AR. In vitro, however, agonist-occupied chick M2 muscarinic cholinergic and human $\alpha_{2A}$-adrenergic receptors are excellent substrates for phosphorylation by $\beta$ARK (125, 126). $\beta$ARK also phosphorylates light-activated rhodopsin, and rhodopsin kinase phosphorylates agonist-occupied $\beta_2$-AR, but each enzyme displays a much higher $V_{max}$ for its natural substrate (127). Although amino acid sequences in general are poorly conserved among the cytoplasmic domains of 7-TMS receptors, nearly all have clusters of Ser and Thr residues within cytoplasmic domains near their C-terminus or in loop V-VI. The $\alpha_{2A}$-AR, for example, is probably phosphorylated in vitro by $\beta$ARK at sites within cytoplasmic loop V-VI (20, 126), rather than near its C-terminus which lacks serine and threonine residues altogether. It is not known if all of these receptors are normally substrates for $\beta$ARK and/or rhodopsin kinase, or if there are additional receptor-specific kinases. Indeed, cDNAs encoding $\beta$ARK and at least one $\beta$ARK homolog have been cloned, but little is known about how the $\beta$ARK-like product might differ functionally from $\beta$ARK (124).

The deduced amino acid sequence of $\beta$ARK suggests a structure consisting of a presumed catalytic domain of 239 amino acids, flanked by unique N- and C-terminal domains of comparable lengths (124). The function of these flanking regions is unknown. The catalytic domain of $\beta$ARK is homologous to catalytic domains of other protein kinases, with greatest similarity to the PKA and PKC families. Interestingly, $\beta$ARK, PKA, and PKC have in common the fact that they mediate or modulate the activity of G protein–coupled receptors.

To study the relative contributions of the homologous and heterologous pathways to receptor desensitization, Lohse et al have recently developed a method for permeabilizing whole cells in a manner that allows the entry of selective inhibitors of PKA or $\beta$ARK (PKI and heparin, respectively) yet leaves the cAMP-generating and desensitization machinery intact (128). PKA was found to be necessary and sufficient for desensitization following exposure to low agonist concentrations. Following exposure to high agonist concentrations, however, either PKA or $\beta$ARK could mediate the desensitization response, although maximal attenuation required the action of both kinases (128).

Receptor dephosphorylation is also presumed to play an important role in regenerating active rhodopsin and $\beta_2$-AR, although it has received relatively little attention. Several groups have reported dephosphorylation of rhodopsin by a phospho-protein phosphatase 2A in rod outer segments (129, 130). Moreover, arrestin inhibits dephosphorylation of photoactivated rhodopsin, but not of non-activated rhodopsin or of an irrelevant substrate (129). Thus, the ability of arrestin to attenuate specifically dephosphorylation of photoactivated rhodopsin may represent a novel mechanism that confers substrate specificity and temporal regulation to a phosphatase with an otherwise broad substrate specificity. Paradoxically, arrestin does not prevent further phosphorylation by rhodopsin kinase of rhodopsin that has been phosphorylated initially to low stoichiometry (119). In other words, arrestin binding to rhodopsin appears to block the action of the phosphatase, but not that of the kinase.

Treatment of cells with low doses of agonist has also been reported to result in rapid and reversible sequestration of the $\beta_2$-AR within the cell (101, 131). Sequestration is distinct from internalization, which occurs over a longer period of time and is associated with receptor down-regulation (degradation). Sequestration takes place after uncoupling has already occurred, however, and it can be completely blocked (with concanavalin A) with no detrimental effect to the overall desensitization process (131, 133). Nonetheless, there is evidence that sequestered $\beta_2$-AR becomes associated with an as-yet-poorly characterized phosphatase that rapidly dephosphorylates the receptor (101). Thus, although receptor sequestration may not be an essential step in the desensitization process per se, it might play a role in accelerating recovery from short-term desensitization.

## THE APPARENT PARADOX OF TWO KINASES THAT MEDIATE DESENSITIZATION

The existence of two kinases that mediate rapid $\beta_2$-AR phosphorylation and desensitization following exposure to agonists is as yet unexplained. While $\beta$ARK is very specific for the agonist-occupied receptor, PKA is just as effective and at much lower agonist concentrations. One role for separate enzymes may be in the temporal requirements for desensitization in highly innervated versus non-innervated tissues. The rapid release of high concentrations of a neurotransmitter at a synapse might require a more rapid turn off mechanism than is necessary in non-innervated tissues. Like rhodopsin kinase (which is expressed exclusively in the retina), $\beta$ARK phosphorylation could mediate a rapid-response mechanism that directly targets the agonist-occupied receptor. In support of this model, mRNA analysis indicates relatively high levels of $\beta$ARK and $\beta$arrestin transcripts in highly innervated tissues such as brain, spleen, heart, and lung (122, 124). In contrast, the PKA-dependent process requires several steps, including G protein subunit dissociation, adenylyl cyclase activation, diffusion and binding of four molecules of cAMP to the regulatory subunit of PKA, dissociation of the catalytic and regulatory subunits, and ultimately receptor phosphorylation (134). This sequence of events may be too slow to mediate effectively the rapid desensitization required during neuronal signaling, but might be adequate in peripheral tissues.

## ADDITIONAL COMPONENTS INVOLVED IN RECEPTOR DESENSITIZATION

Recent studies suggest that light adaptation may require components in addition to rhodopsin kinase and arrestin. By monitoring activation and inactivation of transducin in rod outer segment membranes, a soluble protein extract from rod cells was observed to greatly accelerate the arrestin-dependent inactivation of rhodopsin signaling (135, 136). Moreover, in a reconstituted system consisting of purified transducin and rhodopsin, no diminution of rhodopsin-mediated activation of transducin was detected in the presence of arrestin, regardless of the extent of phosphorylation of rhodopsin (137), suggesting that the present model of competition between arrestin and transducin for binding to rhodopsin might be inadequate. An additional soluble factor, perhaps acting in synergy with arrestin, might be required for the rapid desensitization observed in vivo following phosphorylation of rhodopsin. Such a component might be analogous to the *S. cerevisiae CDC36, CDC39*, or *SST2* gene products, which have been identified as negative regulators of receptor–G protein action through the use of genetic analysis to study receptor signaling in yeast (see below).

## YEAST PHEROMONE SIGNALING: SIMILARITIES TO MAMMALIAN SYSTEMS

*S. cerevisiae* can exist in any of three distinct cell types: the haploid mating types *MAT*a or *MAT*α, as well as the diploid *MAT*a/*MAT*α. Prior to mating, *MAT*a and *MAT*α cells both secrete specific peptide pheromones that bind to G protein–coupled receptors on cells of the opposite mating type. Ligand binding triggers a program of developmental processes required for fusion, including transcriptional changes and ensuing morphological changes, as well as growth arrest in the G1 phase of the cell cycle (5, 138).

While the mechanisms involved in G protein–mediated signal transduction in mammalian systems have been elucidated biochemically, the identification of components of the mating signal transduction pathway of yeast were originally identified as mutations that disrupt pheromone responsiveness. Many of these mutations define positive controlling elements, because they cause an unresponsive and nonmating (sterile) phenotype. These lesions led to the identification of genes encoding the α- and a- mating factor receptors (*STE2* and *STE3*, respectively) (139–141), as well as the β (*STE4*) and γ

(*STE18*) subunits of the relevant G protein (142). *STE2* and *STE3* encode typical 7-TMS receptors (Figure 1); however, *STE2* and *STE3* are unique among G protein–coupled receptors because comparison of their sequences with one another and with other 7-TMS receptors reveals little similarity.

The nature of the effector(s) or second messenger(s) in the yeast pheromone response pathway are not yet known. Nonetheless, direct evidence for the functional role of several of the gene products has been obtained. Radioligand binding and affinity labeling methods have been applied to the Ste2 protein, and reveal characteristics predicted of a cell surface receptor for $\alpha$-factor; binding is specific, saturable, of high affinity, and dependent on a wild-type copy of the *STE2* gene (143, 144). The *STE2* gene product appears to be solely responsible for ligand recognition, because characteristic binding is observed when this receptor is expressed in *Xenopus* oocytes (145). Moreover, haploid yeast cells induced to express the receptor of the opposite mating type display autocrine responsiveness to their own secreted pheromone (146, 147).

A number of mutations in negative controlling elements have also been identified. These lesions result in a constitutive signal leading to arrest of cell growth. In this way the G protein $\alpha$-subunit [*GPA1* (148), also designated *SCG1* (149), *CDC70* (150), and *DAC1* (151)] involved in mediating pheromone signaling was identified. Interestingly, analysis of the epistasis relationships among mutations in the three G protein subunits has revealed that it is the $\beta\gamma$ moiety, not $\alpha$, that acts on the downstream effector, in contrast to the mammalian paradigm (142, 152, 153). Rather than blocking the mating signal transduction pathway, disruption or transcriptional repression of the yeast $\alpha$-subunit gene leads to constitutive signaling, arrest, and mating (148, 149), whereas loss of either the $\beta$- or $\gamma$-subunits disrupts pheromone responsiveness and suppresses the arrest phenotype of G$\alpha$ mutants (153).

## DESENSITIZATION IN YEAST

Desensitization in yeast is thought to play an important physiological role if mating is unsuccessful, because attenuation of pheromone signaling is presumably necessary for recovery from cell cycle arrest. Similarly, even if mating is successful, desensitization in the resulting diploid might facilitate reentry into the mitotic cycle.

Genetic and biochemical approaches have been used to determine the extent to which the loss of $\alpha$-factor responsiveness in yeast is analogous to the hormone desensitization observed in animal cells. $\alpha$-Factor receptors undergo rapid endocytosis following exposure to pheromone (154, 155). After receptor internalization, new receptors appear at the cell surface that, while competent to bind ligand, remain refractory to pheromone action (154). This observation suggests that newly synthesized receptors are still uncoupled from the intracellular effector(s) responsible for receptor-mediated signal transduction. It is not known whether it is the receptor or some downstream component that is desensitized. Like mammalian G protein–coupled receptors, however, the $\alpha$-factor-occupied receptor is rapidly phosphorylated on Ser and Thr residues within the C-terminal domain (156).

Deletion mutagenesis of the receptor has confirmed that an intact C-terminal domain is required for the normal cycle of desensitization and recovery after pheromone treatment, consistent with the possibility that desensitization is triggered by phosphorylation (156, 157). However, the overall process of desensitization appears to be distinct from internalization. While several C-terminal truncation mutants of the $\alpha$-factor receptor were shown to be defective in their ability to desensitize, only one of these mutations (which completely removed the C-terminal domain) was significantly defective in internalization (156). Other known functions of the $\alpha$-factor receptor, such as ligand binding and activation of the mating response, are relatively unaffected by even the complete C-terminal truncation.

In addition to mutations that affect positive and negative elements of the pheromone signaling pathway, a third class of mutations alter the extent or duration of pheromone responsiveness. This genetic approach has led to the identification of several genes that are required for efficient recovery from pheromone-induced arrest. One of these, the *SST1* gene product, has been shown to encode an extracellular protease that degrades $\alpha$-factor (158). Loss of another gene, *SST2*, results in cells that are severely defective in the recovery process (159, 160). The function of *SST2*, however, has remained elusive. The deduced sequence of the *SST2* gene product bears no obvious similarity to sequences of kinases or of any other known protein (161), and unlike $\beta$ARK or rhodopsin kinase, Sst2 does not appear to act through the C-terminal domain of the $\alpha$-factor receptor (156, 157).

Recent evidence indicates that negative regulation of pheromone signaling also involves the *CDC36* and *CDC39* gene products (162, 163). Temperature-sensitive mutants of *CDC36* and *CDC39* will, at the restrictive temperature, constitutively activate the pheromone response pathway and promote mating (162, 163). Like *SST2*, however, the role of these components in regulating pheromone responsiveness is not yet known. The response mediated by these mutants requires functional $\beta$- and $\gamma$-subunits, and is suppressed by overexpression of the $\alpha$-subunit, suggesting that all three elements operate at the level of the G protein (162, 163).

Homologues of *SST2*, *CDC36*, and *CDC39* have not yet been identified in mammalian cells. An interesting question is whether these components, identified through genetic analysis in yeast, might also play a role in desensitization of the $\beta_2$-AR and other G protein–coupled receptors in higher eukaryotes. Indeed, as was discussed above, a novel component has been implicated in the adaptation pathway for rhodopsin, in addition to rhodopsin kinase and arrestin.

## USE OF HETEROLOGOUS EXPRESSION SYSTEMS

With a growing number of receptor subtypes, G proteins, and effectors, characterization of ligand binding and G protein recognition properties of receptors is an emerging and important challenge. It has long been known that multiple receptors can couple to a single G protein and, as in the case of epinephrine binding to $\beta_2$- and $\alpha_2$-ARs, a single ligand can bind to multiple functionally distinct receptor subtypes. Moreover, G proteins with similar receptor and effector coupling specificities have also been identified. For example, three species of human $G_i$ have been cloned (164), and alternate mRNA splicing has been shown to result in multiple variants of $G_s$ (165).

Reconstitution experiments were the first to show that receptors can, with varying degrees of specificity, couple to multiple (and in some cases functionally distinct) G proteins (166–170). Cloning and overproduction of the muscarinic and $\alpha_2$-ARs led to the demonstration that a single receptor subtype, when expressed at high levels in the cell, will couple to more than one type of G protein. For each of these receptors, agonist treatment led to both inhibition of adenylyl cyclase and stimulation of phosphoinositide metabolism (171, 172). Finally, individual G protein species have been shown to stimulate more than one effector. $G_s$, for example, has been reported to regulate calcium channels, in addition to adenylyl cyclase (34–36). Given this complexity and apparent degeneracy of function, a question of fundamental importance is how, and under what circumstances, can G proteins organize signals from multiple receptors and direct them to the appropriate effector(s) (173, 173a)? A traditional approach has been to reconstitute the purified receptor and G protein components in vitro. Unfortunately, purification schemes have been successful for only a very limited number of receptor subtypes and their cognate G-proteins. Alternatively, heterologous expression systems may be of more general usefulness in the characterization of cloned receptors and in elucidating receptor–G protein coupling specificity (174–176).

One such system was recently developed in yeast cells, in which the genes for a mammalian $\beta_2$-AR and $G_s$ $\alpha$-subunit were coexpressed (176). Expression of the $\beta_2$-AR to levels several hundred-fold higher than in any human tissue was attained, and ligand binding was shown to be of the appropriate affinity, specificity, and stereoselectivity. Moreover, a $\beta_2$-AR-mediated activation of the pheromone signal transduction pathway was demonstrated (176) by several criteria, including imposition of growth arrest, morphological changes, and induction of a pheromone-responsive promoter *(FUS1)* fused to the *Escherichia coli lacZ* gene (encoding $\beta$-galactosidase).

The ability to control the yeast pheromone response pathway by expression of the $\beta_2$-AR and $G_s\alpha$ has the potential to greatly facilitate structural and functional characterization of the $\beta_2$-AR. By scoring for growth arrest or $\beta$-galactosidase induction, the properties of mutant receptors can now be tested rapidly. Moreover, as additional genes for putative G protein–coupled receptors are isolated, a series of ligands can be conveniently screened to identify those with activity toward the unidentified gene product. Finally, expression of a single receptor in the absence of other related subtypes is often impossible to achieve in mammalian cells. Thus, expression in a microorganism that has no such endogenous receptors can be useful for screening and evaluating subtype-selective drugs (174–176).

## THE CONUNDRUM OF RECEPTOR SUBTYPES

The discovery through molecular cloning of additional receptor subtypes has raised some important questions. The foremost issue is what advantage, if any, is conferred by multiple receptors with apparently similar pharmacological and signaling properties? It was recognized long ago that $\beta$-adrenergic receptors were of at least two subtypes, both having similar ligand-binding specificities and both mediating stimulation of adenylyl cyclase. Recently a third subtype, the $\beta_3$-AR, was cloned by cross-hybridization with the genes encoding the $\beta_1$- and $\beta_2$-ARs (19). The apparent redundancy of receptor subtypes might reflect differences in their regulation of expression in various tissues or over the course of development. Subtle differences in their ability to recognize endogenous ligands or G proteins needs closer examination, and is one area where heterologous expression methods might be useful. Another speculation is that multiple receptor subtypes might differ in their relative abilities to undergo desensitization (3). Whereas the $\beta_1$- and $\beta_2$-ARs both have 11 Ser and Thr residues in their extreme C-terminal domains, the analogous domain of the $\beta_3$-AR is shorter and has only three Ser residues. Moreover, the number of canonical PKA sites in the $\beta_1$-, $\beta_2$-, and $\beta_3$-receptors are one, two, and none, respectively.

## VIRALLY ENCODED 7-TMS RECEPTORS

Recently, the family of 7-TMS receptors has been extended to include several virally encoded receptors. In sequencing the genome of human cytomegalovirus, a pathogen of the herpes virus group, three open reading frames likely to encode 7-TMS receptors were identified (177). By analogy with viral oncogenes, the cytomegalovirus gene products could have been acquired from the genome of a host cell and might even have become constitutively active in the absence of extracellular signals. However, it has not been demonstrated that these genes are expressed, much less what ligands or G proteins, if any, they might recognize. It is also not known if closely related genes exist in the human genome, and an association of cytomegalovirus with malignancy is unclear. Another question is how cytomegalovirus receptors, if expressed, participate in the viral life cycle. The possibility that infected cells express a unique receptor subtype could make them potential targets for specific anti-viral agents. The application of heterologous gene expression methods might be useful in addressing some of these issues.

## REGULATION OF CELL GROWTH AND DIFFERENTIATION BY 7-TMS RECEPTORS

The proliferative effects of chronic activation of a number of different G protein–coupled receptors is well documented. These include the serotonin (5HT1c), angiotensin, and adrenergic receptors (178–180), which stimulate both the inositol phosphate and the cAMP second messenger pathways. Moreover, in the case of some pituitary tumors, a defect in the stimulatory G protein ($G_s$) component has been identified and shown to lead to constitutive activation of adenylyl cyclase (181). These results have implicated G protein–mediated signaling in the regulation of growth control, and G proteins as potential oncogenes.

A constitutively active form of a G protein–coupled receptor would also be of interest. Despite many examples of mutations leading to activation of proto-oncogenes, even oncogenes that resemble receptors, the nature of the intracellular signal or second messenger is not known for most such oncogene products. Given the precedent of oncogenic G proteins, and the potential for amplification of this signaling cascade at the level of the receptor, it seems likely that a receptor with transforming potential might be identified. An example of how this might work is provided by a recently described mutation of the $\alpha_1$-AR, in which substitutions in the C-terminal segment of loop V-VI lead to agonist-independent stimulation of inositol metabolism (60a).

One candidate for a transforming G protein–coupled receptor is the human *mas* oncogene, which appears to bind angiotensin. However, tumorigenic activity may depend on inappropriate expression, rather than a mutation leading to constitutive activation of the receptor (182, 183). Similarly, transfection of a serotonin receptor (5HT1c) in a non-neural cell line (where it may regulate growth and differentiation rather than ion channel activity) results in the formation of transformed foci (179). Transformation requires the addition of serotonin and is blocked by an antagonist. However, injection of transformed cells into nude mice results in tumors that, when cultured, will form foci even in the absence of serotonin. These observations suggest a potential role for serotonin receptors in tumor initiation rather than tumor progression, consistent with the "multistage" model of carcinogenesis.

## CONCLUSIONS

Considerable progress in the study of G protein–mediated receptor signaling has been made over the past few years. Particularly satisfying has been the discovery of unifying relationships among vastly different systems. Many questions remain, however, particularly with regard to the way in which cells direct environmental signals through multiple receptors, G proteins, and effector pathways. This formidable and fascinating problem will continue to challenge researchers in fields as diverse as pharmacology, biochemistry, molecular biology, genetics, and even virology.

ACKNOWLEDGMENTS

The authors are grateful to Drs. M. Cyert, M. Hnatowich, D. Ma, and C. Williams for their critical reading of the manuscript. Supported by NIH grants HL16037 (R. J. L. & M. G. C.) and GM21841 (J. T.), and a postdoctoral fellowship from the Jane Coffin Childs Fund for Medical Research (H. G. D.).

*Literature Cited*

1. Dohlman, H. G., Caron, M. G., Lefko-witz, R. J. 1987. *Biochemistry* 26:2657–64
2. Benovic, J. L., Bouvier, M., Caron, M. G., Lefkowitz, R. J. 1988. *Annu. Rev. Cell Biol.* 4:405–28
3. Hausdorff, W. P., Caron, M. G., Lefko-witz, R. J. 1990. *FASEB J.* 4:2881–89
4. Kuhn, H. 1984. *Prog. Retinal Res.* 3:123–56
5. Blumer, K. J., Thorner, J. 1991. *Annu. Rev. Physiol.* 53:37–57
6. Gilman, A. G. 1987. *Annu. Rev. Biochem.* 56:615–49
7. Stryer, L., Bourne, H. R. 1986. *Annu. Rev. Cell Biol.* 2:391–419
8. Birnbaumer, L. 1990. *Annu. Rev. Pharmacol. Toxicol.* 30:675–705
9. Strathmann, M., Simon, M. I. 1991. *Science.* In press
10. Robison, G. A., Butcher, R. W., Sutherland, E. W. 1967. *Ann. NY Acad. Sci.* 139:703–23
11. Orly, J., Schramm, M. 1976. *Proc. Natl. Acad. Sci. USA* 73:4410–14
12. Limbird, L. E., Lefkowitz, R. J. 1977. *J. Biol. Chem.* 252:799–802
13. Rodbell, M., Birnbaumer, L., Pohl, S. L., Krans, H. M. 1971. *J. Biol. Chem.* 246:1877–82
14. Rodbell, M. 1980. *Nature* 284:17–22
15. Moss, J., Vaughan, M. 1988. *Adv. Enzymol.* 61:303–79
16. Dixon, R. A. F., Kobilka, B. K., Strader, D. J., Benovic, J. L., Dohlman, H. G., et al. 1986. *Nature* 321:75–79
17. Yarden, Y., Rodriguez, H., Wong, S. K. F., Brandt, D. R., May, D. C., et al. 1986. *Proc. Natl. Acad. Sci. USA* 83:6795–99
18. Frielle, T., Collins, S., Daniel, K. W., Caron, M. G., Lefkowitz, R. J., Kobilka, B. K. 1987. *Proc. Natl. Acad. Sci. USA* 84:7920–24
19. Emorine, L. J., Marullo, S., Briend-Sutren, M. M., Patey, G., Tate, K., et al. 1989. *Science* 245:1118–21
20. Kobilka, B. K., Matsui, H., Kobilka, T. S., Yang-Feng, T. L., Francke, U., et al. 1987. *Science* 238:650–56
21. Regan, J. W., Kobilka, T. S., Yang-Feng, T. L., Caron, M. G., Lefkowitz, R. J., Kobilka, B. K. 1988. *Proc. Natl. Acad. Sci. USA* 85:6301–5

22. Brown, A. R., Birnbaumer, L. 1990. *Annu. Rev. Physiol.* 52:197–213
23. Cotecchia, S., Schwinn, D. A., Randall, R. R., Lefkowitz, R. J., Caron, M. G., Kobilka, B. K. 1988. *Proc. Natl. Acad. Sci. USA* 85:7159–63
24. Schwinn, D. A., Lomasney, J. W., Lorenz, W., Szklut, P. J., Fremeau, R. T. Jr., et al. 1990. *J. Biol. Chem.* 265:8183–89
25. Lomasney, J. W., Lorenz, W., Allen, L. F., King, K., Regan, J. W., et al. 1990. *Proc. Natl. Acad. Sci. USA* 87:5094–98
26. Hubbard, R., Kropf, A. 1958. *Proc. Natl. Acad. Sci. USA* 44:130–39
27. Wald, G. 1968. *Nature* 219:800–7
28. Wheeler, G. L., Bitensky, M. W. 1977. *Proc. Natl. Acad. Sci. USA* 74:4238–42
29. Fung, B. K.-K., Stryer, L. 1980. *Proc. Natl. Acad. Sci. USA* 77:2500–4
30. Nathans, J. 1987. *Annu. Rev. Neurosci.* 10:163–94
31. Jelsema, C. L., Axelrod, J. 1987. *Proc. Natl. Acad. Sci. USA* 84:3623–27
32. Brown, D. A. 1990. *Annu. Rev. Physiol.* 52:215–42
33. Szabo, G., Otero, A. S. 1990. *Annu. Rev. Physiol.* 52:293–305
34. Dolphin, A. C. 1990. *Annu. Rev. Physiol.* 52:243–55
35. Trautwein, W., Hescheler, J. 1990. *Annu. Rev. Physiol.* 52:257–74
36. Schultz, G., Rosenthal, W., Hescheler, J., Trautwein, W. 1990. *Annu. Rev. Physiol.* 52:275–92
37. Lefkowitz, R. J., Stadel, J. M., Caron, M. G. 1983. *Annu. Rev. Biochem.* 52:159–86
38. Benovic, J. L., Shorr, R. G. L., Caron, M. G., Lefkowitz, R. J. 1984. *Biochemistry* 23:4510–18
39. Lomasney, J. W., Leeb-Lundberg, L. M., Cotecchia, S., Regan, J. W., DeBernardis, J. F., et al. 1986. *J. Biol. Chem.* 261:7710–16
40. Regan, J. W., Nakata, H., DeMarinis, R. M., Caron, M. G., Lefkowitz, R. J. 1986. *J. Biol. Chem.* 261:3894–900
41. Repaske, M. G., Nunnari, J. M., Limbird, L. E. 1987. *J. Biol. Chem.* 262:12381–86
42. Cerione, R. A., Strulovici, B., Benovic, J. L., Lefkowitz, R. J., Caron, M. G. 1983. *Nature* 306:526–66
43. Asano, T., Katada, T., Gilman, A. G., Ross, E. M. 1984. *J. Biol. Chem.* 259:9351–54
44. Ovchinnikov, Yu. A. 1982. *FEBS Lett.* 148:179–91
45. Hargrave, P. A. 1982. *Prog. Retinal Res.* 1:1–51
46. Kyte, J., Doolittle, R. F. 1982. *J. Mol. Biol.* 157:105–32
47. Henderson, R., Unwin, P. N. T. 1975. *Nature* 257:28–32
48. Henderson, R., Baldwin, J. M., Ceska, T. A., Zemlin, F., Beckmann, E., Downing, K. H. 1990. *J. Mol. Biol.* 213:899–929
49. Ovchinnikov, Yu. A. 1987. *Trends. Biol. Sci.* 12:434–38
50. Mullen, E., Akhtar, M. 1983. *Biochem. J.* 211:45–54
51. Barclay, P. L., Findlay, J. B. C. 1984. *Biochem. J.* 220:75–84
52. Dohlman, H. G., Bouvier, M., Benovic, J. L., Caron, M. G., Lefkowitz, R. J. 1987. *J. Biol. Chem.* 262:14282–88
53. Wang, H.-Y., Lipfert, L., Malbon, C. C., Bahouth, S. 1989. *J. Biol. Chem.* 264:14424–31
54. Rands, E., Candelore, M. R., Cheung, A. H., Hill, W. S., Strader, C. D., Dixon, R. A. F. 1990. *J. Biol. Chem.* 265:10759–64
55. Hargrave, P. A. 1977. *Biochim. Biophys. Acta* 492:83–94
56. Nathans, J., Hogness, D. S. 1983. *Cell* 34:807–14
57. Nathans, J., Thomas, D., Hogness, D. S. 1986. *Science* 232:193–202
58. Nathans, J., Piantanida, T. P., Eddy, R. L., Shows, T. B., Hogness, D. S. 1986. *Science* 232:203–10
59. Kobilka, B. K., Kobilka, T. S., Daniel, K., Regan, J. W., Caron, M. G., Lefkowitz, R. J. 1988. *Science* 240:1310–16
60. Marsh, L., Herskowitz, I. 1988. *Proc. Natl. Acad. Sci. USA* 85:3855–59
60a. Cotecchia, S., Exum, S., Caron, M. G., Lefkowitz, R. J. 1990. *Proc. Natl. Acad. Sci. USA* 87:2896–900
61. Kubo, T., Bujo, H., Akiba, I., Nakai, J., Mishina, M., Numa, S. 1988. *FEBS Lett.* 241:119–25
62. Wong, S. K., Parker, E. M., Ross, E. M. 1990. *J. Biol. Chem.* 265:6219–24
63. Strader, C. D., Dixon, R. A. F., Cheung, A. H., Candelore, M. R., Blake, A. D., Sigal, I. S. 1987. *J. Biol. Chem.* 262:16439–43
64. O'Dowd, B. F., Hnatowich, M., Regan, J. W., Leader, W. M., Caron, M. G., Lefkowitz, R. J. 1988. *J. Biol. Chem.* 263:15985–92
65. Ovchinnikov, Yu. A., Abdulaev, N. G., Bogachuk, A. S. 1988. *FEBS Lett.* 230:1–5
66. O'Dowd, B. F., Hnatowich, M., Caron, M. G., Lefkowitz, R. J., Bouvier, M. 1989. *J. Biol. Chem.* 264:7564–69
67. Palm, D., Munch, G., Malek, D., Dees, C., Hekman, M. 1990. *FEBS Lett.* 261:294–98

68. Konig, B., Arendt, A., McDowell, J. H., Kahlert, M., Hargrave, P. A., Hofmann, K. P. 1989. *Proc. Natl. Acad. Sci. USA* 86:6878–82
69. Higashijima, T., Uzu, S., Nakajima, T., Ross, E. M. 1988. *J. Biol. Chem.* 263:6491–94
70. Segrest, J. P., De Loof, H., Dohlman, J. G., Brouillette, C. G., Anantharamaiah, G. M. 1990. *Proteins: Struct. Funct. Genet.* 8:103–17
71. Strittmatter, S. M., Valenzuela, D., Kennedy, T. E., Neer, E. J., Fishman, M. C. 1990. *Nature* 344:836–41
72. Bownds, K. D. 1967. *Nature* 216:1178–81
73. Oseroff, A. R., Callender, R. H. 1974. *Biochemistry* 13:4243–48
74. Wang, J. K., McDowell, J. H., Hargrave, P. A. 1980. *Biochemistry* 19:5111–17
75. Findlay, J. B. C., Brett, M., Pappin, D. J. C. 1981. *Nature* 293:314–16
76. Thomas, D. D., Stryer, L. 1982. *J. Mol. Biol.* 154:145–57
77. Frielle, T., Daniel, K. W., Caron, M. G., Lefkowitz, R. J. 1988. *Proc. Natl. Acad. Sci. USA* 85:9494–98
78. Marullo, S., Emorine, L. J., Strosberg, A. D., Delavier-Klutchko, C. 1990. *EMBO J.* 9:1471–76
79. Dohlman, H. G., Caron, M. G., Strader, C. D., Amlaiky, N., Lefkowitz, R. J. 1988. *Biochemistry* 27:1813–17
80. Wong, S. K., Slaughter, C., Ruoho, A. E., Ross, E. M. 1988. *J. Biol. Chem.* 263:7925–28
81. Matsui, H., Lefkowitz, R. J., Caron, M. G., Regan, J. W. 1989. *Biochemistry* 28:4125–30
82. Curtis, C. A., Wheatley, M., Bansal, S., Birdsall, N. J., Eveleigh, P., et al. 1989. *J. Biol. Chem.* 264:489–95
83. Braiman, M., Bubis, J., Doi, T., Chen, H. B., Flitsch, S. L., et al. 1988. *Cold Spring Harbor Symp. Quant. Biol.* 53:355–64
84. Mukherjee, M. G., Caron, M. G., Mullikin, D., Lefkowitz, R. J. 1976. *Mol. Pharmacol.* 12:16–31
85. Strader, C. D., Candelore, M. R., Hill, W. S., Sigal, I. S., Dixon, R. A. F. 1989. *J. Biol. Chem.* 264:13572–78
86. Strader, C. D., Sigal, I. S., Candelore, M. R., Rands, E., Hill, W. S., Dixon, R. A. F. 1988. *J. Biol. Chem.* 263:10267–71
87. Sakmar, T. P., Franke, R. R., Khorana, H. G. 1989. *Proc. Natl. Acad. Sci. USA* 86:8309–13
88. Zhukovsky, E. A., Oprian, D. D. 1990. *Science* 246:928–30
89. Strader, C. D., Candelore, M. R., Hill, W. S., Dixon, R. A. F., Sigal, I. S. 1989. *J. Biol. Chem.* 264:16470–77
90. Honig, B., Greenberg, A. D., Dinur, U., Ebrey, T. G. 1976. *Biochemistry* 15:4593–99
91. Nathans, J. 1990. *Biochemistry* 29:937–42
92. Dohlman, H. G., Caron, M. G., DeBlasi, A., Frielle, A., Lefkowitz, R. J. 1990. *Biochemistry* 29:2335–42
93. Dixon, R. A. F., Sigal, I. S., Candelore, M. R., Register, R. B., Scattergood, W., et al. 1987. *EMBO J.* 6:3269–75
94. Fraser, C. M. 1989. *J. Biol. Chem.* 264:9266–70
95. Karnik, S. S., Sakmar, T. P., Chen, H.-B., Khorana, H. G. 1988. *Proc. Natl. Acad. Sci. USA* 85:8459–63
96. Hoffman, B. B., Lefkowitz, R. J. 1990. In *Pharmacological Basis of Therapeutics*, ed. A. G. Gilman, T. W. Rall, A. S. Nies, P. Taylor, pp. 187–220. New York: Permagon. 8th ed.
97. Green, D. A., Clark, R. B. 1981. *J. Biol. Chem.* 256:2105–8
98. Bouvier, M., Hausdorff, W. P., DeBlasi, A., O'Dowd, B. F., Kobilka, B. K., et al. 1988. *Nature* 333:370–73
99. Cheung, A. H., Sigal, I. S., Dixon, R. A. F., Strader, C. D. 1989. *Mol. Pharmacol.* 34:132–38
100. Strader, C. D., Sigal, I. S., Blake, A. D., Cheung, A. H., Register, R. B., et al. 1987. *Cell* 49:855–63
101. Sibley, D. R., Peters, J. R., Nambi, P., Caron, M. G., Lefkowitz, R. J. 1984. *J. Biol. Chem.* 259:9742–49
102. Strulovici, B., Cerione, R. A., Kilpatrick, B. F., Caron, M. G., Lefkowitz, R. J. 1984. *Science* 225:837–40
103. Nambi, P., Peters, J. R., Sibley, D. R., Lefkowitz, R. J. 1985. *J. Biol. Chem.* 260:2165–71
104. Benovic, J. L., Pike, L. J., Cerione, R. A., Staniszewski, C., Yoshimasa, T., et al. 1985. *J. Biol. Chem.* 260:7094–101
105. Sibley, D. R., Benovic, J. L., Caron, M. G., Lefkowitz, R. J. 1987. *Cell* 48:913–22
106. Hausdorff, W. P., Bouvier, M., O'Dowd, B. F., Irons, G. P., Caron, M. G., Lefkowitz, R. J. 1989. *J. Biol. Chem.* 264:12657–63
107. Clark, R. B., Friedman, J., Dixon, R. A. F., Strader, C. D. 1989. *Mol. Pharmacol.* 36:343–48
108. Clark, R. B., Kunkel, M. W., Friedman, J., Goka, T. J., Johnson, J. A. 1988. *Proc. Natl. Acad. Sci. USA* 85:1442–46
109. Benovic, J. L., Strasser, R. H., Caron,

M. G., Lefkowitz, R. J. 1986. *Proc. Natl. Acad. Sci. USA* 83:2797–801

110. Benovic, J. L., Kuhn, H., Weyand, I., Codina, J., Caron, M. G., Lefkowitz, R. J. 1987. *Proc. Natl. Acad. Sci. USA* 84:8879–82

111. Sale, G. J., Towner, D., Akhtar, M. 1978. *Biochem. J.* 175:421–30

112. Thompson, P., Findlay, J. B. C. 1984. *Biochem. J.* 220:773–80

113. Shichi, H., Yamamoto, K., Somers, R. L. 1984. *Vision Res.* 24:1523–31

114. Wilden, U., Kuhn, H. 1982. *Biochemistry* 21:3014–22

115. Sitaramayya, A. 1986. *Biochemistry* 25:5460–68

116. Kuhn, H., Hall, S. W., Wilden, U. 1984. *FEBS Lett.* 176:473–78

117. Miller, J. L., Fox, D. A., Litman, B. J. 1986. *Biochemistry* 25:4983–88

118. Wilden, U., Hall, S. W., Kuhn, H. 1986. *Proc. Natl. Acad. Sci. USA* 83:1174–78

119. Bennett, N., Sitaramayya, A. 1988. *Biochemistry* 27:1710–15

120. Kelleher, D. J., Johnson, G. L. 1988. *Mol. Pharmacol.* 34:452–60

121. Shinohara, T., Dietzschold, B., Craft, C. M., Wistow, G., Early, J. J., et al. 1987. *Proc. Natl. Acad. Sci. USA* 84:6975–79

122. Lohse, M. J., Benovic, J. L., Codina, J., Caron, M. G., Lefkowitz, R. J. 1990. *Science* 248:1547–50

123. Liggett, S. B., Bouvier, M., Hausdorff, W. P., O'Dowd, B. F., Caron, M. G., Lefkowitz, R. J. 1989. *Mol. Pharmacol.* 36:343–48

124. Benovic, J. L., DeBlasi, A., Stone, W. C., Caron, M. G., Lefkowitz, R. J. 1990. *Science* 246:235–40

125. Kwatra, M. M., Benovic, J. L., Caron, M. G., Lefkowitz, R. J., Hosey, M. M. 1989. *Biochemistry* 28:4543–47

126. Benovic, J. L., Regan, J. W., Matsui, H., Mayor, F. Jr., Cottechia, S., et al. 1987. *J. Biol. Chem.* 262:17251–53

127. Benovic, J. L., Mayor, F. Jr., Somers, R. L., Caron, M. G., Lefkowitz, R. J. 1986. *Nature* 322:869–72

128. Lohse, M. J., Benovic, J. L., Caron, M. G., Lefkowitz, R. J. 1990. *J. Biol. Chem.* 265:3202–9

129. Palczewski, K., McDowell, J. H., Jakes, S., Ingebritsen, T. S., Hargrave, P. A. 1989. *J. Biol. Chem.* 264:15770–73

130. Fowles, C., Akhtar, M., Cohen, P. 1989. *Biochemistry* 28:9385–91

131. Waldo, G. L., Northup, J. K., Perkins, J. P., Harden, T. K. 1983. *J. Biol. Chem.* 258:13900–8

132. Deleted in proof

133. Hertel, C., Coulter, S. J., Perkins, J. P. 1983. *J. Biol. Chem.* 260:12547–53

134. Taylor, S. S. 1990. *Annu. Rev. Biochem.* 59:971–1006

135. Wagner, R., Ryba, N., Uhl, R. 1988. *FEBS Lett.* 234:44–48

136. Wagner, R., Ryba, N., Uhl, R. 1988. *FEBS Lett.* 235:103–8

137. Fukuda, Y., Yoshizawa, T., Saito, T., Ohguro, H., Akino, T. 1990. *FEBS Lett.* 261:419–22

138. Herskowitz, I. 1988. *Microbiol. Rev.* 52:536–53

139. Burkholder, A. C., Hartwell, L. H. 1985. *Nucleic Acids Res.* 13:8463–75

140. Nakayama, N., Miyajima, A., Arai, K. 1985. *EMBO J.* 4:2643–48

141. Hagen, D. C., McCaffrey, G., Sprague, G. F. Jr. 1986. *Proc. Natl. Acad. Sci. USA* 83:1418–22

142. Whiteway, M., Hougan, L., Dignard, D., Thomas, D. Y., Bell, L., et al. 1989. *Cell* 56:467–77

143. Jenness, D. D., Burkholder, A. C., Hartwell, L. H. 1986. *Mol. Cell. Biol.* 6:318–20

144. Blumer, K. J., Reneke, J. E., Thorner, J. 1988. *J. Biol. Chem.* 263:10836–42

145. Yu, L., Blumer, K. J., Davidson, N., Lester, H. A., Thorner, J. 1989. *J. Biol. Chem.* 264:20847–50

146. Bender, A., Sprague, G. F. Jr. 1986. *Cell* 47:929–37

147. Nakayama, N., Miyajima, A., Arai, K. 1987. *EMBO J.* 4:2643–48

148. Miyajima, I., Nakafuku, M., Nakayama, N., Brenner, C., Miyajima, A., et al. 1987. *Cell* 50:1011–19

149. Dietzel, C., Kurjan, J. 1987. *Cell* 50:1001–10

150. Jahng, K. Y., Ferguson, J., Reed, S. I. 1988. *Mol. Cell. Biol.* 8:2484–93

151. Fujimura, H. A. 1989. *Mol. Cell. Biol.* 9:152–58

152. Blinder, D., Bouvier, S., Jenness, D. D. 1989. *Cell* 56:479–86

153. Nakayama, N., Kaziro, Y. U., Arai, K., Matsumoto, K. 1988. *Mol. Cell. Biol.* 8:3777–83

154. Jenness, D. D., Spatrick, P. 1986. *Cell* 46:345–53

155. Chvatchko, Y., Howald, I., Riezman, H. 1986. *Cell* 46:355–64

156. Reneke, J. E., Blumer, K. J., Courschesne, W., Thorner, J. 1988. *Cell* 55:221–34

157. Konopka, J. B., Jenness, D. D., Hartwell, L. H. 1988. *Cell* 54:609–20

158. MacKay, V. L., Welch, S. K., Insley, M. Y., Manney, T. R., Holly, J., et al. 1988. *Proc. Natl. Acad. Sci. USA* 85:55–59

159. Chan, R. K., Otte, C. A. 1982. *Mol. Cell. Biol.* 2:11–20
160. Chan, R. K., Otte, C. A. 1982. *Mol. Cell. Biol.* 2:21–29
161. Dietzel, C., Kurjan, J. 1987. *Mol. Cell. Biol.* 7:4169–77
162. Nieman, A. M., Chang, F., Komachi, K., Herskowitz, I. 1990. *Cell Regul.* 1:391–401
163. de Barros Lopes, M., Ho, J.-Y., Reed, S. I. 1990. *Mol. Cell. Biol.* 10:2966–72
164. Itoh, H., Toyama, R., Kozasa, T., Tsukamoto, T., Matsuoka, M., Kaziro, Y. 1988. *J. Biol. Chem.* 263:6656–64
165. Kozasa, T., Itoh, H., Tsukamoto, T., Kaziro, Y. 1988. *Proc. Natl. Acad. Sci. USA* 85:2081–85
166. Kanaho, Y., Tsai, S.-C., Adamik, R., Hewlett, E. L., Moss, J., Vaughan, M. 1984. *J. Biol. Chem.* 259:7378–81
167. Tsai, S.-C., Adamik, R., Kanaho, Y., Hewlett, E. L., Moss, J. 1984. *J. Biol. Chem.* 259:15320–23
168. Cerione, R. A., Staniszewski, C., Benovic, J. L., Lefkowitz, R. J., Caron, M., et al. 1985. *J. Biol. Chem.* 260:1493–500
169. Senogles, S. E., Spiegel, A. M., Padrell, E., Iyengar, R., Caron, M. G. 1990. *J. Biol. Chem.* 265:4507–11
170. Kurose, H., Regan, J. W., Caron, M. G., Lefkowitz, R. J. 1991. Submitted
171. Peralta, E. G., Ashkenazi, A., Winslow, J. W., Ramachandran, J., Capon, D. J. 1988. *Nature* 334:434–37
172. Cotecchia, S., Kobilka, B. K., Daniel, K. W., Nolan, R. D., Lapetina, E. Y., et al. 1990. *J. Biol. Chem.* 265:63–69
173. Ross, E. M. 1989. *Neuron* 3:141–52
173a. Ercolani, L., Stow, J. L., Boyle, J. F., Holtzman, E. J., Lin, H., et al. 1990. *Proc. Natl. Acad. Sci. USA* 87:4635–39
174. Marullo, S., Delavier-Klutchko, C., Eshdat, Y., Strosberg, A. D., Emorine, L. J. 1988. *Proc. Natl. Acad. Sci. USA* 85:7551–55
175. Payette, P., Gossard, F., Whiteway, M., Dennis, M. 1990. *FEBS Lett.* 266:21–25
176. King, K., Dohlman, H. G., Thorner, J., Caron, M. G., Lefkowitz, R. J. 1990. *Science* 250:121–23
177. Chee, M. S., Satchwell, S. C., Preddie, E., Weston, K. M., Barrell, B. G. 1990. *Nature* 344:774–77
178. Hen, R., Axel, R., Obici, S. 1989. *Proc. Natl. Acad. Sci. USA* 86:4785–88
179. Julius, D., Livelli, T. J., Jessell, T. M., Axel, R. 1989. *Science* 244:1057–62
180. Allen, J. M., Baetge, E. E., Abrass, I. B., Palmiter, R. D. 1988. *EMBO J.* 7:133–38
181. Landis, C. A., Masters, S. B., Spada, A., Pace, A. M., Bourne, H. R., Vallar, L. 1989. *Nature* 340:692–96
182. Young, D., Waitches, G., Birchmeier, C., Fasano, O., Wigler, M. 1986. *Cell* 45:711–19
183. Jackson, T. R., Blair, L. A. C., Marshall, J., Goedert, M., Hanley, M. R. 1988. *Nature* 335:437–40
184. Machida, C. A., Bunzow, J. R., Searles, R. P., Van Tol, H., Tester, B., et al. 1990. *J. Biol. Chem.* 265:12960–65
185. Kobilka, B. K., Dixon, R. A. F., Frielle, T., Dohlman, H. G., Bolanowski, M. A., et al. 1987. *Proc. Natl. Acad. Sci. USA* 84:46–50
186. Schofield, P. R., Rhee, L. M., Peralta, E. G. 1987. *Nucleic Acids Res.* 15:3636
187. Chung, F. Z., Lentes, K. U., Gocayne, J., FitzGerald, M., Robinson, D., et al. 1987. *FEBS Lett.* 211:200–6
188. Emorine, L. J., Marullo, S., Delavier-Klutchko, C., Kaveri, S. V., Durieu-Trautmann, O., Strosberg, A. D. 1987. *Proc. Natl. Acad. Sci. USA* 84:6995–99
189. Allen, J. M., Baetge, E. E., Abrass, I. B., Palmiter, R. D. 1988. *EMBO J.* 7:133–38
190. Gocayne, J., Robinson, D. A., FitzGerald, M. G., Chung, F. Z., Kerlavage, A. R., et al. 1987. *Proc. Natl. Acad. Sci. USA* 84:8296–300
191. Buckland, P. R., Hill, R. M., Tidmarsh, S. F., McGuffin, P. 1990. *Nucleic Acids Res.* 18:682
192. Voigt, M. M., Kispert, J., Chin, H. M. 1990. *Nucleic Acids Res.* 18:1053
193. Fraser, C. M., Arakawa, S., McCombie, W. R., Venter, J. C. 1989. *J. Biol. Chem.* 264:11754–61
193a. Chalberg, S. C., Duda, T., Rhine, J. A., Sharma, R. K. 1990. *Mol. Cell. Biochem.* 97:161–72
193b. Guyer, C. A., Horstman, D. A., Wilson, A. L., Clark, J. D., Cragoe, E. J. Jr., Limbird, L. E. 1990. *J. Biol. Chem.* 265:17307–17
194. Zeng, D. W., Harrison, J. K., D'Angelo, D. D., Barber, C. M., Tucker, A. L., et al. 1990. *Proc. Natl. Acad. Sci. USA* 87:3102–6
195. Kobilka, B. K., Frielle, T., Collins, S., Yang-Feng, T., Kobilka, T. S., et al. 1987. *Nature* 329:75–79
196. Fargin, A., Raymond, J. R., Lohse, M. J., Kobilka, B. K., Caron, M. G., Lefkowitz, R. J. 1988. *Nature* 335:358–60
197. Albert, P. R., Zhou, Q. Y., Van Tol, H. H. M., Bunzow, J. R., Civelli, O. 1990. *J. Biol. Chem.* 265:5825–32

198. Julius, D., MacDermott, A. B., Axel, R., Jessell, T. M. 1988. *Science* 241: 558–64
199. Pritchett, D. B., Bach, A. W., Wozny, M., Taleb, O., Dal Toso, R., et al. 1988. *EMBO J.* 7:4135–40
200. Julius, D., Huang, K. N., Livelli, T. J., Axel, R., Jessell, T. M. 1990. *Proc. Natl. Acad. Sci. USA* 87:928–32
201. Kubo, T., Fukuda, K., Mikami, A., Maeda, A., Takahashi, H., et al. 1986. *Nature* 323:411–16
202. Peralta, E. G., Ashkenazi, A., Winslow, J. W., Smith, D. H., Ramachandran, J., Capon, D. J. 1987. *EMBO J.* 6:3923–29
203. Allard, W. J., Sigal, I. S., Dixon, R. A. F. 1987. *Nucleic Acids Res.* 15:10604
204. Bonner, T. I., Buckley, N. J., Young, A. C., Brann, M. R. 1987. *Science* 237:527–32
205. Shapiro, R. A., Scherer, N. M., Habecker, B. A., Subers, E. M., Nathanson, N. M. 1988. *J. Biol. Chem.* 263:18397–403
206. Kubo, T., Maeda, A., Sugimoto, K., Akiba, I., Mikami, A., et al. 1986. *FEBS Lett.* 209:367–72
207. Peralta, E. G., Winslow, J. W., Peterson, G. L., Smith, D. H., Ashkenazi, A., et al. 1987. *Science* 236:600–5
208. Braun, T., Schofield, P. R., Shivers, B. D., Pritchett, D. B., Seeburg, P. H. 1987. *Biochem. Biophys. Res. Commun.* 149:125–32
209. Akiba, I., Kubo, T., Maeda, A., Bujo, H., Nakai, J., et al. 1988. *FEBS Lett.* 235:257–61
210. Bonner, T. I., Young, A. C., Brann, M. R., Buckley, N. J. 1988. *Neuron* 1:403–10
211. Liao, C. F., Themmen, A. P. N., Joho, R., Barberis, C., Birnbaumer, M., Birnbaumer, L. 1989. *J. Biol. Chem.* 264:7328–37
212. Dearry, A., Gingrich, J. A., Falardeau, P., Fremeau, R. T. Jr., Bates, M. D., Caron, M. G. 1990. *Nature* 347:72–75
213. Sunahara, R. K., Niznik, H. B., Weiner, D. M., Stormann, T. M., Brann, M. R., et al. 1990. *Nature* 347:80–82
214. Zhou, Q.-Y., Grandy, D. K., Thambi, L., Kushner, J. A., Van Tol, H. H. M., et al. 1990. *Nature* 347:76–80
214a. Monsma, F. J., Mahan, L. C., McVittie, L. D., Gerfen, C. R., et al. 1990. *Proc. Natl. Acad. Sci. USA* 87:6723–27
214b. Monsma, F. J., Mahan, L. C., McVittie, L. D., Gerfen, C. R., Sibley, D. R. 1990. *Proc. Natl. Acad. Sci. USA* 87:6723–27
215. O'Dowd, B. F., Nguyen, T., Tirosk, A., Jorvie, K. R., Israel, Y., et al. 1990. *FEBS Lett.* 347:8–12
216. Bunzow, J. R., Van Tol, H. H. M., Grandy, D. K., Albert, P., Salon, J., et al. 1988. *Nature* 336:783–87
217. Todd, R. D., Khurana, T. S., Sajovic, P., Stone, K. R., O'Malley, K. L. 1989. *Proc. Natl. Acad. Sci. USA* 86:10134–38
217a. Stormann, T. M., Gdula, D. C., Weiner, D. M., Brann, M. R. 1990. *Mol. Pharmacol.* 37:1–6
218. Grandy, D. K., Marchionni, M. A., Makam, H., Stofko, R. E., Alfano, M., et al. 1989. *Proc. Natl. Acad. Sci. USA* 86:9762–66
219. Monsma, F. J. Jr., McVittie, L. D., Gerfen, C. R., Mahan, L. C., Sibley, D. R. 1989. *Nature* 342:926–29
220. Giros, B., Sokoloff, P., Martres, M. P., Riou, J. F., Emorine, L. J., Schwartz, J. C. 1989. *Nature* 342:923–26
221. Eidne, K. A., Taylor, P. L., Zabavnik, J., Saunders, P. T., Inglis, J. D. 1989. *Nature* 342:865
222. Chio, C. L., Hess, G. F., Graham, R. S., Huff, R. M. 1990. *Nature* 343:266–69
223. Dal Toso, R., Sommer, B., Ewert, M., Herb, A., Pritchett, D. B., et al. 1989. *EMBO J.* 8:4025–34
224. O'Malley, K. L., Mack, K. J., Gandelman, K. Y., Todd, R. D. 1990. *Biochemistry* 29:1367–71
225. Rao, D. D., McKelvy, J., Kebabian, J., MacKenzie, R. G. 1990. *FEBS Lett.* 263:18–22
226. Miller, J. C., Wang, Y., Filer, D. 1990. *Biochem. Biophys. Res. Commun.* 166:109–12
227. Sokoloff, P., Giros, B., Marties, M. P., Boutheret, M. C., Schwartz, J. C. 1990. *Nature* 347:146–51
228. Masu, Y., Nakayama, K., Tamaki, H., Harada, Y., Kuno, M., Nakanishi, S. 1987. *Nature* 329:836–38
229. Sasai, Y., Nakanishi, S. 1989. *Biochem. Biophys. Res. Commun.* 165:695–702
229a. Gerard, N. P., Eddy, R. L., Shows, T. B., Gerard, C. 1990. *J. Biol. Chem.* 265:20455–62
230. Shigemoto, R., Yokota, Y., Tsuchida, K., Nakanishi, S. 1990. *J. Biol. Chem.* 265:623–28
231. Yokota, Y., Sasai, Y., Tanaka, K., Fujiwara, T., Tsuchida, K., et al. 1989. *J. Biol. Chem.* 264:17649–52
232. Hershey, A. D., Krause, J. E. 1990. *Science* 247:958–62
232a. Thomas, K. M., Pyun, H. Y., Navar-

688    DOHLMAN ET AL

ro, J. 1990. *J. Biol. Chem.* 265:20061–64

233. Parmentier, M., Libert, F., Maenhaut, C., Lefort, A., Gerard, C., et al. 1989. *Science* 246:1620–22

234. Libert, F., Parmentier, M., Maenhaut, C., Lefort, A., Gerard, C., et al. 1990. *Mol. Cell. Endocrinol.* 68:R15–17

235. Libert, F., Lefort, A., Gerard, C., Parmentier, M., Perret, J., et al. 1989. *Biochem. Biophys. Res. Commun.* 165:1250–55

235a. Misrahi, M., Loosfelt, H., Atger, M., Sar, S., Guiochon-Mantel, A., Milgrom, E. 1990. *Biochem. Biophys. Res. Commun.* 166:394–403

236. Nagayama, Y., Kaufman, K. D., Seto, P., Rapoport, B. 1989. *Biochem. Biophys. Res. Commun.* 165:1184–90

236a. Akamizu, T., Ikuyama, S., Saji, M., Kosugi, S., Kozak, C., et al. 1990. *Proc. Natl. Acad. Sci. USA* 87:5677–81

237. McFarland, K. C., Sprengel, R., Phillips, H. S., Kohler, M., Rosemblit, N., et al. 1989. *Science* 245:494–99

238. Loosfelt, H., Misrahi, M., Atger, M., Salesse, R., Vu, M. T., et al. 1989. *Science* 245:525–28

238a. Minegish, T., Nakamura, K., Takakura, Y., Miyamoto, K., Hasegawa, Y., et al. 1990. *Biochem. Biophys. Res. Commun.* 172:1049–54

238b. Arai, H., Hori, S., Aramori, I., Ohkubo, H., Nakanishi, S. 1990. *Nature* 348:730–32

238c. Sakurai, T., Yanagisawa, M., Takuwa, Y., Miyazaki, H., Kimura, S., et al. 1990. *Nature* 348:732–35

239. Young, D., O'Neill, K., Jessell, T., Wigler, M. 1988. *Proc. Natl. Acad. Sci. USA* 85:5339–42

240. Nathans, J., Hogness, D. S. 1984. *Proc. Natl. Acad. Sci. USA* 81:4851–55

241. Baehr, W., Falk, J. D., Bugra, K., Triantafyllos, J. T., McGinnis, J. F. 1988. *FEBS Lett.* 238:253–56

242. Matsuda, L. A., Lolait, S. J., Brownstein, M. J., Young, A. C., Bonner, T. I. 1990. *Nature* 346:561–64

243. Libert, F., Parmentier, M., Lefort, A., Dinsart, C., Van Sande, J., et al. 1989. *Science* 244:569–72

244. Hla, T., Maciag, T. 1990. *J. Biol. Chem.* 265:9308–13

245. Ross, P. C., Figler, R. A., Corjay, M. H., Barber, C. M., Adam, N., et al. 1990. *Proc. Natl. Acad. Sci. USA* 87:3052–56

245a. Witz, P., Amlaiky, N., Plassat, J. L., Maroteaux, L., Borrelli, E., Hen, R. 1990. *Proc. Natl. Acad. Sci. USA* 87:8940–44

246. Tietje, K. M., Goldman, P. S., Nathanson, N. M. 1990. *J. Biol. Chem.* 265:2828–34

247. Shapiro, R. A., Wakimoto, B. T., Subers, E. M., Nathanson, N. M. 1989. *Proc. Natl. Acad. Sci. USA* 86:9039–43

248. Onai, T., FitzGerald, M. G., Arakawa, S., Gocayne, J. D., Urquhart, D. A., et al. 1989. *FEBS Lett.* 255:219–25

249. O'Tousa, J. E., Baehr, W., Martin, R. L., Hirsh, J., Pak, W. L., Applebury, M. L. 1985. *Cell* 40:839–50

250. Zuker, C. S., Cowman, A. F., Rubin, G. M. 1985. *Cell* 40:851–58

251. Cowman, A. F., Zuker, C. S., Rubin, G. M. 1986. *Cell* 44:705–10

252. Zuker, C. S., Montell, C., Jones, K., Laverty, T., Rubin, G. M. 1987. *J. Neurosci.* 7:1550–57

253. Fryxell, K. J., Meyerowitz, E. M. 1987. *EMBO J.* 6:443–51

254. Montell, C., Jones, K., Zuker, C., Rubin, G. 1987. *J. Neurosci.* 7:1558–66

255. Ovchinnikov, Yu. A., Abdulaev, N. G., Zolotarev, A. S., Artamonov, I. D., Bespalov, I. A., et al. 1988. *FEBS Lett.* 232:69–72

256. Takao, M., Yasui, A., Tokunaga, F. 1988. *Vision Res.* 28:471–80

257. Arakawa, S., Gocayne, J. D., McCombie, W. R., Urquhart, D. A., Hall, L. M., et al. 1990. *Neuron* 4:343–54

258. Klein, P. S., Sun, T. J., Saxe, C. L. III, Kimmel, A. R., Johnson, R. L., Devreotes, P. N. 1988. *Science* 241:1467–72

Annu. Rev Biochem. 1991. 60:689–715

# RNA POLYMERASE II

*Richard A. Young*

Whitehead Institute for Biomedical Research, Nine Cambridge Center, Cambridge, Massachusetts 02142

and

Department of Biology, Massachusetts Institute of Technology, Cambridge, Massachusetts 02139

KEY WORDS:    transcription, mRNA synthesis.

## CONTENTS

0066-4154/91/0701-0689$02.00

## PERSPECTIVES AND SUMMARY

The control of transcription in eukaryotes requires a greater repertoire of proteins than it does in prokaryotes. In prokaryotes, a single core RNA polymerase enzyme composed of three subunits is responsible for all transcription. Selective promotor recognition and other regulatory controls are conferred by the association of one or more additional polypeptides with the core polymerase (1–8). By contrast, eukaryotic cells contain three nuclear DNA–dependent RNA polymerases that transcribe different sets of genes. RNA polymerase I (or A) synthesizes rRNA precursors, RNA polymerase II (or B) transcribes pre-mRNA, and RNA polymerase III (or C) transcribes 5S rRNA and tRNA genes. The three eukaryotic nuclear RNA polymerases are each composed of 8–14 polypeptides. Each of these enzymes requires the aid of a variety of additional factors for selective promoter recognition and regulated transcription initiation (9–16). The relative number and complexity of the eukaryotic RNA polymerases and their transcription factors probably reflects the need for more elaborate controls on transcription in eukaryotes.

Dramatic progress has been made during the past decade in our understanding of the components of the mRNA transcription apparatus that are important for regulated transcription initiation. The picture that has developed, in outline, is one in which RNA polymerase II associates with a set of TATA-associated general transcription factors at the promoter; gene-specific transcription factors bind to upstream elements (enhancers and Upstream Activating Sequences) and influence the rate of transcription initiation by interacting with components of the TATA-associated transcription complex. The TATA-associated transcription complex has at least five components, TFIIA, TFIIB, TFIID, RNA polymerase II, and TFIIE (17, 18, reviewed in 19), and is competent to initiate transcription accurately in vitro in the absence of regulatory factors. Upstream promoter elements and the transcription factors that bind them are largely responsible for the regulation of transcription initiation (10–15). How signals to begin RNA synthesis are transmitted to RNA polymerase II is not yet clear, despite some recent clues (20-27).

Interest in the dialogue between transcription factors and RNA polymerase II has stimulated renewed interest in the polymerase itself. Several excellent reviews address the subject of eukaryotic nuclear RNA polymerases (19, 28–32). However, considerable new information has come from recent molecular genetic analysis of RNA polymerase II subunit genes isolated from a variety of eukaryotic organisms. This is especially true for yeast, where the isolation of genes encoding all 11 RNA polymerase II subunits is now complete. In addition, new information has recently emerged on the function of the unusual carboxy terminal repeat domain (CTD) of the large subunit.

The conservation of many features of eukaryotic RNA polymerase II subunit structure and function makes the yeast enzyme a good model for eukaryotic RNA polymerase II. The picture of RNA polymerase II that has emerged from molecular genetic analysis of the 11 *S. cerevisiae* RNA polymerase II subunits is reviewed herein. Structural and functional studies of yeast RNA polymerase II genes and their products have revealed that the three largest subunits are related to the prokaryotic core RNA polymerase subunits and are largely responsible for RNA catalysis. Three smaller subunits are essential components shared by all three nuclear RNA polymerases. Two of the remaining small subunits form a dissociable subcomplex that influences the efficiency of transcription initiation in vitro. The structure and function of the RNA polymerase II large subunit CTD is also reviewed here. Recent evidence indicates that the CTD is involved in the regulation of transcription initiation.

## DEFINING THE ENZYME

### RNA Polymerase Purification

Three types of DNA-dependent RNA polymerases are found in the nucleus of all eukaryotic cells, each responsible for the synthesis of different classes of RNA. Roeder & Rutter (33–35) first separated the three mammalian enzymes by DEAE-Sephadex chromatography, and named them I, II, and III according to their order of elution by increasing concentrations of ammonium sulfate. RNA polymerase I is responsible for transcription of rDNA, RNA polymerase II transcribes protein-coding genes, and RNA polymerase III transcribes the genes for small stable RNAs such as tRNA and 5S rRNA. RNA polymerases I, II, and III are sometimes also called RNA polymerases A, B, and C, respectively (28, 29).

RNA polymerase II has been studied somewhat more extensively than the other two nuclear RNA polymerases. Column chromatography has been used to purify RNA polymerase II from a large number of eukaryotes, including *Saccharomyces* (32), *Neurospora* (36), *Acanthamoeba* (37), *Dictyostelium* (38), *Physarum* (39), *Caenorhabditis* (40), *Drosophila* (41), *Xenopus* (42), mouse (43), calf thymus (44), human (45), and a variety of plants (46–50). Descriptions of the standard purification methods can be found in Sentenac (32), and modifications that effectively minimize proteolysis have been reported (44, 51). Immunological methods have been developed recently that permit rapid purification of enzyme that is highly active in transcription assays in vitro (52–54).

Multiple chromatographic and electrophoretic forms of RNA polymerase II are detected in purified preparations of the enzyme. These various forms generally differ in the integrity of the CTD and in the extent to which the CTD is phosphorylated (55–57). The RNA polymerase II large subunit CTD is discussed in detail in a later section.

## General Properties

The enzymatic properties of the RNA polymerases have been reviewed extensively (19, 29, 31, 32). Purified RNA polymerase II is capable of template-dependent synthesis of RNA in the absence of additional factors, but is not capable of precise selective initiation at promoters. Purified RNA polymerase II can be directed to "initiate" transcription at specific sites by using oligonucleotide primers, and the ability of the enzyme to extend the 3' ends of primers has been exploited to investigate its intrinsic elongation and termination properties (reviewed in 19).

The degree to which RNA polymerase II itself contributes to DNA binding, selection of the precise start site, and the response to gene-specific transcription factors is not yet clear. However, certain mutations in RNA polymerase II have been observed to affect transcription in a gene-specific manner (25, 58, 59), and other mutations have been found that alter the precise transcription start site (D. Hekmatpanah, R. A. Young, unpublished). These results suggest that RNA polymerase II does not play an entirely passive role in selective initiation.

RNA polymerase II alone can catalyze RNA chain elongation. The average elongation rate in vitro has been estimated to be between 60 and 600 nucleotides per minute (60–62). The rate of elongation can be influenced significantly by transcriptional pausing (61). Pausing by RNA polymerase II appears to be involved in transcription termination (63). Additional factors may have a role in regulating the rate of pausing, and thus the rate of elongation and termination.

## The Subunit Problem

Eukaryotic RNA polymerases have traditionally been defined as the set of proteins that copurify with transcriptional activity in assays of nonspecific initiation and elongation with heterologous templates. The prokaryotic RNA polymerases, composed of three core subunits ($\beta'$, $\beta$, and $\alpha$) and a specificity factor ($\sigma$), have been defined by reconstitution from purified subunits and by genetic analysis (1–3). The eukaryotic enzymes have not yet been reconstituted from purified subunits, and it is not clear whether all of the polypeptides associated with the purified enzymes are genuine subunits, as opposed to adventitiously associated polypeptides. However, molecular genetic studies of *Saccharomyces* RNA polymerase II subunit structure and function, described below, indicate that most of the polypeptides associated with the conventionally purified enzyme have a role in transcription, and are by this criterion genuine components of the transcription apparatus.

## Molecular Architecture of RNA Polymerase II

Three features of RNA polymerase II are conserved in eukaryotes (Figure 1). First, RNA polymerase II is generally composed of 10 ± 2 subunits. Second, the enzyme always contains two large subunits with molecular sizes of approximately 220 and 140 kDa. Finally, RNA polymerase II contains three subunits of 14–28 kDa that are also found associated with RNA polymerases I and III in all eukaryotes in which all three nuclear enzymes have been studied. These polypeptides are called "common" or "shared" subunits.

The information presented in Figure 1 may exaggerate differences in the actual number and size of the smaller RNA polymerase II subunits found in different eukaryotic organisms. The subunit compositions of the RNA polymerases were ascertained in independent laboratories, using different gel electrophoresis methods and size markers. In addition, efforts to visualize the entire range of subunit sizes on SDS-PAGE sometimes resulted in poor resolution of the smaller subunits. As is discussed in more detail below, it is likely that most of the small subunits of yeast RNA polymerase II have conserved counterparts in the other RNA polymerase II enzymes.

*S. cerevisiae* RNA polymerases have been more intensively studied than other eukaryotic RNA polymerases, in part because of the relative ease of genetic and biochemical investigation. The conservation of many features of eukaryotic RNA polymerase II subunit structure and function indicate that the yeast enzyme is a good prototype for higher eukaryotic RNA polymerase II, and the ability to combine powerful molecular genetic and biochemical tools with *Saccharomyces* greatly facilitates investigation of the molecular mechanisms involved in gene expression.

## YEAST RNA POLYMERASE II

*Saccharomyces* RNA polymerase II is composed of 11 polypeptides with apparent molecular weights that range from 220 to 10 kDa. RNA polymerase II purified by conventional techniques (32) or by immunoprecipitation with monoclonal antibodies directed against specific subunits (53, 54) contains 10 polypeptides that are readily resolved by SDS-PAGE, and an 11th protein that comigrates with the 9th largest (12-kDa) subunit. A good deal is now known about this enzyme, both from extensive biochemical characterization and from molecular genetic analysis of RNA polymerase II subunit genes.

### Eleven Unique Subunit Genes

The genes that encode all 11 yeast RNA polymerase II subunits have been isolated and sequenced (55, 64–72; N. Woychik and R. Young, unpublished). These genes, *RPB1–RPB11,* were isolated by screening recombinant DNA libraries of *S. cerevisiae* genomic DNA with a *Drosophila* large subunit

# RNA POLYMERASE II

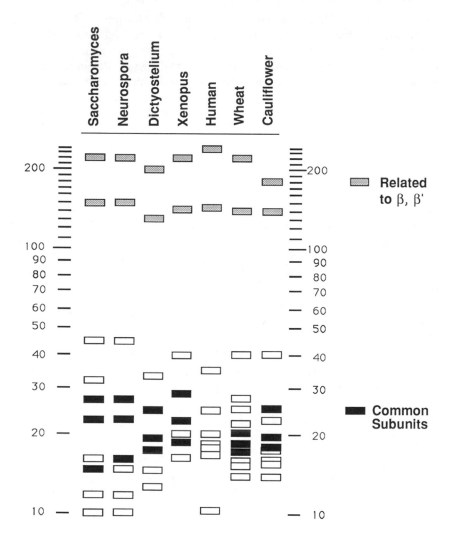

*Figure 1* Comparison of RNA polymerase II subunit compositions among selected eukaryotes. Evidence that the two largest subunits are related to the *Escherichia coli* RNA polymerase subunits β and β' comes from a combination of sequence analysis (in the case of *Saccharomyces*) and antibody cross-reactivity (see text for details). Molecular genetic studies and antigenic analysis have demonstrated that three *Saccharomyces* RNA polymerase II subunits are common to the three nuclear RNA polymerases (32, 70). RNA polymerase II enzymes from *Neurospora*, *Dictyostelium*, *Xenopus*, cauliflower, and wheat contain three subunits of 14–28 kDa that have counterparts of identical SDS gel mobility in RNA polymerases I and III. Subunit information was from Refs. 32 *(Saccharomyces)*, 36 *(Neurospora)*, 38 *(Dictyostelium)*, 42 *(Xenopus)*, 45 (human), and 49 (wheat and cauliflower).

cDNA (73), subunit-specific antibodies (74, 75), and oligonucleotide probes whose sequence was predicted from amino acid sequences of individual subunits. Each of the 11 RNA polymerase II subunit genes occurs in single copy in the haploid genome and, with one possible exception, all are necessary for yeast cells to grow at wild-type rates and for survival throughout the normal temperature range for cell growth (Table 1).

## Subunit Features and Functions

Clues to the functions of the subunits have been obtained from sequence similarities with prokaryotic relatives, from studies of DNA and nucleotide binding in vitro, and from the analysis of mutant defects. The yeast RNA polymerase II subunits can be grouped into three general types that are summarized in Figure 2. The three largest subunits are related to prokaryotic core RNA polymerase components and are primarily responsible for RNA catalysis, three smaller subunits are shared with RNA polymerases I and III, and two other small subunits form a subcomplex that tends to dissociate from the enzyme and may have a role in initiation. The three smallest subunits so far lack any known function.

THE "CORE" SUBUNITS: RPB1, RPB2, AND RPB3    The two largest yeast RNA polymerase II subunits, RPB1 and RPB2, are clearly related to the two largest subunits of *E. coli* RNA polymerase, $\beta'$ and $\beta$, respectively (55, 67) (Table 2). The sequence similarity occurs in 8–9 segments that span the two large subunits (67, 76, reviewed in 77) (Figure 3). The two large subunits of yeast RNA polymerase I and III are also homologues of the two large *E. coli* subunits (55, 78). In every case where the two large RNA polymerase

**Table 1**    Yeast RNA polymerase II subunit genes

| Gene | SDS-PAGE mobility $(M_r \times 10^{-3})$ | Gene copy number | Chromosomal location | Deletion viability | Refs. |
|------|------|------|------|------|------|
| *RPB1*[a] | 220 | 1 | IV[b] | inviable | 55, 64, 65 |
| *RPB2* | 150 | 1 | XV[b] | inviable | 67 |
| *RPB3* | 45 | 1 | IX[b] | inviable | 68 |
| *RPB4* | 32 | 1 | X | conditional | 69 |
| *RPB5* | 27 | 1 | II | inviable | 70 |
| *RPB6* | 23 | 1 | XVI | inviable | 70 |
| *RPB7* | 17 | 1 | | | |
| *RPB8* | 14 | 1 | XV | inviable | 70 |
| *RPB9* | 13 | 1 | VII | conditional | 72 |
| *RPB10* | 10 | 1 | XV | inviable | 71 |
| *RPB11* | 13 | 1 | XV[b] | inviable | |

[a] Also known as *RPO21, RPB220*
[b] Precise map position determined

# YEAST RNA POLYMERASE II

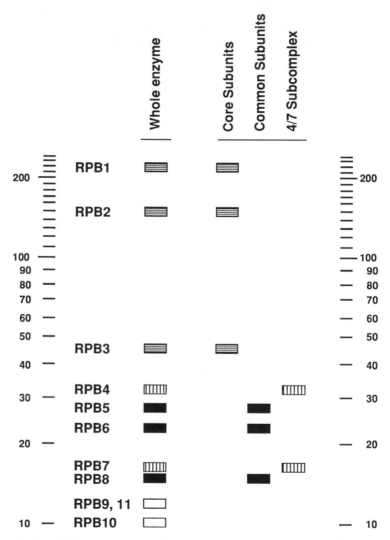

*Figure 2* Yeast RNA polymerase II subunits can be grouped into three general types. Core subunits are relatives of the *E. coli* RNA polymerase core subunits $\beta'$, $\beta$, and $\alpha$. Common subunits are proteins associated with all three nuclear RNA polymerases. THe RPB4/RPB7 subcomplex can readily dissociate from the purified enzyme, and appears to influence the efficiency of transcription initiation. RPB9 and RPB11 frequently comigrate on SDS-PAGE.

**Table 2**  Yeast RNA polymerase II subunits

| Subunit | SDS-PAGE mobility ($M_r \times 10^{-3}$) | Protein mass ($M_r \times 10^{-3}$) | Stoichiometry | Modifications | Prokaryotic homologue | Notable sequences |
|---|---|---|---|---|---|---|
| RPB1 | 220 | 190 | 1.1 | phosphate | $\beta'$ | carboxy terminal (YSPTSPS)$_{27}$<br>$C-X_2-C-X_6-C-X_2-H-X_2-H-X_{23}-C-X_2-C$ |
| RPB2 | 150 | 140 | 1.0 | phosphate | $\beta$ | $C-X_2-C-X_{15}-C-X_2-C$ |
| RPB3 | 45 | 35 | 2.1 | | $(\alpha)$[a] | |
| RPB4 | 32 | 25 | 0.5 | | | |
| RPB5 | 27 | 25 | 2.0 | | | |
| RPB6 | 23 | 18 | 0.9 | phosphate | | |
| RPB7 | 17 | — | 0.5 | | | |
| RPB8 | 14 | 17 | 0.8 | | | $C-X_2-C-X_{18}-C-X_2-C$ |
| RPB9 | 13 | 14 | 2.0 | | | $C-X_2-C-X_{24}-C-X_2-C$ |
| RPB10 | 10 | 5.4 | 0.9 | | | $C-X_2-C-G$ |
| RPB11 | 13 | 14 | — | | | |

[a] limited sequence similarity

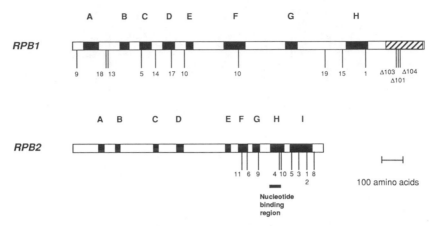

*Figure 3*  Segments of RPB1 and RPB2 that are similar in sequence to the prokaryotic RNA polymerase subunits β' and β, respectively. The eight regions of substantial homology between RPB1 protein and the β' subunit of *E. coli* RNA polymerase (76, 77) and the nine regions conserved between RPB2 and the β subunit of *E. coli* RNA polymerase (67, 77) are shown as black boxes. The box with diagonal lines represents the heptapeptide repeat domain. The position of *RPB1* and *RPB2* conditional mutations that affect RNA polymerase II function or assembly are also indicated; these mutations generally affect amino acid residues that are invariant among homologous subunits from other eukaryotes (148). A portion of the RPB2 subunit that is involved in nucleotide binding has been mapped to region H (98, 99).

subunits of eukaryotes other than yeast have been investigated, whether by immunological cross-reactivity (75, 79, 80), or by gene isolation and sequence analysis (76, 81–86), the two large subunits are closely related to the corresponding subunits of yeast RNA polymerase II.

The two large subunits of prokaryotic and eukaryotic RNA polymerases are functional homologues. Each of the two large subunits may contain portions of the "catalytic site" for RNA synthesis. The *E. coli* RNA polymerase β' subunit alone can bind to DNA, suggesting that this subunit is directly involved in the interaction of RNA polymerase with template DNA (2, 3, 87). The largest subunit of eukaryotic RNA polymerase II also appears to be involved in binding DNA, as suggested by blotting and UV crosslinking experiments (88–90) and the ability of monoclonal antibodies to inhibit binding (91). Template DNA and the nascent RNA chain can be UV crosslinked to both of the largest eukaryotic subunits (88, 92); the nascent RNA chain is also crosslinked to the two largest prokaryotic subunits (93). The second largest subunit of *E. coli* RNA polymerase (β) and its counterpart in eukaryotic RNA polymerases can be tagged with nucleoside triphosphate substrates (94–99), and a nucleotide affinity label has been mapped to the highly conserved homology region H of yeast RPB2 (98, 99) (Figure 3). Highly conserved lysine and histidine residues in this region may be involved in nucleotide binding, and this region thus appears to contain a portion of the catalytic site for RNA synthesis (98, 99).

Several features of RPB3 (68), the third largest yeast RNA polymerase II subunit, and the *E. coli* RNA polymerase $\alpha$ subunit (100) suggest that the two polypeptides may be partial homologues. The two subunits are similar in size (68, 100), in sequence (over a portion of the molecule), in stoichiometry (both polypeptides appear to exist in two copies in the RNA polymerase molecule) (2, 54), and in their apparent roles in assembly (mutations in both subunits result in the accumulation of similar subunit subcomplexes) (2, 101). RPB3 and $\alpha$ are 318 and 330 amino acids long, respectively. When the two proteins are aligned, there is a 25-amino-acid long segment [residues 29–48 in RPB3 (68)] in which 32% of the residues are identical and all but three residues are conserved; other aligned segments of the two proteins are also conserved, albeit less strikingly. Subcomplexes of RNA polymerase II subunits that accumulate in RPB1, RPB2, and RPB3 assembly mutants correspond to those found in $\beta'$, $\beta$, and $\alpha$ assembly mutants (2, 101). The resemblance in size and stoichiometry, the presence of limited amino acid sequence similarity, and the shared assembly mutant phenotypes together suggest that RPB3 and $\alpha$ may play similar roles in eukaryotic and prokaryotic RNA polymerases. One of these roles may be to nucleate assembly of the enzyme.

The third largest subunit of human RNA polymerase II (Figure 1) is a homologue of the yeast RPB3 protein (102). Analysis of a cDNA clone encoding the human subunit (hRPB33) revealed that nearly half of the amino acid residues of human and yeast RPB3 subunits are identical.

Although the three large eukaryotic RNA polymerase II subunits share many features with the three prokaryotic core subunits, the RPB1 subunit has an unusual feature that is not shared with its prokaryotic homologue. RPB1 contains a CTD that consists of multiple heptapeptide repeats of the consensus sequence Tyr-Ser-Pro-Thr-Ser-Pro-Ser (55, 56, 76, 81, 86, reviewed in 57). The CTD is essential for cell viability (103 106), is highly phosphorylated in a substantial portion of the RNA polymerase II molecules in the cell (51, 54, 107), and appears to be involved in the response to transcription activation at some promoters (24, 25). The CTD is discussed in more detail below.

THE COMMON SUBUNITS: RPB5, RPB6, AND RPB8    RPB5, RPB6, and RPB8 are essential components of all three nuclear RNA polymerases (70). The RPB5, RPB6, and RPB8 proteins, with apparent molecular sizes deduced from SDS-PAGE of 27, 23, and 14.5 kDa (32), have molecular sizes predicted from gene sequences of 25, 18, and 16 kDa, respectively (70) (Table 2). The aberrant mobility of the RPB6 protein is apparently due to the fact that it is phosphorylated (54, 108). Whether isolated from RNA polymerase I, II, or III, the subunits RPB5, RPB6, and RPB8 are indistinguishable by SDS-polyacrylamide gel mobility, fingerprint patterns (109, 110), isoelectric point (109), and antigenic recognition (74, 75, 111, 112). Although their sequences have not provided clues to the functions of the common subunits, these proteins could be involved in nuclear localization, transcriptional efficiency, or the coordinate regulation of rRNA, mRNA, and tRNA synthesis. A role in DNA binding has been suggested for RPB6, but the evidence for this is indirect (74, 112). Conditional mutations already constructed in two of the common subunit genes may help elucidate the role of these subunits in RNA polymerases I, II, and III (112a; N. Woychik, S.-M. Liao, R. A. Young, unpublished).

RNA polymerase II contains three subunits of 14–28 kDa that comigrate with subunits associated with RNA polymerases I and III in all eukaryotes in which all three nuclear enzymes have been carefully studied (36–38, 42, 49, 70) (Figure 1). Antibodies have been used to confirm that, in plants, the three common subunits are in fact identical in RNA polymerases I, II, and III (49).

THE RPB4/RPB7 SUBCOMPLEX    The RPB4 and RPB7 proteins, with apparent molecular sizes deduced from SDS-PAGE of 32 and 16 kDa (32), have molecular sizes predicted from gene sequences of 25 and 18 kDa, respectively (69; N. Woychik and R. Young, unpublished) (Table 2). These two proteins appear to form a subcomplex within RNA polymerase II that can dissociate from purified RNA polymerase II under partially denaturing conditions (113–115). The RPB4 and RPB7 subunits are not essential for mRNA synthesis in vivo, as cells lacking either or both of these proteins are viable, at least at moderate temperatures (69). RPB4-deficient cells grow slowly and are temperature sensitive, suggesting that RNA polymerase II requires the RPB4 subunit for maximal efficiency (69). Both RPB4 and RPB7 are missing from RNA polymerase II immunoprecipitated from cells lacking RPB4 (54, 115), suggesting that RPB4 provides a crucial link between the RPB4/RPB7 complex and the rest of the polymerase subunits.

Purified RNA polymerase II lacking the RPB4/RPB7 subcomplex is indistinguishable from the normal enzyme in promoter-independent initiation/ chain elongation activity in vitro (115). However, enzyme lacking the RPB4/ RPB7 subcomplex exhibits only limited activity in Gal4-VP16-stimulated, promoter-directed transcription initiation assays in vitro. The addition of equimolar amounts of purified RPB4/RPB7 subcomplex to enzyme lacking these subunits substantially increases the levels of selective initiation activity (115). These results suggest that the RPB4/RPB7 subcomplex influences the efficiency of selective transcription initiation.

The central 102 amino acids in RPB4 are 30% identical to residues in a portion of the prokaryotic RNA polymerase $\sigma^{70}$ subunit (69). The function of this portion of $\sigma^{70}$ is not yet defined and so the implications of this sequence similarity are not yet clear.

THE THREE SMALL SUBUNITS: RPB9–RPB11 The RPB9 subunit has an apparent molecular size deduced from SDS-PAGE of 12 kDa (32) and a molecular size predicted from gene sequences of 14 kDa (72) (Table 2). RPB9 is not necessary for mRNA catalysis but is essential for cell growth at temperature extremes (72), and probably has a role in fine tuning the efficiency of the transcription apparatus. RPB9 is unusual among RNA polymerase II subunits in that it contains two cysteine repeat motifs, $C-X_2-C-X_{18}-C-X_2-C$ and $C-X_2-C-X_{24}-C-X_2-C$, which may bind some of the zinc that is associated with the enzyme (31, 116, 117).

The RPB10 subunit has an apparent molecular size deduced from SDS-PAGE of 10 kDa (32) and a predicted molecular size of 5.4 kDa (71) (Table 2). RPB10 is vital for the yeast cell despite contributing only 46 amino acid residues to an enzyme that consists of a total of 4329 residues. Antigenic evidence suggests that RPB10 may be a fourth common subunit; however, the RPB10 protein used to prepare antibodies was not characterized for its purity, and this data is difficult to interpret (112). The functional contribution of RPB10 to the transcription apparatus is not yet understood.

The observation that the 12.5-kDa SDS-PAGE band can occasionally be resolved as a doublet (53) led to the discovery that RPB9 comigrates with a second polypeptide, and this second 12.5-kDa subunit was designated RPB11 (N. Woychik and R. Young, unpublished). RPB11 has a molecular size predicted from the gene sequence of 14 kDa. Unlike RPB9, RPB11 is essential for yeast cell viability.

## Subunit Stoichiometry and Phosphorylation

The isolation of the genes for yeast RNA polymerase II subunits has allowed investigators to confirm the subunit composition of the enzyme, to estimate the relative stoichiometry of each subunit, and to ascertain which subunits are phosphorylated. Ten polypeptides are immunoprecipitated from radiolabeled yeast cell extracts containing an epitope-tagged subunit (54); as mentioned previously, the 11th subunit (RPB11) frequently comigrates with RPB9. The 10 polypeptides that are isolated by this method are identical in size and number to those obtained for yeast RNA polymerase II purified by conventional column chromatography. Similar results are obtained when the enzyme is isolated using a monoclonal antibody directed against the CTD of the largest subunit (53).

The relative stoichiometry of the yeast RNA polymerase II subunits has been estimated by combining knowledge of the methionine content of the subunits deduced from sequence analysis and the relative extents of labeling with $^{35}$S-methionine (54, 101) (Table 2). The average RNA polymerase II molecule appears to be composed of one copy each of RPB1, 2, 6, 8, 10, and 11, and two copies of RPB3, 6, and 9. The RPB4 and 7 subunits are present at less than one copy per enzyme molecule.

The in vivo phosphorylation state of RNA polymerase II subunits has also been investigated using the immunoprecipitation approach (54). The RPB1 and RPB6 subunits are clearly phosphorylated. RPB1 appears to be present in both phosphorylated and unphosphorylated forms in vivo, with approximately half of the RNA polymerase II molecules containing one form, and half the other form. Most or all of the phosphate on RPB1 appears to be associated with the CTD (101). It is not yet clear whether the phosphorylated form of RPB1 is a single species or whether there is some heterogeneity in the degree of phosphorylation of the RPB1 CTD. Relatively low levels of phosphate can be detected on RPB2 (54); it is conceivable that this represents phosphate that is transiently bound to the subunit as an intermediate during hydrolysis of nucleoside triphosphates. Possible roles for RPB1 phosphorylation are discussed below.

## Subunit Interactions and RNA Polymerase Structure

The results of genetic studies in *Saccharomyces* and in *Drosophila* indicate that RPB1 and RPB2 interact with one another, probably at multiple sites (118, 119). In yeast, suppressors of a temperature-sensitive mutation *(rpb1-1)* in conserved region H of RPB1 (Figure 3) are found clustered in conserved region I of RPB2 (118). Conversely, suppressors of a temperature-sensitive mutation *(rpb2-2)* in region I of RPB2 are found both in region H and in two other regions of RPB1 (118). Genetic analysis of the two large subunits of *Saccharomyces* RNA polymerase I has revealed that one of the interactions between the two large subunits of all three nuclear RNA polymerases may occur through two putative zinc-binding domains that are highly conserved among RNA polymerases I, II, and III (120). One of these putative zinc-binding domains is located in conserved region A of the largest subunit, and the other immediately follows conserved region I of the second largest subunit (Figure 3). While suppressor analysis should be interpreted cautiously, it is possible that both regions A and H of the largest subunit interact with region I of the second largest subunit.

Although crystals of RNA polymerase II suitable for X-ray analysis have not yet been obtained, a low-resolution structure of yeast RNA polymerase II has been determined using electron diffraction of two-dimensional crystals (53, 121). Two-dimensional crystals of single RNA polymerase II molecules were formed on charged lipid layers; the crystals gave electron diffraction patterns extending to about 30 angstroms (53). Much greater resolution (to 16 angstroms) of RNA polymerase II in crystals was obtained with enzyme lacking the RPB4/RPB7 subcomplex (121), presumably because the RNA polymerase molecules in the original crystals were not homogeneous, as only a portion of the enzyme molecules contain the RPB4/RPB7 subcomplex. RNA polymerase II lacking the RPB4/RPB7 subcomplex has a 25 angstrom groove that bears a striking resemblance to the *E. coli* RNA polymerase groove that is thought to be the DNA-binding channel (122). Future analysis of two-dimensional crystals of RNA polymerase containing modified subunits may reveal the positions of individual subunits in the enzyme, and may thus provide additional clues to subunit functions.

# HIGHER EUKARYOTIC RNA POLYMERASE II

## Conserved Features of Subunit Architecture

RNA polymerase II enzymes purified from higher eukaryotes appear to be closely related to the *S. cerevisiae* enzyme. Purified by conventional column chromatography, the eukaryotic enzymes are composed of 10 ± 2 subunits (Figure 1). The two largest subunits have molecular sizes of approximately 220 and 140 kDa and, where sequences are available, the two large subunits are always closely related to the yeast RPB1 and RPB2 subunits. In addition, where the subunit composition of all three nuclear RNA polymerases has been investigated, three subunits of molecular sizes 14–28 kDa appear to be shared by all three enzymes.

## Highly Conserved Subunit Sequences

RPB1    Nearly 40% of the amino acid residues of the largest RNA polymerase II subunit (RPB1) of *S. cerevisiae* (55) are identical in *Caenorhabditis elegans* (81), *Drosophila melanogaster* (76), and the mouse (82), and a large fraction of the remaining residues are conserved. This subunit is a homologue of the *E. coli* RNA polymerase $\beta'$ subunit, and the regions that are most conserved between the yeast and the *E. coli* large subunit are also the most conserved when any eukaryotic RPB1 subunit is compared with its prokaryotic counterpart. With one exception noted below, all of the eukaryotic RPB1 subunits contain the unusual CTD that consists of repeats of the heptapeptide consensus sequence Tyr-Ser-Pro-Thr-Ser-Pro-Ser. The CTD repeat length tends to increase with the complexity of the eukaryote; yeast has 26 or 27 repeats, depending on the strain, and man has 52.

RPB2    The gene encoding the second largest subunit of *D. melanogaster* RNA polymerase II has been cloned and sequenced (85, 123). More than a third of the amino acid residues of the *S. cerevisiae* and *D. melanogaster* RPB2 subunits are identical.

RPB3    The isolation and sequence analysis of a cDNA clone encoding the third largest human RNA polymerase II subunit (hRPB33) has revealed that 45% of the amino acid residues of human and yeast RPB3 subunits are identical (102).

COMMON SUBUNITS (RPB5, 6, AND 8)    The subunit composition of all three nuclear RNA polymerases has been determined for *Saccharomyces* (32), *Neurospora* (36), *Acanthamoeba* (37), *Dictyostelium* (38), *Xenopus* (42), wheat, and cauliflower (49). In each case, three polypeptides of RNA polymerases I, II, and III have identical gel mobility (Figure 1). These apparently shared polypeptides range from 14 to 28 kDa. A cDNA clone encoding a human homologue of *Saccharomyces* RPB5 has been isolated and sequenced (124). This protein, called hRPB23 or HP-23, is 30% identical to yeast RPB5.

RPB4/RPB7 SUBCOMPLEX    The *Saccharomyces* RPB4 and RPB7 subunits form a subcomplex that tends to dissociate from the enzyme and appears to influence the efficiency of initiation. It is not yet clear whether higher eukaryotic RNA polymerase II enzymes have components related to RPB4 and RPB7. Purified human RNA polymerase II does not contain detectable levels of an RPB4-like subunit; the fourth largest RNA polymerase II subunit of human RNA polymerase II (hRPB23) is homologous to the *Saccharomyces* common subunit RPB5 (124).

RPB9    The yeast RPB9 subunit appears to have a homologue in *D. melanogaster,* which is encoded by DNA upstream of the suppressor of Hairy Wing gene (72, 125); almost 50% of the amino acid residues of the yeast and *Drosophila* proteins are identical.

## THE LARGE SUBUNIT CARBOXY TERMINAL DOMAIN

This section reviews current knowledge about the structure and function of the RNA polymerase II carboxy terminal domain. The CTD is an essential component of the enzyme that appears to be necessary for normal responses to activation signals at some promoters.

### Conserved Structure

The RNA polymerase II large subunit CTD consists of multiple repeats of the consensus sequence Tyr-Ser-Pro-Thr-Ser-Pro-Ser. This domain does not exist in the large subunits of RNA polymerases I and III (55, 78), nor is it found in eubacterial or archaebacterial RNA polymerases (126, 127) or a viral RNA polymerase (128). The CTD consensus sequence is repeated 17 times in *Plasmodium* RNA polymerase II (86), 26–27 times in *Saccharomyces* (55, 103), 32 times in *Caenorhabditis* (81), 40 times in *Arabadopsis* (128a), about 45 times in *Drosophila* (104, 106), and 52 times in hamster (104) and mouse (56) (Figure 4). The CTD has an essential role in transcription; deletion mutations that remove most or all of

| | Mouse | Drosophila | Arabadopsis | Caenorhabditis | Yeast | Plasmodium |
|---|---|---|---|---|---|---|
| 1 | YSPTSPA | YSPTSPN | YSPSSPA | LSPRTPS | FGVSSPG | YSPS |
| 2 | YEPRSPGG | YTASSPG | YSPTSPG | YGGMSPGV | FSPTSPT | YSPTSPT |
| 3 | YTPQSPS | GASPN | YSPTSPG | YSPSSPQ | YSPTSPA | YNANNAY |
| 4 | YSPTSPS | YSPSSPN | YSPTSPG | FSMTSPH | YSPTSPS | YSPTSPK |
| 5 | YSPTSPS | YSPTSPL | YSPTSPT | YSPSSPQ | YSPTSPS | NQNDQMNVNSQ |
| 6 | YSPTSPN | YA  SPR | YSPSSPG | YSPTSPS | YSPTSPS | YNVMSPV |
| 7 | YSPTSPS | YASTTPN | YSPTSPA | YSPTSPAAG | YSPTSPS | YSVTSPK |
| 8 | YSPTSPS | FNPNSTG | YSPTSPS | QSPVSPS | YSPTSPS | YSPTSPK |
| 9 | YSPTSPS | YSPSSSG | YSPTSPS | YSPTSPS | YSPTSPS | YSPTSPK |
| 10 | YSPTSPS | YSPTSPV | YSPTSPS | YSPTSPS | YSPTSPS | YSPTSPK |
| 11 | YSPTSPS | YSPTVQ | YSPTSPS | YSPTSPS | YSPTSPS | YSPTSPK |
| 12 | YSPTSPS | FGSSPS | YSPTSPS | YSPTSPS | YSPTSPS | YSPTSPK |
| 13 | YSPTSPS | FAGSGSNI | YSPTSPS | YSPTSPS | YSPTSPS | YSPTSPK |
| 14 | YSPTSPS | YSPGN A | YSPTSPS | YSPTSPS | YSPTSPS | YSPTSPK |
| 15 | YSPTSPS | YSPSSSN | YSPTSPA | YSPSSPS | YSPTSPS | YSPTSPVA |
| 16 | YSPTSPS | YSPNSPS | YSPTSPA | YSPSSPS | YSPTSPS | QNIASPN |
| 17 | YSPTSPS | YSPTSPS | YSPTSPA | YSPSSPR | YSPTSPS | YSP |
| 18 | YSPTSPS | YSPSSPS | YSPTSPS | YSPTSPT | YSPTSPA | YSITSPK |
| 19 | YSPTSPS | YSPTSPC | YSPTSPS | YSPTSPT | YSPTSPS | FSPTSPA |
| 20 | YSPTSPS | YSPTSPS | YSPTSPS | YSPTSPT | YSPTSPS | YSISSPV  (+68 AA) |
| 21 | YSPTSPS | YSPTSPN | YSPTSPS | YSPTSPT | YSPTSPS | |
| 22 | YSPTSPN | YTPVTPS | YSPTSPS | YSPTSPS | YSPTSPS | |
| 23 | YSPTSPN | YSPTSPN | YSPTSPA | YESGGG | YSPTSPN | |
| 24 | YTPTSPS | YS ASPQ | YSPTSPG | YSPSSPK | YSPTSPS | |
| 25 | YSPTSPS | YSPASPA | YSPTSPS | YSPSSPT | YSPTSPG | |
| 26 | YSPTSPN | YSQTGVK | YSPTSPS | YSPTSPS | YSPGSPA | |
| 27 | YTPTSPN | YSPTSPT | YGPTSPS | YSPTSPQ | YSPKQDEQKHNENENSR | |
| 28 | YSPTSPS | YSPPSPSDG | YNPQSAK | YSPTSPQ | | |
| 29 | YSPTSPS | YSPGSPQ | YSP SIA | YSPSSPT | | |
| 30 | YSPTSPS | YTPGSPQ | YSPSNAR | YSPTSPT | | |
| 31 | YSPSSPR | YSPASPK | LSPASP | YSPTSPS | | |
| 32 | YTPQSPT | YSPTSPL | YSPTSPN | YTPSSPQ | | |
| 33 | YTPSSPS | YSPSSPQ | YSPTSPS | YSPTSPT | | |
| 34 | YSPSSPS | HSPS SQ | YSPTSPS | YTP SPSEQPGTSAQVDFFYSKLNF | | |
| 35 | YSPTSPK | YSPTGST | YSPSSPT | | | |
| 36 | YTPTSPS | YSPTSPR | YSPSSPY | | | |
| 37 | YSPSSPE | YSPNMSI | SSGASPD | | | |
| 38 | YTPASPK | YSPSSTK | YSPSAG | | | |
| 39 | YSPTSPK | YSPTSPT | YSPTLPG | | | |
| 40 | YSPTSPK | YTPTARN | YTPSSTGQ | | | |
| 41 | YSPTSPT | YSPTSPM | YTPHEGDKKDKTGKKDASKDDKGNP | | | |
| 42 | YSPTTPK | YSPTAPSH | | | | |
| 43 | YSPTSPT | YSPTSPA | | | | |
| 44 | YSPTSPV | YSPSSPTFEESED | | | | |
| 45 | YTPTSPK | | | | | |
| 46 | YSPTSPT | | | | | |
| 47 | YSPTSPK | | | | | |
| 48 | YSPTSPT | | | | | |
| 49 | YSPTSPKGST | | | | | |
| 50 | YSPTSPG | | | | | |
| 51 | YSPTSPT | | | | | |
| 52 | YSLTSPAISPDDSDEEN | | | | | |

*Figure 4*    RNA polymerase II CTDs. CTD sequences are shown for *Plasmodium* (86), *Saccharomyces* (103), *Caenorhabditis* (81), *Arabadopsis* (128a), *Drosophila* (104, 106), and mouse (56). Different yeast strains have somewhat different CTD sequences (55, 103); and that from strain S288C (103) is shown here.

the CTD are lethal to yeast (103, 104), *Drosophila* (106), and murine (105) cells. The CTD is highly phosphorylated in a substantial portion of the RNA polymerase II molecules in the cell (54, 107), and protein kinases have been purified that appear to be involved in CTD phosphorylation (129–131).

Both the CTD repeat length and the degree of heptapeptide conservation vary from one organism to another (Figure 4). The length of the CTD generally increases with the genomic complexity of the organism. The precise heptapeptide consensus sequence Tyr-Ser-Pro-Thr-Ser-Pro-Ser occurs in 30–60% of the CTD repeats in *Saccharomyces*, *Caenorhabditis*, *Arabadopsis*, and mouse. The consensus sequence is much less well conserved in *Drosophila* and in *Plasmodium*. Certain amino acid residues in the repeat are highly conserved in all of these organisms; in particular, the tyrosines are almost invariant. The repeat form of the CTD does not occur in all eukaryotes. The protozoan parasites *Trypanosoma brucei* and *Crithidia fasciculata* do not contain a heptapeptide repeat structure at the carboxyl terminus of the large subunit (83, 84, 132), but are instead rich in the amino acid residues found in the CTDs of higher eukaryotes (serine, threonine, proline, and tyrosine).

Modification and proteolysis of the CTD almost certainly accounts for most of the various forms of RNA polymerase II that have been purified from eukaryotes (51, 54–56, 107, 130, 131, 133; reviewed in 57). The largest subunit of RNA polymerase II can be resolved into three major forms in SDS-PAGE. In order of increasing mobility, these are called $II_o$, $II_a$, and $II_b$. The $II_o$ form of the subunit contains a highly phosphorylated CTD that can be generated in vivo or in vitro (51, 107, 108, 131, 134). The $II_a$ form of the subunit appears to be an unmodified primary translation product (55, 56); the $II_a$ form can be generated by phosphatase treatment of the $II_o$ form of the subunit (107). The $II_b$ form lacks most or all of the CTD (55, 56, 106), and is probably a proteolytic artifact of purification (44, 54). The CTD of the mammalian $II_o$ form of the subunit contains phosphoserine and phosphothreonine, but not detectable amounts of phosphotyrosine (107). The $II_o$ form of the subunit is thought to predominate in HeLa cells (44), but similar amounts of the $II_o$ form and the $II_a$ form occur in yeast RNA polymerase II isolated with a rapid immunoprecipitation procedure (54).

## Essential Function

Truncation of the *S. cerevisiae* RNA polymerase II CTD reduces the efficiency of a function that is essential for cell viability (103, 104, 135). RNA polymerase II large subunit genes encoding fewer than 10 complete heptapeptide repeats are unable to sustain cell viability, those encoding 10 to 12 complete repeats produce cells that are temperature sensitive and cold sensitive, and those encoding at least 13 complete repeats are sufficient for wild-type growth (103). Yeast cells containing CTDs of 10 to 12 complete repeats are also inositol auxotrophs (25, 135), and are unable to use pyruvate as a carbon source (D. Chao, R. A. Young, unpublished). The phenotypes associated with these CTD mutations are not a consequence of instability of the large subunit, as CTD truncation does not appear to alter the levels of RNA polymerase II in the cell (103); rather, these phenotypes appear to reflect some functional deficiency of the enzyme. The conditional phenotypes and the inositol auxotrophy suggest that the expression of some genes is more sensitive to certain CTD partial deletion mutations than are others.

Truncation of the mouse RNA polymerase II CTD, which contains 52 repeats, also appears to affect the efficiency of a function that is essential for cell viability (105). The effects of CTD deletion mutations were assayed by measuring the ability of mutant genes containing an $\alpha$-amanitin resistance mutation to confer $\alpha$-amanitin resistance on transfected rodent cells. RNA polymerase II large subunit genes containing the $\alpha$-amanitin resistance mutation and encoding between 36 and 78 heptapeptide repeats were able to confer $\alpha$-amanitin resistance on transfected cells while those encoding fewer than 25 repeats were not. An intermediate effect was observed when the polymerase gene encoded 29, 31, or 32 repeats; the number and size of transfected $\alpha$-amanitin-resistant cells were reduced. Like the yeast enzyme, mouse RNA polymerase II is partially defective when the CTD is reduced to approximately half its normal size.

The RNA polymerase II CTD has also been shown to be essential in *Drosophila*. A P-element insertion that truncates the *Drosophila* CTD from 42 to 20 repeats is lethal when the allele is in the homozygous state (106).

The CTD appears to be a functionally redundant structure whose normal size is important for functional efficiency. Although a considerable portion of the CTD can be deleted in yeast and in mammalian cells without gross alterations in cell viability, the elimination of smaller portions of the CTD can have subtle but important effects on gene expression. For example, the ability to induce *INO1* expression in yeast is reduced by the loss of only nine heptapeptide repeats (25). It appears that different mammalian cell types may have different requirements for CTD length or for certain amino acid sequences within specific heptapeptide repeats (105). Thus, while portions of the CTD are not required for cell viability in simple growth assays, it seems likely that the entire CTD is necessary for maximal transcriptional efficiency in each eukaryote.

## Evidence for a Role in Initiation

At most well-characterized promoters, positive and negative regulatory factors strike a complex balance to regulate the rate of transcription initiation. Although the role of the RNA polymerase II CTD in transcription is not yet firmly established, the simplest general model that accounts for all of the available data is that the RNA polymerase II CTD is one of several components of the transcription apparatus that contribute to the response to activating signals from factors associated with the enhancer. The contribution of the CTD to the response to activating signals relative to the contribution of other mechanisms involved in activation varies at different promoters.

GENETIC STUDIES    Yeast cells containing RNA polymerase II with a series of CTD partial deletion mutations have been constructed to study the effect of CTD truncation on the ability to transcribe specific genes (25, 103). RNA polymerase II CTD mutations can affect the ability of the enzyme to respond to regulatory signals from Upstream Activating Sequences (UASs). Cells containing RNA polymerase II with a large subunit CTD of 11 3/7 repeats are highly defective in derepression of *INO1* mRNA synthesis, partially defective in derepression of *GAL10* mRNA synthesis, and unaffected in derepression of *HIS4* transcription. The relative ability of an RNA polymerase II CTD mutant to transcribe *INO1, GAL10,* and *HIS4* upon induction correlates with its relative ability to utilize the UAS elements of these three genes when they substitute for the *CYC1* UAS in vivo. Investigation of the effect of various repeat lengths on utilization of the *INO1, GAL10,* and *HIS4* UASs confirms that different UASs vary in their sensitivity to CTD repeat length, and demonstrates that there is a rapid transition between repeat lengths that are responsive to UAS-specific activating signals and those that are not. Inadequate *INO1* UAS utilization appears to be a direct result of RNA polymerase II CTD dysfunction at the *INO1* promoter, rather than an indirect result of a lack of *INO1* transcription factors. Finally, inadequate responses to the GAL4-binding site UAS can account for the inadequate response of CTD mutants to the *GAL10* UAS. Although the level of *GAL4* mRNA is reduced in CTD mutant cells, restoration of *GAL4* mRNA to wild-type levels does not diminish the defect in the response to the *GAL10* UAS in CTD mutants. These results implicate the CTD in the response to activation signals from the UAS elements of certain genes.

The inability of RNA polymerase II CTD truncation mutants to respond fully to signals from some transactivating proteins could be a consequence of several direct or indirect effects. The CTD mutations might affect the ability of the CTD to interact directly with some transactivation factors or their accessory factors. CTD truncation could affect the ability of RNA polymerase to interact appropriately with general transcription factors, which themselves may respond directly or indirectly to transactivation factors. CTD truncation could alter the conformation of other portions of RNA polymerase, which are normally important for the enzyme's response to transactivating signals.

Support for the notion that the RNA polymerase CTD may help mediate the response to signals from transcription factors also comes from investigating the effect of RNA polymerase II with substantially different CTD sizes on transcription from a *GAL1* promoter (the *GAL1* and *GAL10* promoters share the *GAL1–10* UAS) in the presence of a variety of *GAL4* transcriptional activation domain mutations (24). In the *GAL4* mutants, increases in transcription from the *GAL1* promoter were observed as the RNA polymerase II CTD length increased from 13 to 38 repeats. In contrast, there was no difference in the ability of RNA polymerase II with CTD lengths of 13, 26, and 38 repeats to express $\beta$-galactosidase in the presence of wild-type GAL4. The observation that 13 repeats is close to the threshold number of repeats necessary for wild-type responses to the *GAL1–10* UAS in the presence of wild-type GAL4 (25) may explain why no substantial differences in *GAL1–10* UAS responses were observed with CTD lengths of 13, 26, and 38 repeats (24).

The removal of negative regulators can partially restore the ability of yeast RNA polymerase II with short CTDs to respond to activation signals at specific promoters. The negative regulatory factor SPT2 (also called SIN1) represses transcription of the yeast *HO* and *INO1* genes (136), among others (137). Deletion of the *SPT2* gene partially restores the ability of RNA polymerase II with a CTD of 10 heptapeptide repeats to respond to induction at the *INO1* promoter, and reduces the severity of the cold-sensitive phenotype of cells containing the RNA polymerase mutation (136). Since CTD truncation severely reduces the ability of RNA polymerase II to respond to induction at promoters regulated by SPT2 (*HO* and *INO1*), and because the absence of SPT2 partially restores the response, one of the roles of the CTD may be to antagonize the negative regulatory activity of SPT2. It is not yet clear whether the CTD and SPT2 act through the same molecular mechanism. Evidence that the CTD and SPT2 both bind to DNA nonspecifically in vitro led Peterson et al (136) to suggest that the CTD and SPT2 may compete for DNA binding. It is also possible that the antagonism is more indirect; for example, if the CTD facilitates interaction between upstream transcription factors and the TATA-associated initiation complex, nonspecific DNA-binding by SPT2 could curtail that interaction by restricting DNA looping.

BIOCHEMICAL STUDIES    The results of in vitro transcription experiments thus far indicate that the CTD is required for efficient promoter-dependent transcription at some promoters but not others, and that the CTD is not involved in RNA chain elongation in the presence of naked DNA templates. Monoclonal antibodies directed against the CTD inhibit promoter-dependent transcription from the adenovirus-2 major late promoter and the murine dihydrofolate reductase promoter in HeLa cell nuclear extracts, but do not appear to affect elongation (52, 138, 139). In addition, synthetic peptides containing the consensus repeat sequence and conjugated to bovine serum albumin (BSA) inhibit specific transcription initiation but not elongation in vitro (52). Purified calf thymus RNA polymerase II in which the CTD has been removed by proteolysis is not capable of initiation at the murine dihydrofolate reductase promoter (52). However, purified calf thymus RNA polymerase II lacking the CTD can initiate transcription from the adenovirus-2 major late promoter (52, 139, 140). Similarly, purified *Drosophila* RNA polymerase II lacking the CTD is able to initiate promoter-dependent transcription of *Drosophila* actin 5C and histone H3 and H4 templates (106). Together, these data suggest that the importance of the CTD in transcription initiation varies from one promoter to another, consistent with studies of the ability of RNA polymerase CTD mutants to respond to signals from various UAS elements in vivo (25). In addition, these in vitro transcription experiments argue against a general role for the CTD in RNA chain elongation.

Nuclear extracts of yeast RNA polymerase II CTD truncation mutants are fully competent in promoter-independent initiation and elongation assays and contain normal levels of RNA polymerase II, but are strikingly defective in their response to GAL4-VP16 stimulated, promoter-dependent transcription in vitro (S.-M. Liao, R. A. Young, unpublished). Progressive CTD truncation results in a progressive loss of GAL4-VP16 stimulated promoter-dependent transcription using a template containing a GAL4-binding site UAS. Promoter-dependent transcription can be restored by adding purified wild-type RNA polymerase II to the CTD mutant nuclear extracts, indicating that the defect is not due to the absence of necessary transcription factors. These results confirm that the CTD is important for normal responses to some upstream activating factors.

At two promoters where RNA polymerase II lacking the CTD can initiate transcription in vitro, the CTD-less enzyme can respond to stimulation by at least two transcription factors. The calf thymus RNA polymerase CTD is not required to obtain three- to fourfold stimulation of transcription from the adenovirus-2 major late promoter in vitro with the adenovirus Major Late Transcription Factor (140). In addition, the factor Sp1 stimulates transcription threefold from a hybrid HSP70 promoter in a *Drosophila* extract in vitro independent of the presence or the absence of the CTD (141). These results suggest that transcription activation by these factors does not involve the CTD, but it is also possible that these assays do not detect contributions that are normally made by this domain during initiation in vivo.

Loss of the CTD does not appear to affect the ability of RNA polymerase II to assemble with general transcription factors on TATA-containing promoter DNA (140). When the CTD is removed from mammalian RNA polymerase II by proteolysis, the CTD-less RNA polymerase and wild-type RNA polymerase are indistinguishable in their ability to assemble with TATA-associated factors on adenovirus-2 major late promoter DNA.

## General Models for CTD Function

Few aspects of transcription have been left untouched in models that have been proposed for CTD function. These proposed functions include involvement in transcription initiation, enzyme localization, DNA binding, removal of chromatin proteins from DNA, and general regulation of enzyme activity (55–57, 82, 103–105, 136, 138, 142, 143). The simplest interpretation of all of the available data is that the RNA polymerase II CTD is one of several components of the transcription apparatus that contributes to the response to activating signals from the enhancer. The importance of this contribution relative to the contribution of other mechanisms involved in activation varies at different promoters.

It seems likely that the CTD acts to facilitate interactions between upstream activating factors (either directly or through additional factors) and components of the mRNA transcription complex (the general transcription factors and/or RNA polymerase). For example, the hydroxyl-rich CTD could provide an initial target for an upstream transcription factor whose ultimate target is a component of the mRNA transcription complex. Transient weak interactions between heptapeptide repeat sequences and the upstream factor might increase the rate with which the factor finds the ultimate target in the mRNA transcription complex.

Phosphorylation of the RNA polymerase II CTD may play a role in transcription initiation, providing a switch between steps during initiation (129, 133, 144; reviewed in 145). The relatively unphosphorylated form of RNA polymerase ($II_a$) becomes phosphorylated during initiation in vitro. It is difficult to distinguish between a requirement for phosphorylation and coincidental phosphorylation during this initiation reaction in vitro, but this is a very interesting model. The postulated interaction between the acidic activating domain of a transcription factor and the hydroxyl-rich CTD (143) could be disrupted by phosphorylation of the CTD, permitting a transition between initiation and elongation. If the CTD interacts with DNA in vivo (136, 142), that interaction would be disrupted by CTD phosphorylation, permitting a similar type of transition.

## ACTIVATION OF RNA POLYMERASE II

The mechanisms by which factors deliver signals to initiate transcription to RNA polymerase II are not yet clear. Signals could be delivered directly through interactions between upstream factors and RNA polymerase, or indirectly through intermediates such as the TATA-associated transcription factors or other cofactors. Whatever the mechanism(s), RNA polymerase itself must have the capacity to respond to activation signals. The large subunit CTD is one component of RNA polymerase II that appears to be involved in the response to activating signals. Genetic evidence suggests that other portions of the enzyme may also be involved in this response. For example, an intragenic suppresor of partial CTD truncation mutations occurs within segment H of RPB1 (135), suggesting that this portion of the large subunit interacts with the CTD or is involved in the same function as the CTD. In addition, some mutations in RPB2 affect the transcription of some genes but not others in vivo, and exhibit some but not all of the phenotypes associated with CTD truncations (59). In this context, it is interesting that the *E. coli* RNA polymerase homologue of RPB2 ($\beta$) probably interacts with the $\sigma$ subunit; the $\beta$ and $\sigma$ subunits can be crosslinked to DNA sequences only five nucleotides apart in the *lac* UV5 promoter (146), and mutations in $\beta$ can affect $\sigma$ binding to the core enzyme (147). How signal transduction to RNA polymerase occurs remains one of the more interesting unknown mechanisms in the control of transcription.

## CONCLUDING REMARKS

A well-defined RNA polymerase II enzyme provides an important starting point for genetic and biochemical experiments that should improve our understanding of transcription and its regulation. Conditional mutations in many of the yeast RNA polymerase II subunit genes have been isolated (59, 68, 69, 112a, 148–150) and are being used in combination with in vitro transcription (151) and other biochemical assays to obtain clues to subunit functions. This potent combination of genetics and biochemistry will surely yield valuable new insights into the molecular mechanisms that govern transcription.

ACKNOWLEDGMENTS

I am especially grateful to Richard Burgess, Steve Buratowski, David Chao, Gerry Fink, Peter Kolodziej, Charles Scafe, and Nancy Woychik for insights, criticism, and stimulating discussions. I thank the many investigators who shared with me their data and thoughts, including David Baltimore, Ekkehard Bautz, Jean-Marie Buhler, Richard Burgess, Steve Buratowski, Pierre Chambon, Jeffry Corden, A. Cornelissen, Michael Dahmus, Aled Edwards, Arno Greenleaf, Leonard Guarente, Ira Herskowitz, James Ingles, Carl Mann, Mark Johnston, Roger Kornberg, Mark Mortin, Masayasu Nomura, Craig Peterson, Michel Riva, Alan Sachs, Andre Sentenac, Phil Sharp, Robert Tjian, and Sherman Weissman. I am also grateful to Carolyn C. Carpenter for preparing the manuscript and to David Rothstein for assistance with literature research. This work was supported by Public Health Service Grant GM34365 and by a Burroughs Wellcome Molecular Parasitology Award.

*Literature Cited*

1. Chamberlin, M. J. 1976. *RNA Polymerase*, ed. R. Losick, M. Chamberlin, pp. 17–67. Cold Spring Harbor, NY: Cold Spring Harbor Lab. 899 pp.
2. Yura, T., Ishihama, A. 1979. *Annu. Rev. Genet.* 13:59–97
3. Chamberlin, M. J. 1982. *Enzymes* 15: 61–86
4. McClure, W. R. 1985. *Annu. Rev. Biochem.* 54:171–204
5. Schleif, R. 1988. *Science* 241:1182–87
6. Gralla, J. D. 1989. *Cell* 57:193–95
7. Helmann, J. D., Chamberlin, M. J. 1988. *Annu. Rev. Biochem.* 57:839–72
8. Platt, T. 1986. *Annu. Rev. Biochem.* 55:339–72
9. Bell, S. P., Learned, R. M., Jantzen, H.-M., Tjian, R. 1988. *Science* 241: 11192–97
10. Guarente, L. 1987. *Annu. Rev. Genet.* 21:425–52
11. Johnson, P. F., McKnight, S. L. 1989. *Annu. Rev. Biochem.* 58:799–839
12. Struhl, K. 1989. *Annu. Rev. Biochem.* 58:1051–77
13. Mitchell, P. J., Tjian, R. 1989. *Science* 245:371–78
14. Guarente, L. 1988. *Cell* 52:303–5
15. Ptashne, M. 1988. *Nature* 335:683–89
16. Geiduschek, E. P. 1988. *Annu. Rev. Biochem.* 57:873–914
17. Van Dyke, M. W., Roeder, R. G., Sawadogo, M. 1988. *Science* 241:1335–38
18. Buratowski, S., Hahn, S., Guarente, L., Sharp, P. A. 1989. *Cell* 56:549–61
19. Sawadogo, M., Sentenac, A. 1990. *Annu. Rev. Biochem.* 59:711–54
20. Pugh, B. F., Tjian, R. 1990. *Cell* 61:1187–97
21. Berger, S. L., Cress, W. D., Cress, A., Triezenberg, S. J., Guarente, L. 1990. *Cell* 61:1199–208
22. Kelleher, R. J., Flanagan, P. M., Kornberg, R. D. 1990. *Cell* 61:1209–15
23. Stringer, K. F., Ingles, C. J., Greenblatt, J. 1990. *Nature* 345:783–86
24. Allison, L. A., Ingles, C. J. 1989. *Proc. Natl. Acad. Sci. USA* 86:2794–98
25. Scafe, C., Chao, D., Lopes, J., Hirsch, J. P., Henry, S., Young, R. A. 1990. *Nature* 347:491–94
26. Lewin, B. 1990. *Cell* 61:1161–64
27. Ptashne, M., Gann, A. A. F. 1990. *Nature* 346:329–31
28. Chambon, P. 1975. *Annu. Rev. Biochem.* 44:613–38
29. Roeder, R. G. 1976. See Ref. 1, pp. 285–329
30. Paule, M. R. 1981. *Trends Biochem. Sci.* 6:128–31
31. Lewis, M. K., Burgess, R. R. 1982. *Enzymes* 15:109–53
32. Sentenac, A. 1985. *Crit. Rev. Biochem.* 18:31–91
33. Roeder, R. G., Rutter, W. J. 1969. *Nature* 224:234
34. Roeder, R. G., Rutter, W. J. 1970. *Proc. Natl. Acad. Sci. USA* 65:675
35. Roeder, R. G., Rutter, W. J. 1970. *Biochemistry* 9:2543
36. Armaleo, D., Gross, S. R. 1985. *J. Biol. Chem.* 260:16169–73
37. D'Alessio, J. M., Perna, P. J., Paule, M. R. 1979. *J. Biol. Chem.* 254:11282–87
38. Renart, M. F., Sastre, L., Diaz, V., Sebastian, J. 1985. *Mol. Cell. Biochem.* 66:21–29
39. Smith, S. S., Braun, R. 1981. *FEBS Lett.* 125:107–10
40. Sanford, T., Prenger, J. P., Golomb, M. 1985. *J. Biol. Chem.* 260:8064–69
41. Greenleaf, A. L., Bautz, E. K. F. 1975. *Eur. J. Biochem.* 60:169–79
42. Engelke, D. R., Shastry, B. S., Roeder, R. G. 1983. *J. Biol. Chem.* 258:1921–31
43. Schwartz, L. B., Roeder, R. G. 1975. *J. Biol. Chem.* 250:3221–28
44. Kim, W-Y., Dahmus, M. 1988. *J. Biol. Chem.* 263:18880–85
45. Freund, E., McGuire, P. M. 1986. *J. Cell. Physiol.* 127:432–38
46. Jendrisak, J. J., Petranyi, P. W., Burgess, R. 1976. See Ref. 1, pp. 779–92
47. Hodo, H. G. III, Blatti, S. P. 1977. *Biochemistry* 16:2334–43
48. Jendrisak, J., Guilfoyle, T. J. 1978. *Biochemistry* 17:1322–27
49. Guilfoyle, T. J., Hagen, G., Malcolm, S. 1984. *J. Biol. Chem.* 259:640–48
50. Guilfoyle, T. J. 1980. *Biochemistry* 19: 5966–72
51. Kim, W. Y., Dahmus, M. E. 1986. *J. Biol. Chem.* 261:14219–25
52. Thompson, N. E., Steinberg, T. H., Aronson, D. B., Burgess, R. R. 1989. *J. Biol. Chem.* 264:11511–20
53. Edwards, A. M., Darst, S. A., Feaver, W. J., Thompson, N. E., Burgess, R. R., Kornberg, R. D. 1990. *Proc. Natl. Acad. Sci. USA* 87:2122–26
54. Kolodziej, P. A., Woychik, N. A., Liao, S.-M., Young, R. A. 1990. *Mol. Cell. Biol.* 10:1915–20

55. Allison, L. A., Moyle, M., Shales, M., Ingles, C. J. 1985. *Cell* 42:599–610
56. Corden, J. L., Cadena, D. L., Ahearn, J. M., Dahmus, M. E. 1985. *Proc. Natl. Acad. Sci. USA* 82:7934–38
57. Corden, J. L. 1990. *Trends. Biochem. Sci.* 15:383–87
58. Arndt, K., Styles, C., Fink, G. R. 1989. *Cell* 56:527–37
59. Scafe, C., Nonet, M., Young, R. A. 1990. *Mol. Cell. Biol.* 10:1010–16
60. Kadesch, T. R., Chamberlin, M. J. 1982. *J. Biol. Chem.* 257:5286–95
61. Sluder, A. E., Price, D. H., Greenleaf, A. L. 1987. *J. Biol. Chem.* 263:9917–25
62. Coulter, D. E., Greenleaf, A. L. 1985. *J. Biol. Chem.* 260:13190–98
63. Proudfoot, N. J. 1989. *Trends Biochem. Sci.* 14:105–10
64. Young, R. A., Davis, R. W. 1983. *Science* 222:778–82
65. Ingles, C. J., Himmelfarb, H. J., Shales, M., Greenleaf, A. L., Friesen, J. D. 1984. *Proc. Natl. Acad. Sci. USA* 81:2157–61
66. Riva, M., Memet, S., Micouin, J. Y., Huet, J., Treich, I., et al. 1986. *Proc. Natl. Acad. Sci. USA* 83:1554–58
67. Sweetser, D., Nonet, M., Young, R. A. 1987. *Proc. Natl. Acad. Sci. USA* 84:1192–96
68. Kolodziej, P. A., Young, R. A. 1989. *Mol. Cell Biol.* 9:5387–94
69. Woychik, N. A., Young, R. A. 1989. *Mol. Cell Biol.* 9:2854–59
70. Woychik, N. A., Liao, S.-M., Kolodziej, P., Young, R. A. 1990. *Genes Dev.* 4:313–23
71. Woychik, N. A., Young, R. A. 1990. *J. Biol. Chem.* 265:17816–19
72. Woychik, N. A., Lane, W. S., Young, R. A. Submitted
73. Searles, L. L., Jokerst, R. S., Bingham, P. M., Voelker, R. A., Greenleaf, A. L. 1982. *Cell* 31:585–92
74. Breant, B., Huet, J., Sentenac, A., Fromageot, P. 1983. *J. Biol. Chem.* 258:11968–73
75. Buhler, J.-M., Huet, J., Davies, K. E., Sentenac, A., Fromageot, P. 1980. *J. Biol. Chem.* 255:9949–54
76. Jokerst, R. S., Weeks, J. R., Zehring, W. A., Greenleaf, A. L. 1989. *Mol. Gen. Genet.* 215:266–75
77. Cornelissen, A. W. C. A., Evers, R., Kock, J. 1988. *Oxford Surv. Eukaryotic Genes* 5:91–131
78. Memet, S., Gouy, M., Marck, C., Sentenac, A., Buhler, J.-M. 1988. *J. Biol. Chem.* 262:2830–39
79. Huet, J., Sentenac, A., Fromageot, P. 1982. *J. Biol. Chem.* 257:2613–18
80. Weeks, J. R., Coulter, D. E., Greenleaf, A. L. 1982. *J. Biol. Chem.* 257:5884–91
81. Bird, D. M., Riddle, D. L. 1989. *Mol. Cell. Biol.* 9:4119–30
82. Ahearn, J. M. Jr., Bartolomei, M. S., West, M. L., Cisek, L. J., Corden, J. L. 1987. *J. Biol. Chem.* 262:10695–705
83. Evers, R., Hammer, A., Kock, J., Jess, W., Borst, P., et al. 1989. *Cell* 56:585–97
84. Smith, J. L., Levin, J. R., Ingles, C. J., Agabian, N. 1989. *Cell* 56:815–27
85. Falkenburg, D., Dworniczak, B., Faust, D., Bautz, E. K. F. 1987. *J. Mol. Biol.* 195:929–37
86. Li, W.-B., Bzik, D. J., Gu, H., Tanaka, M., Fox, B. A., Inselburg, J. 1989. *Nucleic Acids Res.* 17:9621–36
87. Zillig, W., Palm, P., Heil, A. 1976. See Ref. 1, pp. 101–25
88. Gundelfinger, E. D. 1983. *FEBS Lett.* 157:133–38
89. Horikoshi, M., Tamura, H. O., Sekimizu, K., Obinata, M., Natori, S. 1983. *J. Biochem.* 94:1761–67
90. Chuang, R. Y., Chuang, L. F. 1987. *Biochem. Biophys. Res. Commun.* 145:73–80
91. Carroll, S. B., Stollar, B. D. 1983. *J. Mol. Biol.* 170:777–90
92. Bartholomew, B., Dahmus, M. E., Meares, C. F. 1986. *J. Biol. Chem.* 261:14226–31
93. Hanna, M. M., Meares, C. F. 1983. *Proc. Natl. Acad. Sci. USA* 80:4238–42
94. Grachev, M. A., Hartmann, G. R., Maximova, T. G., Mustaev, A. A., Schaffner, A. R., et al. 1986. *FEBS Lett.* 200:287–90
95. Grachev, M. A., Kolocheva, T., Lukhtanov, E. A., Mustaev, A. A. 1987. *Eur. J. Biochem.* 163:113–21
96. Riva, M., Schaffner, A., Sentenac, A., Hartmann, G., Mustaev, A., et al. 1987. *J. Biol. Chem.* 262:14377–80
97. Grachev, M. A., Lukhtanov, E. A., Mustaev, A. A., Zaychikov, E. F., Abdukayumov, M. N., et al. 1989. *Eur. J. Biochem.* 180:577–85
98. Berghofer, B., Krockel, L., Kortner, C., Truss, M., Schallenberg, J., Klein, A. 1988. *Nucleic Acids Res.* 16:8113–28
99. Riva, M., Carles, C., Sentenac, A., Grachev, M. A., Mustaev, A. A., Zaychikov, E. F. 1990. *J. Biol. Chem.* 265:16498–503
100. Ovchinnikov, Y., Lipkin, V. M., Modyanov, N. N., Chertov, O. Y., Smirnov, Y. V. 1977. *FEBS Lett.* 76:108–11
101. Kolodziej, P. 1991. PhD thesis, MIT
102. Pati, U. K., Weissman, S. M. 1990. *J. Biol. Chem.* 265:8400–3

103. Nonet, M., Sweetser, D., Young, R. A. 1987. *Cell* 50:909–15
104. Allison, L. A., Wong, J. K.-C., Fitzpatrick, V. D., Moyle, M., Ingles, C. J. 1988. *Mol. Cell. Biol.* 8:321–29
105. Bartolomei, M. S., Halden, N. F., Cullen, C. R., Corden, J. L. 1988. *Mol. Cell. Biol.* 8:330–39
106. Zehring, W. A., Lee, J. M., Weeks, J. R., Jokerst, R. S., Greenleaf, A. L. 1988. *Proc. Natl. Acad. Sci. USA* 85:3698–702
107. Cadena, D., Dahmus, M. 1987. *J. Biol. Chem.* 262:12468–74
108. Buhler, J.-M., Iborra, F., Sentenac, A., Fromageot, P. 1976. *FEBS Lett.* 71:37–41
109. Buhler, J.-M., Iborra, F., Sentenac, A., Fromageot, P. 1976. *J. Biol. Chem.* 251:1712–17
110. Valenzuela, P., Bell, G. I., Weinburg, F., Rutter, W. J. 1976. *Biochem. Biophys. Res. Commun.* 71:1319–25
111. Huet, J., Phalente, L., Buttin, G., Sentenac, A., Fromageot, P. 1982. *EMBO J.* 1:1193–98
112. Huet, J., Riva, M., Sentenac, A., Fromageot, P. 1985. *J. Biol. Chem.* 260:15304–10
112a. Archambault, J., Schappert, K. T., Friesen, J. D. 1990. *Mol. Cell. Biol.* 10:6123–31
113. Dezelee, S., Wyers, F., Sentenac, A., Fromageot, P. 1976. *Eur. J. Biochem.* 65:543–52
114. Ruet, A., Sentenac, A., Fromageot, P., Winsor, B., Lacroute, F. 1980. *J. Biol. Chem.* 255:6450–55
115. Edwards, A. M., Kane, C. M., Young, R. A., Kornberg, R. D. 1991. *J. Biol. Chem.* 266:71–75
116. Solaiman, D., Wu, F. Y.-H. 1984. *Biochemistry* 23:6369–77
117. Lattke, H., Weser, U. 1976. *FEBS Lett.* 65:288–91
118. Martin, C., Okamura, S., Young, R. A. 1990. *Mol. Cell. Biol.* 10:1908–14
119. Mortin, M. A. 1990. *Proc. Natl. Acad. Sci. USA* 87:4864–68
120. Yano, R., Nomura, M. 1991. *Mol. Cell. Biol.* 11:754–64
121. Darst, S., Edwards, A., Kubalek, E., Kornberg, R. 1990. Submitted
122. Darst, S. A., Kubalek, E. W., Kornberg, R. D. 1989. *Nature* 340:730–32
123. Faust, D. M., Renkawitz-Pohl, R., Falkenburg, D., Gasch, A., Bailojan, S., et al. 1986. *EMBO J.* 5:741–46
124. Pati, U. K., Weissman, S. M. 1989. *J. Biol. Chem.* 264:13114–21
125. Parkhurst, S. M., Harrison, D. A., Remington, M. P., Spana, C., Kelly, R. L., et al. 1988. *Genes Dev.* 2:1205–15
126. Puhler, G., Leffers, H., Gropp, F., Palm, P., Klenk, H.-P., et al. 1989. *Proc. Natl. Acad. Sci. USA* 86:4569–73
127. Ovchinikov, Y. A., Monastyrskaya, G. S., Gubanov, V. V., Guryev, S. O., Salomatina, I. S., et al. 1982. *Nucleic Acids Res.* 10:4035–44
128. Broyles, S. S., Moss, B. 1986. *Proc. Natl. Acad. Sci. USA* 83:3141–45
128a. Dietrich, M. A., Prenger, J. P., Guilfoyle, T. J. 1990. *Plant Mol. Biol.* 15:207–23
129. Payne, J. M., Laybourn, P. J., Dahmus, M. E. 1989. *J. Biol. Chem.* 264:19621–29
130. Cisek, L., Corden, J. 1989. *Nature* 339:679–84
131. Lee, J. M., Greenleaf, A. L. 1989. *Proc. Natl. Acad. Sci. USA* 86:3624–28
132. Evers, R., Hammer, A., Cornelissen, A. W. C. A. 1989. *Nucleic Acids Res.* 17:3404–13
133. Laybourn, P. J., Dahmus, M. E. 1989. *J. Biol. Chem.* 264:6693–98
134. Bell, G. I., Valenzuela, P., Rutter, W. J. 1977. *J. Biol. Chem.* 252:3082–91
135. Nonet, M., Young, R. A. 1989. *Genetics* 123:715–24
136. Peterson, C., Kruger, W., Herskowitz, I. 1991. *Cell.* In press
137. Roeder, G. S., Beard, C., Smith, M., Keranen, S. 1985. *Mol. Cell. Biol.* 5:1543–53
138. Moyle, M., Lee, J. S., Anderson, W. F., Ingles, C. J. 1989. *Mol. Cell. Biol.* 9:5750–53
139. Kim, W.-Y., Dahmus, M. E. 1989. *J. Biol. Chem.* 264:3169–76
140. Buratowski, S., Sharp, P. 1990. *Mol. Cell. Biol.* 10:5562–64
141. Zehring, W. A., Greenleaf, A. L. 1990. *J. Biol. Chem.* 265:8351–53
142. Suzuki, M. 1990. *Nature* 344:562–65
143. Sigler, P. 1988. *Nature* 333:210–12
144. Laybourn, P. J., Dahmus, M. E. 1990. *J. Biol. Chem.* 265:13165–73
145. Dahmus, M. E., Dynan, W. S. 1991. *Transcriptional Regulation*, ed. K. Yamamoto, R. Tjian. Cold Spring Harbor, NY: Cold Spring Harbor Lab. In press
146. Simpson, R. B. 1979. *Cell* 18:277–85
147. Glass, R. E., Honda, A., Ishihama, A. 1986. *Mol. Gen. Genet.* 203:492–95
148. Scafe, C., Martin, C., Nonet, M., Podos, S., Okamura, S., Young, R. A. 1990. *Mol. Cell. Biol.* 10:1270–75
149. Nonet, M., Scafe, C., Sexton, J., Young, R. A. 1987. *Mol. Cell. Biol.* 7:1601–11
150. Himmelfarb, H. J., Simpson, E. M., Friesen, J. D. 1987. *Mol. Cell. Biol.* 7:2155–64
151. Lue, N. F., Kornberg, R. D. 1987. *Proc. Natl. Acad. Sci. USA* 84:8839–43

Annu. Rev. Biochem. 1991. 60:717–55

# TRANSLATIONAL CONTROL IN MAMMALIAN CELLS

*John W. B. Hershey*

Department of Biological Chemistry, School of Medicine, University of California, Davis, California 95616

KEY WORDS:   protein synthesis, phosphorylation, initiation factor, elongation factor, ribosome.

## CONTENTS

## PERSPECTIVES AND SUMMARY

Protein synthesis is an integral part of the pathway of gene expression and makes important contributions to modulation of the expression of specific genes. As a major consumer of energy in the cell, protein synthesis must be integrated into the overall metabolic activity of the cell by controlling the rates

717

0066-4154/91/0701-0717$02.00

of global protein synthesis. In order to understand how translation is regulated at the molecular level, knowledge of the mechanism of protein synthesis is required. In the five years since protein synthesis was reviewed in this series (1), a more detailed understanding of the mechanism of mRNA binding to ribosomes and the catalytic function of initiation factors has been gained. In particular, the ribosome scanning model for initiation has been well established and an alternate mechanism involving ribosome binding directly into internal regions of mRNAs has been demonstrated. A plausible working model of the pathway of protein synthesis is postulated, but a detailed molecular description of how mRNA and aminoacyl-tRNAs bind and interact on ribosomes and how soluble factors promote these interactions is not yet possible. Nor is it certain that all of the components of the translational apparatus have been identified.

The rate-limiting step of translation usually but not always occurs during the initiation phase. Methods are available to measure changes in rates of initiation and elongation for global or specific protein synthesis on polysomes, and to detect changes in the distribution of mRNAs between active polysomes and messenger ribonucleoprotein particles (mRNPs). The intrinsic activity or strength of most mRNAs is determined first by accessibility of the capped 5'-terminus to initiation factors, then by the ability of the 40S ribosomal subunit to bind and scan the mRNA distally, and finally by the frequency of recognition of the initiator codon and surrounding context. The mobilization of mRNAs from repressed mRNPs into active polysomes is controlled separately from ribosome initiation on polysomes and is an important regulator of the expression of specific mRNAs. The physical basis for mRNP repression is poorly understood in most cases, but insights into mRNP structure and activity are beginning to emerge. In particular, the repression of ferritin mRNA translation is explained by the binding of a heme-regulated repressor protein to a specific cis-acting RNA sequence in the 5'-untranslated region of the mRNA.

The dominant mechanism of control of global protein synthesis is phosphorylation/dephosphorylation of translational components, primarily initiation and elongation factors. In the stimulation of protein synthesis, eIF-4F (the mRNA cap-binding protein complex) is the major phosphorylation target, although phosphorylation of other proteins such as eIF-3, eIF-4B, and ribosomal protein S6 may be important also. The fact that overexpression of the cDNA encoding the $\alpha$-subunit of eIF-4F is oncogenic, whereas synthesis of a mutant form that resists phosphorylation is not, serves to emphasize the importance of translation and its regulation through phosphorylation in growth control. Phosphorylation also causes repression of protein synthesis, primarily by modifying the $\alpha$-subunit of initiation factor eIF-2 or the elongation factor eEF-2. Phosphorylation of eIF-2$\alpha$ appears to be a general mechanism for inhibiting initiation of translation, whereas eEF-2 phosphorylation affects elongation rates and may be used less frequently. The kinases thus far identified as responsible for the phosphorylation of translational components are usually highly specific, whereas the phosphatases exhibit a broad range of specificity. Their regulation is mostly obscure and represents one of the larger challenges for the future. A complex net of interactions, some redundant or overlapping in effect, may serve to tie translation to the overall metabolic state of the cell.

A wide variety of control mechanisms operate on classes of mRNAs or specific mRNAs. The poly(A) tail at the 3'-terminus of most mRNAs influences both mRNA stability and translational efficiency. *Trans*-acting proteins that bind to special *cis*-acting regions of an mRNA may be a general way to regulate the synthesis of specific proteins. Ribosomal frameshifting and translational readthrough of stop codons also contribute to the regulated expression of specific genes, especially the reverse transcriptase gene in retroviruses. In a growing number of examples, gene expression is regulated at the level of translation by unknown mechanisms. We can anticipate confronting a great variety of new and exciting regulatory mechanisms whose elucidation presents a challenge for future studies.

## PATHWAY AND MECHANISM OF PROTEIN SYNTHESIS

Protein synthesis is divided into three phases: initiation; elongation; and termination. The reactions in each phase are promoted by soluble protein factors (see Table 1) that transiently interact with the ribosome, mRNA, and aminoacyl-tRNAs. A striking feature of the pathway is the plethora of noncovalent binding interactions among the protein and RNA components and the dearth of covalent bond-making and -breaking reactions. Thus, during the initiation phase, only the hydrolysis of pyrophosphate bonds in ATP and GTP occurs, whereas a great many macromolecular binding and dissociation reactions occur. A thorough understanding of the pathway and mechanism requires knowledge of the structures of the various intermediates and of the rules governing protein and nucleic acid interactions. Neither aspect is known to a satisfactory degree, but a working model of the pathway may be proposed (Figure 1) to assist in our discussion of the reactions involved. The model is based primarily on in vitro studies with purified components, where stable putative intermediate complexes are detected and then arranged in a logical order. Such a model should be used with caution and skepticism, since in vivo evidence and kinetic confirmation are largely lacking. Most of the work

**Table 1**  Mammalian soluble factors[a]

| Factor | Other names | Functions | Factor mass (kDa) | Subunits mass (kDa) | Cloned cD species (R |
|---|---|---|---|---|---|
| **Initiation factors** | | | | | |
| eIF-1 | | stimulates formation of 40S preinitiation complexes | 15 | | |
| eIF-1A | eIF-4C | ribosome dissociation; stimulates 40S preinitiation complexes | 17.6 | | |
| eIF-2 | | GTP-dependent Met-tRNA binding to 40S ribosome | 130 | $\alpha$  36.1<br>$\beta$  38.4<br>$\gamma$  55 | human, rat (<br>human (234) |
| eIF-2A | | AUG-dependent Met-tRNA binding to 40S ribosome | 65 | | |
| eIF-2B | GEF | GTP:GDP exchange | 272 | $\alpha$  26<br>$\beta$  39<br>$\gamma$  58<br>$\delta$  67<br>$\epsilon$  82 | |
| eIF-2C | Co-eIF-2A | stabilizes ternary complex | 94 | | |
| eIF-3 | | stimulates formation of 40S preinitiation complex | 550 | $\alpha$  35<br>$\beta$  36<br>$\gamma$  40<br>$\delta$  44<br>$\epsilon$  47<br>$\zeta$  67<br>$\eta$ 115<br>$\theta$ 170 | |
| eIF-3A | eIF-6 | ribosome dissociation; binds 60S ribosome | 25 | | |
| eIF-4A | | RNA-dependent ATPase; helicase; stimulates mRNA binding | 44.4 | | mouse (23: |
| eIF-4B | | helicase; stimulates mRNA binding | 80 | | human (23 |
| eIF-4F | CBP-II<br>eIF-4E<br><br>eIF-4A<br>p220 | cap recognition; helicase; stimulates mRNA binding<br><br><br>cleaved by poliovirus | 270 | $\alpha$  25<br><br>$\beta$  44.4<br>$\gamma$  220 | human (23<br>mouse (23 |
| eIF-5 | | stimulates ribosome junction | 150 | | |
| eIF-5A | eIF-4D | stimulates 1st peptide bond formation | 16.7 | | human (2: |
| eEF-1$\alpha$ | | binds aa-tRNA; GTPase | 51 | | human (2<br>mouse (2 |

**Table 1**   (*Continued*)

| Factor | Other names | Functions | Factor mass (kDa) | Subunits mass (kDa) | Cloned cDNA species (Ref.) |
|---|---|---|---|---|---|
| eEF-1$\beta$ | | GTP:GDP exchange on eEF-1$\alpha$ | 23 | | *Artemia* (242) |
| eEF-1$\gamma$ | | GTP:GDP exchange with eEF-1$\beta$ | 49 | | *Artemia* (243) |
| eEF-2 | | stimulates translocation; GTPase | 100 | | rat, hamster (244) |
| Termination factor | | | | | |
| eRF | | recognizes stop codons, stimulates peptidyl-tRNA cleavage and release | 54.0 | | rabbit (245) |

[a] Information in the table was obtained in part from (4), where further references may be found.

leading to the model was conducted prior to 1985 and therefore has been reviewed in detail in this series (1). An extensive, up-to-date treatment of eukaryotic translation also is available (2) as well as descriptions of the structure, function, and genetics of ribosomes (3). In the following sections, I merely sketch the outlines of the various reactions and concentrate on aspects relevant to translational control.

## Initiation

During the initiation phase, two important tasks are achieved: (*a*) an mRNA among many available candidates is selected for translation by the initiating ribosome; and (*b*) the ribosome identifies the initiator codon and begins translation in the appropriate reading frame. The various binding reactions are promoted by at least 10 initiation factors (abbreviated eIF), whose physical characteristics are listed in Table 1. These proteins have been highly purified, and a number of their cDNAs or genes have been cloned and sequenced. Throughout this review, I employ the revised nomenclature for soluble factors as specified by the Nomenclature Committee of the International Union of Biochemistry (4). Previous names are included in Table 1 to facilitate the reader's transition to the new nomenclature.

We identify four major steps in the initiation pathway: 1. ribosome dissociation into 40S and 60S subunits; 2. Met-tRNA$_i$ binding to the 40S ribosomal subunit to form a 40S preinitiation complex; 3. mRNA binding to the 40S preinitiation complex to form a 40S initiation complex; and 4. junction of the 40S initiation complex with the 60S subunit to form an 80S initiation complex. I concentrate primarily on steps 2 and 3, and limit comments to descriptions of our current understanding of the pathway.

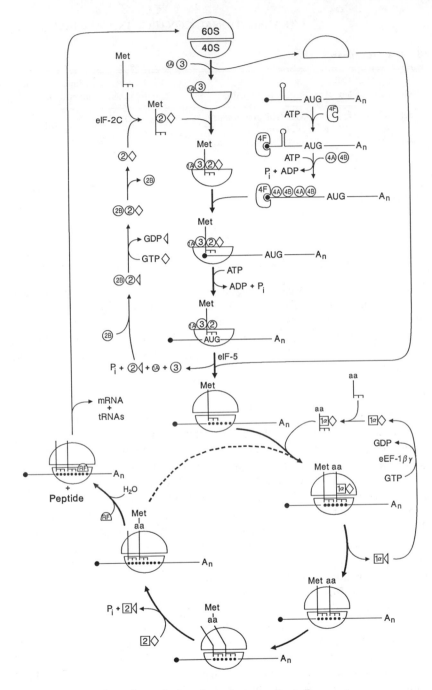

*Figure 1*　Model of protein synthesis pathway in mammalian cells

DISSOCIATION OF 80S RIBOSOMES    At $Mg^{2+}$ concentrations thought to be physiological ($>1$ mM), 80S ribosomes predominate in active equilibrium with dissociated subunits. Two initiation factors, eIF-1A and eIF-3, shift the equilibrium towards dissociation by binding to 40S subunits and preventing their association with 60S subunits (5). Another factor, eIF-3A, binds to 60S subunits and prevents subunit association in vitro (6), although the role of this protein in translation is not well established. The mechanism of ribosome dissociation has been reviewed in detail (1); little additional work has been reported since then. Since the levels of eIF-3 and possibly eIF-1A are nearly equal to those of ribosomes, one would expect extensive 80S dissociation in translationally repressed cells, yet mostly 80S ribosomes rather than predominantly subunits are observed. Either the in vitro studies do not reflect the process of dissociation in vivo accurately, or other components interact with 80S ribosomes to favor their association.

METHIONYL-tRNA BINDING AND eIF-2 FUNCTION    The binding of the initiator methionyl-tRNA (Met-tRNA$_i$) to 40S ribosomal subunits is a step common to the translation of all mRNAs. eIF-2, in a binary complex with GTP, binds Met-tRNA$_i$ to form a ternary complex. The ternary complex is an obligatory intermediate that can be readily identified and isolated. Ternary complexes are formed in high yield with purified components at physiological concentrations of factor and tRNA (ca. 1 $\mu$M) (7, 8). At lower concentrations ($<0.1$ $\mu$M), stabilization by ancillary factors has been observed. One of these, eIF-2C (formally called Co-eIF-2A$^{80}$; a 94 kDa protein), acts stoichiometrically to stabilize the complex and prevent its disruption when naked mRNA is added (9). Another is eIF-3, which also stimulates ternary complex formation at low concentrations of eIF-2 and Met-tRNA$_i$ (10). In the case of eIF-3, disruption by mRNA is not prevented, however. It is not clear if the interactions of eIF-2C and eIF-3 with the ternary complex actually occur in vivo off the ribosome or whether they reflect subsequent interactions on the surface of the 40S ribosomal subunit. Since eIF-2 binds GDP 400-fold more tightly than GTP (11), the proportion of eIF-2 in potentially active binary complexes with GTP will be sensitive to the energy charge of the cell (see also discussion of eIF-2 recycling below).

The ternary complex binds to the 40S ribosomal subunit to form a 40S preinitiation complex that is sufficiently stable to be isolated by sucrose density gradient centrifugation (12). Such complexes are stabilized by eIF-3 and eIF-1A, but do not require the presence of mRNA (13, 14). Furthermore, GTP hydrolysis does not occur and nonhydrolyzable GTP analogues can function in these steps. Since stable mRNA·40S complexes are not detected in the absence of Met-tRNA$_i$, the proposed pathway of initiation places Met-tRNA$_i$ binding prior to mRNA binding. An important caveat is that other pathways involving unstable intermediates are possible. Initiation complexes are most frequently analyzed by sucrose density gradient centrifugation, which usually requires hours of centrifugation to separate ribosome-bound and free components. Complexes are detected only if their dissociation rates are slower than the centrifugation times, yet many physiologically relevant complexes may dissociate in seconds rather than hours. Thus we cannot rule out a pathway that includes initial binding of mRNA to 40S ribosomal subunits in the absence of eIF-2 and Met-tRNA$_i$ (see section below). A kinetic analysis of the possible reactions is required to define the pathway more precisely.

mRNA BINDING    An initiating 40S ribosomal subunit interacts with mRNA in the form of either a free mRNP particle or a polysome with associated polysomal RNP proteins and translating ribosomes. It is generally believed that the 40S preinitiation complex with Met-tRNA$_i$, eIF-2, eIF-3, and eIF-1A is the active form of the ribosomal subunit (14), but this view is not firmly established. Either of two modes of mRNA binding occur: the ribosome binds to the 5'-terminus of the mRNA, then scans linearly down the mRNA until it recognizes the initiator codon (the scanning model); or the ribosome binds to an internal region of the mRNA, either directly at the initiator codon, or somewhat upstream followed by a scanning process (internal initiation model). In either case, formation of the 40S initiation complex is thought to be promoted by ATP hydrolysis and by a number of initiation factors. A tentative pathway for the scanning model is depicted in Figure 1.

The scanning model accounts for initiation of the large majority of cellular mRNAs (15). Important features of mRNA structure [reviewed in (15, 16)] include the presence of a 7-methyl-guanylic acid "cap" at the 5'-terminus, aspects of secondary and tertiary structure, the initiator codon (nearly always AUG) and surrounding sequences (context), and the poly-adenylate [poly(A)] tail at the 3'-terminus (discussed in greater detail below).

The role of initiation factors in the scanning mechanism is only beginning to be understood [reviewed in (17, 18)]. A working hypothesis is that eIF-4F binds to the cap through its $\alpha$-subunit (eIF-4E), which is known as the cap-binding protein because it binds and crosslinks to the cap structure. eIF-4A and eIF-4B may then join the complex. Highly purified eIF-4A or eIF-4F in combination with eIF-4B exhibits an ATP-dependent RNA helicase activity (19–21). eIF-4A is an RNA-dependent ATPase (22) and is homologous to other "D-E-A-D box" proteins with helicase activity (23); eIF-4B contains a consensus RNA-binding domain present in a number of other RNA-binding proteins (236). It is not yet known if eIF-4A and eIF-4B continue their helicase activity until the entire 5'-untranslated region (5'-

UTR) of the mRNA is melted up to the initiator codon, with the 40S subunit passively following, or if the 40S ribosome joins these factors earlier, and is an essential contributor to the melting and movement towards the AUG. In either case, ATP hydrolysis is essential for the unwinding process but may or may not be directly involved in 40S ribosomal subunit migration. It is possible that higher-order initiation factor complexes are involved in cap recognition. Both eIF-4F and eIF-4B bind to eIF-3, and evidence for such super-factor complexes has been reported (25, 26). Since eIF-3 binds tightly to 40S ribosomal subunits and also to mRNA, it may be the link between the 40S ribosomal subunit and the mRNA-factor complexes. Alternatively, the α-subunit of eIF-4F alone may interact first with the cap, followed by the binding of the other factors and the ribosome, as suggested by Rhoads (16). Clearly, additional biochemical studies are needed to elucidate how the initiation factors contribute to the scanning mechanism and how the 40S ribosomal subunit slides distally from the cap along the mRNA.

Following recognition of the cap structure by the initiation factors and the melting out of secondary structure near the 5'-terminus, the 40S preinitiation complex binds to the cap-proximal region of the mRNA. The 40S ribosomal subunit then begins the scanning process in search of the initiator codon. The scanning model predicts that the 5'-proximal initiator codon should be preferred, and this is true for >90% of cellular mRNAs (27). A modulating feature is the context in which the AUG occurs. Important elements in a strong context include purines at positions $-3$ and $+4$ (where the A of AUG is $+1$) (28). When a scanning ribosome encounters an AUG with a weak consensus sequence, it may pass over the AUG and continue to scan the mRNA. What recognizes the AUG and consensus sequences? The anticodon of Met-tRNA$_i$ surely interacts with the AUG, as has been shown conclusively in yeast (29). Genetic studies also implicate eIF-2 in the recognition process, where mutations in the α- and β-subunits of eIF-2 alter codon selection in yeast (30, 31). However, the macromolecules that recognize and interact with the surrounding consensus sequences remain a mystery. It is difficult to see how Met-tRNA$_i$ can play this role, nor has evidence been generated for rRNA involvement. Specific recognition of AUG-ribosome binding sites by eIF-2 (32) and eIF-4B (33) has been proposed, but needs further experimental verification. Whatever the mechanism, the 40S initiation complex presumably pauses at AUGs with strong consensus sequences, thereby enabling the 60S junction reaction to occur. mRNA secondary structure properly placed downstream of the initiator AUG promotes initiation, presumably by enhancing pausing near the AUG and impeding further scanning (24).

mRNA binding to ribosomes by the "internal initiation" mechanism has been demonstrated for poliovirus (34) and encephalomyocarditis (EMC) virus RNAs (35). The long 5'-UTRs and the presence of numerous upstream AUGs between the uncapped 5'-terminus and initiator codon strain the credibility of the scanning mechanism for these RNAs. Proof for an internal initiation mechanism has been accomplished by constructing di-cistronic genes with the ribosome entry region of the poliovirus or EMC virus 5'-UTRs inserted between the cistrons (34, 35). Enhanced in vivo expression of the downstream cistron in transfected cells, even when translation of the upstream cistron is inhibited, indicates ribosomal entry between the cistrons. The possibility that translation of the second cistron is due to partial RNA cleavage has been ruled out by Northern blot analyses of polysomes actively translating only the second cistron. Whereas most evidence for internal initiation has been obtained with viral RNAs, there is a report of a cellular mRNA encoding the glucose-regulated protein 78/immunoglobulin heavy chain binding protein (GRP78/BiP) that utilizes a cap-independent process and may initiate by the internal binding mechanism (36). Cellular mRNAs with long 5'-UTRs that contain numerous AUG's [e.g. human c-*abl* mRNA (37)], also are good candidates for translation by the cap-independent mechanism. Whereas the internal initiation mechanism likely is a minor contributor to initiation in mammalian cells, it may play an important role in enabling the translation of a small class of mRNAs when the cap-dependent scanning mechanism is repressed.

A detailed molecular mechanism of internal initiation is not yet available, nor have essential initiation factors been identified. It appears likely that the same initiation factors are involved in both mechanisms, except that internal initiation does not involve cap recognition and therefore may not require a functional eIF-4F. An attractive hypothesis is that eIF-4F is non-essential and that eIF-4A and eIF-4B bind directly to the internal region and begin their helicase activity. An added complexity may be the binding of mRNA-specific proteins to specific structures in these 5'-UTRs. Poliovirus-specific proteins of 52 kDa (38) and 50 kDa (39) and a 57 kDa EMC virus-specific protein (40) have been identified that may assist in directing ribosome binding. In the case of poliovirus, ribosome entry appears to occur upstream of the initiator AUG, since secondary structure introduced between the ribosome entry site and the initiator AUG inhibits translation (34). In contrast, with EMC viral RNA, the ribosome appears to bind directly at the 11th AUG, the authentic initiator codon (41).

80S INITIATION COMPLEX FORMATION AND INITIATION FACTOR RECYCLING    The 60S ribosomal subunit binds to the 40S subunit carrying mRNA and Met-tRNA$_i$ positioned at the initiator codon. The reaction requires the function of eIF-5 and the hydrolysis of the GTP molecule bound to eIF-2. The GTPase reaction is promoted by eIF-5 in the absence of 60S subunits and results in the ejection of eIF-2·GDP and other bound factors such as eIF-1A and eIF-3 (14, 42, 43). Subsequent binding of the 60S subunit may be rapid, since the stability of bound Met-tRNA$_i$ to 40S ribosomes in the absence of eIF-2 is low. The physical nature of eIF-5 remains controversial. Initial purification and characterization of the factor identified a protein of 150–170 kDa [reviewed in (1)], whereas a 62 kDa protein with qualitatively comparable activity has been isolated (44) and shown not to be a proteolytic fragment of the larger form (45).

The eIF-2·GDP binary complex must exchange the GDP for GTP in order for eIF-2 to catalyze another round of initiation. The rate of guanine nucleotide exchange is slow at $Mg^{2+}$ concentrations in the physiological range and the reaction thus requires catalysis. eIF-2B, frequently called GEF (for guanine nucleotide exchange factor), interacts with eIF-2·GDP to enhance the exchange rate by a mechanism that remains controversial (11, 46). A separate issue is the thermodynamics of the reaction: eIF-2 binds GDP ca. 400-fold more avidly than GTP. The high GTP/GDP ratio (energy charge) and the removal of the eIF-2·GTP product of the equilibrium reaction by ternary complex formation contribute to a reaction flux towards the eIF-2·GTP complex.

Following 80S initiation complex formation, the elongation phase of protein synthesis commences. However, formation of the first peptide bond has unique features that distinguish it from subsequent peptide bond formation reactions: the Met-tRNA$_i$ in the ribosomal P site (donor site) has a charged $\alpha$-amino group, whereas in subsequent rounds of the elongation cycle this aminoacyl-tRNA derivative is acylated as peptidyl-tRNA. The initiation factor eIF-5A appears to promote formation of the first peptide bond, usually measured by methionyl-puromycin synthesis (47). The factor is uniquely modified on Lys50 by the polyamine, spermidine, and subsequently is hydroxylated to form a hypusine residue [$N^{\epsilon}$ (2-hydroxy-4-aminobutyl)lysine] (48). The hypusine (or deoxyhypusine) modification is essential for eIF-5A activity in the in vitro methionyl-puromycin synthesis assay (49) and is required for eIF-5A function in vivo in yeast (J. Schnier, H. G. Schwelberger, Z. Smit-McBride, H. Kang, J. W. B. Hershey, manuscript submitted). How eIF-5A and its hypusine residue function in the reaction is not known.

## Elongation and Termination

The elongation phase of protein synthesis is a cyclic process that adds one amino acid residue to the C-terminal end of the nascent polypeptide chain per turn of the cycle. It is promoted by four proteins, called elongation factors (abbreviated eEF), which have been purified and whose cDNAs have been cloned and sequenced (Table 1). The pathway (Figure 1) involves four major steps: 1. binding of aminoacyl-tRNA to the A site of the ribosome, catalyzed by eEF-1$\alpha$; 2. GTP hydrolysis and ejection of eEF-1$\alpha$·GDP; 3. formation of the peptide bond, catalyzed by the peptidyl transferase center in the 60S ribosomal subunit; and 4. translocation, promoted by eEF-2. Considerable energy is expended during each cycle: at least one high-energy pyrophosphate bond in GTP is cleaved at both the binding and translocation steps; and two additional high-energy bonds are required to synthesize the aminoacyl-tRNA that is consumed. The process is rapid, a ribosome incorporating up to six amino acids per second, although not so rapid as in bacteria (15–18 residues/ sec). Key features are high fidelity, i.e. the ability to match properly the aminoacyl-tRNA and the codon in the mRNA template, and processivity, i.e. the ability to synthesize long polypeptides without premature dissociation of the peptidyl-tRNA. The termination phase, in contrast, is relatively simple. When a termination codon is positioned in the A site of the ribosome, a release factor (eRF) (Table 1) promotes cleavage of the completed peptidyl-tRNA, releasing the protein.

PATHWAY AND MECHANISM    Our knowledge of the molecular events of elongation and termination in mammalian cells is primarily by analogy to mechanisms elucidated in bacteria, where some important advances have been made recently. The components and pathway have been reviewed by Moldave in this series (1), and more currently by Slobin (50). Aminoacyl-tRNA forms a ternary complex with eEF-1$\alpha$ and GTP prior to binding to the A site of the ribosome. Following ribosome binding but before peptide bond formation can occur, the GTP is hydrolyzed by the factor, and eEF-1$\alpha$ dissociates as a binary complex with GDP. The ejection of eEF-1$\alpha$-GDP may be the rate-limiting step of elongation, as it appears to be in bacteria (51), and is thought to provide an additional opportunity for incorrectly bound aminoacyl-tRNA to dissociate from the ribosome (proofreading). In order for eEF-1$\alpha$ to promote another binding reaction, the GDP must be exchanged for GTP. Reminiscent of the situation with eIF-2, the release of GDP from eEF-1$\alpha$ is slow [$14 \times 10^{-3}$ s$^{-1}$ for the *Artemia* protein (52)], and requires catalysis by a factor complex called eEF-1$\beta\gamma$. eEF-1$\alpha$ is a highly abundant protein in cells, and is posttranslationally modified by methylation of a number of lysine residues (53, 54). It also is uniquely modified on two glutamic acid residues by formation of an amide linkage to glycerylphosphorylethanolamine (55, 56). However, the role of these modifications in vivo is unclear; they do not appear to be required for factor activity in vitro (W. C. Merrick, personal communication).

Little is known about the catalytic entities on the 60S ribosomal subunit that promote peptide bond formation. In contrast, significant advances in our understanding of the translocation reaction have been made for the *Escherichia coli* system. Using RNA modification methods and refined models of the bacterial ribosome, Noller and coworkers were able to distinguish at the structural level between bound peptidyl-tRNA before and after translocation (57). In the pretranslocation state, the anticodon end of the peptidyl-tRNA is bound in the A site, but the peptidyl end already is located in the P site. The translocation reaction actually occurs prior to GTP hydrolysis (i.e. with the GMP-PNP analogue) (58). It therefore is likely that eEF-2·GTP binding alone promotes the movement of the mRNA and the anticodon portion of the peptidyl-tRNA into the P-site. How eEF-2 promotes the movement of mRNA and peptidyl-tRNA on the ribosomal surface is unclear. eEF-2 also is uniquely modified by conversion of a specific histidine residue to diphthamide. Diphthamide in turn is the target of toxins from either *Corynebacterium diphtheriae* or *Pseudomonas aeruginosa,* which ADP-ribosylate the factor and inhibit its activity (59, 60). Transfection of mammalian cells with mutant forms of the cDNA encoding eEF-2, where the target His is altered to Lys or Arg, does not result in toxin resistance (61). It is possible that the mutant factor lacking diphthamide does not function in protein synthesis, or that endogenous eEF-2 when ADP-ribosylated inhibits dominantly. However, yeast mutants lacking enzymes in the biosynthetic pathway for diphthamide are viable, indicating that the His residue fully modified to diphthamide is not essential for protein synthesis (59). Since the diphthamide modification reactions are reversible, a regulatory role for this modification remains a possibility.

## Reinitiation

Nearly all cellular mRNAs are monocistronic, i.e. only a single translational start and stop signal are recognized and utilized. This is in contrast to protein synthesis in bacteria, where polycistronic mRNAs are common and are translated either independently or by coupling translation of the downstream cistron to that of the upstream cistron. Recent work suggests that eukaryotic ribosomes also are capable of termination at one open reading frame (ORF) and initiating at another downstream ORF, a process we call *reinitiation.* The following observations indicate that reinitiation occurs at low frequency in mammalian cells. An mRNA with an upstream AUG and strong consensus sequence inserted in the 5'-UTR translates the major protein coding sequence poorly, but does so more efficiently when a termination codon is introduced in-frame and distal to the upstream AUG (62). Translated upstream ORFs are usually short (63), although low-level reinitiation after a major ORF has been reported for the xanthine-guanine phosphoribosyl transferase (64) and dihydrofolate reductase (65) coding regions. Placement of the upstream "minicistron" affects the efficiency of reinitiation; increased efficiency is seen as the distance separating the two ORFs becomes greater (66). The mechanism of reinitiation remains unclear: after termination, either 80S ribosomes or 40S ribosomal subunits resume scanning in a downstream or upstream direction.

Translation of the small upstream ORF may result in incomplete dissociation of soluble factors following termination, so that reinitiation is facilitated. Alternatively, sufficient distance separating the ORFs may provide enough time for Met-tRNA$_i$ and initiation factors to bind to the scanning ribosome. Reinitiation plays a key role in regulating the translational efficiency of the yeast *GCN4* mRNA, where four upstream minicistrons are found [reviewed in (67, 68)]. Elucidation of this complicated control mechanism with the assistance of yeast genetics is providing insight into the mechanisms of initiation, termination, and reinitiation.

# APPROACHES TO STUDYING TRANSLATIONAL CONTROL

## Determinants of Protein Synthesis Rates

In order to change the rate of protein synthesis, a cell must alter the kinetic parameters of the step that is overall rate limiting. Changing the rates of other steps will not lead to translational regulation so long as the rate-limiting step remains constant. Therefore it is pertinent to ask: What limits the rate of protein synthesis in cells? The answer varies, depending on the cell type and its physiological state. A priori, rates depend on the level of the translational machinery and its specific activity. For specific protein synthesis, the limiting components are the particular mRNA's cytoplasmic concentration and specific activity, the latter determined by the mRNA's intrinsic activity (i.e. "strength") and by *trans*-acting elements. For global protein synthesis, the limiting parameters are more difficult to identify. We examine here those features of the translational machinery that contribute to determining the rates of specific and global protein synthesis.

mRNA LEVELS AND STRUCTURE    The level of an mRNA is determined by its rate of synthesis (transcription), the efficiency of posttranscriptional processing and transport out of the nucleus, and its degradation rate in the cytoplasm. These issues lie outside the scope of this review and are not considered further. We note only that in terms of global protein synthesis, mRNAs seem not to be limiting. In cells such as hepatocytes following feeding, or cultured cells in exponential growth, where at least 90% of the ribosomes are active in polysomes, there remains a sizeable portion of cytoplasmic mRNA in the non-active mRNP compartment.

The intrinsic efficiency whereby a mRNA is translated depends on the structure of the specific RNA [reviewed in detail by Kozak (15) and by Rhoads (16)]. Some mRNAs are translated efficiently, whereas others are translated inefficiently. What determines such efficiency? Four features appear to be important for mRNAs translated by the scanning mechanism.

1. The presence and accessibility of a $5'$-$m^7G$ cap structure enhances mRNA translation, as demonstrated in vitro. All mammalian cellular mRNAs are thought to be capped, but uncapped mRNAs are found in some virus-infected cells. The accessibility of the $5'$-cap structure to eIF-4F correlates with mRNA strength; higher-order structures that bury the cap reduce translational efficiency (69, 70).

2. Secondary and tertiary structures of mRNAs play an important role, depending on their stability and position. Secondary structure in the $5'$-UTR may affect cap accessibility and may impede the scanning 40S ribosomal subunit if sufficiently stable (>50 kcal/mol) (15, 71, 72). However, when placed just downstream from the initiator codon, such structures cause the scanning 40S ribosomal subunit to pause at the AUG and thereby enhance initiation (28). If the cap is accessible, the translational machinery can melt out moderately stable structures during scanning. This is in contrast to bacterial initiation, where efficiency is greatly influenced by secondary structure near the initiator AUG (73, 74). Secondary structure in the coding region of mammalian mRNAs, even when very stable, does not block or greatly impede the elongating ribosome (75), nor do long cDNAs annealed to this region of the mRNA (76). This likely is due to partial opening (breathing) of such structures, allowing the ribosome to advance a codon at a time, thereby eventually melting out the entire structure, regardless of its overall stability. A problem in evaluating the role of secondary and tertiary structures in translational efficiency is that these structures are difficult to determine experimentally or to predict with confidence from the mRNA sequence. Sequences in the coding or $3'$-untranslated region can bind to the $5'$-UTR by long-range interactions, thereby affecting initiation. A further complication is that mRNAs are presented to ribosomes, not as naked RNAs, but as mRNP particles. Very little is known about how the proteins in mRNPs influence higher-order structures and thus the process of translation (see below).

3. The placement and context of the AUG initiator codon affect translational efficiency as discussed earlier. The use of non-AUG initiator codons, while rare, reduces initiation strength (24).

4. Finally, the presence and length of poly(A) tails at the $3'$-terminus contribute to translational efficiency, as described in a later section of this review. The poly(A) tail length for a mRNA may vary with time, thereby affecting its translation.

In summary, mRNA strength is a measure of the frequency of binding of a 40S ribosomal subunit to the AUG initiator codon and initiation of translation there. Cap accessibility is critical for opening up the $5'$-terminus of mRNAs to allow 40S preinitiation complexes to bind. Bound 40S subunits must migrate along the mRNA towards the initiator AUG, during which secondary structure and/or premature dissociation from the mRNA may influence the frequency of reaching the AUG. The initiator codon and its context then influence the efficiency of productive initiation events. It is obvious that one cannot predict the innate strength of an mRNA on the basis of its primary structure or initiator codon context alone.

LEVEL OF TRANSLATIONAL MACHINERY    Ribosome levels appear to define the overall capacity of a cell to synthesize proteins. Tissues relatively active in protein synthesis, such as liver and brain, contain large amounts of ribosomes and their associated soluble protein factors, whereas the opposite is true for many other cell types. Shortly after feeding, essentially all liver ribosomes are engaged in translation, and are functioning at their maximum elongation rate (77). Thus in this situation, ribosome levels limit protein synthesis. However, some hours later, only a fraction of available ribosomes are functioning; some other component becomes rate limiting. It is generally believed that the rate-limiting step of protein synthesis under most physiological conditions is the initiation phase (78). This view is based in part on analyses of polysome size, where ribosomes are usually spaced along the mRNA at intervals of 80–100 nucleotide residues. Were initiation rates fast relative to elongation, ribosomes could bind as close as one every 30 nucleotides, as is observed when ribosomes stack up behind pause sites in the mRNA (75). Another method to distinguish initiation vs elongation as the rate-limiting step is to treat cells with low concentrations of the elongation inhibitor, cycloheximide (79, 80); mRNAs whose translation is elongation rate-limited are highly sensitive to the inhibitor, whereas mRNAs with relatively weak initiation rates are more resistant.

Kinetic studies of translation indicate that both the 40S preinitiation complex (81) and the initiation factor eIF-4F (82–84) contribute to limiting the overall rate of translation. However, polysome profiles show the presence of "half-mers," where a 40S ribosomal subunit is bound to a polysome, implying that the 60S junction step also may be relatively slow and near-limiting. It is likely that a number of different steps may be close to rate-limiting, so that an increase in the level of a single limiting component may cause little overall effect. The cellular concentrations of the soluble factors are likely coordinately regulated with ribosome levels, as has been shown for some of the initiation factors in lymphosarcoma cells down-regulated by treatment with glucocorticoids (85). On the other hand, the ratio of eIF-2B to eIF-2 varies substantially, depending on the cell type; the ratio for rabbit reticulocytes is ca. 0.15 whereas that for Ehrlich ascites cells is ca. 0.5 (86). Similarly, the relative and absolute levels of the two forms of eIF-4A, as deduced from mRNA concentrations, vary appreciably in different tissues (87). Such variability raises the possibility that a component limiting in one cell type or physiological state may not be limiting in another. Unfortunately, a comprehensive analysis of the levels of translational components is lacking.

SPECIFIC ACTIVITY OF TRANSLATIONAL MACHINERY    Translation rates may be regulated by the activity of mRNAs, ribosomes, and their associated soluble factors. Regulation of specific activities rather than levels allows the cell to alter protein synthesis rates rapidly, and to reverse the inhibition or activation equally rapidly. The translational control of ferritin synthesis described below is an example of the cell's need to respond quickly to an environmental change, namely the presence of iron. Posttranslational covalent modification of protein components of the translational machinery, such as phosphorylation or methylation, is known to occur. Phosphorylation/dephosphorylation appears to be the most prevalent mechanism of translational control, especially at the global level. The synthesis or regulation of *trans*-acting factors specific for one or a small class of mRNAs also may occur. How the activity of the translational machinery is regulated is the major focus of the remainder of this review.

## Measuring Protein Synthesis Rates

The rate of protein synthesis is most frequently determined by measuring the incorporation of radioactive amino acid precursors into protein. Its application to estimating global protein synthesis rates is obvious; the major caveat is that intracellular amino acid pools may vary, thereby altering the specific radioactivities of the precursors. It appears that in most cases the specific radioactivity of amino acids in the intracellular pool equilibrates within one to two minutes with that of the extracellular pool (88, 89), but exceptions have been reported (90). Measurement of rates of synthesis of specific proteins requires their separation from the bulk of proteins, usually accomplished by immunoprecipitation or fractionation by two-dimensional isoelectric focussing–SDS polyacrylamide gel electrophoresis. The methods provide a measure of the number of amino acids incorporated per unit time, from which the number of protein molecules synthesized per unit time can be calculated if the specific or average size and amino acid composition of the protein(s) are known.

An alternative method for estimating protein synthesis rates measures the number of ribosomes actively engaged in protein synthesis together with the average rate of amino acid incorporation by the ribosomes (91). For bulk protein synthesis, the number of active ribosomes is determined by sucrose density gradient centrifugation, which separates active ribosomes (polysomes) from nontranslating 80S particles and ribosomal subunits. For specific proteins, the average number of ribosomes per mRNA and the number of specific mRNAs in polysomes may be determined from gradient analyses by hybridization techniques, yielding the total number of ribosomes engaged in protein synthesis on the specific mRNA. The rate of amino acid incorporation by active ribosomes (elongation rate) is determined by measuring the ribosome transit time, the time required for a ribosome to translate a mRNA. Ribosome transit times are measured for bulk or specific protein synthesis by in vivo labeling with amino acid precursors followed by separation of released proteins from ribosome-bound (nascent) protein by centrifugation (89, 92). The method is not sensitive to variations in amino acid pools or specific activities, but requires that the synthesis of similar-sized products be compared. The elongation rate is equal to the number of amino acid residues encoded by the mRNA divided by the ribosome transit time. Thus:

$$\text{protein synthesis rate} = \frac{[\text{\# active ribosomes}] \times [\text{\# amino acid residues}]}{\text{ribosome transit time}}$$

Values obtained usually fall in the range of 3 to 6 amino acids incorporated per ribosome per second at 37°C.

A priori, a third method is to measure the absolute rate of initiation of specific or bulk proteins. Since the rate of protein synthesis can be expressed as the number of proteins synthesized per unit time, this value under steady-state conditions is equal to the number of initiation events irrespective of the elongation rate (assuming that each initiation event results in a completed protein product, a condition that holds approximately). No general method exists for measuring initiation events in vivo directly, so this approach is not useful. An absolute initiation rate, however, can be calculated from polysome size and elongation rate measurements (91):

$$\text{\# initiation events/min} = \frac{[\text{\# ribosomes in polysome}]}{\text{ribosome transit time (min)}}$$

An efficiently translated mRNA at 37°C initiates protein synthesis once every 5 to 6 sec.

## Targets of Translational Control

Regulation of protein synthesis requires altering the rates of one or more of the phases of protein synthesis. To increase clarity, the initiation phase needs to be further divided and defined. We restrict the use of the term *initiation* to the process of ribosome binding to a preexisting polysome. The binding of the first ribosome to an mRNP to begin the formation of a polysome is quite distinct (see below) and is called *mobilization*. When a ribosome terminates translation on one cistron and then initiates translation on a distal cistron of the same mRNA, the process is called *reinitiation* (as defined earlier).

MOBILIZATION OF mRNPs INTO POLYSOMES   Cytoplasmic mRNAs are found in two distinctly different functional states: those actively being translated by ribosomes, in polysomes; and those that are not active, in mRNP particles. Regulation of the proportion of active to non-active mRNAs constitutes a type of translational control that is readily distinguished from changes in the initiation rate on polysomes. A substantial amount of total cytoplasmic mRNA is found as free mRNPs in a wide variety of cells. For example, 30% of mRNAs are in mRNP particles in active muscle cells (93). An example of a shift of mRNA from polysomes into mRNPs, with little change in polysome size, is the inhibition of translation of ribosomal protein mRNAs upon treatment of mouse lymphosarcoma cells with dexamethasone (94). The amount of mRNA decreases in polysomes and increases in mRNPs, but the number of ribosomes present per mRNA changes very little. A shift of mRNA from mRNPs into polysomes is a dominant event in early development, especially in sea urchins and in the maturation of *Xenopus* oocytes (95). Many specific mRNAs are present simultaneously in free mRNPs and in large polysomes. For example, the distribution of ferritin mRNA is bimodal in iron-depleted cells (96). Insulin mRNA is found primarily in mRNPs in islet cells treated with low glucose, but is mobilized into polysomes in the presence of high glucose (97). The fact that mRNPs (which have zero bound ribosomes) fall well outside the Poisson distribution for the number of ribosomes bound to a specific mRNA means that the binding of the first ribosome to an mRNP is kinetically different from the initiation rates that help maintain a polysome size. One plausible explanation is that mRNPs are repressed or masked by a *trans*-acting element, blocking initiation. Another is that initiation by a ribosome on an mRNP may differ mechanistically from initiation on a polysome.

What regulates the inability of mRNPs to be translated (i.e. their apparent repression)? mRNPs comprise about 75% protein and 25% RNA, with up to 30 different protein species ranging in mass from 25 to 150 kDa. Few of these proteins have been well characterized except the 72 kDa poly(A) binding protein (PABP) (98, 99). Most of the proteins appear to be present in the RNPs of a wide range of mRNA species, whereas it is expected (but not yet widely demonstrated) that some proteins may be restricted to a given RNA or class of RNAs. Lacking are determinations of where the proteins bind (except for PABP), their stoichiometry in the mRNP particle, and their precise functions. Some of the proteins may remain bound as the mRNA is mobilized into polysomes, since polysomal mRNA stripped of ribosomes still is complexed with proteins. The most abundant of the polysomal RNP proteins are the PABP and a 50 kDa protein.

Recent work begins to provide some insight into how mRNPs are repressed and activated. The rate of recruitment of ribosomes into polysomes was studied in a nuclease-treated reticulocyte lysate with exogenous globin mRNA (100). The first ribosome bound to the mRNA rapidly, but subsequent ribosome binding was slower, requiring about 15 minutes to reach full-sized polysomes. This result is unexpected, since in vivo analyses suggest that binding of the first ribosome is a slow step. A possible explanation is that once a ribosome translates an mRNA and then terminates, it preferentially and efficiently reinitiates protein synthesis on the same mRNA. The introduction of new ribosomes from the free pool of ribosomes would appear to be a relatively slow process in this study. In another study, the ability of mRNAs to be translated in vitro, either as naked RNA or in RNP complexes, was compared in rabbit reticulocyte and wheat germ lysates (101). In the reticulocyte lysate, high concentrations of naked mRNA but not mRNPs inhibit protein synthesis, perhaps explained by the fact that naked RNA binds and sequesters initiation and elongation factors essential for translation (102). The wheat germ lysate does not translate the mRNPs well, indicating that it lacks the ability to unmask the particles. Treatment of mRNPs with high salt or with the ribosomal high salt wash from reticulocyte lysates activates the mRNPs for translation in the wheat germ system, apparently by removing a masking factor (103). A 50 kDa protein, purified from mRNPs, inhibits translation of naked mRNAs in wheat germ lysates (104) and may be the putative masking factor. Standart and coworkers (105) have developed a "competitive unmasking" assay with clam oocyte mRNPs and a rabbit reticulocyte lysate. Translation is stimulated by addition of antisense RNAs complementary to relatively short regions in the 3'-UTR of the mRNAs. This suggests that the small RNAs interfere with the binding to the 3'-UTR of a repressor protein(s) that is responsible for masking the activity of the mRNA. The latter two studies point towards a role for *trans*-acting factors that specifically repress one or a small class of mRNAs. The repression of ferritin mRNA described in detail below is another example of masking by a *trans*-acting protein. Additional studies of the mechanism of mRNP mobilization are required to understand better how this process is controlled in cells.

ALTERATION OF INITIATION RATES    Changes in initiation rates are reflected in polysome size, which is proportional to the initiation rate (and coding length of the mRNA) and inversely proportional to the elongation rate. Since the elongation rate can be determined from ribosome transit time measurements, a change in the relative rate of initiation can be calculated (as discussed above) (91). At constant elongation rate, an increased rate of initiation results in larger polysomes, whereas a decreased rate results in smaller polysomes. Severe inhibition of initiation results in essentially complete run-off of ribosomes from mRNAs and the disappearance of polysomes. A change in the rate of global initiation can be detected by shifts in the polysome profile measured by absorption at 260 nm of sucrose density gradient fractions. Examples are the mild inhibition of protein synthesis observed during mitosis in HeLa cells (92), or more severe inhibitions seen upon heat shock (106) or serum deprivation (107, 108). A change in the rate of initiation of a specific protein is most readily detected by measuring polysome size by hybridization with specific DNA or RNA probes.

MODULATION OF ELONGATION/TERMINATION RATES    We are concerned here with temporal modulations of elongation and termination rates, as opposed to intrinsic differences in elongation efficiencies caused by codon usage or mRNA secondary structure. A priori, translational control of the elongation or termination phases seems unlikely for a number of reasons: 1. the initiation phase generally is rate-limiting, so modest changes in the elongation rate should have little or no effect on the overall rate of protein synthesis; 2. the elongation and termination phases occur by the same basic mechanism for all mRNAs and are thought to be rather insensitive to mRNA sequence or structure, so regulation of specific mRNAs is difficult to envision; 3. stimulation of elongation rates can only occur to a limited extent since most cells are functioning at near-maximal rates (ca. 5 amino acids/ribosome/sec); and 4. inhibition of elongation of specific mRNAs would result in an inefficient use of ribosomes, since they would be unavailable for the translation of other mRNAs. Nevertheless, small changes in elongation rate are seen and often accompany changes in initiation rates. For example, HeLa cells deprived of serum for 24 h exhibit an elongation rate that is half that of serum-fed cells (89). Severe inhibition of elongation has been reported for *Drosophila* mRNAs following heat shock (109) [although the postulated block at elongation is disputed (110)] and for mRNAs in HeLa cells stressed by amino acid analogues (111). The viral S1 mRNA in reovirus-infected cells also appears to be regulated at the level of elongation (112). Perhaps the best documented example of such regulation concerns translation of the HSP 70 mRNA in chicken reticulocytes (113). In this case, the mRNA in nonstressed cells is present in polysomes but is translated very slowly, whereas following heat stress, the elongation block is removed and the efficiency of HSP 70 synthesis increases 10-fold.

During elongation, ribosomes may shift the reading frame, or upon reaching a termination codon, may read through the stop signal. These events are usually regarded as translational errors and they occur at low frequencies. It is now apparent, however, that the frequency of such errors can be increased substantially by *cis*-acting sequences in the mRNA, and can be utilized as a means to control gene expression. Regulation through frameshifting and readthrough is developed in more detail below.

The yield of protein product from an mRNA is usually thought to be directly proportional to the number of translational initiation events. This is not strictly true, however, since not all ribosomes that initiate actually complete synthesis of the protein. As has been shown in prokaryotic cells, there is a measurable frequency whereby ribosomes abort synthesis by falling off the mRNA before reaching the termination codon. This rate, approximately $2 \times 10^{-4}$ abortive reactions per codon translated (114), is substantial but may be less in mammalian cells given that extremely large proteins ($>500$ kDa) are known to be synthesized. The ability of a ribosome to retain the peptidyl-tRNA at each step depends on the affinity of peptidyl-tRNA binding. However, high-affinity binding is expected to result in slow rates of tRNA release at steps such as translocation. Thus processivity and rate of protein synthesis are diametrically opposed. It is tempting to speculate that the need to synthesize very long proteins is the cause for the threefold slower overall rate of elongation in mammalian cells compared to prokaryotes. Little or nothing is known about ribosome processivity in mammalian cells, or about if this aspect of the elongation phase is regulated under special circumstances.

## GLOBAL CONTROLS BY PHOSPHORYLATION

Phosphorylation/dephosphorylation of numerous protein components of the translational apparatus appears to play an important role in controlling the overall rate of protein synthesis in mammalian cells [reviewed in (115)]. A large number of phosphoproteins have been identified: at least 13 initiation factor subunits; 3 of the 4 proteins comprising the elongation factors; 3 ribosomal proteins; and a number of aminoacyl-tRNA synthetases. Phosphorylation of a subset of the components correlates with inhibition of translation, whereas phosphorylation of others may cause stimulation. Current research in this area seeks to identify the precise sites of in vivo phosphorylation and the protein kinases and phosphatases involved, to demonstrate in vitro a change in activity of the protein due to phosphorylation, to elucidate how the extent of phosphorylation is controlled, and to establish that phosphorylation actually causes modulation of translation rates in vivo.

### Stimulation of Protein Synthesis and Cell Growth

The phosphorylation of a number of initiation factors, namely eIF-2B, eIF-3, eIF-4B, and eIF-4F, as well as ribosomal protein S6, correlates with activation of protein synthesis in vivo and/or with increased factor activity in vitro (115). We discuss those factors implicated in mRNA binding here, and postpone consideration of eIF-2B to the following section on repression of translation.

eIF-4F    The best-studied initiation factor in this group is the relatively low-abundance cap-binding protein, eIF-4Fα (eIF-4E), which is phosphorylated in vivo primarily at Ser53 (116). The same site is phosphorylated in vitro by protein kinase C (117, 118), where the preferred substrate is the eIF-4F complex rather than the free eIF-4Fα subunit (119). Another kinase specific for eIF-4Fα has been identified (120), but is not yet well characterized. Phosphorylation of eIF-4Fα does not appear to alter its affinity for m$^7$G cap structures; however, numerous correlative data link phosphorylation and the active state of the factor. The factor is dephosphorylated in cells during mitosis (121) or when treated at high temperatures (25), conditions that inhibit protein synthesis, and is phosphorylated by treatment of quiescent cells with serum, insulin, tumor necrosis factor α, or the mitogen, phorbol ester (TPA) (118, 122, 123). These results suggest that phosphorylation of eIF-4Fα activates the factor. In vitro studies with radiolabeled protein expressed from eIF-4Fα cDNAs demonstrate that the protein binds to 40S initiation complexes, where it is mainly in the phosphorylated state (124). In contrast, a mutant form, where Ala is substituted for Ser53 thereby preventing phosphorylation at that site, does not bind to the complexes. A startling finding is that the wild-type eIF-4Fα cDNA, when overexpressed in murine 3T3 cells or in rat 2 fibroblasts, causes the cells to undergo malignant transformation and become tumorigenic (125). Furthermore, microinjection of eIF-4F or eIF-4Fα into serum-starved NIH 3T3 cells results in the stimulation of DNA synthesis and a transient morphological transformation (126). A role for phosphorylation in these events is implied by the fact that overexpression or microinjection of the Ser53-Ala mutant form of eIF-4Fα is not oncogenic. The results above implicate the factor as a key target in the phosphorylation cascades that lead to control of cell growth through signal transduction mechanisms and protein kinase C.

The γ-subunit of eIF-4F, p220, also is phosphorylated in 3T3-L1 cells treated by insulin and phorbol esters (127). Furthermore, protein kinase C and protease-activated kinase I and II phosphorylate eIF-4Fγ in vitro (119). When eIF-4F is treated with protein kinase C, there is a fivefold increase in the factor's activity in a cell-free globin synthesis assay (128). Since the α-subunit also is phosphorylated by protein kinase C, it is not yet clear if phosphorylation of either or both the α- and γ-subunits causes enhanced activity. The site(s) of phosphorylation of eIF-4Fγ have not yet been determined, nor is it certain that protein kinase C is responsible for the protein's phosphorylation in vivo.

RIBOSOMAL PROTEIN S6    S6 is phosphorylated in vivo on up to five serine residues near the C-terminus of the protein [reviewed in (129)]. Phosphorylation correlates with activation of protein synthesis caused by mitogens and growth factors, but no compelling evidence has been generated to show that S6 phosphorylation enhances the activity of 40S ribosomal subunits in translation. Protein kinases relatively specific for S6 have been purified from a variety of cell types (130–132), and a cDNA encoding a mouse S6 kinase has been cloned recently (133). The mammalian S6 kinase itself is stimulated upon phosphorylation on serine/threonine by an unidentified kinase (134), although MAP-2 kinase has been implicated in the *Xenopus* system (135). Protein kinase C also is implicated in the pathway, since treatment of cells with phorbol esters increases S6 phosphorylation (136). Phosphoprotein phosphatase type-1 dephosphorylates S6 and is differentially regulated by insulin and epidermal growth factor (137). Thus, a complex net of kinases and phosphatases appears to be involved in establishing the phosphorylation level of S6. A role for S6 in the regulation of protein synthesis may be better established now that its kinase has been cloned.

OTHER INITIATION FACTORS    eIF-4B, like S6, is phosphorylated at multiple sites, namely on eight or more serine residues. The hyperphosphorylation of eIF-4B in vivo correlates with activation of protein synthesis, and partial dephosphorylation occurs upon repression of translation caused by heat shock and serum deprivation (106, 138). eIF-4B is phosphorylated in vivo when cells are treated with serum or phorbol esters (127) and in vitro by the mammalian or *Xenopus* S6 kinases (R. Duncan, unpublished results) as well as by protein kinase C and a variety of other kinases (117), but it remains to be shown whether or not the sites are identical to those phosphorylated in vivo. There is as yet no evidence that the phosphorylation state of eIF-4B affects the factor's activities in vitro or in vivo. The difficulty in demonstrating activity changes due to phosphorylation stems from the low activities of highly fractionated assays, and to rapid changes in the phosphorylation states in more active, crude assay systems. Another problem may be that simultaneous phosphorylation of a number of different translational components is required to effect a measurable change in protein synthesis.

eIF-3 is another initiation factor whose phosphorylation on the $\eta$-subunit (115–120 kDa) is enhanced by insulin and phorbol esters (127). eIF-3 comprises at least eight different polypeptide subunits (139), and evidence for in vivo phosphorylation on the $\delta$-, $\epsilon$-, $\xi$-, $\eta$-, and $\theta$-subunits exists (139, 140). No reports implicate phosphorylation of these subunits in translational control, however.

The simultaneous stimulation of the phosphorylation on eIF-4F, eIF-4B, eIF-3, and ribosomal protein S6 by phorbol esters and growth factors suggests that there is coordinate regulation of these translational components. Too little is known about the kinases involved in modifying these proteins to speculate how the apparent coordination is achieved, but it seems safe to postulate that protein kinase C is a part of at least one of the pathways. eIF-4F seems to be a particularly important target for control of cell growth and translation rates. It remains to be established whether or not phosphorylation of the other proteins is necessary for, adds only marginally to, or is entirely gratuitous with respect to changes in rates of protein synthesis.

## Repression of Translation

Repression of global protein synthesis is seen in cells deprived of serum, starved for amino acids, and subjected to stress by heat, among other things. In each of these cases, the extent of phosphorylation of eIF-4F$\alpha$ and eIF-4B is decreased (106, 138, 141) and may contribute to reduced protein synthesis activity, as described above. Thus increased phosphatase action or decreased kinase activity on any of the proteins whose phosphorylation is implicated in stimulation of protein synthesis is expected to contribute to translational repression. In addition to this kind of control, inhibition of protein synthesis may be caused by enhanced phosphorylation of two other soluble factors: eIF-2 and eEF-2.

eIF-2 AND eIF-2B    The phosphorylation of the $\alpha$-subunit of eIF-2 was first detected (142) in rabbit reticulocyte lysates deprived of hemin [the reader is referred to Jackson's recent review (143) of the extensive literature on this subject]. The absence of hemin results in an activation of a highly specific protein kinase, called the hemin regulated inhibitor (HRI, or HCR). The phosphorylation of eIF-2$\alpha$ in many other cell types also correlates with inhibition of the initiation rate, suggesting that this is a common mechanism for controlling protein synthesis (115). At least one other highly specific eIF-2$\alpha$ kinase, an interferon-induced protein called the double-stranded RNA activated inhibitor (DAI, or dsI), phosphorylates the same site, namely Ser51 (144, 145). The phosphate on Ser51 turns over rapidly, and broad-specificity phospho-protein phosphatases of type-1 and type-2A are implicated. Phosphorylation of only 25–30% of the factor is sufficient to cause a strong inhibition of the initiation phase in reticulocyte lysates. Phosphorylation does not directly inhibit formation of the ternary complex or ribosomal initiation complexes, as these reactions proceed efficiently in vitro with completely phosphorylated factor (146). Rather, it impedes the guanine nucleotide exchange reaction that enables the factor to recycle and promote multiple rounds of initiation (Figure 1) [reviewed in (147)]. Phosphorylated eIF-2·GDP binds more tightly to eIF-2B, but does not exchange GDP for GTP (11, 46). Since the cellular level of eIF-2B is thought to be lower than that of eIF-2, only a portion of eIF-2 need be phosphorylated in order to inhibit eIF-2B activity. In effect, eIF-2B is sequestered, thereby preventing its catalysis of GTP exchange on nonphosphorylated eIF-2. Although this mechanism appears to explain the inhibition caused by HRI or DAI, a number of issues remain unresolved. Evidence suggests that the guanine exchange reaction may occur on the surface of the ribosome, and that phosphorylation may result in a failure of eIF-2·GDP to be ejected, thereby inhibiting 80S initiation complex formation (148–150). Elucidation of the molecular mechanisms that control HRI and DAI is incomplete, and still other eIF-2$\alpha$ kinases may be involved.

Although the phosphorylation of eIF-2$\alpha$ correlates with inhibition of protein synthesis and results in loss of factor activity in vitro, these findings do not prove that eIF-2$\alpha$ phosphorylation actually causes translational repression in vivo. To address this point, mutant forms of cDNA encoding eIF-2$\alpha$ altered at the site of phosphorylation were transfected and expressed in mammalian cells. Expression of eIF-2$\alpha$ with the Ser51 residue substituted with Asp, which resembles phosphoserine, results in severe inhibition of global protein synthesis following accumulation of small amounts of the protein (151). The mutant form of eIF-2$\alpha$ exchanges into the endogenous eIF-2 complex (S.-Y. Choi, unpublished results) and presumably mimics phosphorylated eIF-2. Phosphorylation therefore appears to be sufficient to cause inhibition in cells under the growth conditions used in this study. The mutant factor, eIF-2$\alpha$(Ser51-Ala), stimulates the translation of both plasmid-derived mRNAs in transiently transfected COS-1 cells (151) and viral mRNAs in human 293 cells infected with a mutant form of adenovirus lacking the *VA-I* gene (152). In these cells, DAI kinase is activated (153, 154) but the repressive effects of the kinase are prevented by the mutant eIF-2$\alpha$ protein, which is not phosphorylated. The results implicate the importance of eIF-2$\alpha$ phosphorylation by the DAI kinase and indicate that eIF-2$\alpha$ phosphorylation is the cause of translational repression in intact cells.

eIF-2 phosphorylation is perhaps the best-characterized translational control mechanism in mammalian cells. A large body of evidence points to its involvement in many instances of translational repression, although not every kind of inhibition involves this mechanism (155). A complication in the interpretation of the above results is the observation that inhibition of protein synthesis does not always occur when eIF-2$\alpha$ is phosphorylated. For example, 2-aminopurine prevents host mRNA shut-off following adenovirus infection, although DAI remains active and eIF-2$\alpha$ is phosphorylated to a relatively high extent (156). A possible explanation is that eIF-2B may be modified to suppress the effects of eIF-2$\alpha$ phosphorylation. Dholakia & Wahba (157) have shown that eIF-2B is phosphorylated in vitro on the $\epsilon$-subunit (82 kDa) by casein kinase II, resulting in a fivefold stimulation of its specific activity in the guanine nucleotide exchange reaction. They also argue that NADPH binding to the protein alters its function (158). It is therefore possible, but not yet demonstrated, that repression by eIF-2$\alpha$ phosphorylation may be modulated by modifications of eIF-2B structure and/or activity.

eEF-2    A second phosphorylation pathway for inhibition of protein synthesis involves the elongation factor, eEF-2. Evidence for the phosphorylation of this factor comes from analyses of reticulocyte lysates incubated with [$\gamma$-$^{32}$P]ATP (159) and of mammalian fibroblasts treated with serum or growth factors (160). A highly specific kinase, called the $Ca^{2+}$/calmodulin-dependent protein kinase III or eEF-2 kinase, has been identified and shown to phosphorylate primarily Thr56 (161, 162), but Thr53 and Thr58 are also labeled on prolonged treatment. These Thr residues reside in a region that may be involved in eEF-2 binding to ribosomes (163). Phosphorylated eEF-2 binds to 80S ribosomes but does not promote in vitro the translocation reaction with GTP (164). Thus phosphorylation inactivates eEF-2 and inhibits the elongation phase of protein synthesis. This is consistent with the findings that cAMP stimulation of reticulocyte lysates correlates with dephosphorylation of eEF-2 (165), whereas okadaic acid inhibition correlates with its phosphorylation (166). What is the physiological role of eEF-2 phosphorylation? Celis and coworkers (167) reported that in transformed human amnion cells, eEF-2 becomes phosphorylated during mitosis, when a transient increase in $Ca^{2+}$ concentration occurs. The view that elongation is inhibited during mitosis may not apply generally, however, as there is contradictory evidence for a decreased initiation rate with no change in elongation rate in mitotic HeLa cells (92). It is perplexing that eEF-2 is transiently phosphorylated in fibroblasts treated with serum or growth factors (160), circumstances that cause a stimulation of protein synthesis. Since treatment of cells with cycloheximide causes ribosomes to shift into polysomes (79), a transient inhibition of the elongation rate by phosphorylation of eEF-2 also may result in the mobilization of ribosomes and mRNPs. The widespread occurrence of eEF-2 phosphorylation and its proven effects on protein synthesis in vitro likely will attract considerable interest in elucidating the physiological roles played by this phosphorylation mechanism.

## Other Phosphorylation Targets

The phosphorylation of a variety of other protein components of the translational apparatus has been described but not yet studied extensively. Although no clear role in translational control is indicated at this time, some of these modifications may turn out to be especially significant. The $\beta$-subunit of eIF-2 is phosphorylated in vivo, but there are only minor changes in its phosphorylation status when protein synthesis is either inhibited or stimulated (106, 168). eIF-2$\beta$ is phosphorylated on Ser2 by casein kinase II and on Ser13 by protein kinase C (169, 170). eIF-5 is phosphorylated at multiple sites in vivo and by protein kinase II in vitro (140), but changes in the extent of phosphorylation have not been determined. Phosphorylated eEF-1$\alpha$ has been reported (171) but not characterized. Potentially interesting is eEF-1$\beta\gamma$, proteins involved in the guanine nucleotide exchange reaction with eEF-1$\alpha$, which are substrates for P34$^{cdc32}$ protein kinase (172). Here, phosphorylation may contribute to inhibition of protein synthesis during mitosis, together with the dephosphorylation of eIF-4F$\alpha$ (121) and the phosphorylation of eEF-2 (167). *Artemia* eEF-1$\beta$ is phosphorylated on Ser89 in vitro, resulting in a decrease in its GTP exchange activity (173). Besides ribosomal protein S6, the large subunit acidic proteins P1 and P2 are phosphoproteins (174), and the small subunit proteins S2 and S13 are phosphorylated following infection of cells by vaccinia virus (175). Whether or not the phosphorylation status of any of these ribosomal proteins affects translation rates remains to be determined, although phosphorylation of P1-P2 is required for their in vitro assembly into 60S particles (176). Finally, up to eight different aminoacyl-tRNA synthetases are phosphorylated (177), although there appears to be no change in their capacities to charge tRNAs [see the recent discussion by Clemens (178)].

In summary, there exists a large number of phosphorylation targets among the proteins in the translational apparatus. It is possible that many of these modifications are gratuitous, without functional effects. Yet is also is likely that regulation by phosphorylation proceeds by multiple, redundant mechanisms, thereby ensuring proper control of this crucial pathway.

## OTHER CONTROL MECHANISMS

Control of global rates of protein synthesis is important in coordinating translation with the overall metabolism of the cell, whereas regulation of a single mRNA or a class of mRNAs may be employed to provide temporal control of specific gene expression. Specific translational controls are known to operate in bacteria, where the synthesis of ribosomal proteins (179) and a variety of other proteins (180) are under autogenous control. That is, the gene product, upon reaching a given critical concentration in the cell, binds to its mRNA and inhibits its own synthesis. Likewise, bacteria and their phages regulate the synthesis of specific or groups of proteins with translational repressors (180). It is therefore likely that mammalian cells would employ translational control mechanisms to regulate the synthesis of specific proteins. Such regulation is expected also because control at the level of transcription and RNA processing involves a time delay of 15 minutes or more in the response, whereas translational control can be much more rapid. A number of cases involving regulation of specific mRNAs have been described and a detailed mechanism has been elucidated for ferritin mRNA (see below). We focus on a few examples of control mechanisms involving whole classes or individual mRNAs with the intent to provide some insight into the kinds of regulation that can occur.

## Polyadenylation of mRNAs

Most mammalian mRNAs are polyadenylated soon after transcription in the nucleus, and carry 200–250 adenylate residues at their 3'-termini as they are transported into the cytoplasm. The poly(A) tails are further metabolized in the cytoplasm, where in general they are gradually shortened to about 50–70 residues, but also can be lengthened by cytoplasmic enzymes. Besides possible functions in nuclear processing and transport, the poly(A) tails are thought to affect both mRNA stability and translational efficiency. The role of poly(A) tails in mRNA degradation and translation has been reviewed recently (181). We shall limit our discussion to their influence on the efficiency of translation.

A role for poly(A) is indicated by in vitro studies of rabbit reticulocyte lysates, where mRNAs carrying poly(A) tails are translated more efficiently than their deadenylated counterparts (182). The requirement of a poly(A) tail is not absolute, as considerable translation occurs with poly(A)$^-$ mRNAs. Further in vitro evidence comes from the observation that addition of oligo(A) molecules inhibits protein synthesis (182, 183). Reversal of inhibition by the poly(A) binding protein (PABP) suggests that the oligo(A) molecules compete for the PABP and implies that the PABP and poly(A) tail together are positive effectors of translation. A detailed investigation showed that poly(A)$^-$ mRNAs are recruited less efficiently into polysomes compared to poly(A)$^+$ mRNAs, possibly through a defect in the junction step of the 60S subunit with the 40S initiation complex (184). A compelling case for poly(A) stimulation in vivo comes from studies of the tissue plasminogen activator (TPA) mRNA (185). Its poly(A) tail is lengthened to 400–600 adenylate residues during mouse oocyte maturation, resulting in a strong activation of translation. Polyadenylation and translational activation also occur upon microinjection of an in vitro chimeric transcript carrying the 3'-UTR of the TPA mRNA inserted next to a reporter gene.

Experiments involving microinjection of mammalian mRNAs into *Xenopus* oocytes provide evidence in support of the view that polyadenylation affects translational efficiency. mRNAs with poly(A) tails are translated much more efficiently than those lacking poly(A), a result that is not attributable to differences in mRNA stability (186, 187). In a recent study of the mobilization of maternal mRNAs into polysomes, Richter, Dworkin, and coworkers found that the act of cytoplasmic polyadenylation, not the size of the poly(A) tail per se, is required for activation of the mRNA (188). However, it is uncertain whether or not results from *Xenopus* oocytes can be extrapolated to mammalian systems.

Finally, genetic studies in yeast provide some insight into the role of poly(A) and the PABP in translation. Deletion of the PABP gene is lethal and second site revertants have been obtained, one of which is a 60S ribosomal protein (189). Careful examination of the polysome profiles from PABP-deficient cells indicates that the amount of RNA in polysomes is decreased, but not the number of ribosomes per mRNA. This suggests that poly(A) and PABP are involved in establishing the distribution of mRNA between mRNPs and polysomes rather than in the initiation rate on polysomes.

## Poliovirus-Infected Cells

When poliovirus infects a mammalian cell, host mRNA translation is strongly inhibited, whereas poliovirus mRNA is translated vigorously [for a comprehensive review, see (190)]. The mechanism for selectively inhibiting the cell's mRNA translation involves proteolytic cleavage of the $\gamma$-subunit of eIF-4F (191), thereby inactivating the factor and reducing translation of capped mRNAs by the scanning mechanism. Poliovirus mRNAs are not capped and initiate by the internal initiation mechanism, thereby evading the effects of eIF-4F$\gamma$ cleavage. Apparently the infection also activates an eIF-2$\alpha$ kinase, leading to the phosphorylation of eIF-2 (192). It is proposed that cleavage of eIF-4F$\gamma$ results in only a partial inhibition of capped mRNA translation, and that eIF-2$\alpha$ phosphorylation further inhibits these mRNAs. It is unclear how the internal initiation mechanism would be less sensitive to repression by eIF-2 phosphorylation.

The poliovirus genome encodes two proteases, named 2A and 3C, which are involved in the processing of the polyprotein translation product. Protease 2A is implicated in eIF-4F$\gamma$ cleavage, yet appears not to cleave the initiation factor directly, but to activate a cellular protease. It has been shown recently that both protease 2A and eIF-3 are required to generate the eIF-4F$\gamma$ protease activity (193). Possibly eIF-3 itself carries the catalytic site for eIF-4F$\gamma$ cleavage, or eIF-3 alters the substrate or protease conformation to promote cleavage. It is interesting to speculate that there may be cellular mechanisms not involving poliovirus protease 2A that activate eIF-4F$\gamma$ cleavage and cause inhibition of protein synthesis. Indeed, specific cleavage of eIF-4F$\gamma$ comparable to that seen in poliovirus-infected cells, is occasionally observed. In this regard, it has been shown recently that eIF-4F$\gamma$ is a substrate for multiple $Ca^{2+}$-dependent enzymes (194). It remains to be shown whether or not such enzymes play a role in regulating translation by cleavage of eIF-4F$\gamma$.

## Ferritin mRNA

Ferritins are ubiquitous intracellular proteins that bind and store iron molecules, thereby protecting cells from the toxic effects of free iron. Expression of the two highly homologous genes, H and L, is dramatically regulated at the translational level by the extracellular iron supply [for a review, see (195)]. Ferritin mRNA is repressed in iron-deficient cells, but the rate of ferritin synthesis is stimulated 10–40-fold when cells are shifted to an iron-containing medium. Whereas a part of the induction may be attributed to enhanced transcription in some cells (196), most is due to an increase in translational efficiency of preexisting mRNAs. Analysis of *cis*-acting mRNA sequences identified a conserved 28-nucleotide region in the 5'-UTR of all ferritin mRNAs that is necessary for iron regulation (197, 198) and that confers such regulation when inserted into other mRNAs. The sequence, called the iron responsive element (IRE), may fold into a stem-loop structure. Mutations affecting the strength of the putative stem-loop structure alter the regulatory properties of the IRE (199). It is interesting to note that transferrin receptor mRNA contains five similar IREs, but in contrast to ferritin mRNA, they are situated in the 3'-UTR. The transferrin receptor IREs play a role in protecting the mRNA from degradation when iron levels are low.

A *trans*-acting ferritin repressor protein (FRP) has been purified to homogeneity from mammalian cells or tissue on the basis of its binding to RNA sequences containing the IRE (200–202). The protein inhibits ferritin mRNA translation in a wheat germ cell-free system (203), leading to the conclusion that the FRP is a repressor of ferritin mRNA translation in the absence of iron. Thach and coworkers have shown that hemin, not iron itself, reverses the inhibition of in vitro ferritin synthesis by FRP (204). They argue that hemin may be a physiological inducer in mammalian cells (205). An alternative, though not necessarily a mutually exclusive, view is that the redox state of the FRP is critical. When the FRP is in an oxidized state, its affinity for the IRE is low ($K_d = 5$ nM); upon reduction, a high-affinity ($K_d = \sim 20$ pM) FRP form is generated (206). Klausner and coworkers propose a disulfide–sulfhydryl switch regulated by iron (207). The position of the IRE in the 5' UTR is critical; no inhibition by the FRP occurs when the IRE lies more than 67 nucleotides distal from the cap (208). This suggests that the IRE-FRP complex inhibits cap recognition or the initial 40S ribosomal sub-unit binding at the 5'-terminus rather than the scanning process itself. The recent cloning of the cDNA encoding the FRP (209) should provide a tool for better evaluating the mechanism of mRNA binding and derepression.

## *Frameshifting and Readthrough*

A translating ribosome normally advances along the mRNA by precisely three nucleotides at each turn of the elongation cycle, then terminates synthesis when a nonsense codon is encountered in the A site. However, at a frequency estimated to be less than $1 \times 10^{-3}$, a mistake (other than incorporation of a wrong amino acid at a sense codon) is made and the ribosome either shifts its mRNA reading frame or binds an aminoacyl-tRNA at the nonsense codon and continues translation. In both prokaryotic and eukaryotic cells, the structures of some specific mRNAs have evolved so as to increase the tendency to frameshift or read through termination codons. By increasing the frequency of such errors, the ribosome is able to translate a more distal part of the mRNA at low but significant levels. In addition, bacterial ribosomes can "hop" over a significant number of nucleotides, leaving an internal section of the mRNA untranslated, as seen in the bacteriophage T4 DNA topoisomerase gene 60 mRNA (210, 211). No evidence for hopping in mammalian cells has been found, however. We examine here a number of examples of how frameshifting and readthrough are employed to express downstream regions in mammalian mRNAs [reviewed recently in (212)].

FRAMESHIFTING     Frameshifting is usually caused by slippage of a tRNA derivative from one codon to an overlapping codon in the −1 or +1 reading frame. For example, tRNA$^{Leu}$, which binds initially to CUU in the sequence CUU UGA of the RF-2 mRNA of *E. coli*, slips to UUU, thereby establishing a new +1 reading frame (213). Runs of homopolymers facilitate frameshifting and thereby create potential shift sites. In mammalian systems, frameshifting is frequently encountered in the *gag-pol* region of retroviruses, where a −1 shift is invariably seen. The frameshift enables the virus to express at low levels the *pol* gene, which lies in a different translational reading frame from the upstream *gag* gene. In many cases, two adjacent tRNAs shift in tandem, for example at the sequences A AAA AAG in mouse mammary tumor virus (MMTV) (214), and at U UUA AAC in the coronavirus IBV (215). The frequency of the frameshift event is increased by other elements in the mRNA sequence, particularly those that cause pausing of the ribosome over the shifty sequence. Examples of stimulators of frameshifting are: secondary structure elements, particularly pseudoknots, located about seven nucleotides downstream from the shift site, seen in Rous sarcoma virus (216) and coronavirus IBV (217); an adjacent stop codon, as in the *E. coli* RF-2 mRNA (213); and (perhaps only in bacteria) the upstream Shine/Dalgarno region of the RF-2 mRNA that interacts with 16S rRNA (218). Frameshift frequencies range from a few percent to as high as 23% (for MMTV), quite sufficient to synthesize adequate amounts of the *pol* gene product.

READTHROUGH     Readthrough enables the ribosome to continue to translate a downstream, in-frame region of an mRNA that is distal to a stop codon. The readthrough phenomenon is seen in the murine leukemia virus, where the *gag* and *pol* genes are separated by a UAG stop codon that is read at 10% frequency by tRNA$^{Gln}$ to generate a fusion protein (219). Other examples are readthrough of UGA in Sindbis virus (220) and UAG in feline leukemia virus (221). As in frameshifting, secondary structure elements may enhance the frequency of such readthrough events (222, 223). The insertion of selenocysteine into glutathione oxidase in mammalian cells is coded by a UGA codon (224), and therefore represents a special class of the readthrough mechanism. Selenocysteinyl-tRNA is synthesized from a minor seryl-tRNA that recognizes the UGA codon. The mechanism of mammalian cell recognition of the specific UGA codons where selenocysteine is incorporated is not known. However, in the synthesis of formate dehydrogenase in *E. coli,* nucleotides 3' to the UGA are critical (225) and a specific EF-Tu–like elongation factor is involved (226). It will be interesting to elucidate further the molecular mechanisms that lead to specific incorporation of selenocysteine in both prokaryotic and eukaryotic proteins.

## *Miscellaneous Control Mechanisms*

A variety of additional translational controls are not addressed due to limitations of space. Global controls by intracellular pH and energy charge have been observed, and $Ca^{2+}$ levels are reported to affect protein synthesis [for a recent review, see (227)]. Cells subjected to nutrient deprivation, heat shock, or heavy metals are translationally repressed [reviewed in (228)], likely by mechanisms involving the phosphorylation/dephosphorylation of translational components. In contrast, hypertonic shock of cells causes an inhibition of the initiation phase (229) by an unknown mechanism not involving changes in eIF-2 or eIF-4B phosphorylation (138). A sizeable body of evidence points to an involvement of the redox potential in the cell, and sugar phosphates or other small metabolite molecules may affect the translational machinery [see discussion by Jackson (143)]. Hormones influence protein synthesis, likely in large part through the phosphorylation mechanisms discussed earlier, but other pathways of control may occur as well. Infection of mammalian cells by viruses frequently leads to inhibition of host protein synthesis, and a number of different mechanisms are involved [reviewed in (154)]. Small cytoplasmic RNAs may specifically inhibit translation, in some cases through complementary interactions (230). Polysomal mRNAs appear to be attached in some way to the cytoskeleton (231), yet little is known about the nature of the attachment or its importance in translation. Finally, degradation of mRNAs is sometimes coupled to their translation. The best-studied example is the tubulin system, where translation of tubulin mRNA is required for mRNA degradation by a mechanism that recognizes the nascent polypeptide product (232). The main topics of this review and the above examples do not exhaust the list of translational control possibilities, but indicate the richness and variety of ways to regulate gene expression at the translational level. As our knowledge of ribosome structure and the protein synthesis pathway is refined, it should be possible to describe these translational control mechanisms in detailed molecular terms.

ACKNOWLEDGMENTS

I thank Drs. E. Ehrenfeld, H. Ernst, R. J. Kaufman, W. C. Merrick, N. Sonenberg, and R. E. Thach together with members of my laboratory group for helpful comments on the text, and Drs. J. Bag, M. Clemens, R. Jackson, R. Rhoads, and L. Slobin for sharing manuscripts prior to publication. I am especially indebted to Drs. R. F. Duncan and R. J. Jackson for numerous stimulating discussions and insightful criticisms. The work from the author's laboratory was supported by USPHS grant GM22135 from the N.I.H.

*Literature Cited*

1. Moldave, K. 1985. *Annu. Rev. Biochem.* 54:1109–49
2. Trachsel, H. 1990. *Translation in Eukaryotes.* Caldwell, NJ: Telford
3. Wool, I. G., Endo, Y., Chan, Y.-L., Glück, A. 1990. In *The Ribosome: Structure, Function, and Evolution,* ed. W. Hill, A. Dahlberg, R. Garrett, P. Moore, D. Schlessinger, J. Warner, pp. 203–14. Washington, DC: Am. Soc. Microbiol.
4. Safer, B. 1989. *Eur. J. Biochem.* 186:1–3
5. Goumans, H., Thomas, A., Verhoeven, A., Voorma, H. O., Benne, R. 1980. *Biochim. Biophys. Acta* 608:39–46
6. Russell, D. W., Spremulli, L. L. 1979. *J. Biol. Chem.* 254:8796–800
7. Benne, R., Amesz, H., Hershey, J. W. B., Voorma, H. O. 1979. *J. Biol. Chem.* 254:3201–5
8. Konieczny, A., Safer, B. 1983. *J. Biol. Chem.* 258:3402–8
9. Roy, A. L., Chakrabarti, D., Datta, B., Hileman, R. E., Gupta, N. K. 1988. *Biochemistry* 27:8203–9
10. Gupta, N. K., Roy, A. L., Nag, M. K., Kinzy, T. G., MacMillan, S., et al. 1991. In *Post-Transcriptional Control of Gene Expression,* ed. J. McCarthy, pp. 521–26. Berlin: Springer-Verlag. In press
11. Rowlands, A. G., Panniers, R., Henshaw, E. C. 1988. *J. Biol. Chem.* 263:5526–33
12. Schreier, M. H., Staehelin, T. 1973. In *Regulation of Translation in Eukaryotes,* ed. E. Bautz, pp. 335–49. Berlin/New York: Springer-Verlag
13. Trachsel, H., Erni, B., Schreier, M. H., Staehelin, T. 1977. *J. Mol. Biol.* 116:755–67
14. Benne, R., Hershey, J. W. B. 1978. *J. Biol. Chem.* 253:3078–87
15. Kozak, M. 1989. *J. Cell. Biol.* 108:229–41
16. Rhoads, R. E. 1991. See Ref. 2
17. Sonenberg, N. 1988. *Prog. Nucleic Acids Res. Mol. Biol.* 35:173–207
18. Rhoads, R. E. 1988. *Trends Biochem. Sci.* 13:52–56
19. Ray, B. K., Lawson, T. G., Kramer, J. C., Cladaras, M. H., Grifo, J. A., et al. 1985. *J. Biol. Chem.* 260:7651–58
20. Lawson, T. G., Lee, K. A., Maimone, M. M., Abramson, R. D., Dever, T. E., et al. 1989. *Biochemistry* 28:4729–34
21. Rozen, F., Edery, I., Meerovitch, K., Dever, T. E., Merrick, W. C., Sonenberg, N. 1990. *Mol. Cell. Biol.* 10:1134–44
22. Grifo, J. A., Abramson, R. D., Satler, C. A., Merrick, W. C. 1984. *J. Biol. Chem.* 259:8648–54
23. Linder, P., Lasko, P. F., Ashburner, M., Leroy, P., Nielsen, P. J., et al. 1989. *Nature* 337:121–22
24. Kozak, M. 1990. *Proc. Natl. Acad. Sci. USA* 87:8301–5
25. Duncan, R., Milburn, S. C., Hershey, J. W. B. 1987. *J. Biol. Chem.* 262:380–88
26. Etchison, D., Smith, K. 1990. *J. Biol. Chem.* 265:7492–500
27. Kozak, M. 1987. *J. Mol. Biol.* 196:947–50
28. Kozak, M. 1989. *Mol. Cell. Biol.* 9:5073–80
29. Cigan, A. M., Feng, L., Donahue, T. F. 1988. *Science* 242:93–97
30. Donahue, T. F., Cigan, A. M., Pabich, E. K., Valavicius, B. C. 1988. *Cell* 54:621–32
31. Cigan, A. M., Pabich, E. K., Feng, L., Donahue, T. F. 1989. *Proc. Natl. Acad. Sci. USA* 86:2784–88
32. Kaempfer, R., Emmelo, J. V., Fiers, W. 1981. *Proc. Natl. Acad. Sci. USA* 78:1542–46
33. Goss, D. J., Woodley, C. L., Wahba, A. J. 1987. *Biochemistry* 26:1551–56

34. Pelletier, J., Sonenberg, N. 1988. *Nature* 334:320–25
35. Jang, S. K., Davies, M. V., Kaufman, R. J., Wimmer, E. 1989. *J. Virol.* 63:1651–60
36. Sarnow, P. 1989. *Proc. Natl. Acad. Sci. USA* 86:5795–99
37. Bernards, A., Rubin, C. M., Westbrook, C. A., Paskind, M., Baltimore, D. 1987. *Mol. Cell. Biol.* 7:3231–36
38. Meerovitch, K., Pelletier, J., Sonenberg, N. 1989. *Genes Dev.* 3:1026–34
39. Najita, L., Sarnow, P. 1990. *Proc. Natl. Acad. Sci. USA* 87:5846–50
40. Jang, S. K., Wimmer, E. 1990. *Genes Dev.* 4:1560–72
41. Kaminski, A., Howell, M. T., Jackson, R. J. 1990. *EMBO J.* 8:3753–59
42. Peterson, D. T., Safer, B., Merrick, W. C. 1979. *J. Biol. Chem.* 254:7730–35
43. Raychaudhuri, P., Chaudhuri, A., Maitra, U. 1985. *J. Biol. Chem.* 260:2140–45
44. Raychaudhuri, P., Chaudhuri, A., Maitra, U. 1985. *J. Biol. Chem.* 260:2132–39
45. Ghosh, S., Chevesich, J., Maitra, U. 1989. *J. Biol. Chem.* 264:5134–40
46. Dholakia, J. N., Wahba, A. J. 1989. *J. Biol. Chem.* 264:546–50
47. Kemper, W. M., Berry, K. W., Merrick, W. C. 1976. *J. Biol. Chem.* 251:5551–57
48. Park, M. H., Liberato, D. J., Yergey, A. L., Folk, J. E. 1984. *J. Biol. Chem.* 259:12123–27
49. Smit-McBride, Z., Schnier, J., Kaufman, R. J., Hershey, J. W. B. 1989. *J. Biol. Chem.* 264:18527–30
50. Slobin, L. I. 1991. See Ref. 2
51. Thompson, R. C. 1988. *Trends Biochem. Sci.* 13:91–93
52. Janssen, G. M. C., Moller, W. 1988. *J. Biol. Chem.* 263:1773–78
53. Dever, T. E., Costello, C. E., Owens, C. L., Rosenberry, T. L., Merrick, W. C. 1989. *J. Biol. Chem.* 264:20518–25
54. Coppard, N. J., Clark, B. F. C., Cramer, F. 1983. *FEBS Lett.* 164:330–34
55. Whiteheart, S. W., Shenbagamurthi, P., Chen, L., Cotter, R. J., Hart, G. W. 1989. *J. Biol. Chem.* 264:14334–41
56. Rosenberry, T. L., Krall, J. A., Dever, T. E., Hass, R., Louvard, D., Merrick, W. C. 1989. *J. Biol. Chem.* 264:7096–99
57. Moazed, D., Noller, H. F. 1989. *Nature* 342:142–48
58. Inoue-Yokosawa, N., Ishikawa, C., Kaziro, Y. 1974. *J. Biol. Chem.* 249:4321–23
59. Bodley, J. W., Veldman, S. A. 1990. In *ADP-Ribosylating Toxins and G Proteins,* ed. J. Moss, M. Vaughan, pp. 21–30. Washington, DC: Am. Soc. Microbiol.
60. Wick, M. J., Iglewski, B. H. 1990. See Ref. 59, pp. 31–43
61. Omura, F., Kohno, K., Uchida, T. 1989. *Eur. J. Biochem.* 180:1–8
62. Kozak, M. 1984. *Nucleic Acids Res.* 12:3873–93
63. Hackett, P. B., Peterson, R. B., Hensel, C., Albericio, F., Gunderson, S. I., Palmenberg, A. C. 1986. *J. Mol. Biol.* 190:45–57
64. Peabody, D. S., Berg, P. 1986. *Mol. Cell. Biol.* 6:2695–703
65. Kaufman, R. J., Murtha, P., Davies, M. V. 1987. *EMBO J.* 6:187–93
66. Kozak, M. 1987. *Mol. Cell. Biol.* 7:3438–45
67. Hinnebusch, A. G. 1988. *Microbiol. Rev.* 52:248–73
68. Hinnebusch, A. G. 1990. *Trends Biochem. Sci.* 15:148–52
69. Lawson, T. G., Ray, B. K., Dodds, J. T., Grifo, J. A., Abramson, R. D., et al. 1986. *J. Biol. Chem.* 261:13979–89
70. Lawson, T. G., Cladaras, M. H., Ray, B. K., Lee, K. A., Abramson, R. D., et al. 1988. *J. Biol. Chem.* 263:7266
71. Pelletier, J., Sonenberg, N. 1985. *Cell* 40:515–26
72. Kozak, M. 1986. *Proc. Natl. Acad. Sci. USA* 83:2850–54
73. de Smit, M. H., van Duin, J. 1990. *Prog. Nucleic Acids Res. Mol. Biol.* 38:1–35
74. de Smit, M. H., van Duin, J. 1990. *Proc. Natl. Acad. Sci. USA* 87:7668–72
75. Wolin, S. L., Walter, P. 1988. *EMBO J.* 7:3559–69
76. Minshull, J., Hunt, T. 1986. *Nucleic Acids Res.* 14:6433–51
77. Pronczuk, A. W., Rogers, Q. R., Munro, H. N. 1970. *J. Nutr.* 100:1249–58
78. Jagus, R., Anderson, W. F., Safer, B. 1981. *Prog. Nucleic Acids Res. Mol. Biol.* 25:127–85
79. Perlman, S., Hirsch, M., Penman, S. 1972. *Nature New Biol.* 238:143–44
80. Walden, W. E., Godefroy-Colburn, T., Thach, R. E. 1981. *J. Biol. Chem.* 256:11739–46
81. Lodish, H. F. 1974. *Nature* 251:385–88
82. Godefroy-Colburn, T., Thach, R. E. 1981. *J. Biol. Chem.* 256:11762–73
83. Ray, B. K., Brendler, T. G., Adya, S., Daniels-McQueen, S., Kelvin-Miller, J., et al. 1983. *Proc. Natl. Acad. Sci. USA* 80:663–67
84. Sarkar, G., Edery, I., Gallo, R., Sonenberg, N. 1984. *Biochim. Biophys. Acta* 783:122–29

85. Huang, S., Hershey, J. W. B. 1989. *Mol. Cell. Biol.* 9:3679–84
86. Rowlands, A. G., Montine, K. S., Henshaw, E. C., Panniers, R. 1988. *Eur. J. Biochem.* 175:93–99
87. Nielsen, P. J., Trachsel, H. 1988. *EMBO J.* 7:2097–105
88. van Venrooji, W. J., Moonen, H., van Loon-Klassen, L. 1974. *Eur. J. Biochem.* 50:297–304
89. Nielsen, P. J., McConkey, E. H. 1980. *J. Cell Physiol.* 104:269–81
90. Hargreaves-Wall, K. M., Buciak, J. L., Pardridge, W. M. 1990. *J. Cereb. Blood Flow Metab.* 10:162–69
91. Palmiter, R. D. 1975. *Cell* 4:189–97
92. Fan, H., Penman, S. 1970. *J. Mol. Biol.* 50:655–70
93. Bag, J., Sarkar, S. 1975. *Biochemistry* 14:3800–7
94. Meyuhas, O., Thompson, E. A., Perry, R. P. 1987. *Mol. Cell. Biol.* 7:2691–99
95. Davidson, E. H. 1986. *Gene Activity in Early Development.* New York: Academic
96. Aziz, N., Munro, H. N. 1986. *Nucleic Acids Res.* 14:915–27
97. Welsh, M., Scherberg, N., Gilmore, R., Steiner, D. F. 1986. *Biochem. J.* 235:459–67
98. Dreyfuss, G. 1986. *Annu. Rev. Cell Biol.* 2:459–98
99. Bag, J. 1991. See Ref. 2
100. Nelson, E. M., Winkler, M. M. 1987. *J. Biol. Chem.* 262:11501–6
101. Minich, W. B., Korneyeva, N. L., Ovchinnikov, L. P. 1989. *FEBS Lett.* 257:257–59
102. Ovchinnikov, L. P., Spirin, A. S., Erni, B., Staehelin, T. 1978. *FEBS Lett.* 88:21–26
103. Minich, W. B., Korneyeva, N. L., Berezin, Y. V., Ovchinnikov, L. P. 1989. *FEBS Lett.* 258:227–29
104. Minich, W. B., Volyanik, E. V., Korneyeva, N. L., Berezin, Y. V., Ovchinnikov, L. P. 1990. *Mol. Biol. Rep.* 14:65–67
105. Standart, N., Dale, M., Stewart, E., Hunt, T. 1990. *Genes Dev.* 4:2157–68
106. Duncan, R., Hershey, J. W. B. 1984. *J. Biol. Chem.* 259:11882–89
107. Duncan, R., McConkey, E. H. 1982. *Eur. J. Biochem.* 123:539–44
108. Thomas, G., Martin-Perez, J., Siegmann, M., Otto, A. M. 1982. *Cell* 30:235–42
109. Ballinger, D. G., Pardue, M. L. 1983. *Cell* 33:103–14
110. Sirkin, E. R., Lindquist, S. 1985. In *Sequence Specificity in Transcription and Translation*, pp. 669–79. New York: Liss
111. Thomas, G. P., Mathews, M. B. 1984. *Mol. Cell. Biol.* 4:1063–72
112. Fajardo, J. E., Shatkin, A. J. 1990. *Proc. Natl. Acad. Sci. USA* 87:328–32
113. Theodorakis, N. G., Banerji, S. S., Morimoto, R. I. 1988. *J. Biol. Chem.* 263:14579–85
114. Tsung, K., Inouye, S., Inouye, M. 1989. *J. Biol. Chem.* 264:4428–33
115. Hershey, J. W. B. 1989. *J. Biol. Chem.* 264:20823–26
116. Rychlik, W., Russ, M. A., Rhoads, R. E. 1987. *J. Biol. Chem.* 262:10434–37
117. Tuazon, P. T., Merrick, W. C., Traugh, J. A. 1989. *J. Biol. Chem.* 264:2773–77
118. Morley, S. J., Traugh, J. A. 1989. *J. Biol. Chem.* 264:2401–4
119. Tuazon, P. T., Morley, S. J., Dever, T. E., Merrick, W. C., Rhoads, R. E., Traugh, J. A. 1990. *J. Biol. Chem.* 265:10617–21
120. McMullin, E. L., Haas, D. W., Abramson, R. D., Thach, R. E., Merrick, W. C., Hagedorn, C. A. 1988. *Biochem. Biophys. Res. Commun.* 153:340–46
121. Bonneau, A. M., Sonenberg, N. 1987. *J. Biol. Chem.* 262:11134–39
122. Marino, M. W., Pfeffer, L. M., Guidon, P. T., Donner, D. B. 1989. *Proc. Natl. Acad. Sci. USA* 86:8417–21
123. Kaspar, R. L., Rychlik, W., White, M. W., Rhoads, R. E., Morris, D. R. 1990. *J. Biol. Chem.* 265:3619–22
124. Joshi-Barve, S., Rychlik, W., Rhoads, R. E. 1990. *J. Biol. Chem.* 265:2979–83
125. Lazaris-Karatzas, A., Montine, K. S., Sonenberg, N. 1990. *Nature* 345:544–47
126. Smith, M. R., Jaramillo, M., Liu, Y., Dever, T. E., Merrick, W. C., et al. 1990. *New Biol.* 2:648–54
127. Morley, S. J., Traugh, J. A. 1990. *J. Biol. Chem.* 265:10611–16
128. Morley, S., Dever, T. E., Etchison, D., Traugh, J. A. 1991. *J. Biol. Chem.* 266: In press
129. Kozma, S. C., Ferrari, S., Thomas, G. 1989. *Cell. Signal.* 1:219–25
130. Erikson, E., Maller, J. L. 1986. *J. Biol. Chem.* 261:350–55
131. Jeno, P., Ballou, L. M., Novak-Hofer, I., Thomas, G. 1988. *Proc. Natl. Acad. Sci. USA* 85:406–10
132. Tabarini, D., DeHerreros, A. G., Heinrich, J., Rosen, O. M. 1987. *Biochem. Biophys. Res. Commun.* 144:891–900
133. Kozma, S. C., Ferrari, S., Bassand, P., Siegmann, M., Totty, N., Thomas, G. 1990. *Proc. Natl. Acad. Sci. USA* 87:7365–69
134. Ballou, L. M., Siegmann, M., Thomas,

G. 1988. *Proc. Natl. Acad. Sci. USA* 85:7154–58
135. Sturgill, T. W., Ray, L. B., Erikson, E., Maller, J. L. 1988. *Nature* 334:715–18
136. Blenis, J., Erikson, R. L. 1986. *Proc. Natl. Acad. Sci. USA* 83:1733–37
137. Oliver, A. R., Ballou, L. M., Thomas, G. 1988. *Proc. Natl. Acad. Sci. USA* 85:4720–24
138. Duncan, R. F., Hershey, J. W. B. 1987. *Mol. Cell. Biol.* 7:1293–95
139. Milburn, S. C., Duncan, R. F., Hershey, J. W. B. 1990. *Arch. Biochem. Biophys.* 276:6–11
140. Benne, R., Edman, J., Traut, R. R., Hershey, J. W. B. 1978. *Proc. Natl. Acad. Sci. USA* 75:108–12
141. Duncan, R., Hershey, J. W. B. 1986. In *Current Communications in Molecular Biology: Translational Control*, pp. 47–51. New York: Cold Spring Harbor Lab.
142. Farrell, P. J., Hunt, T., Jackson, R. J. 1977. *Eur. J. Biochem.* 89:517–21
143. Jackson, R. J. 1991. See Ref. 2
144. Pathak, V. K., Schindler, D., Hershey, J. W. B. 1988. *Mol. Cell. Biol.* 8:993–95
145. Colthurst, D. R., Campbell, D. G., Proud, C. G. 1987. *Eur. J. Biochem.* 166:357–63
146. Trachsel, H., Staehelin, T. 1978. *Proc. Natl. Acad. Sci. USA* 75:204–8
147. Pain, V. M. 1986. *Biochem. J.* 235:625–37
148. Thomas, N. S. B., Matts, R. L., Levin, D. H., London, I. M. 1985. *J. Biol. Chem.* 260:9860–66
149. Gross, M., Redman, R., Kaplansky, D. A. 1985. *J. Biol. Chem.* 260:9491–500
150. Gross, M., Wing, M., Rundquist, C., Rubino, M. S. 1987. *J. Biol. Chem.* 262:6899–907
151. Kaufman, R. J., Davies, M. V., Pathak, V. K., Hershey, J. W. B. 1989. *Mol. Cell. Biol.* 9:946–58
152. Davies, M. V., Furtado, M., Hershey, J. W. B., Thimmappaya, B., Kaufman, R. J. 1989. *Proc. Natl. Acad. Sci. USA* 86:9163–67
153. Kaufman, R. J., Murtha, P. 1987. *Mol. Cell. Biol.* 7:1568–71
154. Schneider, R. J., Shenk, T. 1987. *Annu. Rev. Biochem.* 56:317–32
155. Duncan, R. F., Hershey, J. W. B. 1987. *Arch. Biochem. Biophys.* 256:651–61
156. Huang, J., Schneider, R. J. 1990. *Proc. Natl. Acad. Sci. USA* 87:7115–19
157. Dholakia, J. N., Wahba, A. J. 1988. *Proc. Natl. Acad. Sci. USA* 85:51–54

158. Dholakia, J. N., Mueser, T. C., Woodley, C. L., Parkhurst, L. J., Wahba, A. J. 1986. *Proc. Natl. Acad. Sci. USA* 83:6746–50
159. Ryazanov, A. G. 1987. *FEBS Lett.* 214:331–34
160. Nairn, A. C., Palfrey, H. C. 1987. *J. Biol. Chem.* 262:17299–303
161. Price, N. T., Redpath, N. T., Severinov, K. V., Campbell, D. G., Russell, R. J. M., Proud, C. G. 1990. *FEBS Lett.*
162. Ovchinnikov, L. P., Motuz, L. P., Natapov, P. G., Averbuch, L. J., Wettenhall, R. E. H., et al. 1990. *FEBS Lett.* 275:209–12
163. Peter, M. E., Schirmer, N. K., Reiser, C. O. A., Sprinzl, M. 1990. *Biochemistry* 29:2876–84
164. Ryazanov, A. G., Davydova, E. K. 1989. *FEBS Lett.* 251:187–90
165. Sitikov, A. S., Simonenko, P. N., Shestakova, E. A., Ryazanov, A. G., Ovchinnikov, L. P. 1988. *FEBS Lett.* 228:327–31
166. Redpath, N., Proud, C. G. 1989. *Biochem. J.* 262:69–75
167. Celis, J. E., Madsen, P., Ryazanov, A. G. 1990. *Proc. Natl. Acad. Sci. USA* 87:4231–35
168. Duncan, R., Hershey, J. W. B. 1985. *J. Biol. Chem.* 260:5493–97
169. Clark, S. J., Ashford, A. J., Price, N. T., Proud, C. G. 1989. *Biochim. Biophys. Acta* 1010:377–80
170. Clark, S. J., Colthurst, D. R., Proud, C. G. 1988. *Biochim. Biophys. Acta* 968:211–19
171. Davydova, E. K., Sitikov, A. S., Ovchinnikov, L. P. 1984. *FEBS Lett.* 176:401–5
172. Belle, E., Derancourt, J., Poulhe, R., Capony, J.-P., Ozon, R., Mulner-Lorillon, O. 1989. *FEBS Lett.* 255:101–4
173. Janssen, G. M. C., Maessen, G. D. F., Amons, R., Moeller, W. 1988. *J. Biol. Chem.* 263:11063–66
174. Arpin, M., Madjar, J. J., Reboud, J.-P. 1978. *Biochim. Biophys. Acta* 519:537–41
175. Beaud, G., Masse, T., Madjar, J.-J., Leader, D. P. 1989. *FEBS Lett.* 259:10–14
176. Lavergne, J.-P., Conquet, F., Reboud, J.-P., Reboud, A.-M. 1987. *FEBS Lett.* 216:83–88
177. Dang, C. V., Traugh, J. A. 1989. *J. Biol. Chem.* 264:5861–65
178. Clemens, M. J. 1990. *Trends Biochem. Sci.* 15:172–75
179. Lindahl, L., Zengel, J. M. 1986. *Annu. Rev. Genet.* 20:297–326

180. Gold, L. 1988. *Annu. Rev. Biochem.* 57:199–233
181. Jackson, R. J., Standart, N. 1990. *Cell* 62:15–24
182. Jacobson, A., Favreau, M. 1983. *Nucleic Acids Res.* 18:6353–68
183. Lemay, G., Millward, S. 1986. *Arch. Biochem. Biophys.* 249:191–98
184. Munroe, D., Jacobson, A. 1990. *Gene* 91:151–58
185. Vassalli, J.-D., Huarte, J., Belin, D., Gubler, P., Vassalli, A., et al. 1989. *Genes Dev.* 3:2163–71
186. Drummond, D. R., Armstrong, J., Colman, A. 1985. *Nucleic Acids Res.* 13:7375–592
187. Galili, G., Kawata, E. E., Smith, L. D., Larkins, B. A. 1988. *J. Biol. Chem.* 263:5764–70
188. McGrew, L. L., Dworkin-Rastl, E., Dworkin, M. B., Richter, J. D. 1989. *Genes Dev.* 3:803–15
189. Sachs, A. B., Davis, R. W. 1990. *Science* 247:1077–79
190. Sonenberg, N. 1987. *Adv. Virus Res.* 33:175–204
191. Etchison, D., Milburn, S. C., Edery, I., Sonenberg, N., Hershey, J. W. B. 1982. *J. Biol. Chem.* 257:14806–10
192. O'Neill, R. E., Racaniello, V. R. 1989. *J. Virol.* 63:5069–75
193. Wyckoff, E. E., Hershey, J. W. B., Ehrenfeld, E. 1990. *Proc. Natl. Acad. Sci. USA* 87:9529–33
194. Wyckoff, E. E., Croall, D. E., Ehrenfeld, E. 1990. *Biochemistry* 29:10055–61
195. Theil, E. C. 1987. *Annu. Rev. Biochem.* 56:289–315
196. White, K., Munro, H. N. 1988. *J. Biol. Chem.* 263:8938–42
197. Hentze, M. W., Caughman, S. W., Rouault, T. A., Barriocanal, J. G., Dancis, A., et al. 1987. *Science* 238:1570–73
198. Aziz, N., Munro, H. N. 1987. *Proc. Natl. Acad. Sci. USA* 84:8478–82
199. Rouault, T. A., Hentze, M. W., Caughman, S. W., Harford, J. B., Klausner, R. D. 1988. *Science* 241:1207–10
200. Walden, W. E., Patino, M. M., Gaffield, L. 1989. *J. Biol. Chem.* 264:13765–69
201. Walden, W. E., Daniels-McQueen, S., Brown, P. H., Gaffield, L., Russell, D. A., et al. 1988. *Proc. Natl. Acad. Sci. USA* 85:9503–7
202. Rouault, T. A., Hentze, M. W., Haile, D. J., Harford, J. B., Klausner, R. D. 1989. *Proc. Natl. Acad. Sci. USA* 86:5768–72
203. Brown, P. H., Daniels-McQueen, S., Walden, W. E., Patino, M. M., Gaffield, L., et al. 1989. *J. Biol. Chem.* 264:13383–86
204. Lin, J.-J., Daniels-McQueen, S., Patino, M. M., Gaffield, L., Walden, W. E., Thach, R. E. 1990. *Science* 247:74–77
205. Lin, J.-J., Daniels-McQueen, S., Gaffield, L., Patino, M. M., Walden, W. E., Thach, R. E. 1990. *Biochim. Biophys. Acta* 1050:146–50
206. Hentze, M. W., Rouault, T. A., Harford, J. B., Klausner, R. D. 1989. *Science* 244:357–59
207. Klausner, R. D., Harford, J. B. 1989. *Science* 246:870–72
208. Goossen, B., Caughman, S. W., Harford, J. B., Klausner, R. D., Hentze, M. W. 1990. *EMBO J.* 9:4127–33
209. Rouault, T. A., Tang, C. K., Kaptain, S., Burgess, W. H., Haile, D. J., et al. 1990. *Proc. Natl. Acad. Sci. USA* 87:7958–62
210. Huang, W. M., Ao, S.-Z., Casjens, S., Orlandi, R., Zeikus, R., et al. 1988. *Science* 239:1005–12
211. Weiss, R. B., Huang, W. M., Dunn, D. M. 1990. *Cell* 62:117–26
212. Atkins, J. F., Weiss, R. B., Gesteland, R. F. 1990. *Cell* 62:413–23
213. Craigen, W. J., Cook, R. G., Tate, W. P., Caskey, C. T. 1985. *Proc. Natl. Acad. Sci. USA* 82:3616–20
214. Jacks, T., Townsley, K., Varmus, H. E., Majors, J. 1987. *Proc. Natl. Acad. Sci. USA* 84:4298–302
215. Brierley, I., Boursnell, M. E. G., Binns, M. M., Bilimoria, B., Blok, V. C., et al. 1987. *EMBO J.* 6:3779–85
216. Jacks, T., Madhani, H. D., Masiarz, F. R., Varmus, H. E. 1988. *Cell* 55:447–58
217. Brierley, I., Digard, P., Inglis, S. C. 1989. *Cell* 57:537–47
218. Weiss, R. B., Dunn, D. M., Dahlberg, A. E., Atkins, J. F., Gesteland, R. F. 1988. *EMBO J.* 7:1503–7
219. Yoshinaka, Y., Katoh, I., Copeland, T. D., Oroszlan, S. 1985. *Proc. Natl. Acad. Sci. USA* 82:1618–22
220. Strauss, E. G., Rice, C. M., Strauss, J. H. 1983. *Proc. Natl. Acad. Sci. USA* 80:5271–75
221. Yoshinaka, Y., Katoh, I., Copeland, T. D., Oroszlan, S. 1985. *J. Virol.* 55:870–73
222. Jones, D. S., Nemoto, F., Kuchino, Y., Masuda, M., Yoshikura, H., Nishimura, S. 1989. *Nucleic Acids Res.* 17:5933–45
223. ten Dam, E. B., Pleij, C. W. A., Bosch, L. 1990. *Virus Genes* 4:121–36

224. Chambers, I., Frampton, J., Goldfarb, P., Affara, N., McBain, W., Harrison, P. R. 1986. *EMBO J.* 5:1221–27
225. Zinoni, F., Heider, J., Böck, A. 1990. *Proc. Natl. Acad. Sci. USA* 87:4660–64
226. Forchhammer, K., Rucknagel, K. P., Böck, A. 1990. *J. Biol. Chem.* 265:9346–50
227. Brostrom, C. O., Brostrom, M. A. 1990. *Annu. Rev. Physiol.* 52:577–90
228. Pain, V. M., Clemens, M. J. 1991. See Ref. 2
229. Saborio, J. L., Pong, S.-S., Koch, G. 1974. *J. Mol. Biol.* 85:195–211
230. van der Krol, A. R., Mol, J. N. M., Stuitje, A. R. 1989. *Biotechniques* 6:958–76
231. Nielsen, P., Goelz, S., Trachsel, H. 1983. *Cell Biol. Int. Rep.* 7:245–54
232. Gay, D. A., Sisodia, S. S., Cleveland, D. W. 1989. *Proc. Natl. Acad. Sci. USA* 86:5763–67
233. Ernst, H., Duncan, R. F., Hershey, J. W. B. 1987. *J. Biol. Chem.* 262:1206–12
234. Pathak, V. K., Nielsen, P. J., Trachsel, H., Hershey, J. W. B. 1988. *Cell* 54:633–39
235. Nielsen, P. J., McMaster, G. K., Trachsel, H. 1985. *Nucleic Acids Res.* 13:6867–80

236. Milburn, S. C., Hershey, J. W. B., Davies, M. V., Kelleher, K., Kaufman, R. J. 1990. *EMBO J.* 9:2783–90
237. Rychlik, W., Domier, L. L., Gardner, P. R., Hellmann, G. M., Rhoads, R. E. 1987. *Proc. Natl. Acad. Sci. USA* 84:945–49
238. McCubbin, W. D., Edery, I., Altmann, M., Sonenberg, N., Kay, C. M. 1988. *J. Biol. Chem.* 263:17663–71
239. Smit-McBride, Z., Dever, T. E., Hershey, J. W. B., Merrick, W. C. 1989. *J. Biol. Chem.* 264:1578–83
240. Brands, J. H. G. M., Maassen, J. A., van Hemert, F. J., Amons, R., Möller, W. 1986. *Eur. J. Biochem.* 155:167–71
241. Xiang, L., Werner, D. 1989. *Nucleic Acids Res.* 17:442
242. Maessen, G. D. F., Amons, R., Maassen, J. A., Möller, W. 1986. *FEBS Lett.* 208:77–83
243. Maessen, G. D. F., Amons, R., Zeelen, J. P., Möller, W. 1987. *FEBS Lett.* 223:181–86
244. Kohno, K., Uchida, T., Ohkubo, H., Nakanishi, S., Nakanishi, T., et al. 1986. *Proc. Natl. Acad. Sci. USA* 83:4978–82
245. Lee, C. C., Craigen, W. J., Muzny, D. M., Harlow, E., Caskey, C. T. 1990. *Proc. Natl. Acad. Sci. USA* 87:3508–12

*Annu. Rev. Biochem. 1991. 60:757–94*
*Copyright © 1991 by Annual Reviews Inc. All rights reserved*

# STRUCTURE AND FUNCTION OF HEXOSE TRANSPORTERS

## Mel Silverman

MRC Group in Membrane Biology, Department of Medicine, University of Toronto, Toronto, Ontario, M5S 1A8 Canada

KEY WORDS:   glucose transport, glucose carriers, transport of sugars.

## CONTENTS

## INTRODUCTION

In this review I focus exclusively on a single example of carrier-mediated transport: sugar movement across cell membranes.

757

0066-4154/91/0701-0757$02.00

Important discoveries have taken place within the past five years, and progress has accelerated within the past 36 months. Occasionally, to put some of the new developments in perspective, I cite references from the older literature, but I emphasize recent work. I have relied heavily on three excellent overview articles on the subject of facilitative glucose transport that have appeared in the past 6–12 months (1–3).

## PERSPECTIVES AND SUMMARY

All living creatures, prokaryotes and eukaryotes alike, control their internal milieu by a continuous exchange of physical and chemical information with the external environment. Information transfer is mediated by means of specific transmembrane signalling pathways initiated by cell surface receptor occupancy (e.g. by hormones), which then triggers a cascade of intramembrane and cytoplasmic transducing events. But by far the most important means of regulating molecular traffic across cell membranes is through cooperative action of specialized intramembrane channels and carrier protein complexes linked to each other functionally via a complex network of transmembrane electrochemical gradients. Many impressive advances have occurred in the field of membrane transport (especially within the past few years) relating to identification, isolation, and purification of various carrier proteins, yet elucidation of their molecular mechanism of action remains an unsolved problem. Nevertheless, the rapid advances being made in our understanding of transporter structure will undoubtedly soon provide valuable insights into function.

Several reasons justify selecting sugars as a prototype substrate with which to explore general concepts of transporter function. For example, sugars are relatively simple uncharged compounds, whose configurational and conformational properties are well described in the chemical literature. This greatly simplifies interpretations of structure-activity considerations, both from the point of view of substrate-solvent and substrate-membrane interactions.

Sugar transport is also a vital area of research in human biology and pathophysiology. Mammalian species have a variety of sugar carriers, each specialized to individual cellular needs. Perhaps this specialization is not too surprising, since the crucial importance of sugars as metabolic substrates for the brain necessitates precise regulation of blood glucose concentrations. Thus not only are sugar carriers ubiquitous in nature, but in addition interesting functional variations have evolved that endow the sugar transport system with intriguing specificity and kinetic characteristics.

Beginning in the 1930s, physiologic, biochemical, and biophysical studies provided an increasingly detailed picture of hexose transport. By the early-to-mid-1980s, a three-category functional classification was available for mammalian sugar transport. Carriers in the first category behaved as facilitated (equilibrative) mechanisms; the most intensively studied prototype is the human red cell. Hormone-sensitive carriers, such as the insulin-modulated system in adipocytes, constituted a second group. The third category consisted of $Na^+$-coupled cotransport mechanisms, of which the best examples were found in renal and intestinal epithelium.

In general, purification of sugar transporters (like other membrane proteins) was difficult to achieve using traditional protein separation methods because of the relatively low density of these transporters in the plasma membranes. One exception was the human red cell glucose carrier, which represents about 5% of the erythrocyte-to-membrane protein. Nevertheless, even for the case of this transport protein, it was difficult to obtain sufficient material for detailed sequencing and structural analysis.

During the past five years the intensive application of molecular biological techniques to clone cDNAs of membrane transporters has dramatically altered the whole field and has clarified the molecular basis for the functional differences of the three categories of sugar transporter discussed above ($Na^+$-independent, insulin sensitive, and $Na^+$-dependent). We now know that there exists a family of glucose transporter genes, located on different chromosomes, that code for the sugar transporter in all cells. So far at least five different isoforms of the facilitative glucose transporter have been characterized and had their amino acid sequences deduced from cloned cDNA (these isoforms have been designated GLUT 1 to 5 based on the chronological order of cloning of their genes). *GLUT 1* is the gene that encodes the red cell and brain glucose carrier, and *GLUT 4* encodes the insulin-dependent glucose carrier in adipocytes and striated muscles. *GLUT 2* encodes the glucose transporter in liver and the $\beta$-cell of pancreatic islets. *GLUT 3* and *GLUT 5* encode glucose carriers distributed in various tissues, including brain and the small intestine. The sizes of the mammalian facilitative glucose carriers vary from 492 to 524 amino acids. There is 39–65% identity and 50–76% similarity between the amino acid sequences of the different human isoforms. For all the five forms, there is a high degree of sequence conservation across species.

The strong homologies that exist within the family of human facilitative glucose carriers are reflected in the fact that they have almost identical topological organization in the plasma membrane based on hydropathic and secondary structure predictions. The essential features of this structure are the 12 transmembrane $\alpha$-helical segments, which form the polypeptide backbone. The N and C termini, which vary in length and/or sequence in the different isoforms, are located on the cytoplasmic face; there is a highly charged endofacial domain between M6 and M7 consisting of 65 amino acids and a smaller exofacial domain between M1 and M2 of 30 amino acids (GLUT 1,

GLUT 3, and GLUT 5), 37 amino acids (GLUT 4), and 67 amino acids (GLUT 2). This exofacial segment contains a potential site for asparagine-linked glycosylation. The structural differences between members of the GLUT family of proteins presumably account for their unique behavior.

Strong structural homologies exist not only within the family of mammalian facilitative glucose transporters, but also between mammalian and certain prokaryotic sugar carriers such as the yeast glucose transporter, and the *Escherichia coli* symporters, $H^+$-xylose and $H^+$-arabinose. All of these carrier proteins also have in common the fact that they are inserted posttranslationally into microsomes (such insertion is unusual for mammalian membrane proteins, most of which are inserted cotranslationally). The proteins encoded by the *GLUT* genes have a molecular weight of around 55,000, and this protein subunit appears to be sufficient to carry out sugar transport activity. Radiation inactivation studies suggest that the native erythrocyte glucose transporter is a homotetramer. It is still uncertain, however, whether or not the behavior in the intact membrane reflects the cooperative interaction of the oligomeric structure.

The details of the chemical mechanism of action of the facilitative glucose carrier remain unknown, but detailed kinetic analysis of tracer fluxes in intact erythrocytes and measurement of changes in intrinsic tryptophan fluorescence associated with ligand binding in purified glucose transporter support the following sequence of events during the transmembrane transport process. At 37°C under pre-equilibrated conditions of 5 mM D-glucose, the glucose carriers fall into four groups: (*a*) outward-facing conformer, unloaded (~35%), (*b*) outward-facing conformer, glucose-loaded (~15%), (*c*) inward-facing conformer, unloaded (~35%), and (*d*) inward-facing conformer, glucose-loaded (~15%). The sugar ligand "docks" in its binding site by exchanging hydrogen bonds from water to protein residues. The resulting exclusion of water lowers the free energy of the ligand-receptor complex by ~11 kcal/mole, and this is associated with a sugar-induced conformational change that reorients the carrier and results in translocation of solute.

A sixth gene, designated *SGLT 1*, has been cloned that encodes an apparently totally different sugar transporter, the human $Na^+$-D-glucose intestinal cotransporter. *SGLT 1* codes for a 664-amino-acid protein of $M_r$ ~73,000 that is almost identical to the $Na^+$-D-glucose transporter from rabbit intestine. The predicted structure for SGLT 1 has 11 transmembrane segments with N and C termini on the endo- and exofacial surfaces respectively. Extensive homology exists between SGLT 1 and the renal brush border $Na^+$/D-glucose cotransporter, and there is also strong similarity to the Na-proline symporter from *E. coli*. Thus the $Na^+$-D-glucose symporters represent a distinct class of carrier proteins from the facilitative glucose transporters.

At the molecular level, the mechanism of action for the $Na^+/D$-glucose cotransporter is very similar to that of the facilitative carriers, but is complicated by the interaction of $Na^+$ with the transport protein. Dynamic studies of the isolated purified transporter as well as kinetic flux studies carried out using isolated membrane vesicles indicate that the cotransport of $Na^+$ and D-glucose via SGLT 1 occurs as follows: $Na^+$ binds to its site on the 73-kDa protein and induces a conformational change that drastically increases the affinity of binding of D-glucose to its site. The binding of glucose then induces a second conformational change that results in translocation. There is evidence that in the native state the $Na^+/D$-glucose cotransporter exists as a homotetramer, but it is not known whether or not the biologic activity in the normal membrane requires the cooperative interaction of all four subunits.

Ultimately, the structure-function correlations that determine the biologic activity of the hexose transport proteins will require three-dimensional crystallography studies of the loaded and unloaded carrier. In the meantime, we must be content with all available indirect approaches, realizing that we may have reached a point predicted by D. D. Brown some years ago (4) where "we can imagine the paradoxical situation of having in hand a set of fully sequenced genes with completely characterized products [i.e. those of carrier molecules] all known to have important and interrelated development function but without the slightest notion of what these products do [or how they work!] in the cell."

# FACILITATIVE GLUCOSE TRANSPORTERS

## The Red Cell/Brain Glucose Transporter—GLUT 1

By the early 1980s, the human red cell glucose transporter had been identified as a 55-kDa glycoprotein running as band 4.5 on SDS-PAGE gels of the erythrocyte membrane. Treatment with N-glycanase, which removes N-linked oligosaccharides, resolved the broad 55-kDa species to a single 46-kDa polypeptide (reviewed in 5, 6).

The primary structure of this red cell glucose carrier was definitively established in 1985 when Mueckler et al (7), using a polyclonal antibody against the purified red cell carrier, screened a complementary DNA (cDNA) library from the human Hep G2 hepatoma cell line. The cDNA isolated from Hep G2 was found to encode a 492-amino-acid glucose transport protein, the amino acid sequence of which was found to be in good agreement with that obtained from partial sequence data of the purified erythrocyte glucose carrier (8).

Later, it was independently confirmed that the cDNA isolated from Hep G2 coded for a glucose transport protein. This confirmation was accomplished by using a bacterial expression system consisting of an *E. coli* mutant strain deficient in three principal proteins that catalyze glucose uptake (9, 10). Through use of this system, it was demonstrated that the glucose transport induced in the bacterial membrane exhibited stereospecificity for D- but not L-glucose and was inhibited by cytochalasin B and mercury chloride. These results correlated with the well-known characteristics of the glucose transporter in intact human erythrocyte.

The consensus in the field, therefore, was that the amino acid sequence derived from the Hep G2 cDNA clone truly represents the red cell glucose carrier, and it was designated GLUT 1.

A similar glucose transporter was isolated from a rat brain expression library (11) and found to be 98% identical in amino acid sequence to the human red cell glucose transporter. This so-called erythroid/brain glucose carrier has also been detected in fetal tissue, the placenta, and in adult heart, kidney, fibroblasts, adipose cells, and every transformed cell line tested. Most importantly, it was found to be expressed in high levels in brain microvessels, suggesting that it is an important component of the blood-brain capillary barrier (12).

A model for the two-dimensional orientation of GLUT 1 in the plasma membrane was proposed (7) based on hydropathy plots of the deduced amino acid sequence (Figure 1). The essential features of this structure are that the transporter is organized into three domains: (*a*) 12 α-helices, which each span the membrane with the N and C termini of the protein located on the cytoplasmic side of the plasma membrane, (*b*) a highly charged intracellular (endofacial) domain, which consists of 65 hydrophilic amino acids (between M6 and M7), and (*c*) a smaller 33-amino-acid extracellular (exofacial) segment (between M1 and M2), which contains a single site for an asparagine-linked oligosaccharide, at asparagine 45.

The topological features of the model structure depicted in Figure 1 have been largely confirmed by vectorial proteolytic digestion and sequence-specific antibody studies (13–16). Trypsin only cleaves at the cytoplasmic surface of the transporter to yield two large membrane-bound fragments, one of which bears the N-linked oligosaccharide and the other, the cytochalasin B binding site (see Figure 1 and Ref. 14). Infrared and circular dichroism spectroscopic studies have confirmed that the transporter is predominantly composed of α-helices, but with some random coil and β turn structure also reported (17, 18). The extramembranous domains have some α-helical structure (15).

The endo- and exo-facial surfaces of the transporter have been covalently labelled with the high-affinity transport inhibitors cytochalasin B and ASA-BMPA [2-N-(4-azidosalicoyl)-1,3-bis(D-mannose-4'-yloxy) propyl-2-amine], respectively (19). Proteolytic digestion of the labelled transporter

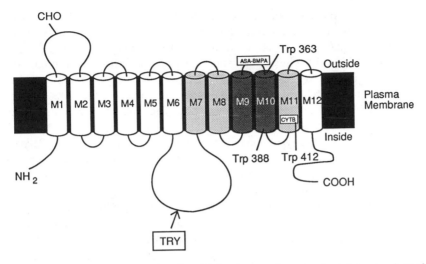

*Figure 1*   Schematic respresentation of model showing two-dimensional membrane topology of the prototypical facilitative glucose transporter GLUT 1. Putative membrane-spanning α-helices are labelled M1–M12. M7, M8, and M11 have been putatively assigned as the hydrophilic channel and M9 and M10 as the hydrophobic clefts. Sites of inhibitor binding for cytochalasin B and ASA-BMPA are indicated as well as potential glucose-binding sites near tryptophan residues 363, 388, and 412. CHO represents the location of asparagine-linked glycosylation and TRY shows the endofacial cleavage location of trypsin (adapted from Ref. 3).

with either chymotrypsin or subtilisin yields a 6–7-kDa fragment that can be labelled with either ligand. The locations of the ligand-binding sites ASA-BMPA and cytochalasin B between helices nine and eleven (see Figure 1) were deduced from the fragmentation produced by chemical cleavage at cysteine and tryptophan residues (using 2-nitro-5-thiocyanobenzoic acid and *N*-bromosuccinimide, respectively) (15, 19).

It must be emphasized that the structure depicted schematically in Figure 1 is a two-dimensional model of the proposed intramembrane topology of GLUT 1. The three-dimensional structure is unknown. Radiation inactivation studies of intact red cell membranes, however, suggest that in its native state GLUT 1 may exist as a homotetramer (20). The implication of this oligomeric structure in terms of transporter function is not yet determined.

## The Liver Glucose Transporter—GLUT 2

Biochemical and molecular biological studies suggested that the protein responsible for glucose transport in liver is distinct from the erythrocyte-type transporter described in the previous section. The binding constant for cytochalasin B (a potent competitive inhibitor of glucose transport) and the apparent $K_m$ for glucose of the liver glucose carrier were observed to be about

10-fold greater than those of the erythrocyte glucose transporter (21, 22). The very low levels of *GLUT 1* mRNA in adult liver were also consistent with the notion that hepatocytes expressed a novel glucose transporter (11, 23).

Assuming that the glucose transporter expressed in the liver might be structurally related to GLUT 1, liver cDNA libraries for clones that cross-hybridized to the *GLUT 1* cDNA were screened under conditions of low stringency (24, 25), and cDNA clones were isolated that encoded human and rat proteins of 524 and 522 amino acids, respectively. The amino acid sequence of this liver-type glucose transporter (designated GLUT 2) had 55.5% identity with that of GLUT 1, and there is 82% identity between the amino acid sequences of human and rat GLUT 2. Analysis of the predicted primary structure indicated that the topological organization of GLUT 2 protein in the plasma membrane was identical to that predicted for GLUT 1 (Figure 1). Expression of this transporter in an *E. coli* strain defective in glucose transport indicated that the protein was indeed a glucose transporter with the expected stereospecificity (24). Analysis of protein and mRNA distribution indicated that GLUT 2 was preferentially expressed in the liver, kidney, small intestine, and insulin-producing $\beta$-cell of the endocrine pancreas. By immunofluorescence, both GLUT 2 and GLUT 1 were localized to the basolateral surface of the kidney and intestinal epithelium (26). The presence of GLUT 2 in the kidney and small intestine suggests that it may be responsible for the transport of glucose across the serosal surface of the epithelial cells that line the intestine and nephron. (As is discussed below, luminal surfaces of these cells have a unique $Na^+$-dependent glucose carrier or symporter, the sequence of which is unrelated to that of any of the facilitative glucose transporter family.) The expression of GLUT 2 in islets may be of considerable importance in the maintenance of blood glucose levels, perhaps acting in tandem with glucokinase to regulate the control of insulin secretion from $\beta$-cells.

## The Brain Glucose Transporter—GLUT 3

RNA blotting studies indicated that skeletal muscle, which is responsible for disposal of about 90% of the glucose after a meal, contains very low levels of *GLUT 1* mRNA and no *GLUT 2* mRNA, suggesting that neither of these transporters is responsible for glucose uptake by this tissue. Using the low-stringency-hybridization strategy with a *GLUT 1* cDNA probe, Kayano et al (27) isolated a cDNA from a human fetal skeletal muscle library that encodes a protein of 496 amino acids having 64% and 52% identity with GLUT 1 and GLUT 2, respectively. Analysis of the amino acid sequence suggested that the membrane topology of GLUT 3 is similar to that of GLUT 1 (Figure 1). RNA for this transporter isoform is present in all tissues, but is most abundant in adult brain, kidney, and placenta. As *GLUT 3* mRNA is present in fibroblasts and smooth muscle, it is possible that the apparent ubiquitous distribution of this transporter isoform is due to the presence in all tissues of fibroblasts and blood vessels. *GLUT 3* mRNA levels are relatively low in adult muscle, suggesting that another facilitative glucose transporter isoform exists in this tissue.

## The Insulin-Responsive Glucose Transporter—GLUT 4

Insulin causes a 20–30-fold increase in the rate of glucose transport across the plasma membrane of adipocytes. The effects of insulin on glucose transport are acute, occurring with a half-time of two or three minutes, and in the absence of new protein synthesis (28, 29). This insulin-stimulated increase in glucose transport is due, at least in part, to the translocation of a latent pool of glucose transporters from an intracellular location to the plasma membrane (30, 31). Antibodies raised against the erythrocyte glucose transporter (GLUT 1) and to peptides predicted by the *GLUT 1* cDNA have been used to demonstrate that in both rat adipocytes and 3T3-L1 adipocytes, GLUT 1 undergoes insulin-stimulated translocation (32–34). A series of experiments using 3T3-L1 adipocytes showed that upon insulin challenge, the numbers of GLUT 1 proteins at the cell surface increased approximately threefold. In contrast, the rate of deoxyglucose transport measured under identical conditions was stimulated 12–15-fold, indicating that the translocation of GLUT 1 is itself insufficient to account for the full stimulation of transport by insulin, unless insulin also increased the intrinsic activity of the GLUT 1 after translocation to the plasma membrane (28, 32–35).

This issue was resolved with the identification of a unique glucose transporter (GLUT 4) expressed in insulin-responsive tissues, which was localized predominantly in an intracellular location in the basal state and which translocated to the plasma membrane in response to insulin. In rat adipocytes, this protein accounts for ~90% of the total detectable glucose transporter present (36, 37). GLUT 4 was first identified using a monoclonal antibody prepared against a low-density microsomal fraction of rat adipocytes (38). Recently, rat, human, and mouse cDNA clones encoding this protein have been isolated and characterized (39–43). GLUT 4 is a 509-amino-acid protein having 65%, 54%, and 58% identity with GLUT 1, 2, and 3, respectively; there is 95% identity between human and rat proteins. GLUT 4 has been shown to represent a functional glucose transporter by expression of synthetic GLUT 4 mRNA in *Xenopus* oocytes (40). The two-dimensional orientation of GLUT 4 in the membrane is similar to the model proposed for the other isoforms of the facilitative glucose transporter (Figure 1).

RNA blotting studies have shown that the highest levels of *GLUT 4* mRNA are in insulin-responsive tissues such as cardiac and skeletal muscle and brown and white fat (39–43). GLUT 4 protein has also been demonstrated in these tissues by immunoblotting (39, 40), and is expressed in 3T3-L1 adipocytes, but only after differentiation of the fibroblastic cell line into adipocytes (39, 40, 43).

## Other Facilitative Glucose Transporter-like Proteins— GLUT 5

In addition to the proteins described above, cDNA clones encoding a 501-amino-acid protein, which is expressed at very high levels in human small intestine, have been isolated and characterized (44). This protein has 42%, 40%, 39%, and 42% identity with GLUT 1, 2, 3, and 4 respectively. It is unknown at this time whether GLUT 5 transports glucose or is a carrier for another sugar or other small polar molecule.

The main features of the family of facilitative glucose transporters so far identified are summarized in Tables 1 and 2 taken from Refs. 1 and 45.

## Conservation of Facilitative Sugar Transporter Structure During Evolution

All mammalian facilitative glucose transporters share the same arrangement of hydrophobic and hydrophilic segments as is found in the erythroid/brain carrier (GLUT 1). The yeast high-affinity glucose carrier has a similar structure except for an additional COOH-terminal extension of 360 amino acids (46). In addition, two E. coli symporters, $H^+$-xylose and $H^+$-arabinose, have the same structure and share 20–25% sequence homology with mammalian glucose transporters (47, 48).

In addition to the strong structural homologies between mammalian and bacterial glucose transporter, both types of protein are inserted posttrans-

**Table 1**  Human facilitative glucose transporters

| Designation (common name) | Size (no. of amino acids) | Major sites of expression | Chromosomal location |
|---|---|---|---|
| Facilitative glucose transporters | | | |
| GLUT 1 (Erythrocyte, Hep G2, brain) | 492 | Fetal tissues, brain, kidney, and colon | 1 |
| GLUT 2 (Liver) | 524 | Liver, $\beta$-cell, kidney, and small intestine | 3 |
| GLUT 3 (Fetal muscle) | 496 | Many tissues including brain, placenta, and kidney | 12 |
| GLUT 4 (Muscle/adipocyte, insulin-regulatable) | 509 | Skeletal muscle, heart, and adipocytes | 17 |
| GLUT 5 (Small intestine) | 501 | Small intestine | 1 |

Reprinted from Ref. 1

**Table 2** Facilitated-diffusion glucose transporters in mammalian tissues

| Nomenclature | Tissue from which gene for transporter was cloned | Tissues in which transporter is expressed | Regulatory factors | |
|---|---|---|---|---|
| | | | In vitro | In vivo |
| GLUT 1 | Hepatoma cells (Hep G2), brain | Many tissues; abundant in human erythrocytes, blood-brain barrier, placenta, transformed cells in culture | Oncogenes, phorbol esters, growth factors, insulin, cAMP, protein kinase C, vanadate, sulfonylureas, glucose, dexamethasone, cellular differentiation | Fasting, refeeding high-fat feeding, genetic obesity (Zucker rats), insulin,[a] exercise[a] developmental stage |
| GLUT 2 | Liver, pancreatic $\beta$-cells | Liver, pancreatic $\beta$-cells, kidney, intestine (basolateral membrane) | | Fasting,[a] refeeding,[a] STZ-D[a,b] |
| GLUT 3 | Fetal skeletal muscle | Many tissues; abundant in brain, placenta, kidney. Low levels in adult skeletal muscle | | Developmental stage in muscle |
| GLUT 4 | Heart, adipose cells, skeletal | White adipose cells, brown fat, red and white muscle heart, smooth muscle (?) | Cellular differentiation | Insulin, STZ-D, fasting, refeeding high-fat feeding, genetic obesity (Zucker rats), exercise[a] |

[a] mRNA levels are altered, but protein levels are unchanged or not studied
[b] STZ-D, streptozocin-induced diabetes
Reprinted from Ref. 45

lationally (49) into microsomes, which could explain the ability of mammalian glucose carrier to be expressed functionally in bacteria.

## Structural Comparison of Facilitative Glucose Transporters

Comparisons indicated several regions in which the sequences of all five mammalian glucose transporters are identical (see Figure 1). These conserved regions presumably serve a common function in each isoform, and it seems likely that several of the conserved amino acids form the channel or pore through which the glucose moves. Of the transmembrane segments, the sequences of M4, M5, and M6 are the most highly conserved (38%, 48%, and 38% of the amino acids are identical), whereas for example only 14% of the residues are identical in M9. The sequence conservation in the hydrophobic membrane-spanning regions includes not only nonpolar-hydrophobic amino acids, which are a feature of such regions, but also polar residues, which may be involved in the transport of glucose. Perhaps, the hydroxyl and amino groups of these amino acids make up part of the channel or pore through which glucose moves. In particular, the sequences of M5 and M8 when plotted as $\alpha$-helices are notably amphipathic, and this feature is conserved in all the transporters including the bacterial and yeast proteins. The polar groups in M7 and M8 would be located on one surface of the $\alpha$-helix and thus could form a polar channel through which the sugar is transported.

Other significant features of the conserved regions include the clusters of aromatic residues in M11 and M12, and the region in M10 that is rich in glycine and proline. The clusters of aromatic amino acids might indicate that M11 and M12 are in close proximity in the protein, with planar interactions between the aromatic side chains of these amino acids stabilizing the structure. The glycine/proline-rich domain in M10 might be expected to impart considerable flexibility to this portion of M10, perhaps indicating a role for this region in the inward- to outward-facing conformational change that accompanies the transport of glucose. Indeed, membrane-buried proline residues have been proposed to be key players in such conformational changes (50). The precise boundaries of these putative transmembrane-spanning domains, however, remain to be established. As a final point, a recent study demonstrated that more than 80% of the protein backbone of the erythrocyte transporter (GLUT 1) is accessible to deuterium exchange, suggesting that much more of the protein is accessible to water than might be expected based on the hydrophobic nature of the primary sequence (51). This observation is consistent with the idea that the transmembrane domains form a pore in the membrane through which glucose is transported. Presumably such a pore is accessible to extracellular water and can thus account for the high level of solvent water contact reported.

In addition to the membrane-associated domains of the facilitative glucose transporters, there are several other regions in which the sequence identity is quite striking. In particular, there is substantial sequence identity in the short exofacial loops connecting the transmembrane domains, which is in marked contrast to the endofacial loops. The conserved residues in the exofacial loops are primarily charged amino acids, suggesting that they may participate in ionic interactions between different regions of the transporter itself or with cytoplasmic proteins such as hexokinase/glucokinase that are involved in the metabolism of glucose once it has been transported into the cell.

While several of the membrane-spanning segments exhibit the highest degree of identity, the N- and C-termini and the extracellular loop between M1 and M2 are the regions that display the greatest divergence, both in primary structure and in size. The central loop between M6 and M7 is intermediate in sequence divergence, and 17 out of 65 residues (26%) in this domain are conserved. The sequences that are responsible for isoform-specific properties such as kinetic parameters and modes of regulation (e.g. insulin-stimulated translocation, or catecholamine-induced inhibition of transport) have not been defined, nor is it known whether there is a sequence-specific motif in GLUT 4 that is responsible for targeting the GLUT 4 protein to an intracellularly localized vesicle fraction whereas the other transporters are directed to the plasma membrane. Expressing the different transporter isoforms and chimeric transporters in various systems, including insulin-responsive and nonresponsive cells, will elucidate the importance of individual regions of this new and functionally important family of proteins (52).

## Structure-Function Correlations

To gain insights into the precise chemical mechanism(s) underlying the transmembrane transport of glucose, it will ultimately be necessary to visualize the sugar substrate bound within its transporter by X-ray crystallography. Until such time as this is possible, we must be satisfied with indirect approaches.

As already mentioned, hydrogen exchange studies have shown that a large proportion of the protein is readily accessible to the aqueous environment, possibly indicating that certain of the $\alpha$-helical transmembrane segments are arranged in a water-filled channel (27, 29). However, in order to prevent leakage of small ions, the structure must be constricted by a tight association of the helices. Helices 7 and 8 are likely candidates for the channel since they contain many serine and glutamine residues capable of hydrogen bonding to glucose.

Since the transporter is fixed in the plane of the lipid bilayer (i.e. there is no evidence of physical displacement of carrier in the transmembrane direction), translocation of sugar occurs by passage of glucose through the protein; and since, as just discussed, it is likely that glucose movement occurs via a water-filled channel contained within the carrier protein, it is relevant to examine the chemistry of sugar:water interaction.

The conformation of monosaccharides in aqueous solution has been extensively studied by Reeves (53) using cupraammonium complexes. He proposed that there are two rigid chair forms, $^4C_1$ and $^1C_4$, and an infinite number of flexible 'boat forms,' which may be described in terms of six specific conformers whose interconversion involves little ring strain (ring atoms that lie above the reference plane are written as superscripts, those below as subscripts).

Which of the two chair forms predominates and whether the $\alpha$ or $\beta$ anomer is more stable depend upon nonbonded and other interactions among the ring substituents and the ring oxygen.

In all tissues exhibiting specialized sugar transport systems, the "preferred" substrate conformation for carrier-mediated transport is that of a pyranose chair in the $^4C_1$ conformation D-glucose configuration. Presumably this preference is a manifestation of the stability of sugars in this particular conformation in aqueous solution, reflecting a state of maximal hydrogen bonding between sugar and water molecules.

The monosaccharides can be assigned instability ratings, which depend upon the presence of axial groups and the relative weighting factors attached to them. The greater the instability rating, the greater the conformational variations, and presumably, the weaker the solute:water interactions. The importance of these conformational variations in determining partitioning of monosaccharides between different phases was borne out by the work of Marsden (54). He found that a low instability rating was associated with exclusion from a gel phase. In other words, the greater the sugar:water interaction, the less the partitioning between solute and gel.

These and other data suggest a rather strong sugar:water interaction. In fact, nuclear magnetic resonance evidence (55) reveals that each sugar hydroxyl group forms at least two hydrogen bonds with water. There is also theoretical and experimental evidence for cooperativity in $H^+$-bonding (56). Hydroxyl groups for which the oxygen atom accepts a $H^+$ bond form stronger bonds at greater frequency, leading to H-bonded chains . . . OH. . . . OH. . . . OH . . . Oxygen atoms that act as donors, however, produce weak $H^+$ bonds. Thus, if we imagine a lattice of water and sugar molecules as part of the external bathing medium adjacent to a membrane phase, and if there is any tendency for $H^+$ bonding into chains on the part of water, then the sugar molecules will tend to act as chain "stoppers."

The biologic implication is that sugar:water, sugar:membrane interactions that occur through $H^+$ bonding will tend to involve multiple weak bonds.

There is important experimental evidence to support the view that networks of hydrogen bonds determine the sugar-transporter binding reaction and that sugar binding induces a protein conformational change. The data are derived from many sources and reviewed in Ref. 57. In particular, the binding of L-arabinose within the arabinose-binding protein (ABP) exhibits extensive networks of hydrogen bonds radiating from the sugar molecule and extending to form shell-like structures with six residues in the vicinity of the sugar (Asp, Glu, and Asn are hydrogen bond acceptors, and Lys, Asn, and Arg are hydrogen bond donors). The sugar ring oxygen is almost tetrahedrally coordinated, and two sequestered water molecules that form hydrogen bonds with $O_2$ and $O_5$ of the sugar mediate the interactions between sugars and ligand site residues as well as provide additional noncovalent linkages at the opening of the binding cleft. Finally, the bound L-arabinose and essentially all of the hydrogen-bonding residues are buried and inaccessible to the bulk solvent.

Two categories of changes in protein conformation may occur upon sugar binding (a) small local changes affecting only residues in and around the binding site region and (b) large changes requiring movement between domains of bilobal proteins and enzymes.

Of most relevance to a discussion of the mechanism of action of sugar transporters is the possible role that sugar-induced protein conformational changes play in causing reorientation of the sugar carrier within the membrane. For example, the binding cleft of hexokinase has been observed to close upon glucose binding, a movement that causes the glucose molecule to become buried and presumably excludes water molecules from the binding site (58). This type of internalization movement has a modest energy requirement compared to that needed to introduce hydrophilic regions into a hydrophobic domain, and therefore this is a more plausible mechanism to account for the catalysis of transmembrane translocation by transport proteins.

## Hydrogen Bonding Specificity of GLUT 1

The hydrogen-bonding specificity of the red cell glucose transporter has emerged from a long series of studies using intact erythrocytes. In 1958 LeFevre & Marshall (59) carried out a comparative study of 14 aldoses in which the dissociation constants of the carrier-sugar complex were derived from kinetic measurements and then correlated with the degree of competitive inhibition by phloretin. It was found that within this homologous series of sugars, the ranked order of affinity of sugars for the carrier paralleled the sequence of increasing stability in the $^4C_1$ chair conformation. There seemed to be no special requirement for particular side group arrangements.

Using intact red cell membranes, Kahlenberg & Dolansky (60) carried out a systematic investigation of the structural requirements of D-glucose for its binding to the carrier. The data indicated that no specific hydroxyl group on the D-glucose molecule could be held responsible for the binding event. But the evidence highlighted the fact that several of the sugar's hydroxyl groups might be involved in a concerted fashion. For example, the OH groups at C-2 and C-6 were demonstrated not to be involved in binding, while removal or derivatization of the anomeric or C-1 hydroxyl was more critical. Kahlenberg & Dolansky (60) went on to point out that the most likely interaction between sugar and carrier was via H-bonding.

Later Barnett (61) carried out further detailed studies to determine the structural requirements for binding to the glucose red cell carrier. Derivatives in which a hydroxyl group in the D-glucose configuration was inverted, or replaced by a hydrogen atom at C-1, C-2, C-3, C-4 or C-6, all bound to the carrier. This observation confirmed the previous findings of LeFevre & Marshall (59) and of Kahlenberg & Dolansky (60), that no single hydroxyl group is essential for binding. Nevertheless, it was evident that binding of sugars in the $\beta$-pyranose form was enhanced.

By comparing the transport of deoxy sugars with their fluorinated derivatives, Barnett showed that C-1-, C-3-, and C-6-hydroxyls of glucose are involved in binding to the transporter (62). Other work using glucose analogs with bulky hydrophobic substituents in the C-1, C-6, and C-4 positions suggested that there was a hydrophobic cleft adjacent to the C-4 and C-6 positions of bound glucose, and that it is the C-1 and C-6 ends of the sugar that initially interact with the inward- and outward-facing transporter, respectively. Based on these studies, it was proposed that when a glucose molecule approaches the transporter from the extracellular medium, the transport protein opens up around the C-1 end of the sugar and closes behind the C-6 end (Figure 2).

The hydrogen bonds involved in holding the glucose in the binding site may largely be provided by a conjunction of helices seven, eight, and eleven (see Figure 1), while the hydrophobic cleft could be formed by helices nine and ten. In support of this supposition, hydrogen-exchange experiments indicate that ligand binding collapses the water-filled channel.

## Biochemical Asymmetry of GLUT 1

Considerable data has accumulated pointing to the asymmetry of the red cell glucose carrier. Almost all of these studies have been carried out on intact membranes. It is now possible, however, to interpret the transporter biochemistry in terms of the cloned carrier structure depicted in Figure 1.

Several lines of evidence indicate that the transporter can adopt mutually exclusive inward- and outward-facing conformations. For example, thermolysin cleavage of *GLUT 1* is promoted by cytochalasin B binding but retarded by ASA-BMPA binding (63). A possible interpretation of these effects is that

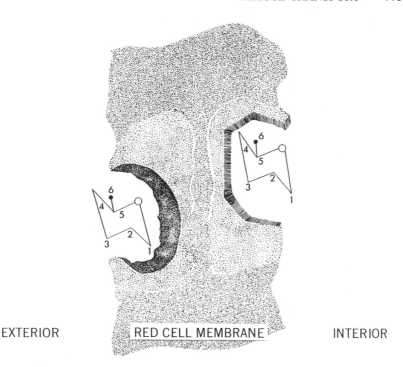

EXTERIOR          RED CELL MEMBRANE          INTERIOR

*Figure 2*  Diagram depicting docking of pyranose substrate in hydrophobic cleft of red cell glucose transporter (GLUT 1). For details see text. (Reprinted with permission.) (Ref. 95)

cytochalasin B sequesters the transporter in the inward-facing conformation, which is more susceptible to cleavage than the outward-facing conformation stabilized by ASA-BMPA. Cytochalasin B is thought to bind to the inward-facing conformation of GLUT 1 because it is a competitive inhibitor of glucose efflux but a noncompetitive inhibitor of influx. Moreover, the inward- and outward-facing conformations of GLUT 1 are thought to be mutually exclusive since the binding of ligands (such as ethylidene glucose) to the exofacial surface of the transporter precludes cytochalasin B binding (64, 65). On the other hand, ASA-BMPA is an impermeable competitive inhibitor of glucose influx. Consequently, ASA-BMPA is thought to bind to the outward-facing conformation. Similar conclusions have been drawn from kinetic analyses of the simultaneous action of inhibitors that bind to opposite surfaces of the transporter (66).

Studies involving the inactivation of the transporter with irreversible inhibitors also support this view of mutually exclusive conformations. For example, inactivation of the transporter with tetrathionate is accelerated by maltose and phloretin (binding to the outward-facing conformation), whereas cytochalasin B retards inactivation (67). Interestingly, inactivation of 1-fluoro-2,4-dinitrobenzene (FDNB) is accelerated by transported sugars, but ligands that stabilize either conformation of the transporter protect against inactivation (68). This finding suggests that as the transporter changes conformation, a FDNB-sensitive site is exposed. A cysteine residue is implicated, since chloronitrobenzoxadiazole and ethyl-maleimide inactivation are also accelerated by transported sugars (69). The cysteine exposed during the conformational change might be at position 347 or 421, since these are located in the areas designated as putative binding sites (located towards the middle of helix nine and the exofacial end of helix eleven, respectively). Exposure of Cys327 seems more likely, however, since the fluorescence emission spectrum of the chloronitrobenzoxadiazole-alkylated transporter indicates that the cysteine residue involved is located in a hydrophobic environment (69). All of the transporter cysteines, apart from Cys347 on M9, are located towards the termini of a particular $\alpha$-helix, where one would expect them to be in an aqueous environment. Therefore it has been suggested that the conformational change involves movement of helix nine, which is presumed to form a hydrophobic cleft in the substrate-binding site. On the other hand, Pawagi & Deber (70) have recently provided evidence that helix 10 contains the dynamic segment of the glucose transporter close to Trp388.

The binding of ligands (e.g. D-glucose, cytochalasin B, and ethylidene glucose) to the transporter has been shown to quench the intrinsic fluorescence of the transport protein, while causing a slight blue shift in the wavelength of maximum fluorescence emission (70, 71). These effects are indicative of a conformational change in the transporter such that one or more tryptophan residues move from a hydrophilic to a hydrophobic environment. As discussed in Ref. 70, this may be attributable to ligand binding near to either Trp363, Trp388, or Trp412, but of all of these, Trp388 seems to be closest to the dynamic segment of the carrier protein.

A stopped-flow method has been used to demonstrate a fluorescent transient (72) associated with the binding of ethylidene glucose to the transporter (reconstituted into leaky vesicles). The existence of this transient has been attributed to reorientation of the transporter from the inward- to the outward-facing conformation, since ethylidene glucose preferentially binds exofacially and does not undergo translocation. The observed transient cannot be reconciled with a simple binding process, since binding occurs with an apparent first order rate constant, which decreases with increasing ligand concentration. This decline probably reflects ligand binding to the inward-facing conformation, which has a low affinity for ethylidene glucose. Furthermore, ethylidene glucose binding causes a slight change in the infrared spectrum of the transporter, indicating a conformational change.

## Kinetics of Red Cell (GLUT 1) Sugar Transport

The classic "mobile" carrier hypothesis to explain facilitative glucose transport was proposed by Widdas (73) to account for the rapidity of sugar entry into cells as well as the phenomena of saturation kinetics and trans stimulation of substrate transport. The model envisioned an alternating mechanism in which the transporter was exposed to first one and then the other side of the membrane. As originally conceived, the sugar-transporter binding interaction at both inside (i) or outside (o) surfaces was considered to be symmetrical, but as already discussed it is now known that the sugar specificities of transport at the cytoplasmic and external surfaces are different. In addition, the rate of exchange flux for D-glucose across red cells is faster than net flux. This latter finding implies that return of unloaded (free) carrier from one side to the other side must be rate limiting for net transport.

Currently, two competing models exist to account for the molecular mechanism of facilitative glucose transport. In the first model, an alternating conformation is described in which the transporter protein oscillates between two conformations—an extracellular or intracellular conformer state—with sugars binding in a fixed orientation (see Figure 3). When the substrate-binding site is occupied, the conformational change results in the translocation of the bound glucose across the membrane. In this model, the kinetic analyses are based on the assumption that reorientation of carrier is slow compared to glucose binding and release, i.e. that the substrate-carrier interaction achieves rapid equilibration relative to translocation (so-called simple asymmetric carrier behavior). Lowe & Walmsley (74) have recently measured zero trans infux, zero trans efflux, and equilibrium exchange of glucose transport in human erythrocytes and have calculated the kinetic constants for the reaction scheme depicted in Figure 3 at 10°C. Appleman & Lienhard (75) have made direct measurements of the rates of interconversion of the transporter conformation using a reconstituted purified glucose carrier and have also analyzed their data according to the assumption of rapid equilibration of substrate and transporter. There is reasonably good agreement between these last two studies, demonstrating that the purified reconstituted GLUT 1 transporter functions very similarly to the transporter in the intact erythrocyte.

Based on the data of Walmsley (2), the relative populations of the conformational states of the transporter at 37°C, equilibrated in 5 mM glucose, would be 28.1 and 17.7% for the outward-facing unloaded and glucose-loaded transporter, respectively, and 42.3 and 11.8% for the inward-facing unloaded and glucose-loaded transporter, respectively. Considering these features, the transport system is apparently poised under physiological con-

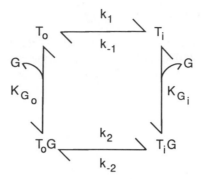

*Figure 3*    Schematic representation of kinetic model of facilitative glucose transporter. The $k$s represent rate constants, $K$s represent substrate-receptor dissociation constants, $G$ and $T$ refer to glucose and transporter concentration, respectively.

ditions for the rapid equilibration of glucose across the membrane in either direction. This may be of importance in rapidly loading the red cells in order ultimately to supply the brain with glucose.

An interesting result of the study by Appleman & Lienhard (75) is the finding that the activation energy for conformer interconversion is much less when glucose is bound than for unliganded transporter (28 compared to 17 kcal/mol). This lower energy value is presumably due to preclusion of water and transient stabilization of the binding site by glucose occupancy.

A second model to account for the molecular mechanism of Glu1 operation proposes two substrate-binding sites simultaneously present (76). This model requires a more complex kinetic analysis and may imply that the sugar-transporter reaction is not of the rapid equilibration type, i.e. that the association dissociation rates are of the same order of magnitude as transporter reorientation rates. It should also be noted that if the facilitative glucose carriers, such as GLUT 1, exist in their native state as oligomers, then more complicated kinetic behavior may result.

## Chemical Mechanism of Facilitative Glucose Transport

In summary then, the structure of GLUT 1 combined with the kinetic analysis of the functional carrier leads us to the following picture: as glucose binds, some water molecules dissociate from both glucose and the binding site of the transporter as glucose effectively exchanges hydrogen bonds with water for a similar number of hydrogen bonds with the transporter. Assuming that each OH group on a glucose molecule participates in two hydrogen bonds with the carrier (57) and that the average energy of a hydrogen bond is $\sim 3$ kcal/mole, then the extrusion of two or three water molecules from the sugar-protein binding cleft would reflect a change in entropy of $\sim 12$–$18$ kcal/mole. This calculation is consistent with the measured estimate of the difference in activation energy for loaded vs unloaded carrier (75) of ~11 kcal/mole and supports the concept that the glucose preclusion of water could transiently stabilize the binding site and thereby reduce the energy barrier to reorientation of the glucose-loaded transporter. Moreover, given the magnitude of the entropy changes involved in sugar binding, the thermodynamics of reorientation would not suggest any major rearrangement of the transport protein structure. Instead, the small free energy changes associated with reorientation would indicate that the transporter can readily oscillate between conformational states, so that under sugar gradient conditions glucose bound at one surface is rapidly transferred to the other side of the membrane through the reorientation of the loaded carrier.

Of course, these thermodynamic considerations provide no detailed information about what is actually taking place at the atomic level within the carrier protein. It would seem likely that the tertiary (quaternary) structure of all transport proteins is optimized to transform a substrate-induced conformational change into a transmembrane translocation event. As more membrane protein genes are cloned, it may become feasible to discover the rules governing their intramembrane three-dimensional organization and thereby deduce the structural features that allow them to function as membrane transporters.

# NA$^+$-DEPENDENT GLUCOSE COTRANSPORTERS

There is now evidence that the Na$^+$-dependent D-glucose cotransporters localized to the luminal (brush border) surface of the intestine and renal tubular epithelium arise from a different gene family than do the facilitative glucose transporters.

## The Intestinal Cotransporter

Hediger et al (77) first reported the cloning of the gene for the sodium-dependent D-glucose transporter from the rabbit intestinal brush border membrane in 1987. These workers injected poly A$^+$-RNA from rabbit intestine into *Xenopus laevis* oocytes and observed the expression in the oocytes of sodium-dependent phlorizin-sensitive α-methyl-D-glucopyranoside uptake. After size selection of poly A$^+$-RNA, the mRNA coding for the glucose transporter was found to be highly enriched in a 2.3-kilobase fraction. A cDNA library was prepared from this fraction using an in vitro expression vector. Clones were screened by injecting mRNA synthesized in vitro into oocytes and then assaying for sodium-dependent α-methyl-D-glucoside uptake. The specificity characteristics, sodium dependence and phlorizin sensitivity, both support the conclusion that the authors had cloned the cDNA for the 'classical' intestinal brush border sodium-dependent D-glucose transporter. The resulting DNA sequence yields a polypeptide containing 662 amino acids with a molecular weight of 73,080.

Based on the analysis of hydrophobicity plots, Hediger et al (77) have constructed a model for the intestinal sodium-dependent D-glucose carrier consisting of 11 membrane-spanning sequences (Figure 4), which are presumed to be $\alpha$-helical. The N and C terminals are on opposite sides of the membrane, endo- and exofacial, respectively. Two highly charged surface loops connect M7 to M8 and M10 to M11. The protein is glycosylated at an asparagine residue in the small exofacial loop connecting M5 to M6. The molecular weight of the protein cloned by Hediger et al (77) is very close to that previously associated with the intestinal brush border membrane sodium-dependent D-glucose transporter by photoaffinity labelling with 4 azido phlorizin (78), by immunoprecipitation with monoclonal antibodies (79), by covalent labelling with fluorescein isothiocyanate (80), or 10-$N$-(bromo-acetyl)amino-1-decyl-$\beta$-D-glucopyranoside (81).

The gene for the cloned rabbit intestinal Na$^+$/glucose cotransporter has been expressed in *Xenopus* oocytes (82), and the kinetic and specificity properties of this cloned protein were very similar to those of the native Na$^+$/glucose cotransporter. The gene for the intestinal Na$^+$/glucose cotransporter was also expressed and characterized in COS-7 cells (83). When this mammalian cell line, which does not express the Na$^+$-dependent cotransporter, was transfected with the cDNA coding for the transporter, the COS-7 cells exhibited the same general sugar kinetic characteristics of the intestinal brush border membrane. Tunicamycin, which prevents asparagine-linked glycosylation, markedly inhibited expression of Na$^+$/D-glucose cotransport

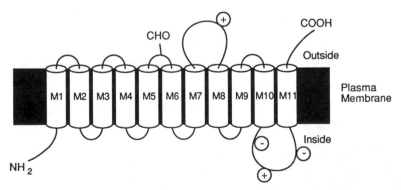

*Figure 4*   Schematic representation model showing two-dimensional membrane topology of the Na$^+$/D-glucose cotransporter SGLT 1 from rabbit intestinal epithelium. Putative membrane-spanning $\alpha$-helices are labelled M1–M11. Endo- and exofacial hydrophilic loops with highly charged residues are putative glucose-binding sites (adapted from Ref. 3).

in COS-7 cells, indicating that N-linked glycosylation is important for functional expression of this membrane protein (83).

The published sequence of the sodium-dependent D-glucose carrier is totally different from that of the sodium-independent glucose carrier GLUT 1 cloned by Mueckler et al (7), which has an estimated molecular weight of 55,000. Because of the absence of any homology with the sodium-independent carrier, Hediger et al (77) proposed that their 73-kDa sodium-dependent sugar transporter represents a new class of transport proteins.

The gene for the human $Na^+$/glucose cotransporter has also been cloned (84), and has been designated *SGLT 1*. The gene has been localized to the 22q11.2 → qter region of chromosome 22. SGLT 1 is composed of 664 amino acids. There is 85% identity and 94% similarity between the rabbit and human clones.

Interestingly, although no other mammalian transporters (other than the renal $Na^+$/D-glucose cotransporter discussed below) have been found with homologies to SGLT 1, a significant homology has been found with the *E. coli* $Na^+$/proline cotransporter (84), suggesting an evolutionary link between bacterial and human cotransport proteins and further supporting the concept that the $Na^+$/D-glucose symporters represent a unique family of transport proteins.

## The Kidney Cotransporter

The cDNA encoding the $Na^+$/D-glucose cotransporter from rabbit jejunum was used as a probe in Northern blots of mRNA extracted from rabbit kidney cortex and medulla (85). The probe hybridized with mRNA of ~2.2 kb from both regions. The renal medulla contained ~20% of the amount of transcript found in the cortex. Using the same cDNA probe, a 1.6 kb partial-length clone encoding 484 amino acids was isolated from a rabbit renal cortex (whole) cDNA library. There was >99% identity between the cDNA sequences and 100% identity between the amino acid sequences of the renal clone and the rabbit intestinal $Na^+$/D-glucose cotransporter. Therefore the rabbit kidney cortex contains a $Na^+$/D-glucose cotransporter that is virtually identical to the intesinal $Na^+$/D-glucose cotransporter. It is possible that the same gene may be involved. A polyclonal antibody raised against the hydrophilic loop between M7 and M8 specifically recognizes a 76-kDa band in LLC-PK1 brush borders (a pig renal cell line) (86). Using this same antibody (kindly provided by E. Wright), our laboratory has also shown a specific band at ~75-kDa in Western analysis in brush border membranes from dog and human kidney (M. Silverman, unpublished results).

Even before the availability of antibody probes against the intestinal $Na^+$/D-glucose cotransport, substantial progress had already been made over the past five years in the isolation and purification of this protein. Through use of a covalently D-glucose analog 10-*N*-(bromoacetyl)amino-1-decyl-β-D-glucopyranoside (BADG) and monoclonal antibodies directed against the $Na^+/D$-glucose cotransporter (81), 75-kDa and 47-kDa polypeptides were both identified as components of the $Na^+/D$-glucose cotransporter (87, 88). In other studies, correlation was found between progressive enrichment and purification of a ~66-kDa polypeptide from dog renal brush border membranes and $Na^+$-dependent binding of $^3H$-phlorizin (89). Finally, monoclonal antibodies that bind the renal $Na^+$/glucose symporter were shown to recognize a 75-kDa antigen in brush border membranes from LLC PK1 cells (90).

There is both kinetic and physical evidence for the existence of two $Na^+$/glucose cotransporters in the rabbit kidney. Vesicles prepared from the outer portion of the renal cortex contain a low-affinity transporter with an apparent Michaelis constant ($K_m$) for glucose of 6 mM (at 40 mM NaCl). Whole vesicles from the outer medulla contain a high-affinity transporter with a $K_m$ of 0.35 mM (91). A preparation of vesicles from whole cortex contains ~80% low-affinity ("cortical") and 20% high-affinity ("medullary") $Na^+$/glucose cotransporters (91). Furthermore, polyclonal antibodies raised to two separate regions of the cloned intestinal transporter (residues 402–419 and 604–615) recognized two broad bands at 56 and 64 kDa in Western analysis of rabbit renal cortical brush borders (see Ref. 85). The fact that hybridization of renal mRNA to the intestinal $Na^+$/glucose cotransporter cDNA was greater in renal cortex than in medulla suggests that the partial renal clone, obtained from a renal cortex cDNA library, encodes the low-affinity (i.e. cortical) $Na^+$/glucose cotransporter. It is more likely, however, that the renal clone encodes the medullary (i.e. high-affinity) cotransporter for the following reasons. First, the library from which the clone was isolated probably contained mRNA encoding both transporters, since it was prepared from whole cortex. Second, the cloned intestinal transporter (92) exhibits high affinity and 2:1 $Na:D$-glucose stoichiometry, transport properties which are more closely related to the $Na^+$/glucose cotransporter in the proximal straight tubule. Finally, it is possible that the signal seen on the Northern blot represents hybridization of the intestinal $Na^+$/glucose cotransporter cDNA with more than one species of renal mRNA of the same size.

## Structure-Function Correlations of the $Na^+/D$-glucose Cotransporter

The specificity characteristics of sugar transport in the intestinal and kidney brush border membranes are different from each other and also diverge from the facilitative carrier systems described earlier. But there are also some important similarities in terms of the sugar-transporter binding reaction that emerge when we compare the hydrogen-binding specificity of, for example, the $Na^+/D$-glucose transporter from the kidney, with that of the red cell. The specificity characteristics of the early proximal tubule $Na^+/D$-glucose cotransporter are remarkably species invariant as determined for the rat, rabbit, dog, and man (94–96). Through systematic testing of a series of homologous monosaccharides, each with different substituents at the various carbon positions in the pyranose structure, it has been possible to "map out" the chemical and steric determinants of sugar-carrier interaction at the exo-facial region of the renal brush border membrane. The cumulative evidence suggests that substrate "docking" requires formation of hydrogen bonds between sugar molecules in the $^4C_1$ pyranose chair conformation D-glucose configuration and specific amino acid side chains at the active site of the transporter. This bonding apparently involves the participation of six different positions on the pyranose substrate including the oxygen in the pyranose ring and the hydroxyl groups at $C_1$, $C_2$, $C_3$, $C_4$, and $C_6$. The participation of OH groups at $C_3$ and $C_6$ appears to be critical for the transport event. These findings imply that the $C_4$, $C_6$ end of the sugar molecule is tightly constrained within a hydrophilic pocket of carrier-binding site, whereas the $C_1$ end of the molecule probably has more freedom from steric hindrances, allowing for hydrophobic interactions. This description of sugar occupancy of transport-binding site is almost identical to that of sugar binding to the red cell carrier depicted in Figure 2 and suggests a similar functional paradigm of sugar-induced conformational changes in the transporter.

The lysine residues in the highly charged loops connecting M7 to M8 and M10 to M11 (see Figure 4) have been postulated to be involved in glucose binding based on two observations: (a) that there are lysyl residues at the glucose/phlorizin binding site of the intestinal $Na^+$/D-glucose cotransporter (97, 98), and (b) that in sugar-binding proteins of prokaryotes such as the arabinose-binding protein (99), extensive use is made of charged residues including lysine to form hydrogen bonds with the sugar.

Studies in intact intestinal brush border membranes also have provided evidence that a tyrosine residue is located at or near the $Na^+$-binding site (100); that the $Na^+$ active site is 30 to 40 Å away from the glucose sites; and that $Na^+$ binding to the transporter produces a long-range conformational change that increases the affinity of the glucose site for substrates but is not a transport-committing step. It is noteworthy that there are 27 tyrosine residues in the published sequence of the $Na^+$/D-glucose cotransporter, but as yet there is no clue as to which tyrosine(s) participates in $Na^+$ binding to the carrier.

Recently (101), substrate-induced conformational changes of the intestinal $Na^+$/D-glucose cotransporter have been reexamined by measuring changes in tryptophan fluorescence of the purified symporter in response to $Na^+$, glucose, and phlorizin. It appears that in addition to the conformational changes brought about when $Na^+$ binds (i.e. increasing the efficiency of glucose binding), a second conformational change occurs that is specific for $Na^+$ and D-glucose and involves more than binding to the glucose site because phlorizin cannot substitute for D-glucose. The implication is that the cotransporter undergoes a second conformational change upon binding D-glucose, which shifts the protein into a transport-competent conformation.

Some experimental data suggests that the brush border $Na^+$/D-glucose cotransport carrier is asymmetrical. The evidence is based on observed differences in phlorizin-binding characteristics after exposure of brush border membrane vesicles to different chemical and proteolytic perturbances before and after deoxycholate treatment (102). The assumption made in these studies is that deoxycholate only acts to increase vesicle permeability (at least under the conditions employed), thereby providing free access of the cytoplasmic face of the membrane to the action of various perturbing chemical agents. This assumption may not be entirely justified, because deoxycholate even in low concentrations (i.e. .01%) causes solubilization of both peripheral and integral membrane proteins.

Finally, there is very recent evidence based on radiation inactivation studies that the functional unit molecular size of the intestinal brush border membrane $Na^+$/glucose cotransporter is 290 kDa, representing a homotetramer comprising four independent 73-kDa subunits (103).

## Kinetics and Chemical Mechanisms of $Na^+$/D-glucose Cotransport

During the 1950s, evidence accumulated documenting the $Na^+$-dependence of sugar transport. In the early 1960s, Crane (104) developed the $Na^+$ gradient hypothesis. According to this model, $Na^+$ and glucose are obligatorily cotransported by the same carrier. Since the flux of $Na^+$ and glucose is coupled, this means that movement of one species uphill against a concentration gradient can be driven by a favorable downhill gradient of the other chemical species. Therefore, the $Na^+$/D-glucose symport mechanism is a secondary active transport system in the sense that energy derived for uphill transport is not directly linked to a metabolic reaction, e.g. as is the case for the $Na^+$ pump.

By the mid-1970s, experimental data favoring the $Na^+$ gradient hypothesis became overwhelming. The information derived from isolated vesicle studies was especially important because transport could be completely dissociated from the metabolic activity of the cell. For example, in 1974 Murer & Hopfer (105) showed not only that $Na^+$ and D-glucose were cotransported but also that their movement involved the transfer of electrical charge (i.e. was electrogenic) and was directly sensitive to favorable transmembrane electrical potential diffusion.

In developing a kinetic analysis of secondary active cotransport system, the electrochemical potential for all transported species across the membrane is set equal to zero (otherwise, the system would be primarily active). This assumption then leads to thermodynamic constraints on the parameters of the model. But still the mechanisms for cotransport are considerably more complicated than for facilitative diffusion. Nevertheless, much of the kinetic behavior of cotransport systems is completely analogous to that described for sugar transport in $Na^+$-independent systems, such as the red cell, i.e. saturation, competitive inhibition, counterflow, and accelerated exchange. But because of the presence of two chemical species (activator $Na^+$ and substrate sugar), several new features emerge. For example, it becomes of importance to ask if there is any ordering in the initial binding interaction with the carrier system and following this, in what sequence the transported species are released at the other side of the membrane.

It has long been known that phlorizin is a reversible competitive inhibitor of the $Na^+$/D-glucose cotransport system located at the luminal surface of both the kidney and intestine. Within the past decade it has also been possible to demonstrate the existence of a single set of high-affinity $Na^+$-dependent phlorizin receptor sites at the mucosal surface of many epithelia. There is now overwhelming evidence that the high-affinity phlorizin receptor is identical to the sugar carrier in the brush border membrane (106).

The interaction of phlorizin and $Na^+$ with the phlorizin receptor exhibits $1:1$ stoichiometry; the same as is found for $Na^+$/D-glucose cotransport (107). A critical finding is that phlorizin is a nontransported substrate of the $Na^+$-dependent D-glucose carrier. In other words, phlorizin binds to the external face of the carrier but is not translocated across the membrane. It is this fact that makes the phlorizin molecule such a useful ligand with which to probe the kinetic mechanism of the $Na^+$/D-glucose cotransport system. The evidence justifying this assumption is summarized for the kidney in Refs. 107 and 108.

$Na^+$/D-glucose cotransport is electrogenic and does not involve cotransport of another anion nor countertransport of a cation (105). Using brush border membrane vesicles, Aronson demonstrated that inside negative potentials stimulate phlorizin binding (108). Since phlorizin binds but does not cross the membrane, this observation complements the results of Ref. 105 in the sense that the electrogenicity of D-glucose transport (as defined by the response to inside negative membrane potential) can now be accounted for by assigning a net negative charge to the free carrier.

Turner & Silverman (107) have shown that the binding of inhibitor (phlorizin) and activator ($Na^+$) to renal brush border membrane vesicles was random. Later Turner & Silverman (109) and Turner (110) developed a formulation to analyze a family of cotransport models, including the possibilities of random binding of sugar and $Na^+$, as well as ordered binding. For the latter situation, the symmetry of the binding events at the two membrane faces was also taken into consideration. Starting with the assumption that translocation is rate limiting, the expression for the usual kinetic parameters $K_m$ and $V_{max}$ were derived for different experimental cases (e.g. zero trans, infinite cis, equilibrium exhange, etc) and the resulting formulas were used to devise different tests to distinguish between various modes of kinetic behavior, i.e. ordered vs random binding, etc.

If the equilibrium model is applied to the small intestinal $Na^+$/glucose cotransporter, then experiments by Kessler & Semenza (111) concerning the moderating effect of negative inside membrane potential on the transinhibition of glucose influx by intravesicular glucose lead to the conclusion that the transporter carries one or more negative charges that move across the membrane barrier as part of the catalytic (binding) cycle. This view proposes that there is a negatively charged carboxylate group at the active site, and that this negative charge explains the voltage dependence of the phlorizin on rate and the voltage independence of the off rate.

Meanwhile, Hopfer had taken a completely different approach and was studying exclusively equilibrium exchange kinetics of $Na^+$/D-glucose cotransport in intestinal brush border vesicles (112–114). He found that influx of either $Na^+$ or D-glucose plotted as a function of increasing concentration of its cotransported portion yielded a biphasic response. It turns out that this biphasic behavior can only be predicted if translocation is not considered as rate limiting. Moreover, these activation curves provide a kinetic test for distinguishing between various modes of carrier mechanism. In contrast to the findings of Turner & Silverman (107), Hopfer's laboratory reported that the $Na^+$/D-glucose carrier system exhibited glide symmetry (first in first out), rather than mirror symmetry (first in last out) (see Figure 5). Another important feature of Hopfer's analysis is that the interaction of the cotransporter with the $Na^+$ and sugar substrates is ordered rather than random. This conclusion is based on experimental data, which were interpreted as showing "back to zero" inhibition in either $Na^+$ activation by D-glucose or D-glucose activation by $Na^+$. It is not clear, however, whether the data are really good enough to distinguish between the case where the ordinate becomes vanishingly small and the case where it approaches a nonzero asymptote. Which situation actually prevails is quite critical when it comes to deducing the correct kinetic mechanism. Only for true "back to zero" conditions is it appropriate to infer that a glide mechanism is the correct scheme. Otherwise the data imply a random mechanism.

Harrison et al (115) studied a completely general scheme of cotransport using computer simulations and then fitted the theoretical curves to available experimental data from Hopfer's laboratory. The random model clearly does the best job, but qualitatively both glide and random schemes behave identically. A strategy was then adopted that involved detailed parametrization of the glide model followed by computer simulation to predict quantitative as well as qualitative behavior of cotransport, i.e. the biphasic nature of the $Na^+$

vs D-glucose activation curves. It was reasoned that if a unique set of microscopic rate constants could not be found to fit the experimental observations satisfactorily, then the glide model should be rejected. Unfortunately, there is still inadequate experimental data upon which to make a crucial test simulation.

A synthesis of current information derived from kidney and intestine kinetic vesicle data as well as from the biochemical and structural information obtained with purified transporter leads us to construct the following picture: Na$^+$ and D-glucose interact randomly with a carrier molecule at either face (internal or external) of the membrane. Once the initial binding event has taken place, two possibilities exist: either glide symmetry is preserved during translocation so that whatever binds first comes off first, or alternatively the mechanism is completely random so that regardless of which order Na$^+$ and D-glucose enter the carrier, either can exit first. It should be pointed out that even in a completely random scheme, there can be a marked preference for one or the other pathway, and both in the case of the intestinal and the renal cotransport systems, Na$^+$ even at low concentrations ($\sim$20 mM) improves the efficiency of glucose binding so dramatically that for all intents and purposes the reaction scheme is ordered—first Na$^+$, then glucose binding. Thus we visualize that first Na$^+$ binds, leading to a conformational charge that en-

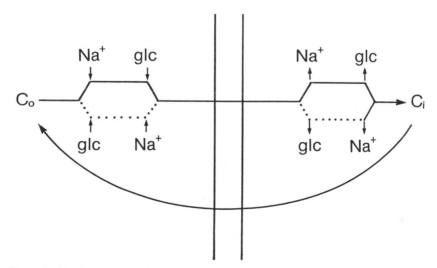

*Figure 5*   Kinetic model of Na$^+$/glucose cotransport with 1 : 1 stoichiometry. Solid line indicates preferred pathway or binding. The model is written as net transport under physiological conditions; however, all steps are assumed to be freely reversible. In Cleland's nomenclature, the mechanism is iso-random-bi-bi with preference for a particular ordered transport sequence and Na$^+$ binding on the inside (solid line) at high [Na$^+$] (Reprinted with permission). (Ref. 123)

hances glucose binding. Once the glucose site is also occupied, a second conformation occurs that facilitates reorientation of the carrier and results in solute translocation. Although each 73-kDa subunit can independently bind $Na^+$ and sugar, it is possible as proposed in Ref. 103 that all four subunits work cooperatively.

The implication of these findings is that the most plausible mechanistic model to explain transmembrane sugar transport in intestinal and renal brush border vesicles is that of a "mobile gate." The concept of a "mobile gate" envisions a slight physical displacement of critical residues with binding site, causing the sugar ligand to switch from an outward- to an inward-facing orientation (or vice versa). Certainly this is the most likely scheme capable of preserving glide symmetry along with the other kinetic features that are characteristic of the $Na^+/D$-glucose cotransport system.

An alternative two-substrate site model for the $Na^+/D$-glucose cotransporter has recently been proposed by Koepsell (116) after careful kinetic analysis of phlorizin binding and D-glucose transport measurements. It is also possible, however, that this new data can be explained by the presence of equal amounts of separate populations of high- and low-affinity $Na^+/D$-glucose cotransporters.

In a similar vein, the kinetics of $Na^+$-dependent $\alpha$-methyl glucoside uptake into intestinal brush border vesicles are curvilinear and can be interpreted as either reflecting the presence of two independent saturable transporters—a high- and a low-affinity system—or alternatively different manifestations of a single transporter. Interestingly, the kinetics of $\alpha$-methyl-D-glucoside uptake in *Xenopus* oocytes injected with the cloned $Na^+/D$-glucose cotransporter reflect the activity of a single carrier species only (82).

## Heterogeneity of $Na^+/D$-glucose Cotransporters

So far no family of Na/glucose cotransport genes has been cloned as has been the case for the facilitative sugar transporters. Nevertheless, there is reasonable experimental evidence, alluded to previously, to suggest that such a family of symporters does exist with even the possibility of extensive homology to other $Na^+$ cotransporters such as those mediating amino acid uptake in epithelia.

Early studies of $Na^+$-dependent D-glucose transport in brush border membrane vesicles prepared from dog and human kidney showed curvilinear uptake/[glucose] versus glucose uptake (Eadie-Hofstee) plots (117, 118). These curvilinear Eadie-Hofstee plots cannot be accounted for by a single transporter obeying Michaelis-Menten kinetics. Although more complex coupled transport models can explain the observed deviations from linearity,

Turner & Silverman provided experimental evidence (117, 118) indicating that this effect may be due to the presence of two (or more) $Na^+$-dependent D-glucose transporters in the renal brush border membrane. These authors further suggested that the existence of multiple glucose carriers may reflect heterogeneity of transport sites along the length of the proximal tubule. This conjecture later received support from the work of Barfuss & Schafer (119) discussed above. Even these studies, however, did not conclusively demonstrate the existence of multiple transporters, because the kinetic parameters measured by Barfuss & Schafer are characteristic of transepithelial rather than trans brush border membrane fluxes. Thus, for example, the observed changes in kinetic properties from segment to segment may reflect differences in the intracellular electrochemical potential difference for sodium or in transport properties at the antiluminal membrane rather than actual differences at the brush border.

Turner & Moran (91) subsequently directly established the existence of proximal tubular brush border membrane D-glucose transport heterogeneity. These authors compared the D-glucose transport properties of brush border membrane vesicles prepared from two regions of the rabbit kidney: the outer cortex and the outer medulla. Owing to the structural organization of nephrons in the kidney, the outer cortical preparation is expected to contain predominantly brush border membranes from proximal convoluted tubules of superficial nephrons (early proximal tubule from S1 and some S2 cell types), whereas the outer medullary preparation is expected to contain predominantly brush border membranes from the late proximal straight tubule (from S3 cell types). Eadie Hofstee plots of the initial rate of sodium-dependent D-glucose transport into the two preparations revealed linear plots with dramatically different slopes for outer cortical and outer medullary vesicles. The linearity of these plots indicates that the data fit the Michaelis-Menten equation and thus are consistent with the existence of a single transporter in each preparation. Fitting these data to the Michaelis-Menten equation yields $K_m = 6$ mM in the outer cortex and $K_m = 0.4$ mM in the outer medulla. Thus D-glucose reabsorption is mediated by a low-affinity system in the early proximal tubule and by a high-affinity system in the late proximal tubule. These two transporters also show different specificity properties and different sensitivities to inhibition by phlorizin (91, 93). Also, it is the low-affinity site that is associated with the $Na^+$-dependent high-affinity phlorizin-binding site commonly found in renal brush border membrane preparations (93, 120).

In additional studies (93, 120), it was demonstrated that the sodium : glucose stoichiometries (i.e. the number of sodium ions transported per glucose molecule) was different for the two transporters. In the outer cortical preparation, one sodium ion was translocated per glucose molecule, whereas in the outer medullary preparation, two sodiums were translocated per glucose. The possible physiological significance of these observations is discussed below.

## Functional Organization of Sugar Transporters in Epithelia

Through use of the information derived from cloning and tissue localization studies, it is possible to specify the transport molecules responsible for reabsorption of sugars across epithelia. As shown schematically in Figure 6, the glucose carrier molecules are distributed asymmetrically with the $Na^+$/D-glucose cotransporters localized apically and a facilitative carrier localized at the serosal surface. The energetics are determined by the activity of serosal $Na^+$-$K^+$-ATPase, which actively extrudes $Na^+$ from the epithelial cells, thus establishing a transmucosal electrochemical gradient for $Na^+$ and allowing the secondary active reabsorption of sugars.

In the intestine, the constituents of the glucose reabsorptive mechanism would be apical SGLT 1 and basolateral GLUT 2. In the kidney epithelium, the situation is more complicated (as discussed previously). In the proximal straight portion of the tubule, there is a carrier identical to SGLT 1 with GLUT 1 situated basolaterally. In the proximal convoluted tubule, we predict the existence of a low-affinity $Na^+$/D-glucose carrier at the apex, partly homologous to SGLT 1, which we designate SGLT 2, and GLUT 2 at the opposite pole (basolateral). Distally in the medullary and papillary segments, immunolocalization studies reveal the presence of basolateral GLUT 1. Glucose transport in this region of the nephron is probably related to the metabolic requirements and not substrate conservation.

It is interesting that in the kidney, the localization of GLUT 1 parallels the distribution of glycolytic activity, and colocalizes with hexokinase. This distribution of facilitative glucose transporters appears to coincide with a hypothesis that transport proteins are "reactive site directed" proposed years ago by Berlin (121). He suggested that since the hexokinase reaction involves phosphorylation at the carbon 6 position, that this $C_6$ position would also be a critical requirement for recognition by the transport protein so that it could be "protected." Indeed as we have already seen, the specificity characteristics of both the $Na^+$/D-glucose cotransporter and GLUT 1 are consistent with a sugar-carrier interaction that causes the protein to open up around the $C_1$ end of the sugar and close behind the $C_6$ position, thus "protecting" the $C_6$ position during translocation. By contrast, in a segment that is dominated by gluconeogenesis (proximal convoluted tubule) and where hexokinase is absent, it is worth pointing out that the specificity requirements for sugar entry across the basolateral membrane are critically dependent upon interaction between the $C_1$ position and carrier (106), implying that in this segment the $C_6$ position on glucose is not protected during the translocation process between blood and cell.

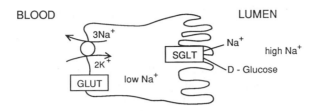

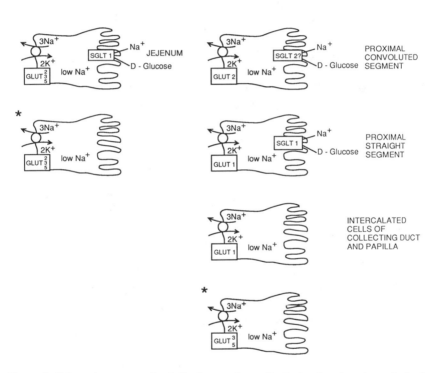

*Figure 6* Schematic representation indicating membrane distribution (basolateral or apical) of cloned glucose transporters in epithelia based on immunolocalization studies. Orientation permits net reabsorption (secondary active transport) from lumen to blood. Upper figure shows general format. Diagrams marked with a * are tentative assignments based on mRNA hybridization studies. See text (and Table 2) for details.

## Functional Consequences of $Na^+/D$-glucose Cotransporter Heterogeneity

Just as the tissue localization exhibited by GLUT 1 and 2 appears to correlate, respectively, with predominant glycolytic (red cell, renal medulla) vs gluconeogenic (liver) activity, the apparent heterogeneity of $Na^+/D$-glucose cotransporters along the renal tubule can also be rationalized into a functional paradigm that enables the renal (and possibly the intestinal) epithelium to conserve sugar substrate completely.

To understand the systematics of D-glucose transport along the length of the proximal tubule, it is first necessary to review the thermodynamic consequences of Crane's Gradient Hypothesis. Consider a proximal tubular cell where the intracellular sodium concentration is maintained at some fixed value by Na/K ATPase. Glucose will be driven into this cell down the extracellular to intracellular sodium electrochemical gradient via the $Na^+$-dependent transporter and will accumulate against its concentration gradient. Equilibrium thermodynamics predicts (122) that the ratio of the intracellular to extracellular glucose concentrations at steady state will be given by the equation

$$G_i/G_o \leq [(Na_o/Na_i)exp(-F\delta\psi/RT)]^n$$

where i and o denote intracellular and extracellular concentrations, respectively, $\delta\psi = \psi_i - \psi_o$ is the intracellular potential, $n$ is the sodium:glucose stoichiometry. The $\leq$ sign in this equation becomes an equality in the absence of D-glucose leak pathways. In a proximal tubular cell, the main leak pathway is expected to be via the basolateral transporter. Thus the cell will never reach the maximum intracellular glucose concentration predicted by the above equation and a pump-leak situation will exist. The actual value of $G_i/G_o$ will depend on the magnitude of the leak pathways and the kinetic properties of the coupled transporter (122).

Consider now the functioning of the proximal tubular cells in situ. Recall that the D-glucose transporter located in the early proximal tubule brush border membrane has a sodium:glucose stoichiometry of 1:1, while the transporter in the late proximal tubule brush border has a sodium:glucose stoichiometry of 2:1 (93, 120). In the early proximal tubule, the luminal glucose concentration will be approximately equal to that in plasma. Glucose will thus be concentrated in the proximal tubular cell above the level in plasma by the 1:1 coupled system and will leak down its concentration gradient back into the blood via the basolateral transporter. As glucose is reabsorbed along the length of the tubule, however, the intracellular glucose concentration will fall along with the urine concentration. Eventually a point will be reached where the electrochemical gradient for sodium is too low for a 1:1 system to drive the intracellular glucose concentration to a value higher than that found in the blood. At this point glucose reabsorption would cease (in fact, some glucose reabsorption would continue due to fluid reabsorption in the proximal tubule, but this does not affect the arguments given here and is, therefore, neglected). Before this point, however, the 2:1 system is presumably already present in the luminal membrane. Since the concentrating capacity of a coupled transporter increases as the power of the coupling stoichiometry (see above equation), this transporter can use the same sodium electrochemical gradient to produce substantially higher intracellular glucose concentrations than the 1:1 system; thus it can continue to drive D-glucose reabsorption in the late proximal tubule. In this way the bulk of the D-glucose load is thought to be reabsorbed in the early proximal tubule by the 1:1 coupled transporter at an energetic cost of one sodium ion per glucose molecule (the energy expended in a cotransport event is simply the energy required to pump the cotransported sodium ions back out of the cell). The last traces of D-glucose are then pumped from the urine at the higher energetic cost of two sodium ions per glucose molecule. This arrangement of transporters in series along the proximal tubule leads to a more energy-efficient reabsorptive mechanism for D-glucose than could be achieved by either of the transporters acting alone. The distribution of basolateral $Na^+$-independent glucose carriers, low-affinity GLUT 2 in the early proximal segment, and higher-affinity GLUT 1 (see Figure 6), would also be consistent with the colocalization of low-affinity SGLT 1 and high-affinity ?SGLT 2 at the luminal surface of the corresponding segment.

## CONCLUSION

Remarkable progress has been achieved in identification and biochemical characterization of the putative carrier proteins responsible for $Na^+$-dependent, $Na^+$-independent, and insulin-responsive sugar transport in mammalian cells. The amino acid sequences of these carrier proteins are now known. The dynamics of sugar:transporter interactions are under active investigation using either the purified proteins or cells that can exhibit functional expression of cloned transporter genes. New information is rapidly emerging about the chemical and kinetic mechanisms that permit the glucose transporter to catalyze transmembrane translocation of sugars. A detailed understanding of the molecular basis for carrier-mediated transport, however, will require three-dimensional crystallographic studies of the purified carriers in the bound and unbound state.

ACKNOWLEDGMENTS

This work was supported by a grant from the Medical Research Council of Canada to the MRC Group in Membrane Biology. The author thanks R. Reithmeier for helpful suggestions.

## Literature Cited

1. Gould, G. W., Bell, G. I. 1990. *Trends Biochem. Sci.* 15:18–23
2. Walmsley, A. R. 1988. *Trends Biochem. Sci.* 13:226–31
3. Bell, G. I., Kayano, T., Buse, J. B., Burant, C. F., Takeda, J., et al. 1990. *Diabetes Care* 13:198–208
4. Brown, D. D. 1981. *Science* 211:667–74
5. Wheeler, T. J., Henkle, P. C. 1985. *Annu. Rev. Physiol.* 47:503–17
6. Baldwin, S. A., Lienhard, G. E. 1981. *Trends Biochem. Sci.* 6:208–11
7. Mueckler, M., Caruso, C., Baldwin, S. A., Panico, M., Blench, I., et al. 1985. *Science* 229:942–45
8. Baldwin, S. A., Baldwin, J. M., Lienhard, G. E. 1982. *Biochemistry* 21:3836–42
9. Tabor, S., Richardson, C. C. 1985. *Proc. Natl. Acad. Sci. USA* 82:1074–78
10. Bouma, C. L., Meadow, N. D., Stover, E. W., Roseman, S. 1987. *Proc. Natl. Acad. Sci. USA* 84:930–34
11. Birnbaum, M. J., Haspel, H. C., Rosen, D. M. 1986. *Proc. Natl. Acad. Sci. USA* 83:5784–88
12. Thorens, B., Charron, M. J., Lodish, H. F. 1990. *Diabetes Care* 13:209–18
13. Shanahan, M. F., D'Artel-Ellis, J. 1984. *J. Biol. Chem.* 259:13878–84
14. Cairns, M. T., Elliot, D. A., Scudder, P. R., Baldwin, S. A. 1984. *Biochem. J.* 225:179–88
15. Cairns, M. T., Alvarez, J., Panico, M., Gibbs, A. F., Morris, H. R., et al. 1987. *Biochim. Biophys. Acta* 905:295–310
16. Davies, A., Meeran, K., Cairns, M. T., Baldwin, S. A. 1987. *J. Biol. Chem.* 262:9347–52
17. Alvarez, J., Lee, D. C., Baldwin, S. A., Chapman, D. 1987. *J. Biol. Chem.* 262:3502–9
18. Chin, J. J., Jung, E. K. Y., Chen, V., Jung, C. Y. 1987. *Proc. Natl. Acad. Sci. USA* 84:4113–16
19. Holman, G. D., Rees, W. D. 1987. *Biochim. Biophys. Acta* 897:395–405
20. Cuppoletti, J., Jung, C. Y., Green, F. A. 1981. *J. Biol. Chem.* 256:1305–6
21. Axelrod, J. D., Pilch, P. F. 1983. *Biochemistry* 22:2222–27
22. Maenz, D. D., Cheeseman, C. I. 1987. *J. Membr. Biol.* 97:259–66
23. Flier, J. S., Mueckler, M., McCall, A. L., Lodish, H. F. 1987. *J. Clin. Invest.* 79:657–61
24. Thorens, B., Sarkar, H. K., Kaback, H. R., Lodish, H. F. 1988. *Cell* 55:281–90
25. Fukumoto, H., Seino, S., Imura, H., Seino, Y., Eddy, R. L., et al. 1988. *Proc. Natl. Acad. Sci. USA* 85:5434–38
26. Thorens, B., Lodish, H. F., Brown, D. 1990. *Am. J. Physiol.* 259:286–94
27. Kayano, T., Fukumoto, H., Eddy, R. L., Fan, Y-S., Byers, M. G., et al. 1988. *J. Biol. Chem.* 263:15245–48
28. Gibbs, E. M., Lienhard, G. E., Gould, G. W. 1988. *Biochemistry* 27:6681–85
29. Karnelli, E., Zarnowski, M. J., Hissin, P. J., Simpson, I. A., Salans, L. B., et al. 1981. *J. Biol. Chem.* 256:4772–77
30. Cushman, S. W., Wardzala, L. J. 1980. *J. Biol. Chem.* 255:4758–62
31. Suzuki, K., Kono, T. 1980. *Proc. Natl. Acad. Sci. USA* 79:3777–79
32. Blok, J., Gibbs, E. M., Lienhard, G. E., Slot, J. W., Gueze, H. J. 1988. *J. Cell. Biol.* 106:69–76
33. Calderhead, D. M., Lienhard, G. E. 1988. *J. Biol. Chem.* 263:12171–74
34. Lienhard, G. E., Kim, H. H., Ransome, K. J., Gorga, J. C. 1982. *Biochem. Biophys. Res. Commun.* 105:1150–56
35. Joost, H. G., Weber, T. M., Cushman, S. W. 1988. *Biochem. J.* 249:155–61
36. Oka, Y., Asano, T., Shibasaki, Y., Kasuga, M., Kanazawa, Y., Takaku, F. 1988. *J. Biol. Chem.* 263:13432–39
37. Zorzano, A., Wilkinson, W., Kotliar, N., Thoidis, G., Wadzinkski, B. E., et al. 1989. *J. Biol. Chem.* 264:12358–63
38. James, D. E., Brown, R., Navarro, J., Pilch, P. F. 1988. *Nature* 333:183–85
39. James, D. E., Strube, M., Mueckler, M. 1989. *Nature* 338:83–87
40. Birnbaum, M. J. 1989. *Cell* 57:305–15
41. Charron, M. J., Brosius, F. C., Alper, S. L., Lodish, H. F. 1989. *Proc. Natl. Acad. Sci. USA* 86:2535–39
42. Fukumoto, H., Kayano, T., Buse, J. B., Edwards, Y., Pilch, P. F., et al. 1989. *J. Biol. Chem.* 264:7776–79
43. Kaestner, K. H., Christy, R. J., McLenithan, J. C., Braiterman, L. T., Cornelius, P., et al. 1989. *Proc. Natl. Acad. Sci. USA* 86:3150–54
44. Kayano, T., Burant, C. F., Fukumoto, H., Gould, G. W., Fan, Y-S., et al. 1990. *J. Biol. Chem.* 265:13276–82
45. Kahn, B. B., Flier, J. S. 1990. *Diabetes Care* 13:548–64
46. Cilenza, J. L., Marshall-Carlson, L., Carlson, M. 1988. *Proc. Natl. Acad. Sci. USA* 85:2130–34
47. Maiden, M. C. J., Davis, E. O., Baldwin, S. A., Moore, D. C. M., Henderson, P. J. F. 1987. *Nature* 325:641–43

48. Szkutnicka, K., Tschopp, J. F., Andrews, L., Cirillo, V. P. 1989. *J. Bacteriol.* 171:4486–93
49. Mueckler, M., Lodish, H. F. 1986. *Cell* 44:629–37
50. Brandl, C. J., Deber, C. M. 1986. *Proc. Natl. Acad. Sci. USA* 83:917–21
51. Alvarez, J., Lee, D. C., Baldwin, S. A., Chapman, D. 1987. *J. Biol. Chem.* 262:3502–9
52. Gould, G. W., Derechin, V., James, D. E., Tordjman, K., Ahern, S., et al. 1989. *J. Biol. Chem.* 264:2180–84
53. Reeves, R. E. 1951. *Advances in Carbohydrate Chemistry,* Vol. 6. New York: Academic. 264 pp.
54. Marsden, N. V. B. 1965. *Ann. NY Acad. Sci.* 125:428
55. Harvey, J. M., Symons, M. C. R., Naftalin, R. J. 1976. *Nature* 261:435–36
56. Jeffrey, G. A., Lewis, L. 1978. *Carbohydrate Res.* 60:179–82
57. Quiocho, F. A. 1986. *Annu. Rev. Biochem.* 55:287–316
58. Bennett, W. S., Steitz, T. A. 1980. *J. Mol. Biol.* 140:183–209
59. LeFevre, P. G., Marshall, J. K. 1958. *Am. J. Physiol.* 194:333
60. Kahlenberg, A., Dolansky, D. 1972. *Can. J. Biochem.* 50:638
61. Barnett, J. E. G., Holman, G. D., Chalkley, R. A., Munday, K. A. 1975. *Biochem. J.* 145:417–29
62. Barnett, J. E. G., Holman, G. D., Munday, K. A. 1973. *Biochem. J.* 131:211–21
63. Holman, G. D., Rees, W. D. 1987. *Biochim. Biophys. Acta* 897:395–405
64. Deves, R., Krupka, R. M. 1978. *Biochim. Biophys. Acta* 510:339–48
65. Gorga, F. R., Lienhard, G. E. 1981. *Biochemistry* 20:5108–13
66. Krupka, R. M., Deves, R. 1981. *J. Biol. Chem.* 256:5410–16
67. Krupka, R. M. 1985. *J. Membr. Biol.* 84:35–43
68. Krupka, R. M. 1971. *Biochemistry* 10:1143–48
69. Rampal, A. L., Jung, C. Y. 1987. *Biochim. Biophys. Acta* 896:287–94
70. Pawagi, A. B., Deber, C. M. 1990. *Biochemistry* 29:950–55
71. Gorga, F. R., Lienhard, G. E. 1985. *Biochemistry* 21:1905–8
72. Appleman, J. R., Lienhard, G. E. 1985. *J. Biol. Chem.* 260:4757–78
73. Widdas, W. F. 1952. *J. Physiol.* 118:23
74. Lowe, A. G., Walmsley, A. R. 1986. *Biochim. Biophys. Acta* 857:146–54
75. Appleman, J. R., Lienhard, G. E. 1989. *Biochemistry* 28:8221–29
76. Carruthers, A. 1986. *Biochemistry* 25:3592–602
77. Hediger, M. A., Coady, M. J., Ikeda, T. S., Wright, E. M. 1987. *Nature* 330:379–81
78. Hosang, M., Gibbs, E. M., Diedrich, D. F., Semenza, G. 1981. *FEBS Lett.* 130:244–48
79. Schmidt, U. M., Eddy, B., Fraser, C. M., Venter, J. C., Semenza, G. 1983. *FEBS Lett.* 161:279–83
80. Peerce, B. E., Wright, E. M. 1984. *Proc. Natl. Acad. Sci. USA* 81:2223–26
81. Neeb, M., Fasold, H., Koepsell, H. 1985. *FEBS Lett.* 182:139–44
82. Ikeda, T. S., Want, H., Coady, M. J., Hirayama, B. A., Hediger, M. A., Wright, E. M. 1989. *J. Membr. Biol.* 110:87–95
83. Birnir, B., Lee, H-S., Hediger, M. A., Wright, E. M. 1990. *Biochim. Biophys. Acta* 1048:100–4
84. Hediger, M. A., Turk, E., Wright, E. M. 1989. *Proc. Natl. Acad. Sci. USA* 86:5748–52
85. Coady, M. J., Pajor, A. M., Wright, E. M. 1990. *Am. J. Physiol.* 259:C605–C610
86. Wright, E. M., Hediger, M. A., Coady, M. J., Hirayama, B., Turk, E. 1989. *Biochem Soc. Trans.* 17:810–11
87. Neeb, M., Kunz, U., Koepsell, H. 1987. *J. Biol. Chem.* 262:10718–27
88. Koepsell, H., Korn, K., Raszeja-Specht, A., Bernotat-Danielowski, S., Ollig, D. 1988. *J. Biol. Chem.* 263:18419–29
89. Silverman, M., Speight, P. 1986. *J. Biol. Chem.* 261:13820–26
90. Wu, J-S. R., Lever, J. E. 1987. *Biochemistry* 26:5783–90
91. Turner, R. J., Moran, A. 1982. *Am. J. Physiol.* 242:F406–F414
92. Ikeda, T. S., Hwang, F-S., Coady, M. J., Hirayama, B. A., Hediger, M. A., Wright, E. M. 1989. *J. Membr. Biol.* 110:87–95
93. Turner, R. J., Moran, A. 1982. *J. Membr. Biol.* 70:37–45
94. Silverman, M. 1976. *Biochim. Biophys. Acta* 457:303–51
95. Silverman, M. 1981. *Can. J. Physiol. Pharmacol.* 59:209–24
96. Ullrich, K. J., Rumrich, G., Kloss, S. 1974. *Pfluegers Arch.* 351:35–48
97. Peerce, B. E., Wright, E. M. 1987. *J. Biol. Chem.* 259:14105–12
98. Peerce, B. E., Wright, E. M. 1987. *Biochemistry* 26:4272–76
99. Quiocho, F. A. 1984. *Nature* 310:381–86
100. Peerce, B. E., Wright, E. M. 1985. *J. Biol. Chem.* 260:6026–31

101. Peerce, B. E., Wright, E. M. 1990. *J. Biol. Chem.* 265:1737–41
102. Klip, A., Grinstein, S., Biber, J., Semenza, G. 1980. *Biochim. Biophys. Acta* 598:100–14
103. Stevens, B. R., Fernandez, A., Hirayama, B., Wright, E. M., Kempner, E. S. 1990. *Proc. Natl. Acad. Sci. USA* 87:1456–60
104. Crane, R. K. 1962. *Fed. Proc.* 21:891
105. Murer, H., Hopfer, U. 1974. *Proc. Natl. Acad. Sci. USA* 71:484–88
106. Silverman, M., Turner, R. J. 1990. *Handb. Physiol.* In press
107. Turner, R. J., Silverman, M. 1981. *J. Membr. Biol.* 58:43–55
108. Aronson, P. S. 1978. *J. Membr. Biol.* 42:81–98
109. Turner, R. J., Silverman, M. 1980. *Biochim. Biophys. Acta* 596:272–91
110. Turner, R. J. 1981. *Biochim. Biophys. Acta* 649:269–80
111. Kessler, M., Semenza, G. 1983. *J. Membr. Biol.* 76:27–56
112. Hopfer, U. 1977. *J. Supramol. Struct.* 7:1–13
113. Hopfer, U., Groseclose, R. 1980. *J. Biol. Chem.* 255:4453–62
114. Hopfer, U., Liedtke, C. M. 1981. *Membr. Biochem.* 4:11–29
115. Harrison, D., Rowe, G., Lumsden, C. J., Silverman, M. 1984. *Biochim. Biophys. Acta* 774:1–10
116. Koepsell, H., Fritzsch, G., Korn, K., Madrala, A. 1990. *J. Membr. Biol.* 114:113–32
117. Turner, R. J., Silverman, M. 1977. *Proc. Natl. Acad. Sci. USA* 75:2825–29
118. Turner, R. J., Silverman, M. 1978. *Biochim. Biophys. Acta* 511:470–86
119. Barfuss, D. W., Schafer, J. A. 1981. *Am. J. Physiol.* 241:F322–F332
120. Turner, R. J., Moran, A. 1982. *J. Membr. Biol.* 67:73–80
121. Berlin, R. D. 1970. *Science* 168:1539
122. Turner, R. J. 1985. *Ann. NY Acad. Sci.* 456:10–25
123. Hopfer, U. 1987. *Physiology of the Gastrointestinal Tract*, ed. L. R. Johnson, pp. 1499–1526. New York: Raven. 2nd ed.

Annu. Rev. Biochem. 1991. 60:795–825

# DENATURED STATES OF PROTEINS

*Ken A. Dill*

Department of Pharmaceutical Chemistry, University of California, San Francisco, California 94143

*David Shortle*

Department of Biological Chemistry, The Johns Hopkins University School of Medicine, Baltimore, Maryland 21205

KEY WORDS: unfolded states, conformational change, molten globules, acid denaturation, solvent denaturation.

## CONTENTS

## PERSPECTIVES AND OVERVIEW

Because the biological properties of proteins arise primarily from their native conformations, considerably more attention during the past 60 years has been focused on the native conformations than on the non-native conformations. In the past few years, however, there has been increasing interest in non-native and denatured states of proteins since these less ordered states play important roles in at least three important phenomena: 1. *Protein folding and stability*. The denatured states of proteins are equal in importance to the native states in determining the stability of a protein, since stability is defined as the difference in free energies between native and denatured states. In cells and

0066-4154/91/0701-0795$02.00

tissues the native states are in dynamic equilibrium with the denatured conformations (1–3). 2. *Transport across membranes*. To cross a lipid bilayer after translation and folding, some proteins must first be converted into a partially folded or denatured conformation, either spontaneously or with the help of "unfoldases" (4–6). 3. *Proteolysis and protein turnover*. Numerous studies have implicated the denatured state as the primary target for degradative enzymes, both in states of stress such as heat shock, and under physiological conditions (7–9).

In this review, we summarize results from theory and experiment that bear on the following questions: What are the denatured conformations of proteins? How are they affected by external conditions, such as solvent composition, temperature, pH, and ionic strength? In what way do these denatured conformations depend on the amino acid composition and specific sequence of the protein? A brief review of the forces that determine conformations in polymers leads to the concepts of "good," "poor," and "theta" solvents. The denatured state is not a single entity microscopically or macroscopically. Rather, theory and experiments indicate that the denatured state should be viewed as a distribution of many microstates that change with the conditions of solution and with the protein sequence.

Although classical experiments by Tanford (1) showed that several proteins in 6M guanidine hydrochloride (GuHCl) exhibit the hydrodynamic properties of random coils, this result has often been misconstrued to imply that the denatured state is a random coil under all conditions. There is now evidence that, even in 6M GuHCl, proteins can have significant amounts of internal structure. Under the much broader range of conditions that are typically used to denature proteins, recent theory and experimental data suggest that denatured proteins are often very compact, with persistent hydrophobic clustering and considerable residual secondary structure. The traditional view holds that the native state is determined by the amino acid sequence and that the denatured states are not, but current evidence indicates that the primary sequence is also a major determinant of denatured conformations. Perhaps the most important denatured state, defined here as $D_0$, is that which is in equilibrium with the native molecule under physiological conditions. Although no experiment has directly observed this state, model systems consisting of peptides and shortened proteins may lead to a greater understanding of the properties of $D_0$.

## INTRODUCTION

Whereas the native state of a protein has a relatively well defined set of atomic coordinates, the denatured states do not, and must be described within a different framework, involving conformational ensembles. We therefore believe it is important to begin a discussion of denatured states with definitions of some terminology. First, we distinguish between "denatured" and "unfolded" states as follows. The term "denatured" was first given an operational definition (1) to designate " . . . a major change from the original native structure . . . " that is noncovalent, cooperative, and reversible, in principle if not in practice. ["Denaturation" is sometimes used to refer to a variety of poorly understood irreversible alterations in protein structure that include aggregation, disulfide bond rearrangement, and backbone cleavage (10), but we do not use that meaning here.] We use "unfolded" state to refer to a specific subset of denatured states, namely conformations that are highly open and solvent-exposed, with little or no residual structure; such states are generally obtained only under strongly denaturing conditions. A second subset of denatured states is referred to here as "compact denatured" states, obtained under weaker denaturing conditions. Whereas "denatured" and "unfolded" have often been used interchangeably, we believe experimental resolution has now progressed to the stage that it is useful to divide the category of denatured states into two parts: unfolded and compact.

In this review, the term "state" has its usual thermodynamic meaning; it refers to an ensemble of the many different configurations of individual molecules, the average of which corresponds to a given macroscopically observable quantity. We denote this macroscopic parameter, or "reaction coordinate," as $a$ from the native state to the denatured state. For example, $a$ might be the molecular radius as measured by intrinsic viscosity or by scattering experiments. We usually use the term "state" more narrowly as an abbreviation to mean "stable equilibrium state," i.e. an ensemble of configurations at a minimum of free energy. For example, consider the radius $r$ as the reaction coordinate. The observed radius in an equilibrium population (see Figure 1) represents a large number of different configurations (i.e. an "ensemble") of the many molecules in solution, because: ($a$) a chain can configure in many different ways to have radius $r = r^*$, corresponding to the equilibrium state, and ($b$) other chain configurations with $r \neq r^*$, will also be populated in accord with the Boltzmann distribution law. A "two-state system" has two minima along the reaction coordinate. It follows that some sort of free energy barrier must separate these two minima. This review is not concerned with the nature of this barrier, or with the kinetics of folding; for a recent review of these topics, see Kim & Baldwin (12). If the conformational free energy, $\Delta G\ (a,\ x)$, of a protein chain also depends on an external parameter, $x$, such as temperature or concentration of some denaturing agent, then external changes of $x$ can lead to a transition, as shown in Figure 1. By "cooperative," we mean that the state of the protein (as represented by the average $<a>$) is observed to have a sigmoidal dependence on $x$ (for example, see $<a>$ vs $x$ in Figures 1A–D). By sigmoidal, we refer only to the S-shaped

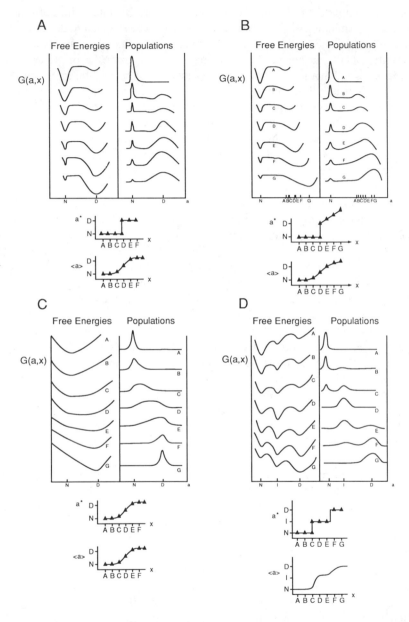

*Figure 1*   Thermodynamic models for transitions such as protein denaturation. The axis labelled *a* on the upper figures represents a property, such as the radius of the molecule, that varies from the native to denatured states. The y-axis shows the free energy. The deepest minimum corresponds to the stable state. The populations are given by the Boltzmann distribution law [proportional to exp (-$\Delta G/RT$)]. The series of curves, $x = x_A, x_B, \ldots$ represent increasing quantities

nature of a curve and not to any specific mathematical functional form. One of the major purposes in measuring transition behavior is to unravel the molecular balance of forces. In that regard, "cooperative" is a very nonspecific term; it can refer to many different types of molecular behavior. Figures 1A–D show four examples. (A) *"Fixed" two-state model.* The free energy behavior shown in Figure 1A is referred to as a "first-order," or "two-state" transition (11); there is a free energy barrier separating two stable states, N and D, which are "fixed" insofar as they do not change with $x$. Because the free energy is high between states N and D, the populations of the intermediate states will be small. (B) *Variable two-state model.* In this case too, there is a free energy barrier separating two states, but one state $(D)$ changes with $x$ (13, 14). (C) *Higher-order transition.* In this type of transition, there would be no free energy barrier between states N and D. Therefore at the transition, significant populations of intermediate states would be observed. An example of a higher-order transition is the critical point of the liquid-gas phase change. (D) *Classical three-state model.* In this model, there are three stable states, N, D, and an intermediate, I, separated by two free energy barriers.

To ascertain the molecular balance of forces in protein folding transitions, it is necessary to determine which of these (or other) types of free energy behavior govern the transition. It is clear from Figure 1 that the simple observation of sigmoidal behavior is not sufficient to prove or disprove the two-state model or any one specific transition model. All of the models shown in Figures 1A–C lead to sigmoidal behavior. What evidence then is required to prove or disprove the various types of molecular models of transition behavior? First, if a physical system undergoes fixed two-state behavior, it is shown in the appendix that two different measured properties must have coincident sigmoidal curves. Therefore, if the measurement of two different properties leads to noncoincident sigmoidal curves, it follows that the transition will not be of the fixed two-state type. However, it cannot then be inferred, as is sometimes done, that noncoincidence of two sigmoidal curves

---

of some denaturing agent. The set of curves in each of Figures A–D is summarized by two figures below each set that describe the experimental observables: $a*$ vs $x$ or $<a>$ vs $x$: for example, molecular radius vs amount of denaturant. Two types of property are shown. 1. $a*$ is a type of property (state-distinguishable) in which both states (N and D) are simultaneously observable; the * indicates whichever state has the highest population. 2. Another type of property (state-averaged), such as intrinsic viscosity, only detects an average, $<a>$. The latter will show sigmoidal behavior vs $x$. Four different transition models are shown: (A) *Fixed Two-State* (first order): free energy barrier separates two fixed states, (B) *Variable Two-State:* two-state but the denatured state varies with denaturant, (C) *Higher Order:* no barrier between the states, (D) *Fixed Three-State:* two barriers separate three stable states.

implies three-state or multistate transition behavior. It is shown in the appendix that the noncoincidence of the sigmoidal curves of two different physical properties can also be consistent with the variable two-state model. The second, and most direct, type of evidence that can confirm the existence of a free energy barrier, and thus a first-order transition, is to show that populations of the states intermediate between N and D are small. Third, the traditional evidence for a first-order transition is the observation of a latent heat of transition (30). In addition, the ratio of the calorimetric enthalpy of transition to the van't Hoff enthalpy (which is based on assuming a two-state model) is useful evidence for or against two-state behavior, for example in contrast to a simple three-state model.

The molecular transition behaviors of proteins may differ for different proteins and for transitions caused by different external denaturing agents. For some simple, single-domain proteins, the existence of a single free energy barrier between N and D is now well-established by: (a) the coincidence of sigmoidal curves for multiple physical properties (110; 111), (b) the lack of evidence of significant intermediate populations (12), and (c) the existence of a latent heat equal to the van't Hoff enthalpy (30). Nevertheless, other protein denaturations are observed to be more complex than two-state (23, 12, 60).

Figure 1 shows two different types of experimental property. (a) A "state-distinguishable" property is one that identifies individual states: for example two different peaks in a NMR spectroscopy experiment—one monitors N and the other monitors D. The quantity $a^*$ refers to which ever state has the highest population (lowest free energy) as the quantity of denaturing agent, $x$, is changed. In a two-state transition, this type of property shows a sharp transition as a function of $x$. (b) A "state-averaged" property is one that monitors only the average over all the different molecular species in solution. For example the intrinsic viscosity monitors the Boltzmann-averaged radius (averaged over all molecules, native and denatured). A state-averaged property shows a smooth sigmoidal transition of native and denatured molecules. Even though the radii may be very different, their populations change smoothly with $x$ and this causes the average, $<a>$, to change smoothly.

## THEORETICAL FRAMEWORK

What forces determine the conformations of denatured proteins? The balance of forces that contributes to polymer collapse and expansion was first described by Flory (15, 16); subsequent theoretical developments have recently been reviewed (17). The free energy of a polymer chain in solution can be divided into three components, chain elasticity, excluded volume, and solvent interactions, which are briefly summarized below.

1. CHAIN ELASTICITY    A characteristic feature of polymer conformations in solution is their statistical nature. For long chains composed of bonds that have relatively free rotations (i.e. small energy differences among rotational isomers), the number of conformations accessible to a chain molecule is large and grows exponentially with chain length. The properties of such chain molecules are governed by the statistical distribution of these conformations. These properties can be approximated by assuming the chain is a random flight: a sequence of completely independent bonds, each with a random orientation relative to its predecessor. If a random-flight chain is caused to undergo either radial expansion or contraction relative to its "unperturbed" state of maximum conformational freedom, then fewer conformations are accessible to it. Hence, such a perturbation leads to a decrease in entropy and increase in the free energy. A perturbed chain will then experience a retractive force toward its unperturbed radius in order to reduce its configurational free energy. This entropic force is the basis for rubber elasticity and acts to oppose the collapse of a polymer chain to a compact state, as when a protein folds to its native state. This component of the conformational free energy of a chain molecule is due to "local" forces; i.e. this form of "ordering" is of each bond in the chain relative to its immediately neighboring bonds in the sequence. For peptides, this ordering involves changes in $\phi$, $\psi$, and side-chain angles.

2. EXCLUDED VOLUME ENTROPY    Another type of chain configurational entropy also opposes the collapse of a polymer molecule to compact states. This is the "excluded volume" entropy due to "nonlocal" factors, where nonlocal refers to the effect of one monomer on another that is far apart in the sequence. Because the atoms that make up a chain molecule occupy volume, fewer configurations are possible for a chain that occupies a small volume than for a chain configured to occupy a larger volume; see Figure 2. The number of configurations available to a chain molecule as compact as in Figures 2C, D, and E is many tens of orders of magnitude fewer than are available to a molecule with the compactness of Figure 2A. These steric constraints are nonlocal in the sense that the volume occupied within any region of space can be due to monomers quite distant in the sequence. Excluded volume contributions to the entropy are large for a chain molecule that becomes as extraordinarily compact as the native state of a globular protein. Protein folding is strongly opposed by this entropy (14, 17).

3. POLYMER-SOLVENT CONTACT INTERACTIONS    Even though entropic effects 1 and 2 oppose the collapse of polymers and proteins to their compact states, polymers can be driven to compactness if they are dissolved in an unfavorable solvent. In a solvent in which the monomer-monomer attractions

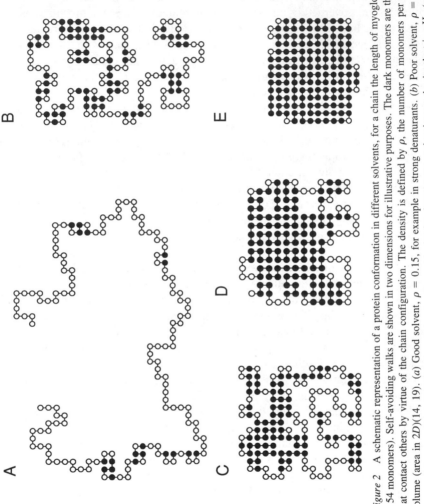

*Figure 2*  A schematic representation of a protein conformation in different solvents, for a chain the length of myoglobin (154 monomers). Self-avoiding walks are shown in two dimensions for illustrative purposes. The dark monomers are those that contact others by virtue of the chain configuration. The density is defined by $\rho$, the number of monomers per unit volume (area in 2D)(14, 19). (*a*) Good solvent, $\rho = 0.15$, for example in strong denaturants. (*b*) Poor solvent, $\rho = 0.5$ corresponds approximately to chains at denaturation midpoint by denaturants or heating at the isoelectric pH. (*c*, *d*) Compact denatured states for proteins in weak denaturants, $\rho = 0.7$ and 0.8, (*e*) Native state typical compactness, $\rho = 0.92$.

are stronger than monomer-solvent attractions, a chain will contract because contraction increases the number of monomer-monomer contacts, and this free energy can be sufficient to overcome the effects of chain elasticity and excluded volume.

## Good Solvents, Poor Solvents, and Theta Solvents

One of the main results of quantitative theories that describe the balance of forces in long-chain homopolymers (i.e. polymers composed of a single type of monomer) is the concept that solvents for a given chain can be divided into three classes: "good," "theta," and "poor" (reviewed in 17). This classification is made on the basis of how the ensemble-averaged (rms) radius ($r$) of the dissolved chain increases as a function of the number of amino acids in the chain ($n$).

Chain solvation by a "theta" solvent is somewhat unfavorable, causing the chain to contract just enough to cancel the chain expansion caused by excluded volume. In a theta solvent, the net sum of nonlocal forces is zero, so the conformations are governed by local forces and random-flight statistics, and the rms radius $r$ will increase in proportion to $n^{1/2}$ where $n$ is the number of monomers in the chain, i.e. $r = (constant)n^{1/2}$. For a given type of polymer chain, the observation of this dependence of radius on chain length implies the chain is in a theta solvent.

A "good" solvent is one that interacts favorably with the monomeric units of the chain. In general, for systems governed by dispersion interactions, the best possible solvent is composed of molecules that have a structure identical to a chain monomer (18, 16). This is an "inert" solvent: there is no free energy difference between a polymer-solvent contact and a polymer intrachain contact. In a good solvent, the contact free energy of the polymer with the solvent is favorable (zero for an inert solvent), and the chain expands relative to the random-flight dimensions. For a given type of polymer, a good solvent is that in which the rms radius scales as $r = (constant)n^{3/5}$.

A "poor" solvent is defined as one that is more unfavorable than a theta solvent. In a poor solvent, a chain contracts and becomes relatively compact, $r = (constant)n^{1/3}$, because the intrachain attractions that collapse the polymer are stronger than the excluded volume repulsions that expand it.

Thus in order to establish whether a solvent is good, theta, or poor for a given type of polymer, measurements must be made of the exponent for the dependence of molecular radius on the chain length. In a theta solvent this exponent is ½, whereas in a good solvent it is ⅗, and in a poor solvent it is ⅓.

## Heteropolymer Model

The theory described above applies to homopolymers. The three solvent types are defined in terms of a single quantity, $\chi$, the free energy of interaction between the solvent and the single type of chain monomer (16) (divided by *RT*, gas constant times temperature, so $\chi$ is a dimensionless quantity). For heteropolymers such as proteins composed of different monomer types, a more general theory is required. We describe below the results of one such theory for protein stability and denatured states; we refer to it as the hetero-polymer theory (14, 17, 19). Other theoretical models also exist (reviewed in reference 17). It is not our purpose here to compare theories, but rather to analyze denatured conformations under different solution conditions. At present, the heteropolymer model is the only theory that has been applied to predictions for specific conditions of temperature, denaturant concentrations, charge, and protein composition.

For a heteropolymer, there will be different $\chi$ values for the interactions of solvent with each of the 20 different monomer types. However, if the interactions of nonpolar monomers with water are dominant, then the $\chi$ corresponding to this type of interaction will be more important than the others, and the behavior of the chain will be dominated by the aversion of nonpolar amino acids for water. For a homopolymer $\chi < \frac{1}{2}$ defines a good solvent, $\chi = \frac{1}{2}$ defines a theta solvent, and $\chi > \frac{1}{2}$ defines a poor solvent. For nonpolar amino acids in water, it is estimated from oil/water partition experiments that $\chi = 3.37$ at T = 25°C (19, 47). Hence water is an extremely poor solvent for polypeptide chains which, like most naturally occurring proteins, contain from 25 to 50% nonpolar amino acids. For a heteropolymer, it is not $\chi$ alone that predicts whether the solvent will expand or contract the polymer. Instead the quantity $\chi\phi^2$, where $\phi$ is the fraction of residues in the chain that are nonpolar, is the approximate predictor of chain expansion or contraction. The net result is that, even though 6M GuHCl reduces $\chi$ by only about 25% (20, 47), this is sufficient to cause an aqueous solution of 6M GuHCl to be a good solvent for most proteins. Significantly lower concentrations of GuHCl (i.e. 0–4 M) will be poor solvents for most proteins (47). Therefore, for a given set of conditions, proteins with many nonpolar residues will have denatured states that are more compact than will proteins with few nonpolar residues. The heteropolymer theory predicts that a protein at its isoelectric pH should denature if $\chi$ exceeds a particular value, $\chi_0$, irrespective of whether $\chi$ is caused to increase by temperature or by denaturants. Thus at the midpoint temperature, the conformations of a thermally denatured molecule should closely resemble those of a molecule denatured by GuHCl or urea (at the midpoint denaturant concentration).

Proteins are typically denatured by GuHCl in the range of 1 to 4 M; for most proteins these are poor solvent conditions. Therefore, in the absence of charge effects, denaturation either by heating to high temperatures, on the one hand, or by GuHCl or urea, on the other hand, should lead to a relatively compact denatured state (at the transition midpoint) compared to a chain in a theta solvent. Moreover, further increase in denaturant (i.e. either of temperature or of denaturing agent) beyond the denaturation midpoint should lead to further gradual increase of the radius as the solvent becomes "better," as shown in Figures 3 and 4. Figures 3 and 4 also show that the radius of the denatured states at the denaturation midpoint depends on the fraction of nonpolar residues in the chain. The theory predicts that weakly denatured proteins, either by temperature or by low denaturant concentrations, have considerable numbers of contacts among hydrophobic residues.

Another property is predicted by theory for polymer chains in poor solvents: compact chains should have some internal structure: helices, sheets, and turns (21, 22). The amount of structure should increase with compactness (21, 22) and will also depend on specific interactions. Structure arises because

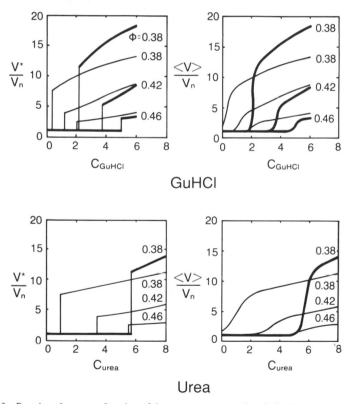

*Figure 3*  Protein volume as a function of denaturant concentration C (for GuHCl or urea), from the heteropolymer model (14, 19). The volume $<V>$ is the average volume, $V_n$ is the native volume, and $V^*$ is the volume of the most stable species, N or D. The volume continues to increase with denaturant beyond the denaturation midpoint because the solvent becomes better. The details of denaturation are dependent on the amino acid composition; $\phi$ is the fraction of residues that are nonpolar. Light lines are for $n = 100$ monomer chains; heavy lines are for $n = 200$ monomer chains.

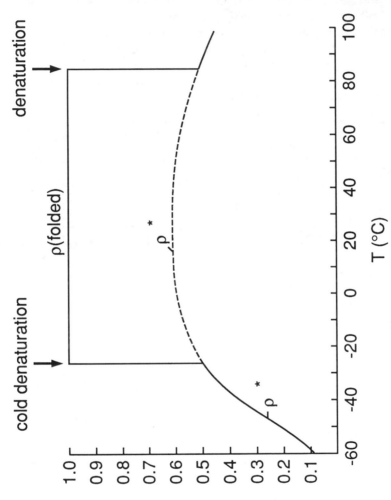

*Figure 4* The density ($\rho$) of the protein (inversely proportional to the volume) as a function of temperature, at the isoelectric pH, predicted from the heteropolymer model (19). $\rho^*$ indicates the density of the denatured state. At the thermal denaturation midpoint, (about 85°C in this case) the protein is predicted to be about half as compact as the native protein. The volume then further increases (density decreases) with further increase in temperature.

the severe steric constraints in relatively compact chains are selective for certain configurations, including helices, sheets, and turns. "Local" interactions among neighboring residues in the chain also contribute to structure in denatured states.

# EXPERIMENTS

## Solvent Denatured States

As emphasized above, the properties of denatured states depend on the solvent conditions. The denatured state under physiological conditions, $D_0$, is the most relevant to an understanding of protein stability, since stability is the free energy difference, $\Delta G = G_{native} - G_{denatured} = -RT \ln [N]/[D_0]$. Unfortunately, since most proteins under native conditions are stable by $+5$ to $+20$ kcal/mole, then only one out of every $10^4$ to $10^{15}$ protein chains is in the $D_0$ state. Consequently, experimental studies have concentrated on D states under conditions quite different than aqueous buffers at neutral pH and 20° to 37°C. By analyzing how the properties of denatured states vary under these more extreme conditions, estimates of the properties of $D_0$ can often be made by extrapolation back to physiological conditions.

Experimental studies of denatured states have also been hampered by two major practical obstacles. The most serious of these is that the denaturing conditions used in an experiment are often those that also lead to limited solubility, so it is often necessary (but very difficult) to distinguish between denatured and precipitated protein. In addition, until recently, the physical methods for characterizing the structure of denatured proteins have only provided information at low levels of resolution. Intrinsic viscosity, gel filtration, and quasi-elastic light scattering can determine the molecular radius (23–25), and various types of spectroscopy can monitor the amount of secondary structure and the environments around tyrosine and tryptophan residues (23, 26, 27). Important as they are for characterizing low-resolution structure, these methods provide relatively little insight into the molecular details of those conformations. Fortunately, high-resolution NMR spectroscopy has enormous potential for providing this type of structural information (for a recent review, see reference 28).

A problem of long-standing interest has been whether the denatured states that arise from heating as opposed to the addition of denaturants are different (29, 30, 1). Heat denaturation may be something of a misnomer, since heating a protein to high temperatures at its isoelectric pH has not been widely used for reversible denaturation experiments because it often leads to irreversible covalent changes (1, 10). Thus thermal denaturation generally refers to heating to moderate temperatures at low (or high) pH. In these cases, the role of electrostatics is probably very important. Therefore, these processes are considered in the next section on pH denaturation rather than in the present section.

The most widely used strong denaturants are GuHCl and urea, which lower protein stability in proportion to their concentrations (3, 31, 32). Although the chemistry of interaction between proteins and aqueous solutions of urea and guanidine is not fully understood, most available evidence indicates that these compounds cause water to become a better solvent for the nonpolar amino acids (33, 34); i.e. they weaken the hydrophobic interaction.

As mentioned above, Tanford observed that 6M GuHCl is a good solvent for proteins and it is "universal" in the sense that the radius, as determined by intrinsic viscosity and light scattering measurements, depends only on chain length, $r = (\text{constant})n^{0.67}$, and not on the amino acid sequence, for more than 10 different proteins (1, 2); see Figure 5. Early evidence supported the conclusion that, in high concentrations of urea or guanidine hydrochloride, denatured proteins are highly unfolded and have little residual structure (35–39, 1). On this basis, denatured states of proteins have often been assumed to resemble random coil conformations.

However, computer simulations show that the radius of random-flight chains is not a particularly sensitive measure of partial clustering of the chain (40). Therefore the Tanford experiment does not rule out the possibility that denatured proteins may have partial structure in 6M GuHCl. Recent studies on a number of proteins using spectroscopic methods have turned up unambiguous evidence of residual structure, in some cases even at very high denaturant concentrations. For example, Bierzynski & Baldwin (41) used NMR spectroscopy to characterize GuHCl-denatured ribonuclease A at low pH and found evidence that the $\alpha$-helix that corresponds to the C-peptide is present a significant fraction of the time. Haas et al (42, 43) used fluorescence energy transfer between two extrinsic probes to measure the distance between amino acids 1 and (49 + 53) in reduced ribonuclease A in 6M GuHCl. They observed an average distance significantly less than that predicted for a random coil polypeptide. A similar fluorescence transfer study by Amir & Haas (44) of reduced bovine pancreatic trypsin inhibitor (BPTI) in 6M GuHCl suggests that approximately 50% of molecules exhibit native distances between residues 1 and 26, while in the other 50% of the molecules these two residues are separated by a much greater distance. Since this persistent structure was greatly diminished in 5.4M GuSCN, Amir & Haas proposed that the highly hydrophobic chain segment between the donor-acceptor pair forms structure that persists in 6M GuHCl but is nearly eliminated under still stronger denaturing conditions.

More specific evidence from high-resolution NMR also shows the persistence of structure in strong denaturants. Wishart (45) labeled the alpha

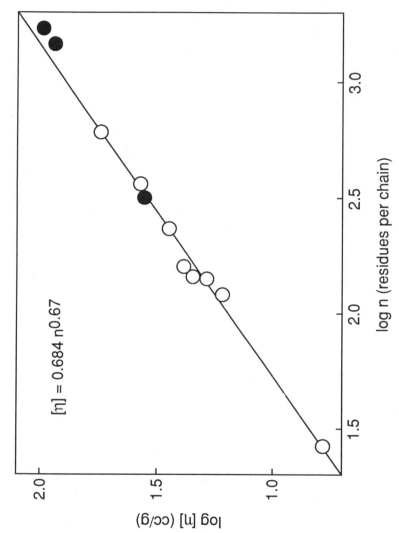

*Figure 5* Experiments of Tanford (1, 2) showing that for several different proteins, the intrinsic viscosity $[\eta]$ and hence the radius, $r$, is proportional to $n^{0.67}$ in 6M GuHCl; thus this is a good solvent.

carbons of the nine glycines of *Escherichia coli* thioredoxin with $C^{13}$. While all nine peaks could be resolved in the native state, five or six peaks were still resolved in 9M urea at room temperature. Since in 2.5M GuHCl at 65°C all the resonances collapsed into three broad bands, the authors concluded that significant local structure must persist in 9M urea to give rise to the observed chemical shift differences. Similarly, the proton NMR spectrum of BPTI in 8M urea shows clear evidence for some residual structure (46).

Another probe of denatured conformations is the amount of solvent exposure of the amino acid residues, which can be estimated from the dependence of protein stability on denaturant concentration. Thermodynamic and statistical mechanical models (34, 47) indicate that $m$, the rate of change of the free energy of denaturation with respect to the denaturant concentration, is approximately proportional to the difference in solvent accessible surface area $A$ between the D and N states. If the proportionality constant for this relationship were known, $A_d$ (exposed nonpolar area of the denatured state) could be determined from $m$, since $A_n$ (exposed area of the native state) can be calculated from the X-ray crystal structure. The heteropolymer theory (47) and experimental evidence (2, 32) suggest that $A_d$ for a typical protein is considerably less than that predicted for a fully exposed chain. A large number of experiments on a variety of proteins (reviewed in reference 3) show that $m$ is almost invariably a constant, displaying little if any dependence on denaturant concentration over the small concentration ranges that are typically accessible experimentally; see Figure 6. An approximate linearity over narrow ranges of denaturant concentration is expected from theory (34, 47); but whether there are small deviations from linearity has not yet been fully resolved.

Recent data strongly support the idea that replacing single amino acid residues can affect the denatured states of a protein. For different mutant forms, $m$ can either increase or decrease relative to the wild-type value (32, 48–55). Small changes in $m$ of either sign can be accounted for in part by significant changes in the area exposed in the denatured state, $A_d$ (47). However, some single-site replacements are observed to lead to very large changes in $m$ (48) that cannot be accounted for by the heteropolymer theory (47). For staphylococcal nuclease, a multiple mutant has been identified that increases $m_{GuHCl}$ by almost a factor of two (D. Shortle, A. M. Meeker, unpublished observations), suggesting that the wild-type D state must retain a large number of residues in a solvent-excluded configuration.

For several proteins, increasing the charge on the protein by increasing or decreasing pH can also lead to very significant increases in $m_{urea}$ (56, 57). Presumably, the fraction of buried surface exposed upon denaturation is increased by electrostatic repulsion when the polypeptide chain is highly charged.

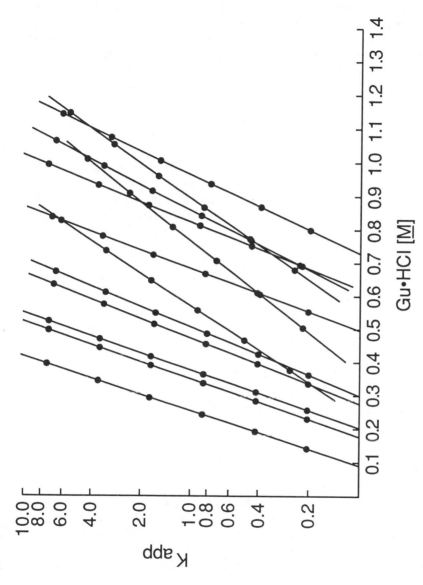

*Figure 6* The logarithm of $K_{app}$ (the apparent equilibrium constant for denaturation) versus the concentration of GuHCl for a series of 11 mutant forms and the wild-type of staphylococcal nuclease. The slope of each line is $m_{GuHCl}$. See reference 48.

To analyze the contribution of the large hydrophobic residues of staphylococcal nuclease to the stability of the native state and to the residual structure that persists in the denatured state, Shortle et al (58) mutated individually each of the leucine, valine, isoleucine, phenylalanine, tyrosine, and methionine residues to both alanine and glucine. After overexpressing and purifying 83 single mutant proteins, GuHCl denaturation was used to determine $m_{GuHCl}$ and $\Delta G_{H_2O}$, the stability in the absence of denaturant. Perhaps their most important finding was the existence of a statistically significant correlation between the absolute value of the change in $m_{GuHCl}$ away from that of wild-type and a loss in stability. From this result, they made the argument that part of the stability loss for some mutants must be a consequence of the altered D state structure reflected in the change in $m_{GuHCl}$.

On analyzing the changes in stability in this collection of mutant nucleases, Shortle et al (58) found it very difficult to account for various patterns in their data solely in terms of changes in native state interactions. For example, if the principal effects of shortening aliphatic side chains were to decrease the free energy of hydrophobic group transfer on folding and/or to generate cavities in the native structure, one might expect that the stability loss for a series of substitutions at one position would show a linear relationship with the number of carbon atoms (or methylene equivalents) removed. But at more than half of the 29 positions analyzed, a twofold or greater variation was found among the changes in stability that accompanied removal of methylenes from leucine, isoleucine, valine, and mathionine. At some positions, the gamma, delta, and epsilon- carbons made a greater contribution to stability than did the beta carbon, whereas at other positions the reverse was found. In addition, at several positions the stability loss that accompanied removal of the beta carbon exceeded the 2.0 to 2.5 kcal/mole upper limit estimated for the contribution of methyl groups to the cohesion of hydrocarbon solids. Shortle et al proposed that some of these unexpected effects could be a consequence of disruption of the hydrophobic clustering in the denatured state predicted on the basis of changes in $m_{GuHCl}$. For mutant nucleases with a less structured denatured state, part of the stability loss would then be accounted for by the larger value of the entropy gain on denaturation.

Flanagan et al (59) have recently used small-angle scattering of both X-rays and neutrons to monitor the denaturation of staphylococcal nuclease in urea. Both the radius of gyration and the density of local clustering of the polypeptide chain in the denatured state have been determined. Perhaps their most surprising finding was the bilobed shape of the denatured state around 3M urea (conditions where only 10% of molecules remain in the native state). This structure gradually expanded and became more nearly spherical with increasing denaturant concentrations. Data collected on several mutant nucleases confirmed earlier suggestions that single amino acid substitutions can significantly alter the residual structure in the denatured state (48, 23).

A few relatively unstable proteins, such as alpha-lactalbumin (76), carbonic anhydrase (62), bovine growth hormone (62), and beta-lactamase from *Staphylococcus aureus* (63), which are denatured in low concentrations of GuHCl, show transitions more complex than those between two simple fixed states. Although such behavior has been attributed to three states—the native state, the unfolded state, and a compact denatured state (60)—it may also be consistent with a two-state mechanism in which the denatured state changes continuously from a compact form to an unfolded form with increasing denaturant concentration (see appendix and Ref. 13). Additional evidence is needed to unequivocally establish the existence of a discrete third species, or equivalently to establish that there is a free energy barrier between the compact and unfolded denatured states for these transitions in GuHCl.

Whereas urea and GuHCl are among the most widely used agents for denaturation of proteins, nearly all organic solvents that are miscible with water also act as protein denaturants. At low concentrations (5–20%), simple aliphatic alcohols promote the breakdown of protein structure, presumably by direct association of the hydrophobic component of the solvent with exposed hydrophobic residues of the denatured protein (64, 65). The efficacy of alcohols and similar compounds in promoting denaturation is markedly dependent on both their concentration and the temperature. Elevated temperatures increase denaturation, whereas low concentrations of some compounds, such as methanol and ethanol, can stabilize the native state around 0°C (64). At higher concentrations of organic solvent, which lead to aggregation and precipitation of many proteins, the appearance of a more structured state has been noted (66). Interpretation of the optical rotary dispersion (ORD) and circular dichroism (CD) spectra of this state suggests the polypeptide chain is driven into a conformation with large amounts of alpha-helical structure (66). For example, trifluoroethanol is commonly used to promote alpha helix formation in small peptides (67).

## pH and Thermally Denatured States

Since proteins invariably contain basic and acidic residues, they possess a net charge which increases as the solution becomes more acidic or more basic relative to the isoelectric pH. In the absence of ion-pairing, the electrostatic free energy of a charged protein is reduced upon denaturation because charge repulsion is reduced as charges are distributed over the larger volume of the denatured molecule.

Unlike the case with solvent denatured states, the persistence of large amounts of residual structure in acid-denatured states has been simple to demonstrate, especially in the presence of salt (68–71). The most direct evidence was that of Aune et al (36), who performed two parallel experiments. In the first, they denatured three different proteins (ribonuclease, lysozyme, and chymotrypsinogen) at pHs from 1.65 to 2.18 using temperatures between 56°C and 60°C and then tested the nature of the resultant denatured state by further treatment with increasing GuHCl concentrations. They observed a small cooperative transition, which they interpreted as evidence that the low-pH thermal denaturation process had not fully unfolded the protein. In a parallel experiment, they denatured the protein in 6M GuHCl (at pH 4.4 and 6.7), followed by heating. In this experiment, they observed no cooperative transition and therefore concluded that denaturation in 6M GuHCl leads to a less structured denatured state than thermal denaturation at low pH.

More recent experiments have confirmed that thermal unfolding under acidic conditions often leads to a compact denatured state, sometimes referred to as the A state (72–74). At pH below 4.0, dramatic changes are observed in the environment of buried aromatic residues in alpha-lactalbumin as monitored by CD spectroscopy and UV absorption, whereas indicators of secondary structure such as CD in the far ultraviolet and infrared spectroscopy detect only minor structural changes (75, 76) (see Figure 7A). This acid-denatured form, termed the A state, remains quite compact, with gel chromatography, small-angle X-ray studies, and quasi-elastic light scattering all suggesting that the polypeptide chain has increased its radius from approximately 18 Å in the N state to 20 Å in the A state (29, 77).

Perhaps the most convincing evidence that the A state is not a relatively subtle alteration in the N state comes from studies of its thermal stability. No cooperative transition, only a monotonic decrease, was observed in the CD signal as the temperature was raised for molecules in the A state (76, 78). In addition, no appreciable heat absorption was detected by scanning calorimetry (79).

Recent NMR studies by Baum et al (80) revealed that the proton spectrum of the A state is very different compared to either the native or the urea-denatured states. For some of the resolvable protons (shown in Figure 7B), tentative assignments were made by using magnetization transfer from the native state under conditions where N and A are in rapid exchange. The four aromatic residues in the N state with the largest chemical shifts away from the random coil values also had large chemical shifts in the A state, suggesting that some of the residual structure may be native-like. This conclusion was supported by hydrogen-exchange measurements that revealed that some of the slowly exchanging protons in the A state are also among the most slowly exchanging in the N state. In particular, amide protons of residues 89 to 96 exchanged 50 times more slowly than expected for a random coil. Since this chain segment forms an alpha helix in the N state, it may also be helical in the A state. Because protection from hydrogen exchange was not found for other amide protons that form alpha helices in the N state, Baum et al proposed that a hydrophobic kernel is the most prominent structure in the A state (80).

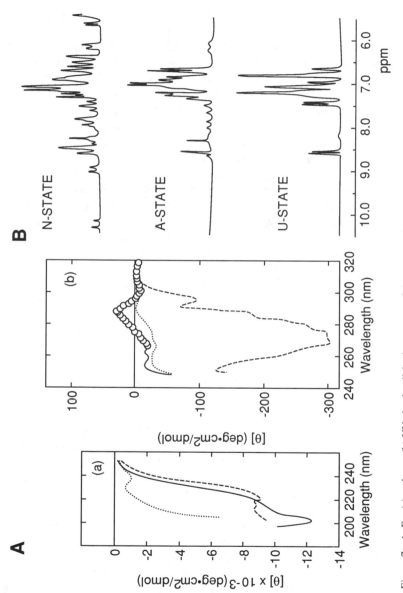

*Figure 7*  *A.* Far (a) and near (b) UV circular dichroism spectra of three states of alpha-lactalbumin: N(– – –), A (– – –), and D (····), where A is the acid-denatured form and D is the urea-denatured form (109). *B.* Low field regions of the proton NMR spectra for the three states of alpha-lactalbumin (80).

Carbonic anhydrase (81, 62), beta-lactamase (82), cytochrome c (83, 84), apomyoglobin (84), staphylococcal nuclease (85), and several other proteins have all been shown to give rise to compact states on acid denaturation. Kuwajima has listed six properties characteristic of the A state of α-lactalbumin (60): (a) secondary structure similar to the native structure, (b) solvent inaccessibility of some residues (tryptophans) that are also buried in the native structure, (c) small expansion of the radius relative to the native state less than or equal to 10% (i.e. a 30% increase in volume), (d) propensity to aggregate, perhaps due to exposure of hydrophobic residues, (e) no thermal transition upon heating (although there is some cooperativity upon GuHCl denaturation), (f) slow but extensive intramolecular fluctuations. Kuwajima refers to this as the "molten globule" state.

The term "molten globule" originated with Ohgushi & Wada (83). Although it originally referred only to the acid-denatured A state, the term molten globule has now come to refer to compact states that may also arise from other conditions of denaturation, for example by thermal or GuHCl denaturation (60, 74, 77). "Globule" refers to the native-like compactness, and "molten" refers to the higher enthalpy and entropy relative to the native structure, as in the melting of a solid, and to the larger structural fluctuations, particularly in the side chains (77), relative to the native structure. Because there are some differences in definition of molten globule among groups, we favor the recommendation of Kim & Baldwin (12) that the term "molten globule" be used to designate the theoretical model of non-native states of proteins that have "native-like" topology described by Ptitsyn (74) and Shakh-novich & Finkelstein (86, 87); that "compact denatured states" refer to experimentally observed compact non-native states; and that the A state refer specifically to acid-denatured compact states. Fink et al (73, 82, 88) have shown, particularly by CD spectra taken in the near and far UV, that the acid-denatured state may not be unique, and that for different proteins, these states may have little, moderate, or large amounts of residual structure that may or may not change continuously with salt concentration.

Recently, Fink and colleagues (88) have provided important new information on acid-denatured states for several different proteins. Characterization of the efficacy of different salts in stabilizing the A state relative to the unfolded state clearly suggests that charge neutralization, not indirect solvent effects, is the dominant role of added anions (88). In addition, proteins that denature directly to the A state at low pH without added salt and proteins that do not denature even at pHs below 1.0 develop a salt requirement for stability of the A state when urea is added. The data of Fink et al can be explained in terms of a balance between electrostatic repulsion among residues of like charge and the forces that act to constrain and structure the polypeptide chain. Some proteins (i.e. alpha-lactalbumin) tend to require little or no salt to collapse into the A state, whereas others (i.e. beta-lactamase) often require high concentrations of salt to first reduce electrostatic repulsion. An extreme example of the latter effect can be seen in histone H1, which spontaneously denatures at very low salt concentrations even at neutral pH (89).

The existence of two stable populations of denatured states at low pH, separated by a free energy barrier, has been predicted by addition of electrostatics contributions to the heteropolymer theory (90). Low pH denaturation can lead to both an open unfolded state, and a compact denatured state. As shown in Figure 8, the predictions are in good agreement with the pH-ionic strength phase diagram of myoglobin of Goto & Fink (73). According to the theory, the appearance of two stable denatured states is a consequence of a balance of four forces. The open denatured state is stabilized by the electrostatic force, which leads to expansion, balanced by chain elasticity, which opposes expansion. Similarly, the compact denatured state is stabilized by the hydrophobic interaction, which leads to contraction, balanced by the excluded volume entropy, which opposes contraction. States intermediate in radius are slightly less stable, because either charge repulsion drives them to expand or hydrophobic interactions drive them to contract.

The heteropolymer model predicts two very different mechanisms of denaturation. On the one hand, denaturation of charged molecules can lead to a transition from N to either of two stable denatured states in equilibrium: one unfolded, and one compact with a small free-energy barrier between them. On the other hand, for denaturation near the isoelectric pH either by heating or by denaturants, the heteropolymer model predicts only a single sharp transition from native to a compact denatured state, which then gradually unfolds with further increase in temperature or denaturant. In support of this theory, experiments of Pfeil (29) show some similarity between temperature- and solvent-denatured alpha-lactalbumin compared to the acid-induced denatured state.

## Experimental Models of Denatured States Under Physiological Conditions

To develop better experimental models for $D_0$, the denatured state in equilibrium with the native structure under physiological conditions, it is necessary to characterize molecules at the margin of stability. Only unstable molecules have significant populations of denatured conformations. One approach is to study the properties of short peptides, which can often be analyzed in aqueous solutions at neutral pH and ambient temperature, and therefore promise to be more representative of the $D_0$ state.

Early efforts to detect persistent secondary structure in short peptides with sequences corresponding to alpha helices or beta turns of known proteins (91–93) were unsuccessful in all cases but one (94). More recent work,

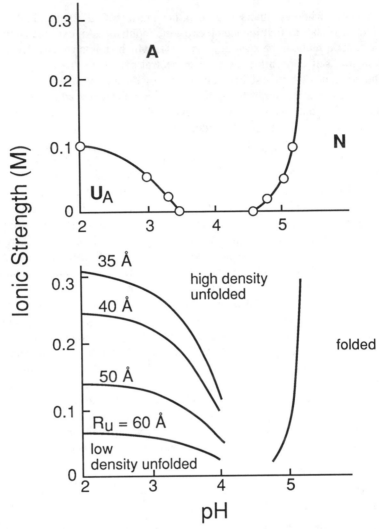

*Figure 8* Phase diagram of apomyoglobin vs pH and ionic strength. (*a*) Experiments of Goto & Fink (73) showing three states: native (N), acid unfolded (U$_A$), and compact denatured (A). (*b*) Predictions of the heteropolymer theory with electrostatics (90). Iso-radius lines on the left identify the possible phase boundary between U$_A$ and A.

however, has provided a number of examples in which synthetic peptides, free in solution, can adopt the same secondary structure as a segment of a known protein with the same amino acid sequence. The most extensively studied example is the C-peptide, which corresponds to an alpha helix formed by the first 13 residues of ribonuclease A. Generated by cyanogen bromide cleavage, this protein fragment exhibits approximately 30% helical structure at 0°C by CD (94). Baldwin and colleagues have studied extensively synthetic peptides with altered amino acid sequences in order to determine the origins of the unusually high stability of this short helix (95, 96). They conclude that, in addition to the favorable helix propensities of the residues, salt bridge formation and an aromatic interaction between side chains, plus favorable interactions of terminal charged groups with the helix dipole, all contribute to helix stability (97).

Similarly, synthetic peptides corresponding to an alpha helix in BPTI (98), the C helix of myohemerythrin (99), residues 132–153 of sperm whale myoglobin (100), residues 50–61 of ribonuclease A (101), and a proteolytic fragment of bovine growth hormone (102) all form alpha helices in solution. None of these helices, however, are as stable as that formed by the C- peptide. A short synthetic peptide corresponding to a highly immunogenic segment of the influenza hemagglutinin molecule has been demonstrated to form a relatively stable type II beta turn in solution, although this sequence in the native protein does not form a turn (103). In several short peptides containing three or more hydrophobic residues, the protons on the hydrophobic side chains have been found by NMR spectroscopy to exhibit significant chemical shift dispersion, a result consistent with aggregation or clustering of these neighboring hydrophobic groups (104–106). It appears that isolated peptides in solution under native conditions can have a propensity to form native-like structures.

This propensity to form structure can be further enhanced if protein substructures are appropriately constrained by disulfide bonds or hydrophobic contacts. One example is a peptide model of parts of BPTI studied by Oas & Kim (107). A disulfide-bonded pair of peptides corresponding to 30 of the 56 residues of BPTI forms a structure in solution that is similar to that of native BPTI (107). One peptide corresponds to a two-turn helix and a short beta strand, whereas the other has the sequence of parts of two antiparallel beta strands. The CD spectra and the aromatic region of the proton NMR spectrum clearly demonstrate the large changes in structure that occur as a function of the state of oxidation of the disulfide and at different temperatures. Thus, isolated pieces of BPTI held together by a native disulfide bond are induced to form native-like architecture.

Fry et al (108) have characterized a synthetic 45-mer that corresponds in sequence to the amino terminus of adenylate kinase. This long peptide binds ATP with the same affinity as intact enzyme, and 2D-NMR studies suggest that, some fraction of the time, it assumes the correct native structure consisting of two alpha helices and two beta strands (108).

Mutant proteins that are unable to stably enter the N state in aqueous buffer at neutral pH may also exhibit some of the properties of $D_0$. Shortle & Meeker

(23) have made short deletions by removing a small number of residues from either the amino-terminal or carboxyterminal ends of staphylococcal nuclease for such model studies. For this protein and many others, the ends assume a peripheral position in the native structure, and thus this modification of the amino acid sequence may not grossly perturb most of the potential interactions that occur in the D state. Analysis of the CD spectra of two large nonsense fragments of staphylococcal nuclease suggested that the wild-type $D_0$ may exhibit approximately 50% of native secondary structure. Extrapolation of the gel filtration behavior of these fragments to normal chain length suggested that this state may occupy a volume only 30–50% larger than the N state (23). In addition, all of the mutants with increased values of $m_{GuHCl}$ exhibited significantly less residual structure, supporting the conclusion discussed above that changes in $m_{GuHCl}$ reflect an alteration in D state structure.

## SUMMARY

The denatured "state" of a protein is a distribution of many different molecular conformations, the averages of which are measured by experiments. The properties of this ensemble depend sensitively on the solution conditions. There is now considerable evidence that even in strong denaturants such as 6M GuHCl and 9M urea, some structure may remain in protein chains. Under milder or physiological conditions, the denatured states of most proteins appear to be highly compact with extensive secondary structure. Both theoretical and experimental studies suggest that hydrophobic interactions, chain conformational entropies, and electrostatic forces are dominant in determining this structure.

The denaturation reaction of many proteins in GuHCl or urea can be most simply modelled as a two-state transition between the native structure and a relatively compact denatured state, which then undergoes a gradual increase in radius on further addition of denaturant. However, when a protein acquires a large net charge in acids or bases, it can have two stable denatured populations, one compact and the other more highly unfolded.

The prediction and elucidation of the structural details of the non-native states of proteins may ultimately prove to be as difficult as predicting the native structures, particularly for $D_0$, the denatured state under physiological conditions. Just as with the native state, the structure of this biologically important denatured state appears to depend on the amino acid sequence. The development of synthetic peptide and protein fragment models of the denatured state and the recent progress in NMR spectroscopy provide bases for optimism that new insights will be gained into this poorly understood realm of protein biochemistry.

ACKNOWLEDGMENTS

We thank R. L. Baldwin, Linda De Young, Tony Fink, and Oleg Ptitsyn for helpful comments and K. Kuwajima, C. Dobson, C. R. Matthews, P. E. Wright, P. Kim, Elisha Haas, Fred Richards, and C. N. Pace for helpful communications. Research in the laboratories of the authors has been supported by grants from the National Institutes of Health (NIH) and the University Research Initiative of the Defense Advanced Research Projects Agency (DARPA) (K. A. D.).

# APPENDIX

In this appendix, we show that the observation of noncoincident sigmoidal curves of two different properties for the denaturation of a protein does not constitute proof of the existence of an intermediate state of the type shown in Figure 1D. Noncoincidence of sigmoidal denaturation curves is also consistent with the variable two-state model shown in Figure 1B. Suppose the fraction of protein molecules in the denatured state, $f_D(x)$, has the sigmoidal dependence on denaturing agent $x$ shown in Figure 9. Suppose one property, $a$, has the value $a_N$ for pure native protein and the value $a_D$ for pure denatured protein; for example see Figure 1. Then the averaged value, $<a(x)>$, will be the population-weighted sum:

$$<a(x)> = a_N f_N(x) + a_D f_D(x) \qquad\qquad 1.$$
$$= a_N(1 - f_D(x)) + a_D f_D(x) \qquad\qquad 2.$$
$$= a_N + (a_D - a_N)f_D(x) \qquad\qquad 3.$$

Without loss of generality, suppose $a_N = 0$ for simplicity of illustration. Then Eq. 3 becomes

$$<a(x)> = a_D f_D(x) \qquad\qquad 4.$$

Thus when a protein denaturation involves two fixed states, then the measured curve, $<a(x)>$, scaled by the value $a_D$, will superimpose on the actual denaturation curve. It follows from Eq. 3 that for a classical two-state transition, any property will superimpose on this curve, irrespective of the values of the qualities $a_D$ and $a_N$ provided it is expressed in the form:

$$\frac{<a(x)> - a_N}{a_D - a_N} = f_D(x) \qquad\qquad 5.$$

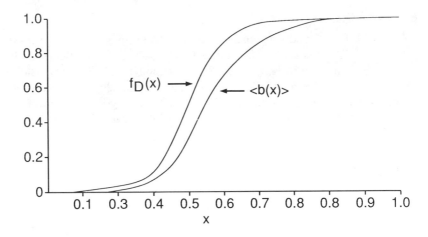

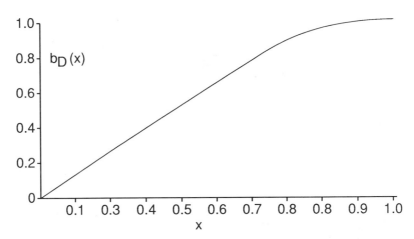

*Figure 9*   The denaturation curve $f_D(x)$ vs denaturing agent $x$ does not coincide with an experimental measure of it, $<b(x)>$, if the property $b$ detects variations in the denatured state with $x$. For illustration, $b_D(x)$ is assumed to increase linearly then reach a plateau (bottom figure), as is predicted to be the case for the volume of a denatured molecule (see Figure 3).

Now consider a second property $b$ that has a single value $b_N$ when all the molecules in solution are native, but which changes with $x$ as the denatured state expands, $b_D(x)$. Since $b_D$ depends on $x$, the measured property $<b(x)>$ will not be coincident with the denaturation curve, $f_D(x)$ because Eq. 4 now becomes:

$$<b(x)> = b_D(x)f_D(x) \qquad\qquad\qquad 6.$$

To demonstrate the principle, we assume a simplest functional form, motivated by the heteropolymer theory prediction that the volume of the denatured state increases with denaturant and ultimately reaches a plateau beyond which it does not change further (see Figure 3). Therefore for the purpose of illustration, we let $b_D(x)$ be linear in $x$ and saturate as shown in Figure 9. In this case, the measured curve $<b(x)>$ retains a sigmoidal shape but is shifted in $x$ relative to the actual transition curve, $f_D(x)$.

## Literature Cited

1. Tanford, C. 1968. *Adv. Protein Chem.* 23:121–282
2. Tanford, C. 1970. *Adv. Protein Chem.* 24:1–95
3. Pace, C. N. 1975. *Crit. Rev. Biochem.* 3:1–43
4. Rothman, J. E., Kornberg, R. D. 1986. *Nature* 322:209
5. Verner, K., Schatz, G. 1987. *EMBO J.* 6:2449
6. Deshaies, R. J., Koch, B. D., Schekman, R. 1988. *Trends Biochem. Sci.* 13:384–88
7. McLendon, G., Radany, E. 1978. *J. Biol. Chem.* 253:6335–37
8. Holzer, H., Heinrich, P. C. 1980. *Annu. Rev. Biochem.* 49:63–92
9. Parsell, D. A., Sauer, R. T. 1989. *J. Biol. Chem.* 264:7590–95
10. Zale, S. E., Klibanov, A. M. 1986. *Biochemistry* 25:5432–44
11. Lumry, R., Biltonen, R., Brandts, J. F. 1966. *Biopolymers* 4:917–44
12. Kim, P. S., Baldwin, R. L. 1990. *Annu. Rev. Biochem.* 59:631–60
13. Shortle, D. 1987. *Protein Structure, Folding, and Design*, pp. 353–61. New York: Liss
14. Dill, K. A. 1985. *Biochemistry* 24:1501
15. Flory, P. J. 1949. *J. Chem. Phys.* 17:303
16. Flory, P. J. 1953. In *Principles of Polymer Chemistry*. Ithaca, New York: Cornell Univ.
17. Chan, H. S., Dill, K. A. 1991. *Annu. Rev. Biophys. Biophys. Chem.* 20:447–90
18. Hildebrand, J. H., Scott, R. L. 1950. In *The Solubility of Nonelectrolytes*. New York: Reinhold
19. Dill, K. A., Alonso, D. O. V., Hutchinson, K. 1989. *Biochemistry* 28:5439
20. Nokazi, Y., Tanford, C. 1970. *J. Biol. Chem.* 245:1698
21. Chan, H. S., Dill, K. A. 1989. *Macromolecules* 22:4559
22. Chan, H. S., Dill, K. A. 1990. *Proc. Natl. Acad. Sci. USA* 87:6388
23. Shortle, D., Meeker, A. K. 1989. *Biochemistry* 28:936–44
24. Goto, Y., Calciano, L. J., Fink, A. L. 1990. *Proc. Natl. Acad. Sci. USA* 87:573–77
25. Gast, K., Zirwer, D., Welfle, H., Bychkova, V. E., Ptitsyn, O. B. 1986. *Int. J. Biol. Macromol.* 8:231–36
26. Kronman, M. J., Hoffman, W. B., Jeroszko, J., Sage, G. W. 1974. *Biochem. Biophys. Acta* 285:124–44
27. Kuwajima, K., Ogawa, Y., Sugai, S. 1979. *Biochemistry* 18:878–82
28. Roder, H. 1989. *Methods Enzymol.* 176:446–73
29. Pfeil, W. 1987. *Biochim. Biophys. Acta* 911:114–16
30. Privalov, P. L. 1979. *Adv. Protein Chem.* 33:167–241
31. Santoro, M. M., Bolen, D. W. 1988. *Biochemistry* 27:8063–68
32. Shortle, D., Meeker, A. K., Gerring, S. L. 1989. *Arch. Biochem. Biophys.* 272:103–13
33. Lee, J. C., Timasheff, S. N. 1974. *Biochemistry* 13:257–65
34. Schellman, J. A., 1978. *Biopolymers* 17:1305–22
35. Greene, R. F. Jr., Pace, C. N. 1974. *J. Biol. Chem.* 249:5388
36. Aune, K. C., Salahuddin, A., Zarlengo, M. H., Tanford, C. 1967. *J. Biol. Chem.* 242:4486
37. Bradbury, J. H., King, N. L. R., O'Shea, J. M. 1972. *Int. J. Pept. Prot. Res.* 4:257
38. Bradbury, J. H., Norton, R. S. 1973. *Biochim. Biophys. Acta* 10:10
39. McDonald, C. C., Phillips, W. D. 1969. *J. Am. Chem. Soc.* 91:1513

40. Miller, W. G., Goebel, C. V. 1968. *Biochemistry* 7:3925–35
41. Bierzynski, A., Baldwin, R. L. 1982. *J. Mol. Biol.* 162:173–86
42. McWherter, C. A., Haas, E., Leed, A. R., Scheraga, H. A. 1986. *Biochemistry* 25:1951–63
43. Haas, E., McWherter, C. A., Scheraga, H. A., 1988. *Biopolymers* 27:1–21
44. Amir, D., Haas, E. 1988. *Biochemistry* 27:8889–93
45. Wishart, D. 1990. PhD thesis. Univ. Alberta, Edmonton, Canada
46. Karplus, S., Snyder, G. H., Sykes, B. D. 1973. *Biochemistry* 12:1323–29
47. Alonso, D. O. V., Dill, K. A. 1991. *Biochemistry*. In press
48. Shortle, D., Meeker, A. K. 1986. *Proteins: Struct. Funct. Genet.* 1:81–89
49. Sondek, J. E., Shortle, D. 1990. *Proteins: Struct. Funct. Genet.* 7:299–305
50. Perry, K. M., Onuffer, J. J., Touchette, N. A., Herndon, C. S., Gittelman, M. S., et al. 1987. *Biochemistry* 26:2674–82
51. Villafranca, J. E., Howell, E. E., Oatley, S. J., Xuong, N. H., Kraut, J. 1987. *Biochemistry* 26:2182–89
52. Garvey, E. P., Matthews, C. R. 1989. *Biochemistry* 29:2083–2093
53. Shirley, B. A., Stanssens, P., Steyaert, J., Pace, C. N. 1989. *J. Biol. Chem.* 264:11621–25
54. Kellis, J. T. Jr., Nyberg, K., Fersht, A. R. 1989. *Biochemistry* 28:4914–22
55. Hughson, F. M., Baldwin, R. L. 1989. *Biochemistry* 28:4415–22
56. Pace, C. N., Laurents, D. V., Thomson, J. A. 1990. *Biochemistry* 29:2564–72
57. Pace, C. N., Vanderburg, K. E. 1979. *Biochemistry* 18:288–92
58. Shortle, D., Stites, W. E., Meeker, A. K. 1990. *Biochemistry* 29:8033–41
59. Flanagan, J. M., Kataoka, M., Shortle, D., Engleman, D. M. 1991. Submitted
60. Kuwajima, K. 1989. *Proteins: Struct. Funct. Genet.* 6:87–103
61. Brazhnikov, E. V., Chirgadze, Y. N., Dolgikh, D. A., Ptitsyn, O. B. 1985. *Biopolymers* 24:1899–907
62. Brems, D. N., Plaisted, S. M., Havel, H. A., Kaufman, E. W., Stodola, J. D., et al. 1985. *Biochemistry* 24:7662
63. Robson, B., Pain, R. H. 1976. *Biochem. J.* 155:331–44
64. Brandts, J. F., Hunt, L. 1967. *J. Amer. Chem. Soc.* 89:4826–38
65. Eagland, D. 1975. *Water, A Comprehensive Treatise*, Vol. 4, ed. F. Franks. New York: Plenum
66. Conio, G., Patrone, G., Brighetti, S. 1970. *J. Biol. Chem.* 245:3335

67. Nelson, J. W., Kallenbach, N. R. 1986. *Proteins* 1:211
68. Biltonen, R. L., Lumry, R., Madison, V., Parker, H. 1965. *Proc. Natl. Acad. Sci. USA* 54:1412
69. Brandts, J. F., Lumry, R. 1963. *J. Phys. Chem.* 67:1484
70. Brandts, J. F. 1964a. *J. Am. Chem. Soc.* 86:4291–302
71. Brandts, J. F., 1964b. *J. Am. Chem. Soc.* 86:4302
72. Dolgikh, D. A., Gilmanshin, R. I., Brazhnikov, E. V., Bychkova, V. E., Semisotnov, G. V., et al. 1981. See Ref. 76
73. Goto, Y., Fink, A. L. 1990. *J. Mol. Biol.* 214:803
74. Ptitsyn, O. B., 1987. *J. Prot. Chem.* 6:273–93
75. Kronman, M. J., Jeroszko, J., Sage, G. W. 1972. *Biochim. Biophys. Acta* 285:145–66
76. Dolgikh, D. A., Gilmanshin, R. I., Brazhnikov, E. V., Bychkova, V. E., Semisotnov, G. V., et al. 1981. *FEBS Lett.* 136:311–15
77. Dolgikh, D. A. et al. 1985. *Eur. Biophys. J.* 13:109–21
78. Kuwajima, K., Kiraoka, Y., Ikeguchi, M., Sugai, S. 1985. *Biochemistry* 24:874–81
79. Pfeil, W., Bychkova, V. E., Ptitsyn, O. B. 1986. *FEBS Lett.* 198;287–91
80. Baum, J., Dobson, C. M., Evans, P. A., Hanley, C. 1989. *Biochemistry* 28:7–13
81. Wong, K. P., Hamlin, L. M. 1974. *Biochemistry* 13:2678–83
82. Goto, Y., Fink, A. L. 1989. *Biochemistry* 28:945–52
83. Ohgushi, M., Wada, A. 1983. *FEBS Lett.* 164:21–24
84. Goto, Y., Calciano, L. J., Fink, A. L. 1990. *Proc. Natl. Acad. Sci. USA* 87:573–77
85. Epstein, H. F., Schecter, A. N., Chen, R. F., Anfinsen, C. B. 1971. *J. Mol. Biol.* 60:499–508
86. Shakhnovich, E. I., Finkelstein, A. V. 1989. *Biopolymers* 28:1667–80
87. Finkelstein, A. V., Shakhnovich, E. I. 1989. *Biopolymers* 28:1681–94
88. Goto, Y., Takahashi, N., Fink, A. L. 1990. *Biochemistry* 29:3480–88
89. Libertini, L. J., Small, E. W. 1985. *Biophys. J.* 47:765–72
90. Stigter, D., Alonso, D. O. V., Dill, K. A. 1991. *Proc. Natl. Acad. Sci. USA*. In press
91. Epand, R. M., Scheraga, H. A. 1968. *Biochemistry* 7:2864–72
92. Taniuchi, H., Anfinsen, C. B. 1969. *J. Biol. Chem.* 244:3864–75

93. Howard, J. C., Ali, A., Scheraga, H. A., Momany, F. A. 1975. *Macromolecules* 8:607–22
94. Brown, J. E., Klee, W. A. 1971. *Biochemistry* 10:470–76
95. Bierzynski, A., Kim, P. S., Baldwin, R. L. 1982. *Proc. Natl. Acad. Sci. USA* 79:2470–74
96. Kim, P. S., Baldwin, R. L. 1984. *Nature* 307:329–34
97. Osterhout, J. J. Jr., Baldwin, R. L., York, E. J., Steward, J. M., Dyson, H. J., Wright, P. E. 1989. *Biochemistry* 28:7059–64
98. Goodman, E. M., Kim, P. S. 1989. *Biochemistry* 28:4343–47
99. Dyson, H. J., Rance, M., Houghten, R. A., Wright, P. E., Lerner, R. A. 1988. *J. Mol. Biol.* 201:201–7
100. Waltho, J. P., Feher, V. A., Lerner, R. A., Wright, P. E. 1989. *FEBS Lett.* 250:400–4
101. Jimenez, M. A., Nieto, J. L., Herranz, J., Rico, M., Santoro, J. 1987. *FEBS Lett.* 221:320–24

102. Gooley, P. R., Mackenzie, N. E. 1988. *Biochemistry* 27:4032–40
103. Dyson, H. J., Rance, M., Houghten, R. A., Lerner, R. A., Wright, P. E. 1988. *J. Mol. Biol.* 201:161–200
104. Bundi, A., Andreatta, R. H., Wuthrich, K. 1978. *Eur. J. Biochem.* 91:201–8
105. Wright, P. E., Dyson, H. J., Lerner, R. A. 1988. *Biochemistry* 27:7167–75
106. Haas, E., Montelione, G. T., McWherter, C. A., Scheraga, H. A. 1987. *Biochemistry* 26:1672–83
107. Oas, T. G., Kim, P. S. 1988. *Nature* 336:42–48
108. Fry, D. C., Byler, D. M., Susi, H., Brown, E. M., Kuby, S. A., Mildvan, A. S. 1988. *Biochemistry* 27:3588–98
109. Kuwajima, K., Ogawa, Y., Sugai, S. 1981. *J. Biochem.* 89:759–70
110. Ginsberg, A., Carroll, W. R. 1965. *Biochemistry* 4:2159
111. Thomson, J. A., Shirley, B. A., Grimsley, R., Pace, C. N. 1989. *J. Biol. Chem.* 264:11614

Annu. Rev. Biochem. 1991. 60:827–61

# THE REGULATION OF HISTONE SYNTHESIS IN THE CELL CYCLE

*M. A. Osley*

Program in Molecular Biology, Sloan Kettering Cancer Center, New York, New York 10021

KEY WORDS:   histone RNAs: transcription, processing, turnover.

CONTENTS

## SUMMARY AND PERSPECTIVE

The recent focus on the eukaryotic cell cycle has renewed interest in genes whose expression is regulated during the division cycle. The most prominent and best-studied examples of cell cycle–regulated genes are those encoding histones, proteins that are synthesized during S phase. Multiple levels of control are involved in restricting the synthesis of histones to this phase of the cell cycle, and their combined effect is to regulate the production of histone

827

0066-4154/91/0701-0827$02.00

mRNAs. Three major pathways contribute to this regulation, and they act at the levels of transcription, pre-mRNA processing, and mRNA stability. During the past 10 years, each pathway has been subjected to detailed analysis; cis-acting regulatory elements and trans-acting regulatory factors have been identified, and in some instances, the mechanisms of regulation have begun to be defined. This review summarizes the major developments underlying the contribution of each pathway to the production of histone mRNAs in the eukaryotic cell cycle. Because significant differences are found between higher and lower eukaryotes in the regulation of histone genes, these two groups of organisms are considered separately.

Although this review emphasizes the control of histone synthesis during the cell cycle, histone synthesis is also regulated during development, and the interested reader is referred to (1–5) for discussions of this specialized aspect of regulation.

## ORGANIZATION OF HISTONE GENES—STRUCTURE OF HISTONE mRNAs

Two classes of histone genes are found in most eukaryotes. The most frequent class contains the replication-dependent histone genes whose expression is regulated in the cell cycle, while a second, less abundant class contains genes that encode the minor histone variants and whose expression occurs at a basal level throughout the cell cycle (6). In all eukaryotes the replication-dependent core *(H2A, H2B, H3, H4)* histone genes are members of a multigene family, and in higher eukaryotes the linker *(H1)* histone genes are also repeated. The core histone gene family has as few as one to two members each, as exemplified by fungi (7–14), to as many as several hundred per core histone gene in invertebrates (2, 3, 15, 16). The genomic organization of the histone genes differs widely among organisms. In invertebrates the genes are part of a reiterated tandem repeat encompassing the five histone genes (3, 16). This arrangement is not found in vertebrates or lower eukaryotes, where the genomic organization of histone genes is largely dispersive, albeit with some clustering (2, 17–23). For example, in humans the *H4* gene occurs in heterogeneous clusters, either unlinked to other histone genes or associated with as many as three other core histone genes (17, 19, 23). In budding yeast and other fungi, the genes encoding *H2A* and *H2B* are often linked and genetically separable from linked pairs of *H3* and *H4* genes (7, 8, 10, 13, 14).

A dispersive genomic organization implies that the histone genes are not regulated by far upstream enhancer elements. The regulatory sequences of these genes are in fact very compactly arranged. The 5' sequences responsible for transcriptional control are usually found within several hundred bases upstream of the mRNA initiation site, and the 3' processing signals in vertebrate histone genes are positioned close to the translation termination site (24). Moreover, vertebrate histone genes containing short 5' and 3' flanking sequences are correctly regulated following transfection into tissue culture cells (25–29), and yeast histone genes remain cell cycle regulated when placed in ectopic chromosomal locations (30).

In vertebrates, two remarkable features distinguish the replication-dependent histone genes from both the replication-independent histone genes and the majority of genes transcribed by RNA polymerase II. The first is the absence of introns (24). The second is the structure of the 3' terminus of mature histone RNAs. While most pol II transcripts are polyadenylated in vertebrate cells, the mRNAs produced by the replication-dependent histone genes do not contain poly(A) sequences and terminate with a highly conserved stem-loop structure (24, 31).

In most lower eukaryotes the histone genes are also free of introns (7, 11, 12), except for the histone genes of *Neurospora* (14), *Aspergillus* (13, 32), and *Physarum* (33). Another contrast between higher and lower eukaryotes is seen in the structure of their mature histone RNAs. The histone mRNAs of some lower eukaryotes (fungi and ciliates) do not terminate with the stem-loop structure that is characteristic of most eukaryotic histone mRNAs, but instead are polyadenylated (7, 34–36). It is intriguing that the replication-independent histone genes of vertebrates generally contain introns and are transcribed into polyadenylated mRNAs, since it is clear that neither of these features precludes the regulation of histone genes in the cell cycle of some lower eukaryotes.

## OVERVIEW OF HISTONE SYNTHESIS IN THE CELL CYCLE

The coupling of histone biosynthesis to the cell cycle was noted more than two decades ago by Robbins & Borun, who reported that histones were synthesized only in S phase cells (37). Subsequent studies showed that this pattern of histone protein synthesis is the direct consequence of the restriction of histone mRNA synthesis to the period of DNA replication (6, 38, 39). With the isolation of individual histone genes, the tools became available to study in detail the molecular basis of the coupling between histone biosynthesis and the DNA synthetic phase. What has emerged is a complicated picture of cell cycle regulation in which a combination of transcriptional and posttranscriptional controls regulates the levels of histone mRNAs.

Two general approaches have been taken to study histone mRNA metabolism in the cell cycle. In the first, synchronized cell cultures have been used to evaluate at what point histone mRNAs accumulate during an uninterrupted cell cycle. In the second, inhibitors of DNA chain elongation have been employed to test whether ongoing DNA synthesis is a prerequisite for the continued production of histone mRNAs. The major conclusions of a large body of work are that histone mRNA production is temporally restricted to the S phase of the cell cycle and dependent on the presence of replicating DNA. During the cell cycle of most eukaryotes, histone mRNAs accumulate to maximal levels only during S phase. As cells pass from G1 into S phase, the levels of these mRNAs increase 15–30-fold, with peak accumulation occurring in mid S phase (13, 18, 23, 40a–51). When DNA replication is interrupted, histone mRNAs rapidly disappear from the cytoplasm (6, 40a, 41, 52–54).

Three pathways regulate the production of histone mRNAs in the cell cycle of higher eukaryotes. The first regulates transcription at the G1-S phase boundary: as cells enter S phase, nascent histone mRNA synthesis increases three to five fold over basal levels of transcription characteristic of G1 phase cells (40a, 40b, 41–43, 52). The two remaining pathways regulate histone mRNA production at a posttranscriptional level. These latter pathways are responsible for the remaining five- to sixfold increase in histone mRNA levels that occurs during S phase and for the disappearance of histone mRNAs from the cytoplasm when chromosome replication is blocked.

The two pathways of posttranscriptional regulation affect histone mRNA levels very differently in the cell cycle. The first acts in the cytoplasm to control the half-life of histone mRNAs. The major role of this pathway is to degrade histone mRNAs when DNA synthesis is inhibited (41, 52–55). The half-life of histone mRNAs in S phase has been estimated to be 30–60 minutes by several different techniques (40a, 40b). When chromosome replication is blocked with inhibitors of DNA chain elongation, the apparent half-life of these mRNAs drops to 10–15 minutes (40a, 41, 52–54, 56). Numerous studies have asked whether this pathway might also operate during the uninterrupted cell cycle. One point at which it might act is upon entry into S phase, where histone mRNAs show maximal levels of accumulation. Here, the pathway would be used to stabilize histone mRNAs in order to accumulate these transcripts. A compelling argument that changes in histone mRNA stability can take place during S phase arises from the observation that histone mRNA half-life can be modulated both rapidly and reversibly in this phase of the cell cycle. The destabilization of histone mRNAs that results from a replication block is abrogated when the block is removed; histone mRNAs quickly reaccumulate in the cytoplasm as the consequence of a lengthened half-life (40a, 52–54, 56). These results reinforce the notion that during an unperturbed cell cycle, histone mRNAs synthesized in late G1-early S phase are stabilized when cells enter S phase. Support for this view has come from the analysis of *H3* mRNA stability during the HeLa cell cycle when transcriptional and posttranscriptional controls were uncoupled (40c). *H3* mRNA was almost three times more stable in early S phase cells than in G2-M-G1 phase cells. This proposition, however, has been challenged by the results of a study on the relative half-life of histone mRNAs in the G1 and S phases of synchronized CHO cells (40b). During both of these cell cycle stages, histone mRNAs had a half-life of 40–45 minutes, suggesting that the intrinsic stability of histone messengers may not be significantly different between unperturbed G1 and S phase cells. In contrast, the degradation pathway apparently does act at the natural completion of chromosome replication. At this point in the cell cycle, the rate of transcription is very reduced (57), and the half-life of histone mRNAs has been estimated to be only 10–20 minutes (40b, 40c).

The second pathway of posttranscriptional regulation in higher eukaryotes acts in the nucleus to process histone pre-mRNAs. After their transcription, histone pre-mRNAs are processed by endonucleolytic cleavage at their 3' termini to produce the mature cytoplasmic mRNA species (58, 59). The processing pathway has been postulated to act primarily at the G1-S phase boundary, the same period of the cell cycle in which transcription is activated. This was first demonstrated with a temperature-sensitive mouse mastocytoma mutant that could be blocked in G1 phase at the restrictive temperature. In such blocked cells, transcription of histone genes continued at almost maximal levels, but the nascent histone transcripts were processed inefficiently and few mature histone mRNAs were found in the cytoplasm (40b, 60). Processing was activated when the cells were released from the G1 block, and the subsequent entry into S phase was correlated with an almost 10-fold increase in the levels of mature histone RNAs (40b). The temporal relationship between efficient nuclear processing and the cytoplasmic accumulation of histone mRNAs suggests that the processing pathway may contribute substantially to the increase in histone mRNA levels as cells enter S phase. The observation that histone transcripts that could be correctly processed but not degraded continued to accumulate in S phase is consistent with this view (40b, 61). It is therefore possible that the regulation of histone pre-mRNA processing is the major posttranscriptional pathway utilized in higher eukaryotes upon entry into S phase, while the regulation of histone mRNA stability may be a specialized response either to an interrupted S phase or at the natural completion of S.

The discovery that three major pathways (one transcriptional and two posttranscriptional) regulate the levels of histone mRNAs in the vertebrate cell cycle does not necessarily extend to all eukaryotes. Since fungi and ciliates have polyadenylated histone mRNAs, these organisms lack the posttranscriptional pathway that regulates pre-mRNA processing. The two remaining pathways are apparently common to all eukaryotes, although in some lower eukaryotes, transcription figures more prominently than messenger turnover in regulating the levels of histone mRNAs. In budding yeast, for example, transcription of the histone genes, which is hardly detectable in early G1 phase cells, is activated 10–20-fold in late G1 phase cells (62). This degree of stimulation approximates the magnitude of accumulation of these mRNAs during S phase (46, 62). Moreover, when DNA synthesis is blocked in yeast, the major effect is not to destabilize histone mRNAs but to turn off transcription of the histone genes (63). The contribution of posttranscriptional regulation to the accumulation of histone mRNAs in the cell cycle of lower eukaryotes has been examined systematically only in budding yeast. When several different yeast histone genes were transcribed constitutively, the levels of their mRNAs showed significant fluctuations during the cell cycle, so that maximal accumulation of these mRNAs continued to occur in S phase (63–65). Since yeast histone H2B mRNA can be degraded selectively under some circumstances (66), these fluctuations are believed to represent differential effects on histone mRNA stability during the G1 and G2 phases of the cell cycle.

Among the major forms of gene regulation found in eukaryotes, only translational control does not make a significant contribution to the synthesis of histones in the cell cycle. An exception to this rule, however, is found in the lower eukaryote, *Physarum*. Histone mRNAs synthesized in G2 (the equivalent of G1 phase in other eukaryotes) are transported to the cytoplasm and stored as a translationally inactive RNP particle (67). Translation commences only when cells enter S phase (67). It is likely that proteins associated with histone mRNAs in the RNP particles mask translation, since these same mRNAs can be efficiently translated in vitro after the particles have been deproteinized (68). Similar translational regulation of histone biosynthesis hitherto has been noted only during embryogenesis (4, 69, 70).

## TRANSCRIPTIONAL REGULATION

### Higher Eukaryotes

In vertebrates, where most of the studies to be described have been performed, the histone genes are transcribed by RNA polymerase II, and the sequences that regulate transcription occur 5' to the site of transcription initiation. A number of approaches have been taken to identify discrete regulatory elements within these sequences and the proteins with which they interact. An initial approach has been to search the 5' flanking regions of histone genes to identify DNA sequences that are either common to the histone gene family or specific to a particular class of histone gene. A second has been to identify regulatory elements directly through functional analysis of upstream sequences. This analysis has been facilitated by the discovery that exogenous histone genes—either chimeric reporter genes containing histone-specific 5' sequences (25, 27–29) or heterologous histone genes (26, 71)—are correctly cell cycle regulated when they are transfected into cells in culture. With these genes it has been possible to study the effects of deletion, linker substitution, and point mutations in upstream regulatory sequences in vivo. A third avenue has employed techniques that detect specific protein-DNA contacts in vitro (72–79) and in vivo (80–82) to identify factors that interact with transcriptional regulatory elements. Finally, the development of a soluble transcription system that faithfully mimics S phase activation of histone genes in vitro (55) has permitted the assignment of critical regulatory roles to factors that interact with upstream regulatory elements.

With the possible exception of a human *H4* gene (83, 84), vertebrate histone genes do not contain enhancer-like regulatory sequences that act from a distance. Most of the regulatory elements in the promoters of these genes are located within several hundred base pairs upstream of a highly conserved mRNA capping site (24, 85). In many respects, the promoters of histone genes are not organized very differently from the promoters of other differentially regulated genes. They are generally modular in nature, and contain a number of discrete, independently functioning sequence elements that contribute in toto to transcriptional activity. Among the regulatory elements found in vertebrate histone genes are several general promoter elements common to genes transcribed by RNA polymerase II, e.g. the TATA box (15, 24, 86, 87) and CCAAT motif (15, 24, 88), elements found in a subset of genes [SP1 sites (24, 89, 90)], elements that are specific to all classes of replication-dependent histone genes, and elements that occur only in particular classes of histone genes. The histone family–specific element is a hexameric sequence (GACTTC) that is analogous to a pentameric sequence (GATCC) first noted in sea urchin histone genes (15, 24, 91). Histone gene- or subtype-specific elements (24) occur in the promoters of *H1* (92, 93), *H2A* (91, 93–95), *H2B* (93, 96), and *H4* (93, 97) genes.

Because the transcription of the five vertebrate histone genes is coordinately activated upon entry into S phase, a unifying model to account for their simultaneous regulation would be that a unique transcription factor interacts with an element common to all histone genes. The likely candidate for such a element is the histone family–specific hexamer sequence. The surprising conclusion from a large body of work is that this is an overly reductionist view of cell cycle regulation. Indeed, no known cell cycle–related function yet has been assigned to the hexamer element, which acts primarily to maintain maximal levels of transcription. Rather, each kind of vertebrate histone gene harbors a different S phase regulatory element that binds a distinct regulatory factor with specific activation functions. Thus, although each of these disparate transcription factors acts independently, they all must be regulated coordinately to elicit their S phase–specific functions.

Extensive analysis has been performed on the transcriptional regulatory sequences and proteins of four of the five replication-dependent histone genes of human, rodent, and chicken origin. Data on the regulation of each of these genes is presented below and summarized in Figure 1.

*H1*   Five distinct sequence elements have been found in the promoters of *H1* genes examined to date (Figure 1). The most proximal is a TATA box motif located approximately $-30$ bp from the mRNA cap site (24, 98, 99). A G-rich element (GGGCGGG) that is conserved across species lines and shows a good fit to the consensus SP1-binding site (G/TG/AGGCGG/TG/AG/AC/T) is also found in most *H1* genes at $-70$ bp from the cap site (24, 98, 99). The two remaining sequence elements are ubiquitous in *H1* promoters. The first, the AC box, having a consensus sequence of AAACACA, is the most distal element in *H1* promoters ($-100$ bp from the cap site) (24, 92). The second is more proximal, interposed between the G-rich element and TATA box (24, 99, 100). Although the latter element contains a general CCAAT box motif, this motif is actually part of a more extended, *H1*-specific homology box of GCACCAATCACAGCGCGC (78, 99, 100). Finally, the most distal promoter element is a copy of the histone family–specific hexamer sequence (100).

The contribution of each of these elements to maximal or S phase–activated transcription has been determined by analysis of *H1* promoter mutations in vivo following transfection of mutant genes (79, 98, 100) and in an in vitro transcription assay (100). Progressive 5' deletions of a human *H1* promoter resulted in an incremental loss in transcriptional efficiency, suggesting a positive and additive role for each element in maximal gene expression (100). This theme is reiterated in the regulation of each of the different histone genes. The only element with a demonstrated role in cell cycle activation is the *H1*-specific AC box. Deletion of the AC element from a chicken *H1* promoter or 4 bp substitution mutations in the same motif greatly reduced the overall level of *H1* transcription and specifically abolished S phase activation (79). No equivalent cell cycle function has yet been assigned to the more proximal *H1*-specific element, although this element and the TATA box motif are juxtaposed, a positional context that has been noted for the S phase–specific activation elements of other classes of histone genes (Figure 1).

The potential significance of the AC box to transcriptional regulation is underscored by the observation that a site-specific DNA-binding protein [H1-SF in chickens (74); H1TF1 in mammals (100)] interacts with this element. H1-SF DNA complexes were 12-fold more abundant in S phase extracts than in G1 phase extracts (101), supporting a role for these complexes in cell cycle regulation. However, it has not yet been demonstrated by an in vitro transcription assay that the H1-SF factor can positively regulate the synthesis of *H1* mRNA, so its precise role in S phase activation remains unknown (J. Wells, personal communication).

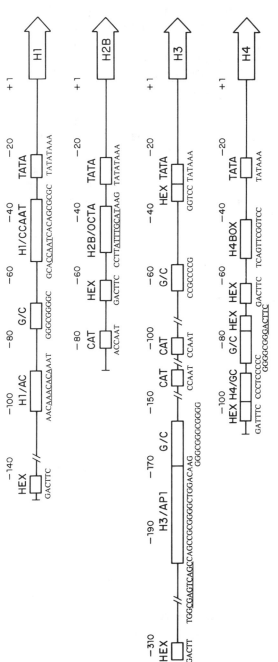

*Figure 1*  Regulatory elements in vertebrate histone gene promoters.

Regulatory elements that are characteristic of four different vertebrate histone gene promoters are indicated as boxes, with pertinent sequences indicated below each box. Underlined sequences point to specialized motifs within certain elements. Data for each gene have been adapted from the following references: *H1*: human gene (100); *H2B*: human gene (76, 138); *H3*: hamster gene (29, 108); *H4*: human gene (144, 147, 149). The coordinates are relative to the mRNA cap site.

Additional protein-DNA interactions have been described in mammalian *H1* promoters. One of the most interesting is a factor that binds specifically to the more proximal *H1*-specific element (Figure 1), an element that is required for maximal levels of transcription but has no identifiable role in cell cycle activation. This factor, called H1TF2 (100) or HiNFB (82), requires a functional CCAAT box for its binding, although it protects the more extended sequence in which the CCAAT motif is located (82, 100). HiNFB is a constitutive DNA-binding protein that interacts with the *H1*-specific CCAAT element throughout the cell cycle (78). The H1TF2 binding activity is conferred by a single polypeptide of 47 kDa (100), while the HiNFB binding activity appears to be composed of two components (78, 82). The binding of H1TF2 to DNA is correlated with a positive role in transcription, since partially pure factor activated transcription in vitro specifically through the CCAAT element (100). Numerous CCAAT box–binding factors with specific roles in the transcriptional activation of a large number of genes now have been identified (102–106). The H1TF2/HiNFB factor, by a number of physical, biochemical, and DNA-binding properties, is apparently a novel CCAAT-binding factor (78, 82, 100).

A second site-specific binding factor also has been noted in some mammalian *H1* genes. This factor, HiNFA, interacts with A/T-rich repeats present in a human *H1* promoter, and may be the same factor that forms complexes with similar repeated sequences in human *H3* and *H4* promoters (82). Its significance to transcriptional regulation is unclear, however, since these repeats have no reported role in the expression of the *H1* gene.

*H2A*    Little is known of the promoter elements or factors that regulate transcription of *H2A* genes in the cell cycle. Human and chicken *H2A* genes contain canonical TATA and CCAAT box motifs at appropriate positions upstream of the mRNA cap site [Figure 1; (24)], and a conserved, species-specific element (general consensus CACTRRYRCATTTRTG) has been noted in the promoters of sea urchin (94), frog (93), and human *H2A* genes (91). The functional significance of this sequence has been demonstrated only for the sea urchin *H2A* gene; point mutations in the element abolished transcription of the *H2A* gene in a frog oocyte assay system (95). In vertebrates, many *H2A* genes are closely linked to *H2B* genes, often forming divergent transcription units. In one chicken transcription unit, deletion or base substitution mutations in the conserved *H2B* subtype-specific element (see below) were found to reduce transcription of both the *H2A* and *H2B* genes in vivo (107). It was not reported, however, whether these same mutations also abrogated the S phase activation function associated with the *H2B* element.

*H2B*    This histone gene is a paradigm for studies on cell cycle–regulated transcription in vertebrates. S phase activation is achieved through the function of a ubiquitous and well-studied transcription factor that binds to a conserved, *H2B*-specific element. The promoter of the typical vertebrate *H2B* gene has a modular architecture, containing both canonical sequence elements (the proximal TATA box and more distal CCAAT box) and elements unique to either histone genes in general or *H2B* genes in particular (Figure 1). The general histone element is the hexamer sequence that is common to many histone gene promoters, while the *H2B*-specific element (consensus CCTTATTTGCATAAG) is found exclusively in the promoters of *H2B* genes across species lines (24, 96). As noted for the *H1* gene, each promoter element contributes a positive and incremental role to the transcription of *H2B* genes; as assayed both in vivo and in vitro, 5' deletions progressively attenuated transcriptional efficiency and base change mutations in individual sequence motifs reduced the maximal level of transcription (29, 108). The *H2B*-specific box has been identified as the element essential for S phase activation: when human 293 cells were transfected with an *H2B-CAT* gene whose transcription was driven by the *H2B* box and TATA element, the reporter gene was activated 4–6-fold when cells entered S phase; point mutations in the H2B box abolished this activation but did not alter the basal level of transcription (108). The H2B box therefore acts independently of the other promoter elements (with the exception of the TATA element, which is essential for transcription), and the latter elements are likely to be responsible for maintaining the overall levels of transcription.

The most intriguing feature of the *H2B* box is that it contains an octamer (OCTA) subsequence [ATTTGCAT (108)] found in the promoters of a number of genes whose transcription is not cell cycle–regulated. These genes include some pol II genes that are ubiquitously transcribed, such as the *U1* and *U2* snRNA genes (109, 110) as well as several genes transcribed by RNA polymerase III [the *U6* (110–112) and *7SK* RNA genes (113)]. It also is found in the SV40 enhancer (114), the HSV immediate early promoter (115, 116), and the nuclear factor III (NFIII)–binding site, which has a role in adenovirus DNA replication (117). A related octamer sequence (ATGCAAAT) also occurs in the promoters/enhancers of immunoglobulin genes, whose transcription is not ubiquitous but is restricted to cells of lymphoid origin (118, 119). The octamer sequence in the *H2B* box is absolutely required for the S phase activation function of this element. Point mutations in the OCTA motif destroyed this activity of the *H2B* box both in vivo and in vitro (29, 108). Similar studies with other genes that contain the OCTA motif have revealed the essential function of this sequence in the specific activation of these genes as well (113, 118, 120, 121).

Two proteins that interact specifically with the OCTA sequence have been purified, and the genes that encode them have been isolated. The first protein, Oct1 (Otf1/OPBF), is a ubiquitous 100 kDa protein (73, 122–124) and the second, Oct2 (Otf2), is a B cell–specific factor that may be a family of polypeptides (125–128). Both Oct1 and Oct2 bind to the octamer element in an apparently indistinguishable manner (109, 125, 129), and analysis of the two proteins has revealed that the DNA-binding domains are highly conserved at the amino acid sequence level (123, 127, 130). These domains fall within a 150–160-amino-acid-long region called the POU (Pit-OCTA-Unc) domain, which is present in a number of transcriptional activators that act in development (123, 126, 130, 131). Oct1 and Oct2 share 87% identity over the entire POU domain, which has been further subdivided into two subdomains. DNA-binding specificity is conferred by subdomain 1, which is characterized by the presence of a homeobox domain with a typical helix-turn-helix motif (126, 127, 130–132). The remaining portions of the two proteins have diverged, but both are characterized by a high concentration of glutamic acid residues in their amino termini and clusters of serine- and threonine-rich residues in their carboxy termini (123, 127, 130).

Two critical questions are raised by the observation that Oct1 and Oct2 share identical DNA-binding specificities [(131, 133, 134) for reviews]. The first is: What determines the selective activation functions of the two factors so that Oct2 activates immunoglobulin genes while Oct1 activates a larger subset of genes? The second is: How can a ubiquitous protein like Oct1 discriminate among the array of pol II- and pol III-transcribed genes with which it interacts so that when cells enter S phase it only activates the *H2B* gene? Oct1 is apparently bound to the *H2B* subtype element throughout the cell cycle (73, 101), although in cultured diploid hamster cells the octamer DNA binding activity has exhibited cell cycle–dependent fluctuations (135). If Oct1 is indeed constitutively bound to the H2B box, then it might have to be modified to exert its selective activation function. Such a mechanism is apparently used to permit Oct1 to discriminate among other non-B-cell promoters. Transcription of the HSV immediate early promoter, for example, requires both Oct1 and the viral transactivator VP16 (116, 136, 137), and it has been postulated that the interaction between VP16 and Oct1 converts Oct1 into an HSV-specific activator. Thus, the S phase activation function of Oct1 might be uncovered by protein-protein interaction between Oct1 and a VP16-like factor that is itself cell cycle regulated. This model, however, does not necessarily explain why periodic modification of Oct1 would not also lead to periodic activation of the wide array of genes regulated by this factor. It is perhaps noteworthy that the OCTA sequences in the *H2B* box are part of a larger unique motif (29, 96, 108). The flanking sequences might therefore function to recruit an *H2B*-specific transactivator to the *H2B* box. Although no other protein-DNA contacts have been reported in the *H2B* box, additional proteins may form a complex with Oct1 bound to this element; DNaseI analysis has revealed a 20 bp footprint that overlaps the *H2B* box and the juxtaposed TFIID-binding site (73).

*H3*   No gene-specific consensus element has been discerned from the sequence analysis of *H3* genes. Otherwise the 5' regions of these genes mirror the promoters of other histone genes in their modular organization (Figure 1). The hamster *H3.2* promoter, for example, contains several copies of the histone-specific hexamer element, canonical TATA and CCAAT motifs, and G+C-rich repeats that are putative SP1-binding sites (138). Progressive 5' deletions of the *H3.2* promoter gradually lowered the transcriptional efficiency of an *H3-neo* gene in vivo (138), again consistent with the view that vertebrate histone genes are positively regulated. Despite the absence of an *H3*-specific promoter element, an S phase activation site has been identified in the hamster gene. This site maps to a 32 bp region approximately 200 bp upstream from the mRNA cap site; 5' deletions into this region lowered transcriptional efficiency 10-fold and specifically eliminated S phase activation (138). The critical role of this region in cell cycle regulation recently has been confirmed by the observation that point mutations abolished its S phase activation function in vivo (A. Lee, personal communication).

Protein DNA contacts have been mapped in vivo and in vitro in the promoter sequences of both human (81) and hamster [(76) A. Lee, personal communication) *H3* genes. In the human gene, two binding sites have been identified in vivo, both of which are occupied throughout the cell cycle. Since the elements regulating transcription of the human gene have not been identified, the significance of these protein contacts to maximal gene expression or S phase activation cannot be fully evaluated. An important protein-DNA contact is found in the hamster *H3.2* promoter, however, because it occurs in the 32 bp region implicated in cell cycle control (76). This region contains a site (CTGACTCG) that has significant homology to the AP1 consensus site [C/GTGACTC/AA (139, 140)], although AP1 itself does not interact with the *H3* site (76). Instead, the two proteins that bind to this region appear to be members of the *jun* oncogene family (141, 142). Since Jun proteins regulate genes involved in cell cycle control (142, 143), the observation that they may also regulate some histone genes provides a possible mechanism to link the histone gene transcription pathway to the progression of the cell cycle.

*H4*   The *H4* genes, like the other histone genes, contain both proximal and distal promoter elements that are required for transcriptional activation (Figure 1). Deletion analysis of the human *H4a* gene has shown that the most proximal element, the TATA box motif, is essential for transcription, and that sequences important for maximal transcriptional efficiency reside in a more distal region 70–100 nucleotides upstream of the mRNA cap site (144). This region contains two copies of the histone-specific hexamer element as well as several G+C-rich, Sp1-like repeats. An *H4* subtype-specific element [TCAN$_4$GGTCC (145)] occurs just distal to the TATA motif, but unlike the analogous *H1*- and *H2B*-specific elements, it has no clearly defined role in the activation of transcription during S phase (144).

The promoter of a second human *H4* gene (FO108) has also been subjected to deletion analysis (27, 146). Many of the same sequence motifs that characterize the *H4a* promoter are also present, and they are each important for establishing maximal levels of transcription (27). The promoter of the FO108 gene may be more complex, however, since sequences farther upstream also contribute to maximal transcriptional efficiency. A putative enhancer element, the only such example reported in histone genes, has been identified 6.5 kb upstream from the mRNA cap site; its significance remains unknown (83, 84).

Despite the apparent absence of a recognizable S phase regulatory element, nuclear extracts prepared from S phase cells preferentially recognized the distal promoter region in an in vitro transcription assay (55, 144). Two significant protein-DNA contacts have been mapped in vitro in this region— one by a factor, H4TF-2 (HiNFD), that binds to the subtype-specific element, and the second by a factor, H4TF-1 (HiNFC), that binds to the more distal G+C-rich sequences (72, 77, 147). Both sets of contacts persist in vivo throughout the cell cycle (77, 80), although in nontransformed cell lines HiNFD only showed significant binding activity in extracts prepared from S phase cells (148). H4TF-2 is a single 65 kDa protein species that requires zinc for its specific binding activity (147, 149). H4TF-1 is a doublet of 105 kDa and 110 kDa proteins and is also a zinc-dependent DNA-binding protein (77, 147, 149). Although H4TF-1 and transcription factor Sp1 have similar characteristics, they apparently are distinct transcription factors (147).

The significance of these two sites of protein-DNA interaction to the transcription of the *H4* gene has been shown by the dependence of an in vitro transcription assay on the addition of purified H4TF-1 or H4TF-2. In this assay, transcription was stimulated fourfold by H4TF-1 and two- to threefold by H4TF-2 (147). Since the target of H4TF-2 is the *H4* subtype-specific element, the latter effect confirmed the role of this element in transcription of the H4 gene. Unlike the interaction of Oct1 with the *H2B*-specific element, however, it is not yet known whether this interaction is sufficient to account for S phase activation.

It is convenient to consider the activation of histone genes in the cell cycle as part of a signal transduction pathway. Many of the downstream elements in this pathway are now known for the vertebrate histone genes; cis-acting elements that regulate S phase activation have been defined and proteins that bind to these elements have been identified. These proteins are often constitutively bound to DNA throughout the cell cycle, and they function as transcriptional activators. One of the surprises of this analysis has been the discovery that although the five vertebrate histone genes are coordinately activated, each class of genes has a unique set of S phase regulatory elements and proteins. Thus, there must be more upstream steps in the transcription pathway that transduce a common signal when cells reach the G1-S phase boundary.

## Lower Eukaryotes

In contrast to vertebrates, some lower eukaryotes utilize a common promoter element to regulate the transcription of histone genes in the cell cycle. In both budding and fission yeast, short, homologous repeats have been noted in the 5' sequences that precede each of the histone genes (12, 30, 150). The significance of these repeats to the regulation of transcription has been analyzed most extensively in budding yeast, *Saccharomyces cerevisiae*. The histone genes in this organism have a promoter structure that is typical of the majority of yeast genes transcribed by RNA polymerase II (Figure 2). The proximal promoter sequences contain consensus TATA box motifs, and the distal promoter sequences, which occur several hundred base pairs from the start site of transcription, contain elements that selectively regulate transcription. Two or three copies of a conserved 16 bp sequence (consensus GCGAAAAANTNNGAAC) are found within the distal promoter regions of each of the four histone gene loci. Deletion and promoter substitution analyses performed in vivo with histone-*lacZ* reporter genes derived from either the *HTA1-HTB1* [encoding H2A-1 and H2B-1 (8, 11)] or *HHT2-HHF2* [encoding H3-2 and H4-2 (10, 11)] locus have identified this sequence as an upstream activation (UAS) element [(30); M. Smith, personal communication]. This element has a S phase–specific function as well because three copies of the 16 bp repeat can activate the transcription of the normally constitutive *CYC1* gene at the G1-S phase boundary (30). It therefore is likely that the 16 bp UAS element is responsible for the coordinated activation of all four histone gene loci in late G1.

A number of other genes are also transcribed during the G1 phase of the yeast cell cycle [(150, 151) for reviews]. Among these genes are those encoding products involved in DNA metabolism *(CDC8, CDC9, CDC17, CDC21, RNR1, RAD6)* or mating type switching *(HO)* (150–160). Two different activation elements, each distinct from the consensus histone UAS, have been implicated in the periodic transcription of these genes (150, 151). The *HO* element has the sequence $CACGA_4$ (152), and the sequence TPuACGCGTN(T/A) (Mlu1 motif) has been correlated with cell cycle activation in the remaining periodic genes (151, 159a, 159b). The existence of

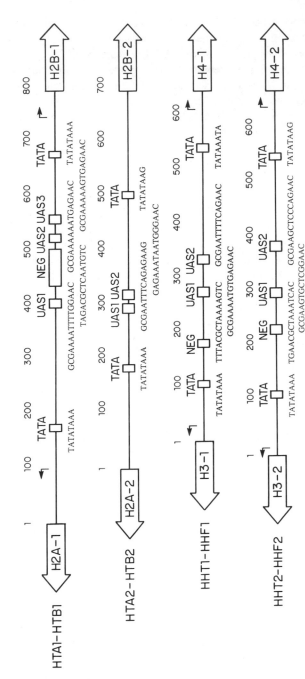

*Figure 2*  Regulatory elements in *Saccharomyces cerevisiae* histone gene promoters.

Regulatory elements present in the promoters of the four divergently transcribed histone gene loci are indicated as boxes, with relevant sequences shown below each box. In the *HTA1-HTB1* promoter, UASI, UAS3, and the 15 bp negative site consensus sequence are present on the *H2A-1* coding strand and UAS2 is present on the *H2B-1* coding strand. In the *HTA2-HTB2* promoter, UAS1 is present on the *H2A-2* coding strand and UAS2 is present on the *H2B-2* coding strand. In both the *HHT1-HHF1* and *HHT2-HHF2* promoters, UAS1 is present on the *H3* coding strand and UAS2 and the negative site consensus sequence are present on the *H4* coding strand. The coordinates are relative to the appropriate ATG codon in each locus. Small arrows indicate the start sites of transcription, where they have been precisely mapped (M. A. Osley, unpublished, and Ref. 47).

multiple, G1-specific activation elements may partly reflect when the three groups of genes are transcribed in G1. The histone genes are the last to be transcribed in this phase of the cell cycle and are not expressed until the *CDC4*-dependent step has been completed (see below); in contrast, the other periodic genes are transcribed early in G1, independently of the *CDC4* step (150, 151, 153).

The *HTA1-HTB1* locus in budding yeast contains a second regulatory site that also affects periodic transcription (Figure 2). This site negatively regulates transcription, and it constitutes the only known example of this kind of control in cell cycle–regulated transcription. The site first was recognized by deletions in the *HTA1-HTB1* promoter that resulted in high-level, constitutive transcription of histone-*lacZ* reporter genes in vivo (30). It was additionally noted that an isolated negative site could bring the constitutive *CYC1* UAS under cell cycle control, although the heterologous gene was activated almost 10–20 minutes earlier in G1 than the histone genes. The negative site has been localized to a 67 bp region in the *HTA1-HTB1* promoter that is characterized by several sequence motifs, including direct and inverted repeats, and it contains a 15 bp sequence (consensus TNNACGCTNAANGNC) also found in the *HHT1-HHF1* and *HHT2-HHF2* promoters, but not in the *HTA2-HTB2* promoter (150). It is not known whether this sequence acts as an independent regulatory element, although deletions that remove it lead to derepressed, constitutive transcription (30). Because isolated histone UAS elements are sufficient by themselves to activate transcription periodically, the negative site apparently provides a redundant function in cell cycle–regulated transcription. A possible role for this site in periodic transcription is discussed below.

A genetic approach has been taken to study the factors that regulate transcription of the *S. cerevisiae* histone genes. This approach has the potential to identify both upstream and downstream steps in the cell cycle transduction pathway. It has been known for some time that two *CDC* (cell division cycle) genes (161) required for the G1 to S phase transition also affect the periodic transcription of the histone genes: a functional *CDC4* gene product is required to turn on transcription (153) and the *CDC7* gene product is required to turn off transcription (62). These gene products might affect transcription directly by modifying or interacting with regulatory factors such as site-specific DNA binding proteins. Thus, it is interesting that the *CDC7* gene product encodes a protein kinase (162). Alternatively, the *CDC* genes could be more upstream regulators in a signal transduction cascade that affects both cell cycle progression and histone gene transcription. A further link between transcription and a cell cycle regulatory step has recently been shown by the ability of alpha factor [which arrests cells at the G1 control point, "Start" (161)] to inhibit the transcription of histone genes when cells exit from stationary phase (163). Normally, these genes cannot be activated until "Start" has been completed.

Genes that regulate transcription more directly have been sought using histone-*lacZ* (65) or histone-*neo* (M. Grunstein, personal communication) reporter genes to screen for mutants that constitutively transcribe the histone genes. Two classes of mutants (*hir* for **hi**stone **r**egulation) have been isolated by this approach. One class contains a single dominant mutant (*HIR9⁻*) in which the regulation of only the *HTA1-HTB1* locus is affected (M. Spector and M. A. Osley, unpublished observations). The phenotype of this mutant suggests that it may contain an altered product that is normally involved in activating transcription; however, the exact nature of its defect is still not known. The second, larger class contains recessive mutants that are derepressed for transcription because of specific effects on genes whose products act through the negative regulatory site (65). In these *hir* mutant strains, transcription of each histone locus except the *HTA2-HTB2* locus (which does not contain a negative site) is derepressed throughout the cell cycle. The mutations do not entirely abolish cell cycle control, however, since they do not affect the ability of isolated UAS elements to activate transcription periodically.

Other data have shown that in addition to their effects on cell cycle–regulated transcription, the products of the *HIR* genes are involved in the autoregulated transcription of the *HTA1-HTB1* locus by histones H2A and H2B. This phenomenon was first identified as a response to an altered copy number of the genes encoding H2A and H2B (164); transcription of an *HTA1-lacZ* reporter gene was repressed 10–20-fold when the copy number of *HTA* and *HTB* genes was increased and derepressed 4–14-fold when the copy number of these same genes was reduced by one half. Three factors are required for this regulation. The first two are the *HIR* gene products and the negative site through which these products act; *hir* mutant strains or an *HTA1-HTB1* locus from which the negative site had been deleted showed no regulation by *HTA* and *HTB* copy number. The third is the intracellular signal that transduces the effect of altered histone gene copy number. This signal is apparently the histone proteins themselves; a frameshift mutation in *HTB1* eliminated the ability of the *HTA1-HTB1* locus to repress transcription when it was present in high copy number. In addition, H2A and H2B together form the regulatory signal. Individual *HTA* or *HTB* genes failed to repress transcription when overexpressed; these genes had to be present in the same cell to act as a functional regulatory signal.

The major role of autoregulation may be to maintain the balanced synthesis of histones, since the unbalanced production of these proteins has been shown to affect processes as diverse as chromosome transmission (165) and transcription (166). The system of autoregulation may be used during the cell cycle as well, thereby providing a specific role for the apparently redundant function of negative regulation in periodic transcription. Two observations have suggested that histones may act as negative regulators of their own genes in the cell cycle. Strains with reduced levels of H2A and H2B showed derepressed transcription of the *HTA1-HTB1* locus during S phase (L. Moran and M. A. Osley, unpublished observations) and failed to turn off transcription when chromosome replication was blocked with hydroxyurea (164).

Additional genes affecting the transcription of the histone loci have been identified in unrelated screens for mutants that show suppression of Ty insertion mutations in the *HIS4* and *LYS2* loci (167). Recessive mutations in three *SPT* (suppressor of Ty) genes, *SPT1* (167), *SPT10* (168), and *SPT21* (G. Natsoulis and J. Boeke, personal communication), alter the regulation of histone genes; in these strains, transcription of the *HTA1-HTB1* locus continues in the presence of a replication block (P. Sherwood and M. A. Osley, unpublished observations). The s*pt1* and *hir2* mutations are allelic; *SPT10* and *SPT21* are distinct from the remaining *HIR* genes even though mutations in all of the *HIR* genes confer a Spt⁻ phenotype (P. Sherwood, M. Spector, and M. A. Osley, unpublished observations). Mutations in *SPT10* and *SPT21*, unlike mutations in other *HIR* genes, reduce the levels of mRNAs transcribed from the *HTA2-HTB2* locus by almost 10-fold (F. Winston, G. Natsoulis, and J. Boeke, personal communication). It is not known whether the *SPT10* or *SPT21* products might therefore act through the consensus UAS elements in this locus (Figure 2). Because of the locus-specific effects of mutations in the two genes, the *SPT10* and *SPT21* products would have to interact selectively with a factor(s) that recognizes the conserved UAS element or act through an as-yet-unidentified promoter element that is unique to the *HTA2-HTB2* locus.

Thus, although the yeast histone genes have comparatively simple promoters, a large array of genes affects their transcription, and it is likely that their regulation will be complex. It will be important to identify the products of these regulatory genes and to determine where they act in the cell cycle transduction pathway. To initiate such an analysis, several *HIR* genes have now been isolated by complementation of the Spt⁻ phenotype that is conferred by mutations in these genes (P. Sherwood, M. Spector, and M. A. Osley, unpublished observations).

A variety of approaches is used by other lower eukaryotes to regulate the transcription of histone genes. Fission yeast, *Schizosaccharomyces pombe,* like budding yeast, harbors a short consensus sequence (ATCAC/AAACCCTAACCCT) upstream of each of its histone genes (12). The accumulation of histone mRNAs in the cell cycle of fission yeast is in many respects similar to the pattern observed in budding yeast: histone mRNAs accumulate in late G1 in the absence of DNA replication (48). It is therefore likely that the conserved element plays an analogous role to the unrelated budding yeast UAS to activate transcription at the G1-S phase boundary. This has not yet been directly tested, however, since histone mRNA accumulation, not transcription, has been measured during the cell cycle of this organism. In contrast to the yeasts, other fungi such as *Aspergillis* and *Neurospora* have no identifiable 5' sequence motifs common to their histone genes (13, 14), and an element associated with S phase activation has not been defined. In the ciliated protozoan, *Tetrahymena,* a putative regulatory element (TATCCAATCARA) has been noted in the promoters of the *H3* and *H4* genes, but this element is not unique to the histone genes and precedes several other genes in this organism (169). It may instead be a general regulatory element since it contains a motif similar to the CCAAT box motif of vertebrate cells.

In summary, eukaryotes use widely different strategies to regulate the transcription of histone genes in the cell cycle. One extreme is exemplified by an organism like *Aspergillus,* which has no identifiable histone family– or histone gene–specific regulatory elements. In such an organism, general transcription factors may be utilized differentially to control periodic transcription. An organism like budding yeast uses a second strategy. Its histone genes have promoters that contain two kinds of elements with specific roles in cell cycle–regulated transcription. One is an activation element, present in the promoters of all the histone genes, which is likely to bind a common transcription factor. The second is a negative element, found only in the promoters of some histone genes, that may regulate transcription through feedback repression by the histones themselves. A third strategy is provided by vertebrates, whose histone genes also contain specific elements that activate transcription in the cell cycle. Unlike budding yeast, however, the cell cycle regulatory elements in these eukaryotes show no sequence conservation, and each class of histone gene contains a different S phase activation element that binds a distinct transcription factor. This latter strategy is surprising considering the evolution of histones. These proteins are among the most highly conserved in nature, probably reflecting their fundamental roles in nucleosome structure and function. Yet, the sequences that regulate transcription clearly have diverged among the five kinds of vertebrate histone genes. Why should constraints have been placed on the evolution of coding sequences but not regulatory sequences? It is now clear that unless the synthesis of the four core histones is balanced in vivo, the cell will suffer deleterious effects (165). The diversity of regulatory elements in vertebrate histone promoters may therefore reflect a way to maintain balanced histone synthesis in a large variety of cell types if it is assumed that the abundance or activity of particular regulatory factors varies among tissues (170, 171).

# POSTTRANSCRIPTIONAL REGULATION

## *Higher Eukaryotes*

Following transcription of the histone genes at the G1-S phase boundary, the histone RNAs of vertebrates are processed to their mature form, transported to the cytoplasm and translated throughout S, and then degraded at the natural completion of S phase. Several of these steps—nuclear pre-mRNA processing and cytoplasmic mRNA turnover—are tightly regulated in an apparently evolutionarily conserved process [(172–174) for reviews].

HISTONE RNA PROCESSING    Mature histone mRNAs are formed in the nucleus by the endonucleolytic cleavage of a longer precursor RNA (58, 59) to yield transcripts whose 3' termini contain a highly conserved stem-loop structure (GGCU/CCUUUUCAGA/GGCC) followed by the sequence ACCA (24, 175). Two types of assay systems have been utilized to reveal many of the details of how histone pre-mRNAs are processed: a frog oocyte system (176, 177) that can process either injected cloned DNAs or RNAs synthesized in vitro, and a soluble processing extract derived from HeLa cell nuclei (59, 178).

The sequences that are required for processing have been well studied (Figure 3). The pre-mRNA processing signal is bipartite, and includes both the conserved stem-loop structure and a purine-rich element lying 13–17 nucleotides downstream of the 3' end of mature histone mRNAs (15, 24, 172, 175). In mammals, the sequence of the second element can be quite flexible (PuAAAGACU vertebrate consensus), but its position relative to the stem-loop structure is fixed (24). In contrast, in sea urchins, where it was first noted, the second element is strictly conserved in both sequence (CAAGAAAGA) and position (15, 24, 175, 177). Mutational analysis has demonstrated that both sequence motifs are important for histone pre-mRNA processing. In mammals, the downstream purine-rich element is absolutely essential for processing and the stem-loop structure contributes to the overall efficiency of this event (179), while in sea urchins, both elements are indispensable (176, 177, 180, 181). The maintenance of a discrete distance between the two processing signals is also a critical parameter in processing (180).

Although an endonuclease that cleaves histone mRNA precursors has not been identified, at least three distinct factors have roles in processing. Two of the factors have been shown to interact directly with the conserved processing elements. The best studied is U7-snRNP, a rare member of the group of cellular U-snRNPs, many of which have established roles in the splicing of introns from pre-mRNAs [(182, 183) for reviews] and the processing of ribosomal RNA (183, 184) for reviews]. U7-snRNP was first identified in sea

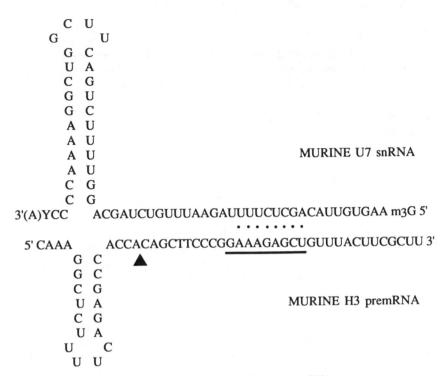

*Figure 3* Sequences involved in the processing of histone pre-mRNAs.

The 3' terminus of a mouse *H3* pre-mRNA is shown to indicate the two conserved processing elements (sequence adapted from Ref. 179). The proximal palindromic sequence is shown as a stem-loop structure and the downstream element is underlined. The arrowhead indicates the position of endonucleolytic cleavage. The interaction between murine U7 snRNA (sequence adapted from Ref. 189) and *H3* pre-mRNA during processing is shown to indicate potential base pair contacts between 5' sequences of the U7 RNA moiety and the downstream element in *H3* pre-mRNA (dots).

urchins as a factor that promoted efficient 3' processing of exogenous histone RNAs in frog oocytes (185, 186). It was subsequently found to be precipitated by human Sm antisera, which recognize a family of U-snRNPs (178, 186, 187). The U7 RNA moiety has been isolated from sea urchins (188) and mammals (178, 189, 190), and its nucleotide sequence has been determined from corresponding cDNAs. At its 5' end U7 RNA contains a region of extensive sequence complementarity to the purine-rich processing element (31, 178, 189, 190), and two lines of evidence have suggested that U7 interacts with this most downstream processing element through specific base-pairing contacts (Figure 3). First, oligonucleotides complementary to the 5' sequences of U7 RNA inhibited processing in vitro (189, 190). Second, a mutation in the downstream element that led to defective processing in the frog oocyte system could be abrogated by coinjection of U7 RNAs that contained a compensatory mutation to restore base pairing (191).

A second factor, the hairpin-binding factor (HBF), was identified in nuclear extracts that were active in processing as a factor that bound specifically to the highly conserved stem-loop element (178, 192). The hairpin-binding factor is also precipitable with anti-Sm antibodies (178, 187, 192); however, it can still be detected in fractionated extracts that lack U7-snRNPs (192). Thus, the hairpin-binding factor is distinct from U7-snRNP, although it could represent a protein component of this snRNP.

The third component in the processing reaction is a heat-labile factor, first identified in HeLa cell nuclear extracts proficient for processing; when these extracts were subjected to a mild heat treatment, processing was abolished (193). Since processing proficiency could be restored to heat-treated extracts by the addition of extracts depleted of U7-snRNP and HBF, the heat-labile activity must be distinct from either of these two factors (193). Unlike HBF or U7-snRNA, the heat-labile factor does not interact directly with either of the two conserved processing elements (193). This factor might therefore regulate processing or possibly be the endonuclease itself.

A number of studies suggest that histone pre-mRNA processing is a regulated event in the cell cycle, specifically in G1 phase. The first indication that processing may be under cell cycle control came from the examination of histone mRNA synthesis in 21-TB cells, a temperature-sensitive mouse mastocytoma cell line that can be blocked at an undefined point in G1 at the restrictive temperature (60, 194). In G1-arrested cells, two effects were seen on the expression of an endogenous *H4* gene: its transcription was repressed threefold, but the levels of its mature mRNA were reduced approximately 200-fold by posttranscriptional regulation (60). As few as 80 bp of 3' terminal sequences from the *H4* gene were sufficient to confer posttranscriptional regulation in the mutant cell line (194). Since these terminal sequences contained the two conserved processing signals in *H4* pre-mRNA, the results suggested that the failure to accumulate histone mRNAs in the cytoplasm was the consequence of a failure to process primary transcripts in the nucleus. However, the same stem-loop structure required for nuclear processing is also present in mature histone mRNAs, where it serves as a cytoplasmic signal for degradation (see below). Thus, the possibility that decreased cytoplasmic stability of *H4* mRNAs contributed to the reduced levels of *H4* transcripts could not be eliminated. Nonetheless, further evidence supports the view that histone pre-mRNA processing is regulated in G1 phase cells. First, nuclear precursors of *H4* mRNAs, which are undetectable in exponentially dividing cells, accumulate in G1-arrested cells (195). Second, a constitutively synthesized histone RNA that could be correctly processed but was insensitive to half-life regulation was still subject to posttranscriptional regulation in G1 cells (61). Finally, the most compelling evidence comes from the observation that the heat-labile processing factor is deficient in cells arrested in either G1 or G0, two states where processing is abolished (61, 195). In these cells, neither U7-snRNP nor HBF was limiting (61).

The mechanism by which processing is regulated in the cell cycle is unknown, although U7-snRNP may be the primary target of regulation. One study has shown that in S phase cells, the 5' sequences of U7 RNA were freely accessible to base pair with complementary sequences in the purine-rich processing element, but in serum-starved, G0 arrested cells, occlusion of these same U7 sequences inhibited their base pairing with the distal processing signal (196). As cells enter S phase, processing might therefore be regulated through changes in the accessibility of U7 RNA 5' sequences. A second study suggests that the availability of U7 RNA to base pair with the downstream element is unaltered throughout the cell cycle; instead, the electrophoretic mobility of U7-snRNP complexes appears to change when cells are arrested in G1 (D. Schumperli, personal communication). Regardless of the nature of the U7 modification in G1 cells, it is significant that the heat-labile factor is also absent from these same cells. This factor might therefore be required to modify and thereby activate the essential U7-snRNP processing factor as cells enter S phase. Thus, the regulation of processing in the cell cycle would ultimately depend on the availability or activity of the heat-labile factor.

HISTONE mRNA TURNOVER     Once they are processed, mature histone mRNAs are transported from the nucleus to the cytoplasm where their stability is differentially regulated. Differential turnover is seen most dramatically when S phase is interrupted; the half-lives of histone mRNAs in such cells drop from 30–60 minutes to 10–15 minutes. Histone mRNA stability is not obligatorily linked to DNA replication, however, because the degradation of these mRNAs can be uncoupled from a replication block by the simultaneous inhibition of protein synthesis. Under these conditions, histone mRNAs are stabilized selectively compared to bulk cellular mRNAs (41, 52–54, 56, 197).

Chimeric mRNAs containing histone gene–specific 5' or 3' sequences have been used to localize the sequences in histone mRNA required for its destabilization in vivo. Only mRNAs that contained histone 3' sequences were degraded following a replication block (198–202), although in one exceptional case, histone 5' sequences were also reported to confer instability (203). The structural features of the 3' sequences have been studied in detail. The major determinant is the highly conserved stem-loop structure described above that is present at the terminus of all replication-dependent histone mRNAs. When this structure was mutated and replaced with sequences containing polyadenylation signals, histone mRNAs remained stable when DNA replication was inhibited (200, 204). Moreover, the histone mRNA terminus is sufficient to confer instability; a normally stable globin mRNA whose polyadenylation site was deleted showed replication-dependent instability when it ended with only 30 nucleotides of histone 3' sequences (including the stem-loop structure) (202). The position of the stem-loop structure is, however, a critical factor in its role in mRNA turnover. This structure must be at the extreme terminus of histone mRNA to act as an instability element. When a strong polyadenylation signal was placed downstream of the stem-loop sequence, histone mRNAs became polyadenylated and remained stable when DNA synthesis was inhibited (200, 202, 205, 206). The observation that several replication-dependent histone genes contain cryptic polyadenylation signals downstream of the stem-loop structure (204, 207) may therefore provide a way to uncouple the production of some histone mRNAs from chromosome replication.

Additional requirements must be met to regulate histone mRNA degradation in vivo. First, histone mRNAs must be translated (199). This effect does not require the presence of specific coding sequences since a histone gene that could encode only a short peptide was wild-type for degradation (198, 199). Second, the translation termination codon must be positioned at a fixed distance (<300 nucleotides) from the terminal stem-loop structure (199); histone mRNAs were stabilized when translation was prematurely terminated by the introduction of stop codons (199) or if translation was allowed to continue into 3' sequences that are not normally translated (198).

Two other factors aside from the structure of histone mRNAs are involved in the degradation pathway. One of these is the activity that degrades histone mRNAs. Following the inhibition of DNA replication in vivo, these mRNAs are apparently degraded by a 3' exonuclease that initiates at the stem-loop structure (208). The first step in this reaction is the cleavage of 5–15 nucleotides from the 3' terminus, followed by the degradation of the body of the mRNA in a 3' to 5' direction (208). Although an exonuclease that acts on histone mRNAs in vivo has not been identified, an exonucleolytic activity that degrades these mRNA with the same polarity in vitro has been identified in a crude mRNA decay system (209). This exonuclease is loosely associated with ribosomes and requires divalent cations for its activity (210). It is a nonspecific nuclease, however, because it degrades a number of other polysome-associated mRNAs aside from histone mRNAs, including mRNAs that are polyadenylated (209, 211). Nonetheless, it shows fidelity because the relative turnover rates of most mRNAs in the in vitro system accurately reflect their stabilities in vivo (209). This fidelity also extends to the histone mRNAs; polyadenylated histone transcripts were 10-fold more stable in vitro (mimicking the in vivo stability of such transcripts), while unadenylated histone mRNAs were rapidly turned over (211). There is still no direct evidence that this nuclease is the same activity that acts on histone mRNAs in vivo. In order for it to be that activity, it must acquire specificity for histone mRNAs when DNA replication is blocked; this has not yet been demonstrated for the in vitro exonuclease (209).

The second and most elusive factor in histone mRNA destabilization is the intracellular signal that activates the degradation pathway. Two possibilities have been suggested. One is that the regulatory signal is related to deoxynucleotide metabolism, in particular to the levels of guanosine nucleotide in the cell (174). This is prompted by the observation that most inhibitors of DNA replication alter the pools of this nucleotide (53). A second is that histones themselves constitute the intracellular signal by autogenously regulating the turnover of their mRNAs (212, 213). This is consistent with earlier data showing that cytoplasmic pools of histones rose transiently when chromosome replication was interrupted (214). It has some additional experimental support from studies performed in vitro with the crude mRNA decay system; the addition of exogenous histones to factors present in a S100 supernatant fraction preferentially stimulated histone mRNA turnover (212). The appeal of this proposal is that it provides a way for the regulatory signal to exhibit selectivity.

Two general models have been advanced to explain how histone mRNAs might be specifically degraded following the inhibition of chromosome replication (174). In the first model, histone mRNA degradation is postulated to be coupled directly to the translation of these mRNAs (199). It is suggested that a ribosome-associated nuclease, which is activated by a replication block, recognizes the 3' terminus of histone mRNA after the termination of translation, and degradation then ensues. By coupling degradation to translation, this model offers an explanation for the stabilization of histone mRNAs when DNA replication and protein synthesis are simultaneously inhibited; during the inhibition of protein synthesis, histone mRNAs would be released from polysomes and therefore inaccessible to the nuclease.

In the second model, translation of histone mRNA per se is not a prerequisite for degradation, and histones themselves regulate histone mRNA turnover by an autoregulatory feedback mechanism (212, 213). Thus, during the inhibition of DNA replication, a rise in free histone pools (a possible consequence of the failure to assemble nucleosomes at stalled replication forks) might trigger the decay of histone mRNAs by activating a nuclease; simultaneous inhibition of protein synthesis would prevent the accumulation of histones, and histone mRNAs would therefore be stabilized.

In both models, the implicit target of the intracellular activating signal is the nuclease that degrades histone mRNAs. This nuclease could be specific for histone mRNAs, and exist in a latent state until activated by the inhibition of DNA replication. Alternatively, it could be a nonspecific nuclease like the one identified in the in vitro mRNA decay system that is loosely associated with polysomes. This latter nuclease would then have to acquire specificity towards histone mRNAs following a replication block. A second target, however, might be the stem-loop structure at the histone mRNA 3' terminus, which is the primary substrate of a nuclease. It is intriguing that this structure is required both for the degradation of mature histone mRNAs in the cytoplasm and as part of the bipartite signal for histone pre-mRNA processing in the nucleus. If the hairpin-binding factor or a similar factor binds to the 3' terminus of cytoplasmic histone mRNAs, it might protect the terminus from exonucleolytic attack. Displacement of the factor upon the inhibition of DNA replication could therefore make the terminus susceptible to a nuclease. If, for example, histones were the intracellular regulatory signal, they might displace the hairpin-binding factor by competing with it for association with histone mRNAs (212).

Does regulation of histone mRNA stability play any role in the uninterrupted vertebrate cell cycle? Histone mRNAs rapidly disappear from the cytoplasm at the termination of S phase, consistent with the notion that these mRNAs are degraded in G2-M phase (43). In this phase of the cell cycle, the half-life of histone mRNAs has been found to be under 20 minutes, compared to a 40–60-minute half-life in G1-S phase (40b, 40c). The requirements for this turnover are unknown, although the degradation systems activated upon the inhibition of DNA synthesis and at the natural completion of chromosome replication might be identical.

## Lower Eukaryotes

Any consideration of how posttranscriptional regulation affects histone synthesis in lower eukaryotes must take into account the observation that the 3' termini of histone mRNAs in many of these organisms (e.g. fungi and ciliates) do not end in the stem-loop structure but are polyadenylated. Since most of the studies relating to the posttranscriptional regulation of polyadenylated histone mRNAs have been conducted with budding yeast, *S. cerevisiae,* the discussion is limited to this organism.

Studies performed in vitro have shown that the 3' terminus of yeast mRNAs, including those of histone mRNAs, is formed from the endonucleolytic cleavage of a longer precursor molecule to which poly(A) is subsequently added (215). Unlike the formation of polyadenylated transcripts in higher eukaryotes, poly(A) addition in yeast is not associated with the presence of a conserved AAUAAA sequence (216–219). A loose consensus sequence that might be associated with cleavage has been noted in the 3' untranslated region of many different yeast genes (220). The histone genes also contain this sequence (11), but there are no data to suggest that histone mRNAs are differentially regulated at the level of 3' end processing. Thus, it is most likely that the major form of posttranscriptional regulation acting on histone RNAs occurs at the level of mRNA turnover.

Two different kinds of studies have indicated that the turnover of yeast histone mRNAs can be differentially regulated. The most direct example comes from the results of gene dosage experiments. When the copy number of *HTA1-HTB1* genes (Figure 2) was doubled, the half-life of *HTB1* mRNA decreased by one half (66). In contrast, the amount of fusion mRNA produced by multiple copies of an *HTA1-lacZ* fusion gene that contained only histone 5' sequences, including the first 13 amino acids of coding sequence, was commensurate with gene dosage (D. Lycan and M. A. Osley, unpublished observations). These data point to a requirement for 3' coding or untranslated sequences in regulating histone mRNA turnover in response to gene dosage alterations. They also suggest that histone mRNA degradation might be autoregulated, although there is as yet no direct evidence to support this idea.

A second example is provided by the analysis of histone mRNA accumulation in the cell cycle when the contribution of transcriptional control was eliminated. When full-length *HTA* or *HTB* genes were constitutively transcribed, significant, periodic fluctuations were seen in the levels of their mRNAs during the cell cycle, consistent with the notion that the histone mRNAs could be degraded in G1 and G2 (63, 65). The 3' region of histone mRNA is again implicated in turnover; deletion of 3' coding and untranslated sequences from these same genes resulted in the continuous accumulation of the histone mRNAs (63). This conclusion is supported by the observation that 3' sequences from the *HTB1* gene could confer cell cycle regulation to a chimeric *GAL10-neo* gene (64). In this study, 17 amino acids of *HTB1* 3' coding sequences plus the entire *HTB1* 3' untranslated region were sufficient to cause periodic accumulation of the heterologous mRNAs.

If an intact 3' terminus is a prerequisite for histone mRNA turnover, the polyadenylation of these transcripts might be a factor in their degradation. The role of polyadenylation in messenger stability is controversial [(221–223) for reviews]. In higher cells, poly(A) tract shortening appears to precede mRNA degradation. If the length of the poly(A) tract is related to the stability of yeast histone mRNAs, one potential target of regulation might be the conserved poly(A)-binding protein (PAB) that is bound to these tracts (224, 225). The addition of exogenous PAB to an in vitro mRNA decay system derived from HeLa cells stabilized a number of different polyadenylated mRNAs of vertebrates, presumably by protecting the poly(A) tracts from nucleases (226). Thus, it is possible that the first step in yeast histone mRNA degradation might be the removal of PAB from these mRNAs in order to expose their poly(A) sequences to exonucleolytic attack. A compelling argument against this interpretation is the observation that yeast strains deleted of

PAB showed longer poly(A) tracts on the bulk of their mRNAs (227). Further evidence that polyadenylation per se may not be directly responsible for the stability of histone mRNAs comes from an analysis of the parameters involved in yeast mRNA stability in vivo; poly(A) metabolism is not significantly different between stable and unstable mRNAs (228). It is now becoming clear that additional features of eukaryotic mRNAs (so called instability elements), which may be part of coding sequences, also contribute to the stability of a particular mRNA species (221, 223, 228, 229). Thus, an as-yet-undefined instability element present either in 3' coding or untranslated sequences might be responsible for regulating yeast histone mRNA turnover.

The observation that yeast histone gene expression is predominantly regulated at the level of transcription (63) probably reflects the fact that histone mRNAs in this organism have half-lives of only 5–15 minutes (66, 228). Thus, transcriptional control may be sufficient to provide rapid adjustments in histone mRNA levels, e.g. to decrease the levels of histone mRNAs in response to a block in DNA replication. What role then does posttranscriptional regulation, specifically mRNA turnover, play in the biosynthesis of histones in yeast? From the examples described above, it has at least two roles. One is apparently to adjust the intracellular levels of histone mRNAs as part of a dosage compensation mechanism. The second is the potential to regulate the levels of histone mRNAs in the cell cycle, which was only noted when transcription was constitutive. Under such conditions, it is estimated that the extent of posttranscriptional regulation in the yeast cell cycle ranges from 3- (63) to 12-fold (64) effects on the levels of histone mRNAs. Nothing is known about the signal that regulates turnover in the cell cycle. One intriguing observation, however, has shown that posttranscriptional regulation may operate on a clock that is independent of, but in synchrony with, the cell cycle (63). When a synchronized cell population was blocked in S, and histone mRNA synthesis was altered to be constitutive, the histone transcripts continued to show the same periodic fluctuations as they did in unperturbed cells.

## MULTIPLE FORMS OF REGULATION MODULATE HISTONE mRNA LEVELS IN THE CELL CYCLE

Why should so many different levels of regulation—transcriptional, pre-mRNA processing, and mRNA degradation—be used to control the production of histones in the eukaryotic cell cycle? In mitotically dividing somatic cells little or no free pools of histones are observed because newly synthesized histones are rapidly assembled into chromatin at replication forks (230). Replicating DNA could in fact be thought of as a sink for new histone molecules. A growing number of studies have suggested that the overproduction of histones relative to the rate of chromosome replication results in deleterious effects to the cell, including an increased frequency of chromo-some loss (165). Presumably many of these effects occur because of gross alterations to chromatin structure. To avoid such catastrophe, a number of different regulatory strategies might be required to control histone levels as a function of DNA replication. For example, in vertebrate cells the interruption of DNA replication is associated with rapid degradation of histone mRNAs; since the half-lives of these mRNAs are 30–60 minutes, turnoff of transcrip-tion would not be sufficient to deplete the cell of translatable histone mRNAs rapidly. Similarly, the increased frequency of transcription at the G1-S phase boundary might be required to provide enough histone mRNA to keep pace with the need for new chromatin synthesis during S phase. The role of the highly conserved histone pre-mRNA processing system in cell cycle regula-tion is more problematic. Both transcription and processing are regulated at the G1-S phase boundary. Perhaps the regulation of processing is a backup system to prevent the premature accumulation of new histone transcripts in the cytoplasm before significant DNA synthesis has occurred (173). An alternative possibility is that processing is used as a necessary adjunct to transcriptional control to provide sufficient levels of histone mRNAs as cells enter S phase, where their requirement for new histone synthesis is greatest.

If it is indeed critical for the cell to maintain a fixed relationship between the level of histone synthesis and the rate of chromosome replication, it is of interest that a number of replication-independent histone genes are found in most eukaryotes. These genes are minor members of the histone gene family and are generally not expressed to the same levels as the majority of the replication-dependent histone genes. Their expression outside of the normal period of chromosome replication is presumably a way to remodel chromatin structure at specific loci, with accordant effects on gene expression or other cellular processes.

## FUTURE PROSPECTS

The enormous body of research performed on the histone genes during the past decade has been fruitful on two fronts. It has led to the identification of three major pathways that regulate the levels of histone mRNAs in the cell cycle. In addition, it has provided a wealth of detail on many of the regulatory factors that are important in each pathway. The nature of the signal(s) transduced from the cell cycle and how this signal coordinates these factors into each of the regulatory pathways are the most fundamental questions that remain. It is hoped that the next 10 years will see the resolution of these critical issues.

ACKNOWLEDGMENTS

Arthur Lustig, Lorraine Moran, Peter Sherwood, Stewart Shuman, and Mona Spector are thanked for critically reading this manuscript, and Dennis Ballinger is gratefully acknowledged for his help with computer programs. Thanks also go to numerous colleagues for reprints and communication of unpublished data, and to members of my laboratory for their contributions to the field of yeast histone gene regulation. Supported by N.I.H. grant GM40118.

## Literature Cited

1. Busslinger, M., Schumperli, D., Birnstiel, M. L. 1985. *Cold Spring Harbor Symp. Quant. Biol.* 50:665–70
2. Maxson, R., Cohn, R., Kedes, L. 1983. *Annu. Rev. Genet.* 17:239–77
3. Kedes, L. H. 1979. *Annu. Rev. Biochem.* 48:837–70
4. Angerer, L. M., DeLeon, D. V., Angerer, R. C., Showman, R. M., Wells, D. E., Raff, R. A. 1984. *Dev. Biol.* 101:477–84
5. Anderson, K. V., Lengyel, J. A. 1984. *Histone Genes: Structure, Organization and Regulation,* ed. G. Stein, J. Stein, W. F. Marzluff, pp. 135–62. New York: Wiley
6. Wu, R. S., Bonner, W. M. 1981. *Cell* 27:321–30
7. Choe, J., Schuster, T., Grunstein, M. 1985. *Mol. Cell. Biol.* 5:3261–69
8. Hereford, L., Fahrner, K., Woolford, J., Rosbash, M., Kaback, D. B. 1979. *Cell* 18:1261–71
9. Smith, M. M., Murray, K. 1983. *J. Mol. Biol.* 169:641–61
10. Smith, M. M., Andresson, O. S. 1983. *J. Mol. Biol.* 169:663–90
11. Smith, M. M. 1984. See Ref. 5, pp. 3–33
12. Matsumoto, S., Yanagida, M. 1985. *EMBO J.* 4:3531–38
13. Ehinger, A., Denison, S., May, G. 1990. *Mol. Gen. Genet.* 222:416–24
14. Woudt, L. P., Pastink, A., Kempers-Veenstra, A. E., Jansen, A. E. M., Mager, W. H., Planta, R. J. 1983. *Nucleic Acids Res.* 11:5347–60
15. Hentschel, C. C., Birnstiel, M. L. 1981. *Cell* 25:301–13
16. Lifton, R. P., Goldberg, M. L., Karp, R. W., Hogness, D. S. 1977. *Cold Spring Harbor Symp. Quant. Biol.* 42:1047–51
17. Heintz, N., Zernik, M., Roeder, R. G. 1981. *Cell* 24:661–68
18. Marzluff, W. F., Graves, R. A. 1984. See Ref. 5, pp. 281–315

19. Sierra, F., Lichtler, A., Marashi, F., Rickles, R., Van Dyke, T., et al. 1982. *Proc. Natl. Acad. Sci. USA* 79:1795–99
20. Sittman, D., Chiu, I.-M., Pan, C.-I., Cohn, R. H., Kedes, L. H., Marzluff, W. F. 1981. *Proc. Natl. Acad. Sci. USA* 78:4078–82
21. Engel, J. D., Dodgson, J. B. 1981. *Proc. Natl. Acad. Sci. USA* 78:2856–60
22. Engel, J. D. 1984. See Ref. 5, pp. 263–80
23. Stein, G. S., Sierra, F., Stein, J. L., Plumb, M., Marashi, F., et al. 1984. See Ref. 5, pp. 397–455
24. Wells, D. E. 1986. *Nucleic Acids Res.* 14:R119–49
25. Artishevsky, A., Grafsky, A., Lee, A. S. 1985. *Science* 230:1061–63
26. Green, L., Schlaffer, I., Wright, K., Moreno, M. L., Berand, D., et al. 1986. *Proc. Natl. Acad. Sci. USA* 83:2315–19
27. Kroeger, P., Stewart, C., Schaap, T., van Wijnen, A., Hirshman, J., et al. 1987. *Proc. Natl. Acad. Sci. USA* 84:3982–86
28. Seiler-Tuyns, A., Paterson, B. M. 1987. *Mol. Cell. Biol.* 7:1048–54
29. La Bella, F., Sive, H. L., Roeder, R. G., Heintz, N. 1988. *Genes Dev.* 2:32–39
30. Osley, M. A., Gould, J., Kim, S. Y., Kane, M., Hereford, L. 1986. *Cell* 45:537–44
31. Birnstiel, M. L., Busslinger, M., Strub, K. 1985. *Cell* 41:349–59
32. May, G. S., Morris, N. R. 1987. *Gene* 58:59–66
33. Wilhelm, M. L., Wilhelm, F. X. 1987. *Nucleic Acids Res.* 15:5478
34. Bannon, G. A., Calzone, F. J., Bowen, J. K., Allis, C. D., Gorovsky, M. A. 1983. *Nucleic Acids Res.* 11:3903–17
35. Fahrner, K., Yarger, J., Hereford, L. 1980. *Nucleic Acids Res.* 8:5725–39
36. Horowitz, S., Bowen, J. K., Bannon,

G. A., Gorovsky, M. A. 1987. *Nucleic Acids Res.* 15:141–60
37. Robbins, E., Borun, T. W. 1967. *Proc. Natl. Acad. Sci. USA* 57:409–16
38. Borun, T. W., Gabrielli, F., Ajiro, K., Zweidler, A., Baglioni, C. 1975. *Cell* 4:59–67
39. Detke, S., Lichtler, A., Phillips, I., Stein, J., Stein, G. 1979. *Proc. Natl. Acad. Sci. USA* 76:4995–99
40a. Heintz, N., Sive, H. L., Roeder, R. G. 1983. *Mol. Cell. Biol.* 3:539–50
40b. Harris, M. E., Bohni, R., Schneiderman, M. H., Ramamurthy, L., Schumperli, D., Marzluff, W. F. 1991. *Mol. Cell. Biol.* In press
40c. Morris, T. D., Weber, L. A., Hickey, E., Stein, G. S., Stein, J. L. 1991. *Mol. Cell. Biol.* 11:544–53
41. Baumbach, L. L., Stein, G. S., Stein, J. L. 1987. *Biochemistry* 26:6178–87
42. Artishevsky, A., Delegeane, A. M., Lee, A. S. 1984. *Mol. Cell. Biol.* 4:2364–69
43. Plumb, M., Stein, J., Stein, G. 1983. *Nucleic Acids Res.* 11:2391–410
44. Coffino, P., Stimac, E., Groppi, V. E., Bieber, D. 1984. See Ref. 5, pp. 317–38
45. Wu, R. S., West, M. H. P., Bonner, W. M. 1984. See Ref. 5, pp. 457–83
46. Hereford, L. M., Osley, M. A., Ludwig, J. R., McLaughlin, C. S. 1981. *Cell* 24:367–75
47. Cross, S. L., Smith, M. M. 1988. *Mol. Cell. Biol.* 8:945–54
48. Matsumoto, S., Yanagida, M., Nurse, P. 1987. *EMBO J.* 6:1093–97
49. Yu, S.-M., Horowitz, S., Gorovsky, M. A. 1987. *Genes Dev.* 1:683–92
50. Wilhelm, M. L., Toublan, B., Jalouzot, R., Wilhelm, F. X. 1984. *EMBO J.* 3:2659–62
51. Carrino, J. J., Laffler, T. G. 1986. *J. Cell Biol.* 102:1666–70
52. DeLisle, A. J., Graves, R. A., Marzluff, W. F., Johnson, L. F. 1983. *Mol. Cell. Biol.* 3:1920–29
53. Graves, R. A., Marzluff, W. F. 1984. *Mol. Cell. Biol.* 4:351–57
54. Sittman, D. B., Graves, R. A., Marzluff, W. F. 1983. *Proc. Natl. Acad. Sci. USA* 80:1849–53
55. Heintz, N., Roeder, R. G. 1984. *Proc. Natl. Acad. Sci. USA* 81:2713–17
56. Sive, H. L., Heintz, N., Roeder, R. G. 1984. *Mol. Cell. Biol.* 4:2723–34
57. Plumb, M., Marashi, F., Green, L., Zimmerman, A., Zimmerman, S., et al. 1984. *Proc. Natl. Acad. Sci. USA* 81:434–38
58. Krieg, P. A., Melton, D. A. 1984. *Nature* 308:203–6

59. Gick, O., Kramer, A., Keller, W., Birnstiel, M. L. 1986. *EMBO J.* 5:1319–26
60. Luscher, B., Stauber, C., Schindler, R., Schumperli, D. 1985. *Proc. Natl. Acad. Sci. USA* 82:4389–93
61. Stauber, C., Schumperli, D. 1988. *Nucleic Acids Res.* 16:9399–414
62. Hereford, L., Bromley, S., Osley, M. A. 1982. *Cell* 30:305–10
63. Lycan, D. E., Osley, M. A., Hereford, L. M. 1987. *Mol. Cell. Biol.* 7:614–21
64. Xu, H. X., Johnson, L., Grunstein, M. 1990. *Mol. Cell. Biol.* 10:2687–94
65. Osley, M. A., Lycan, D. 1987. *Mol. Cell. Biol.* 7:4204–10
66. Osley, M. A., Hereford, L. M. 1981. *Cell* 24:377–84
67. Wilhelm, M. L., Toublan, B., Fujita, R. A., Wilhelm, F. X. 1988. *Biochem. Biophys. Res. Commun.* 153:162–71
68. Laffler, T. G., Carrino, J. 1987. *J. Bacteriol.* 169:2291–93
69. Ruddell, A., Jacobs-Lorena, M. 1985. *Proc. Natl. Acad. Sci. USA* 82:3316–19
70. Wells, D. E., Showman, R. M., Klein, W. H., Raff, R. A. 1981. *Nature* 292:477–78
71. Capasso, O., Heintz, N. 1985. *Proc. Natl. Acad. Sci. USA* 82:5622–26
72. Dailey, L., Hanly, S. M., Roeder, R. G., Heintz, N. 1986. *Proc. Natl. Acad. Sci. USA* 83:7241–45
73. Fletcher, C., Heintz, N., Roeder, R. G. 1987. *Cell* 51:773–81
74. Dalton, S., Wells, J. R. E. 1988. *Mol. Cell. Biol.* 8:4576–78
75. van Wijnen, A. J., Stein, J. L., Stein, G. S. 1987. *Nucleic Acids Res.* 15:1679–97
76. Sharma, A., Bos, T. J., Pekkala-Flagan, A., Vogt, P. K., Lee, A. S. 1989. *Proc. Natl. Acad. Sci. USA* 86:491–95
77. van Wijnen, A. J., Wright, K. L., Lian, J. B., Stein, J. L., Stein, G. S. 1989. *J. Biol. Chem.* 264:15034–42
78. van Wijnen, A. J., Massung, R. F., Stein, J. L., Stein, G. S. 1988. *Biochemistry* 27:6534–41
79. Dalton, S., Wells, J. R. E., 1988. *EMBO J.* 7:49–56
80. Pauli, U., Chrysogelos, S., Stein, G., Stein, J., Nick, H. 1987. *Science* 236:1308–11
81. Pauli, U., Chrysogelos, S., Nick, H., Stein, G., Stein, J. 1989. *Nucleic Acids Res.* 17:2333–50
82. van Wijnen, A. J., Wright, K. L., Massung, R. F., Gerretsen, M., Stein, J. L., Stein, G. S. 1988. *Nucleic Acids Res.* 16:571–92
83. Shiels, A., Marashi, F., Stein, G.,

Stein, J. 1985. *Biochem. Biophys. Res. Commun.* 127:239–46
84. Helms, S. R., van Wijnen, A. J., Kroeger, P., Shiels, A., Stewart, C., et al. 1987. *J. Cell. Physiol.* 132:552–58
85. Busslinger, M., Portmann, R., Irminger, J. C., Birnstiel, M. L. 1983. *Nucleic Acids Res.* 8:957–77
86. Gannon, F., O'Hare, K., Perrin, F., LePennec, J. P., Benoist, C., et al. 1979. *Nature* 278:428–34
87. Breathnach, R., Chambon, P. 1981. *Annu. Rev. Biochem.* 50:349–83
88. Efstradiatis, A., Posakony, J. W., Maniatis, T., Lawn, R. M., O'Connell, C., et al. 1980. *Cell* 21:653–68
89. Dynan, W. S., Tjian, R. 1983. *Cell* 35:79–87
90. Dynan, W. S., Tjian, R. 1983. *Cell* 32:669–80
91. Zhong, R., Roeder, R. G., Heintz, N. 1983. *Nucleic Acids Res.* 11:7409–25
92. Coles, L. S., Wells, J. R. E. 1985. *Nucleic Acids Res.* 13:585–94
93. Perry, M., Thomsen, G. H., Roeder, R. G. 1985. *J. Mol. Biol.* 185:479–99
94. Busslinger, M., Portmann, R., Irminger, J., Birnstiel, M. 1983. See Ref. 85
95. Grosschedl, R., Machler, M., Rohrer, U., Birnstiel, M. L. 1983. *Nucleic Acids Res.* 11:8123–36
96. Harvey, R. P., Robins, A. J., Wells, J. R. E. 1982. *Nucleic Acids Res.* 10:7851–63
97. Sugarman, B. J., Dodgson, J. B., Engel, J. D. 1983. *J. Biol. Chem.* 258:9005–16
98. Younghusband, H. B., Sturm, R., Wells, J. R. E. 1986. *Nucleic Acids Res.* 14:635–44
99. Coles, L. S., Robins, A. J., Madley, L. K., Wells, J. R. E. 1987. *J. Biol. Chem.* 262:9656–63
100. Gallinari, P., La Bella, F., Heintz, N. 1989. *Mol. Cell. Biol.* 9:1566–75
101. Dalton, S., Wells, J. R. E. 1988. *Mol. Cell. Biol.* 8:4576–78
102. Dorn, A., Bollekens, J., Staub, A., Benoist, C., Mathis, D. 1987. *Cell* 50:863–72
103. Chodosh, L. A., Baldwin, A. S., Carthew, R. W., Sharp, P. A. 1988. *Cell* 53:11–24
104. Cohen, R. B., Sheffery, M., Kim, C. G. 1986. *Mol. Cell Biol.* 6:821–32
105. Knight, G. B., Gudas, J. M., Pardee, A. B. 1987. *Proc. Natl. Acad. Sci. USA* 84:8350–54
106. Raymondjean, M., Cereghini, S., Yaniv, M. 1988. *Proc. Natl. Acad. Sci. USA* 85:757–61
107. Sturm, R. A., Dalton, S., Wells, J. R. E. 1988. *Nucleic Acids Res.* 16:8571–86
108. Sive, H. L., Heintz, N., Roeder, R. G. 1986. *Mol. Cell. Biol.* 6:3329–40
109. Sive, H. L., Roeder, R. G. 1986. *Proc. Natl. Acad. Sci. USA* 83:6382–86
110. Carbon, P., Murgo, S., Ebel, J.-P., Krol, A., Tebb, G., Mattaj, I. W. 1987. *Cell* 51:71–79
111. Bark, C., Weller, P., Zabielski, J., Janson, L., Pettersson, U. 1987. *Nature* 328:356–59
112. Sollner-Webb, B. 1988. *Cell* 52:153–54
113. Murphy, S., Pierani, A., Scheidereit, C., Melli, M., Roeder, R. G. 1989. *Cell* 59:1071–80
114. Falkner, F. G., Zachau, H. G. 1984. *Nature* 310:71–74
115. Mackem, S., Roizman, B. 1982. *J. Virol.* 44:939–49
116. O'Hare, P., Goding, C. R. 1988. *Cell* 52:435–45
117. Pruijn, G. J. M., van Driel, W., van der Vliet, P. C. 1986. *Nature* 322:656–59
118. Bergman, Y., Rice, D., Grosschedl, R., Baltimore, D. 1984. *Proc. Natl. Acad. Sci. USA* 81:7041–45
119. Parslow, T. G., Blair, D. L., Murphy, W. J., Granner, D. K. 1984. *Proc. Natl. Acad. Sci. USA* 81:2650–54
120. Wirth, T., Staudt, L., Baltimore, D. 1987. *Nature* 329:174–78
121. Muller, M. M., Ruppert, S., Schaffner, W., Matthias, P. 1988. *Nature* 336:544–51
122. Sturm, R., Baumruker, T., Franza, B. R., Herr, W. 1987. *Genes Dev.* 1:1147–60
123. Sturm, R. A., Das, G., Herr, W. 1988. *Genes Dev.* 2:1582–99
124. O'Neill, E. A., Fletcher, C., Burrow, C. R., Heintz, N., Roeder, R. G., Kelly, T. J. 1988. *Science* 241:1210–13
125. Scheidereit, C., Heguy, A., Roeder, R. G. 1987. *Cell* 51:783–93
126. Scheidereit, C., Cromlish, J. A., Gerster, T., Kawakami, K., Balmaceda, C.-G., et al. 1988. *Nature* 336:551–57
127. Clerc, R. G., Corcoran, L. M., LeBowitz, J. H., Baltimore, D., Sharp, P. A. 1988. *Genes Dev.* 2:1570–81
128. Schreiber, E., Matthias, P., Muller, M. M., Schaffner, W. 1988. *EMBO J.* 7:4221–29
129. Poellinger, L., Roeder, R. G. 1989. *Mol. Cell. Biol.* 9:747–56
130. Herr, W., Sturm, R. A., Clerc, R. G., Corcoran, L. M., Baltimore, D., et al. 1988. *Genes Dev.* 2:1513–16
131. Garcia-Blanco, M. A., Clerc, R. G.,

Sharp, P. A. 1989. *Genes Dev.* 3:739–45

132. Sturm, R. A., Herr, W. 1988. *Nature* 336:601–4

133. Schaffner, W. 1989. *Trends Genet.* 5:37–39

134. Berk, A., Schmidt, M. 1990. *Genes Dev.* 4:151–55

135. Ito, M., Sharma, A., Lee, A. S., Maxson, R. 1989. *Mol. Cell. Biol.* 9:869–73

136. Gerster, T., Roeder, R. G. 1988. *Proc. Natl. Acad. Sci. USA* 85:6347–51

137. Kristie, T. M., LeBowitz, J. H., Sharp, P. A. 1989. *EMBO J.* 8:4229–38

138. Artishevsky, A., Wooden, S., Sharma, A., Resendez, E., Lee, A. S. 1987. *Nature* 328:823–27

139. Angel, P., Imagawa, M., Chiu, R., Stein, B., Imbra, R. J., et al. 1987. *Cell* 49:729–39

140. Lee, W., Mitchell, P., Tjian, R. 1987. *Cell* 49:741–52

141. Bohmann, D., Bos, T. J., Admon, A., Nishimura, T., Vogt, P. K., Tjian, R. 1987. *Science* 238:1386–92

142. Ryder, K., Lau, L. F., Nathans, D. 1988. *Proc. Natl. Acad. Sci. USA* 85:1487–91

143. Lau, L. F., Nathans, D. 1987. *Proc. Natl. Acad. Sci. USA* 84:1182–86

144. Hanly, S. M., Bleecker, G. C., Heintz, N. 1985. *Mol. Cell. Biol.* 5:380–89

145. Clerc, R. G., Bucher, P., Strub, K., Birnstiel, M. L. 1983. *Nucleic Acids Res.* 11:8641–57

146. Sierra, F., Stein, G., Stein, J. 1983. *Nucleic Acids Res.* 11:7069–86

147. Dailey, L., Roberts, S. B., Heintz, N. 1988. *Genes Dev.* 2:1700–12

148. Holthuis, J., Owen, T. A., van Wijnen, A., Wright, K. L., Ramsey-Ewing, A., et al. 1990. *Science* 247:1454–57

149. Dailey, L., Roberts, S. B., Heintz, N. 1987. *Mol. Cell. Biol.* 7:4582–84

150. Breeden, L. 1988. *Trends Genet.* 4:249–53

151. Andrews, B., Herskowitz, I. 1990. *J. Biol. Chem.* 265:14057–60

152. Nasmyth, K. 1985. *Cell* 42:225–35

153. White, J. H. M., Green, S. R., Barker, D. G., Dumas, L. B., Johnston, L. H. 1987. *Exp. Cell. Res.* 171:223–31

154. Johnston, L. H., White, J. H. M., Johnson, A. L., Lucchini, G., Plevani, P. 1987. *Nucleic Acids Res.* 15:5017–30

155. Peterson, T. A., Prakash, L., Prakash, S., Osley, M. A., Reed, S. I. 1985. *Mol. Cell. Biol.* 5:226–35

156. Storms, R. K., Ord, R. W., Greenwood, M. T., Mirdamadi, B., Chu, F. K., Belfort, M. 1984. *Mol. Cell. Biol.* 4:2858–64

157. Barker, D. G., White, J. H. M., Johnston, L. H. 1985. *Nucleic Acids Res.* 13:8323–37

158. Kupiec, M., Simchen, G. 1986. *Mol. Gen. Genet.* 203:538–43

159a. McIntosh, E. M., Ord, R. W., Storms, R. K. 1988. *Mol. Cell. Biol.* 8:4616–24

159b. McIntosh, E. M., Atkinson, T., Storms, R. K., Smith, M. 1991. *Mol. Cell Biol.* 11:329–37

160. Elledge, S. J., Davis, R. W. 1990. *Genes Dev.* 4:740–51

161. Hartwell, L. H. 1974. *Bacteriol. Rev.* 38:164–98

162. Hollingsworth, R., Sclafani, R. 1990. *Proc. Natl. Acad. Sci. USA* 87:6272–76

163. Drebot, M. A., Veinot-Drebot, L., Singer, R. A., Johnston, G. C. 1990. *Mol. Cell. Biol.* 10:6356–61

164. Moran, L., Norris, D., Osley, M. A. 1990. *Genes Dev.* 4:752–63

165. Meeks-Wagner, D., Hartwell, L. H. 1986. *Cell* 44:43–52

166. Clark-Adams, C. D., Norris, D., Osley, M. A., Fassler, J. S., Winston, F. 1988. *Genes Dev.* 2:150–59

167. Winston, F., Chaleff, D. T., Valent, B., Fink, G. R. 1984. *Genetics* 107:179–97

168. Fassler, J. S., Winston, F. 1988. *Genetics* 118:203–12

169. Brunk, C. F., Sadler, L. A. 1990. *Nucleic Acids Res.* 18:323–29

170. La Bella, F., Heintz, N. 1988. *J. Biol. Chem.* 263:2115–18

171. Collart, D. G., Wright, K. L., van Wijnen, A. J., Ramsey, A. L., Lian, J., et al. 1988. *J. Biol. Chem.* 263:15860–63

172. Birnstiel, M. L., Schaufele, F. 1988. *Structure and Function of Major and Minor Small Nuclear Ribonucleoprotein Particles*, ed. M. L. Birnstiel, pp. 155–82. New York/London/Paris/Tokyo: Springer-Verlag

173. Schumperli, D. 1988. *Trends Genet.* 4:187–91

174. Marzluff, W. F., Pandey, N. B. 1988. *Trends Biochem. Sci.* 13:49–52

175. Busslinger, M., Portmann, R., Birnstiel, M. L. 1979. *Nucleic Acids Res.* 6:2997–3008

176. Birchmeier, C., Grosschedl, R., Birnstiel, M. L. 1982. *Cell* 28:739–45

177. Birchmeier, C., Folk, W., Birnstiel, M. L. 1983. *Cell* 35:433–40

178. Mowry, K. L., Steitz, J. A. 1987. *Science* 238:1682–87

179. Mowry, K. L., Oh, R., Steitz, J. A. 1989. *Mol. Cell. Biol.* 9:3105–8

180. Georgiev, O., Birnstiel, M. L. 1985. *EMBO J.* 4:481–89

181. Birchmeier, C., Schumperli, D., Scon-

zo, G., Birnstiel, M. L. 1984. *Proc. Natl. Acad. Sci. USA* 81:1057–61
182. Guthrie, C., Patterson, B. 1988. *Annu. Rev. Genet.* 22:387–419
183. Steitz, J. A., Black, D., Gerke, V., Parker, K., Kramer, A., et al. 1988. See Ref. 172, pp. 115–54
184. Reddy, R., Busch, H. 1988. See Ref. 172, pp. 1–37
185. Galli, G., Hofstetter, H., Stunnenberg, H. G., Birnstiel, M. L. 1983. *Cell* 34:823–28
186. Strub, K., Birnstiel, M. 1986. *EMBO J.* 5:1675–82
187. Mowry, K. L., Steitz, J. A. 1987. *Mol. Cell. Biol.* 7:1663–72
188. Strub, K., Galli, G., Busslinger, M., Birnstiel, M. L. 1984. *EMBO J.* 3:2801–7
189. Soldati, D., Schumperli, D. 1988. *Mol. Cell. Biol.* 8:1518–24
190. Cotten, M., Gick, O., Vasserot, A., Schaffner, G., Birnstiel, M. L. 1988. *EMBO J.* 7:801–8
191. Schaufele, F., Gilmartin, G. M., Bannwarth, W., Birnstiel, M. L. 1986. *Nature* 323:777–81
192. Vasserot, A. P., Schaufele, F. J., Birnstiel, M. L. 1989. *Proc. Natl. Acad. Sci. USA* 86:4345–49
193. Gick, O., Kramer, A., Vasserot, A., Birnstiel, M. L. 1987. *Proc. Natl. Acad. Sci. USA* 84:8937–40
194. Stauber, C., Luscher, B., Eckner, R., Lotscher, E., Schumperli, D. 1986. *EMBO J.* 5:3297–303
195. Luscher, B., Schumperli, D. 1987. *EMBO J.* 6:1721–26
196. Hoffmann, I., Birnstiel, M. L. 1990. *Nature* 346:665–68
197. Stimac, E., Groppi, V. E., Coffino, P. 1984. *Mol. Cell. Biol.* 4:2082–90
198. Capasso, O., Bleecker, G. C., Heintz, N. 1987. *EMBO J.* 6:1825–31
199. Graves, R. A., Pandey, N. B., Chodchoy, N., Marzluff, W. F. 1987. *Cell* 48:615–26
200. Levine, B. J., Chodchoy, N., Marzluff, W. F., Skoultchi, A. I. 1987. *Proc. Natl. Acad. Sci. USA* 84:6189–93
201. Levine, B. J., Liu, T.-J., Marzluff, W. F., Skoultchi, A. I. 1988. *Mol. Cell. Biol.* 8:1887–95
202. Pandey, N. B., Marzluff, W. F. 1987. *Mol. Cell. Biol.* 7:4557–59
203. Morris, T., Marashi, F., Weber, L., Hickey, E., Greenspan, D., et al. 1986. *Proc. Natl. Acad. Sci. USA* 83:981–85
204. Chodchoy, N., Levine, B. J., Sprecher, C., Skoultchi, A. I., Marzluff, W. F. 1987. *Mol. Cell. Biol.* 7:1039–47
205. Alterman, R. B. M., Sprecher, C., Graves, R., Marzluff, W. F., Skoultchi, A. I. 1985. *Mol. Cell. Biol.* 5:2316–24
206. Whitelaw, E., Coates, A., Proudfoot, N. J. 1986. *Nucleic Acids Res.* 14:7059–70
207. Cheng, G. H., Nandi, A., Clerk, S., Skoultchi, A. I. 1989. *Proc. Natl. Acad. Sci. USA* 86:7002–6
208. Ross, J., Peltz, S. W., Kobs, G., Brewer, G. 1986. *Mol. Cell. Biol.* 6:4362–71
209. Ross, J., Kobs, G. 1986. *J. Mol. Biol.* 188:579–93
210. Ross, J., Kobs, G., Brewer, G., Peltz, S. 1987. *J. Biol. Chem.* 262:9374–81
211. Peltz, S. W., Brewer, G., Kobs, G., Ross, J. 1987. *J. Biol. Chem.* 262:9382–88
212. Peltz, S. W., Ross, J. 1987. *Mol. Cell. Biol.* 7:4345–56
213. Sariban, E., Wu, R. S., Erickson, L. C., Bonner, W. M. 1985. *Mol. Cell. Biol.* 5:1279–86
214. Butler, W. B., Mueller, G. C. 1973. *Biochim. Biophys. Acta* 294:481–96
215. Butler, J. S., Sadhale, P. P., Platt, T. 1990. *Mol. Cell. Biol.* 10:2599–605
216. Fitzgerald, M., Shenk, T. 1981. *Cell* 24:251–60
217. Wickens, M., Stephenson, P. 1984. *Science* 226:1045–51
218. Conway, L., Wickens, M. 1985. *Proc. Natl. Acad. Sci. USA* 82:3949–53
219. Green, T. L., Hart, R. P. 1988. *Mol. Cell. Biol.* 8:1839–41
220. Zaret, K. S., Sherman, F. 1982. *Cell* 28:563–73
221. Bernstein, P., Ross, J. 1989. *Trends Biochem. Sci.* 14:373–77
222. Jackson, R. J., Standart, N. 1990. *Cell* 62:15–24
223. Brawerman, G. 1987. *Cell* 48:5–6
224. Sachs, A. B., Bond, M. W., Kornberg, R. D. 1986. *Cell* 45:827–35
225. Sachs, A. B., Davis, R. W., Kornberg, R. D. 1987. *Mol. Cell. Biol.* 7:3268–76
226. Bernstein, P., Peltz, S. W., Ross, J. 1989. *Mol. Cell. Biol.* 9:659–70
227. Sachs, A. B., Davis, R. W. 1989. *Cell* 58:857–67
228. Herrick, D., Parker, R., Jacobson, A. 1990. *Mol. Cell. Biol.* 10:2269–84
229. Raghow, R. 1987. *Trends Biochem. Sci.* 12:358–60
230. Stillman, B. 1986. *Cell* 45:555–65

# AUTHOR INDEX

# SUBJECT INDEX

## A

*Acanthamoeba* spp.
RNA polymerase II and, 691,
704
Acetate
bacterial chemotaxis and,
419–20
α-*N*-Acetylgalactosaminidase
lysosomal storage disorders
and, 265
*N*-Acetylglucosamine 6-sulfatase
lysosomal storage disorders
and, 264
Acetylglucosaminidase
lysosomal storage disorders
and, 264
Acid phosphatase
lysosomal storage disorders
and, 259, 262, 264–65
Acquired immunodeficiency syndrome (AIDS)
antiviral therapy for, 619–21
history of, 579–80
vaccine for, 618–19, 621
see also Human im-
munodeficiency virus
Acyclovir
DNA polymerase gene muta-
tions and, 532
Acyl-CoA cholesterol acyltransferase (ACAT)
lipid metabolism and, 92
Adenosine diphosphate (ADP)
GC-A protein and, 567
protein phosphorylation and,
410, 412
Adenosine monophosphate
(AMP)
phage φ29 DNA replication
and, 48, 52, 54
phage Cp-1 DNA replication
and, 56
Adenosine triphosphate (ATP)
GC-A protein and, 567
GTP-binding proteins and,
352, 353
history of, 10–13, 35
molecular chaperones and,
327, 335, 337, 339–43
phage λ DNA maturation
and, 127, 130
phage φ29 DNA replication
and, 48, 50, 52, 54
phage Cp-1 DNA replication
and, 56

protein phosphorylation and,
402, 410, 412, 425, 427
SecA protein and, 102, 108–
9, 112–14, 116–20
terminases and, 133–35
translational control and, 725
Adenovirus
HIV activation and, 611
protein-priming mechanism of
DNA replication and,
40–47
Adenylate kinase
history of, 14
Adenylyl cyclases
molecular cloning of guanylyl
cyclases and, 562–63
transmembrane signaling and,
654–56
Adrenal corticotropic hormone
(ACTH)
steroidogenesis and, 93
β₂-Adrenergic receptors
desensitization and, 671–74
heterologous receptor systems
and, 679–80
ligand binding determinants
and, 667–71
phosphorylation and, 672–
74
schematic representation of,
662
subtypes of, 680
transmembrane signaling and,
655, 657–60
Affinity labeling
rRNA and, 198–99, 201
*Agaricus bitorquis*
linear plasmids and, 57
*Agrobacterium tumefaciens*
phage HB-3 and, 56
AIDS
see Acquired im-
munodeficiency syn-
drome
Alcohol dehydrogenase
conformation of, 28–29
Aldosterone secretion inhibitory
factor (ASIF)
classification of, 234
Alpha accessory factor (AAF)
DNA polymerases and, 527
Alpha factor receptor
ligand binding determinants
and, 670–71
transmembrane signaling and,
664

Amantadine
HIV entry and, 585
Ammonium chloride
HIV entry and, 585
Angiotensin
transmembrane signaling and,
661
Antibiotics
rRNA interactions with, 209–
11, 214, 218
Antibodies
against genome-linked proteins of RNA viruses,
62
against phospholipid transfer
proteins, 76, 78, 89–90
Antisense RNA
biological regulation and,
637–49
naturally occurring, 646–48
RNA-RNA interaction and,
632–37, 649
Antithrombin
proteoglycans and, 458–59,
461–62, 464, 467
AP1 transcription factor
HIV gene expression and,
589
Aphidicolin
DNA polymerase gene muta-
tions and, 532, 545
phage φ29 DNA polymerase
inhibition and, 50
Apomyoglobin
acid denaturation and, 816
*aQ* gene
antisense RNA and, 646
*Arabidopsis thaliana*
G-proteins and, 365, 378–79,
385
RNA polymerase II and, 704–
6
Arabinose-binding protein (ABP)
hydrogen bonds and, 771
*Arbacia punctulata*
guanylyl cyclase and, 566,
568
*ARF* genes
GTP binding and, 351
*Artemia* spp.
protein synthesis and, 728,
744
Arylsulfatases
lysosomal storage disorders
and, 264–66, 269, 272–
73

# CUMULATIVE INDEXES

## CONTRIBUTING AUTHORS, VOLUMES 56–60

# CHAPTER TITLES, VOLUMES 56–60

938

# ANNUAL REVIEWS INC.

## A NONPROFIT SCIENTIFIC PUBLISHER

 4139 El Camino Way
P.O. Box 10139
Palo Alto, CA 94303-0897 ● USA

---

Annual Reviews Inc. publications may be ordered directly from our office; through booksellers and subscription agents, worldwide; and through participating professional societies. Prices subject to change without notice.                                               ARI Federal I.D. #94-1156476

- **Individuals:** Prepayment required on new accounts by check or money order (in U.S. dollars, check drawn on U.S. bank) or charge to credit card — American Express, VISA, MasterCard.
- **Institutional buyers:** Please include purchase order.
- **Students:** $10.00 discount from retail price, per volume. Prepayment required. Proof of student status must be provided (photocopy of student I.D. or signature of department secretary is acceptable). Students must send orders direct to Annual Reviews. Orders received through bookstores and institutions requesting student rates will be returned. You may order at the Student Rate for a maximum of 3 years.
- **Professional Society Members:** Members of professional societies that have a contractual arrangement with Annual Reviews may order books through their society at a reduced rate. Check with your society for information.
- **Toll Free Telephone orders:** Call 1-800-523-8635 (except from California) for orders paid by credit card or purchase order and customer service calls only. California customers and all other business calls use 415-493-4400 (not toll free). Hours: 8:00 AM to 4:00 PM, Monday-Friday, Pacific Time. **Written confirmation** is required on purchase orders from universities before shipment.
- **FAX: 415-855-9815      Telex: 910-290-0275**
- **We do not ship on approval.**

Regular orders: Please list below the volumes you wish to order by volume number.
Standing orders: New volume in the series will be sent to you automatically each year upon publication. Cancellation may be made at any time. Please indicate volume number to begin standing order.
Prepublication orders: Volumes not yet published will be shipped in month and year indicated.
California orders: Add applicable sales tax. Canada: Add GST tax.
Postage paid (4th class bookrate/surface mail) by Annual Reviews Inc. UPS domestic ground service available (except Alaska and Hawaii) at $2.00 extra per book. Airmail postage or UPS air service also available at prevailing costs. UPS must have street address. P.O. Box, APO or FPO not acceptable.

---

| ANNUAL REVIEWS SERIES | Prices postpaid, per volume USA & Canada / elsewhere | | Regular Order Please Send | Standing Order Begin With |
|---|---|---|---|---|
| | Until 12-31-90 | After 1-1-91 | Vol. Number: | Vol. Number: |
| Annual Review of **ANTHROPOLOGY** | | | | |
| Vols. 1-16     (1972-1987) ........... | $31.00/$35.00 | $33.00/$38.00 | | |
| Vols. 17-18     (1988-1989) ........... | $35.00/$39.00 | $37.00/$42.00 | | |
| Vol. 19     (1990) .............. | $39.00/$43.00 | $41.00/$46.00 | | |
| Vol. 20     (avail. Oct. 1991) ....... | $41.00/$46.00 | $41.00/$46.00 | Vol(s). _____ | Vol. _____ |
| Annual Review of **ASTRONOMY AND ASTROPHYSICS** | | | | |
| Vols. 1, 5-14     (1963, 1967-1976) | | | | |
| 16-20     (1978-1982) .......... | $31.00/$35.00 | $33.00/$38.00 | | |
| Vols. 21-27     (1983-1989) .......... | $47.00/$51.00 | $49.00/$54.00 | | |
| Vol. 28     (1990) .............. | $51.00/$55.00 | $53.00/$58.00 | | |
| Vol. 29     (avail. Sept. 1991) ...... | $53.00/$58.00 | $53.00/$58.00 | Vol(s). _____ | Vol. _____ |
| Annual Review of **BIOCHEMISTRY** | | | | |
| Vols. 30-34, 36-56     (1961-1965, 1967-1987) .. | $33.00/$37.00 | $35.00/$40.00 | | |
| Vols. 57-58     (1988-1989) .......... | $35.00/$39.00 | $37.00/$42.00 | | |
| Vol. 59     (1990) .............. | $39.00/$44.00 | $41.00/$47.00 | | |
| Vol. 60     (avail. July 1991) ....... | $41.00/$47.00 | $41.00/$47.00 | Vol(s). _____ | Vol. _____ |
| Annual Review of **BIOPHYSICS AND BIOPHYSICAL CHEMISTRY** | | | | |
| Vols. 1-11     (1972-1982) .......... | $31.00/$35.00 | $33.00/$38.00 | | |
| Vols. 12-18     (1983-1989) .......... | $49.00/$53.00 | $51.00/$56.00 | | |
| Vol. 19     (1990) .............. | $53.00/$57.00 | $55.00/$60.00 | | |
| Vol. 20     (avail. June 1991) ...... | $55.00/$60.00 | $55.00/$60.00 | Vol(s). _____ | Vol. _____ |

| ANNUAL REVIEWS SERIES | Prices postpaid, per volume USA & Canada / elsewhere | | Regular Order Please Send Vol. Number: | Standing Order Begin With Vol. Number: |
|---|---|---|---|---|
| | Until 12-31-90 | After 1-1-91 | | |

### Annual Review of CELL BIOLOGY

| | | | |
|---|---|---|---|
| Vols. 1-3 | (1985-1987) .......... $31.00/$35.00 | $33.00/$38.00 | |
| Vols. 4-5 | (1988-1989) .......... $35.00/$39.00 | $37.00/$42.00 | |
| Vol. 6 | (1990) ............... $39.00/$43.00 | $41.00/$46.00 | |
| Vol. 7 | (avail. Nov. 1991) ....... $41.00/$46.00 | $41.00/$46.00 | Vol(s). _____ Vol. _____ |

### Annual Review of COMPUTER SCIENCE

| | | | |
|---|---|---|---|
| Vols. 1-2 | (1986-1987) .......... $39.00/$43.00 | $41.00/$46.00 | |
| Vols. 3-4 | (1988, 1989-1990) ...... $45.00/$49.00 | $47.00/$52.00 | Vol(s). _____ Vol. _____ |

Series suspended until further notice. SPECIAL OFFER: Volumes 1-4 are available at the special promotional price of $100.00 USA & Canada / $115.00 elsewhere, when all 4 volumes are purchased at one time. Orders at the special price must be prepaid.

### Annual Review of EARTH AND PLANETARY SCIENCES

| | | | |
|---|---|---|---|
| Vols. 1-10 | (1973-1982) .......... $31.00/$35.00 | $33.00/$38.00 | |
| Vols. 11-17 | (1983-1989) .......... $49.00/$53.00 | $51.00/$56.00 | |
| Vol. 18 | (1990) ............... $53.00/$57.00 | $55.00/$60.00 | |
| Vol. 19 | (avail. May 1991) ....... $55.00/$60.00 | $55.00/$60.00 | Vol(s). _____ Vol. _____ |

### Annual Review of ECOLOGY AND SYSTEMATICS

| | | | |
|---|---|---|---|
| Vols. 2-18 | (1971-1987) .......... $31.00/$35.00 | $33.00/$38.00 | |
| Vols. 19-20 | (1988-1989) .......... $34.00/$38.00 | $36.00/$41.00 | |
| Vol. 21 | (1990) ............... $38.00/$42.00 | $40.00/$45.00 | |
| Vol. 22 | (avail. Nov. 1991) ....... $40.00/$45.00 | $40.00/$45.00 | Vol(s). _____ Vol. _____ |

### Annual Review of ENERGY

| | | | |
|---|---|---|---|
| Vols. 1-7 | (1976-1982) .......... $31.00/$35.00 | $33.00/$38.00 | |
| Vols. 8-14 | (1983-1989) .......... $58.00/$62.00 | $60.00/$65.00 | |
| Vol. 15 | (1990) ............... $62.00/$66.00 | $64.00/$69.00 | |
| Vol. 16 | (avail. Oct. 1991) ....... $64.00/$69.00 | $64.00/$69.00 | Vol(s). _____ Vol. _____ |

### Annual Review of ENTOMOLOGY

| | | | |
|---|---|---|---|
| Vols. 10-16, 18 | (1965-1971, 1973) | | |
| 20-32 | (1975-1987) .......... $31.00/$35.00 | $33.00/$38.00 | |
| Vols. 33-34 | (1988-1989) .......... $34.00/$38.00 | $36.00/$41.00 | |
| Vol. 35 | (1990) ............... $38.00/$42.00 | $40.00/$45.00 | |
| Vol. 36 | (avail. Jan. 1991) ....... $40.00/$45.00 | $40.00/$45.00 | Vol(s). _____ Vol. _____ |

### Annual Review of FLUID MECHANICS

| | | | |
|---|---|---|---|
| Vols. 2-4, 7 | (1970-1972, 1975) | | |
| 9-19 | (1977-1987) .......... $32.00/$36.00 | $34.00/$39.00 | |
| Vols. 20-21 | (1988-1989) .......... $34.00/$38.00 | $36.00/$41.00 | |
| Vol. 22 | (1990) ............... $38.00/$42.00 | $40.00/$45.00 | |
| Vol. 23 | (avail. Jan. 1991) ....... $40.00/$45.00 | $40.00/$45.00 | Vol(s). _____ Vol. _____ |

### Annual Review of GENETICS

| | | | |
|---|---|---|---|
| Vols. 1-21 | (1967-1987) .......... $31.00/$35.00 | $33.00/$38.00 | |
| Vols. 22-23 | (1988-1989) .......... $34.00/$38.00 | $36.00/$41.00 | |
| Vol. 24 | (1990) ............... $38.00/$42.00 | $40.00/$45.00 | |
| Vol. 25 | (avail. Dec. 1991) ....... $40.00/$45.00 | $40.00/$45.00 | Vol(s). _____ Vol. _____ |

### Annual Review of IMMUNOLOGY

| | | | |
|---|---|---|---|
| Vols. 1-5 | (1983-1987) .......... $31.00/$35.00 | $33.00/$38.00 | |
| Vols. 6-7 | (1988-1989) .......... $34.00/$38.00 | $36.00/$41.00 | |
| Vol. 8 | (1990) ............... $38.00/$42.00 | $40.00/$45.00 | |
| Vol. 9 | (avail. April 1991) ....... $41.00/$46.00 | $41.00/$46.00 | Vol(s). _____ Vol. _____ |

### Annual Review of MATERIALS SCIENCE

| | | | |
|---|---|---|---|
| Vols. 1, 3-12 | (1971, 1973-1982) ...... $31.00/$35.00 | $33.00/$38.00 | |
| Vols. 13-19 | (1983-1989) .......... $66.00/$70.00 | $68.00/$73.00 | |
| Vol. 20 | (1990) ............... $70.00/$74.00 | $72.00/$77.00 | |
| Vol. 21 | (avail. Aug. 1991) ....... $72.00/$77.00 | $72.00/$77.00 | Vol(s). _____ Vol. _____ |

FROM

NAME _____

ADDRESS _____

_____

ZIP CODE _____

serving science · Annual
since
1932
Reviews · Annual

# Annual Reviews Inc.

A NONPROFIT
SCIENTIFIC PUBLISHER

4139 El Camino Way
P.O. Box 10139
Palo Alto, CA 94303-0897

PLACE
STAMP
HERE

| ANNUAL REVIEWS SERIES | Prices postpaid, per volume USA & Canada / elsewhere | | Regular Order Please Send | Standing Order Begin With |
|---|---|---|---|---|
| | Until 12-31-90 | After 1-1-91 | Vol. Number: | Vol. Number: |

**Annual Review of PSYCHOLOGY**

| | | | |
|---|---|---|---|
| Vols. 4, 5, 8, 10, 13-24 | (1953, 1954, 1957, 1959, 1962-1973) | | |
| 26-30, 32-38 | (1975-1979, 1981-1987) . . | $31.00/$35.00 | $33.00/$38.00 |
| Vols. 39-40 | (1988-1989) . . . . . . . . . . | $34.00/$38.00 | $36.00/$41.00 |
| Vol. 41 | (1990) . . . . . . . . . . . . . . | $38.00/$42.00 | $40.00/$45.00 |
| Vol. 42 | (avail. Feb. 1991) . . . . . . . | $40.00/$45.00 | $40.00/$45.00 | Vol(s). _____ Vol. _____ |

**Annual Review of PUBLIC HEALTH**

| | | | |
|---|---|---|---|
| Vols. 1-8 | (1980-1987) . . . . . . . . . . | $31.00/$35.00 | $33.00/$38.00 |
| Vols. 9-10 | (1988-1989) . . . . . . . . . . | $39.00/$43.00 | $41.00/$46.00 |
| Vol. 11 | (1990) . . . . . . . . . . . . . . | $43.00/$47.00 | $45.00/$50.00 |
| Vol. 12 | (avail. May 1991) . . . . . . . | $45.00/$50.00 | $45.00/$50.00 | Vol(s). _____ Vol. _____ |

**Annual Review of SOCIOLOGY**

| | | | |
|---|---|---|---|
| Vols. 1-13 | (1975-1987) . . . . . . . . . . | $31.00/$35.00 | $33.00/$38.00 |
| Vols. 14-15 | (1988-1989) . . . . . . . . . . | $39.00/$43.00 | $41.00/$46.00 |
| Vol. 16 | (1990) . . . . . . . . . . . . . . | $43.00/$47.00 | $45.00/$50.00 |
| Vol. 17 | (avail. Aug. 1991) . . . . . . . | $45.00/$50.00 | $45.00/$50.00 | Vol(s). _____ Vol. _____ |

# Note: Volumes not listed are out of print.

| SPECIAL PUBLICATIONS | Prices postpaid, per volume USA & Canada / elsewhere | Regular Order Please Send |
|---|---|---|

**The Excitement and Fascination of Science**

| | | | | |
|---|---|---|---|---|
| Volume 1 | (published 1965) | Clothbound . . . . . . . . . . | $25.00/$29.00 | _____ Copy(ies). |
| Volume 2 | (published 1978) | Softcover . . . . . . . . . . . . | $15.00/$19.00 | _____ Copy(ies). |
| Volume 3 | (published 1990) | Clothbound . . . . . . . . . . | $90.00/$95.00 | _____ Copy(ies). |

(Volume 3 is published in two parts with complete indexes for Volumes 1, 2, and both parts of Volume 3. Sold in two-part set only.)

**Intelligence and Affectivity:**
**Their Relationship During Child Development,** by Jean Piaget

| | | | | |
|---|---|---|---|---|
| (published 1981) | Hardcover . . . . . . . . . . . . . . . . . . . . . . . . . . . . . | $8.00/$9.00 | _____ Copy(ies). |

TO:  ANNUAL REVIEWS INC., a nonprofit scientific publisher
4139 El Camino Way ● P.O. Box 10139
Palo Alto, CA 94303-0897 USA

Please enter my order for the publications checked above. **California orders, please add applicable sales tax.**
Prices subject to change without notice.

☐ Proof of student status enclosed.

Institutional purchase order No. _____

Amount of remittance enclosed $ _____

Charge my account  ☐ VISA

☐ MasterCard  ☐ American Express

INDIVIDUALS: Prepayment required in U.S. funds or charge to bank card below. Include card number, expiration date, and signature.

Acct. No. _____

Exp. Date _____  _____
                              Signature

Name _____
        Please print

Address _____
          Please print

_____ Zip Code _____ Date _____

_____ Send free copy of current **Prospectus** ☐
Area(s) of Interest                                    ARI Federal I.D. #94-1156476